Stafford Library
Columbia College
1001 Rogers Street
Columbia, Missouri 65216

WITHDRAWN

WITHDRAWN

# Handbook of Elliptic and Hyperelliptic Curve Cryptography

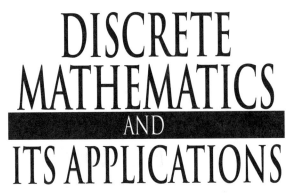

# DISCRETE MATHEMATICS AND ITS APPLICATIONS

Series Editor
**Kenneth H. Rosen, Ph.D.**

*Juergen Bierbrauer*, Introduction to Coding Theory

*Kun-Mao Chao and Bang Ye Wu*, Spanning Trees and Optimization Problems

*Charalambos A. Charalambides*, Enumerative Combinatorics

*Henri Cohen, Gerhard Frey, et al.*, Handbook of Elliptic and Hyperelliptic Curve Cryptography

*Charles J. Colbourn and Jeffrey H. Dinitz*, The CRC Handbook of Combinatorial Designs

*Steven Furino, Ying Miao, and Jianxing Yin*, Frames and Resolvable Designs: Uses, Constructions, and Existence

*Randy Goldberg and Lance Riek*, A Practical Handbook of Speech Coders

*Jacob E. Goodman and Joseph O'Rourke*, Handbook of Discrete and Computational Geometry, Second Edition

*Jonathan Gross and Jay Yellen*, Graph Theory and Its Applications

*Jonathan Gross and Jay Yellen*, Handbook of Graph Theory

*Darrel R. Hankerson, Greg A. Harris, and Peter D. Johnson*, Introduction to Information Theory and Data Compression, Second Edition

*Daryl D. Harms, Miroslav Kraetzl, Charles J. Colbourn, and John S. Devitt*, Network Reliability: Experiments with a Symbolic Algebra Environment

*Derek F. Holt with Bettina Eick and Eamonn A. O'Brien*, Handbook of Computational Group Theory

*David M. Jackson and Terry I. Visentin*, An Atlas of Smaller Maps in Orientable and Nonorientable Surfaces

*Richard E. Klima, Ernest Stitzinger, and Neil P. Sigmon*, Abstract Algebra Applications with Maple

*Patrick Knupp and Kambiz Salari*, Verification of Computer Codes in Computational Science and Engineering

*William Kocay and Donald L. Kreher*, Graphs, Algorithms, and Optimization

*Donald L. Kreher and Douglas R. Stinson*, Combinatorial Algorithms: Generation Enumeration and Search

*Charles C. Lindner and Christopher A. Rodgers*, Design Theory

*Alfred J. Menezes, Paul C. van Oorschot, and Scott A. Vanstone*, Handbook of Applied Cryptography

## Continued Titles

*Richard A. Mollin,* Algebraic Number Theory

*Richard A. Mollin,* Codes: The Guide to Secrecy from Ancient to Modern Times

*Richard A. Mollin,* Fundamental Number Theory with Applications

*Richard A. Mollin,* An Introduction to Cryptography

*Richard A. Mollin,* Quadratics

*Richard A. Mollin,* RSA and Public-Key Cryptography

*Kenneth H. Rosen,* Handbook of Discrete and Combinatorial Mathematics

*Douglas R. Shier and K.T. Wallenius,* Applied Mathematical Modeling: A Multidisciplinary Approach

*Jörn Steuding,* Diophantine Analysis

*Douglas R. Stinson,* Cryptography: Theory and Practice, Second Edition

*Roberto Togneri and Christopher J. deSilva,* Fundamentals of Information Theory and Coding Design

*Lawrence C. Washington,* Elliptic Curves: Number Theory and Cryptography

DISCRETE MATHEMATICS AND ITS APPLICATIONS
Series Editor KENNETH H. ROSEN

# HANDBOOK OF ELLIPTIC AND HYPERELLIPTIC CURVE CRYPTOGRAPHY

HENRI COHEN

GERHARD FREY

ROBERTO AVANZI, CHRISTOPHE DOCHE, TANJA LANGE,
KIM NGUYEN, AND FREDERIK VERCAUTEREN

Chapman & Hall/CRC
Taylor & Francis Group

Boca Raton   London   New York   Singapore

Published in 2006 by
Chapman & Hall/CRC
Taylor & Francis Group
6000 Broken Sound Parkway NW, Suite 300
Boca Raton, FL 33487-2742

© 2006 by Taylor & Francis Group, LLC
Chapman & Hall/CRC is an imprint of Taylor & Francis Group

No claim to original U.S. Government works
Printed in the United States of America on acid-free paper
10 9 8 7 6 5 4 3 2

International Standard Book Number-10: 1-58488-518-1 (Hardcover)
International Standard Book Number-13: 978-1-58488-518-4 (Hardcover)
Library of Congress Card Number 2005041841

This book contains information obtained from authentic and highly regarded sources. Reprinted material is quoted with permission, and sources are indicated. A wide variety of references are listed. Reasonable efforts have been made to publish reliable data and information, but the author and the publisher cannot assume responsibility for the validity of all materials or for the consequences of their use.

No part of this book may be reprinted, reproduced, transmitted, or utilized in any form by any electronic, mechanical, or other means, now known or hereafter invented, including photocopying, microfilming, and recording, or in any information storage or retrieval system, without written permission from the publishers.

For permission to photocopy or use material electronically from this work, please access www.copyright.com (http://www.copyright.com/) or contact the Copyright Clearance Center, Inc. (CCC) 222 Rosewood Drive, Danvers, MA 01923, 978-750-8400. CCC is a not-for-profit organization that provides licenses and registration for a variety of users. For organizations that have been granted a photocopy license by the CCC, a separate system of payment has been arranged.

**Trademark Notice:** Product or corporate names may be trademarks or registered trademarks, and are used only for identification and explanation without intent to infringe.

---

**Library of Congress Cataloging-in-Publication Data**

Handbook of elliptic and hyperelliptic curve cryptography / Scientific editors, Henri Cohen & Gerard Frey ; authors, Roberto M Avanzi ... [et. al.].
    p. cm. – (Discrete mathematics and its applications)
Includes bibliographical references and index.
ISBN 1-58488-518-1 (alk. paper)
1.Curves, Elliptic – Handbooks, manuals, etc. 2. Cryptography – mathematics -- handbooks, manuals, etc. 3. Machine theory – Handbooks, manuals etc. I. Cohen, Henri. II. Frey, Gerhard, 1994- III. Avanzi, Roberto M. IV. Series.

QA567.2.E44H36 2005
516.'52 – dc22
                                                2005041841

---

Taylor & Francis Group
is the Academic Division of T&F Informa plc.

Visit the Taylor & Francis Web site at
http://www.taylorandfrancis.com

and the CRC Press Web site at
http://www.crcpress.com

Dr. Henri Cohen is Professor of Mathematics at the University of Bordeaux, France. His research interests are number theory in general, and computational number theory in particular.

Dr. Gerhard Frey holds a chair for number theory at the Institute for Experimental Mathematics at the University of Duisburg-Essen, Germany. His research interests are number theory and arithmetical geometry as well as applications to coding theory and cryptography.

Dr. Christophe Doche is lecturer at Macquarie University, Sydney, Australia. His research is focused on analytic and algorithmic number theory as well as cryptography.

Dr. Roberto M. Avanzi is currently Junior Professor at the Ruhr-University, Bochum. His research interests include arithmetic and algorithmic aspects of curve-based cryptography, integer recodings and addition chains, side-channel analysis, and diophantine analysis.

Dr. Tanja Lange is Associate Professor of Mathematics at the Technical University of Denmark in Copenhagen. Her research covers mathematical aspects of public-key cryptography and computational number theory with focus on curve cryptography.

Dr. Kim Nguyen received a Ph.D. in number theory and cryptography in 2001 at the University of Essen. His first position outside academia was with the Cryptology Competence Center of Philips Semiconductors Hamburg. He currently works for the Bundesdruckerei GmbH in Berlin, Germany.

Dr. Frederik Vercauteren is a Post-Doc at the Katholieke Universiteit Leuven, Belgium. His research interests are computational algebraic geometry and number theory, with applications to cryptography.

**Scientific Editors:** *Henri Cohen and Gerhard Frey*

**Executive Editor:** *Christophe Doche*

**Authors:** *Roberto M. Avanzi, Henri Cohen, Christophe Doche, Gerhard Frey, Tanja Lange, Kim Nguyen, and Frederik Vercauteren*

**Contributors:** *Bertrand Byramjee, Jean-Christophe Courrège, Sylvain Duquesne, Benoît Feix, Reynald Lercier, David Lubicz, Nicolas Thériault, and Andrew Weigl*

Roberto M. Avanzi
Faculty of Mathematics and
Horst Görtz Institute for IT-Security
Ruhr University Bochum, Germany
Roberto.Avanzi@ruhr-uni-bochum.de

Henri Cohen
Université Bordeaux I
Laboratoire A2X, France
Henri.Cohen@math.u-bordeaux1.fr

Christophe Doche
Macquarie University
Department of Computing, Australia
doche@ics.mq.edu.au

Benoît Feix
CEACI, Toulouse, France
Benoit.Feix@cnes.fr

Tanja Lange
Technical University of Denmark
Department of Mathematics
t.lange@mat.dtu.dk

David Lubicz
Centre d'ÉLectronique de l'ARmement
France
david.lubicz@math.univ-rennes1.fr

Nicolas Thériault
University of Waterloo,
Department of Combinatorics
and Optimization, Canada
ntheriau@uwaterloo.ca

Andrew Weigl
University of Bremen
ITEM, Germany
a.s.weigl@ieee.org

Bertrand Byramjee
bbyramjee@libertysurf.fr

Jean-Christophe Courrège
CEACI, Toulouse, France
Jean-Christophe.Courrege@cnes.fr

Sylvain Duquesne
Université Montpellier II
Laboratoire I3M, France
duquesne@math.univ-montp2.fr

Gerhard Frey
University of Duisburg-Essen
IEM, Germany
frey@iem.uni-due.de

Reynald Lercier
Centre d'ÉLectronique de l'ARmement
France
Reynald.Lercier@m4x.org

Kim Nguyen
nguyen.kim@web.de

Frederik Vercauteren
Katholieke Universiteit Leuven
COSIC - Electrical Engineering
Belgium
fvercaut@esat.kuleuven.be

# Table of Contents

**List of Algorithms** . . . . . . . . . . . . . . xxiii

**Preface** . . . . . . . . . . . . . . . xxix

**1 Introduction to Public-Key Cryptography** . . . . . . . . . 1
   1.1 Cryptography . . . . . . . . . . . . . . 2
   1.2 Complexity . . . . . . . . . . . . . . 2
   1.3 Public-key cryptography . . . . . . . . . . . 5
   1.4 Factorization and primality . . . . . . . . . . 6
      1.4.1 Primality . . . . . . . . . . . . . 6
      1.4.2 Complexity of factoring . . . . . . . . . 6
      1.4.3 RSA . . . . . . . . . . . . . . 7
   1.5 Discrete logarithm systems . . . . . . . . . . 8
      1.5.1 Generic discrete logarithm systems . . . . . . 8
      1.5.2 Discrete logarithm systems with bilinear structure . . 9
   1.6 Protocols . . . . . . . . . . . . . . 9
      1.6.1 Diffie–Hellman key exchange . . . . . . . 10
      1.6.2 Asymmetric Diffie–Hellman and ElGamal encryption . 10
      1.6.3 Signature scheme of ElGamal-type . . . . . . 12
      1.6.4 Tripartite key exchange . . . . . . . . . 13
   1.7 Other problems . . . . . . . . . . . . . 14

## I Mathematical Background

**2 Algebraic Background** . . . . . . . . . . . . 19
   2.1 Elementary algebraic structures . . . . . . . . . 19
      2.1.1 Groups . . . . . . . . . . . . . 19
      2.1.2 Rings . . . . . . . . . . . . . 21
      2.1.3 Fields . . . . . . . . . . . . . 23
      2.1.4 Vector spaces . . . . . . . . . . . 24
   2.2 Introduction to number theory . . . . . . . . . 24
      2.2.1 Extension of fields . . . . . . . . . . 25
      2.2.2 Algebraic closure . . . . . . . . . . 27
      2.2.3 Galois theory . . . . . . . . . . . 27
      2.2.4 Number fields . . . . . . . . . . . 29
   2.3 Finite fields . . . . . . . . . . . . . . 31
      2.3.1 First properties . . . . . . . . . . . 31
      2.3.2 Algebraic extensions of a finite field . . . . . . 32
      2.3.3 Finite field representations . . . . . . . . 33
      2.3.4 Finite field characters . . . . . . . . . 35

# 3 Background on $p$-adic Numbers . . . . . . . . . . . . 39
- 3.1 Definition of $\mathbb{Q}_p$ and first properties . . . . . . . . . . 39
- 3.2 Complete discrete valuation rings and fields . . . . . . . 41
  - 3.2.1 First properties . . . . . . . . . . . . . . 41
  - 3.2.2 Lifting a solution of a polynomial equation . . . . . 42
- 3.3 The field $\mathbb{Q}_p$ and its extensions . . . . . . . . . . . 43
  - 3.3.1 Unramified extensions . . . . . . . . . . . . 43
  - 3.3.2 Totally ramified extensions . . . . . . . . . . 43
  - 3.3.3 Multiplicative system of representatives . . . . . . 44
  - 3.3.4 Witt vectors . . . . . . . . . . . . . . . 44

# 4 Background on Curves and Jacobians . . . . . . . . . 45
- 4.1 Algebraic varieties . . . . . . . . . . . . . . . 45
  - 4.1.1 Affine and projective varieties . . . . . . . . . 46
- 4.2 Function fields . . . . . . . . . . . . . . . . 51
  - 4.2.1 Morphisms of affine varieties . . . . . . . . . 52
  - 4.2.2 Rational maps of affine varieties . . . . . . . . 53
  - 4.2.3 Regular functions . . . . . . . . . . . . . 54
  - 4.2.4 Generalization to projective varieties . . . . . . . 55
- 4.3 Abelian varieties . . . . . . . . . . . . . . . 55
  - 4.3.1 Algebraic groups . . . . . . . . . . . . . 55
  - 4.3.2 Birational group laws . . . . . . . . . . . . 56
  - 4.3.3 Homomorphisms of abelian varieties . . . . . . . 57
  - 4.3.4 Isomorphisms and isogenies . . . . . . . . . 58
  - 4.3.5 Points of finite order and Tate modules . . . . . . 60
  - 4.3.6 Background on $\ell$-adic representations . . . . . . . 61
  - 4.3.7 Complex multiplication . . . . . . . . . . . 63
- 4.4 Arithmetic of curves . . . . . . . . . . . . . . 64
  - 4.4.1 Local rings and smoothness . . . . . . . . . 64
  - 4.4.2 Genus and Riemann–Roch theorem . . . . . . . 66
  - 4.4.3 Divisor class group . . . . . . . . . . . . 76
  - 4.4.4 The Jacobian variety of curves . . . . . . . . . 77
  - 4.4.5 Jacobian variety of elliptic curves and group law . . . 79
  - 4.4.6 Ideal class group . . . . . . . . . . . . . 81
  - 4.4.7 Class groups of hyperelliptic curves . . . . . . . 83

# 5 Varieties over Special Fields . . . . . . . . . . . . 87
- 5.1 Varieties over the field of complex numbers . . . . . . . 87
  - 5.1.1 Analytic varieties . . . . . . . . . . . . . 87
  - 5.1.2 Curves over $\mathbb{C}$ . . . . . . . . . . . . . . 89
  - 5.1.3 Complex tori and abelian varieties . . . . . . . . 92
  - 5.1.4 Isogenies of abelian varieties over $\mathbb{C}$ . . . . . . . 94
  - 5.1.5 Elliptic curves over $\mathbb{C}$ . . . . . . . . . . . 95
  - 5.1.6 Hyperelliptic curves over $\mathbb{C}$ . . . . . . . . . 100
- 5.2 Varieties over finite fields . . . . . . . . . . . . 108
  - 5.2.1 The Frobenius morphism . . . . . . . . . . 109
  - 5.2.2 The characteristic polynomial of the Frobenius endomorphism . 109
  - 5.2.3 The theorem of Hasse–Weil for Jacobians . . . . . . 110
  - 5.2.4 Tate's isogeny theorem . . . . . . . . . . . 112

## 6 Background on Pairings . . . . . . . . . . . . . . 115
6.1 General duality results . . . . . . . . . . . . 115
6.2 The Tate pairing . . . . . . . . . . . . . . . 116
6.3 Pairings over local fields . . . . . . . . . . . . 117
    6.3.1 The local Tate pairing . . . . . . . . . . 118
    6.3.2 The Lichtenbaum pairing on Jacobian varieties . . . . 119
6.4 An explicit pairing . . . . . . . . . . . . . . 122
    6.4.1 The Tate–Lichtenbaum pairing . . . . . . . . 122
    6.4.2 Size of the embedding degree . . . . . . . . . 123

## 7 Background on Weil Descent . . . . . . . . . . . . 125
7.1 Affine Weil descent . . . . . . . . . . . . . . 125
7.2 The projective Weil descent . . . . . . . . . . . . 127
7.3 Descent by Galois theory . . . . . . . . . . . . 128
7.4 Zariski closed subsets inside of the Weil descent . . . . . 129
    7.4.1 Hyperplane sections . . . . . . . . . . . 129
    7.4.2 Trace zero varieties . . . . . . . . . . . 130
    7.4.3 Covers of curves . . . . . . . . . . . . 131
    7.4.4 The GHS approach . . . . . . . . . . . . 131

## 8 Cohomological Background on Point Counting . . . . . . 133
8.1 General principle . . . . . . . . . . . . . . 133
    8.1.1 Zeta function and the Weil conjectures . . . . . . 134
    8.1.2 Cohomology and Lefschetz fixed point formula . . . 135
8.2 Overview of $\ell$-adic methods . . . . . . . . . . . 137
8.3 Overview of $p$-adic methods . . . . . . . . . . . 138
    8.3.1 Serre–Tate canonical lift . . . . . . . . . . 138
    8.3.2 Monsky–Washnitzer cohomology . . . . . . . . 139

# II Elementary Arithmetic

## 9 Exponentiation . . . . . . . . . . . . . . . . . 145
9.1 Generic methods . . . . . . . . . . . . . . . 146
    9.1.1 Binary methods . . . . . . . . . . . . 146
    9.1.2 Left-to-right $2^k$-ary algorithm . . . . . . . . 148
    9.1.3 Sliding window method . . . . . . . . . . 149
    9.1.4 Signed-digit recoding . . . . . . . . . . . 150
    9.1.5 Multi-exponentiation . . . . . . . . . . . 154
9.2 Fixed exponent . . . . . . . . . . . . . . . 157
    9.2.1 Introduction to addition chains . . . . . . . . 157
    9.2.2 Short addition chains search . . . . . . . . . 160
    9.2.3 Exponentiation using addition chains . . . . . . 163
9.3 Fixed base point . . . . . . . . . . . . . . . 164
    9.3.1 Yao's method . . . . . . . . . . . . . 165
    9.3.2 Euclidean method . . . . . . . . . . . . 166
    9.3.3 Fixed-base comb method . . . . . . . . . . 166

## 10 Integer Arithmetic . . . . . . . . . . . . . . **169**

- 10.1 Multiprecision integers. . . . . . . . . . . . . 170
  - 10.1.1 Introduction . . . . . . . . . . . . . 170
  - 10.1.2 Internal representation . . . . . . . . . . . 171
  - 10.1.3 Elementary operations. . . . . . . . . . . . 172
- 10.2 Addition and subtraction . . . . . . . . . . . . 172
- 10.3 Multiplication . . . . . . . . . . . . . . . 174
  - 10.3.1 Schoolbook multiplication . . . . . . . . . . 174
  - 10.3.2 Karatsuba multiplication . . . . . . . . . . 176
  - 10.3.3 Squaring . . . . . . . . . . . . . . 177
- 10.4 Modular reduction . . . . . . . . . . . . . . 178
  - 10.4.1 Barrett method. . . . . . . . . . . . . 178
  - 10.4.2 Montgomery reduction. . . . . . . . . . . 180
  - 10.4.3 Special moduli. . . . . . . . . . . . . 182
  - 10.4.4 Reduction modulo several primes . . . . . . . . 184
- 10.5 Division . . . . . . . . . . . . . . . . 184
  - 10.5.1 Schoolbook division . . . . . . . . . . . 185
  - 10.5.2 Recursive division . . . . . . . . . . . . 187
  - 10.5.3 Exact division . . . . . . . . . . . . . 189
- 10.6 Greatest common divisor . . . . . . . . . . . . 190
  - 10.6.1 Euclid extended gcd . . . . . . . . . . . 191
  - 10.6.2 Lehmer extended gcd . . . . . . . . . . . 192
  - 10.6.3 Binary extended gcd . . . . . . . . . . . 194
  - 10.6.4 Chinese remainder theorem . . . . . . . . . 196
- 10.7 Square root . . . . . . . . . . . . . . . 197
  - 10.7.1 Integer square root . . . . . . . . . . . . 197
  - 10.7.2 Perfect square detection . . . . . . . . . . 198

## 11 Finite Field Arithmetic . . . . . . . . . . . . . . **201**

- 11.1 Prime fields of odd characteristic. . . . . . . . . . . 201
  - 11.1.1 Representations and reductions . . . . . . . . 202
  - 11.1.2 Multiplication . . . . . . . . . . . . . 202
  - 11.1.3 Inversion and division . . . . . . . . . . . 205
  - 11.1.4 Exponentiation. . . . . . . . . . . . . 209
  - 11.1.5 Squares and square roots . . . . . . . . . . 210
- 11.2 Finite fields of characteristic $2$ . . . . . . . . . . . 213
  - 11.2.1 Representation . . . . . . . . . . . . . 213
  - 11.2.2 Multiplication . . . . . . . . . . . . . 218
  - 11.2.3 Squaring . . . . . . . . . . . . . . 221
  - 11.2.4 Inversion and division . . . . . . . . . . . 222
  - 11.2.5 Exponentiation . . . . . . . . . . . . . 225
  - 11.2.6 Square roots and quadratic equations . . . . . . . 228
- 11.3 Optimal extension fields . . . . . . . . . . . . 229
  - 11.3.1 Introduction . . . . . . . . . . . . . 229
  - 11.3.2 Multiplication . . . . . . . . . . . . . 231
  - 11.3.3 Exponentiation. . . . . . . . . . . . . 231
  - 11.3.4 Inversion . . . . . . . . . . . . . . 233
  - 11.3.5 Squares and square roots . . . . . . . . . . 234
  - 11.3.6 Specific improvements for degrees $3$ and $5$ . . . . . . 235

## 12 Arithmetic of $p$-adic Numbers . . . . . . . . . . . . . . . 239
  12.1 Representation . . . . . . . . . . . . . . . . 239
    12.1.1 Introduction . . . . . . . . . . . . . . . 239
    12.1.2 Computing the Teichmüller modulus . . . . . . . . 240
  12.2 Modular arithmetic . . . . . . . . . . . . . . . 244
    12.2.1 Modular multiplication . . . . . . . . . . . . 244
    12.2.2 Fast division with remainder . . . . . . . . . . 244
  12.3 Newton lifting . . . . . . . . . . . . . . . . 246
    12.3.1 Inverse . . . . . . . . . . . . . . . . 247
    12.3.2 Inverse square root . . . . . . . . . . . . . 248
    12.3.3 Square root . . . . . . . . . . . . . . . 249
  12.4 Hensel lifting . . . . . . . . . . . . . . . . 249
  12.5 Frobenius substitution . . . . . . . . . . . . . . 250
    12.5.1 Sparse modulus . . . . . . . . . . . . . . 251
    12.5.2 Teichmüller modulus . . . . . . . . . . . . 252
    12.5.3 Gaussian normal basis . . . . . . . . . . . . 252
  12.6 Artin–Schreier equations . . . . . . . . . . . . . 252
    12.6.1 Lercier–Lubicz algorithm . . . . . . . . . . . 253
    12.6.2 Harley's algorithm . . . . . . . . . . . . . 254
  12.7 Generalized Newton lifting . . . . . . . . . . . . 256
  12.8 Applications . . . . . . . . . . . . . . . . . 257
    12.8.1 Teichmüller lift . . . . . . . . . . . . . . 257
    12.8.2 Logarithm . . . . . . . . . . . . . . . 258
    12.8.3 Exponential . . . . . . . . . . . . . . . 259
    12.8.4 Trace . . . . . . . . . . . . . . . . 260
    12.8.5 Norm . . . . . . . . . . . . . . . . 261

# III Arithmetic of Curves

## 13 Arithmetic of Elliptic Curves . . . . . . . . . . . . . . 267
  13.1 Summary of background on elliptic curves . . . . . . . 268
    13.1.1 First properties and group law . . . . . . . . . 268
    13.1.2 Scalar multiplication . . . . . . . . . . . . 271
    13.1.3 Rational points . . . . . . . . . . . . . . 272
    13.1.4 Torsion points . . . . . . . . . . . . . . 273
    13.1.5 Isomorphisms . . . . . . . . . . . . . . 273
    13.1.6 Isogenies . . . . . . . . . . . . . . . 277
    13.1.7 Endomorphisms . . . . . . . . . . . . . 277
    13.1.8 Cardinality . . . . . . . . . . . . . . . 278
  13.2 Arithmetic of elliptic curves defined over $\mathbb{F}_p$ . . . . . . . 280
    13.2.1 Choice of the coordinates . . . . . . . . . . . 280
    13.2.2 Mixed coordinates . . . . . . . . . . . . . 283
    13.2.3 Montgomery scalar multiplication . . . . . . . . 285
    13.2.4 Parallel implementations . . . . . . . . . . . 288
    13.2.5 Compression of points . . . . . . . . . . . 288
  13.3 Arithmetic of elliptic curves defined over $\mathbb{F}_{2^d}$ . . . . . . 289
    13.3.1 Choice of the coordinates . . . . . . . . . . . 291
    13.3.2 Faster doublings in affine coordinates . . . . . . . 295

|  |  |  |
|---|---|---|
| 13.3.3 | Mixed coordinates | 296 |
| 13.3.4 | Montgomery scalar multiplication | 298 |
| 13.3.5 | Point halving and applications | 299 |
| 13.3.6 | Parallel implementation | 302 |
| 13.3.7 | Compression of points | 302 |

## 14 Arithmetic of Hyperelliptic Curves — 303
- 14.1 Summary of background on hyperelliptic curves — 304
  - 14.1.1 Group law for hyperelliptic curves — 304
  - 14.1.2 Divisor class group and ideal class group — 306
  - 14.1.3 Isomorphisms and isogenies — 308
  - 14.1.4 Torsion elements — 309
  - 14.1.5 Endomorphisms — 310
  - 14.1.6 Cardinality — 310
- 14.2 Compression techniques — 311
  - 14.2.1 Compression in odd characteristic — 311
  - 14.2.2 Compression in even characteristic — 313
- 14.3 Arithmetic on genus $2$ curves over arbitrary characteristic — 313
  - 14.3.1 Different cases — 314
  - 14.3.2 Addition and doubling in affine coordinates — 316
- 14.4 Arithmetic on genus $2$ curves in odd characteristic — 320
  - 14.4.1 Projective coordinates — 321
  - 14.4.2 New coordinates in odd characteristic — 323
  - 14.4.3 Different sets of coordinates in odd characteristic — 325
  - 14.4.4 Montgomery arithmetic for genus $2$ curves in odd characteristic — 328
- 14.5 Arithmetic on genus $2$ curves in even characteristic — 334
  - 14.5.1 Classification of genus $2$ curves in even characteristic — 334
  - 14.5.2 Explicit formulas in even characteristic in affine coordinates — 336
  - 14.5.3 Inversion-free systems for even characteristic when $h_2 \neq 0$ — 341
  - 14.5.4 Projective coordinates — 341
  - 14.5.5 Inversion-free systems for even characteristic when $h_2 = 0$ — 345
- 14.6 Arithmetic on genus $3$ curves — 348
  - 14.6.1 Addition in most common case — 348
  - 14.6.2 Doubling in most common case — 349
  - 14.6.3 Doubling on genus $3$ curves for even characteristic when $h(x) = 1$ — 351
- 14.7 Other curves and comparison — 352

## 15 Arithmetic of Special Curves — 355
- 15.1 Koblitz curves — 355
  - 15.1.1 Elliptic binary Koblitz curves — 356
  - 15.1.2 Generalized Koblitz curves — 367
  - 15.1.3 Alternative setup — 375
- 15.2 Scalar multiplication using endomorphisms — 376
  - 15.2.1 GLV method — 377
  - 15.2.2 Generalizations — 380
  - 15.2.3 Combination of GLV and Koblitz curve strategies — 381
  - 15.2.4 Curves with endomorphisms for identity-based parameters — 382
- 15.3 Trace zero varieties — 383
  - 15.3.1 Background on trace zero varieties — 384
  - 15.3.2 Arithmetic in $G$ — 385

## 16 Implementation of Pairings . . . . . . . . . . . . . . 389
### 16.1 The basic algorithm. . . . . . . . . . . . . . . 389
#### 16.1.1 The setting . . . . . . . . . . . . . . . 390
#### 16.1.2 Preparation . . . . . . . . . . . . . . . 391
#### 16.1.3 The pairing computation algorithm . . . . . . . . . . 391
#### 16.1.4 The case of nontrivial embedding degree $k$ . . . . . . . 393
#### 16.1.5 Comparison with the Weil pairing . . . . . . . . . 395
### 16.2 Elliptic curves . . . . . . . . . . . . . . . . . 396
#### 16.2.1 The basic step . . . . . . . . . . . . . . . 396
#### 16.2.2 The representation . . . . . . . . . . . . . 396
#### 16.2.3 The pairing algorithm . . . . . . . . . . . . 397
#### 16.2.4 Example . . . . . . . . . . . . . . . . 397
### 16.3 Hyperelliptic curves of genus $2$ . . . . . . . . . . . . 398
#### 16.3.1 The basic step . . . . . . . . . . . . . . 399
#### 16.3.2 Representation for $k > 2$ . . . . . . . . . . . . 399
### 16.4 Improving the pairing algorithm . . . . . . . . . . . 400
#### 16.4.1 Elimination of divisions . . . . . . . . . . . . 400
#### 16.4.2 Choice of the representation . . . . . . . . . . . 400
#### 16.4.3 Precomputations . . . . . . . . . . . . . . 400
### 16.5 Specific improvements for elliptic curves . . . . . . . . . 400
#### 16.5.1 Systems of coordinates . . . . . . . . . . . . 401
#### 16.5.2 Subfield computations . . . . . . . . . . . . . 401
#### 16.5.3 Even embedding degree . . . . . . . . . . . . 402
#### 16.5.4 Example . . . . . . . . . . . . . . . . 403

# IV Point Counting

## 17 Point Counting on Elliptic and Hyperelliptic Curves . . . . . . 407
### 17.1 Elementary methods . . . . . . . . . . . . . . . 407
#### 17.1.1 Enumeration . . . . . . . . . . . . . . . 407
#### 17.1.2 Subfield curves . . . . . . . . . . . . . . 409
#### 17.1.3 Square root algorithms . . . . . . . . . . . . 410
#### 17.1.4 Cartier–Manin operator . . . . . . . . . . . . 411
### 17.2 Overview of $\ell$-adic methods . . . . . . . . . . . . 413
#### 17.2.1 Schoof's algorithm . . . . . . . . . . . . . 413
#### 17.2.2 Schoof–Elkies–Atkin's algorithm . . . . . . . . . 414
#### 17.2.3 Modular polynomials . . . . . . . . . . . . 416
#### 17.2.4 Computing separable isogenies in finite fields of large characteristic . 419
#### 17.2.5 Complete SEA algorithm . . . . . . . . . . . . 421
### 17.3 Overview of $p$-adic methods . . . . . . . . . . . . 422
#### 17.3.1 Satoh's algorithm . . . . . . . . . . . . . 423
#### 17.3.2 Arithmetic–Geometric–Mean algorithm . . . . . . . 434
#### 17.3.3 Kedlaya's algorithm . . . . . . . . . . . . . 449

**18 Complex Multiplication** . . . . . . . . . . . . . **455**
    18.1 CM for elliptic curves . . . . . . . . . . . . 456
        18.1.1 Summary of background . . . . . . . . . . 456
        18.1.2 Outline of the algorithm . . . . . . . . . . 456
        18.1.3 Computation of class polynomials . . . . . . . . 457
        18.1.4 Computation of norms. . . . . . . . . . . 458
        18.1.5 The algorithm . . . . . . . . . . . . . 459
        18.1.6 Experimental results . . . . . . . . . . . 459
    18.2 CM for curves of genus $2$ . . . . . . . . . . . . 460
        18.2.1 Summary of background . . . . . . . . . . 462
        18.2.2 Outline of the algorithm . . . . . . . . . . 462
        18.2.3 CM-types and period matrices . . . . . . . . 463
        18.2.4 Computation of the class polynomials . . . . . . 465
        18.2.5 Finding a curve . . . . . . . . . . . . 467
        18.2.6 The algorithm . . . . . . . . . . . . . 469
    18.3 CM for larger genera . . . . . . . . . . . . 470
        18.3.1 Strategy and difficulties in the general case . . . . 470
        18.3.2 Hyperelliptic curves with automorphisms . . . . . 471
        18.3.3 The case of genus $3$ . . . . . . . . . . . 472

# V Computation of Discrete Logarithms

**19 Generic Algorithms for Computing Discrete Logarithms** . . . **477**
    19.1 Introduction . . . . . . . . . . . . . . . 478
    19.2 Brute force . . . . . . . . . . . . . . . . 479
    19.3 Chinese remaindering . . . . . . . . . . . . 479
    19.4 Baby-step giant-step . . . . . . . . . . . . . 480
        19.4.1 Adaptive giant-step width . . . . . . . . . 481
        19.4.2 Search in intervals and parallelization . . . . . . 482
        19.4.3 Congruence classes . . . . . . . . . . . 483
    19.5 Pollard's rho method . . . . . . . . . . . . 483
        19.5.1 Cycle detection . . . . . . . . . . . . 484
        19.5.2 Application to DL . . . . . . . . . . . 488
        19.5.3 More on random walks. . . . . . . . . . . 489
        19.5.4 Parallelization . . . . . . . . . . . . . 489
        19.5.5 Automorphisms of the group . . . . . . . . 490
    19.6 Pollard's kangaroo method . . . . . . . . . . . 491
        19.6.1 The lambda method . . . . . . . . . . . 492
        19.6.2 Parallelization . . . . . . . . . . . . . 493
        19.6.3 Automorphisms of the group . . . . . . . . 494

**20 Index Calculus** . . . . . . . . . . . . . . . **495**
    20.1 Introduction. . . . . . . . . . . . . . . . 495
    20.2 Arithmetical formations . . . . . . . . . . . . 496
        20.2.1 Examples of formations . . . . . . . . . . 497
    20.3 The algorithm . . . . . . . . . . . . . . . 498
        20.3.1 On the relation search . . . . . . . . . . . 499
        20.3.2 Parallelization of the relation search . . . . . . . 500

|     |       | 20.3.3 On the linear algebra | 500 |
|---|---|---|---|
|     |       | 20.3.4 Filtering | 503 |
|     |       | 20.3.5 Automorphisms of the group | 505 |
|     | 20.4  | An important example: finite fields | 506 |
|     | 20.5  | Large primes | 507 |
|     |       | 20.5.1 One large prime | 507 |
|     |       | 20.5.2 Two large primes | 508 |
|     |       | 20.5.3 More large primes | 509 |
| **21** | **Index Calculus for Hyperelliptic Curves** | | **511** |
|     | 21.1  | General algorithm | 511 |
|     |       | 21.1.1 Hyperelliptic involution | 512 |
|     |       | 21.1.2 Adleman–DeMarrais–Huang | 512 |
|     |       | 21.1.3 Enge–Gaudry | 516 |
|     | 21.2  | Curves of small genus | 516 |
|     |       | 21.2.1 Gaudry's algorithm | 517 |
|     |       | 21.2.2 Refined factor base | 517 |
|     |       | 21.2.3 Harvesting | 518 |
|     | 21.3  | Large prime methods | 519 |
|     |       | 21.3.1 Single large prime | 520 |
|     |       | 21.3.2 Double large primes | 521 |
| **22** | **Transfer of Discrete Logarithms** | | **529** |
|     | 22.1  | Transfer of discrete logarithms to $\mathbb{F}_q$-vector spaces | 529 |
|     | 22.2  | Transfer of discrete logarithms by pairings | 530 |
|     | 22.3  | Transfer of discrete logarithms by Weil descent | 530 |
|     |       | 22.3.1 Summary of background | 531 |
|     |       | 22.3.2 The GHS algorithm | 531 |
|     |       | 22.3.3 Odd characteristic | 536 |
|     |       | 22.3.4 Transfer via covers | 538 |
|     |       | 22.3.5 Index calculus method via hyperplane sections | 541 |

# VI Applications

| **23** | **Algebraic Realizations of DL Systems** | | **547** |
|---|---|---|---|
|     | 23.1  | Candidates for secure DL systems | 547 |
|     |       | 23.1.1 Groups with numeration and the DLP | 548 |
|     |       | 23.1.2 Ideal class groups and divisor class groups | 548 |
|     |       | 23.1.3 Examples: elliptic and hyperelliptic curves | 551 |
|     |       | 23.1.4 Conclusion | 553 |
|     | 23.2  | Security of systems based on $\mathrm{Pic}_C^0$ | 554 |
|     |       | 23.2.1 Security under index calculus attacks | 554 |
|     |       | 23.2.2 Transfers by Galois theory | 555 |
|     | 23.3  | Efficient systems | 557 |
|     |       | 23.3.1 Choice of the finite field | 558 |
|     |       | 23.3.2 Choice of genus and curve equation | 560 |
|     |       | 23.3.3 Special choices of curves and scalar multiplication | 563 |
|     | 23.4  | Construction of systems | 564 |

   23.4.1 Heuristics of class group orders . . . . . . . . 564
   23.4.2 Finding groups of suitable size . . . . . . . . 565
  23.5 Protocols . . . . . . . . . . . . . . 569
   23.5.1 System parameters . . . . . . . . . . 569
   23.5.2 Protocols on $\mathrm{Pic}_C^0$ . . . . . . . . . . . 570
  23.6 Summary . . . . . . . . . . . . . 571

## 24 Pairing-Based Cryptography   573
  24.1 Protocols . . . . . . . . . . . . . . 573
   24.1.1 Multiparty key exchange . . . . . . . . . 574
   24.1.2 Identity-based cryptography . . . . . . . . 576
   24.1.3 Short signatures . . . . . . . . . . . 578
  24.2 Realization . . . . . . . . . . . . . 579
   24.2.1 Supersingular elliptic curves . . . . . . . . 580
   24.2.2 Supersingular hyperelliptic curves . . . . . . . 584
   24.2.3 Ordinary curves with small embedding degree . . . . 586
   24.2.4 Performance . . . . . . . . . . . . 589
   24.2.5 Hash functions on the Jacobian . . . . . . . 590

## 25 Compositeness and Primality Testing – Factoring   591
  25.1 Compositeness tests . . . . . . . . . . . . 592
   25.1.1 Trial division . . . . . . . . . . . . 592
   25.1.2 Fermat tests . . . . . . . . . . . . 593
   25.1.3 Rabin–Miller test . . . . . . . . . . . 594
   25.1.4 Lucas pseudoprime tests . . . . . . . . . 595
   25.1.5 BPSW tests . . . . . . . . . . . . 596
  25.2 Primality tests . . . . . . . . . . . . . 596
   25.2.1 Introduction . . . . . . . . . . . . 596
   25.2.2 Atkin–Morain ECPP test . . . . . . . . . 597
   25.2.3 APRCL Jacobi sum test . . . . . . . . . 599
   25.2.4 Theoretical considerations and the AKS test . . . . 600
  25.3 Factoring . . . . . . . . . . . . . . 601
   25.3.1 Pollard's rho method . . . . . . . . . . 601
   25.3.2 Pollard's $p - 1$ method . . . . . . . . . . 603
   25.3.3 Factoring with elliptic curves . . . . . . . . 604
   25.3.4 Fermat–Morrison–Brillhart approach . . . . . . 607

# *VII Realization of Discrete Logarithm Systems*

## 26 Fast Arithmetic in Hardware .   617
  26.1 Design of cryptographic coprocessors . . . . . . . . 618
   26.1.1 Design criterions . . . . . . . . . . . 618
  26.2 Complement representations of signed numbers . . . . . . 620
  26.3 The operation $XY + Z$ . . . . . . . . . . . 622
   26.3.1 Multiplication using left shifts . . . . . . . . 623
   26.3.2 Multiplication using right shifts . . . . . . . 624
  26.4 Reducing the number of partial products . . . . . . . 625
   26.4.1 Booth or signed digit encoding . . . . . . . 625

  26.4.2 Advanced recoding techniques . . . . . . . 626
 26.5 Accumulation of partial products . . . . . . . . 627
  26.5.1 Full adders . . . . . . . . . . . 627
  26.5.2 Faster carry propagation . . . . . . . . 628
  26.5.3 Analysis of carry propagation . . . . . . . 631
  26.5.4 Multi-operand operations . . . . . . . . 633
 26.6 Modular reduction in hardware . . . . . . . . 638
 26.7 Finite fields of characteristic $2$ . . . . . . . . . 641
  26.7.1 Polynomial basis . . . . . . . . . . 642
  26.7.2 Normal basis . . . . . . . . . . . 643
 26.8 Unified multipliers . . . . . . . . . . . 644
 26.9 Modular inversion in hardware. . . . . . . . . 645

## 27 Smart Cards . . . . . . . . . . . . 647
 27.1 History. . . . . . . . . . . . . . 647
 27.2 Smart card properties . . . . . . . . . . 648
  27.2.1 Physical properties . . . . . . . . . 648
  27.2.2 Electrical properties . . . . . . . . . 650
  27.2.3 Memory . . . . . . . . . . . . 651
  27.2.4 Environment and software . . . . . . . . 656
 27.3 Smart card interfaces . . . . . . . . . . 659
  27.3.1 Transmission protocols . . . . . . . . . 659
  27.3.2 Physical interfaces . . . . . . . . . . 663
 27.4 Types of smart cards . . . . . . . . . . 664
  27.4.1 Memory only cards (synchronous cards) . . . . . 664
  27.4.2 Microprocessor cards (asynchronous cards) . . . . 665

## 28 Practical Attacks on Smart Cards . . . . . . . 669
 28.1 Introduction. . . . . . . . . . . . . 669
 28.2 Invasive attacks . . . . . . . . . . . . 670
  28.2.1 Gaining access to the chip . . . . . . . . 670
  28.2.2 Reconstitution of the layers . . . . . . . . 670
  28.2.3 Reading the memories . . . . . . . . . 671
  28.2.4 Probing . . . . . . . . . . . . 671
  28.2.5 FIB and test engineers scheme flaws . . . . . . 672
 28.3 Non-invasive attacks . . . . . . . . . . 673
  28.3.1 Timing attacks . . . . . . . . . . . 673
  28.3.2 Power consumption analysis . . . . . . . . 675
  28.3.3 Electromagnetic radiation attacks . . . . . . . 682
  28.3.4 Differential fault analysis (DFA) and fault injection attacks . . 683

## 29 Mathematical Countermeasures against Side-Channel Attacks . . 687
 29.1 Countermeasures against simple SCA . . . . . . . 688
  29.1.1 Dummy arithmetic instructions . . . . . . . 689
  29.1.2 Unified addition formulas . . . . . . . . 694
  29.1.3 Montgomery arithmetic . . . . . . . . . 696
 29.2 Countermeasures against differential SCA . . . . . . 697
  29.2.1 Implementation of DSCA . . . . . . . . 698
  29.2.2 Scalar randomization . . . . . . . . . 699
  29.2.3 Randomization of group elements . . . . . . . 700

|        | 29.2.4 Randomization of the curve equation . . . . . . 700 |
|--------|---|
| 29.3   | Countermeasures against Goubin type attacks . . . . 703 |
| 29.4   | Countermeasures against higher order differential SCA . . . . 704 |
| 29.5   | Countermeasures against timing attacks. . . . . . . 705 |
| 29.6   | Countermeasures against fault attacks . . . . . . 705 |
|        | 29.6.1 Countermeasures against simple fault analysis . . . . 706 |
|        | 29.6.2 Countermeasures against differential fault analysis . . . 706 |
|        | 29.6.3 Conclusion on fault induction . . . . . . 708 |
| 29.7   | Countermeasures for special curves . . . . . . . 709 |
|        | 29.7.1 Countermeasures against SSCA on Koblitz curves . . . 709 |
|        | 29.7.2 Countermeasures against DSCA on Koblitz curves . . . 711 |
|        | 29.7.3 Countermeasures for GLV curves . . . . . . 713 |

## 30 Random Numbers – Generation and Testing . . . . . 715

- 30.1 Definition of a random sequence. . . . . . . . 715
- 30.2 Random number generators . . . . . . . . 717
  - 30.2.1 History . . . . . . . . . . 717
  - 30.2.2 Properties of random number generators . . . . 718
  - 30.2.3 Types of random number generators . . . . . 718
  - 30.2.4 Popular random number generators . . . . . 720
- 30.3 Testing of random number generators . . . . . . 722
- 30.4 Testing a device . . . . . . . . . . 722
- 30.5 Statistical (empirical) tests . . . . . . . . 723
- 30.6 Some examples of statistical models on $\Sigma^n$ . . . . . 725
- 30.7 Hypothesis testings and random sequences . . . . . 726
- 30.8 Empirical test examples for binary sequences. . . . . 727
  - 30.8.1 Random walk . . . . . . . . . 727
  - 30.8.2 Runs . . . . . . . . . . . 728
  - 30.8.3 Autocorrelation . . . . . . . . 728
- 30.9 Pseudorandom number generators . . . . . . . 729
  - 30.9.1 Relevant measures . . . . . . . . 730
  - 30.9.2 Pseudorandom number generators from curves . . . 732
  - 30.9.3 Other applications . . . . . . . . 735

**References** . . . . . . . . . . . . . 737

**Notation Index** . . . . . . . . . . . . 777

**General Index** . . . . . . . . . . . . 785

# List of Algorithms

| | | |
|---|---|---|
| 1.15 | Key generation. | 11 |
| 1.16 | Asymmetric Diffie–Hellman encryption. | 11 |
| 1.17 | ElGamal encryption. | 11 |
| 1.18 | ElGamal signature | 12 |
| 1.20 | Signature verification | 13 |
| 1.21 | Three party key exchange | 14 |
| 9.1 | Square and multiply method. | 146 |
| 9.2 | Right-to-left binary exponentiation | 146 |
| 9.5 | Montgomery's ladder | 148 |
| 9.7 | Left-to-right $2^k$-ary exponentiation | 148 |
| 9.10 | Sliding window exponentiation | 150 |
| 9.14 | NAF representation. | 151 |
| 9.17 | Koyama–Tsuruoka signed-binary recoding | 152 |
| 9.20 | $NAF_w$ representation | 153 |
| 9.23 | Multi-exponentiation using Straus' trick | 155 |
| 9.27 | Joint sparse form recoding | 156 |
| 9.41 | Exponentiation using addition chain | 164 |
| 9.43 | Multi-exponentiation using vectorial addition chain | 164 |
| 9.44 | Improved Yao's exponentiation | 165 |
| 9.47 | Euclidean exponentiation. | 166 |
| 9.49 | Fixed-base comb exponentiation. | 167 |
| 10.3 | Addition of nonnegative multiprecision integers | 173 |
| 10.5 | Subtraction of nonnegative multiprecision integers. | 173 |
| 10.8 | Multiplication of positive multiprecision integers | 174 |
| 10.11 | Karatsuba multiplication of positive multiprecision integers | 176 |
| 10.14 | Squaring of a positive multiprecision integer. | 177 |
| 10.17 | Division-free modulo of positive multiprecision integers | 179 |
| 10.18 | Reciprocation of positive multiprecision integers. | 179 |
| 10.22 | Montgomery reduction REDC of multiprecision integers | 181 |
| 10.25 | Fast reduction for special form moduli. | 183 |
| 10.28 | Short division of positive multiprecision integers | 185 |
| 10.30 | Schoolbook division of positive multiprecision integers. | 185 |
| 10.35 | Recursive division of positive multiprecision integers | 188 |
| 10.39 | Exact division of positive multiprecision integers in base $b = 2^\ell$. | 189 |
| 10.42 | Euclid extended gcd of positive integers | 191 |
| 10.45 | Lehmer extended gcd of multiprecision positive integers | 192 |
| 10.46 | Partial gcd of positive multiprecision integers in base $b = 2^\ell$ | 193 |
| 10.49 | Extended binary gcd of positive integers | 195 |
| 10.52 | Chinese remainder computation | 196 |

| | | |
|---|---|---:|
| 10.55 | Integer square root | 198 |
| 11.1 | Interleaved multiplication-reduction of multiprecision integers | 203 |
| 11.3 | Multiplication in Montgomery representation | 204 |
| 11.6 | Plus-minus inversion method | 206 |
| 11.9 | Prime field inversion. | 207 |
| 11.12 | Montgomery inverse in Montgomery representation | 208 |
| 11.15 | Simultaneous inversion modulo $p$ | 209 |
| 11.17 | Binary exponentiation using Montgomery representation | 210 |
| 11.19 | Legendre symbol | 210 |
| 11.23 | Tonelli and Shanks square root computation | 212 |
| 11.26 | Square root computation. | 213 |
| 11.31 | Division by a sparse polynomial | 214 |
| 11.34 | Multiplication of polynomials in $\mathbb{F}_2[X]$ | 218 |
| 11.37 | Multiplication of polynomials in $\mathbb{F}_2[X]$ using window technique | 219 |
| 11.41 | Euclid extended polynomial gcd | 222 |
| 11.44 | Inverse of an element of $\mathbb{F}_{2^d}^*$ in polynomial representation | 223 |
| 11.47 | Inverse of an element of $\mathbb{F}_{q^d}^*$ using Lagrange's theorem | 224 |
| 11.50 | Modular composition of Brent and Kung | 226 |
| 11.53 | Shoup exponentiation algorithm | 227 |
| 11.66 | OEF inversion | 234 |
| 11.69 | Legendre–Kronecker–Jacobi symbol | 235 |
| 12.2 | Teichmüller modulus | 242 |
| 12.3 | Teichmüller modulus increment | 242 |
| 12.5 | Polynomial inversion | 245 |
| 12.6 | Fast division with remainder. | 245 |
| 12.9 | Newton iteration | 246 |
| 12.10 | Inverse. | 247 |
| 12.12 | Inverse square root ($p = 2$) | 248 |
| 12.15 | Hensel lift iteration | 249 |
| 12.16 | Hensel lift. | 250 |
| 12.18 | Artin–Schreier root square multiply | 253 |
| 12.19 | Artin–Schreier root I. | 254 |
| 12.21 | Artin–Schreier root II | 255 |
| 12.23 | Generalized Newton lift | 257 |
| 12.24 | Teichmüller lift | 258 |
| 12.33 | Norm I | 261 |
| 12.35 | Norm II. | 262 |
| 13.6 | Sliding window scalar multiplication on elliptic curves | 271 |
| 13.35 | Scalar multiplication using Montgomery's ladder | 287 |
| 13.42 | Repeated doublings | 295 |
| 13.45 | Point halving | 300 |
| 13.48 | Halve and add scalar multiplication | 301 |
| 14.7 | Cantor's algorithm | 308 |
| 14.19 | Addition ($g = 2$ and $\deg u_1 = \deg u_2 = 2$) | 317 |
| 14.20 | Addition ($g = 2$, $\deg u_1 = 1$, and $\deg u_2 = 2$) | 318 |
| 14.21 | Doubling ($g = 2$ and $\deg u = 2$) | 319 |

| | | |
|---|---|---|
| 14.22 | Addition in projective coordinates ($g=2$ and $q$ odd) | 321 |
| 14.23 | Doubling in projective coordinates ($g=2$ and $q$ odd) | 322 |
| 14.25 | Addition in new coordinates ($g=2$ and $q$ odd) | 324 |
| 14.26 | Doubling in new coordinates ($g=2$ and $q$ odd) | 325 |
| 14.30 | Montgomery scalar multiplication for genus $2$ curves | 331 |
| 14.41 | Doubling on Type Ia curves ($g=2$ and $q$ even) | 337 |
| 14.42 | Doubling on Type Ib curves ($g=2$ and $q$ even) | 338 |
| 14.43 | Doubling on Type II curves ($g=2$ and $q$ even) | 339 |
| 14.44 | Doubling on Type III curves ($g=2$ and $q$ even) | 340 |
| 14.45 | Doubling in projective coordinates ($g=2$, $h_2 \neq 0$, and $q$ even) | 341 |
| 14.47 | Addition in new coordinates ($g=2$, $h_2 \neq 0$, and $q$ even) | 342 |
| 14.48 | Doubling in new coordinates ($g=2$, $h_2 \neq 0$, and $q$ even) | 343 |
| 14.49 | Doubling in projective coordinates ($g=2$, $h_2 = 0$, and $q$ even) | 346 |
| 14.50 | Addition in recent coordinates ($g=2$, $h_2 = 0$, and $q$ even) | 346 |
| 14.51 | Doubling in recent coordinates ($g=2$, $h_2 = 0$, and $q$ even) | 347 |
| 14.52 | Addition on curves of genus $3$ in the general case | 348 |
| 14.53 | Doubling on curves of genus $3$ in the general case | 350 |
| 14.54 | Doubling on curves of genus $3$ with $h(x)=1$ | 351 |
| 15.6 | $\tau$NAF representation | 359 |
| 15.9 | Rounding-off of an element of $\mathbb{Q}[\tau]$ | 361 |
| 15.11 | Division with remainder in $\mathbb{Z}[\tau]$ | 361 |
| 15.13 | Reduction of $n$ modulo $\delta$ | 362 |
| 15.17 | $\tau$NAF$_w$ representation | 364 |
| 15.21 | Recoding in $\tau$-adic joint sparse form | 365 |
| 15.28 | Expansion in $\tau$-adic form | 370 |
| 15.34 | Representation of $\delta = (\tau^k - 1)/(\tau - 1)$ in $\mathbb{Z}[\tau]$ | 372 |
| 15.35 | Computation of $n$-folds using $\tau$-adic expansions | 373 |
| 15.41 | GLV representation | 379 |
| 16.5 | Relative prime representation | 391 |
| 16.8 | Tate–Lichtenbaum pairing | 392 |
| 16.11 | Tate–Lichtenbaum pairing if $k > g$ | 394 |
| 16.12 | Tate–Lichtenbaum pairing for $g=1$ if $k>1$ | 397 |
| 16.16 | Tate–Lichtenbaum pairing for $g=1$ if $k>1$ and $k=k_1 e$ | 402 |
| 16.17 | Tate–Lichtenbaum pairing for $g=1$ and $k=2e$ | 402 |
| 17.25 | SEA | 421 |
| 17.31 | Lift $j$-invariants | 425 |
| 17.36 | Lift kernel | 428 |
| 17.38 | Satoh's point counting method | 430 |
| 17.54 | Elliptic curve AGM | 439 |
| 17.58 | Univariate elliptic curve AGM | 441 |
| 17.71 | Hyperelliptic curve AGM | 446 |
| 17.80 | Kedlaya's point counting method for $p \geqslant 3$ | 452 |
| 18.4 | Cornacchia's algorithm | 458 |
| 18.5 | Construction of elliptic curves via CM | 459 |
| 18.12 | Construction of genus $2$ curves via CM | 469 |

| | | |
|---|---|---|
| 19.5 | Shanks' baby-step giant-step algorithm | 480 |
| 19.7 | BJT variant of the baby-step giant-step algorithm | 481 |
| 19.9 | Terr's variant of the baby-step giant-step algorithm | 482 |
| 19.10 | Baby-step giant-step algorithm in an interval | 482 |
| 19.12 | Floyd's cycle-finding algorithm | 484 |
| 19.13 | Gosper's cycle finding algorithm | 485 |
| 19.14 | Brent's cycle-finding algorithm | 485 |
| 19.16 | Brent's improved cycle-finding algorithm | 486 |
| 19.17 | Nivash' cycle-finding algorithm | 488 |
| 20.3 | Index calculus | 498 |
| 21.1 | Divisor decomposition | 511 |
| 21.5 | Computing the factor base | 513 |
| 21.6 | Smoothing a prime divisor | 513 |
| 21.7 | Finding smooth principal divisors | 514 |
| 21.8 | Decomposition into unramified primes | 514 |
| 21.9 | Adleman–DeMarrais–Huang | 515 |
| 21.12 | Relation search | 517 |
| 21.16 | Single large prime | 520 |
| 21.18 | Double large primes | 521 |
| 21.19 | Full graph method | 522 |
| 21.21 | Simplified graph method | 524 |
| 21.24 | Concentric circles method | 525 |
| 22.7 | Weil descent of $C$ via GHS | 535 |
| 23.2 | Cantor's algorithm | 552 |
| 23.12 | Elliptic curve AGM | 566 |
| 23.13 | Construction of elliptic curves via CM | 567 |
| 23.14 | HECDSA – Signature generation | 570 |
| 23.15 | HECDSA – Signature verification | 571 |
| 24.2 | Tripartite key exchange | 575 |
| 24.3 | Identity-based encryption | 577 |
| 24.4 | Private-key extraction | 577 |
| 24.5 | Identity-based decryption | 577 |
| 24.6 | Signature in short signature scheme | 578 |
| 24.7 | Signature verification in short signature scheme | 578 |
| 24.10 | Triple and add scalar multiplication | 581 |
| 24.12 | Tate–Lichtenbaum pairing in characteristic 3 | 582 |
| 24.15 | Construction of elliptic curves with prescribed embedding degree | 587 |
| 25.7 | Pollard's rho with Brent's cycle detection | 603 |
| 25.12 | Fermat–Morrison–Brillhart factoring algorithm | 608 |
| 25.14 | Sieve method | 609 |
| 25.15 | The number field sieve | 613 |
| 26.9 | Sequential multiplication of two integers using left shift | 623 |
| 26.11 | Sequential multiplication of two integers using right shift | 624 |

| | | |
|---|---|---|
| 26.14 | Recoding the multiplier in signed-digit representation | 625 |
| 26.17 | Recoding the multiplier in radix-4 SD representation | 626 |
| 26.22 | Addition using full adders | 628 |
| 26.49 | Least-significant-digit-first shift-and-add multiplication in $\mathbb{F}_{2^d}$ | 643 |
| 28.1 | Modular reduction | 674 |
| 28.2 | Modular reduction against timing attacks | 674 |
| 28.3 | Double and add always | 676 |
| 29.1 | Atomic addition-doubling formulas | 691 |
| 29.2 | Elliptic curve doubling in atomic blocks | 692 |
| 29.3 | Elliptic curve addition in atomic blocks | 693 |
| 29.4 | Montgomery's ladder | 697 |
| 29.6 | Right to left binary exponentiation | 707 |
| 29.7 | SSCA and DSCA by random assignment | 710 |
| 29.8 | Randomized GLV method | 714 |
| 30.4 | A pseudorandom number generator | 719 |

# *Preface*

Information security is of the greatest importance in a world in which communication over open networks and storage of data in digital form play a key role in daily life. The science of cryptography provides efficient tools to secure information. In [MEOO⁺ 1996] one finds an excellent definition of cryptography as "*the study of mathematical techniques related to aspects of information security such as confidentiality, data integrity, entity identification, and data origin authentication*" as well as a thorough description of these information security objectives and how cryptography addresses these goals.

Historically, cryptography has mainly dealt with methods to transmit information in a confidential manner such that a third party (called the *adversary*) cannot read the information, even if the transmission is done through an insecure channel such as a public telephone line. This should not be confused with *coding theory*, the goal of which is, on the contrary, to add some redundant information to a message so that even if it is slightly garbled, it can still be decoded correctly.

To achieve secure transmission, one can use the oldest, and by far the fastest, type of cryptography, *secret-key* cryptography, also called *symmetric-key* cryptography. This is essentially based on the sharing of a *secret key* between the people who want to communicate. This secret key is used both in the encryption process, in which the ciphertext is computed from the cleartext and the key, and also in the decryption process. This is why the method is called symmetric. The current standardized method of this type is the AES symmetric-key cryptosystem. Almost all methods of this type are based on bit manipulations between the bits of the message (after the message has been translated into binary digits) and the bits of the secret key. Decryption is simply done by reversing these bit manipulations. All these operations are thus very fast. The main disadvantages of symmetric-key cryptography are that one *shared* secret key per pair must be exchanged beforehand in a secure way, and that key management is more tricky in a large network.

At the end of the 1970's the revolutionary new notion of *public-key* cryptography, also called *asymmetric* cryptography, appeared. It emerged from the the pioneering work of Diffie and Hellman published 1976 in the paper "New directions in cryptography" [DIHE 1976].

Public-key cryptography is based on the idea of *one-way functions*: in rough terms these are functions, whose inverse functions cannot be computed in any reasonable amount of time. If we use such a function for encryption, an adversary will in principle not be able to decrypt the encrypted messages. This will be the case even if the function is public knowledge. In fact, having the encryption public has many advantages such as enabling protocols for authentication, signature, etc., which are typical of public-key cryptosystems; see Chapter 1.

All known methods for public-key cryptography are rather slow, at least compared to secret-key cryptography, and the situation will probably always stay that way. Thus public-key cryptography is used as a complement to secret-key cryptography, either for signatures or authentication, or for key exchange, since the messages to be transmitted in all these cases are quite short.

There remains the major problem of finding suitable public-key cryptosystems. Many proposals have been made in the 25 years since the invention of the concept, but we can reasonably say that only two types of methods have survived. The first and by far the most widely used methods are variants of the RSA cryptosystem. These are based on the asymmetrical fact that it is very easy to create at will, quite large prime numbers (for instance 768 or 1024-bit is a typical cryptographic size for reasonable security), but it is totally impossible (at the time this book is being written), except by some incredible stroke of luck (or bad choice of the primes) to factor the product of two primes of this size, in other words a 1536 or 2048-bit number. The present world record for factoring an RSA type number is the factoring of the 576-bit RSA challenge integer. The remarkable fact is that this result was obtained *without* an enormous effort of thousands of PC's interconnected via the Internet. Using today's algorithms and computer technologies it seems possible to factor 700-bit numbers. Of course, this is much smaller than the number of bits that we mention. But we will see in Chapter 25 that there are many *subexponential* algorithms for factoring which, although not as efficient as polynomial time algorithms, are still much faster than naïve factoring approaches. The existence of these subexponential algorithms explains the necessity of using 768 or 1024-bit primes as mentioned above, in other words very large keys.

The second type of method is based on the discrete logarithm problem (DLP) in cyclic groups of prime order that are embedded in elliptic curves or more generally Jacobians of curves (or even general abelian varieties) over finite fields. In short, if $G$ is a group, $g \in G$, $k \in \mathbb{Z}$ and $h = g^k$, the DLP in $G$ consists in computing $k$ knowing $g$ and $h$. In the case where $G = (\mathbb{Z}/n\mathbb{Z})^*$ it can be shown that the subexponential methods used for factoring can be adapted to give subexponential methods for the DLP in $G$, so the security of such methods is analogous to the security of RSA, and in particular one needs very large keys. On the other hand for elliptic curves no subexponential algorithm is known for the DLP, and this is also the case for Jacobians of curves of small genus. In other words the only attacks known for the DLP working on all elliptic curves are generic (see Chapter 19). This is bad from an algorithmic point of view, but is of course very good news for the cryptographer, since it means that she can use much smaller keys than in cryptosystems such as RSA for which there exist subexponential attacks. Typically, to have the same security as 2048-bit RSA one estimates that an elliptic curve (or the Jacobian of a curve of small genus) over a finite field should have a number of points equal to a small multiple of a prime of 224 or perhaps 256 bits. Even though the basic operations on elliptic curves and on Jacobians are more complicated than in $(\mathbb{Z}/n\mathbb{Z})^*$, the small key size largely compensates for this, especially in restricted environments such as smart cards (cf. Chapter 27) where silicon space is small.

## Aim of the book

The goal of this book is to explain in great detail the theory and algorithms involved in elliptic and hyperelliptic curve cryptography. The reader is strongly advised to read carefully what follows before reading the rest of the book, otherwise she may be discouraged by some parts.

The intended audience is broad: our book is targeted both at students and at professionals, whether they have a mathematical, computer science, or engineering background. It is *not* a textbook, and in particular contains very few proofs. On the other hand it is reasonably self-contained, in that essentially all of the mathematical background is explained quite precisely. This book contains many algorithms, of which some appear for the first time in book form. They have been written in such a way that they can be immediately implemented by anyone wanting to go as fast as possible to the bottom line, without bothering about the detailed understanding of the algorithm or its mathematical background, when there is one. This is why this book is a *handbook*. On the other hand, it is *not* a cookbook: we have not been content with giving the algorithms, but for the more mathematically minded ones we have given in great detail all the necessary definitions, theorems, and justifications for the understanding of the algorithm.

Thus this book can be read at several levels, and the reader can make her choice depending on her interests and background.

The reader may be primarily interested in the mathematical parts, and how some quite abstract mathematical notions are transformed into very practical algorithms. For instance, particular mention should be made of point counting algorithms, where apparently quite abstract cohomology theories and classical results from explicit class field theory are used for efficient implementations to count points on hyperelliptic curves. A most striking example is the use of $p$-adic methods. Kedlaya's algorithm is a very practical implementation of the Monsky–Washnitzer cohomology, and Satoh's algorithm together with its successors are very practical uses of the notion of canonical $p$-adic liftings. Another example is the Tate duality of abelian varieties which provides the most efficient realization of a bilinear structure opening new possibilities for public-key protocols. Thus the reader may read this book to learn about the mathematics involved in elliptic and hyperelliptic curve cryptography.

On the other hand the reader may be primarily interested in having the algorithms implemented as fast as possible. In that case, she can usually implement the algorithms directly as they are written, even though some of them are quite complex. She even can use the book to look for appropriate solutions for a concrete problem in data security, find optimal instances corresponding to the computational environment and then delve deeper into the background.

To achieve these aims we present almost all topics at different levels which are linked by numerous references but which can be read independently (with the exception that one should at least have some idea about the principles of public-key cryptography as explained in Chapter 1 and that one knows the basic algebraic structures described in Chapter 2).

## Mathematical background

The *first level* is the mathematical background concerning the needed tools from algebraic geometry and arithmetic. This constitutes the first part of the book. We define the elementary algebraic structures and the basic facts on number theory including finite fields and $p$-adic numbers in Chapters 2 and 3. The basic results about curves and the necessary concepts from algebraic geometry are given in Chapter 4. In Chapter 5 we consider the special cases in which the ground field is a finite field or equal to the complex numbers. In Chapters 6, 7, and 8 we explain the importance of Galois theory and especially of the Frobenius automorphism for the arithmetic of curves over finite fields and develop the background for pairings, Weil descent, and point counting.

On the one hand a mathematically experienced reader will find many topics well-known to her. On the other hand some chapters in this part may not be easy to grasp for a reader not having a sufficient mathematical background. In both cases we encourage the reader to skip (parts of) Part I on first reading and we hope that she will come back to it after being motivated by applications and implementations.

This skipping is possible since in later parts dealing with implementations we always repeat the crucial notions and results in a summary at the beginning of the chapters.

## Algorithms and their implementation

In Part II of the book, we treat exponentiation techniques in Chapter 9. Chapters 10, 11, and 12 present in a very concrete way the arithmetic in the ring of integers, in finite fields and in $p$-adic fields. The algorithms developed in these chapters are amongst the key ingredients if one wants to have fast algorithms on algebraic objects like polynomial rings and so their careful use is necessary for implementations of algorithms developed in the next part.

In Part III, we give in great detail the algorithms that are necessary for addition in groups consisting of points of elliptic curves and of Jacobian varieties of hyperelliptic curves. Both generic and special cases are treated, and the advantages and disadvantages of different coordinate systems are discussed. In Chapter 13 this is done for elliptic curves, and in Chapter 14 one finds the results concerning hyperelliptic curves. Chapter 15 is devoted to elliptic and hyperelliptic curves, which have an extra structure like fast computable endomorphisms. The reader who plans to implement a system based on discrete logarithms should read these chapters very carefully. The tables counting the necessary field operations will be of special interest for her, and in conjunction with the arithmetic of finite fields she will be able to choose an optimal system adapted to the given or planned environment. (At this place we already mention that security issues must not be neglected, and as a rule, special structures will allow more attacks. So Part V or the conclusions in Chapter 23 should be consulted.) Since all necessary definitions and results from the mathematical background are restated in Part III it can be read *independently* of the chapters before. But of course we hope that designers of systems will become curious to learn more about the foundations.

The same remarks apply to the next Chapter, 16. It is necessary if one wants to use protocols based on bilinear structures like tripartite-key exchange or identity-based cryptography. It begins with a down-to-earth definition of a variant of the Tate pairing under the conditions that make it practicable nowadays and then describe algorithms that become very simple in concrete situations. Again it should be interesting not only for mathematicians which general structures are fundamental and so a glimpse at Chapter 6 is recommended.

In Part IV of the book, it is more difficult to separate the background from the implementation. Nevertheless we get as results concrete and effective algorithms to count the number of points on hyperelliptic curves and their Jacobian varieties over finite fields. We present a complete version of the Schoof–Elkies–Atkin algorithm, which counts the number of rational points on random *elliptic* curves in Section 17.2 in the most efficient version known today. The $p$-adic methods for *elliptic and hyperelliptic* curves over fields with small characteristics are given in Section 17.3 and the method using complex multiplication is described in Chapter 18 for the relevant cases of curves of genus 1, 2, and 3. In the end of this part, one finds algorithms in such a detailed manner that it should be possible to implement point counting without understanding all mathematical details, but some experience with computational number theory is necessary to get efficient algorithms, and for instance in Chapter 18 it is advised that for some precomputations like determining class polynomials a published list should be used. For readers who do not want to go into these details a way out could be to use standard curves instead of generating their own. For mathematicians interested in computational aspects this part of the book should provide a very interesting lecture after they have read the mathematical background part. All readers should be convinced by Part IV that there are efficient algorithms that provide in abundance instances for discrete logarithm systems usable for public-key cryptography.

Until now the constructive point of view was in the center. Now we have to discuss security issues and investigate how hard it is to compute discrete logarithms. In Part V we discuss these methods in detail. The most important and fastest algorithms rely on the index calculus method which is in abstract form presented in Chapter 20. In Chapter 21 it is implemented in the most efficient way known nowadays for hyperelliptic curves. Contrary to the algorithms considered in the constructive part it has a high complexity. But the computation is feasible even nowadays in wide ranges and its subexponential complexity has as a consequence that it will become applicable to much larger instances in the near future — if one does not avoid curves of genus $\geqslant 4$ and if one is not careful about the choice of parameters for curves of genus 3. In Chapter 22 we describe how results from the mathematical background like Weil descent or pairings can result in algorithms that transfer the discrete logarithm from seemingly secure instances to the ones endangered by index calculus methods. Though the topics of Part V are not as easily accessible as the results and algorithms in Part III, every designer of cryptosystems based on the discrete logarithm problem should have a look at least at the type of algorithms used and the complexity obtained by index calculus methods, and to make this task easier we have given a digest of all the results obtained until now in the *third* level of the book.

## Applications

In Part VI of the book we reach the third level and discuss how to find cryptographic primitives that can be used in appropriate protocols such that the desired level of security and efficiency is obtained. In Chapter 23 this is done for DL systems, and Chapter 24 deals with protocols using bilinear structures. Again these chapters are self-contained but necessarily in the style of a digest with many references to earlier chapters. One finds the mathematical nature of the groups used as cryptographic primitives, how to compute inside of them, a security analysis, and how to obtain efficient implementations. Moreover, it is explained how to transfer the abstract protocols, for instance, signature schemes, from Chapter 1 to real protocols using elliptic and hyperelliptic curves in a most efficient and secure way. The summary at the end of Chapter 23 is a scenario as to how this book could be used as an aid in developing a system. At the same time it could give one (a bit extreme) way how to read the book. Begin with Section 23.6 and take it as hints for reading the previous sections of this chapter. If an algorithm is found to be interesting or important go to the corresponding chapters and sections cited there; proceed to chapters in Parts II, III, IV and V but not necessarily in this order and not necessarily in an exhaustive way. And then, just for fun and better understanding, read the background chapters on which the implementations rely.

If the reader is interested in applications involving bilinear structures she can use the same recipe as above but beginning in Chapter 24.

A not so direct but important application of elliptic curves and (at a minor level) hyperelliptic curves in cryptography is their use in algorithms for primality testing and factoring. Because of the importance of these algorithms for public-key cryptography we present the state of the art information on these topics in Chapter 25.

## Realization of Discrete Logarithm systems

Until now our topics were totally inside of mathematics with a strong emphasis on computational aspects and some hints coming from the need of protocols. But on several occasions we have mentioned already the importance of the computational environment as basis for optimal choices of DL systems and their parameters. To understand the conditions and restrictions enforced by the physical realization of the system one has to understand its architecture and the properties of the used hardware. In Part VII, mathematics is at backstage and in the foreground are methods used to implement discrete logarithm systems based on elliptic and hyperelliptic curves in *hardware* and especially in restricted environments such as *smart cards*.

In Chapter 26 it is shown how the arithmetic over finite fields is realized in hardware. In Chapter 27 one finds a detailed description of a restricted environment in which DL systems will have their most important applications, namely smart cards. The physical realization of the systems opens new lines of attacks, so called side-channel attacks, not computing the discrete logarithm to break the system but relying, for instance, on the analysis of timings and power consumption during the computational processes. The background for such attacks is explained in Chapter 28 while Chapter 29 shows how mathematical methods can be used to develop countermeasures.

A most important ingredient in all variants of protocols used in public-key cryptography is randomization. So one has to have (pseudo-)random number generators of high quality at hand. This problem and possible solutions are discussed in Chapter 30. As a side-result it is shown that elliptic or hyperelliptic curves can be used with good success.

## Acknowledgments

We would like to thank the following people for their great help:

For Chapter 1, we thank David Lubicz. For Chapter 2, we thank Jean-Luc Stehlé. For Chapter 11, we thank Guerric Meurice de Dormale. For Chapter 14, we thank Jan Pelzl and Colin Stahlke. For Chapter 19, we thank Edlyn Teske. For Chapter 22, we thank Claus Diem and Jasper Scholten. For Chapter 25, we thank Scott Contini. For Chapter 26, we thank Johann Großschädl. For Chapter 27, we thank Benoît Feix, Christian Gorecki, and Ralf Jährig. For Chapter 28, we thank Vanessa Chazal, Mathieu Ciet, and Guillaume Poupard. For Chapter 30, we thank Peter Beelen, Igor Shparlinski, and Johann Großschädl.

Photos in Figure 28.2 are courtesy of Oliver Kömmerling and Markus Kuhn.

We would like to thank Alfred J. Menezes and Igor Shparlinski for their useful advice.

Roberto M. Avanzi expresses his gratitude to the Institute of Experimental Mathematics (IEM) at the University of Duisburg-Essen and the Communication Security Group (COSY) at the Ruhr-University Bochum for providing very simulating environments. The present book was written while he was affiliated with the IEM, then with COSY.

Christophe Doche would like to thank the A2X department at the University of Bordeaux for providing excellent working conditions. Some parts of this book were written while he was affiliated in Bordeaux. He is also grateful to Alejandra M. O'Donnell for careful proofreading.

Tanja Lange would like to thank the Faculty of Mathematics at the Ruhr-University Bochum for creating a good working atmosphere. This book was written while she was affiliated with Bochum.

Frederik Vercauteren would like to thank his former colleagues at the CS department of the University of Bristol for their encouragement and support. This book was written while he was affiliated with Bristol.

Andrew Weigl would like to thank Professor Walter Anheier and the Institute for Electromagnetic Theory and Microelectronics at the University of Bremen for their support on this project.

Jean-Christophe Courrège and Benoît Feix would like to thank people from the CEACI, the CNES and the DCSSI.

This book grew out of a report for the European Commission in the scope of the AREHCC (Advanced Research in Elliptic and Hyperelliptic Curves Cryptography) project. The work described in this book has been supported by the Commission of the European Communities through the IST Programme under Contract IST-2001-32613 (see http://www.arehcc.com). As the momentum around this project built up, some new people joined the core team as contributors.

The information in this document is provided as is, and no guarantee or warranty is given or implied that the information is fit for any particular purpose. The user thereof uses the information at its sole risk and liability. The views expressed are those of the authors and do not represent an official view or position of the AREHCC project (as a whole).

Henri Cohen and Gerhard Frey

# Chapter 1

# Introduction to Public-Key Cryptography

Roberto M. Avanzi and Tanja Lange

## Contents in Brief

| | | |
|---|---|---|
| 1.1 | **Cryptography** | 2 |
| 1.2 | **Complexity** | 2 |
| 1.3 | **Public-key cryptography** | 5 |
| 1.4 | **Factorization and primality** | 6 |
| | Primality • Complexity of factoring • RSA | |
| 1.5 | **Discrete logarithm systems** | 8 |
| | Generic discrete logarithm systems • Discrete logarithm systems with bilinear structure | |
| 1.6 | **Protocols** | 9 |
| | Diffie–Hellman key exchange • Asymmetric Diffie–Hellman and ElGamal encryption • Signature scheme of ElGamal-type • Tripartite key exchange | |
| 1.7 | **Other problems** | 14 |

In this chapter we introduce the basic building blocks for cryptography based on the *discrete logarithm problem* that will constitute the main motivation for considering the groups studied in this book. We also briefly introduce the *RSA cryptosystem* as for use in practice it is still an important public-key cryptosystem.

Assume a situation where two people, called Alice and Bob in the sequel (the names had been used since the beginning of cryptography because they allow using the letters A and B as handy abbreviations), want to communicate via an insecure channel in a secure manner. In other words, an eavesdropper Eve (abbreviated as E) listening to the encrypted conversation should not be able to read the cleartext or change it. To achieve these aims one uses *cryptographic primitives* based on a problem that should be easy to set up by either Alice, or Bob, or by both, but impossible to solve for Eve. Loosely speaking, infeasibility means computational infeasibility for Eve if she does not have at least partial access to the secret information exploited by Alice and Bob to set up the problem.

Examples of such primitives are RSA, cf. [PKCS], which could be solved if the *integer factorization problem* was easy, i.e., if one could find a nontrivial factor of a composite integer $n$, and the discrete logarithm problem, i.e., the problem of finding an integer $k$ with $[k]P = Q$ where $P$ is a generator of a cyclic group $(G, \oplus)$ and $Q \in G$. These primitives are reviewed in Sections 1.4.3

and 1.5. They are applied in a prescribed way given by protocols. We will only briefly state the necessary problems and hardness assumptions in Section 1.6 but not go into the details.

Then we go briefly into issues of primality proving and integer factorization. The next section is devoted to discrete logarithm systems. This is the category of cryptographic primitives in which elliptic and hyperelliptic curves are applied. Finally, we consider protocols, i.e., algorithms using the cryptographic primitive to establish a common key, encrypt a message for a receiver, or sign electronically.

## 1.1 Cryptography

In ancient times, the use of cryptography was restricted to a small community essentially formed by the military and the secret service. The keys were distributed secretly by a courier and the same key allowed to encipher and, later, decipher the messages. These *symmetric systems* include the ancient Caesar's cipher, Enigma, and other rotor machines. Today's standard symmetric cipher is the AES (Advanced Encryption Standard) [FIPS 197]. Symmetric systems still are by far the fastest means to communicate secretly – provided that a joint key is established.

In order to thrive, e-commerce requires the possibility of secure transactions on an electronically connected global network. Therefore, it is necessary to rely on mechanisms that allow a key exchange between two or more parties that have not met each other before. One of the main features of public-key cryptography is to relax the security requirement of the channel used to perform a key exchange: in the case of symmetric cryptography it must be protected in integrity and confidentiality though integrity suffices in public-key cryptography. It allows for building easier-to-set-up and more scalable secured networks. It also provides cryptographic services like signatures with non-repudiation, which are not available in symmetric cryptography. The security of public-key cryptography relies on the evaluation of the computational difficulty of some families of mathematical problems and their classification with respect to their complexity.

## 1.2 Complexity

The aim of complexity theory is to define formal models for the processors and algorithms that we use in our everyday computers and to provide a classification of the algorithms with respect to their memory or time consumption.

Surprisingly all the complex computations carried out with a computer can be simulated by an automaton given by a very simple mathematical structure called a Turing machine. A *Turing machine* is defined by a finite set of states: an initial state, a finite set of symbols, and a transition function. A Turing machine proceeds step-by-step following the rules given by the transition function and can write symbols on a memory string. It is then easy to define the execution time of an algorithm as the number of steps between its beginning and end and the memory consumption as the number of symbols written on the memory string. For convenience, in the course of this book we will use a slightly stronger *model of computation*, called a *Random Access Machine* because it is very close to the behavior of our everyday microprocessors. Then determining the execution time of an algorithm boils down to counting the number of basic operations on machine words needed for its execution. For more details, the reader should refer to [PAP 1994].

The security of protocols is often linked to the assumed hardness of some problems. In the theory of computation a problem is a set of finite-length questions (strings) with associated finite-length answers (strings). In our context the input will usually consist of mathematical objects like integers or group elements coded with a string. The problems can be loosely classified into problems to compute something, e.g., a further group element and problems that ask for a *yes or no* answer.

**Definition 1.1** A problem is called a *decision problem* if the problem is to *decide* whether a statement about an input is true or false.

A problem is called a *computation problem* if it asks to *compute* an output maybe more elaborate than true or false on a certain set of inputs.

One can formulate a computation problem from a decision problem. Many protocols base their security on a decision problem rather than on a computation problem.

**Example 1.2** The problem to compute the square root of 16 is a computation problem whereas the question, whether 4 is a square root of 16, is a decision problem. Here, the decision problem can be answered by just computing the square $4^2 = 16$ and comparing the answers.

A further decision problem in this context is also to answer whether 16 is a square. Clearly this decision problem can be answered by solving the above computation problem.

**Example 1.3** A second important example that we will discuss in the next section is the problem to decide whether a certain integer $m$ is a prime. This is related to the computation problem of finding the factorization of $m$.

Given a model of computation, one can attach a certain function $f$ to an algorithm that bounds a certain resource used for the computations given the length of the input called the *complexity parameter*. If the resource considered is the execution time (resp. the memory consumption) of the algorithm, $f$ measures its *time complexity* (resp. *space complexity*). In fact, in order to state the complexity independently from the specific processor used it is convenient to express the cost of an algorithm only "up to a constant factor." In other words, what is given is not the exact operation count as a function of the input size, but the growth rate of this count.

The Schoolbook multiplication of $n$-digit integers, for example, is an "$n^2$ algorithm." By this it is understood that, in order to multiply two $n$-digit integers, no more than $c\,n^2$ single-digit multiplications are necessary, for some real constant $c$ — but we are not interested in $c$. The "big-O" notation is one way of formalizing this "sloppiness," as [GAGE 1999] put it.

**Definition 1.4** Let $f$ and $g$ be two real functions of $s$ variables. The function $g(N_1, \ldots, N_s)$ is of order $f(N_1, \ldots, N_s)$ denoted by $O\big(f(N_1, \ldots, N_s)\big)$ if for a positive constant $c$ one has

$$|g(N_1, \ldots, N_s)| \leqslant c f(N_1, \ldots, N_s),$$

with $N_i > N$ for some constant $N$. Sometimes a finite set of values of the tuples $(N_1, \ldots, N_s)$ is excluded, for example those for which the functions $f$ and $g$ have no meaning or are not defined.

Additional to this *"big-O" notation* one needs the *"small-o"notation*.

The function $g(N_1, \ldots, N_s)$ is of order $o\big(f(N_1, \ldots, N_s)\big)$ if one has

$$\lim_{N_1,\ldots,N_s \to \infty} \frac{g(N_1, \ldots, N_s)}{f(N_1, \ldots, N_s)} = 0.$$

Finally we write $f(n) = \tilde{O}\big(g(n)\big)$ as a shorthand for $f(n) = O\big(g(n) \lg^k g(n)\big)$ for some $k$.

Note that we denote by lg the logarithm to base 2 and by ln the natural logarithm. As these expressions differ only by constants the big-$O$ expressions always contain the binary logarithm. In case other bases are needed we use $\log_a b$ to denote the logarithm of $b$ to base $a$. This must not be confused with the discrete logarithm introduced in Section 1.5 but the meaning should be clear from the context.

**Example 1.5** Consider $g(N) = 10N^2 + 30N + 5000$. It is of order $O(N^2)$ as for $c = 5040$ one has $g(N) \leqslant cN^2$ for all $N$. We may write $g(N) = O(N^2)$. In addition $g(N)$ is $o(N^3)$.

**Example 1.6** Consider the task of computing the $n$-fold of some integer $m$. Instead of computing $n \times m = m + m + \cdots + m$ ($n$-times) we can do much better reducing the complexity of scalar multiplication from $O(n)$ to $O(\lg n)$. We make the following observation: we have $4m = 2(2m)$ and a doubling takes about the same time as an addition of two distinct elements. Hence, the number of operations is reduced from 3 to 2. This idea can be extended to other scalars: $5m = 2(2m) + m$ needing 3 operations instead of 4. In more generality let $n = \sum_{i=0}^{l-1} n_i 2^i, n_i \in \{0, 1\}$ be the binary expansion of $n$ with $l - 1 = \lfloor \lg n \rfloor$. Then

$$n \times m = 2(2(\cdots 2(2(2m + n_{l-2}m) + n_{l-3}m) + \cdots + n_2 m) + n_1 m) + n_0 m.$$

This way of computing $n \times m$ needs $l - 1$ doublings and $\sum_{i=0}^{l-1} n_i \leqslant l$ additions. Hence, the algorithm has complexity $O(\lg n)$. Furthermore, we can bound the constant $c$ from above by 2. Algorithms achieving a smaller constant are treated in Chapter 9 together with a general study of scalar multiplication.

An algorithm has running time *exponential in* $N$ if its running time can be bounded from above and below by $e^{f(N)}$ and $e^{g(N)}$ for some polynomials $f, g$. In particular, its running time is of order $O(e^{f(N)})$. Its running time is *polynomial in* $N$ if it is of order $O(f(N))$. Algorithms belonging to the first category are computationally hard, those of the second are easy. Note that the involved constants can imply that for a certain chosen small $N$ an exponential-time algorithm may take less time than a polynomial-time one. However, the growth of $N$ to achieve a certain increase in the running time is smaller in the case of exponential running time.

**Definition 1.7** For the complexity of algorithms depending on $N$ we define the shorthand

$$L_N(\alpha, c) := \exp\big((c + o(1))(\ln N)^\alpha (\ln \ln N)^{1-\alpha}\big)$$

with $0 \leqslant \alpha \leqslant 1$ and $c > 0$. The $o(1)$ refers to the asymptotic behavior of $N$. If the second parameter is omitted, it is understood that it equals $1/2$.

The parameter $\alpha$ is the more important one. Depending on it, $L_N(\alpha, c)$ interpolates between polynomial complexity for $\alpha = 0$ and exponential complexity for $\alpha = 1$. For $\alpha < 1$ the complexity is said to be *subexponential*.

One might expect a cryptographic primitive to be efficient while at the same time difficult to break. This is why it is important to classify the hardness of a problem — and to find instances of hard problems. Note that for cryptographic purposes, we need problems that are *hard on average*, i.e., it should be rather easy to construct really hard instances of a given problem. (The classification of problems **P** and not **P** must therefore be considered with care, keeping in mind that a particular problem in **NP** can be easy to solve in most cases that can be constructed in practice, and there need to be only some hard instances for the problem itself to let it be in **NP**.)

In practice, we often measure the hardness of a problem by the complexity of the best known algorithm to solve it. The complexity of an algorithm solving a particular problem can only be an upper bound for the complexity of solving the problem itself, hence security is always only "to our best knowledge." For some problems it is possible to give also lower bounds showing that an algorithm needs at least a certain number of steps. It is not our purpose to give a detailed treatment of complexity here. The curious reader will find a broader and deeper discussion in [GRKN+ 1994, Chapter 9], [SHP 2003], and [BRBR 1996, Chapter 3].

## 1.3 Public-key cryptography

In public-key cryptography, each participant possesses two keys — a *public key* and a *private key*. These are linked in a unique manner by a *one-way function*.

**Definition 1.8** Let $\Sigma^*$ be the set of binary strings and $f$ be a function from $\Sigma^*$ to $\Sigma^*$. We say that $f$ is a one-way function if

- the function $f$ is one-on-one and for all $x \in \Sigma^*$, $f(x)$ is at most polynomially longer or shorter than $x$
- for all $x \in \Sigma^*$, $f(x)$ can be computed in polynomial time
- there is no polynomial-time algorithm which for all $y \in \Sigma^*$ returns either "no" if $y \notin \text{Im}(f)$ or $x \in \Sigma^*$ such that $y = f(x)$.

**Remark 1.9** So far, there is no proof of the existence of a one-way function. In fact, it is easy to see that it would imply that $\mathbf{P} \neq \mathbf{NP}$, which is a far reaching conjecture of complexity theory. This means in particular that the security of public-key cryptosystems always relies on the unproven hypothesis of the hardness of some computational problem. But in the course of this book, we are going to present some families of functions that are widely believed to be good candidates for being a one-way function and we now give an example of such a function.

**Example 1.10** Let $\Sigma = \{0, 1, \square\}$ be the alphabet with three letters. Note that even if a computer manipulates only bits of information, that is an alphabet with two letters, it is easy by grouping the bits by packets of $\lceil \lg |\Sigma| \rceil$ to simulate computations on $\Sigma$. The character $\square$ will serve as separator in our data structure. Let $s$ be the function that assigns to each integer $n \in \mathbb{N}$ its binary representation. It is then possible to define a function $f$ from $\Sigma^*$ to $\Sigma^*$ such that for all couples of prime numbers $(p, q)$ the function evaluates as $f\big(s(p) \,\|\, \square \,\|\, s(q)\big) = s(pq)$ with $\|$ the concatenation. As there exists a polynomial-time algorithm to multiply two integers, $f$ satisfies clearly the two first conditions of a one-way function. It is also widely believed that there is no polynomial-time algorithm to invert $f$ but up to now there is no proof of such an assumption. The next section gives some details on the complexity of the best algorithms known in order to invert this function.

Given a one-way function one can choose as the private key an $a \in \Sigma^*$ and obtains the public key $f(a)$. This value can be published since it is computationally infeasible to defer $a$ from it. Complexity theory considered in the previous section gives a mathematical measure to define what is meant by "computationally infeasible." For some applications it will be necessary to have a special class of computational one-way functions that can be inverted if one possesses additional information. These functions are called *trapdoor one-way functions*.

If an encrypted message is transmitted, the cryptographic framework has to ensure that no other party can obtain any information on the message or change it without being noticed. To map the paper and pencil based world to electronic processes, other issues have to be solved in order to allow electronic commerce and contracts. Just as a handwritten signature is bound to the signer by the uniqueness of his handwriting, the message as signature and text are linked by physical means, and any change of the message would be visible, an electronic signature needs to guarantee the following:

- *Reliability*: The signature is bound uniquely to the signer.
- *Non-repudiation*: The signer cannot deny his signature.
- *Unforgeability*: The signature is bound to the signed text.

More details for all material treated in this section can be found in [MEOO$^+$ 1996, STI 1995, SCH 1996].

## 1.4 Factorization and primality

This section is devoted to prime numbers. We need them to construct finite fields and the cryptosystems in which we are interested are designed around groups of prime order. The prime number theorem gives an estimate on the number $\pi(x)$ of primes less than $x$.

**Theorem 1.11 (Prime number theorem)** *Asymptotically there exist*

$$\pi(x) \sim \int_2^x \frac{dy}{\ln y}$$

*prime numbers less than $x$. A slightly worse estimate that is easier to remember is*

$$\pi(x) \sim \frac{x}{\ln x}.$$

### 1.4.1 Primality

In the applications we envision we must be sure that a given integer $N$ is prime. The most obvious way is to try for all integers $n \leqslant \sqrt{N}$ whether $N \equiv 0 \pmod{n}$ in which case one even has found a divisor of $N$. However, this method requires $O(\sqrt{N})$ modular reductions, which is far too large for the size of $N$ encountered in practice.

In fact, it is too time-consuming to prove primality, or for that matter compositeness, of a given integer by failing or succeeding to find a proper factor of it. The best primality test algorithms described in Chapter 25 will only prove $N$ to be prime and will not output any divisor in case it is not. Most algorithms will be probabilistic in nature in the sense that one output is always true while the other is only true with a certain probability. Iterating this algorithm allows us to enlarge the probability that the answer that was given only with a certain probability actually holds true. These techniques offer quite good performance.

To prove primality using probabilistic algorithms one usually starts with some iterations of an algorithm whose output "nonprime" is always correct while the output "prime" is true only with a certain probability. After passing some rounds one uses an algorithm that is correct when it outputs "prime."

The reason for this order is that usually the algorithms of the first type have a shorter running time and allow us to detect composite integers very efficiently. Factoring algorithms, on the other hand, are usually much slower.

### 1.4.2 Complexity of factoring

Even though we shall return to this matter in Chapter 25 we briefly recall the complexity of finding factors of composite numbers.

By brute force one can check divisibility by 2, 3, 5, 7, 11, and so on in succession. Even if $N$ is 1000 decimal digits long (about 3222 bits), it takes only a few seconds on a modern personal computer to divide it by all integers up to $10^7$. Checking whether a large number $N$ is divisible by $n$ (where $N$ is much larger than $n$) by trial division requires at most $O\big(\lg(n)\lg(N)\big)$ operations assuming simple techniques and

$$O\left(\lg(n)^{1+\epsilon}\frac{\lg(N)}{\lg(n)}\right) = O\big(\lg(n)^{\epsilon}\lg(N)\big)$$

asymptotically.

The *elliptic curve method (ECM)* of factorization given in Section 25.3.3 has expected complexity $L_p(\frac{1}{2}, \sqrt{2})$ for finding the smallest prime factor $p$ of $N$. It is expected that in the near future this method will be able to find 60-digits, i.e., 200-bit factors.

Boosted by RSA challenges [RSA] (see below) there has been a lot of research on the *number field sieve* methods for factoring integers. In theory the number field sieve should be able to factor any number in heuristic expected time

$$L_N\left(\tfrac{1}{3}, \left(\tfrac{64}{9}\right)^{1/3}\right) \text{ where } \left(\tfrac{64}{9}\right)^{1/3} \approx 1.923.$$

This algorithm was long thought not to be practical, but recent years have seen tremendous success in its implementation and its improvements. The largest RSA modulus factored so far is the 200-digit RSA challenge integer, a feat achieved by Franke and Kleinjung [WEI 2005, CON 2005].

### 1.4.3 RSA

We give only the schoolbook method here. It is understood that one should not implement RSA this way. The RSA cryptosystem [RISH+ 1978] uses an integer $N = pq$ which is the product of two primes $p$ and $q$. The system uses the modular relation $m^{(p-1)(q-1)} \equiv 1 \pmod{pq}$, which holds for every $m$ coprime to $N$. This relation will be proved in Section 2.1.2 in a more general context.

Alice's public and secret keys are two integers $e$ and $d$ such that $ed \equiv 1 \pmod{(p-1)(q-1)}$ (this alone implies that $e$ must be relatively prime to $p-1$ and $q-1$). If Bob wants to send a message $m$ to Alice, where we assume that $m$ is a natural number smaller than $pq$, he computes $c = m^e \bmod n$. To recover the message, Alice simply computes

$$c^d = m^{ed} = m^{1+k(p-1)(q-1)} \equiv m \pmod{N}$$

by the relation above. In the RSA cryptosystem, the one-way function is thus $m \mapsto m^e \bmod N$.

The security is based on the *RSA assumption*, namely the assumption that given $e, m^e \bmod N$ and $N$ one cannot recover $m$. This would be easy if one were able to compute $d$ from the given information, e.g., if $(p-1)(q-1)$ would be known. Apparently this could be done if one could factor $N$. However, factoring is not easy as shown in the previous section.

The primes $p$ and $q$ are usually chosen of similar bit length, in order to prevent as much as possible attacks arising from future development of algorithms whose complexity depends on the smallest prime. At the same time $p$ and $q$ cannot be too close, otherwise if $p = \lfloor\sqrt{N}\rfloor + a, q = \lfloor\sqrt{N}\rfloor - b$ for some small $a, b$ one has $N = pq = \lfloor\sqrt{N}\rfloor^2 + (a-b)\lfloor\sqrt{N}\rfloor - ab$. From this one gets $N - \lfloor\sqrt{N}\rfloor^2 = (a-b)\lfloor\sqrt{N}\rfloor - ab$. Performing a division with remainder by $\lfloor\sqrt{N}\rfloor$ one obtains $ab$ and $a - b$ provided they are smaller than $\lfloor\sqrt{N}\rfloor$.

It is not clear whether the *RSA problem*, which can be loosely formulated as inverting the function $m \mapsto m^e \bmod N$, is equivalent to the factorization problem. It is possible, in theory, that there is some way of computing $m$ from $m^e$ that does not involve determining $p$ and $q$. In the original RSA paper, the authors say:

> "It may be possible to prove that any general method of breaking our scheme yields an efficient factoring algorithm. This would establish that any way of breaking our scheme must be as difficult as factoring. We have not been able to prove this conjecture, however."

However, as of today, if $p$ and $q$ are large enough, the cryptosystem can be considered secure. Large enough currently means that the two numbers are of at least 500 bits and that their product has about 1024 bits.

## 1.5 Discrete logarithm systems

In this section we introduce a second kind of problem where hard instances can be constructed. In practice a *discrete logarithm (DL) system* is usually based on a cyclic group of prime order. For some protocols considered later on, a commutative semigroup would be enough.

In this book, we are mainly concerned with cryptographic use of elliptic and hyperelliptic curves and there one works in a group, hence, we restrict our attention to this case here as well. For the definition of a group and the examples used in the sequel, we refer to Chapter 2.

### 1.5.1 Generic discrete logarithm systems

Let $(G, \oplus)$ be a cyclic group of prime order $\ell$ and let $P$ be a generator of $G$. The map

$$\varphi : \mathbb{Z} \to G$$
$$n \to [n]P = \underbrace{P \oplus P \oplus \cdots \oplus P}_{n\text{-times}}$$

has kernel $\ell \mathbb{Z}$, thus $\varphi$ leads to an isomorphism between $(G, \oplus)$ and $(\mathbb{Z}/\ell\mathbb{Z}, +)$. The problem of computing the inverse map is called the *discrete logarithm problem (DLP) to the base of $P$*. It is the problem given $P$ and $Q$ to determine $k \in \mathbb{Z}$ such that $Q = [k]P$, i.e., to find $k \in \mathbb{N}$ such that $\varphi(k) = Q$. The *discrete logarithm of $Q$ to the base of $P$* is denoted by $\log_P(Q)$. Note, that it is unique only modulo the group order $\ell$. The complexity of this problem depends on the choice of $G$ and $\oplus$. To show the dependency on the generator $P$ of $G$ we speak of the DL system $(G, \oplus, P)$.

**Example 1.12** Let $(G, \oplus) = (\mathbb{Z}/\ell\mathbb{Z}, +)$ with generator $1 + \ell\mathbb{Z}$. The discrete logarithm of $n + \ell\mathbb{Z}$ is simply given by $n$. Also, if the generator is chosen to be $a + \ell\mathbb{Z}$ for some integer $a$, this problem is easy to solve as it is nothing but computing the inverse modulo $\ell$. In Chapter 10 this is shown to have complexity polynomial in the size of the operands, i.e., in $\lg \ell$.

Hence, this group *cannot* be used in cryptographic applications.

**Example 1.13** Choose a prime $p$ such that $\ell$ divides $p - 1$. Choose $\zeta \neq 1$ in $\mathbb{Z}/p\mathbb{Z}$ with $\zeta^\ell = 1$ (i.e., $\zeta$ is a primitive $\ell$-th root of unity). Then $(G, \oplus) = (\langle \zeta \rangle, \times)$ and $\varphi(n) = \zeta^n$.

In Chapter 19, we will show that this DLP is of subexponential complexity, thus harder than in the previous example but not optimal.

An obvious generalization is to work in extension fields $\mathbb{F}_q$, with $q = p^d, \ell \mid p^d - 1$ for $p$ prime. To represent the finite field $\mathbb{F}_{p^d}$ one fixes an irreducible polynomial $m(X) \in \mathbb{F}_p[X]$ of degree $d$ and uses the isomorphism $\mathbb{F}_{p^d} \simeq \mathbb{F}_p[X]/(m(X))$. For an introduction to finite fields and their arithmetic see Chapters 2 and 11.

Systems based on the DLP in the multiplicative group of finite fields are easy to construct. Starting with a prime $\ell$ of appropriate size one searches for $p$ and $d$ such that $\ell \mid p^d - 1$. For appropriately chosen subgroups, compression methods based on traces such as LUC [SMSK 1995] and XTR [LEVE 2000] can be used to represent the subgroup elements. These groups additionally allow faster group operations.

The groups associated to elliptic and hyperelliptic curves of small genus that will be studied in the sequel of the book are believed to have a DLP of *exponential complexity*.

To efficiently implement a DL system one needs to find good instances of groups in which the DLP is hard: in order to put aside easy instances this implies in particular that the group size can be efficiently computed to ensure that there exists a large prime order subgroup $(G, \oplus, P)$. We also need to have a short representation of group elements needing $O(\lg \ell)$ space and the group operation $Q \oplus R$ needs to be performed efficiently for any input $Q, R \in G$. The complexity of computing the DL is studied in Chapter 19. Note that the computation of the order of $P$ is a special instance for the DLP, namely $\mathrm{ord}(P) = \log_P 1$, where 1 is the neutral element in $G$. Techniques for scalar multiplication are studied in Chapter 9. For groups based on elliptic and hyperelliptic curves, see Chapters 13 and 14.

### 1.5.2 Discrete logarithm systems with bilinear structure

Some groups have an additional structure that can either be considered a weakness, as it allows transfers (see below), or an advantage, as it can be used constructively in special protocols (cf. Section 24.1.2).

**Definition 1.14** Assume that a DL system is given by $(G, \oplus)$ a group of prime order $\ell$ and that there is a group $(G', \oplus')$ of the same order $\ell$ in which we can compute "as fast" as in $G$. Assume moreover that $(H, \boxplus)$ is another DL system and that a map

$$e : G \times G' \to H$$

satisfies the following requirements:

- the map $e$ is computable in polynomial time (this includes that the elements in $H$ need only $O(\lg \ell)$ space),
- for all $n_1, n_2 \in \mathbb{N}$ and random elements $(P_1, P_2') \in G \times G'$ we have

$$e([n_1]P_1, [n_2]P_2') = [n_1 n_2] e(P_1, P_2'),$$

- the map $e$ is nondegenerate. Hence, for random $P' \in G'$ we have $e(P_1, P') = e(P_2, P')$ if and only if $P_1 = P_2$.

Then we call $(G, e)$ a *DL system with bilinear structure*.

There are two immediate consequences:

- Assume that $G = G'$ and hence
$$e(P, P) \neq 0.$$
Then for all triples $(P_1, P_2, P_3) \in \langle P \rangle^3$ one can decide in time polynomial in $\lg \ell$ whether
$$\log_P(P_3) = \log_P(P_1) \log_P(P_2).$$

- The DL system $G$ is at most as secure as the system $H$. Even if $G \neq G'$ one can transfer the DLP in $G$ to a DLP in $H$, provided that one can find an element $P' \in G'$ such that the map $P \to e(P, P')$ is injective, i.e., it induces an injective homomorphism of $G$ into a subgroup of $H$. Hence, instead of solving the DLP in $G$ one transfers the problem to $H$ where it might be easier to solve.

## 1.6 Protocols

This book is concerned with cryptographic applications of elliptic and hyperelliptic curves. So far we have described DL systems in an abstract setting. In this section we motivate that groups are a good choice and show how two or more parties can agree on a joint secret key, exchange encrypted data, and sign electronically. We also show how the identity of a participant can be used to form his public key.

But this book is *not* mainly concerned with protocols. We just show the bare-bones schoolbook protocols. Their use is twofold: they should motivate the reader to consider DL systems in more detail and at the same time highlight which computational problems need to be solved in order to get fast cryptographic systems. For an actual implementation one needs to take care not to weaken the system in applying a flawed protocol. For a great overview consider [MEOO$^+$ 1996].

### 1.6.1 Diffie–Hellman key exchange

The publication of Diffie and Hellman's seminal paper *New directions in cryptography* [DIHE 1976] can be seen as the start of public-key cryptography in public. We describe the *Diffie–Hellman (DH)* protocol for an abstract group $(G, \oplus)$. In their paper they proposed the multiplicative group of a finite prime field (cf. Example 1.13).

The two parties Alice (A) and Bob (B) have the public parameters $(G, \oplus, P)$ and want to agree on a joint key that is a group element. Once they are in the possession of such a joint secret $P_k$ they can use a *key derivation function* to derive a bit-string useful as a key in a symmetric system. To this aim A secretly and randomly chooses $a_A \in_R \mathbb{N}$ ($\in_R$ means choosing at random) and computes $P_A = [a_A]P$ while B ends up with $P_B = [a_B]P$. They publicly exchange these intermediate results. If the DLP (cf. Section 1.5) is hard in $G$ one cannot extract $a_A$ from $P_A$ or $a_B$ from $P_B$. Upon receiving $P_B$, A computes $P_k = [a_A]P_B = [a_A a_B]P$. Now B can obtain the same result as $[a_B]P_A = [a_B a_A]P$, thus they are both in possession of a group element $P_k$, which should not be computable from the public values $P_A$ and $P_B$.

Clearly, this last assumption does not hold if the DLP in $(G, \oplus)$ is easy. The problem of computing $[a_A a_B]P$ given $[a_A]P$ and $[a_B]P$ is called the *computational Diffie–Hellman problem (CDHP)*.

Maurer and Wolf [MAWO 1999] study the equivalence of the CDHP and the DLP. An important tool in their proof are elliptic curves of split group order. They show that if such curves can be found, then an oracle to solve the CDHP can be used to solve DLP in polynomially many steps. For groups related to elliptic curves this question is studied in [MUSM$^+$ 2004].

In most DL systems it is also hard to verify whether a proposed solution to the CDHP is correct. The problem given $[a_A]P, [a_B]P$ and $[c]P$ to decide whether $[c]P = [a_A a_B]P$ is called the *decision Diffie–Hellman problem (DDHP)*. Obviously, it is no harder than CDHP. For the DLP a decision version is not useful to consider as one could simply try the pretended solution.

If $(G, \oplus)$ is a DL system with bilinear structure (Section 1.5.2), the DDHP can easily be solved by comparing $e([a_A]P, [a_B]P) = [a_A a_B](e(P, P))$ to $e(P, [c]P) = [c](e(P, P))$. Groups in which the CDHP is assumed to be hard while the DDHP is easy are called *Gap-Diffie–Hellman groups*.

As usual the presented version is *not ready to implement*. An eavesdropper Eve (E) could intercept the communication and act as Bob for Alice sending $[a_E]P$ to her and as Alice for Bob sending her key to him as well. Then she would have a joint key $P_{k,A}$ with Alice and one with Bob $P_{k,B}$. Hence, she can decipher any message from Alice intended for Bob and encipher it again for Bob using $P_{k,B}$. This way, no party would notice her presence and she could read any message. This attack is called the *man-in-the-middle* attack.

### 1.6.2 Asymmetric Diffie–Hellman and ElGamal encryption

The Diffie–Hellman key exchange requires that both parties to work online, i.e., they are both active at the same time. The following two protocols are asymmetric also in the sense that the sender and the receiver perform different steps and that there are two different keys — a private key and a public key.

## § 1.6 Protocols

If the DLP is hard in $(G, \oplus)$ then Alice could just as well publish her public key $P_A = [a_A]P$ in a directory. The process of computing the public and private key pair is called *key generation*. The systems described in this section require that the receiver of the message has already set up and published his public key. The problem of how to make accessible this data and put confidence in the link between A and her public key is considered in the literature on public-key infrastructure (PKI).

---
**Algorithm 1.15** Key generation
---
INPUT: The public parameters $(G, \oplus, P)$.
OUTPUT: The public key $P_A$ and private key $a_A$.

1. $a_A \in_R \mathbb{N}$                                                                                                            [choose $a$ at random in $\mathbb{N}$]
2. $P_A \leftarrow [a_A]P$
3. **return** $P_A$ and $a_A$

---

The random choice should be done by the computing device to avoid biases introduced by humans, like choosing small numbers to facilitate the computations. In Chapter 30 we deal with random number generators.

If Bob wants to send the message $m$ to Alice, he looks up her public key in a directory. He can perform an asymmetric version of the Diffie–Hellman key exchange if there is a map $\psi : G \to \mathcal{K}$ from the group to the *keyspace* $\mathcal{K}$ and a symmetric cipher $E_\kappa$ depending on the key $\kappa$. The decryption function, i.e., the inverse of $E_\kappa$, is denoted by $D_\kappa$.

---
**Algorithm 1.16** Asymmetric Diffie–Hellman encryption
---
INPUT: A message $m$, the public parameters $(G, \oplus, P)$ and the public key $P_A \in G$.
OUTPUT: The encrypted message $(Q, c)$.

1. $k \in_R \mathbb{N}$
2. $Q \leftarrow [k]P$
3. $P_k \leftarrow [k]P_A$
4. $\kappa \leftarrow \psi(P_k)$
5. $c \leftarrow E_\kappa(m)$
6. **return** $(Q, c)$

---

To decrypt, Alice computes $P_k = [a_A]Q$, using her private key $a_A$, from which she determines $\kappa = \psi(P_k)$. She recovers the plaintext as $m = D_\kappa(c)$.

The randomly chosen *nonce* $k \in_R \mathbb{N}$ makes this a *randomized encryption*.

If there is an invertible map $\varphi$ from the message space $\mathcal{M}$ to $G$ one can also use *ElGamal encryption*.

---
**Algorithm 1.17** ElGamal encryption
---
INPUT: A message $m$, the public parameters $(G, \oplus, P)$ and the public key $P_A \in G$.
OUTPUT: The encrypted message $(Q, c)$.

1. $k \in_R \mathbb{N}$
2. $Q \leftarrow [k]P$
3. $P_k \leftarrow [k]P_A$

4. $R \leftarrow P_k \oplus \varphi(m)$
5. **return** $(Q, R)$

To decrypt, Alice uses $P_k = [a_A]Q$ and computes $m = \varphi^{-1}(R \ominus P_k)$.

Note that by this method one can only encrypt messages of size at most $\lg \ell$, where $\ell$ is the order of $G$. It is possible to encrypt longer messages using a mode of operation making multiple calls to the Algorithm 1.17; however, this is hardly ever done because of the relative slowness of this encryption scheme. Instead the transmitted message $m$ will act as a secret key in some following symmetric encryption.

### 1.6.3 Signature scheme of ElGamal-type

An electronic signature should bind the signer to the content of the signed message. A *hash function* (see [MEOo$^+$ 1996]) is a map $h : S \to T$ between two sets $S, T$, where usually $|S| > |T|$, e.g., the input is a bit-string of arbitrary length and the output has fixed length.

Additional properties are required from *cryptographic hash functions*:

- *Preimage resistant*: for essentially all outputs $t \in T$ it is computationally infeasible to find any $s \in S$ such that $t = h(s)$.
- *2nd-preimage resistant*: for any given $s_1 \in S$ it is computationally infeasible to find a different $s_2 \in S$ such that $h(s_1) = h(s_2)$.
- *Collision resistant*: it is computationally infeasible to find any distinct inputs $s_1, s_2$ such that $h(s_1) = h(s_2)$.

For practical use one requires the signature to be of fixed length no matter how long the signed message is. Therefore, one only signs the hash of the message. The hash function should be collision resistant as otherwise a malicious party could ask to sign some innocent message $m_1$ and use the signature, which only depends on $h(m_1)$, as a signature for a different message $m_2$ with $h(m_1) = h(m_2)$. We shall also apply hash functions to elements of the group $G$. Here, we assume that these are represented via a bit-string and thus write $h(Q)$ for $Q \in G$.

To compute an electronic signature, Alice must have performed Algorithm 1.15 in advance.

**Algorithm 1.18** ElGamal signature

INPUT: A message $m$, the public parameters $(G, \oplus, P)$ with $|G| = \ell$ and the private key $a_A \in G$.
OUTPUT: The signature $(Q, s)$ on $m$.

1. $k \in_R \mathbb{N}$
2. $Q \leftarrow [k]P$
3. $s \leftarrow \left(k^{-1}\bigl(h(m) - a_A h(Q)\bigr)\right) \bmod \ell$
4. **return** $(Q, s)$

## Remarks 1.19

(i) In the signature scheme the short term secret, i.e., the random nonce $k$, must be kept secret as otherwise the long-term secret, the private key $a_A$, can be recovered as

$$a_A \equiv h(Q)^{-1}(h(m) - sk) \pmod{\ell}.$$

(ii) There are many variants of ElGamal signature schemes. Some have the advantage that one need not invert $k$ modulo the group order. This is especially interesting if one is concerned with restricted environments (cf. Chapters 27 and 26) as this way one avoids implementing modular arithmetic for two different moduli (finite field arithmetic for the group arithmetic and computations modulo $\ell$). An overview of different schemes is provided in [MEOO$^+$ 1996, Note 11.6]; e.g., the signature can also be given by

$$s = \bigl(kh(m) + a_A h(Q)\bigr) \bmod \ell$$

with notations as above.

A signature can be verified by everybody.

---
**Algorithm 1.20** Signature verification

INPUT: A message $m$, its signature $(Q, s)$ from Algorithm 1.18, the public parameters $(G, \oplus, P)$ where $|G| = \ell$, and the public key $P_A$.
OUTPUT: Acceptance or rejection of signature.

1. $R_1 \leftarrow [h(Q)]P_A \oplus [s]Q$
2. $R_2 \leftarrow [h(m)]P$
3. **if** $R_1 = R_2$ **return** "acceptance" **else return** "rejection"

---

The algorithm is valid as a correct signature gets accepted. Namely,

$$R_1 = [h(Q)]P_A \oplus [s]Q = [a_A h(Q)]P \oplus [ks]P = [a_A h(Q) + h(m) - a_A h(Q)]P = [h(m)]P = R_2.$$

In Line 1 one can apply simultaneous multiplication techniques (cf. Chapter 9).

Depending on the special properties of the group it might be possible to transmit only some part of $Q$. The standard for digital signatures (DSA) works in a subgroup of the multiplicative group of a finite field. For elliptic curves the standard is called the *elliptic curve digital signature algorithm* (ECDSA) [ANSI X9.62], an adaption of Algorithm 1.18. The German standard GECDSA avoids inversions modulo the group order. So far, there is no standard for hyperelliptic curves. A version analogous to the ECDSA can be found in [AVLA 2005].

### 1.6.4 Tripartite key exchange

We now give an example of how the additional structure of a DL system with bilinear structure can be used in protocols. We come back to this study in Chapter 24 where we apply special bilinear maps for elliptic and hyperelliptic curves. Here, we show how three persons can agree on a joint secret group element using using two DL systems, $(G, \oplus)$ and $(H, \boxplus)$, and $e$, a bilinear map from $G \times G$ into $H$ and needing only one round [JOU 2000]. Note that there are other protocols based on the usual DH protocol to solve this for arbitrarily many group members, but in two rounds [BUDE 1995, BUDE 1997], and clearly the protocol as such is a schoolbook version that can easily be attacked by a man-in-the-middle attack.

The following algorithm shows the computations done by person A.

---
**Algorithm 1.21** Three party key exchange
---
INPUT: The public parameters $(G, \oplus, H, \boxplus, P, e)$ with the bilinear map $e$.
OUTPUT: The joint key $K \in H$.

---
1. $a_\text{A} \in_R \mathbb{N}$
2. $P_\text{A} \leftarrow [a_\text{A}]P$
3. send $P_\text{A}$ to B, C
4. receive $P_\text{B}, P_\text{C}$ from B, C
5. $K \leftarrow [a_\text{A}]\bigl(e(P_\text{B}, P_\text{C})\bigr)$
6. **return** $K$
---

Applying this algorithm, A, B, and C obtain the same element of $H$ as

$$[a_\text{A}]\bigl(e(P_\text{B}, P_\text{C})\bigr) = [a_\text{A} a_\text{B} a_\text{C}]\bigl(e(P, P)\bigr) = [a_\text{B}]\bigl(e(P_\text{A}, P_\text{C})\bigr) = [a_\text{C}]\bigl(e(P_\text{A}, P_\text{B})\bigr).$$

Obviously, this protocol can be extended to more parties by the same methods as the DH protocol.

## 1.7 Other problems

Few problems have been investigated as thoroughly as the RSA and discrete logarithm problem. Furthermore, these problems are suitable for designing schemes that address most of today's needs: key exchange, en- and decryption, and generation and verification of digital signatures. After more than 20 years of research, they still stand as hard ones, with discrete logarithms in the Jacobians of curves offering the advantage of shorter keys compared to RSA and the gap becomes larger as the security demands increase. In a certain sense, curves offer *balanced* systems: they allow to design protocols for all applications, offering good performance.

Besides RSA and DLP some other computationally hard problems have been proposed as a basis for cryptosystems. Some of the systems designed around these problems require an extremely careful choice of parameters in order to attain security: the parameters had to be redefined several times as new attacks have been discovered. Therefore, we do not go into the details here but only list them with some references.

- The *Knapsack* (subset sum) problem is to determine a subset of a given set of integers such that the sum of the elements equals a given integer $s$. The corresponding decision problem is to decide whether there exists such a sum leading to $s$.
  Cryptosystems based on the knapsack problem [SAH 1975, IBKI 1975] were very well received when they were created, because they were the first alternative to RSA. But the fact that in a cryptosystem the hard problem should be easy to set up, implies that not all types of subset-sum problems are suitable for cryptography. In particular, the first proposals using knapsacks of low density have been broken in polynomial time [ODL 1990]. There exist some proposals that have not been broken so far but they are hardly ever applied.
- The *NTRU encryption* [HOPI+ 1998] and the NTRU-Sign [HOHO+ 2003] signature system, see [NTRU], are systems based on the problem of recovering a sparse polynomial that is a factor of a polynomial modulo $X^N - 1$ in the polynomial ring of some

finite field $\mathbb{F}_q$. This problem can be transferred to the setting of lattices where Coppersmith and Shamir [CoSh 1997] showed that this problem can be reduced to a shortest vector problem. Even though the latter is in general a hard problem, many instances of it arising from NTRU systems have been proved easy to solve. Since the NTRU system possesses a remarkable speed, there is a lot of interest in finding the parameters that make it secure. An excellent survey on the subject is [GeSm 2003].

- In the last two decades, several public-key schemes based on the difficulty of solving *Multivariate Quadratic equations* (MQ) have been proposed [MaIm 1988, Pat 1996, KiPa$^+$ 1999, KiPa$^+$ 2003]. As it often happens, even though the general MQ problem is **NP**-complete, many instances that have been proposed for designing cryptosystems have been revealed solvable. In spite of the recent progress in Gröbner bases computation [FaJo 2003] there are some signature schemes based on enhanced versions of the *hidden field equations* (HFE) system [Pat 1996] that are believed to be secure. It is still an active area of research to devise provably secure variants of HFE cryptosystems.

- McEliece proposed the first system based on algebraic coding theory [McE 1978]. His intent was to take advantage of the very efficient encoding and decoding algorithms for the binary Goppa codes to propose a very fast asymmetric encryption scheme. The security was related to the difficulty of decoding in a linear code [BeMc$^+$ 1978]: this problem is **NP**-hard. In 1986 Niederreiter proposed a dual version [Nie 1986] of the system, with an equivalent security [LiDe$^+$ 1994]. A specialized version of the latter based on Reed–Muller code was also proposed in 1994 [Sid 1994]. The first digital signature scheme using codes was presented in 2001 [CoFi$^+$ 2001].

- *Braid groups* $B_m$ are a special class of noncommutative groups. This allows us to define the *conjugacy problem*, namely the problem of finding an $a \in B_m$ for given $x, y \in B_m$ such that $y = axa^{-1}$. The systems based on this problem [KoLe$^+$ 2000, AnAn$^+$ 1999] have been broken completely in [ChJu 2003] by showing a polynomial-time algorithm to determine the secret information from the data made public for key exchange and encryption.

Noncommutative groups still receive some interest, but at the moment we are not aware of any proposed cryptoscheme.

# Mathematical Background

# Chapter 2

# Algebraic Background

### Christophe Doche and David Lubicz

**Contents in Brief**

| | | |
|---|---|---|
| 2.1 | **Elementary algebraic structures** <br> Groups • Rings • Fields • Vector spaces | 19 |
| 2.2 | **Introduction to number theory** <br> Extension of fields • Algebraic closure • Galois theory • Number fields | 24 |
| 2.3 | **Finite fields** <br> First properties • Algebraic extensions of a finite field • Finite field representations • Finite field characters | 31 |

In the first part we state definitions and simple properties of the algebraic structures we shall use constantly in the remainder of the book. More details can be found in [LAN 2002a].

The next section deals with number theory. We shall give at this occasion an introduction to extension of fields, including the algebraic closure, Galois theory, and number fields. We refer mainly to [LAN 2002a] and [FRTA 1991] for this part.

Finally, we conclude with an elementary theory of finite fields that are of crucial importance for elliptic and hyperelliptic curve cryptography. Finite fields are extensively discussed in [LINI 1997].

## 2.1 Elementary algebraic structures

We shall recall here basic properties of groups, rings, fields, and vector spaces.

### 2.1.1 Groups

**Definition 2.1** Given a set $S$, a *composition law* $\times$ *of $S$ into itself* is a mapping from the Cartesian product $S \times S$ to $S$. Common notations for the image of $(x, y)$ under this mapping are $x \times y$, $x.y$ or simply $xy$. When the law is *commutative*, i.e., when the images of $(x, y)$ and $(y, x)$ under the composition law are the same for all $x, y \in S$, it is customary to denote it by $+$.

**Definition 2.2** A *group* $G$ is a set with a composition law $\times$ such that

- $\times$ is *associative*, that is for all $x, y, z \in G$ we have $(xy)z = x(yz)$
- $\times$ has a *unit element* $e$, i.e., for all $x \in G$ we have $xe = ex = x$
- for every $x \in G$ there exists $y$, an *inverse* of $x$ such that $xy = yx = e$.

**Remarks 2.3**

(i) The group $G$ is said to be *commutative* or *abelian*, if the composition law is commutative. As previously mentioned, the law is often denoted by $+$ or $\oplus$ and the unit element by 0 in this case.

(ii) The unit of a group $G$ is necessarily unique as well as the inverse of an element $x$ that is denoted by $x^{-1}$. If $G$ is commutative the inverse of $x$ is usually denoted by $-x$.

(iii) The cardinality of a group $G$ is also called its *order*. The group $G$ is *finite* if its order is finite.

**Definition 2.4** Let $G$ be a group. A *subgroup* $H$ of $G$ is a subset of $G$ containing the unit element $e$ and such that

- for all $x, y \in H$ one has $xy \in H$
- if $x \in H$ then also $x^{-1} \in H$.

**Example 2.5** Let $x \in G$. The set $\{x^n \mid n \in \mathbb{Z}\}$ is the *subgroup of $G$ generated by* $x$. It is denoted by $\langle x \rangle$.

**Definition 2.6** Let $G$ be a group. An element $x \in G$ is of *finite order* if $\langle x \rangle$ is finite. In this case, the *order of* $x$ is $|\langle x \rangle|$, that is, the smallest positive integer $n$ such that $x^n = e$. Otherwise, $x$ is of *infinite order*.

**Definition 2.7** A group $G$ is *cyclic* if there is $x \in G$ such that $\langle x \rangle = G$. If such an element $x$ exists, it is called a *generator* of $G$.

**Remark 2.8** Every subgroup of a cyclic group $G$ is also cyclic. More precisely, if the order of $G$ is $n$, then for each divisor $d$ of $n$, $G$ contains exactly one cyclic subgroup of order $d$.

**Definition 2.9** Let $G$ be a group and $H$ be a subgroup of $G$. For all $x, y \in G$, the relation $x \sim y \in H$, if and only if $x^{-1}y \in H$, respectively $x \sim y$ if and only if $yx^{-1} \in H$, is an equivalence relation. An equivalence class for this relation is denoted by $xH = \{xh \mid h \in H\}$, respectively $Hx = \{hx \mid h \in H\}$ and are called respectively left and right cosets of $H$. The numbers of classes for both relations are the same. This invariant is called the *index* of $H$ in $G$ and is denoted by $[G : H]$.

**Theorem 2.10 (Lagrange)** Let $G$ be a finite group and $H$ be a subgroup of $G$. Then the order of $H$ divides the order of $G$. As a consequence, the order of every element also divides the order of $G$.

Since all the classes modulo $H$ have the same cardinality $|H|$ and form a partition of $G$, we have the more precise result $|G| = [G : H]|H|$.

**Definition 2.11** Let $G$ be a group. A subgroup $H$ is *normal* if for all $x \in G$, $xH = Hx$. In this case $G/H$ can be endowed with a group structure such that $(xH)(yH) = xyH$.

For example, the group $G = (\mathbb{Z}, +)$ is abelian. Hence the group of multiples of $n$, called $n\mathbb{Z}$ is a normal subgroup of $G$ for every integer $n$, and one can consider the *quotient group* $\mathbb{Z}/n\mathbb{Z} = \{x + n\mathbb{Z} \mid x \in \mathbb{Z}\}$. An element of $\mathbb{Z}/n\mathbb{Z}$ is a *class modulo* $n$. Two integers $x$ and $y$ are *congruent*

*§ 2.1 Elementary algebraic structures*

modulo $n$ if they belong to the same class modulo $n$, i.e., if and only if $x - y \in n\mathbb{Z}$. In this case, we write $x \equiv y \pmod{n}$.

For every integer $x$, there is a unique integer $r$ in the interval $[0, n-1]$, which belongs to the class of $x$. This integer $r$ is called the *canonical representative of* $x$ and we write $r = x \bmod n$. Therefore we have

$$\mathbb{Z}/n\mathbb{Z} = \{r + n\mathbb{Z} \mid r \in [0, n-1]\}.$$

But other choices are possible. For example, to minimize the absolute value of the representatives, we write $x \bmod s\, n$ for the unique integer in $[\lfloor -n/2 \rfloor + 1, \lfloor n/2 \rfloor]$ congruent to $x$ modulo $n$.

**Definition 2.12** Let $G$ and $G'$ be two groups with respective laws $\times$ and $\otimes$ and units $e$ and $e'$.

- A *group homomorphism* $\psi$ between $G$ and $G'$ is a map from $G$ to $G'$ such that for all $x, y \in G$, $\psi(x \times y) = \psi(x) \otimes \psi(y)$.
- The *kernel of* $\psi$ is $\ker \psi = \{x \in G \mid \psi(x) = e'\}$.

**Remark 2.13** The kernel of $\psi$ is never empty as it is easy to see that $\psi(e) = e'$. In addition, $\ker \psi$ is always a subgroup of $G$, which is in addition normal.

**Definition 2.14** Let $S$ be a set and $G$ be a group. The group $G$ *acts on* $S$ if there is a map $\sigma$ from $G \times S$ into $S$ such that

- $\sigma(e, t) = t$, for all $t \in S$
- $\sigma\big(x, \sigma(y, t)\big) = \sigma(xy, t)$, for all $t \in S$ and for all $x, y \in G$.

### 2.1.2 Rings

**Definition 2.15** A *ring* $R$ is a set together with two composition laws $+$ and $\times$ such that

- $R$ is a commutative group with respect to $+$
- $\times$ is associative and has a unit element $1$, which is different from $0$, the unit of $+$
- $\times$ is *distributive over* $+$, that is for all $x, y, z \in R$, $x(y + z) = xy + xz$ and $(y + z)x = yx + zx$.

**Remarks 2.16**

(i) The ring $R$ is said to be *commutative*, if the law $\times$ is commutative.
(ii) A commutative ring $R$ such that for all $x, y \in R$, the equality $xy = 0$ implies that $x = 0$ or $y = 0$ is called an *integral domain*.

**Example 2.17** The set $\mathbb{Z}$ of integers together with the usual addition and multiplication is a ring. The set $\mathbb{Z}[X]$ of polynomials with coefficients in $\mathbb{Z}$ together with the addition and multiplication of polynomials is a ring.

**Definition 2.18** Let $R$ and $R'$ be two rings with the respective operations $+, \times$ and $\oplus, \otimes$. A ring homomorphism $\psi$ is an application from $R$ to $R'$ such that for all $x, y \in R$

- $\psi(x + y) = \psi(x) \oplus \psi(y)$
- $\psi(x \times y) = \psi(x) \otimes \psi(y)$
- $\psi(1) = 1$.

**Definition 2.19** Let $R$ be a ring, $I$ is an *ideal of* $R$ if it is a nonempty subset of $R$ such that

- $I$ is a subgroup of $R$ with respect to the law $+$
- for all $x \in R$ and all $y \in I$, $xy \in I$ and $yx \in I$.

The ideal $I \subsetneq R$ is *prime* if for all $x, y \in R$ with $xy \in I$ one obtains $x \in I$ or $y \in I$.
The ideal $I \subsetneq R$ is *maximal* if for any ideal $J$ of $R$ the inclusion $I \subset J$ implies $J = I$ or $J = R$.
Two ideals $I$ and $J$ of $R$ are *coprime* if $I + J = \{i + j \mid i \in I \text{ and } j \in J\}$ is equal to $R$.

**Remark 2.20** It is easy to prove that a maximal ideal is also prime. The converse is not true in general.

**Definition 2.21** An ideal $I$ of a ring $R$ is *finitely generated* if there are elements $a_1, \ldots, a_n$ such that every $x \in I$ can be written $x = x_1 a_1 + \cdots + x_n a_n$ with $x_1, \ldots, x_n \in R$.

The ideal $I$ is *principal* if $I = aR$ and $R$ is a *principal ideal domain (PID)* if it is an integral domain and if every ideal of $R$ is principal.

**Example 2.22** The integer ring $\mathbb{Z}$ and the polynomial ring $K[X]$ where $K$ is a field are principal ideal domains.

**Theorem 2.23 (Chinese remainder theorem)** Let $I_1, \ldots, I_k$ be pairwise coprime ideals of $R$. Then
$$R / \prod_{i=1}^{k} I_i \simeq \prod_{i=1}^{k} R / I_i.$$

**Corollary 2.24** Let $n_1, \ldots, n_k$ be pairwise coprime integers, i.e., such that $\gcd(n_i, n_j) = 1$ for $i \neq j$. Then, for any integers $x_i$, there exists an integer $x$ such that
$$\begin{cases} x \equiv x_1 \pmod{n_1} \\ x \equiv x_2 \pmod{n_2} \\ \vdots \\ x \equiv x_k \pmod{n_k}. \end{cases}$$
Furthermore, $x$ is unique modulo $\prod_{i=1}^{k} n_i$.

**Remark 2.25** See Algorithm 10.52 for an efficient method to compute $x$ given the $x_i$'s.

Next we define an important arithmetic invariant. Let $R$ be a ring and let $\psi$ be the natural ring homomorphism from $\mathbb{Z}$ to $R$. So
$$\psi(n) = \begin{cases} 1 + \cdots + 1 & n \text{ times if } n \geqslant 0 \\ -(1 + \cdots + 1) & -n \text{ times otherwise.} \end{cases} \tag{2.1}$$

The kernel of $\psi$ is an ideal of $\mathbb{Z}$ and if the multiples of 1 are all different then $\ker \psi = \{0\}$. Otherwise, for example if $R$ is finite, some multiples of 1 must be zero. In other words, the kernel of $\psi$ is generated by a positive integer $m$.

**Definition 2.26** Let $R$ be a ring and $\psi$ defined as above. The kernel of $\psi$ is of the form $m\mathbb{Z}$, for some nonnegative integer $m$, which is called the *characteristic of $R$* and is denoted by $\mathrm{char}(R)$.

**Remark 2.27** In a commutative ring $R$ of prime characteristic $p$, the binomial formula simplifies to
$$(\alpha + \beta)^{p^n} = \alpha^{p^n} + \beta^{p^n} \quad \text{for all } \alpha, \beta \in R \text{ and } n \in \mathbb{N}. \tag{2.2}$$

**Definition 2.28** Let $R$ be a ring. An element $x \in R$ is said to be *invertible* if there is an element $y$ satisfying $xy = yx = 1$. Such an *inverse* $y$, also called a *unit*, is necessarily unique and is denoted by $x^{-1}$. The set of all the invertible elements is a group under multiplication denoted by $R^*$.

**Example 2.29** Take a positive integer $N$ and consider the ring $\mathbb{Z}/N\mathbb{Z}$ obtained as the quotient of the usual integer ring $\mathbb{Z}$ by the ideal $N\mathbb{Z}$. The invertible elements of $\mathbb{Z}/N\mathbb{Z}$ are in one-to-one correspondence with the canonical representatives coprime with $N$. The inverse of an element is given by an extended gcd computation, cf. Section 10.6.

**Definition 2.30** Let $N \geqslant 1$ and let us denote $|(\mathbb{Z}/N\mathbb{Z})^*|$ by $\varphi(N)$. The function $\varphi$ is called the *Euler totient function* and one has $\varphi(N) = |\{x \mid 1 \leqslant x \leqslant N, \gcd(x, N) = 1\}|$.

From Lagrange's Theorem 2.10, it is easy to prove the following.

**Theorem 2.31 (Euler)** Let $N$ and $x$ be integers such that $x$ is coprime to $N$, then

$$x^{\varphi(N)} \equiv 1 \pmod{N}.$$

This result was first proved by Fermat when the modulus $N$ is a prime $p$. In this case, Theorem 2.31 reduces to $x^{p-1} \equiv 1 \pmod{p}$ for $x$ prime to $p$. Therefore this restricted version if often referred to as *Fermat's little theorem*.

The ring $\mathbb{Z}/p\mathbb{Z}$ has many other marvelous properties. In particular, every nonzero element has an inverse, which means that $\mathbb{Z}/p\mathbb{Z}$ is a field.

### 2.1.3 Fields

**Definition 2.32** A *field* $K$ is a commutative ring such that every nonzero element is invertible.

**Example 2.33** The set of rational numbers $\mathbb{Q}$ with the usual addition and multiplication law is a field. The quotient set $\mathbb{Z}/p\mathbb{Z}$ with the induced integer addition and multiplication is also a field for any prime number $p$.

An easy consequence of Definition 2.32 is that a field is an integral domain. Now, quotienting $K$ by the kernel of $\psi$ as defined by (2.1), we see that $K$ contains a field isomorphic to $\mathbb{Z}/\operatorname{char}(K)\mathbb{Z}$. These two facts imply the following result.

**Proposition 2.34** The characteristic of a field is either 0 or a prime number $p$.

As a corollary, a field $K$ contains a subfield which is isomorphic to $\mathbb{Q}$ or $\mathbb{Z}/p\mathbb{Z}$.

Given an integral domain $R$, a common way to obtain a field is to add to $R$ the formal inverses of all the elements of $R$. The set obtained is the *field of fractions of $R$*. For instance, $K(X)$ is the field of fractions of the polynomial ring $K[X]$. Next proposition is also very much used in practice to construct fields.

**Proposition 2.35** Let $R$ be a ring and $I$ an ideal of $R$. Then the quotient set $R/I$ is a field if and only if $I$ is maximal.

**Definition 2.36** Let $K$ and $L$ be fields. A *homomorphism of fields* is a ring homomorphism between $K$ and $L$.

We remark that a homomorphism of fields is always injective, for it is immediate that its kernel is reduced to $\{0\}$.

### 2.1.4 Vector spaces

In the remainder of this part, $K$ will denote a field.

**Definition 2.37** A *vector space* $V$ over $K$ is an abelian group for a first operation denoted by $+$, together with a *scalar multiplication* from $K \times V$ into $V$, which sends $(\lambda, x)$ on $\lambda x$ and such that for all $x, y \in V$, for all $\lambda, \mu \in K$ we have

- $\lambda(x + y) = \lambda x + \lambda y$
- $(\lambda + \mu)x = \lambda x + \mu x$
- $(\lambda \mu)x = \lambda(\mu x)$
- $1x = x$.

An element $x$ of $V$ is a *vector* whereas an element $\lambda$ of $K$ is called a *scalar*.

**Definition 2.38** A $K$-*basis* of a vector space $V$ is a subset $S \subset V$ which

- is *linearly independent over* $K$, i.e., for any finite subset $\{x_1, \ldots, x_n\} \subset S$ and any $\lambda_1, \ldots, \lambda_n \in K$, one has that

$$\sum_{i=1}^{n} \lambda_i x_i = 0 \text{ implies that } \lambda_i = 0 \text{ for all } i$$

- *generates $V$ over $K$*, i.e., for all $x \in V$ there exist finitely many vectors $x_1, \ldots, x_n$ and scalars $\lambda_1, \ldots, \lambda_n$ such that

$$x = \sum_{i=1}^{n} \lambda_i x_i.$$

**Theorem 2.39** Let $V$ be a vector space over $K$. If $V$ is different from $\{0\}$ then $V$ has a $K$-basis.

**Definition 2.40** Two bases of a vector space $V$ over $K$ have the same cardinality. This invariant is called the $K$-*dimension of $V$* or simply the *dimension of $V$*. Note that the dimension is allowed to be infinite.

**Example 2.41** The set of complex numbers $\mathbb{C}$ together with the usual addition and coefficient wise multiplication with elements of $\mathbb{R}$ is a vector space over $\mathbb{R}$ of dimension 2. A real basis is for instance $\{1, i\}$.

**Example 2.42** The set $K[X]$ of polynomials in one variable over a field $K$ is an infinite dimensional vector space with the usual addition of polynomials and multiplications with elements from $K$. A basis is given by $\{1, X, X^2, \ldots, X^n, \ldots\}$

**Remark 2.43** When the field $K$ is replaced by a ring $R$, the axioms of Definition 2.37 give rise to a *module over the ring $R$*.

## 2.2 Introduction to number theory

We refer to Section 2.1 for an elementary presentation of groups, rings, and fields. More details can be found in [LAN 2002a].

In this section, we review the construction of an extension of a field $K$ by formally adding some elements to it. Then we describe some properties of algebraic extensions of fields in order to be able to state the main result of Galois theory. We conclude with a brief presentation of number fields.

## 2.2.1 Extension of fields

**Definition 2.44** Let $K$ and $L$ be fields, we say that $L$ is an *extension field* of $K$ if there exists a field homomorphism from $K$ into $L$. Such an extension field is denoted by $L/K$.

**Remark 2.45** As said before, a field homomorphism is always injective, so we shall identify $K$ with the corresponding subfield of $L$ when considering $L/K$.

**Example 2.46** Let $\mathbb{R}$ be the field of real numbers with usual addition and multiplication. Obviously, $\mathbb{R}$ is an extension of $\mathbb{Q}$. Now, let us describe a less trivial example. Consider the element $\sqrt{2} \in \mathbb{R}$ and the subset of $\mathbb{R}$ of the elements of the form $a + \sqrt{2}b$ with $a, b \in \mathbb{Q}$. If we put for $a + \sqrt{2}b$ and $a' + \sqrt{2}b'$

$$(a + \sqrt{2}b) + (a' + \sqrt{2}b') = a + a' + \sqrt{2}(b + b')$$

and

$$(a + \sqrt{2}b) \times (a' + \sqrt{2}b') = aa' + 2bb' + \sqrt{2}(ab' + a'b),$$

it is easy to see that we obtain a field denoted by $\mathbb{Q}(\sqrt{2})$, which is an extension of $\mathbb{Q}$.

**Definition 2.47** Let $L$ and $L'$ be two extension fields of $K$ and $\sigma$ a field isomorphism from $L$ to $L'$. One says that $\sigma$ is a *$K$-isomorphism* if $\sigma(x) = x$ for all $x \in K$.

**Definition 2.48** Let $L/K$ be a field extension then $L$ can be considered as a $K$-vector space. The dimension of $L/K$ is called the *degree* of $L/K$ denoted by $[L : K]$ or $\deg(L/K)$. If the degree of $L/K$ is finite then we say that the extension $L/K$ is *finite*.

The following result is straightforward.

**Proposition 2.49** Let $K \subset L \subset F$ be a tower of extension fields then

$$\deg(F/K) = \deg(F/L) \deg(L/K).$$

Now, let $L/K$ be a field extension and let $x$ be an element of $L$. There is a unique ring homomorphism $\psi : K[X] \to L$ such that $\psi(X) = x$ and for all $z \in K$, $\psi(z) = z$. We can consider the kernel of this homomorphism which is either $\{0\}$ or a maximal ideal $I$ of $K[X]$.

**Definition 2.50** Suppose that $I$ as defined above is nonzero. As $K[X]$ is a principal ideal domain, there exists a unique monic irreducible polynomial

$$m(X) = X^d + a_{d-1}X^{d-1} + \cdots + a_0$$

such that $I = m(X)K[X]$. We say that $x$ is an *algebraic element* of $L$ of degree $d$ and that $m(X)$ is the *minimal polynomial* of $x$.

Quotienting by $\ker \psi$, one sees that $\psi$ gives rise to a field inclusion $\overline{\psi}$ of $K[X]/(m(X)K[X])$ into $L$. Let $K[x] = \{f(x) \mid f(X) \in K[X]\}$ be the image of $\psi$ in $L$. It is an extension field of $K$ and the monic polynomial $m(X)$ is an invariant of the extension: in fact, if there exists $y \in L$ such that $K[x] \simeq K[y]$ then by construction $x$ and $y$ have the same minimal polynomials.

**Definition 2.51** If every element of $L$ is algebraic over $K$, we say that $L$ is an *algebraic extension* of $K$.

It is easy to see that $K[x]/K$ is a finite extension and $d = \deg(m)$ is equal to its degree: in fact $\overline{\psi}$ gives a bijective map from the $K$-vector space of polynomials with coefficients in $K$ of degree less than $d$ to $K[x]$. This bijection $\overline{\psi}$ is called a *polynomial representation* of $K[x]/K$.

Not all algebraic extensions are finite, but a finite extension is always algebraic, and if $L/K$ is a finite extension then there exists a finite sequence of elements $x_1, \ldots, x_n \in L$ such that $L = K[x_1, \ldots, x_n]$. If $L = K[x]$ we say that $L$ is a *monogenic extension* of $K$.

**Definition 2.52** Let $L/K$ be a finite algebraic extension and let $x \in L$. The application of right multiplication by $x$ from $L$ to $L$ considered as a $K$-vector space is linear. The trace and the norm of this endomorphism of $L$ are called respectively the *trace* and *norm* of $x$ and denoted by $\text{Tr}_{L/K}(x)$ and $\text{N}_{L/K}(x)$. We use the notations $\text{Tr}(x)$ and $\text{N}(x)$ when no confusion is likely to arise.

If $x$ is a generating element of $L/K$ with minimal polynomial $m(X) = X^d + a_{d-1} X^{d-1} + \cdots + a_0$ then $\text{Tr}(x) = -a_{d-1}$ and $\text{N}(x) = (-1)^d a_0$.

The trace and norm are both maps of $L$ to $K$. We have the basic properties:

**Lemma 2.53** Let $L/K$ be a degree $d$ finite algebraic extension. For $x, y \in L$ and $a \in K$ we have

$$\begin{aligned}
\text{Tr}(x+y) &= \text{Tr}(x) + \text{Tr}(y), & \text{N}(xy) &= \text{N}(x)\,\text{N}(y) \\
\text{Tr}(a) &= da, & \text{N}(a) &= a^d \\
\text{Tr}(ax) &= a\,\text{Tr}(x), & \text{N}(x) = 0 &\Rightarrow x = 0.
\end{aligned}$$

Let $K \subset L \subset F$ be a tower of finite algebraic extensions, let $x$ be an element of $F$ then

$$\text{Tr}_{F/K}(x) = \text{Tr}_{L/K}\big(\text{Tr}_{F/L}(x)\big) \quad \text{and} \quad \text{N}_{F/K}(x) = \text{N}_{L/K}\big(\text{N}_{F/L}(x)\big).$$

When $x$ is not a root of any polynomial equation with coefficients in $K$ one needs a new notion.

**Definition 2.54** If the kernel of $\psi$ is equal to $\{0\}$ we say that $x$ is a *transcendental element* of $L$. If every element of $L$ is transcendental over $K$, we say that $L$ is a *pure transcendental extension* over $K$. More generally, if there exists an element of $L$ which is not algebraic over $K$, then $L/K$ is a *transcendental extension* of $K$.

In this case, one can extend $\psi$ to an inclusion $\widetilde{\psi}$ of the fraction field $K(X)$ into $L$ by setting $\widetilde{\psi}(1/X) = 1/x$. Let $K(x)$ be the image of $\widetilde{\psi}$ in $L$. If $L$ is not an algebraic extension over $K(x)$ then putting $x_1 = x$ we can find $x_2$, a transcendental element of $L/K(x_1)$ which is not in $K(x_1)$, and build in the same way an inclusion of $K(X_1, X_2)$ into $L$. Iterating this process we can find $n \in \mathbb{N} \cup \{\infty\}$ the maximum number such that $K(x_1, \ldots, x_n)$ is a subfield of $L$ isomorphic to $K(X_1, \ldots, X_n)$. It can be shown that $n$ is independent of the sequence of transcendental elements $x_1, \ldots, x_n$ of $L$ over $K$ chosen.

**Definition 2.55** The number $n$ defined above is called the *transcendence degree* of $L$ over $K$.

It is quite clear from the previous discussion that every extension $K \to L$ can be written as the composition $K \to K_{\text{trans}} \to K_{\text{alg}}$ where $K_{\text{trans}}$ is a pure transcendental extension of $K$ and $K_{\text{alg}}$ is an algebraic extension over $K_{\text{trans}}$.

**Example 2.56** Let $\mathbb{Q}(X)$ be the field of rational functions over $\mathbb{Q}$; then obviously $\mathbb{Q}(X)/\mathbb{Q}$ is a pure transcendental extension of $\mathbb{Q}$ of transcendence degree equal to 1. Now let $\overline{\mathbb{Q}}$ be the algebraic closure of $\mathbb{Q}$, then by definition $\overline{\mathbb{Q}}/\mathbb{Q}$ is an algebraic extension but it is not a finite extension. Finally $\mathbb{Q}(X, \sqrt{2})/\mathbb{Q}$ is a transcendental extension that can be written as $\mathbb{Q} \to \mathbb{Q}(X) \to \mathbb{Q}(X, \sqrt{2})$.

## 2.2.2 Algebraic closure

Let $K$ be a field and consider a monogenic algebraic extension $K[x]$ of $K$ defined by a polynomial $m(X)$ irreducible over $K$. The polynomial $m(X)$ can be written as a product $\prod m_i(X)$ of irreducible polynomials over $K[x]$. As by construction $x$ is a root of $m(X)$ in $K[x]$ then $(X - x)$ is an irreducible factor of $m(X)$ and as a consequence for each $i$, $\deg m_i < \deg m$. If the $m_i(X)$'s are all of degree 1 then we say that $m(X)$ *splits completely* in $K[x]$.

If $m(X)$ does not split completely over $K[x]$ then there exists $m_{i_1}(X)$, an irreducible polynomial of degree greater than or equal to 2 and we can consider the extension $K[x,y]$ over $K[x]$ defined by $m_{i_1}(X)$. Repeating this process, one can build recursively an extension field over which $m(X)$ splits completely.

**Definition 2.57** The smallest extension of $K$ over which $m(X)$ completely splits is called the *splitting field* of $m(X)$. It is unique up to a $K$-isomorphism.

It is well known that every polynomial with coefficients in $\mathbb{R}$ splits completely in $\mathbb{C}$. More generally, if $K$ is a field, we would like to consider a maximal algebraic extension of $K$ in which every algebraic extension of $K$ could be embedded. Such an extension has the property that every polynomial of $K[X]$ splits completely in $\overline{K}$. The following theorem [STE 1910] asserts its existence.

**Theorem 2.58 (Steinitz)** There exists a unique algebraic extension of $K$ in which every polynomial $m(X) \in K[X]$ splits completely. This extension called the *algebraic closure of* $K$ and denoted by $\overline{K}$ is unique up to a $K$-isomorphism.

Next, we review some basic properties of algebraic extensions in order to state the main theorem of Galois theory.

## 2.2.3 Galois theory

For most parts of the book we consider finite algebraic extension fields. Therefore we restrict the discussion of Galois theory to this important case.

**Definition 2.59** An extension $L$ over $K$ is said to be *normal* if every irreducible polynomial over $K$ that has a root in $L$ splits completely in $L$.

As an immediate consequence every field automorphism of $\overline{L}$ fixing $K$ leaves $L$ invariant. Let $K$ be a field, $\overline{K}$ its algebraic closure, $\sigma$ an embedding of $K$ into $\overline{K}$ and for $x \in \overline{K}$, $K[x]$ an algebraic monogenic extension of $K$ defined by a polynomial $m(X)$ of degree $d$. Let $x = x_1, x_2 \ldots, x_s$ be the different roots of $m(X)$ in $\overline{K}$. Then for $i = 1, \ldots, s$, it is possible to define the unique field inclusion $\sigma_i$ of $K[x]$ into $\overline{K}$ imposing that the restriction of $\sigma_i$ on $K$ is $\sigma$ and $\sigma_i(x) = x_i$. The $\sigma_i$'s are all the inclusion homomorphisms of $K[x]$ in $\overline{K}$, the restriction of which is given by $\sigma$ on $K$. We remark that $s$ is always less than or equal to $d$, the degree of $K[x]/K$. This integer $s$ is called the degree of separability of $K[x_1]$ over $K$ or the degree of separability of $x_1$. More generally, we have:

**Definition 2.60** Let $L$ be a finite algebraic extension of $K$, $\overline{K}$ be the algebraic closure of $K$ and $\sigma$ an inclusion of $K$ into $\overline{K}$. Then the *degree of separability* of $L$ over $K$ denoted by $\deg_s(L/K)$ is the number $s$ of different field inclusions $\sigma_i$, $i = 1, \ldots, s$ of $L$ into $\overline{K}$ restricting to $\sigma$ over $K$. If $\deg_s(L/K) = \deg(L/K)$, we say that $L/K$ is *separable*.

If $x \in L$, the elements $\sigma_i(x) \in \overline{K}$ are called the *conjugates* of $x$.

An immediate consequence of the definition and the preceding discussion is:

**Lemma 2.61** A monogenic algebraic extension $K[x]$ defined by a minimal polynomial $m(X)$ is separable if and only if $m(X)$ is prime to its derivative $m'(X)$.

Concerning the composition of degree of separability we have

**Proposition 2.62** Let $L/K$ and $F/K$ be a tower of extension fields, then

$$\deg_s(F/K) = \deg_s(F/L)\deg_s(L/K).$$

We have the basic fact

**Fact 2.63** Let $F/L$ and $L/K$ be field extensions and let $x \in F$ be separable over $K$; then it is separable over $L$.

From the previous proposition, we deduce that an algebraic finite extension is separable if and only if it can be written as the composition of monogenic separable extensions. From this and the preceding fact, we can state

**Proposition 2.64** A finite algebraic extension $L/K$ is separable if and only if every $x \in L$ is separable over $K$.

Then the criterion of Proposition 2.61 tells us that every algebraic extension over a field of characteristic 0 is separable. We have the following definition

**Definition 2.65** A field over which every algebraic extension is separable is called a *perfect field*.

A field is perfect if and only if every irreducible polynomial is prime to its derivative. We saw that every field of characteristic zero is perfect. More generally, we have

**Proposition 2.66** A field $K$ is a perfect field if and only if one the the following conditions is realized

- $\text{char}(K) = 0$,
- $\text{char}(K) = p$ and $K^p = K$.

As a consequence of this proposition, we shall see that every finite field is perfect. The following theorem shows that every finite algebraic separable extension is in fact monogenic.

**Theorem 2.67** If $L/K$ is a separable finite algebraic extension of $K$ then $L/K$ is monogenic, i.e., there exists $x \in L$ such that $L = K[x]$ and $x$ is called a *defining element*.

**Definition 2.68** An extension $L/K$ is a *Galois extension* if it is normal and separable. We define the *Galois group of L over K* denoted by $G_{L/K}$ or $\text{Gal}(L/K)$ to be the group of $K$-automorphisms of $L$. There is a natural action of $G_{L/K}$ on $L$ defined for $g \in G_{L/K}$ and $x \in L$ by $g \cdot x = g(x)$. By its very definition, this action leaves the elements of $K$ invariant.

If $H$ is a subgroup of $G$, we denote by $L^H$ the set of elements of $L$ invariant under the action of $H$. It is easy to see that $L^H$ is a subfield of $L$. Moreover, $L^H$ is a Galois extension of $K$ if and only if $H$ is a normal subgroup of $G$.

Obviously, the separability condition of Galois extensions implies that the order of the Galois group of $L/K$ is equal to the degree of this extension. Then, we have the following result, which is the starting point of Galois theory

**Theorem 2.69** Let $L/K$ be a finite Galois extension. Then there is a one-to-one correspondence between the set of subfields of $L$ containing $K$ and the subgroups of $G_{L/K}$. To a subgroup $H$ of $G_{L/K}$ this correspondence associates the field $L^H$.

## § 2.2 Introduction to number theory

### 2.2.4 Number fields

**Definition 2.70** A *number field* $K$ is an algebraic extension of $\mathbb{Q}$ of finite degree. An element of $K$ is called an *algebraic number*.

**Remarks 2.71**

(i) There are number fields of any degree $d$, since, for instance, the polynomial $X^d - 2$ is irreducible over $\mathbb{Q}$ for every positive integer $d$.
(ii) As a consequence of Theorem 2.67, for every number field $K$, there is an algebraic number $\theta \in K$ such that $K = \mathbb{Q}(\theta)$.
(iii) The degree $d$ of $K/\mathbb{Q}$ is equal to the number of field homomorphisms $\sigma_i$ from $K$ to $\mathbb{C}$. Thus if $K = \mathbb{Q}(\theta)$, the degree $d$ is equal to the number of conjugates $\sigma_i(\theta)$ of $\theta$ and corresponds to the degree of the minimal polynomial of $\theta$.

**Definition 2.72** Let $\psi$ be a homomorphism from $K$ to $\mathbb{C}$. If the image of $\psi$ is in fact included in $\mathbb{R}$ then $\psi$ is a *real homomorphism*. Otherwise $\psi$ is called a *complex homomorphism*.

**Definition 2.73** The numbers of real and complex homomorphisms of $K/\mathbb{Q}$, respectively denoted by $r_1$ and $2r_2$, satisfy $d = r_1 + 2r_2$. If $r_2 = 0$ then $K/\mathbb{Q}$ is said to be *totally real*. In case $r_1 = 0$ then $K/\mathbb{Q}$ is *totally complex*. The ordered pair $(r_1, r_2)$ is called the *signature of $K/\mathbb{Q}$*.

**Fact 2.74** It is clear that a Galois extension must be totally real or totally complex.

**Remark 2.75** The signature of $K/\mathbb{Q}$ can be found easily. Let us write $K = \mathbb{Q}(\theta)$ and let $m(X)$ be the minimal polynomial of $\theta$. Then $r_1$ and $2r_2$ are respectively the numbers of real and nonreal roots of $m(X)$.

**Example 2.76** The signature of the totally real field $\mathbb{Q}(\sqrt{2})$ of degree 2 is $(2, 0)$.
The extension $\mathbb{Q}(i)$ generated by $X^2 + 1$ is totally complex and its signature is $(0, 1)$.
The signature of $\mathbb{Q}(\theta)$ where $\theta$ is the unique real root of the polynomial $X^3 - X - 1$ is $(1, 1)$.

**Proposition 2.77** Let $K/\mathbb{Q}$ be a number field of degree $d$, let $\sigma_1, \ldots, \sigma_d$ be the field homomorphisms of $K$ to $\mathbb{C}$ and let $\alpha$ be an algebraic number in $K$. Following Definition 2.52, the *trace* and *norm* of $\alpha$ are explicitly given by

$$\mathrm{Tr}_{K/\mathbb{Q}}(\alpha) = \sum_{i=1}^{d} \sigma_i(\alpha) \text{ and } \mathrm{N}_{K/\mathbb{Q}}(\alpha) = \prod_{i=1}^{d} \sigma_i(\alpha).$$

**Definition 2.78** Let $K/\mathbb{Q}$ be a number field. An algebraic number $\alpha$ is called *integral over* $\mathbb{Z}$ or an *algebraic integer* if $\alpha$ is a zero of a monic polynomial with coefficients in $\mathbb{Z}$.

The set of all the algebraic integers of $K$ under the addition and the multiplication of $K$ is a ring, called the *integer ring of $K$* and is denoted by $\mathcal{O}_K$.

**Remarks 2.79**

(i) If $K = \mathbb{Q}(\theta)$ is of degree $d$ then the ring $\mathcal{O}_K$ is a $\mathbb{Z}$-module having an *integral basis*, that is a set of integral elements $\{\alpha_1, \ldots, \alpha_d\}$ such that every $\alpha \in \mathcal{O}_K$ can be written as

$$\alpha = a_1 \alpha_1 + \cdots + a_d \alpha_d \text{ for some } a_i \in \mathbb{Z}.$$

(ii) The ring $\mathbb{Z}[\theta] = \{f(\theta) \mid f(X) \in \mathbb{Z}[X]\}$ is a subring of $\mathcal{O}_K$. In general it is different from $\mathcal{O}_K$.

**Example 2.80** Let $\alpha$ be an algebraic integer such that $\alpha^2 = n$ with $n \in \mathbb{Z}$ different from 0, 1, and squarefree. Then $\mathcal{O}_K$ is explicitly determined.

- If $n \equiv 1 \pmod 4$ then $\mathcal{O}_K = \mathbb{Z}\left[\frac{1+\alpha}{2}\right]$.
- If $n \equiv 2, 3 \pmod 4$ then $\mathcal{O}_K = \mathbb{Z}[\alpha]$.

**Definition 2.81** Let $K$ be a number field of degree $d$. An *order of* $K$ is a subring of $\mathcal{O}_K$ of finite index which contains an integral basis of length $d$. The ring $\mathcal{O}_K$ is itself an order known as the *maximal order of* $K$.

**Theorem 2.82** The ring $\mathcal{O}_K$ is a *Dedekind ring*, in other words

- it is *Noetherian*, i.e., every ideal $\mathfrak{a}$ of $\mathcal{O}_K$ is finitely generated
- it is *integrally closed*, that is, the set of all the roots of polynomials with coefficients in $\mathcal{O}_K$ is equal to $\mathcal{O}_K$ itself
- every nonzero prime ideal is maximal in $\mathcal{O}_K$.

In a Dedekind ring, an element has not necessarily a unique factorization. For instance, in the field $\mathbb{Q}(i\sqrt{5})$ the ring of integers is $\mathbb{Z}[i\sqrt{5}]$ and one has

$$\begin{aligned} 21 &= 3 \times 7 \\ &= (1 + 2i\sqrt{5})(1 - 2i\sqrt{5}). \end{aligned}$$

However, if we consider ideals instead, we have unique factorization. First, we define the *product of two ideals* $\mathfrak{a}$ and $\mathfrak{b}$ by

$$\mathfrak{a}\mathfrak{b} = \left\{ \sum_i a_i b_i \mid a_i \in \mathfrak{a}, b_i \in \mathfrak{b} \right\}.$$

**Theorem 2.83** Let $\mathcal{O}$ be a Dedekind ring and $\mathfrak{a}$ be an ideal of $\mathcal{O}$ different from $(0)$ and $(1)$. Then $\mathfrak{a}$ admits a factorization

$$\mathfrak{a} = \mathfrak{p}_1 \ldots \mathfrak{p}_r$$

where the $\mathfrak{p}_i$'s are nonzero prime ideals. The factorization is unique up to the order of the factors.

**Definition 2.84** Let $K$ be a number field and let an order $\mathcal{O}$ be a Dedekind ring. A *fractional ideal of* $K$ is a submodule of $K$ over $\mathcal{O}$.

**Remark 2.85** An $\mathcal{O}$-submodule $\mathfrak{a}$ is a fractional ideal of $K$ if and only if there exists $c \in \mathcal{O}$ such that $c\mathfrak{a} \subset \mathcal{O}$.

The fractional ideals form a group $J_K$ under the product defined above and the *inverse of a fractional ideal* $\mathfrak{a}$ in $J_K$ is the fractional ideal

$$\mathfrak{a}^{-1} = \{x \in K \mid x\mathfrak{a} \subset \mathcal{O}\}.$$

The fractional principal ideals, i.e., fractional ideals of the form $a\mathcal{O}$ for $a \in K^*$ form a subgroup $P_K$ of $J_K$.

**Definition 2.86** Let $K$ be a field. The *class group* of $\mathcal{O}_K$ is the quotient group $\mathrm{Cl}_K = J_K/P_K$.

**Theorem 2.87** For any number field $K$, the class group $\mathrm{Cl}_K$ is a finite abelian group. Its cardinality, called the *class number* is denoted by $h(K)$.

**Theorem 2.88** Let $K/\mathbb{Q}$ be a number field whose signature is $(s,t)$, let $\mathcal{O}_K$ be its integer ring and let $r = s + t - 1$. Then there are units $\varepsilon_0, \varepsilon_1, \ldots, \varepsilon_r$ such that

- the unit $\varepsilon_0$ is of finite order $w(K)$ and generates the group of roots of unity in $K$
- every unit $\varepsilon \in \mathcal{O}_K^*$ can be written in a unique way as

$$\varepsilon = \prod_{0 \leqslant i \leqslant r} \varepsilon_i^{n_i}$$

with $0 \leqslant n_0 < w(K)$ and $n_i \in \mathbb{Z}$ for $1 \leqslant i \leqslant r$.

A family $(\varepsilon_1, \ldots, \varepsilon_r)$ as above is called a basis of *fundamental units*.

**Definition 2.89** Let $K/\mathbb{Q}$ be a number field of degree $d$ let $p$ be a prime integer. Then the ideal decomposition of $p\mathcal{O}_K$ is of the form

$$p\mathcal{O}_K = \prod_{i=1}^{g} \mathfrak{p}_i^{e_i} \text{ with } e_i \geqslant 1.$$

The ideals $\mathfrak{p}_i$ are *above the ideal* $p\mathbb{Z}$. We say that

- $p$ is *ramified in* $K$ if there is $i$ such that $e_i > 1$. The corresponding ideal $\mathfrak{p}_i$ is said to be *ramified* as well.
- $p$ *splits* in $K$ if $p\mathcal{O}_K$ is the product of distinct prime ideals. In this case $g = d$ and $e_i = 1$ for all $i$.
- $p$ *is inert in* $K$, if $p\mathcal{O}_K$ is again a prime ideal, i.e., $g = 1$ and $e_1 = 1$.

## 2.3 Finite fields

Finite fields are central objects in cryptography, because they enjoy very special properties. For instance, their multiplicative group is cyclic and their Galois structure is remarkably simple. Initially they were the core of cryptosystems such as ElGamal's which relies on the difficulty to solve the discrete logarithm problem in the group of units of a well chosen finite field. For our purpose they serve as elementary blocks since elliptic and hyperelliptic curves used in cryptography are always defined over finite fields. We refer mainly to [LINI 1997] for this section.

### 2.3.1 First properties

**Definition 2.90** A *finite field* is a field whose order is finite. Finite fields are also referred to as *Galois fields*.

Let $K$ be a finite field. From Proposition 2.34, we know that the characteristic of $K$ is necessarily a prime number $p$, since otherwise $K$ would be of characteristic 0 and would contain $\mathbb{Q}$. The cardinality of $K$ is precisely determined as well. Indeed, if we factorize $\psi$ as defined by (2.1), we see that $K$ contains a subfield we shall identify with $\mathbb{Z}/p\mathbb{Z}$. This implies that it is a $\mathbb{Z}/p\mathbb{Z}$-vector space of finite dimension $d$. In particular, the order of $K$ is equal to $p^d$. The following theorem classifies all the finite fields.

**Theorem 2.91** For any prime $p$ and any positive integer $d$ there exists a finite field with $q = p^d$ elements. This field is unique up to isomorphism and is denoted by $\mathbb{F}_q$ or $GF(q)$.

This result can be proved using the uniqueness of the splitting field of $X^{p^d} - X$ in the algebraic closure of $\mathbb{Z}/p\mathbb{Z}$.

**Definition 2.92** A finite field that does not contain any proper subfield is called a *prime field*.

It is clear that $\mathbb{F}_p$ is the only prime field of characteristic $p$. More generally, there is a bijection between the subfields of $\mathbb{F}_{p^d}$ and the divisors of $d$ since

$$\mathbb{F}_{p^c} \subset \mathbb{F}_{p^d} \text{ if and only if } c \mid d.$$

As a consequence

$$\mathbb{F}_{p^{d_1}} \cap \mathbb{F}_{p^{d_2}} = \mathbb{F}_{p^{\gcd(d_1,d_2)}} \text{ and } \mathbb{F}_{p^{d_1}}.\mathbb{F}_{p^{d_2}} = \mathbb{F}_{p^{\operatorname{lcm}(d_1,d_2)}}.$$

The *multiplicative group* of nonzero elements of $\mathbb{F}_q$ is, as usual, denoted by $\mathbb{F}_q^*$. Lagrange's theorem shows that

$$\alpha^{q-1} = 1 \text{ for every } \alpha \in \mathbb{F}_q^*.$$

This generalizes Fermat's theorem and has many important consequences. For example, it implies that no finite field is algebraically closed. Indeed the polynomial $X^q - X + 1 \in \mathbb{F}_q[X]$ has no roots in $\mathbb{F}_q$. Using the property that a polynomial of degree $n$ with coefficients in $\mathbb{F}_q$ has at most $n$ roots in $\mathbb{F}_q$ one can show the following.

**Theorem 2.93** Let $\mathbb{F}_q$ be a finite field. The group $\mathbb{F}_q^*$ is cyclic.

A generator $\gamma$ of $\mathbb{F}_q^*$, i.e., an element such that $\mathbb{F}_q^* = \langle \gamma \rangle$, is called a *primitive element*.

**Remark 2.94** It is easy to see that there are $\varphi(q-1)$ primitive elements, where $\varphi$ is the Euler's totient function, as given in Definition 2.30. More generally, if $e \mid q-1$ then there are exactly $\varphi(e)$ elements of order $e$ in $\mathbb{F}_q^*$. Note also that the map $f(\alpha) = \alpha^e$ is a bijection if and only if $\gcd(e, q-1) = 1$.

Finite fields do have transcendental extensions. For example $\mathbb{F}_q(X)$ is an extension field of $\mathbb{F}_q$, which is not algebraic over $\mathbb{F}_q$. However, in the remainder we shall only describe algebraic extensions of a finite field.

## 2.3.2 Algebraic extensions of a finite field

There are algebraic extensions of $\mathbb{F}_q$ of infinite degree, the first example being $\overline{\mathbb{F}}_q$, the algebraic closure of $\mathbb{F}_q$. Concerning finite extensions, the field $\mathbb{F}_q$ being perfect, see Definition 2.65 and Proposition 2.66, this implies that $\mathbb{F}_{q^k}/\mathbb{F}_q$ can always be written $\mathbb{F}_q(\theta)$ where $\theta$ is algebraic of degree $k$ over $\mathbb{F}_q$, cf. Theorem 2.67.

The polynomial representation of an extension of $\mathbb{F}_q$ gives a practical way to construct $\mathbb{F}_{q^k}$. Since there is a unique finite field of cardinality $q^k$, it is sufficient to find a polynomial of degree $k$ irreducible over $\mathbb{F}_q$.

Gauß established the equality

$$X^{q^k} - X = \prod_{j \mid k} \prod_{P \in \mathcal{I}_j} P(X) \qquad (2.3)$$

where $\mathcal{I}_j$ is the set of all the irreducible monic polynomials of degree $j$ in $\mathbb{F}_q[X]$. It follows that the number of monic irreducible polynomials of degree $k$ over $\mathbb{F}_q$ is given by the formula

$$\frac{1}{k} \sum_{j \mid k} \mu(j) q^{k/j} \qquad (2.4)$$

where $\mu$ is the Möbius function. As a consequence, there is at least one irreducible polynomial of degree $k$ over $\mathbb{F}_q$, for all $k \geqslant 1$.

Relation (2.2) shows that the map that sends $\alpha$ to $\alpha^p$ is an automorphism of any field of characteristic $p$. Furthermore $\alpha^p = \alpha$ for every $\alpha \in \mathbb{F}_p$.

**Definition 2.95** Let $\alpha \in \mathbb{F}_{q^k}$. The map $\phi_p$, which sends $\alpha$ to $\alpha^p$, is a $\mathbb{F}_p$-automorphism called the *absolute Frobenius automorphism* of $\mathbb{F}_{q^k}$. More generally, the application

$$\begin{aligned} \phi_q : \mathbb{F}_{q^k} &\to \mathbb{F}_{q^k} \\ \alpha &\mapsto \alpha^q \end{aligned}$$

is an $\mathbb{F}_q$-automorphism of $\mathbb{F}_{q^k}$ called the *relative Frobenius automorphism of* $\mathbb{F}_{q^k}/\mathbb{F}_q$.

The following important result was proved by Galois.

**Theorem 2.96** Every finite extension $\mathbb{F}_{q^k}/\mathbb{F}_q$ is Galois and $\mathrm{Gal}(\mathbb{F}_{q^k}/\mathbb{F}_q)$ is a cyclic group of order $k$ generated by $\phi_q$.

If $m(X) \in \mathbb{F}_q[X]$ is an irreducible polynomial of degree $k$ then it splits completely in $\mathbb{F}_{q^k}$. If $\alpha$ is a root of $m(X)$ in $\mathbb{F}_{q^k}$, one sees by direct calculation that the conjugates of $\alpha$ are the distinct elements $\alpha, \alpha^q, \alpha^{q^2}, \ldots, \alpha^{q^{k-1}}$ of $\mathbb{F}_{q^k}$. In case $\alpha \in \mathbb{F}_{q^k}$ is not of degree $k$, the conjugates of $\alpha$ are no longer distinct since $\alpha^{q^j} = \alpha$, for some $j$ dividing $k$. In any case, we have the following result similar to Proposition 2.77.

**Proposition 2.97** Let $\alpha \in \mathbb{F}_{q^k}$. The trace and the norm of $\alpha$ are given by the formulas

$$\mathrm{Tr}_{\mathbb{F}_{q^k}/\mathbb{F}_q}(\alpha) = \sum_{i=1}^{k} \alpha^{q^i} \quad \text{and} \quad \mathrm{N}_{\mathbb{F}_{q^k}/\mathbb{F}_q}(\alpha) = \prod_{i=1}^{k} \alpha^{q^i}.$$

### 2.3.3 Finite field representations

In Section 2.3.1, we have seen that the multiplicative group of a finite field is cyclic. This allows us to easily describe the multiplication in $\mathbb{F}_q^*$. Likewise, considering $\mathbb{F}_q$ as a vector space over its prime field $\mathbb{F}_p$ with respect to some basis allows us to add efficiently in $\mathbb{F}_q$. However, the interplay between these structures needs to be investigated.

The notion of Zech's logarithm can be used for prime fields as well as for extension fields. It relies on the multiplicative structure of $\mathbb{F}_q^*$. Let $\gamma$ be a primitive element of $\mathbb{F}_q$. For $\alpha = \gamma^n$ put $\log_\gamma \alpha = n$, the discrete logarithm of $\alpha$ to the base of $\gamma$, where $n$ is defined modulo $q-1$ and $\log_\gamma 0 = \infty$. This representation is well adapted to products since

$$\log_\gamma \alpha_1 \alpha_2 = \log_\gamma \alpha_1 + \log_\gamma \alpha_2 = n_1 + n_2,$$

but computing $\log_\gamma(\alpha_1 + \alpha_2) = n_1 \log_\gamma(1 + \gamma^{n_2 - n_1})$ is not straightforward. For $n \in \mathbb{N}$, one defines *Zech's logarithm* $Z(n)$ of $\gamma^n$ to be the discrete logarithm of $1 + \gamma^n$, namely

$$1 + \gamma^n = \gamma^{Z(n)}.$$

Note that $Z(0) = \infty$ if $q$ is even and $Z((q-1)/2) = \infty$ if $q$ is odd. In practice, we have to precompute Zech's logarithm of each element of the field. So this representation is only useful for fields of small order.

Concerning prime fields, an element of $\mathbb{F}_p$ is usually represented by an integer between 0 and $p-1$ and computations are done modulo $p$. More details and efficiency considerations can be found in Section 11.1.1.

Concerning extension fields, elements of $\mathbb{F}_{q^n}$ are represented using the $\mathbb{F}_q$-vector space structure with respect to some basis. Additions are performed coefficient wise. However, to multiply one needs to know about the dependencies between the elements of this basis. For an efficiency focused discussion, we refer to Section 11.2.1. The following two bases are the most common ones.

### 2.3.3.a Polynomial representation

An element $\alpha \in \mathbb{F}_{q^k}$ can be represented as a polynomial with coefficients in $\mathbb{F}_q$ modulo an irreducible polynomial $m(X) \in \mathbb{F}_q[X]$ of degree $k$. If $\theta$ is a root of $m(X)$ then $\{1, \theta, \theta^2, \ldots, \theta^{k-1}\}$ is a basis of $\mathbb{F}_{q^k}$ over $\mathbb{F}_q$. Such a basis is called a *polynomial basis*; see Section 11.2.2.a. Note that one deduces from (2.4) that there are approximately $q^k/k$ irreducible monic polynomials of degree $k$ in $\mathbb{F}_q[X]$. Addition, subtraction, and multiplication are made modulo $m(X)$. Usually, inversion is obtained with an extended gcd computation in $\mathbb{F}_q[X]$. As the field polynomial $m(X)$ is irreducible and the polynomial $a(X)$ representing the field element $\alpha$ is of degree less than $k$, one can find a polynomials $u(X)$ and $v(X)$ of degree less than $k$ such that $a(X)u(X) + m(X)v(X) = 1$. Accordingly, $u(X) = a(x)^{-1} \mod m(X)$. In some cases we can also use the identity $\alpha^{q^k} = 1$.

Any irreducible polynomial of degree $k$ can be used to define $\mathbb{F}_{q^k}$ but in practice polynomials with special properties are chosen. One one hand, it can be useful to consider an irreducible polynomial having a primitive element, that is a generator of $\mathbb{F}_{q^k}^*$, among its roots. Such a polynomial is called a *primitive polynomial* and there are exactly $\varphi(q^k - 1)/k$ monic primitive polynomials of degree $k$ in $\mathbb{F}_q[X]$. On the other hand, a *sparse polynomial*, that is a polynomial with only a few nonzero coefficients, allows a fast reduction; see Algorithm 11.31. So irreducible binomials, trinomials, and pentanomials are commonly used to define extensions of a finite field, cf. Section 11.2.1.a and Definition 11.60. Some polynomials enjoy these two properties, e.g., the trinomial $X^{167} + X^6 + 1$ in $\mathbb{F}_2[X]$ is both primitive and sparse.

Finally, note that in some cases it can be more efficient to use a reducible polynomial, that is to embed the field $\mathbb{F}_{q^k}$ into a ring where the computations are done. This variant gives a so-called *redundant polynomial basis*. The representation of a field element is no longer unique but the reduction can be cheaper, see Section 11.2.1.b.

### 2.3.3.b Normal basis representation

**Definition 2.98** The element $\alpha \in \mathbb{F}_{q^k}$ is said to be *normal over* $\mathbb{F}_q$ if $\alpha, \alpha^q, \ldots, \alpha^{q^{k-1}}$ are linearly independent over $\mathbb{F}_q$. In this case $\{\alpha, \phi_q(\alpha), \ldots, \phi_q^{k-1}(\alpha)\}$ is a basis of $\mathbb{F}_{q^k}/\mathbb{F}_q$ which is called a *normal basis of* $\mathbb{F}_{q^k}$ *over* $\mathbb{F}_q$.

The element $\alpha$ is normal if and only if

$$\gcd\big(X^k - 1, \alpha X^{k-1} + \phi_q(\alpha) X^{k-2} + \cdots + \phi_q^{k-2}(\alpha) X + \phi_q^{k-1}(\alpha)\big) = 1.$$

Hensel proved that there always exists a normal basis of $\mathbb{F}_{q^k}$ over $\mathbb{F}_q$. For $\beta \in \mathbb{F}_{q^k}$, the computation of $\beta^q$ is very easy with this representation since it is simply a cyclic shift of the coordinates of $\beta$ represented with respect to $\{\alpha, \phi_q(\alpha), \ldots, \phi_q^{k-1}(\alpha)\}$. This is especially interesting when $q = 2$ because, then in the usual exponentiation using square and multiply, the squarings can simply be replaced by these far less expensive shiftings. Multiplying two elements is more painful and requires precomputing a table, cf. Section 11.2.2.b. To actually build a normal basis, Gauß periods are very convenient.

**Proposition 2.99** Let $r = kn + 1$ be a prime number, let $\mathcal{K}$ be the unique subgroup of order $n$ of $(\mathbb{Z}/r\mathbb{Z})^*$ and $\zeta$ be a primitive $r$-th root of unity. Then the *Gauß period of type* $(n, k)$

$$\alpha = \sum_{a \in \mathcal{K}} \zeta^a$$

## § 2.3 Finite fields

generates a normal basis $\left(\alpha, \alpha^q, \ldots, \alpha^{q^{k-1}}\right)$ of $\mathbb{F}_{q^k}$ over $\mathbb{F}_q$ if and only if $kn/e$ is prime to $k$, where $e$ is the order of $q$ in $(\mathbb{Z}/r\mathbb{Z})^*$. For $q = 2^d$ and a given $k$ this holds true for some $n \in \mathbb{N}$ if and only if $\gcd(k,d) = 1$ and $8 \nmid k$, [GAO 2001]. For $q = 2$, the obtained normal basis is called a *Gaussian normal basis of type* $n$ and leads to optimal normal basis when $n = 1$ or 2, cf. Section 11.2.1.c.

There is also a definition of general Gauß periods of type $(k, \mathcal{K})$ for any integer $r$ [FEGA⁺ 1999, NÖC 2001]. In this case $\mathcal{K}$ is an explicit subgroup of $(\mathbb{Z}/r\mathbb{Z})^*$ of order $n$ with $\varphi(r) = kn$.

Let $r = p_1^{v_1} \ldots p_t^{v_t}$ where the $p_i$'s are prime and $p_i \neq p_j$ for $i \neq j$. We denote by $\zeta_r$ a primitive $r$-th root of unity and let $r = r_1 r_2$ where $r_1$ is the squarefree part of $r$, i.e., $r = p_1 \ldots p_t$.

**Definition 2.100** With these settings, the *general Gauß period of type* $(k, \mathcal{K})$ is

$$\alpha = \sum_{a \in \mathcal{K}} \left( \zeta_r^{ar_2} \prod_{p_i | r_2} \sum_{1 \leq s < v_i} \zeta_r^{ar/p_i^s} \right).$$

Now this element $\alpha$ is normal in $\mathbb{F}_{q^k}/\mathbb{F}_q$ if and only if $\langle q, \mathcal{K} \rangle = (\mathbb{Z}/r\mathbb{Z})^*$ [FEGA⁺ 1999].

See [SHP 1999, Section 4.1] for more details on normal bases and Section 11.2.2.b for efficiency considerations.

### 2.3.3.c Dual basis representation

Two bases $\{\alpha_1, \ldots, \alpha_k\}$ and $\{\beta_1, \ldots, \beta_k\}$ of $\mathbb{F}_{q^k}/\mathbb{F}_q$ are *dual* or *complementary* with respect to the trace if $\text{Tr}(\alpha_i \beta_j) = \delta_{ij}$, where $\delta_{ij}$ is Kronecker's symbol. The dual basis of $\{\alpha_1, \ldots, \alpha_k\}$ is uniquely determined.

The use of dual bases leads to very efficient hardware implementation [BER 1982, OMMA 1986] to multiply an element $\alpha$ expressed with respect to some basis by $\beta$ expressed with respect to its dual basis. The result is obtained in dual coordinates.

As several bases are used one needs efficient conversion techniques to change bases. To overcome this difficulty, one can use a *self-dual basis*, i.e., a basis equal to its own dual basis. The field $\mathbb{F}_{q^k}$ has a self-dual basis over $\mathbb{F}_q$ if and only if $q$ is even or both $q$ and $k$ are odd [SELE 1980].

In some very particular and interesting cases a self-dual basis is also normal. Such a basis is traditionally called a *self-complementary normal basis* and it exists [BLGA⁺ 1994b] in every extension of $\mathbb{F}_{q^k}/\mathbb{F}_q$ where

- $q$ is even and $k$ is not a multiple of 4, or
- $q$ and $k$ are both odd.

Jungnickel et al. [JUME⁺ 1990] determined the total number of self-dual and self-complementary normal bases of $\mathbb{F}_{q^k}$ over $\mathbb{F}_q$ but their proof is not constructive. In [GAGA⁺ 2000], it is proved that a normal basis generated by a Gauß period of type $(k, n)$ with $k > 2$ is self-dual if and only if $n$ is even and divisible by the characteristic of $\mathbb{F}_q$. See also [BLGA⁺ 1994b] for an explicit construction in one of the following cases:

- $k$ is equal to the characteristic of $\mathbb{F}_q$, or
- $k \mid (q - 1)$ and $k$ is odd, or
- $k \mid (q + 1)$ and $k$ is odd.

### 2.3.4 Finite field characters

**Definition 2.101** A *character* of the finite field $\mathbb{F}_q$ is a group homomorphism from $\mathbb{F}_q^*$ into $\mathbb{C}^*$.

In this part, we assume that the characteristic of $\mathbb{F}_q$ is an odd prime number $p$ and we shall only describe two examples of characters, namely the Legendre symbol, and its generalization to extension fields, the Legendre–Kronecker–Jacobi symbol.

### 2.3.4.a The Legendre symbol

Take an integer $a$ and let us consider the equation

$$x^2 \equiv a \pmod{p}. \tag{2.5}$$

If $a \equiv 0 \pmod{p}$ then $x = 0$ is the only solution. When $a \not\equiv 0 \pmod{p}$, $a$ is said to be a *quadratic nonresidue* if equation (2.5) has no solution. Otherwise, there are two solutions and $a$ is a *quadratic residue*. Obviously, there are $(p-1)/2$ quadratic residues in $\mathbb{F}_p$ and the same number of quadratic nonresidues.

**Definition 2.102** The *Legendre symbol* $\left(\frac{a}{p}\right)$ is precisely the number of solutions of the above equation minus 1. Namely

$$\left(\frac{a}{p}\right) = \begin{cases} -1 & \text{if } a \text{ is a quadratic nonresidue} \\ 0 & \text{if } a \equiv 0 \pmod{p} \\ 1 & \text{if } a \text{ is a quadratic residue.} \end{cases}$$

**Theorem 2.103** The Legendre symbol satisfies the following properties

$$\left(\frac{a}{p}\right) \equiv a^{(p-1)/2} \pmod{p}$$

$$\left(\frac{ab}{p}\right) = \left(\frac{a}{p}\right)\left(\frac{b}{p}\right)$$

$$\left(\frac{-1}{p}\right) = (-1)^{(p-1)/2}$$

$$\left(\frac{2}{p}\right) = (-1)^{(p^2-1)/8}.$$

If $p$ and $q$ are both odd primes then one has the *quadratic reciprocity law*

$$\left(\frac{p}{q}\right)\left(\frac{q}{p}\right) = (-1)^{(p-1)(q-1)/4}. \tag{2.6}$$

The Legendre symbol can be extended to the *Kronecker–Jacobi* symbol $\left(\frac{a}{b}\right)$ where $a, b \in \mathbb{Z}$. We shall only use it when $b = \prod_i p_i^{\nu_i}$ is odd. Its main feature is

$$\left(\frac{a}{b}\right) = \prod_i \left(\frac{a}{p_i}\right)^{\nu_i}.$$

With these settings the reciprocity law (2.6) can be extended to any odd integers $p$ and $q$ and leads to an efficient method to compute the Legendre symbol, cf. Algorithm 11.19.

### 2.3.4.b The Legendre–Kronecker–Jacobi symbol

The case of extension fields of odd characteristic is very similar to prime fields of odd characteristic. Let $m(X)$ be an irreducible polynomial of degree $d$ such that $\mathbb{F}_p[X]/(m(X))$ is isomorphic to $\mathbb{F}_q$. There is a generalization of the Legendre symbol for an element $f(X) \in \mathbb{F}_p[X]$ denoted by

$$\left(\frac{f(X)}{m(X)}\right)$$

## § 2.3 Finite fields

which is equal to 0 if $m(X) \mid f(X)$, 1 if $f(X)$ is a nonzero square mod $m(X)$ and $-1$ if it is not a square mod $m(X)$. The Legendre symbol can be extended to the Kronecker–Jacobi symbol when $m(X)$ is not irreducible. This is useful because of an analogue of the reciprocity law (2.6) independently discovered by Kühne [KÜH 1902], Schmidt [SCH 1927], and Carlitz [CAR 1932].

**Theorem 2.104** Let $f(X)$ and $m(X)$ be monic polynomials of $\mathbb{F}_p[X]$. Then

$$\left(\frac{f(X)}{m(X)}\right) = \begin{cases} \left(\frac{m(X)}{f(X)}\right) & \text{if } q \equiv 1 \pmod{4} \text{ or if } \deg m(X) \text{ or } \deg f(X) \text{ is even} \\ -\left(\frac{m(X)}{f(X)}\right) & \text{otherwise.} \end{cases} \quad (2.7)$$

Algorithm 11.69 makes use of the law (2.7) to compute the Legendre–Kronecker–Jacobi symbol.

# Chapter 3

## *Background on p-adic Numbers*

### David Lubicz

**Contents in Brief**

| | | |
|---|---|---|
| 3.1 | **Definition of $\mathbb{Q}_p$ and first properties** | 39 |
| 3.2 | **Complete discrete valuation rings and fields** | 41 |
| | First properties • Lifting a solution of a polynomial equation | |
| 3.3 | **The field $\mathbb{Q}_p$ and its extensions** | 43 |
| | Unramified extensions • Totally ramified extensions • Multiplicative system of representatives • Witt vectors | |

The $p$-adic numbers play an important role in algebraic number theory. Many of the fruitful properties they enjoy stem from Hensel's lemma that allows one to lift the modulo $p$ factorization of a polynomial. As a consequence, although $\mathbb{Q}_p$ is a characteristic zero field, its absolutely unramified extensions reflect the same structure as the algebraic extensions of the finite field $\mathbb{F}_p$. On the other hand, the completion of the algebraic closure of $\mathbb{Q}_p$ can be embedded as a field, but not as a valuation field, into $\mathbb{C}$. Consequently, $p$-adic numbers are used to bridge the gap between finite field algebraic geometry and complex algebraic geometry by the use of the so-called Lefschetz principle.

In this chapter, we review the definition and basic properties of $p$-adic numbers. More details can be found in the excellent book by Serre [SER 1979].

## 3.1 Definition of $\mathbb{Q}_p$ and first properties

First, we introduce the notion of inverse limit of a directed family, which is useful in the construction of the $p$-adic numbers.

**Definition 3.1** Let $I$ be a set with a partial ordering relation $\geqslant$, i.e., for all $i, j, k \in I$ we have

- $i \geqslant i$,
- if $i \geqslant j$ and $j \geqslant k$ then $i \geqslant k$,
- if $i \geqslant j$ and $j \geqslant i$ then $i = j$.

Then $I$ is a *directed set* if for all $i, j$ there exists a $k \in I$ such that $k \geqslant i$ and $k \geqslant j$.

**Definition 3.2** A *directed family of groups* is given by

- a directed set $I$
- for each $i \in I$ a group $A_i$ and for $i, j \in I$, $i \geqslant j$ a morphism of groups $p_{ij} : A_i \to A_j$ satisfying the compatibility relation: for all $i, j, k \in I$ with $i \geqslant j \geqslant k$, $p_{jk} \circ p_{ij} = p_{ik}$.

We denote by $(A_i, \{p_{ij}\}_{j \in I})$ such a directed family.

**Definition 3.3** Let $(A_i, \{p_{ij}\}_{j \in I})$ be a directed family of groups and let $A$ be a group, together with a set of morphisms $\{p_i : A \to A_i\}_{i \in I}$ compatible with the $p_{ij}$, i.e., $p_j = p_{ij} \circ p_i$ for $i \geqslant j$, that satisfies the following universal property: let $B$ be a group and let $\{\phi_i\}_{i \in I}$ be a set of morphisms $\phi_i : B \to A_i$ such that the following diagram commutes for $i \geqslant j$

then there is a morphism $\phi$ such that for all $i, j \in I$, the following diagram

$$B \xrightarrow{\phi} A$$
$$\phi_j \searrow \quad \downarrow p_j$$
$$A_j$$

is commutative. The group $A$ is called the *inverse limit* of $(A_i, \{p_{ij}\}_{j \in I})$ and is denoted by $\varprojlim A_i$. The universal property implies that $\varprojlim A_i$ is unique up to isomorphism.

**Proposition 3.4** Let $(A_i, \{p_{ij}\}_{j \in I})$ be a directed family of groups with $I = \mathbb{N}^*$. Let $A = \prod A_i$ be the product of the family. Note that $A$ itself is a group where the group law is defined componentwise. Consider the subset $\Gamma$ of $A$ consisting of all elements $(a_i)$ with $a_i \in A_i$ and for $i \geqslant j$, $p_{ij}(a_i) = a_j$. It is easily verified that $\Gamma$ is a subgroup of $A$ which is isomorphic to $\varprojlim A_i$ where the projections $p_k$ for $k \in \mathbb{N}^*$ are given by $p_k : (a_i) \mapsto a_k$. In particular, the inverse limit of a directed family of groups always exists. Moreover, if the $A_i$ are rings, then $\varprojlim A_i$ is a ring, where the ring operations are defined componentwise.

**Definition 3.5** Let $p$ be a prime number and $I = \mathbb{N}^*$. For $i \geqslant j \in I$, let $p_{ij} : \mathbb{Z}/p^i\mathbb{Z} \to \mathbb{Z}/p^j\mathbb{Z}$ be the natural projections given by reduction modulo $p^j$, then $(\mathbb{Z}/p^i\mathbb{Z}, \{p_{ij}\}_{j \in I})$ is a directed family. Its inverse limit $\varprojlim \mathbb{Z}/p^i\mathbb{Z}$, denoted by $\mathbb{Z}_p$, is called the *ring of p-adic integers*.

The natural morphism of rings $\psi : \mathbb{Z} \to \mathbb{Z}_p$ with $\psi(1_\mathbb{Z}) = 1_{\mathbb{Z}_p}$ is injective, which implies that $\operatorname{char} \mathbb{Z}_p = 0$. The invertible elements in $\mathbb{Z}_p$ are characterized by the following proposition.

**Proposition 3.6** An element $z \in \mathbb{Z}_p$ is invertible if and only if $z$ is not in the kernel of $p_1$. For every nonzero element $z \in \mathbb{Z}_p$ there exists a unique $v_p(z) \in \mathbb{N}$ such that $z = u\psi(p)^{v_p(z)}$ with $u$ an invertible element in $\mathbb{Z}_p$. The integer $v_p(z)$ is called the *p-adic valuation* of $z$ and we extend the map $v_p$ to $\mathbb{Z}_p$ by defining $v_p(0) = -\infty$.

**Definition 3.7** Let $\mathcal{R}$ be a ring and let $v : \mathcal{R} \to \mathbb{Z}$ be a map such that for all $x, y \in \mathcal{R}$,

- $v(xy) = v(x)v(y)$;
- $v(x + y) \geqslant \min(v(x), v(y))$ with equality when $v(x) \neq v(y)$.

A map with the above properties is called a *discrete valuation*.

**Lemma 3.8** The map $v_p$ is a discrete valuation.

The ring $\mathbb{Z}_p$ is an integral domain and $\mathbb{Q}_p$ denotes its field of fractions. The $p$-adic valuation $v_p$ can be extended to $\mathbb{Q}_p$ by defining $v_p(1/x) = -v_p(x)$ for $x \in \mathbb{Z}_p$. Similarly, the natural embedding $\psi : \mathbb{Z} \to \mathbb{Z}_p$ can also be extended to the embedding $\mathbb{Q} \to \mathbb{Q}_p$ by defining $\psi(1/x) = 1/\psi(x)$, for $x \in \mathbb{Z}$.

The valuation of $\mathbb{Q}_p$ induces a map defined by $|x|_p = p^{-v_p(x)}$ for $x \in \mathbb{Q}_p$. The properties of $v_p$ imply that $|\cdot|_p$ is a norm on $\mathbb{Q}_p$, which is called the *p-adic norm*. The $p$-adic norm also defines a norm on $\mathbb{Z}$ and on $\mathbb{Q}$ via the map $\psi$. For $x \in \mathbb{Z}$, the norm is given by $|x|_p = p^{-\nu}$ with $\nu$ the power of $p$ in the prime factorization of $x$. For $x/y \in \mathbb{Q}$, the norm is given by $|x/y|_p = |x|_p/|y|_p$. The set $\psi(\mathbb{Q})$ is a dense subset of $\mathbb{Q}_p$ for $|\cdot|_p$. In fact, $\mathbb{Q}_p$ can also be defined as the completion of $\mathbb{Q}$ with respect to $|\cdot|_p$. This definition is similar to the definition of $\mathbb{R}$ as the completion of $\mathbb{Q}$ endowed with the usual archimedean norm.

## 3.2 Complete discrete valuation rings and fields

### 3.2.1 First properties

**Definition 3.9** A field $K$ is a *complete discrete valuation field* if

- $K$ is endowed with a discrete valuation $v_K$
- the valuation induces a norm $|\cdot|_K$ on $K$ by defining $|x|_K = \lambda^{-v_K(x)}$ with $\lambda > 1$
- every sequence in $K$ which is Cauchy for $|\cdot|_K$ has a limit in $K$.

**Remark 3.10** The topology induced by the norm $|x|_K = \lambda^{-v_K(x)}$ does not depend on $\lambda$.

It is easy to see that the subset $\mathcal{R} = \{x \in K \mid |x|_K \leq 1\}$ is a ring. This ring is an integral domain which is integrally closed, i.e., if $x \in K$ is a zero of a monic polynomial with coefficients in $\mathcal{R}$ then $x \in \mathcal{R}$. The ring $\mathcal{R}$ is called the *valuation ring* of $K$. Clearly, $\mathcal{M} = \{x \in \mathcal{R} \mid |x|_K < 1\}$ is the unique maximal ideal of $\mathcal{R}$. The field $\mathcal{K} = \mathcal{R}/\mathcal{M}$ is called the *residue field* of $K$. In the remainder of this chapter, we will assume that the residue field is finite.

**Proposition 3.11** An element $x \in \mathcal{R}$ is invertible in $\mathcal{R}$ if and only if $x$ is not in $\mathcal{M}$.

Note that if $K$ is a complete discrete valuation field with valuation ring $\mathcal{R}$ and maximal ideal $\mathcal{M}$, then the rings $A_i = \mathcal{R}/\mathcal{M}^i$ together with the natural projections $p_{ij} : A_i \to A_j$ for $i \geq j$ form a directed family of rings. It is easy to see that $\mathcal{R}$ is isomorphic to $\varprojlim A_i$.

**Example 3.12** The field $\mathbb{Q}_p$ is a complete discrete valuation field with residue field $\mathbb{F}_p$.

**Definition 3.13** An element $\pi \in \mathcal{R}$ is called a *uniformizing element* if $v_K(\pi) = 1$. Let $p_1$ be the canonical projection from $\mathcal{R}$ to $\mathcal{K}$. A map $\omega : \mathcal{K} \to \mathcal{R}$ is a *system of representatives* of $\mathcal{K}$ if for all $x \in \mathcal{K}$ we have $p_1(\omega(x)) = x$.

**Definition 3.14** An element $x \in \mathcal{R}$ is called a *lift of an element* $x_0 \in \mathcal{K}$ if $p_1(x) = x_0$. Consequently, for all $x \in \mathcal{K}$, $\omega(x)$ is a lift of $x$.

Now, let $\pi$ be a uniformizing element, $\omega$ a system of representatives of $\mathcal{K}$ in $\mathcal{R}$ and $x \in \mathcal{R}$. Let $n = v_K(x)$, then $x/\pi^n$ is an invertible element of $\mathcal{R}$ and there exists a unique $x_n \in \mathcal{K}$ such that $v_K(x - \pi^n \omega(x_n)) = n + 1$. Iterating this process and using the Cauchy property of $K$ we obtain the existence of the unique sequence $(x_i)_{i \geq 0}$ of elements of $\mathcal{K}$ such that

$$x = \sum_{i=0}^{\infty} \omega(x_i) \pi^i.$$

The following theorem classifies the complete discrete valuation fields.

**Theorem 3.15** Let $K$ be a complete discrete valuation field with valuation ring $\mathcal{R}$ and residue field $\mathcal{K}$, assumed finite of characteristic $p$. If char $K = p$ then $\mathcal{R}$ is isomorphic to a power series ring $\mathcal{K}[[X_1, X_2, \ldots]]$. If char $K = 0$ then $K$ is an algebraic extension of $\mathbb{Q}_p$.

From now on, we restrict ourselves to complete discrete valuation ring of characteristic 0 with a finite residue field. By Theorem 3.15, any such ring can be viewed as the valuation ring of an algebraic extension of $\mathbb{Q}_p$.

## 3.2.2 Lifting a solution of a polynomial equation

Let $K$ be a complete discrete valuation field with norm $|\cdot|_K$ and let $\mathcal{R}$ be its valuation ring. Let $\mathcal{R}[X]$ denote the univariate polynomial ring over $\mathcal{R}$. The main result of this section is Newton's algorithm which provides an efficient way to compute a zero of a polynomial $f \in \mathcal{R}[X]$ to arbitrary precision starting from an approximate solution.

**Proposition 3.16** Let $K$ be a complete discrete valuation field with valuation ring $\mathcal{R}$ and norm $|\cdot|_K$. Let $f \in \mathcal{R}[X]$ and let $x_0 \in \mathcal{R}$ be such that

$$|f(x_0)|_K < |f'(x_0)|_K^2$$

then the sequence

$$x_{n+1} = x_n - \frac{f(x_n)}{f'(x_n)} \tag{3.1}$$

converges quadratically towards a zero of $f$ in $\mathcal{R}$.

The quadratic convergence implies that the precision of the approximation nearly doubles at each iteration. More precisely, let $k = v_K(f'(x_0))$ and let $x$ be the limit of the sequence (3.1). Suppose that $x_i$ is an approximation of $x$ to precision $n$, i.e., $(x - x_i) \in \mathcal{M}^n$, then $x_{i+1} = x_i - f(x_i)/f'(x_i)$ is an approximation of $x$ to precision $2n - k$. Very closely related to the problem of lifting the solution of a polynomial equation is Hensel's lemma that enables one to lift the factorization of a polynomial.

**Lemma 3.17 (Hensel)** Let $f, A_k, B_k, U, V$ be polynomials with coefficients in $\mathcal{R}$ such that

- $f \equiv A_k B_k \pmod{\mathcal{M}^k}$,
- $U(X)A_k(X) + V(X)B_k(X) = 1$, with $A_k$ monic and $\deg U(X) < \deg B_k(X)$ and $\deg V(X) < \deg A_k(X)$

then there exist polynomials $A_{k+1}$ and $B_{k+1}$ satisfying the same conditions as above with $k$ replaced by $k + 1$ and

$$A_{k+1} \equiv A_k \pmod{\mathcal{M}^k}, \qquad B_{k+1} \equiv B_k \pmod{\mathcal{M}^k}.$$

Iterating this lemma, we obtain an algorithm to compute a factor of a polynomial over $\mathcal{R}$ given a factor modulo $\mathcal{M}$.

**Corollary 3.18** With the notation of Proposition 3.16, let $f \in \mathcal{R}[X]$ be a polynomial and $x_0 \in \mathcal{K}$ such that $x_0$ is a simple zero of the polynomial $f_0 = p_1(f)$. Then there exists an element $x \in \mathcal{R}$ such that $p_1(x) = x_0$ and $x$ is a zero of $f$.

## 3.3 The field $\mathbb{Q}_p$ and its extensions

Let $K$ be a finite algebraic extension of $\mathbb{Q}_p$ defined by an irreducible polynomial $m \in \mathbb{Q}_p[X]$. It can be shown that there exists a unique norm $|\cdot|_K$ on $K$ extending the p-adic norm on $\mathbb{Q}_p$. Let $\mathcal{R} = \{x \in K \mid |x|_K \leq 1\}$ denote the valuation ring of $K$ and let $\mathcal{M} = \{x \in \mathcal{R} \mid |x|_K < 1\}$ be the unique maximal ideal of $\mathcal{R}$. Then $\mathcal{K} = \mathcal{R}/\mathcal{M}$ is an algebraic extension of $\mathbb{F}_p$, the degree of which is called the *inertia degree* of $K$ and is denoted by $f$. The *absolute ramification index* of $K$ is the integer $e = v_K(\psi(p))$. The extension degree $[K : \mathbb{Q}_p]$, the inertia degree $f$, and the ramification index satisfy the following fundamental relation.

**Theorem 3.19** Let $d$ be the degree of $K/\mathbb{Q}_p$, then $d = ef$.

### 3.3.1 Unramified extensions

**Definition 3.20** Let $K/\mathbb{Q}_p$ be a finite algebraic extension, then $K$ is called *absolutely unramified* if $e = 1$. An absolutely unramified extension of degree $d$ is denoted by $\mathbb{Q}_q$ with $q = p^d$ and its valuation ring by $\mathbb{Z}_q$.

**Proposition 3.21** Denote by $P_1$ the reduction morphism $\mathcal{R}[X] \to \mathcal{K}[X]$ induced by $p_1$ and let $\overline{m}$ be the irreducible polynomial defined by $P_1(m)$. The extension $K/\mathbb{Q}_p$ is absolutely unramified if and only if $\deg m = \deg \overline{m}$. Let $d = \deg \overline{m}$ and $\mathbb{F}_q = \mathbb{F}_{p^d}$ the finite field defined by $\overline{m}$, then we have $p_1(\mathcal{R}) = \mathbb{F}_q$. Let $K_1$ and $K_2$ be two unramified extensions of $\mathbb{Q}_p$ defined respectively by $m_1$ and $m_2$ then $K_1 \simeq K_2$ if and only if $\deg m_1 = \deg m_2$.

As a consequence, every unramified extension of $\mathbb{Q}_p$ is isomorphic to $\mathbb{Q}_p[X]/(m(X))$ with $m$ being an arbitrary degree $d$ lift of an irreducible polynomial over $\mathbb{F}_p$ of degree $d$. Let $\omega : \mathbb{F}_q \to \mathbb{Z}_q$ be a system of representatives of $\mathbb{F}_q$; every element $x$ of $\mathbb{Z}_q$ can be written as a power series $x = \sum_{i=0}^{\infty} \omega(x_i) p^i$ with $(x_i)_{i \geq 0}$ a sequence of elements of $\mathbb{F}_q$.

**Proposition 3.22** An unramified extension of $\mathbb{Q}_p$ is Galois and its Galois group is cyclic generated by an element $\Sigma$ that reduces to the Frobenius morphism on the residue field. We call this automorphism the *Frobenius substitution on $K$*.

### 3.3.2 Totally ramified extensions

**Definition 3.23** Let $K/\mathbb{Q}_p$ be a finite algebraic extension, then $K$ is called totally ramified if $f = 1$.

**Definition 3.24** A monic degree $d$ polynomial $P(X) = \sum_{i=0}^{d} a_i X^i$ in $\mathbb{Q}_p[X]$ is an *Eisenstein polynomial* if it satisfies

- $v_p(a_0) = 1$,
- $v_p(a_i) > 1$, for $i = 1, \ldots, d-1$.

Such a polynomial is irreducible.

**Proposition 3.25** Let $K$ be a totally ramified extension of $\mathbb{Q}_p$; then there exists an Eisenstein polynomial $P$ such that $K$ is isomorphic to $\mathbb{Q}_p[X]/(P)$.

### 3.3.3 Multiplicative system of representatives

Let $K$ be a complete discrete valuation field of characteristic zero, with valuation ring $\mathcal{R}$ and residue field $\mathcal{K}$, assumed finite of characteristic $p$. Then $K$ is isomorphic to an algebraic extension of $\mathbb{Q}_p$.

**Proposition 3.26** There exists a unique system of representatives $\omega$ which commutes with $p$-th powering, i.e., for all $x \in \mathcal{K}$, $\omega(x^p) = \omega(x)^p$. This system $\omega$ is multiplicative in the following way: for all $x, y \in \mathcal{K}$ we have $\omega(xy) = \omega(x)\omega(y)$.

Such a system can be obtained as follows. Let $x_0 \in \mathcal{K}$. Since $\mathcal{K}$ is a perfect field, for each $r \in \mathbb{N}$ there exists $x_r \in \mathcal{K}$ such that $x_r^{p^r} = x_0$. Set $X_r = \{x \in \mathcal{R} \mid p_1(x) = x_r\}$ and let $Y_r$ be the set $\{x^{p^r} \mid x \in X_r\}$. It is easy to see that for all $y \in Y_r$, $p_1(y) = x_0$. Moreover, we have for all $x, y \in Y_r$, $v_K(x - y) \geqslant r$. This means by the Cauchy property that there exists a unique element $z \in \mathcal{R}$ such that $z \in Y_r$, for all $r$. Then simply define $\omega(x_0) = z$. The system of representatives defined in this way is exactly the unique system that commutes with $p$-th powering.

Let $\pi$ be a uniformizing element of $\mathcal{R}$ and let $\omega$ be the multiplicative system of representatives of $\mathcal{K}$ in $\mathcal{R}$ that commutes with $p$-th powering. Write $x \in \mathcal{R}$ as $x = \sum_{i=0}^{\infty} \omega(x_i) \pi^i$ with $(x_i)_{i \geqslant 0}$ the unique sequence of elements of $\mathcal{K}$ as defined in Section 3.2. Let $\Sigma$ be the Frobenius substitution on $K$; then we have

$$\Sigma(x) = \sum_{i=0}^{\infty} \omega(x_i)^p \pi^i.$$

### 3.3.4 Witt vectors

**Definition 3.27** Let $p$ be a prime number and $(X_i)_{i \in \mathbb{N}}$ a sequence of indeterminates. The *Witt polynomials* $W_n \in \mathbb{Z}[X_0, \ldots, X_n]$ are defined as

$$\begin{aligned} W_0 &= X_0, \\ W_1 &= X_0^p + pX_1, \\ W_n &= \sum_{i=0}^{n} p^i X_i^{p^{n-i}}. \end{aligned}$$

**Theorem 3.28** Let $(Y_i)_{i \in \mathbb{N}}$ be a sequence of indeterminates, then for every $\Phi(X, Y) \in \mathbb{Z}[X, Y]$ there exists a unique sequence $(\phi_i)_{i \in \mathbb{N}} \in \mathbb{Z}[X_0, X_1, \ldots; Y_0, Y_1, \ldots]$ such that for all $n \geqslant 0$

$$W_n(\phi_0, \ldots, \phi_n) = \Phi(W_n(X_0, \ldots, X_n), W_n(Y_0, \ldots, Y_n)).$$

Let $(S_i)_{i \in \mathbb{N}}$, resp. $(P_i)_{i \in \mathbb{N}}$, be the sequence of polynomials $(\phi_i)_{i \in \mathbb{N}}$ associated via Theorem 3.28 with the polynomials $\Phi(X, Y) = X + Y$, resp. $\Phi(X, Y) = X \times Y$. Then for any commutative ring $\mathcal{R}$, we can define two composition laws on $\mathcal{R}^{\mathbb{N}}$: let $a = (a_i)_{i \in \mathbb{N}} \in \mathcal{R}^{\mathbb{N}}$ and $b = (b_i)_{i \in \mathbb{N}} \in \mathcal{R}^{\mathbb{N}}$, then

$$a + b = \big(S_i(a, b)\big)_{i \in \mathbb{N}} \quad \text{and} \quad a \times b = \big(P_i(a, b)\big)_{i \in \mathbb{N}}.$$

**Definition 3.29** The set $\mathcal{R}^{\mathbb{N}}$ endowed with the two previous composition laws is a ring called the ring of *Witt vectors* with coefficients in $\mathcal{R}$ and is denoted by $W(\mathcal{R})$.

The relation with $p$-adic numbers is the following. Let $\mathbb{F}_q$ with $q = p^d$ be a finite field of characteristic $p$, then $W(\mathbb{F}_q)$ is canonically isomorphic to the valuation ring of the unramified extension of degree $d$ of $\mathbb{Q}_p$. Via this isomorphism, the map $F : W(\mathbb{F}_q) \to W(\mathbb{F}_q)$ given by $F\big((a_i)_{i \in \mathbb{N}}\big) = (a_i^p)_{i \in \mathbb{N}}$ corresponds to the Frobenius substitution $\Sigma$.

# Chapter 4

# *Background on Curves and Jacobians*

*Gerhard Frey and Tanja Lange*

### Contents in Brief

| | | |
|---|---|---|
| 4.1 | **Algebraic varieties** | 45 |
| | Affine and projective varieties | |
| 4.2 | **Function fields** | 51 |
| | Morphisms of affine varieties • Rational maps of affine varieties • Regular functions • Generalization to projective varieties | |
| 4.3 | **Abelian varieties** | 55 |
| | Algebraic groups • Birational group laws • Homomorphisms of abelian varieties • Isomorphisms and isogenies • Points of finite order and Tate modules • Background on $\ell$-adic representations • Complex multiplication | |
| 4.4 | **Arithmetic of curves** | 64 |
| | Local rings and smoothness • Genus and Riemann–Roch theorem • Divisor class group • The Jacobian variety of curves • Jacobian variety of elliptic curves and group law • Ideal class group • Class groups of hyperelliptic curves | |

This chapter introduces the main characters of this book — curves and their Jacobians. To this aim we give a brief introduction to algebraic and arithmetic geometry. We first deal with arbitrary varieties and abelian varieties to give the general definitions in a concise way. Then we concentrate on Jacobians of curves and their arithmetic properties, where we highlight elliptic and hyperelliptic curves as main examples. The reader not interested in the mathematical background may skip the complete chapter as the chapters on implementation summarize the necessary mathematical properties. For full details and proofs we refer the interested reader to the books [CAFL 1996, FUL 1969, LOR 1996, SIL 1986, STI 1993, ZASA 1976].

Throughout this chapter let $K$ denote a *perfect field* (cf. Chapter 2) and $\overline{K}$ its algebraic closure. Let $L$ be an extension field of $K$. Its absolute Galois group $\text{Aut}_L(\overline{L})$ is denoted by $G_L$.

## 4.1 Algebraic varieties

We first introduce the basic notions of algebraic geometry in projective and affine spaces.

### 4.1.1 Affine and projective varieties

Before we can define curves we need to introduce the space where they are defined and it is also useful to have coordinates at hand.

#### 4.1.1.a Projective space

We shall fix a field $K$ as above. As a first approximation of the $n$-dimensional projective space $\mathbb{P}^n/K := \mathbb{P}^n$ over $K$ we describe its set of $\overline{K}$-rational points as the set of $(n+1)$-tuples

$$\mathbb{P}^n(\overline{K}) := \{(X_0 : X_1 : \ldots : X_n) \mid X_i \in \overline{K}, \text{ at least one } X_i \text{ is nonzero}\}/\sim$$

where $\sim$ is the equivalence relation

$$(X_0 : X_1 : \ldots : X_n) \sim (Y_0 : Y_1 : \ldots : Y_n) \iff \exists \lambda \in \overline{K} \; \forall i : X_i = \lambda Y_i.$$

The coordinates are called *homogeneous coordinates*. The equivalence classes are called *projective points*. Next we endow this set with a $K$-rational structure by using Galois theory.

**Definition 4.1** Let $L$ be an extension field of $K$ contained in $\overline{K}$. Its absolute Galois group $G_L$ operates on $\mathbb{P}^n(\overline{K})$ via the action on the coordinates. Obviously, this preserves the equivalence classes of $\sim$. The set of $L$-rational points $\mathbb{P}^n(L)$ is defined to be equal to the subset of $\mathbb{P}^n$ fixed by $G_L$. In terms of coordinates this means:

$$\mathbb{P}^n(L) := \{(X_0 : \ldots : X_n) \in \mathbb{P}^n \mid \exists \lambda \in \overline{K} \; \forall i : \lambda X_i \in L\}.$$

Note that in this definition for an $L$-rational point one does not automatically have $X_i \in L$. However, if $X_j \neq 0$ then $\forall i : X_i/X_j \in L$.

Let $P \in \mathbb{P}^n(\overline{K})$. The smallest extension field $L$ of $K$ such that $P \in \mathbb{P}^n(L)$ is denoted by $K(P)$ and called the *field of definition* of $P$. One has

$$K(P) = \bigcap_{G_L \cdot P = P} L.$$

Let $S \subset \mathbb{P}^n(\overline{K})$ and $L$ be a subfield of $\overline{K}$ containing $K$. Then $S$ is called *defined over* $L$ if and only if for all $P \in S$ the field $K(P)$ is contained in $L$, or, equivalently, $G_L \cdot S = S$.

**Remark 4.2** Let $L$ be any extension field of $K$, not necessarily contained in $\overline{K}$. We can define points in the $n$-dimensional projective space over $L$ in an analogous way and an embedding of $\overline{K}$ into $\overline{L}$ induces a natural inclusion of points of the projective space over $K$ to the one over $L$. This is a special case of *base change*.

To be more rigorous, one should not only look at the points of $\mathbb{P}^n$ over extension fields of $K$ as sets, but endow $\mathbb{P}^n$ with the structure of a topological space with respect to the Zariski topology. This will explain the role of the base field $K$ much better.

First recall that a polynomial $f(X_0, \ldots, X_n) \in K[X_0, \ldots, X_n]$ is called *homogeneous of degree* $d$ if it is the sum of monomials of the same degree $d$. This is equivalent to requiring that $f(\lambda X_0, \ldots, \lambda X_n) = \lambda^d f(X_0, \ldots, X_n)$ for all $\lambda \in \overline{K}$. Especially, this implies that the set

$$D_f(L) := \{P \in \mathbb{P}^n(L) \mid f(P) \neq 0\}$$

is well defined.

One defines a topology on $\mathbb{P}^n(\overline{K})$ by taking the sets $D_f(\overline{K}) =: D_f$ as basic *open sets*. The $L$-rational points are denoted by $D_f(L) = \mathbb{P}^n(L) \bigcap D_f$. To describe *closed sets* we need the notion

## § 4.1 Algebraic varieties

of homogeneous ideals. An *ideal* $I \subseteq K[X_0, X_1, \ldots, X_n]$ *is homogeneous* if it is generated by homogeneous polynomials. For $I \neq \langle X_0, \ldots, X_n \rangle$, define

$$V_I := \{P \in \mathbb{P}^n(\overline{K}) \mid f(P) = 0, \forall f \in I\}$$

and $V_I(L) = V_I \cap \mathbb{P}^n(L)$. One sees immediately that $V_I$ is well defined. So a subset $S \subset \mathbb{P}^n(\overline{K})$ is closed with respect to the Zariski topology attached to the projective space over $K$ if it is the set of simultaneous zeroes of homogeneous polynomials lying in $K[X_0, \ldots, X_n]$.

**Example 4.3** The set of points of the projective $n$-space $\mathbb{P}^n$ and the empty set $\emptyset$ are closed sets as they are the roots of the constant polynomials 0 and 1. By the same argument they are also open sets.

**Example 4.4** Let $f \in K[X_0, X_1, \ldots, X_n]$ be a homogeneous polynomial. The closed set $V_{(f)}$ is called a *hypersurface*.

**Example 4.5** Define $U_i := D_{X_i}$, thus

$$U_i(L) = \{(X_0 : X_1 : \ldots : X_n) \in \mathbb{P}^n(L) \mid X_i \neq 0\}$$

and let $W_i := V_{(X_i)}$ with

$$W_i(L) = \{(X_0 : X_1 : \ldots : X_n) \in \mathbb{P}^n(L) \mid X_i = 0\}.$$

The $U_i$ are open sets, the $W_i$ are closed.

**Example 4.6** Let $(k_0, \ldots, k_n) \in K^{n+1}$ and not all $k_i = 0$. Take $f_{ij}(X_0, \ldots, X_n) := k_j X_i - k_i X_j$ and $I = (\{f_{ij} \mid 0 \leq i, j \leq n\})$. Obviously, $I$ is a homogeneous ideal and taking $(k_0 : \ldots : k_n)$ as a homogeneous point, $I$ is independent of the representative. Then $V_I(L) = \{(k_0 : \ldots : k_n)\}$, $\forall L$. This shows that $K$-rational points are closed with respect to the Zariski topology. This is not true if $P$ is not defined over $K$. The smallest closed set containing $P$ is the $G_K$-orbit $G_K \cdot P$.

From now on we write $X$ for $(X_0, \ldots, X_n)$. If $T \subset K[X]$ is a finite set of homogeneous polynomials we define $V(T)$ to be the intersection of the $V_{(f_i)}, f_i \in T$. Let $I = (T)$ be the ideal generated by the $f_i$. Then $V(T) = V_I$.

### 4.1.1.b Affine space

As in the projective space we begin with the set of $\overline{K}$-rational points of the *affine space of dimension $n$ over $K$* given by the set of $n$-tuples

$$\mathbb{A}^n := \{(x_1, \ldots, x_n) \mid x_i \in \overline{K}\}.$$

The set of $L$-rational points is given by

$$\mathbb{A}^n(L) := \{(x_1, \ldots, x_n) \mid x_i \in L\}$$

which is the set of $G_L$-invariant points in $\mathbb{A}^n(\overline{K})$ under the natural action on the coordinates.

As in the projective case one has to consider $\mathbb{A}^n$ as a topologic space with respect to the Zariski topology, defined now in the following way: For $f \in K[x_1, \ldots, x_n]$ let

$$D_f(L) := \{P \in \mathbb{A}^n(L) \mid f(P) \neq 0\}$$

and take these sets as base for the open sets.

Closed sets are given in the following way: for an ideal $I \subseteq K[x_1, \ldots, x_n]$ let

$$V_I(L) = \{P \in \mathbb{A}^n(L) \mid f(P) = 0, \forall f \in I\}.$$

A set $S \subset \mathbb{A}^n$ is closed if there is an ideal $I \subseteq K[x_1, \ldots, x_n]$ with $S = V_I$.

**Example 4.7** Let $(k_1, \ldots, k_n) \in \mathbb{A}^n(K)$ and put $f_i = x_i - k_i$ and $I = (\{f_i \mid 1 \leq i \leq n\})$. Then $V_I = \{(k_1, \ldots, k_n)\}$. Hence, the $K$-rational points are closed.

Please note, if $P \in \mathbb{A}^n \smallsetminus \mathbb{A}^n(K)$ the set $\{P\}$ is not closed.

**Remark 4.8** For closed $S \subset \mathbb{A}^n$ assume that $S = V_I$. The ideal $I$ is not uniquely determined by $S$. Obviously there is a maximal choice for such an ideal, and it is equal to the *radical ideal* (cf. [ZASA 1976, pp. 164]) defined as

$$\sqrt{I} = \{f \in K[x_1, \ldots, x_n] \mid \exists k \in \mathbb{N} \text{ with } f^k \in I\}.$$

As in the projective case we take $x$ as a shorthand for $(x_1, \ldots, x_n)$.

### 4.1.1.c Varieties and dimension

To define varieties we use the definition of irreducible sets. A subset $S$ of a topological space is called *irreducible* if it cannot be expressed as the union $S = S_1 \cup S_2$ of two proper subsets closed in $S$. We additionally define that the empty set is not irreducible.

**Definition 4.9** Let $V$ be an affine (projective) closed set. One calls $V$ an *affine (projective) variety* if it is irreducible.

**Example 4.10** The affine 1-space $\mathbb{A}^1$ is irreducible because $K[x_1]$ is a principal ideal domain and so every closed set is the set of zeroes of a polynomial in $x_1$. Therefore, any closed set is either finite or equal to $\mathbb{A}^1$. Since $\mathbb{A}^1$ is infinite it cannot be the union of two proper closed subsets.

From commutative algebra we get a criterion for when a closed set is a variety.

**Proposition 4.11** A subset $V$ of $\mathbb{A}^n$ (resp. $\mathbb{P}^n$) is an affine (projective) variety if and only if $V = V_I$ with $I$ a (homogeneous) prime ideal in $K[x]$ (resp. $K[X]$).

We recall that the Zariski topology is defined relative to the ground field $K$. For extension fields $L$ and given embeddings $\sigma$ of $\overline{K}$ into $\overline{L}$ fixing $K$ we have induced embeddings of $\mathbb{P}^n/K \to \mathbb{P}^n/L$. Due to the obvious embedding of $K[X]$ into $L[X]$ and as the topology depends on these polynomial rings, we can try to compare the Zariski topologies of affine and projective spaces over $K$ with corresponding ones over $L$.

If $L$ is arbitrary, a closed set in the space over $K$ may not remain closed in the space over $L$.

But if $L$ is algebraic over $K$ and if $S$ is closed in the affine (projective) space over $K$ then its embedding $\sigma \cdot S$ is closed over $L$. Namely, if $S = V_I$ with $I \subseteq K[x]$ (resp. $K[X]$) then $\sigma \cdot S = V_{I \cdot L[x]}$ (resp. $\sigma \cdot S = V_{I \cdot L[X]}$).

But varieties over $K$ do not have to be varieties over $L$ since for prime ideals $I$ in $K[x]$ it may not be true that $I \cdot L[x]$ is a prime ideal.

**Example 4.12** Consider $I = (x_1^2 - 2x_2^2) \subseteq \mathbb{Q}[x_1, x_2]$. Over $\mathbb{Q}(\sqrt{2})$ the variety $V_I$ splits because $x_1^2 - 2x_2^2 = (x_1 - \sqrt{2}x_2)(x_1 + \sqrt{2}x_2)$. Therefore, the property of a closed set being a variety depends on the field of consideration.

**Example 4.13** Let $V$ be an affine variety, i.e., a closed set in some $\mathbb{A}^n$ for which the defining ideal $I$ is prime in $K[x]$. The $m$-fold Cartesian product $V^m$ is also a variety, embedded in the affine space $\mathbb{A}^{nm}$. For affine coordinates choose $(x_1^1, \ldots, x_n^1, \ldots, x_1^m, \ldots, x_n^m)$, define $I_i \subseteq K[x^i]$ obtained from $I$ by replacing $x_j$ by $x_j^i$. Then the ideal of $V^m$ is given by $\langle I_1, \ldots, I_m \rangle$.

**Definition 4.14** A variety $V$ of the affine (projective) space $\mathbb{A}^n$ ($\mathbb{P}^n$) over $K$ is called *absolutely irreducible* if it is irreducible as closed set with respect to the Zariski topology of the corresponding spaces over $\overline{K}$.

## § 4.1 Algebraic varieties

**Example 4.15**

(i) The $n$-dimensional spaces $\mathbb{A}^n$ and $\mathbb{P}^n$ are absolutely irreducible varieties as they correspond to the prime ideal $(0)$.

(ii) The sets $V_{(f)}$ and $V_{(F)}$ with $f \in K[x]$ and $F \in K[X]$ are absolutely irreducible if and only if $f$ and $F$ are absolutely irreducible polynomials, i.e., they are irreducible over $\overline{K}$.

(iii) Let $S$ be a *finite* set in an affine or projective space over $K$. The set $S$ is absolutely irreducible if and only if it consists of one ($K$-rational) point.

**Example 4.16** Let $f(x_1, x_2) = x_2^2 - x_1^3 - a_4 x_1 - a_6 \in K[x_1, x_2]$. This polynomial is absolutely irreducible, hence $V_{(f)}$ is an irreducible variety over $K$ and over any extension field of $K$ contained in $\overline{K}$.

The affine and the projective $n$-spaces are *Noetherian*, which means that any sequence of closed subsets $S_1 \supseteq S_2 \supseteq \ldots$ will eventually become stationary, i.e., there exists an index $r$ such that $S_r = S_{r+1} = \ldots$. This holds true as any closed set corresponds to an ideal of $K[x]$ or $K[X]$, respectively, and these rings are Noetherian.

**Definition 4.17** Let $V$ be an affine (projective) variety. The *dimension* $\dim(V)$ is defined to be the supremum on the lengths of all chains $S_0 \supset S_1 \supset \cdots \supset S_n$ of distinct irreducible closed subspaces $S_i$ of $V$. A variety is called a *curve* if it is a variety of dimension 1.

**Example 4.18** The dimension of $\mathbb{A}^1$ is 1 as the only irreducible subsets correspond to nonzero irreducible polynomials in 1 variable. In general, $\mathbb{A}^n$ and $\mathbb{P}^n$ are varieties of dimension $n$.

**Example 4.19** Let $0, 1 \neq f \in K[x_1, x_2]$ be absolutely irreducible. Then $V_{(f)}$ is an affine curve as the only proper subvarieties are points $P \in \mathbb{A}^2$ satisfying $f(P) = 0$.

**Example 4.20** Let $V$ be an affine variety of dimension $d$. Then the Cartesian product (cf. Example 4.13) $V^m$ has dimension $md$ by concatenating the chains of varieties.

### 4.1.1.d Relations between affine and projective space

Here we show how the topologies introduced for $\mathbb{P}^n$ and $\mathbb{A}^n$ are made compatible. For both spaces we defined open and closed sets via polynomials and ideals, respectively.

Let $F \in K[X_0, X_1, \ldots, X_n]$ be a homogeneous polynomial of degree $d$. The process of replacing

$$F(X_0, X_1, \ldots, X_n) \text{ by } F_i := F(x_1, \ldots, x_i, 1, x_{i+1}, \ldots, x_n) \in K[x_1, \ldots, x_n]$$

is called *dehomogenization with respect to* $X_i$. The reverse process takes a polynomial $f \in K[x]$ and maps it to

$$f_i := X_i^d f(X_0/X_i, X_1/X_i, \ldots, X_{i-1}/X_i, X_{i+1}/X_i, \ldots, X_n/X_i),$$

where $d$ is minimal such that $f_i$ is a polynomial in $K[X]$. By applying these transformations, we relate homogeneous (prime) ideals in $K[X]$ to (prime) ideals in $K[x]$ and conversely. So we can expect that we can relate affine spaces with projective spaces including properties of the Zariski topologies.

**Example 4.21** The open sets $U_i = D_{X_i} \subset \mathbb{P}^n$ are mapped to $\mathbb{A}^n$ by dehomogenizing their defining polynomial $X_i$ with respect to $X_i$. The inverse mappings are given by

$$\begin{aligned} \phi_i : \mathbb{A}^n &\to U_i \\ (x_1, \ldots, x_n) &\mapsto (x_1 : \ldots : x_i : 1 : x_{i+1} : \ldots : x_n) \end{aligned}$$

Therefore, for any $0 \leq i \leq n$ we have a canonical bijection between $U_i$ and $\mathbb{A}^n$ which is a homeomorphism as it maps closed sets of $U_i$ to closed sets in $\mathbb{A}^n$.

The sets $U_0, \ldots, U_n$ cover the projective space $\mathbb{P}^n$. This covering is called the *standard covering*. The maps $\phi_i$ can be seen as inclusions $\mathbb{A}^n \subset \mathbb{P}^n$.

If $V$ is a projective closed set such that $V = V_{I(V)}$ with homogeneous ideal $I(V) \subseteq \overline{K}[X_0, \ldots, X_n]$ we denote by $V_i$ the set $\phi_i^{-1}(V \cap U_i)$ for $0 \leq i \leq n$. The resulting set is a closed affine set with ideal obtained by dehomogenizing all polynomials in $I(V)$ with respect to $X_i$. This way, $V$ is covered by the $n+1$ sets $\phi_i(V_i)$.

For the inverse process we need a further definition:

**Definition 4.22** Let $V_I \subseteq \mathbb{A}^n$ be an affine closed set. Using one of the $\phi_i$, embed $V_I$ into $\mathbb{P}^n$ by

$$V_I \subset \mathbb{A}^n \xrightarrow{\phi_i} \mathbb{P}^n.$$

The *projective closure* $\overline{V}_I$ of $V_I$ is the closed projective set defined by the ideal $\overline{I}$ generated by the homogenized polynomials $\{f_i \mid f \in I\}$.

The points added to get the projective closure are called *points at infinity*. Note that in the definition we need to use the ideal *generated* by the $f_i$'s, a set of generators of $I$ does not automatically homogenize to a set of generators of $\overline{I}$. These processes lead to the following lemma that describes the relation between affine and projective varieties.

**Lemma 4.23** We choose one embedding $\phi_i$ from $\mathbb{A}^n$ to $\mathbb{P}^n$ and identify $\mathbb{A}^n$ with its image. Let $V \subseteq \mathbb{A}^n$ be an affine variety, then $\overline{V}$ is a projective variety and

$$V = \overline{V} \cap \mathbb{A}^n.$$

Let $V \subseteq \mathbb{P}^n$ be a projective variety, then $V \cap \mathbb{A}^n$ is an affine variety and either

$$V \cap \mathbb{A}^n = \emptyset \text{ or } V = \overline{V \cap \mathbb{A}^n}.$$

If $V$ is a projective variety defined over $K$ then $V \cap \mathbb{A}^n$ is empty or an affine variety defined over $K$. There is always at least one $i$ such that $V \cap \phi_i \mathbb{A}^n =: V_{(i)}$ is nonempty. We call $V_{(i)}$ a *nonempty affine part of* $V$.

For example, let $C \subset \mathbb{P}^n$ be a projective curve. The intersections $C \cap U_i$ lead to affine curves $C_{(i)}$. Starting from an affine curve $C_a \subset \mathbb{A}^n$ one can embed the points of $C_a$ into $\mathbb{P}^n$ via $\phi_i$. The result will not be closed in the Zariski topology of $\mathbb{P}^n$ so one needs to include points from $\mathbb{P}^n \setminus U_i$ to obtain the projective closure $\overline{C}_a$.

**Example 4.24** Consider the projective line $\mathbb{P}^1$. It is covered by two copies of the affine line $\mathbb{A}^1$. When embedding $\mathbb{A}^1$ in $\mathbb{P}^1$ via $\phi_0$ we miss a single point $(0:1)$ which is called the *point at infinity* denoted by $\infty$.

**Example 4.25** Let $V$ be a projective variety embedded in $\mathbb{P}^n$. To define the $m$-fold Cartesian product one uses the construction for affine varieties (cf. Example 4.13) for affine parts $V_a$ and "glues them together."

*Warning:* it is not possible to embed $V^m$ in $\mathbb{P}^{mn}$ in general. One has to use constructions due to Segre [HAR 1977, pp. 13] and ends up in a higher dimensional space.

## 4.2 Function fields

**Definition 4.26** Let $V$ be an affine variety in the $n$-dimensional space $\mathbb{A}^n$ over $K$ with corresponding prime ideal $I$. Denote by
$$K[V] := K[x_1, \ldots, x_n]/I$$
the quotient ring of $K[x_1, \ldots, x_n]$ modulo the ideal $I$. This is an integral domain, called the coordinate ring of $V$. The *function field $K(V)$ of $V$* is the quotient field
$$K(V) := \mathrm{Quot}(K[V]).$$
The maximal algebraic extension of the field $K$ contained in $K(V)$ is called the *field of constants of $K(V)/K$*.

**Definition 4.27** Let $V$ be a projective variety over $K$. Let $V_a \subseteq \mathbb{A}^n$ be a nonempty affine part of $V$. Then the *function field $K(V)$* is defined as $K(V_a)$.

One can check that $K(V)$ is independent of the choice of the affine part $V_a$. Thus, the notation $K(V)$ makes sense. But note that $K[V_a]$ depends on the choice of $V_a$.

Obviously, the elements $f \in K(V)$ can be represented by fractions of polynomials $f = g/h$, $f, g \in K[x_1, \ldots, x_n]$ or as fractions of homogeneous polynomials of the same degree $f = g/h$, $f, g \in K[X_0, X_1, \ldots, X_n]$. Then functions $f_1 = g_1/h_1$ and $f_2 = g_2/h_2$ are equal if $g_1 h_2 - g_2 h_1 \in I(V)$.

In Example 4.12, the splitting was induced by an algebraic extension of the ground field. We can formulate a criterion for $V$ to be absolutely irreducible:

**Proposition 4.28** A variety $V$ is absolutely irreducible if and only if $K$ is algebraically closed in $K(V)$, i.e., $K$ is the full constant field of $K(V)$ (cf. [STI 1993, Cor. III.6.7]).

**Example 4.29** Consider $\mathbb{A}^n$ as affine part of $\mathbb{P}^n$. Its coordinate ring $K[\mathbb{A}^n] = K[x_1, \ldots, x_n]$ is the polynomial ring in $n$ variables. The function field of $\mathbb{P}^n$ is the field of rational functions in $n$ variables.

From now on, we assume that $V$ is absolutely irreducible.

Let $L$ be an algebraic extension field of $K$. As pointed out above the set $V$ is closed under the Zariski topology related to the new ground field $L$ and again irreducible by assumption. We denote this variety by $V_L$. We get

**Proposition 4.30** If $V$ is affine then $K[V_L] = K[V] \cdot L$. If $V$ is affine or projective then $K(V_L) = K(V) \cdot L$.

The proof of this proposition follows immediately from the fact that for affine $V$ with corresponding prime ideal $I$ we get $V_L = V_{I \cdot L[x]}$.

**Example 4.31** Consider the projective curve $C = \mathbb{P}^1$ and the affine part $C_a = \mathbb{A}^1$. For any field $K \subseteq L \subseteq \overline{K}$ the coordinate ring of $C_a$ is the polynomial ring in one variable $L[C_a] = L[x_1]$ and the function field is the function field in one variable $L(x_1)$.

A function field $K(V)$ of a projective variety $V$ is finitely generated. Since $K$ is perfect the extension is also separably generated. Therefore, the transcendence degree of $K(V)/K$ is finite.

**Lemma 4.32** Let $K(V)$ be the function field corresponding to the projective variety $V$. The dimension of $V$ is equal to the transcendence degree of $K(V)$.

### 4.2.1 Morphisms of affine varieties

We want to define maps between affine varieties that are continuous with respect to the Zariski topologies. We shall call such maps *morphisms*. We begin with $V = \mathbb{A}^n$.

**Definition 4.33** A *morphism $\varphi$ from $\mathbb{A}^n$ to the affine line $\mathbb{A}^1$* is given by a polynomial $f(x) \in K[x]$ and defined by

$$\varphi : \mathbb{A}^n \to \mathbb{A}^1$$
$$P = (a_1, \ldots, a_n) \mapsto f\big((a_1, \ldots, a_n)\big) =: f(P).$$

One sees immediately that $f$ is uniquely determined by $\varphi$.

To ease notation we shall identify $f$ with $\varphi$. Hence the set of morphisms from $\mathbb{A}^n$ to the affine line is identified with $K[x]$. In fact we can make the set of morphisms to a $K$-algebra in the usual way by adding and multiplying values. As $K$-algebra it is then isomorphic to $K[x]$.

As desired, the map $f$ is continuous with respect to the Zariski topology. It maps closed sets to closed sets, varieties to varieties, and for extension fields $L$ of $K$ we get $f\big(\mathbb{A}^n(L)\big) \subset \mathbb{A}^1(L)$.

**Definition 4.34** A *morphism $\varphi$ from $\mathbb{A}^n$ to $\mathbb{A}^m$* (for $n, m \in \mathbb{N}$) is given by an $m$-tuple

$$\big(f_1(x), \ldots, f_m(x)\big)$$

of polynomials in $K[x]$ mapping $P \in \mathbb{A}^n$ to $\big(f_1(P), \ldots, f_m(P)\big)$.

Since $\varphi$ is determined by $f_1, \ldots, f_m$, the set of morphisms from $\mathbb{A}^n$ to $\mathbb{A}^m$ can be identified with $K[x]^m$. Again one checks without difficulty that morphisms are continuous with respect to the Zariski topology and map varieties to varieties.

Let $V$ be an affine variety in $\mathbb{A}^n$ with corresponding prime ideal $I \subset K[x]$.

**Definition 4.35** A *morphism from $V \subset \mathbb{A}^n$ to a variety $W \subset \mathbb{A}^m$* is given by the restriction to $V$ of a morphism from $\mathbb{A}^n$ to $\mathbb{A}^m$ with image in $W$.

We denote the set of morphisms from $V$ to $W$ by $\mathrm{Mor}_K(V, W)$.

**Example 4.36** As basic example take $W = \mathbb{A}^1$. For $V = \mathbb{A}^1$ we already have $\mathrm{Mor}_K(\mathbb{A}^1, \mathbb{A}^1) = K[x]$. For an arbitrary variety $V = V_I$ one has that $\mathrm{Mor}_K(V, \mathbb{A}^1)$ is as $K$-algebra isomorphic to $K[V] = K[x]/I$.

**Remark 4.37** Take $\varphi \in \mathrm{Mor}_K(V, W)$ and $f \in \mathrm{Mor}_K(W, \mathbb{A}^1) = K[W]$. Then $f \circ \varphi$ is an element of $\mathrm{Mor}_K(V, \mathbb{A}^1) = K[V]$, and so we get an *induced $K$-algebra morphism*

$$\varphi^* : K[W] \to K[V].$$

The morphism $\varphi^*$ is injective if and only if $\varphi$ is surjective. If $\varphi^*$ is surjective then $\varphi$ is injective.

**Definition 4.38** The map $\varphi$ is an isomorphism if and only if $\varphi^*$ is an isomorphism. This means that the inverse map of $\varphi$ is again a morphism, i.e., given by polynomials.

Two varieties $V$ and $W$ are called isomorphic if there exists an isomorphism from $V$ to $W$, and we have seen that this is equivalent to the fact that $K[V]$ is isomorphic to $K[W]$ as $K$-algebra.

**Example 4.39** Assume that $\mathrm{char}(K) = p > 0$. Then the exponentiation with $p$ is an automorphism $\phi_p$ of $K$ since $K$ is assumed to be perfect. The map $\phi_p$ is called the (absolute) Frobenius automorphism of $K$ (cf. Section 2.3.2).

## § 4.2 Function fields

We can extend $\phi_p$ to points of projective spaces over $K$ by sending the point $(X_0, \ldots, X_n)$ to $(X_0^p, \ldots, X_n^p)$. We apply $\phi_p$ to polynomials over $K$ by applying it to the coefficients.

If $V$ is a projective variety over $K$ with ideal $I$ we can apply $\phi_p$ to $I$ and get a variety $\phi_p(V)$ with ideal $\phi_p(I)$. The points of $V$ are mapped to points on $\phi_p(V)$.

The corresponding morphism from $V$ to $\phi_p(V)$ is called the *Frobenius morphism* and is again denoted by $\phi_p$. We note that $\phi_p$ is *not* an isomorphism as the polynomial rings $K[V]/K[\phi_p(V)]$ form a proper inseparable extension.

### 4.2.2 Rational maps of affine varieties

Let $V \subset \mathbb{A}^n$ be an affine variety with ideal $I = I(V)$ and take $\varphi \in K[V]$ with representing element $f \in K[x]$.

By definition, the set $D_f$ consists of the points $P$ in $\mathbb{A}^n$ in which $f(P) \neq 0$. It is open in the Zariski topology of $\mathbb{A}^n$, and hence $U_\varphi := D_f \cap V$ is open in $V$. Its complement $V_\varphi$ in $V$ is the zero locus of $\varphi$. It is not equal to $V$ if and only if $U_\varphi$ is not empty, and this is equivalent to $f \notin I$.

We assume now that $f \notin I$. For $P \in U_\varphi$ define $(1/\varphi)(P) := f(P)^{-1}$.

**Definition 4.40** Assume that $U$ is a nonempty open set of an affine variety $V$ and let the map $r_U$ be given by

$$\begin{aligned} r_U : U &\to \mathbb{A}^1 \\ P &\mapsto (\psi/\varphi)(P) \end{aligned}$$

for some $\psi, \varphi \in K[V]$ and $U \subset U_\varphi$. Then $r_U$ is a *rational map from $V$ to $\mathbb{A}^1$* with *definition set* $U$.

We introduce an equivalence relation on rational maps: for given $V$ the rational map $r_U$ is equivalent to $r'_{U'}$ if for all points $P \in U \cap U'$ we have: $r_U(P) = r'_{U'}(P)$.

**Definition 4.41** The equivalence class of a rational map from $V$ to $\mathbb{A}^1$ is called a *rational function* on $V$.

**Proposition 4.42** Let $V$ be an affine variety. The set of rational functions on $V$ is equal to $K(V)$. The addition (resp. multiplication) in $K(V)$ corresponds to the addition (resp. multiplication) of rational functions defined by addition (resp. multiplication) of the values.

Let $V \subset \mathbb{A}^n$. As in the case of morphisms we can extend the notion of rational maps from the case $W = \mathbb{A}^1$ to the general case that $W \subset \mathbb{A}^m$ is a variety:

**Definition 4.43** A *rational map $r$ from $V$ to $W$* is an $m$-tuple of rational functions $(r_1, \ldots, r_m)$ with $r_i \in K(V)$ having representatives $R_i$ defined on a nonempty open set $U \subset V$ with $R(U) := (R_1(U), \ldots, R_m(U)) \subset W$.

A rational map $r$ from $V$ to $W$ is *dominant* if (with the notation from above) $R(U)$ is dense in $W$, i.e., if the smallest closed subset in $W$ containing $R(U)$ is equal to $W$.

A rational map $r : V \to W$ is *birational* if there exists an inverse rational map $r' : W \to V$ such that $r' \circ r$ is equivalent to $\mathrm{Id}_V$ and $r \circ r'$ is equivalent to $\mathrm{Id}_W$.

If there exists a birational map from $V$ to $W$ the varieties are called *birationally equivalent*.

**Example 4.44** Consider the rational maps

$$r^{ij} : \mathbb{A}^n \to \mathbb{A}^n,\ r^{ij} = (r_1^{ij}, \ldots, r_n^{ij}),$$

where (for $i \leqslant j$)

$$r_k^{ij}(x_1, \ldots, x_n) := \begin{cases} x_k/x_j, & k < i \\ 1/x_j, & k = i \\ x_{k-1}/x_j, & i < k \leqslant j \\ x_k/x_j, & j < k. \end{cases}$$

The case $i > j$ works just the same. For fixed $j$ and arbitrary $i$ the maps $r^{ij}$ are defined on $D_{x_j}$.

Using the embeddings $\phi_i$ of $\mathbb{A}^n$ into $\mathbb{P}^n$ one has the description

$$r^{ij} = \phi_j^{-1} \circ \phi_i.$$

The inverse map is just $r^{ji}$ and so $r^{ij}$ represents a birational map regular on $D_{x_j} \cap D_{x_i}$. It describes the coordinate transition of affine coordinates with respect to $\phi_i$ to affine coordinates with respect to $\phi_j$ on $\mathbb{P}^n$.

**Proposition 4.45** Assume that the rational map $r$ from $V$ to $W$ is dominant. Then the composition of $r$ with elements in $K(W)$ induces a field embedding $r^*$ of $K(W)$ into $K(V)$ fixing elements in $K$, generalizing the definition made for morphisms in Remark 4.37.

If $r$ is birational then $K(V)$ is isomorphic to $K(W)$ as $K$-algebra.

**Example 4.46** A projective curve $C$ corresponds to a function field of transcendence degree 1. Since $K$ is perfect, there are elements $x_1, x_2 \in K(C)$ and an irreducible polynomial $f(x_1, x_2)$ such that $K(C) = \mathrm{Quot}\bigl(K[x_1, x_2]/(f(x_1, x_2))\bigr)$. Hence, $C$ (and every affine part of dimension 1 of $C$) is birationally equivalent to the plane curve $V_{(f)}$ and of course to its projective closure $\overline{V}_{(f)} \subset \mathbb{P}^2$.

**Example 4.47** We consider again the Frobenius morphism $\phi_p$ from Example 4.39. The map

$$\phi_p^* : K\bigl(\phi_p(V)\bigr) \to K(V)$$

has as its image $K(V)^p$ since the coordinate functions of $V$ are exponentiated by $p$ under the map $\phi_p$.

### 4.2.3 Regular functions

We continue to assume that $V$ is an affine variety.

**Definition 4.48** A rational function $f \in K(V)$ is *regular at a point* $P \in V$ if $f$ has as representative a rational map $\tilde{f}$ with set of definition $U$ containing $P$.

In other words $f$ is regular at $P \in V$ if there is an open neighborhood $U$ of $P$ where $f_{|U} = (g/h)_{|U}$ for $g, h \in K[x]$ and $P \in D_h$. If this is the case we say that $f$ is defined at $P$ with value $f(P) = g(P)/h(P)$.

**Definition 4.49** For two varieties $V \subset \mathbb{A}^n$, $W \subset \mathbb{A}^m$ a rational map $r : V \to W$ *is regular at $P$* if there is a nonempty open set $U$ of $V$ containing $P$ such that the restriction of $r$ to $U$ is given by an $m$-tuple of rational maps defined on $U$.

In other words: a map $r$ is regular if locally it can be represented via $m$-tuples of quotients of polynomials in $K[x]$ which are defined at $P \in U$.

### 4.2.4 Generalization to projective varieties

We want to generalize the definitions of morphisms to projective varieties.

**Definition 4.50** Let $V \subset \mathbb{P}^n$ and $W \subset \mathbb{P}^m$ be projective varieties. Let $\varphi$ be a map from $V$ to $W$ such that the following holds:

(i) The set $V = \bigcup_{i=1}^{n} V_i$ with $V_i = V \cap U_i$ the standard affine parts of $V$.
(ii) The morphism $\varphi_i := \varphi_{|V_i}$ is an affine morphism to an affine part $W_i$ of $W$.
(iii) The polynomials $\left(f_1^i(x), \ldots, f_m^i(x)\right)$ describing $\varphi$ on $V_i$ with respect to the standard affine coordinates are transformed into the polynomials $\left(f_1^j(x), \ldots, f_m^j(x)\right)$ describing $\varphi_j$ in the standard affine coordinates related to $V_j$ under the coordinate transformation considered in Example 4.44.

Then $\varphi$ *is a morphism from $V$ to $W$*: $\varphi \in \mathrm{Mor}_K(V, W)$.

The notions of rational functions of projective varieties $V$ and of regularity in a point $P$ of such functions are easier to define since they are local definitions.

We define rational maps as equivalence classes of rational maps defined on the affine parts of $V$ compatible with the transition maps on intersections of standard affine pieces $U_i$ (cf. Example 4.44). To define regularity at $P$ we first choose an affine part $V_i$ of $V$ containing $P$ and then require that there is an open neighborhood $U$ of $P$ in $V_i$ such that the rational map obtained by restriction is defined on $U$.

A rational map from $W$ to $\mathbb{P}^1$ is called a *rational function of $V$*.

**Proposition 4.51** The set of rational functions on a projective variety $V$ forms a field isomorphic to $K(V)$ which is equal to the field of rational functions on a nonempty affine part of $V$.

A function $f : V \to K$ is regular at $P \in V$ if there is an open neighborhood $U$ of $P$ where $f = g/h$ for homogeneous polynomials $g, h \in K[X]$ of the same degree and $h(Q) \neq 0$, $\forall Q \in U$.

## 4.3 Abelian varieties

We want to use the concepts introduced above for a structure that will become most important for the purposes of the book.

**Remark 4.52** Already in the definition we shall restrict ourselves to the cases that will be considered in the sequel. So we shall assume throughout the whole section that all varieties are defined over $K$ and are *absolutely irreducible*.

### 4.3.1 Algebraic groups

We combine the concept of groups with the concept of varieties in a functorial way.

**Definition 4.53** An (absolutely irreducible) *algebraic group $\mathcal{G}$ over a field $K$* is an (affine or projective) absolutely irreducible variety defined over $K$ together with three additional ingredients:

(i) the addition, i.e., a morphism
$$m : \mathcal{G} \times \mathcal{G} \to \mathcal{G},$$

(ii) the inverse, i.e., a morphism
$$i : \mathcal{G} \to \mathcal{G},$$

(iii) the neutral element, i.e., a $K$-rational point

$$0 \in \mathcal{G}(K),$$

satisfying the usual group laws:

$$m \circ (\mathrm{Id}_\mathcal{G} \times m) = m \circ (m \times \mathrm{Id}_\mathcal{G}) \text{ (associativity)},$$

$$m_{|\{0\} \times \mathcal{G}} = p_2,$$

where $p_2$ is the projection of $\mathcal{G} \times \mathcal{G}$ on the second argument, and

$$m \circ (i \times \mathrm{Id}_\mathcal{G}) \circ \delta_\mathcal{G} = c_0,$$

where $\delta_\mathcal{G}$ is the diagonal map from $\mathcal{G}$ to $\mathcal{G} \times \mathcal{G}$ and $c_0$ is the map which sends $\mathcal{G}$ to $0$.

Let $L$ be an extension field of $K$. Let $\mathcal{G}(L)$ denote the set of $L$-rational points. The set $\mathcal{G}(L)$ is a group in which the sum and the inverse of elements are computed by evaluating morphisms that are defined over $K$, that do not depend on $L$, and in which the neutral element is the point $0$.

A surprising fact is that if $\mathcal{G}$ is a projective variety the group law $m$ is necessarily commutative.

**Definition 4.54** Projective algebraic groups are called *abelian varieties*.

From now on we shall require $m$ to be *commutative*. We can use a classification theorem which yields that $\mathcal{G}$ is an extension of an abelian variety by an affine (i.e., the underlying variety is affine) algebraic group. So, for cryptographic purposes we can assume that $\mathcal{G}$ is either affine or an abelian variety as by Theorem 2.23 one is interested only in (sub)groups of prime order.

Affine commutative group schemes that are interesting for cryptography are called *tori*. The reader can find the definition and an interesting discussion on how to use them for DL systems in [SIRU 2004].

**Remark 4.55** To make the connection with abelian groups more obvious we replace "$m(P, Q)$" with the notation $P \oplus Q$ for $P, Q \in \mathcal{G}(\overline{K})$ and $i(P)$ by $-P$.

We shall concentrate on abelian varieties from now on and shall use as standard notation $\mathcal{A}$ instead of $\mathcal{G}$.

### 4.3.2 Birational group laws

Assume that we are given an abelian variety $\mathcal{A}$. Since we want to use $\mathcal{A}$ for DL systems we shall not only need structural properties of $\mathcal{A}$ but explicitly compute with its points. In general this seems to be hopeless. Results of Mumford [MUM 1966] and Lange–Ruppert [LARU 1985] show that the number of coordinate functions and the degree of the addition formulas both grow exponentially with the dimension of the abelian variety. Therefore, we have to use special abelian varieties on which we can describe the addition at least on open affine parts.

By definition $\mathcal{A}$ can be covered by affine subvarieties $V_i$. Choose one such $V := V_i$. For $l$ depending on $V$ one finds coordinate functions $X_1, \ldots, X_l$ defining $V$ by polynomial relations

$$\{f_1(X_1, \ldots, X_l), \ldots, f_k(X_1, \ldots, X_l)\}.$$

The $L$-rational points $V(L) \subset \mathcal{A}(L)$ are the elements $(x_1, \ldots, x_l) \in L^l$, where the polynomials $f_i$ vanish simultaneously. The addition law can be restricted to $V \times V$ and induces a morphism

$$m_V : V \times V \to \mathcal{A}.$$

## § 4.3 Abelian varieties

For generic points of $V \times V$ the image of $m_V$ is again contained in $V$. So $m_V$ is given by *addition functions* $R_i \in K(X_1, \ldots, X_l; Y_1, \ldots, Y_l)$ such that for pairs of $L$-rational points in $V \times V$ we get

$$((x_1, \ldots, x_l) \oplus (y_1, \ldots, y_l)) = (R_1(x_1, \ldots, x_l; y_1, \ldots, y_l), \ldots, R_l(x_1, \ldots x_l; y_1, \ldots, y_l)).$$

**Remark 4.56** This is a birational description of the addition law that is true outside proper closed subvarieties of $V \times V$. The set of points where this map is not defined is of small dimension and hence with high probability one will not run into it by chance. But it can happen that we use pairs of points on purpose (e.g., lying on the diagonal in $V \times V$) for which we need an extra description of $m$.

We shall encounter examples of abelian varieties with birational description of the group law in later chapters. In fact it will be shown that one can define abelian varieties from elliptic and hyperelliptic curves — they constitute even the main topics of this book.

### 4.3.3 Homomorphisms of abelian varieties

We assume that $\mathcal{A}$ and $\mathcal{B}$ are abelian varieties over $K$ with addition laws $\oplus$ (resp. $\oplus'$). Let $\varphi$ be an element of $\mathrm{Mor}_K(\mathcal{A}, \mathcal{B})$.

**Example 4.57** Let $P \in \mathcal{A}(K)$ and define

$$\begin{aligned} t_P : \mathcal{A} &\to \mathcal{A} \\ Q &\mapsto P \oplus Q. \end{aligned}$$

Here, $t_P$ is called the *translation by* $P$ and lies in $\mathrm{Mor}_K(\mathcal{A}, \mathcal{A})$.

A surprising fact is that for all $\varphi \in \mathrm{Mor}_K(\mathcal{A}, \mathcal{B})$ we have

$$\varphi(P \oplus Q) = \varphi(P) \oplus' \varphi(Q)$$

for all points $P, Q$ of $\mathcal{A}$ if and only if $\varphi(0)$ is the neutral element of $\mathcal{B}$. In other words every morphism from $\mathcal{A}$ to $\mathcal{B}$ is a *homomorphism* with respect to the addition laws up to the translation map $t_{-(\varphi(0))}$ in $\mathcal{B}$. The set of homomorphisms from $\mathcal{A}$ to $\mathcal{B}$ is denoted by $\mathrm{Hom}_K(\mathcal{A}, \mathcal{B})$.

Let $L$ be an extension field of $K$ and take $\varphi \in \mathrm{Hom}_K(\mathcal{A}, \mathcal{B})$. We get a group homomorphism $\varphi_L : \mathcal{A}(L) \to \mathcal{B}(L)$ which is given by evaluating polynomials with coefficients in $K$. An important observation is that $\varphi_L$ commutes with the action of the Galois group $G_K$ of $K$.

The set of homomorphisms $\mathrm{Hom}_K(\mathcal{A}, \mathcal{B})$ becomes a $\mathbb{Z}$-module in the usual way: for $\varphi_1, \varphi_2 \in \mathrm{Hom}_K(\mathcal{A}, \mathcal{B})$ and points $P$ of $\mathcal{A}$ define

$$(\varphi_1 + \varphi_2)(P) := (\varphi_1(P) \oplus \varphi_2(P)).$$

In many cases it is useful to deal with vector spaces instead of modules, and so we define

$$\mathrm{Hom}_K(\mathcal{A}, \mathcal{B})^0 := \mathrm{Hom}_K(\mathcal{A}, \mathcal{B}) \otimes_{\mathbb{Z}} \mathbb{Q}.$$

In the next chapter we shall see that $\mathrm{Hom}_K(\mathcal{A}, \mathcal{B})^0$ is a finite dimensional vector space over $\mathbb{Q}$.

**Remark 4.58** Homomorphisms of abelian varieties behave in a natural way under base change: let $L$ be an extension field of $K$ and let $\mathcal{A}_L, \mathcal{B}_L$ be the abelian varieties obtained by scalar extension to $L$, $\mathrm{Hom}_L(\mathcal{A}, \mathcal{B}) := \mathrm{Hom}_L(\mathcal{A}_L, \mathcal{B}_L)$. The Galois group $G_L$ acts in a natural way on $\mathrm{Mor}_L(\mathcal{A}_L, \mathcal{B}_L)$ and hence on $\mathrm{Hom}_L(\mathcal{A}, \mathcal{B})$.

**Lemma 4.59** With the notations from above we get

(i) Let $L_0$ be the algebraic closure of $K$ in $L$. Then $\mathrm{Hom}_L(\mathcal{A},\mathcal{B}) = \mathrm{Hom}_{L_0}(\mathcal{A},\mathcal{B})$.

(ii) For $L$ contained in $\overline{K}$ we get $\mathrm{Hom}_L(\mathcal{A},\mathcal{B}) = \mathrm{Hom}_{\overline{K}}(\mathcal{A},\mathcal{B})^{G_L}$.

Because of the next results we can think of abelian varieties as behaving like abelian groups.

**Proposition 4.60** Take $\varphi \in \mathrm{Hom}_K(\mathcal{A},\mathcal{B})$.

(i) The image $\mathrm{Im}(\varphi)$ of $\varphi$ is a subvariety of $\mathcal{B}$, which becomes an abelian variety by restricting the addition law from $B$, i.e., it is an abelian subvariety of $\mathcal{B}$.

(ii) The kernel $\ker(\varphi)$ of $\varphi$ is by definition the inverse image of $0_\mathcal{B}$. It is closed (in the Zariski topology) in $\mathcal{A}$. Its points consist of all points in $\mathcal{A}(\overline{K})$ that are mapped to $0_\mathcal{B}$ by $\varphi_{\overline{K}}$ and hence form a subgroup of $\mathcal{A}(\overline{K})$.

(iii) The kernel $\ker(\varphi)$ contains a maximal absolutely irreducible subvariety $\ker(\varphi)^0$ containing $0_\mathcal{A}$. This subvariety is called the *connected component of the unity of* $\ker(\varphi)$. It is an abelian subvariety of $\mathcal{A}$.

(iv) For the dimension one has
$$\dim\bigl(\mathrm{Im}(\varphi)\bigr) + \dim\bigl(\ker(\varphi)^0\bigr) = \dim(\mathcal{A}).$$

**Remark 4.61** Warning: in general it is not true that the sequence of abelian groups
$$0 \to \ker(\varphi)(L) \to \mathcal{A}(L) \to \mathrm{Im}\bigl(\varphi(L)\bigr) \to 0$$
is exact. This holds, however, if $L = \overline{K}$.

## 4.3.4 Isomorphisms and isogenies

To study abelian varieties it is (as usual) important to have an insight into isomorphisms between them. Very closely related to isomorphisms are homomorphisms which preserve the dimension of the abelian variety. They are intensively used both in theory and in applications to cryptography.

**Definition 4.62** We assume that $\mathcal{A},\mathcal{B}$ are abelian varieties over $K$.

(i) The map $\varphi \in \mathrm{Hom}_K(\mathcal{A},\mathcal{B})$ is an *isogeny* if and only if $\mathrm{Im}(\varphi) = \mathcal{B}$ and $\ker(\varphi)$ is finite.

(ii) The morphism $\varphi$ is an *isomorphism* if and only if there is a $\psi \in \mathrm{Hom}_K(\mathcal{B},\mathcal{A})$ with $\varphi \circ \psi = \mathrm{Id}_\mathcal{B}$ and $\psi \circ \varphi = \mathrm{Id}_\mathcal{A}$. So necessarily one has $\ker(\varphi) = \{0_\mathcal{A}\}$.

(iii) The variety $\mathcal{A}$ *is isogenous to* $\mathcal{B}$ ($\mathcal{A} \sim \mathcal{B}$) if and only if there exists an isogeny in $\mathrm{Hom}_K(\mathcal{A},\mathcal{B})$.

(iv) The variety $\mathcal{A}$ *is isomorphic to* $\mathcal{B}$ ($\mathcal{A} \simeq \mathcal{B}$) if and only if there exists an isomorphism in $\mathrm{Hom}_K(\mathcal{A},\mathcal{B})$.

Let $\varphi \in \mathrm{Hom}_K(\mathcal{A},\mathcal{B})$ be dominant. By mapping $f \in K(\mathcal{B})$ to $f \circ \varphi$ we get an injection $\varphi^*$ of $K(\mathcal{B})$ into $K(\mathcal{A})$ (cf. Remark 4.45).

**Proposition 4.63** The homomorphism $\varphi \in \mathrm{Hom}_K(\mathcal{A},\mathcal{B})$ is an isogeny if and only if $\dim(\mathcal{A}) = \dim(\mathcal{B})$ and $\dim\bigl(\ker(\varphi)^0\bigr) = 0$.

Equivalently we have that $\varphi$ is dominant and $K(\mathcal{A})$ is a finite algebraic extension of $\varphi^*\bigl(K(\mathcal{B})\bigr)$.

The relations $\simeq$ and $\sim$ (cf. Corollary 4.76) are equivalence relations between abelian varieties. Hence we can speak about ($K$-)isogeny classes (resp. ($K$-)isomorphism classes) of abelian varieties over $K$.

## § 4.3 Abelian varieties

**Definition 4.64** Let $\varphi$ be an isogeny from $\mathcal{A}$ to $\mathcal{B}$. The *degree of* $\varphi$ is defined as $[K(\mathcal{A}) : \varphi^*(K(\mathcal{B}))]$.

The isogeny $\varphi$ is called *separable* if and only if $K(\mathcal{A})/\varphi^*(K(\mathcal{B}))$ is a separable extension. It is called *(purely) inseparable* if $K(\mathcal{A})/\varphi^*(K(\mathcal{B}))$ is purely inseparable. In this case $\ker(\varphi) = \{0\}$ but nevertheless $\varphi$ is not an isomorphism in general.

As for abelian groups we can describe the abelian varieties that are isomorphic to a homomorphic image of $\mathcal{A}$.

Let $\mathcal{C}$ be a closed set (with respect to the Zariski topology) in $\mathcal{A}$ with $\mathcal{C}(\overline{K})$ a subgroup of $\mathcal{A}(\overline{K})$. Then there exists an (up to $K$-isomorphisms unique) abelian variety $\mathcal{B}$ defined over $K$ and a unique $\pi := \pi_\mathcal{C} \in \mathrm{Hom}_K(\mathcal{A}, \mathcal{B})$ such that

- $\mathrm{Im}(\pi) = \mathcal{B}$,
- $\ker(\pi) = \mathcal{C}$ and
- $K(\mathcal{A})/\pi^*(K(\mathcal{B}))$ is separable.

**Definition 4.65** With the notation from above we call $\mathcal{B} =: \mathcal{A}/\mathcal{C}$ the *quotient of* $\mathcal{A}$ *modulo* $\mathcal{C}$. Hence, $\mathcal{B}(\overline{K}) = \mathcal{A}(\overline{K})/\mathcal{C}(\overline{K})$.

For general homomorphisms $\varphi \in \mathrm{Hom}_K(\mathcal{A}, \mathcal{B})$ we get:

$$\varphi = \psi \circ \pi_{\ker(\varphi)}$$

where $\psi$ is a purely inseparable isogeny from $\mathcal{A}/\ker(\varphi)$ to $\mathrm{Im}(\varphi)$.

Hence we can classify all abelian varieties defined over $K$ that are separably isogenous to $\mathcal{A}$ up to isomorphisms:

**Proposition 4.66** The $K$-isomorphism classes of abelian varieties that are $K$-separably isogenous to $\mathcal{A}$ correspond one-to-one to the finite subgroups $C \subseteq \mathcal{A}(\overline{K})$ that are invariant under the action of $G_K$. They are isomorphic to $\mathcal{A}/C$. The field $K(\mathcal{A})$ is a separable extension of $\pi^*(K(\mathcal{A}/C))$, which is a Galois extension with Galois group canonically isomorphic to $C$: the automorphisms of $K(\mathcal{A})$ fixing $\pi^*(K(\mathcal{A}/C))$ are induced by translation maps $t_P$ with $P \in \ker(\pi)$, i.e., $\pi^*(K(\mathcal{A}/C))$ consists of the functions on $\mathcal{A}$ that are invariant under translations of the argument by points in $C$.

To describe all abelian varieties that are $K$-isogenous to $\mathcal{A}$ we have to compose separable isogenies with purely inseparable ones. As seen the notion of finite subgroups of $\mathcal{A}(\overline{K})$ is not sufficient for this; we would have to go to the category of *group schemes* to repair this deficiency. This is beyond the scope of this introduction. For details see e.g., [MUM 1974, pp. 93].

For our purposes there is a most prominent inseparable isogeny, the *Frobenius homomorphism*. Assume that $\mathrm{char}(K) = p > 0$. We recall that we have defined the Frobenius morphism $\phi_p$ (cf. Example 4.39) for varieties over $K$. It is easily checked that $\phi_p(\mathcal{A})$ is again an abelian variety over $K$. By the description of $\phi_p^*$ it follows at once that $\phi_p$ is a purely inseparable isogeny of degree $p^{\dim(\mathcal{A})}$ and its kernel is $\{0\}$.

Now we consider the special case that $\mathcal{B} = \mathcal{A}$.

**Definition 4.67** The homomorphisms $\mathrm{End}_K(\mathcal{A}) := \mathrm{Hom}_K(\mathcal{A}, \mathcal{A})$ are the *endomorphisms of* $\mathcal{A}$.

The set $\mathrm{End}_K(\mathcal{A})$ is a ring with composition as multiplicative structure.

**Example 4.68** Assume that $K = \mathbb{F}_p$. Then $\phi_p$ induces the identity map on polynomials over $K$ and so $\phi_p(\mathcal{A}) = \mathcal{A}$. Therefore, $\phi_p \in \mathrm{End}_K(\mathcal{A})$. It is called the *Frobenius endomorphism*.

A slight but important generalization is to consider $K = \mathbb{F}_q$ with $q = p^d$. Then $\phi_q := \phi_p^d$ is the relative Frobenius automorphism fixing $K$ element wise. We can apply the considerations made above to $\phi_q$ and get a totally inseparable endomorphism of $\mathcal{A}$ of degree $p^{d \dim(\mathcal{A})}$ which is called the (relative) Frobenius endomorphism of $\mathcal{A}$.

To avoid quotient groups we introduce a further definition.

**Definition 4.69** An abelian variety is *simple* if and only if it does not contain a proper abelian subvariety.

Assume that $\mathcal{A}$ is simple. It follows that $\varphi \in \mathrm{Hom}_K(\mathcal{A}, \mathcal{B})$ is either the zero map or has a finite kernel, hence its image is isogenous to $\mathcal{A}$.

**Proposition 4.70** If $\mathcal{A}$ is simple then $\mathrm{End}_K(\mathcal{A})$ is a ring without zero divisors and $\mathrm{End}_K(\mathcal{A})^0 := \mathrm{End}_K(\mathcal{A}) \otimes \mathbb{Q}$ is a skew field.

One proves by induction with respect to the dimension that every abelian variety is isogenous to the direct product of simple abelian varieties. So we get

**Corollary 4.71** $\mathrm{End}_K(\mathcal{A})^0$ is isomorphic to a product of matrix rings over skew fields.

### 4.3.5 Points of finite order and Tate modules

We come to most simple but important examples of elements in $\mathrm{End}_K(\mathcal{A})$.

For $n \in \mathbb{N}$ define
$$[n] : \mathcal{A} \to \mathcal{A}$$
as the $(n-1)$-fold application of the addition $\oplus$ to the point $P \in \mathcal{A}$. For $n = 0$ define $[0]$ as zero map, and for $n < 0$ define $[n] := -[|n|]$. By identifying $n$ with $[n]$ we get an injective homomorphism of $\mathbb{Z}$ into $\mathrm{End}_K(\mathcal{A})$.

By definition $[n]$ commutes with every element in $\mathrm{End}_K(\mathcal{A})$ and with $G_K$ and so lies in the center of the $G_K$-module $\mathrm{End}_K(\mathcal{A})$.

The kernel of $[n]$ is finite if and only if $n \neq 0$. Hence $[n]$ is an isogeny for $n \neq 0$. It is an isomorphism if $|n| = 1$.

**Definition 4.72** Let $n \in \mathbb{N}$.

(i) The kernel of $[n]$ is denoted by $\mathcal{A}[n]$.
(ii) The points in $\mathcal{A}[n]$ are called *n-torsion points*.
(iii) There exists a homogeneous ideal defined over $K$ such that $\mathcal{A}[n](\overline{K})$ is the set of points on $\mathcal{A}(\overline{K})$ at which the ideal vanishes. It is called the *n-division ideal*.

For the latter we recall that $\mathcal{A}$ is a projective variety with a fixed embedding into a projective space.

For elliptic curves (cf. Section 4.4.2.a) which are abelian varieties of dimension one the $n$-division ideal is a principal ideal. The generating polynomial is called the *n-division polynomial*. We will give the division polynomials explicitly in Section 4.4.5.a together with a recursive construction.

A fundamental result is

**Theorem 4.73** The degree of $[n]$ is equal to $n^{2\dim(\mathcal{A})}$. The isogeny $[n]$ is separable if and only if $n$ is prime to $\mathrm{char}(K)$. In this case $\mathcal{A}[n](\overline{K}) \simeq (\mathbb{Z}/n\mathbb{Z})^{2\dim(\mathcal{A})}$. If $n = p^s$ with $p = \mathrm{char}(K)$ then $\mathcal{A}[p^s](\overline{K}) = \mathbb{Z}/p^{ts}\mathbb{Z}$ with $t \leq \dim(\mathcal{A})$ independent of $s$.

A proof of these facts can be found in [MUM 1974, p. 64].

**Definition 4.74** Let $p = \mathrm{char}(K)$.

(i) The variety $\mathcal{A}$ is called *ordinary* if $\mathcal{A}[p^s](\overline{K}) = \mathbb{Z}/p^{ts}\mathbb{Z}$ with $t = \dim(\mathcal{A})$.

## § 4.3 Abelian varieties

(ii) If $\mathcal{A}[p^s](\overline{K}) = \mathbb{Z}/p^{ts}\mathbb{Z}$ then the abelian variety $\mathcal{A}$ has *p-rank* $t$.

(iii) If $\mathcal{A}$ is an elliptic curve, i.e., an abelian variety of dimension 1 (cf. Section 4.4.2.a), it is called *supersingular* if it has p-rank 0.

(iv) The abelian variety $\mathcal{A}$ is supersingular if it is isogenous to a product of supersingular elliptic curves.

**Remark 4.75** If an abelian variety $\mathcal{A}$ is supersingular then it has p-rank 0. The converse is only true for abelian varieties of dimension $\leqslant 2$.

**Corollary 4.76** Let $\varphi$ be an isogeny from $\mathcal{A}$ to $\mathcal{B}$ of degree $n = \prod_{i=1}^{t} \ell_i^{k_i}$ with $\ell_i$ primes.

(i) There is a sequence $\varphi_i$ of isogenies from abelian varieties $\mathcal{A}_i$ to $\mathcal{A}_{i+1}$ with $\mathcal{A}_1 = \mathcal{A}$ and $\mathcal{A}_{t+1} = \mathcal{B}$ with $\deg(\varphi_i) = \ell_i^{k_i}$ and $\varphi = \varphi_t \circ \varphi_{t-1} \circ \cdots \circ \varphi_1$.

(ii) Assume that $n$ is prime to $\mathrm{char}(K)$. Let $B_n = \varphi(\mathcal{A}[n](\overline{K}))$. Then $B_n$ is $G_K$-invariant, $\mathcal{B}/B_n$ is isomorphic to $\mathcal{A}$ and $\pi_{B_n} \circ \varphi = [n]$.

(iii) If $\mathcal{A}$ is ordinary then taking $\varphi = \phi_p$

$$[p] = \phi_p \circ \pi_{\mathcal{A}[p]} = \pi_{\phi_p(\mathcal{A}[p])} \circ \phi_p.$$

**Example 4.77** Assume that $K = \mathbb{F}_p$ and that $\mathcal{A}$ is an ordinary abelian variety over $K$. Then there is a uniquely determined separable endomorphism $\overline{V}_p \in \mathrm{End}_K(\mathcal{A})$ called *Verschiebung* with

$$[p] = \phi_p \circ \overline{V}_p = \overline{V}_p \circ \phi_p$$

of degree $p^{\dim(\mathcal{A})}$, where $\phi_p$ is the absolute Frobenius endomorphism

**Corollary 4.78** Let $\ell$ be a prime different from $\mathrm{char}(K)$ and $k \in \mathbb{N}$. Then

$$[\ell]\mathcal{A}[\ell^{k+1}] = \mathcal{A}[\ell^k].$$

We can interpret this result in the following way: the collection of groups

$$\ldots \mathcal{A}[\ell^{k+i}], \ldots, \mathcal{A}[\ell^k], \ldots$$

forms a projective system with connecting maps $[\ell^i]$ and so we can form their *projective limit* $\varprojlim \mathcal{A}[\ell^k]$. The reader should recall that the system with groups $\mathbb{Z}/\ell^k\mathbb{Z}$ has as projective limit the $\ell$-adic integers $\mathbb{Z}_\ell$ (cf. Chapter 3). In fact there is a close connection:

**Definition 4.79** Let $\ell$ be a prime different from $\mathrm{char}(K)$. The *$\ell$-adic Tate module of $\mathcal{A}$* is

$$T_\ell(\mathcal{A}) := \varprojlim \mathcal{A}[\ell^k].$$

**Corollary 4.80** The Tate module $T_\ell(\mathcal{A})$ is (as $\mathbb{Z}_\ell$-module) isomorphic to $\mathbb{Z}_\ell^{2\dim(\mathcal{A})}$.

### 4.3.6 Background on $\ell$-adic representations

The torsion points and the Tate modules of abelian varieties are used to construct most important representations. The basic fact provided by Proposition 4.73 is that for all $n \in \mathbb{N}$ prime to $\mathrm{char}(K)$ the groups $\mathcal{A}[n](\overline{K})$ are free $\mathbb{Z}/n\mathbb{Z}$-modules and for primes $\ell$ different from $\mathrm{char}(K)$ the Tate modules $T_\ell(\mathcal{A})$ are free $\mathbb{Z}_\ell$-modules each of them having rank equal to $2\dim(\mathcal{A})$. Hence $\mathrm{Aut}_K(\mathcal{A}[n])$, respectively $\mathrm{Aut}_{\mathbb{Z}_\ell}(T_\ell(\mathcal{A}))$, can be identified (by choosing bases) with the group of invertible $2\dim(\mathcal{A}) \times 2\dim(\mathcal{A})$-matrices over $\mathbb{Z}/n\mathbb{Z}$, respectively $\mathbb{Z}_\ell$. Likewise one can identify $\mathrm{End}_K(\mathcal{A}[n])$ and $\mathrm{End}_{\mathbb{Z}_\ell}(T_\ell(\mathcal{A}))$ with the $2\dim(\mathcal{A}) \times 2\dim(\mathcal{A})$-matrices over $\mathbb{Z}/n\mathbb{Z}$ or $\mathbb{Z}_\ell$, respectively.

The first type of these representations relates the arithmetic of $K$ to the arithmetic of $\mathcal{A}$ via Galois theory. This will become very important in the case that $K$ is a finite field or a finite algebraic extension of either $\mathbb{Q}$ or a $p$-adic field $\mathbb{Q}_p$.

The Galois action of $G_K$ on $\mathcal{A}(\overline{K})$ maps $\mathcal{A}[\ell^k]$ into itself and extends in a natural way to $T_\ell(\mathcal{A})$. Hence both the groups of points in $\mathcal{A}[n]$ and in $T_\ell(\mathcal{A})$ carry the structure of a $G_K$-module and so they give rise to representations of $G_K$.

**Definition 4.81** For a natural number $n$ prime to $\mathrm{char}(K)$ the *representation induced by the action of $G_K$ on $\mathcal{A}[n]$* is denoted by $\rho_{\mathcal{A},n}$.

For primes $\ell$ prime to $\mathrm{char}(K)$ the representation induced by the action on $T_\ell(\mathcal{A})$ is denoted by $\tilde{\rho}_{\mathcal{A},\ell}$ and is called the *$\ell$-adic Galois representation attached to $\mathcal{A}$*.

Second, we take $\varphi \in \mathrm{End}_K(\mathcal{A})$. It commutes with $[n]$ for all natural numbers and so it operates on $T_\ell(\mathcal{A})$ continuously with respect to the $\ell$-adic topology. Let $T_\ell(\varphi)$ denote the corresponding element in $\mathrm{End}_{\mathbb{Z}_\ell}(T_\ell(\mathcal{A}))$.

With the results about abelian varieties we have mentioned already, it is not difficult to see that the set of points of $\ell$-power order in $\mathcal{A}(\overline{K})$ is Zariski-dense, i.e., the only Zariski-closed subvariety of $\mathcal{A}$ containing all points of $\ell$-power order is equal to $\mathcal{A}$ itself.

It follows that $T_\ell(\varphi) = 0$ if and only if $\varphi = 0$ and so we get an injective homomorphism $T_\ell$ from $\mathrm{End}_K(\mathcal{A})$ into $\mathrm{End}_{\mathbb{Z}_\ell}(T_\ell(\mathcal{A}))$.

Much deeper and stronger is the following result:

**Theorem 4.82** We use the notation from above. The Tate module, $T_\ell$ induces a continuous $\mathbb{Z}_\ell$-module monomorphism, again denoted by $T_\ell$, from $\mathrm{End}_{\mathbb{Z}_\ell}(T_\ell(\mathcal{A})) \otimes_{\mathbb{Z}} \mathbb{Z}_\ell$ into $\mathrm{End}_{\mathbb{Z}_\ell}(T_\ell(\mathcal{A}))$.

For the proof see [MUM 1974, pp. 176].

It follows that $\mathrm{End}_K(\mathcal{A})$ is a free finitely generated $\mathbb{Z}$-module of rank $\leqslant (2\dim(\mathcal{A}))^2$ and so $\mathrm{End}_K(\mathcal{A}) \otimes \mathbb{Q}$ is a finite dimensional semisimple algebra over $\mathbb{Q}$. There is an extensive theory about such algebras and a complete classification. For more details see again [MUM 1974, pp. 193].

Moreover we can associate to $\varphi$ the characteristic polynomial

$$\chi(T_\ell(\varphi))(T) := \det(T - T_\ell(\varphi)),$$

which is a monic polynomial of degree $2\dim(\mathcal{A})$ with coefficients in $\mathbb{Z}_\ell$ by definition.

But here another fundamental result steps in:

**Theorem 4.83** The characteristic polynomial $\chi(T_\ell(\varphi))(T)$ does not depend on the prime number $\ell$ and has coefficients in $\mathbb{Z}$, hence it is a monic polynomial of degree $2\dim(\mathcal{A})$ in $\mathbb{Z}[T]$.

For the proof see [MUM 1974, p. 181].

Because of this result the following definition makes sense.

**Definition 4.84** Let $\varphi$ be an element in $\mathrm{End}_K(\mathcal{A})$. The characteristic polynomial $\chi(\varphi)_{\mathcal{A}}(T)$ is equal to the characteristic polynomial of $T_\ell(\varphi)$ for any $\ell$ different from $\mathrm{char}(K)$.

**Corollary 4.85** We get

$$\deg(\varphi) = \chi(\varphi)_{\mathcal{A}}(0)$$

## § 4.3 Abelian varieties

and more generally
$$\deg([n] - \varphi) = \chi(\varphi)_{\mathcal{A}}(n).$$

For $n$ prime to $\operatorname{char}(K)$ the restriction of $\varphi$ to $\mathcal{A}[n]$ has the characteristic polynomial
$$\chi(\varphi)_{\mathcal{A}}(T) \pmod{n}.$$

There is an important refinement of Theorem 4.82 taking into account the Galois action. As seen $G_K$ is mapped into $\operatorname{End}_{\mathbb{Z}_\ell}(T_\ell(\mathcal{A}))$ by the representation $\tilde{\rho}_{\mathcal{A},\ell}$. By definition the image of $T_\ell$ commutes with the image of $\tilde{\rho}_{\mathcal{A},\ell}$ and hence we get:

**Corollary 4.86** Moving to the Tate module $T_\ell$ induces a map from $\operatorname{End}_K(\mathcal{A})$ to $\operatorname{End}_{\mathbb{Z}_\ell[G_K]}(T_\ell(\mathcal{A}))$.

**Example 4.87** Let $K = \mathbb{F}_p$ and let $\mathcal{A}$ be an abelian variety defined over $K$.

The Frobenius automorphism of $K$ has a Galois $\ell$-adic representation $\tilde{\rho}_{\mathcal{A},\ell}(\phi_p)$ and a representation as endomorphism of $\mathcal{A}$ in $\operatorname{End}_{\mathbb{Z}_\ell}(T_\ell(\mathcal{A}))$. By the very definition both images in $\operatorname{End}_{\mathbb{Z}_\ell}(T_\ell(\mathcal{A}))$ coincide.

It follows that the endomorphism $\phi_p$ attached to the Frobenius automorphism of $K$ commutes with every element in $\operatorname{End}_K(\mathcal{A})$.

Moreover its characteristic polynomial $\chi(\phi_p)_{\mathcal{A}}(T)$ is equal to the characteristic polynomial of $\tilde{\rho}_{\mathcal{A},\ell}(\phi_p)$ for all $\ell$ prime to $\operatorname{char}(K)$.

For $n$ prime to $\operatorname{char}(K)$, the kernel of $[n] - \phi_p$ has order $\chi(\phi_p)_{\mathcal{A}}(n)$.

### 4.3.7 Complex multiplication

The results of the section above are the key ingredients for the study of $\operatorname{End}_K(\mathcal{A})$.

For instance, it follows for simple abelian varieties $\mathcal{A}$ that a maximal subfield $F$ of $\operatorname{End}_K(\mathcal{A}) \otimes \mathbb{Q}$ is a number field of degree at most $2\dim(\mathcal{A})$ over $\mathbb{Q}$, cf. [MUM 1974, p. 182].

**Definition 4.88** A simple abelian variety $\mathcal{A}$ over $K$ has *complex multiplication* if $\operatorname{End}_K(\mathcal{A}) \otimes \mathbb{Q}$ contains a number field $F$ of degree $2\dim(\mathcal{A})$ over $\mathbb{Q}$.

If an $F$ of this maximal degree exists then it has to be a field of *CM-type*. That means that it is a quadratic extension of degree 2 of a totally real field $F_0$ (i.e., every embedding of $F_0$ into $\mathbb{C}$ lies in $\mathbb{R}$), and no embedding of $F$ into $\mathbb{C}$ is contained in $\mathbb{R}$. Therefore, $F = \operatorname{End}_K(\mathcal{A}) \otimes \mathbb{Q}$.
If $K$ is a field of characteristic 0 we get more:

**Proposition 4.89** Let $K$ be a field of characteristic 0. Let $\mathcal{A}$ be a simple abelian variety defined over $K$ with complex multiplication. Then $\operatorname{End}_K(\mathcal{A}) \otimes \mathbb{Q}$ is equal to a number field $F$ of degree $2\dim(\mathcal{A})$ which is of CM-type. The ring $\operatorname{End}_K(\mathcal{A})$ is an order (cf. Definition 2.81) in $F$.

**Example 4.90** Let $K$ be a field of characteristic 0 and let $E$ be an elliptic curve over $K$ (cf. Section 4.4.2.a). Then either $\operatorname{End}_K(E) = \{[n] \mid n \in \mathbb{Z}\}$ or $E$ has complex multiplication and $\operatorname{End}_K(E)$ is an order in an imaginary quadratic field. In either case the ring of endomorphisms of $E$ is commutative.

The results both of the proposition and of the example are wrong if $\operatorname{char}(K) > 0$.

For instance, take a supersingular curve $E$ defined over a finite field $\mathbb{F}_{p^2}$ (cf. Definition 4.74). Then the center of $\operatorname{End}_{\mathbb{F}_{p^2}}(E) \otimes \mathbb{Q}$ is equal to $\mathbb{Q}$, and $\operatorname{End}_{\mathbb{F}_{p^2}}(E) \otimes \mathbb{Q}$ is a quaternion algebra over an imaginary quadratic number field $F$ (in fact there are infinitely many such quadratic number fields). Hence $E$ has complex multiplication but $\operatorname{End}_{\mathbb{F}_{p^2}}(E) \otimes \mathbb{Q}$ is *not* commutative and *not* an order in $F$.

**Remark 4.91** For elliptic curves (cf. Section 4.4.2.a) it is a strong requirement to have complex multiplication. If $K$ has characteristic $0$ we shall see that $E$ has to be defined over a number field $K_0$, and its absolute invariant $j_E$ which will be defined in Corollary 4.118 has to be an algebraic integer in $K_0$ (satisfying more conditions as we shall see in Theorem 5.47).

If $\mathrm{char}(K) = p > 0$ a necessary condition is that $j_E$ lies in a finite field. After at most a quadratic extension of $K$ this condition becomes sufficient.

## 4.4 Arithmetic of curves

From now on we concentrate on curves.

### 4.4.1 Local rings and smoothness

**Definition 4.92** Let $P$ be a point on an affine curve $C$. The set of rational functions that are regular at $P$ form a subring $\mathcal{O}_P$ of $K(C)$.

In fact, $\mathcal{O}_P$ is a local ring with maximal ideal

$$\mathfrak{m}_P = \{f \in \mathcal{O}_P \mid f(P) = 0\}.$$

It is called the *local ring of* $P$.

The *residue field of* $P$ is defined as $\mathcal{O}_P/\mathfrak{m}_P$.

One has $K(P) = \mathcal{O}_P/\mathfrak{m}_P$, hence, $\deg(P) = [K(P) : K]$.

For $S \subset C$ define $\mathcal{O}_S := \bigcap_{P \in S} \mathcal{O}_P$. It is the *ring of regular functions on* $S$. If $S$ is closed then $\mathcal{O}_S$ is the localization of $K[C]$ with respect to the ideal defining $S$.

A rational function $r$ on $C$ is a morphism if and only if $r \in \mathcal{O}_C = K[C]$.

For a projective curve, the ring of rational functions on $C$ that are regular at $P$ is equal to $\mathcal{O}_P$, the local ring of $P$ in a nonempty affine part of $C$.

**Definition 4.93** Let $P \in C$ for a projective curve $C$. The point $P$ is *nonsingular* if $\mathcal{O}_P$ is integrally closed in $K(C)$. Otherwise the point is called *singular*. A curve is called *nonsingular* or *smooth* if every point of $C(\overline{K})$ is nonsingular.

A smooth curve satisfies that $K[C_a]$ is integrally closed in $K(C)$ for any choice of $C_a$. If $C$ is projective but not smooth we take an affine covering $C_i$ and define $\widetilde{C}_i$ as affine curve corresponding to the integral closure of $K[C_i]$. By the uniqueness of the integral closure we can glue together the curves $\widetilde{C}_i$ to a projective curve $\widetilde{C}$ called the *desingularization of the curve* $C$. Note that in general even for $C$ a plane curve, $\widetilde{C}$ shall not be plane.

There is a morphism $\varphi : \widetilde{C} \to C$ that is a bijection on the nonsingular points of $C$. Hence projective smooth curves that are birationally equivalent are isomorphic.

Therefore, irreducible projective nonsingular curves are in one-to-one correspondence to function fields of dimension 1 over $K$.

To have a criterion for smoothness that can be verified more easily we restrict ourselves to affine parts of curves.

**Lemma 4.94 (Jacobi criterion)** Let $C_a \subseteq \mathbb{A}^n$ be an affine curve, let $f_1, \ldots, f_d \in K[x]$ be generators of $I(C_a)$, and let $P \in C_a(\overline{K})$. If the rank of the matrix $\left((\partial f_i/\partial x_j)(P)\right)_{i,j}$ is $n - 1$ then the curve is nonsingular at $P$.

## § 4.4 Arithmetic of curves

Using this lemma one can show that there are only finitely many singular points on a curve.

For a nonsingular point $P$ the dimension of $\mathfrak{m}_P/\mathfrak{m}_P^2$ is one. Therefore, the local ring $\mathcal{O}_P$ is a discrete valuation ring.

**Definition 4.95** Let $C$ be a curve and $P \in C$ be nonsingular. The *valuation at $P$* on $\mathcal{O}_P$ is given by

$$v_P : \mathcal{O}_P \to \{0, 1, 2, \ldots\} \cup \{\infty\}, \quad v_P(f) = \max\{i \in \mathbb{Z} \mid f \in \mathfrak{m}_P^i\}.$$

The valuation is extended to $K(C)$ by putting $v_P(g/h) = v_P(g) - v_P(h)$. The value group of $v_P$ is equal to $\mathbb{Z}$.

The valuation $v_P$ is a non-archimedean discrete normalized valuation (cf. Chapter 3).

A function $t$ with $v(t) = 1$ is called *uniformizer for $C$ at $P$*.

Let $P_1$ and $P_2$ be nonsingular points. Then $v_{P_1} = v_{P_2}$ if and only if $P_1 \in G_K \cdot P_2$.

**Example 4.96** Let $C = \mathbb{P}^1/K$ and choose $P \in \mathbb{A}^1$. Let $f \in K(x)$. The value $v_P(f)$ of $f$ at $P = (a) \in \overline{K}$ equals the multiplicity of $a$ as a root of $f$. If $a$ is a pole of $f$, the pole-multiplicity is taken with negative sign as it is the zero-multiplicity of $1/f$.

This leads to a correspondence of Galois orbits of nonsingular points of $C$ to normalized valuations of $K(C)$ that are trivial on $K$. For a nonsingular curve this is even a bijection. Namely, to each valuation $v$ of $K(C)$ corresponds a local ring defined by $\mathcal{O}_v := \{f \in K(C) \mid v(f) \geq 0\}$ with maximal ideal $\mathfrak{m}_v$. If $C$ is smooth, there exists a maximal ideal $M_v \subset K[C_a]$, where $C_a$ is chosen such that $K[C_a] \subset \mathcal{O}_v$, satisfying $\mathcal{O}_v = \mathcal{O}_\mathfrak{p}$. Over the algebraic closure there exist points $P_1, \ldots, P_d$ such that $\mathcal{O}_v$ equals $\mathcal{O}_{P_i}$ and the $P_i$ form an orbit under $G_K$. The degree of $M_v$ is $[K[C_a]/M_v : K] = [\mathcal{O}_v/\mathfrak{m}_v : K]$. It is equal to the order of $G_K \cdot P_i$ of one of the corresponding points on $C$.

Two valuations of $v_1, v_2$ of $K(C)$ are called *equivalent* if there exists a number $c \in \mathbb{R}_{>0}$ with $v_1 = cv_2$.

**Definition 4.97** The equivalence class of a valuation $v$ of $K(C)$ which is trivial on $K$ is called a *place* $\mathfrak{p}$ of $K(C)$. The set of places of $F/K$ is denoted by $\Sigma_{F/K}$.

In every place there is one valuation with value group $\mathbb{Z}$. It is called the *normalized valuation of* $\mathfrak{p}$ and denoted by $v_\mathfrak{p}$.

We have seen:

**Lemma 4.98** Let $F/K$ be a function field and let $C/K$ be a smooth projective absolute irreducible curve such that $F \simeq K(C)$ with an isomorphism $\varphi$ fixing each element of $K$.

There is a natural one-to-one correspondence induced by $\varphi$ between the places of $F/K$ and the Galois orbits of points on $C$.

**Example 4.99** Consider the function field $K(x_1)$ with associated smooth curve $\mathbb{P}^1/K$ and affine coordinate ring $K[x_1]$. The normalized valuations in $\Sigma_{K(C)/K}$ for which the valuation ring contains $K[x_1]$ correspond one-to-one to the irreducible monic polynomials in $K[x_1]$. There is one additional valuation with negative value at $x_1$, called $v_\infty$, which is equal to the negative degree valuation, corresponding to the valuation at $p_\infty(t) = t$ in $K[t] = K[1/x_1]$. Geometrically $v_\infty$ corresponds to $\mathbb{P}^1 \smallsetminus \mathbb{A}^1$.

## 4.4.2 Genus and Riemann–Roch theorem

We want to define a group associated to the points of a curve $C$.

**Definition 4.100** Let $C/K$ be a curve. The *divisor group* $\mathrm{Div}_C$ of $C$ is the free abelian group over the places of $K(C)/K$. An element $D \in \mathrm{Div}_C$ is called a *divisor*. It is given by

$$D = \sum_{\mathfrak{p}_i \in \Sigma_{K(C)/K}} n_i \mathfrak{p}_i,$$

where $n_i \in \mathbb{Z}$ and $n_i = 0$ for almost all $i$.

The divisor $D$ is called a *prime divisor* if $D = \mathfrak{p}$ with $\mathfrak{p}$ a place of $K(C)/K$.

The *degree* $\deg(D)$ *of a divisor* $D$ is given by

$$\begin{aligned}\deg : \mathrm{Div}_C &\to \mathbb{Z} \\ D &\mapsto \deg(D) = \sum_{\mathfrak{p}_i \in \Sigma_{K(C)/K}} n_i \deg(\mathfrak{p}_i).\end{aligned}$$

A divisor is called *effective* if all $n_i \geq 0$. By $E \geq D$ one means that $E - D$ is effective.

For $D \in \mathrm{Div}_C$ put

$$D_0 = \sum_{\substack{\mathfrak{p}_i \in \Sigma_{K(C)/K} \\ n_i \geq 0}} n_i \mathfrak{p}_i \quad \text{and} \quad D_\infty = \sum_{\substack{\mathfrak{p}_i \in \Sigma_{K(C)/K} \\ n_i \leq 0}} -n_i \mathfrak{p}_i,$$

thus $D = D_0 - D_\infty$.

Recall that over $\overline{K}$ each place $\mathfrak{p}_i$ corresponds to a Galois orbit of points on the projective nonsingular curve attached to $K(C)$. Thus, $D$ can also be given in the form

$$D = \sum_{P_i \in C} n_i P_i$$

with $n_i \in \mathbb{Z}$, almost all $n_i = 0$ and $n_i = n_j$ if $P_i \in P_j \cdot G_K$.

Assume now that $C$ is absolutely irreducible. Then we can make a base change from $K$ to $\overline{K}$. As a result we get again an irreducible curve $C \cdot \overline{K}$ (given by the same equations as $C$ but interpreted over $\overline{K}$) with function field $K(C) \cdot \overline{K}$.

Applying the results from above we get

$$\mathrm{Div}_{C \cdot \overline{K}} = \left\{ \sum_{P_i \in C} n_i P_i \right\}$$

with $n_i \in \mathbb{Z}$ and almost all $n_i = 0$. For all fields $L$ between $K$ and $\overline{K}$ the Galois group $G_L$ operates by linear extension of the operation on points.

**Proposition 4.101** Assume that $C/K$ is a projective nonsingular absolutely irreducible curve. Let $L$ be a field between $K$ and $\overline{K}$ and denote by $\mathrm{Div}_{C \cdot L}$ the group of divisors of the curve over $L$ obtained by base change from $K$ to $L$. Then

$$\mathrm{Div}_{C \cdot L} = \{ D \in \mathrm{Div}_{C \cdot \overline{K}} \mid \sigma(D) = D, \text{ for all } \sigma \in G_L \}.$$

Especially: $\mathrm{Div}_C = \mathrm{Div}_{C \cdot \overline{K}}^{G_K}$.

## § 4.4 Arithmetic of curves

Important examples of divisors of $C$ are associated to functions. We use the relation between normalized valuations of $K(C)$ which are trivial on $K$ and prime divisors.

**Definition 4.102** Let $C/K$ be a curve and $f \in K(C)^*$. The *divisor* $\mathrm{div}(f)$ *of* $f$ is given by

$$\begin{aligned} \mathrm{div}: K(C) &\to \mathrm{Div}_C \\ f &\mapsto \mathrm{div}(f) = \sum_{\mathfrak{p}_i \in \Sigma_{K(C)/K}} v_{\mathfrak{p}_i}(f)\mathfrak{p}_i. \end{aligned}$$

A divisor associated to a function is called a *principal divisor*. The set of principal divisors forms a group $\mathrm{Princ}_C$.

We have a presentation of $\mathrm{div}(f)$ as difference of effective divisors as above:

$$\mathrm{div}(f) = \mathrm{div}(f)_0 - \mathrm{div}(f)_\infty.$$

The points occurring in $\mathrm{div}(f)_0$ (resp. in $\mathrm{div}(f)_\infty$) with nonzero coefficient are called *zeroes (resp. poles) of $f$*.

**Example 4.103** Recall the setting of Example 4.99 for the curve $C = \mathbb{P}^1$. Since polynomials of degree $d$ over fields have $d$ zeroes (counted with multiplicities) over $\overline{K}$ we get immediately from the definition:

$$\deg(f) = 0, \text{ for all } f \in K(x_1)^*.$$

Now let $C$ be arbitrary. Take $f \in K(C)^*$. For constant $f \in K^*$ the divisor is $\mathrm{div}(f) = 0$. Otherwise $K(f)$ is of transcendence degree 1 over $K$ and can be interpreted as function field of the projective line (with affine coordinate $f$) over $K$. By commutative algebra (cf. [ZASA 1976]) we learn about the close connection between valuations in $K(f)$ and $K(C)$, the latter being a finite algebraic extension of $K(f)$. Namely, $\mathrm{div}(f)_\infty$ is the conorm of the negative degree valuation on $K(f)$ and hence has degree $[K(C):K(f)]$ (cf. [STI 1993, p. 106]).

Since $\mathrm{div}(f)_0 = \mathrm{div}(f^{-1})_\infty$ we get:

**Proposition 4.104** Let $C$ be an absolutely irreducible curve with function field $K(C)$ and $f \in K(C)^*$.

(i) $\deg\bigl(\mathrm{div}(f)_0\bigr) = 0$ if and only if $f \in K^*$.
(ii) If $f \in K(C) \smallsetminus K$ then $[K(C):K(f)] = \deg\bigl(\mathrm{div}(f)_\infty\bigr) = \deg\bigl(\mathrm{div}(f)_0\bigr)$.
(iii) For all $f \in K(C)^*$ we get: $\deg\bigl(\mathrm{div}(f)\bigr) = 0$.

So the principal divisors form a subgroup of the group $\mathrm{Div}_C^0$ of degree zero divisors.

To each divisor $D$ we associate a vector space consisting of those functions with pole order at places $\mathfrak{p}_i$ bounded by the coefficients $n_i$ of $D$.

**Definition 4.105** Let $D \in \mathrm{Div}_C$. Define

$$L(D) := \{f \in K(C) \mid \mathrm{div}(f) \geqslant -D\}.$$

It is not difficult to see that $L(D)$ is a finite dimensional $K$-vector space. Put $\ell(D) = \dim_K\bigl(L(D)\bigr)$.

The *Theorem of Riemann–Roch* gives a very important connection between $\deg(D)$ and $\ell(D)$.

We give a simplified version of this theorem, which is sufficient for our purposes. The interested reader can find the complete version in [STI 1993, Theorem I.5.15].

**Theorem 4.106 (Riemann–Roch)** Let $C/K$ be an absolutely irreducible curve with function field $K(C)$. There exists an integer $g \geqslant 0$ such that for every divisor $D \in \mathrm{Div}_C$

$$\ell(D) \geqslant \deg(D) - g + 1.$$

For all $D \in \mathrm{Div}_C$ with $\deg(D) > 2g - 2$ one even has equality $\ell(D) = \deg(D) - g + 1$.

**Definition 4.107** The number $g$ from Theorem 4.106 is called the *genus of $K(C)$* or the *geometric genus of $C$*. If $C$ is projective nonsingular then $g$ is called the *genus of $C$*.

The Riemann–Roch theorem guarantees the existence of functions with prescribed poles and zeroes provided that the number of required zeroes is at most $2g - 2$ less than the number of poles. Namely, if $n_i > 0$ at $\mathfrak{p}_i$ then $f \in L(D)$ is allowed to have a pole of order at most $n_i$ at $\mathfrak{p}_i$. Vice versa a negative $n_i$ requires a zero of multiplicity at least $n_i$ at $\mathfrak{p}_i$.

As an important application we get:

**Lemma 4.108** Let $C/K$ be a nonsingular curve and let $D = \sum n_i \mathfrak{p}_i$ be a $K$-rational divisor of $C$ of degree $\geqslant g$. Then there is a function $f \in K(C)$ which has poles of order at most $n_i$ (hence zeroes of order at least $-n_i$ if $n_i < 0$) in the points $P_i \in C$ corresponding to $\mathfrak{p}_i$ and no poles elsewhere. In other words: the divisor $D + (f)$ is effective.

**Example 4.109** For the function field $K(x_1)$, Lagrange interpolation allows to find quotients of polynomials for any given zeroes and poles. This leads to $\ell(D) = \deg(D) + 1$. The curve $\mathbb{P}^1/K$ has genus 0.

The *Hurwitz genus formula* relates the genus of algebraic extensions $F'/F/K$. It is given in a special case in the following theorem (cf. [STI 1993, Theorem III.4.12] for the general case).

**Theorem 4.110 (Hurwitz Genus Formula)** Let $F'/F$ be a tame finite separable extension of algebraic function fields having the same constant field $K$. Let $g$ (resp. $g'$) denote the genus of $F/K$ (resp. $F'/K$). Then

$$2g' - 2 = [F' : F](2g - 2) \sum_{\mathfrak{p} \in \Sigma_{F/K}} \sum_{\mathfrak{p}'|\mathfrak{p}} \big(e(\mathfrak{p}'|\mathfrak{p}) - 1\big) \deg(\mathfrak{p}').$$

One of the most important applications of the Riemann–Roch theorem is to find affine equations for a curve with given function field. We shall demonstrate this in two special cases which will be the center of interest later on.

#### 4.4.2.a Elliptic curves

**Definition 4.111** A nonsingular absolutely irreducible projective curve defined over $K$ of genus 1 with at least one $K$-rational point is called an *elliptic curve*.

Let $C$ be such a smooth absolutely irreducible curve of genus 1 with at least one $K$-rational point $P_\infty$ and let $F/K$ be its function field. As $\ell(P_\infty) = 1$ we have thus $L(P_\infty) = K$.

Theorem 4.106 guarantees $\ell(2P_\infty) = 2$, hence there exists a function $x \in F$ such that $\{1, x\}$ is a basis of $L(2P_\infty)$ over $K$. There also exists $y \in F$ such that $\{1, x, y\}$ is a basis of $L(3P_\infty)$ over $K$. We easily find that $\{1, x, y, x^2\}$ is a basis of $L(4P_\infty)$ and that $L(5P_\infty)$ has basis $\{1, x, y, x^2, xy\}$.

The space $L(6P_\infty) \supset \langle \{1, x, y, x^2, xy, x^3, y^2\} \rangle$ has dimension six, hence there must be a linear dependence between these seven functions. In this relation $y^2$ has to have a nontrivial coefficient $a$. By multiplying the relation with $a$ and by replacing $y$ by $a^{-1}y$ we can assume that $a = 1$. The

## § 4.4 Arithmetic of curves

function $x^3$ has to appear nontrivially, too, with some coefficient $b$. Multiply the relation by $b^2$ and replace $x$ by $b^{-1}x$, $y$ by $b^{-1}y$. Then the coefficients of $y^2$ and $x^3$ are equal to 1 and we get a relation

$$y^2 + a_1 xy + a_3 y - x^3 - a_2 x^2 - a_4 x - a_6, \; a_i \in K.$$

This is the equation of an absolutely irreducible plane affine curve. It is a fact (again obtained by the use of the theorem of Riemann–Roch) that this curve is smooth.

The projective closure $\overline{C}$ of $C$ is given by

$$Y^2 Z + a_1 XYZ + a_3 YZ^2 - X^3 - a_2 X^2 Z - a_4 XZ^2 - a_6 Z^3, \; a_i \in K$$

with plane projective coordinates $(X : Y : Z)$. One sees at once that $\overline{C} \smallsetminus C = \{(0 : 1 : 0)\}$ and that $P_\infty := (0 : 1 : 0)$ is smooth. Hence $C$ is a nonsingular absolutely irreducible plane projective curve of genus 1.

Again by using the Riemann–Roch theorem one can prove that the converse holds, too. The projective curve given by

$$Y^2 Z + a_1 XYZ + a_3 YZ^2 - X^3 - a_2 X^2 Z - a_4 XZ^2 - a_6 Z^3, \; a_i \in K$$

is a curve of genus 1 if and only if it is smooth.

We have seen that the Riemann–Roch theorem yields

**Theorem 4.112** A function field $F/K$ of genus 1 with a prime divisor of degree 1 is the function field of an elliptic curve $E$. This curve is isomorphic to a smooth plane projective curve given by a *Weierstraß equation*

$$E : Y^2 Z + a_1 XYZ + a_3 YZ^2 - X^3 - a_2 X^2 Z - a_4 XZ^2 - a_6 Z^3, \; a_i \in K.$$

A plane nonsingular affine part $E_a$ of $E$ is given by

$$y^2 + a_1 xy + a_3 y - x^3 - a_2 x^2 - a_4 x - a_6, \; a_i \in K.$$

$E \smallsetminus E_a$ consists of one point with homogeneous coordinates $(0 : 1 : 0)$.

Conversely nonsingular curves given by equations

$$E : Y^2 Z + a_1 XYZ + a_3 YZ^2 - X^3 - a_2 X^2 Z - a_4 XZ^2 - a_6 Z^3, \; a_i \in K$$

have function fields of genus 1 with at least one prime divisor of degree 1 and so are elliptic curves.

In the remainder of the book $E$ will be a standard notation for an elliptic curve given by a Weierstraß equation, and we shall often abuse notation and denote by $E$ the affine part $E_a$, too. Since elliptic curves are one of the central topics of this book we use the opportunity to study their equations in more detail.

### Short normal forms and invariants

Let $E$ be an elliptic curve defined over $K$ with affine Weierstraß equation

$$E : y^2 + a_1 xy + a_3 y = x^3 + a_2 x^2 + a_4 x + a_6.$$

We shall simplify this equation under assumptions about the characteristic of $K$. To achieve this we shall map $(x, y)$ to $(x', y')$ by invertible linear transformations. These transformations correspond to morphisms of the affine part of $E$ to the affine part of another elliptic curve $E'$, and since the infinite point remains unchanged we get an isomorphism between $E$ and $E'$. Having done the transformation we change notation and denote the transformed curve by $E$ with coordinates $(x, y)$ again.

First assume that the characteristic of $K$ is odd. We make the following transformations

$$x \mapsto x' = x \quad \text{and} \quad y \mapsto y' = y + \frac{1}{2}\left(a_1 x + \frac{a_3}{2}\right).$$

The equation of $E$ expressed in the coordinates $(x', y')$ and then, following our convention to change notation and to write $x$ for $x'$ and $y$ for $y'$ is:

$$E : y^2 = x^3 + \frac{b_2}{4}x^2 + \frac{b_4}{2}x + \frac{b_6}{4}$$

where $b_2 = a_1^2 + 4a_2$, $b_4 = 2a_4 + a_1 a_3$ and $b_6 = a_3^2 + 4a_6$.

Now we assume in addition that the characteristic of $K$ is prime to 6. We transform

$$x \mapsto x' = x + \frac{b_2}{12} \quad \text{and} \quad y \mapsto y' = y$$

and — applying our conventions — get the equation

$$E : y^2 = x^3 - \frac{c_4}{48}x - \frac{c_6}{864},$$

where $c_4$ and $c_6$ are expressed in an obvious way in terms of $b_2, b_4, b_6$ as

$$c_4 = b_2^2 - 24b_4 \quad \text{and} \quad c_6 = -b_2^3 + 36b_2 b_4 - 216b_6.$$

So if $\text{char}(K)$ is prime to 6, we can always assume that an elliptic curve is given by a *short Weierstraß equation* of the type

$$y^2 = x^3 + a_4 x + a_6.$$

Next we have to decide which Weierstraß equations define isomorphic elliptic curves. We can and will restrict ourselves to isomorphisms that fix the point at infinity, i.e., we fix one rational point on $E$ or equivalently we fix one place of degree 1 in the function field of $E$, which is the place $P_\infty$ used when we derived the equation in Section 4.4.2.a.

To make the discussion not too complicated we shall continue to assume that the characteristic of $K$ is prime to 6 and so we have to look for invertible transformations of the affine coordinates for which the transformed equation is again a short Weierstraß equation.

So let $E$ be given by

$$E : y^2 = x^3 + a_4 x + a_6.$$

One sees immediately that the conditions imposed on the transformations imply

$$x \mapsto x' = u^{-2} x \quad \text{and} \quad y \mapsto y' = u^{-3} y$$

with $u \in K^*$, and that the resulting equation is

$$E' : y'^2 = x'^3 + u^4 a_4 x + u^6 a_6.$$

**Proposition 4.113** Assume that the characteristic of $K$ is prime to 6 and let $E$ be an elliptic curve defined over $K$. Let $E$ be given by a short Weierstraß equation

$$E : y^2 = x^3 + a_4 x + a_6.$$

- If $a_4 = 0$ then the coefficient of $x$ is equal to 0 in all short Weierstraß equations for $E$, and $a_6$ is determined up to a sixth power in $K^*$.
- If $a_6 = 0$ then the absolute term in all short Weierstraß equations for $E$ is equal to 0 and $a_4$ is determined up to a forth power in $K^*$.
- If $a_4 a_6 \neq 0$ then $a_6/a_4$ is determined up to a square in $K^*$.

Conversely:

- If $a_4 = 0$ then $E$ is isomorphic to $E'$ if in a short Weierstraß form of $E'$ the coefficient $a'_4$ of $x$ is equal to 0 and $a'_6/a_6$ is a sixth power in $K^*$.
- If $a_6 = 0$ then $E$ is isomorphic to $E'$ if in a short Weierstraß form of $E'$ the absolute term is equal to 0 and $a'_4/a_4$ is a fourth power in $K^*$.
- If $a_4 a_6 \neq 0$ then $E$ is isomorphic to $E'$ if in a short Weierstraß form of $E'$ we have: there is an element $v \in K^*$ with $a'_4 = v^2 a_4$ and $a'_6 = v^3 a_6$.

**Corollary 4.114** Assume that the characteristic of $K$ is prime to 6 and let $E$ be given by a short Weierstraß equation
$$E : y^2 = x^3 + a_4 x + a_6.$$

- If $a_4 = 0$ then for every $a'_6 \in K^*$ the curve $E$ is isomorphic to
$$E' : y^2 = x^3 + a'_6 \quad \text{over} \quad K\big((a_6/a'_6)^{1/6}\big).$$

- If $a_6 = 0$ then for every $a'_4 \in K^*$ the curve $E$ is isomorphic to
$$E' : y^2 = x^3 + a'_4 x \quad \text{over} \quad K\big((a_4/a'_4)^{1/4}\big).$$

- If $a_4 a_6 \neq 0$ then for every $v \in K^*$ the curve $E$ is isomorphic to
$$\widetilde{E}_v : y^2 = x^3 + a'_4 x + a'_6 \quad \text{with} \quad a'_4 = v^2 a_4 \text{ and } a'_6 = v^3 a_6 \quad \text{over} \quad K(\sqrt{v}).$$

The curves occurring in the Corollary are called *twists of E*. The curves $\widetilde{E}_v$ are called *quadratic twists* of $E$. Note that $E$ is isomorphic to $\widetilde{E}_v$ over $K$ if and only if $v$ is a square in $K^*$. Therefore up to isomorphisms there is only one quadratic twist of a curve with $a_4 a_6 \neq 0$.

We want to translate the results of the proposition and of the lemma into "invariants" of $E$ that can be read off from any Weierstraß equation.

Recall that a crucial part of the definition of elliptic curves was that the affine part has no singular points. This is translated into the condition that the discriminant of the equation of $E$ is not equal to 0. This discriminant is a polynomial in the coefficients $a_i$, which is particularly easy to write down if we have a short Weierstraß equation. So let $E$ be given by
$$E : y^2 = x^3 + a_4 x + a_6 := f(x).$$

**Definition 4.115** The discriminant $\Delta_E$ of $E$ is equal to the polynomial discriminant of $f(x)$ which is (up to a sign) the product of the differences of the zeroes of $f(x)$, which we endow with a constant for historical reasons:
$$\Delta_E = -16(4 a_4^3 + 27 a_6^2).$$

We note that this definition is to be taken with caution: it depends on the chosen Weierstraß equation and not only on the isomorphism class of $E$. To make the discriminant well defined we have to consider it modulo 12-th powers in $K^*$.

To get an invariant of the isomorphism class of $E$ we use the transformations of $a_4, a_6$, and $\Delta_E$ under transformations of Weierstraß forms.

**Definition 4.116** The absolute invariant (sometimes called $j$-invariant) $j_E$ of $E$ is defined by
$$j_E = 12^3 \frac{-4 a_4^3}{\Delta_E}.$$

**Lemma 4.117** Assume that the characteristic of $K$ is prime to 6 and let $E$ be given by a short Weierstraß equation
$$E: y^2 = x^3 + a_4 x + a_6.$$
The absolute invariant $j_E$ depends only on the isomorphism class of $E$.

(i) We have $j_E = 0$ if and only if $a_4 = 0$.
(ii) We have $j_E = 12^3$ if and only if $a_6 = 0$.
(iii) If $j \in K$ is not equal to $0, 12^3$ then $E$ is a quadratic twist of the elliptic curve
$$E_j : y^2 = x^3 - \frac{27j}{4(j - 12^3)} x + \frac{27j}{4(j - 12^3)}.$$

**Corollary 4.118** Assume that the characteristic of $K$ is prime to 6. The isomorphism classes of elliptic curves $E$ over $K$ are, up to twists, uniquely determined by the absolute invariants $j_E$, and for every $j \in K$ there exists an elliptic curve with absolute invariant $j$.

If $K$ is algebraically closed then the isomorphism classes of elliptic curves over $K$ correspond one-to-one to the elements in $K$ via the map $E \mapsto j_E$.

Of course it is annoying that we have to restrict ourselves to fields whose characteristic is prime to 6. In fact this is not necessary at all; completely analogous discussions can be done for characteristics 2 and 3 and can be found in [SIL 1986] and also in Chapter 13.

We give a very short sketch of the discussions there.

We start with a general Weierstraß equation for $E$ over a field with odd characteristic.
$$E: y^2 + a_1 xy + a_3 y = x^3 + a_2 x^2 + a_4 x + a_6,$$
and recall the definitions of $b_2 = a_1^2 + 4a_2$, $b_4 = 2a_4 + a_1 a_3$, $b_6 = a_3^2 + 4a_6$, and $c_4 = b_2^2 - 24b_4$.

In addition we define
$$b_8 = a_1^2 a_6 - a_1 a_3 a_4 + 4 a_2 a_6 + a_2 a_3^2 - a_4^2.$$

**Definition 4.119** The discriminant of $E$ is
$$\Delta_E := -b_2^2 b_8 - 8 b_4^3 - 27 b_6^2 + 9 b_2 b_4 b_6$$
and the absolute invariant of $E$ is
$$j_E = c_4^3 / \Delta_E.$$

If the characteristic of $K$ is equal to 2 one also finds normal forms for $E$ (cf. Section 13.3). Either $a_1 = 0$ and then $j_E := 0$. Otherwise we can find an equation for $E$ with $a_3 = a_4 = 0$ and $a_1 = 1$. Then $j_E = a_6^{-1}$.

We summarize the definitions in Table 4.1. Using these extra considerations one easily sees that the conclusions of Corollary 4.118 hold without any restrictions about the characteristic of the ground field.

**Theorem 4.120** Let $K$ be a field. The isomorphism classes of elliptic curves $E$ over $K$ are, up to twists, uniquely determined by the absolute invariants $j_E$, and for every $j \in K$ there exists an elliptic curve $E$ with absolute invariant $j_E = j$.

If $K$ is algebraically closed then the isomorphism classes of elliptic curves over $K$ correspond one-to-one to the elements in $K$ via the map $E \mapsto j_E$.

## § 4.4 Arithmetic of curves

**Table 4.1** Short Weierstraß equations.

| char $K$ | Equation | $\Delta$ | $j$ |
|---|---|---|---|
| $\neq 2, 3$ | $y^2 = x^3 + a_4 x + a_6$ | $-16(4a_4^3 + 27a_6^2)$ | $1728 a_4^3 / 4\Delta$ |
| 3 | $y^2 = x^3 + a_4 x + a_6$ | $-a_4^3$ | 0 |
| 3 | $y^2 = x^3 + a_2 x^2 + a_6$ | $-a_2^3 a_6$ | $-a_2^3 / a_6$ |
| 2 | $y^2 + a_3 y = x^3 + a_4 x + a_6$ | $a_3^4$ | 0 |
| 2 | $y^2 + xy = x^3 + a_2 x^2 + a_6$ | $a_6$ | $1/a_6$ |

### 4.4.2.b Hyperelliptic curves

**Definition 4.121** A nonsingular curve $C/K$ of genus $g > 1$ is called a *hyperelliptic curve* if the function field $K(C)$ is a separable extension of degree 2 of the rational function field $K(x)$ for some function $x$. Let $\omega$ denote the nontrivial automorphism of this extension. It induces an involution $\omega_*$ on $C$ with quotient $\mathbb{P}^1$. The fixed points $P_1, \ldots, P_{2g+2}$ of $\omega_*$ are called *Weierstraß points*.

From a geometrical point of view, $C$ is a hyperelliptic curve if there exists a generically étale morphism $\pi$ of degree 2 to $\mathbb{P}^1$. The Weierstraß points are exactly the points in which $\pi$ is ramified.

Classically, elliptic curves are not subsumed under hyperelliptic curves. The main difference is that for $g > 1$ the rational subfield of index 2 is unique. That implies that the function $x$ is uniquely determined up to transformations

$$x \mapsto \frac{ax+b}{cx+d} \quad \text{with } a, b, c, d \in K \text{ and } ad - bc \neq 0.$$

For elliptic curves this is wrong. If, for instance, $K$ is algebraically closed then there exist infinitely many rational subfields of index 2. In this book we will often consider elliptic curves as hyperelliptic curves of genus one since most of the arithmetic properties we are interested in are the same.

We now use the Riemann–Roch theorem to find an equation describing a plane affine part of $C$.

The definition implies that there exists a divisor $D$ of degree 2, which is the conorm of the negative degree valuation on $K(x)$ (cf. [STI 1993, p. 106]).

From the construction we have that $L(D)$ has basis $\{1, x\}$ and, hence, $\ell(D) = 2$. For $1 \leqslant j \leqslant g$ we have that $\ell(jD) \geqslant 2j$ and the elements $\{1, x, \ldots, x^j\}$ are linearly independent in $L(jD)$. As $\deg((g+1)D) = 2(g+1) > 2g - 2$, Theorem 4.106 implies that

$$\ell((g+1)D) = \deg((g+1)D) - g + 1 = g + 3.$$

Hence, besides the $g+2$ elements $1, x, \ldots, x^{g+1}$ there must be a $(g+3)$-th function $y \in L((g+1)D)$ independent of the powers of $x$.

Therefore, $y \notin K[x]$. The space $L(2(g+1)D)$ has dimension $3g + 3$. It contains the $3g + 4$ functions

$$1, x, \ldots, x^{g+1}, y, x^{g+2}, xy, \ldots, x^{2(g+1)}, x^{g+1} y, y^2.$$

Therefore there must exist a linear combination defined over $K$ among them. In this relation $y^2$ has to have some nontrivial coefficient $a$ as $y \notin K[x]$. By multiplying the relation with $a$ and by replacing $y$ by $a^{-1} y$ we can assume that $a = 1$.

This leads to an equation

$$y^2 + h(x) y = f(x), \quad h(x), f(x) \in K[x],$$

where $\deg(h) \leqslant g + 1$ and $\deg(f) \leqslant 2g + 2$.

To determine the exact degrees we use the *Hurwitz genus formula* stated in Theorem 4.110. In our case $[K(C) : K(x)] = 2$ and thus $e(\mathfrak{p}'|\mathfrak{p}) \leqslant 2$. To simplify we shall assume that the characteristic of $K$ is odd. After applying the usual transformation $y \mapsto y - h(x)/2$ we can assume that $h(x) = 0$. Then the fixed points of $\omega_*$ are points with $y$-coordinate equal to 0 or are points lying over $x = \infty$. The latter case occurs if and only if $D$ is a divisor of the form $2P_\infty$, i.e., if there is only one point $P_\infty$ lying over $\infty$ on the nonsingular curve with function field $K(C)$. Moreover the $x$-coordinates of these points correspond to the places of $K(x)$ which ramify in the extension $K(C)/K(x)$.

By the genus formula the number of the ramified points has to be equal to $2g + 2$. Hence $f(x)$ has to have $2g + 2$ different zeroes if $\infty$ is not ramified, and $2g + 1$ different zeroes if $\infty$ is ramified. As a result we get: the degree of $f(x)$ is equal to $2g + 2$ if $D = P_1 + P_2$ with different $P_1, P_2$ and equal to $2g + 1$ if $D = 2P_\infty$, and $f(x)$ has no double zeroes.

Moreover the affine curve given by the equation

$$C_a : y^2 + h(x)y = f(x), \; h(x), f(x) \in K[x]$$

is nonsingular.

**Theorem 4.122** A function field $F/K$ of genus $g > 1$ with an automorphism $\omega^*$ of order 2 with rational fixed field is the function field of a plane affine curve given by an equation

$$C : y^2 + h(x)y = f(x), \; h(x), f(x) \in K[x], \tag{4.1}$$

where $2g + 1 \leqslant \deg f \leqslant 2g + 2$ and $\deg h \leqslant g + 1$ without singularities.

Conversely the nonsingular projective curve birationally isomorphic to an affine nonsingular curve given by an equation of this type is a hyperelliptic curve of genus $g$.

The homogenized equation has a singularity at infinity exactly if there is a single point in $\pi^{-1}(\infty)$ and then the degree of $f(x)$ is equal to $2g + 1$. In this case we can achieve a monic $f$. Let $b$ be the leading coefficient. Multiplying the equation by $b^{2g}$ and replacing $y \mapsto y/b^g, x \mapsto x/b^2$ we obtain

$$C_a : y^2 + h(x)y = f(x) \text{ with } h(x), f(x) \in K[x], \deg(f) = 2g + 1, \deg(h) \leqslant g \text{ and } f \text{ monic.}$$

In the sequel we shall always characterize hyperelliptic curves by their affine plane parts and assume them given by equations of the form (4.1).

### Short Weierstraß equations

Later on we shall concentrate on the case that $\deg(f) = 2g + 1$, i.e., curves having a $K$-rational Weierstraß point. In this case we can simplify the equations analogously to the case of elliptic curves. We distinguish between the case of $K$ having odd or even characteristic.

Let $C$ be a hyperelliptic curve of genus $g$ defined over a field of characteristic $\neq 2$ by an equation of the form (4.1) with $\deg(f) = 2g+1$. The transformation $y \mapsto y - h(x)/2$ leads to an isomorphic curve given by

$$C : y^2 = f(x), f \in K[x] \text{ and } \deg(f) = 2g + 1. \tag{4.2}$$

The Jacobi criterion (cf. Lemma 4.94) states that $C$ is nonsingular if and only if no point on the curve satisfies both partial derivative equations $2y = 0$ and $f'(x) = 0$. The points with $y = 0$ are just the points $P_i = (x_i, 0)$, where $f(x_i) = 0$. The second condition shows that the singular points are just the Weierstraß points for which the first coordinate is a multiple root of $f$. Therefore, a necessary and sufficient criterion for (4.2) to be nonsingular is that $f$ has only simple roots over the algebraic closure.

Let

$$f(x) = x^{2g+1} + \sum_{i=0}^{2g} f_i x^i.$$

## § 4.4 Arithmetic of curves

If additionally $\mathrm{char}(K)$ is coprime to $2g$, the transformation $x \mapsto x - f_{2g}/(2g)$ allows to give

$$f(x) = x^{2g+1} + f_{2g-1}x^{2g-1} + \cdots + f_1 x + f_0 \text{ with } f_i \in K.$$

Let $C$ be a hyperelliptic curve of genus $g$ over a field of characteristic 2. Assume first that $h(x) = 0$, i.e., $y^2 = f(x)$ like above.

The partial derivatives are $2y = 0$ and $f'(x)$. Any of the $2g+1$ roots $x_P$ of $f'$ can be extended to a point $(x_P, y_P)$ satisfying $y_P^2 = f(x_P)$ and both partial derivatives. Hence, $h(x) = 0$ immediately leads to a singular point and so we must have $h(x) \neq 0$.

### 4.4.2.c Differentials

We shall now give another application of the theorem of Riemann–Roch. For this we have to introduce differentials. We shall do this in the abstract setting of function fields. If the ground field $K$ is equal to $\mathbb{C}$ this concept coincides with the "usual" notion of differentials known from calculus.

Let $K(C)$ be the function field of a curve $C$ defined over $K$. To every $f \in K(C)$ we attach a symbol $df$, the *differential* of $f$ lying in a $K(C)$-vector space $\Omega(K(C))$, which is the free vector space generated by the symbols $df$ modulo the following relations.

For $f, g \in K(C)$ and $\lambda \in K$ we have

(R1) $$d(\lambda f + g) = \lambda df + dg$$

(R2) $$d(fg) = f dg + g df.$$

Recall that a *derivation* of $K(C)$ is a $K$-linear map

$$D : K(C) \to K(C)$$

vanishing on $K$ with

$$D(fg) = D(f)g + D(g)f.$$

Let $x \in K(C)$ be such that $K(C)$ is a finite separable extension of $K(x)$. Then there is exactly one derivation $D$ of $K(C)$ with $D(x) = 1$ (cf. [ZASA 1976]) call this derivation the *partial derivative with respect to $x$* and denote the image of $f \in K(C)$ under this derivation by $\partial f/\partial x$.

The relation between derivations and differentials is given by the *chain rule*.

**Lemma 4.123 (Chain rule)** Let $x$ be as above and $f \in K(C)$. Then $df = (\partial f/\partial x) dx$.

**Corollary 4.124** The $K(C)$-vector space of differentials $\Omega(K(C))$ has dimension 1.
It is generated by $dx$ for any $x \in K(C)$ for which $K(C)/K(x)$ is finite separable.

Let $P$ be a point on $C$. Take a uniformizer for $C$ at $P$, i.e., a function $t_P \in K(C)$ that generates the maximal ideal $M_P$ of the place associated to $P$. So $t_P$ is a function that vanishes at $P$ with multiplicity 1. It follows that $K(C)/K(t_P)$ is finite separable.

Let $\omega \in \Omega(K(C))$ be a differential of $C$. We attach a divisor $\mathrm{div}(\omega) = \sum_{P \in C} n_P P$ given by the following recipe: for $P \in C$ choose a uniformizer $t_P$ and express $\omega$ by $\omega = f_P dt_P$ with $f_P \in K(C)$. Then

$$n_P = v_P(f_P).$$

**Lemma 4.125** The sum $\mathrm{div}(\omega) = \sum_{P \in C} n_P P$ defines a divisor of $C$ that is independent of the choice of the uniformizers $t_P$. The degree of $\mathrm{div}(\omega)$ is equal to $2g - 2$.

For a proof of the lemma see [STI 1993].

**Definition 4.126** A differential $\omega$ is *holomorphic* if $\operatorname{div}(\omega)$ is an effective divisor.

The set of holomorphic differentials of $C$ forms a $K$-vector space $\Omega^0(K(C))$.

A consequence of the Riemann–Roch theorem is:

**Theorem 4.127** The $K$-vector space $\Omega^0(K(C))$ has dimension $g$.

**Example 4.128** Let $E$ be an elliptic curve defined over $K$ and given by an affine Weierstraß equation $G(x, y) = 0$, where
$$G(x, y) = y^2 + a_1 xy + a_3 y - x^3 - a_2 x^2 - a_4 x - a_6, \quad \text{with } a_i \in K.$$

The differential $1/(\partial G(x,y)/\partial y)dx$ is holomorphic, and up to a multiplicative constant it is the unique holomorphic differential of $E$. Note that it has neither poles nor zeroes.

## 4.4.3 Divisor class group

In this section we shall attach an abelian group to each nonsingular curve starting from the group of divisors as defined in Section 4.4.2. This construction will give us an intimate relation between the arithmetic of curves and abelian varieties.

Let $C/K$ be an absolutely irreducible smooth projective curve. Let $\operatorname{Div}_C^0$ denote the group of $K$-rational divisors of $C$ of degree 0.

Recall that principal divisors have degree zero and form a subgroup $\operatorname{Princ}_C \subseteq \operatorname{Div}_C^0$.

**Definition 4.129** The *divisor class group* $\operatorname{Pic}_C^0$ of $C$ of degree zero is the quotient of the group of degree zero divisors $\operatorname{Div}_C^0$ by the principal divisors. It is also called the *Picard group of $C$*.

Hence, two divisors $D_1$ and $D_2$ are in the same class if there exists an $f \in K(C)$ with $\operatorname{div}(f) = D_1 - D_2$.

**Example 4.130** Let $\omega$ and $\omega'$ be two differentials of $C$ that are not equal to 0. Then $\operatorname{div}(\omega)$ is in the same class as $\operatorname{div}(\omega')$.

In contrast to the group of divisors, the divisor class group has many torsion elements. If the field $K$ is finite, it is even a finite group.

We now give an example of how to deal with torsion elements.

### 4.4.3.a Divisor classes of order equal to char($K$)

We assume that $K$ is a field of characteristic $p > 0$. Let $C$ be a projective absolutely irreducible nonsingular curve of genus $g$ defined over $K$. Let $\operatorname{Pic}_C^0[p]$ be the group of divisor classes of $C$ with order dividing $p$.

In [SER 1958] we find the following result.

**Proposition 4.131** There is a monomorphism $\alpha$ from $\operatorname{Pic}_C^0[p]$ into $\Omega^0(C)$, the $K$-vector space of holomorphic differentials on $C$ given by the following rule: choose a $K$-rational divisor $D$ with $pD = \operatorname{div}(f)$ where $f$ is a function on $C$. Then the divisor class $\overline{D}$ of $D$ is mapped under $\alpha$ to the holomorphic differential $(1/f)df$.

Note that $(1/f)df$ is holomorphic since the multiplicity of the zeroes of $f$ is divisible by $p = \operatorname{char}(K)$. Next we choose a point $P_0 \in C(K)$. Let $t$ be a uniformizer of $C$ at $P_0$ (i.e., $t \in K(C)$ with $t(P_0) = 0$ and $\partial f/\partial t(P_0) \neq 0$). We take $(1/f)df$ as in the proposition and express it via the chain rule in the form $((\partial f/\partial t)/f)dt$. Let $(a_0, a_1, \ldots, a_{2g-2})(f)$ be the tuple whose coordinates are first coefficients of the power series expansion of $(\partial f/\partial t)/f$ at $P_0$ and assume that there is another holomorphic differential $hdt$ with $h$ having the same power series expansion as $(\partial f/\partial t)/f$ modulo $t^{2g-1}$. Then $(1/f)df - hdt$ is a holomorphic differential whose divisor has a coefficient $\geqslant 2g-1$ at $P_0$ and so its degree is $\geqslant 2g-1$. But this implies that it is equal to 0 and so $(1/f)df = hdt$.

Hence we get

**Proposition 4.132** Let $K$ be a field of characteristic $p > 0$ and $C$ a curve of genus $g$ defined over $K$. For divisor classes $\overline{D} \in \mathrm{Pic}_C^0[p]$ choose a divisor $D \in \overline{D}$ and take $f \in K(C)$ with $pD = \mathrm{div}(f)$.

The map

$$\begin{aligned} \Phi : \mathrm{Pic}_C^0[p] &\to K^{2g-1} \\ \overline{D} &\mapsto (a_0, a_1, \ldots, a_{2g-2})(f) \end{aligned}$$

is an injective homomorphism.

**Remark 4.133** For applications later on we shall be interested mostly in the case that $K$ is a finite field $\mathbb{F}_q$. Moreover, computational aspects will become important. If we want to use Proposition 4.132 in practice to identify $\mathrm{Pic}_C^0[p](\mathbb{F}_q)$ with a subgroup of $\mathbb{F}_q^{2g-1}$ we must be able to compute the first coefficients of the power series expansion of $(1/f)df$ at $P_0$ fast. The problem is that the degree of $f$ can be very large. Nevertheless this can be done in polynomial time (depending on $g$ and $\lg q$). The method is similar to the one we shall use later on for computing the Tate pairing (see Chapter 16) and so we refer here to [RÜC 1999] for details.

### 4.4.4 The Jacobian variety of curves

We come back to a projective absolutely irreducible curve $C$ defined over the field $K$ and the study of its divisor class group.

A first and easily verified observation is that $G_K$ acts in a natural way on $\mathrm{Pic}_{C \cdot \overline{K}}^0$ and that

$$\mathrm{Pic}_C^0 = (\mathrm{Pic}_{C \cdot \overline{K}}^0)^{G_K}$$

where $\mathrm{Pic}_{C \cdot \overline{K}}^0$ is the divisor class group of degree 0 of the curve over $\overline{K}$ obtained by base change from $C$.

More generally, take any field $L$ between $K$ and $\overline{K}$. Then $\mathrm{Pic}_{C \cdot L}^0 = (\mathrm{Pic}_{C \cdot \overline{K}}^0)^{G_L}$.

In the language of categories this means that for a fixed curve $C$, the Picard groups corresponding to curves obtained from $C$ by base change define a functor $\mathrm{Pic}^0$ from the set of intermediate fields $L$ between $K$ and $\overline{K}$ to the category of abelian groups.

It is very important that this functor can be represented by an absolutely irreducible smooth projective variety $J_C$ defined over $K$. For all fields $L$ between $K$ and $\overline{K}$ we have that the functors of sets $L \mapsto J_C(L)$, the set of $L$-rational points of $J_C$, can be identified in a natural way with $L \mapsto \mathrm{Pic}_{C \cdot L}^0$.

But even more is true: $J_C$ has the structure of an *algebraic group*. Since $J_C$ is projective and absolutely irreducible this means that *$J_C$ is an abelian variety*.

In particular, this implies that $J_C(L)$ is a group in which the group composition $\oplus$ is given by the evaluation of rational functions (if one takes affine coordinates) or polynomials (in projective coordinates) with coefficients in $K$ on pairs $(P_1, P_2) \in J_C(L)^2$.

As a result we can introduce coordinates for elements in $\mathrm{Pic}_C^0$ and compute by using algebraic formulas.

**Definition 4.134** The variety $J_C$ is called the *Jacobian (variety) of* $C$.

By using Theorem 4.106 we can give a birational description of $J_C$, which (essentially) proves its existence and makes it accessible for computations. It is based on the following lemma.

**Lemma 4.135** Let $C/K$ be a nonsingular, projective, absolutely irreducible curve of genus $g$ with a $K$-rational point $P_\infty$ corresponding to the place $\mathfrak{p}_\infty$. For every $K$-rational divisor class $\overline{D}$ of degree $0$ of $C$ there exists an effective divisor $D$ of degree $\deg(D) = g$ such that $D - g\,\mathfrak{p}_\infty \in \overline{D}$.

**Proof.** Take any $D' \in \overline{D}$ with $D' = D_1 - D_2$ as difference of two effective $K$-rational divisors. In the first step we choose $l$ large enough so that $l - \deg(D_2) > g$ and by Lemma 4.108 find a function $f_1$ such that $-D_2 + (f_1) + l\,\mathfrak{p}_\infty$ is effective.

By replacing $D'$ by $D' + (f_1)$ we can assume that $D' = D - k\,\mathfrak{p}_\infty$ with $D$ effective and $k = \deg(D)$. If $k > g$ (otherwise we are done) we apply Lemma 4.108 to the divisor $D - (k-g)\,\mathfrak{p}_\infty$ and find a function $f$ such that $D - (k-g)\,\mathfrak{p}_\infty + (f) := D_0$ is effective and therefore $D + (f) - k\,\mathfrak{p}_\infty = D_0 - g\,\mathfrak{p}_\infty$ is an element of $\overline{D}$ of the required form. $\square$

Let $C$ be as in Lemma 4.135 and take the $g$-fold Cartesian product $C^g$ of $C$. As per Example 4.25, $C^g$ is a projective variety of dimension $g$. Recall that an affine part of it is given in the following way:
Take $C_a$ as a nonempty affine part of $C$ in some affine space $\mathbb{A}^n$ with affine coordinates $(x_1, \ldots, x_n)$ and denote by $C^{(i)}$ an isomorphic copy of $C$ with coordinates $(x_1^i, \ldots, x_n^i)$. Then $C_a^g$ can be embedded into the affine space $\mathbb{A}^{gn}$ with coordinates $(x_1^1, \ldots, x_n^1, \ldots, x_1^g, \ldots, x_n^g)$.

Let $S_g$ be the symmetric group acting on $\{1, \ldots, g\}$. It acts in a natural way on $C^g$ by permuting the factors. On the affine part described above this action is given by permuting the sections $(x_1^i, \ldots, x_n^i)$. The action of $S_g$ defines an equivalence relation on $C^g$. We denote the quotient by $C^g/S_g$.

It is not difficult to see that $C^g/S_g$ is a projective variety defined over $K$. On the affine part $C_a^g/S_g$ one proves this as follows: take the ring of polynomials $K[x^1, \ldots, x^g]$ where $x^j$ is shorthand for the $n$ variables $x_1^j, \ldots, x_n^j$. On this ring, the group $S_g$ acts by permuting $\{x^1, \ldots, x^g\}$. The polynomials fixed under $S_g$ are symmetric and form a ring $R = K[x^1, \ldots, x^g]^{S_g}$. By a theorem of Noether (cf. [ZASA 1976]) there is a number $m$ and an ideal $I$ in $K[Y_1, \ldots, Y_m]$ with $R = K[Y_1, \ldots, Y_m]/I$. Hence, $C_a^g/S_g$ is isomorphic to $V_I \subset \mathbb{A}^m$.

Let $\underline{P}$ be a point in $C^g/S_g(L)$ for a field $L$ between $K$ and $\overline{K}$. Then $\underline{P}$ is the equivalence class of a $g$-tuple $(P_1, \ldots, P_g)$ with $P_i \in C$ and for all $\sigma \in G_L$ we get: there is a permutation $\pi \in S_g$ such that $(\sigma P_1, \ldots, \sigma P_g) = (P_{\pi(1)}, \ldots, P_{\pi(g)})$.

This means that for any $P_i$ the tuple $(P_1, \ldots, P_g)$ contains the whole Galois orbit $G_L \cdot P_i$. Assume that it consists of $k$ disjoint $G_L$-orbits, each of them corresponding to a place $\mathfrak{p}_1, \ldots, \mathfrak{p}_k$ of $K(C) \cdot L$. Hence the formal sum $P_1 + \cdots + P_g$ corresponds to the $L$-rational divisor $\mathfrak{p}_1 + \cdots + \mathfrak{p}_k$ which is positive and of degree $g$.

This way we get a map $\phi_L$ from $C^g(L)$ to $\mathrm{Pic}^0_{C \cdot L}$ defined by

$$\phi_L(\underline{P}) \mapsto (\mathfrak{p}_1 + \cdots + \mathfrak{p}_k - g\,\mathfrak{p}_\infty),$$

where $(\mathfrak{p}_1 + \cdots + \mathfrak{p}_k - g\,\mathfrak{p}_\infty)$ is the divisor class of degree zero associated to $\mathfrak{p}_1 + \cdots + \mathfrak{p}_k$.

Using the alternative description of $L$-rational divisors as sums of points on $C$ that consist of Galois orbits under $G_L$ we get a more elegant description of $\phi_L$: let $(P_1, \ldots, P_g) \in C^g$ be a representative of $\underline{P} \in C^g/S_g$. Define $\phi(\underline{P})$ as the divisor class of $P_1 + \cdots + P_g - g P_\infty$ in $\mathrm{Pic}^0_{C \cdot \overline{K}}$. Then $\phi_L$ is the restriction of $\phi$ to $\mathrm{Pic}^0_{C \cdot L}$.

**Theorem 4.136** Assume that $C$ is a curve of genus $g > 0$ with a $K$-rational point $P_\infty$. Then $C^g/S_g$ is birationally isomorphic to $J_C$, and the map $\phi$ defined above represents a birational part of the functorial isomorphism between $J_C(L)$ and $\mathrm{Pic}^0_{C \cdot L}$. It maps the symmetry class $\underline{P}_\infty$ of the point $(P_\infty, \ldots, P_\infty)$ to the zero class and so $\underline{P}_\infty$ corresponds to the neutral element of the algebraic group $J_C$.

## 4.4.5 Jacobian variety of elliptic curves and group law

We come back to elliptic curves as introduced in Definition 4.111 and make concrete all of the considerations discussed above.

Assume that $E$ is an elliptic curve with function field $K(E)$. Hence, $E$ can be given as plane projective cubic without singularities and with (at least) one $K$-rational point $P_\infty$. Clearly $E^1/S_1 = E$.

Let $\overline{D} \in \mathrm{Pic}^0_E$ be a divisor class of degree 0, $D \in \overline{D}$ a $K$-rational divisor. By the Riemann–Roch theorem 4.106 the space $L(D + P_\infty)$ has dimension 1. So there is an effective divisor in the class of $D + P_\infty$, and since this divisor has degree 1 it is a prime divisor corresponding to a point $P \in E(K)$, and $\phi_K(P) = \overline{D}$. So, $E(K)$ is mapped bijectively to $\mathrm{Pic}^0_E$, the preimage of a divisor class $\overline{D}$ is the point $P$ on $E$ corresponding to the uniquely determined prime divisor in the class of $D + P_\infty$ with $D \in \overline{D}$.

This implies that $E$ is isomorphic to its Jacobian as projective curve. So $E(K)$ itself is an abelian group with the chosen point $P_\infty$ as neutral element, and the addition of two points is given by rational functions in the coordinates in the points.

Hence $E$ is an abelian variety of dimension 1 (and vice versa) and we can apply all the structural properties of abelian varieties discussed above to study the structure of $E(K)$ (in dependence of $K$). This and the description of the addition with respect to carefully chosen equations for $E$ will be among the central parts of the algorithmic and applied parts of the book (cf. Chapter 13).

Let $P = (x_1, y_1)$ and $Q = (x_2, y_2)$ be two points with $x_1 \neq x_2$ of the affine curve

$$E : y^2 + a_1 xy + a_3 y = x^3 + a_2 x^2 + a_4 x + a_6.$$

The isomorphism maps them to divisor classes with representatives $D_P = P - P_\infty$ and $D_Q = Q - P_\infty$ of degree 0. The space $L(D_P + D_Q + P_\infty)$ has dimension 1 by Riemann–Roch. Hence, there exists a function passing through $P$ and $Q$ and having a pole of order at most 1 in $P_\infty$. Such a function is given by the line $l(x, y) = y - \lambda x - \mu = 0$ connecting $P$ and $Q$. It has

$$\lambda = \frac{y_2 - y_1}{x_2 - x_1} \quad \text{and} \quad \mu = y_1 - \lambda x_1.$$

As $D_P + D_Q + P_\infty = P + Q - P_\infty$ has degree 1 and $l \in L(D_P + D_Q + P_\infty)$, there exists an effective divisor in this class that we denote by $R$ and which is a prime divisor. Hence, in the divisor class group we have $\overline{D}_P + \overline{D}_Q = \overline{R} + \overline{P}_\infty$, which is equivalent to $P \oplus Q = R$ on $E$ using the isomorphism from above.

Choosing $P \neq Q$ with $x_1 = x_2$ we apply the same geometric construction and get as connecting line the parallel to the $y$-axis $x = x_1$. Hence, the third intersection point has to be interpreted as the point $P_\infty$. This associates to each point $P \in E$ an inverse point $-P$ which has the same $x$-coordinate.

In the remaining case $P = Q$ one can use the considerations above. The function $l \in L(2P - P_\infty)$ passes through $P$ with multiplicity 2, i.e., it is the tangent line to the curve at $P$. In formulas this means

$$\lambda = \frac{3x_1^2 + 2a_2 x_1 + a_4 - a_1 y_1}{2y_1 + a_1 x_1 + a_3} \quad \text{and} \quad \mu = y_1 - \lambda x_1.$$

### 4.4.5.a  Division polynomials

By Theorem 4.73 we know the structure of the group of $n$-torsion points on $E$. In that context we showed that for each $n$ there exists a polynomial $\psi_n$ such that the $x$-coordinates of $n$-torsion points are the roots of $\psi_n$. These polynomials are called *division polynomials*.

If $\operatorname{char}(K) \neq 2$, put

$$f_0(x) = 0, \; f_1(x) = 1, \; f_2(x) = 1,$$
$$f_3(x) = 3x^4 + b_2 x^3 + 3b_4 x^2 + 3b_6 x + b_8,$$
$$f_4(x) = 2x^6 + b_2 x^5 + 5b_4 x^4 + 10 b_6 x^3 + 10 b_8 x^2 + (b_2 b_8 - b_4 b_6) x + (b_4 b_8 - b_6^2)$$

where the $b_i$'s are defined as in Section 4.4.2.a and for $n \geqslant 5$

$$f_{2n} = f_n(f_{n+2} f_{n-1}^2 - f_{n-2} f_{n+1}^2),$$
$$f_{2n+1} = \begin{cases} \tilde{f}^2 f_{n+2} f_n^3 - f_{n-1} f_{n+1}^3 & \text{if } n \text{ is even,} \\ f_{n+2} f_n^3 - \tilde{f}^2 f_{n-1} f_{n+1}^3 & \text{otherwise.} \end{cases}$$

with $\tilde{f}(x) = 4x^3 + b_2 x^2 + 2 b_4 x + b_6$.

If $\operatorname{char}(K) = 2$ and $E : y^2 + xy = x^3 + a_2 x^2 + a_6$ then set

$$f_0(x) = 0, \; f_1(x) = 1, \; f_2(x) = x,$$
$$f_3(x) = x^4 + x^3 + a_6, \; f_4(x) = x^6 + x^2 a_6.$$

Otherwise $E : y^2 + a_3 y = x^3 + a_4 x + a_6$ and put

$$f_0(x) = 0, \; f_1(x) = 1, \; f_2(x) = a_3,$$
$$f_3(x) = x^4 + a_3^2 x + a_4^2, \; f_4(x) = a_3^5.$$

For $n \geqslant 5$, they can be computed recursively in both cases with the formulas

$$f_{2n+1} = f_{n+2} f_n^3 - f_{n-1} f_{n+1}^3,$$
$$f_2 f_{2n} = f_{n+2} f_n f_{n-1}^2 - f_{n-2} f_n f_{n+1}^2.$$

Now if $P = (x_1, y_1) \in E(\overline{K})$ is not a 2-torsion point then $P \in E[n]$ if and only if $P = P_\infty$ or $f_n(x) = 0$.

In addition there are explicit formulas for $[n]$ when $\operatorname{char}(K)$ is different from 2, namely

$$[n] : E \to E$$
$$P \mapsto [n]P = \begin{cases} P_\infty & \text{if } P \in E[n], \\ \left( \dfrac{\phi_n(x,y)}{\psi_n^2(x,y)}, \dfrac{\omega_n(x,y)}{\psi_n^3(x,y)} \right) & \text{if } P \in E(\overline{K}) \smallsetminus E[n]. \end{cases}$$

where

$$\psi_n = \begin{cases} (2y + a_1 x + a_3) f_n & \text{if } n \text{ is even,} \\ f_n & \text{otherwise} \end{cases}$$

and

$$\phi_n = x \psi_n^2 - \psi_{n-1} \psi_{n+1} \text{ and } 2 \psi_n \omega_n = \psi_{2n} - \psi_n^2 (a_1 \phi_n + a_3 \psi_n^2).$$

Note that in general these formulas are not used to compute $[n]P$ for given $n$ and $P$.

## 4.4.6 Ideal class group

The divisor class group relies on the projective curve and leads to points on an abelian variety. For computational reasons it is sometimes easier to work with affine parts and the arithmetic of corresponding affine algebras $\mathcal{O}$.

The objects corresponding to divisors are ideals of $\mathcal{O}$ and the objects corresponding to divisor classes are ideal classes of $\mathcal{O}$. The purpose of this section is to discuss the relation between these groups.

Let $C$ be an affine smooth curve with function field $K(C)$ and coordinate ring $\mathcal{O} = K[C]$.

We recall that $\mathcal{O}$ is a Dedekind ring and so every ideal $\neq (0)$ is a product of powers of maximal ideals $M$ in a unique way and every maximal ideal $M$ corresponds to a place $\mathfrak{p}_M$ of $K(C)$.

The ideals $\neq 0$ of $\mathcal{O}$ form a semigroup freely generated by the maximal ideals. To get a group one generalizes $\mathcal{O}$-ideals:

**Definition 4.137** The set $B \subset K(C)$ is a *fractional $\mathcal{O}$-ideal* if there exists a function $f \in K(C)^*$ such that $fB$ is an ideal of $\mathcal{O}$. For a maximal ideal $M \subset \mathcal{O}$ define $v_M(B) := \max\{k \in \mathbb{Z} \mid B \subset M^k\}$. Then

$$B = \prod_{M \text{ maximal in } \mathcal{O}} M^{v_M(B)}$$

and $B \subset \mathcal{O}$ if and only if all $v_M(B) \geqslant 0$.

To form the ideal class group we let two $\mathcal{O}$-ideals $B_1$ and $B_2$ be *equivalent* if and only if there exists a function $f \in K(C)$ with $v_M(B_1) = v_M(B_2) + v_M((f))$ for all maximal ideals $M$ of $\mathcal{O}$.
The group of $\mathcal{O}$-ideal classes is denoted by $\text{Cl}(\mathcal{O})$.

### 4.4.6.a Relation between divisor and ideal class groups

Here we want to explain the relation between ideal class groups of rings of regular functions of affine parts of absolutely irreducible smooth projective curves $C$ and the divisor class group of $C$ (hence points of the Jacobian $J_C$).

For the simplicity of our presentation we shall assume that there is a $K$-rational point $P_\infty$. Let $x_1$ be a nonconstant function on $C$ with pole divisor

$$\text{div}(x_1)_\infty = m_\infty P_\infty + \sum_{2 \leqslant j \leqslant t} m_j P_{\infty_j},$$

and $t \geqslant 0$, $m_\infty > 0$, $m_j > 0$ and $P_{\infty_j} \in C(\overline{K})$. Put $P_{\infty_1} = P_\infty$. Let $\mathcal{O}$ be the ring of functions on $C$ that are regular outside of the points $P_{\infty_j}$. So $\mathcal{O}$ is the intersection of the valuation rings $\mathcal{O}_\mathfrak{p}$ of all places $\mathfrak{p}$ of $K(C)$ with $v_\mathfrak{p}(x_1) \geqslant 0$.

It follows that $\mathcal{O}$ is the integral closure of the polynomial ring $K[x_1]$ in $K(C)$. It is the coordinate ring of the affine curve $C_\mathcal{O}$ with $C_\mathcal{O}(\overline{K}) = C(\overline{K}) \setminus \{P_{\infty_1}, \ldots, P_{\infty_t}\}$.

The inclusion $K[x_1] \to \mathcal{O}$ corresponds to a morphism $C_\mathcal{O} \to \mathbb{A}^1$, which extends to a map $\pi : C \to \mathbb{P}^1$ with $\pi^{-1}(\infty) = \{P_{\infty_1}, \ldots, P_{\infty_t}\}$. To describe a relation between points on $J_C$ and elements of $\text{Cl}(\mathcal{O})$, we state that every place of $K(C)$ is either equal to $\mathfrak{p}_M$ for some maximal ideal $M$ of $\mathcal{O}$ or to an extension of the infinite place on $\mathbb{P}^1$ to $C$.

Hence, there is a one-to-one correspondence between proper ideals $A \subset \mathcal{O}$ and effective $K$-rational divisors $D$ of $C$ in which only points of $C_\mathcal{O}$ occur, given by

$$\sum n_i \mathfrak{p}_i \leftrightarrow \prod M_{\mathfrak{p}_i}^{n_i},$$

where the $\mathfrak{p}_i$ are not extensions of $\mathfrak{p}_\infty$. If $A$ corresponds to $D$ then $\deg(D) = \deg(A)$. This correspondence extends naturally to fractional ideals and arbitrary divisors.

Now we apply the theorem of Riemann–Roch to ideal classes of $\mathcal{O}$ to get

**Lemma 4.138** With notation as above let $C$ be a curve of genus $g$. Let $c$ be an element of $\mathrm{Cl}(\mathcal{O})$. Then $c$ contains an ideal $A \subset \mathcal{O}$ with $\deg(A) \leqslant g$.

**Proof.** Let $A' \in c$ be an $\mathcal{O}$-ideal and assume that $\deg(A') > g$. Take the effective divisor $D_{A'}$ associated to $A'$ and a function $f$ such that $D' := (f) + D_{A'} - (\deg(A') - g)P_\infty$ is effective of degree $g$. Let $D''$ be the divisor obtained from $D'$ by removing points in $\pi^{-1}(\infty)$ and let $A$ be the ideal obtained from $D''$. Then $A \in c$ and $\deg(A) \leqslant g$. $\square$

Note that principal divisors are mapped to principal ideals. Therefore, one can consider the correspondence between divisor classes and ideal classes. We are now ready to define a homomorphism from $J_C$ to the ideal class group $\mathrm{Cl}(\mathcal{O})$.

Define $\phi : J_C(K) \to \mathrm{Cl}(\mathcal{O})$ by the following rule: in the divisor class $c$ take a representative $D'$ of the form $D' = D - gP_\infty$, $D$ effective. Remove from $D$ all points in $\pi^{-1}(\infty)$ and define $A$ as ideal in $\mathcal{O}$ like above. Then $\phi(c)$ is the class of $A$ in $\mathrm{Cl}(\mathcal{O})$. By Lemma 4.138 $\phi$ is surjective.

For applications one is usually interested in the case that the kernel of $\phi$ is trivial, i.e., in choices for $C$ and $\mathcal{O}$ such that $\mathrm{Cl}(\mathcal{O}) \simeq \mathrm{Pic}_C^0$. This allows us to use the interpretation via ideal classes of polynomial orders $\mathcal{O}$ for the computations whereas the interpretation as points on the Jacobian of $C$ is used for the structural background.

So let us describe the kernel of $\phi$: let $c$ be a divisor class of degree 0 represented by the divisor $D = D_1 + D_2 - gP_{\infty_1}$, where $D_i$ are effective divisors and $D_1 = \sum n_i P_i$ with $P_i \notin \pi^{-1}\{\infty\}$ and $D_2 = \sum m_j P_{\infty_j}$ with $P_{\infty_j} \in \pi^{-1}\{\infty\}$. If $\phi(c) = 0$ then $\prod M_{P_i}^{n_i}$ is a principal ideal $(f)$ with $f \in \mathcal{O}$. Hence, all prime divisors occurring in the pole divisor of $f$ correspond to points in $\pi^{-1}\{\infty\}$ and we can replace $D$ by an equivalent divisor $D - (f)$ of degree 0, which is a sum of prime divisors all corresponding to points in $\pi^{-1}\{\infty\}$.

**Proposition 4.139** We use the notation from above. The homomorphism

$$\phi : J_C(K) \to \mathrm{Cl}(\mathcal{O})$$

is surjective.

The kernel of $\phi$ is equal to the divisor classes of degree 0 in

$$\left\{ \sum m_j P_{\infty_j} \mid \sum m_j = 0 \text{ and all } P_{\infty_j} \in \pi^{-1}\{\infty\} \right\}.$$

**Proposition 4.140** Assume that there is a cover

$$\varphi : C \to \mathbb{P}^1,$$

in which one point $P_\infty$ is totally ramified and induces the place $v_\infty$ in the function field $K(x_1)$ of $\mathbb{P}^1$. Let $\mathcal{O}$ be the integral closure of $K[x_1]$ in the function field of $C$. Then $\phi$ is an isomorphism and, hence, the ideal class group of $\mathcal{O}$ is (in a natural way) isomorphic to the divisor class group of $C$.

This gives a very nice relation of the projective algebraic geometry and the ideal theory in Dedekind rings. Due to the isomorphism the ideal class group can be used for arithmetic while the divisor class group setting provides structural background.

**Definition 4.141** A nonsingular curve $C/K$ for which there exists a cover $\varphi$ in which one $K$-point $P_\infty \in C$ is totally ramified is called a $C_{ab}$-curve.

If the functions $x$ and $y$ have pole divisor $aP_\infty$ and $bP_\infty$, respectively, one finds an equation over $K$ given by

$$C : \alpha_{b,0} x^b + \alpha_{0,1} + \sum_{ia+jb<ab} \alpha_{i,j} x^i y^j, \text{ with } \alpha_{i,j} \in K,$$

with $\alpha_{b,0}, \alpha_{0,1} \neq 0$.

§ 4.4 Arithmetic of curves

In particular, hyperelliptic curves are $C_{ab}$ curves if they have a $K$-rational Weierstraß point and if we take as affine part the curve given by a Weierstraß equation (4.1). This relation is the topic of the following section.

**Example 4.142** An interesting class of curves are the *Picard curves* of genus 3. Over a field of characteristic $\mathrm{char}(K) \neq 3$ containing the third roots of unity they can by given by an equation of the form
$$y^3 = f(x),$$
where $f(x) \in K[x]$ is monic of degree 4 and has only simple roots over $\overline{K}$.

### 4.4.7 Class groups of hyperelliptic curves

The type of hyperelliptic curves $C$ we consider in this book additionally satisfies that there exists one $K$-rational Weierstraß point of $C$. This point is totally ramified under a cover $\phi$ and is denoted by $P_\infty$. By the considerations of the previous paragraph these curves satisfy that the ideal class group and the divisor class group are isomorphic. In Chapter 14 on the arithmetic of hyperelliptic curves we will use the ideal class group for the efficient computations inside the group. To fix notation we still speak of divisor classes usually implying this isomorphism. In Section 4.4.2.b we have shown how one can use the definition and the Riemann–Roch theorem to derive an affine plane equation. The $K$-rational point $P_\infty$ allows us to use the divisor of degree one in the construction.

Recall that a hyperelliptic curve over $K$ of genus $g$ with at least one $K$-rational Weierstraß point can be given by a *Weierstraß equation*

$$y^2 + h(x)y = f(x), \quad \text{with } h(x) \text{ and } f(x) \in K[x], \tag{4.3}$$

where $f$ is monic of degree $2g + 1$ and $\deg(h) \leq g$. By abuse of language we denote affine curves given by such an equation as *imaginary quadratic curves*.

We use the equation of such curves $C$ to describe explicitly their ideal class group.

**Theorem 4.143** Let $C/K$ be an imaginary quadratic hyperelliptic curve of genus $g$ and let $\omega$ denote the nontrivial automorphism of $K(C)$ over $K(x)$ with a $K$-rational Weierstraß point $P_\infty$ lying over the place $x_\infty$ of $K(x)$. Let $\mathcal{O} = K[x,y]/(y^2 + h(x)y - f(x))$.

(i) In every nontrivial ideal class $c$ of $\mathrm{Cl}(\mathcal{O})$ there is exactly one ideal $I \subseteq \mathcal{O}$ of degree $t \leq g$ with the property: the only prime ideals that could divide both $I$ and $\omega(I)$ are those resulting from Weierstraß points.

(ii) Let $I$ be as above. Then $I = K[x]u(x) + K[x](v(x) - y)$ with $u(x), v(x) \in K[x]$, $u$ monic of degree $t$, $\deg(v) < t$ and $u$ divides $v^2 + h(x)v - f(x)$.

(iii) The polynomials $u(x)$ and $v(x)$ are uniquely determined by $I$ and hence by $c$. So $[u, v]$ can be used as coordinates for $c$.

**Proof.** Since for every ideal $J$ we get that $J \cdot \omega(J)$ is a principal ideal we can reduce $I$ repeatedly until the condition in (i) is satisfied without changing its class. After this process we call $J$ reduced.

Now assume that $\deg(I) \leq g$, $\deg(J) \leq g$, with $I, J$ reduced and that $I \sim J$. Then $I \cdot \omega(J)$ is a principal ideal in $\mathcal{O}$ and so it is equal to $(b)$ with $b \in K(C)$ having only one pole of order $\leq 2g$ in $P_\infty$. By Riemann–Roch all such functions lie in a $K$-vector space of dimension $g + 1$ and a basis of this space is given by $\{1, x, x^2, \ldots, x^g\}$. So $b \in K[x]$ and $I \cdot \omega(J)$ is the conorm of an ideal in $K[x]$. Since $I$ and $J$ are reduced this means that $I = J$ and (i) is proved.

(ii). Let $I \in O$ be an ideal of degree $t$. Recall that $\{1, y\}$ is a basis of $\mathcal{O}$ as $K[x]$-module. We choose any basis $\{w_1 = f_1(x) + f_2(x)y, w_2 = g_1(x) + g_2(x)y\}$ of $I$ as $K[x]$-module. We find relatively prime polynomials $h_1, h_2$ with $f_2 h_1 - g_2 h_2 = 0$ and choose $u_1, u_2 \in K[x]$ with $u_1 h_1 - u_2 h_2 = 1$. Now take $u' := h_1 w_1 + h_2 w_2, w_2' = u_2 w_1 + u_1 w_2$. Since the determinant of this transformation is 1 the pair $\{u'(x), w_2' = v_1(x) + v_2(x)y\}$ is again a basis of $I$. Since the rank of $I$ is 2, $v_2(x)$ is not equal to 0. So $I \cap K[X]$ is generated by $u'$. Since $I$ is reduced, the degree of $I$ is equal to the degree of $u'$ and we can and will take $u$ monic. Now write $v_1 = a u + v$ with $\deg v < t$. By replacing $w_2'$ by $w_2' - a u$ we get a basis $\{u(x), v(x) + v_2(x)y\}$ of $I$. Since the degree of $I$ is equal to $u(x)v_2(x)$ we get: $v_2(x)$ is constant and so we can assume $v_2(x) = -1$. The element $(v+y)(v-y) = v^2 + h(x)y - f(x) = \left(v^2 + h(x)v - f(x)\right) - h(x)(v-y)$ lies in $I$ and so the last claim of (ii) follows.

(iii). From the proof of (ii) we have that $u(x)$ is determined by $I$ as monic generator of $I \cap K[x]$. Now assume that $v' - y \in I$ with $\deg(v') < t$. Then $v' - v \in I \cap K[x]$ and hence $v' - v = 0$. □

**Remark 4.144** We are in a very similar situation as in the case of class groups of imaginary quadratic fields. In fact, Artin has generalized the theory of ideal classes of imaginary quadratic number fields, due to Gauß, to hyperelliptic function fields connecting ideal classes of $\mathcal{O}$ with reduced quadratic forms of discriminant $f(x)$ and the addition $\oplus$ with the composition of such forms. Theorem 4.143 and its proof can easily be translated into this language.

We are now in a position to use the results obtained in the previous section and describe the divisor class group of $C$ using the ideal class group of the affine part.

**Theorem 4.145 (Mumford representation)**
Let $C$ be a genus $g$ hyperelliptic curve with affine part given by $y^2 + h(x)y - f(x)$, where $h, f \in K[x]$, $\deg f = 2g+1$, $\deg h \leq g$. Each nontrivial group element $\overline{D} \in \text{Pic}_C^0$ can be represented via a unique pair of polynomials $u(x)$ and $v(x)$, $u, v \in K[x]$, where

(i) $u$ is monic,

(ii) $\deg v < \deg u \leq g$,

(iii) $u \mid v^2 + vh - f$.

Let $\overline{D}$ be uniquely represented by $D = \sum_{i=1}^{r} P_i - r P_\infty$, where $P_i \neq P_\infty, P_i \neq -P_j$ for $i \neq j$ and $r \leq g$. Put $P_i = (x_i, y_i)$. Then the corresponding polynomials are defined by

$$u = \prod_{i=1}^{r}(x - x_i)$$

and the property that if $P_i$ occurs $n_i$ times then

$$\left(\frac{d}{dx}\right)^j \left[v(x)^2 + v(x)h(x) - f(x)\right]_{|x=x_i} = 0, \text{ for } 0 \leq j \leq n_i - 1.$$

A divisor with at most $g$ points in the support satisfying $P_i \neq P_\infty, P_i \neq -P_j$ for $i \neq j$ is called a *reduced divisor*. The first part states that each class can be represented by a reduced divisor. The second part of the theorem means that for all points $P_i = (x_i, y_i)$ occurring in $D$ we have $u(x_i) = 0$ and the third condition guarantees that $v(x_i) = y_i$ with appropriate multiplicity.

Like for elliptic curves (cf. Section 4.4.5) one can make explicit the group operations in the ideal class group. Consider the classes represented by $[u_1(x), v_1(x)]$ and $[u_2(x), v_2(x)]$ and assume them in general position. The product of the representatives is generated by

$$\langle u_1 u_2, u_1(y - v_2), u_2(y - v_1), (y - v_1)(y - v_2) \rangle.$$

By Hermite reduction from the generating system we obtain a basis $\{u_3'(x), v_3'(x) + w_3'(x)y\}$. This ideal lies in the class of the product of the ideal classes but is usually not yet reduced. To reduce it one recursively applies the fact that $u \mid v^2 + hv - f$. This procedure is formalized and applied to arbitrary inputs in Cantor's algorithm, which we state in Chapter 14 on the arithmetic of hyperelliptic curves.

# Chapter 5

# *Varieties over Special Fields*

### *Gerhard Frey and Tanja Lange*

### Contents in Brief

| | | |
|---|---|---|
| 5.1 | **Varieties over the field of complex numbers** | 87 |
| | Analytic varieties • Curves over $\mathbb{C}$ • Complex tori and abelian varieties • Isogenies of abelian varieties over $\mathbb{C}$ • Elliptic curves over $\mathbb{C}$ • Hyperelliptic curves over $\mathbb{C}$ | |
| 5.2 | **Varieties over finite fields** | 108 |
| | The Frobenius morphism • The characteristic polynomial of the Frobenius endomorphism • The theorem of Hasse–Weil for Jacobians • Tate's isogeny theorem | |

In the previous chapter we dealt with algebraic and geometric objects over arbitrary fields. In this chapter we explain additional properties of these objects when considered over special fields. We concentrate on varieties over the complex numbers and finite fields.

## 5.1 Varieties over the field of complex numbers

In the whole section we take ground field $K$ as the field of complex numbers $\mathbb{C}$. Since $\mathbb{C}$ is algebraically closed we can identify the affine space $\mathbb{A}^n$ (respectively the projective space $\mathbb{P}^n$) with the set of points in $\mathbb{C}^n$ (respectively the homogeneous classes of $(n+1)$-tuples in $\mathbb{C}^{n+1}$) together with the topological structure induced by the Zariski topology. Recall that closed sets are given as zeroes of polynomial equations.

The absolute value $|\cdot|$ makes $\mathbb{C}^n$ to a metric space and hence induces a "natural" topology. Since polynomial functions are continuous in this topology it follows that Zariski closed sets are also closed in this topology.

### 5.1.1 Analytic varieties

First we shall describe very briefly the *analytic structure* on $\mathbb{A}^n$ (respectively $\mathbb{P}^n$): the key notions are *holomorphic functions*. Locally, holomorphic functions are given by power series converging in an open ball.

For open sets $U \subset \mathbb{P}^n$ one can globalize to get holomorphic functions by gluing together the local "germs." So a holomorphic function $f$ on $U$ is a complex valued function defined on $U$ such that for all $P \in U$ there is an open ball around $P$ on which $f$ is given by a convergent power series. Examples for holomorphic functions are polynomials (for $U = \mathbb{A}^n$) and rational functions (for $U$ equal to the set of definition).

*Meromorphic functions* on $U$ are defined as quotients of holomorphic functions. Locally they are given by Laurent series with finite negative part. To a meromorphic function $f$ on $U$ and to any point $P \in U$ we can associate the order of vanishing $n_P(f)$ of $f$ at $P$. It is negative if $f$ has a pole of order $|n_P(f)|$ at $P$, and positive if $f$ has a zero of order $n_P(f)$ at $P$. If $n_P(f) = 0$ there is a neighborhood of $P$ in $U$ such that the restriction of $f$ is invertible in this neighborhood as a holomorphic function. In particular, it follows that the set of zeroes and poles of meromorphic functions on $U$ does not have a limit point in $U$. The (analytic) divisor $\mathrm{div}_{\mathrm{an}}(f)$ is equal to the formal sum $\mathrm{div}_{\mathrm{an}}(f) = \sum_{P \in U} n_P(f) P$.

One can differentiate and integrate holomorphic and meromorphic functions and as usual one has meromorphic differentials $\omega$ on $U$. Locally at a point $P \in U$ they are of the form $f_P(x)dx_1 \cdots dx_n$ with $f_P$ meromorphic and $(x_1, \ldots, x_n)$ a local system of coordinates (mapping the chosen neighborhood to an open ball in $\mathbb{C}^n$ with 0 as image of $P$). Their divisor is $\mathrm{div}_{\mathrm{an}}(\omega) = \sum_{P \in U} n_P(f_P) P$. One sees that $\omega$ is holomorphic on $U$ if and only if the divisor of $\omega$ has only nonnegative coefficients.

In the sequel we shall need a further concept, namely *analytic varieties*. For the notion of analytic varieties (without boundary) in projective spaces we refer to [GRHA 1978].

One essential property of analytic varieties $V_{\mathrm{an}} \subseteq \mathbb{P}^n$ is that there exists a number $d \leqslant n$ such that $V_{\mathrm{an}}$ is locally isomorphic to a ball in an affine space $\mathbb{A}^d$, or equivalently: every point $P \in V_{\mathrm{an}}$ has an open neighborhood $U_P$ (with respect to the topology on $V_{\mathrm{an}}$ induced by the restriction of the topology on $\mathbb{P}^n$ to $V_{\mathrm{an}}$) and local coordinate functions (holomorphic in $U_P$) which map $U_P$ bijectively to a ball in $\mathbb{C}^d$ with 0 as image of $P$.

Using this local analytic structure one defines holomorphic functions on open subsets of $V_{\mathrm{an}}$, meromorphic functions on $V_{\mathrm{an}}$, holomorphic (respectively meromorphic) differentials and holomorphic (respectively meromorphic) maps between two analytic varieties. The number $d$ is the *dimension of $V_{\mathrm{an}}$*.

Now assume that $V$ is an (affine or projective) algebraic variety of algebraic dimension $d$ embedded in $\mathbb{P}^n$. First of all the underlying set is closed. Next, all points of $V$ are nonsingular, and the Jacobi criterion (cf. Lemma 4.94) for the local (algebraic) coordinate functions together with the implicit function theorem ensures that this set satisfies the conditions of analytic varieties being locally isomorphic to $\mathbb{C}^d$. So we can give $V$ the structure of an analytic variety of dimension $d$ denoted by $V_{\mathrm{an}}$. Note that rational functions on $V$ are meromorphic functions on $V_{\mathrm{an}}$. Of course, the converse does not have to hold true.

But there is a very important special case. Assume that $V$ is a projective algebraic variety. Then the underlying set is *compact* in $\mathbb{P}^n$. It follows that meromorphic functions on $V_{\mathrm{an}}$ have only finitely many zeroes and poles. Therefore, they are rational functions on $V$. So the field of meromorphic functions on $V_{\mathrm{an}}$ is equal to $K(V)$ and has transcendental degree $d$ over $\mathbb{C}$.

The converse of this remark is true, too. So we can state the following fundamental result.

**Theorem 5.1** Let $V_{\mathrm{an}}$ be a compact analytic variety in $\mathbb{P}^n$ of dimension $d$. There is an algebraic projective variety $V \subset \mathbb{P}^n$ such that the induced analytic variety is equal to $V_{\mathrm{an}}$, and the field of meromorphic functions on $V_{\mathrm{an}}$ has transcendence degree $d$ over $\mathbb{C}$ and hence is equal to $K(V)$.

The next lemma gives a slight generalization of the above facts about functions on varieties.

**Lemma 5.2** Let $V, W$ be projective algebraic varieties. Then the set of holomorphic maps from $V_{\mathrm{an}}$ to $W_{\mathrm{an}}$ is (in a natural way) identical with $\mathrm{Hom}_{\mathbb{C}}(V, W)$.

§ 5.1 Varieties over the field of complex numbers

As a consequence of these comparison results, we can use the full power of complex analysis to get purely algebraic properties of objects related to varieties defined over the complex numbers.

Before discussing the two examples that are the most important for us we will conclude this section with a remark.

**Remark 5.3** It is well-known in number theory that the interpretation of number fields $K$ as subfields of $\mathbb{C}$ is in a most fruitful way generalized to the study of number fields as subfields of $p$-adic fields. The same is true if we want to study objects of algebraic geometry by analytic methods. As counterpart of $\mathbb{C}$ one uses the completion of the algebraic closure of $\mathbb{Q}_p$. Over these fields we have the highly developed machinery of rigid analytic geometry. In Chapter 17 we shall have to use parts of this theory as *background* for discussing $p$-adic point counting methods on curves over finite fields, which have become important in recent years.

### 5.1.2 Curves over $\mathbb{C}$

Analytic curves $C_{\mathrm{an}}$ are one-dimensional analytic varieties embedded into a projective space over $\mathbb{C}$. From now on we shall assume that $C_{\mathrm{an}}$ is compact. Then there is a nonsingular projective irreducible curve $C$ such that $C_{\mathrm{an}}$ is the corresponding analytic curve. Hence, from an abstract point of view the rational functions on $C$ cannot be distinguished from the meromorphic functions on $C_{\mathrm{an}}$.

One uses this to produce functions on $C$ by analytic methods: locally there are many more meromorphic functions given by converging Laurent series, and by the gluing process we can hope to get global meromorphic functions that turn out to be *algebraic*.

In the previous section we introduced the notion of divisors on analytic curves $C_{\mathrm{an}}$ in a way analogous to the algebraic case. The finiteness condition for coefficients not equal to 0 is replaced by the condition that poles and zeroes have no limit point. But since we have assumed that $C_{\mathrm{an}}$ is compact this is exactly the same condition as in the algebraic case. Therefore, analytic divisors can be identified with algebraic divisors in a canonical way. The same is true for divisor class groups. (Note again that the situation changes totally if we go to affine parts of $C$.)

We introduced differentials for algebraic curves in Section 4.4.2.c. We now look at them from the analytic point of view. Here the usual calculus methods are used to construct the meromorphic differentials. Again we get:

**Proposition 5.4** Meromorphic (respectively holomorphic) differentials on $C_{\mathrm{an}}$ can be identified with meromorphic (respectively holomorphic) differentials on $C$.

We have defined the genus $g$ of $C$ with the help of the theorem of Riemann–Roch (cf. Theorem 4.106). This theorem also holds for the divisor theory of $C_{\mathrm{an}}$. (In fact its original version was proved in this context.) One of its consequences is that the holomorphic differentials $\Omega_C$ on $C_{\mathrm{an}}$ (or on $C$) form a $g$-dimensional vector space over $\mathbb{C}$ and that these differentials can be identified with algebraic differentials with effective divisors. Let us choose a basis $\{\omega_1, \ldots, \omega_g\}$. To get the full power of analytic methods we have to go one step further and go to real surfaces.

#### Digression: the easiest example of Weil descent

Next we use an additional special property of $\mathbb{C}$: it is a two-dimensional vector space over the field of real numbers $\mathbb{R}$ with basis $\{1, i\}$ where as usual $i^2 = -1$.

Replacing a complex variable $z$ by two real variables $x, y$ using $z = x + iy$ identifies the metric vector space $\mathbb{C}^n$ with the usual Euclidean space $\mathbb{R}^{2n}$. By this process we lose the analytic structure but have a *differentiable structure* from usual real calculus again compatible with the Zariski topology. Applying this to algebraic varieties $V$ of dimension $d$ in $\mathbb{A}^n_{\mathbb{C}}$ we find in a natural way an affine variety $W_{\mathbb{R}} \subset \mathbb{A}^{2n}_{\mathbb{R}}$ of dimension $2d$ with $W(\mathbb{R}) = V(\mathbb{C})$: we replace the $n$ complex affine coordinates $(X_1, \ldots, X_n)$ by the real coordinates $(U_1, V_1, \ldots, U_n, V_n)$ with $X_j = U_j + iV_j$, plug them into the equations $(f_1(X), \ldots, f_m(X))$ defining $V$ and separate the resulting polynomials into their real and imaginary part $f_k(U,V) = g_k(U,V) + ih_k(U,V)$, where $g_k$ and $h_k$ are defined over $\mathbb{R}$. Then $W$ is the variety defined by $(g_k, h_k)$.

By a gluing process we can apply this procedure to *projective* algebraic varieties. So we attach to every affine or projective variety $V$ of dimension $d$ defined over $\mathbb{C}$ an affine (respectively projective) algebraic variety $W_V$ of dimension $2d$ defined over $\mathbb{R}$ with $W_V(\mathbb{R}) = V(\mathbb{C})$. It is a nice exercise to show that $W_V \cdot \mathbb{C}$ is isomorphic to $V \times V$ as algebraic variety over $\mathbb{C}$.

What we have sketched above is the most simple example of scalar (or Weil-) restriction of varieties defined over a finite algebraic extension field $L$ of a field $K$ to varieties over $K$. This construction will play an important role later (cf. Chapter 7 on Weil descent).

### Riemann surfaces

We apply Weil descent to irreducible nonsingular projective curves $C$ defined over $\mathbb{C}$ and get an irreducible two-dimensional projective variety $W_C$ defined over $\mathbb{R}$. The analytic structure of $C$ induces a differentiable real structure that makes $W_C$ locally isomorphic to a unit ball in $\mathbb{R}^2$. That means that for every $P \in W_C(\mathbb{R})$ there is an open neighborhood $U_P \in W_C$ and real differentiable functions $f_1, f_2$ defined on $U_P$ mapping $U_P$ to the open unit disc in $\mathbb{R}^2$ and sending $P$ to $(0,0)$.

Since $C(\mathbb{C})$ is compact, it follows that $W_C$ is compact in the real topology.

As result we get that the projective curve $C$ carries in a natural way the structure of a *compact Riemann surface*. We remark that the converse is true, too: every compact Riemann surface is the Weil descent of a projective nonsingular irreducible curve defined over $\mathbb{C}$.

Riemann surfaces $R$ are classical and very well studied objects in geometry. One key ingredient is the study of paths on them up to homology (cf. [GRHA 1978]). They can be used to define the topological genus $g_{\text{top}}$ of $R$. Namely fixing a base point $P_0$ and composing closed paths in a natural way we turn the set of points $\mathcal{P}$ into a group. By identifying homologous paths we get the *fundamental group* $\Pi_R$ of $R$ as quotient of $\mathcal{P}$. It is generated by $2g_{\text{top}}$ paths satisfying one relation which lies in the commutator subgroup of the fundamental group. This implies that the maximal abelian factor group of $\Pi_R$, the first homology group $H_1(R, \mathbb{Z})$, is a free abelian group with $2g_{\text{top}}$ generators $(\alpha_1, \ldots, \alpha_{2g_{\text{top}}})$.

We come back to the case that $R = W_C$ with $C$ a projective curve over $\mathbb{C}$.

**Proposition 5.5** The genus $g$ of $C$ is equal to the topological genus $g_{\text{top}}$ of $W_C$.

Using well-known results from (real and complex) calculus we do integration on $W_C$ using holomorphic differentials $\omega$ on $C$ and closed paths $\alpha$ on $W_C$. As above we choose a base point $P_0$ on $W_C$ and get the group $\mathcal{P}$ by composing closed (continuous) paths beginning in $P_0$.

**Lemma 5.6** We have a map
$$\langle \cdot, \cdot \rangle_0 : \mathcal{P} \times \Omega_C \to \mathbb{C}$$
defined by $\langle \alpha, \omega \rangle_0 := \int_\alpha \omega$ where $\int_\alpha$ is the line integral along the path $\alpha$.

Moreover, $\langle \cdot, \cdot \rangle_0$ is independent of the homology class of $\alpha$ and vanishes when restricted to the commutator subgroup of $\mathcal{P}$ in the first component.

**Corollary 5.7** The map $\langle \cdot, \cdot \rangle_0$ induces a pairing, that is denoted by $\langle \cdot, \cdot \rangle$, between the $\mathbb{Z}$-module $H_1(W_C, \mathbb{Z})$ and the $\mathbb{C}$-vector space $\Omega_C$.

## § 5.1 Varieties over the field of complex numbers

Recall that we have chosen a basis $(\omega_1, \ldots, \omega_g)$ of the space of holomorphic differentials on $C$. We define the map

$$\begin{aligned} \phi : H_1(W_C, \mathbb{Z}) &\longrightarrow \mathbb{C}^g \\ \tau &\longmapsto \left( \int_\alpha \omega_1, \ldots, \int_\alpha \omega_g \right) \end{aligned}$$

where $\alpha$ is a path in the class of $\tau$.

**Proposition 5.8** The image $\Lambda_C$ of $\phi$ is a full lattice in $\mathbb{C}^g$, i.e., a discrete free $\mathbb{Z}$-module of rank $2g$ in $\mathbb{C}^g$.

By this proposition we can associate a full rank lattice to each curve over $\mathbb{C}$. The following lemma describes what quotients of lattices look like.

**Lemma 5.9** Let $\Lambda$ be a lattice of full rank in $\mathbb{C}^g$ and let $\mathbb{C}^g/\Lambda$ be the quotient group with quotient topology. Then $\mathbb{C}^g/\Lambda$ is compact and locally isomorphic (as topological space) to the unit ball in $\mathbb{C}^g$.

**Corollary 5.10** The set $\mathbb{C}^g/\Lambda_C$ is a compact topological space with respect to the quotient topology inherited from $\mathbb{C}^g$. It is locally homeomorphic to the unit ball in $\mathbb{C}^g$.

**Definition 5.11** The lattice $\Lambda_C$ is called the *period lattice of* $C$ (with respect to the basis $\{\omega_1, \ldots, \omega_g\}$ of the holomorphic differentials).

We are now ready to define the Abel–Jacobi map. We fix the base point $P_0 \in C(\mathbb{C})$. For $P \in C(\mathbb{C})$ choose a path $\gamma$ from $P_0$ to $P$ and define $J_\gamma(P) := \left( \int_\gamma \omega_1, \ldots, \int_\gamma \omega_n \right) \in \mathbb{C}^g$. The tuple $J_\gamma(P)$ will — in general — depend on the choice of $\gamma$. If $\gamma'$ is another path from $P_0$ to $P$ then $\gamma$ and $\gamma'$ differ by a closed path beginning in $P_0$ and so $J_\gamma(P) - J_{\gamma'}(P)$ is an element of $\Lambda_C$.

**Definition 5.12** The *Abel–Jacobi map* is defined by

$$\begin{aligned} J : C(\mathbb{C}) &\to \mathbb{C}^g/\Lambda_C \\ P &\mapsto J_\gamma(P) + \Lambda_C. \end{aligned}$$

We can generalize this definition to divisors on $C$ by linear extension. We denote the corresponding map again by $J$.

**Theorem 5.13 (Abel–Jacobi)**

(i) Let $D$ be a principal divisor of $C$. Then $J(D) = 0$. So $J$ induces a map $\bar{J}$ from $\mathrm{Pic}^0_C$ to $\mathbb{C}^g/\Lambda_C$.
(ii) The map $\bar{J}$ is a group isomorphism.

By Lemma 5.9 the group $\mathbb{C}^g/\Lambda_C$ carries an analytic structure since it is locally homeomorphic to the unit ball in $\mathbb{C}^g$. The group $\mathrm{Pic}^0_C$ carries the structure of an abelian variety, namely the Jacobian variety $J_C$ of $C$ (see Definition 4.134). Hence, it has an analytic structure, too. The theorem of Abel–Jacobi includes that $\bar{J}$ is an *analytic isomorphism*.

So the structure of $J_C$ as analytic variety is rather simple and described by $\mathbb{C}^g/\Lambda$.

### 5.1.3 Complex tori and abelian varieties

An important part of the introduction to the objects relevant for cryptography were the connected *projective* algebraic groups called abelian varieties (cf. Section 4.3). The analytic counterpart are connected *compact* complex Lie groups (cf. e.g., Lang [LAN 2002a]).

We give the most simple example of a complex Lie group: take $\mathbb{C}^d$ with the usual complex structure and with vector addition $+$ as group composition. It is obvious that the addition $+$ as well as the inversion $-$ are holomorphic. The group $\mathbb{C}^d$ is not compact but we can easily find quotients that are compact.

For this we choose a lattice (always assumed to be of full rank) $\Lambda \subset \mathbb{C}^d$, i.e., there is a basis $\{\mu_1, \ldots, \mu_{2d}\}$ of $\mathbb{C}^d$ as real vector space such that

$$\Lambda = \left\{ \sum_{j=1}^{2d} z_j \mu_j \mid z_j \in \mathbb{Z} \right\}.$$

Equivalently we have: $\Lambda$ is a $\mathbb{Z}$-submodule of $\mathbb{C}^d$ of rank $2d$ which is discrete, i.e., in every bounded subset of $\mathbb{C}^d$ there are only finitely many elements of $\Lambda$.

We can endow $\mathbb{C}^d/\Lambda$ with an analytic structure in a natural way. Let the $U_j$ be open sets covering $\mathbb{C}^d/\Lambda$ such that each $U_j$ is homeomorphic via bijective continuous maps $\varphi_j$ to balls $B_j$ in $\mathbb{C}^d$. The maps $\varphi_j$ are assumed to be compatible with restrictions to intersections of the sets $U_j$. We define holomorphic functions on $U_j$ as functions $f_j : U_j \to \mathbb{C}$ such that $f_j \circ \varphi_j^{-1}$ are holomorphic on $B_j$ and come to global functions by gluing local holomorphic functions. Meromorphic functions are defined as quotients of holomorphic functions. It is an immediate consequence from these definitions that $\mathbb{C}^d/\Lambda$ carries the structure of a complex connected Lie group that is a quotient (as Lie group) of $\mathbb{C}^d/\Lambda$.

**Definition 5.14** A complex Lie group isomorphic to $\mathbb{C}^d/\Lambda$ is called a *complex $d$-dimensional torus*.

A fundamental result is:

**Proposition 5.15** Let $X$ be a connected compact complex Lie group of dimension $d$. Then $X$ is isomorphic to a torus $T := \mathbb{C}^d/\Lambda$.

For the proof see [MUM 1974, p. 2].

We apply this to an abelian variety $\mathcal{A}$ of dimension $d$ defined over $\mathbb{C}$. The associated analytic variety $\mathcal{A}_{\text{an}}$ is connected and compact. Since addition and inversion on $\mathcal{A}$ are given by polynomials, $\mathcal{A}_{\text{an}}$ is a torus and, hence, is isomorphic to the Lie group $\mathbb{C}^d/\Lambda$ for some lattice $\Lambda$. Note that by this isomorphism the addition on $\mathcal{A}$ is transferred into a very easy form. It is just the vector addition in $\mathbb{C}^d$ modulo $\Lambda$.

Next we shall study the converse. We want to decide whether $T$ is the analytic companion of an algebraic variety. By Chow's theorem this is equivalent to the question whether we can embed $T$ into a projective space such that the analytic structures are compatible.

If this is possible we shall find $d$ algebraically independent meromorphic functions on $T$. By standard methods of algebraic theory (the key word is "ample line bundle") one sees that the converse is true, too. So one has to *construct* meromorphic functions on $T$, or equivalently, meromorphic functions on $\mathbb{C}^d$ which are *periodic* with respect to $\Lambda$. There are well-known methods for this (for $d=1$ one uses results like the Weierstraß product theorem or the Mittag–Leffler theorem). In general the main ingredients are theta functions attached to $\Lambda$. We shall need them later on (cf. Chapter 18) and then deal explicitly with the case that is most interesting for us, and so we do not give a formal definition here.

In [MUM 1974, pp. 24-35] one finds the discussion what additional properties $\Lambda$ has to have in order to have enough periodic functions.

## § 5.1 Varieties over the field of complex numbers

First recall that a Hermitian form $H$ on $\mathbb{C}^d \times \mathbb{C}^d$ can be decomposed as
$$H(x,y) = E(ix,y) + iE(x,y)$$
where $E$ is a skew symmetric real form on $\mathbb{C}^d$ satisfying $E(ix, iy) = E(x, y)$. The form $E$ is called the imaginary part $\Im m(H)$ of $H$. (Since these notations are rather standard we find it convenient not to change them though the letter $E$ is used for elliptic curves in most cases. We hope that this does not give rise to confusion.)

**Theorem 5.16** *The torus $T = \mathbb{C}^d/\Lambda$ can be embedded into a projective space and, hence, equals the analytic variety attached to an abelian variety if and only if there exists a positive definite Hermitian form $H$ on $\mathbb{C}^d$ with $E = \Im m(H)$ such that $E$ restricted to $\Lambda \times \Lambda$ has values in $\mathbb{Z}$.*

We use the structure theorems for Hermitian forms and get

**Corollary 5.17** *Let $T = \mathbb{C}^d/\Lambda$ be a complex torus attached to an abelian variety $\mathcal{A}$. Then $\Lambda$ is isomorphic to $\mathbb{Z}^d \oplus \Omega \cdot \mathbb{Z}^d$, where the $(d \times d)$-matrix $\Omega$ is symmetric and has a positive definite imaginary part, i.e., lies in the Siegel upper half plane $\mathbb{H}_g$.*

**Corollary 5.18** *Assume that $d = 1$, i.e., $\mathcal{A}$ is an elliptic curve $E$. Then the torus associated to $E$ is isomorphic to $\mathbb{C}/(\mathbb{Z} + \tau\mathbb{Z})$ where $\tau$ is a complex number with positive imaginary part.*

**Definition 5.19** We call $\Omega$ the *period matrix of $\mathcal{A}$*.

We continue to assume that $T = \mathcal{A}_{\mathrm{an}}$ with Hermitian form $H$ and $E = \Im m(H)$.

With the help of $E$ we can define a dual lattice $\widehat{\Lambda}$ given by
$$\widehat{\Lambda} := \{x \in \mathbb{C}^d \mid E(x,y) \in \mathbb{Z}, \text{ for all } y \in \Lambda\}.$$

The lattice $\widehat{\Lambda}$ contains $\Lambda$ and $\widehat{\Lambda}/\Lambda$ is finite. Furthermore, $\widehat{\Lambda}$ belongs to a torus $\widehat{T}$, which is attached to an abelian variety $\widehat{\mathcal{A}}$. In fact we have just constructed the *dual abelian variety* to $\mathcal{A}$ by analytic methods over the complex numbers (see [MUM 1974], 82-86). There it is also shown how this dual abelian variety can be constructed by purely algebraic methods over any ground field.

For us a special case is most important. Assume that $\widehat{\Lambda} = \Lambda$ and so $\mathcal{A}$ is equal to its dual.

**Definition 5.20** If $\widehat{\Lambda} = \Lambda$ then $\mathcal{A}$ is called *principally polarized*.

**Corollary 5.21** *Let $\mathcal{A}$ be a principally polarized abelian variety over $\mathbb{C}$ with lattice $\Lambda$, Hermitian form $H$ and $E = \Im m(H)$. Then there exists a basis $\{\mu_1, \ldots, \mu_{2d}\}$ of $\Lambda$ such that*
$$\left[E(\mu_i, \mu_j)\right]_{1 \leq i,j \leq 2d} = \begin{bmatrix} 0 & I_d \\ -I_d & 0 \end{bmatrix}.$$

Now we come back to the theme of this book, namely projective irreducible nonsingular curves $C$ and their Jacobians $J_C$.

By the theorem of Abel–Jacobi (cf. Theorem 5.13) we have found an isomorphism from $\mathrm{Pic}_C^0$, the divisor class group of degree 0 of $C$ to $\mathbb{C}^g/\Lambda_C$ by integrating a basis of holomorphic differentials along paths that form a basis of the first homology group of $C$. So the period lattice of $C$ is attached to the isomorphism class of $(J_C)_{\mathrm{an}}$.

**Definition 5.22** The period matrix $\Omega_C$ of $\Lambda_C$ is called the *period matrix of $C$*. The form $E(x,y)$ is called the *Riemann form*.

**Lemma 5.23** *The period matrix $\Omega_C$ can be computed by integrating a basis of holomorphic differentials along paths on the Riemann surface corresponding to $C$.*

By duality theorems about differentials and paths on Riemann surfaces one sees:

**Proposition 5.24** *The Jacobian of a projective irreducible nonsingular curve is a principally polarized abelian variety.*

### 5.1.4 Isogenies of abelian varieties over $\mathbb{C}$

We can use the torus representation of abelian varieties to find the algebraic results about torsion points, isogenies and endomorphisms. So assume that $\mathcal{A}$ is analytically given by $T = \mathbb{C}^d/\Lambda$.

First we find a result given previously.

**Proposition 5.25** Let $n$ be a natural number. The points of order dividing $n$ of $\mathcal{A}$, $\mathcal{A}[n]$, are isomorphic to the subgroup $\frac{1}{n}\Lambda/\Lambda \subset \mathbb{C}^d/\Lambda$ and hence isomorphic to $(\mathbb{Z}/n\mathbb{Z})^{2d}$.

Let $G$ be a subgroup of $\frac{1}{n}\Lambda/\Lambda$. The inverse image of $G$ in $\mathbb{C}^d$ is a lattice $\Lambda_G$ that contains $\Lambda$, and hence we get a quotient map from $T$ to $\mathbb{C}^d/\Lambda_G = T/G$ with kernel isomorphic to $G$. This quotient map is, by definition of the analytic structure on tori, an analytic map. The Hermitian structure on $T$ induces one on $T_G$ that satisfies the condition from Theorem 5.16 and hence $T_G$ corresponds to an abelian variety $\mathcal{A}_G$.

By Lemma 5.2 the quotient map comes from an algebraic morphism that is an isogeny from $\mathcal{A}$ to $\mathcal{A}_G$ with kernel corresponding to $G$.

**Proposition 5.26** Let $\mathcal{A}$ be an abelian variety defined over $\mathbb{C}$ with lattice $\Lambda \subset \mathbb{C}^d$. The isogenies $\eta$ of degree $n$ of $\mathcal{A}$ are, up to isomorphisms, in one-to-one correspondence with lattices $\Lambda_\eta$ which contain $\Lambda$ and satisfying $[\Lambda_\eta : \Lambda] = n$. The kernel of $\eta$ is isomorphic to $\Lambda_\eta/\Lambda$.

Of special interest are isogenies with image isomorphic to $\mathcal{A}$. For simplicity and since it is in the center of our interest we restrict the discussion to *simple abelian varieties*.

We know that in this case the ring $\text{End}_{\mathbb{C}}(\mathcal{A})$ of endomorphisms of $\mathcal{A}$ is a skew field and that all endomorphisms different from the zero map are isogenies.

We want to use the results of Proposition 5.26 but look at them from a slightly different point of view. In the proposition we interpreted isogenies as quotient maps of the identity map on $\mathbb{C}^d$ with changing lattices. Now we shall *fix the lattice* $\Lambda$ and study holomorphic additive maps $\alpha : \mathbb{C}^d \to \mathbb{C}^d$. Such a map $\alpha$ induces an endomorphism of $\mathcal{A}$ if and only if it is well defined modulo $\Lambda$, i.e., $\alpha(\Lambda) \subset \Lambda$.

**Example 5.27** We give the most simple example to explain this. Look at the endomorphism $[n]$ obtained by scalar multiplication with $n$.

In the first interpretation we take as lattice of the image the lattice $\frac{1}{n}\Lambda$ and take the quotient map from $\mathbb{C}^d/\Lambda$ to $\mathbb{C}^d/\left(\frac{1}{n}\Lambda\right)$.

In the second interpretation we multiply elements in $\mathbb{C}^d$ by $n$ and so the subset $\frac{1}{n}\Lambda$ is mapped to $\Lambda$ and hence to the zero element of the torus associated to $\mathcal{A}$.

From the condition imposed on $\alpha$ (it has to be continuous) it follows that $\alpha$ is a linear invertible map on the real vector space of dimension $2d$ attached to $\mathbb{C}^d$. Hence (after having chosen a basis $\{\mu_1, \ldots, \mu_{2d}\}$ of $\Lambda$) we can describe $\alpha$ by a real invertible $(2d \times 2d)$-matrix $B$ with the additional condition that $\alpha$ maps $\Lambda$ into itself. But this is equivalent to the condition that $B$ has integers as coefficients. Hence the characteristic polynomial $\chi(\alpha)_\mathcal{A}(T)$ of $\alpha$ is a monic polynomial of degree $2d$ with integers as coefficients.

**Remark 5.28** The reader should recall that we have described endomorphisms $\alpha$ in the algebraic setting by using Tate modules to produce $\ell$-adic representations. One of the crucial results due to Weil is that the characteristic polynomials do not depend on the prime $\ell$.

Here we use the period lattice to produce an integral representation again of dimension $2d$. It plays the role of Tate modules in the analytic setting. The resulting characteristic polynomial $\chi_\alpha(T)$ is *the same* as the corresponding $\ell$-adic polynomial. This remark will become important for point counting algorithms.

## § 5.1 Varieties over the field of complex numbers

Until now we have only looked at linear algebra and continuity. But we have to take into account the analytic structure that yields holomorphy conditions for $\alpha$.

We shall explain this in the simplest case.

**Example 5.29** Let $\mathcal{A} =: E$ be an elliptic curve. The associated analytic variety is isomorphic to $\mathbb{C}/(\mathbb{Z} + \tau\mathbb{Z})$ with $\tau \notin \mathbb{R}$. A holomorphic additive map $\alpha$ is given by a matrix

$$B = \begin{bmatrix} n_1 & n_2 \\ n_3 & n_4 \end{bmatrix}$$

over $\mathbb{Z}$ if we take $\{1, \tau\}$ as basis, it also represents a multiplication by a complex number $\beta$ that is determined by $\alpha(1) =: \beta = n_1 + n_2\tau$. Hence it maps $\tau$ to $\beta\tau = n_1\tau + n_2\tau^2 = n_3 + n_4\tau$.

Now assume that $n_2 \neq 0$ or, equivalently, that $\beta \notin \mathbb{Z}$. Then $\tau$ satisfies the equation

$$n_2\tau^2 + (n_1 - n_4)\tau - n_3 = 0$$

and, hence, $\mathbb{Q}(\tau)$ is an imaginary quadratic field $K$.

The lattice $\mathbb{Z} + \tau\mathbb{Z}$ is an ideal $A_\tau$ of an order of $K$, and the isogenies correspond to numbers $n_1 + n_2\mathbb{Z}$ that map $A_\tau$ into itself. But this means that $\text{End}_\mathbb{C}(E)$ is an order (cf. Definition 2.81) in $K$ and that $E$ has complex multiplication.

For higher dimensional abelian varieties $\mathcal{A}$, analogous but more complicated considerations lead to the CM-theory mentioned already in the algebraic part. Again one gets that the lattice of abelian varieties with complex multiplication is very special and that the period matrix has an algebraic structure. This combined with class field theory is the key of the CM method used to construct abelian varieties over finite fields with known number of points. We shall be more precise in the next sections in the case of elliptic and hyperelliptic curves and come to algorithmic details in Chapter 18.

### 5.1.5 Elliptic curves over $\mathbb{C}$

In this section we shall apply the theory of curves and their Jacobians over $\mathbb{C}$ for elliptic curves $E$.

#### 5.1.5.a The complex theory of elliptic curves

We recall Corollary 5.18 that the Jacobian variety of $E$ and hence $E$ itself is analytically isomorphic to $\mathbb{C}/(\mathbb{Z} + \tau\mathbb{Z})$ where $\tau$ is a complex number with a positive imaginary part.

Let $E$ be given by an affine Weierstraß equation

$$E : y^2 = x^3 + a_4 x + a_6 \quad \text{with} \quad a_4, a_6 \in \mathbb{C}.$$

As a consequence of the theorem of Abel–Jacobi 5.13 we get: there is an analytic isomorphism between the groups $E(\mathbb{C})$ and $\mathbb{C}/\Lambda_E$ where $\Lambda_E$ is a lattice $\mathbb{Z}\omega_1 + \mathbb{Z}\omega_2$ in $\mathbb{C}$.

We want to describe explicitly this isomorphism. For this we *begin* with the lattice $\Lambda = \mathbb{Z}\omega_1 + \mathbb{Z}\omega_2$ and then construct the elliptic curve corresponding to it. We shall follow closely [COH 2000, Chapter 7] and [SIL 1986, Chapter VI, section 3]. The first step is to find the meromorphic functions.

**Definition 5.30** Let $\omega_1, \omega_2 \in \mathbb{C}$ be linearly independent over $\mathbb{R}$. An *elliptic function with periods* $\{\omega_1, \omega_2\}$ is a meromorphic function $f(x)$ on $\mathbb{C}$ such that for all $x \in \mathbb{C}$ one has $f(x + \omega_1) = f(x + \omega_2) = f(x)$.

We shall fix $\omega_1, \omega_2$ as well as the lattice $\Lambda$ spanned by them in the following. Elliptic functions will always be periodic with respect to $\Lambda$.

The task is to construct nonconstant elliptic functions. It was solved by Weierstraß.

**Definition 5.31** The Weierstraß $\wp$-function is defined by the series

$$\wp(z, \Lambda) = \frac{1}{z^2} + \sum_{\omega \in \Lambda \setminus \{0\}} \left( \frac{1}{(z+\omega)^2} - \frac{1}{\omega^2} \right). \tag{5.1}$$

This series converges uniformly on every compact subset of $\mathbb{C} \setminus \Lambda$. The function $\wp := \wp(z, \Lambda)$ defined by (5.1) is a meromorphic function on $\mathbb{C}$ with poles (of order 2) in $\Lambda$. It is an even function, i.e., $\wp(z, \Lambda) = \wp(-z, \Lambda)$ for all $z \in \mathbb{C} \setminus \Lambda$.

The proofs are straightforward applications of the basics of complex analysis, see e.g., [SIL 1986, Chapter VI, Theorem 3.1].

As usual we denote by $\wp' := \wp'(z, \Lambda)$ the derivative of $\wp$. Again it is an elliptic function. It can be computed easily by using the series defining $\wp$; the result is again a series whose first term is $-\frac{2}{z^3}$. It follows immediately that $\wp'$ is an odd function, i.e., $\wp'(z) = -\wp'(-z)$.

Define the *Eisenstein series* $G_n := G_n(\Lambda)$ of weight $n$ for $\Lambda$ by

$$G_n(\Lambda) := \sum_{\omega \in \Lambda \setminus \{0\}} \omega^{-n}.$$

The fundamental observation is:

**Theorem 5.32** The elliptic functions $\wp$ an $\wp'$ satisfy the equation

$$\wp'(z)^2 = 4\wp(z)^3 - 60G_4 \wp(z) - 140G_6.$$

This is the affine equation for an elliptic curve $E_\Lambda$ with function field $\mathbb{C}(\wp, \wp')$. The map

$$\Phi : \mathbb{C}/\Lambda \to E_\Lambda \subset \mathbb{P}^2$$
$$z \mapsto \begin{cases} (\wp(z) : \wp'(z) : 1) & \text{for } z \notin \Lambda \\ \Phi(\Lambda) = (0 : 1 : 0) \end{cases}$$

is an isomorphism of Riemann surfaces, which is a group homomorphism (using the induced natural additive group structure on $\mathbb{C}/\Lambda$ and the elliptic curve group structure on $E_\Lambda$).

Hence $E_\Lambda$ is the abelian variety attached to the torus $\mathbb{C}/\Lambda$, and we can interpret the map $\Phi$ as the inverse of the Abel–Jacobi map from $E_\Lambda$ as curve to its Jacobian variety which is isomorphic to $E_\Lambda$.

**Remark 5.33** The equation defining $E_\Lambda$ is not quite in the standard Weierstraß form. We obtain it if we replace $\wp'$ by $y = 1/2 \wp'$ and set $x = \wp$, $g_2 := g_2(\Lambda) := 15G_4(\Lambda)$ and $g_3 := g_3(\Lambda) := 35G_6(\Lambda)$. The resulting equation is

$$E_\Lambda : y^2 = x^3 - g_2 x - g_3.$$

We have seen that for every lattice $\Lambda$ we can use $g_2(\Lambda)$ and $g_3(\Lambda)$ to obtain the equation of the corresponding elliptic curve $E_\Lambda$. The first question is now to describe in terms of the two lattices $\Lambda, \Lambda'$ what it means that $E_\Lambda$ is isomorphic to $E_{\Lambda'}$.

As we have seen in Lemma 5.2 this is equivalent to the question under which conditions $\mathbb{C}/\Lambda$ is analytically isomorphic to $\mathbb{C}_{\Lambda'}$.

## § 5.1 Varieties over the field of complex numbers

**Example 5.34** Take $\alpha \in \mathbb{C}^*$ and define the map $t_\alpha$ from $\mathbb{C}$ to $\mathbb{C}$ by $z \mapsto \alpha z$. Define $\Lambda' := \alpha \Lambda$. Then $t_\alpha$ induces an analytic isomorphism

$$h_\alpha : \mathbb{C}/\Lambda \to \mathbb{C}/\alpha\Lambda.$$

Motivated by the example we define that two lattices $\Lambda_1$ and $\Lambda_2$ are *homothetic* if there is an $\alpha \in \mathbb{C}^*$ such that $\alpha \Lambda_1 = \Lambda_2$.

**Theorem 5.35** There is a canonical isomorphism between the set of $\mathbb{C}$-isomorphism classes of elliptic curves and the set of homothety classes of lattices in $\mathbb{C}$.

**Corollary 5.36** Let $\Lambda$ be a lattice in $\mathbb{C}$ with basis $\{\omega_1, \omega_2\}$. We can assume (by replacing $\omega_1$ by $-\omega_1$ if necessary) that $\tau := \omega_2/\omega_1$ is a complex number with positive imaginary part. Let $\Lambda_\tau$ be the lattice $\mathbb{Z} + \tau \mathbb{Z}$. Then the elliptic curve $E_\Lambda$ is isomorphic to $E_{\Lambda_\tau}$.

By this result we have attached to every (isomorphism class of) elliptic curves over $\mathbb{C}$ a unique lattice $\Lambda_\tau := \mathbb{Z} + \tau \mathbb{Z}$ such that $E$ is isomorphic to $E_{\mathbb{Z}+\tau\mathbb{Z}} =: E_\tau$ with $\tau \in \mathbb{C}$ with imaginary part $\Im m(\tau) > 0$. But $\tau$ is not uniquely determined by $\Lambda_\tau$.

**Lemma 5.37** Let $\tau, \tau'$ be complex numbers with positive imaginary part. Then $\Lambda_\tau = \Lambda_{\tau'}$ if and only if there exist integers $a, b, c, d$ with $ad - bc = 1$ and $\tau' = \frac{a\tau+b}{c\tau+d}$.

**Definition 5.38** A complex function $f$ which is holomorphic on the upper half plane

$$\mathcal{H} := \{\tau \in \mathbb{C} \mid \Im m(\tau) > 0)\}$$

and which satisfies

$$f(\tau) = f\left(\frac{a\tau+b}{c\tau+d}\right)$$

for all integers $a, b, c, d$ with $ad - bc = 1$ is called a *modular function*.

The set of modular functions forms a field $F_1$.

**Example 5.39** Define

$$\begin{aligned} j : \mathcal{H} &\to \mathbb{C} \\ \tau &\mapsto j(\tau) := 1728 \frac{g_2(\Lambda_\tau)^3}{4g_2(\Lambda_\tau)^3 - 27g_3(\Lambda_\tau)^2}. \end{aligned}$$

Then $j \in F_1$.

**Theorem 5.40**

(i) The field of modular functions is equal to $\mathbb{C}(j)$.
(ii) The elliptic curve $E_\tau$ is isomorphic to $E_{\tau'}$ if and only if $j(\tau) = j(\tau')$.
(iii) Let $E$ be an elliptic curve defined over $\mathbb{C}$ with absolute invariant $j_E$ (cf. Corollary 4.118) Then there is a $\tau \in \mathcal{H}$ with $j(\tau) = j_E$ and $E$ is isomorphic to $E_\tau$.

Since $j \in F_1$ we have $j(\tau + 1) = j(\tau)$. (Take $a = 1, b = 1, c = 0, d = 1$.) We can use this identity to develop $j$ into a Laurent series "at $\infty$."

Define $q := e^{2\pi i \tau}$ and $j^*(q) := j(\tau)$. Observe that $q$ approaches $0$ when $\Im m(\tau)$ becomes large. It turns out that $j^*$ can be extended to a meromorphic function with a pole in $0$ of order $1$. Its Laurent series has integer coefficients. It is called the *q-expansion of the j-function*.

**Proposition 5.41** The $q$-expansion of the $j$-function is given by

$$j(q) = \frac{(1 + 240 \sum_{n=1}^{\infty} \sigma_3(n) q^n)^3}{q \prod_{n \in \mathbb{N}} (1 - q^n)^{24}}.$$

For a proof we refer to [SIL 1994, Chapter I, Remark 7.4.2].

After having an explicit description of isomorphism classes of elliptic curves over $\mathbb{C}$ we now determine the *isogeny classes* again by using the theory of complex tori (see Section 5.26) applied to elliptic curves and get:

**Proposition 5.42** Let $E$ and $E'$ be two elliptic curves defined over $\mathbb{C}$ with lattices $\Lambda$ (respectively $\Lambda'$).

Then $E$ is isogenous to $E'$ if and only if there exists an $\alpha \in \mathbb{C}^*$ with $\alpha \Lambda \subset \Lambda'$. If so denote by $\eta_\alpha$ the isogeny from $E$ to $E'$. Then the kernel of $\eta_\alpha$ is canonically isomorphic to $\alpha^{-1} \Lambda' / \Lambda$.

**Corollary 5.43** Assume that $E$ is an elliptic curve over $\mathbb{C}$ with $j_E = j(\tau)$. Then

$$\mathrm{End}_{\mathbb{C}}(E) = \{\alpha \in \mathbb{C} \mid \alpha \Lambda_\tau \subset \Lambda_\tau\}.$$

### 5.1.5.b Elliptic curves with complex multiplication

The ring

$$\mathrm{End}_{\mathbb{C}}(E) = \{\alpha \in \mathbb{C} \mid \alpha \Lambda_\tau \subset \Lambda_\tau\}$$

always contains and in general will be equal to $\mathbb{Z}$.

We reformulate the definition of complex multiplication (cf. Definition 4.88) applied to elliptic curves $E$ over $\mathbb{C}$.

**Definition 5.44** The elliptic curve $E$ has complex multiplication if and only if $\mathrm{End}_{\mathbb{C}}(E)$ is larger than $\mathbb{Z}$.

In Example 4.90 we have already discussed that this implies:

**Corollary 5.45** Let $E$ be an elliptic curve defined over $\mathbb{C}$ with period $\tau$. Then $\tau$ is a *nonrational integer in an imaginary quadratic field* $K_\tau$ and $\mathrm{End}_{\mathbb{C}}(E)$ is the *order corresponding to* $\mathbb{Z} + \tau \mathbb{Z}$ in $K_\tau$.

The converse is true as well.

**Proposition 5.46** Let $K$ be an imaginary quadratic field, let $\mathcal{O}$ be an order of $K$, and let $A$ be an ideal of $\mathcal{O}$. Then $A \subset \mathbb{C}$ is a lattice, the elliptic curve $E_A := \mathbb{C}/A$ is an elliptic curve with complex multiplication and $\mathrm{End}_{\mathbb{C}}(E_A) = \mathcal{O}$. For two ideals $A$, $A'$ of $\mathcal{O}$ we get: $E_A$ is isomorphic to $E_{A'}$ over $\mathbb{C}$ (i.e., the absolute $j$-invariants are equal) if and only if $A$ and $A'$ are in the same ideal class.

So elliptic curves with complex multiplication have *algebraic* periods $\tau$. But even more important we get that the absolute invariant $j(\tau)$ is a very special algebraic integer, i.e., it is the zero of a monic polynomial over $\mathbb{Z}$, and is obtained as $j$-invariant of an ideal in an imaginary quadratic field. The exact statement is the key result of class field theory of imaginary quadratic fields.

**Theorem 5.47** Assume that $E$ is defined over $\mathbb{C}$ and has complex multiplication. Let $\tau$ be its period. Then $\mathbb{Q}(\tau)$ is an imaginary quadratic field, $\mathrm{End}_{\mathbb{Q}(\tau)}(E) = \mathrm{End}_{\mathbb{C}}(E)$ is an order $\mathcal{O}_E$ in $\mathbb{Q}(\tau)$ and the absolute invariant $j(\tau)$ is an algebraic integer that lies in the ring class field $H_{\mathcal{O}_E}$ over $\mathbb{Q}(\tau)$. The invariant $j(\tau)$ is the $j$-function evaluated at an ideal of $\mathcal{O}_E$.

## § 5.1 Varieties over the field of complex numbers

Recall that the ring class field of $\mathcal{O}_E$ is an abelian extension of $\mathbb{Q}(\tau)$ whose Galois group is isomorphic in a canonical way to the ideal class group of $\mathcal{O}_E$. The most important case for us will be that $\mathcal{O}_E$ is the ring of integers $\mathcal{O}$ of $\mathbb{Q}(\tau)$. Then $H_{\mathcal{O}_E}$ is the *Hilbert class field* $H$ of $\mathbb{Q}(\tau)$, the maximal Galois extension of $\mathbb{Q}(\tau)$, which is unramified and has an abelian Galois group. In particular, we get that the degree of $H$ over $\mathbb{Q}(\tau)$ is equal to the order of $\mathrm{Cl}(\mathcal{O})$, which is called the class number of $\mathbb{Q}(\tau)$.

On the other side it follows easily from Theorem 5.35 that $j_{E_A}$ depends only on the ideal class group of $A$ in $\mathrm{Cl}(\mathcal{O})$ and class field theory tells us that all the algebraic numbers $j_{E_A}$ are conjugates under the action of the Galois group of $H$ over $\mathbb{Q}(\tau)$. From this we get:

**Corollary 5.48** Let $K = \mathbb{Q}(\sqrt{-d})$ be an imaginary quadratic field with ring of integers $\mathcal{O}$. Let $E$ be an elliptic curve with $\mathrm{End}_{\mathbb{C}}(E) = \mathcal{O}$. Then the minimal polynomial of $j_E$ is the *Hilbert class polynomial*

$$H_d(x) := \prod_{i=1}^{h_d}(x - j(A_i))$$

where $j(A_i)$ is the $j$-invariant of the elliptic curve corresponding to $A_i$, the number $h_d$ is the order of the ideal class group of $K$ and $A_i$ are representatives of elements of the class group of $\mathcal{O}$. The coefficients of the Hilbert class polynomial are rational numbers. As $j_E$ is an algebraic integer, they are integers.

For the proof of Theorem 5.47 and Corollary 5.48, see for example [SIL 1986, Appendix C, Theorem 11.2], or [LAN 1973, Chapter 10, Theorem 1].

### Reduction of elliptic curves with complex multiplication

In Section 5.2 below we shall discuss elliptic curves over finite fields. The determination of the order of the rational points will be one of the most important topics in this part. Here we can give a bridge from elliptic curves over number fields to elliptic curves over finite fields.

The class polynomial

$$H_d(x) := \prod_{i=1}^{h_d}(x - j(A_i))$$

can be reduced modulo a prime $p$ to a polynomial $H_d(x)_p$ defined over $\mathbb{F}_p$, and it has simple roots if $p$ does not divide $d$.

Let $\mathbb{F}_{p^r}$ be the smallest field that contains a root $j_p$ of $H_d(x)_p$. It is the reduction modulo $p$ of one of the invariants $j(A_i)$. As the elements $j(A_i)$ are conjugate it follows that all roots of $H_d(x)_p$ are in this field.

By the algebraic theory of elliptic curves we know that there are elliptic curves $E_p$ defined over $\mathbb{F}_{p^r}$ with absolute invariant $j_p$. The curve $E_p$ is determined up to twists, and if $j_p \neq 0, 12^3$ there is exactly one twist of $E_p$.

For the sake of simplicity we shall assume now that $r = 1$ and that the prime number $p$ is decomposed in $\mathbb{Q}(\sqrt{-d})$. Class field theory of imaginary quadratic fields gives the following remarkable result.

**Theorem 5.49** There is an integer $\pi \in \mathbb{Q}(\sqrt{-d})$ such that $\pi\bar{\pi} = p$ and $|p + 1 - (\pi + \bar{\pi})|$ is the number of $\mathbb{F}_p$-rational points on either $E_p$ or one of its twists.

To understand this theorem one needs the theory of elliptic curves over finite fields and in particular of the Frobenius endomorphism and its related characteristic polynomial (cf. Example 4.87) made explicit in Example 5.83. The theorem then states that the algebraic integer $\pi$, interpreted

### 5.1.6 Hyperelliptic curves over $\mathbb{C}$

#### 5.1.6.a Periods and invariants

Let $C$ be a hyperelliptic curve of genus $g$ defined over $\mathbb{C}$ with Jacobian variety $J_C$. As we know $J_C$ is as analytic variety isomorphic to a torus $\mathbb{C}^g/\Lambda_C$. Since $J_C$ is principally polarized $\Lambda_C$ can be chosen in the form $\mathbb{Z}^d \oplus \Omega \cdot \mathbb{Z}^d$, where the $(g \times g)$-matrix $\Omega$ is symmetric and has a positive definite imaginary part, i.e., lies in the Siegel upper half plane $\mathbb{H}_g$.

The matrix $\Omega$ is the *period matrix of* $C$. It can be computed by integrating a basis of holomorphic differentials along paths on the Riemann surface corresponding to $C$. Since such a basis is explicitly known for hyperelliptic curves (see Chapter 17) it is in principle possible to compute it. For elliptic curves $E$ this gives the period $\tau$, a complex number with a positive imaginary part.

The next step for elliptic curves was to determine the isomorphism class when the period is known. This task was solved by the $j$-function whose value at $\tau$ is the absolute invariant of $E$. To construct $j$ we used Eisenstein series as special functions on lattices, i.e., modular forms.

Analogous to the elliptic curve case we define values of complex functions to lattices that are now *Siegel modular forms*: let $\Omega \in \mathbb{H}_g$ the period matrix of a principally polarized abelian variety and let $z \in \mathbb{C}^g$ be a column vector. The Riemann theta function is given by

$$\theta(z, \Omega) = \sum_{\mathbf{n} \in \mathbb{Z}^g} \exp\bigl(\pi i (\mathbf{n}^t \Omega \mathbf{n} + 2\mathbf{n}^t z)\bigr).$$

This function is $\mathbb{C}$-valued, holomorphic and symmetric, i.e., $\theta(z, \Omega) = \theta(-z, \Omega)$.

For fixed $\Omega \in \mathbb{H}_g$ we get a function from $\mathbb{C}^g$ to $\mathbb{C}$ and we define the *Riemann theta divisor* by

$$\Theta^{(\Omega)} := \{z \bmod \Lambda \mid \theta(z, \Omega) = 0\}.$$

Recall that $\tau$ and $\tau'$ define isomorphic elliptic curves if and only if $\tau' = \frac{a\tau + b}{c\tau + d}$ with $a, b, c, d \in \mathbb{Z}$ and $ad - bc = 1$, i.e., if $\tau$ and $\tau'$ are equivalent under the action of $\mathrm{SL}_2(\mathbb{Z})$, the group of invertible $(2 \times 2)$-matrices over $\mathbb{Z}$ with determinant 1.

An analogous result holds for arbitrary dimension. We define $\mathrm{Sp}(2g, \mathbb{Z})$ to be the *symplectic group* of dimension $g$ over $\mathbb{Z}$. (For $g = 1$ this is $\mathrm{SL}_2(\mathbb{Z})$.) It acts on $\mathbb{H}_g$ in a natural way (cf. [LAN 1982]).

**Theorem 5.50** *Two period matrices $\Omega, \Omega'$ define isomorphic principally polarized abelian varieties if and only if they lie on the same orbit under the operation of the symplectic group $\mathrm{Sp}(g, \mathbb{Z})$ on $\mathbb{H}_g$.*

For a proof see [LAN 1982].

The theta divisors of two equivalent period matrices $\Omega, \Omega'$ do not have to be equal. But if they are equivalent then there exists an $a \in \Omega(\frac{1}{2}\mathbb{Z}^g) + \frac{1}{2}\mathbb{Z}^g$ such that $\Theta^{(\Omega')} = \Theta_a^{(\Omega)}$ where $\Theta_a^{(\Omega)}$ denotes the translation of $\Theta^{(\Omega)}$ by $a$.

This motivates the introduction of *theta characteristics*

$$\theta\begin{bmatrix}\delta\\\epsilon\end{bmatrix}(z, \Omega) = \sum_{\mathbf{n} \in \mathbb{Z}^g} \exp\left(\pi i \left(\mathbf{n} + \frac{1}{2}\delta\right)^t \Omega \left(\mathbf{n} + \frac{1}{2}\delta\right) + 2\left(\mathbf{n} + \frac{1}{2}\delta\right)^t \left(z + \frac{1}{2}\epsilon\right)\right) \qquad (5.2)$$

with column vectors $\delta$ and $\epsilon \in (\mathbb{Z}/2\mathbb{Z})^g$. If we fix $\delta, \epsilon$ and set $z = 0$, we obtain functions on $\mathbb{H}_g$, called the *theta constants*. A theta constant is even, if $\delta^t \epsilon \equiv 0 \pmod{2}$, and odd otherwise. All odd theta constants vanish for principally polarized varieties. There are $2^{g-1}(2^g + 1)$ even theta constants.

**Theorem 5.51** The complete set of theta constants uniquely determines the isomorphism class of a principally polarized abelian variety of dimension $g$.

The proof is given in [IGU 1960].

**Example 5.52** For $g = 2$ there are 10 even theta constants. A list of the vectors $\delta, \epsilon \in (\mathbb{Z}/2\mathbb{Z})^2$ used to get them is found in [WEN 2003]. For $g = 3$ there are 32 even theta constants.

By the theta constants we have found a complete system of invariants for isomorphism classes of principally polarized abelian varieties of dimension $g \geqslant 2$. But two things are disturbing. First these invariants are not "independent." Secondly and worse they are defined analytically. But they define points in an algebraic variety, the *moduli space of isomorphism classes of principally polarized abelian varieties of dimension g*. So we would like to have algebraically defined invariants.

For $g = 2, 3$ we can make this precise. Due to results of Weil [WEI 1957] we know that every principally polarized abelian variety $\mathcal{A}$ of dimension $g \leqslant 3$ is the Jacobian variety of a curve $C$. Because of the famous theorem of Torelli, the isomorphism class of $\mathcal{A}$ *with its polarization* is determined uniquely by the isomorphism class of $C$. So the invariants have to be algebraic expressions in the coefficients of the equation defining $C$. Recall that for elliptic curves $E$ we can express $j_E$ by the coefficients $g_2, g_3$ of a Weierstraß equation.

In fact we can find these algebraic invariants for hyperelliptic curves of genus 2 and 3 using work of Igusa [IGU 1960] and Shioda [SHI 1967].

### 5.1.6.b Hyperelliptic curves of genus 2

For every principally polarized abelian variety $\mathcal{A}$ of dimension two there exist three absolute invariants $j_1, j_2$, and $j_3$ called *Igusa invariants*, which determine its isomorphism class. They can be expressed in terms of the theta constants. The explicit formulas can be found in [WEN 2003, Section 5].

Let $C : y^2 = x^5 + f_4 x^4 + f_3 x^3 + f_2 x^2 + f_1 x + f_0$ be the curve with $J_C = \mathcal{A}$. Then the invariants $j_i$ of the Jacobian of $C$ can be expressed by

$$j_1 = I_2^5/I_{10}, \quad j_2 = I_2^3 I_4/I_{10} \text{ and } j_3 = I_2^2 I_6/I_{10}, \tag{5.3}$$

where the $I_i$'s are given below expressed in the coefficients of the curve.

By Spallek [SPA 1994, p. 71] the absolute invariants $I_i$ are given in terms of the coefficients $f_j$ as

$$\begin{aligned}
I_2 =& \; 6f_3^2 - 16f_4 f_2 + 40f_1, \\
I_4 =& \; 4\big(f_4^2 f_2^2 - 3f_3 f_2^2 - 3f_4^2 f_3 f_1 + 9f_3^2 f_1 + f_4 f_2 f_1 - 20f_1^4 + 12f_4^3 f_0 - 45f_4 f_3 f_0 + 75f_2 f_0\big), \\
I_6 =& \; -2\big(-4f_4^2 f_3^2 f_2^2 + 12f_3^3 f_2^2 + 12f_4^3 f_2^3 - 38f_4 f_3 f_2^3 + 18f_2^4 + 12f_4^2 f_3^3 f_1 - 36f_3^4 f_1 \\
& -38f_4^3 f_3 f_2 f_1 + 119f_4 f_3^2 f_2 f_1 - 14f_4^2 f_2^2 f_1 - 13f_3 f_2^2 f_1 + 18f_4^4 f_1^2 - 13f_4^2 f_3 f_1^2 - 88f_3^2 f_1^2 \\
& -32f_4 f_2 f_1^2 + 160 f_1^4 - 30f_4^3 f_3^2 f_0 + 99 f_4 f_3^3 f_0 + 80 f_4^4 f_2 f_0 - 246 f_4^2 f_3 f_2 f_0 - 165 f_3^2 f_2 f_0 \\
& + 320 f_4 f_2^2 f_0 - 308 f_4^3 f_1 f_0 + 930 f_4 f_3 f_1 f_0 - 800 f_2 f_1 f_0 + 450 f_4^2 f_0^2 - 1125 f_3 f_0^2\big), \\
I_{10} =& \; f_4^2 f_3^2 f_2^2 f_1^2 - 4 f_3^3 f_2^2 f_1^2 - 4 f_4^3 f_2^3 f_1^2 + 18 f_4 f_3 f_2^3 f_1^2 - 27 f_2^4 f_1^2 - 4 f_4^2 f_3^3 f_1^3 + 16 f_3^4 f_1^3 \\
& + 18 f_4^3 f_3 f_2 f_1^3 - 80 f_4 f_3^2 f_2 f_1^3 - 6 f_4^2 f_2^2 f_1^3 + 144 f_3 f_2^2 f_1^3 - 27 f_4^4 f_1^4 + 144 f_4^2 f_3 f_1^4 \\
& - 128 f_3^2 f_1^4 - 192 f_4 f_2 f_1^4 + 256 f_1^5 - 4 f_4^2 f_3^2 f_2^3 f_0 + 16 f_3^3 f_2^3 f_0 + 16 f_4^3 f_2^4 f_0 - 72 f_4 f_3 f_2^4 f_0 \\
& + 108 f_2^5 f_0 + 18 f_4^3 f_3^2 f_2 f_1 f_0 - 72 f_3^3 f_2 f_1 f_0 - 80 f_4^3 f_3 f_2^2 f_1 f_0 + 356 f_4 f_3^2 f_2^2 f_1 f_0 \\
& + 24 f_4^2 f_3^3 f_1 f_0 - 630 f_3 f_2^3 f_1 f_0 - 6 f_4^3 f_2^2 f_1^2 f_0 + 24 f_4 f_3^2 f_1^2 f_0 + 144 f_4^4 f_2 f_1^2 f_0 \\
& - 746 f_4^2 f_3 f_2 f_1^2 f_0 + 560 f_3^2 f_2 f_1^2 f_0 + 1020 f_4 f_2^2 f_1^2 f_0 - 36 f_4^3 f_1^3 f_0 + 160 f_4 f_3 f_1^3 f_0
\end{aligned}$$

$$
\begin{aligned}
&- 1600 f_2 f_1^3 f_0 - 27 f_4^2 f_3^4 f_0^2 + 108 f_3^5 f_0^2 + 144 f_4^3 f_3^2 f_2 f_0^2 - 630 f_4 f_3^3 f_2 f_0^2 - 128 f_4^4 f_2^3 f_0^2 \\
&+ 560 f_4^2 f_3 f_2^2 f_0^2 + 825 f_3^2 f_2^2 f_0^2 - 900 f_4 f_2^3 f_0^2 - 192 f_4^4 f_3 f_1 f_0^2 + 1020 f_4^2 f_3^2 f_1 f_0^2 \\
&- 900 f_3^3 f_1 f_0^2 + 160 f_4^3 f_2 f_1 f_0^2 - 2050 f_4 f_3 f_2 f_1 f_0^2 + 2250 f_2^2 f_1 f_0^2 - 50 f_4^2 f_1^2 f_0^2 \\
&+ 2000 f_3 f_1^2 f_0^2 + 256 f_4^5 f_0^3 - 1600 f_4^3 f_3 f_0^3 + 2250 f_4 f_3^2 f_0^3 + 2000 f_4^2 f_2 f_0^3 - 3750 f_3 f_2 f_0^3 \\
&- 2500 f_4 f_1 f_0^3 + 3125 f_0^4.
\end{aligned}
$$

Hence we can compute the invariants $j_i$ of the curve $C$ if we know either its period matrix or the curve equation. Conversely, from the invariants we get a system of polynomial equations for the coefficients of an equation defining $C$ and we can solve this system in principle, e.g., by applying Buchberger's algorithm.

But there is a much more efficient way due to Mestre [MES 1991].

To use it we have to define new invariants, which we call *Mestre's invariants*. In his paper, Mestre introduces the invariants $A, B, C, D$ [MES 1991] and invariants $j_1', j_2', j_3'$ with

$$j_1' = A^5/D, \; j_2' = A^3 B/D \text{ and } j_3' = A^2 C/D$$

which satisfy

$$j_1' = \frac{-j_1}{120^5}, \; j_2' = \frac{720 j_1'}{6750} - \frac{j_2}{(120^3 \times 6750)}, \; j_3' = \frac{j_3}{120^2 \times 2025100} + \frac{1080 j_2'}{2025} - \frac{16 j_1'}{375}.$$

In addition we need

$$\alpha = \frac{-1}{4556250}\left(\frac{1}{j_1'} + 62208\right) + \frac{16 j_2'}{75 j_1'} + \frac{16 j_3'}{45 j_1'} - 2\frac{j_2'^2}{3 j_1'^2} - \frac{4 j_2' j_3'}{3 j_1'^2}$$

which relates Mestre's invariant $D$ with Igusa's discriminant $\Delta$ by $\alpha = \frac{D}{\Delta}$.

Next one defines a conic $\mathcal{Q}(j_1, j_2, j_3)$ by the equation

$$\sum_{1 \leqslant i, k \leqslant 3} Q_{ik} x_i x_k = 0$$

with

$$
\begin{aligned}
Q_{11} &= \frac{6 j_3' + j_2'}{3 j_1'}, \\
Q_{12} &= Q_{21} = \frac{2(j_2'^2 + j_1' j_3')}{3 j_1'^2}, \\
Q_{13} &= Q_{31} = Q_{22} \alpha \\
Q_{23} &= Q_{32} \frac{1}{j_1'^2}\left(\frac{j_2'^3}{3 j_1'} + \frac{4 j_2' j_3'}{9} + \frac{2 j_3'^2}{3}\right), \\
Q_{33} &= \frac{1}{j_1'^2}\left(\frac{j_1' j_2' \alpha}{2} + \frac{2 j_2'^2 j_3'}{9 j_1'} + \frac{2 j_3'^2}{9}\right).
\end{aligned}
$$

This conic is intersected with a cubic $\mathcal{H}(j_1, j_2, j_3)$ given by the equation

$$\sum_{1 \leqslant i, k, l \leqslant 3} H_{ikl} x_i x_k x_l$$

where

$$H_{111} = \frac{2\left(j_1'j_3' - 6j_2'j_3' + 9j_1'^2\alpha\right)}{9j_1'^2},$$

$$H_{112} = H_{211} = \frac{2j_2'^3 + 4j_1'j_2'j_3' + 12j_1'j_3'^2 + j_1'^3\alpha}{9j_1'^3},$$

$$H_{113} = H_{311} = H_{131} = H_{122} = \frac{j_2'^3 + 4/3\,j_1'j_2'j_3' + 4j_2'^2 j_3' + 6j_1'j_3'^2 + 3j_1'^2 j_2'\alpha}{9j_1'^3},$$

$$H_{123} = \frac{1}{18j_1'^3}\left(\frac{2j_2'^4}{j_1'} + 4j_2'^2 j_3' + \frac{4j_1'j_3'^2}{3} + 4j_2'j_3'^2 + 3j_1'^2 j_2'\alpha + 12j_1'^2 j_3'\alpha\right),$$

$$H_{133} = H_{313} = H_{331} = \frac{1}{18j_1'^3}\left(\frac{j_2'^4}{j_1'} + \frac{4j_2'^2 j_3'}{3} + \frac{16j_2'^3 j_3'}{3j_1'} + \frac{26j_2'j_3'^2}{3} + 8j_3'^3 + 3j_1'j_2'^2\alpha + 2j_1'^2 j_3'\alpha\right),$$

$$H_{222} = \frac{1}{9j_1'^3}\left(\frac{3j_2'^4}{j_1'} + 6j_2'^2 j_3' + \frac{8j_1'j_3'^2}{3} + 2j_2'j_3'^2 - 3j_1'^2 j_3'\alpha\right),$$

$$H_{223} = H_{232} = H_{322} = \frac{1}{18j_1'^3}\left(-\frac{2j_2'^3 j_3'}{3j_1'} - \frac{4j_2'j_3'^2}{3} - 4j_3'^3 + 9j_1'j_2'^2\alpha + 8j_1'^2 j_3'\alpha\right),$$

$$H_{233} = H_{323} = H_{332} = \frac{1}{18j_1'^3}\left(\frac{j_2'^5}{j_1'^2} + \frac{2j_2'^3 j_3'}{j_1'} + \frac{8j_2'j_3'^3}{9} + \frac{2j_2'^2 j_3'^2}{3j_1'} - j_1'j_2'j_3'\alpha + 9j_1'^3\alpha^2\right),$$

$$H_{333} = \frac{1}{36j_1'^3}\left(-\frac{2j_2'^4 j_3'}{j_1'^2} - \frac{4j_2'^2 j_3'^2}{j_1'} - \frac{16j_3'^3}{9} - \frac{4j_2'j_3'^3}{j_1'} + 9j_2'^3\alpha + 12j_1'j_2'j_3'\alpha + 20j_1'j_3'^2\alpha\right).$$

Note that this is easily done if the conic has a rational point. Then the set of points on the conic can be parameterized by a parameter $t$. So in the worst case we have to go to a quadratic extension of $K$ to perform this step. The intersection consists of six points that are the zeroes of a polynomial $f(t)$ of degree 6 in the parameter $t$.

**Lemma 5.53 (Mestre)** The curve $C$ with Igusa invariants $\{j_1, j_2, j_3\}$ can be given by the equation

$$y^2 = f(x).$$

where $f$ is the polynomial of degree 6 constructed above.

We note that this is not the standard form for an equation of genus 2. But we can transform one of the zeroes of $f(x)$ to be the infinite point on $C$ and then find an equation

$$y^2 = \bar{f}(x)$$

with $\bar{f}$ a polynomial of degree 5 for the curve $C$.

Until now we have done all computations over $\mathbb{C}$. But Mestre's result is a purely algebraic one, and so we get:

**Theorem 5.54** Let $\mathcal{A}$ be a principally polarized abelian variety defined over $\mathbb{C}$ with Igusa invariants $\{j_1, j_2, j_3\}$. Let $K_0 \subset \mathbb{C}$ be a field containing these invariants and such that the conic $\mathcal{Q}(j_1, j_2, j_3)$ has a $K_0$-rational point. Then $\mathcal{A}$ is the Jacobian variety of a curve $C$ of genus 2 defined over $K_0$. Its equation is

$$y^2 = f(x),$$

where $f(x)$ is the polynomial of degree 6 from Lemma 5.53.

Let $K$ be an extension field of $K_0$ such that $f(x)$ has a zero $x_0$ in $K$. Then $C$ as curve over $K$ can be given by the equation

$$y^2 = \bar{f}(x)$$

which is obtained by transforming the point $(x_0, 0)$ to infinity.

### 5.1.6.c  Hyperelliptic curves of genus 3

Let $\mathcal{A}$ be a principally polarized abelian variety of dimension 3. We assume that we know its period matrix. So we know the theta constants and, by using a theorem of Mumford–Poor, we can decide whether it is the Jacobian of a hyperelliptic curve (cf. [WEN 2001a, Theorem 4.3]. If so we want to find the equation of the corresponding curve given in the form

$$C : y^2 = f(x)$$

where $f(x)$ is a polynomial of degree 7.

In principle we can proceed as in the case of genus 2. Only things become more complicated. One way proposed in [WEB 1997] is as follows. First one computes the *Rosenhain model*

$$y^2 = x(x - \lambda_1) \cdots (x - \lambda_7)$$

of $C$ where the complex numbers $\lambda_i$ are rational expressions in theta constants. Having this equation one computes the *Shioda invariants* $j_1, j_3, j_5, j_7, j_9$, which determine the isomorphism class of $C$ as curve over $\mathbb{C}$.

Then a variant of Mestre's method allows us to find an equation for $C$ that is defined over field of degree $\leqslant 2$ over $\mathbb{Q}(j_1, j_3, j_5, j_7, j_9)$. For details we refer to [WEB 1997] and [WEN 2001a].

**Remarks 5.55**

(i) In [WEB 1997] the theoretical results and the algorithms to compute curves are given for hyperelliptic curves of genus $\leqslant 5$.

(ii) In the elliptic case we went further. By using the Weierstraß $\wp$ function and its derivative we were able to make (the inverse of) the Abel–Jacobi map explicit. In [KAM 1991] it is shown that an analogous definition of Weierstraß functions and its higher derivatives can be used to achieve this for hyperelliptic curves of any genus.

### 5.1.6.d  Hyperelliptic curves of genus 2 and 3 with CM

In the last section we have seen that the knowledge of the period matrix of a hyperelliptic curve $C$ of genus 2 or 3 makes it possible to compute its invariants and then to determine its equation in an algebraic way.

We shall discuss now how the theory of CM-fields makes it possible to determine the invariants in an algebraic way if $J_C$ has complex multiplication. Though the ideas are quite analogous to those that occurred in the case of complex multiplication of elliptic curves we need considerably more technical details. The key ingredients were developed in the important book of Taniyama–Shimura [SHTA 1961]. The reader who is interested in this deep and beautiful theory is encouraged to use this book as reference for the whole section.

We shall begin by giving a very rough sketch of the general CM theory and then we shall apply it to the special case of Jacobian varieties of hyperelliptic curves of genus 2 and 3.

## § 5.1 Varieties over the field of complex numbers

### Abelian varieties to CM-types

A number field $K$ with $[K : \mathbb{Q}] = 2g$ is called *CM-field* if $K$ is an imaginary quadratic extension of a totally real number field $K_0$.

Let $\varphi_i, 1 \leqslant i \leqslant 2g$ be the $2g$ distinct embeddings from $K$ into $\mathbb{C}$. A tuple

$$(K, \Phi) := \big(K, \{\varphi_1, \varphi_2, \ldots, \varphi_g\}\big)$$

is called CM-type, if all embeddings $\varphi_i$ are distinct and no two of them are complex conjugate to each other.

Let $\mathcal{A} \simeq \mathbb{C}^g / \Lambda_\mathcal{A}$ be an abelian variety over $\mathbb{C}$ with $\mathrm{End}(\mathcal{A}) \otimes \mathbb{Q} \simeq K$. Hence $\mathcal{A}$ has complex multiplication with ring of endomorphisms being an order $\mathcal{O} \subset K$. We have to make the identification of $\mathcal{O}$ with $\mathrm{End}(\mathcal{A})$ more explicit.

**Definition 5.56** Assume that the operation of $\alpha \in \mathcal{O}$ on $\mathcal{A}$ is given by the action of

$$\begin{bmatrix} \varphi_1(\alpha) & & \\ & \ddots & \\ & & \varphi_g(\alpha) \end{bmatrix}.$$

on $\mathbb{C}^g$. Then $\mathcal{A}$ is an abelian variety of CM-type $(K, \Phi) = \big(K, \{\varphi_1, \cdots, \varphi_g\}\big)$.

**Proposition 5.57** For every abelian variety $\mathcal{A}$ with $\mathrm{End}(\mathcal{A}) \otimes \mathbb{Q} \simeq K$ there exists a CM-type $(K, \Phi) = \big(K, \{\varphi_1, \cdots, \varphi_g\}\big)$.

To ease things we restrict ourselves (as in the case of elliptic curves) to the case that $\mathrm{End}(\mathcal{A}) = \mathcal{O}_K$, the ring of integers in $K$.

**Theorem 5.58** Let $\mathfrak{A}$ be an ideal in $\mathcal{O}_K$ and let $(K, \Phi)$ be a CM-type. Take

$$\Phi(\mathfrak{A}) := \Big\{ \big(\varphi_1(\alpha), \ldots, \varphi_g(\alpha)\big)^t \mid \alpha \in \mathfrak{A} \Big\}$$

in $\mathbb{C}^g$.

Then $\Phi(\mathfrak{A})$ is a lattice in $\mathbb{C}^g$ and the torus $\mathbb{C}^g / \Phi(\mathfrak{A})$ is an abelian variety $\mathcal{A}_{\mathfrak{A}, \Phi}$ which has complex multiplication by $\mathcal{O}_K$.

The action of $\mathcal{O}_K$ on $\mathbb{C}^g / \Phi(\mathfrak{A})$ is given by the action of the $g$-tuple

$$\begin{bmatrix} \varphi_1(\gamma) & & \\ & \ddots & \\ & & \varphi_g(\gamma) \end{bmatrix} \text{ with } \gamma \in \mathcal{O}_K$$

on $\mathbb{C}^g$. Conversely every abelian variety $\mathcal{A}$ of CM-type $(K, \Phi)$ with complex multiplication by $\mathcal{O}_K$ is isomorphic to an abelian variety $\mathcal{A}_{\mathfrak{A}, \Phi}$.

The proof of this theorem can be found in [SHTA 1961].

### Principal polarizations

We are interested in Jacobian varieties and corresponding curves with complex multiplications and so we need a finer structure: we want to construct principally polarized abelian varieties and we have to determine isomorphism classes of abelian varieties *with principal polarizations*. For this it is convenient to make an *additional assumption* that is very often satisfied: the maximal real subfield $K_0$ of $K$ has class number 1, i.e., the ring of integers $\mathcal{O}_{K_0}$ is principal.

**Lemma 5.59** Assume that the maximal real subfield $K_0$ in the CM-field $K$ has class number 1. Let $(K, \Phi)$ be a CM-type, $\mathfrak{A}$ an ideal of $\mathcal{O}_K$ and $\mathcal{A}_{\mathfrak{A},\Phi}$ the abelian variety attached to these data. There exists a basis $\{\alpha_1, \ldots, \alpha_{2g}\}$ of $\Phi(\mathfrak{A})$ such that the Riemann form 5.22 is

$$[E(\alpha_i, \alpha_j)]_{1 \leq i,j \leq 2n} = \begin{bmatrix} 0 & I_g \\ -I_g & 0 \end{bmatrix}.$$

Hence the period matrix of $\mathcal{A}_{\mathfrak{A},\Phi}$ lies in the Siegel upper half plane $\mathbb{H}_g$ and we can endow $\mathcal{A}_{\mathfrak{A},\Phi}$ with a principal polarization determined by an element $\gamma$ in $K$ (cf. [LAN 1982]).

For a proof see [WEN 2001b].

**Definition 5.60** We take the notations as in the Lemma 5.59. We shall write $(\mathcal{A}_{\mathfrak{A},\Phi}, \gamma)$ for the abelian variety corresponding to the ideal $\mathfrak{A}$, the CM-type $\Phi$ and the polarization attached to $\gamma$.

As in the case of elliptic curves, we now need a characterization of *isomorphism classes* of abelian varieties with principal polarization that correspond to a given CM-type $(K, \Phi)$.

For this we need some notation. Let $K$ be a CM-field with CM-type $\Phi$. We assume that the maximal totally real subfield has class number 1.

Let $U^+$ denote the totally positive units of $K_0$ (i.e., units $u$ in $\mathcal{O}_K$ such that for all $1 \leq j \leq g$ we have $\varphi_i(u)$ is a positive real number). Let $U_1$ be the image of the norm map from $K$ to $K_0$ applied to the units in $\mathcal{O}_K$. We denote by $\epsilon_1, \ldots, \epsilon_d$ a system of representatives for $U^+/U_1$. Note that the complex conjugation $^-$ generates the Galois group of $K$ over $K_0$. Using our assumption that the class number of $K_0$ is 1 we get that for any ideal $\mathfrak{A}$ of $K$ the ideal $\mathfrak{A}\overline{\mathfrak{A}}$ can be interpreted as a principal ideal $(\alpha)$ of $K_0$.

**Definition 5.61** The subgroup $\mathrm{Cl}'(\mathcal{O}_K)$ of the class group $\mathrm{Cl}(\mathcal{O}_K)$ consists of the ideal classes $c$ that contain an ideal $\mathfrak{A}$ with $\mathfrak{A}\overline{\mathfrak{A}} = \alpha \mathcal{O}_K$ with $\varphi_i(\alpha)$ totally positive, i.e., $\varphi_i(\alpha)$ is a real positive number for every $1 \leq i \leq g$. The order of $\mathrm{Cl}(\mathcal{O}_K)'$ is denoted by $h'_K$.

We have the following theorem:

**Theorem 5.62** Let $(\mathcal{A}_{\mathcal{O}_K,\Phi}, \gamma)$ be a principally polarized abelian variety attached to $\mathcal{O}_K$. Let $\mathfrak{A}_1, \ldots, \mathfrak{A}_{h'_K}$ be a system of representatives for $\mathrm{Cl}(\mathcal{O}_K)'$ with $\mathfrak{A}_i \overline{\mathfrak{A}_i} = (\alpha_i)$ and $\alpha_i$ totally positive. There are $h'_k d$ isomorphism classes of principally polarized abelian varieties with complex multiplication by $\mathcal{O}_K$ of CM-type $(K, \Phi)$.

Let $K_\Phi = \bigcup_{l=1}^{d} K_\Phi^l$ with

$$K_\Phi^l = \left\{ (\mathcal{A}_{\mathfrak{A}_i}, \epsilon_l (\alpha_i \gamma)^{-1}) \mid i = 1, \ldots, h'_k \right\}.$$

The set $K_\Phi$ is a set of representatives of the isomorphism classes of principally polarized abelian varieties of CM-type $(K, \Phi)$.

*Warning:* principally polarized abelian varieties of different CM-types can be isomorphic.

**Example 5.63** Consider the case where the principally polarized abelian variety has dimension two. Here, the CM-field is an imaginary quadratic extension of a real quadratic field $K_0$.

If $K$ is Galois, we get all isomorphism classes of principally polarized abelian varieties with complex multiplication with $\mathcal{O}_K$ by choosing one CM-type.

If $K$ is non-normal, we need two CM-types to get all isomorphism classes of principally polarized abelian varieties.

## Class polynomials for hyperelliptic curves of genus 2 and 3

Recall from the previous paragraph that for elliptic curves with complex multiplication by $\mathcal{O}_K$ the $j$-invariant lies in the Hilbert class field of the imaginary quadratic field $K$. Again the situation is analogous but more complicated in the higher dimensional case.

We need the notion of the *reflex CM-field* $\widehat{K}$ ([SHI 1998]), which for $g = 1$ is equal to $K$ and in general different from $K$. We shall not need the explicit definition of the reflex CM-field but use the arithmetic information from class field theory to determine minimal polynomials for invariants.

**Theorem 5.64** Let $K$ be a CM-field of degree 4 over $\mathbb{Q}$.

(i) The *Igusa invariants* $j_1(C), j_2(C), j_3(C)$ for hyperelliptic curves $C$ of genus 2 with complex multiplication with the ring of integers $\mathcal{O}_K$ of $K$ are algebraic numbers that lie in a class field over the reflex CM-field $\widehat{K}$.

(ii) For hyperelliptic curves $C$ and $C'$ with complex multiplication with $\mathcal{O}_K$ we get that for $k \in \{1, 2, 3\}$ the invariants $j_k(C)$ and $j_k(C')$ are Galois conjugates.

(iii) Let $\{C_1, \ldots, C_s\}$ be a set of representatives of isomorphism classes of curves of genus 2 whose Jacobian varieties have complex multiplication with endomorphism ring $\mathcal{O}_K$. We denote by $j_k(i)$ the $k$-th Igusa invariant belonging to the curve $C_i$.
The three *class polynomials*

$$H_{K,k}(X) = \prod_{i=1}^{s}(X - j_k^{(i)}), k = 1, \ldots, 3.$$

have coefficients in $\mathbb{Q}$.

For hyperelliptic curves of genus 3 we get a completely analogous result.

**Theorem 5.65** Let $K$ be a CM-field of degree 6 over $\mathbb{Q}$.

(i) The *Shioda invariants* $j_1(C), j_3(C), j_5(C), j_7(C), j_9(C)$ for hyperelliptic curves $C$ of genus 3 with complex multiplication with the ring of integers $\mathcal{O}_K$ of $K$ are algebraic numbers that lie in a class field over the reflex CM-field $\widehat{K}$.

(ii) For hyperelliptic curves $C$ and $C'$ with complex multiplication with $\mathcal{O}_K$ we get that for $k \in \{1, 3, 5, 7, 9\}$ the invariants $j_k(C)$ and $j_k(C')$ are Galois conjugate.

(iii) Let $\{C_1, \ldots, C_s\}$ be a set of representatives of isomorphism classes of curves of genus 3 whose Jacobian varieties have complex multiplication with endomorphism ring $\mathcal{O}_K$. We denote by $j_k(i)$ the $k$-th Igusa invariant belonging to the curve $C_i$.
The five *class polynomials*

$$H_{K,k}(X) = \prod_{i=1}^{s}(X - j_k^{(i)}), \ \ k \in \{1, 3, 5, 7, 9\}.$$

have coefficients in $\mathbb{Q}$.

## Denominators in the class polynomials

The careful reader will have remarked that — contrary to the elliptic case — we did not claim in Theorems 5.64 and 5.65 that the class polynomials have integer coefficients. In fact this is wrong.

There are two reasons for this. First, small primes occur (for $g = 2$ up to 5 and for $g = 3$ up to 7) because we did not normalize the invariants in a careful enough way. But much more serious is the second reason: it may happen that the Jacobian of a curve has good reduction modulo a place p of the field over which it is defined but the curve does not have good reduction. The curve may become reducible modulo p.

There are famous conjectures about the arithmetic of curves over number fields related to the $ABC$-conjecture that this should occur only for places with moderate norm.

In practice this is confirmed. So to compute the coefficients of the class polynomial one computes a real approximation with high precision and then determines the denominator using the continued fraction algorithm.

### Reduction of hyperelliptic curves of genus 2 and 3 with complex multiplication

The invariants of a hyperelliptic curves of genus 2 or 3 with complex multiplication with a CM-field $K$ are zeroes of polynomials over $\mathbb{Q}$. Let us choose a prime $p$ that does not divide the denominator of the coefficients of these polynomials. Then we can reduce the class polynomials modulo $p$.

We can factor the resulting polynomials over $\mathbb{F}_p$ and find zeroes in an extension field $\mathbb{F}_q$. By Galois theory we see that the class polynomials will split in linear factors over $\mathbb{F}_q$. Combining "related" zeroes we get systems of invariants for which the resulting curves $C_q$ have a Jacobian variety with ring of endomorphisms containing an isomorphic copy of $\mathcal{O}_K$.

So, we have very explicit information about the endomorphisms of the Jacobian variety of $C_q$, which are defined over (possibly a quadratic extension of) $\mathbb{F}_q$. Class field theory of CM-fields can be used to identify the Frobenius endomorphism.

We explain the easiest case, which is the most important one for practical use: we assume that the genus of $C_q$ is equal to 2 and that $q = p$.

**Theorem 5.66** Let $K$ be a CM-field of degree 4 and assume that $p$ is a prime $\geqslant 7$, which does not divide the denominator of the class polynomials $H_{K,k}(X) =: H_k(X)$.

- For every $w \in \mathcal{O}_K$ with $w\overline{w} = p$ the polynomials $H_k(X)$ have a linear factor over $\mathbb{F}_p$ corresponding to $w$.
- Let $\overline{j_k}$ be a zero of $H_k(X)$ modulo $p$. There are two $\mathbb{F}_p$-isomorphism classes $\mathcal{A}_{p,1}$ and $\mathcal{A}_{p,2}$ of principally polarized abelian varieties over $\mathbb{F}_p$ with Igusa invariants $\overline{j_k}$.
- The principally polarized abelian varieties $\mathcal{A}_{p,1}$ and $\mathcal{A}_{p,2}$ have complex multiplication by $\mathcal{O}_K$.
- The number of $\mathbb{F}_p$-rational points of $\mathcal{A}_{p,m}, m = 1, 2$ is given by

$$\prod_{i=1}^{4}\left(1 + (-1)^m w_i\right)$$

  where $w = w_1$ and $w_i$ are conjugates of $w$.
- The equation $w\overline{w} = p$ with $w \in \mathcal{O}_K$ has (up to conjugacy and sign) at most two different solutions, i.e., for every CM-field of degree 4 there are at most four different possible orders of groups of $\mathbb{F}_p$-rational points of principally polarized abelian varieties, defined over $\mathbb{F}_p$ with complex multiplication by $\mathcal{O}_K$.

For genus 3 an analogous result holds. We refer the interested reader to Weng [WEN 2001a].

## 5.2 Varieties over finite fields

In this section we shall deal with varieties defined over finite fields. We assume that the ground field $K$ is equal to $\mathbb{F}_q$ with $q = p^d$.

## 5.2.1 The Frobenius morphism

In this section, we consider two extension fields of $\mathbb{F}_p$. We assume $K = \mathbb{F}_q$ with $q = p^d$, and consider an arbitrary power $\phi_p^k$ of the absolute Frobenius automorphism, which fixes the elements of $\mathbb{F}_{p^k} \subset \overline{\mathbb{F}}_p$. We recall the definition of the Frobenius endomorphism and its action on varieties over $\mathbb{F}_q$ given in Example 4.39, which we shall need in a slightly more general way.

Take $k \in \mathbb{N}$ and let $\phi_{p^k}$ be the Frobenius automorphism of the field $\mathbb{F}_{p^k}$, sending $\alpha \in \mathbb{F}_{p^k}$ to $\pi_k(\alpha) = \alpha^{p^k}$. We can extend $\phi_{p^k}$ to points of projective spaces over $\mathbb{F}_{q^k}$ by sending points $(X_0, \ldots, X_n)$ to $(X_0^{p^k}, \ldots X_n^{p^k})$. We apply $\phi_{p^k}$ to polynomials with coefficients in the algebraic closure $\overline{\mathbb{F}}_p$ of $\mathbb{F}_p$ by applying it to the coefficients.

If $V$ is a projective variety over $\mathbb{F}_q$ with ideal $I$ we can apply $\phi_{p^k}$ to $I$ and get a variety $\phi_{p^k}(V)$ with ideal $\phi_{p^k}(I)$. The points of $V$ are mapped to points on $\phi_{p^k}(V)$.

The corresponding morphism from $V$ to $\phi_{p^k}(V)$ is called the *Frobenius morphism* with respect to the field $\mathbb{F}_{p^k}$ and is again denoted by $\phi_{p^k}$. It is the $k$-th power of the *absolute Frobenius* $\phi_p$.

We note that though $\phi_{p^k}$ is by definition a Galois group element it induces a *morphism* from $V$ to $\phi_{p^k}(V)$. We recall that in the language of function fields the corresponding rational map $\phi_{p^k}^*$ from $K(\phi_{p^k}(V))$ is given as follows: choose an open affine part of $V$ and affine coordinate functions $x_1, \ldots, x_n$; then the image of $\phi_{p^k}^*$ in $K(V)$ is generated by $x_1^{p^k}, \ldots, x_n^{p^k}$.

It follows that this rational map is purely inseparable of degree $p^{k \dim(V)}$.

In general $\phi_{p^k}(V)$ will not be isomorphic to $V$. But if $d$ divides $k$ then $V = \phi_{p^k}(V)$ since then $\phi_{p^k}(I) = I$.

**Proposition 5.67** Let $s$ be a natural number such that $ks$ is divisible by $d$. Put $V_0 := V$ and for $i = 1, \ldots, s-1$ define $V_i := \phi_{p^k}(V_{i-1})$.

Then we get the chain of morphisms

$$V = V_0 \xrightarrow{\phi_{p^k}} V_1 \xrightarrow{\phi_{p^k}} \cdots \xrightarrow{\phi_{p^k}} V_{s-1} \xrightarrow{\phi_{p^k}} V_s = V$$

each being purely inseparable of degree $p^{k \dim(V)}$.

The composite of the morphisms is $\phi_{p^{ks}}$.

Hence for $k = 1$ we get a decomposition of $\phi_q$ into a chain in which the absolute Frobenius endomorphism occurs.

## 5.2.2 The characteristic polynomial of the Frobenius endomorphism

We assume now that $C$ is a projective absolutely irreducible nonsingular curve over $\mathbb{F}_q$ of genus $g \geqslant 1$. As seen above the Frobenius endomorphism operates on the rational functions on $C$, on the points of $C$ and — by linear continuation — on the divisors of $C$. It maps principal divisors to principal divisors and preserves the degree of divisors. So it operates in a natural way on $\mathrm{Pic}^0_{C_{\overline{\mathbb{F}}_q}}$, the divisor class group of degree 0 of the curve $C$ over $\overline{\mathbb{F}}_q$.

From the results in the last paragraph and from the fact that the Galois group of $\mathbb{F}_q$ is (topologically) generated by $\phi_q$ we get:

**Proposition 5.68** The Frobenius morphism induces a homomorphism of $\mathrm{Pic}^0_{C_{\overline{\mathbb{F}}_q}}$ and hence an endomorphism, also denoted by $\phi_q$, of the Jacobian variety $J_C$ defined over $\mathbb{F}_q$.

This endomorphism is an isogeny that is purely inseparable of degree $q^g$.

The elements fixed by $\phi_q$ in $J_{C(\overline{\mathbb{F}}_q)}$ are $J_C(\mathbb{F}_q) = \mathrm{Pic}^0_C$. Hence $J_C(\mathbb{F}_q)$ is the kernel of $\mathrm{Id}_{J_C} - \phi_q$ and $|\mathrm{Pic}^0_C| = \deg(\mathrm{Id}_{J_C} - \phi_q)$.

Now recall that for primes $\ell$ different from $p$ we have attached a Galois $\ell$-adic representation $\tilde{\rho}_{J_C,\ell}$ induced by the action of $G_{\mathbb{F}_q}$ on points of order $\ell^k$ of $J_C$ Theorem 4.82. In fact, we have to replace the field $\mathbb{F}_p$ by $\mathbb{F}_q$ and the absolute Frobenius endomorphism by the relative one $\phi_q$ but all the results about $\ell$-adic representations of Galois elements and endomorphisms remain true after this change.

We associate to $\phi_q$ the characteristic polynomial $\chi\bigl(T_\ell(\phi_q)\bigr)_{J_C}(T)$ of $\tilde{\rho}_{J_C,\ell}(\phi_q)$, which is a monic polynomial of degree $2g$ with coefficients in $\mathbb{Z}$ and it is *independent* of the choice of $\ell$.

**Definition 5.69** The polynomial $\chi(\phi_q)_{J_C}(T) := \chi\bigl(T_\ell(\phi_q)\bigr)_{J_C}(T)$ is the *characteristic polynomial of the Frobenius endomorphism* $\phi_q$ on $C$ and of $J_C$. To simplify notation we also use $\chi(\phi_q)_{\overline{C}}(T)$ to denote it.

Since we know that $\deg([1] - \phi_q) = \chi(\phi_q)_C(1)$ we get:

**Corollary 5.70** The order of $\mathrm{Pic}^0_C$, or equivalently, of $J_C(\mathbb{F}_q)$ is equal to $\chi(\phi_q)_C(1)$.

Hence the determination of the number of elements in $\mathrm{Pic}^0_C$ is easy if we can compute the characteristic polynomial of the Frobenius endomorphism on $C$.

The following remark is very useful if we want to compute this polynomial.

**Lemma 5.71** For $n$ prime to $p$ the restriction of $\phi_q$ to $J_C[n]$ has the characteristic polynomial $\chi(\phi_q)_C(T) \pmod{n}$.

**Corollary 5.72** The endomorphism $\chi(\phi_q)_C(\phi_q)$ is equal to the zero map on $J_C$.

There are two distinguished coefficients of the characteristic polynomial of a linear map: the absolute coefficient, which is (up to a sign) the determinant of the map, and the second highest coefficient, which is the negative of the sum of the eigenvalues and is called the trace of the map.

In our case we know that $\chi(\phi_q)_C(0) = q^g$ since the degree of $\phi_q$ as endomorphism on $J_C$ is $q^{\dim(J_C)}$.

The trace of $\chi(\phi_q)_C(T)$ is called the *trace of the Frobenius endomorphism on* $C$ and denoted by $\mathrm{Tr}(\phi_q)$.

**Example 5.73** Let $E$ be an elliptic curve over $\mathbb{F}_q$. Then $\chi(\phi_q)_E(T) = T^2 - \mathrm{Tr}(\phi_q)T + q$, and so

$$|E(\mathbb{F}_q)| = q + 1 - \mathrm{Tr}(\phi_q).$$

### 5.2.3 The theorem of Hasse–Weil for Jacobians

The following results are true for arbitrary abelian varieties over finite fields. We shall state them only for Jacobians of curves $C$ of genus $g > 0$.

**Definition 5.74** The zeroes $\lambda_1, \ldots, \lambda_{2g}$ of $\chi(\phi_q)_C(T)$ are called the *eigenvalues of the Frobenius* $\phi_q$ on $C$ and on $J_C$.

By definition the eigenvalues of $\phi_q$ are algebraic integers lying in a number field of degree $\leqslant g$.

The product is equal to

$$\prod_{i=1}^{2g} \lambda_i = q^g.$$

Because of the duality on Jacobian varieties (or as a consequence of the theorem of Riemann–Roch [STI 1993]) one can make a finer statement.

## § 5.2 Varieties over finite fields

**Proposition 5.75** We can arrange the eigenvalues of $\phi_q$ on $C$ such that

$$\text{for all } i = 1, \ldots, g \text{ we have } \lambda_i \lambda_{i+g} = q.$$

But there is a much deeper result. It is the analogue of the famous *Riemann hypothesis* for the Riemann $\zeta$-function and it says that *the absolute value of each eigenvalue $\lambda_i$ interpreted as a complex number is, for every curve $C$ of arbitrary positive genus $g$, equal to $\sqrt{q}$.*

This result was proved by Hasse for elliptic curves and by Weil for abelian varieties. A generalization for arbitrary varieties over finite fields was formulated by Weil. One of the greatest achievements of mathematics in the twentieth century was the proof of these Weil conjectures by Deligne.

The general philosophy is that the number of rational points on varieties over finite fields should not differ "too much" from the number of points of the projective spaces of the same dimension, and the difference is expressed in terms of the size of the trace of the Frobenius endomorphism acting on attached vector spaces like Tate modules, or more generally, cohomology groups.

Let us come back to our situation and resume what we know.

**Theorem 5.76** Let $C$ be a projective absolutely irreducible nonsingular curve of genus $g > 0$ over $\mathbb{F}_q$. Let $\lambda_1, \ldots, \lambda_{2g}$ be the eigenvalues of the Frobenius endomorphism on $C$.

(i) Each $\lambda_i$ is an algebraic integer of degree $\leqslant 2g$.

(ii) We can numerate the eigenvalues such that for $1 \leqslant i \leqslant g$ we have

$$\lambda_i \lambda_{i+g} = q.$$

(iii) For $1 \leqslant i \leqslant 2g$ take any embedding of $\lambda_i$ into $\mathbb{C}$. Then the complex absolute value $|\lambda_i|$ is equal to $\sqrt{q}$.

For the proof of these fundamental results about the arithmetic of curves and abelian varieties we refer to [STI 1993] or, in a more general frame, to [MUM 1974, pp. 203–207].

**Corollary 5.77** Let $C/\mathbb{F}_q$ be a curve of genus $g$. If

$$J_C(\mathbb{F}_q)[n] \supseteq (\mathbb{Z}/n\mathbb{Z})^t$$

for some $t > g$ then

$$n \mid q - 1.$$

**Proof.** We find $t$ linear independent $\overline{D}_1, \ldots, \overline{D}_t$ elements in $J_C(\mathbb{F}_q)[n]$, which lie in the eigenspace $\rho_{J_C,n}(\phi_q)$ with eigenvalue 1 (mod $n$). Hence, there is a $1 \leqslant i \leqslant g$ such that $\lambda_i$ and $\lambda_{i+g}$ are both equivalent to 1 modulo $n$. Since $\lambda_i \lambda_{i+g} = q$ we have $q \equiv 1 \pmod{n}$. □

We can combine this corollary with Theorem 4.73 to get the following proposition.

**Proposition 5.78** Let $C/\mathbb{F}_q$ be a curve of genus $g$. For the structure of the group of $\mathbb{F}_q$-rational points on the Jacobian we have

$$J_C(\mathbb{F}_q)[n] \simeq \mathbb{Z}/n_1\mathbb{Z} \times \mathbb{Z}/n_2\mathbb{Z} \times \cdots \times \mathbb{Z}/n_{2g}\mathbb{Z},$$

where $n_i \mid n_{i+1}$ for $1 \leqslant i < 2g$ and for all $1 \leqslant i \leqslant g$ one has $n_i \mid q - 1$.

From the Theorem 5.76 we obtain bounds on the number of points on the curve and its Jacobian.

**Corollary 5.79** Let $C$ be as in Theorem 5.76.
Then
$$\left||\text{Pic}_C^0| - q^g\right| = \left|\prod_{i=1}^{2g}(1 - \lambda_i) - q^g\right| = O\left(q^{g-1/2}\right).$$

Take $k \in \mathbb{N}$. Since $\phi_{q^k} = \phi_q^k$ we can extend this result:

**Corollary 5.80** The number $N_k$ of $\mathbb{F}_{q^k}$-rational points of $J_C$, or equivalently, the number of elements in $\text{Pic}_{C \cdot \mathbb{F}_{q^k}}^0$ is estimated by
$$\left|N_k - q^{gk}\right| = \left|\prod_{i=1}^{2g}(1 - \lambda_i^k) - q^{gk}\right| = O\left(q^{k(g-1/2)}\right).$$

This corollary can be used to compute the $\zeta$-function of the curve $C$ [STI 1993] and to get a bound for the number of rational points on $C$.

**Corollary 5.81** Let $C$ be as above.
Then
$$||C(\mathbb{F}_q)| - q - 1| \leqslant 2g\sqrt{q}.$$

The estimates for the number of elements of $\text{Pic}_C^0$ and of $C(\mathbb{F}_q)$ are called the *Hasse–Weil bounds*.

In fact the Serre bound gives the sharper estimate
$$||C(\mathbb{F}_q)| - q - 1| \leqslant g\lfloor 2\sqrt{q}\rfloor.$$

When one wants to compute the characteristic polynomial of the Frobenius endomorphism it is very important that one has *ad hoc* estimates for the size of the coefficients of this polynomial. Again Theorem 5.76 can be used in an obvious way to get

**Corollary 5.82** The characteristic polynomial of $\phi_q$ has a very symmetric shape given by
$$\chi(\phi_q)_C(T) = T^{2g} + a_1 T^{2g-1} + \cdots + a_g T^g + \cdots + a_1 q^{g-1} T + q^g,$$
where $a_i \in \mathbb{Z}, 1 \leqslant i \leqslant g$.

The absolute value of the $i$-th coefficient of $\chi(\phi_q)_C(T)$ is bounded by $\binom{2g}{i} q^{(2g-i)/2}$.

**Example 5.83** Let $E$ be an elliptic curve over $\mathbb{F}_q$. The eigenvalues $\lambda_1$ and $\lambda_2$ of $\phi_q$ on $E$ are algebraic integers of degree $\leqslant 2$ with absolute value $|\lambda_i| = \sqrt{q}$ and $\lambda_1 \lambda_2 = q$. The number of points in $E(\mathbb{F}_q)$ is estimated by
$$||E(\mathbb{F}_q)| - q - 1| \leqslant 2\sqrt{q}.$$

The interval $[-2\sqrt{q} + q + 1, 2\sqrt{q} + q + 1]$ is called the *Hasse–Weil interval*. All elliptic curves defined over $\mathbb{F}_q$ are forced to have their number of rational points lying in this interval.

### 5.2.4 Tate's isogeny theorem

We end this section by stating deep results due to Tate and Tate–Honda [TAT 1966], which demonstrate the importance of characteristic polynomials of Frobenius endomorphisms.

**Theorem 5.84**

(i) Let $\mathcal{A}$ and $\mathcal{A}'$ be abelian varieties over $\mathbb{F}_q$. Then $\mathcal{A}$ is isogenous to $\mathcal{A}'$ over $\mathbb{F}_q$ if and only if $\chi(\phi_q)_{\mathcal{A}}(T) = \chi(\phi_q)_{\mathcal{A}'}(T)$.

(ii) Assume that $\lambda_1, \ldots, \lambda_{2g}$ are algebraic integers lying in a number field of degree $\leqslant 2g$ and satisfying the properties of eigenvalues of Frobenius endomorphism as stated in Corollary 4.118. Then there is an abelian variety $\mathcal{A}$ over $\mathbb{F}_q$ such that $\lambda_1, \ldots, \lambda_{2g}$ are the eigenvalues of the Frobenius endomorphism on $\mathcal{A}$.

Note that this abelian variety need not be principally polarized, and if it is, it need not to be a Jacobian of a curve.

Maisner and Nart [MANA 2002] study the problem to decide whether $\lambda_1, \ldots, \lambda_{2g}$ belong to a hyperelliptic curve.

# Chapter 6

# Background on Pairings

*Sylvain Duquesne and Gerhard Frey*

### Contents in Brief

| | | |
|---|---|---|
| 6.1 | General duality results | 115 |
| 6.2 | The Tate pairing | 116 |
| 6.3 | Pairings over local fields | 117 |
| | The local Tate pairing • The Lichtenbaum pairing on Jacobian varieties | |
| 6.4 | An explicit pairing | 122 |
| | The Tate–Lichtenbaum pairing • Size of the embedding degree | |

## 6.1 General duality results

Let $C$ be a curve of genus $g$ defined over a field $K$ with Jacobian variety $J_C$.

As mentioned in Section 5.1.3 there is a duality theory of abelian varieties, and Jacobians are self-dual. Hence, the points of order $n$ form an abelian group $V = J_C[n]$ endowed with a natural bilinear map.

We have already treated the case when the characteristic of $K$ is $p$ and $n = p$ (although without emphasizing the duality theory behind it) and we have seen in Proposition 4.132 that we can identify $V = J_C[p]$ with a subgroup of the additive group $K^{2g-1}$.

We shall therefore assume from now on that $n$ is a natural number prime to $p$. In this case the duality of $J_C$ induces a nondegenerate pairing on $J_C[n]$ with values in the group of $n$-th roots of unity $\mu_n$, so $J_C[n]$ becomes self-dual with respect to this pairing. In fact this pairing was well-known already in "classical times" for $K = \mathbb{C}$. It is related to Riemann forms and is discussed in [MUM 1974, Section 20]. It is defined for arbitrary principally polarized abelian varieties and given in an explicit form on p. 187 of that book. Using this explicit form it is easy to generalize the pairing in the abstract setting of abelian varieties by the same formulas. In the case of elliptic curves this can be found in [SIL 1986]. The resulting pairing is called the *Weil pairing*.

The Weil pairing has very nice properties that can all be found in [MUM 1974]. It is defined as pairing on the group of torsion points of order $n$ prime to $p$ and it is compatible with the natural maps from $J_C[n]$ to $J_C[n']$ for $n'$ dividing $n$. Hence taking $n = \ell^k$ it can be extended to a pairing of Tate modules $T_\ell(J_C)$.

Moreover, the pairing is compatible with the action of the absolute Galois group $G_K$ of $K$. A useful consequence is that $J_C[n](\overline{K}) = J_C(K)[n]$ implies that $K$ contains the $n$-th roots of unity. (If $K$ is a finite field we find another argument in Corollary 5.77 to see this).

One disadvantage is that the Weil pairing is skew symmetric. This implies that there are subspaces of dimension $g$ in $J_C[n]$ such that the restriction of the pairing vanishes identically ("maximal isotropic subspaces"). So we need the rationality of many $n$-torsion points in order to get subspaces of $J_C[n]$ on which the Weil pairing is nondegenerate.

Nevertheless the Weil pairing is a possibility for constructing a bilinear structure on DL systems inside abelian varieties. In this book however we shall prefer a derived pairing which, for our purposes, has nicer properties both from the theoretical and from the computational point of view: *the Tate pairing*.

## 6.2 The Tate pairing

We shall begin by giving the general background for the construction of the Tate pairing. It involves Galois cohomology, hence requires some knowledge of nonelementary algebra. We shall then specialize more and more, and in the end we shall obtain a description of the pairing in the cases that are of interest to us. At this point, it is quite elementary and does not use any advanced ingredients. We advise the reader who is only interested in applications to skip as much of the following as he (dis-)likes and in the worst case to begin reading Section 6.4.

After this warning we begin with the theory. We are interested in Jacobians but to define the Tate pairing we use results that are true for arbitrary abelian varieties $A$ defined over $K$. For simplicity we shall assume that $A$ is principally polarized and so it is isomorphic to its dual variety (as is true for Jacobians). The assumption that $n$ is prime to $p$ implies that the *Kummer sequence*

$$0 \to A\left(\overline{K}\right)[n] \to A\left(\overline{K}\right) \xrightarrow{[n]} A\left(\overline{K}\right) \to 0 \tag{6.1}$$

is an exact sequence of $G_K$-modules.

We can therefore apply Galois cohomology and obtain the exact sequence

$$0 \to A(K)/nA(K) \xrightarrow{\delta} H^1\left(G_K, A\left(\overline{K}\right)[n]\right) \xrightarrow{\alpha} H^1\left(G_K, A\left(\overline{K}\right)\right)[n] \to 0.$$

Since this is an important sequence both in theory and in practice we explain it in more detail.

Recall that for a $G_K$-module $M$ the group $H^n(G_K, M)$ is a quotient of the group of $n$-cocycles (i.e., of maps $c$ from the $n$-fold Cartesian product $G_K^n$ to $M$ satisfying a combinatorial condition) modulo the subgroup of $n$-coboundaries. Whenever needed we shall give an explicit description of the cohomology groups that occur.

**Example 6.1** Obviously 1-cocycles are maps

$$c : G_K \to M$$

such that for all $\sigma, \tau \in G_K$ we have

$$c(\sigma\tau) = c(\sigma) + \sigma c(\tau)$$

and 1-coboundaries are maps

$$c : G_K \to M$$

such that there exists an element $m \in M$ with

$$c(\sigma) = \sigma \cdot m - m$$

for all $\sigma \in G_K$.

Let $P \in A(K)$. There exists a point $Q \in A\left(\overline{K}\right)$ such that $[n]Q = P$. Define

$$\delta'(P) : G_K \to A\left(\overline{K}\right)[n]$$
$$\sigma \mapsto \sigma \cdot Q - Q.$$

We easily check that $\delta'(P)$ is a 1-cocycle with image in $A\left(\overline{K}\right)[n]$ and that another choice of $Q'$ with $[n]Q' = P$ changes this cocycle by a coboundary and so we get a well defined map from $A(K)$ to $H^1\left(G_K, A\left(\overline{K}\right)[n]\right)$. Another immediate check shows that the kernel of this map is exactly $[n]A(K)$. This explains the first part of the Kummer sequence.

We now use the injection of $A\left(\overline{K}\right)[n]$ into $A\left(\overline{K}\right)$ to interpret cocycles with values in $A\left(\overline{K}\right)[n]$ as cocycles with values in $A\left(\overline{K}\right)$. Going to the quotient modulo coboundaries gives the map $\alpha$. Since the arguments of the induced cocycles are points of order $n$ it follows that the image of $\alpha$ is contained in the subgroup of $H^1\left(G_K, A\left(\overline{K}\right)\right)$, which is annihilated by the map "multiplication by $n$."

We can check, either directly or by using properties of cohomology, that $\alpha$ is surjective and that the kernel of $\alpha$ is equal to the image of $\delta$.

Next we use that $A\left(\overline{K}\right)[n]$ is self-dual as a $G_K$-module under the Weil pairing, which we denote by $W_n$. We obtain a cup product

$$\cup : H^1\left(G_K, A\left(\overline{K}\right)[n]\right) \times H^1\left(G_K, A\left(\overline{K}\right)[n]\right) \to H^2(G_K, \overline{K}^*)[n]$$

in the following way:

Represent $\zeta_1, \zeta_2 \in H^1\left(G_K, A\left(\overline{K}\right)[n]\right)$ by cocycles $c_1, c_2$. Then $\zeta_1 \cup \zeta_2$ is the cohomology class of the 2-cocycle

$$c : G_K \times G_K \to \overline{K}^*$$

given by

$$c(\sigma_1, \sigma_2) := W_n\bigl(c_1(\sigma_1), c_2(\sigma_2)\bigr).$$

This map $\cup$ is bilinear.

We can apply this to a point $P \in A(K)$ and a cohomology class $\gamma \in H^1\left(G_K, A\left(\overline{K}\right)\right)[n]$ to define the *Tate pairing*

$$\langle \cdot, \cdot \rangle_{T,n} : A(K)/nA(K) \times H^1\left(G_K, A\left(\overline{K}\right)\right)[n] \to H^2(G_K, \overline{K}^*)[n]$$

by

$$\langle P + nA(K), \gamma \rangle_{T,n} = \delta\bigl(P + nA(K)\bigr) \cup \alpha^{-1}(\gamma).$$

It is routine to check that $\langle \cdot, \cdot \rangle_{T,n}$ is well defined and bilinear.

**Remark 6.2** The Tate pairing relates three very interesting groups occurring in Arithmetic Geometry: the Mordell–Weil group of $A$, the first cohomology group of $A$, which can be interpreted as group of principally homogeneous spaces over $A$, and $H^2(G_K, \overline{K}^*)$, the *Brauer group* of the ground field $K$, which can be interpreted as group of classes of central simple algebras with center $K$ with the class of full matrix groups as neutral element.

## 6.3 Pairings over local fields

We now assume that $K$ is a *local field* (e.g., a $p$-adic field) with finite residue field $\mathbb{F}_q$.

### 6.3.1 The local Tate pairing

We have the beautiful result of Tate [TAT 1958]:

**Theorem 6.3** Let $A$ be a principally polarized abelian variety over $K$, e.g., $A = J_C$.
Let
$$\langle \cdot, \cdot \rangle_{T,n} : A(K)/nA(K) \times H^1\left(G_K, A\left(\overline{K}\right)\right)[n] \to H^2(G_K, \overline{K}^*)[n]$$
be the Tate pairing as defined above. Then $\langle \cdot, \cdot \rangle_{T,n}$ is a nondegenerate $\mathbb{Z}$-bilinear map.

It is thus certainly worthwhile to study in more detail the groups that are involved.

To simplify the situation we shall assume that $A$ has good reduction, i.e., we find equations for $A$ with coefficients in the ring of integers of $K$ whose reductions modulo the valuation ideal of $K$ again define an abelian variety over $\mathbb{F}_q$. This situation is typical for the applications that we have in mind. In fact we shall begin with an abelian variety over $\mathbb{F}_q$ and then lift it to an abelian variety over $K$. This motivates a change of notation: $A \mapsto \tilde{A}$ and the reduction of $\tilde{A}$ is now denoted by $A$.

Let us consider the first group occurring in Tate duality. Using Hensel's lemma we get
$$\tilde{A}(K)/n\tilde{A}(K) \simeq A(\mathbb{F}_q)/nA(\mathbb{F}_q).$$

**Remark 6.4** If we assume that $n = \ell$ is a prime and that $A(\mathbb{F}_q)$ has no points of order $\ell^2$ then $A(\mathbb{F}_q)/\ell A(\mathbb{F}_q)$ is isomorphic to $A(\mathbb{F}_q)[\ell]$ in a natural way.

We now come to the discussion of $H^1\left(G_K, A\left(\overline{K}\right)\right)[n]$. Since unramified extensions of $K$ do not split elements in this group we can use a well-known inflation-restriction sequence to change our base field from $K$ to the maximal unramified extension $K^{\mathrm{ur}}$ of $K$, compute the cohomology group over this larger field, and look for elements that are invariant under the Galois group of $K^{\mathrm{ur}}/K$ which is topologically generated by (a canonical lift of) the Frobenius automorphism $\phi_q$ of $\mathbb{F}_q$. Note that this automorphism acts both on $G\left(\overline{K}/K^{\mathrm{ur}}\right)$ and on $A[n] = A(K^{\mathrm{ur}})[n]$.

Let $K_{\mathrm{tame}}$ be the unique cyclic extension of $K^{\mathrm{ur}}$ of degree $n$ (which has to be fully ramified). We obtain

**Proposition 6.5** The first cohomology group $H^1\left(G_K, A\left(\overline{K}\right)\right)[n]$ is equal to the group of elements in
$$\mathrm{Hom}(G(K_{\mathrm{tame}}/K^{\mathrm{ur}}), A[n]),$$
which are invariant under the natural action of the Frobenius automorphism.

After fixing a generator $\tau$ of $G(K_{\mathrm{tame}}/K^{\mathrm{ur}})$ we can identify
$$\psi \in \mathrm{Hom}(G(K_{\mathrm{tame}}/K^{\mathrm{ur}}), A[n])$$
with
$$\psi(\tau) =: P_\tau \in A[n]$$
and hence $\mathrm{Hom}(G(K_{\mathrm{tame}}/K^{\mathrm{ur}}), A[n])$ with $A[n]$.

*Warning:* The identification of $\mathrm{Hom}(G(K_{\mathrm{tame}}/K^{\mathrm{ur}}), A[n])$ with $A[n]$ is, in general, not compatible with Galois actions. Here the cyclotomic character becomes important: over $K(\mu_n)$ we can realize a ramified cyclic extension $K_n$ of degree $n$ by choosing an $n$-th root $t$ of a uniformizing element $\pi$ of $K$. Since $\tau$ maps $t$ to $\zeta_n t$ for some $n$-th root of unity $\zeta_n$ and the Frobenius automorphism $\phi_q$ maps $\zeta_n$ to $\zeta_n^q$ we deduce that $\phi_q$ operates on $\langle \tau \rangle$ by conjugation, sending $\tau$ to $\tau^{-q}$.

**Corollary 6.6** The first cohomology group $H^1\left(G_K, A\left(\overline{K}\right)\right)[n]$ can be identified with the subgroup $A_0$ of points in $A[n]$ defined by
$$A_0 = \{P \in A[n] \mid \phi_q(P) = [q]P\}.$$

## § 6.3 Pairings over local fields

We summarize what we have found up to now in the case where $n$ is a prime $\ell$.

**Proposition 6.7** Let $K$ be a local field with residue field $\mathbb{F}_q$, let $A$ be an abelian variety defined over $\mathbb{F}_q$, let $\ell$ be a prime number not dividing $q$, and assume that $A(\mathbb{F}_q)$ contains no elements of order $\ell^2$. Define $A_0 = \{P \in A[\ell] \mid \phi_q(P) = [q]P\}$.

The Tate pairing induces a nondegenerate pairing
$$\langle \cdot, \cdot \rangle_{T,\ell} : A(\mathbb{F}_q)[\ell] \times A_0 \to \mathrm{Br}(K)[\ell].$$

**Corollary 6.8** Under the assumptions of the propositions we get that $A_0$ is (as an abelian group) isomorphic to $A(\mathbb{F}_q)[\ell]$.

**Example 6.9** Assume that $A[\ell](\mathbb{F}_q)$ is cyclic of order $\ell$ and generated by $P$.

1. Assume that $\ell \mid (q-1)$. Then $A_0 = A[\ell](\mathbb{F}_q)$, and $\langle P, P \rangle_{\mathbb{F}_q, \ell} \neq 0$.
2. Assume that $\ell \nmid q-1$. Then $\phi$ with $\phi(\tau) = P$ is not in $H^1\left(G_K, A\left(\overline{K}\right)\right)[\ell]$. In particular "$\langle P, P \rangle_{T,\ell}$" is not defined. Of course we can extend the ground field until the pairing over these larger fields permits the argument $(P, P)$. But the value will then necessarily be equal to 0.

**Example 6.10** If $A[\ell](\mathbb{F}_q)$ is not cyclic and $\ell \mid q-1$ then for all points $P, Q \in A[\ell](\mathbb{F}_q)$ we can form $\langle P, Q \rangle_{T,\ell}$ but it is not clear whether there is a $P$ with $\langle P, P \rangle_{T,\ell} \neq 0$.

We now come to the discussion of the Brauer group. First one has

**Theorem 6.11** The Brauer group of $K$ is (canonically) isomorphic to $\mathbb{Q}/\mathbb{Z}$.

More precisely, there is a map, the invariant map $\mathrm{inv}_K$, such that for all $n \in \mathbb{N}$ we have an isomorphism
$$\mathrm{inv} : \mathrm{Br}(K)[n] \to \mathbb{Z}/n\mathbb{Z}.$$

Thus computations in $\mathrm{Br}(K)$ boil down to the computation of the invariant map. A further study of the theory of local fields shows that this is closely related to the computation of the discrete logarithm in $\mathbb{F}_q^*$ (see [NGU 2001]).

This becomes more obvious if we replace $\mathbb{F}_q$ by $\mathbb{F}_q(\mu_n) = \mathbb{F}_{q^k}$ with $k$ minimal such that $n \mid (q^k - 1)$. Put $K_1 = K(\mu_n)$ and let $K_n$ be a cyclic ramified extension of degree $n$ of $K_1$. The value of the Tate pairing is then in $H^2(G(K_n/K), K_n^*)[n]$, and elementary computations with cohomology groups yield that this group is isomorphic (canonically after the choice of $\tau$) to $\mathbb{F}_{q^k}^* / \left(\mathbb{F}_{q^k}^*\right)^n$.

**Proposition 6.12** Let $n$ be equal to a prime number $\ell$, and let $k$ be as above. Assume that $A(\mathbb{F}_q)$ contains no points of order $\ell^2$.

The Tate pairing induces a pairing
$$\langle \cdot, \cdot \rangle_{T,\ell} : A[\ell](\mathbb{F}_q) \times A[\ell](\mathbb{F}_{q^k}) \to \mathbb{F}_{q^k}^* / \left(\mathbb{F}_{q^k}^*\right)^\ell$$

which is nondegenerate on the left, i.e., if $\langle P, Q \rangle_{T,\ell} = 0$ for all $Q \in A[\ell](\mathbb{F}_{q^k})$ then $P = 0$.

### 6.3.2 The Lichtenbaum pairing on Jacobian varieties

In Proposition 6.12 we have described a pairing that can, in principle, be used to transfer the discrete logarithm from $A[\ell](\mathbb{F}_q)$ to $\mathbb{F}_{q^k}^* / \left(\mathbb{F}_{q^k}^*\right)^\ell$. However, looking at the conditions formulated in Section 1.5.2 we see that one crucial ingredient is missing: we must be able to compute the pairing very fast.

The next two sections are devoted to this goal in the case that we are interested in.

We come back from the general theory of abelian varieties to the special theory of Jacobian varieties $J_C$ of projective curves $C$ of genus $g$ over finite ground fields $\mathbb{F}_q$.

We want to use the duality theory over local fields and so we lift $C$ to a curve $\widetilde{C}$ of genus $g$ over the local field $K$ with corresponding abelian variety $J_{\widetilde{C}}$. We remark that the reduction of $J_{\widetilde{C}}$ is $J_C$.

The central part of the construction of the Tate pairing was the construction of a 2-cocycle from $G_K^2$ into $\overline{K}^*$. For a point $P \in J_{\widetilde{C}}(K)$ and an element $\gamma \in H^1\big(G_K, J_{\widetilde{C}}(\overline{K})\big)[n]$ we choose a 1-cocycle $c$ in $\alpha^{-1}(\gamma)$ and define

$$c(\sigma_1, \sigma_2) = W_n\big(\delta'(P)(\sigma_1), c(\sigma_2)\big)$$

where $\delta'$ and $\alpha$ are the maps defined in Section 6.3.1 and $W_n$ is the Weil pairing.

In his paper [LIC 1969] Lichtenbaum used sequences of divisor groups of curves to define a pairing in the following way.

Put $\overline{C} = \widetilde{C} \times \overline{K}$. We have discussed the group of divisor classes of degree 0 of $\overline{C}$ together with the action of $G_K$ on this group in Section 4.4.4. As a consequence we get the exact sequence of $G - K$-modules (see 6.1)

$$1 \to \mathrm{Princ}_{\overline{C}} \to \mathrm{Div}_{\overline{C}}^0 \to \mathrm{Pic}_{\overline{C}}^0 \to 0.$$

We can apply cohomology theory to this sequence and obtain a map

$$\delta_1 : H^1(G_K, \mathrm{Pic}_{\overline{C}}^0) \to H^2(G_K, \mathrm{Princ}_{\overline{C}})$$

which associates to $\gamma \in H^1(G_K, \mathrm{Pic}_{\overline{C}}^0)$ a 2-cocycle from $G_K^2$ to $\mathrm{Princ}_{\overline{C}}$.

In other words, given $\gamma$ we find for each pair $(\sigma_1, \sigma_2) \in G_K$ a function $f_{\sigma_1, \sigma_2} \in \overline{K}(\overline{C})$ such that the class of $(f_{\sigma_1, \sigma_2}; (\sigma_1, \sigma_2) \in G_K)$ is equal to $\delta_1(\gamma)$.

**Definition 6.13** The notations are as above. Let $c \in \mathrm{Pic}_{\overline{C}}^0$ be a $K$-rational divisor class of degree 0 with divisor $D \in c$.

The Lichtenbaum pairing

$$\langle \cdot, \cdot \rangle_L : \mathrm{Pic}_{\overline{C}}^0 \times H^1(G_K, \mathrm{Pic}_{\overline{C}}^0) \to H^2(G_K, \overline{K}^*)$$

maps $(c, \gamma)$ to the class in $H^2(G_K, \overline{K}^*)$ of the cocycle

$$G_K^2 \to \overline{K}^*$$

given by

$$(\sigma_1, \sigma_2) \mapsto f_{\sigma_1, \sigma_2}(D) .$$

(Here $D$ has to be chosen such that it is prime to the set of poles and zeroes of $f_{\sigma_1, \sigma_2}$, which is always possible.)

Since we have seen that $J_C(\overline{K}) = \mathrm{Pic}_{\overline{C}}^0$ we can compare Tate's pairing with $\langle \cdot, \cdot \rangle_L$. It is shown in [LIC 1969, pp. 126-127], that the two pairings are the "same" in the following sense.

**Proposition 6.14** For all natural numbers $n$ denote by $\langle \cdot, \cdot \rangle_{L,n}$ the pairing induced by $\langle \cdot, \cdot \rangle_L$ on $\mathrm{Pic}_{\overline{C}}^0 / n\mathrm{Pic}_{\overline{C}}^0 \times H^1(G_K, \mathrm{Pic}_{\overline{C}}^0)[n]$.

Then $\langle \cdot, \cdot \rangle_{L,n}$ is equal (up to sign) to the Tate pairing $\langle \cdot, \cdot \rangle_{T,n}$ applied to the abelian variety $J_{\widetilde{C}}$.

In fact, Lichtenbaum uses this result to prove nondegeneracy of his pairing for a local field $K$.

The importance of Lichtenbaum's result for our purposes is that we have a description of the Tate pairing related to Jacobian varieties that only uses objects directly defined by the curve $\widetilde{C}$. In particular the Weil pairing has completely disappeared.

We can now use the general considerations given in Section 6.3.1 and come to the final version of the pairing in the situation that we are interested in.

Thus we come back to a curve $C$ defined over $\mathbb{F}_q$, choose a lifting to $\widetilde{C}$ over $K$, and compute the groups occurring in the pairing. As a result we shall obtain a pairing that only involves the curve $C$ itself.

The first group can be identified with $\operatorname{Pic}_C^0/n\operatorname{Pic}_C^0$. To avoid trivial cases we shall assume that this group is nontrivial, i.e., that $\operatorname{Pic}_C^0$ contains elements of order $n$.

The group $H^1(G_K, \operatorname{Pic}_{\widetilde{C}}^0)[n]$ can be identified with the subgroup $J_0$ of points in $\operatorname{Pic}_{\widetilde{C}}^0[n]$ defined by
$$J_0 = \left\{ c \in \operatorname{Pic}_{\widetilde{C}}^0[n] \mid \phi_q(c) = [q]c \right\}.$$

A first application of these results is that we can describe the Lichtenbaum pairing by Galois cohomology groups related to a finite extension of $K$. We first enlarge $K$ to a field $K_1$ that is unramified and such that over its residue field all elements of $J_0$ are rational. Automatically $K_1$ contains the $n$-th roots of unity. It follows that there exists a cyclic ramified extension $K_n$ of degree $n$ of $K_1$.

The image of the pairing $\langle \cdot, \cdot \rangle_{L,n}$ will be contained in $H^2(G(K_n/K_1), K_n^*)$. Let us fix a generator $\tau$ of the Galois group of $K_n/K_1$. We identify $\gamma$ in $H^1(G_K, \operatorname{Pic}_{\widetilde{C}}^0)[n]$ with the class of the cocycle $\zeta$ from $G(K_n/K_1)$ given by
$$\zeta(\tau^i) = [i]\tilde{c} \text{ for } 0 \leqslant i \leqslant n-1,$$
where $\tilde{c} \in \operatorname{Pic}_{\widetilde{C} \times K_n}^0[n]$ is a lift of $c \in J_0$.

The image of $\gamma$ under the map $\delta_1$ will be a 2-cocycle mapping from $\langle \tau \rangle \times \langle \tau \rangle$ to $\operatorname{Princ}_{\widetilde{C} \times K_n}$ which we must describe. Choose a divisor $D \in \tilde{c}$ rational over $K_1$. Thus $iD \in \zeta(\tau^i)$.

By definition
$$\delta_1(\gamma)(\tau^i, \tau^j) = \tau^i j D - r_{i+j} D + jD$$
for $0 \leqslant i, j \leqslant n-1$, where $r_{i+j}$ is the smallest nonnegative residue of $i+j$ modulo $n$.

Since $\tau D = D$ it follows that $\delta_1(\gamma)(\tau^i, \tau^j) = 0$ for $i+j < n$ and $\delta_1(\gamma)(\tau^i, \tau^j) = nD$ for $i+j \geqslant n$. Since $D$ is a divisor of degree 0 in a class of order $n$ the divisor $nD$ is the divisor of a function $f_D$ (which is defined over $K_1$).

Now choose $\tilde{c}_1 \in \operatorname{Pic}_{\widetilde{C}}^0$ and a divisor $E \in \tilde{c}_1$ such that $E$ is prime to $D$. Then
$$\langle \tilde{c}_1 + n\operatorname{Pic}_{\widetilde{C}}^0, \gamma \rangle_{L,n}(\tau^i, \tau^j) = 1$$
if $i+j < n$ and
$$\langle \tilde{c}_1 + n\operatorname{Pic}_{\widetilde{C}}^0, \gamma \rangle_{L,n}(\tau^i, \tau^j) = f_D(E)$$
if $i+j \geqslant n$.

Thanks to this result we can immediately identify the element in the Brauer group of $K_1$ which corresponds to $\langle \tilde{c}_1 + n\operatorname{Pic}_{\widetilde{C}}^0, \gamma \rangle_{L,n}$: it is the cyclic algebra split by $K_n$ corresponding to the class of the pair $(\tau, f_D(E))$ (cf. [NGU 2001]). If we fix $\tau$ the class is uniquely determined by the norm class $f_D(E) \operatorname{N}_{K_n/K_1}(K_n^*)$.

Since $K_n/K_1$ is totally ramified we can compute $\operatorname{N}_{K_n/K_1}(K_n^*)$ to be (canonically) isomorphic to $\mathbb{F}_{q^k}^* / \left(\mathbb{F}_{q^k}^*\right)^n$.

Hence $\langle \tilde{c}_1 + n\operatorname{Pic}_{\widetilde{C}}^0, \gamma \rangle_{L,n}$ is uniquely determined by the image of $f_D(E)$ in the residue field $\mathbb{F}_{q^k}$ modulo $\left(\mathbb{F}_{q^k}^*\right)^n$.

We can obtain this image directly: choosing $c_1 \in \operatorname{Pic}_C^0$, $E \in c_1$, $c \in J_0$, $D \in c$, and $f_D$ a function on $C$ defined over $\mathbb{F}_{q^k}$ with no zeroes and poles in divisors of $E$, we have
$$\langle \tilde{c}_1 + n\operatorname{Pic}_{\widetilde{C}}^0, \gamma \rangle_{L,n} = f_D(E)\left(\mathbb{F}_{q^k}^*\right)^n.$$

## 6.4 An explicit pairing

### 6.4.1 The Tate–Lichtenbaum pairing

We have done all the necessary work so as to be able to describe the Tate pairing in the Lichtenbaum version for abelian varieties, which are Jacobian varieties of projective curves, in a very elementary and explicit manner as promised in the beginning of this section. In order to use it, no knowledge of the preceding sections is necessary. In fact, even the proof of the following theorem can be given without using Galois cohomology and liftings to local fields by a clever application of Kummer theory to function fields over finite fields (cf. [HES 2004]). Nevertheless we think that it may be interesting for some readers to see the structural background involved.

**Theorem 6.15** Let $\mathbb{F}_q$ be the field with $q$ elements, let $n$ be a number prime to $q$, and let $k \in \mathbb{N}$ be minimal with $n \mid (q^k - 1)$. Let $C$ be a projective irreducible nonsingular curve of genus $g$ defined over $\mathbb{F}_q$ with a rational point.
Define
$$J_0 := \{c \in \operatorname{Pic}^0_{C \times \bar{\mathbb{F}}_q}[n] \mid \phi_q(c) = [q]c\}.$$
There exists a nondegenerate bilinear map
$$T_n : \operatorname{Pic}^0_C/[n]\operatorname{Pic}^0_C \times J_0 \to \mathbb{F}^*_{q^k}/\left(\mathbb{F}^*_{q^k}\right)^n$$
defined in the following way:
For $c_1 \in \operatorname{Pic}^0_C$ and $c_2 \in J_0$ we choose divisors $E \in c_1$ and $D \in c_2$ rational over $\mathbb{F}_q$ respectively $\mathbb{F}_{q^k}$ such that no point $P$ of $C$ occurs both in $E$ and in $D$. Let $f_D$ be a function on $C$ rational over $\mathbb{F}_{q^k}$ with principal divisor $nD$.
Then
$$T_n(c_1, c_2) = f_D(E)\left(\mathbb{F}^*_{q^k}\right)^n.$$

**Definition 6.16** The *Tate–Lichtenbaum pairing* is the bilinear map
$$T_n : \operatorname{Pic}^0_C/[n]\operatorname{Pic}^0_C \times J_0 \to \mathbb{F}^*_{q^k}/\left(\mathbb{F}^*_{q^k}\right)^n$$
described in Theorem 6.15.

Let us consider this pairing from a computational point of view. It is not difficult to find divisors $D$ and $E$ for given $c_1, c_2$. However, for large $n$ it is not obvious how to find $f_D$ and how to evaluate it at $E$. In Chapter 16 we shall give a polynomial-time algorithm to perform this computation. Here we describe very briefly the theoretical background of this algorithm.

We must solve the following task: Let $C$ be a curve of genus $g$ defined over some ground field $K$ with a $K$-rational point $P_0$, let $E$ be a $K$-rational divisor of degree 0 on $C$ and $c$ a $K$-rational divisor class of degree 0 and of order $n$ on $C$. Let $D_1 = A_1 - gP_0 \in c$ be a divisor, where $A_1$ is an effective divisor of degree $g$. Any multiple $[i]c$ can be represented in a similar way by $D_i = A_i - gP_0$. We assume that the support of $E$ is prime to the support of all divisors $D_i$. In particular the divisor $nD_1$ is the principal divisor of a function $f$ on $C$, which has no poles and zeroes at the points of the support of $E$. Hence $c(E) = f(E)$ is a well defined element of $K^*$.

We want to compute this element fast, and we follow an idea that for elliptic curves has been described by Miller in an unpublished manuscript [MIL 1986] (now published in [MIL 2004]), and which, in the general case, is inspired by Mumford's theory of Theta groups of abelian varieties.

The *basic step* for the computation is the following. For given positive divisors $A, A'$ of degree $g$ find a positive divisor $B$ of degree $g$ and a function $h$ on $C$ such that $A + A' - B - gP_0 = \operatorname{div}(h)$.

## § 6.4 An explicit pairing

Define the following *group law* on $\langle c \rangle \times K^*$:

$$(ic, a_1) \circ (jc, a_2) := \big((i+j)c, a_1 a_2 h_{i,j}(E)\big),$$

with $A_i + A_j - A_{i+j} - gP_0 = \operatorname{div}(h_{i,j})$. The assumptions on $E$ guarantee that each $h_{i,j}(E) \in K^*$, and the degree of $h_{i,j}$ is at most equal to $g$. It can be easily seen by induction that $m \cdot (c, 1) = \big(mc, h_{m-1}(E)\big)$, where $h_{m-1}$ is a function on $C$ satisfying $mA - A_{m-1} - (m-1)gP_0 = \operatorname{div}(h_{m-1})$. It follows that repeating $n$ times this process gives the result $\big(0, f(E)\big)$, where $f$ is a function on $C$ such that $\operatorname{div}(f) = nD_1$.

We can now use the group structure on $\langle c \rangle \times K^*$ and apply the square and multiply algorithm to evaluate $f$ at $E$ in $O(\lg n)$ basic steps.

**Corollary 6.17** The Tate–Lichtenbaum pairing $T_n$ can be computed in $O(\lg n)$ basic steps over $\mathbb{F}_{q^k}$.

**Remark 6.18** By a clever choice of $P_0$ we can accelerate the computation. For instance, with hyperelliptic curves, we shall choose the point at infinity $P_\infty$.

### 6.4.2 Size of the embedding degree

Recall that $n$ is a number prime to $q$. The *embedding degree* (with respect to $n$) is the smallest number $k$ such that $n \mid q^k - 1$.

The above result is of practical importance only if $k$ is small.

In general, the necessary conditions for $C$ such that $\operatorname{Pic}^0_C$ has elements of order $\ell$ rational over $\mathbb{F}_q$ with $\ell$ in a cryptographically interesting range, *and* the conditions for $q$ that for a small $k$ the field $\mathbb{F}_{q^k}$ contains $\ell$-th roots of unity, will not be satisfied at the same time.

To see this we look at $\chi(\phi_{q^k})_C(T)$, the characteristic polynomial of $\phi_{q^k}$. Its zeroes $(\lambda_1, \ldots, \lambda_{2g})$ are integers in a number field $K$ and we order them so that $\lambda_i \lambda_{g+i} = q$ for $1 \leqslant i \leqslant g$, which is always possible (cf. Proposition 5.75).

Since $\operatorname{Pic}^0_C$ has elements of order $\ell$ there exists an eigenvalue $\lambda_i$ of $\phi_q$ such that a prime ideal $\mathfrak{l}$ of $K$ dividing $(\ell)$ divides $(1 - \lambda_i)$. Of course this implies that for all natural numbers $d$ the ideal $\mathfrak{l}$ divides $(1 - \lambda_i^d)$.

Now assume that $\mathbb{F}_{q^k}$ contains the $\ell$-th roots of unity and hence that $q^k \equiv 1 \pmod{\ell}$. Since $\lambda_i^k \lambda_{g+i}^k = q^k \equiv 1 \pmod{\mathfrak{l}}$ we get that the prime ideal $\mathfrak{l}$ divides *simultaneously* $(1 - \lambda_i^k)$ and $(1 - \lambda_{g+i}^k)$ and so

$$\lambda_i^k + \lambda_{i+g}^k \equiv 2 \pmod{\mathfrak{l}}.$$

For elliptic curves this yields

**Proposition 6.19** Let $E$ be an elliptic curve defined over $\mathbb{F}_q$ and $\ell$ a prime such that $\ell$ divides $|E(\mathbb{F}_q)|$. Let $\phi_q$ be the Frobenius endomorphism acting on $E[\ell]$. The corresponding discrete logarithm in $E(\mathbb{F}_q)[\ell]$ can be reduced to the discrete logarithm in $\mathbb{F}_{q^k}^*[\ell]$ by the use of the Tate–Lichtenbaum pairing if and only if the characteristic polynomial of the endomorphism $\phi_q^k$ on $E$ is congruent to $T^2 - 2T + 1$ modulo $\ell$.

*Avoiding* elliptic curves with small $k$ is easy. For randomly chosen elliptic curves $E$ we can expect that $k$ will be large.

But there is an important class of special elliptic curves for which $k$ is always small: the *supersingular elliptic curves*. The crucial facts that we use are that the characteristic $p$ of $\mathbb{F}_q$ divides the trace of the Frobenius acting on supersingular elliptic curves $E$ and that their absolute invariant $j_E$ lies either in $\mathbb{F}_p$ or in $\mathbb{F}_{p^2}$ [LAN 1973].

Let us discuss the easiest case in detail. We assume that $p > 3$ and $j_E \in \mathbb{F}_p$. Let $E_0$ be an elliptic curve defined over $\mathbb{F}_p$ with invariant $j_E$. Let $T^2 - aT + p$ be the characteristic polynomial of $\phi_p$ on $E_0$. We know that $a = \lambda p$ with $\lambda \in \mathbb{Z}$. The estimate

$$|1 - \lambda p + p| \leqslant 2\sqrt{p} + (p+1) \quad \text{with} \quad \lambda \in \mathbb{Z}$$

implies that $\lambda = 0$, hence the eigenvalues $\lambda_1, \lambda_2$ of $\phi_p$ acting on $E_0$ satisfy

$$\lambda_1 = -\lambda_2 \quad \text{and} \quad \lambda_1 \lambda_2 = p$$

hence $\lambda_i = \pm \sqrt{-p}$.

Assume now that $q = p^d$. Since $E$ becomes isomorphic to $E_0$ over $\mathbb{F}_{q^2}$ the characteristic polynomial of the Frobenius endomorphism on $E$ over $\mathbb{F}_{q^2}$ is equal to

$$T^2 - (\lambda_1^{2d} + \lambda_2^{2d}) + q^2 = T^2 - 2\lambda_1^{2d} + \lambda_1^{4d} = (T - \lambda_1^{2d})^2.$$

Since by assumption $E(\mathbb{F}_q)$ has elements of order $\ell$, we obtain that $\ell$ divides $(1 - \lambda_1^{2d})$.

Since $\lambda_1^2 = -\lambda_1 \lambda_2 = -p$ it follows that $\ell$ divides $1 - (-p)^d$. But this implies that $k = 1$ if $d$ is even, and $k = 2$ if $d$ is odd.

The other cases can be treated by similar considerations. As a result we obtain

**Proposition 6.20** Let $E$ be a supersingular elliptic curve over $\mathbb{F}_q$ with $q = p^d$. Assume that $E$ has a $\mathbb{F}_q$-rational point of order $\ell$. Let $k$ be the smallest natural number such that $\ell \mid q^k - 1$. Then

- in characteristic 2 we have $k \leqslant 4$,
- in characteristic 3 we have $k \leqslant 6$,
- over prime fields $\mathbb{F}_p$ with $p \geqslant 5$ we have $k \leqslant 2$,

and these bounds are attained.

In general we have

**Theorem 6.21** There is an integer $k(g)$ such that for all finite fields $\mathbb{F}_q$ and for all supersingular abelian varieties of dimension $g$ over $\mathbb{F}_q$ we have $k \leqslant k(g)$.

The number $k(g)$ can be found in [GAL 2001a] and in Section 24.2.2.

# Chapter 7

# Background on Weil Descent

*Gerhard Frey and Tanja Lange*

**Contents in Brief**

| | | |
|---|---|---|
| 7.1 | Affine Weil descent | 125 |
| 7.2 | The projective Weil descent | 127 |
| 7.3 | Descent by Galois theory | 128 |
| 7.4 | Zariski closed subsets inside of the Weil descent | 129 |
| | Hyperplane sections • Trace zero varieties • Covers of curves • The GHS approach | |

*Weil descent* — or, as it is alternatively called — *scalar restriction*, is a well-known technique in algebraic geometry. It is applicable to all geometric objects like curves, differentials, and Picard groups, if we work over a separable field $L$ of degree $d$ of a ground field $K$.

It relates $t$-dimensional objects over $L$ to $td$-dimensional objects over $K$. As guideline the reader should use the theory of algebraic curves over $\mathbb{C}$, which become surfaces over $\mathbb{R}$. This example, detailed in Section 5.1.2, already shows that the structure of the objects after scalar restriction can be much richer: the surfaces we get from algebraic curves carry the structure of a Riemann surface and so methods from topology and Kähler manifolds can be applied to questions about curves over $\mathbb{C}$.

This was the reason to suggest that Weil descent should be studied with respect to (constructive and destructive) applications for DL systems [FRE 1998]. We shall come to such applications in Sections 15.3 and 22.3.

In the next two sections we give a short sketch of the mathematical properties of Weil descent. The purpose is to provide a mathematical basis for the descent and show how to construct it. For a thorough discussion in the frame of algebraic geometry and using the language of schemes, we refer to [DIE 2001].

## 7.1 Affine Weil descent

We begin with the easiest case. Let $V$ be an affine variety in the affine space $\mathbb{A}_L^n$ over $L$ defined by $m$ equations

$$F_i(x_1,\ldots,x_n) = 0;\ i = 1,\ldots,m$$

with $F_i(x) \in L[x_1, \ldots, x_n]$.

We want to find an affine variety $W_{L/K}(V)$ defined over $K$ with the following properties:

(W1) For any field $K' \subset \overline{K}$ for which the degree of $L \cdot K'$ over $K'$ is equal to $d$ (i.e., $K'$ is linearly disjoint from $L$ over $K$) we have a natural identification of $W_{L/K}(V)(K')$ with $V(L \cdot K')$.

(W2) The variety $W_{L/K}(V)_L$ obtained from $W_{L/K}(V)$ by base extension from $K$ to $L$ is isomorphic to $V^d$, the $d$-fold Cartesian product of $V$ with itself.

To achieve this we choose a basis $\{u_1, \ldots, u_d\}$ of $L$ as $K$-vector space. Then we define the $nd$ variables $y_{i,j}$ by

$$x_i = u_1 y_{1,i} + \cdots + u_d y_{d,i}, \text{ for } i = 1, \ldots, n.$$

We replace the variables $x_i$ in the relations defining $V$ by these expressions.

Next we write the coefficients of the resulting relations as $K$-linear combinations of the basis $\{u_1, \ldots, u_d\}$ and order these relations according to this basis. As result we get $m$ equations of the form

$$G_i(y) = g_{i,1}(y) u_1 + \cdots + g_{i,d}(y) u_d = 0$$

with $g_{i,j} \in K[y_{1,1}, \ldots, y_{n,d}]$. Because of the linear independence of the elements $u_i$ and because of condition W1 we see that we have to define $W$ as the Zariski closed subset in $\mathbb{A}^{nd}$ given by the $md$ equations

$$g_{i,j}(y) = 0.$$

**Proposition 7.1** Let $V$ and $W$ be as above. Then $W$ is an affine variety defined over $K$ satisfying the conditions W1 and W2. So $W$ is the Weil descent $W_{L/K}(V)$ of $V$.

**Example 7.2** Let $V$ be equal to the affine space of dimension $n$ over $L$ with coordinate functions $x_1, \ldots, x_n$.

Then $W_{L/K}(V) = \mathbb{A}^{nd}$ with coordinate functions $y_{i,j}$ defined by

$$x_i = u_1 y_{1,i} + \cdots + u_d y_{d,i}.$$

As a special case, take $L = \mathbb{C}$ and $K = \mathbb{R}$, $n = 1$, and take as complex coordinate function the variable $z$ and as basis of $\mathbb{C}/\mathbb{R}$, the elements $1, i$ with $i^2 = -1$.

As usual we choose real variables $x, y$ satisfying the identity

$$z = x + iy.$$

A polynomial or more generally a rational function $G(z)$ in $z$ gives rise to a function in $G_{\mathbb{R}}(x, y)$ that we can interpret as a function from $\mathbb{R}^2$ to $\mathbb{C}$. We separate its real and imaginary part and get

$$G(z) = g_1(x, y) + i g_2(x, y).$$

**Example 7.3** Assume that $L = K(\alpha)$ with $\{1, \alpha, \alpha^2\}$ a basis of $L/K$ and $\alpha^3 = b \in K$ and assume that $\text{char}(K) \neq 3$.

Take the affine part of the elliptic curve given by the equation

$$E_a : x_1^2 - x_2^3 - 1 = 0.$$

Replace $x_i$ by $y_{1,i} + \alpha y_{2,i} + \alpha^2 y_{3,i}$ to get the equation

$$\left(y_{1,1} + \alpha y_{2,1} + \alpha^2 y_{3,1}\right)^2 - \left(y_{1,2} + \alpha y_{2,2} + \alpha^2 y_{3,2}\right)^3 - 1 = 0.$$

This yields the following system of equations

$$y_{1,1}^2 + 2by_{2,1}y_{3,1} - y_{2,1}^3 - b^2 y_{3,2}^3 - by_{2,3}^3 - 6by_{1,2}y_{2,2}y_{3,2} - 1 = 0$$
$$by_{1,3}^2 + 2y_{1,1}y_{2,1} - 3y_{1,2}^2 y_{2,2} - 3by_{2,2}^2 y_{3,2} - 3by_{1,2}y_{3,2}^2 = 0$$
$$y_{1,2}^2 + 2y_{1,1}y_{3,1} - 3y_{1,2}^2 y_{3,2} - 3y_{1,2}y_{2,2}^2 - 3by_{2,2}y_{3,2}^2 = 0$$

which defines the Weil descent $W_{L/K}(E_a)$ of $E_a$.

**Remark 7.4** The example is interesting since it is an open affine part of an abelian variety of dimension 3 defined over $K$, whose rational points are in a natural way equal to the $L$-rational points of the elliptic curve $E$.

## 7.2 The projective Weil descent

Having defined the Weil descent for affine varieties we proceed in the usual way to define it for projective varieties $V$ defined over $L$, which are embedded in some projective space $\mathbb{P}_L^n$.

We cover $V$ by affine subvarieties $V_i$ and apply the restriction of scalars to the $V_i$ to get a collection of affine varieties $W_{L/K}(V_i) =: W_i$ over $K$. The varieties $V_i$ are intersecting in Zariski open parts of $V$ and there are rational maps from $V_i$ to $V_j$ induced by the rational maps between the different embeddings of the affine space $\mathbb{A}_L^n$ into $\mathbb{P}_L^n$ (cf. Example 4.44). By using the functoriality properties of the Weil descent (or by a direct computation in the respective coordinates as in the examples) one concludes that the affine varieties $W_i$ can be glued together in a projective space (which is the Weil descent of $\mathbb{P}_L^n$). If we take the coverings fine enough we get as a result of the gluing process a projective variety $W_{L/K}(V)$.

*Warning.* Not every cover of $V$ by affine subvarieties $V_i$ has the property that the varieties $W_{L/K}(V_i)$ cover $W_{L/K}(V)$. For instance let $E$ be a plane projective elliptic curve given by the equation

$$E : Y^2 Z = X^3 + a_4 X Z^2 + a_6 Z^3.$$

Then $E$ is covered by the affine curves $E_1$ and $E_2$ one gets by intersecting $\mathbb{P}_L^2$ with the open parts for which $Z \neq 0$ (respectively $Y \neq 0$) holds. But we also need $E_3$, which is the intersection of $E$ with the open part of $\mathbb{P}^2$ defined by $X \neq 0$ to get $W_{L/K}(E)$ by the gluing procedure described above.

There is another complication if we want to describe the projective variety $W_{L/K}(V)$ explicitly as a subvariety of the projective space $\mathbb{P}^N$: the dimension of this space can become rather large. Here is an estimate for this dimension:

**Lemma 7.5** Let $V$ be a projective variety embedded into $\mathbb{P}_L^n$. Then $W_{L/K}(V)$ can be embedded (in a canonical way) into $\mathbb{P}^{(n+1)^d - 1}$.

This lemma follows from the construction via affine covers and the application of the Segre map (cf. Examples 4.13 and 4.25) of products of projective spaces into a projective space.

We can summarize our results and get the following theorem:

**Theorem 7.6** Let $L/K$ be a finite separable field extension of degree $d$. Let $V$ be an affine or a projective variety defined over $L$. The Weil restriction $W_{L/K}(V)$ satisfies the properties W1 and W2. If $V$ is affine (respectively projective) and has dimension $t$ then $W_{L/K}(V)$ is an affine (respectively projective) variety defined over $K$ of dimension $td$.

Again by functoriality properties one can conclude that the Weil restriction of an algebraic group is again an algebraic group. Hence we get:

**Corollary 7.7** The Weil restriction of an abelian variety $\mathcal{A}$ over $L$ is an abelian variety $W_{L/K}(\mathcal{A})$ over $K$.

Let $\mathcal{A}_1$ be a Zariski-open nonempty affine subvariety of $\mathcal{A}$. Then $W_{L/K}(\mathcal{A}_1)$ is an affine Zariski-open nonempty subvariety of $W_{L/K}(\mathcal{A})$ and hence it is birationally equivalent to $W_{L/K}(\mathcal{A})$.

This corollary justifies Remark 7.4.

## 7.3 Descent by Galois theory

In the last sections we have introduced an explicit method to construct the Weil descent of varieties by using affine coordinates. The advantage of this approach is the explicit definition of the Weil descent by equations. The disadvantage is that the number of variables and the number of relations grow and so the description becomes very complicated. This is especially striking if we want to apply the descent to projective varieties or if the degree of $L/K$ is not small. For many purposes it is enough to have the Weil descent and its properties as background. Then we apply it using definitions by Galois theory as this is much more elegant.

This approach becomes most natural if we assume that *$L/K$ is a Galois extension with relative Galois group $G(L/K) = G$*. Note that for us the most important case is that $K$ and $L$ are finite fields and then this assumption is always satisfied.

Let $V$ be a variety defined over $L$ and let $\sigma \in G$ be an automorphism of $L$ fixing $K$.

We want to define the image of $V$ under $\sigma$. We assume that $V$ is affine. If $V$ is projective one can proceed in a completely analogous way.

We choose affine coordinate functions $x = (x_1, \ldots, x_n)$ of $\mathbb{A}^n_L$ and define the points on $V$ as the set of zeroes of the equations defining $V$ as usual. Let $I$ be the prime ideal generated by these equations in $L[x]$. We apply $\sigma$ to the coefficients of rational functions $F$ in $L(x)$ and denote by $\sigma \cdot F$ the image.

The ideal $I_\sigma := \sigma \cdot I$ is again a prime ideal in $L[x]$ and so it defines an affine variety $V^\sigma$ over $L$. Let us extend $\sigma$ to an automorphism $\tilde{\sigma}$ of $\overline{K}$. By definition we get $\tilde{\sigma} \cdot I = \sigma \cdot I$ and so $V^{\tilde{\sigma}} = V^\sigma$ does not depend on the choice of the extension. Let $P$ be a point in $V(\overline{K})$. Then $\tilde{\sigma}(P)$ is a point in $V^\sigma(\overline{K})$ and conversely. So $V^\sigma(\overline{K}) = \tilde{\sigma} \cdot V(\overline{K})$.

For all points $Q \in V^\sigma(\overline{K})$ and $f \in L(V)$ we get the identity

$$(\sigma \cdot f)(Q) = \tilde{\sigma}\big(f\big(\tilde{\sigma}^{-1}(Q)\big)\big).$$

In particular, it follows that we can interpret $\sigma \cdot f$ as rational function on $V^\sigma$.

We apply this to the functions $x_i$. To clarify what we mean, we denote by $x_{i,\sigma}$ the function on $V^\sigma$ induced by the coordinate function $x_i$, i.e., $x_{i,\sigma}$ is the image of $x_i$ in $L[x]/(\sigma \cdot I)$. We get: let $P$ be a point in $V$ and let $x_i(P)$ be the value of the $i$-th coordinate function on $V$ applied to $P$. Let $x_{i,\sigma}$ be the $i$-th coordinate function on $V^\sigma$. Then $x_{i,\sigma} = \sigma \cdot x_i$ and the value of $x_{i,\sigma}$ applied to $\tilde{\sigma}(P)$ is equal to $\tilde{\sigma}(x_i(P))$.

All these considerations are near to tautological statements but they allow us to define an action of the absolute Galois group $G_K$ of $K$ on the variety

$$W := \prod_{\sigma \in G} V^\sigma.$$

Indeed let $P := (\ldots, P_\sigma, \ldots)_{\sigma \in G}$, with $P_\sigma \in V^\sigma(\overline{K})$, be a point in $W(\overline{K})$. Let $\tilde{\tau}$ be an element of $G_K$ whose restriction to $L$ is equal to $\tau$. Then

$$\tilde{\tau}(P) := (\ldots, Q_\sigma, \ldots)_{\sigma \in G} \text{ with } Q_\sigma = \tilde{\tau}(P_{\tau^{-1} \circ \sigma}).$$

**Theorem 7.8** The variety $W$ is equal to the Weil restriction $W_{L/K}(V)$.

**Proof.** To prove this theorem we check the properties characterizing the Weil descent.

First $W$ is a variety defined over $L$. As we have seen its set of points is invariant under the action of $G_K$. So $W$ is a variety defined over $K$.

A point $P \in W$ is $K$-rational if and only if it is $L$-rational and for all $\tau \in G$ we have

$$\tau(P_{\tau^{-1}\circ\sigma}) = P_\sigma.$$

Taking $\tau = \sigma^{-1}$ this means that $P_\sigma = \sigma^{-1} P_{\mathrm{Id}}$ for all $\sigma \in G$ with $P \in V(L)$. It follows that $W(K) = V(L)$.

Next we extend the ground field $K$ to $L$ and look at $W_L$. On the Galois theoretic side this means that we restrict the Galois action of $G_K$ on $W$ to an action of $G_L$. But this group leaves each $V^\sigma$ invariant and so $W_L$ is isomorphic to $\prod_{\sigma \in G} V = V^d$. $\square$

We shall be interested in the special case that $K = \mathbb{F}_q$ and $L = \mathbb{F}_{q^d}$.

**Corollary 7.9** Let $V$ be a (projective or affine) variety defined over $\mathbb{F}_{q^d}$ of dimension $t$. For $i = 0, \ldots, d-1$ let $V_i$ be the image of $V$ with respect to $\phi_q^i$ (cf. Proposition 5.67).
Then

$$W(V) := \prod_{i=0}^{d-1} V_i$$

is a variety defined over $\mathbb{F}_q$ of dimension $td$ which is $K$-isomorphic to $W_{\mathbb{F}_{q^d}/\mathbb{F}_q}(V)$.

If $V$ is affine (respectively projective) then $W(V)$ is an affine (respectively projective) variety defined over $K$.

If $V$ is an abelian variety over $\mathbb{F}_{q^d}$ then $W(V)$ is an abelian variety over $\mathbb{F}_q$.

The action of $\phi$ on $W(V)$ is given as follows: Let $P = (\ldots, P_i, \ldots)$ be a point in $W(V)(\overline{K})$. Then $\phi_q(P) = (\ldots, Q_i, \ldots)$ with $Q_i = \phi_q(P)_{(i-1 \bmod d)}$.

**Remark 7.10** In general the Weil restriction of a Jacobian variety is not a Jacobian variety.

## 7.4 Zariski closed subsets inside of the Weil descent

As mentioned already, one main application of the Weil descent method is that in $W_{L/K}$ there are Zariski closed subsets which cannot be defined in $V$.

In the following we shall describe strategies to find such subsets.

### 7.4.1 Hyperplane sections

To simplify the discussion we assume that $V$ is affine with coordinate functions $x_1, \ldots, x_n$ and we take the description of $W_{L/K}(V)$ given in Proposition 7.1. There we have introduced $nd$ coordinates functions $y_{i,j}$ for $W_{L/K}(V)$ by

$$x_i = u_1 y_{1,i} + \cdots + u_d y_{d,i}, \text{ for } i = 1, \ldots, n,$$

where $\{u_1, \ldots, u_d\}$ is a basis of $L/K$. Take $J \subset \{1, \ldots, d\} \times \{1, \ldots, n\}$ and adjoin the equations $y_{i,j} = 0$ for $(i,j) \in J$ to the equations defining $W_{L/K}(V)$.

The resulting Zariski closed set inside of $W_{L/K}(V)$ is denoted by $W_J$. It is the intersection of the Weil restriction of $V$ with the affine hyperplanes defined by $y_{i,j} = 0$; $(i,j) \in J$.

"In general" we can expect that $W_J$ is again a variety over $K$ of dimension $td - |J|$.

**Example 7.11** Let $E$ be an elliptic curve defined over $L$ given by a Weierstraß equation

$$E : x_1^2 + a_1 x_1 x_2 + a_3 x_1 = f(x_2),$$

where $f$ is monic of degree 3. Take $1 \leqslant m \leqslant d-1$ and $J = \{1,\ldots,m\} \times \{2\}$.

Then $W_J(K)$ consists of all points in $E(L)$ whose $x_2$-coordinate is a $K$-linear combination of the elements $u_1, \ldots, u_m$.

**Remark 7.12** This example is the mathematical background of a subexponential attack to the discrete logarithm in elliptic curves over nonprime fields found recently by Gaudry and Diem (cf. Section 22.3.5).

### 7.4.2 Trace zero varieties

We assume for simplicity that $L = \mathbb{F}_{q^d}$ and $K = \mathbb{F}_q$ and we use the Galois theoretic description of the Weil descent.

Let $V$ be a variety defined over $K$. So we get $V^{\phi_q} = V$. Note that nevertheless $W_{\mathbb{F}_{q^d}/\mathbb{F}_q}(V)$ is *not* $\mathbb{F}_q$-isomorphic to $V^d$ because of the twisted Galois operation. But we can embed $V$ into $W_{\mathbb{F}_{q^d}/\mathbb{F}_q}(V)$ as diagonal:

Map the point $P \in V(\overline{K})$ to the point $(\ldots, \phi_q^i(P), \ldots) \in \prod_{i=0}^{d-1} V$. By this map we can identify $V$ with a subvariety of $W_{\mathbb{F}_{q^d}/\mathbb{F}_q}(V)$.

Now assume in addition that $V = \mathcal{A}$ is an abelian variety. Then we find a complementary abelian subvariety to $\mathcal{A}$ inside of $W_{\mathbb{F}_{q^d}/\mathbb{F}_q}(\mathcal{A})$.

We use the existence of an automorphism $\pi$ of order $d$ of $W_{\mathbb{F}_{q^d}/\mathbb{F}_q}(\mathcal{A})$ defined by

$$P = (\ldots, P_i, \ldots) \mapsto \pi(P) = (\ldots, Q_i, \ldots) \text{ with } Q_i = P_{i-1 \bmod d}.$$

The map $\pi$ is obviously an automorphism over $\mathbb{F}_{q^d}$. To prove that $\pi$ is defined over $\mathbb{F}_q$ we have to show that $\pi$ commutes with the action of $\phi_q$. But

$$\pi(\phi_q(P)) = (\ldots, Q_i', \ldots) \text{ with } Q_i' = \phi_q(Q_{i-2 \bmod d})$$

and this is equal to $\phi_q(\pi(P))$.

Denote by $\mathcal{A}_0$ the kernel of the endomorphism $\sum_{i=0}^{d-1} \pi^i$. It is an abelian subvariety of $\mathcal{A}$ and it is called the *trace zero subvariety* of $\mathcal{A}$. Note that the intersection set of $\mathcal{A}$ — embedded as diagonal into $W_{\mathbb{F}_{q^d}/\mathbb{F}_q}(\mathcal{A})$ — with $\mathcal{A}_0$ consists of the points of $\mathcal{A}$ of order dividing $d$, and the $\mathbb{F}_q$-rational points of $\mathcal{A}_0$ are the points $P$ in $\mathcal{A}(\mathbb{F}_{q^d})$ with $\text{Tr}(\phi_q)(P) = 0$.

To see that $\mathcal{A}$ and $\mathcal{A}_0$ generate $W_{\mathbb{F}_{q^d}/\mathbb{F}_q}(\mathcal{A})$ we use that $\mathcal{A}$ is the kernel of $\pi - \text{Id}$ and that $\mathcal{A}_0$ contains $(\pi - \text{Id})(W_{\mathbb{F}_{q^d}/\mathbb{F}_q}(\mathcal{A}))$.

We summarize:

**Proposition 7.13** Let $\mathcal{A}$ be an abelian variety defined over $\mathbb{F}_q$. We use the product representation of $W_{\mathbb{F}_{q^d}/\mathbb{F}_q}(\mathcal{A})$ and define $\pi$ as automorphism induced by a cyclic permutation of the factors. Then we have the following results:

1. $\mathcal{A}$ can be embedded (as diagonal) into $W_{\mathbb{F}_{q^d}/\mathbb{F}_q}(\mathcal{A})$. Its image under this embedding is the kernel of $\pi - \text{Id}$.
2. The image of $\pi - \text{Id}$ is the trace zero subvariety $\mathcal{A}_0$.
3. The $\mathbb{F}_q$-rational points of $\mathcal{A}_0$ are the images of points $P \in \mathcal{A}(\mathbb{F}_{q^d})$ with $\text{Tr}(\pi)(P) = 0$.
4. Inside of $W_{\mathbb{F}_{q^d}/\mathbb{F}_q}(\mathcal{A})$ the subvarieties $\mathcal{A}$ and $\mathcal{A}_0$ intersect in the group of points of $\mathcal{A}$ of order dividing $d$.

For an example with $\mathcal{A} = E$ an elliptic curve and $d = 3$ we refer to [FRE 2001]; for $\mathcal{A} = J_C$ being the Jacobian of a hyperelliptic curve $C$, see [LAN 2004c]. We further investigate these constructive applications of Weil descent in Section 15.3.

### 7.4.3 Covers of curves

Let $C$ be a curve defined over $\mathbb{F}_{q^d}$ with Jacobian variety $J_C$. We want to apply Weil descent to get information about $\text{Pic}^0_C$ from $W_{\mathbb{F}_{q^d}/\mathbb{F}_q}(J_C)(\mathbb{F}_q)$.

Here we investigate the idea of looking for curves $C'$ defined over $\mathbb{F}_q$ that are embedded into $W_{\mathbb{F}_{q^d}/\mathbb{F}_q}(J_C)$. Then the Jacobian of $C'$ has $W_{\mathbb{F}_{q^d}/\mathbb{F}_q}(J_C)$ as a factor and we can use information about $\text{Pic}^0_{C'}$ to study $\text{Pic}^0_C$. Of course this is only a promising approach if the genus of $C'$ is not too large.

One can try to construct $C'$ directly, for instance, by using hyperplane sections. But it is very improbable that this will work if we are not in very special situations. Hence, it is not clear whether this variant can be used in practice. But this approach leads to interesting mathematical questions:

- Which abelian varieties have curves of small genus as sub-schemes?
- Which curves can be embedded into Jacobian varieties of modular curves?
- Which curves have the scalar restriction of an abelian variety (e.g., an elliptic curve) as Jacobian?

In [BODI+ 2004] one finds families of curves for which the last question is answered positively.

### 7.4.4 The GHS approach

In practice another approach is surprisingly successful. *A priori* it has nothing to do with Weil descent, but as a background and in order to prove results the Weil descent method is useful.

Let $L$ be a Galois extension of the field $K$. In our applications we shall take $L = \mathbb{F}_{q^d}$ and $K = \mathbb{F}_q$. Assume that $C$ is a projective irreducible nonsingular curve *defined over* $L$, and $D$ is a projective irreducible nonsingular curve *defined over* $K$.

Let
$$\varphi : D_L \to C$$

be a nonconstant morphism defined over $L$. As usual we denote by $\varphi^*$ the induced map from $\text{Pic}^0_C$ to $\text{Pic}^0_{D_L}$. It corresponds to the conorm map of divisors in the function fields $\varphi^*(L(C)) \subset L(D_L)$. Next we use the inclusion $K(D) \subset L(D_L)$ to define a correspondence map on divisor classes

$$\psi : \text{Pic}^0(C) \to \text{Pic}^0(D)$$

given by
$$\psi := \text{N}_{L/K} \circ \varphi^*,$$

where $\text{N}_{L/K}$ is the norm of $L/K$.

Assume that we are interested in a subgroup $G$ (for instance, of large prime order $\ell$) in $\text{Pic}^0_C$ and assume that we can prove that $G \cap \ker(\psi) = \{0\}$. Then we have transferred the study of $G$ as subgroup of a Jacobian variety over $L$ to the study of a subgroup of a Jacobian variety over $K$ which may be easier.

The relation with the Weil descent method is that by the Weil descent of the cover map $\varphi$ we get an embedding of $D$ into $W_{L/K}(J_C)$. This method is the background of the so-called GHS algorithm. We shall come to this in more detail in 22.3.2.

The mathematically interesting aspect of this method is that it relates the study of Picard groups of curves to the highly interesting theory of fundamental groups of curves over non-algebraically closed ground fields.

# Chapter 8

## Cohomological Background on Point Counting

*David Lubicz and Frederik Vercauteren*

### Contents in Brief

| | | |
|---|---|---|
| 8.1 | **General principle** <br> Zeta function and the Weil conjectures • Cohomology and Lefschetz fixed point formula | 133 |
| 8.2 | **Overview of $\ell$-adic methods** | 137 |
| 8.3 | **Overview of $p$-adic methods** <br> Serre–Tate canonical lift • Monsky–Washnitzer cohomology | 138 |

## 8.1 General principle

Let $p$ be a prime and $\mathbb{F}_q$ a finite field with $q = p^d$ elements. Consider a projective nonsingular curve $C$ defined over $\mathbb{F}_q$, e.g., an elliptic or hyperelliptic curve. Let $\mathbb{F}_{q^k}$ be a finite extension of $\mathbb{F}_q$ of degree $k$, then a point on $C$ is called $\mathbb{F}_{q^k}$-rational if a representative of its homogeneous coordinates is defined over $\mathbb{F}_{q^k}$. Let $P$ be a point in $C(\overline{\mathbb{F}}_q)$ and denote with $\phi_q$ the Frobenius morphism, then $\phi_q(P) = P$ if and only if $P$ is $\mathbb{F}_q$-rational. More generally, the number of $\mathbb{F}_{q^k}$-rational points on $C$ is the number of fixed points of $\phi_q^k$. A first natural question to ask is thus: how to efficiently compute $|C(\mathbb{F}_{q^k})|$ for any positive $k$.

As described in Section 4.4.4, we can embed $C$ in a projective group variety over $\mathbb{F}_q$, called the Jacobian variety of $C$ and denoted by $J_C$. The Frobenius morphism $\phi_q$ then induces an isogeny of $J_C$, also denoted by $\phi_q$, and clearly $J_C(\mathbb{F}_q) = \ker(\phi_q - [1])$. A second natural question to ask is thus: how to efficiently compute $|J_C(\mathbb{F}_{q^k})|$ for any positive $k$.

This chapter shows that the above questions are in fact closely related, and introduces different approaches to solving both of them. The general strategy is based on ideas introduced by Weil, Serre, Grothendieck, Dwork, etc. in order to prove the Weil conjectures (see Section 8.1.1). The main idea is the following: the number of $\mathbb{F}_{q^k}$-rational points on $C$ or $J_C$ is the number of fixed points of $\phi_q^k$. In the complex setting, there exists a general formula due to Lefschetz for the number of fixed points of an analytic map.

**Theorem 8.1 (Lefschetz Fixed Point Theorem)** Let $M$ be a compact complex analytic manifold and $f : M \to M$ an analytic map. Assume that $f$ only has isolated nondegenerate fixed points; then

$$\left|\{P \in M \mid f(P) = P\}\right| = \sum_i (-1)^i \operatorname{Tr}\bigl(f^*; H^i_{DR}(M)\bigr).$$

The $H^i_{DR}(M)$ in the above theorem are the de Rham cohomology groups of $M$ and are finite dimensional vector spaces over $\mathbb{C}$ on which $f$ induces a linear map $f^*$. The number of fixed points of $f$ is thus the alternating sum of the traces of the linear map $f^*$ on the vector spaces $H^i_{DR}(M)$.

The dream of Weil was to mimic this situation for varieties over finite fields, i.e., construct a good cohomology theory (necessarily over a characteristic zero field) such that the number of fixed points of the Frobenius morphism is given by a Lefschetz fixed point formula.

The different approaches described in this chapter all fit in the following slightly more general framework: construct vector spaces over some characteristic zero field together with an action of the Frobenius morphism $\phi_q$ that provides information about the number of fixed points of $\phi_q$ and thus the number of $\mathbb{F}_q$-rational points on $C$ or $J_C$.

### 8.1.1 Zeta function and the Weil conjectures

Let $\mathbb{F}_q$ be a finite field with $q = p^d$ and $p$ prime. For any algebraic variety $X$ defined over $\mathbb{F}_q$, let $N_k$ denote the number of $\mathbb{F}_{q^k}$-rational points on $X$.

**Definition 8.2** The *zeta function* $Z(X/\mathbb{F}_q; T)$ of $X$ over $\mathbb{F}_q$ is the generating function

$$Z(X/\mathbb{F}_q; T) = \exp\left(\sum_{k=1}^\infty \frac{N_k}{k} T^k\right).$$

The zeta function should be interpreted as a formal power series with coefficients in $\mathbb{Q}$. In 1949, Weil [WEI 1949] stated the following conjectures, all of which have now been proven.

**Theorem 8.3 (Weil Conjectures)** Let $X$ be a smooth projective variety of dimension $n$ defined over a finite field with $q$ elements.

1. Rationality: $Z(X/\mathbb{F}_q; T) \in \mathbb{Q}[[T]]$ is a rational function.
2. Functional equation: $Z(T) = Z(X/\mathbb{F}_q; T)$ satisfies

$$Z\left(\frac{1}{q^n T}\right) = \pm q^{nE/2} T^E Z(T),$$

with $E$ equal to the Euler–Poincaré characteristic of $X$, i.e., the intersection number of the diagonal with itself in the product $X \times X$.

3. Riemann hypothesis: there exist polynomials $P_i(T) \in \mathbb{Z}[T]$ for $i = 0, \ldots, 2n$, such that

$$Z(X/\mathbb{F}_q; T) = \frac{P_1(T) \cdots P_{2n-1}(T)}{P_0(T) \cdots P_{2n}(T)}$$

with $P_0(T) = 1 - T$, $P_{2n}(T) = 1 - q^n T$ and for $1 \leqslant r \leqslant 2n - 1$

$$P_r(T) = \prod_{i=1}^{\beta_r} (1 - \alpha_{r,i} T)$$

where the $\alpha_{r,i}$ are algebraic integers of absolute value $q^{r/2}$.

## § 8.1 General principle

Weil [WEI 1948] proved these conjectures for curves and abelian varieties. The rationality of the zeta function of any algebraic variety was settled in 1960 by Dwork [DWO 1960] using $p$-adic methods. Soon after, the Grothendieck school developed $\ell$-adic cohomology and gave another proof of the rationality and the functional equation. Finally, in 1973, Deligne [DEL 1974] proved the Riemann hypothesis.

Let $\phi_q$ be the Frobenius endomorphism of $J_C$, then the elements fixed by $\phi_q$ are exactly $J_C(\mathbb{F}_q)$ or $\ker(\phi_q - [1]) = J_C(\mathbb{F}_q)$. As introduced in Section 5.2.2, we can associate to $\phi_q$ its characteristic polynomial $\chi(\phi_q)_C$, which is a monic polynomial of degree $2g$ with coefficients in $\mathbb{Z}$. Furthermore, by Corollary 5.70 we have that $|J_C(\mathbb{F}_q)| = \chi(\phi_q)_C(1)$.

The relation between $\chi(\phi_q)_C$ and the zeta function of the smooth projective curve $C$ is as follows:

**Proposition 8.4** Let $C$ be a smooth projective curve of genus $g$ and let $\chi(\phi_q)_C(T) \in \mathbb{Z}[T]$ be the characteristic polynomial of $\phi_q$. Define the *L-polynomial* of $C$ by

$$L(T) = T^{2g} \chi(\phi_q)_C\left(\frac{1}{T}\right),$$

then the zeta function of $C$ is given by

$$Z(C/\mathbb{F}_q; T) = \frac{L(T)}{(1-T)(1-qT)}.$$

Let $L(T) = a_0 + a_1 T + \cdots + a_{2g} T^{2g}$, then the functional equation shows that $a_{2g-i} = q^{g-i} a_i$ for $i = 0, \ldots, g$. If we write $L(T) = \prod_{i=1}^{2g}(1 - \alpha_i T)$, then the Riemann hypothesis implies $|\alpha_i| = \sqrt{q}$ and again by the functional equation, we can label the $\alpha_i$ such that $\alpha_i \alpha_{i+g} = q$ for $i = 0, \ldots, g$. This shows that Theorem 5.76 immediately follows from the Weil conjectures.

Taking the logarithm of both expressions for the zeta function leads to

$$\ln Z(C/\mathbb{F}_q; T) = \sum_{k=1}^{\infty} \frac{N_k}{k} T^k = \sum_{i=1}^{2g} \ln(1 - \alpha_i T) - \ln(1-T) - \ln(1-qT).$$

Since $\ln(1 - sT) = -\sum_{i=1}^{\infty} \frac{(sT)^k}{k}$, we conclude that for all positive $k$

$$N_k = q^k + 1 - \sum_{i=1}^{2g} \alpha_i^k.$$

The zeta function $Z(C/\mathbb{F}_q; T)$ of a curve $C$ contains important geometric information about $C$ and its Jacobian $J_C$. For example, Stichtenoth [STI 1979] proved the following theorem.

**Theorem 8.5** Let $L(T) = a_0 + \cdots + a_{2g} T^{2g}$, then the $p$-rank of $J_C$ is equal to

$$\max\{i \mid a_i \not\equiv 0 \pmod{p}\}$$

Furthermore, Stichtenoth and Xing [STXI 1995] showed that $J_C$ is supersingular, i.e., isogenous over $\overline{\mathbb{F}}_q$ to a product of supersingular elliptic curves, if and only if $p^{\lceil dk/2 \rceil} \mid a_k$ for all $1 \leq k \leq g$.

### 8.1.2 Cohomology and Lefschetz fixed point formula

In this section we indicate how the Weil conjectures, except for the Riemann hypothesis, almost immediately follow from a good cohomology theory. Let $X$ be a projective, smooth algebraic variety of dimension $n$ over a finite field $\mathbb{F}_q$ of characteristic $p$ and let $\overline{X} \otimes_{\mathbb{F}_q} \overline{\mathbb{F}}_q$ denote the corresponding variety over the algebraic closure $\overline{\mathbb{F}}_q$ of $\mathbb{F}_q$.

Let $\ell$ denote a prime different from $p$ and let $\mathbb{Q}_\ell$ be the field of $\ell$-adic numbers. Grothendieck introduced the $\ell$-adic cohomology groups $H^i(\overline{X}, \mathbb{Q}_\ell)$ (see [SGA 4]), which he used to prove the rationality and functional equation of the zeta function. The description of these cohomology groups is far beyond the scope of this book and we will simply state their main properties. However, for $X$, a smooth projective curve, we have the following theorem.

**Theorem 8.6** Let $C$ be a smooth projective curve over a finite field $\mathbb{F}_q$ of characteristic $p$ and let $\ell$ be a prime different from $p$; then there exists an isomorphism
$$H^1(C, \mathbb{Z}_\ell) \simeq T_\ell(J_C).$$

To prove the rationality of the zeta function and the factorization of its numerator and denominator, we only need the following two properties:

- The $\ell$-adic cohomology groups $H^i(\overline{X}, \mathbb{Q}_\ell)$ are finite dimensional vector spaces over $\mathbb{Q}_\ell$ and $H^i(\overline{X}, \mathbb{Q}_\ell) = 0$ for $i < 0$ and $i > 2n$.
- Let $f : \overline{X} \to \overline{X}$ be a morphism with isolated fixed points and suppose moreover that each fixed point has multiplicity 1. Then the number $N(f, X)$ of fixed points of $f$ is given by a Lefschetz fixed point formula:
$$N(f, X) = \sum_{i=0}^{2n} (-1)^i \operatorname{Tr}\left(f^*; H^i(\overline{X}, \mathbb{Q}_\ell)\right).$$

Recall that the number $N_k$ of $\mathbb{F}_{q^k}$-rational points on $X$ equals the number of fixed points of $\phi_q^k$ with $\phi_q$ the Frobenius morphism. By the Lefschetz fixed point formula, we have
$$N_k = \sum_{i=0}^{2n} (-1)^i \operatorname{Tr}\left(\phi_q^{k*}; H^i(\overline{X}, \mathbb{Q}_\ell)\right).$$

Substituting this in the definition of the zeta function proves the following theorem.

**Theorem 8.7** Let $X$ be a projective, smooth algebraic variety over $\mathbb{F}_q$ of dimension $n$, then
$$Z(X/\mathbb{F}_q, T) = \frac{P_1(T) \ldots P_{2n-1}(T)}{P_0(T) \ldots P_{2n}(T)}$$
with
$$P_i(T) = \det(1 - \phi_q^* T; H^i(\overline{X}, \mathbb{Q}_\ell)).$$

The above theorem constitutes the first cohomological approach to computing the zeta function of a projective, smooth algebraic variety: construct a basis for the $\ell$-adic cohomology groups $H^i(\overline{X}, \mathbb{Q}_\ell)$ and compute the characteristic polynomial of the representation of $\phi_q$ on $H^i(\overline{X}, \mathbb{Q}_\ell)$. Unfortunately, the definition of the $H^i(\overline{X}, \mathbb{Q}_\ell)$ is very abstract and thus useless from an algorithmic point of view. For curves not all is lost, since by Theorem 8.6 we have the isomorphism $H^1(C, \mathbb{Z}_\ell) \simeq T_\ell(J_C)$.

The second cohomological approach constructs $p$-adic cohomology groups defined over the unramified extension $\mathbb{Q}_q$ of $\mathbb{Q}_p$. Several different theories that satisfy a Lefschetz fixed point formula exist, e.g., Monsky–Washnitzer cohomology [MOWA 1968, MON 1968, MON 1971], Lubkin's $p$-adic cohomology [LUB 1968], crystalline cohomology by Grothendieck [GRO 1968] and Berthelot [BER 1974], and finally, rigid cohomology by Berthelot [BER 1986]. The main algorithmic advantage over the $\ell$-adic cohomology theory is the existence of comparison theorems that provide an isomorphism with the algebraic de Rham cohomology, i.e., modules of differentials modulo exact differentials. The algebraic de Rham cohomology itself is very computational in nature and thus more amenable to computations than $\ell$-adic cohomology. The main disadvantage of the $p$-adic approach is that the complexity of the resulting algorithms is exponential in $p$.

## 8.2 Overview of $\ell$-adic methods

Let $\mathbb{F}_q$ be a finite field of characteristic $p$ and let $C$ be a smooth, projective curve defined over $\mathbb{F}_q$ of genus $g$. The Jacobian variety $J_C$ of $C$ then is an abelian variety over $\mathbb{F}_q$ of dimension $g$. Let $\chi(\phi_q)_C(T) \in \mathbb{Z}[T]$ be the characteristic polynomial of the Frobenius endomorphism $\phi_q$ acting on the Tate module $T_\ell(J_C)$ for $\ell$ a prime different from $p$, then we can write

$$\chi(\phi_q)_C(T) = \sum_{i=0}^{2g} a_{2g-i} T^i \text{ with } a_0 = 1 \text{ and } a_{2g} = q^g.$$

By the functional equation of the zeta function we have $a_{2g-i} = q^{g-i} a_i$ for $i = 0, \ldots, g$, so it suffices to determine the $a_i$ for $i = 0, \ldots, g$.

The main idea of the $\ell$-adic methods is to approximate $T_\ell(J_C)$ by the $\ell$-torsion points $J_C[\ell]$. Recall that since $\ell \neq p$, the $\ell$-torsion is a $2g$ dimensional vector space over $\mathbb{Z}/\ell\mathbb{Z}$ and the restriction of $\phi_q$ to $J_C[\ell]$ is a linear transformation of this vector space. Let $P_\ell(T)$ denote the characteristic polynomial of this restriction, then by Lemma 5.71 we have $P_\ell(T) \equiv \chi(\phi_q)_C(T) \pmod{\ell}$.

Furthermore, by the Riemann hypothesis or Corollary 5.82, the coefficients $a_0, \ldots, a_g$ are bounded by

$$|a_i| \leqslant \binom{2g}{i} q^{i/2} \leqslant \binom{2g}{g} q^{g/2}.$$

Using the Chinese remainder theorem, we can therefore uniquely recover $\chi(\phi_q)_C$ from $P_\ell(T)$ for primes $\ell \leqslant H \ln q$ with $H$ a constant such that

$$\prod_{\substack{\text{primes } \ell \leqslant H \ln q \\ \gcd(\ell, q) = 1}} \ell > 2 \binom{2g}{g} q^{g/2}.$$

The constant $H$ only depends on the genus $g$ and the prime number theorem implies that $H$ is linear in $g$.

For a given prime $\ell$, the polynomial $P_\ell$ can be computed as follows. Assume that $J_C$ is embedded in $\mathbb{P}^N$ and that an affine part is defined by a system of polynomial equations $F_1, \ldots, F_s \in \mathbb{F}_q[X]$ with $X = (X_1, \ldots, X_N)$. Furthermore, assume that the addition law is explicitly given by an $N$-tuple of rational functions $(G_1(X, Y), \ldots, G_N(X, Y))$ with $Y = (Y_1, \ldots, Y_N)$. Using the double and add method, we can compute a set of polynomials $Q_1^\ell, \ldots, Q_{k_\ell}^\ell$ generating the ideal of the subvariety of $\ell$-torsion points of $J_C$. Let $I_\ell$ be the radical ideal of $\langle F_1, \ldots, F_s, Q_1^\ell, \ldots, Q_{k_\ell}^\ell \rangle$; to recover $P_\ell$ we have to find integers $0 \leqslant a_i < \ell$ for $i = 0, \ldots, 2g$ such that

$$\sum_{i=0}^{2g} [a_{2g-i}](X_1^{q^i}, \ldots, X_N^{q^i}) \in I_\ell,$$

where the addition in the above equation is the group law on $J_C$ and $[m]$ denotes multiplication by $m$ on $J_C$. The resulting algorithm has complexity $O\bigl((\lg q)^\Delta\bigr)$, where $\Delta$ only depends on $g$, the dimension of the embedding space $\mathbb{P}^N$, the number and degrees of the defining equations of $J_C$ and the group law. More details about this approach can be found in [PIL 1990].

**Remark 8.8** Although the above algorithm has a polynomial time complexity in $\lg q$, it is currently only practical for elliptic curves and hyperelliptic curves of genus 2. The reason for this is that the degrees of the polynomials $Q_1^\ell, \ldots, Q_{k_\ell}^\ell$ grow as $O(\ell^{2g})$, since $J_C[\ell] \simeq (\mathbb{Z}/\ell\mathbb{Z})^{2g}$. However, for elliptic curves, the above algorithm can be improved substantially by restricting to a subgroup of $J_C[\ell]$, which is the kernel of an isogeny of degree $\ell$. Further details can be found in Section 17.2.

## 8.3 Overview of $p$-adic methods

The best known application of $p$-adic methods in algebraic geometry is undoubtedly Dwork's ingenious proof of the rationality of the zeta function [DWO 1960]. Although Dwork's proof can be transformed easily in an algorithm to compute the zeta function of any algebraic variety, nobody seemed to realize this and for more than a decade only $\ell$-adic algorithms were used.

At the end of 1999, Satoh [SAT 2000] introduced the $p$-adic approach into computational algebraic geometry by describing a $p$-adic algorithm to compute the number of points on an ordinary elliptic curve over a finite field. Following this breakthrough development, many existing $p$-adic theories were used as the basis for new algorithms:

- Dwork's $p$-adic analytic methods by Lauder and Wan [LAWA 2002b]
- Serre–Tate canonical lift by Satoh [SAT 2000], Mestre [MES 2000b], etc.
- Monsky–Washnitzer cohomology by Kedlaya [KED 2001]
- Dwork–Reich cohomology by Lauder and Wan [LAWA 2002a, LAWA 2004]
- Dwork's deformation theory by Lauder [LAU 2004].

Finally, we note that the use of $p$-adic methods as the basis for an algorithm to compute the zeta function of an elliptic curve already appeared in the work of Kato and Lubkin [KALU 1982].

In this section we will only review the two $p$-adic theories that are most important for practical applications, namely the Serre–Tate canonical lift and Monsky–Washnitzer cohomology.

### 8.3.1 Serre–Tate canonical lift

Let $\overline{\mathcal{A}}$ be an abelian variety defined over $\mathbb{F}_q$ with $q = p^d$ and $p$ a prime. Let $\mathbb{Q}_q$ be an unramified extension of $\mathbb{Q}_p$ of degree $d$ with valuation ring $\mathbb{Z}_q$ and residue field $\mathbb{Z}_q/(p\mathbb{Z}_q) \simeq \mathbb{F}_q$. Consider an arbitrary lift $\mathcal{A}$ of $\overline{\mathcal{A}}$ defined over $\mathbb{Z}_q$, i.e., $\mathcal{A}$ reduces to $\overline{\mathcal{A}}$ modulo $p$, then in general there will not exist an endomorphism $\mathcal{F} \in \mathrm{End}(\mathcal{A})$ that reduces to the $q$-th power Frobenius endomorphism $\phi_q \in \mathrm{End}(\overline{\mathcal{A}})$.

**Definition 8.9** A *canonical lift* of an abelian variety $\overline{\mathcal{A}}$ over $\mathbb{F}_q$ is an abelian variety $\mathcal{A}$ over $\mathbb{Q}_q$ such that $\mathcal{A}$ reduces to $\overline{\mathcal{A}}$ modulo $p$ and the ring homomorphism $\mathrm{End}(\mathcal{A}) \longrightarrow \mathrm{End}(\overline{\mathcal{A}})$ induced by reduction modulo $p$ is an isomorphism.

This definition implies that if $\overline{\mathcal{A}}$ admits a canonical lift $\mathcal{A}_c$, then there exists a lift $\mathcal{F} \in \mathrm{End}(\mathcal{A}_c)$ of the Frobenius endomorphism $\phi_q \in \mathrm{End}(\overline{\mathcal{A}})$. In fact, the reverse is also true: let $\mathcal{A}$ be a lift of $\overline{\mathcal{A}}$ and assume that $\mathcal{F} \in \mathrm{End}(\mathcal{A})$ reduces to $\phi_q \in \mathrm{End}(\overline{\mathcal{A}})$, then $\mathcal{A}$ is a canonical lift of $\overline{\mathcal{A}}$. Deuring [DEU 1941] proved that for an ordinary elliptic curve, a canonical lift always exists and is unique up to isomorphism. The question of existence and uniqueness of the canonical lift for general abelian varieties was settled by Lubin, Serre and Tate [LUSE$^+$ 1964].

**Theorem 8.10 (Lubin–Serre–Tate)** Let $\overline{\mathcal{A}}$ be an *ordinary* abelian variety over $\mathbb{F}_q$. Then there exists a canonical lift $\mathcal{A}_c$ of $\overline{\mathcal{A}}$ over $\mathbb{Z}_q$ and $\mathcal{A}_c$ is unique up to isomorphism.

## § 8.3 Overview of $p$-adic methods

Recall that an abelian variety $\overline{\mathcal{A}}$ is ordinary if it has maximal $p$-rank, i.e., $\overline{\mathcal{A}}[p] = (\mathbb{Z}/p\mathbb{Z})^{\dim(\overline{\mathcal{A}})}$.

The construction of a $p$-adic approximation of $\mathcal{A}_c$ given $\overline{\mathcal{A}}$ proceeds as follows: let $\mathcal{A}_0$ be a lift of $\overline{\mathcal{A}}$ to $\mathbb{Z}_q$ and denote with $\pi : \mathcal{A}_0 \to \overline{\mathcal{A}}$ reduction modulo $p$. Consider the subgroup $\mathcal{A}_0[p]^{\text{loc}} = \mathcal{A}_0[p] \cap \ker(\pi)$, i.e., the $p$-torsion points on $\mathcal{A}_0$ that reduce to the neutral element of $\overline{\mathcal{A}}$. As shown by Carls [CAR 2003], $\mathcal{A}_1 = \mathcal{A}_0/\mathcal{A}_0[p]^{\text{loc}}$ is again an abelian variety such that its reduction is ordinary and there exists an isogeny $I_0 : \mathcal{A}_0 \longrightarrow \mathcal{A}_1$, which reduces to the $p$-th power Frobenius morphism $\sigma : \overline{\mathcal{A}} \longrightarrow \overline{\mathcal{A}}^\sigma$. Repeating this construction we can define $\mathcal{A}_i = \mathcal{A}_{i-1}/\mathcal{A}_{i-1}[p]^{\text{loc}}$ for $i$ positive and we get a sequence of abelian varieties and isogenies

$$\mathcal{A}_0 \xrightarrow{I_0} \mathcal{A}_1 \xrightarrow{I_1} \mathcal{A}_2 \xrightarrow{I_2} \mathcal{A}_3 \xrightarrow{I_3} \ldots$$

Clearly we have that $\mathcal{A}_{kd}$ for $k \in \mathbb{N}$ reduces to $\overline{\mathcal{A}}$ modulo $p$; furthermore, the sequence $\{\mathcal{A}_{kd}\}_{k \in \mathbb{N}}$ converges to the canonical lift $\mathcal{A}_c$ and the convergence is linear.

Let $C$ be a smooth projective curve defined over $\mathbb{F}_q$ of genus $g$, with Jacobian variety $J_C$. Assuming that $J_C$ is ordinary, we can consider its canonical lift $\mathcal{A}_c$. Note that $\mathcal{A}_c$ itself does not have to be the Jacobian variety of a curve [OOTS 1986]. Since $\text{End}(\mathcal{A}_c)$ is isomorphic to $\text{End}(J_C)$, there exists a lift $\mathcal{F}$ of the Frobenius endomorphism $\phi_q$.

To recover the characteristic polynomial of $\phi_q$, we proceed as follows: let $\mathfrak{D}_0(\mathcal{A}_c, \mathbb{Q}_q)$ denote the space of holomorphic differential forms of degree 1 on $\mathcal{A}_c$ defined over $\mathbb{Q}_q$, then we have $\dim(\mathfrak{D}_0(\mathcal{A}_c, \mathbb{Q}_q)) = g$, since $\dim(J_C) = g$. Given a basis $B$ of $\mathfrak{D}_0(\mathcal{A}_c, \mathbb{Q}_q)$, every endomorphism $\lambda \in \text{End}_{\mathbb{Q}_q}(\mathcal{A}_c)$ can be represented by a $g \times g$ matrix $M$ defined over $\mathbb{Q}_q$ by considering the action of $\lambda^*$ on $B$, i.e., $\lambda^*(B) = MB$. The link with the characteristic polynomial of Frobenius $\chi(\phi_q)_C$ is then given by the following proposition.

**Proposition 8.11** Let $\mathcal{F} \in \text{End}_{\mathbb{Q}_q}(\mathcal{A}_c)$ be the lift of the Frobenius endomorphism $\phi_q \in \text{End}_{\mathbb{F}_q}(J_C)$ and let $M_\mathcal{F}$ be the matrix through which $\phi_q^*$ acts on $\mathfrak{D}_0(\mathcal{A}_c, \mathbb{Q}_q)$. If $P(T) \in \mathbb{Z}_q(T)$ is the characteristic polynomial of $M_\mathcal{F} + qM_\mathcal{F}^{-1}$, then the characteristic polynomial $\chi(\phi_q)_C$ is given by

$$\chi(\phi_q)_C(T) = T^g P\left(T + \frac{q}{T}\right). \tag{8.1}$$

Note that we can also write $\chi(\phi_q)_C(T) = P_1(X)P_2(X)$ with $P_1$ the characteristic polynomial of $M_\mathcal{F}$ and $P_2$ the characteristic polynomial of $qM_\mathcal{F}^{-1}$.

The point-counting algorithms based on the canonical lift thus proceed in two stages: in the first stage, a sufficiently precise approximation of the canonical lift of $J_C$ (or its invariants) is computed and in the second stage, the action of the lifted Frobenius endomorphism $\mathcal{F}$ is computed on $\mathfrak{D}_0(\mathcal{A}_c, \mathbb{Q}_q)$.

### 8.3.2 Monsky–Washnitzer cohomology

In this section we will specialize the formalism of Monsky–Washnitzer cohomology as described in the seminal papers by Monsky and Washnitzer [MOWA 1968, MON 1968, MON 1971], to smooth affine plane curves. Further details can be found in the lectures by Monsky [MON 1970] and in the survey by van der Put [PUT 1986].

Let $\overline{C}$ be a smooth affine plane curve over a finite field $\mathbb{F}_q$ with $q = p^d$ elements, and let $\mathbb{Q}_q$ be a degree $d$ unramified extension of $\mathbb{Q}_p$ with valuation ring $\mathbb{Z}_q$, such that $\mathbb{Z}_q/p\mathbb{Z}_q = \mathbb{F}_q$. The aim of Monsky–Washnitzer cohomology is to express the zeta function of the curve $\overline{C}$ in terms of a Frobenius operator $\mathcal{F}$ acting on $p$-adic cohomology groups $H^i(\overline{C}, \mathbb{Q}_q)$ defined over $\mathbb{Q}_q$ associated to $\overline{C}$. Note that it is necessary to work over a field of characteristic 0; otherwise, it would only be possible to obtain the zeta function modulo $p$. For smooth curves, most of these groups are zero as illustrated in the next proposition.

**Proposition 8.12** Let $\overline{C}$ be a nonsingular affine curve over a finite field $\mathbb{F}_q$, then the only nonzero Monsky–Washnitzer cohomology groups are $H^0(\overline{C}, \mathbb{Q}_q)$ and $H^1(\overline{C}, \mathbb{Q}_q)$.

In the remainder of this section, we introduce the cohomology groups $H^0(\overline{C}, \mathbb{Q}_q)$ and $H^1(\overline{C}, \mathbb{Q}_q)$ and review their main properties.

Since $\overline{C}$ is plane, $\overline{C}$ can be given by a bivariate polynomial equation $\overline{g}(x, y) = 0$ with $\overline{g} \in \mathbb{F}_q[x, y]$. Let $\overline{A} = \mathbb{F}_q[x, y]/(\overline{g}(x, y))$ be the coordinate ring of $\overline{C}$. Take an arbitrary lift $g(x, y) \in \mathbb{Z}_q[x, y]$ of $\overline{g}(x, y)$ and let $C$ be the curve defined by $g(x, y) = 0$ with coordinate ring $A = \mathbb{Z}_q[x, y]/(g(x, y))$. To compute the zeta function of $\overline{C}$ in terms of a Frobenius operator, we need to lift the Frobenius endomorphism $\phi_q$ on $\overline{A}$ to the $\mathbb{Z}_q$-algebra $A$, but as illustrated in the previous section, this is almost never possible. Furthermore, the $\mathbb{Z}_q$-algebra $A$ depends essentially on the choices made in the lifting process as the following example illustrates.

**Example 8.13** Consider $\overline{C} : xy - 1 = 0$ over $\mathbb{F}_p$ with coordinate ring $\overline{A} = \mathbb{F}_p[x, 1/x]$, and consider the two lifts
$$g_1(x, y) = xy - 1 \qquad g_2(x, y) = x(1 + px)y - 1$$
then we have that $A_1 = \mathbb{Z}_p[x, 1/x]$ and $A_2 = \mathbb{Z}_p[x, 1/(x(1 + px))]$, which are not isomorphic.

A first attempt to remedy both difficulties is to work with the $p$-adic completion $A^\infty$ of $A$, which is unique up to isomorphism and does admit a lift of $\phi_q$ to $A^\infty$. But then a new problem arises since the de Rham cohomology of $A^\infty$, which provides the vector spaces we are looking for, is too big.

**Example 8.14** Consider the affine line over $\mathbb{F}_q$, then $A = \mathbb{Z}_q[x]$ and $A^\infty$ is the ring of power series
$$\sum_{i=0}^{\infty} r_i x^i \text{ with } r_i \in \mathbb{Z}_q \text{ and } \lim_{i \to \infty} r_i = 0.$$

We would like to define $H^1(\overline{A}, \mathbb{Q}_q)$ as $A^\infty \, dx/d(A^\infty) \otimes_{\mathbb{Z}_q} \mathbb{Q}_q$, but this turns out to be infinite dimensional. For example, it is clear that each term in the differential form $\sum_{i=0}^{\infty} p^i x^{p^i-1} dx$ is exact but its sum is not, since $\sum_{i=0}^{\infty} x^{p^i}$ is not in $A^\infty$. The fundamental problem is that $\sum_{i=0}^{\infty} p^i x^{p^i-1}$ does not converge fast enough for its integral to converge as well.

Monsky and Washnitzer therefore work with a subalgebra $A^\dagger$ of $A^\infty$, whose elements satisfy growth conditions.

**Definition 8.15** Let $A = \mathbb{Z}_q[x, y]/(g(x, y))$, then the *dagger ring* or *weak completion* $A^\dagger$ is defined as $A^\dagger = \mathbb{Z}_q\langle x, y\rangle^\dagger/(g(x, y))$, where $\mathbb{Z}_q\langle x, y\rangle^\dagger$ is the ring of overconvergent power series
$$\left\{\sum r_{i,j} x^i y^j \in \mathbb{Z}_q[[x, y]] \mid \exists \delta, \varepsilon \in \mathbb{R}, \varepsilon > 0, \forall (i, j) : v_p(r_{i,j}) \geqslant \varepsilon(i + j) + \delta\right\}.$$

The ring $A^\dagger$ satisfies $A^\dagger/(pA^\dagger) = \overline{A}$ and depends up to $\mathbb{Z}_q$-isomorphism only on $\overline{A}$. Furthermore, Monsky and Washnitzer show that if $\overline{\varphi}$ is an $\mathbb{F}_q$-endomorphism of $\overline{A}$, then there exists a $\mathbb{Z}_q$-endomorphism $\varphi$ of $A^\dagger$ lifting $\overline{\varphi}$. In particular, we can lift the Frobenius endomorphism $\phi_q$ on $\overline{A}$ to a $\mathbb{Z}_q$-endomorphism $\mathcal{F}$ on $A^\dagger$.

To each element $s \in A^\dagger$ we can associate the differential $ds$ such that the usual Leibniz rule applies: for $s, t \in A^\dagger : d(st) = s \, dt + t \, ds$, which implies that $d(a) = 0$ for $a \in \mathbb{Z}_q$. The set of all these differentials clearly is a module over $A^\dagger$ and is denoted by $D^1(A^\dagger)$. The following lemma gives a precise description of this module.

**Lemma 8.16** The universal module $D^1(A^\dagger)$ of differentials satisfies
$$D^1(A^\dagger) = \left(A^\dagger \, dx + A^\dagger \, dy\right) \Big/ \left(A^\dagger \left(\frac{\partial g}{\partial x} dx + \frac{\partial g}{\partial y} dy\right)\right).$$

## § 8.3 Overview of $p$-adic methods

Taking the total differential of the equation $g(x,y) = 0$ gives $\frac{\partial g}{\partial x} dx + \frac{\partial g}{\partial y} dy = 0$, which explains the module $A^\dagger(\frac{\partial g}{\partial x} dx + \frac{\partial g}{\partial y} dy)$ in the above lemma. The map $d: A^\dagger \longrightarrow D^1(A^\dagger)$ is a well defined derivation, so it makes sense to consider its kernel and cokernel.

**Definition 8.17** The cohomology groups $H^0(\overline{A}, \mathbb{Q}_q)$ and $H^1(\overline{A}, \mathbb{Q}_q)$ are defined by

$$H^0(\overline{A}, \mathbb{Q}_q) = \ker(d) \otimes_{\mathbb{Z}_q} \mathbb{Q}_q \quad \text{and} \quad H^1(\overline{A}, \mathbb{Q}_q) = \operatorname{coker}(d) \otimes_{\mathbb{Z}_q} \mathbb{Q}_q. \tag{8.2}$$

By definition we have $H^1(\overline{A}, \mathbb{Q}_q) = \left(D^1(A^\dagger)/d(A^\dagger)\right) \otimes_{\mathbb{Z}_q} \mathbb{Q}_q$; the elements of $d(A^\dagger)$ are called *exact differentials*. One can prove that $H^0(\overline{A}, \mathbb{Q}_q)$ and $H^1(\overline{A}, \mathbb{Q}_q)$ are well defined, only depend on $\overline{A}$, and are finite dimensional vector spaces over $\mathbb{Q}_q$.

**Proposition 8.18** Let $\overline{C}$ be a nonsingular affine curve of genus $g$, then $\dim\left(H^0(\overline{A}, \mathbb{Q}_q)\right) = 1$ and $\dim\left(H^1(\overline{A}, \mathbb{Q}_q)\right) = 2g + m - 1$, where $m$ is the number of points needed to complete $\overline{C}$ to a smooth projective curve.

Let $\mathcal{F}$ be a lift of the Frobenius endomorphism $\phi_q$ to $A^\dagger$, then $\mathcal{F}$ induces an endomorphism $\mathcal{F}^*$ on the cohomology groups. The main theorem of Monsky–Washnitzer cohomology is the following Lefschetz fixed point formula.

**Theorem 8.19 (Lefschetz fixed point formula)** Let $\overline{C}/\mathbb{F}_q$ be a nonsingular affine curve over $\mathbb{F}_q$, then the number of $\mathbb{F}_{q^k}$-rational points on $\overline{C}$ is equal to

$$|\overline{C}(\mathbb{F}_{q^k})| = \operatorname{Tr}\left(q^k \mathcal{F}^{-k*}; H^0(\overline{C}, \mathbb{Q}_q)\right) - \operatorname{Tr}\left(q^k \mathcal{F}^{-k*}; H^1(\overline{C}, \mathbb{Q}_q)\right).$$

Since $H^0(\overline{C}, \mathbb{Q}_q)$ is a one-dimensional vector space on which $\mathcal{F}^*$ acts as the identity, we conclude that $\operatorname{Tr}\left(q^k \mathcal{F}^{-k*}; H^0(\overline{C}, \mathbb{Q}_q)\right) = q^k$. To count the number of $\mathbb{F}_{q^k}$-rational points on $\overline{C}$, it thus suffices to compute the action of $\mathcal{F}^*$ on $H^1(\overline{C}, \mathbb{Q}_q)$.

The algorithms based on Monsky–Washnitzer cohomology thus also proceed in two stages: in the first stage, a sufficiently precise approximation of the lift $\mathcal{F}$ is computed and in the second stage, a basis of $H^1(\overline{C}, \mathbb{Q}_q)$ is constructed together with reduction formulas to express any differential form on this basis. More algorithmic details can be found in Section 17.3.

# Elementary Arithmetic

# Chapter 9

# *Exponentiation*

*Christophe Doche*

### Contents in Brief

| | | |
|---|---|---|
| **9.1** | **Generic methods** | **146** |
| | Binary methods • Left-to-right $2^k$-ary algorithm • Sliding window method • Signed-digit recoding • Multi-exponentiation | |
| **9.2** | **Fixed exponent** | **157** |
| | Introduction to addition chains • Short addition chains search • Exponentiation using addition chains | |
| **9.3** | **Fixed base point** | **164** |
| | Yao's method • Euclidean method • Fixed-base comb method | |

Given an element $x$ of a group $(G, \times)$ and an integer $n \in \mathbb{Z}$ one describes in this chapter efficient methods to perform the *exponentiation* $x^n$. Only positive exponents are considered since $x^n = (1/x)^{-n}$ but nothing more is assumed especially regarding the structure and the properties of $G$. See Chapter 11 for specific improvements concerning finite fields. Two elementary operations are used, namely multiplications and squarings. The distinction is made for performance reasons since squarings can often be implemented more efficiently; see Chapters 10 and 11 for details. In the context of elliptic and hyperelliptic curves, the computations are done in an abelian group denoted additively $(G, \oplus)$. The equivalent of the exponentiation $x^n$ is the *scalar multiplication* $[n]P$. All the techniques described in this chapter can be adapted in a trivial way, replacing multiplication by addition and squaring by doubling. See Chapter 13 for additional details concerning elliptic curves and Chapter 14 for hyperelliptic curves.

Exponentiation is a very important operation in algorithmic number theory. For example, it is intensively used in many primality testing and factoring algorithms. Therefore efficient methods have been studied over centuries. In cryptosystems based on the discrete logarithm problem (cf. Chapter 1) exponentiation is often the most time-consuming part, and thus determines the efficiency of cryptographic protocols like key exchange, authentication, and signature.

Three typical situations occur. The *base point* $x$ and the *exponent* $n$ may both vary from one computation to another. Generic methods will be used to get $x^n$ in this case. If the same exponent is used several times a closer study of $n$, especially the search of a short addition chain for $n$, can lead to substantial improvements. Finally, if different powers of the same element are needed, some precomputations, whose cost can be neglected, give a noticeable speedup.

Most of the algorithms described in the remainder of this chapter can be found in [MEO0+ 1996, GOR 1998, KNU 1997, STA 2003, BER 2002].

## 9.1 Generic methods

In this section both $x$ and $n$ may vary. Computing $x^n$ naïvely requires $n-1$ multiplications, but much better methods exist, some of them being very simple.

### 9.1.1 Binary methods

It is clear that $x^{2^k}$ can be obtained with only $k$ squarings, namely $x^2, x^4, x^8, \ldots, x^{2^k}$. Building upon this observation, the following method, known for more than 2000 years, allows us to compute $x^n$ in $O(\lg n)$ operations, whatever the value of $n$.

---

**Algorithm 9.1** Square and multiply method

INPUT: An element $x$ of $G$ and a nonnegative integer $n = (n_{\ell-1} \ldots n_0)_2$.
OUTPUT: The element $x^n \in G$.

1. $y \leftarrow 1$ and $i \leftarrow \ell - 1$
2. **while** $i \geqslant 0$
3. $\quad y \leftarrow y^2$
4. $\quad$ **if** $n_i = 1$ **then** $y \leftarrow x \times y$
5. $\quad i \leftarrow i - 1$
6. **return** $y$

---

This method is based on the equality

$$x^{(n_{\ell-1}\ldots n_{i+1} n_i)_2} = \left(x^{(n_{\ell-1}\ldots n_{i+1})_2}\right)^2 \times x^{n_i}.$$

As the bits are processed from the most to the least significant one, Algorithm 9.1 is also referred to as the *left-to-right binary method*.

There is another method relying on

$$x^{(n_i n_{i-1} \ldots n_0)_2} = x^{n_i 2^i} \times x^{(n_{i-1}\ldots n_0)_2}$$

which operates from the right to the left.

---

**Algorithm 9.2** Right-to-left binary exponentiation

INPUT: An element $x$ of $G$ and a nonnegative integer $n = (n_{\ell-1} \ldots n_0)_2$.
OUTPUT: The element $x^n \in G$.

1. $y \leftarrow 1, z \leftarrow x$ and $i \leftarrow 0$
2. **while** $i \leqslant \ell - 1$ **do**
3. $\quad$ **if** $n_i = 1$ **then** $y \leftarrow y \times z$

## § 9.1 Generic methods

    4.          $z \leftarrow z^2$

    5.          $i \leftarrow i + 1$

    6.  **return** $y$

**Remarks 9.3**

(i) Algorithm 9.1 is related to Horner's rule, more precisely computing $x^n$ is similar to evaluating the polynomial $\sum_{i=0}^{\ell-1} n_i X^i$ at $X = 2$ with Horner's rule.

(ii) Further enhancements may apply to the products $y \times x$ in Algorithm 9.1 since one of the operands is fixed during the whole computation. For example, if $x$ is well chosen the multiplication can be computed more efficiently. Such an improvement is impossible with Algorithm 9.2 where different terms of approximately the same size are involved in the products $y \times z$.

(iii) In Algorithm 9.2, whatever the value of $n$, the extra variable $z$ contains the successive squares $x^2, x^4, \ldots$ which can be evaluated in parallel to the multiplication.

The next example provides a comparison of Algorithms 9.1 and 9.2.

**Example 9.4** Let us compute $x^{314}$. One has $314 = (100111010)_2$ and $\ell = 9$.

| Algorithm 9.1 | | | | | | | | | |
|---|---|---|---|---|---|---|---|---|---|
| $i$ | 8 | 7 | 6 | 5 | 4 | 3 | 2 | 1 | 0 |
| $n_i$ | 1 | 0 | 0 | 1 | 1 | 1 | 0 | 1 | 0 |
| $y$ | $x$ | $x^2$ | $x^4$ | $x^9$ | $x^{19}$ | $x^{39}$ | $x^{78}$ | $x^{157}$ | $x^{314}$ |

| Algorithm 9.2 | | | | | | | | | |
|---|---|---|---|---|---|---|---|---|---|
| $i$ | 0 | 1 | 2 | 3 | 4 | 5 | 6 | 7 | 8 |
| $n_i$ | 0 | 1 | 0 | 1 | 1 | 1 | 0 | 0 | 1 |
| $z$ | $x$ | $x^2$ | $x^4$ | $x^8$ | $x^{16}$ | $x^{32}$ | $x^{64}$ | $x^{128}$ | $x^{256}$ |
| $y$ | 1 | $x^2$ | $x^2$ | $x^{10}$ | $x^{26}$ | $x^{58}$ | $x^{58}$ | $x^{58}$ | $x^{314}$ |

The number of squarings is the same for both algorithms and equal to the bit length $\ell \sim \lg n$ of $n$. In fact this will be the case for all the methods discussed throughout the chapter.

The number of required multiplications directly depends on $\nu(n)$ the *Hamming weight* of $n$, i.e., the number of nonzero terms in the binary expansion of $n$. On average $\frac{1}{2} \lg n$ multiplications are needed.

Further improvements introduced below tend to decrease the number of multiplications, leading to a considerable speedup. Many algorithms also require some precomputations to be done. In the case where several exponentiations with the same base have to be performed in a single run, these precomputations need to be done only once per session, and if the base is fixed in a given system, they can even be stored, so that their cost might become almost negligible. This is the case considered in Algorithms 9.7, 9.10, and 9.23. Depending on the bit processed, a single squaring or a multiplication and a squaring are performed at each step in both Algorithms 9.1 and 9.2. This implies that it can be possible to retrieve each bit and thus the value of the exponent from an analysis of the computation. This has serious consequences when the exponent is some secret key. See Chapters 28 and 29 for a description of side-channel attacks. An elegant technique, called *Montgomery's ladder*, overcomes this issue. Indeed, this variant of Algorithm 9.1 performs a squaring and a multiplication at each step to compute $x^n$.

**Algorithm 9.5** Montgomery's ladder

INPUT: An element $x \in G$ and a positive integer $n = (n_{\ell-1} \ldots n_0)_2$.
OUTPUT: The element $x^n \in G$.

1. $x_1 \leftarrow x$ and $x_2 \leftarrow x^2$
2. **for** $i = \ell - 2$ **down to** $0$ **do**
3.     **if** $n_i = 0$ **then**
4.         $x_1 \leftarrow x_1^2$ and $x_2 \leftarrow x_1 \times x_2$
5.     **else**
6.         $x_1 \leftarrow x_1 \times x_2$ and $x_2 \leftarrow x_2^2$
7. **return** $x_1$

**Example 9.6** To illustrate the method, let us compute $x^{314}$ using, this time, Algorithm 9.5. Starting from $(x_1, x_2) = (x, x^2)$, the next values of $(x_1, x_2)$ are given below.

| $i$ | 7 | 6 | 5 | 4 | 3 | 2 | 1 | 0 |
|---|---|---|---|---|---|---|---|---|
| $n_i$ | 0 | 0 | 1 | 1 | 1 | 0 | 1 | 0 |
| $(x_1, x_2)$ | $(x^2, x^3)$ | $(x^4, x^5)$ | $(x^9, x^{10})$ | $(x^{19}, x^{20})$ | $(x^{39}, x^{40})$ | $(x^{78}, x^{79})$ | $(x^{157}, x^{158})$ | $(x^{314}, x^{315})$ |

See also Chapter 13, for a description of Montgomery's ladder in the context of scalar multiplication on an elliptic curve.

### 9.1.2 Left-to-right $2^k$-ary algorithm

The general idea of this method, introduced by Brauer [BRA 1939], is to write the exponent on a larger base $b = 2^k$. Some precomputations are needed but several bits can be processed at a time.

In the following the function $\sigma$ is defined by $\sigma(0) = (k, 0)$ and $\sigma(m) = (s, u)$ where $m = 2^s u$ with $u$ odd.

**Algorithm 9.7** Left-to-right $2^k$-ary exponentiation

INPUT: An element $x$ of $G$, a parameter $k \geqslant 1$, a nonnegative integer $n = (n_{\ell-1} \ldots n_0)_{2^k}$ and the precomputed values $x^3, x^5, \ldots, x^{2^k-1}$.
OUTPUT: The element $x^n \in G$.

1. $y \leftarrow 1$ and $i \leftarrow \ell - 1$
2. $(s, u) \leftarrow \sigma(n_i)$                                                                                            [$n_i = 2^s u$]
3. **while** $i \geqslant 0$ **do**
4.     **for** $j = 1$ **to** $k - s$ **do** $y \leftarrow y^2$
5.     $y \leftarrow y \times x^u$
6.     **for** $j = 1$ **to** $s$ **do** $y \leftarrow y^2$

## § 9.1 Generic methods

7. $\quad i \leftarrow i - 1$
8. **return** $y$

---

**Remarks 9.8**

(i) Lines 4 to 6 compute $y^{2^k} x^{n_i}$ i.e., the exact analogue of $y^2 x^{n_i}$ in Algorithm 9.1. To reduce the amount of precomputations, note that $(y^{2^{k-s}} x^u)^{2^s} = y^{2^k} x^{n_i}$ is actually computed.

(ii) The number of elementary operations performed is $\lg n + \lg n \bigl(1 + o(1)\bigr) \lg n / \lg \lg n$.

(iii) For optimal efficiency, $k$ should be equal to the smallest integer satisfying

$$\lg n \leqslant \frac{k(k+1) 2^{2k}}{2^{k+1} - k - 2}.$$

See [COH 2000] for details. This leads to the following table, which gives for all intervals of bit lengths the appropriate value for $k$.

| $k$ | 1 | 2 | 3 | 4 | 5 | 6 | 7 |
|---|---|---|---|---|---|---|---|
| No. of binary digits | $[1, 9]$ | $[10, 25]$ | $[26, 70]$ | $[70, 197]$ | $[197, 539]$ | $[539, 1434]$ | $[1434, 3715]$ |

**Example 9.9** Take $n = 11957708941720303968251$ whose binary representation is

$(10100010000111010100011000001111111010110010111011100000001111111111011)_2$.

As its binary length is 74, take $k = 4$. The representation of $n$ in radix $2^4$ is

$(\underset{2\times 1}{2}\ \underset{8\times 1}{8}\ \underset{8\times 1}{8}\ 3\ \underset{2\times 5}{10}\ \underset{8\times 1}{8}\ \underset{4\times 3}{12}\ 1\ 15\ 13\ \underset{2\times 3}{6}\ 5\ \underset{2\times 7}{14}\ \underset{2\times 7}{14}\ 0\ 1\ 15\ 15\ 11)_{2^4}$.

Thus the successive values of $y$ are $1, x, x^2, x^4, x^5, x^{10}, x^{20}, x^{40}, x^{80}, x^{81}, x^{162}, x^{324}, \ldots, x^n$. Let us denote a multiplication by M and a squaring by S. Then the precomputations cost $7M + S$ and additionally one needs $17M + 72S$, i.e., 97 elementary operations in total. By way of comparison, Algorithm 9.1 needs 112 operations, $39M + 73S$.

### 9.1.3 Sliding window method

The $2^k$-ary method consists of slicing the binary representation of $n$ into pieces using a window of length $k$ and to process the parts one by one. Letting the window slide allows us to skip strings of consecutive zeroes. For instance, let $n = 334 = (101001110)_2$. Take a window of length 3 or, in other words, precompute $x^3, x^5$ and $x^7$ only. The successive values of $y$ computed by Algorithm 9.7 are $1, x^5, x^{10}, x^{20}, x^{40}, x^{41}, x^{82}, x^{164}, x^{167}$ and $x^{334}$ as reflected by

$$334 = (\underset{5}{101}\ \underset{1}{001}\ \underset{2\times 3}{110})_2.$$

But one could compute $1, x^5, x^{10}, x^{20}, x^{40}, x^{80}, x^{160}, x^{167}, x^{334}$ instead. This saves one multiplication and amounts to allowing non-adjacent windows

$$334 = (\underset{5}{101}\,00\,\underset{7}{111}\,0)_2$$

where the strings of many consecutive zeroes are ignored.

Here is the general algorithm.

---

**Algorithm 9.10** Sliding window exponentiation

INPUT: An element $x$ of $G$, a nonnegative integer $n = (n_{\ell-1} \ldots n_0)_2$, a parameter $k \geq 1$ and the precomputed values $x^3, x^5, \ldots, x^{2^k-1}$.
OUTPUT: The element $x^n \in G$.

1.    $y \leftarrow 1$ and $i \leftarrow \ell - 1$
2.    **while** $i \geq 0$ **do**
3.        **if** $n_i = 0$ **then** $y \leftarrow y^2$ and $i \leftarrow i - 1$
4.        **else**
5.            $s \leftarrow \max\{i - k + 1, 0\}$
6.            **while** $n_s = 0$ **do** $s \leftarrow s + 1$
7.            **for** $h = 1$ **to** $i - s + 1$ **do** $y \leftarrow y^2$
8.            $u \leftarrow (n_i \ldots n_s)_2$        $[n_i = n_s = 1 \text{ and } i - s + 1 \leq k]$
9.            $y \leftarrow y \times x^u$        [$u$ is odd so that $x^u$ is precomputed]
10.       $i \leftarrow s - 1$
11.    **return** $y$

---

**Remarks 9.11**

(i) In Line 6 the index $i$ is fixed, $n_i = 1$ and the while loop finds the longest substring $(n_i \ldots n_s)$ of length less than or equal to $k$ such that $n_s = 1$. So $u = (n_i \ldots n_s)_2$ is odd and belongs to the set of precomputed values.

(ii) Only the values $x^u$ occurring in Line 9 actually need to be precomputed and not all the values $x^3, x^5, \ldots, x^{2^k-1}$.

(iii) In certain cases it is possible to skip some squarings at the beginning, at the cost of an additional multiplication. For the sake of clarity assume that $k = 5$ and that the binary expansion of $n$ is $(1000000)_2$. Then Algorithm 9.10 computes $x, x^2, x^4, x^8, x^{16}, x^{32}, x^{64}$. But one could perform $x^{31} \times x$ instead to obtain $x^{32}$ directly, taking advantage of the precomputations. However, this trick is interesting only if the first value of $u$ is less than $2^{k-1}$.

**Example 9.12** With $n = 1195770894172030396825$1 and $k = 4$ the sliding window method makes use of the following decomposition

$(\underbrace{101}_{5} 000 \underbrace{1}_{1} 00000 \underbrace{111}_{7} 0 \underbrace{101}_{5} 000 \underbrace{11}_{3} 00000 \underbrace{1111}_{15} \underbrace{111}_{7} 0 \underbrace{1011}_{11} 00 \underbrace{1011}_{11} \underbrace{1101}_{13} \underbrace{11}_{3} 00000000 \underbrace{1111}_{15} \underbrace{1111}_{15} \underbrace{1101}_{13} \underbrace{1}_{1})_2.$

The successive values of $y$ are $1, x^5, x^{10}, x^{20}, x^{40}, x^{80}, x^{81}, x^{162}, x^{324}, x^{648}, \ldots, x^n$. In this case 93 operations are needed, namely $21M + 72S$, precomputations included.

---

## 9.1.4 Signed-digit recoding

When inversion in $G$ is fast (or when $x$ is fixed and $x^{-1}$ precomputed) it can be very efficient to multiply by either $x$ or $x^{-1}$. This can be used to save additional multiplications on the cost of allowing negative coefficients and hence using the inverse of precomputed values. The extreme

*§ 9.1 Generic methods*

example is the computation of $x^{2^k-1}$. With the binary method, cf. Algorithm 9.1, one needs $k-1$ squarings and $k-1$ multiplications. But one could also perform $k$ squarings to get $x^{2^k}$ followed by a multiplication by $x^{-1}$. This remark leads to the following concept.

**Definition 9.13** A *signed-digit representation of an integer* $n$ *in radix* $b$ is given by

$$n = \sum_{i=0}^{\ell-1} n_i b^i \text{ with } |n_i| < b.$$

A *signed-binary* representation corresponds to the particular choice $b = 2$ and $n_i \in \{-1, 0, 1\}$.

It is denoted by $(n_{\ell-1} \ldots n_0)_s$ and usually obtained by some *recoding* technique. The representation is said to be in *non-adjacent form*, NAF for short, if $n_i n_{i+1} = 0$, for all $i \geqslant 0$. It is denoted by $(n_{\ell-1} \ldots n_0)_{\text{NAF}}$.

For example, take $n = 478$ and let $\bar{1} = -1$. Then $(10\bar{1}1100\bar{1}10)_s$ is a signed-binary representation of $n$. The first recoding technique was proposed by Booth [BOO 1951]. It consists of replacing each string of $i$ consecutive 1 in the binary expansion of $n$ by 1 followed by a string of $i-1$ consecutive 0 and then $\bar{1}$. For $478 = (111011110)_2$ it gives $(100\bar{1}1000\bar{1}0)_s$. Obviously, the signed-binary representation of $n$ is not unique. However, the NAF of a given $n$ is unique and its Hamming weight is minimal among all signed-digit representations of $n$. For example, the NAF of 478 is equal to $(1000\bar{1}000\bar{1}0)_{\text{NAF}}$. On average the number of nonzero terms in an NAF expansion of length $\ell$ is equal to $\ell/3$. See [BOS 2001] for a precise analysis of the NAF density. There is a very simple algorithm to compute it [REI 1962, MOOL 1990].

---

**Algorithm 9.14** NAF representation

INPUT: A positive integer $n = (n_\ell n_{\ell-1} \ldots n_0)_2$ with $n_\ell = n_{\ell-1} = 0$.
OUTPUT: The signed-binary representation of $n$ in non-adjacent form $(n'_{\ell-1} \ldots n'_0)_{\text{NAF}}$.

1. $c_0 \leftarrow 0$
2. **for** $i = 0$ **to** $\ell - 1$ **do**
3. $\quad c_{i+1} \leftarrow \lfloor (c_i + n_i + n_{i+1})/2 \rfloor$
4. $\quad n'_i \leftarrow c_i + n_i - 2c_{i+1}$
5. **return** $(n'_{\ell-1} \ldots n'_0)_{\text{NAF}}$

---

**Remarks 9.15**

(i) Algorithm 9.14 subtracts $n$ from $3n$ with the rule $0 - 1 = \bar{1}$ and discards the least significant digit of the result. For each $i$, $c_i$ is the carry occurring in the addition $n + 2n$. Let $s_i = c_i + n_i + n_{i+1} - 2c_{i+1}$ so that the binary expansion of $3n$ is equal to $(s_{\ell-1} \ldots s_0 n_0)_2$. Now $n'_i = c_i + n_i - 2c_{i+1} \in \{\bar{1}, 0, 1\}$. The following observation ensures the non-adjacent property of the expansion [JOYE 2000]. If $n'_i \neq 0$, we have $c_i + n_i = 1$, which implies that $c_{i+1} = n_{i+1}$. So $n'_{i+1} = 2(n_{i+1} - c_{i+2})$ must be zero.

(ii) Finding a signed-binary representation in non-adjacent form can be done by table lookup. Indeed $c_{i+1}$ and $n'_i$, computed in Lines 3 and 4, only depend on $n_{i+1}$, $n_i$ and $c_i$ giving just eight cases.

(iii) Algorithm 9.14 operates from the right to the left. Since most of the exponentiation algorithms presented so far process the bits from the left to the right, the signed-binary representation must first be computed and stored. To enable "on the fly" recoding, which is particularly interesting for hardware applications, cf. Chapter 26, Joye and Yen designed a left-to-right signed-digit recoding algorithm. The result is not necessarily in non-adjacent form but its Hamming weight is still minimal [JOYE 2000].

(iv) Algorithms 9.1, 9.2, 9.7, and 9.10 can be updated in a trivial way to deal with signed-binary representation.

(v) A generalization of the NAF is presented below; see Algorithm 9.20.

**Example 9.16** Again take $n = 1195770894172030396825$1. Algorithm 9.14 gives

$$n = (10100010000100\bar{1}01010010\bar{1}0000100000001\bar{0}\bar{1}0\bar{1}010\bar{1}0000\bar{1}00\bar{1}0000000100000000000\bar{1}0\bar{1})_{\text{NAF}}.$$

Now one can combine this representation to a sliding window algorithm of length 4 to get the following decomposition

$$(\underbrace{101}_{5}\,000\,\underbrace{1}_{1}\,0000\,\underbrace{100\bar{1}}_{7}\,0\,\underbrace{101}_{5}\,00\,\underbrace{10\bar{1}}_{3}\,0000\,\underbrace{1}_{1}\,0000000\,\underbrace{\bar{1}0\bar{1}}_{-5}\,0\,\underbrace{\bar{1}01}_{-3}\,0\,\underbrace{\bar{1}}_{-1}\,0000\,\underbrace{\bar{1}00\bar{1}}_{-9}\,0000000\,\underbrace{1}_{1}\,0000000000\,\underbrace{\bar{1}0\bar{1}}_{-5})_{\text{NAF}}.$$

The number of operations, precomputations included, is 90, namely $18\text{M} + 72\text{S}$.

Koyama and Tsuruoka [KOTS 1993] designed another transformation, getting rid of the condition $n_i n_{i+1} = 0$ but still minimizing the Hamming weight. Its average length of zero runs is 1.42 against 1.29 for the NAF.

---

**Algorithm 9.17** Koyama–Tsuruoka signed-binary recoding

INPUT: The binary representation of $n = (n'_{\ell-1} \ldots n'_0)_2$.
OUTPUT: The signed-binary representation $(n_\ell \ldots n_0)_s$ of $n$ in Koyama–Tsuruoka form.

1. $m \leftarrow 0, i \leftarrow 0, j \leftarrow 0, u \leftarrow 0, v \leftarrow 0, w \leftarrow 0, y \leftarrow 0$ and $z \leftarrow 0$
2. **while** $i < \lfloor \lg n \rfloor$ **do**
3.     **if** $n'_i = 1$ **then** $y \leftarrow y + 1$ **else** $y \leftarrow y - 1$
4.     $i \leftarrow i + 1$
5.     **if** $m = 0$ **then**
6.         **if** $y - z \geqslant 3$ **then**
7.             **while** $j < w$ **do** $n_j = b_j$ and $j \leftarrow j + 1$
8.             $n_j \leftarrow -1, j \leftarrow j + 1, v \leftarrow y, u \leftarrow i$ and $m \leftarrow 1$
9.         **else if** $y < z$ **then** $z \leftarrow y$ and $w \leftarrow i$
10.     **else**
11.         **if** $v - y \geqslant 3$ **then**
12.             **while** $j < u$ **do** $n_j = b_j - 1$ and $j \leftarrow j + 1$
13.             $n_j \leftarrow 1, j \leftarrow j + 1, z \leftarrow y, w \leftarrow i$ and $m \leftarrow 0$
14.         **else if** $y > v$ **then** $v \leftarrow y$ and $u \leftarrow i$
15. **if** $m = 0$ **or** ($m = 1$ **and** $v \leqslant y$) **then**
16.     **while** $j < i$ **do** $n_j = b_j - m$ and $j \leftarrow j + 1$
17.     $n_j \leftarrow 1 - m$ and $n_{j+1} \leftarrow m$
18. **else**
19.     **while** $j < u$ **do** $n_j = b_j - 1$ and $j \leftarrow j + 1$
20.     $n_j \leftarrow 1$ and $j \leftarrow j + 1$

## § 9.1 Generic methods

21.              **while** $j < i$ **do** $n_j = b_j$ and $j \leftarrow j+1$

22.              $n_j \leftarrow 1$ and $n_{j+1} \leftarrow 0$

23. **return** $(n_\ell \ldots n_0)_s$

---

This approach gives good results when combined with the sliding window method.

**Example 9.18** For the same $n = 11957708941720303968251$, a sliding window exponentiation of length 4 based on the expansion given by Algorithm 9.17 corresponds to

$$(\underbrace{101}_{5}\,000\,\underbrace{1}_{1}0000\,\underbrace{1}_{1}000\,\underbrace{\bar{1}0\bar{1}\bar{1}}_{-11}\,000\,\underbrace{11}_{3}0000\,\underbrace{1}_{1}0000000\,\underbrace{\bar{1}0\bar{1}}_{-5}\,00\,\underbrace{\bar{1}\bar{1}0\bar{1}}_{-13}\,0000\,\underbrace{\bar{1}00\bar{1}}_{-9}\,0000000\,\underbrace{1}_{1}0000000000\,\underbrace{\bar{1}0\bar{1}}_{-5})_s.$$

In total 89 operations are necessary, i.e., $17M + 72S$, including the precomputations.

Now one introduces a generalization of the NAF, which combines window and signed methods as suggested in [MOOL 1990] and explained in [COMI$^+$ 1997, COH 2005].

**Definition 9.19** Let $w$ be a parameter greater than 1. Then every positive integer $n$ has a unique signed-digit expansion

$$n = \sum_{i=0}^{\ell-1} n_i 2^i$$

where

- each $n_i$ is zero or odd
- $|n_i| < 2^{w-1}$
- among any $w$ consecutive coefficients at most one is nonzero.

An expansion of this particular form is called *width-$w$ non-adjacent form*, $\text{NAF}_w$ for short, and is denoted by $(n_{\ell-1} \ldots n_0)_{\text{NAF}_w}$.

In [AVA 2005a], Avanzi shows that the $\text{NAF}_w$ is optimal, in the sense that it is a recoding of smallest weight among all those with coefficients smaller in absolute value than $2^{w-1}$. See also [MUST 2004] for a similar result.

A generalization of Algorithm 9.14 allows us to compute the $\text{NAF}_w$ of any number $n > 0$.

---

**Algorithm 9.20** $\text{NAF}_w$ representation

INPUT: A positive integer $n$ and a parameter $w > 1$.
OUTPUT: The $\text{NAF}_w$ representation $(n_{\ell-1} \ldots n_0)_{\text{NAF}_w}$ of $n$.

1. $i \leftarrow 0$
2. **while** $n > 0$ **do**
3.     **if** $n$ is odd **then**
4.         $n_i \leftarrow n \bmod s\, 2^w$
5.         $n \leftarrow n - n_i$
6.     **else** $n_i \leftarrow 0$
7.     $n \leftarrow n/2$ and $i \leftarrow i+1$
8. **return** $(n_{\ell-1} \ldots n_0)_{\text{NAF}_w}$

**Remarks 9.21**

(i) The function mods used in Line 4 of Algorithm 9.20 returns the smallest residue in absolute value. Hence, $n$ mods $2^w$ belongs to $[-2^{w-1}+1, 2^{w-1}]$.

(ii) For $w=2$ the $\text{NAF}_w$ corresponds to the classical NAF, cf. Definition 9.13.

(iii) The length of the $\text{NAF}_w$ of $n$ is at most equal to $\lfloor \lg n \rfloor + 1$. The average density of the $\text{NAF}_w$ expansion of $n$ is $1/(w+1)$ as $n$ tends to infinity. For a precise analysis, see [COH 2005].

(iv) A left-to-right variant to compute an $\text{NAF}_w$ expansion of an integer can be found both in [AVA 2005a] and in [MUST 2005]. The result may differ from the expansion produced by Algorithm 9.20 but they have the same digit set and the same optimal weight.

(v) Let $w > 1$ and precompute the values $x^{\pm 3}, \ldots, x^{\pm (2^{w-1}-1)}$. Then in Algorithm 9.1 it is sufficient to replace the statement

    4.   **if** $n_i = 1$ **then** $y \leftarrow x \times y$

by

    4.   **if** $n_i \neq 0$ **then** $y \leftarrow x^{n_i} \times y$

to take advantage of the $\text{NAF}_w$ expansion of $n = (n_{\ell-1} \ldots n_0)_{\text{NAF}_w}$ to compute $x^n$.

(vi) See [MÖL 2003] for a further generalization called the *signed fractional window method*, where only a subset of $\{x^{\pm 3}, \ldots, x^{\pm (2^{w-1}-1)}\}$ is actually precomputed.

**Example 9.22** For $n = 11957708941720303968251$ and $w = 4$ one has

$$n = (500010000000700050000300001000000\bar{1}000\bar{5}0003000\bar{1}000700000001000000000000\bar{5})_{\text{NAF}_w}$$

where $\bar{n}_i = -n_i$. With this representation and the modification of Algorithm 9.1 explained above $x^n$ can be obtained with $3\text{M}+\text{S}$ for the precomputations and then $12\text{M}+69\text{S}$, that is 85 operations in total.

### 9.1.5 Multi-exponentiation

The group $G$ is assumed to be abelian in this section.

It is often needed in cryptography, for example during a signature verification, cf. Chapter 1, to evaluate $x_0^{n_0} x_1^{n_1}$ where $x_0, x_1 \in G$ and $n_0, n_1 \in \mathbb{Z}$. Instead of computing $x_0^{n_0}$ and $x_1^{n_1}$ separately and then multiplying these terms, it is suggested in [ELG 1985] to adapt Algorithm 9.1 in the following way, in order to get $x_0^{n_0} x_1^{n_1}$ in one round. Indeed, start from $y \leftarrow 1$. Scan the bits of $n_0$ and $n_1$ simultaneously from the left to the right and do $y \leftarrow y^2$. Then if the current bits of $\substack{n_0 \\ n_1}$ are $\substack{1 \\ 0}, \substack{0 \\ 1}$ or $\substack{1 \\ 1}$ multiply by $x_0$, $x_1$ or $x_0 x_1$ accordingly.

For example, to compute $x_0^{51} x_1^{166}$ write the binary expansion of 51 and 166

$$51 = (00110011)_2$$
$$166 = (10100110)_2$$

and apply the rules above so that the successive values of $y$ are at each step $1, x_1, x_1^2, x_0 x_1^5, x_0^3 x_1^{10}$, $x_0^6 x_1^{20}, x_0^{12} x_1^{41}, x_0^{25} x_1^{83}$, and finally $x_0^{51} x_1^{166}$.

This trick is often credited to Shamir although it is a special case of an idea of Straus [STR 1964] described below. Note that the binary coefficients of $n_j$ are denoted by $n_{j,k}$. If necessary, the expansion of $n_j$ is padded with zeroes in order to be of length $\ell$.

*§ 9.1 Generic methods*

---

**Algorithm 9.23** Multi-exponentiation using Straus' trick

INPUT: The elements $x_0, \ldots, x_{r-1} \in G$ and the $\ell$-bit positive exponents $n_0, \ldots, n_{r-1}$. For each $i = (i_{r-1} \ldots i_0)_2 \in [0, 2^r - 1]$, the precomputed value $g_i = \prod_{j=0}^{r-1} x_j^{i_j}$.
OUTPUT: The element $x_0^{n_0} \cdots x_{r-1}^{n_{r-1}}$.

1. $y \leftarrow 1$
2. **for** $k = \ell - 1$ **down to** $0$ **do**
3. $\quad y \leftarrow y^2$
4. $\quad i \leftarrow \sum_{j=0}^{r-1} n_{j,k} 2^j$ $\qquad\qquad [i = (n_{r-1,k} \ldots n_{0,k})_2]$
5. $\quad y \leftarrow y \times g_i$
6. **return** $y$

---

**Remarks 9.24**

(i) Computing $x_0^{n_0} \ldots x_{r-1}^{n_{r-1}}$ in a naïve way requires $r\ell$ squarings and $r\ell/2$ multiplications on average. With Algorithm 9.23, precomputations cost $2^r - r - 1$ multiplications, then only $\ell$ squarings and $(1 - 1/2^r)\ell$ multiplications are necessary on average. However one needs to store $2^r - r$ values.

(ii) One can use Algorithm 9.23 to compute $x^n$. To do so, write $n = (n_{\ell-1} \ldots n_0)_b$ in base $b$, then set $x_i = x^{b^i}$ and compute $x^n = \prod_{i=0}^{\ell-1} x_i^{n_i}$. This approach can be seen as a baby-step giant-step algorithm, where the giant steps $x^{b^i}$ are computed first.

(iii) All the improvements of Algorithm 9.1 described previously apply to Algorithm 9.23 as well. In particular the use of parallel sliding window leads to a faster method; see [AVA 2002, AVA 2005b, BER 2002] for a general overview on multi-exponentiation.

**Example 9.25** Let us compute $x^{31021}$. One has $31021 = (7\ 36\ 45)_{64}$, so that $x^n = x_0^{45} x_1^{36} x_2^7$ where $x_0 = x$, $x_1 = x^{64}$ and $x_2 = x^{64^2}$. First precompute the $g_i$'s

| $i$ | 0 | 1 | 2 | 3 | 4 | 5 | 6 | 7 |
|---|---|---|---|---|---|---|---|---|
| $g_i$ | 1 | $x_0$ | $x_1$ | $x_0 x_1$ | $x_2$ | $x_0 x_2$ | $x_1 x_2$ | $x_0 x_1 x_2$ |

Then one gets

| $k$ | 5 | 4 | 3 | 2 | 1 | 0 |
|---|---|---|---|---|---|---|
| $n_{2,k}$ | 0 | 0 | 0 | 1 | 1 | 1 |
| $n_{1,k}$ | 1 | 0 | 0 | 1 | 0 | 0 |
| $n_{0,k}$ | 1 | 0 | 1 | 1 | 0 | 1 |
| $i$ | 3 | 0 | 1 | 7 | 4 | 5 |
| $y$ | $x_0 x_1$ | $x_0^2 x_1^2$ | $x_0^5 x_1^4$ | $x_0^{11} x_1^9 x_2$ | $x_0^{22} x_1^{18} x_2^3$ | $x_0^{45} x_1^{36} x_2^7$ |

To improve Straus' method in case of a double exponentiation within a group where inversion can be performed efficiently, Solinas [SOL 2001] made signed-binary expansions come back into play.

**Definition 9.26** The *joint sparse form*, JSF for short, of the $\ell$-bit integers $n_0$ and $n_1$ is a representation of the form

$$\begin{pmatrix} n_0 \\ n_1 \end{pmatrix} = \begin{pmatrix} n_{0,\ell} \ldots n_{0,0} \\ n_{1,\ell} \ldots n_{1,0} \end{pmatrix}_{\text{JSF}}$$

such that

- of any three consecutive positions, at least one is a zero column, that is for all $i$ and all positive $j$ one has $n_{i,j+k} = n_{1-i,j+k} = 0$, for at least one $k$ in $\{0, \pm 1\}$
- adjacent terms do not have opposite signs, i.e., it is never the case that $n_{i,j}n_{i,j+1} = -1$
- if $n_{i,j+1}n_{i,j} \neq 0$ then one has $n_{1-i,j+1} = \pm 1$ and $n_{1-i,j} = 0$.

The *joint Hamming weight* is the number of positions different from a zero column.

Solinas also gives an algorithm to compute the JSF of two integers.

---

**Algorithm 9.27** Joint sparse form recoding

INPUT: Nonnegative $\ell$-bit integers $n_0$ and $n_1$ not both zero.
OUTPUT: The joint sparse form of $n_0$ and $n_1$.

1. $j \leftarrow 0$, $S_0 = ()$, $S_1 = ()$, $d_0 \leftarrow 0$ and $d_1 \leftarrow 0$
2. **while** $n_0 + d_0 > 0$ **or** $n_1 + d_1 > 0$ **do**
3.     $\ell_0 \leftarrow d_0 + n_0$ and $\ell_1 \leftarrow d_1 + n_1$
4.     **for** $i = 0$ **to** 1 **do**
5.         **if** $\ell_i \equiv 0 \pmod{2}$ **then** $r_i \leftarrow 0$
6.         **else**
7.             $r_i \leftarrow \ell_i \text{ mods } 4$
8.             **if** $\ell_i \equiv \pm 3 \pmod{8}$ **and** $\ell_{1-i} \equiv 2 \pmod{4}$ **then** $r_i \leftarrow -r_i$
9.         $S_i \leftarrow r_i \parallel S_i$                       [$r_i$ prepended to $S_i$]
10.     **for** $i = 0$ **to** 1 **do**
11.         **if** $2d_i = 1 + n'_i$ **then** $d_i \leftarrow 1 - d_i$
12.         $n_i \leftarrow \lfloor n_i/2 \rfloor$
13.     $j \leftarrow j + 1$
14. **return** $S_0$ and $S_1$

---

### Remarks 9.28

(i) The joint sparse form of $n_0$ and $n_1$ is unique. The joint Hamming weight of the JSF of $n_0$ and $n_1$ is equal to $\ell/2$ on average and the JSF is optimal in the sense that it has the smallest joint Hamming weight among all joint signed representations of $n_0$ and $n_1$ [SOL 2001].

(ii) The naïve computation of $x_0^{n_0} x_1^{n_1}$ involving NAF representations of $n_0$ and $n_1$ requires $2\ell$ squarings and $2\ell/3$ multiplications on average. Only $\ell$ squarings and $\ell/2$ multiplications are necessary with the JSF, neglecting the cost of the precomputations of $x_0 x_1$ and $x_0/x_1$. Applying Straus' trick with two integers in NAF results in a Hamming weight of $5\ell/9$ on average.

(iii) In [GRHE+ 2004], Grabner et al. introduce the *simple joint sparse form* whose joint Hamming weight is also minimal but which can be obtained in an easier way.

**Example 9.29** Let us compute $x_0^{51} x_1^{166}$. The joint NAF expansion of 51 and 166 is

$$51 = (010\bar{1}010\bar{1})_{\text{NAF}}$$
$$166 = (10101 0\bar{1}0)_{\text{NAF}}.$$

Its joint Hamming weight is 8. The JSF of 51 and 166, as given by Algorithm 9.27, is

$$\binom{51}{166} = \binom{0010\bar{1}0011}{10\bar{1}0\bar{1}\bar{1}0\bar{1}0}_{\text{JSF}}$$

with a joint Hamming weight equal to 6.

The next section is devoted to the case when several exponentiations to the same exponent $n$ have to be performed.

## 9.2 Fixed exponent

The methods considered in this section essentially give better algorithms when the exponent $n$ is fixed. They rely on the concept of addition chains. However, the computation of a short addition chain for a given exponent can be very costly. But if the exponent is to be used several times it is probably a good investment to carry out this search.

In the following, different kinds of addition chains are discussed, then efficient methods to actually find short addition chains are introduced before related exponentiation algorithms are described.

### 9.2.1 Introduction to addition chains

**Definition 9.30** An *addition chain* computing an integer $n$ is given by two sequences $v$ and $w$ such that

$$\begin{aligned} v &= (v_0, \ldots, v_s), \ v_0 = 1, \ v_s = n \\ v_i &= v_j + v_k \ \text{for all} \ 1 \leqslant i \leqslant s \ \text{with respect to} \\ w &= (w_1, \ldots, w_s), \ w_i = (j, k) \ \text{and} \ 0 \leqslant j, k \leqslant i - 1. \end{aligned} \quad (9.1)$$

The *length* of the addition chain is $s$.

A *star addition chain* satisfies the additional property that at each step $v_i = v_{i-1} + v_k$ for some $k$ such that $0 \leqslant k \leqslant i - 1$.

Note that one should write $v_i = v_{j(i)} + v_{k(i)}$ since the indexes depend on $i$. They are omitted for the sake of simplicity. Sometimes only $v$ is given since it is easy to retrieve $w$ from $v$. For example $v = (1, 2, 3, 6, 7, 14, 15)$ is an addition chain for 15 of length 6. It is implicit in the computation of $x^{15}$ by Algorithm 9.1. In fact binary or window methods can be seen as methods producing and using special classes of addition chains but it is often possible to do better, that is to find a shorter chain. For instance $(1, 2, 3, 6, 12, 15)$ computes also 15 and is of length 5.

For a given $n$, the smallest $s$ for which there exists an addition chain of length $s$ computing $n$ is denoted by $\ell(n)$. It is not hard to see that $\ell(15) = 5$ but the determination of $\ell(n)$ can be a difficult problem even for rather small $n$.

As complexities of squarings and multiplications are usually slightly different, note that a carefully chosen complexity measure should take into account not only the length of the chain but also the respective numbers of squarings and multiplications involved. For example $(1, 2, 4, 5, 6, 11)$ and

$(1, 2, 3, 4, 8, 11)$ compute 11 and have the same length 5. However the first chain needs 3 multiplications whereas the latter requires only 2.

There is an abundant literature about addition chains. It is known [SCH 1975, BRA 1939] that

$$\lg(n) + \lg(\nu(n)) - 2.13 \leqslant \ell(n) \leqslant \lg(n) + \lg(n)(1 + o(1))/\lg(\lg(n)),$$

where $\nu(n)$ is the Hamming weight of $n$.

As said before, finding an addition chain of the shortest length can be very hard. To make this process easier, it seems harmless to restrict the search to star addition chains. But Hansen [HAN 1959] proved that for some $n$, the smallest being $n = 12509$, there is no star addition chain of minimal length $\ell(n)$. The shortest length $\ell(n)$ has been determined for all $n$ up to $2^{20}$, pruning trees to speed up the search [BLFL]. See also Thurber's algorithm, which is able to find all the addition chains for a given $n$ [THU 1999]. The hardness of this search depends primarily on $\nu(n)$, so that it is longer to find the minimal length of $191 = (10111111)_2$ than $1048577 = (10000000000000000001)_2$, but the running time can be quite long, even for small integers with a rather low density.

The concept of addition chain can be extended in at least three different ways.

**Definition 9.31** An *addition-subtraction chain* is similar to an addition chain except that the condition $v_i = v_j + v_k$ is replaced by $v_i = v_j + v_k$ or $v_i = v_j - v_k$.

For example, an addition chain for 314 is $v = (1, 2, 4, 8, 9, 19, 38, 39, 78, 156, 157, 314)$. The addition–subtraction chain $v = (1, 2, 4, 5, 10, 20, 40, 39, 78, 156, 157, 314)$ is one term shorter.

**Definition 9.32** An *addition sequence* for the set of integers $S = \{n_0, \ldots, n_{r-1}\}$ is an addition chain $v$ that contains each element of $S$. In other words, for all $i$ there is $j$ such that $n_i = v_j$.

For example, an addition sequence computing $\{47, 117, 343, 499, 933, 5689\}$ is

$(1, 2, 4, 8, 10, 11, 18, 36, \underline{47}, 55, 91, 109, \underline{117}, 226, \underline{343}, 434, 489, \underline{499}, \underline{933}, 1422, 2844, 5688, \underline{5689})$.

In [YAO 1976], it is shown that the shortest length of an addition sequence computing the set of integers $\{n_0, \ldots, n_{r-1}\}$ is less than

$$\lg N + cr \lg N / \lg \lg N,$$

where $N = \max_i \{n_i\}$ and $c$ is some constant.

**Definition 9.33** Let $k$ and $s$ be positive integers. A *vectorial addition chain* is a sequence $V$ of $k$-dimensional vectors of nonnegative integers $v_i$ for $-k + 1 \leqslant i \leqslant s$ together with a sequence $w$, such that

$$\begin{aligned}
v_{-k+1} &= [1, 0, 0, \ldots, 0, 0] \\
v_{-k+2} &= [0, 1, 0, \ldots, 0, 0] \\
&\vdots \\
v_0 &= [0, 0, 0, \ldots, 0, 1] \\
v_i &= v_j + v_k \text{ for all } 1 \leqslant i \leqslant s \text{ with } -k + 1 \leqslant j, k \leqslant i - 1 \\
v_s &= [n_0, \ldots, n_{r-1}] \\
w &= (w_1, \ldots, w_s), \ w_i = (j, k).
\end{aligned} \quad (9.2)$$

For example, a vectorial addition chain for $[45, 36, 7]$ is

$$\begin{aligned}
V = \ (&[1, 0, 0], [0, 1, 0], [0, 0, 1], [1, 1, 0], [2, 2, 0], [4, 4, 0], [5, 4, 0], [10, 8, 0], \\
&[11, 9, 0], [11, 9, 1], [22, 18, 2], [22, 18, 3], [44, 36, 6], [45, 36, 6], [45, 36, 7])
\end{aligned}$$

## § 9.2 Fixed exponent

$$w = ((-2, -1), (1, 1), (2, 2), (-2, 3), (4, 4), (1, 5), (0, 6), (7, 7), (0, 8), (9, 9), (-2, 10), (0, 11)).$$

Since $k = 3$, the first three terms of $V$ are $v_{-2} = [1, 0, 0]$, $v_{-1} = [0, 1, 0]$, and $v_0 = [0, 0, 1]$. This chain is of length 12 and is implicitly produced by Algorithm 9.23.

Addition sequences and vectorial addition chains are in some sense dual. We refer the interested reader to [BER 2002, STA 2003] for details and to [KNPA 1981] for a more general approach. In [OLI 1981] Olivos describes a procedure to transform an addition sequence computing $\{n_0, \ldots, n_{r-1}\}$ of length $\ell$ into a vectorial addition chain of length $\ell + r - 1$ for $[n_0, \ldots, n_{r-1}]$.

To illustrate his method let us deduce a vectorial addition chain for $[45, 36, 7]$ from the addition sequence $v = (1, 2, 4, 6, 7, 9, 18, 36, 45)$ computing $\{7, 36, 45\}$. Let $\{e_j \mid 0 \leq j \leq k\}$ be the canonical basis of $\mathbb{R}^{k+1}$. The idea is then to build an array by induction, starting in the lower right corner with a 2-by-2 array, and then processing the successive elements $v_h$ of the addition sequence, following two rules:

- if $v_h = 2v_i$ then the line to be added on top is the double of line $i$ and the new two columns on the left are $2e_h$ and $2e_h + e_i$
- if $v_h$ satisfies $v_h = v_i + v_j$ then the new line on top is the sum of lines $i$ and $j$ and the two columns on the left are $e_h + e_i$ and $e_h + e_j$.

The expression of $v_h$ in terms of the $v_i$'s is written on the right. The first steps are:

| 2 | 2 | 2 = 1 + 1 |
|---|---|---|
| 0 | 1 | 1 |

$\implies$

| 2 | 2 | 4 | 4 | 4 = 2 + 2 |
|---|---|---|---|---|
| 0 | 1 | 2 | 2 | 2 = 1 + 1 |
| 0 | 0 | 0 | 1 | 1 |

$\implies$

| 1 | 1 | 2 | 3 | 6 | 6 | 6 = 4 + 2 |
|---|---|---|---|---|---|---|
| 1 | 0 | 2 | 2 | 4 | 4 | 4 = 2 + 2 |
| 0 | 1 | 0 | 1 | 2 | 2 | 2 = 1 + 1 |
| 0 | 0 | 0 | 0 | 0 | 1 | 1 |

At the end one has

| 1 | 1 | 2 | 2 | 4 | 5 | 5 | 5 | 5 | 5 | 10 | 10 | 20 | 40 | 45 | 45 = 36 + 9 ← |
|---|---|---|---|---|---|---|---|---|---|----|----|----|----|----|---|
| 1 | 0 | 2 | 2 | 4 | 4 | 4 | 4 | 4 | 4 | 8  | 8  | 16 | 32 | 36 | 36 = 18 + 18 ← |
| 0 | 0 | 0 | 1 | 2 | 2 | 2 | 2 | 2 | 2 | 4  | 4  | 8  | 16 | 18 | 18 = 9 + 9 |
| 0 | 1 | 0 | 0 | 0 | 1 | 1 | 1 | 1 | 1 | 2  | 2  | 4  | 8  | 9  | 9 = 7 + 2 |
| 0 | 0 | 0 | 0 | 0 | 0 | 1 | 0 | 1 | 1 | 1  | 2  | 3  | 6  | 7  | 7 = 6 + 1 ← |
| 0 | 0 | 0 | 0 | 0 | 0 | 0 | 0 | 1 | 1 | 1  | 2  | 3  | 6  | 6  | 6 = 4 + 2 |
| 0 | 0 | 0 | 0 | 0 | 0 | 0 | 0 | 0 | 0 | 1  | 0  | 2  | 4  | 4  | 4 = 2 + 2 |
| 0 | 0 | 0 | 0 | 0 | 0 | 0 | 1 | 0 | 0 | 0  | 1  | 2  | 2  | 2  | 2 = 1 + 1 |
| 0 | 0 | 0 | 0 | 0 | 0 | 0 | 0 | 0 | 1 | 0  | 0  | 0  | 1  | 1  | 1 |

Then discard all the lines except the ones marked by an arrow and corresponding to 7, 36, and 45. Consider the columns from the left to the right, eliminate redundancies and finally add the canonical vectors of $\mathbb{R}^r$ so that a vectorial addition sequence for $[45, 36, 7]$ is

$$([1, 0, 0], [0, 1, 0], [0, 0, 1], [1, 1, 0], [2, 2, 0], [4, 4, 0], [5, 4, 0],$$
$$[5, 4, 1], [10, 8, 1], [10, 8, 2], [20, 16, 3], [40, 32, 6], [45, 36, 7]).$$

Conversely the procedure to get an addition sequence from a vectorial addition exists as well [OLI 1981].

Before explaining how to find short chains, let us remark that the set of vectors $(n_0, \ldots, n_{r-1}, \ell)$ such that there is an addition sequence of length $\ell$ containing $n_0, \ldots, n_{r-1}$ has been shown to be **NP**-complete [DOLE⁺ 1981]. This does not imply, as it is sometimes claimed, that finding a shortest addition chain for $n$ is **NP**-complete. However, we have seen that dedicated algorithms to find a shortest addition chain are in practice limited to small exponents.

## 9.2.2 Short addition chains search

In the following, different heuristics to find short addition chains are discussed. They are rather efficient but do not necessarily find a shortest possible chain. Most of the methods described here use the concept of dictionary.

**Definition 9.34** Given an integer $n$, a *dictionary* $\mathcal{D}$ for $n$ is a set of integers such that

$$n = \sum_{i=0}^{k} b_{i,d} d 2^i, \text{ with } b_{i,d} \in \{0,1\} \text{ and } d \in \mathcal{D}.$$

Note that all the algorithms introduced in Section 9.1 can be used to produce addition chains and implicitly use a dictionary. For example, the dictionary associated to window methods of length $k$ is the set $\{1, 3, \ldots, 2^k - 1\}$. For the $\text{NAF}_w$ it is $\{\pm 1, \pm 3, \ldots, \pm (2^{k-1} - 1)\}$.

In [GoHA$^+$ 1996, O'Co 2001] the dictionary is simply made of the elements $2^i - 1$ for $0 < i \leqslant w$, for some fixed parameter $w$.

The power tree method [KNU 1997] is quite simple to implement but it does not always return an optimal addition chain, the first counter example being $n = 77$. Like other algorithms exploring trees it cannot be used for exponents of cryptographic relevance, as the size grows too fast in the bit size of the exponent and is too large for the required sizes.

A more sophisticated method is described in [KUYA 1998] and is related to the Tunstall method; see [TUN 1968]. Namely, choose a parameter $k$, let $p$ be the number of zeroes in the expansion of $n$ divided by its length $\ell$ and let $q = 1 - p$. If the expansion is signed let $\hat{q} = (1-p)/2$. Then form a tree having a root of weight 1 and while the number of leaves is less than $k+1$ add leaves to this tree according to the following procedure. Take the leaf of highest weight $w$ and create two children with weight $wp$ and $wq$, labeled respectively by 0 and 1. If the expansion is signed create three children with weight $wp$, $w\hat{q}$, and $w\hat{q}$, labeled respectively by 0, $\bar{1}0$, and 10, instead. At the end read the labels from the root to the leaves and concatenate 1 (10 if signed) at the beginning of each sequence. The dictionary $\mathcal{D}$ is the set of odd integers obtained by removing the zeroes at the end of each sequence. The result is a function of $\ell$ and of the number of zeroes in the signed-binary expansion of $n$. The best choice for the size of the dictionary depends on $\ell$ and can be as large as 20 for 512-bit exponents [KUYA 1998].

**Example 9.35** Take $n = 587257$ and $k = 4$. The signed-binary recoding Algorithm 9.14 gives $n = (1001000\bar{1}0\bar{1}00000\bar{1}001)_{\text{NAF}}$. One has $p = 7/10$ and $\hat{q} = 3/20$. After two iterations, there are five leaves and $\mathcal{D} = \{(00), (010), (0\bar{1}0), (10), (\bar{1}0)\}$ as shown below

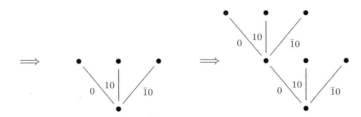

Then concatenate (10) at the beginning of each sequence of $\mathcal{D}$ and remove all the zeroes at the end to finally get $\mathcal{D} = \{1, 3, 5, 7, 9\}$. From this one can compute $n = 2^{19} + 2^{16} - 5 \times 2^9 - 7$.

Yacobi suggests a completely different approach [YAC 1998], namely to use the well-known Lempel–Ziv compression algorithm [ZILE 1977, LEZI 1978] to get the dictionary.

At the beginning the dictionary is empty and new elements are added while the binary expansion of the exponent is scanned from the right to the left. Take the longest word in the dictionary that

## § 9.2 Fixed exponent

fits as prefix of the unscanned part of the exponent and concatenate it to the next unscanned digit of the exponent to form a new element of the dictionary. Repeat the process until all the digits are processed. There is also a sliding version that skips strings of zeroes.

**Example 9.36** For $n = 587257$ one gets $(10\underline{00}\ \underline{1111}\ \underline{010}\ \underline{111}\ \underline{11}\ \underline{10}\ 0\ \underline{1})_2$ and the dictionary $\{1, 0, 2, 3, 7, 2, 15, 0, 2\}$, which actually gives rise to $\mathcal{D} = \{1, 3, 7, 15\}$. One has $n = 1 \times 2^{19} + 15 \times 2^{12} + 1 \times 2^{10} + 7 \times 2^6 + 3 \times 2^4 + 1 \times 2^3 + 1$ so that an addition chain for $n$ is

$$(1, 2, 3, 4, 7, 8, 15, 16, 32, 64, 128, 143, 286, 572, 573, 1146, 2292, 4584,$$
$$9168, 9175, 18350, 36700, 36703, 73406, 73407, 146814, 293628, 587256, 587257)$$

whose length is 28.

The sliding version returns $\mathcal{D} = \{1, 3, 5, 7\}$ from the decomposition

$$(1000\,\underline{0111}\,\underline{101}\,\underline{01}\,\underline{111}\,\underline{11}\,00\,\underline{1})_2.$$

In this case $n = 1 \times 2^{19} + 7 \times 2^{13} + 5 \times 2^{10} + 1 \times 2^8 + 7 \times 2^6 + 3 \times 2^3 + 1$ and an addition chain for $n$ is

$$(1, 2, 3, 5, 7, 8, 16, 32, 64, 71, 142, 284, 568, 573, 1146, 2292, 2293, 4586,$$
$$9172, 18344, 18351, 36702, 73404, 73407, 146814, 293628, 587256, 587257)$$

of length 27.

This method can also be used with signed-digit representations and is particularly efficient when the number of zeroes is small.

In [BEBE$^+$ 1989] continued fractions and the Euclid algorithm are used to produce short addition chains. First let $\otimes$ and $\oplus$ be two simple operations on addition chains, defined as follows. If $v = (v_0, \ldots, v_s)$ and $w = (w_0, \ldots, w_t)$ then

$$v \otimes w = (v_0, \ldots, v_s, v_s w_0, \ldots, v_s w_t)$$

and if $j$ is an integer

$$v \oplus j = (v_0, \ldots, v_s, v_s + j).$$

Now let $1 < k < n$ be some integer. Then

$$(1, \ldots, k, \ldots, n) = (1, \ldots, n \bmod k, \ldots, k) \otimes (1, \ldots, \lfloor n/k \rfloor) \oplus (n \bmod k). \tag{9.3}$$

The point is to choose the best possible $k$. If $k = \lfloor n/2 \rfloor$ then the addition chain is equal to the one obtained with binary methods. Instead the authors propose a *dichotomic strategy*, that is to take

$$k = \gamma(n) = \left\lfloor \frac{n}{2^{\lceil \lfloor \lg n \rfloor / 2 \rceil}} \right\rfloor.$$

The rule (9.3) is then applied recursively in `minchain(n)`, which uses the additional procedure `chain(n, k)`,

```
minchain(n)
  1. if n = 2^ℓ then return (1, 2, 4, ..., 2^ℓ)
  2. if n = 3 then return (1, 2, 3)
  3. return chain(n, 2^⌊lg n/2⌋)
```

and

```
chain(n, k)
  1. q ← ⌊n/k⌋ and r ← n mod k
  2. if r = 0 then return (minchain(k) ⊗ minchain(q))
  3. else return chain(k, r) ⊗ minchain(q) ⊕ r
```

Note that these algorithms are able to find short addition sequences as well.

**Example 9.37** For $n = 87$, one has $k = \lfloor 87/8 \rfloor = 10$ and the successive calls are

$$\text{chain}(87, 10)$$
$$\text{chain}(10, 7) \otimes \text{minchain}(8) \oplus 7$$
$$(\text{chain}(7, 3) \otimes \text{minchain}(1) \oplus 3) \otimes \text{minchain}(8) \oplus 7$$
$$((\text{minchain}(3) \otimes \text{minchain}(2) \oplus 1) \otimes \text{minchain}(1) \oplus 3) \otimes \text{minchain}(8) \oplus 7$$

so that the final result is the optimal addition chain $(1, 2, 3, 6, 7, 10, 20, 40, 80, 87)$.

In [BEBE⁺ 1994], the authors generalize this approach, introducing new strategies to determine a set of possible values for $k$. So the choice of $k$ is no longer deterministic and it is necessary to backtrack the best possible $k$. For the *factor method*, see also [KNU 1997], one has

$$\begin{cases} k \in \{n-1\} & \text{if } n \text{ is prime} \\ k \in \{n-1, p\} & \text{if } p \text{ is the smallest prime dividing } n. \end{cases}$$

For the *total strategy*, $k \in \{2, 3, \ldots, n-1\}$. For the *dyadic strategy*,

$$k \in \left\{ \left\lfloor \frac{n}{2^j} \right\rfloor, j = 1, \ldots \right\}.$$

Note that only *Fermat's strategy*, where

$$k \in \left\{ \left\lfloor \frac{n}{2^{2^j}} \right\rfloor, j = 0, 1, \ldots \right\}$$

has a reasonable complexity and is well suited for large exponents.

**Example 9.38** The corresponding results for 87 are all optimal and given in the next table.

| Strategy | Initial $k$ | Result |
|---|---|---|
| Factor | 3 | $(1, 2, 3, 6, 12, 24, 48, 72, 84, 87)$ |
| Total | 17 | $(1, 2, 4, 8, 16, 17, 34, 68, 85, 87)$ |
| Dyadic | 2 | $(1, 2, 4, 6, 10, 20, 40, 80, 86, 87)$ |
| Fermat | 5 | $(1, 2, 3, 5, 10, 20, 40, 80, 85, 87)$ |

In [BOCO 1990] Bos and Coster use similar ideas to produce an addition sequence. See also [COSTER]. Starting from 1, 2 and the requested numbers $\ldots, f_2, f_1, f$ they replace at each step the last term by new elements produced by one of four different methods. A weight function helps

## § 9.2 Fixed exponent

to decide which rule should be used at each stage. Here is a brief description of these strategies with some examples.

### Approximation

Condition  $0 \leqslant f - (f_i + f_j) = \varepsilon$ with $f_i \leqslant f_j$ and $\varepsilon$ small
Insert  $f_i + \varepsilon$
Example  49 67 85 117 → 49 $\underline{50}$ 67 85 (because 117 - (49 + 67) = 1)

### Division

Condition  $f$ is divisible by $p \in \{3, 5, 9, 17\}$. Let $(1, 2, \ldots, \alpha_r = p)$ be an addition chain for $p$
Insert  $f/p, 2f/p, \ldots, \alpha_{r-1}f/p$
Example  17 48 → $\underline{16}$ 17 $\underline{32}$ (because 48/3 = 16 and (1,2,3) computes 3)

### Halving

Condition  $f/f_i \geqslant 2^u$ and $\lfloor f/2^u \rfloor = k$
Insert  $k, 2k, \ldots, k2^u$
Example  14 382 → 14 $\underline{23}$ $\underline{46}$ $\underline{92}$ $\underline{184}$ $\underline{368}$ (because 382/14 $\geqslant 2^4$ and $\lfloor 382/2^4 \rfloor = 23$)

### Lucas

Condition  $f$ and $f_i$ belong to a Lucas series ($f_i = u_0$, $f = u_k$, $k \geqslant 3$ and $u_{i+1} = u_i + u_{i-1}$)
Insert  $u_1, u_2, \ldots, u_{k-1}$
Example  4 23 → 4 $\underline{5}$ $\underline{9}$ $\underline{14}$ (because 4,5,9,14,23 is a Lucas series)

For faster results use only **Approximation** and **Halving** steps. The choice is simpler and does not require any weight function.

**Example 9.39** This method applied to $\{1, 2, 47, 117, 343, 499, 933, 5689\}$ returns

$(1, 2, 4, 8, 10, 11, 18, 36, \underline{47}, 55, 91, 109, \underline{117}, 226, \underline{343}, 434, 489, \underline{499}, \underline{933}, 1422, 2844, 5688, \underline{5689})$.

The method of Bos and Coster when combined to a sliding window of big length allows us to compute $x^n$ with a dictionary of small size and no precomputation. The following example is taken from [BoCo 1990].

**Example 9.40** Let $n = 2623594742895366318191$ and take a window of length 10 (except for the first digit corresponding to a window of length 13). Then

$n = (\underbrace{1011000111001}_{5689}\,0000001\underbrace{110100101}_{933}\,00\underbrace{1110101}_{117}\,000000\underbrace{101111}_{47}\,00000\underbrace{111110011}_{499}\,00\underbrace{101010111}_{343})_2$

so that the dictionary is $\mathcal{D} = \{47, 117, 343, 499, 933, 5689\}$ and the corresponding addition chain built with $\mathcal{D}$ is of length 89. By way of comparison, Algorithms 9.1 and 9.10 need respectively 110 and 93 operations.

Finally, see [NEMA 2002] for techniques related to genetic algorithms.

### 9.2.3 Exponentiation using addition chains

Once an addition chain for $n$ is found it is straightforward to deduce $x^n$.

---
**Algorithm 9.41** Exponentiation using addition chain

INPUT: An element $x$ of $G$ and an addition chain computing $n$ i.e., $v$ and $w$ as in (9.1).
OUTPUT: The element $x^n$.

1. $x_1 \leftarrow x$
2. **for** $i = 1$ **to** $s$ **do** $x_i \leftarrow x_j \times x_k$      $[w(i) = (j,k)]$
3. **return** $x_s$

---

**Example 9.42** Let us compute $x^{314}$, from the addition chain for 314 given below

| $i$ | 0 | 1 | 2 | 3 | 4 | 5 | 6 | 7 | 8 | 9 | 10 | 11 | 12 |
|---|---|---|---|---|---|---|---|---|---|---|---|---|---|
| $v_i$ | 1 | 2 | 4 | 8 | 9 | 18 | 19 | 38 | 39 | 78 | 156 | 157 | 314 |
| $w_i$ | — | (0,0) | (1,1) | (2,2) | (3,0) | (4,4) | (5,0) | (6,6) | (7,0) | (8,8) | (9,9) | (10,0) | (11,11) |
| $x_i$ | $x$ | $x^2$ | $x^4$ | $x^8$ | $x^9$ | $x^{18}$ | $x^{19}$ | $x^{38}$ | $x^{39}$ | $x^{78}$ | $x^{156}$ | $x^{157}$ | $x^{314}$ |

Vectorial addition chains are well suited to multi-exponentiation. Here again $G$ is assumed to be abelian.

---
**Algorithm 9.43** Multi-exponentiation using vectorial addition chain

INPUT: Elements $x_0, \ldots, x_{r-1}$ of $G$ and a vectorial addition chain of dimension $r$ computing $[n_0, \ldots, n_{r-1}]$ as in (9.2).
OUTPUT: The element $x_0^{n_0} \cdots x_{r-1}^{n_{r-1}}$.

1. **for** $i = -k + 1$ **to** $0$ **do** $y_i \leftarrow x_{i+k-1}$
2. **for** $i = 1$ **to** $s$ **do** $y_i \leftarrow y_j \times y_k$      $[w(i) = (j,k)]$
3. **return** $y_s$

---

The vectorial addition chain for $[45, 36, 7]$ implicitly produced by Algorithm 9.23 is of length 12. A careful search reveals a chain of length 10 as it can be seen in the next table, which displays the execution of Algorithm 9.43 while computing $x_0^{45} x_1^{36} x_2^7$ with it. Recall that $y_{-2} = x_0$, $y_{-1} = x_1$ and $y_0 = x_2$.

| $i$ | 1 | 2 | 3 | 4 | 5 | 6 | 7 | 8 | 9 | 10 |
|---|---|---|---|---|---|---|---|---|---|---|
| $w_i$ | $(-2,-1)$ | $(1,1)$ | $(2,2)$ | $(-2,3)$ | $(0,4)$ | $(4,5)$ | $(0,6)$ | $(6,7)$ | $(8,8)$ | $(5,9)$ |
| $y_i$ | $x_0 x_1$ | $x_0^2 x_1^2$ | $x_0^4 x_1^4$ | $x_0^5 x_1^4$ | $x_0^5 x_1^4 x_2$ | $x_0^{10} x_1^8 x_2$ | $x_0^{10} x_1^8 x_2^2$ | $x_0^{20} x_1^{16} x_2^3$ | $x_0^{40} x_1^{32} x_2^6$ | $x_0^{45} x_1^{36} x_2^7$ |

## 9.3 Fixed base point

In some situations the element $x$ is always the same whereas the exponent varies. Precomputations are the key point here.

## 9.3.1 Yao's method

A simpler version of Algorithm 9.44, which can be seen as the dual of the $2^k$-ary method, was first described in [YAO 1976]. A slightly improved form is presented in [KNU 1981, answer to exercise 9, Section 4.6.3]. Note that it is identical to the patented BGMW's algorithm [BRGO$^+$ 1993].

Let $n$, $n_i$, $b_i$, $\ell$ and $h$ be integers. Suppose that

$$n = \sum_{i=0}^{\ell-1} n_i b_i \text{ with } 0 \leqslant n_i < h \text{ for all } i \in [0, \ell-1]. \tag{9.4}$$

Let $x_i = x^{b_i}$. The method relies on the equality

$$x^n = \prod_{i=0}^{\ell-1} x_i^{n_i} = \prod_{j=1}^{h-1} \left[ \prod_{n_i = j} x_i \right]^j.$$

---

**Algorithm 9.44** Improved Yao's exponentiation

INPUT: The element $x$ of $G$, an exponent $n$ written as in (9.4) and the precomputed values $x^{b_0}, x^{b_1}, \ldots, x^{b_{\ell-1}}$.
OUTPUT: The element $x^n$.

1. $y \leftarrow 1, u \leftarrow 1$ and $j \leftarrow h - 1$
2. **while** $j \geqslant 1$ **do**
3.     **for** $i = 0$ **to** $\ell - 1$ **do if** $n_i = j$ **then** $u \leftarrow u \times x^{b_i}$
4.     $y \leftarrow y \times u$
5.     $j \leftarrow j - 1$
6. **return** $y$

---

**Remarks 9.45**

(i) The term $\left[\prod_{n_i=j} x_i\right]^j$ is computed by repeated iterations in Line 4.
One obtains the correct powers $x_i^{n_i}$ as in each round the result is multiplied with $u$ and $x_i$ is included in $u$ for $n_i$ rounds.

(ii) The choice of $h$ and of the $b_i$'s is free. One can set $h = 2^k$ and $b_i = h^i$ so that the $n_i$'s are simply the digits of $n$ in base $h$.

(iii) One needs $\ell + h - 2$ multiplications and $\ell + 1$ elements must be stored to compute $x^n$.

**Example 9.46** Let us compute $x^{2989}$. Set $h = 4$, $b_i = 4^i$ then $2989 = (2\,3\,2\,2\,3\,1)_4$ and $\ell = 6$. Suppose that $x, x^4, x^{16}, x^{64}, x^{256}$ and $x^{1024}$ are precomputed and stored.

| $j$ | 3 | 2 | 1 |
|---|---|---|---|
| $u$ | $x^4 x^{256} = x^{260}$ | $x^{260} x^{16} x^{64} x^{1024} = x^{1364}$ | $x^{1364} x = x^{1365}$ |
| $y$ | $x^{260}$ | $x^{260} x^{1364} = x^{1624}$ | $x^{1624} x^{1365} = x^{2989}$ |

## 9.3.2 Euclidean method

The Euclidean method was first introduced in [Roo 1995], see also [SEM 1983]. Algorithm 9.47 computes $x^n$ generalizing a method to compute the double exponentiation $x_0^{n_0} x_1^{n_1}$ discussed by Bergeron et al. [BEBE+ 1989] which is similar to the technique introduced in Section 9.2.2 to find short addition chains. The idea is to recursively use the equality

$$x_0^{n_0} x_1^{n_1} = (x_0 x_1^q)^{n_0} \times x_1^{(n_1 \bmod n_0)} \text{ where } q = \lfloor n_1/n_0 \rfloor.$$

---

**Algorithm 9.47** Euclidean exponentiation

INPUT: The element $x$ of $G$, an exponent $n$ as in (9.4) and the precomputed values $x_0 = x^{b_0}, x_1 = x^{b_1}, \ldots, x_{\ell-1} = x^{b_{\ell-1}}$.
OUTPUT: The element $x^n$.

1. **while true do**
2.     Find $M$ such that $n_M \geqslant n_i$ for all $i \in [0, \ell-1]$
3.     Find $N \neq M$ such that $n_N \geqslant n_i$ for all $i \in [0, \ell-1]$, $i \neq M$
4.     **if** $n_N \neq 0$ **then**
5.         $q \leftarrow \lfloor n_M/n_N \rfloor, x_N \leftarrow x_M{}^q \times x_N$ and $n_M \leftarrow n_M \bmod n_N$
6.     **else break**
7. **return** $x_M^{n_M}$

---

**Example 9.48** Take the same exponent $2989 = (2\,3\,2\,2\,3\,1)_4$ and let us evaluate $x^{2989}$.

| $n_5$ | $n_4$ | $n_3$ | $n_2$ | $n_1$ | $n_0$ | $M$ | $N$ | $q$ | $x_5$ | $x_4$ | $x_3$ | $x_2$ | $x_1$ | $x_0$ |
|---|---|---|---|---|---|---|---|---|---|---|---|---|---|---|
| — | — | — | — | — | — | — | — | — | $x^{1024}$ | $x^{256}$ | $x^{64}$ | $x^{16}$ | $x^4$ | $x$ |
| 2 | 3 | 2 | 2 | 3 | 1 | 4 | 1 | 1 | $x^{1024}$ | $x^{256}$ | $x^{64}$ | $x^{16}$ | $x^{260}$ | $x$ |
| 2 | 0 | 2 | 2 | 3 | 1 | 1 | 2 | 1 | $x^{1024}$ | $x^{256}$ | $x^{64}$ | $x^{276}$ | $x^{260}$ | $x$ |
| 2 | 0 | 2 | 2 | 1 | 1 | 5 | 3 | 1 | $x^{1024}$ | $x^{256}$ | $x^{1088}$ | $x^{276}$ | $x^{260}$ | $x$ |
| 0 | 0 | 2 | 2 | 1 | 1 | 3 | 2 | 1 | $x^{1024}$ | $x^{256}$ | $x^{1088}$ | $x^{1364}$ | $x^{260}$ | $x$ |
| 0 | 0 | 0 | 2 | 1 | 1 | 2 | 1 | 2 | $x^{1024}$ | $x^{256}$ | $x^{1088}$ | $x^{1364}$ | $x^{2988}$ | $x$ |
| 0 | 0 | 0 | 0 | 1 | 1 | 1 | 0 | 1 | $x^{1024}$ | $x^{256}$ | $x^{1088}$ | $x^{1364}$ | $x^{2988}$ | $x^{2989}$ |
| 0 | 0 | 0 | 0 | 0 | 1 | 0 | 1 | — | $x^{1024}$ | $x^{256}$ | $x^{1088}$ | $x^{1364}$ | $x^{2988}$ | $x^{2989}$ |

## 9.3.3 Fixed-base comb method

This algorithm is a special case of Pippenger's algorithm [BER 2002, PIP 1979, PIP 1980]. It is also often referred to as Lim–Lee method [LILE 1994]. It is essentially a special case of Algorithm 9.23 where the different base points are in fact distinct powers of a single base. Suppose that $n = (n_{\ell-1} \ldots n_0)_2$. Select an integer $h \in [1, \ell]$. Let $a = \lceil \ell/h \rceil$ and choose $v \in [1, a]$. Let $r = \lceil a/v \rceil$ and write the $n_i$'s in an array with $h$ rows and $a$ columns as below (pad the representation of $n$ with zeroes if necessary)

## § 9.3 Fixed base point

| $a-1$ | $\cdots$ | 1 | 0 |
|---|---|---|---|
| $n_{a-1}$ | $\cdots$ | $n_1$ | $n_0$ |
| $n_{2a-1}$ | $\cdots$ | $n_{a+1}$ | $n_a$ |
| $\vdots$ | $\vdots$ | $\vdots$ | $\vdots$ |
| $n_{ah-1}$ | $\cdots$ | $n_{ah-a+1}$ | $n_{ah-a}$ |

For each $s$, the column number $s$ can be read as the binary representation of an integer denoted by $I(s)$. For example the last column $I(0)$ is the binary representation of $I(0) = (n_{ah-a} \ldots n_a n_0)_2$. The algorithm relies on the following equality

$$x^n = \prod_{k=0}^{r-1} \left( \prod_{j=0}^{v-1} G[j, I(jr+k)] \right)^{2^k}$$

where

$$G[j, i] = \left( \prod_{s=0}^{h-1} x^{i_s 2^{as}} \right)^{2^{jr}} \quad \text{for } j \in [0, v-1] \text{ and } i = (i_{h-1} \ldots i_0)_2 \in [0, 2^h - 1].$$

---

**Algorithm 9.49** Fixed-base comb exponentiation

INPUT: The element $x$ of $G$ and an exponent $n$. Let $h, a, v, r$ and $G[j, i]$ be as above.
OUTPUT: The element $x^n$.

1.     $y \leftarrow 1$ and $k \leftarrow r - 1$
2.     **for** $k = r - 1$ **down to** $0$ **do**
3.          $y \leftarrow y^2$
4.          **for** $j = v - 1$ **down to** $0$ **do**
5.              $I \leftarrow \sum_{s=0}^{h-1} n_{as+jr+k} 2^s$      $[I = I(jr+k)]$
6.              $y \leftarrow y \times G[j, I]$
7.     **return** $y$

---

**Example 9.50** Once again, let us compute $x^{2989}$. Set $h = 3$ and $v = 2$ so that $a = 4 = rv$, hence $r = 2$. Form the following array from the digits of $2989 = (101110101101)_2$

| $s$ | 3 | 2 | 1 | 0 |
|---|---|---|---|---|
| $n_s \ldots n_0$ | 1 | 1 | 0 | 1 |
| $n_{a+s} \ldots n_a$ | 1 | 0 | 1 | 0 |
| $n_{2a+s} \ldots n_{2a}$ | 1 | 0 | 1 | 1 |
| $I(s)$ | 7 | 1 | 6 | 5 |

The precomputed values are

| $i$ | 0 | 1 | 2 | 3 | 4 | 5 | 6 | 7 |
|---|---|---|---|---|---|---|---|---|
| $G[0, i]$ | 1 | $x$ | $x^{16}$ | $xx^{16}$ | $x^{256}$ | $xx^{256}$ | $x^{16}x^{256}$ | $xx^{16}x^{256}$ |
| $G[1, i]$ | 1 | $x^4$ | $x^{64}$ | $x^4 x^{64}$ | $x^{1024}$ | $x^4 x^{1024}$ | $x^{64} x^{1024}$ | $x^4 x^{64} x^{1024}$ |

The algorithm proceeds as follows

| $k$ | 1 | 1 | 1 | 0 | 0 | 0 |
|---|---|---|---|---|---|---|
| $j$ | — | 1 | 0 | — | 1 | 0 |
| $jr+k$ | — | 3 | 1 | — | 2 | 0 |
| $I$ | — | 7 | 6 | — | 1 | 5 |
| $G[j,I]$ | — | $x^{1092}$ | $x^{272}$ | — | $x^4$ | $x^{257}$ |
| $y$ | 1 | $x^{1092}$ | $x^{1364}$ | $x^{2728}$ | $x^{2732}$ | $x^{2989}$ |

**Remarks 9.51**

(i) One needs at most $a + r - 2$ multiplications and $v(2^h - 1)$ precomputed values. If squarings can be achieved efficiently or "on the fly" (for example in finite fields of even characteristic represented through normal bases, see Section 11.2.1.c), only $2^h - 1$ values must be stored.

(ii) The adaptation of this method to larger base representations or signed representations such as the non-adjacent form is straightforward.

# Chapter 10

# Integer Arithmetic

*Christophe Doche*

### Contents in Brief

| | | |
|---|---|---|
| 10.1 | **Multiprecision integers** <br> Introduction • Internal representation • Elementary operations | 170 |
| 10.2 | **Addition and subtraction** | 172 |
| 10.3 | **Multiplication** <br> Schoolbook multiplication • Karatsuba multiplication • Squaring | 174 |
| 10.4 | **Modular reduction** <br> Barrett method • Montgomery reduction • Special moduli • Reduction modulo several primes | 178 |
| 10.5 | **Division** <br> Schoolbook division • Recursive division • Exact division | 184 |
| 10.6 | **Greatest common divisor** <br> Euclid extended gcd • Lehmer extended gcd • Binary extended gcd • Chinese remainder theorem | 190 |
| 10.7 | **Square root** <br> Integer square root • Perfect square detection | 197 |

In most of the cases, the integer ring $\mathbb{Z}$ is the fundamental mathematical layer of many cryptosystems. Once it is possible to compute with integers, one can build on top of them finite fields, then curves and even more complicated objects. More generally, rational, real, complex, and $p$-adic numbers, but also polynomials with coefficients in these sets, rely on integers and their arithmetic is greatly influenced by the underlying integer algorithms. That is why integer arithmetic is so important and should be performed as efficiently as possible.

Practical considerations are therefore the core of this chapter. For instance, we first recall some elementary notions on computers to describe how numbers are internally represented and how we can manipulate them. Then we describe the four operations and related items such as square root computations or extended gcd algorithms.

All the techniques for an efficient implementation from scratch are presented in the following chapter. However, there already exist a great number of libraries that are highly optimized and available on the Internet. Most of them are written in C, like GMP [GMP], BigNum [BIGNUM] or FreeLip [FREELIP].

Sometimes, different algorithms are given for the same operation. In this case, the size of the operands mainly determines which method should be used. The algorithms presented here have been directly taken from [MEOO$^+$ 1996], [KNU 1997], [CRPO 2001], and [COH 2000].

## 10.1 Multiprecision integers

Without special software or hardware, computers can only operate on rather small integers. In order to consider larger quantities, we first need to recall some elementary facts on integers and computers.

### 10.1.1 Introduction

Let $b \geqslant 2$ be an integer called the *base* or the *radix*. Every integer $u > 0$ can be written in a unique way as the sum

$$u = u_{n-1}b^{n-1} + \cdots + u_1 b + u_0$$

provided $0 \leqslant u_i < b$ and $u_{n-1} \neq 0$. This is what we will call the *base $b$ representation* of $u$ and will be denoted by $(u_{n-1} \ldots u_0)_b$. When $b$ is understood, it will be usually omitted and $u$ simply written as $u = (u_{n-1} \ldots u_0)$. The $u_i$'s are the *digits* of $u$, $u_{n-1}$ and $u_0$ being respectively the *most significant digit* of $u$ and the *least significant digit* of $u$. The *precision* of $u$ is the largest $i$ such that $u_{i-1} > 0$. It corresponds to the *length* $n$ of $(u_{n-1} \ldots u_0)_b$ and is denoted $|u|_b$.

**Remarks 10.1**

(i) The base $b$ representation of zero is always $(0)_b$.
(ii) It is sometimes useful to add a certain number of zeroes at the beginning of the representation of $u$, i.e., to consider $(0 \ldots 0 \, u_{n-1} \ldots u_0)_b$. This operation, which obviously does not change the value of $u$, is called *padding*.

In a computer, the base $b$ is usually a power of 2 and a number is internally stored as a sequence of 0 and 1 called *bits*. The important elementary operations on bits are the following. Given two bits $x$ and $y$, we can compute the

- *complement* of $x$ denoted by $\overline{x}$, which is equal to 1 if and only if $x$ equals 0
- *conjunction* of $x$ and $y$, $x \wedge y$, which is equal to 1 if and only if $x$ and $y$ both equal 1
- *disjunction* of $x$ and $y$, $x \vee y$, which is equal to 1 if and only if at least one of $x, y$ equals 1
- *exclusive disjunction* of $x$ and $y$, also called XOR. The result $x$ XOR $y$ is equal to 1 if and only if exactly one of the two values $x, y$ equals 1.

Usually computers cannot manipulate bits directly. The smallest quantity of main memory that a computer can address is a *byte*, which is nowadays almost always a sequence of eight bits. A processor can operate on several bytes at the same time by means of a *register*. This important hardware component determines very low-level properties of a processor. The size of a register, called a *word*, is one of the main characteristics of a chip. Modern computers commonly use words of 32 or 64 bits, however it is not unusual, especially in the smart card world, to work with 8 or 16-bit devices.

## 10.1.2 Internal representation

A *single precision integer*, also called a 1-*word integer*, or just a *single* needs only a word to be represented. For 32-bit architectures such integers belong to $[0, 2^{32} - 1]$. The concept of *multiprecision* was introduced in order to manipulate objects that do not fit in a word. In this case, an array of consecutive words is allocated and for instance a *n-word integer* is an integer whose representation requires $n$ words.

There are several ways to store information in the memory of a computer, the order being especially important. Usually a byte comes with the most significant bit first. To describe the ordering of bytes within a word, there is a specific terminology. Namely, in *little endian* format the least significant byte is stored first, whereas in *big endian* representation, the sequence begins with the most significant byte. For instance, on a 32-bit PC, the single $262657 = 2^{18} + 2^9 + 1$ is represented in little endian format by 00000001 00000010 00000100 00000000. The representation of the same integer in big endian, commonly used in RISC architectures, is 00000000 00000100 00000010 00000001. To represent multiple precision integers, the same kind of choice occurs for words as well. Note that when the least significant word comes first, as it is the case for many multiprecision implementations, an integer can have different representations with as many high order zeroes as wanted. To illustrate the exposition, we will give examples more classically written with the most significant word first.

For example, take $u = (1128103691808033)_{10}$ and $b = 2^{32}$. Then $u$ has 51 digits in base 2, namely

$$u = (\underbrace{1000000001000000001}_{262657}\ \underbrace{00011011110100011110110100100001}_{466742561})_2$$

so that $u = (262657\ 466742561)_{2^{32}}$. Thus $u$ is a 2-word integer, also called a *double precision* integer or just a *double*.

We will also have to deal with negative values. They can be represented in two different ways.

- In *signed-magnitude* notation, the sign of an integer is independently coded by a bit, byte, or word. For instance, 0 for positive values and 1, or $b - 1$ for negative ones. Thus $u = \bigl(s, (u_{n-1} \ldots u_0)_b\bigr)$ stands for $u_{n-1}b^{n-1} + \cdots + u_0$ or $-(u_{n-1}b^{n-1} + \cdots + u_0)$ depending on the value of $s$. Therefore 0 has two representations $\bigl(0, (0\ldots 0)_b\bigr)$ and say $\bigl(b-1, (0\ldots 0)_b\bigr)$. As a consequence, the opposite of $u$ is easily obtained since we only need to modify $s$.

- In *complement* format, if the highest bit of the most significant word is 0 then the integer $(u_{n-1} \ldots u_0)_b$ is nonnegative and must be understood as $u_{n-1}b^{n-1} + \cdots + u_0$. Otherwise $(u_{n-1} \ldots u_0)_b$ is equal to $-b^n + u_{n-1}b^{n-1} + \cdots + u_0$ which is always negative. The representation $(0\ldots 0)_b$ of 0 is unique and if $u$ is nonzero, the opposite of $u$ is $(\overline{u_{n-1}} \ldots \overline{u_0})_b + 1$ where $\overline{u_i} = b - 1 - u_i$ is the *bitwise complement* of $u_i$. Computing the opposite is therefore more costly with this representation as each bit needs to be changed. Note also that padding $u$ with zeroes drastically changes the representation of the opposite in this case.

**Example 10.2** With a radix $b$ equal to 4, 152 is represented by $(0, (2120)_4)$ in signed-magnitude format and $-152$ is trivially $(3, (2120)_4)$. In complement notation, $152 = (02120)_4$ and $(31213)_4 + 1$ that is $(31220)_4$ stands for $-4^4 + 4^3 + 2 \times 4^2 + 2 \times 4 = -152$.

In base 2, complement notation is called *two's complement* notation and the highest bit of an integer codes its sign in both representations. Therefore the same sequence of bits corresponds to different values. For instance $(10011000)_2$ is equal to $-24$ in signed-magnitude format whereas it is equal to $-104$ in two's complement notation.

### 10.1.3 Elementary operations

Computers have highly optimized built-in operations for single precision integers. Thus in the sequel we will describe the arithmetic of multiprecision integers, assuming the existence of the following low-level operations:

- *Comparison* of two singles returning a boolean 0 or 1. For instance, the equality $=$ and the natural ordering $<$ of two integers can be directly determined.
- *Bitwise complement* of a single $u$, that is $\bar{u} = b - 1 - u$.
- *Bitwise conjunction*, *disjunction* and *exclusive disjunction* of the singles $u$ and $v$, that is respectively $u \wedge v$, $u \vee v$ and $u$ XOR $v$.
- The *right* and *left shifts* of $t$ bits of the single $u$, respectively denoted by $u \gg t$ and $u \ll t$, corresponding to $\lfloor u/2^t \rfloor$ and $u2^t \bmod b$.
- *Addition* of two singles $u$ and $v$ giving a single $w$ and a *carry bit* $k$ equal to 0 or 1 so that the correct result is $u + v = kb + w$.
- *Subtraction* of a single $v$ from a single $u$, that is $u - v$, giving as a result a single $w$ and a *carry* $k$. If $u \geqslant v$ then $w = u - v$ and $k = 0$, otherwise $w = b + u - v > 0$ and $k = -1$. The nonnegative quantity $-k$ is sometimes called the *borrow bit*.
- *Multiplication* of two singles $u$ and $v$ giving a double $w = u \times v$.
- *Division* of a double $u$ by a single $v$, when the quotient $q = \lfloor u/v \rfloor$ and the remainder $r = u \bmod v$ are both singles. This operation computes $q$ and $r$ simultaneously.

Two remarks should be made on these basic operations. First the more complicated operations such as multiplication and especially double by single division are not always available in modern processors as integer instructions, but only as floating point ones. Second and more importantly usual high-level languages such as C do not allow, at least in their basic versions, to access the carry or borrow bit, or the high part of a multiply instruction, or to have simultaneously the quotient and remainder of a double by single division. If high efficiency is desired it is thus necessary to bypass these restrictions, either by choosing a C compiler and library allowing you to do these things, or to implement them in explicit or inline assembly language programs.

In addition to the above, there are other useful basic operations such as extracting the top bit of a word or counting the parity of the number of bits equal to 1, which do not exist as CPU instructions and should be very carefully implemented. Note also that on some architectures, floating registers provide faster operations than integer ones and should be used instead for multiprecision operations, cf. [ZIM 2001] and [HAME+ 2003, § 5.1.2].

In the following, we present several algorithms for addition, subtraction, multiplication, and division of multiprecision positive integers written in base $b$. Even if they work in general for every radix, in practice $b$ is almost always a power of 2, namely $2^8$, $2^{16}$, $2^{32}$, or $2^{64}$. However, for clarity, algorithms will be often illustrated with $b = 10$. We start with the simplest operations, addition and subtraction.

## 10.2 Addition and subtraction

Algorithm 10.3 is an obvious generalization of the schoolbook method for any radix $b$.

## § 10.2 Addition and subtraction

**Algorithm 10.3** Addition of nonnegative multiprecision integers

INPUT: Two $n$-word integers $u = (u_{n-1} \ldots u_0)_b$ and $v = (v_{n-1} \ldots v_0)_b$.
OUTPUT: The $(n+1)$-word integer $w = (w_n \ldots w_0)_b$ such that $w = u + v$, $w_n$ being 0 or 1.

1. $k \leftarrow 0$      [$k$ is the carry]
2. **for** $i = 0$ **to** $n-1$ **do**
3.      $w_i \leftarrow (u_i + v_i + k) \bmod b$      [$0 \leqslant w_i < b$]
4.      $k \leftarrow \lfloor (u_i + v_i + k)/b \rfloor$      [$k = 0$ or $1$]
5. $w_n \leftarrow k$
6. **return** $(w_n \ldots w_0)_b$

**Example 10.4** Take $b = 10$, $u = (9635)_{10}$ and $v = (827)_{10}$ and let us compute $u + v$ using Algorithm 10.3. The algorithm proceeds as follows

| $i$ | $u_i + v_i$ | $k$ | $w_4$ | $w_3$ | $w_2$ | $w_1$ | $w_0$ |
|---|---|---|---|---|---|---|---|
| 0 | 12 | 0 | — | — | — | — | 2 |
| 1 | 5  | 1 | — | — | — | 6 | 2 |
| 2 | 14 | 0 | — | — | 4 | 6 | 2 |
| 3 | 9  | 1 | — | — | 4 | 6 | 2 |
| — | —  | 1 | 1 | 0 | 4 | 6 | 2 |

and the result is $9635 + 827 = 10462$ as expected.

The subtraction algorithm is very similar to the addition algorithm. Indeed, a simple change of sign in Algorithm 10.3 is enough to get $u - v$ instead of $u + v$ provided $u \geqslant v$.

**Algorithm 10.5** Subtraction of nonnegative multiprecision integers

INPUT: Two $n$-word integers $u = (u_{n-1} \ldots u_0)_b$ and $v = (v_{n-1} \ldots v_0)_b$ such that $u \geqslant v$.
OUTPUT: The $n$-word integer $w = (w_{n-1} \ldots w_0)_b$ such that $w = u - v$.

1. $k \leftarrow 0$      [$k$ is the carry]
2. **for** $i = 0$ **to** $n-1$ **do**
3.      $w_i \leftarrow (u_i - v_i + k) \bmod b$      [$0 \leqslant w_i < b$]
4.      $k \leftarrow \lfloor (u_i - v_i + k)/b \rfloor$      [$k = 0$ or $-1$]
5. **return** $(w_{n-1} \ldots w_0)_b$      [if $k = -1$ then $u < v$]

**Remarks 10.6**

(i) When working with a fixed number of words, the for loops of Algorithms 10.3 and 10.5 should be unrolled by hand for faster results. Also the computation of $w_i$ and $k$, in Lines 3 and 4 should be implemented as a single operation.
(ii) For adding or subtracting integers of different lengths we must first pad the smallest number with as many zeroes as necessary so that they are both of the same length. For efficiency this padding is not done explicitly but implicitly.

(iii) One can apply Algorithm 10.5 without checking if $u \geqslant v$. If $k = -1$ at the end, then the output is related to the complement representation of the negative number $u - v$. To obtain $|u - v|$ one can repeat Algorithm 10.5 with $u = (0\ldots0)_b$ and $v = w$. Note that the second subtraction is actually a computation of the complement of $w$ and it is faster implemented via a simplified version of the algorithm with the value $u = 0$ hardwired.

**Example 10.7** Let us compute $u - v$ when $u$ is smaller than $v$. Take $u = 2165$ and $v = 58646$.

| $i$ | $u_i - v_i$ | $k$ | $w_4$ | $w_3$ | $w_2$ | $w_1$ | $w_0$ |
|---|---|---|---|---|---|---|---|
| 0 | $-1$ | 0 | — | — | — | — | 9 |
| 1 | 2 | $-1$ | — | — | — | 1 | 9 |
| 2 | $-5$ | 0 | — | — | 5 | 1 | 9 |
| 3 | $-6$ | $-1$ | 0 | 3 | 5 | 1 | 9 |
| 4 | $-5$ | $-1$ | 4 | 3 | 5 | 1 | 9 |
| 2-nd execution with $u = (00000)_{10}$ and $v = w$ ||||||||
| 0 | $-9$ | 0 | — | — | — | — | 1 |
| 1 | $-1$ | $-1$ | — | — | — | 8 | 1 |
| 2 | $-5$ | $-1$ | — | — | 4 | 8 | 1 |
| 3 | $-3$ | $-1$ | — | 6 | 4 | 8 | 1 |
| 4 | $-4$ | $-1$ | 5 | 6 | 4 | 8 | 1 |

So $|u - v| = 56481$ and it can be deduced that $u - v = -56481$ since $k = -1$ at the end of the execution. The complement representation of $-56481$ is $943519$.

## 10.3 Multiplication

Multiplication is a very important operation. Its optimization is crucial since it is the most time-consuming part for a wide range of applications. For instance, the efficiency of division algorithms depends to a large extent on the speed of the multiplication that is used. The complexity of a multiplication algorithm is therefore an important parameter for a complete arithmetic system. In the following, the number of elementary operations necessary to multiply two $n$-word integers will be denoted by $\mathrm{M}(n)$. In the remainder, we shall often encounter situations where the best algorithm to be used to perform a given task is chosen in a set and determined by parameters such as the size of the arguments and the adopted computer architecture. This is the case for multiplication. We describe in detail the schoolbook and the Karatsuba multiplication, which are the only interesting ones for the range of integers we consider. However, many other algorithms exist. See [BER 2001a] for a comprehensive presentation of multiplication methods.

### 10.3.1 Schoolbook multiplication

One starts with the simplest method known for at least four millennia.

**Algorithm 10.8** Multiplication of positive multiprecision integers

INPUT: An $m$-word integer $u = (u_{m-1}\ldots u_0)_b$ and an $n$-word integer $v = (v_{n-1}\ldots v_0)_b$.
OUTPUT: The $(m+n)$-word integer $w = (w_{m+n-1}\ldots w_0)_b$ such that $w = uv$.

1. **for** $i = 0$ **to** $n - 1$ **do** $w_i \leftarrow 0$     [see the remark below]

2. **for** $i = 0$ **to** $n - 1$ **do**

## § 10.3 Multiplication

| | | | |
|---|---|---|---|
| 3. | $k \leftarrow 0$ | | [$k$ is the carry] |
| 4. | **if** $v_i = 0$ **then** $w_{m+i} \leftarrow 0$ | | [optional test] |
| 5. | **else** | | |
| 6. |     **for** $j = 0$ **to** $m - 1$ **do** | | |
| 7. |         $t \leftarrow v_i u_j + w_{i+j} + k$ | | $[0 \leqslant t < b^2]$ |
| 8. |         $w_{i+j} \leftarrow t \bmod b$ | | $[0 \leqslant w_{i+j} < b]$ |
| 9. |         $k \leftarrow \lfloor t/b \rfloor$ | | $[0 \leqslant k < b]$ |
| 10. |     $w_{m+i} \leftarrow k$ | | |
| 11. | **return** $(w_{m+n-1} \ldots w_0)_b$ | | |

**Remarks 10.9**

(i) In fact Algorithm 10.8 performs the following *multiply and add* operation

$$(w_{n+m-1} \ldots w_0)_b \leftarrow (u_{n-1} \ldots u_0)_b \times (v_{m-1} \ldots v_0)_b + (w_{n-1} \ldots w_0)_b.$$

(ii) Actually, the schoolbook method consists in computing the intermediate results $uv_i$ before adding them. Here we multiply and add the terms *simultaneously* inside the $j$ loop.

(iii) Checking if $v_i = 0$ in Line 4 is useless unless $b$ is small.

**Example 10.10** Take $u = (9712)_{10}$ and $v = (526)_{10}$. So $m = 4$ and $n = 3$. Let us compute $uv$ with Algorithm 10.8. The table shows the relevant values after execution of Line 9.

| $i$ | $j$ | $v_i u_j$ | $w_{i+j}$ | $t$ | $k$ | $w_6$ | $w_5$ | $w_4$ | $w_3$ | $w_2$ | $w_1$ | $w_0$ |
|---|---|---|---|---|---|---|---|---|---|---|---|---|
| 0 | 0 | 12 | 0 | 12 | 1 | 0 | 0 | 0 | 0 | 0 | 0 | 2 |
|   | 1 | 6  | 0 | 7  | 0 | 0 | 0 | 0 | 0 | 0 | 7 | 2 |
|   | 2 | 42 | 0 | 42 | 4 | 0 | 0 | 0 | 0 | 2 | 7 | 2 |
|   | 3 | 54 | 0 | 58 | 5 | 0 | 0 | 5 | 8 | 2 | 7 | 2 |
| 1 | 0 | 4  | 7 | 11 | 1 | 0 | 0 | 5 | 8 | 2 | 1 | 2 |
|   | 1 | 2  | 2 | 5  | 0 | 0 | 0 | 5 | 8 | 5 | 1 | 2 |
|   | 2 | 14 | 8 | 22 | 2 | 0 | 0 | 5 | 2 | 5 | 1 | 2 |
|   | 3 | 18 | 5 | 25 | 2 | 0 | 2 | 5 | 2 | 5 | 1 | 2 |
| 2 | 0 | 10 | 5 | 15 | 1 | 0 | 2 | 5 | 2 | 5 | 1 | 2 |
|   | 1 | 5  | 2 | 8  | 0 | 0 | 2 | 5 | 8 | 5 | 1 | 2 |
|   | 2 | 35 | 5 | 40 | 4 | 0 | 2 | 0 | 8 | 5 | 1 | 2 |
|   | 3 | 45 | 2 | 51 | 5 | 5 | 1 | 0 | 8 | 5 | 1 | 2 |

At the end of each $i$ loop, is computed respectively $9712 \times 6 = 58272$, $9712 \times 20 + 58272 = 252512$, and $9712 \times 500 + 252512 = 5108512$.

The number of elementary multiplications carried out by the schoolbook method is $nm$. Thus $M(n) = O(n^2)$ for Algorithm 10.8.

## 10.3.2 Karatsuba multiplication

In the mid-1950's, Kolmogorov made the conjecture that the lower estimate of $M(n)$ was of the order of $n^2$, whatever the method used. However, in 1960, one of his students, namely Karatsuba, discovered a method in $O(n^{\lg 3})$, where lg is the logarithm to base 2. The method was later published in [KAOF 1962]. The interesting genesis of the so-called Karatsuba method is explained in [KAR 1995].

For the sake of clarity, set $R = b^n$, $d = 2n$ and let $u = (u_{d-1} \ldots u_0)_b$ and $v = (v_{d-1} \ldots v_0)_b$ be two $d$-word integers. The method relies on the following observation. Split both $u$ and $v$ in two, namely the least and most significant parts, so that $u = U_1 R + U_0$ and $v = V_1 R + V_0$. Then one can check that

$$uv = U_1 V_1 R^2 + \big((U_0 + U_1)(V_0 + V_1) - U_1 V_1 - U_0 V_0\big) R + U_0 V_0.$$

One needs *a priori* four multiplications to compute $uv$ but as a multiplication by $R$ is just a shift, one performs actually some additions and only three multiplications, which are $U_1 V_1$, $(U_1 + U_0)(V_0 + V_1)$, and $U_0 V_0$. Moreover, a recursive approach allows us to reduce the size of the operands until they are sufficiently small so that the schoolbook multiplication is faster. In practice, this holds when $d$ becomes smaller than a *threshold* $d_0$, depending essentially on the processor used. Granlund performed tests with GMP on several architectures to determine the optimal value of $d_0$. Results spread from 8 up to more than 100 [GRA 2004, GMP].

---

**Algorithm 10.11** Karatsuba multiplication of positive multiprecision integers

INPUT: An $n$-word integer $u = (u_{n-1} \ldots u_0)_b$, an $m$-word integer $v = (v_{m-1} \ldots v_0)_b$, the size $d = \max\{m, n\}$, and a threshold $d_0$.
OUTPUT: The $(m+n)$-word integer $w = (w_{m+n-1} \ldots w_0)_b$ such that $w = uv$.

1. **if** $d \leqslant d_0$ **then return** $uv$           [use Algorithm 10.8]
2. $p \leftarrow \lfloor d/2 \rfloor$ and $q \leftarrow \lceil d/2 \rceil$
3. $U_0 \leftarrow (u_{q-1} \ldots u_0)_b$ and $V_0 \leftarrow (v_{q-1} \ldots v_0)_b$
4. $U_1 \leftarrow (u_{p+q-1} \ldots u_q)_b$ and $V_1 \leftarrow (v_{p+q-1} \ldots v_q)_b$     [pad with 0's if necessary]
5. $U_s \leftarrow U_0 + U_1$ and $V_s \leftarrow V_0 + V_1$
6. compute recursively $U_0 V_0$, $U_1 V_1$ and $U_s V_s$     [corresponding sizes being $q$, $p$ and $q$]
7. **return** $U_1 V_1 b^{2q} + \big((U_s V_s - U_1 V_1 - U_0 V_0)\big) b^q + U_0 V_0$

---

**Remark 10.12** The number of elementary operations required by Algorithm 10.11 to multiply two $n$-word integers shall be denoted by $K(n)$. As $K(n) \leqslant 3 K(n/2) + cn/2$ for some constant $c$, one finds by induction that $K(n) = O(n^{\lg 3}) \approx O(n^{1.585})$.

**Example 10.13** Let us multiply $u = 564986$ and $v = 1279871$ by Algorithm 10.11 with $d_0 = 4$. So one has $d = 7$, $q = 4$ and $R = 10^4$. Put

$$U_1 = 56, \quad U_0 = 4986$$
$$V_1 = 127, \quad V_0 = 9871$$
$$U_s = 5042, \quad V_s = 9998.$$

With Algorithm 10.8, compute

$$U_0 V_0 = 49216806, \quad U_1 V_1 = 7112 \quad \text{and} \quad U_s V_s = 50409916$$

to obtain

$$uv = 7112 \times 10^8 + (50409916 - 7112 - 49216806) \times 10^4 + 49216806$$
$$= 723109196806.$$

Other multiplication algorithms, like Toom–Cook or Fast Fourier Transform methods [KNU 1997] based on interpolation are asymptotically more efficient, but the gain occurs only for very large $n$, out of the range of the sizes used nowadays for cryptosystems based on elliptic and hyperelliptic curves.

### 10.3.3 Squaring

It seems simpler to square a number than to multiply two arbitrary integers. This feeling is supported by the existence of a specific algorithm suggested by the formula

$$\left(\sum_{i=0}^{n-1} u_i b^i\right)^2 = \sum_{i=0}^{n-1} u_i^2 b^{2i} + 2 \sum_{i<j} u_i u_j b^{i+j}.$$

Thus a schoolbook squaring takes only $(n^2 + n)/2$ elementary multiplications (against $n^2$ for the general algorithm).

---

**Algorithm 10.14** Squaring of a positive multiprecision integer

INPUT: An $n$-word integer $u = (u_{n-1} \ldots u_0)_b$.
OUTPUT: The $(2n)$-word integer $w = (w_{2n-1} \ldots w_0)_b$ such that $w = u^2$.

1. **for** $i = 0$ **to** $2n - 1$ **do** $w_i \leftarrow 0$
2. **for** $i = 0$ **to** $n - 1$ **do**
3.     $t \leftarrow u_i^2 + w_{2i}$
4.     $w_{2i} \leftarrow t \bmod b$ and $k \leftarrow \lfloor t/b \rfloor$
5.     **for** $j = i+1$ **to** $n - 1$ **do**
6.         $t \leftarrow 2 u_i u_j + w_{i+j} + k$
7.         $w_{i+j} \leftarrow t \bmod b$ and $k \leftarrow \lfloor t/b \rfloor$
8.     $w_{i+n} \leftarrow k$
9. **return** $(w_{2n-1} \ldots w_0)_b$

---

**Remarks 10.15**

(i) In the $j$ loop, one has $0 \leqslant k \leqslant 2(b-1)$. This implies that during the process, namely in $w_{i+n} \leftarrow k$ an overflow could occur. However, at the end all the $w_j$'s are single precision integers.

(ii) In practice, Algorithm 10.14 is about 20% faster than the standard multiplication $u \times u$.

(iii) Consider any commutative ring $\mathcal{R}$ of characteristic different from 2. In the unlikely event that a squaring is more than twice as fast as a multiplication we can use $4uv = (u+v)^2 - (u-v)^2$ to get $uv$. Note that this identity can be used in practice, see [CRPO 2001, exercise 9.6].

**Example 10.16** Let us follow the computation of the square of $u = (769)_{10}$ by Algorithm 10.14. Below are displayed the values of some relevant parameters after Lines 4 and 7 execute.

| $i$ | $j$ | $u_i^2$ | $w_{2i}$ | $2u_iu_j$ | $w_{i+j}$ | $t$ | $k$ | $w_5$ | $w_4$ | $w_3$ | $w_2$ | $w_1$ | $w_0$ |
|---|---|---|---|---|---|---|---|---|---|---|---|---|---|
| 0 | — | 81 | 0 | — | — | 81 | 8 | 0 | 0 | 0 | 0 | 0 | 1 |
|   | 1 | — | — | 108 | 0 | 116 | 11 | 0 | 0 | 0 | 0 | 6 | 1 |
|   | 2 | — | — | 126 | 0 | 137 | 13 | 0 | 0 | 13 | 7 | 6 | 1 |
| 1 | — | 36 | 7 | — | — | 43 | 4 | 0 | 0 | 13 | 3 | 6 | 1 |
|   | 2 | — | — | 84 | 13 | 101 | 10 | 0 | 10 | 1 | 3 | 6 | 1 |
| 2 | — | 49 | 10 | — | — | 59 | 5 | 5 | 9 | 1 | 3 | 6 | 1 |

It is also possible to write a specific squaring procedure inspired by Karatsuba's idea. Indeed, one can check that if $u = U_1 R + U_0$ and $U_s = U_0 + U_1$, as defined in Section 10.3.2, then

$$u^2 = U_1^2 R^2 + (U_s^2 - U_1^2 - U_0^2) R + U_0^2$$

so that intermediate steps only require squarings as well. The minimal word size $d_1$, for which it is faster to use, Karatsuba method against a naïve approach, should be greater than the threshold $d_0$ used for Karatsuba multiplication in Algorithm 10.11. Set approximately $d_1 = 2d_0$ [GRA 2004].

## 10.4 Modular reduction

In many situations, only the remainder of a Euclidean division is required. In the following, one describes two general methods to reduce a number modulo an integer $N$. In practice $N$ will often be prime, and the corresponding *reduction* is an essential operation for prime field arithmetic. We also consider moduli of special form and introduce a reduction method modulo several primes.

The obvious way to obtain $u \bmod N$ consists in dividing $u$ by $N$ and computing the remainder. The following methods allow more efficient execution. Sometimes an almost reduced element which is not minimal is accepted as an intermediate result.

### 10.4.1 Barrett method

If $u$ and $N$ are both real numbers or formal series, there is another method to divide $u$ by $N$. First, compute the inverse of $N$ to a sufficient precision and then multiply it by $u$. The inverse is often computed with Newton method, i.e., the iteration

$$x \leftarrow x - x(Nx - 1)$$

starting from an initial approximation $x_0$. Such a precomputation of $N^{-1}$ proves useful if many reductions modulo $N$ are necessary.

There is a similar technique for integers. Let $N$ be an $n$-word integer. We define the *reciprocal integer* of $N$ as $R(N) = \lfloor b^{2n}/N \rfloor$. Now if $u$ is a $2n$-word integer we see that

$$q = \left\lfloor \frac{u}{N} \right\rfloor \text{ is also equal to } \left\lfloor \frac{\frac{u}{b^{n-1}} \frac{b^{2n}}{N}}{b^{n+1}} \right\rfloor$$

which can be approximated by

$$\hat{q} = \left\lfloor \frac{\lfloor \frac{u}{b^{n-1}} \rfloor R}{b^{n+1}} \right\rfloor.$$

## § 10.4 Modular reduction

In addition we have $q-2 \leqslant \hat{q} \leqslant q$ and it can be shown that $\hat{q} = q$ in about 90% of the cases whereas $\hat{q}$ will be 2 in error in only 1% of the cases. This is the so-called *Barrett method* [BAR 1987]. See [BoGo+ 1994] for further improvements.

For modular reduction this approximation is sufficient as $\hat{u} = u - \hat{q}N$ is in the same residue class modulo $N$ as $u$ and will be minimal if $\hat{q} = q$. If $\hat{u} > N$, one needs at most 2 subtractions by $N$ to obtain the minimal representative.

When a lot of divisions are performed with a fixed divisor, for instance when computing in a finite field, this approach is very efficient. Indeed $R$ is precomputed and the quotient and remainder of a Euclidean division by some power of $b$ can be trivially determined.

---

**Algorithm 10.17** Division-free modulo of positive multiprecision integers

INPUT: A $2n$-word integer $u = (u_{2n-1} \ldots u_0)_b$ and the $n$-word integer $N = (N_{n-1} \ldots N_0)_b$ with $N_{n-1} \neq 0$. The quantity $R = \lfloor b^{2n}/N \rfloor$ is precomputed.
OUTPUT: The $n$-word integer $r = (r_{n-1} \ldots r_0)_b$ such that $u \equiv r \pmod{N}$.

1. $\hat{q} \leftarrow \lfloor \lfloor (u/b^{n-1}) \rfloor R/b^{n+1} \rfloor$      $[q-2 \leqslant \hat{q} \leqslant q]$
2. $r_1 \leftarrow u \bmod b^{n+1}$, $r_2 \leftarrow (\hat{q}N) \bmod b^{n+1}$ and $r \leftarrow r_1 - r_2$
3. **if** $r < 0$ **then** $r \leftarrow r + b^{n+1}$
4. **while** $r \geqslant N$ **do** $r \leftarrow r - N$
5. **return** $r$

---

Of course, Algorithm 10.17 can be trivially modified such that it computes the quotient $q$ as well. Even if $R$ is precomputed and thus the performance is not crucial, we now present an algorithm inspired by the Newton method, which computes it in an efficient way.

---

**Algorithm 10.18** Reciprocation of positive multiprecision integers

INPUT: An $n$-word integer $N = (N_{n-1} \ldots N_0)_b$.
OUTPUT: The $(n+2)$-word integer $R = \lfloor b^{2n}/N \rfloor$.

1. $R \leftarrow b^n$
2. **repeat**
3.      $s \leftarrow R$
4.      $R \leftarrow 2R - \lfloor N \lfloor R^2/b^n \rfloor/b^n \rfloor$      [discrete Newton iteration]
5. **until** $R \leqslant s$
6. $t \leftarrow b^{2n} - NR$
7. **while** $t < 0$ **do** $R \leftarrow R - 1$ and $t \leftarrow t + N$      [performed at most twice]
8. **return** $R$

---

### Remarks 10.19

(i) The complexity of Algorithm 10.17 is at first glance equal to 2 multiplications of size $n$. However, the $n$ lowest words of the product $\lfloor (u/b^{n-1}) \rfloor R$ will be discarded when dividing by $b^{n+1}$ so that it is no use to compute them. In the same manner, only the $n$ lowest words of the product $\hat{q}N$ are required. Therefore it can be shown that computing the remainder of a $2n$-word $u$ modulo the $n$-word $N$ has asymptotically the same complexity as an $n$-word multiplication [MEO0+ 1996, p. 604], cf. also Section 11.1.1.

(ii) The number of iterations through Algorithm 10.18 is $O\bigl(\lg\lg(N+2)\bigr)$.

**Example 10.20** Let $u = 893994278$ and $N = 21987$ and take $b = 2$. First, one computes $R$ with Algorithm 10.18. Since $n = 15$, set $R = 2^{15}$ and the successive values of $R$ are indicated below

| $s$ | $R$ | $t$ |
|---|---|---|
| — | 32768 | — |
| 32768 | 43549 | — |
| 43549 | 48264 | — |
| 48264 | 48829 | — |
| 48829 | 48836 | — |
| 48836 | 48836 | — |
| — | 48836 | $-15308$ |
| — | 48835 | 6679 |

Finally $R$ is 48835. Then the determination of the quotient can be done with Algorithm 10.17. Indeed $n = 15$, $u = (11010101001001010001010010011 0)_2$ and we can see that $\lfloor u/2^{n-1} \rfloor = (1101010100100101)_2 = 54565$. After computing the highest bits of the product of this last term by $R$, another shift gives an approximation of the quotient $\hat{q} = 40659$ whereas the exact value of $q$ is 40660. Then one computes $r_1 = 17702$, $r_2 = 58393$ with another partial product and $r = r_1 - r_2 = -40691$. Since $r < 0$ one sets $r = r + 2^{16} = 24845$ which is larger than $N$ so that the remainder of $u$ modulo $N$ is finally equal to $r - N = 2858$.

## 10.4.2 Montgomery reduction

Montgomery introduced a clever way to represent elements of $\mathbb{Z}/N\mathbb{Z}$ such that arithmetic and especially multiplication becomes easy [MON 1985]. It can be viewed as a generalization of *Hensel odd division* for computing inverses of 2-adic numbers [HEN 1908]. In practice, $N$ will often be a prime number.

**Definition 10.21** Let $R$ be some integer greater than $N$ and coprime with it. The *Montgomery representation* of $x \in [0, N-1]$ is $[x] = (xR) \bmod N$. The *Montgomery reduction* of $u \in [0, RN-1]$ is $\text{REDC}(u) = (uR^{-1}) \bmod N$.

When $R$ is a power of the radix $b$ there is an efficient algorithm to perform the reduction of $u$. Indeed let $N' = (-N^{-1}) \bmod R$ and let $k$ be the unique integer in $[0, N-1]$ such that $k \equiv uN' \pmod{R}$. Then clearly $(u + kN)$ is a multiple of $R$. Let $t = (u + kN)/R$. As $N$ and $R$ are relatively prime, this implies that $t \equiv uR^{-1} \pmod{N}$. Finally, $0 \leqslant u < RN$ by assumption and it can be easily shown that $0 \leqslant t < 2N$ so that $t$ or $t - N$ is equal to the desired result $\text{REDC}(u)$.

The following algorithm makes use of these ideas, with some improvements, to handle multiprecision integers.

## § 10.4 Modular reduction

**Algorithm 10.22** Montgomery reduction REDC of multiprecision integers

INPUT: An $n$-word integer $N = (N_{n-1} \ldots N_0)_b$ such that $\gcd(N,b) = 1$, $R = b^n$, $N' = (-N^{-1}) \bmod b$ and a $2n$-word integer $u = (u_{2n-1} \ldots u_0)_b < RN$.
OUTPUT: The $n$-word integer $t = (t_{n-1} \ldots t_0)_b$ such that $t = \text{REDC}(u) = (uR^{-1}) \bmod N$.

1. $(t_{2n-1} \ldots t_0)_b \leftarrow (u_{2n-1} \ldots u_0)_b$
2. **for** $i = 0$ **to** $n - 1$ **do**
3. $\quad k_i \leftarrow (t_i N') \bmod b$
4. $\quad t \leftarrow t + k_i N b^i$
5. $t \leftarrow t/R$
6. **if** $t \geqslant N$ **then** $t \leftarrow t - N$
7. **return** $t$

**Remarks 10.23**

(i) It is immediate that $[x] = \text{REDC}\big((xR^2) \bmod N\big)$ and that $\text{REDC}([x]) = x$ for all $x \in [0, N-1]$. The value $R^2 \bmod N$ can also be precomputed.

(ii) Algorithm 10.22 requires $n^2 + n$ single precision multiplications to compute Montgomery reduction.

(iii) Since $N'$ is precomputed once $N$ is fixed, the way it is obtained is not crucial for the overall performance of the method. However, Dussé and Kaliski designed a short procedure to efficiently compute it [DUKA 1990]. One can use also an idea of Jebelean who gave an efficient recursive method to compute an inverse modulo some power of $2^8$, see Remark 10.40(ii).

(iv) Classical reduction computes the remainder processing the digits of $u$ from the left to the right. Montgomery reduction is in one sense dual since it operates from the right to the left.

(v) If $u \geqslant RN$ then Algorithm 10.22 does not return $t = uR^{-1} \bmod N$ but $t \equiv uR^{-1} \pmod{N}$. A divisibility criterion makes use of this remark, see Section 10.5.3.

**Example 10.24** Let $N = 2011$ and $b = 2^3$ so that $R = 8^4 = 4096$. Let $u = 8170821 = (37126505)_8$. One can check that $u < RN$. A direct computation gives

$$
\begin{aligned}
N' &= 941 = (1655)_8 \\
uN' &= (37126505)_8 \times (1655)_8 = (71222163241)_8 \\
k &\equiv (3241)_8 \pmod{R} \\
u + kN &= (54140000)_8 \\
t &= (54140000)_8/R = (5414)_8.
\end{aligned}
$$

As $t \geqslant N$, Montgomery reduction of $u$ is $t - N = (5414)_8 - (3733)_8 = (1461)_8 = 817$.

Now let us use Algorithm 10.22 instead. If we consider $N = (3733)_8 = 2011$ as a 4-word integer in base $b = 2^3$, one first sets $N' = (-N^{-1}) \mod 8 = 5$ and the successive steps are

| $i$ | $t_i$ | $k_i$ | $k_i N b^i$ | $t$ |
|---|---|---|---|---|
| — | — | — | — | $(37126505)_8$ |
| 0 | 5 | 1 | $(3733)_8$ | $(37132440)_8$ |
| 1 | 4 | 4 | $(175540)_8$ | $(37330200)_8$ |
| 2 | 2 | 2 | $(766600)_8$ | $(40317000)_8$ |
| 3 | 7 | 3 | $(13621000)_8$ | $(54140000)_8$ |
| — | — | — | — | $(5414)_8$ |
| — | — | — | — | $(1461)_8$ |

In [BoGo+ 1994] the authors give the following table that compares the complexities of classical, Barrett, and Montgomery reduction for $n$-bit integers.

| Algorithm | Classical | Barrett | Montgomery |
|---|---|---|---|
| Multiplications | $n(n+2.5)$ | $n(n+4)$ | $n(n+1)$ |
| Divisions | $n$ | 0 | 0 |
| Precomputations | Normalization of $N$ | $\lfloor b^{2n}/N \rfloor$ | Reduction |
| Restrictions | None | $u < b^{2n}$ | $u < N b^n$ |

For cryptographic applications the Montgomery method is reported to be faster than Barrett, see tests in [BoGo+ 1994] and an interesting discussion of these two methods in [CrPo 2001].

### 10.4.3 Special moduli

Special moduli are usually considered only for arithmetic modulo primes $p$, i.e., finite field arithmetic; however, as previously, the following algorithms depend by no means on this restriction. Note that even if the overall modular arithmetic is performed more efficiently, one must be aware that restricting the range of values for the modulus could also benefit a potential attacker.

The primes $p$ we shall consider can be seen as generalizations of the concept of *Mersenne primes*, that are prime numbers of the form $p = 2^k - 1$. In practice, reduction modulo a Mersenne prime is completely trivial since it requires only one field addition. Indeed, if $0 \leqslant x < p^2$ then $x$ can easily be written as $x = x_1 2^k + x_0$ with $x_0$ and $x_1$ less than $2^k$. As $2^k \equiv 1 \pmod{p}$, it follows that $x \equiv x_1 + x_0 \pmod{p}$. For example, take $p = 2^7 - 1$ and $x = 10905 = (10101010011001)_2$. One has immediately

$$\begin{aligned}
x &\equiv \big((1010101)_2 + (0011001)_2\big) \pmod{127} \\
&\equiv (1101110)_2 \pmod{127} \\
&= 110 \mod 127.
\end{aligned}$$

The indexes $k$ less than 1000 that give a Mersenne prime are $2, 3, 5, 7, 13, 17, 19, 31, 61, 89, 107, 127, 521$, and $607$. In cryptography, they are used to define prime fields or extension fields. We consider such optimal extension fields, [AvMi 2004], in Section 11.3. However, their lack in the interval $[2^{128}, 2^{520}]$, which is of great interest for elliptic and hyperelliptic curve cryptography, led to different kinds of generalizations. First Crandall introduced primes of the form $2^k - c$ with $c > 0$ sufficiently small [Cra 1992]. In fact, it is no more complicated to consider $p = b^k + c$ with $b$ some power of 2 and $|c|$ small. The fundamental relation is

$$x \equiv \big((x \mod b^k) - c \lfloor x/b^k \rfloor\big) \pmod{p}.$$

## § 10.4 Modular reduction

One deduces the following algorithm.

---
**Algorithm 10.25** Fast reduction for special form moduli

INPUT: A positive integer $x$ and a modulus $p = b^k + c$ such that $|c| \leq b^{k-1} - 1$.
OUTPUT: The positive residue $x$ modulo $p$.

1. $q_0 \leftarrow \lfloor x/b^k \rfloor$, $r_0 \leftarrow x - q_0 b^k$, $r \leftarrow r_0$, $i = 0$ and $c' = |c|$
2. **while** $q_i > 0$ **do**
3. $\qquad q_{i+1} \leftarrow \lfloor q_i c'/b^k \rfloor$
4. $\qquad r_{i+1} \leftarrow q_i c' - q_{i+1} b^k$
5. $\qquad r \leftarrow r + (-1)^{i+1} r_i$ and $i \leftarrow i + 1$
6. **while** $r \geq p$ **do** $r \leftarrow r - p$
7. **while** $r < 0$ **do** $r \leftarrow r + p$
8. **return** $r$

---

**Remarks 10.26**

(i) If $x$ is a $2k$-word integer in base $b$ and if $|c| \leq b^{k/2} - 1$ then at most 3 multiplications by $c'$ are required to find the final residue [MEOo+ 1996].

(ii) It is also possible to consider a prime number $p$ that divides $N = b^k + c$. If the cofactor $N/p$ is sufficiently small, the computations can be done modulo $N$ without any additional cost since multiplications are usually performed at a word level. At the end of the whole process the result is reduced modulo $p$.

**Example 10.27** Take $b = 8$, $k = 6$, $c = 3$, and let us reduce $x = 35061808269$, equal to $(405166136215)_8$ in base eight, modulo the prime number $8^6 + 3$. Execution of Algorithm 10.25 is as follows.

| $i$ | $q_i$ | $r_i$ | $r$ |
|---|---|---|---|
| 0 | $(405166)_8$ | $(136215)_8$ | $(136215)_8$ |
| 1 | $(1)_8$ | $(417542)_8$ | $-(261325)_8$ |
| 2 | $(0)_8$ | $(3)_8$ | $-(261322)_8$ |
| — | — | — | $(516461)_8$ |

Finally $171313 = (516461)_8$ and one checks that $x \equiv 171313 \pmod{8^6 + 3}$.

Another possible generalization of the definition of a Mersenne prime [SOL 1999a] is to consider primes $p$ of the form

$$p = 2^{n_k w} \pm 2^{n_{k-1} w} \pm \cdots \pm 2^{n_1 w} \pm 1$$

where $w = 16, 32$ or $64$. These primes are often referred to as *NIST primes* [FIPS 186-2]. The fundamental reduction relation is

$$2^{n_k w} \equiv \mp 2^{n_{k-1} w} \mp \cdots \mp 2^{n_1 w} \mp 1 \pmod{p}.$$

Used recursively, it allows us to reduce $x < p^2$ written as

$$x = x_{2n_k - 1} 2^{(2n_k - 1)w} \mp x_{2n_k - 2} 2^{(2n_k - 2)w} \mp \cdots \mp x_1 2^w \mp x_0.$$

For example, the field $\mathbb{F}_p$ where $p = 2^{192} - 2^{64} - 1$ is recommended by standards for elliptic curve cryptography. If $x < p^2$ then $x$ can be written

$$x = x_5 2^{320} + x_4 2^{256} + x_3 2^{192} + x_2 2^{128} + x_1 2^{64} + x_0$$

and using

$$\begin{align}
2^{320} &\equiv 2^{192} + 2^{128} \pmod{p} \\
2^{256} &\equiv 2^{128} + 2^{64} \pmod{p} \\
2^{192} &\equiv 2^{64} + 1 \pmod{p}
\end{align}$$

one obtains

$$\begin{align}
x &\equiv x_4 2^{256} + (x_5 + x_3) 2^{192} + (x_5 + x_2) 2^{128} + x_1 2^{64} + x_0 \pmod{p} \\
x &\equiv (x_5 + x_3) 2^{192} + (x_5 + x_4 + x_2) 2^{128} + (x_4 + x_1) 2^{64} + x_0 \pmod{p} \\
x &\equiv (x_5 + x_4 + x_2) 2^{128} + (x_5 + x_4 + x_3 + x_1) 2^{64} + (x_5 + x_3 + x_0) \pmod{p}.
\end{align}$$

### 10.4.4 Reduction modulo several primes

Reducing $x$ simultaneously modulo several primes $p_1, \ldots, p_k$ can be done [MOBO 1972] using a *remainder tree* in time $n(\lg n)^{2+o(1)}$ where $n$ is the total number of bits in $x, p_1, \ldots, p_k$. For instance, if one wants to reduce $x$ modulo $p_1, p_2, p_3$ and $p_4$, the idea is first to compute $x \bmod p_1 p_2 p_3 p_4$, second reduce the result modulo $p_1 p_2$, modulo $p_3 p_4$, and finally modulo each $p_i$, according to the following diagram

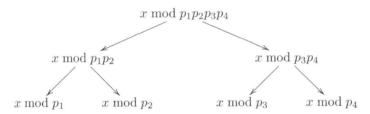

In [BER 2004b], Bernstein introduces the *scaled remainder tree* where computations are done modulo 1. Indeed modular reductions are replaced by real divisions, which are in turn replaced by multiprecision multiplications. This gives an algorithm with the same asymptotical complexity $n(\lg n)^{2+o(1)}$ but with a smaller $o(1)$. Note that similar techniques also apply to polynomials.

## 10.5 Division

Let us divide $u$ by $v$, that is compute the quotient $q = \lfloor u/v \rfloor$ and the remainder $r = u \bmod v$, where $u$ and $v$ are positive integers. Note that when only the remainder is needed, more efficient methods are discussed in Section 10.4.

Before addressing the general case, let us see what can be done when $v$ has special properties, and first of all when $v = 2$.

In many subsequent algorithms, one has to compute the *even part* and the *odd part* of $u$, that is respectively the biggest power $2^k$ that divides $u$ and $u/2^k$. If the processor has no specific instruction to count the number of low zero bits, one can compute $u \wedge (\overline{u} + 1)$ to find the lowest bit of $u$ that is nonzero. However, the simplest method, in this case, is certainly to shift $u$ to the right as many times as necessary. This process can be sped up with some precomputations, as suggested by the following piece of code for 32-bits architecture. The $i$-th entry in array $T$ contains the maximal exponent $r \leqslant 5$ such that $2^r$ divides $i$.

1. $T \leftarrow [5,0,1,0,2,0,1,0,3,0,1,0,2,0,1,0,4,0,1,0,2,0,1,0,3,0,1,0,2,0,1,0]$
2. $i \leftarrow u \wedge 31$
3. $u \leftarrow u \gg T[i]$
4. **while** $u \wedge 1 = 0$ **do** $u \leftarrow u \gg 1$ and $T[i] \leftarrow T[i] + 1$

At the end, the even part is $2^{T[i]}$ and $u$ holds the odd part. Note that the instructions in the while loop are performed only in case $i = 0$ when $2^k$ divides $u$ with $k > 5$.

Another simple case of interest is when $v$ is a single precision integer. The next algorithm will often be used in the sequel, at least implicitly.

---

**Algorithm 10.28** Short division of positive multiprecision integers

INPUT: An $n$-word integer $u = (u_{n-1} \ldots u_0)_b$ and a nonzero single $v = (v_0)_b$.
OUTPUT: The $n$-word integer $q = (q_{n-1} \ldots q_0)_b$ and the single $r = (r_0)_b$ such that $u = vq + r$.

1. $r_0 \leftarrow 0$                                                                              [$r_0$ is the remainder]
2. **for** $i \leftarrow n - 1$ **down to** $0$ **do**
3.     $t \leftarrow (r_0 b + u_i)$
4.     $q_i \leftarrow \lfloor (r_0 b + u_i)/v_0 \rfloor$ and $r_0 \leftarrow t \bmod v_0$                   [$q_{n-1}$ may be 0]
5. **return** $(q, r)$

---

**Example 10.29** Let us divide $(8789)_{10}$ by 7 using Algorithm 10.28.

| $i$ | $u_i$ | $t$ | $q_i$ | $r_0$ |
|---|---|---|---|---|
| 3 | 8 | 8 | 1 | 1 |
| 2 | 7 | 17 | 2 | 3 |
| 1 | 8 | 38 | 5 | 3 |
| 0 | 9 | 39 | 5 | 4 |

At the end, $q = (1255)_{10}$ and $r_0 = 4$ as expected since $8789 = 7 \times 1255 + 4$.

## 10.5.1 Schoolbook division

The next algorithm due to Knuth [KNU 1997] is a refined version of the customary method taught at school for performing division of multidigit numbers.

---

**Algorithm 10.30** Schoolbook division of positive multiprecision integers

INPUT: An $(m + n)$-word integer $u = (u_{m+n-1} \ldots u_0)_b$ and an $n$-word integer $v = (v_{n-1} \ldots v_0)_b$ where $v_{n-1} > 0$ and $n > 1$.
OUTPUT: The $(m+1)$-word integer $q = (q_m \ldots q_0)_b$ and the $n$-word integer $r = (r_{n-1} \ldots r_0)_b$ such that $u = vq + r$.

1. $u_{m+n} \leftarrow 0$ and $d \leftarrow 1$
2. **while** $v_{n-1} < b/2$ **do**
3.     $v \leftarrow 2v$, $u \leftarrow 2u$ and $d \leftarrow 2d$                                           [normalization]

4.   **for** $i = m$ **down to** 0 **do**

5.   $\quad\hat{q} \leftarrow \min(\lfloor(u_{i+n}b + u_{i+n-1})/v_{n-1}\rfloor, b-1)$

6.   $\quad$**while** $\hat{q}(v_{n-1}b + v_{n-2}) > (u_{i+n}b^2 + u_{i+n-1}b + u_{i+n-2})$ **do**

7.   $\quad\quad\hat{q} \leftarrow \hat{q} - 1$

8.   $\quad(u_{i+n}\ldots u_i)_b \leftarrow (u_{i+n}\ldots u_i)_b - \hat{q}(v_{n-1}\ldots v_0)_b$

9.   $\quad$**if** $(u_{i+n}\ldots u_i)_b < 0$ **then** $\hspace{2cm}$ [occurs with probability $2/b$]

10.  $\quad\quad\hat{q} \leftarrow \hat{q} - 1$

11.  $\quad\quad(u_{i+n}\ldots u_i)_b \leftarrow (u_{i+n}\ldots u_i)_b + (0\,v_{n-1}\ldots v_0)_b$

12.  $\quad q_i \leftarrow \hat{q}$

13.  $r \leftarrow u/d$ $\hspace{6cm}$ [unnormalization]

14.  **return** $(q, r)$

**Remarks 10.31**

(i) The for loop computes $q_i = \lfloor (u_{i+n}\ldots u_i)_b/(v_{n-1}\ldots v_0)_b \rfloor$. These steps are similar to the Line 4 of Algorithm 10.28. This determination relies on a guess, i.e., the approximation of $q_i$ by $\hat{q}$, Line 5. One always has $q_i \leqslant \hat{q}$.

(ii) At the beginning, $u$ and $v$ are multiplied by $d$, a suitable power of 2, such that $v_{n-1} \geqslant b/2$. This normalization is particularly well suited when the radix is a power of 2. It ensures that $\hat{q} \leqslant q_i + 2$. Thus the statement $\hat{q} \leftarrow \hat{q} - 1$ in Line 7 is encountered at most twice. The normalization does not change the quotient whereas the remainder must be divided by $d$ at the end.

**Example 10.32** Let us divide $(115923)_{10}$ by $(344)_{10}$. So $n = 3$ and $m = 3$. Because of the normalization, one sets $u = (231846)_{10}$, $v = (688)_{10}$ and $d = 2$.

| $i$ | 3 | 2 | 1 | 0 |
|---|---|---|---|---|
| $(u_{i+n}b + u_{i+n-1})$ | 2 | 23 | 25 | 48 |
| $\hat{q} = \min(\lfloor(u_{i+n}b + u_{i+n-1})/v_{n-1}\rfloor, b-1)$ | 0 | 3 | 4 | 8 |
| $u_{i+n}b^2 + u_{i+n-1}b + u_{i+n-2}$ | 23 | 231 | 254 | 480 |
| $\hat{q}(v_{n-1}b + v_{n-2})$ | 0 | $3 \times 68 = 204$ | $4 \times 68 = 272$ | $8 \times 68 = 544$ |
| $\hat{q}$ in Line 8 | 0 | 3 | 3 | 7 |
| $(u_{i+n}\ldots u_i)_b$ | 231 | 2318 | 2544 | 4806 |
| $\hat{q}(v_{n-1}\ldots v_0)_b$ | 0 | $3 \times 688 = 2064$ | $3 \times 688 = 2064$ | $7 \times 688 = 4816$ |
| $q_i$ | 0 | 3 | 3 | 6 |
| $(u_{i+n}\ldots u_i)_b$ in Line 12 | $(0231)_{10}$ | $(0254)_{10}$ | $(0480)_{10}$ | $(0678)_{10}$ |

Finally $231846 = 688 \times 336 + 678$ so that $115923 = 344 \times 336 + 339$.

## § 10.5 Division

**Remark 10.33** Another normalization allows us to do less elementary divisions. Following an idea of Quisquater [QUI 1990, QUI 1992], let $v'$ be the smallest multiple of $v$ bigger than $b^{n+1}$. Let $q' = u/v'$ and $r' = u \bmod v'$. The approximation of the digits of $q'$ is now trivial since the first two digits of $v'$ are $(10)_b$, but as above, some corrections might be necessary. When $q'$ is determined we easily deduce $q$ and $r$ from $q'$ and $r'$.

When several divisions by the same $v$ are carried out this technique is even more interesting as we compute $v'$ only once. Therefore this idea is useful to speed up prime field reductions; see Section 11.1.2.a.

**Example 10.34** Let $u = (797598)_{10}$ and $v = (983)_{10}$; we have $v' = 11v = (10813)_{10}$. We obtain $q'_1 = 7$ and $u - 70v' = 40688$. Then we should set $q'_0 = 4$ but this approximation is one in excess so that $q'_0 = 3$ and $u - 73v' = 8249$. So

$$u = q'v' + r' \text{ with } q' = 73 \text{ and } r' = 8249.$$

Now $r' = 8v + 385$ so that

$$u = (11 \times 73 + 8)v + 385 = 811v + 385.$$

We perform only 2 true divisions instead of $n$. When $b$ is larger the saving is noticeable [ZIM 2001].

Another simplification is possible since we can suppose that:

the length of $u$ is $2n$, the length of $v$ is $n$, $v$ is normalized, i.e., $b/2 \leqslant v_{n-1} < b$
and $q$ is of length $n$, i.e., $u < b^n v$. (10.1)

Indeed, if $u$ has length $2n$ and $u \geqslant b^n v$, it is sufficient to do $u \leftarrow u - b^n v$ to get $u < b^n v$, $v$ being normalized. Now suppose that $u$ is not of length $2n$ but $n + m$. Two possibilities arise.

- When $m \geqslant n$, one can divide $(u_{n+m-1} \ldots u_{m-n})_b$ of length $2n$ by $(v_{n-1} \ldots v_0)_b$ to get the first digits $q^*$ of $q$ and set $u \leftarrow u - q^* v b^k$ (where $k$ is a suitable power) until the length of $u$ is less than $2n$.
- If $m \leqslant n$ one divides $(u_{n+m-1} \ldots u_{n-m})_b$ of length $2m$ by $(v_{n-1} \ldots v_{n-m})_b$ of length $m$. The result $q^*$ is an approximation by excess of $q$. We easily derive $q$ from $q^*$ if $u - q^* v < 0$.

In the following we shall often assume (10.1).

### 10.5.2 Recursive division

In Algorithm 10.30 the next digit of the quotient is determined by the division of the 3-word number $u_{i+n}b^2 + u_{i+n-1}b + u_{i+n-2}$ by the double precision integer $v_{n-1}b + v_{n-2}$. But this is actually performed by means of a division of the double precision integer $u_{i+n}b + u_{i+n-1}$ by the single one $v_{n-1}$ and some possible corrections. Burnikel and Ziegler [BUZI 1998] apply this idea recursively to blocks of digits. Here we present a slightly different version [HAQU$^+$ 2002] and introduce the *concatenation* of $U_0$ of length $n$ and $U_1$ i.e., $U_1 b^n + U_0$ denoted by $(U_1 \| U_0)_b$.

This method is sometimes referred to as the *Karatsuba division*.

**Algorithm 10.35** Recursive division of positive multiprecision integers

INPUT: A $2n$-word $u = (u_{2n-1} \ldots u_0)_b$, an $n$-word $v = (v_{n-1} \ldots v_0)_b$ as in (10.1). The size $n$ is a parameter whereas the threshold $n_0$ is fixed.
OUTPUT: The $n$-word integers $q = (q_{n-1} \ldots q_0)_b$ and $r = (r_{n-1} \ldots r_0)_b$ such that $u = vq + r$.

1. **if** $n \leqslant n_0$ **then** determine $q$ and $r$ with Algorithm 10.30
2. $p \leftarrow \lceil n/2 \rceil$ and $t \leftarrow \lfloor n/2 \rfloor$
3. $U_1 \leftarrow (u_{2n-1} \ldots u_{2t})_b$ and $U_0 \leftarrow (u_{2t-1} \ldots u_0)_b$  $\quad [u = (U_1 \,\|\, U_0)_b]$
4. $V_1 \leftarrow (v_{n-1} \ldots v_t)_b$ and $V_0 \leftarrow (v_{t-1} \ldots v_0)_b$  $\quad [v = (V_1 \,\|\, V_0)_b]$
5. $Q_1 \leftarrow (q_{n-1} \ldots q_t)_b$ and $Q_0 \leftarrow (q_{t-1} \ldots q_0)_b$  $\quad [q = (Q_1 \,\|\, Q_0)_b]$
6. **if** $U_1 < b^p V_1$ **then**
7. $\quad (Q_1, r) \leftarrow (\lfloor U_1 / V_1 \rfloor, U_1 \bmod V_1)$
8. $\quad r \leftarrow U_0 + b^{2t} r - b^t V_0 Q_1$  $\quad [r \leftarrow (r \,\|\, U_0)_b - b^t V_0 Q_1]$
9. **else**
10. $\quad Q_1 \leftarrow b^p - 1$ and $r \leftarrow u - b^n v + b^t v$  $\quad [r \leftarrow u - b^t v(b^p - 1)]$
11. $\quad$ **while** $r < 0$ **do** $Q_1 \leftarrow Q_1 - 1$ and $r \leftarrow r + b^t v$
12. write $r = R_0 + b^p R_1$ and $V = V_0' + b^p V_1'$
13. **if** $R_1 < b^t V_1'$ **then**
14. $\quad (Q_0, R_1) \leftarrow (\lfloor R_1 / V_1' \rfloor, R_1 \bmod V_1')$
15. **else** $Q_0 \leftarrow b^t - 1$ and $r \leftarrow r - b^t V_1' + b^{n-t}\bigl((\lfloor v / b^{n-t} \rfloor)\bmod b^t\bigr)$
16. $r \leftarrow r - V_0' Q_0$
17. **while** $r < 0$ **do** $q \leftarrow q - 1$ and $r \leftarrow r + v$
18. **return** $(q, r)$

## Remarks 10.36

(i) The computations Lines 7 and 14 are obtained in a recursive way. In the first case, one calls Algorithm 10.35 with parameters $U_1$, $V_1$, $p$, while they are $R_1$, $V_1'$ and $t$ in the second one.

(ii) For optimal results, the threshold $n_0$ should be set after several tests.

(iii) The complexity of Algorithm 10.35 is $2\,\mathrm{K}(n)$ on average.

(iv) It is possible to modify Algorithm 10.35 to compute only the quotient. However this strategy might fail (with a probability less than $1/b$) and the algorithm returns a quantity bigger than $q$ that can be corrected very easily in most of the cases. The complexity is then $1.5\,\mathrm{K}(n)$ on average and only $2\,\mathrm{K}(n)$ in bad cases [HAQU$^+$ 2002] instead of $2.5\,\mathrm{K}(n)$ with the original algorithm of Burnikel and Ziegler.

(v) There is an asymptotically faster method, based on the *recursive middle product* introduced in [HAQU$^+$ 2004]. The complexity to divide $2n$-word by $n$-word integers is about $1.2\,\mathrm{K}(n)$ in this case. However, it is not relevant for sizes used in curve-based cryptography.

**Example 10.37** Set $n_0 = 2$ and let us divide $u = (6541237201)_{10}$ by $v = (65427)_{10}$ using

Algorithm 10.35. The intermediate steps are as follows:

- $n = 5, p = 3, t = 2, U_1 = (654123)_{10}, U_0 = (7201)_{10}, V_1 = (654)_{10}$ and $V_0 = (27)_{10}$.
- $U_1 \geqslant 1000 \times V_1$ so $Q_1 = 999$ and $r = u - 999 \times 100 \times v = 5079\,901$.
- $R_1 = (5079)_{10}, R_0 = (901)_{10}, V_1' = (65)_{10}$ and $V_0' = (427)_{10}$.
- $R_1 < 100 \times 65$ so we get $Q_0 = \lfloor 5079/65 \rfloor = 78$ and $R_1 = 5079 \bmod 65 = 9$ by a recursive call which reduces in this case to a schoolbook division.
- Finally $q = (Q_1 \| Q_0)_{10} = 99978$, $r = (R_1 \| R_0)_{10} = (9 \| 901)_{10} = 9901$ and $r - V_0' Q_0 = -23405 < 0$.

Thus we set $r = r + v = 42022$ and $q = 99977$ and terminate the algorithm. It is easy to check that these are the correct values.

### 10.5.3 Exact division

If $u$ and $v$ are positive numbers, then it is possible to test if $v$ divides $u$ without exactly computing the remainder $u \bmod v$. This is helpful as there are specialized algorithms for exact division.

Indeed, first remark that if $v \mid u$, the largest power of 2 dividing $u$ must be bigger than the one of $v$. So we can assume without loss of generality that $v$ is odd. The idea is now to use Montgomery reduction REDC modulo $v$, see Algorithm 10.22. More precisely, set $t \leftarrow u$ and repeat $t \leftarrow \text{REDC}(t)$ until $t < v$. At the end, $t \equiv u/2^k \pmod{v}$ for some $k$. Obviously, $v$ divides $u$ if and only if $t$ is zero.

**Example 10.38** Let $v$ be the 2-word integer $(573160\ 4090964624)_{2^{32}}$ and $u$ be the 4-word integer $(2242222213\ 1590893749\ 2725169084\ 656228000)_{2^{32}}$. As $v$ is not odd, one computes the even parts of $u$ and $v$, i.e., respectively $2^5$ and $2^4$ and one continues the computations with $u \leftarrow u/2^4$ and $v \leftarrow v/2^4$, which is now odd. With the notations of Section 10.4.2, let $R = 2^{64}$ so that $t = \text{REDC}(u) = (2242205646\ 805464192)_{2^{32}}$. As $t \geqslant v$, one applies another Montgomery reduction which returns 0. This shows that $u \equiv 0 \pmod{v}$.

Now let us introduce Jebelean method [JEB 1993a] to compute the quotient $\lfloor u/v \rfloor$ when it is known that $u \equiv 0 \pmod{v}$ i.e., $\lfloor u/v \rfloor = u/v$.

Once again it is assumed here that the base $b$ is a power of 2. The proposed algorithm relies on the following observation. Let us write $u = Ub + u_0$, $v = Vb + v_0$ and $q = Qb + q_0$, where $0 \leqslant u_0, v_0, q_0 < b$. So $u = vq$ implies that $Ub + u_0 = vQb + Vq_0b + v_0q_0$. Thus $u_0 = v_0q_0 \bmod b$ and $u - vq_0 = vQb$. If $\gcd(v_0, b) = 1$ this shows that $q_0 = (u_0 v_0^{-1}) \bmod b$ and the same arguments work for $(u - vq_0)/b = vQ$ allowing us to find $q_1$ and so on. Moreover only the $(m+1)$ lowest digits of $u$ take part in the computation of $u - vq_0$. So we can perform this subtraction modulo $b^{m+1}$.

---

**Algorithm 10.39** Exact division of positive multiprecision integers in base $b = 2^\ell$

INPUT: An $(m+n)$-word integer $u = (u_{m+n-1} \ldots u_0)_b$ and a $n$-word divisor $v$ of $u$ of the form $v = (v_{n-1} \ldots v_0)_b$.
OUTPUT: The $(m+1)$-word integer $q = (q_m \ldots q_0)_b$ such that $u = vq$.

1. **while** $2 \mid v$ **do** $v \leftarrow v/2$ and $u \leftarrow u/2$
2. $t \leftarrow v_0^{-1} \bmod b$                                                       [see Remark 10.40 (ii)]

3. **for** $i = 0$ **to** $m$ **do**
4. $\qquad q_i \leftarrow (u_0 t) \bmod b$
5. $\qquad u \leftarrow \bigl((u - vq_i) \bmod b^{m+1-i}\bigr)/b$
6. **return** $q$

**Remarks 10.40**

(i) At the end of the first line $\gcd(v_0, 2) = 1$. Since $b = 2^\ell$ this implies that $\gcd(v_0, b) = 1$.
(ii) The quantity $v_0^{-1} \bmod b$ is computed once but is used at each iteration. It can be obtained in a very efficient way. Indeed, if $t = (a^{-1}) \bmod b$ then $a^{-1} = (2t - at^2) \bmod b^2$. This trick used a couple of times after a search in a table of inverses modulo $2^8$ gives good results [JEB 1993a].
(iii) A slightly different procedure allows us to compute the quotient $\lfloor u/v \rfloor$ when the remainder $r = u \bmod v$ is known. Obviously it is enough to replace the last two statements in the for loop by $q_i \leftarrow \bigl((u_0 - r_i)t\bigr) \bmod b$ and $u \leftarrow \bigl((u - r_i - vq_i) \bmod b^{m+1-i}\bigr)/b$.
(iv) Krandick and Jebelean [KRJE 1996] designed a method to compute the highest digits of the quotient without computing the remainder. Their method is well suited to compute the first half of the digits of an exact division, the other half being computed by Algorithm 10.39.

**Example 10.41** One checks that $v = (238019)_{10}$ divides $u = (322413634849)_{10}$. One has $n = m = 6$. Let us find the quotient $q = u/v$ with Algorithm 10.39. The inverse of 9 modulo 10 is $t = 9$. Next table shows the progress of Algorithm 10.39 along the execution of the for loop.

| $i$ | 0 | 1 | 2 | 3 | 4 | 5 | 6 |
|---|---|---|---|---|---|---|---|
| $q_i = (u_0 t) \bmod b$ | 1 | 7 | 5 | 4 | 5 | 3 | 1 |
| $(u - vq_i) \bmod b^{m+1-i}$ | 3396830 | 673550 | 77260 | 55650 | 15470 | 87490 | 70730 |
| $u$ | 339683 | 67355 | 7726 | 5565 | 1547 | 8749 | 7073 |

Finally $q = (1354571)_{10}$.

## 10.6 Greatest common divisor

The following algorithms are described in [COH 2000, LER 1997]. We introduce Euclid, Lehmer and binary methods and give in fact the extended versions of these variants. Indeed, given two integers $x$ and $N$, the algorithms given below not only compute $d = \gcd(x, N)$ but also the integers $u$ and $v$ such that $xu + Nv = d$. Usually, this is the preferred method to compute the inverse of an element in $(\mathbb{Z}/N\mathbb{Z})^*$, see also Section 11.1.3 for specific methods to compute such a modular inverse. Another very important application is linked to the Chinese remainder theorem, cf. Corollary 2.24 and Algorithm 10.52 below.

## 10.6.1 Euclid extended gcd

**Algorithm 10.42** Euclid extended gcd of positive integers

INPUT: Two positive integers $x$ and $N$ such that $x < N$.
OUTPUT: Integers $(u, v, d)$ such that $xu + Nv = d$ with $d = \gcd(x, N)$.

1. $A \leftarrow N, B \leftarrow x, U_A \leftarrow 0$ and $U_B \leftarrow 1$
2. **repeat**
3. $\quad q \leftarrow \lfloor A/B \rfloor$
4. $\quad \begin{bmatrix} A \\ B \end{bmatrix} \leftarrow \begin{bmatrix} 0 & 1 \\ 1 & -q \end{bmatrix} \begin{bmatrix} A \\ B \end{bmatrix}$
5. $\quad \begin{bmatrix} U_A \\ U_B \end{bmatrix} \leftarrow \begin{bmatrix} 0 & 1 \\ 1 & -q \end{bmatrix} \begin{bmatrix} U_A \\ U_B \end{bmatrix}$
6. **until** $B = 0$
7. $d \leftarrow A, u \leftarrow U_A$ and $v \leftarrow (d - xu)/N$
8. **return** $(u, v, d)$

**Remarks 10.43**

(i) If we introduce the variables $V_A$ and $V_B$ such that

$$V_A = 1,\ V_B = 0 \text{ and } \begin{bmatrix} V_A \\ V_B \end{bmatrix} \leftarrow \begin{bmatrix} 0 & 1 \\ 1 & -q \end{bmatrix} \begin{bmatrix} V_A \\ V_B \end{bmatrix}$$

we see that $xU_A + NV_A$ and $xU_B + NV_B$ are constantly equal to respectively $A$ and $B$ during the execution of the algorithm. If the inversion routine is not implemented one can simply add these two variables and update them during each round to avoid the division in Line 7.

(ii) Throughout Algorithm 10.42 $|U_A|, |U_B|$ (resp. $|V_A|, |V_B|$) are less than or equal to $N/A$ (resp. $x/A$).

(iii) The number of necessary steps is $O(\lg N)$ (see [COH 2000] for more precise results). As a consequence, the complexity of Algorithm 10.42 is $O(\lg^2 N)$ when it is carefully implemented.

**Example 10.44** Let us compute $x^{-1} \bmod N$ for $x = 45$ and $N = 127$.

| $q$ | $U_A$ | $V_A$ | $A$ | $xU_A + NV_A$ | $U_B$ | $V_B$ | $B$ | $xU_B + NV_B$ |
|---|---|---|---|---|---|---|---|---|
| — | 0 | 1 | 127 | 127 | 1 | 0 | 45 | 45 |
| 2 | 1 | 0 | 45 | 45 | −2 | 1 | 37 | 37 |
| 1 | −2 | 1 | 37 | 37 | 3 | −1 | 8 | 8 |
| 4 | 3 | −1 | 8 | 8 | −14 | 5 | 5 | 5 |
| 1 | −14 | 5 | 5 | 5 | 17 | −6 | 3 | 3 |
| 1 | 17 | −6 | 3 | 3 | −31 | 11 | 2 | 2 |
| 1 | −31 | 11 | 2 | 2 | 48 | −17 | 1 | 1 |
| 2 | 48 | −17 | 1 | 1 | −127 | 45 | 0 | 0 |

So $48x - 17N = 1$ which implies that $x^{-1} \bmod N = 48$.

Most of the running time is taken by computing the quotient $q \leftarrow \lfloor A/B \rfloor$. Moreover, in 41% of the cases one obtains $q = 1$, which motivates choosing this value in all cases on the cost of more rounds, as is the case for the binary gcd algorithm, cf. Section 10.6.3. Note that Gordon proposed to use an approximation of the quotient by a suitable power of 2 [GOR 1989]. When $x$ and $n$ are multiprecision integers a variant due to Lehmer also avoids a full determination of the quotient.

### 10.6.2 Lehmer extended gcd

Lehmer's idea [LEH 1938] is to approximate $\lfloor A/B \rfloor$ with the most significants digits of $A$ and $B$ and to update them when necessary in computing the matrix product

$$\begin{bmatrix} A \\ B \end{bmatrix} \leftarrow \begin{bmatrix} \alpha & \beta \\ \alpha' & \beta' \end{bmatrix} \begin{bmatrix} A \\ B \end{bmatrix} \tag{10.2}$$

which performs several cumulated steps. The single precision integers $\alpha, \alpha', \beta$ and $\beta'$ are computed by a subalgorithm successively improved by Collins [COL 1980], Jebelean [JEB 1993b] and Lercier [LER 1997]. Here we state the skeleton of this algorithm, the improvements differ in the way Line 3 is performed.

---

**Algorithm 10.45** Lehmer extended gcd of multiprecision positive integers

INPUT: Two $m$-word multiprecision positive integers $x$ and $N$ in base $b = 2^\ell$ such that $x < N$.
OUTPUT: Integers $(u, v, d)$ such that $xu + Nv = d$ with $d = \gcd(x, N)$.

1. $A \leftarrow N, B \leftarrow x, U_A \leftarrow 0$ and $U_B \leftarrow 1$
2. **while** $|B|_2 > \ell$ **do**
3.     compute $\alpha, \alpha', \beta$ and $\beta'$ by **subalgorithm 10.46** with arguments $A$ and $B$
4.     $\begin{bmatrix} A \\ B \end{bmatrix} \leftarrow \begin{bmatrix} \alpha & \beta \\ \alpha' & \beta' \end{bmatrix} \begin{bmatrix} A \\ B \end{bmatrix}$
5.     $\begin{bmatrix} U_A \\ U_B \end{bmatrix} \leftarrow \begin{bmatrix} \alpha & \beta \\ \alpha' & \beta' \end{bmatrix} \begin{bmatrix} U_A \\ U_B \end{bmatrix}$
6. compute $(u, v, d)$ by Algorithm 10.42 with arguments $A$ and $B$
7. $u \leftarrow uU_A + vU_B, v \leftarrow (d - xu)/N$
8. **return** $(u, v, d)$

---

From $\hat{A}$ and $\hat{B}$, i.e., the $\ell$ most significants bits of $A$ and $B$ expressed in base $b = 2^\ell$, Lehmer derives two approximated quotients. Whenever they differ, $A$ and $B$ must be updated as in (10.2). Collins and Jebelean have an equivalent condition to determine $\alpha$, $\alpha'$, $\beta$, and $\beta'$ with only one quotient. Now experiments show that if the size of $\hat{A}$ and $\hat{B}$ is about $b$ then the order of magnitude of $\alpha, \alpha'\beta, \beta'$ is less than $\sqrt{b}$. In order to increase the size of these coefficients Lercier [LER 1997] mixes single and double precision approximations. We present this last improvement now.

### Sub-algorithm 10.46 Partial gcd of positive multiprecision integers in base $b = 2^\ell$

INPUT: Two positive integers $A$ and $B$.
OUTPUT: Integers $\alpha, \alpha', \beta$ and $\beta'$ as above.

1. $\hat{A} \leftarrow \left\lfloor \dfrac{A}{2^{\max(|A|_2 - \ell, 0)}} \right\rfloor$ $\quad\left[\hat{A} \leftarrow \text{the } \ell \text{ most significants bits of } A\right]$
2. $\hat{B} \leftarrow \left\lfloor \dfrac{B}{2^{\max(|A|_2 - \ell, 0)}} \right\rfloor$ $\quad\left[\hat{B} \text{ might be } 0\right]$
3. $\alpha \leftarrow 1, \beta \leftarrow 0, \alpha' \leftarrow 0, \beta' \leftarrow 1$ and $T \leftarrow 0$
4. if $\hat{B} \neq 0$ then $q \leftarrow \lfloor \hat{A}/\hat{B} \rfloor$ and $T \leftarrow \hat{A} \bmod \hat{B}$
5. if $T \geqslant 2^{\ell/2}$ then
6.     while true do
7.         $q' \leftarrow \lfloor \hat{B}/T \rfloor$ and $T' \leftarrow \hat{B} \bmod T$
8.         if $T' < 2^{\ell/2}$ then break
9.         $\hat{A} \leftarrow \hat{B}, \hat{B} \leftarrow T$
10.         $T \leftarrow \alpha - q\alpha', \alpha \leftarrow \alpha'$ and $\alpha' \leftarrow T$
11.         $T \leftarrow \beta - q\beta', \beta \leftarrow \beta'$ and $\beta' \leftarrow T$
12.         $T \leftarrow T'$ and $q \leftarrow q'$
13. if $\beta = 0$ then $\alpha \leftarrow 0, \beta \leftarrow 1, \alpha' \leftarrow 1, \beta' \leftarrow -\lfloor A/B \rfloor$ and return $(\alpha, \beta, \alpha', \beta')$
14. $\hat{A} \leftarrow \left\lfloor \dfrac{A}{2^{\max(|A|_2 - 2\ell, 0)}} \right\rfloor$ $\quad\left[\hat{A} \leftarrow \text{the } 2\ell \text{ most significants bits of } A\right]$
15. $\hat{B} \leftarrow \left\lfloor \dfrac{B}{2^{\max(|A|_2 - 2\ell, 0)}} \right\rfloor$
16. $\begin{bmatrix} \hat{A} \\ \hat{B} \end{bmatrix} = \begin{bmatrix} \alpha & \beta \\ \alpha' & \beta' \end{bmatrix} \begin{bmatrix} \hat{A} \\ \hat{B} \end{bmatrix}$
17. $\hat{A} \leftarrow \left\lfloor \dfrac{\hat{A}}{2^{\max(|\hat{A}|_2 - \ell, 0)}} \right\rfloor$ $\quad\left[\hat{A} \leftarrow \text{the } \ell \text{ most significants bits of } \hat{A}\right]$
18. $\hat{B} \leftarrow \left\lfloor \dfrac{\hat{B}}{2^{\max(|\hat{A}|_2 - \ell, 0)}} \right\rfloor$ and $T \leftarrow 0$
19. if $\hat{B} \neq 0$ then $q \leftarrow \lfloor \hat{A}/\hat{B} \rfloor$ and $T \leftarrow \hat{A} \bmod \hat{B}$
20. if $T \geqslant 2^{\ell/2}$ then
21.     while true do
22.         $q' \leftarrow \lfloor \hat{B}/T \rfloor$ and $T' \leftarrow \hat{B} \bmod T$
23.         if $T' < 2^{\ell/2}$ then return $(\alpha, \beta, \alpha', \beta')$
24.         $\hat{A} \leftarrow \hat{B}, \hat{B} \leftarrow T$
25.         $T \leftarrow \alpha - q\alpha', \alpha \leftarrow \alpha'$ and $\alpha' \leftarrow T$
26.         $T \leftarrow \beta - q\beta', \beta \leftarrow \beta'$ and $\beta' \leftarrow T$
27.         $T \leftarrow T'$ and $q \leftarrow q'$
28. return $(\alpha, \beta, \alpha', \beta')$

**Remark 10.47** With Lehmer's original algorithm [COH 2000] $A$ and $B$ decrease by 13 bits (respectively 29 bits) on average for each iteration when $b = 2^{32}$ (respectively $2^{64}$). With Sub-algorithm 10.46 the gain is of 22 bits (respectively 53 bits) on average.

**Example 10.48** With $b = 2^{16}$ let us compute the extended gcd of $N = 26498041357$ and $x = 8378459450$. With the initial algorithm of Lehmer, the intermediate steps are

| $\begin{bmatrix} \alpha & \beta \\ \alpha' & \beta' \end{bmatrix}$ | $A$ | $B$ | $U_A$ | $U_B$ | $u$ | $v$ | $d$ |
|---|---|---|---|---|---|---|---|
| — | 26498041357 | 8378459450 | 0 | 1 | — | — | — |
| $\begin{bmatrix} 1 & -3 \\ -6 & 19 \end{bmatrix}$ | 1362663007 | 202481408 | $-3$ | 19 | — | — | — |
| $\begin{bmatrix} -4 & 27 \\ 11 & -74 \end{bmatrix}$ | 16345988 | 5668885 | 525 | $-1439$ | — | — | — |
| $\begin{bmatrix} -9 & 26 \\ 17 & -49 \end{bmatrix}$ | 277118 | 106431 | $-42139$ | 79436 | — | — | — |
| $\begin{bmatrix} -3 & 8 \\ 5 & -13 \end{bmatrix}$ | 20094 | 1987 | 761905 | $-1243363$ | — | — | — |
| — | — | — | — | — | 10055119245 | $-3179344757$ | 1 |

With Sub-algorithm 10.46 less steps are needed, namely

| $\begin{bmatrix} \alpha & \beta \\ \alpha' & \beta' \end{bmatrix}$ | $A$ | $B$ | $U_A$ | $U_B$ | $u$ | $v$ | $d$ |
|---|---|---|---|---|---|---|---|
| — | 26498041357 | 8378459450 | 0 | 1 | — | — | — |
| $\begin{bmatrix} -43 & 136 \\ 123 & -389 \end{bmatrix}$ | 54706849 | 38360861 | 136 | $-389$ | — | — | — |
| $\begin{bmatrix} 115 & -164 \\ -291 & 415 \end{bmatrix}$ | 106431 | 64256 | 79436 | $-201011$ | — | — | — |
| — | — | — | — | — | 10055119245 | $-3179344757$ | 1 |

### 10.6.3 Binary extended gcd

To compute the gcd of two integers $A$ and $B$ one can also repeatedly apply the following rules:

- if $A$ and $B$ are both even then $\gcd(A, B) = 2\gcd(A/2, B/2)$
- if $A$ is even and $B$ is odd then $\gcd(A, B) = \gcd(A/2, B)$
- if $A$ and $B$ are both odd then $|A - B|$ is even so that $\gcd(A, B) = \gcd(A, |A - B|/2)$. In addition $|A - B| \leq \max\{A, B\}$.

Since $A$ and $B$ are not necessarily of the same order of magnitude, it is wise to reduce the size of the operands once. So only one division with remainder is required by the following algorithm, which is therefore especially interesting in computing the gcd of multiprecision integers.

## § 10.6 Greatest common divisor

**Algorithm 10.49** Extended binary gcd of positive integers

INPUT: Two positive integers $x$ and $N$ such that $x < N$.
OUTPUT: Integers $(u, v, d)$ such that $xu + Nv = d$ with $d = \gcd(x, N)$.

1. $q \leftarrow \lfloor N/x \rfloor, T \leftarrow N \bmod x, N \leftarrow x$ and $x \leftarrow T$      [reduce size once]
2. **if** $x = 0$ **then** $u \leftarrow 0, v \leftarrow 1, d \leftarrow N$ and **return** $(u, v, d)$
3. $k \leftarrow 0$ and $f \leftarrow 0$      [$f$ is a flag]
4. **while** $N \equiv 0 \pmod 2$ **and** $x \equiv 0 \pmod 2$ **do**
5.      $k \leftarrow k + 1, N \leftarrow N/2$ and $x \leftarrow x/2$
6. **if** $x \equiv 0 \pmod 2$ **then**
7.      $T \leftarrow x, x \leftarrow N, N \leftarrow T$ and $f \leftarrow 1$      [swap $x$ and $N$]
8. $U_B \leftarrow 1, A \leftarrow N, B \leftarrow x$ and $v' \leftarrow x$
9. **if** $N \equiv 1 \pmod 2$ **then** $U_A \leftarrow 0$ and $t' \leftarrow -x$
10. **else** $U_A \leftarrow (1+x)/2$ and $t' \leftarrow n/2$
11. **while** $t' \neq 0$ **do**
12.      **if** $t' > 0$ **then** $U_B \leftarrow U_A$ and $A \leftarrow t'$ **else** $B \leftarrow x - U_A$ and $v' \leftarrow -t'$
13.      $U_A \leftarrow U_B - B$ and $t' \leftarrow A - v'$
14.      **if** $U_A < 0$ **then** $U_A \leftarrow U_A + x$
15.      **while** $t' \equiv 0 \pmod 2$ and $t' \neq 0$
16.          $t' \leftarrow t'/2$
17.          **if** $U_A \equiv 0 \pmod 2$ **then** $U_A \leftarrow U_A/2$ **else** $U_A \leftarrow (U_A + x)/2$
18. $u \leftarrow U_B, v \leftarrow (A - xu)/N$ and $d \leftarrow 2^k A$
19. **if** $f = 1$ **then** $T \leftarrow u, u \leftarrow v$ and $v \leftarrow T$
20. $u \leftarrow u - vq$
21. **return** $(u, v, d)$

We shall not describe here asymptotically faster methods such as the *generalized binary* algorithms [LER 1997], since they become more efficient for integers larger than $2^{600}$. This size is completely out of the range for elliptic and hyperelliptic cryptosystems.

**Remarks 10.50**

(i) On average Euclidean quotients in Algorithm 10.42 are small. It is even possible to show that $q = 1$ is the most probable case. The corresponding probability, defined in a suitable sense, is $0.41504\ldots$, see [COH 2000]. This justifies performing subtractions instead of divisions even if the number of steps needed is greater.

(ii) Traditionally, to compute $(k/x) \bmod N$ one performs the multiplication of $k$ by the inverse of $x$ whereas a direct computation is possible. Indeed it suffices to set $U_B \leftarrow k$ instead of $U_B \leftarrow 1$, in Lines 1, 1 and 8 of Algorithms 10.42, 10.45, and 10.49.

**Example 10.51** Let us compute the extended gcd of $x = 67608$ and $N = 830616$ with Algorithm 10.49.

In Line 1, one performs a classical reduction to obtain variables of comparable size. For these particular values, this step saves many computations since we have $q = \lfloor N/x \rfloor = 12$. Then in the while loop starting Line 4, we remove the highest power of 2 dividing the gcd, that is $k = 3$. After Line 7, $x$ and $N$ are respectively equal to 2415 and 8451.

The next table contains the values of the relevant parameters $A$, $B$, $U_A$, $U_B$, $t'$ and $v'$ before the execution of Line 12.

| $A$ | $B$ | $U_A$ | $U_B$ | $t'$ | $v'$ |
|---|---|---|---|---|---|
| 8451 | 2415 | 0 | 1 | −2415 | 2415 |
| 8451 | 2415 | 604 | 1 | 1509 | 2415 |
| 1509 | 2415 | 302 | 604 | −453 | 2415 |
| 1509 | 2113 | 783 | 604 | 33 | 453 |
| 33 | 2113 | 875 | 783 | −105 | 453 |
| 33 | 1540 | 811 | 783 | −9 | 105 |
| 33 | 1604 | 803 | 783 | 3 | 9 |
| 3 | 1604 | 807 | 803 | −3 | 9 |

We set $u = 803$, $v = (A - Nu)/x = -2810$ and $d = 2^k A = 24$. Finally, the value of $u$ is corrected to take into account the initial reduction, i.e., $u = u - vq = 34523$. We easily check that $ux + vN = 24$.

### 10.6.4 Chinese remainder theorem

Suppose one wants to find a solution to the system

$$\begin{cases} x \equiv x_1 \pmod{n_1} \\ x \equiv x_2 \pmod{n_2} \\ \vdots \\ x \equiv x_k \pmod{n_k} \end{cases}$$

where the $n_i$'s are pairwise coprime integers and the $x_i$'s are fixed integers. Corollary 2.24 ensures that there is a unique solution modulo $N = n_1 n_2 \ldots n_k$ and in fact, such a solution is easy to find. Let $N_i = N/n_i$. Since the $n_i$'s are pairwise coprime one has $\gcd(N_i, n_i) = 1$ for all $i$, and an extended gcd computation gives $a_i$ such that $a_i N_i \equiv 1 \pmod{n_i}$. Clearly, a solution is then given by

$$x = a_1 N_1 x_1 + a_2 N_2 x_2 + \cdots + a_k N_k x_k.$$

This is the idea behind the following algorithm, which performs more efficiently and computes $x$ inductively. At each step, given an integer $x$ such that $x \equiv x_i \pmod{n_i}$ for all $i \leq j$, it finds $x$ satisfying $x \equiv x_i \pmod{n_i}$ for all $i \leq j+1$.

**Algorithm 10.52** Chinese remainder computation

INPUT: Pairwise coprime integers $n_1, \ldots, n_k$ and integers $x_i$ for $1 \leq i \leq k$.
OUTPUT: An integer $x$ such that $x \equiv x_i \pmod{n_i}$, for all $1 \leq i \leq k$.

1. $N \leftarrow n_1$ and $x \leftarrow x_1$
2. **for** $i = 2$ **to** $k$ **do**
3.     compute $u$ and $v$ such that $un_i + vN = 1$     [use an extended gcd algorithm]

§ 10.7 Square root

4.      $x \leftarrow un_i x + vN x_i$

5.      $N \leftarrow N n_i$

6.      $x \leftarrow x \bmod N$

7.      **return** $x$

**Remarks 10.53**

(i) Algorithm 10.52 can be generalized in a straightforward way to the polynomial ring $K[X]$ where $K$ is a field.

(ii) Another variant due to Garner [GAR 1959] involves precomputations and is well suited to solve different systems with fixed $n_i$'s; see for instance [MEOO+ 1996]. This is especially useful for *residue number system arithmetic* where computations modulo $N$ are in fact performed modulo several primes $p_i$ fitting in a word and such that $\prod_i p_i > N^2$. See [BLSE+ 1999] for instance.

**Example 10.54** Let us solve the system

$$\begin{cases} x \equiv 1 \pmod{3} \\ x \equiv 2 \pmod{5} \\ x \equiv 4 \pmod{7} \\ x \equiv 5 \pmod{11} \\ x \equiv 9 \pmod{13}. \end{cases}$$

The moduli $3, 5, 7, 11$, and $13$ prime and thus they are pairwise coprime. Here are the values of relevant parameters before we enter the for loop and at the end of it, Line 6.

| $i$ | $n_i$ | $x_i$ | $N$ | $u$ | $v$ | $x$ | $x \bmod n_i$ |
|---|---|---|---|---|---|---|---|
| 1 | 3 | 1 | 3 | — | — | 1 | 1 |
| 2 | 5 | 2 | 15 | $-1$ | 2 | 7 | 2 |
| 3 | 7 | 4 | 105 | $-2$ | 1 | 67 | 4 |
| 4 | 11 | 5 | 1155 | $-19$ | 2 | 907 | 5 |
| 5 | 13 | 9 | 15015 | $-533$ | 6 | 8992 | 9 |

## 10.7 Square root

First let us describe a simple method to compute the *integer square root* of a positive number $u$, that is $v = \lfloor \sqrt{u} \rfloor$.

### 10.7.1 Integer square root

Newton iteration is a powerful tool to find solutions of the equation $f(x) = 0$. Under suitable conditions, the process $x_{i+1} \leftarrow x_i - f(x_i)/f'(x_i)$ starting from an appropriate approximation, converges quadratically to a root of $f$. Using this method with $f(x) = x^2 - u$ leads to the iteration $x_{i+1} \leftarrow (x_i + u/x_i)/2$ to compute $\sqrt{u}$. Its discrete version is the basis of Algorithm 10.55.

**Algorithm 10.55** Integer square root

INPUT: A positive $n$-word integer $u$.
OUTPUT: The positive integer $v$ such that $v = \lfloor \sqrt{u} \rfloor$.

1. $t \leftarrow 2^{\lceil \lg u/2 \rceil}$ [initial approximation]
2. **repeat**
3. $\quad v \leftarrow t$
4. $\quad t \leftarrow \lfloor (v + \lfloor u/v \rfloor)/2 \rfloor$ [discrete Newton iteration]
5. **until** $t = v$
6. **return** $v$

**Remarks 10.56**

(i) If $u$ fits in a word, it is more efficient to perform a binary search, that is guess the bits of $v$ one by one [BER 1998].
(ii) Algorithm 10.55 needs $O(\lg \lg u)$ iterations to terminate.
(iii) Any integer $t \geqslant \sqrt{u}$ can be chosen as an initial approximation, but $t$ should be as close as possible to $\lfloor \sqrt{u} \rfloor$ for efficiency reasons. To ensure a fast convergence, one possibility is to compute an approximation $\sqrt{u}$, using for instance the most significant word of $u$.
(iv) The working precision should be increased dynamically as the computation progresses.
(v) If Lines 4 and 5 of Algorithm 10.55 are replaced by
   4. $\quad t \leftarrow (v + u/v)/2$
   5. **until** $t - v \leqslant \varepsilon$
   then an approximation up to the precision $\varepsilon$ of the real $\sqrt{u}$ is returned instead.

**Example 10.57** Take $(393419390\ 735536755)_{2^{32}}$ and let us find $v = \lfloor \sqrt{u} \rfloor$ with Algorithm 10.55. Since $u$ is a 61-bit integer, one takes $2^{31}$ as an initial approximation. Then the successive values of $t$ are 1467161214, 1309428509, 1299928331, 1299893617, 1299893616, and again 1299893616 which is the expected result. If the initial value is set to $t \leftarrow \lceil \sqrt{393419390} \rceil \times 2^{16}$ instead, only two iterations are required to get the result.

At present, let us examine a related problem.

## 10.7.2 Perfect square detection

To decide if an integer $u$ is a perfect square or not, one possibility is to compute its integer square root $v = \lfloor \sqrt{u} \rfloor$, with the algorithm above, and to compare $v^2$ to $u$. However, if $u$ is not a square modulo some integer, it is clear that $u$ cannot be a perfect square. Now, if $u$ is a square modulo several integers, for instance 64, 63, 65, and 11, then it is very likely (in fact with a probability bigger than 99%) that $u$ is a square [COH 2000]. Testing more moduli eliminates more integers $u$ that are not a square. For instance, one can choose small odd moduli and pack them into a highly composite integer $N$ fitting in a word. Then one has to reduce $u$ modulo $N$ or, as suggested by Harley [HAR 2002a], use Montgomery reduction, see Section 10.4.2, to derive an appropriate approximation of the residue. Indeed, if $\ell$ is the word size, Montgomery reduction REDC of $u$ modulo $N$, whose cost is very cheap, returns an integer congruent to $u/2^{\ell} \bmod N$. Now if $u$ is an $n$-word integer then $n-1$ successive applications of REDC will give the single $r = u/2^{\ell(n-1)} \bmod N$.

From the Chinese remainder theorem, see Section 2.1.2, and the fact that $\ell$ is even, it is easy to see that $u$ is a square modulo $N$ if and only if $r$ is a square modulo every prime power $p^k$ dividing $N$. As $N$ has only small prime divisors, this can be tested very efficiently since it is sufficient to look at the bit of order $r \bmod p^k$ of a precomputed *mask* $m$, reflecting the residues that are a square modulo $p^k$. If $u$ passes all these tests, Algorithm 10.55 is used to ensure that $u$ is really a square.

**Example 10.58** Take $b = 2^{32}$, $u = (2937606071\ 1090932004\ 1316444929)_{2^{32}}$ and set the value of $N$ to the single $3^2 \times 5^2 \times 7 \times 11 \times 17 \times 19 \times 23 \times 29 = 3732515325$. With the settings of Section 10.4.2, put $R = 2^{32}$ and precompute $(-1/N) \bmod R$. After two successive Montgomery reductions, one obtains $r = 2783108164 = u/2^{64} \bmod N$. Now the squares modulo 9 being $0, 1, 4$ and 7, the value of the mask $m$ is set to $147 = (10010011)_2$. Since $r \bmod 9 = 4$, one looks at the bit of weight 4 of $m$ which is 1. This means that $u$ is a square modulo 9.

This is also the case for all the other prime powers dividing $N$. For instance, the mask modulo 29 is $m = (10011110100010010001011110011)_2$ and the bit of order $r \bmod 29$, that is 1, of $m$ is 1 again. This shows that $u$ has a very high probability (more than 99.54%) to be a square and indeed if one computes $v = \lfloor \sqrt{u} \rfloor$ with Algorithm 10.55 one can check that $v^2 = u$.

Note that a similar idea applies for cubes as well, and more generally for every power $k$. Bernstein [BER 1998] also developed a different method where a real approximation of $v = u^{1/k}$ is first computed before the consistency of the assumption $u = v^k$ is checked on the first few digits of $u$ and $v$.

# Chapter 11

# Finite Field Arithmetic

*Christophe Doche*

### Contents in Brief

| | | |
|---|---|---|
| **11.1** | **Prime fields of odd characteristic** | 201 |
| | Representations and reductions • Multiplication • Inversion and division • Exponentiation • Squares and square roots | |
| **11.2** | **Finite fields of characteristic** 2 | 213 |
| | Representation • Multiplication • Squaring • Inversion and division • Exponentiation • Square roots and quadratic equations | |
| **11.3** | **Optimal extension fields** | 229 |
| | Introduction • Multiplication • Exponentiation • Inversion • Squares and square roots • Specific improvements for degrees 3 and 5 | |

In this chapter, we are mainly interested in performance; see Section 2.3 for a theoretical presentation of finite fields. In the following, we consider three kinds of fields that are of great cryptographic importance, namely prime fields, extension fields of characteristic 2, and optimal extension fields. We will describe efficient methods for performing elementary operations, such as addition, multiplication, inversion, exponentiation, and square roots. The material that we give is implicitly more related to a software approach; see Chapter 26 for a presentation focused on hardware. Efficient finite field arithmetic is crucial in efficient elliptic or hyperelliptic curve cryptosystems and is the subject of abundant literature [JUN 1993, LINI 1997, SHP 1999]. See also the preliminary version of a book written by Shoup and available online [SHO], introducing basic concepts from computational number theory and algebra, and including all the necessary mathematical background.

There are some software packages implementing the algorithms described below, such as ZEN [ZEN], which is a set of optimized C libraries dedicated to finite fields. There are also more general libraries like NTL [NTL] or Lidia [LIDIA]. In addition, several computer algebra systems contain functions for handling finite fields, for example Magma [MAGMA].

## 11.1 Prime fields of odd characteristic

Most of the algorithms detailed here carry through as well to $\mathbb{Z}/N\mathbb{Z}$ for arbitrary moduli $N$, usually with some obvious modifications. However, here we are mainly interested in prime moduli. Methods to find either an industrial-grade prime or a certified prime number $p$ of the desired size are described in Chapter 25.

### 11.1.1 Representations and reductions

For representing finite prime fields we usually use the isomorphism between $\mathbb{F}_p$ and $\mathbb{Z}/p\mathbb{Z}$. Elements of $\mathbb{Z}/p\mathbb{Z}$ are equivalence classes and we have to choose a *representative*, that is a particular element in the class, to perform computations. The most standard choice is to represent $x \in \mathbb{Z}/p\mathbb{Z}$ by the unique integer in $[0, p-1]$, which is in the class of $x$. We can also use other representatives such as the ones belonging to $[-\lfloor p/2 \rfloor, \lfloor p/2 \rfloor]$ or even an *incompletely reduced number*, which is not uniquely determined, for it belongs to an interval of length greater than $p$; see Remark 26.45 (ii) and [YAN 2001, YASA+ 2002].

Given an integer $u$ of arbitrary size we must be able to reduce it, i.e., to find the integer in $[0, p-1]$ which is congruent to $u$ modulo $p$. This *modular reduction* is achieved by computing the remainder of a Euclidean division.

However, since all the reductions are performed modulo the same prime number $p$, there exist several improvements which, for instance, involve some precomputations. The most popular ones are certainly the Montgomery and the Barrett methods; see Section 10.4. In this case the cost of a reduction of a $2n$-word integer modulo an $n$-word integer is asymptotically equal to a size $n$ multiplication.

To compute the remainder faster, other ideas include the choice of a special modulus allowing a fast reduction; see Algorithm 10.25 for the use of a different normalization than the one initially suggested by Knuth in Algorithm 10.30. Quisquater first proposed this method, which speeds up the determination of an approximation of the quotient; see Remark 10.33 and Example 10.34. However, this reduction method will increase the length of the modulus $p$ by at least one digit, resulting in additional multiplications when performing arithmetic in $\mathbb{F}_p$.

In the remainder, we address prime field arithmetic itself. Whatever representation is chosen, prime field addition and subtraction algorithms are straightforward in terms of the corresponding multiprecision algorithms for integers, cf. Algorithms 10.3 and 10.5. For example, with classical representation, if $u, v$ are integers in $[0, p-1]$, then $u + v < 2p$ and the modular addition of $u$ and $v$ is simply $u + v$ or $u + v - p$. In the same way, modular subtraction of $u$ and $v$ is $u - v$ if $u \geq v$ and $u - v + p$ when $u < v$. Montgomery representation is compatible with addition and subtraction as well.

Now let us investigate multiplication algorithms in $\mathbb{F}_p$.

### 11.1.2 Multiplication

Except special methods, like the one explained in [CHCH 1999] that involves precomputations, there are mainly two ways two perform a modular multiplication. The first one consists of a simple integer multiplication with the schoolbook or Karatsuba methods, i.e., one of the Algorithms 10.8 or 10.11, followed by a reduction. The choice of the algorithm depends on the nature and the size of the integer operands as well as on the computer architecture used.

The second one is designed as a single operation. In this case elementary multiplications and reductions are interleaved so that the size of the intermediate results remains bounded. These two options apply to Montgomery representation as well.

#### 11.1.2.a Classical representation

Algorithm 11.1 is a general scheme to compute a modular multiplication. We have

$$uv \equiv \left( \sum_{i=0}^{n-1} u_i v b^i \right) \pmod{p}$$

## § 11.1 Prime fields of odd characteristic

which can be written

$$uv \equiv \left(\left(\ldots\left((u_{n-1}v \bmod p)b + u_{n-2}v \bmod p\right)b + \cdots + u_1 v \bmod p\right)b + u_0 v\right) \pmod{p}.$$

If we set $t_{-1} = 0$ and $t_i = (t_{i-1}b + u_{n-i-1}v) \bmod p$ then $t_{n-1} \equiv uv \pmod{p}$. We deduce the following algorithm:

---

**Algorithm 11.1** Interleaved multiplication-reduction of multiprecision integers

INPUT: Two $n$-word integers $u = (u_{n-1} \ldots u_0)_b$ and $v$.
OUTPUT: An integer $t$ such that $t \equiv uv \pmod{p}$.

1. $t \leftarrow 0$
2. **for** $i = 0$ **to** $n-1$ **do**
3.      $t \leftarrow tb + u_{n-i-1}v$
4.      approximate $q = \lfloor t/p \rfloor$ with $\hat{q}$          [see methods below]
5.      $t \leftarrow t - \hat{q}p$
6. **return** $t$

---

The approximation of $q$ can be achieved by Knuth, Barrett, or Quisquater methods. Knuth's approach has already been explained, cf. Algorithm 10.30. The last two methods are described in detail by Dhem in [DHE 1998]. See also Section 10.4.1 and Remark 10.33. Barrett writes

$$q = \left\lfloor \frac{t}{p} \right\rfloor = \left\lfloor \frac{\frac{t}{2^{n-1}} \frac{2^{2n}}{p}}{2^{n+1}} \right\rfloor$$

where $n$ is the number of bits of $p$, then approximates $q$ by

$$\left\lfloor \frac{\left\lfloor \frac{t}{2^{n-1}} \right\rfloor \left\lfloor \frac{2^{2n}}{p} \right\rfloor}{2^{n+1}} \right\rfloor$$

where we assume that $\left\lfloor \frac{2^{2n}}{p} \right\rfloor$ has been precomputed. Dhem introduced a more general variant, namely

$$\hat{q} = \left\lfloor \frac{\left\lfloor \frac{t}{2^{n+\beta}} \right\rfloor R}{2^{\alpha-\beta}} \right\rfloor \quad \text{with} \quad R = \left\lfloor \frac{2^{n+\alpha}}{p} \right\rfloor.$$

These additional parameters allow us to perform corrections on the remainder only at the end of the whole multiplication process, when they are suitably tuned. Algorithm 11.1 works at a word level and if $b = 2^\ell$ then optimal results are obtained with $\alpha = \ell + 3$ and $\beta = -2$. In this case we have $q - 1 \leqslant \hat{q} \leqslant q$ and the intermediate results grow moderately. More precisely, given $u, v < 2^{n+1}$ then $t \equiv uv \pmod{p}$ returned by Algorithm 11.1 is less than $2^{n+1}$. To get the final result, at most one subtraction is needed. This implies also that an exponentiation or any other long computation can be done with only one correction at the end of the whole process with the same choice of parameters.

Quisquater's method [QUI 1990, QUI 1992] consists in multiplying $p$ by a suitable coefficient $\delta$ such that the reduction modulo $\delta p$ is easy. Set

$$\delta = \left\lfloor \frac{2^{n+\ell+2}}{p} \right\rfloor \quad \text{and get} \quad \hat{q} = \left\lfloor \frac{t}{2^{n+\ell+2}} \right\rfloor.$$

The $(\ell + 2)$ highest bits of $\delta p$ are now equal to 1 and the corresponding quotient is obviously determined, since it is simply equal to the most significants bits of $t$. There is a fast way to compute $\delta$, namely put

$$\hat{\delta} = \left\lfloor \frac{2^{2\ell+6}}{\lfloor \frac{p}{2^{n-\ell-3}} \rfloor} \right\rfloor,$$

which verifies $\delta \leqslant \hat{\delta} \leqslant \delta + 1$, and a simple test gives the correct value. It is also possible to reduce the size of $\delta$; see [DHE 1998, p. 24]. This normalization avoids overflows in Algorithm 11.1 while computing a multiplication or even a modular exponentiation.

Now suppose that one has the result $x \bmod \delta p$ while we still want $x \bmod p$. For this, we could perform $(x \bmod \delta p) \bmod p$ but since an exact division is faster (see Section 10.5.3) it is better to compute

$$x \bmod p = \frac{\delta x \bmod \delta p}{\delta}. \tag{11.1}$$

Note that this method has been used in several smart cards; see for example [QUWA$^+$ 1991] or [FEMA$^+$ 1996].

### 11.1.2.b Montgomery multiplication

Montgomery representation, see Section 10.4.2, was in fact introduced to carry out quick modular multiplications. This property comes from the equality

$$((xR \bmod p)(yR \bmod p)R^{-1} \bmod p) = (xyR) \bmod p$$

which implies that $\text{REDC}([x][y]) = [xy]$. Recall that Montgomery reduction is also useful to convert elements between normal and Montgomery representations. Indeed, $[x] = \text{REDC}(xR')$ where $R' = R^2 \bmod p$ has been precomputed and stored, and $\text{REDC}([x]) = x$.

**Example 11.2** Take $p = 2011$, $b = 2^3$, $R = 4096$ so that $R' = 1454$. Let $x = 45$, $y = 97$. Then

$$\begin{aligned}
[x] &= \text{REDC}(45 \times 1454) = 1319 = (2447)_8 \\
[y] &= \text{REDC}(97 \times 1454) = 1145 = (2171)_8 \\
[x][y] &= 1510255 \\
[xy] &= \text{REDC}(1510255) = 1250 = (2342)_8 \\
xy &= \text{REDC}(1250) = 343.
\end{aligned}$$

One checks that $45 \times 97 \equiv 343 \pmod{2011}$.

Of course, Montgomery method is completely irrelevant when only one product is needed. Instead, operands are converted to and kept in Montgomery representation as long as possible. For instance, if one wants $[x^2y]$, simply perform $\text{REDC}([x][xy])$.

The following algorithm computes directly $\text{REDC}(uv)$ given multiprecision integers $u$ and $v$ in Montgomery representation. It combines Algorithms 10.8 and 10.22.

---

**Algorithm 11.3** Multiplication in Montgomery representation

INPUT: An $n$-word integer $p = (p_{n-1} \ldots p_0)_b$ prime to $b$, $R = b^n$, $p' = -p^{-1} \bmod b$ and two $n$-word integers $u = (u_{n-1} \ldots u_0)_b$ and $v = (v_{n-1} \ldots v_0)_b$ such that $0 \leqslant u, v < p$.
OUTPUT: The $n$-word integer $t = (t_{n-1} \ldots t_0)_b$ equal to $\text{REDC}(uv) = (uvR^{-1}) \bmod p$.

---

1. $t \leftarrow 0$
2. **for** $i = 0$ **to** $n - 1$ **do**
3. $\quad m_i \leftarrow \big((t_0 + u_i v_0)p'\big) \bmod b$

4. $t \leftarrow (t + u_i v + m_i p)/b$

5. **if** $t \geqslant p$ **then** $t \leftarrow t - p$

6. **return** $t$

**Remarks 11.4**

(i) If $R$ is chosen such that $R > 4p$ and if $u$ and $v$ are positive and less than $2p$ then REDC$(uv)$ is bounded by $2p$ as well. This means that it is possible to avoid the subtraction in Line 5 of Algorithm 11.1 during long computations as exponentiations. At the end of the whole process the result is normalized in $\mathbb{Z}/p\mathbb{Z}$ at the cost of a single subtraction [LEN 2002].

(ii) See [KOAC$^+$ 1996] for a comparison of different variations of Montgomery method.

**Example 11.5** Let us perform again the computation of Example 11.2 but at a word level with Algorithm 11.3. Let $u = [45] = 1319 = (2447)_8$ and $v = [97] = 1145 = (2171)_8$. Then

| $i$ | $u_i$ | $t_0$ | $m_i$ | $u_i v$ | $m_i p$ | $t + u_i v + m_i p$ | $t$ |
|---|---|---|---|---|---|---|---|
| — | — | — | — | — | — | — | 0 |
| 0 | 7 | 0 | 3 | $(17517)_8$ | $(13621)_8$ | $(33340)_8$ | $(3334)_8$ |
| 1 | 4 | 4 | 0 | $(10744)_8$ | 0 | $(14300)_8$ | $(1430)_8$ |
| 2 | 4 | 0 | 4 | $(10744)_8$ | $(17554)_8$ | $(32150)_8$ | $(3215)_8$ |
| 3 | 2 | 5 | 3 | $(4362)_8$ | $(13621)_8$ | $(23420)_8$ | $(2342)_8$ |

One obtains REDC$(uv) = (2342)_8 = [xy] = 1250$ as previously.

Concerning modular squaring, the computation of the square of an integer can be achieved faster (see Section 10.3.3); however the reduction takes the same time as in the case of a modular multiplication. Note that there are some dedicated methods like [HOOH$^+$ 1996], which are worth being implemented if modular exponentiation is to be computed, as squarings are a very frequent operation.

## 11.1.3 Inversion and division

To get the inverse of some integer $x$, we can use the multiplicative structure of $\mathbb{F}_p^*$ which implies that $x^{p-2} \times x = x^{p-1} \equiv 1 \pmod{p}$. However, Collins [COL 1969] showed that the average number of arithmetic operations required by this approach is nearly twice as large as for the Euclid extended algorithm, which computes integers $u, v$ such that $xu + pv = 1$. See Section 10.6 for an exhaustive presentation of extended gcd algorithms.

In the following section more specific methods are described, including Montgomery inversion and a useful trick to compute several inverses simultaneously.

### 11.1.3.a Modular inversion

We start with a simplified and improved version of Algorithm 10.6.3, to compute the inverse of $x$ modulo $p$, introduced by Brent and Kung [BRKU 1983] and known as the *plus-minus* method.

**Algorithm 11.6** Plus-minus inversion method

INPUT: An odd modulus $p$ and an integer $x < p$ prime to $p$.
OUTPUT: The integer $x^{-1} \bmod p$.

1.  $A \leftarrow x, B \leftarrow p, U_A \leftarrow 1, U_B \leftarrow 0$ and $\delta \leftarrow 0$
2.  **while** $|A| > 0$ **do**
3.      **while** $A \equiv 0 \pmod 2$ **do**
4.          $A \leftarrow A/2, U \leftarrow (U/2) \bmod p$ and $\delta \leftarrow \delta - 1$
5.      **if** $\delta < 0$ **then**
6.          $T \leftarrow A, A \leftarrow B, B \leftarrow T$
7.          $T \leftarrow U_A, U_A \leftarrow U_B, U_B \leftarrow T$
8.          $\delta \leftarrow -\delta$
9.      **if** $(A + B) \equiv 0 \pmod 4$ **then**
10.         $A \leftarrow (A+B)/2$ and $U_A \leftarrow \big((U_A + U_B)/2\big) \bmod p$
11.     **else** $A \leftarrow (A-B)/2$ and $U_A \leftarrow \big((U_A - U_B)/2\big) \bmod p$
12. **if** $B = 1$ **then** $u \leftarrow U_B$ **else** $u \leftarrow p - U_b$
13. **return** $u$

**Remarks 11.7**

(i) Algorithm 11.6 is based on the observation that if $A$ and $B$ are both odd then either $A+B$ or $A-B$ is divisible by 4. If $A+B \equiv 0 \pmod 4$ then $\gcd(A,B) = \gcd\big((A+B)/2, B\big)$ with $(A+B)/2$ even and $|(A+B)/2| \leqslant \max\{|A|, |B|\}$. Similar results hold if $A-B \equiv 0 \pmod 4$.

(ii) The counter $\delta$ is used to compare $A$ and $B$, as the direct comparison can be time-consuming, especially in hardware. Further improvements are described in [TAK 1998, MEBU$^+$ 2004]. The corresponding algorithms are well suited for hardware realizations and can be implemented in parallel.

**Example 11.8** Take $p = 2^7 - 1 = 127$ and $x = 45$. In the following table are given the values of $\delta, A, B, U_A$, and $U_B$ at the end of the main while loop.

| $\delta$ | $A$ | $B$ | $U_A$ | $U_B$ |
|---|---|---|---|---|
| 0 | 86 | 127 | 64 | 0 |
| 1 | 42 | 43 | 111 | 32 |
| 0 | 32 | 43 | 12 | 32 |
| 5 | 22 | 1 | 40 | 48 |
| 4 | 6 | 1 | 34 | 48 |
| 3 | 2 | 1 | 96 | 48 |
| 2 | 0 | 1 | 0 | 48 |

So, the inverse of 45 modulo 127 is 48.

In a case where the modulus $p$ is prime, one can also use a completely different algorithm due to Thomas et al. [THKE$^+$ 1986] to compute the inverse of $x$ modulo $p$.

## § 11.1 Prime fields of odd characteristic

**Algorithm 11.9** Prime field inversion

INPUT: A prime modulus $p$ and an integer $x$ prime to $p$.
OUTPUT: The integer $x^{-1} \bmod p$.

1. $z \leftarrow x \bmod p$ and $u \leftarrow 1$
2. **while** $z \neq 1$ **do**
3. $\quad q \leftarrow -\lfloor p/z \rfloor$
4. $\quad z \leftarrow p + qz$
5. $\quad u \leftarrow (qu) \bmod p$
6. **return** $u$

**Remarks 11.10**

(i) Algorithm 11.9 is very simple to implement and is reported to be faster than the extended Euclidean algorithm for some types of primes, for example Mersenne primes. Indeed, in this case, the computation of $q$ in Line 3 can be carried out very efficiently. Note that there exists also a dedicated algorithm to compute an inverse modulo a Mersenne prime [CRPO 2001, p. 428].

(ii) In general, the number of iterations needed by Algorithm 11.9 is less than for extended gcd algorithms.

(iii) The modular division $(k/x) \bmod p$ can be directly obtained with Algorithms 11.6 and 11.9. Namely, modify the first line of each algorithm and replace the statements $U_A \leftarrow 1$ and $u \leftarrow 1$ respectively by $U_A \leftarrow k$ and $u \leftarrow k$.

**Example 11.11** Again, take $p = 2^7 - 1 = 127$ and $x = 45$. Here are the values of $q$, $z$, and $u$ at the end of the while loop.

| $q$ | $z$ | $u$ |
|---|---|---|
| $-2$ | 37 | 125 |
| $-3$ | 16 | 6 |
| $-7$ | 15 | 85 |
| $-8$ | 7 | 82 |
| $-18$ | 1 | 48 |

Again, we find that the inverse of 45 modulo 127 is 48. In this case only 5 iterations are needed instead of 7 for Algorithm 11.6, cf. Example 11.8.

### 11.1.3.b Montgomery inversion and division

Montgomery's article also deals with inversions and divisions [MON 1985]. Kaliski [KAL 1995] develops specific algorithms to compute them. Recall the settings of Section 10.4.2 and let $u$ be an integer. Then the *Montgomery inverse* of $u$ is defined as $\text{INV}(u) = (u^{-1} R^2) \bmod p$. So if $u = [x] = xR$, we see that $\text{INV}([x]) = (x^{-1} R) \bmod p = [x^{-1}]$. Thus we have

$$\text{REDC}([x]\,\text{INV}[x]) = R \bmod p = [1].$$

**Algorithm 11.12** Montgomery inverse in Montgomery representation

INPUT: Two $n$-word integers $u$ and $p$ such that $u \in [1, p-1]$. The integer $R = 2^m = b^n$ and the precomputed value $R' = R^2 \bmod p$.
OUTPUT: The $n$-word integer $v$ equal to $\text{INV}(u) = (u^{-1} R^2) \bmod p$.

1. $r \leftarrow u, s \leftarrow 1, t \leftarrow p, v \leftarrow 0$ and $k \leftarrow 0$
2. **while** $r > 0$ **do**
3.     **if** $t \equiv 0 \pmod 2$ **then** $t \leftarrow t/2$ and $s \leftarrow 2s$
4.     **else if** $r \equiv 0 \pmod 2$ **then** $r \leftarrow r/2$ and $v \leftarrow 2v$
5.     **else if** $t > r$ **then** $t \leftarrow (t-r)/2, v \leftarrow v + s$ and $s \leftarrow 2s$
6.     **else** $r \leftarrow (r-t)/2, s \leftarrow v + s$ and $v \leftarrow 2v$
7.     $k \leftarrow k + 1$
8. **if** $v \geqslant p$ **then** $v \leftarrow v - p$
9. $v \leftarrow p - v$
10. **if** $k < m$ **then** $v \leftarrow \text{REDC}(vR')$ and $k \leftarrow k + m$
11. $v \leftarrow \text{REDC}(vR')$
12. $v \leftarrow \text{REDC}(v 2^{2m-k})$
13. **return** $v$

**Remarks 11.13**

(i) Lines 1 to 9 compute the so-called *almost Montgomery inverse* i.e., $(u^{-1} 2^k) \bmod p$ for some $k$ such that $c \leqslant k \leqslant m + c$, where $c$ is the binary length of $p$.

(ii) It is possible to change the end of Algorithm 11.12 in order to compute directly the inverse of $u$, i.e., $u^{-1} \bmod p$ with one or two extra Montgomery multiplications, namely replace Lines 10, 11, and 12 by

    10.  **if** $k > m$ **then** $v \leftarrow \text{REDC}(v)$ and $k \leftarrow k - m$
    11.  $v \leftarrow \text{REDC}(v 2^{m-k})$

(iii) To divide $[x]$ by $[y]$ it suffices to do $\text{REDC}([x] \text{INV}([y]))$ and get $[xy^{-1}]$.

**Example 11.14** With the settings of Example 10.24, let us compute the Montgomery inverse of $[45] = 1319$. Since $p = 2011$ is a 4-word integer in base 8, we have $m = 12$.

- After Line 9, Algorithm 11.12 has computed the almost Montgomery inverse of 1319, which is 1252, and found $k = 17$. This means that $1319^{-1} \times 2^{17} \bmod 2011 = 1252$.
- Lines 10 and 11 compute $\text{REDC}(1252 R') = 142$ and finally $\text{REDC}(142 \times 2^{24-17}) = 1387$, which is the Montgomery inverse of 1319. We check that $\text{REDC}(1319 \times 1387) = 74 \equiv R \pmod p$.
- If we want the inverse of 1319 modulo 2011, we perform $\text{REDC}(1252) = 1267$ and set $k \leftarrow 5$. Then $\text{REDC}(1267 \times 2^{12-5}) = 1485 \equiv 1319^{-1} \pmod{2011}$.

The next section allows us to compute the inverse of several numbers modulo the same number $p$.

### 11.1.3.c  Simultaneous inversion

One needs *a priori* $j$ extended gcd computations to find the inverses of the $j$ elements $a_1, \ldots, a_j$ modulo $p$. Here we present a trick of Montgomery that allows us to do the same with only one extended gcd and $3j - 3$ multiplications modulo $p$. The basic idea is to get the inverse of $\prod_i a_i$ and to multiply it by suitable terms to recover $a_j^{-1}, \ldots, a_1^{-1}$ [COH 2000].

---

**Algorithm 11.15**  Simultaneous inversion modulo $p$

INPUT: A positive integer $p$ and $j$ integers $a_1, \ldots, a_j$ not zero modulo $p$.
OUTPUT: The inverses $b_1, \ldots, b_j$ of the $a_1, \ldots, a_j$ modulo $p$.

1. $c_1 \leftarrow a_1$
2. **for** $i = 2$ **to** $j$ **do** $c_i \leftarrow a_i c_{i-1}$
3. compute $(u, v, d)$ with $u c_j + v p = d$                                           [$d$ is equal to 1]
4. **for** $i = j$ **down to** $2$ **do**
5.       $b_i \leftarrow (u c_{i-1}) \bmod p$ and $u \leftarrow (u a_i) \bmod p$
6. $b_1 \leftarrow u$
7. **return** $(b_1, \ldots, b_j)$

---

**Remarks 11.16**

(i) Let $N$ be a nonprime modulus. If one tries to apply Algorithm 11.15 to the nonzero residues $a_1, \ldots, a_j$ modulo $N$, there are two possibilities. If $a_1, \ldots, a_j$ are all coprime to $N$ then Algorithm 11.15 returns $a_1^{-1}, \ldots, a_j^{-1}$ modulo $N$. If at least one integer is not coprime to $N$ then the gcd computed in Line 3 is different from 1. In this case, if the Lines 4 to 7 of Algorithm 11.15 are replaced by the following statements

    4. **if** $d = N$ **then**
    5.     $i \leftarrow 1$
    6.     **repeat**
    7.         $d \leftarrow \gcd(a_i, N)$ and $i \leftarrow i + 1$
    8.     **until** $d > 1$
    9. **return** $d$

then a nontrivial factor of $N$ is returned.

(ii) This modified algorithm is especially useful for Lenstra's elliptic curve method, cf. Section 25.3.3, where one tries to find factors of $N$ by computing scalar multiples on a curve modulo $N$.

### 11.1.4  Exponentiation

This part deals with specific exponentiation methods for finite fields $\mathbb{F}_p$. The general introduction to the subject can be found in Chapter 9.

#### 11.1.4.a  Ordinary exponentiation

To compute $x^n$, for $x \in \mathbb{F}_p$, one could perform the exponentiation in $\mathbb{Z}$ and then reduce the result. Of course, this approach is completely inefficient even for rather small $n$. However, a systematic reduction after each intermediate step, i.e., a squaring or a multiplication, seems inadequate as well since a modular reduction is quite slow. So, a compromise must be found. One can also use Barrett or Quisquater multiplication algorithms without the remainder correction steps; see Section 11.1.2.a. With appropriate settings, intermediate results are kept bounded such that they still fit in the allocated space, and only at the end of the exponentiation one corrects the result so that it belongs to $[0, p-1]$.

### 11.1.4.b  Montgomery exponentiation

All algorithms presented in Chapter 9 can be adapted to Montgomery representation. The changes are always the same and rather simple: as explained in Section 11.1.2.b, one converts to and from Montgomery representation only for input/output, so any amount of operations can be done in between. These ideas are illustrated in the following adaptation of the classical square and multiply algorithm, cf. Section 9.1.1.

---

**Algorithm 11.17** Binary exponentiation using Montgomery representation

INPUT: An element $x$ of $\mathbb{F}_p$, a positive integer $n = (n_{t-1} \ldots n_0)_2$ such that $n_{t-1} = 1$, the integers $R$ and $R' = R^2 \bmod p$.
OUTPUT: The element $x^n \in \mathbb{F}_p$.

1. $y \leftarrow R \bmod p$ and $t \leftarrow \text{REDC}(xR')$
2. **for** $i = t-1$ **down to** $0$
3. $\quad y \leftarrow \text{REDC}(y^2)$
4. $\quad$ **if** $n_i = 1$ **then** $y \leftarrow \text{REDC}(ty)$
5. **return** $\text{REDC}(y)$

---

**Remark 11.18** Conversion to Montgomery representation is done in Line 1. One has $y = [1]$ and $t = [x]$. In the for loop at each step a Montgomery squaring and possibly a Montgomery multiplication is performed. Finally we come back to the standard representation by a Montgomery reduction. At the end, $y = [x^n]$ so that $\text{REDC}(y) = x^n$, as expected. Note that also here incomplete reduction can be applied.

## 11.1.5  Squares and square roots

Given a nonzero integer $a$ modulo $p$, the Legendre symbol $\left(\frac{a}{p}\right)$ defined in Section 2.3.4 is equal to 1 if and only $a$ is a quadratic residue modulo $p$. From the reciprocity law (2.6) and Theorem 2.103, it is easy to derive an efficient way to compute it.

---

**Algorithm 11.19** Legendre symbol

INPUT: An integer $a$ and an odd prime number $p$.
OUTPUT: The Legendre symbol $\left(\frac{a}{p}\right)$.

1. $k \leftarrow 1$
2. **while** $p \neq 1$ **do**
3. $\quad$ **if** $a = 0$ **then return** $0$
4. $\quad v \leftarrow 0$
5. $\quad$ **while** $a \equiv 0 \pmod 2$ **do** $v \leftarrow v+1$ and $a \leftarrow a/2$

## § 11.1 Prime fields of odd characteristic

6.       **if** $v \equiv 1 \pmod{2}$ **and** $p \equiv \pm 3 \pmod{8}$ **then** $k \leftarrow -k$
7.       **if** $a \equiv 3 \pmod{4}$ **and** $p \equiv 3 \pmod{4}$ **then** $k \leftarrow -k$
8.       $r \leftarrow a, a \leftarrow p \bmod r$ and $p \leftarrow r$
9.       **return** $k$

---

**Remark 11.20** This algorithm is useful, for example, to determine the number of points lying on an elliptic or hyperelliptic curve defined over a finite field of small prime order, cf. Chapter 17.

**Example 11.21** Take the prime $p = 163841$, $a = 109608$ and let us compute $\left(\frac{a}{p}\right)$ with Algorithm 11.19. The next table displays the values of $r$, $a$, $v$ and $k$ after Line 8.

| $r$ | $a$ | $v$ | $k$ |
|---|---|---|---|
| 13701 | 13130 | 3 | 1 |
| 6565 | 571 | 1 | $-1$ |
| 571 | 284 | 0 | $-1$ |
| 71 | 3 | 2 | 1 |
| 3 | 2 | 0 | $-1$ |
| 1 | 0 | 1 | 1 |

These computations reflect the following sequence of equalities

$$\left(\frac{109608}{163841}\right) = \left(\frac{8}{163841}\right)\left(\frac{13701}{163841}\right)$$
$$= \left(\frac{13130}{13701}\right) = \left(\frac{2}{13701}\right)\left(\frac{6565}{13701}\right)$$
$$= -\left(\frac{571}{6565}\right)$$
$$= -\left(\frac{284}{571}\right) = -\left(\frac{4}{571}\right)\left(\frac{71}{571}\right)$$
$$= \left(\frac{3}{71}\right)$$
$$= -\left(\frac{2}{3}\right)$$
$$= 1.$$

So, $109608$ is a quadratic residue modulo $163841$.

When it is known that $a$ is a square, it is often required to determine $x$ such that $x^2 \equiv a \pmod{p}$. For instance, this occurs to actually find a point lying on an elliptic or hyperelliptic curve.

**Lemma 11.22** Given a quadratic residue $a \in \mathbb{F}_p$, there are explicit formulas when $p \not\equiv 1 \pmod{8}$ to determine $x \in \mathbb{F}_p$ such that $x^2 \equiv a \pmod{p}$. Namely,

- $x \equiv \pm a^{(p+1)/4} \pmod{p}$ if $p \equiv 3 \pmod{4}$
- $x \equiv \pm a^{(p+3)/8} \pmod{p}$ if $p \equiv 5 \pmod{8}$ and $a^{(p-1)/4} = 1$
- $x \equiv \pm 2a(4a)^{(p-5)/8} \pmod{p}$ if $p \equiv 5 \pmod{8}$ and $a^{(p-1)/4} = -1$.

When $p \equiv 1 \pmod{8}$ an algorithm of Tonelli and Shanks solves the problem. In fact, this algorithm is correct for all primes.

**Algorithm 11.23** Tonelli and Shanks square root computation

INPUT: A prime $p$ and an integer $a$ such that $\left(\frac{a}{p}\right) = 1$.
OUTPUT: An integer $x$ such that $x^2 \equiv a \pmod{p}$.

1. write $p - 1 = 2^e r$ with $r$ odd                [see the beginning of Section 10.5, p. 185]
2. choose $n$ at random such that $\left(\frac{n}{p}\right) = -1$
3. $z \leftarrow n^r \bmod p, y \leftarrow z, s \leftarrow e$ and $x \leftarrow a^{(r-1)/2} \bmod p$
4. $b \leftarrow (ax^2) \bmod p$ and $x \leftarrow (ax) \bmod p$
5. **while** $b \not\equiv 1 \pmod{p}$
6.      $m \leftarrow 1$
7.      **while** $b^{2^m} \not\equiv 1 \pmod{p}$ **do** $m \leftarrow m + 1$
8.      $t \leftarrow y^{2^{s-m-1}} \bmod p, y \leftarrow t^2 \bmod p$ **and** $s \leftarrow m$
9.      $x \leftarrow (tx) \bmod p$ **and** $b \leftarrow (yb) \bmod p$
10. **return** $x$

## Remarks 11.24

(i) Algorithm 11.23 works in the maximal 2-group of order $2^e$ of $\mathbb{F}_p^*$ generated by some element $z$. If $m = s$ after Line 7, this implies that $a$ is not a quadratic residue modulo $p$. Otherwise $a^r$ is a square in this subgroup and there is an even $k$ less than $e$ such that $a^r z^k \equiv 1 \pmod{p}$. The square root is then given by $x \equiv a^{(r+1)/2} z^{k/2} \pmod{p}$. A variant of Algorithm 11.23 finds $k/2$ by a bit by bit approach [KOB 1994].

(ii) The number of loops performed within the while loop beginning in Line 5 is bounded by $e$ since $s$ is strictly decreasing at each loop.

(iii) The expected running time of Algorithm 11.23 is $O(e^2 \lg^2 p)$.

**Example 11.25** Let us compute the square root of $109608$ modulo $p = 163841$ with Algorithm 11.23. First one sees that $e = 15$ and $r = 5$. The quadratic nonresidue $n$ found at random in Line 2 is $6558$. In the following, we state the values of the principal variables before the while loop in Line 4 and at the end of it Line 9. One can see also that $ab - x^2, y^{2^{s-1}}$ and $b^{2^{s-1}}$ are invariant throughout the execution of the algorithm.

| $m$ | $z$ | $y$ | $x$ | $b$ | $t$ | $ab$ | $x^2$ | $y^{2^{s-1}}$ | $b^{2^{s-1}}$ |
|---|---|---|---|---|---|---|---|---|---|
| — | 12002 | 12002 | 13640 | 100808 | — | 90065 | 90065 | $-1$ | 1 |
| 13 | — | 82347 | 78996 | 68270 | 31765 | 155849 | 155849 | $-1$ | 1 |
| 12 | — | 140942 | 104389 | 56092 | 82347 | 162252 | 162252 | $-1$ | 1 |
| 6 | — | 81165 | 18205 | 57313 | 52992 | 135523 | 135523 | $-1$ | 1 |
| 5 | — | 38297 | 90687 | 101925 | 81165 | 132974 | 132974 | $-1$ | 1 |
| 3 | — | 101925 | 97748 | 39338 | 119418 | 119748 | 119748 | $-1$ | 1 |
| 2 | — | 39338 | 121372 | 163840 | 101925 | 54233 | 54233 | $-1$ | 1 |
| 1 | — | 163840 | 41155 | 1 | 39338 | 109608 | 109608 | $-1$ | 1 |

One checks that $41155^2 \equiv 109608 \pmod{p}$.

When the 2-adic valuation $e$ of $p-1$ is large, as in the previous example, it is better to use another algorithm that works in the quadratic extension $\mathbb{F}_{p^2}$.

**Algorithm 11.26** Square root computation

INPUT: A prime $p$ and an integer $a$ such that $\left(\frac{a}{p}\right) = 1$.
OUTPUT: An integer $x$ such that $x^2 \equiv a \pmod{p}$.

1. choose $b$ at random such that $\left(\frac{b^2-4a}{p}\right) = -1$
2. $f(X) \leftarrow X^2 - bX + a$                                          $[f(X)$ is irreducible over $\mathbb{F}_p]$
3. $x \leftarrow X^{(p+1)/2} \pmod{f(X)}$
4. **return** $x$

**Remarks 11.27**

(i) If $\theta$ is a root of $f(X)$ then $\theta^p$ is the other one and therefore $\theta^{p+1} = a$. So $x$ as defined in Line 3 satisfies $x^2 \equiv a \pmod{f(X)}$. It remains to show that $x$ is in fact an element of $\mathbb{F}_p$. As $a^{(p-1)/2} = 1$ we have $X^{(p^2-1)/2} \equiv 1 \pmod{f(X)}$ so that $x^p \equiv x \pmod{f(X)}$.

(ii) The expected running time of Algorithm 11.26 is $O(\lg^3 p)$.

**Example 11.28** With the same initial values as in Example 11.25, Algorithm 11.26 first finds at random an irreducible polynomial over $\mathbb{F}_p$, in this case, for instance, $f(X) = X^2 + 27249X + 109608$. Then it computes $X^{(p+1)/2}$, which is equivalent to 41155 modulo $f(X)$.

## 11.2 Finite fields of characteristic 2

See Section 2.3.2 for an introduction to algebraic extension of fields. Arithmetic in extension fields of $\mathbb{F}_q$ where $q$ is some power of 2 relies on elementary computer operations like exclusive disjunction and shifts. Note that in general $q$ is simply equal to 2. This allows very efficient implementations, especially in hardware, and gives finite fields of characteristic 2 a great importance in cryptography.

### 11.2.1 Representation

See Section 2.3.3 for a presentation of the different finite field representation systems. In the following we shall focus on efficient implementation techniques used in cryptography. As $\mathbb{F}_{2^d}$ is a vector space of dimension $d$ over $\mathbb{F}_2$, an element can be viewed as a sequence of $d$ coefficients equal to 0 or 1. Therefore it is internally stored as a sequence of bits and the techniques introduced for multiprecision integers apply with some slight modifications. Two kinds of basis are commonly used. In polynomial representation, it is $(1, X, \ldots, X^{d-1})$, whereas with a normal basis it is $(\alpha, \alpha^2, \ldots, \alpha^{2^{d-1}})$, cf. Section 2.3.3. Let us first describe polynomial representation.

#### 11.2.1.a Irreducible polynomial representation

Let $m(X) \in \mathbb{F}_q[X]$ be an irreducible polynomial of degree $d$ and $(m(X))$ the principal ideal generated by $m(X)$. Then $\mathbb{F}_q[X]/(m(X))$ is the finite field with $q^d$ elements. Formula (2.4) proves that there exists an irreducible polynomial of degree $d$ for each positive $d$ but the proof is not constructive. To find such a polynomial, we can consider a random polynomial and test its irreducibility. Since (2.4) shows that the probability for a monic polynomial of degree $d$ to be irreducible is close to $1/d$, we should find one after $d$ attempts on average. There is a variety of polynomial irreducibility tests. For example Rabin [RAB 1980] proved the following.

**Lemma 11.29** Let $m(X) \in \mathbb{F}_q[X]$ of degree $d$ and let $p_1, \ldots, p_k$ be the prime divisors of $d$. Then $m(X)$ is irreducible over $\mathbb{F}_q$ if and only if

- $\gcd(m(X), X^{q^{d/p_i}} - X) = 1$, for $i = 1, \ldots, k$
- $m(X)$ divides $X^{q^d} - X$.

For a deterministic method to find an irreducible polynomial see [SHO 1994b].

Once $m(X)$ has been found, computations are done modulo this irreducible polynomial and reduction is a key operation. For this we need to divide two polynomials with coefficients in a field. Every irreducible polynomial of degree $d$ can be used to build $\mathbb{F}_{q^d}$; however, some special polynomials offer better performance, e.g., monic sparse polynomials are proposed in [SCOR$^+$ 1995].

Usually, one uses *trinomials* or *pentanomials* since binomials and quadrinomials are always divisible by $X + 1$ and so, except for $X + 1$ itself, are never irreducible in $\mathbb{F}_q[X]$. The existence for every $d$ of an irreducible degree $d$ trinomial or pentanomial is still an open question, but this is the case at least for all $d \leqslant 10000$ [SER 1998].

A trinomial $X^d + X^k + 1$ is reducible if both $d$ and $k$ are even as then $X^d + X^k + 1 = (X^{d/2} + X^{k/2} + 1)^2$. Eliminating this trivial case, Swan [SWA 1962] proves the following.

**Lemma 11.30** The trinomial $X^d + X^k + 1$, where at least one of $d$ and $k$ is odd, has an even number of factors if and only if one of the following holds

- $d$ is even, $k$ is odd, $d \neq 2k$ and $\frac{dk}{2} \equiv 0$ or $1 \pmod{4}$
- $d$ is odd, $d \equiv \pm 3 \pmod{8}$, $k$ is even and $k$ does not divide $2d$
- $d$ is odd, $d \equiv \pm 1 \pmod{8}$, $k$ is even and $k$ divides $2d$.

It follows that irreducible trinomials do not exist when $d \equiv 0 \pmod{8}$ and are rather scarce for $d \equiv 3$ or $5 \pmod{8}$. In Table 11.1, we give irreducible polynomials over $\mathbb{F}_2$ of degree less than or equal to 500. More precisely, the coefficients $d, k_1$ in the table stand for the trinomial $X^d + X^{k_1} + 1$. In case there is no trinomial of degree $d$, the sequence $d, k_1, k_2, k_3$ is given for the pentanomial $X^d + X^{k_1} + X^{k_2} + X^{k_3} + 1$. For each $d$ the coefficient $k_1$ is chosen to be minimal, then $k_2$ and so on.

For these sparse polynomials there is a specific reduction algorithm [GANÖ 2005]. The nonrecursive version is given hereafter.

---

**Algorithm 11.31** Division by a sparse polynomial

INPUT: Two polynomials $m(X)$ and $f(X)$ with coefficients in a commutative ring, where $m(X)$ is the sparse polynomial $X^d + \sum_{i=1}^{t} a_i X^{b_i}$ with $b_i < b_{i+1}$ and $b_1 = 0$.
OUTPUT: The polynomials $u$ and $v$ such that $f = um + v$ with $\deg v < d$.

1. $v \leftarrow f$ and $u \leftarrow 0$
2. **while** $\deg(v) \geqslant d$ **do**
3. $\quad k \leftarrow \max\{d, \deg v - d + b_t + 1\}$
4. $\quad$ write $v(X)$ as $u_1(X)X^k + w(X)$ $\qquad$ [$\deg w < k$]
5. $\quad v(X) \leftarrow w(X) - u_1(x)\big(m(X) - X^d\big)X^{k-d}$

6.  $\quad u(X) \leftarrow u_1(X)X^{k-d} + u(X)$
7.  **return** $(u, v)$

**Remarks 11.32**

(i) If $\deg f = d'$ then Algorithm 11.31 needs at most $2t(d' - d + 1)$ field additions to compute $u$ and $v$ such that $f = um + v$. If $d' \leqslant 2d - 2$, as is the case when performing arithmetic modulo $m$, then one needs $4(d-1)$ additions for a reduction modulo a trinomial and $8(d-1)$ additions modulo a pentanomial. The number of loops is at most $\lceil (d' - d + 1)/(d - b_t - 1) \rceil$. Again if $d' \leqslant 2d - 2$, then the number of loops is at most equal to 2 whatever the value of $b_t$, as long as $1 \leqslant b_t \leqslant d/2$.

(ii) To avoid computing the quotient $u$ when it is not required, simply discard Line 6 of Algorithm 11.31.

(iii) When the modulus is fixed, there is in general an even faster algorithm that exploits the form of the polynomial. This is the case for NIST irreducible polynomials [FIPS 186-2], cf. for example [HAME+ 2003, pp. 55–56]

**Example 11.33** Take $m(X) = X^{11} + X^2 + 1$ and $f(X) = X^{20} + X^{16} + X^{15} + X^{12} + X^5 + X^3 + X + 1$, and let us find the quotient and remainder of the division of $f$ by $m$ with Algorithm 11.31.

- First $k = 12$, $u_1(X) = X^8 + X^4 + X^3 + 1$ and $w(X) = X^5 + X^3 + X + 1$.
- The new value of $v(X)$ is $X^{11} + X^9 + X^7 + X^6 + X^4 + 1$ and $u(X) = X^9 + X^5 + X^4 + X$.
- For the next and last loop, $k = 11$, $u_1(X) = 1$ and $w(X) = X^9 + X^7 + X^6 + X^4 + 1$.

Finally, $v(X) = X^9 + X^7 + X^6 + X^4 + X^2$ and $u(X) = X^9 + X^5 + X^4 + X + 1$ and one checks that $f(X) = u(X)m(X) + v(X)$.

Instead of trying to minimize the number of nonzero coefficients of the modulus, another option is to do arithmetic modulo a *sedimentary polynomial* [COP 1984, ODL 1985], that is, a polynomial of the form $X^d + h(X)$ irreducible over $\mathbb{F}_q$ such that the degree of $h(X)$ is minimal. For $q = 2$, it has been shown that for all $d \leqslant 600$ the degree of $h$ is at most 11 [GOMC 1993]. Algorithm 11.31 can be slightly modified to perform reduction modulo a sedimentary polynomial. Namely, replace the statement $k \leftarrow \max\{d, \deg v - d + b_t + 1\}$ by $k \leftarrow \max\{d, \deg v - \deg h\}$.

Tests performed in [GANÖ 2005] indicate that sedimentary polynomials are slightly less efficient than trinomials or pentanomials.

#### 11.2.1.b  Redundant polynomial representation

For some extensions of even degree there is a better choice, namely *all one polynomials*. They are of the form
$$m(X) = X^d + X^{d-1} + \cdots + X + 1.$$

For $d > 1$, such a polynomial is irreducible if and only if $d+1$ is prime and 2 is a primitive element of $\mathbb{F}_{d+1}$. Now it is clear from the definition of $m(X)$ that $m(X)(X+1) = X^{d+1} + 1$. Thus an element of $\mathbb{F}_{2^d}$ can be represented on the basis $(\alpha, \alpha^2, \ldots, \alpha^d)$ where $\alpha$ is a root of $m(X)$. In other words, an element of $\mathbb{F}_{2^d}$ is represented by a polynomial of degree at most $d$ without constant coefficient, 1 being replaced by $X + X^2 + \cdots + X^d$. Alternatively, if the representation does not need to be unique, elements can directly be written on $(1, X, X^2, \ldots, X^d)$. In any case, reductions

**Table 11.1** Irreducible trinomials and pentanomials over $\mathbb{F}_2$.

|  | 2,1 | 3,1 | 4,1 | 5,2 | 6,1 | 7,1 | 8,4,3,1 | 9,1 | 10,3 |
|---|---|---|---|---|---|---|---|---|---|
| 11,2 | 12,3 | 13,4,3,1 | 14,5 | 15,1 | 16,5,3,1 | 17,3 | 18,3 | 19,5,2,1 | 20,3 |
| 21,2 | 22,1 | 23,5 | 24,4,3,1 | 25,3 | 26,4,3,1 | 27,5,2,1 | 28,1 | 29,2 | 30,1 |
| 31,3 | 32,7,3,2 | 33,10 | 34,7 | 35,2 | 36,9 | 37,6,4,1 | 38,6,5,1 | 39,4 | 40,5,4,3 |
| 41,3 | 42,7 | 43,6,4,3 | 44,5 | 45,4,3,1 | 46,1 | 47,5 | 48,5,3,2 | 49,9 | 50,4,3,2 |
| 51,6,3,1 | 52,3 | 53,6,2,1 | 54,9 | 55,7 | 56,7,4,2 | 57,4 | 58,19 | 59,7,4,2 | 60,1 |
| 61,5,2,1 | 62,29 | 63,1 | 64,4,3,1 | 65,18 | 66,3 | 67,5,2,1 | 68,9 | 69,6,5,2 | 70,5,3,1 |
| 71,6 | 72,10,9,3 | 73,25 | 74,35 | 75,6,3,1 | 76,21 | 77,6,5,2 | 78,6,5,3 | 79,9 | 80,9,4,2 |
| 81,4 | 82,8,3,1 | 83,7,4,2 | 84,5 | 85,8,2,1 | 86,21 | 87,13 | 88,7,6,2 | 89,38 | 90,27 |
| 91,8,5,1 | 92,21 | 93,2 | 94,21 | 95,11 | 96,10,9,6 | 97,6 | 98,11 | 99,6,3,1 | 100,15 |
| 101,7,6,1 | 102,29 | 103,9 | 104,4,3,1 | 105,4 | 106,15 | 107,9,7,4 | 108,17 | 109,5,4,2 | 110,33 |
| 111,10 | 112,5,4,3 | 113,9 | 114,5,3,2 | 115,8,7,5 | 116,4,2,1 | 117,5,2,1 | 118,33 | 119,8 | 120,4,3,1 |
| 121,18 | 122,6,2,1 | 123,2 | 124,19 | 125,7,6,5 | 126,21 | 127,1 | 128,7,2,1 | 129,5 | 130,3 |
| 131,8,3,2 | 132,17 | 133,9,8,2 | 134,57 | 135,11 | 136,5,3,2 | 137,21 | 138,8,7,1 | 139,8,5,3 | 140,15 |
| 141,10,4,1 | 142,21 | 143,5,3,2 | 144,7,4,2 | 145,52 | 146,71 | 147,14 | 148,27 | 149,10,9,7 | 150,53 |
| 151,3 | 152,6,3,2 | 153,1 | 154,15 | 155,62 | 156,9 | 157,6,5,2 | 158,8,6,5 | 159,31 | 160,5,3,2 |
| 161,18 | 162,27 | 163,7,6,3 | 164,10,8,7 | 165,9,8,3 | 166,37 | 167,6 | 168,15,3,2 | 169,34 | 170,11 |
| 171,6,5,2 | 172,1 | 173,8,5,2 | 174,13 | 175,6 | 176,11,3,2 | 177,8 | 178,31 | 179,4,2,1 | 180,3 |
| 181,7,6,1 | 182,81 | 183,56 | 184,9,8,7 | 185,24 | 186,11 | 187,7,6,5 | 188,6,5,2 | 189,6,5,2 | 190,8,7,6 |
| 191,9 | 192,7,2,1 | 193,15 | 194,87 | 195,8,3,2 | 196,3 | 197,9,4,2 | 198,9 | 199,34 | 200,5,3,2 |
| 201,14 | 202,55 | 203,8,7,1 | 204,27 | 205,9,5,2 | 206,10,9,5 | 207,43 | 208,9,3,1 | 209,6 | 210,7 |
| 211,11,10,8 | 212,105 | 213,6,5,2 | 214,73 | 215,23 | 216,7,3,1 | 217,45 | 218,11 | 219,8,4,1 | 220,7 |
| 221,8,6,2 | 222,5,4,2 | 223,33 | 224,9,8,3 | 225,32 | 226,10,7,3 | 227,10,9,4 | 228,113 | 229,10,4,1 | 230,8,7,6 |
| 231,26 | 232,9,4,2 | 233,74 | 234,31 | 235,9,6,1 | 236,5 | 237,7,4,1 | 238,73 | 239,36 | 240,8,5,3 |
| 241,70 | 242,95 | 243,8,5,1 | 244,111 | 245,6,4,1 | 246,11,2,1 | 247,82 | 248,15,14,10 | 249,35 | 250,103 |
| 251,7,4,2 | 252,15 | 253,46 | 254,7,2,1 | 255,52 | 256,10,5,2 | 257,12 | 258,71 | 259,10,6,2 | 260,15 |
| 261,7,6,4 | 262,9,8,4 | 263,93 | 264,9,6,2 | 265,42 | 266,47 | 267,8,6,3 | 268,25 | 269,7,6,1 | 270,53 |
| 271,58 | 272,9,3,2 | 273,23 | 274,67 | 275,11,10,9 | 276,63 | 277,12,6,3 | 278,5 | 279,5 | 280,9,5,2 |
| 281,93 | 282,35 | 283,12,7,5 | 284,53 | 285,10,7,5 | 286,69 | 287,71 | 288,11,10,1 | 289,21 | 290,5,3,2 |
| 291,12,11,5 | 292,37 | 293,11,6,1 | 294,33 | 295,48 | 296,7,3,2 | 297,5 | 298,11,8,4 | 299,11,6,4 | 300,5 |
| 301,9,5,2 | 302,41 | 303,1 | 304,11,2,1 | 305,102 | 306,7,3,1 | 307,8,4,2 | 308,15 | 309,10,6,4 | 310,93 |
| 311,7,5,3 | 312,9,7,4 | 313,79 | 314,15 | 315,10,9,1 | 316,63 | 317,7,4,2 | 318,45 | 319,36 | 320,4,3,1 |
| 321,31 | 322,67 | 323,10,3,1 | 324,51 | 325,10,5,2 | 326,10,3,1 | 327,34 | 328,8,3,1 | 329,50 | 330,99 |
| 331,10,6,2 | 332,89 | 333,2 | 334,5,2,1 | 335,10,7,2 | 336,7,4,1 | 337,55 | 338,4,3,1 | 339,16,10,7 | 340,45 |
| 341,10,8,6 | 342,125 | 343,75 | 344,7,2,1 | 345,22 | 346,63 | 347,11,10,3 | 348,103 | 349,6,5,2 | 350,53 |
| 351,34 | 352,13,11,6 | 353,69 | 354,99 | 355,6,5,1 | 356,10,9,7 | 357,11,10,2 | 358,57 | 359,68 | 360,5,3,2 |
| 361,7,4,1 | 362,63 | 363,8,5,3 | 364,9 | 365,9,6,5 | 366,29 | 367,21 | 368,7,3,2 | 369,91 | 370,139 |
| 371,8,3,2 | 372,111 | 373,8,7,2 | 374,8,6,5 | 375,16 | 376,8,7,5 | 377,41 | 378,43 | 379,10,8,5 | 380,47 |
| 381,5,2,1 | 382,81 | 383,90 | 384,12,3,2 | 385,6 | 386,83 | 387,8,7,1 | 388,159 | 389,10,9,5 | 390,9 |
| 391,28 | 392,13,10,6 | 393,7 | 394,135 | 395,11,6,5 | 396,25 | 397,12,7,6 | 398,7,6,2 | 399,26 | 400,5,3,2 |
| 401,152 | 402,171 | 403,9,8,5 | 404,65 | 405,13,8,2 | 406,141 | 407,71 | 408,5,3,2 | 409,87 | 410,10,4,3 |
| 411,12,10,3 | 412,147 | 413,10,7,6 | 414,13 | 415,102 | 416,9,5,2 | 417,107 | 418,199 | 419,15,5,4 | 420,7 |
| 421,5,4,2 | 422,149 | 423,25 | 424,9,7,2 | 425,12 | 426,63 | 427,11,6,5 | 428,105 | 429,10,8,7 | 430,14,6,1 |
| 431,120 | 432,13,4,3 | 433,33 | 434,12,11,5 | 435,12,9,5 | 436,165 | 437,6,2,1 | 438,65 | 439,49 | 440,4,3,1 |
| 441,7 | 442,7,5,2 | 443,10,6,1 | 444,81 | 445,7,6,4 | 446,105 | 447,73 | 448,11,6,4 | 449,134 | 450,47 |
| 451,16,10,1 | 452,6,5,4 | 453,15,6,4 | 454,8,6,1 | 455,38 | 456,18,9,6 | 457,16 | 458,203 | 459,12,5,2 | 460,19 |
| 461,7,6,1 | 462,73 | 463,93 | 464,19,18,13 | 465,31 | 466,14,11,6 | 467,11,6,1 | 468,27 | 469,9,5,2 | 470,9 |
| 471,1 | 472,11,3,2 | 473,200 | 474,191 | 475,9,8,4 | 476,9 | 477,16,15,7 | 478,121 | 479,104 | 480,15,9,6 |
| 481,138 | 482,9,6,5 | 483,9,6,4 | 484,105 | 485,17,16,5 | 486,81 | 487,94 | 488,4,3,1 | 489,83 | 490,219 |
| 491,11,6,3 | 492,7 | 493,10,5,3 | 494,17 | 495,76 | 496,16,5,2 | 497,78 | 498,155 | 499,11,6,5 | 500,27 |

§ 11.2 Finite fields of characteristic 2                                                                                            217

are made modulo $X^{d+1} + 1$ and a squaring is simply a permutation of the coordinates. This idea, first proposed in [ITTS 1989] and rediscovered in [SIL 1999], is known as the *anomalous basis* or the *ghost bit basis* technique.

When $d > 1$ is odd, one can always embed $\mathbb{F}_{2^d}$ into some cyclotomic ring $\mathbb{F}_2[X]/(X^n + 1)$ but only for some $n \geqslant 2d + 1$. So the benefits obtained from a cheap reduction are partially offset by a more expensive multiplication [WUHA⁺ 2002]. For elliptic and hyperelliptic curve cryptography only extensions of prime degree are relevant, cf. Chapter 22, so the best we can hope for with this idea is $n = 2d + 1$.

Adopting the idea of using sparse reducible polynomials with an appropriate irreducible factor, one can use reducible trinomials in case only an irreducible pentanomial exists for some degree $d$. First, we have to find a trinomial $T(X) = X^n + X^k + 1$ with $n$ slightly bigger than $d$ and such that $T(X)$ admits an irreducible factor $m(X)$ of degree $d$. Such a trinomial is called a *redundant trinomial* and the idea is then to embed $\mathbb{F}_{2^d} \sim \mathbb{F}_2[X]/\bigl(m(X)\bigr)$ into $\mathbb{F}_{2^d} \sim \mathbb{F}_2[X]/\bigl(T(X)\bigr)$. In the range $[2, 10000]$, there is no irreducible trinomial in about 50% of the cases (precisely 4853 out of 9999 [SER 1998]) but an exhaustive search has shown that there are redundant trinomials for all the corresponding degrees, see [DOCHE] for a table. In general $n - d$ is small and in more than 85% of the cases the number of 32-bit words required to represent an element of $\mathbb{F}_{2^d}$ are the same with a redundant trinomial of degree $n$ and with an irreducible pentanomial of degree $d$. This implies also that the multiplication has the same cost with both representations, since this operation is usually performed at a word level, cf. Section 11.2.2.a.

From a practical point of view an element of $\mathbb{F}_{2^d}$ is represented by a polynomial of degree less than $n$ and the computations are done modulo $T(X)$. At the end of the whole computation, one can reduce modulo $m(X)$ and this can be done with only $T(X)$ and $\delta(X) = T(X)/m(X)$, since for any polynomial $f(X)$ one has

$$f(X) \bmod m(X) = \frac{f(X)\delta(X) \bmod T(X)}{\delta(X)},$$

as in (11.1). Redundant trinomials can speed up an exponentiation by a factor up to 30%, when compared to irreducible pentanomials, cf. [DOC 2005].

Note that this concept is in fact similar to *almost irreducible trinomials* introduced by Brent and Zimmermann in the context of random number generators in [BRZI 2003]. Similar ideas were also explored by Blake et al. [BLGA⁺ 1994a, BLGA⁺ 1996], and Tromp et al. [TRZH⁺ 1997].

### 11.2.1.c  Normal and optimal normal bases

Another popular way to represent an element of $\mathbb{F}_{q^d}$ over $\mathbb{F}_q$ is to use a normal basis. This is especially true when $q = 2$, since in this case the squaring of an element is just a cyclic shift of its coordinates. However, multiplications are more complicated. As a result only special normal bases, called *optimal normal bases*, ONB for short, are used in practice; see Section 11.2.2.b.

Gauß periods of type $(n, 1)$ and $(n, 2)$, generate optimal normal bases (cf. Section 2.3.3.b and [MUON⁺ 1989]), and it has been proved that all the optimal normal bases can be produced by this construction [GALE 1992].

For $q = 2$, this occurs

1. when $d+1$ is prime and 2 is a primitive element of $\mathbb{F}_{d+1}$. Then the nontrivial $(d+1)$-th roots of unity form an optimal normal basis of $\mathbb{F}_{2^d}$, called a *Type I ONB*.
2. when $2d + 1$ is prime and either
   - 2 is primitive in $\mathbb{F}_{2d+1}$ or
   - 2 generates the quadratic residues in $\mathbb{F}_{2d+1}$, that is $2d + 1 \equiv 3 \pmod{4}$ and the order of 2 in $\mathbb{F}_{2d+1}$ is $d$.

Then there is a primitive $(2d+1)$-th root of unity $\zeta$ in $\mathbb{F}_{2^d}$ and $\zeta+\zeta^{-1}$ is a normal element generating a *Type II ONB*. Such a basis can be written $(\zeta+\zeta^{-1}, \zeta^2+\zeta^{-2}, \ldots, \zeta^d+\zeta^{-d})$ as shown in [BLRO+ 1998].

Note that Type I ONB and anomalous bases are equal up to suitable permutations. So it is possible to enjoy a cheap multiplication and a cheap squaring at the same time. However, as said previously, there is no Type I ONB for an extension of prime degree. The situation is slightly better for Type II ONB. Indeed, in the range $[50, 500]$ there are 80 extension degrees that are prime and among them only 18 have a Type II ONB, namely $53, 83, 89, 113, 131, 173, 179, 191, 233, 239, 251, 281, 293, 359, 419, 431, 443$, and $491$. As a consequence, the use of optimal normal bases for cryptographic purposes is quite constrained in practice.

In the remainder of this section one details the arithmetic itself. First it is clear that addition and subtraction are the same operations in a field of characteristic two. Using polynomial representation or a normal basis one sees that an addition in $\mathbb{F}_{q^d}$ can be carried out with at most $d$ additions in $\mathbb{F}_q$. Ultimately, an addition in $\mathbb{F}_{q^d}$ reduces to a bitwise-XOR hardware operation, which can be performed at a word level. Multiplications are also processed using a word-by-word approach.

### 11.2.2 Multiplication

Again this part mainly deals with software oriented solutions. For a discussion focused on hardware, see Chapter 26.

Montgomery representation for prime fields (see Section 10.4.2) can be easily generalized to extension fields of characteristic 2; see for instance [KOAC 1998]. We shall not investigate this option further but limit ourselves to multiplications using a polynomial basis and a normal basis.

#### 11.2.2.a Polynomial basis

The internal representation of a polynomial is similar to multiprecision integers. Indeed, let $\ell$ be the word size used by the processor. Then a polynomial $u(X)$ of degree less than $d$ will be represented as the $r$-word vector $(u_{r-1} \ldots u_0)$ and the $j$-th bit of the word $u_i$, that is the coefficient of $u(X)$ of degree $i\ell + j$, will be denoted by $u_i[j]$. Many operations on polynomials are strongly related to integer multiprecision arithmetic. For example, polynomials can be multiplied with a slightly modified version of Algorithm 10.8. However in general, we do not have the equivalent of single precision operations. For example, on computers there is usually no hardware multiplication of polynomials in $\mathbb{F}_2[X]$ of bounded degree, even if this operation is simpler than integer multiplication, since there is no carry to handle. Nevertheless, it is possible to perform computations at a word level doing XOR and shifts. Indeed, if $v(X)X^j$ has been already computed then it is easy to deduce $v(X)X^{i\ell+j}$. This is the principle of Algorithm 11.34 introduced in [LÓDA 2000a].

---

**Algorithm 11.34** Multiplication of polynomials in $\mathbb{F}_2[X]$

INPUT: The polynomials $u(X), v(X) \in \mathbb{F}_2[X]$ of degree at most $d-1$ represented as words of size $\ell$ bits.

OUTPUT: The product $w(X) = u(X)v(X)$ of degree at most $2d-2$.

---

1. $w(X) \leftarrow 0$ and $r \leftarrow \lceil \deg u / \ell \rceil$
2. **for** $j = 0$ **to** $\ell - 1$ **do**
3.     **for** $i = 0$ **to** $r - 1$
4.         **if** $u_i[j] = 1$ **then** $w(X) \leftarrow w(X) + v(X)X^{i\ell}$

*§ 11.2 Finite fields of characteristic 2*                                                                                                                        219

5.     **if** $j \neq \ell - 1$ **then** $v(X) \leftarrow v(X)X$

6.     **return** $w$

---

**Remark 11.35** Algorithm 11.34 proceeds the bits of the word $u_i$ from the right to the left. A left-to-right version exists as well, but it is reported to be a bit less efficient [HAME$^+$ 2003].

**Example 11.36** Let $u(X) = X^5+X^4+X^2+X$, $v(X) = X^{10}+X^9+X^7+X^6+X^5+X^4+X^3+1$ and $\ell = 4$. So $u = (0011\ 0110)$, $v = (0110\ 1111\ 1001)$ and $r = 2$. Here are the values of $v(X)$ and $w(X)$ at the end of Line 5 when Algorithm 11.34 executes.

| $j$ | 0 | 1 | 2 | 3 |
|---|---|---|---|---|
| $v$ | (1101 1111 0010) | (0001 1011 1110 0100) | (0011 0111 1100 1000) | (0110 1111 1001 0000) |
| $w$ | (0110 1111 1001 0000) | (1011 1101 0100 0010) | (1010 0110 1010 0110) | (1010 0110 1010 0110) |

Finally $w(X) = X^{15} + X^{13} + X^{10} + X^9 + X^7 + X^5 + X^2 + X$.

Just as for exponentiation algorithms, precomputations and windowing techniques can be very helpful. The next algorithm scans $k$ bits at a time from left to right and accesses intermediate products by table lookup. Usually a good compromise between the speedup and the number of precomputations is to take $k = 4$.

---

**Algorithm 11.37** Multiplication of polynomials in $\mathbb{F}_2[X]$ using window technique

INPUT: The polynomials $u(X), v(X) \in \mathbb{F}_2[X]$ of degree at most $d - 1$ represented as words of size $\ell$ bits. The precomputed products $t(X)v(X)$ for all $t(X)$ of degree less than $k$.
OUTPUT: The product $w(X) = u(X)v(X)$ of degree at most $2d - 2$.

1.   $w(X) \leftarrow 0$ and $r \leftarrow \lceil \deg u/\ell \rceil$
2.   **for** $j = \ell/k - 1$ **down to** $0$ **do**
3.       **for** $i = 0$ **to** $r - 1$
4.           $t(X) \leftarrow t_{k-1}X^{k-1} + \cdots + t_0$ where $t_m = u_i[jk + m]$
5.           $w(X) \leftarrow w(X) + t(X)v(X)X^{i\ell}$            [$t(X)v(X)$ is precomputed]
6.       **if** $j \neq 0$ **then** $w(X) \leftarrow w(X)X^k$
7.   **return** $w$

---

**Remark 11.38** As for prime fields, cf. Algorithm 11.1, it is possible to modify Algorithm 11.37 and interleave polynomial reductions with elementary multiplications in order to get the result in $\mathbb{F}_{2^d}$ directly at the end.

**Example 11.39** To illustrate the way Algorithm 11.37 works, let us take $k = 2$, $u = (0011\ 0110)$, $v = (0110\ 1111\ 1001)$ and $\ell = 4$ as for Example 11.36. The successive values of $t$ come from the bits of $u$ in the following way $(0011\ \underline{01}10)$, $(\underline{00}11\ 0110)$, $(0011\ 01\underline{10})$, and $(00\underline{11}\ 0110)$.

| $j$ | 1 | 1 | 0 | 0 |
|---|---|---|---|---|
| $i$ | 0 | 1 | 0 | 1 |
| $t$ | (01) | (00) | (10) | (11) |
| $w$ | (0110 1111 1001) | (0110 1111 1001) | (0001 0110 0001 0110) | (1010 0110 1010 0110) |

The result is of course the same, i.e., $w(X) = X^{15} + X^{13} + X^{10} + X^9 + X^7 + X^5 + X^2 + X$.

Another idea is to emulate single precision multiplications by storing all the elementary products. However, for 32-bit words the number of precomputed values is far too big. That is why an intermediate approach involving Karatsuba method is often considered instead. In this case, the product of two single precision polynomials of degree less than 32 is computed with 9 multiplications of 8-bit blocks, each elementary product being obtained by table lookup [GAGE 1996].

Karatsuba method can also be applied to perform the whole product directly. In [GANÖ 2005] the crossover degree between *à la* schoolbook and Karatsuba multiplications is reported to be equal to 576. Other more sophisticated techniques like the FFT or Cantor multiplication based on evaluation/interpolation are useful only for even larger degrees. For example, the crossover between Karatsuba and Cantor multiplication is for degree 35840 [GANÖ 2005].

### 11.2.2.b Optimal normal bases

Unlike additions, multiplications are rather involved with normal bases. The standard way to multiply two elements in $\mathbb{F}_{q^d}$ within a normal basis is to introduce the so-called *multiplication matrix* $T_\mathcal{N}$ whose entries $t_{i,h}$ satisfy

$$\alpha^{q^i} \times \alpha = \sum_{h=0}^{d-1} t_{i,h} \alpha^{q^h} \quad \text{so that} \quad \alpha^{q^i} \times \alpha^{q^j} = \sum_{h=0}^{d-1} t_{i-j,h-j} \alpha^{q^h}.$$

So if $u = (u_0, \ldots, u_{d-1})$ and $v = (v_0, \ldots, v_{d-1})$ then the general term $w_h$ of $w = uv$ is

$$w_h = \sum_{0 \leqslant i,j < d} u_i v_j t_{i-j, h-j}.$$

**Example 11.40** The following is taken directly from [OMMA 1986]. Let $\alpha$ be a zero of $m(X) = X^7 + X^6 + 1$. The next equalities are computed mod $m(X)$.

$$\alpha = X \qquad \qquad \alpha^2 = X^2$$
$$\alpha^{2^2} = X^4 \qquad \qquad \alpha^{2^3} = X^6 + X + 1$$
$$\alpha^{2^4} = X^6 + X^5 + X^4 + X^3 + X \qquad \alpha^{2^5} = X^5 + X^4 + X^2 + X + 1$$
$$\alpha^{2^6} = X^4 + X^3 + 1 \qquad \qquad \alpha^{2^7} = \alpha.$$

The products $\alpha^{q^i} \times \alpha$ are

$$\alpha \times \alpha = X^2 \qquad \qquad \alpha^2 \times \alpha = X^3$$
$$\alpha^{2^2} \times \alpha = X^5 \qquad \qquad \alpha^{2^3} \times \alpha = X^6 + X^2 + X + 1$$
$$\alpha^{2^4} \times \alpha = X^5 + X^4 + X^2 + 1 \qquad \alpha^{2^5} \times \alpha = X^6 + X^5 + X^3 + X^2 + X$$
$$\alpha^{2^6} \times \alpha = X^5 + X^4 + X.$$

## § 11.2 Finite fields of characteristic 2

From this and in order to obtain $T_\mathcal{N}$, one introduces the matrix

$$\mathcal{M} = \left[\begin{array}{ccccccc|ccccccc} 0 & 0 & 0 & 1 & 0 & 1 & 1 & 0 & 0 & 0 & 1 & 1 & 0 & 0 \\ 1 & 0 & 0 & 1 & 1 & 1 & 0 & 0 & 0 & 0 & 1 & 0 & 1 & 1 \\ 0 & 1 & 0 & 0 & 0 & 1 & 0 & 1 & 0 & 0 & 1 & 1 & 1 & 0 \\ 0 & 0 & 0 & 0 & 1 & 0 & 1 & 0 & 1 & 0 & 0 & 0 & 1 & 0 \\ 0 & 0 & 1 & 0 & 1 & 1 & 1 & 0 & 0 & 0 & 0 & 1 & 0 & 1 \\ 0 & 0 & 0 & 0 & 1 & 1 & 0 & 0 & 0 & 1 & 0 & 1 & 1 & 1 \\ 0 & 0 & 0 & 1 & 1 & 0 & 0 & 0 & 0 & 0 & 1 & 0 & 1 & 0 \end{array}\right]$$

where the first and last seven columns give respectively the expression of $\alpha^{q^i}$ and of $\alpha^{q^i} \times \alpha$ on the basis $1, X, \ldots, X^6$. To get the identity matrix in the left part of $\mathcal{M}$ one performs a Gaussian elimination, which gives at the same time the transposed matrix of $T_\mathcal{N}$ in the right part of $\mathcal{M}$. Hence

$$T_\mathcal{N} = \begin{bmatrix} 0 & 1 & 0 & 0 & 0 & 0 & 0 \\ 1 & 1 & 0 & 1 & 1 & 1 & 0 \\ 0 & 0 & 0 & 1 & 1 & 0 & 1 \\ 0 & 1 & 0 & 1 & 0 & 0 & 0 \\ 1 & 0 & 0 & 0 & 0 & 1 & 0 \\ 0 & 1 & 1 & 0 & 1 & 0 & 0 \\ 1 & 0 & 1 & 1 & 1 & 0 & 1 \end{bmatrix}.$$

From this matrix, one deduces for instance that $\alpha^2 \times \alpha = \alpha + \alpha^2 + \alpha^{2^3} + \alpha^{2^4} + \alpha^{2^5}$.

The number of nonzero coefficients of the matrix $T_\mathcal{N}$ is denoted by $\delta_\mathcal{N}$ and called the *density of* $T_\mathcal{N}$. It is a crucial parameter for the speed of the system since the multiplication of two elements in $\mathbb{F}_{q^d}$ can be computed with at most $2d\,\delta_\mathcal{N}$ multiplications and $d(\delta_\mathcal{N} - 1)$ additions in $\mathbb{F}_q$. On average the density is about $(q-1)d^2/q$ [BEGE$^+$ 1991] but in fact $\delta_\mathcal{N} \geqslant 2d - 1$ [MUON$^+$ 1989] and this bound is sharp. By definition, an optimal normal basis, cf. Section 11.2.1.c, has such a minimal density.

Concerning $\mathbb{F}_{2^d}$, recall that a multiplication in a Type I ONB can be in fact performed in the corresponding anomalous basis. There is a simple way to transform an element into a polynomial and computations are made modulo $X^{d+1} - 1$. For Type II ONB, there is a similar idea called *palindromic representation* [BLRO$^+$ 1998]. The situation is not as favorable as for Type I ONB since in this case computations must be made modulo $X^{2d+1} - 1$. Optimal normal bases of Type I and II appear as special cases of Gauß periods, cf. Section 2.3.3.b.

### 11.2.3 Squaring

Squaring is a trivial operation for extensions of $\mathbb{F}_2$ in normal basis representation and it is very simple in polynomial representation. The absolute Frobenius $X \mapsto X^2$ being a linear map, one sees that if $u(X) = \sum u_i X^i$ then $u^2(X) = \sum u_i X^{2i}$. Thus, this operation is nothing but inserting 0 bits in the internal representation of $u$ and reducing the result modulo $m(X)$. Precomputing a table of 256 values containing the squares of each byte allows us to speed up the 0-bit insertion process. However, the reduction remains the most time-consuming part of the whole process.

To speed up this process a bit, it is possible to split the square $\sum u_i X^{2i}$ into an even and an odd part so that the number of required bitwise-XOR operations to actually perform the reduction is halved. See [KIN 2001] for details.

## 11.2.4 Inversion and division

There are mainly two ways to compute the inverse of an element $\alpha \in \mathbb{F}_{q^d}$. The first method is to perform an extended gcd computation of the polynomial representing $\alpha$ and the irreducible polynomial defining $\mathbb{F}_{q^d}$. Alternatively, one can exploit the multiplicative structure of the group $\mathbb{F}_{q^d}^*$ with Lagrange's theorem. This is especially useful for normal bases.

### 11.2.4.a Euclid extended gcd

Given two nonzero polynomials $f$ and $m$ in $\mathbb{F}_q[X]$, there are unique polynomials $u$, $v$, and $g$ such that $fu + mv = g$ where $g = \gcd(f, m)$, $\deg u < \deg m$, and $\deg v < \deg f$. In case $m$ is irreducible and $\deg f < \deg m$ we have $g = 1$ so that $u$ is the inverse of $f$ modulo $m$. The following algorithm returns $u$, $v$ and $g$.

---

**Algorithm 11.41** Euclid extended polynomial gcd

INPUT: Two nonzero polynomials $f, m \in \mathbb{F}_q[X]$.
OUTPUT: The polynomials $u, v, g$ in $\mathbb{F}_q[X]$ such that $fu + mv = g$ with $g = \gcd(f, m)$.

1. $u \leftarrow 1, v \leftarrow 0, s \leftarrow m$ and $g \leftarrow f$
2. **while** $s \neq 0$ **do**
3.     compute Euclid division of $g$ by $s$            $[g = qs + r]$
4.     $t \leftarrow u - vq$, $u \leftarrow v$, $g \leftarrow s$, $v \leftarrow t$ and $s \leftarrow r$
5. $v \leftarrow (g - fu)/m$
6. **return** $(u, v, g)$

---

**Remark 11.42** Assuming $\deg f \leqslant d$ and $\deg m \leqslant d$, Algorithm 11.41 requires $O(d^2)$ elementary operations in $\mathbb{F}_q$.

**Example 11.43** Take $m(X) = X^{11} + X^2 + 1$ and $f(X) = X^8 + X^6 + X^5 + X^4 + X + 1$. Algorithm 11.41 proceeds as follows

| $q$ | $r$ | $u$ | $v$ | $g$ |
|---|---|---|---|---|
| (0000) | (0001 0111 0011) | (0000) | (0001) | (1000 0000 0101) |
| (1011) | (1000) | (1011) | (0001 0111 0011) | (0001 0111 0011) |
| (0010 1110) | (0011) | (1011) | (0001 0000 0011) | (1000) |
| (0111) | (0001) | (0001 0000 0011) | (0111 0000 0010) | (0011) |
| (0011) | (0000) | (0111 0000 0010) | (1000 0000 0101) | (0001) |
| — | — | (0111 0000 0010) | (1100 1011) | (0001) |

One deduces that $(X^{10} + X^9 + X^8 + X)f(X) + (X^7 + X^6 + X^3 + X + 1)m(X) = 1$ which implies that $X^{10} + X^9 + X^8 + X$ is the inverse of $f(X)$ in $\mathbb{F}_2[X]/(m(X))$.

## § 11.2 Finite fields of characteristic 2

### 11.2.4.b  Binary inversion

For extensions of $\mathbb{F}_2$, there is also a dedicated algorithm inspired by the binary integer version [BRCU$^+$ 1993].

---

**Algorithm 11.44** Inverse of an element of $\mathbb{F}_{2^d}^*$ in polynomial representation

INPUT: An irreducible polynomial $m(X) \in \mathbb{F}_2[X]$ of degree $d$ and a nonzero polynomial $f(X) \in \mathbb{F}_2[X]$ such that $\deg f < d$.
OUTPUT: The polynomial $u(X) \in \mathbb{F}_2[X]$ such that $fu \equiv 1 \pmod{m}$.

1. $u \leftarrow 1, v \leftarrow 0, s \leftarrow m$ and $\delta \leftarrow 0$
2. **for** $i = 1$ **to** $2d$ **do**
3.    **if** $f_d = 0$ **then**                                               $\bigl[f(X) = f_d X^d + \cdots + f_0\bigr]$
4.       $f(X) \leftarrow Xf(X), u(X) \leftarrow \bigl(Xu(X)\bigr) \bmod m(X)$ and $\delta \leftarrow \delta + 1$
5.    **else**
6.       **if** $s_d = 1$ **then**                                             $\bigl[s(X) = s_d X^d + \cdots + s_0\bigr]$
7.          $s(X) \leftarrow s(X) - f(X)$ and $v(X) \leftarrow \bigl(v(X) - u(X)\bigr) \bmod m(X)$
8.       $s(X) \leftarrow Xs(X)$
9.       **if** $\delta = 0$ **then**
10.          $t(X) \leftarrow f(X), f(X) \leftarrow s(X)$ and $s(X) \leftarrow t(X)$
11.          $t(X) \leftarrow u(X), u(X) \leftarrow v(X)$ and $v(X) \leftarrow t(X)$
12.          $u(X) \leftarrow \bigl(Xu(X)\bigr) \bmod m(X)$
13.          $\delta \leftarrow 1$
14.       **else**
15.          $u(X) \leftarrow \bigl(u(X)/X\bigr) \bmod m(X)$ and $\delta \leftarrow \delta - 1$
16. **return** $u$

---

### Remarks 11.45

(i) The operations $\bigl(Xu(X)\bigr) \bmod m(X)$ and $\bigl(u(X)/X\bigr) \bmod m(X)$ can be very efficiently performed if $X^d \bmod m(X)$ and $X^{-1} \bmod m(X)$ are precomputed.

(ii) Algorithm 11.44, unlike other binary gcd versions (see [HAME$^+$ 2003] for instance) does not require any degree comparison thanks to the use of the counter $\delta$. This idea was first suggested by Brent and Kung for modular inversion (see [BRKU 1983] and Section 11.1.3.a) and gives good performances in both software and hardware.

(iii) A similar algorithm testing least significant bits instead of most significant bits has been recently proposed [WUWU$^+$ 2004].

(iv) It is possible to directly obtain $\bigl(h(X)/f(X)\bigr) \bmod m(X)$ by setting $u(X) \leftarrow h(X)$ instead of $u \leftarrow 1$ in the first line of Algorithms 11.41 and 11.44. In this case, a reduction is almost always needed at the end when the first algorithm is used, whereas the result is already reduced with the second one.

**Example 11.46** With the values of Example 11.43, the successive steps of Algorithm 11.44 are

| $i$ | $u$ | $v$ | $s$ | $\delta$ |
|---|---|---|---|---|
| 1  | (0010) | (0000) | (1000 0000 0101) | 1 |
| 2  | (0100) | (0000) | (1000 0000 0101) | 2 |
| 3  | (1000) | (0000) | (1000 0000 0101) | 3 |
| 4  | (0100) | (1000) | (1110 011 1010) | 2 |
| 5  | (0010) | (1000) | (1110 0111 0100) | 1 |
| 6  | (0001) | (1010) | (1011 1101 1000) | 0 |
| 7  | (0001 0110) | (0001) | (1011 1001 1000) | 1 |
| 8  | (0010 1100) | (0001) | (1011 1001 1000) | 2 |
| 9  | (0101 1000) | (0001) | (1011 1001 1000) | 3 |
| 10 | (1011 0000) | (0001) | (1011 1001 1000) | 4 |
| 11 | (0001 0110 0000) | (0001) | (1011 1001 1000) | 5 |
| 12 | (1011 0000) | (0001 0110 0001) | (0111 0011 0000) | 4 |
| 13 | (0101 1000) | (0001 0110 0001) | (1110 0110 0000) | 3 |
| 14 | (0010 1100) | (0001 0011 1001) | (1100 1100 0000) | 2 |
| 15 | (0001 0110) | (0001 0001 0101) | (1001 1000 0000) | 1 |
| 16 | (1011) | (0001 0000 0011) | (0011 0000 0000) | 0 |
| 17 | (0010 0000 0110) | (1011) | (1000 0000 0000) | 1 |
| 18 | (0100 0000 1100) | (1011) | (1000 0000 0000) | 2 |
| 19 | (0010 0000 0110) | (0100 0000 0111) | (1000 0000 0000) | 1 |
| 20 | (0001 0000 0011) | (0110 0000 0001) | (1000 0000 0000) | 0 |
| 21 | (0110 0000 0001) | (0001 0000 0011) | (1100 0000 0000) | 1 |
| 22 | (0111 0000 0010) | (0111 0000 0010) | (1000 0000 0000) | 0 |

and the final result is the same, that is, the inverse of $f(X)$ is $X^{10} + X^9 + X^8 + X$.

### 11.2.4.c  Inversion based on Lagrange's theorem

It is also possible to use the group structure of $\mathbb{F}_{q^d}^*$ to get the inverse of an element $\alpha$. This method has the same asymptotic complexity as the extended Euclidean one but is reported to be a little faster [NÖC 1996] when a squaring is for free. We know that $|\mathbb{F}_{q^d}^*| = q^d - 1$ with $q$ some power of 2, say $q = 2^k$. So $\alpha^{q^d-2} = 1/\alpha$. Now

$$q^d - 2 = (q^{d-1} - 1)q + q - 2,$$

and we can take advantage of the special expression of $q^{d-1}-1$ in base $q$ and of the Frobenius, which makes the computation of $q$-th powers easier. For better performance, addition chains, presented in Section 9.2.3, are used as well.

---

**Algorithm 11.47** Inverse of an element of $\mathbb{F}_{q^d}^*$ using Lagrange's theorem

INPUT: An element $\alpha \in \mathbb{F}_{q^d}^*$, two addition chains, namely $(a_0, a_1, \ldots, a_{s_1})$ for $q - 2$ and $(b_0, b_1, \ldots, b_{s_2})$ for $d - 1$.
OUTPUT: The inverse of $\alpha$ i.e., $\alpha^{q^d-2} = 1/\alpha$.

1. $y \leftarrow \alpha^{q-2}$  [using $(a_0, a_1, \ldots, a_{s_1})$ and Algorithm 9.41]

## § 11.2 Finite fields of characteristic 2

2. $T[0] \leftarrow \alpha \times y$ and $i \leftarrow 1$
3. **while** $i \leqslant s$ **do**
4. $\quad t \leftarrow T[k]^{q^j}$ where $b_i = b_k + b_j$
5. $\quad T[i] \leftarrow t \times T[j]$ $\hfill \left[T[i] = \alpha^{q^{b_i}-1} \text{ for all } i\right]$
6. $\quad i \leftarrow i+1$
7. $t \leftarrow T[s_2]$ $\hfill [b_{s_2} = d-1]$
8. **return** $yt^q$

**Remarks 11.48**

(i) Note that exchanging $b_k$ and $b_j$, in Line 4, does not alter the correctness of the algorithm. In fact it is better to force $b_k$ to be the maximum of $b_k$ and $b_j$ so that the exponentiation $T[b_k]^{q^{b_j}}$ is simpler.

(ii) One can obtain the inverse of $\alpha \in \mathbb{F}_{q^d}$ with $s_1 + s_2 + 2$ multiplications in $\mathbb{F}_{q^d}$ and $(1 + \sum_i b_j)$ $q$-th power computations where $b_j$ is the integer in $b_i = b_k + b_j$. This last number is equal to $d-1$ when $(b_0, b_1, \ldots, b_{s_2})$ is a star addition chain, cf. Section 9.2.1.

(iii) One of the three methods proposed by Itoh and Tsujii [ITTS 1988] is a special case of Algorithm 11.47 when $q = 2$ and the addition chain computing $d-1$ is derived from the square and multiply method.

(iv) When $q$ is bigger than 2, another option suggested by Itoh and Tsujii is to write $\alpha^{-1}$ as $\alpha^{-r} \times \alpha^{r-1}$ where $r = (q^d - 1)/(q-1) = q^{d-1} + \cdots + q + 1$. As $\alpha^r \in \mathbb{F}_q$, it can be easily inverted. It is the standard way to compute an inverse in an OEF, cf. Section 11.3.

**Example 11.49** Suppose that one wants the inverse of $\alpha \in \mathbb{F}_{2^{19}}$, that is $\alpha^{2^{19}-2}$. Obviously, one has $2^{19} - 2 = 2(2^{18} - 1)$ and an addition chain for 18 is $(1, 2, 3, 6, 12, 18)$.

| $i$ | $b_i = b_k + b_j$ | $T[k]^{q^{b_j}} \times T[j]$ | $T[i]$ |
|---|---|---|---|
| 0 | 1 | — | $\alpha$ |
| 1 | $2 = 1+1$ | $T[0]^{2^1} \times T[0]$ | $\alpha^2 \times \alpha = \alpha^3$ |
| 2 | $3 = 2+1$ | $T[1]^{2^1} \times T[0]$ | $\alpha^6 \times \alpha = \alpha^7$ |
| 3 | $6 = 3+3$ | $T[2]^{2^3} \times T[2]$ | $\alpha^{7 \times 8} \times \alpha^7 = \alpha^{63}$ |
| 4 | $12 = 6+6$ | $T[3]^{2^6} \times T[3]$ | $\alpha^{63 \times 64} \times \alpha^{64} = \alpha^{4095}$ |
| 5 | $18 = 12+6$ | $T[4]^{2^6} \times T[3]$ | $\alpha^{4095 \times 64} \times \alpha^{63} = \alpha^{2^{18}-1}$ |

Finally $T[5]^2 = \alpha^{-1}$.

### 11.2.5 Exponentiation

In polynomial representation, a simple trick can greatly speed up exponentiation. Namely, let $f(X)$, $m(X)$ be polynomials in $\mathbb{F}_q[X]$ and $g(X) = X^{q^r}$. Because of the Frobenius action, it is obvious that $f^{q^r} \equiv f(g) \pmod{m}$. At this point one uses a fast algorithm for modular composition designed by Brent and Kung [BRKU 1978].

The idea, *à la* baby-step giant-step, is to write

$$f(X) = \sum_{0 \leqslant i < k} X^{ki} F_i(X) \text{ with } k = \lceil \deg f \rceil \text{ and } F_i(X) = \sum_{0 \leqslant j < k} f_{ik+j} X^j$$

and to precompute and store $1, g, g^2, \ldots, g^{k-1}$ and $1, g^k, g^{2k}, \ldots, g^{k(k-1)}$ modulo $m$. Here is the complete algorithm:

---
**Algorithm 11.50** Modular composition of Brent and Kung

INPUT: The polynomials $m, f, g \in \mathbb{F}_q[X]$ with $\deg m = d$ and $\deg f, g < d$.
OUTPUT: The polynomial $f(g) \bmod m$.

---

1. $k \leftarrow \lceil \sqrt{d} \rceil$
2. $G[0] \leftarrow 1$
3. **for** $i = 1$ **to** $k$ **do** $G[i] \leftarrow (gG[i-1]) \bmod m$ $\qquad [G[i] = g^i \bmod m]$
4. $P[0] \leftarrow 1$
5. **for** $i = 1$ **to** $k-1$ **do** $P[i] \leftarrow (G[k]P[i-1]) \bmod m$ $\qquad [P[i] = g^{ki} \bmod m]$
6. **for** $i = 0$ **to** $k-1$ **do** $F[i] \leftarrow \sum_{j=0}^{k-1} f_{ik+j} G[j]$ $\qquad [F[i] = F_i(g)]$
7. $R \leftarrow \left(\sum_{i=0}^{k-1} F[i]P[i]\right) \bmod m$
8. **return** $R$

---

**Remark 11.51** With classical arithmetic the complexity of Algorithm 11.50 is $O(d^{5/2})$, but it can be reduced to $O(d^{1/2 + \lg 3})$. Indeed, as shown in [NÖC 1996], the loop in Line 6 can be computed with fast matrix multiplication *à la* Strassen [KNU 1997] and the other multiplications with the Karatsuba method.

**Example 11.52** Let $m(X) = X^{15} + X + 1$ irreducible over $\mathbb{F}_2$, $f(X) = X^{14} + X^{13} + X^8 + X^6 + X^4 + X^3 + 1$ and $g(X) = X^{10} + X^3 + 1$. One has $k = 4$ and

$$f(X) = F_0(X) + X^4 F_1(X) + X^8 F_2(X) + X^{12} F_3(X)$$

with
$$F_0(X) = X^3 + 1, F_1(X) = X^2 + 1, F_2(X) = 1 \text{ and } F_3(X) = X^2 + X.$$

The precomputed values $g^i$ and $g^{ki}$ for $0 \leqslant i \leqslant k$ are respectively stored in the arrays $G$ and $P$ whereas $F[i]$ contains $F_i(g)$.

| $i$ | $G[i]$ | $P[i]$ | $F[i]$ |
|---|---|---|---|
| 0 | (0001) | (0001) | (0101 0100 1011) |
| 1 | (0100 0000 1001) | (0100 0000 0001) | (0010 0000) |
| 2 | (0010 0001) | (0110 0001) | (0001) |
| 3 | (0101 0010 1010) | (0100 0110 0100) | (0100 0100 1000) |

Finally $R = X^{13} + X^{12} + X^{11} + X^9 + X^7 + X^5 + X^3 + X^2 + X + 1$ which is equivalent to $f(g)$ modulo $m$.

Now we present Shoup's algorithm [SHO 1994a, GAGA+ 2000] which is mainly based on the $q^r$-ary method for a well chosen $r$.

## § 11.2 Finite fields of characteristic 2

**Algorithm 11.53** Shoup exponentiation algorithm

INPUT: The polynomials $f, m \in \mathbb{F}_q[X]$ with $\deg m = d$ and $\deg f < d$. A parameter $r$ and an exponent $n = (n_{\ell-1} \ldots n_0)_{q^r}$ such that $0 < n < q^d$.
OUTPUT: The polynomial $f^n \bmod m$.

1. **for** $i = 0$ **to** $\ell - 1$ precompute and store $f^{n_i} \bmod m$
2. $g(X) \leftarrow X^{q^r} \bmod m$ and $y \leftarrow 1$
3. **for** $i = \ell - 1$ **down to** $0$ **do**
4. $\quad y \leftarrow y(g)$ [use Algorithm 11.50]
5. $\quad y \leftarrow (y \times f^{n_i}) \bmod m$
6. **return** $y$

**Remarks 11.54**

(i) The parameter $r$ is usually set to $\lceil d/\log_q d \rceil$ and the precomputations can be done with Yao's method, cf. Algorithm 9.44, as proposed by Gao et al. [GAGA$^+$ 2000].
(ii) Neglecting precomputations, the number of multiplications needed is $O(d/\lg d)$. Its complexity, including the cost of precomputations, is $O(d^3/\lg d + d^2 \lg d)$ with classical arithmetic and $O(d^{1+\lg 3}/\lg d + d^{(1+\lg 7)/2} \lg d)$ with Karatsuba method and *à la* Strassen matrix multiplication techniques for modular composition.
(iii) The number of stored values is $O(d/\lg d)$.
(iv) The for loop starting Line 3 is a *Horner-like scheme*.

**Example 11.55** Take $q = 2, m(X) = X^{15} + X + 1, f(X) = X^{14} + X^{13} + X^8 + X^6 + X^4 + X^3 + 1$ and $n = 23801$. Let us compute $f^n \bmod m$ with Algorithm 11.53. One has $r = \lceil 15/\lg 15 \rceil = 4$, $23801 = (5\ 12\ 15\ 9)_{16}$, and $g(X) \equiv X^2 + X \pmod{m(X)}$. Then for each $i$, $y(g) \equiv y^{16} \pmod m$ and $(y f^{n_i}) \bmod m$ are successively computed. In the following table, we give the corresponding values of $y$ after the execution of Lines 4 and 5 of Shoup's algorithm, as well as the precomputed values $f^{n_i}$ used at each step.

| $i$ | $y$ | $f^{n_i} \bmod m$ |
|---|---|---|
| 3 | 1 | (0001 1001 0011 1000) |
| — | (0001 1001 0011 1000) | — |
| 2 | (0110 1001 0111 1000) | (1001 1101 1100) |
| — | (0010 1001 1001 0000) | — |
| 1 | (0111 0011 1101 0010) | (0111 0100 0000 0011) |
| — | (0100 0101 0111 1110) | — |
| 0 | (0011 1000 1101 1010) | (0100 0001 0111 0011) |
| — | (0001 0111 0001 0101) | — |

### 11.2.6 Square roots and quadratic equations

Every element $\alpha \in \mathbb{F}_{2^d}$ is a square. The square root of $\alpha$ can be easily obtained thanks to the multiplicative structure of $\mathbb{F}_{2^d}^*$, which implies that $\sqrt{\alpha} = \alpha^{2^{d-1}}$. In a normal basis the computation of a square root is therefore immediate. If $\alpha$ is represented by $f(X) = \sum_{i=0}^{d-1} f_i X^i$ on a polynomial basis, it is better to write

$$\sqrt{f(X)} = \sum_{i \text{ even}} f_i X^{i/2} + \sqrt{X} \sum_{i \text{ odd}} f_i X^{\frac{i-1}{2}}$$

where $\sqrt{X}$ has been precomputed modulo $m(X)$. When $m(X)$ is the irreducible trinomial $X^d + X^k + 1$ with $d$ odd, note that $\sqrt{X}$ can be obtained directly. Indeed

$$\sqrt{X} \equiv X^{\frac{d+1}{2}} + X^{\frac{k+1}{2}} \pmod{m(X)}$$

if $k$ is odd and $\sqrt{X} \equiv X^{-\frac{d-1}{2}}(X^{\frac{k}{2}} + 1) \pmod{m(X)}$ otherwise. This technique applies to redundant trinomials as well; see Section 11.2.1.b.

Solving quadratic equations in $\mathbb{F}_{2^d}$ is not as straightforward as computing square roots. Indeed, let us solve the equation $T^2 + aT + b = 0$ in $\mathbb{F}_{2^d}$ where, by the above, $a$ is assumed to be nonzero. The change of variable $T \leftarrow T/a$ yields the simpler equation

$$T^2 + T = c \text{ with } c = b/a^2. \tag{11.2}$$

**Lemma 11.56** Equation (11.2) has a solution in $\mathbb{F}_{2^d}$ if and only if $\text{Tr}(c) = 0$. If $x$ is a solution then $x + 1$ is the other one.

When $d$ is odd, such a solution is given by

$$x = \sum_{i=0}^{(d-3)/2} c^{2^{2i+1}}. \tag{11.3}$$

When $d$ is even, set

$$x = \sum_{i=0}^{d-1} \left( \sum_{j=0}^{i} c^{2^j} \right) y^{2^i} \tag{11.4}$$

where $y \in \mathbb{F}_{2^d}$ is any element of trace 1.

**Proof.** Let $x$ be a solution of (11.2). Then $\text{Tr}(c) = \text{Tr}(x^2 + x) = \text{Tr}(x) + \text{Tr}(x) = 0$. The opposite direction is proved by showing that the proposed solutions actually work. Computing $x^2 + x$, one has in the first case

$$x^2 + x = c + \text{Tr}(c) = c,$$

and in the second one

$$x^2 + x = y \text{Tr}(c) + c \text{Tr}(y) = c.$$

Thus $x$ is always a solution of (11.2) as claimed. □

In practice, several improvements can be considered. First, to check for the existence of a solution and then to actually compute such a solution.

### Remarks 11.57

(i) There is a unique vector $w$ in $\mathbb{F}_{2^d}$ that is orthogonal to all the elements of trace 0. If $w$ is precomputed, it is enough to compute the scalar product $w \cdot c$ in order to deduce the trace of $c$ [KNU 1999].

(ii) When the field $\mathbb{F}_{2^d}$ is defined by an irreducible polynomial of the form

$$X^d + a_{d-1}X^{d-1} + \cdots + a_1 X + a_0 \text{ with } a_j = 0 \text{ for all } j > d/2 \qquad (11.5)$$

the trace of an element can also be obtained very efficiently.
Let $\theta$ be a root of this polynomial. Then using the Newton–Girard formula giving the sum of the conjugates of $\theta^k$ in terms of the $a_i$'s and the linearity of the trace map it is immediate that

$$\text{if } c = \sum_{k=0}^{d-1} c_k \theta^k \text{ then } \text{Tr}(c) = c_0 + \sum_{k=1}^{d-1} k c_k a_{d-k}.$$

As we have seen, moderately large extension fields of characteristic 2 can always be defined by trinomials or pentanomials of the form (11.5), so that the computation of the trace is always simple in practice.

**Example 11.58** In $\mathbb{F}_{2^{233}}$ defined by $X^{233} + X^{74} + 1$ we have

$$\text{Tr}\left(c_{232}\theta^{232} + c_{231}\theta^{231} + \cdots + c_1\theta + c_0\right) = c_0 + c_{159}.$$

**Remark 11.59** Rather than computing a solution using (11.3) or (11.4), it can be faster to use the linearity of the map $\lambda \mapsto \lambda^2 + \lambda$ defined from $\mathbb{F}_{2^d}$ to $\mathbb{F}_{2^d}$. Indeed, precomputing the inverse matrix of this operator gives the result in a straightforward way. Additional tricks can be used to reduce the storage and the amount of computations [KNU 1999].

## 11.3 Optimal extension fields

On the one hand, multiplications in extension fields of characteristic 2 are usually performed less efficiently than in prime fields, due to the lack of a single precision polynomial multiplication on most processors. On the other hand, inversion in prime fields can be a very expensive operation, especially in hardware. To overcome these two difficulties, optimal extension fields have been recently investigated [MIH 1997, BAPA 1998]. They seem to be particularly interesting for smart cards [WOBA+ 2000].

First, we shall briefly introduce optimal extension fields, and give existence criterions and some examples before addressing the arithmetic itself. We conclude with the special cases of extensions of degree 3 and 5.

### 11.3.1 Introduction

Let us take an extension field $\mathbb{F}_{p^d}$ such that

- the characteristic $p$ fits in a machine word and allows a fast reduction in $\mathbb{F}_p$
- the irreducible polynomial defining $\mathbb{F}_{p^d}$ allows a fast polynomial reduction.

This choice leads to the following concept.

**Definition 11.60** An *optimal extension field*, OEF for short, is an extension field $\mathbb{F}_{p^d}$ where

- $p$ is a *pseudo-Mersenne prime*, that is $p = 2^n + c$ with $|c| \leqslant 2^{\lfloor n/2 \rfloor}$
- there is an irreducible binomial $m(X) = X^d - \omega$ over $\mathbb{F}_p$.

If $c = \pm 1$ then the field is said to be of *Type I* and it is of *Type II* when $\omega = 2$.

**Remark 11.61** Generalizing Definition 11.60, cf. [AVMI 2004], it is possible to consider a prime $p$ of another form provided a fast reduction algorithm exists; see Section 10.4.3 for examples.

The cardinality of $\mathbb{F}_{p^d}$ is approximately equal to $2^{nd}$ and in practice, an element $\alpha \in \mathbb{F}_{p^d}$ is represented by the polynomial $a_{d-1}X^{d-1} + \cdots + a_1 X + a_0$ where $a_i \in \mathbb{F}_p$. As suggested before, this implies that computations in OEFs require two kinds of reduction. Intermediate results have to be reduced modulo the binomial $m(X)$, and for this task Algorithm 11.31 is not even required since a reduction modulo $m$ consists simply of replacing $X^d$ by $\omega$. Coefficients of the polynomial also have to be reduced modulo $p$. For Type I OEF, this operation needs one addition in $\mathbb{F}_p$, cf. Section 10.4.3. Otherwise reduction is obtained by Algorithm 10.25 and is more expensive.

OEFs are rather easy to find and their search is simplified by the results below on the irreducibility of $X^d - \omega$ over $\mathbb{F}_p$.

**Theorem 11.62** Let $d \geqslant 2$ be an integer and $\omega \in \mathbb{F}_p^*$. The binomial $m(X) = X^d - \omega$ is irreducible in $\mathbb{F}_p[X]$ if and only if the two following conditions hold

- each prime factor of $d$ divides the order $e$ of $\omega$ but does not divide $(p-1)/e$
- $p \equiv 1 \pmod 4$ if $d \equiv 0 \pmod 4$.

As shown in [JUN 1993] one has the sufficient condition

**Corollary 11.63** If $\omega \in \mathbb{F}_p^*$ is a primitive element and $d \mid (p-1)$ then the polynomial $X^d - \omega$ is irreducible over $\mathbb{F}_p$.

If $d$ is squarefree and $X^d - \omega$ irreducible over $\mathbb{F}_p$ then Theorem 11.62 implies that $p \equiv 1 \pmod d$. This remark is also useful to speed up the search of OEFs.

In Table 11.2 are given all OEFs of Type I, of cryptographic interest sorted with respect to $nd$.

**Table 11.2** Type I OEFs.

| $n$ | $c$ | $d$ | $\omega$ | $nd$ | $n$ | $c$ | $d$ | $\omega$ | $nd$ | $n$ | $c$ | $d$ | $\omega$ | $nd$ | $n$ | $c$ | $d$ | $\omega$ | $nd$ |
|---|---|---|---|---|---|---|---|---|---|---|---|---|---|---|---|---|---|---|---|
| 13 | −1 | 6 | 7 | 78 | 17 | −1 | 5 | 3 | 85 | 13 | −1 | 7 | 3 | 91 | 31 | −1 | 3 | 5 | 93 |
| 7 | −1 | 14 | 3 | 98 | 17 | −1 | 6 | 3 | 102 | 19 | −1 | 6 | 3 | 114 | 13 | −1 | 9 | 7 | 117 |
| 7 | −1 | 18 | 3 | 126 | 8 | 1 | 16 | 3 | 128 | 16 | 1 | 8 | 3 | 128 | 13 | −1 | 10 | 3 | 130 |
| 19 | −1 | 7 | 3 | 133 | 7 | −1 | 21 | 3 | 147 | 17 | −1 | 9 | 3 | 153 | 13 | −1 | 13 | 2 | 169 |
| 17 | −1 | 10 | 3 | 170 | 19 | −1 | 9 | 3 | 171 | 13 | −1 | 14 | 3 | 182 | 61 | −1 | 3 | 5 | 183 |
| 31 | −1 | 6 | 5 | 186 | 7 | −1 | 27 | 3 | 189 | 13 | −1 | 15 | 11 | 195 | 31 | −1 | 7 | 3 | 217 |
| 13 | −1 | 18 | 7 | 234 | 17 | −1 | 15 | 3 | 255 | 8 | 1 | 32 | 3 | 256 | 16 | 1 | 16 | 3 | 256 |
| 19 | −1 | 14 | 3 | 266 | 13 | −1 | 21 | 7 | 273 | 31 | −1 | 9 | 5 | 279 | 17 | −1 | 17 | 2 | 289 |

Table 11.3 contains examples of OEFs of Type II. More precisely, given a size $s$ between 135 and 300, for each $n \geqslant 7$ dividing $s$, the unique parameters $c \in \left[-2^{\lfloor n/2 \rfloor}, 2^{\lfloor n/2 \rfloor}\right]$ and $d$, if any, are given, such that

- $p = 2^n + c$ is prime
- $c$ is minimal in absolute value
- $d = s/n$ and $X^d - 2$ is irreducible over $\mathbb{F}_p$.

Note that when $n = 8, 16, 32$, or $64$ only negative values of $c$ are reported so that elements of $\mathbb{F}_p$ can be represented with a single word on the corresponding commonly used architectures. Since these parameters are of great importance in practice, they appear distinctly in the table.

Concerning arithmetic, additions and subtractions are straightforward and do not enjoy special improvements, unlike other basic operations we shall describe now.

### 11.3.2 Multiplication

Let two elements $\alpha, \beta \in \mathbb{F}_{p^d}$ be represented by $\alpha = \sum_{i=0}^{d-1} a_i X^i$ and $\beta = \sum_{i=0}^{d-1} b_i X^i$ where $a_i$ and $b_i$ are in $\mathbb{F}_p$. Then using the relation $X^d \equiv \omega \pmod{m(X)}$, one has

$$\alpha\beta = c_{d-1} + \sum_{k=0}^{d-2}(c_k + \omega c_{d+k})X^k \quad \text{with} \quad c_k = \sum_{i=0}^{k} a_i b_{k-i}.$$

Instead of reducing $a_i b_{k-i}$ modulo $p$ at each step, it can be faster, especially for OEFs that are not of Type I, to compute $c_k + \omega c_{d+k}$ as a multiprecision integer and to reduce it only once. As shown in [HAME$^+$ 2003], if $p = 2^n + c$ is such that $\lg(1 + \omega(d-1)) + 2\lg|c| \leq n$, then $c_k + \omega c_{d+k}$ can be reduced at once with only two multiplications by $c$.

As suggested in [MIH 2000], one can also use convolutions methods, like the FFT, to multiply $\alpha$ and $\beta$. This is particularly effective when $d$ is close to a power of 2, or close to the product of small primes.

As usual, a squaring should be considered independently and computed with a specific procedure.

### 11.3.3 Exponentiation

The action of the absolute Frobenius $\phi_p$ can be computed very efficiently in OEFs [MIH 2000]. Indeed, since the coefficients of $\alpha = \sum_{j=0}^{d-1} a_j X^j$ are in $\mathbb{F}_p$, one has

$$\alpha^{p^i} = \sum_{j=0}^{d-1} a_j \omega^{\lfloor jp^i/d \rfloor} X^{((jp^i) \bmod d)}.$$

Recall that when $d$ is squarefree, $p \equiv 1 \pmod{d}$ so that $X^{((jp)^i \bmod d)}$ is simply $X^j$. Thus an exponentiation to the power $p^i$ only requires us to multiply each coefficient $a_j$ by some power of $\omega$, which can be precomputed.

Another interesting choice is to take $p = kd + 1$ for a given $d$. In this case, $X^{((jp^i) \bmod d)} = X^j$ as well, and $\omega^{\lfloor jp/d \rfloor} = \zeta^j$ where $\zeta = \omega^{\frac{p-1}{d}} \in \mathbb{F}_p$ is a $d$-th root of unity.

**Example 11.64** Let $p = 2^{16} - 165$, $\mathbb{F}_{p^6} \simeq \mathbb{F}_p[X]/(X^6 - 2)$ and take the random element

$$\alpha = 44048X^5 + 24430X^4 + 54937X^3 + 18304X^2 + 46713X + 63559.$$

One checks that $p - 1 \equiv 0 \pmod{6}$ so that $\zeta = 2^{\lfloor p/d \rfloor}$ is a 6-th primitive root of unity. Precomputing $\zeta, \zeta^2, \zeta^3, \zeta^4$ and $\zeta^5$ modulo $p$, and multiplying $a_j$ by $\zeta^j$, one obtains

$$\alpha^p = 23814X^5 + 34492X^4 + 10602X^3 + 7340X^2 + 40911X + 63559.$$

Using the same set of precomputations, the product of $\alpha$ by the $\zeta^{4j}$'s, componentwise gives

$$\alpha^{p^4} = 41725X^5 + 34492X^4 + 54937X^3 + 7340X^2 + 24628X + 63559.$$

**Table 11.3** Examples of Type II OEFs.

| n | c | d | nd | n | c | d | nd | n | c | d | nd | n | c | d | nd | n | c | d | nd | n | c | d | nd | n | c | d | nd |
|---|---|---|---|---|---|---|---|---|---|---|---|---|---|---|---|---|---|---|---|---|---|---|---|---|---|---|---|
| 15 | −19 | 9 | 135 | 27 | 203 | 5 | 135 | 45 | −55 | 3 | 135 | 17 | 29 | 8 | 136 | 34 | 85 | 4 | 136 | 23 | 11 | 6 | 138 | 46 | −21 | 3 | 138 |
| 10 | 27 | 14 | 140 | 14 | −3 | 10 | 140 | 20 | −3 | 7 | 140 | 28 | −95 | 5 | 140 | 35 | 53 | 4 | 140 | 47 | 5 | 3 | 141 | 11 | −19 | 13 | 143 |
| 9 | −3 | 16 | 144 | 12 | −3 | 12 | 144 | 16 | −15 | 9 | 144 | 18 | −11 | 8 | 144 | 24 | 75 | 6 | 144 | 36 | 117 | 4 | 144 | 48 | 75 | 3 | 144 |
| 29 | 39 | 5 | 145 | 21 | −21 | 7 | 147 | 49 | −139 | 3 | 147 | 37 | 29 | 4 | 148 | 15 | 3 | 10 | 150 | 25 | 35 | 6 | 150 | 30 | 7 | 5 | 150 |
| 50 | −51 | 3 | 150 | 19 | −19 | 8 | 152 | 38 | 13 | 4 | 152 | 17 | −13 | 9 | 153 | 51 | 65 | 3 | 153 | 14 | −15 | 11 | 154 | 22 | −57 | 7 | 154 |
| 31 | 413 | 5 | 155 | 12 | −39 | 13 | 156 | 13 | 29 | 12 | 156 | 26 | −45 | 6 | 156 | 39 | −19 | 4 | 156 | 52 | 21 | 3 | 156 | 53 | 41 | 3 | 159 |
| 10 | −3 | 16 | 160 | 16 | −165 | 10 | 160 | 20 | −3 | 8 | 160 | 32 | −5 | 5 | 160 | 40 | 141 | 4 | 160 | 23 | 11 | 7 | 161 | 9 | 11 | 18 | 162 |
| 18 | 3 | 9 | 162 | 27 | 53 | 6 | 162 | 54 | −33 | 3 | 162 | 41 | −75 | 4 | 164 | 15 | 35 | 11 | 165 | 33 | 29 | 5 | 165 | 55 | 11 | 3 | 165 |
| 14 | −3 | 12 | 168 | 21 | −19 | 8 | 168 | 24 | −63 | 7 | 168 | 28 | 3 | 6 | 168 | 42 | −11 | 4 | 168 | 56 | −57 | 3 | 168 | 13 | −1 | 13 | 169 |
| 10 | −3 | 17 | 170 | 17 | −61 | 10 | 170 | 34 | −113 | 5 | 170 | 19 | −19 | 9 | 171 | 57 | −13 | 3 | 171 | 43 | 29 | 4 | 172 | 29 | −43 | 6 | 174 |
| 58 | −63 | 3 | 174 | 7 | 3 | 25 | 175 | 25 | 41 | 7 | 175 | 35 | 53 | 5 | 175 | 11 | 5 | 16 | 176 | 22 | −3 | 8 | 176 | 44 | 21 | 4 | 176 |
| 59 | −55 | 3 | 177 | 12 | 15 | 15 | 180 | 15 | −19 | 12 | 180 | 18 | −93 | 10 | 180 | 20 | −3 | 9 | 180 | 30 | 3 | 6 | 180 | 36 | −5 | 5 | 180 |
| 45 | −139 | 4 | 180 | 60 | 33 | 3 | 180 | 7 | 3 | 26 | 182 | 13 | 27 | 14 | 182 | 14 | −3 | 13 | 182 | 26 | −45 | 7 | 182 | 61 | −31 | 3 | 183 |
| 23 | −27 | 8 | 184 | 46 | 165 | 4 | 184 | 37 | 9 | 5 | 185 | 31 | 11 | 6 | 186 | 62 | −57 | 3 | 186 | 17 | −61 | 11 | 187 | 47 | 5 | 4 | 188 |
| 21 | −19 | 9 | 189 | 27 | 203 | 7 | 189 | 63 | −25 | 3 | 189 | 19 | −27 | 10 | 190 | 38 | 7 | 5 | 190 | 12 | −3 | 16 | 192 | 16 | −243 | 12 | 192 |
| 24 | −3 | 8 | 192 | 32 | −387 | 6 | 192 | 48 | 21 | 4 | 192 | 64 | −189 | 3 | 192 | 13 | 29 | 15 | 195 | 15 | 71 | 13 | 195 | 39 | 23 | 5 | 195 |
| 14 | −3 | 14 | 196 | 28 | −57 | 7 | 196 | 49 | 69 | 4 | 196 | 11 | 5 | 18 | 198 | 18 | 9 | 11 | 198 | 22 | −3 | 9 | 198 | 33 | 29 | 6 | 198 |
| 10 | −3 | 20 | 200 | 20 | −5 | 10 | 200 | 25 | 69 | 8 | 200 | 40 | 15 | 5 | 200 | 50 | −27 | 4 | 200 | 29 | −3 | 7 | 203 | 17 | 29 | 12 | 204 |
| 34 | −165 | 6 | 204 | 51 | 21 | 4 | 204 | 41 | −21 | 5 | 205 | 23 | 11 | 9 | 207 | 13 | 29 | 16 | 208 | 16 | −15 | 13 | 208 | 26 | −27 | 8 | 208 |
| 52 | 21 | 4 | 208 | 11 | 5 | 19 | 209 | 19 | −27 | 11 | 209 | 10 | −15 | 21 | 210 | 14 | −3 | 15 | 210 | 15 | 21 | 14 | 210 | 21 | −21 | 10 | 210 |
| 30 | 7 | 7 | 210 | 35 | 53 | 6 | 210 | 42 | −33 | 5 | 210 | 53 | 5 | 4 | 212 | 43 | −67 | 5 | 215 | 8 | −15 | 27 | 216 | 12 | 3 | 18 | 216 |
| 18 | 117 | 12 | 216 | 24 | −33 | 9 | 216 | 27 | 29 | 8 | 216 | 36 | 117 | 6 | 216 | 54 | −131 | 4 | 216 | 31 | −85 | 7 | 217 | 20 | 33 | 11 | 220 |
| 22 | 67 | 10 | 220 | 44 | 55 | 5 | 220 | 55 | −67 | 4 | 220 | 13 | −31 | 17 | 221 | 17 | −31 | 13 | 221 | 37 | 269 | 6 | 222 | 14 | −3 | 16 | 224 |
| 16 | −155 | 14 | 224 | 28 | 37 | 8 | 224 | 32 | −17 | 7 | 224 | 56 | −27 | 4 | 224 | 9 | 9 | 25 | 225 | 25 | 35 | 9 | 225 | 45 | 59 | 5 | 225 |
| 19 | −19 | 12 | 228 | 38 | −45 | 6 | 228 | 57 | 141 | 4 | 228 | 10 | −11 | 23 | 230 | 23 | −27 | 10 | 230 | 46 | 127 | 5 | 230 | 21 | −111 | 11 | 231 |
| 33 | 35 | 7 | 231 | 29 | −3 | 8 | 232 | 58 | −27 | 4 | 232 | 13 | −13 | 18 | 234 | 18 | −11 | 13 | 234 | 26 | 15 | 9 | 234 | 39 | −91 | 6 | 234 |
| 47 | −127 | 5 | 235 | 59 | −99 | 4 | 236 | 17 | −115 | 14 | 238 | 34 | −113 | 7 | 238 | 15 | −19 | 16 | 240 | 16 | −15 | 15 | 240 | 20 | −3 | 12 | 240 |
| 24 | 75 | 10 | 240 | 30 | −35 | 8 | 240 | 40 | 141 | 6 | 240 | 48 | −165 | 5 | 240 | 60 | −107 | 4 | 240 | 11 | 21 | 22 | 242 | 22 | 85 | 11 | 242 |
| 9 | 11 | 27 | 243 | 27 | 53 | 9 | 243 | 61 | 21 | 4 | 244 | 35 | −31 | 7 | 245 | 49 | 69 | 5 | 245 | 41 | −133 | 6 | 246 | 13 | 17 | 19 | 247 |
| 19 | 81 | 13 | 247 | 31 | −19 | 8 | 248 | 62 | −171 | 4 | 248 | 10 | −3 | 25 | 250 | 25 | −61 | 10 | 250 | 50 | −113 | 5 | 250 | 12 | 63 | 21 | 252 |
| 14 | −3 | 18 | 252 | 18 | −35 | 14 | 252 | 21 | −19 | 12 | 252 | 28 | 3 | 9 | 252 | 36 | 175 | 7 | 252 | 42 | 75 | 6 | 252 | 63 | 29 | 4 | 252 |
| 23 | 15 | 11 | 253 | 17 | −61 | 15 | 255 | 51 | −237 | 5 | 255 | 16 | −99 | 16 | 256 | 32 | −99 | 8 | 256 | 64 | −59 | 4 | 256 | 43 | −691 | 6 | 258 |
| 37 | 41 | 7 | 259 | 13 | 29 | 20 | 260 | 20 | 57 | 13 | 260 | 26 | 117 | 10 | 260 | 52 | 55 | 5 | 260 | 9 | 11 | 29 | 261 | 29 | −43 | 9 | 261 |
| 11 | 5 | 24 | 264 | 12 | −3 | 22 | 264 | 22 | −3 | 12 | 264 | 24 | 73 | 11 | 264 | 33 | 29 | 8 | 264 | 44 | 21 | 6 | 264 | 53 | −111 | 5 | 265 |
| 14 | 33 | 19 | 266 | 19 | −85 | 14 | 266 | 38 | −45 | 7 | 266 | 10 | 9 | 27 | 270 | 15 | −19 | 18 | 270 | 18 | 87 | 15 | 270 | 27 | 203 | 10 | 270 |
| 30 | 3 | 9 | 270 | 45 | −139 | 6 | 270 | 54 | −33 | 5 | 270 | 16 | −17 | 17 | 272 | 17 | 29 | 16 | 272 | 34 | 85 | 8 | 272 | 13 | 41 | 21 | 273 |
| 21 | −69 | 13 | 273 | 39 | −7 | 7 | 273 | 25 | 35 | 11 | 275 | 55 | 3 | 5 | 275 | 12 | −47 | 23 | 276 | 23 | 29 | 12 | 276 | 46 | −21 | 6 | 276 |
| 31 | 11 | 9 | 279 | 14 | −3 | 20 | 280 | 20 | −3 | 14 | 280 | 28 | −125 | 10 | 280 | 35 | 53 | 8 | 280 | 40 | 27 | 7 | 280 | 56 | 175 | 5 | 280 |
| 47 | 5 | 6 | 282 | 15 | −49 | 19 | 285 | 19 | −67 | 15 | 285 | 57 | −111 | 5 | 285 | 11 | −19 | 26 | 286 | 22 | −87 | 13 | 286 | 26 | 69 | 11 | 286 |
| 41 | −31 | 7 | 287 | 9 | −3 | 32 | 288 | 12 | −3 | 24 | 288 | 16 | −165 | 18 | 288 | 18 | −11 | 16 | 288 | 24 | 117 | 12 | 288 | 32 | −153 | 9 | 288 |
| 36 | 117 | 8 | 288 | 48 | 75 | 6 | 288 | 17 | −1 | 17 | 289 | 29 | 149 | 10 | 290 | 58 | −63 | 5 | 290 | 14 | −3 | 21 | 294 | 21 | −21 | 14 | 294 |
| 42 | −161 | 7 | 294 | 49 | −139 | 6 | 294 | 59 | 273 | 5 | 295 | 37 | 29 | 8 | 296 | 11 | 5 | 27 | 297 | 27 | −39 | 11 | 297 | 33 | 17 | 9 | 297 |
| 23 | 293 | 13 | 299 | 12 | −5 | 25 | 300 | 20 | 435 | 15 | 300 | 25 | 77 | 12 | 300 | 30 | −83 | 10 | 300 | 50 | −51 | 6 | 300 | 60 | 105 | 5 | 300 |

§ 11.3 Optimal extension fields

Accordingly, the use of the Frobenius speeds up an exponentiation to a generic power $n$. Indeed, a traditional approach would require $d \lg p$ squarings to get $\alpha^n$, but writing $n$ in basis $p$, i.e., $n = (n_{\ell-1} \ldots n_0)_p$, it is clear that

$$\alpha^n = \prod_{i=0}^{\ell-1} \phi_p^i(\alpha^{n_i}).$$

Combined with a right-to-left strategy to compute the $\alpha^{n_i}$'s, cf. Section 9.1.1, this idea shows that only $(\lg p - 1)$ squarings are needed, namely $\alpha^2, \alpha^4, \ldots, \alpha^{2^{\lg p - 1}}$. In addition, each term $\alpha^{n_i}$ can be computed in parallel.

**Example 11.65** Let $n = 27071851865689547117393862889$ and let us compute $\alpha^n$. First remark that $n = (22388\ 12209\ 20770\ 63238\ 8078\ 10838)_p$. Using the precomputed values $\alpha^2, \alpha^4, \ldots, \alpha^{2^{15}}$ Algorithm 9.2 gives

$$\begin{aligned}
\alpha^{n_0} &= 13812X^5 + 61164X^4 + 49159X^3 + 1927X^2 + 1781X + 31944 \\
\alpha^{n_1} &= 4807X^5 + 57203X^4 + 62178X^3 + 3283X^2 + 4690X + 33266 \\
\alpha^{n_2} &= 49155X^5 + 5527X^4 + 47396X^3 + 13274X^2 + 13828X + 60304 \\
\alpha^{n_3} &= 21607X^5 + 11848X^4 + 23310X^3 + 30303X^2 + 31752X + 44845 \\
\alpha^{n_4} &= 29730X^5 + 12285X^4 + 27469X^3 + 798X^2 + 9947X + 47295 \\
\alpha^{n_5} &= 54710X^5 + 18029X^4 + 18950X^3 + 23518X^2 + 10120X + 34955
\end{aligned}$$

and with the technique explained above one obtains

$$\begin{aligned}
\alpha^{n_0} &= 13812X^5 + 61164X^4 + 49159X^3 + 1927X^2 + 1781X + 31944 \\
\alpha^{pn_1} &= 60618X^5 + 51323X^4 + 3361X^3 + 4138X^2 + 57415X + 33266 \\
\alpha^{p^2 n_2} &= 8288X^5 + 43479X^4 + 47396X^3 + 53960X^2 + 58194X + 60304 \\
\alpha^{p^3 n_3} &= 43932X^5 + 11848X^4 + 42229X^3 + 30303X^2 + 33787X + 44845 \\
\alpha^{p^4 n_4} &= 46496X^5 + 26171X^4 + 27469X^3 + 28535X^2 + 55771X + 47295 \\
\alpha^{p^5 n_5} &= 37148X^5 + 9908X^4 + 46589X^3 + 49290X^2 + 55179X + 34955
\end{aligned}$$

so that the product of all these values is

$$\alpha^n = 42336X^5 + 42804X^4 + 21557X^3 + 49577X^2 + 22038X + 4278.$$

## 11.3.4 Inversion

Although an inverse could be computed with an extended gcd computation in $\mathbb{F}_{p^d}$, it is much faster to use the Frobenius action and an inversion in $\mathbb{F}_p$ to compute it.

Namely, take

$$r = \frac{p^d - 1}{p - 1} = p^{d-1} + p^{d-2} + \cdots + p + 1.$$

Then $\alpha^{r-1}$ and $\alpha^r$ are easily obtained using the Frobenius and in addition $\alpha^r \in \mathbb{F}_p$ since it is the norm of $\alpha$. So $\alpha^r$ can be easily inverted to obtain

$$\alpha^{-1} = \alpha^{r-1} \times \alpha^{-r}.$$

Further improvements, reminiscent of an addition chain approach, can be applied to compute the term $\alpha^{r-1}$. As an example, let us consider the extension degree $d = 6$, often used in practice with 32-bit architectures. In this case, the successive steps to compute $\alpha^{r-1}$ are

$$\alpha^p, \alpha^{p+1}, \alpha^{p^3+p^2}, \alpha^{p^5+p^4}, \alpha^{p^5+p^4+p^3+p^2} \text{ and } \alpha^{p^5+p^4+p^3+p^2+p}.$$

The entire algorithm is as follows.

**Algorithm 11.66** OEF inversion

INPUT: A nonzero element $\alpha \in \mathbb{F}_{p^d}$.
OUTPUT: The inverse of $\alpha$ in $\mathbb{F}_{p^d}$.

1. $r \leftarrow (p^d - 1)/(p - 1)$
2. $s \leftarrow \alpha^{r-1}$                                           [use an addition-chain-like approach]
3. $t \leftarrow s\alpha$                                                      $[t = \alpha^r \in \mathbb{F}_p]$
4. $u \leftarrow t^{-1}$                                            [compute the inverse of $t$ in $\mathbb{F}_p$]
5. **return** $su$

**Remarks 11.67**

(i) Algorithm 11.66 is in fact a generalization of a method proposed by Itoh and Tsujii [ITTS 1988] for characteristic 2 fields. See also Remark 11.48 (iv).
(ii) Since $t$ belongs to $\mathbb{F}_p$ it is equal to the constant coefficient of the product $s\alpha$. Thus this multiplication needs only $d$ multiplications in $\mathbb{F}_p$ as does the product $su$.
(iii) Let $\nu(k)$ be the Hamming weight of $k$, then $\alpha^{r-1}$ can be computed [HAME$^+$ 2003] with $N_M = \lfloor \lg d - 1 \rfloor + \nu(d-1) - 1$ products in $\mathbb{F}_{p^d}$ and at most $N_{\phi_p}$ Frobenius computations where

$$N_{\phi_p} = \begin{cases} N_M + 1 & \text{if } d \text{ is odd,} \\ \lfloor \lg(d-1) \rfloor + \nu(d) & \text{otherwise.} \end{cases}$$

**Example 11.68** Take $p$, $\mathbb{F}_{p^6}$ and $\alpha$ as defined in Example 11.64 and let us compute the inverse of $\alpha$ by Algorithm 11.66. First $r = p^5 + p^4 + p^3 + p^2 + p + 1$ and $\alpha^{r-1}$ is obtained by computing successively

$$\begin{aligned}
\alpha^p &= 23814X^5 + 34492X^4 + 10602X^3 + 7340X^2 + 40911X + 63559 \\
\alpha^{p+1} &= 27871X^5 + 42246X^4 + 20450X^3 + 8624X^2 + 26549X + 28414 \\
\alpha^{p^3+p^2} &= 47216X^5 + 11126X^4 + 20450X^3 + 50936X^2 + 29251X + 28414 \\
\alpha^{p^5+p^4} &= 55991X^5 + 12167X^4 + 20450X^3 + 5979X^2 + 9739X + 28414 \\
\alpha^{p^5+p^4+p^3+p^2} &= 26086X^5 + 2404X^4 + 35019X^3 + 45382X^2 + 45825X + 22132 \\
\alpha^{r-1} &= 28310X^5 + 14778X^4 + 7889X^3 + 29498X^2 + 2991X + 44851
\end{aligned}$$

with 3 multiplications in $\mathbb{F}_{p^6}$ and 3 applications of $\phi_p$ or $\phi_p^2$. Finally, $t = \alpha^{r-1}\alpha \equiv 42318 \pmod{p}$, $u = t^{-1} \equiv 27541 \pmod{p}$ and

$$\alpha^{-1} = \alpha^{r-1}u = 33766X^5 + 3708X^4 + 9164X^3 + 48513X^2 + 58147X + 27858.$$

### 11.3.5 Squares and square roots

For any extension field $\mathbb{F}_{p^d}$ of odd characteristic, and not only for OEFs, there is a simple method relying on Theorem 2.104 and very similar to Algorithm 11.19 to decide if an element of $\mathbb{F}_{p^d}$ is a square or not.

## § 11.3 Optimal extension fields

---

**Algorithm 11.69** Legendre–Kronecker–Jacobi symbol

INPUT: A polynomial $f(X) \in \mathbb{F}_p[X]$ and an irreducible polynomial $m(X) \in \mathbb{F}_p[X]$.
OUTPUT: The Legendre–Kronecker–Jacobi symbol $\left(\frac{f(X)}{m(X)}\right)$.

1. $k \leftarrow 1$
2. **repeat**
3.     **if** $f(X) = 0$ **then return** $0$
4.     $a \leftarrow$ the leading coefficient of $f(X)$
5.     $f(X) \leftarrow f(X)/a$
6.     **if** $\deg m \equiv 1 \pmod 2$ **then** $k \leftarrow k\left(\frac{a}{p}\right)$
7.     **if** $p^{\deg m} \equiv 3 \pmod 4$ **and** $\deg m \deg f \equiv 1 \pmod 2$ **then** $k \leftarrow -k$
8.     $r(X) \leftarrow f(X)$, $f(X) \leftarrow m(X) \bmod r(X)$ and $m(X) \leftarrow r(X)$
9. **until** $\deg m = 0$
10. **return** $k$

---

**Remark 11.70** Algorithm 11.69 relies on the law (2.7). Since $f(X)$ is not necessarily monic it is first divided by its leading coefficient $a$. Now we remark that $a \in \mathbb{F}_p$ is always a square in an extension of even degree. When the degree is odd $a$ is a quadratic residue if and only if $a = 0$ or $\left(\frac{a}{p}\right) = 1$.

**Example 11.71** Take $p = 7$, let $m(X)$ be the irreducible polynomial $X^9 + 2X^8 + X^7 + 2X^6 + 2X^5 + 4X^2 + 6X + 6$ and $f(X) = X^6 + X^5 + 6X^4 + 2X^3 + 2X^2 + 4X + 1$, both being elements of $\mathbb{F}_7[X]$.

After Line 8 the values of $r$, $f$, $a$ and $k$ are as follows

| $r$ | $f$ | $a$ | $k$ |
|---|---|---|---|
| $X^6 + 4X^5 + 3X^4 + X^3 + X^2 + 2X + 4$ | $4X^5 + 3X^4 + 4X^3 + 3X^2 + 2X + 4$ | 1 | 1 |
| $X^5 + 6X^4 + X^3 + 6X^2 + 4X + 1$ | $4X^3 + 2X^2 + 2X + 6$ | 4 | 1 |
| $X^3 + 4X^2 + 4X + 5$ | $2X^2 + 3X$ | 4 | $-1$ |
| $X^2 + 5X$ | $2X + 5$ | 2 | $-1$ |
| $X + 6$ | $6$ | 2 | $-1$ |
| $1$ | $0$ | 6 | 1 |

So $f(X)$ is a square modulo $m(X)$. Using a trivial generalization of Algorithm 11.26, one finds that $(3X^3 + 6X^2 + 2X + 1)^2 \equiv f(X) \pmod{m(X)}$.

To conclude this part, let us remark that the computation of the trace of an element in an OEF enjoys the same kind of improvements as in characteristic 2, cf. Remark 11.57 (ii).

### 11.3.6 Specific improvements for degrees $3$ and $5$

For some applications, like the implementation of trace zero varieties, cf. Section 15.3, one needs to work in an extension field $\mathbb{F}_{p^d}$ of small degree $d$. In particular, $d = 3$ and $d = 5$ are interesting there. Some specific tricks can be used to make multiplication and inversion more efficient.

We use an explicit description of the field extension as $\mathbb{F}_{p^d} = \mathbb{F}_p[\theta]$, where $\theta$ is a root of an irreducible binomial $X^d - \omega$. Since $d = 3$ or 5 is prime, $X^d - \omega$ is irreducible whenever $p \equiv 1 \pmod{d}$ and $\omega$ is not a $d$-th power in $\mathbb{F}_p$. As $\mathbb{F}_p$ contains a $d$-th root of unity $\zeta$, the roots of $X^d - \omega$ are $\theta, \zeta\theta, \ldots, \zeta^{d-1}\theta$.

**Remark 11.72** It is very likely that there exists $\omega$ of small integer value, which is not a $d$-th power. In fact, by Čebotarev's density theorem, we have with probability $1/d$ that $p \equiv 1 \pmod{d}$ and $X^d - 2$ is irreducible over $\mathbb{F}_p$. With even larger probability, one can find some small $\omega$ such that $X^d - \omega$ is irreducible over $\mathbb{F}_p$ and the multiplication by $\omega$, i.e., the reduction modulo the irreducible binomial, can be computed by additions only.

We shall write all elements of $\mathbb{F}_{p^3}$, respectively $\mathbb{F}_{p^5}$, as polynomials in $\theta$ of degrees at most 2, respectively 4, over $\mathbb{F}_p$. Addition, subtraction, and negation of elements of $\mathbb{F}_{p^d}$ are performed component-wise. If $\omega$ is small we can ignore the costs of reducing modulo $X^d - \omega$.

### 11.3.6.a Multiplication and squaring

Multiplication of elements of $\mathbb{F}_{p^d}$ is split into multiplication of the corresponding polynomials in $\theta$ and then reduction of the result using the fact that $\theta^d = \omega$.

**Multiplication for $d = 3$**

Multiplication in degree 3 extensions is done using Karatsuba's method, which we detail here to have the exact operation count. Let us multiply $\alpha = \sum_{i=0}^{2} a_i \theta^i$ with $\beta = \sum_{i=0}^{2} b_i \theta^i$. We have

$$\begin{aligned}\alpha\beta &= a_0 b_0 + \big((a_0 + a_1)(b_0 + b_1) - a_0 b_0 - a_1 b_1\big)\theta \\ &\quad + \big((a_0 + a_2)(b_0 + b_2) - a_0 b_0 - a_2 b_2 + a_1 b_1\big)\theta^2 \\ &\quad + \big((a_1 + a_2)(b_1 + b_2) - a_1 b_1 - a_2 b_2\big)\theta^3 + (a_2 b_2)\theta^4.\end{aligned} \quad (11.6)$$

It enables us to multiply two degree 2 polynomials by 6 multiplications. By delaying all modular reductions and using incomplete reduction (see [AVMI 2004]), we need to perform 3 modular reductions modulo $p$.

**Multiplication for $d = 5$**

In the degree 5 extension case, to multiply $\alpha = \sum_{i=0}^{4} a_i \theta^i$ with $\beta = \sum_{i=0}^{4} b_i \theta^i$ we put $\xi = \theta^3$ and let $A_0 = \sum_{i=0}^{2} a_i \theta^i$, $A_1 = a_3 + a_4\theta$, $B_0 = \sum_{i=0}^{2} b_i \theta^i$, and $B_1 = b_3 + b_4\theta$. Karatsuba's method is then used to obtain

$$\begin{aligned}\alpha\beta &= (A_0 + A_1\xi)(B_0 + B_1\xi) \\ &= A_0 B_0 + \big((A_0 + A_1)(B_0 + B_1) - A_0 B_0 - A_1 B_1\big)\xi + A_1 B_1 \xi^2.\end{aligned}$$

The product $A_1 B_1$ is computed using Karatsuba's method, and $A_0 B_0$ and $(A_0 + A_1)(B_0 + B_1)$ are both computed using (11.6). Note, however, that having $A_0$, $B_0$ of degree 2 and $A_1$, $B_1$ of degree 1, the coefficients of $\theta^4$ in $A_0 B_0$ and $(A_0 + A_1)(B_0 + B_1)$ are the same, so we can save one $\mathbb{F}_p$-multiplication. The amount of $\mathbb{F}_p$-multiplications needed to multiply two degree 4 polynomials is thus $3 + 2 \times 6 - 1 = 14$. By delaying all modular reductions and using incomplete reduction, we need to compute just 5 modular reductions modulo $p$.

**Squaring**

The squarings are more efficiently carried out using the schoolbook method, since this reduces the number of additions significantly and in several libraries, squarings in $\mathbb{F}_p$ are no cheaper than ordinary multiplications. If this is not the case one should implement both versions (the schoolbook version and the Karatsuba one) and compare their running time.

For $d = 3$, we need 3 squarings and 3 multiplications in $\mathbb{F}_p$ by

$$(a_0 + a_1\theta + a_2\theta^2)^2 = a_0^2 + \omega a_1 a_2 + (\omega a_2^2 + a_0 a_1)\theta + (a_1^2 + a_0 a_2)\theta^2.$$

Likewise, for $d = 5$ we have 5 squarings and 10 multiplications. The number of modular reductions are again 3 and 5 as in the case of multiplications.

### 11.3.6.b  Inversion

For the inversion, the difference from the general OEF approach becomes obvious. To compute the inverse of $\alpha \in \mathbb{F}_{p^d}$, we can consider the multiplication as a linear map and determine a preimage. This method is faster for $d = 3$ and also for $d = 2$ but we do not investigate that case any further. Note that for $d = 5$, the general method made explicit is faster.

### Inversion for $d = 3$

Let $\beta = b_0 + b_1\theta + b_2\theta^2$, with $b_0, b_1, b_2 \in \mathbb{F}_p$, be the inverse of $\alpha = a_0 + a_1\theta + a_2\theta^2 \in \mathbb{F}_{p^3}$ and using $\theta^3 = \omega$, the relation $\alpha\beta = 1$ can be written as:

$$\begin{bmatrix} a_0 & a_2\omega & a_1\omega \\ a_1 & a_0 & a_2\omega \\ a_2 & a_1 & a_0 \end{bmatrix} \begin{bmatrix} b_0 \\ b_1 \\ b_2 \end{bmatrix} = \begin{bmatrix} 1 \\ 0 \\ 0 \end{bmatrix}.$$

Hence

$$\begin{bmatrix} b_0 \\ b_1 \\ b_2 \end{bmatrix} = \begin{bmatrix} a_0 & a_2\omega & a_1\omega \\ a_1 & a_0 & a_2\omega \\ a_2 & a_1 & a_0 \end{bmatrix}^{-1} \begin{bmatrix} 1 \\ 0 \\ 0 \end{bmatrix} = (a_0^3 + \omega a_1^3 + \omega^2 a_2^3 - 3\omega a_0 a_1 a_2)^{-1} \begin{bmatrix} a_0^2 - \omega a_1 a_2 \\ \omega a_2^2 - a_0 a_1 \\ a_1^2 - a_0 a_2 \end{bmatrix}.$$

From this formula we obtain a method for inverting elements in $\mathbb{F}_{p^3}$, which requires (ignoring multiplications by 3 and by $\omega$) just one inversion, 3 squarings, and 9 multiplications in $\mathbb{F}_p$.

This method can be generalized to very small extensions. It is described e.g., in [KOMO$^+$ 1999].

### Inversion for $d = 5$

In this case we use the inversion technique described in Algorithm 11.66 but can save a bit by combining the powers of the Frobenius automorphism, i.e., making the addition chain explicit.

Thus we compute the inversion in $\mathbb{F}_{p^5}$ as:

$$\alpha^{-1} = \frac{(\alpha^p \alpha^{p^2}(\alpha^p \alpha^{p^2})^{p^2}}{\alpha(\alpha^p \alpha^{p^2}(\alpha^p \alpha^{p^2})^{p^2})}.$$

Note that if the result of a $\mathbb{F}_{p^5}$-multiplication is known in advance to be in $\mathbb{F}_p$ (such as the norm), its computation requires just 5 $\mathbb{F}_p$-multiplications. This way we compute inverses in $\mathbb{F}_{p^5}$ by one inversion and 50 multiplications in $\mathbb{F}_p$. This strategy is optimal for $d = 5$ and needs less multiplications than the generalization of the linear algebra approach used for $d = 3$.

# Chapter 12

## Arithmetic of $p$-adic Numbers

### Frederik Vercauteren

**Contents in Brief**

| | | |
|---|---|---|
| **12.1** | **Representation** | 239 |
| | Introduction • Computing the Teichmuller modulus | |
| **12.2** | **Modular arithmetic** | 244 |
| | Modular multiplication • Fast division with remainder | |
| **12.3** | **Newton lifting** | 246 |
| | Inverse • Inverse square root • Square root | |
| **12.4** | **Hensel lifting** | 249 |
| **12.5** | **Frobenius substitution** | 250 |
| | Sparse modulus • Teichmuller modulus • Gaussian normal basis | |
| **12.6** | **Artin–Schreier equations** | 252 |
| | Lercier–Lubicz algorithm • Harley's algorithm | |
| **12.7** | **Generalized Newton lifting** | 256 |
| **12.8** | **Applications** | 257 |
| | Teichmuller lift • Logarithm • Exponential • Trace • Norm | |

This chapter presents efficient algorithms to compute in the ring $\mathbb{Z}_p$ of $p$-adic integers and the valuation ring $\mathbb{Z}_q$ of an unramified extension of $\mathbb{Q}_p$. Many of these algorithms were developed specifically for the $p$-adic point counting algorithms described in Chapter 17, but are by no means limited to this application. Although most of the lifting algorithms remain valid for more general $p$-adic fields, we restrict ourselves to $\mathbb{Z}_p$ and $\mathbb{Z}_q$ due to their importance for practical applications.

## 12.1 Representation

### 12.1.1 Introduction

Let $\mathbb{Q}_q$ be the unramified extension of $\mathbb{Q}_p$ of degree $d$, then Proposition 3.21 implies that $\mathbb{Q}_q$ can be represented as $\mathbb{Q}_p[X]/(M(X))$ with $M \in \mathbb{Z}_p[X]$ of degree $d$ such that $m := P_1(M) \in \mathbb{F}_p[X]$ is irreducible of degree $d$. Since $\deg m = \deg M$, the leading coefficient of $M$ is a unit in $\mathbb{Z}_p$, so without loss of generality we can assume that $M$ is monic.

Let $a \in \mathbb{Q}_q$ be represented by $\sum_{i=0}^{d-1} a_i X^i$ with $a_i \in \mathbb{Q}_p$, then the $p$-adic valuation $v_p$ is given by $v_p(a) = \min_{0 \leqslant i < d} v_p(a_i)$ and the $p$-adic norm by $|a|_p = p^{-v_p(a)}$. This implies that the valuation ring $\mathbb{Z}_q$ of $\mathbb{Q}_q$ can be represented as $\mathbb{Z}_p[X]/(M(X))$. Given a representation of the residue field $\mathbb{F}_q \simeq \mathbb{F}_p[X]/(m(X))$, there exist infinitely many polynomials $M \in \mathbb{Z}_p[X]$ with $P_1(M) = m$. To make reduction modulo $m$ very efficient, $m$ is usually chosen to be sparse.

In practice, one computes with $p$-adic integers up to some precision $N$, i.e., an element $a \in \mathbb{Z}_p$ is approximated by $p_N(a) \in \mathbb{Z}/p^N\mathbb{Z}$ and arithmetic reduces to integer arithmetic modulo $p^N$. For a given precision $N$, each element therefore takes $O(N \lg p)$ space.

This extends in a natural way to $\mathbb{Z}_q$: an element $a \in \mathbb{Z}_q$ represented as $\sum_{i=0}^{d-1} a_i X^i$ with $a_i \in \mathbb{Z}_p$ will be approximated by the polynomial $\sum_{i=0}^{d-1} p_N(a_i) X^i$. Computing in $\mathbb{Z}_q$ up to precision $N$ thus corresponds computing in $(\mathbb{Z}/p^N\mathbb{Z})[X]/(M_N(X))$, with $M_N$ the polynomial obtained by reducing the coefficients of $M$ modulo $p^N$, i.e., $M_N = P_N(M)$. The space required to represent such an element clearly is given by $O(dN \lg p)$.

Given a modulus $m$ defining $\mathbb{F}_q$, there are two common choices for the polynomial $M$ that speed up the arithmetic in $\mathbb{Z}_q$: the first choice preserves the sparse structure of $m$, whereas the second choice is especially suited in case an efficient Frobenius substitution is needed.

- *Sparse modulus representation*: Let $m(X) = \sum_{i=0}^{d} m_i X^i$ with $m_i \in \mathbb{F}_p$ and $m_d = 1$. To preserve the sparseness of $m$, a first natural choice is to define $M(X) = \sum_{i=0}^{d} M_i X^i$ with $M_i$ the unique integer between 0 and $p-1$ such that $M_i \equiv m_i \pmod{p}$. The reduction modulo $M$ of a polynomial of degree $\leq 2(d-1)$ then only takes $d(w-1)$ multiplications of a $\mathbb{Z}_p$-element by a small integer and $dw$ subtractions in $\mathbb{Z}_p$ where $w$ is the number of nonzero coefficients of $m$.

- *Teichmüller modulus representation*: Since $\mathbb{F}_q$ is the splitting field of the polynomial $X^q - X$, the polynomial $m$ divides $X^q - X$. To preserve the simple Galois action on $\mathbb{F}_q$, i.e., $p$-th powering, a second natural choice is to define $M$ as the unique polynomial over $\mathbb{Z}_p$ with $M(X) \mid X^q - X$ and $M(X) \equiv m(X) \pmod{p}$. Every root $\theta \in \mathbb{Z}_q$ of $M(X)$ clearly is a $(q-1)$-th root of unity, therefore also $\Sigma(\theta)$ is a $(q-1)$-th root of unity, with $\Sigma$ the Frobenius substitution of $\mathbb{Z}_q$. Since $\Sigma(\theta) \equiv \theta^p \pmod{p}$, we conclude that $\Sigma(\theta) = \theta^p$ or by abuse of notation $\Sigma(X) = X^p$.

If the finite field $\mathbb{F}_q$ admits a Gaussian normal basis, then Kim et al. [KIPA+ 2002] showed that this basis can be lifted to $\mathbb{Z}_q$.

**Proposition 12.1** Let $p$ be a prime and $d, t$ positive integers such that $dt + 1$ is a prime not equal to $p$. Let $\gamma$ be a primitive $(dt+1)$-th root of unity in some extension field of $\mathbb{Q}_p$. If $\gcd(dt/e, d) = 1$, with $e$ the order of $p$ modulo $dt + 1$, then for any primitive $t$-th root of unity $\tau$ in $\mathbb{Z}/(dt+1)\mathbb{Z}$

$$\theta = \sum_{i=0}^{t-1} \gamma^{\tau^i} \tag{12.1}$$

is a normal element and $[\mathbb{Q}_p(\theta) : \mathbb{Q}_p] = d$. Such a basis is called a *Gaussian normal basis of type t*.

Elements of $\mathbb{Z}_q$ are then represented in a redundant way by computing in the ring

$$\mathbb{Z}_p[X]/(X^{dt+1} - 1).$$

For a given precision $N$, each element therefore requires $O(dNt \lg p)$ space.

## 12.1.2 Computing the Teichmüller modulus

In this section we provide all the details of an algorithm first sketched by Harley in [HAR 2002b], to compute the Teichmüller modulus of the defining polynomial of a finite field. Let $\bar{\theta} \in \mathbb{F}_q$ be such

## § 12.1 Representation

that $\mathbb{F}_q = \mathbb{F}_p[\bar{\theta}]$ and let $\theta \in \mathbb{Z}_q$ be the Teichmüller lift of $\bar{\theta}$, i.e., the unique $(q-1)$-th root of unity that reduces to $\bar{\theta}$. In the previous section we assumed that $\mathbb{Z}_q$ was represented as $\mathbb{Z}_p[X]/(M(X))$ with $M$ the minimal polynomial of $\theta$. Since $\theta$ is a $(q-1)$-th root of unity, we have that $\Sigma(\theta) = \theta^p$ and $M(X) = \prod_{i=0}^{d-1}(X - \theta^{p^i})$. Let $\zeta_p$ be a formal $p$-th root of unity, then

$$M(X^p) = \prod_{i=0}^{p-1} M(\zeta_p^i X). \tag{12.2}$$

Indeed, for $i = 1, \ldots, d-1$ each factor $(X^p - \theta^{p^i})$ of $M(X^p)$ splits as

$$\left(X - \theta^{p^{i-1}}\right)\left(\zeta_p X - \theta^{p^{i-1}}\right) \cdots \left(\zeta_p^{p-1} X - \theta^{p^{i-1}}\right).$$

Write $M$ as $M(X) = \sum_{i=0}^{p-1} M_i(X^p) X^i$ with $M_i \in \mathbb{Z}_q[X]$, then (12.2) can be rewritten as

$$M(X^p) = \prod_{j=0}^{p-1} \left( \sum_{i=0}^{p-1} \zeta_p^{ij} M_i(X^p) X^i \right) = \sum_{k=0}^{p-1} h_k\bigl(M_0(X^p), \ldots, M_{p-1}(X^p)\bigr) X^{pk},$$

with $h_k \in \mathbb{Z}_q[Y_0, \ldots, Y_{p-1}]$ homogeneous polynomials of degree $p$. This implies that $M$ satisfies

$$M(X) = \sum_{k=0}^{p-1} h_k\bigl(M_0(X), \ldots, M_{p-1}(X)\bigr) X^k.$$

The following table lists the first few examples of the polynomials $h_k(Y_0, \ldots, Y_{p-1})$.

| $p$ | $h_k$ |
|---|---|
| 2 | $h_0 = Y_0^2$ |
|   | $h_1 = -Y_1^2$ |
| 3 | $h_0 = Y_0^3$ |
|   | $h_1 = Y_1^3 - 3 Y_0 Y_1 Y_2$ |
|   | $h_2 = Y_2^3$ |
| 5 | $h_0 = Y_0^5$ |
|   | $h_1 = Y_1^5 + 5(Y_0^2 Y_1^2 Y_3 - Y_0^3 Y_1 Y_4 - Y_0^3 Y_2 Y_3 + Y_0^2 Y_1 Y_2^2 - Y_0 Y_1^3 Y_2)$ |
|   | $h_2 = Y_2^5 + 5(Y_0^2 Y_2 Y_4^2 + Y_0^2 Y_3^2 Y_4 + Y_0 Y_1^2 Y_4^2 - Y_0 Y_1 Y_2 Y_3 Y_4 - Y_0 Y_1 Y_3^3$ |
|   | $\quad - Y_0 Y_2^3 Y_4 + Y_0 Y_2^2 Y_3^2 - Y_1^3 Y_3 Y_4 + Y_1^2 Y_2^2 Y_4 + Y_1^2 Y_2 Y_3^2 - Y_1 Y_2^3 Y_3)$ |
|   | $h_3 = Y_3^5 + 5(Y_1 Y_3^2 Y_4^2 - Y_0 Y_3 Y_4^3 - Y_1 Y_2 Y_4^3 + + Y_2^2 Y_3 Y_4^2 - Y_2 Y_3^3 Y_4)$ |
|   | $h_4 = Y_4^5$ |

Assume we know $M_t(X) \equiv M(X) \pmod{p^t}$ and let $\delta_t(X) = \bigl(M(X) - M_t(X)\bigr)/p^t$. Substituting $M_t(X) + p^t \delta_t(X)$ in the equation

$$M(X) - \sum_{k=0}^{p-1} h_k\bigl(M_0(X), \ldots, M_{p-1}(X)\bigr) X^k = 0$$

gives a relation that determines $\delta_t$ modulo $p^t$. For example, consider the case $p = 2$, which leads to

$$\delta_t(X) - 2\bigl(M_{t,0}(X)\delta_{t,0}(X) - X M_{t,1}(X)\delta_{t,1}(X)\bigr) + V_t(X) \equiv 0 \pmod{2^t}, \tag{12.3}$$

with $V_t(X) \equiv \bigl(M_t(X) - M_{t,0}(X)^2 + X M_{t,1}(X)^2\bigr)/2^t \pmod{2^t}$. Assume we have an algorithm to compute $\delta_t \bmod p^{t'}$ with $t' = \lceil t/2 \rceil$, then we can use the same algorithm to compute $\delta_t \bmod p^t$. Indeed, substituting $\delta_t = \delta_{t'} + p^{t'}\Delta_{t'}$ in (12.3) leads to a similar equation modulo $2^{t-t'}$ with $\delta_t$ replaced by $\Delta_{t'}$ and $V_t$ replaced by

$$\frac{V_t(X) + \delta_{t'}(X) - 2\bigl(M_{t,0}(X)\delta_{t',0}(X) - X M_{t,1}(X)\delta_{t',1}(X)\bigr)}{2^{t'}} \bmod 2^{t-t'}.$$

Since $t - t' \leqslant t'$ we can use the same algorithm to find $\Delta_{t'}$ and thus $\delta_t$. This immediately leads to Algorithms 12.2 and 12.3. Note that for odd extension degree $d$, Algorithm 12.2 returns $-M$, so in this case a final negation is necessary.

---

**Algorithm 12.2** Teichmüller modulus

INPUT: A monic irreducible polynomial $m \in \mathbb{F}_2[X]$ of degree $d$ and precision $N$.
OUTPUT: The Teichmüller lift $M \in \mathbb{Z}_2[X]$ up to precision $N$, i.e., $M(X) \mid X^{2^d} - X \bmod 2^N$ and $M(X) \equiv m(X) \pmod 2$.

---

1. **if** $N = 1$ **then**
2. $\quad M(X) \leftarrow m(X) \bmod 2$
3. **else**
4. $\quad N' \leftarrow \lceil \frac{N}{2} \rceil$
5. $\quad M'(X) \leftarrow$ Teichmüller modulus $(m, N')$
6. $\quad M_0(X^2) \leftarrow \bigl(M'(X) + M'(-X)\bigr)/2 \bmod 2^N$
7. $\quad M_1(X^2) \leftarrow \bigl(M'(X) - M'(-X)\bigr)/(2X) \bmod 2^N$
8. $\quad V(X) \leftarrow \bigl(M'(X) - M_0(X)^2 + X M_1(X)^2\bigr)/2^{N'} \bmod 2^{N-N'}$
9. $\quad \delta(X) \leftarrow$ Teichmüller modulus increment $(M_0, M_1, V, N - N')$
10. $\quad M(X) \leftarrow M'(X) + 2^{N'}\delta(X) \bmod 2^N$
11. **return** $M(X)$

---

Algorithm 12.2 relies on a procedure called the *Teichmüller modulus increment*, which returns $\delta$ in $\mathbb{Z}_2[X]$ such that

$$\delta(X) - 2\bigl(M_0(X)\delta_0(X) - X M_1(X)\delta_1(X)\bigr) + V(X) \equiv 0 \pmod{2^N}. \tag{12.4}$$

The algorithm is as follows.

---

**Algorithm 12.3** Teichmüller modulus increment

INPUT: Polynomials $M_0, M_1, V \in \mathbb{Z}_2[X]$ and a precision $N$.
OUTPUT: The polynomial $\delta \in \mathbb{Z}_2[X]$ as in (12.4)

---

1. **if** $N = 1$ **then**
2. $\quad \delta(X) \leftarrow -V(X) \bmod 2$
3. **else**
4. $\quad N' \leftarrow \lceil \frac{N}{2} \rceil$
5. $\quad \delta'(X) \leftarrow$ Teichmüller modulus increment $(M_0, M_1, V, N')$
6. $\quad \delta_0(X^2) \leftarrow \bigl(\delta'(X) + \delta'(-X)\bigr)/2 \bmod 2^N$

## § 12.1 Representation

7.      $\delta_1(X^2) \leftarrow \bigl(\delta'(X) - \delta'(-X)\bigr)/(2X) \bmod 2^N$
8.      $V'(X) \leftarrow \dfrac{V(X) + \delta'(X) - 2\bigl(M_0(X)\delta_0(X) - XM_1(X)\delta_1(X)\bigr)}{2^{N'}} \bmod 2^{N-N'}$
9.      $\Delta(X) \leftarrow$ Teichmüller modulus increment $(M_0, M_1, V', N - N')$
10.    $\delta(X) \leftarrow \delta'(X) + 2^{N'}\Delta(X) \bmod 2^N$
11.    **return** $\delta(X)$

---

The complexity of Algorithm 12.2 is determined by the $O(1)$ multiplications in Line 8 and the call to Algorithm 12.3 in Line 9. The complexity of the latter algorithm is determined by the recursive calls in Lines 5 and 9 and the $O(1)$ multiplications in Line 8. If $T(N)$ is the running time of Algorithm 12.3 for precision $N$, then we have

$$T(N) \leqslant 2T(\lceil N/2 \rceil) + cT_{d,N},$$

for some constant $c$ and $T_{d,N}$ the time to multiply two polynomials in $(\mathbb{Z}/p^N\mathbb{Z})[X]$ of degree less than $d$ assuming $p$ is fixed. The above relation implies by induction that the complexity of Algorithm 12.3 and thus also of Algorithm 12.2 is $O(T_{d,N} \lg N)$.

**Example 12.4** Let $\mathbb{F}_{2^8} \simeq \mathbb{F}_2[X]/(m(X))$ with $m(X) = X^8 + X^4 + X^3 + X^2 + 1$, then on input $(m, 10)$ Algorithm 12.2 computes the following intermediate results:

| | | |
|---|---|---|
| 1 | $M$ | $X^8 + X^4 + X^3 + X^2 + 1$ |
| 2 | $M_0$ | $X^4 + X^2 + X + 1$ |
| | $M_1$ | $X$ |
| | $V$ | $X^6 + X^5 + X^4 + X^2 + X$ |
| | $\delta$ | $X^6 + X^5 + X^4 + X^2 + X$ |
| | $M$ | $X^8 + 2X^6 + 2X^5 + 3X^4 + X^3 + 3X^2 + 2X + 1$ |
| 3 | $M_0$ | $X^4 + 2X^3 + 3X^2 + 3X + 1$ |
| | $M_1$ | $2X^2 + X + 2$ |
| | $V$ | $3X^7 + X^5 + X^3$ |
| | $\delta$ | $X^7 + X^5 + X^3$ |
| | $M$ | $X^8 + 4X^7 + 2X^6 + 6X^5 + 3X^4 + 5X^3 + 3X^2 + 2X + 1$ |
| 5 | $M_0$ | $X^4 + 2X^3 + 3X^2 + 3X + 1$ |
| | $M_1$ | $4X^3 + 6X^2 + 5X + 2$ |
| | $V$ | $2X^7 + X^6 + 3X^4 + X^2$ |
| | $\delta$ | $X^6 + X^4 + 2X^3 + 3X^2 + 2X$ |
| | $M$ | $X^8 + 4X^7 + 10X^6 + 6X^5 + 11X^4 + 21X^3 + 27X^2 + 18X + 1$ |
| 10 | $M_0$ | $X^4 + 10X^3 + 11X^2 + 27X + 1$ |
| | $M_1$ | $4X^3 + 6X^2 + 21X + 18$ |
| | $V$ | $30X^6 + 30X^5 + 24X^4 + 2X^3 + X^2 + 9X$ |
| | $\delta$ | $20X^7 + 26X^6 + 4X^5 + 16X^4 + 31X^2 + 17X$ |
| | $M$ | $X^8 + 644X^7 + 842X^6 + 134X^5 + 523X^4 + 21X^3 + 1019X^2 + 562X + 1$ |

## 12.2 Modular arithmetic

### 12.2.1 Modular multiplication

Let $a \in \mathbb{Z}_p$ and let $N$ be the precision we compute with, then $a$ is approximated by $p_N(a) \in \mathbb{Z}/p^N\mathbb{Z}$. Arithmetic in $\mathbb{Z}/p^N\mathbb{Z}$ is simply integer arithmetic modulo $p^N$ so all methods of Chapter 10 apply. Let $\mu$ be a constant such that multiplication of two $B$-bit integers requires $O(B^\mu)$ bit-operations, for example $\mu = 2$ for schoolbook multiplication, $\mu = \lg 3$ for Karatsuba multiplication [KAOF 1963], and $\mu = 1 + \varepsilon$, with $\varepsilon$ real positive, for Schönhage–Strassen multiplication [SCST 1971]. Since a modular reduction takes the same time as a multiplication, we conclude that modular multiplication up to precision $N$ requires $O(N^\mu)$ bit-operations for fixed $p$.

Let $\mathbb{Z}_q$ be represented as $\mathbb{Z}_p[X]/(M(X))$ with $M \in \mathbb{Z}_p[X]$ a monic, irreducible polynomial of degree $d$, then working up to precision $N$ corresponds to computing in

$$(\mathbb{Z}/p^N\mathbb{Z})[X]/(P_N(M(X))).$$

The modular multiplication proceeds in two steps: multiplication of two polynomials of degree less than $d$ in $(\mathbb{Z}/p^N\mathbb{Z})[X]$ and a modular reduction modulo $P_N(M)$. The former is well known and requires $T_{d,N}$ bit-operations for $p$ fixed, e.g., a combination of schoolbook multiplication in the $p$-adic dimension and Karatsuba multiplication in the polynomial dimension gives $T_{d,N} \in O(d^{\lg 3} N^2)$.

The reduction modulo $P_N(M)$ depends on the choice of representation of $\mathbb{Z}_q$. In the sparse modulus representation, the reduction modulo $P_N(M)$ requires at most $d(w-1)$ multiplications in $\mathbb{Z}/p^N\mathbb{Z}$ with $w$ the number of nonzero coefficients of $P_N(M)$. In the Teichmüller representation, the polynomial $P_N(M(X))$ is no longer sparse and fast reduction methods should be used. The next section shows that this requires $O(1)$ multiplications of polynomials of degree less than $d$. Therefore, we conclude that a modular multiplication in $\mathbb{Z}_q$ to precision $N$ requires $T_{d,N}$ bit-operations.

### 12.2.2 Fast division with remainder

Let $R$ be a commutative ring and $a, b \in R[X]$ polynomials of degree $k$ and $d$ respectively and $b$ monic. Since $b$ is monic, there exist unique polynomials $q, r \in R[X]$ with $a = qb + r$ and $\deg r < \deg b$. Evaluating both sides at $1/X$ and multiplying with $X^k$ gives

$$X^k a\left(\frac{1}{X}\right) = X^{k-d} q\left(\frac{1}{X}\right) X^d b\left(\frac{1}{X}\right) + X^{k-(d-1)} X^{d-1} r\left(\frac{1}{X}\right). \tag{12.5}$$

Note that for a polynomial $P \in R[X]$ of degree $n$, $X^n P(1/X)$ is the polynomial with the coefficients of $P$ reversed. Reducing the above equation modulo $X^{k-(d-1)}$ gives

$$X^k a\left(\frac{1}{X}\right) \equiv X^{k-d} q\left(\frac{1}{X}\right) X^d b\left(\frac{1}{X}\right) \pmod{X^{k-(d-1)}}. \tag{12.6}$$

Since $b$ is monic, $X^d b(1/X)$ has constant coefficient 1 and therefore is coprime to $X^{k-(d-1)}$. Let $c \in R[X]$ be the inverse of $X^d b(1/X)$ modulo $X^{k-(d-1)}$, i.e.,

$$c(X)\left(X^d b(1/X)\right) \equiv 1 \pmod{X^{k-(d-1)}}$$

## § 12.2 Modular arithmetic

then we can recover $q$ from

$$X^{k-d}q\left(\frac{1}{X}\right) \equiv X^k a\left(\frac{1}{X}\right) c(X) \pmod{X^{k-(d-1)}}. \tag{12.7}$$

The polynomial $c$ can be computed using a polynomial version of Algorithm 12.10, i.e., using a Newton iteration to find a "root" of the polynomial $fY - 1 = 0$.

---
**Algorithm 12.5** Polynomial inversion

INPUT: A polynomial $f \in R[X]$ with $f(0) = 1$ and a power $n \in \mathbb{N}$.
OUTPUT: The inverse of $f$ modulo $X^n$.

1. **if** $n = 1$ **then**
2.      $c \leftarrow 1$
3. **else**
4.      $n' \leftarrow \lceil \frac{n}{2} \rceil$
5.      $c \leftarrow$ Polynomial inversion $(f, n')$
6.      $c \leftarrow \bigl(c + c(1 - fc)\bigr) \bmod X^n$
7. **return** $c$

---

Since the degree of the polynomials doubles in each iteration, the complexity of Algorithm 12.5 is determined by the last iteration, which requires $O(1)$ multiplications of polynomials of degree $\leqslant n$ over $R$. Therefore, the complexity amounts to $O(n^\mu)$ operations in $R$.

Once $c$ has been computed, the quotient $q$ follows easily from (12.7) and the remainder is given by $r = a - bq \pmod{X^d}$. This is summarized in Algorithm 12.6.

---
**Algorithm 12.6** Fast division with remainder

INPUT: Polynomials $a, b \in R[X]$ with $b$ monic.
OUTPUT: Polynomials $q, r \in R[X]$ such that $a = qb + r$ and $\deg r < \deg b$.

1. **if** $\deg a < \deg b$ **then**
2.      $q \leftarrow 0$ and $r \leftarrow a$
3. **else**
4.      $n \leftarrow \deg a - \deg b + 1$
5.      $c \leftarrow$ Polynomial inversion $(X^{\deg b} b(1/X), n)$
6.      $\tilde{q} \leftarrow \bigl((X^{\deg a} a(1/X)) c(X)\bigr) \bmod X^n$
7.      $q \leftarrow X^{n-1} \tilde{q}(1/X)$
8.      $r \leftarrow (a - bq) \bmod X^{\deg b}$
9. **return** $q$ and $r$

---

The complexity of Algorithm 12.6 is determined by the call to Algorithm 12.5 in Line 5 and the multiplications in Lines 6 and 8. Since $n = \deg a - \deg b + 1$, we conclude that a fast reduction requires $O\bigl((\deg a - \deg b)^\mu + (\deg b)^\mu\bigr)$ operations in $R$.

**Example 12.7** Let $R = \mathbb{Z}/2^{10}\mathbb{Z}$, then Algorithm 12.6 computes the following intermediate results on input $(a, b)$ (note that $b$ is the modulus computed in Example 12.4):

| | |
|---|---|
| $a$ | $559X^{14} + 781X^{13} + 763X^{12} + 684X^{11} + 133X^{10} + 375X^9 + 922X^8 + 776X^7$ $+452X^6 + 214X^5 + 313X^4 + 148X^3 + 646X^2 + 428X + 168$ |
| $b$ | $X^8 + 644X^7 + 842X^6 + 134X^5 + 523X^4 + 21X^3 + 1019X^2 + 562X + 1$ |
| $c$ | $789X^6 + 747X^5 + 169X^4 + 906X^3 + 198X^2 + 380X + 1$ |
| $\tilde{q}$ | $337X^6 + 755X^5 + 768X^4 + 420X^3 + 673X^2 + 209X + 559$ |
| $q$ | $559X^6 + 209X^5 + 673X^4 + 420X^3 + 768X^2 + 755X + 337$ |
| $r$ | $428X^7 + 728X^6 + 240X^5 + 294X^4 + 10X^3 + 165X^2 + 743X + 855$ |

## 12.3 Newton lifting

Let $f \in \mathbb{Z}_q[X]$ and assume that $a \in \mathbb{Z}_q$ satisfies $v_p(f'(a)) = k$ and $v_p(f(a)) = n + k$ for some $n > k$. Proposition 3.16 then implies that there exists a unique root $b \in \mathbb{Z}_q$ of $f$ with $b \equiv a \pmod{p^n}$. The element $a$ is called an approximate root of $f$ known to precision $n$. Reformulating Proposition 3.16 leads to the following lemma that shows how to compute an approximate root to precision $2n - k$ starting from an approximate root to precision $n$.

**Lemma 12.8** Let $f \in \mathbb{Z}_q[X]$ and assume that $a \in \mathbb{Z}_q$ satisfies $v_p(f'(a)) = k$ and $v_p(f(a)) = n+k$ for some $n > k$. Let $b$ be the unique root of $f$ with $b \equiv a \pmod{p^n}$. Then

$$z = a - \frac{f(a)}{f'(a)} \tag{12.8}$$

satisfies $z \equiv b \pmod{p^{2n-k}}$, $f(z) \equiv 0 \pmod{p^{2n}}$ and $v_p(f'(z)) = k$.

Given an approximate root $a \in \mathbb{Z}_q$ to precision $n$, Algorithm 12.9 computes the unique approximate root $z$ to precision $N$ with $z \equiv a \pmod{n}$. Note that for such $z$, we have $f(z) \equiv 0 \pmod{p^{N+k}}$.

---

**Algorithm 12.9** Newton iteration

INPUT: The polynomial $f \in \mathbb{Z}_q[X]$, an approximate root $a \in \mathbb{Z}_q$, the integer $k$ with $v_p(f'(a)) = k$, precision $n > k$ such that $f(a) \equiv 0 \pmod{p^{n+k}}$ and precision $N$.
OUTPUT: An approximate root $z \in \mathbb{Z}_q$ of $f$ with $z \equiv a \pmod{p^n}$ and $f(z) \equiv 0 \pmod{p^{N+k}}$.

1. **if** $N \leqslant n$ **then**
2. $\quad z \leftarrow a$
3. **else**
4. $\quad N' \leftarrow \lceil \frac{N+k}{2} \rceil$
5. $\quad z \leftarrow$ Newton iteration $(f, a, k, N')$
6. $\quad z \leftarrow z - \frac{f(z)}{f'(z)} \pmod{p^N}$
7. **return** $z$

---

An efficient implementation of the above algorithm requires we always compute with the lowest possible precision. The result in Line 6 has to be correct up to precision $N$. By induction, the numerator $f(z)$ in Line 6 satisfies $f(z) \equiv 0 \pmod{p^{N'+k}}$ and the denominator $f'(z)$ satisfies $v_p(f'(z)) = k$, which implies that $f'(z)/p^k$ is a unit in $\mathbb{Z}_q$. Dividing the numerator by $p^{N'+k}$, we conclude that it suffices to compute $f(z)/p^{N'+k}$ and $f'(z)/p^k$ with precision $N - N'$, which

## § 12.3 Newton lifting

implies that it suffices to compute $f(z) \pmod{p^{N+k}}$ and $f'(z) \pmod{p^{N'+k}}$. Note that also the inverse of $f'(z)/p^k$ and the product with $f(z)/p^{N'+k}$ only has to be computed with precision $N - N'$.

Since in each iteration, the precision we compute with almost doubles, the complexity of Algorithm 12.9 is determined by the last iteration. Let $T_f(N)$ denote the time to evaluate $f$ at precision $N$, then the complexity of Algorithm 12.9 is given by $O(\max\{T_f(N), T_{d,N}\})$, which in most cases reduces to $O(T_f(N))$.

### 12.3.1 Inverse

Let $a \in \mathbb{Z}_q$ be a unit and assume we have computed the inverse of $p_1(a) \in \mathbb{F}_q$ using one of the algorithms given in Chapter 11. Let $z_0$ be any lift of $1/p_1(a)$ to $\mathbb{Z}_q$, then $z_0$ is an approximate root of the polynomial $f(X) = 1 - aX$ to precision 1, since $f'(z_0) = a$ is a unit in $\mathbb{Z}_q$. Applying Lemma 12.8 to $f$ leads to the iteration $z \leftarrow z + (1 - az)/a$. Since the division by $a$ only has to be computed at half precision, we can use $z$ instead of $1/a$ giving the iteration $z \leftarrow z + z(1 - az)$.

---
**Algorithm 12.10** Inverse

INPUT: A unit $a \in \mathbb{Z}_q$ and precision $N$.
OUTPUT: The inverse of $a$ to precision $N$.

1.     **if** $N = 1$ **then**
2.        $z \leftarrow 1/a \bmod p$
3.     **else**
4.        $z \leftarrow$ Inverse $(a, \lceil \frac{N}{2} \rceil)$
5.        $z \leftarrow z + z(1 - az) \bmod p^N$
6.     **return** $z$

---

Let $e_i = 1 - az_i$ be the error in the $i$-th iteration. Then, an easy calculation shows that $e_{i+1} = e_i^2$, so convergence is quadratic, which explains the statement in Line 4. Since evaluating $f$ requires only one multiplication at precision $N$, we conclude that the complexity of Algorithm 12.10 is $O(T_{d,N})$.

**Example 12.11** Assume that $\mathbb{Z}_{2^8}$ up to precision $N = 10$ is represented as $(\mathbb{Z}/2^{10}\mathbb{Z})[X]/(M(X))$ with $M(X) = X^8 + 644X^7 + 842X^6 + 134X^5 + 523X^4 + 21X^3 + 1019X^2 + 562X + 1$, then Algorithm 12.10 computes the following intermediate results on input $(a, 10)$ with

$$a = 982X^7 + 303X^6 + 724X^5 + 458X^4 + 918X^3 + 423X^2 + 650X + 591$$

| $N$ | $z$ |
|---|---|
| 1 | $X^6 + X^3 + X^2 + X$ |
| 2 | $2X^7 + X^6 + 3X^3 + X^2 + X$ |
| 3 | $6X^7 + 5X^6 + 4X^4 + 7X^3 + X^2 + X + 4$ |
| 5 | $22X^7 + 21X^6 + 24X^5 + 4X^4 + 31X^3 + 25X^2 + X + 28$ |
| 10 | $854X^7 + 373X^6 + 760X^5 + 132X^4 + 863X^3 + 697X^2 + 321X + 60$ |

## 12.3.2 Inverse square root

Let $a \in \mathbb{Z}_q$ and consider the polynomial $f(X) = 1 - aX^2$, then if $a$ is an invertible square, $1/\sqrt{a}$ is a root of $f$. Applying Lemma 12.8 to $f$ leads to the iteration $z \leftarrow z + (1 - az^2)/(2az)$, which needs a division by $az$. Note however that this division can be replaced by a multiplication by $z$, giving the final iteration $z \leftarrow z + z(1 - az^2)/2$. Let $e_i = 1 - az_i^2$ be the error in the $i$-th iteration, then an easy calculation shows that $e_{i+1} = (3e_i^2 + e_i^3)/4$. Let $n_i = v_p(e_i) > 0$, then $n_{i+1} = 2n_i$ for $p \geq 5$, $n_{i+1} = 2n_i + 1$ for $p = 3$ and $n_{i+1} = 2n_i - 2$ for $p = 2$. Thus for $p = 2$ we need an approximate root to precision 2 to initialize the Newton iteration. Since every element in a finite field of characteristic 2 is a square, we can always compute $z \equiv 1/\sqrt{p_1(a)} \pmod{2}$ using finite field arithmetic. Let $y = z + 2\Delta$, then $1 - ay^2 \equiv 0 \pmod{8}$ iff $\Delta$ is a solution of $\Delta^2 + z\Delta \equiv (1/a - z^2)/4 \pmod{2}$. If this equation has no solution then $a$ is not a square; otherwise, $z + 2\Delta$ is an approximate root to precision 2.

---

**Algorithm 12.12** Inverse square root ($p = 2$)

INPUT: An invertible square $a \in \mathbb{Z}_q$, an initial approximation $z_0$ to precision 2 and precision $N$.
OUTPUT: The inverse square root of $a$ to precision $N$.

1.  **if** $N \leq 2$ **then**
2.      $z \leftarrow z_0$
3.  **else**
4.      $N' \leftarrow \lceil \frac{N+1}{2} \rceil$
5.      $z \leftarrow$ Inverse square root $(a, x, N')$
6.      $z \leftarrow z + \dfrac{z(1 - az^2)}{2} \bmod 2^N$      $\left[(1 - az^2) \text{ computed modulo } 2^{N+1}\right]$
7.  **return** $z$

---

Since evaluating $f$ requires only $O(1)$ multiplications at precision $N$, we conclude that the complexity of Algorithm 12.12 is $O(T_{d,N})$.

**Example 12.13** Let $\mathbb{Z}_{2^8}$ up to precision $N = 10$ be represented as in Example 12.11. Suppose we want to compute the inverse square root of

$$a = 823X^7 + 707X^6 + 860X^5 + 387X^4 + 663X^3 + 183X^2 + 12X + 354.$$

The initial approximation to precision 2 is given by $z_0 = 2X^7 + X^6 + 3X^3 + X^2 + X$. Since the precision we compute with is $N = 10$, the unique inverse square root $z \equiv z_0 \pmod{4}$ can only be determined up to precision $N - 1$. On input $(a, 9)$, Algorithm 12.12 computes the following intermediate results:

| $N$ | $z$ |
|---|---|
| 2 | $2X^7 + X^6 + 3X^3 + X^2 + X$ |
| 3 | $6X^7 + 5X^6 + 4X^4 + 7X^3 + X^2 + X + 4$ |
| 5 | $22X^7 + 21X^6 + 24X^5 + 4X^4 + 31X^3 + 25X^2 + X + 28$ |
| 9 | $342X^7 + 373X^6 + 248X^5 + 132X^4 + 351X^3 + 185X^2 + 321X + 60$ |

## 12.3.3 Square root

Let $a \in \mathbb{Z}_q$ be a square in $\mathbb{Z}_q$, then $v = v_p(a)$ must be even and $\sqrt{a}$ can be computed as $p^{v/2}\sqrt{a/p^v}$ with $a/p^v$ a unit in $\mathbb{Z}_q$. Therefore, assume that $a$ is an invertible square in $\mathbb{Z}_q$.

A trivial method to obtain $\sqrt{a}$ is simply to compute $c = 1/\sqrt{a}$ using Algorithm 12.12 and to return $\sqrt{a} = ac$, which requires one multiplication at full precision. A trick due to Karp and Markstein [KAMA 1997] can be used to merge this multiplication with the last step of the iteration as follows: compute $b \leftarrow az \bmod p^{N'}$ and replace Line 6 of Algorithm 12.12 in the last iteration by $z \leftarrow b + z(a - b^2)/2 \bmod p^N$. Note that $b$ is only computed at half precision, whereas the trivial method needs a multiplication at full precision.

## 12.4 Hensel lifting

Let $f \in \mathbb{Z}_q[X]$ be a polynomial with integral coefficients and assume that the leading coefficient of $f$ is a unit in $\mathbb{Z}_q$. The reduction of $f$ modulo $p$ is thus a polynomial $P_1(f)$ over $\mathbb{F}_q$ of the same degree as $f$. Given a factorization $f \equiv gh \pmod{p}$ with $g, h \in \mathbb{Z}_q$ such that $g$ and $h$ are coprime modulo $p$, Hensel's lemma 3.17 states that this factorization can be lifted modulo arbitrary powers of $p$. The following lemma is a reformulation of Hensel's lemma, but with quadratic convergence.

**Lemma 12.14** Let $f, g_k, h_k, s_k, t_k \in \mathbb{Z}_q[X]$ be polynomials with integral coefficients such that

$$f \equiv g_k h_k \pmod{p^k} \quad \text{and} \quad s_k g_k + t_k h_k \equiv 1 \pmod{p^k},$$

with $\deg f = \deg P_1(f)$, the leading coefficient of $h_k$ a unit in $\mathbb{Z}_q$, $\deg s_k < \deg h_k$ and $\deg t_k < \deg g_k$. Then there exists polynomials $h_{2k}, g_{2k}, s_{2k}, t_{2k} \in \mathbb{Z}_q[X]$ such that

$$f \equiv g_{2k} h_{2k} \pmod{p^{2k}} \quad \text{and} \quad s_{2k} g_{2k} + t_{2k} h_{2k} \equiv 1 \pmod{p^{2k}}.$$

Furthermore, $g_{2k} \equiv g_k \pmod{p^k}$, $h_{2k} \equiv h_k \pmod{p^k}$, $s_{2k} \equiv s_k \pmod{p^k}$, $t_{2k} \equiv t_k \pmod{p^k}$, $\deg g_{2k} = \deg g_k$, $\deg h_{2k} = \deg h_k$ and $\deg s_{2k} < \deg h_{2k}$ and $\deg t_{2k} < \deg g_{2k}$.

The construction of these polynomials is given as Algorithm 12.15. To illustrate how these formulas are devised, we show that $g_{2k}$ and $h_{2k}$ constructed in Line 3 satisfy the above lemma. Define $\Delta_g$ and $\Delta_h$ by $g_{2k} = g_k + p^k \Delta_g$ and $h_{2k} = h_k + p^k \Delta_h$, then $f \equiv g_{2k} h_{2k} \pmod{p^{2k}}$ implies

$$\frac{f - g_k h_k}{p^k} \equiv e \equiv \Delta_h g_k + \Delta_g h_k \pmod{p^k}. \tag{12.9}$$

Multiplying both sides of the equation $s_k g_k + t_k h_k \equiv 1 \pmod{p^k}$ by $e$, we indeed conclude that $es_k \equiv q_h h_k + \Delta_h \pmod{p^k}$ and $et_k \equiv q_g g_k + \Delta_g \pmod{p^k}$.

---

**Algorithm 12.15** Hensel lift iteration

INPUT: Polynomials $f, g_k, h_k, s_k, t_k \in \mathbb{Z}_q[X]$ as in Lemma 12.14 and precision $k$.
OUTPUT: Polynomials $g_{2k}, h_{2k}, s_{2k}, t_{2k} \in \mathbb{Z}_q[X]$ as in Lemma 12.14.

1. $e \leftarrow (f - g_k h_k)/p^k \bmod p^k$
2. compute $q, r \in \mathbb{Z}_q[X]$ with $\deg r < \deg h_k$ and $es_k \equiv qh_k + r \pmod{p^k}$
3. $g_{2k} \leftarrow g_k + p^k(et_k + qg_k) \bmod p^{2k}$ and $h_{2k} \leftarrow h_k + p^k r \bmod p^{2k}$
4. $e \leftarrow (1 - s_k g_{2k} - t_k h_{2k})/p^k \bmod p^k$
5. compute $q, r \in \mathbb{Z}_q[X]$ with $\deg r < \deg h_{2k}$ and $es_k \equiv qh_{2k} + r \pmod{p^k}$

6. $s_{2k} \leftarrow s_k + p^k r \mod p^{2k}$ and $t_{2k} \leftarrow t_k + p^k(et_k + qg_{2k}) \mod p^{2k}$

7. **return** $g_{2k}, h_{2k}, s_{2k}, t_{2k}$

Using Algorithm 12.15 as a subroutine, we easily deduce an algorithm to lift the factorization of a polynomial modulo $p$ to arbitrarily high powers of $p$.

**Algorithm 12.16** Hensel lift

INPUT: Polynomials $f, g, h, s, t \in \mathbb{Z}_q[X]$ with $f \equiv gh \pmod{p}$ and $sg + th \equiv 1 \pmod{p}$, precision $N$.
OUTPUT: Polynomials $G, H, S, T \in \mathbb{Z}_q[X]$ with $f \equiv GH \pmod{p^N}$ and $SG + TH \equiv 1 \pmod{p^N}$.

1. **if** $N = 1$ **then**
2.     $G \leftarrow g, H \leftarrow h, S \leftarrow s$ and $T \leftarrow t$
3. **else**
4.     $k \leftarrow \lceil \frac{N}{2} \rceil$
5.     $g_k, h_k, s_k, t_k \leftarrow$ Hensel lift $(f, g, h, s, t, k)$
6.     $G, H, S, T \leftarrow$ Hensel lift iteration $(f, g_k, h_k, s_k, t_k, k)$
7. **return** $G, H, S, T$

Since the precision we work with doubles in each iteration, the complexity of Algorithm 12.16 is determined by the last iteration. Let $n = \deg f$, then Algorithm 12.15 requires $O(1)$ multiplications of polynomials of degree less than $n$ and $O(1)$ divisions with remainder, both of which require $O(n^\mu T_{d,N})$ bit-operations.

**Example 12.17** In this example, we illustrate Algorithm 12.16 for polynomials over $\mathbb{Z}_2$. Define

$$f(X) = X^{10} + 321X^9 + 293X^8 + 93X^7 + 843X^6 + 699X^5 + 972X^4 + 781X^3 + 772X^2 + 129X + 376,$$

then clearly $f \equiv gh \pmod{2}$ with $g = X^6 + X^4 + X^3 + 1$ and $h = X^4 + X^3 + X$. Using Euclid extended gcd algorithm in the ring $\mathbb{F}_2[X]$, we compute $s = X^2 + X + 1$ and $t = X^4 + X + 1$ and on input $(f, g, h, s, t, 10)$, Algorithm 12.16 computes the following results:

| | |
|---|---|
| $G$ | $X^6 + 44X^5 + 19X^4 + 165X^3 + 92X^2 + 206X + 529$ |
| $H$ | $X^4 + 277X^3 + 374X^2 + 737X + 504$ |
| $S$ | $660X^3 + 893X^2 + 493X + 345$ |
| $T$ | $364X^5 + 311X^4 + 848X^3 + 618X^2 + 979X + 221$ |

## 12.5 Frobenius substitution

In this section, we present various algorithms to compute the Frobenius substitution $\Sigma$ on $\mathbb{Z}_q$. Depending on the representation of $\mathbb{Z}_q$, different algorithms should be used.

## 12.5.1 Sparse modulus

Let $\mathbb{Z}_q \simeq \mathbb{Z}_p[X]/(M(X))$ and assume that $M$ is sparse, monic of degree $d$ and with coefficients between 0 and $p - 1$. Given any element

$$a = \sum_{i=0}^{d-1} a_i X^i \in \mathbb{Z}_q$$

with $a_i \in \mathbb{Z}_p$ for $0 \leqslant i < d$, then clearly

$$\Sigma(a) = \sum_{i=0}^{d-1} a_i \Sigma(X)^i.$$

Therefore, if we precompute $\Sigma(X)$, we obtain $\Sigma(a)$ by evaluating the polynomial

$$a(Y) = \sum_{i=0}^{d-1} a_i Y^i$$

at $\Sigma(X)$.

Computing $\Sigma(X)$ can be done efficiently using a Newton iteration on $M$ starting from the initial approximation $X^p$. Let $w$ be the number of nonzero coefficients of $M(X)$, then we can clearly evaluate $M$ and $M'$ using $O(w \lg d)$ multiplications in $\mathbb{Z}_q$. Applying Algorithm 12.9 to the polynomial $M$ thus leads to an $O(wT_{d,N} \lg d)$ algorithm to compute $\Sigma(X) \bmod p^N$.

Evaluating the polynomial $a(Y)$ can be done using Horner's rule, which needs $O(d)$ multiplications at precision $N$. This would lead to an $O(dT_{d,N})$ algorithm to compute $\Sigma(a)$. At the expense of storing $O(\sqrt{d})$ elements of $\mathbb{Z}_q$, we can use the Paterson–Stockmeyer algorithm [PAST 1973]: let $B = \lceil \sqrt{d} \rceil$ and precompute $\Sigma(X)^i$ for $0 \leqslant i \leqslant B$ using $O(\sqrt{d})$ multiplications in $\mathbb{Z}_q$. Rewriting $a(Y)$ as

$$a(Y) = \sum_{j=0}^{\lceil n/B \rceil} \left( \sum_{i=0}^{B-1} a_{i+Bj} Y^i \right) Y^{Bj}, \qquad (12.10)$$

with $a_k = 0$ for $k \geqslant d$, we can compute $\Sigma(a)$ using $O(d)$ scalar multiplications and $O(\sqrt{d})$ multiplications in $\mathbb{Z}_q$ or $O(d^2 N^\mu + \sqrt{d} T_{d,N})$ bit-operations for a given precision $N$.

An asymptotically faster method can be obtained by replacing the $O(d)$ scalar multiplications with a matrix product as follows: let $\Gamma$ be the $B \times B$ matrix with $\Gamma[i][j] = a_{i+Bj}$, i.e., the rows of $\Gamma$ contain the scalars of the inner sum in (12.10) and let $\Lambda$ be the $B \times d$ matrix such that

$$\Sigma(X)^i = \sum_{j=0}^{d-1} \Lambda[i][j] X^j, \text{ for } i = 0, \ldots, B - 1.$$

Consider the $B \times d$ matrix $\Delta = \Gamma \Lambda$, then the $j$-th row of $\Delta$ simply contains the coefficients of

$$\sum_{i=0}^{B-1} a_{i+Bj} \Sigma(X)^i.$$

A method by Huang and Pan [HUPA 1998] can be used to compute the rectangular matrix product in $O(B^{3.334})$ ring operations. Therefore, a Frobenius substitution requires $O(d^{1.667} N^\mu + \sqrt{d} T_{d,N})$ bit-operations.

### 12.5.2 Teichmüller modulus

Let $\mathbb{Z}_q \simeq \mathbb{Z}_p[X]/(M(X))$ and assume that $M$ is a Teichmüller modulus of degree $d$. As shown in Section 12.1, this implies that $\Sigma(X) = X^p$. Given an element $a = \sum_{i=0}^{d-1} a_i X^i \in \mathbb{Z}_q$, we can simply compute $\Sigma(a)$ as

$$\Sigma(a) = \sum_{i=0}^{d-1} a_i X^{ip} \pmod{M(X)}.$$

The reduction modulo $M(X)$ takes at most $p - 1$ multiplications over $\mathbb{Z}_q$, thus computing $\Sigma(a)$ $\pmod{p^N}$ for $a \in \mathbb{Z}_q$ requires $O(p\, T_{d,N})$ time.

The inverse Frobenius substitution can also be computed very efficiently as follows:

$$\Sigma^{-1}\left(\sum_{i=0}^{d-1} a_i X^i\right) = \sum_{j=0}^{p-1} \left(\sum_{0 \leqslant pk+j < d} a_{pk+j} X^k\right) C_j(X),$$

where $C_j(X) = \Sigma^{-1}(X^j) = X^{jp^{d-1}} \pmod{M(X)}$. Assuming $C_j$ for $j = 0, \ldots, p-1$ are precomputed, computing $\Sigma^{-1}(a) \pmod{p^N}$ only takes $p - 1$ multiplications over $\mathbb{Z}_q$ or $O(p\, T_{d,N})$ time.

### 12.5.3 Gaussian normal basis

Assume that $\mathbb{Z}_q$ admits a Gaussian normal basis of type $t$, then elements of $\mathbb{Z}_q$ are embedded in the ring $\mathbb{Z}_p[X]/(X^{dt+1} - 1)$. Since $\Sigma^k(X) = X^{p^k}$, we have

$$\Sigma^k(a) = \sum_{i=0}^{dt} a_i X^{ip^k} = a_0 + \sum_{j=1}^{dt} a_{j/p^k \bmod (dt+1)} X^j. \tag{12.11}$$

So, we can compute an arbitrary repeated Frobenius substitution $\Sigma^k(a)$ by a simple permutation of the coefficients of $a$, which only requires $O(dt)$ bit-operations.

## 12.6 Artin–Schreier equations

Recall that if $\mathbb{F}_q$ is a field of characteristic $p$, an Artin–Schreier equation is an equation of the form $X^p - X - \overline{a} = 0$ with $\overline{a} \in \mathbb{F}_q$. The additive version of Hilbert Satz 90 states that such an equation has a solution in $\mathbb{F}_q$ if and only if $\mathrm{Tr}_{\mathbb{F}_q/\mathbb{F}_p}(\overline{a}) = 0$. Since $\sigma(x) = x^p$ for $x \in \mathbb{F}_q$, we can generalize this type of equation to $\mathbb{Z}_q$ by considering

$$\alpha \Sigma(X) + \beta X + \gamma = 0, \tag{12.12}$$

with $\alpha, \beta, \gamma \in \mathbb{Z}_q$ and $\alpha$ a unit in $\mathbb{Z}_q$.

It is easy to see that such an equation always has a unique solution in $\mathbb{Q}_q$ (not necessarily in $\mathbb{Z}_q$), as long as $\mathrm{N}(-\beta/\alpha) \neq 1$. Indeed, let $a_1 = -\beta/\alpha$ and $b_1 = -\gamma/\alpha$, then clearly $\Sigma^1(x) = a_1 x + b_1$ for any solution $x$. Applying $\Sigma$ to both sides, we see that we can recursively define $a_k, b_k \in \mathbb{Z}_q$ such that $\Sigma^k(x) = a_k x + b_k$ for $k = 2, \ldots, d$. Since $\Sigma^d(x) = x$ for all $x \in \mathbb{Q}_q$, we conclude that the unique solution in $\mathbb{Q}_q$ to the above equation is given by $b_d/(1 - a_d)$.

## § 12.6 Artin–Schreier equations

By imposing the condition $v_p(\beta) > 0$, the unique solution to (12.12) is always integral. Indeed, writing out the recursive process explicitly shows that the unique solution is given by

$$x = \frac{\sum_{i=0}^{d-1} \Sigma^i(b_1) \times \prod_{j=i+1}^{d-1} \Sigma^j(a_1)}{1 - N_{\mathbb{Q}_q/\mathbb{Q}_p}(a_1)}.$$

Since $v_p(a_1) = v_p(\beta) > 0$, we conclude that $1 - N_{\mathbb{Q}_q/\mathbb{Q}_p}(a_1)$ is a unit and therefore $x \in \mathbb{Z}_q$.

### 12.6.1 Lercier–Lubicz algorithm

Lercier and Lubicz [LELU 2003] use a simple square and multiply algorithm to compute the $a_k$, $b_k \in \mathbb{Z}_q$ based on the formula

$$\Sigma^{k+n}(x) = \Sigma^n(a_k x + b_k) = \Sigma^n(a_k)(a_n x + b_n) + \Sigma^n(b_k).$$

To find a solution to $\Sigma(x) \equiv ax + b \pmod{p^N}$ we simply call Algorithm 12.18 on input $(a, b, d, N)$, which returns $a_d$ and $b_d$. The complexity of Algorithm 12.18 is determined by Lines 6 and 7 which need $O(1)$ multiplications and $O(1)$ repeated Frobenius substitutions in $\mathbb{Z}_q/p^N \mathbb{Z}_q$.

For fields with a Gaussian normal basis of type $t$, the Frobenius substitution takes $O(dt)$ bit-operations as shown in Section 12.5.3. Since the algorithm needs $O(\lg d)$ recursive calls, the time and space complexity for fields with Gaussian normal basis are $O(T_{d,N} \lg d)$ and $O(dN)$ respectively.

If the field does not admit a Gaussian normal basis, Algorithm 12.18 should be modified to keep track of $\Sigma^{k'}(X)$. This can be achieved by introducing an extra variable $c$ which equals $\Sigma^k(X)$ and is returned in Line 11 together with $a_k$ and $b_k$. The variable $c$ can then be updated by evaluating it at itself, such that $c$ becomes $\Sigma^{2k'}(X)$ and conjugating once if $k$ is odd. Using the Paterson–Stockmeyer trick [PAST 1973], the complexity of Algorithm 12.18 then becomes $O\big((d^2 N^\mu + \sqrt{d}T_{d,N}) \lg d\big)$ and the space complexity is $O(d^{1.5} N)$.

---

**Algorithm 12.18** Artin–Schreier root square multiply

INPUT: Elements $a, b \in \mathbb{Z}_q$, a power $k$ and precision $N$.
OUTPUT: Elements $a_k, b_k \in \mathbb{Z}_q$ such that $\Sigma^k(x) \equiv a_k x + b_k \pmod{p^N}$.

1. **if** $k = 1$ **then**
2.     $a_k \leftarrow a \bmod p^N$ and $b_k \leftarrow b \bmod p^N$
3. **else**
4.     $k' \leftarrow \lfloor \frac{k}{2} \rfloor$
5.     $a_{k'}, b_{k'} \leftarrow$ Artin–Schreier root square multiply $(a, b, k', N)$
6.     $a_k \leftarrow a_{k'} \Sigma^{k'}(a_{k'}) \bmod p^N$
7.     $b_k \leftarrow b_{k'} \Sigma^{k'}(a_{k'}) + \Sigma^{k'}(b_{k'}) \bmod p^N$
8.     **if** $k \equiv 1 \pmod 2$ **then**
9.         $b_k \leftarrow b \Sigma(a_k) + \Sigma(b_k) \bmod p^N$
10.         $a_k \leftarrow a \Sigma(a_k) \bmod p^N$
11. **return** $a_k, b_k$

---

Algorithm 12.19 solves the general Artin–Schreier equation (12.12) and clearly has the same time and space complexity as Algorithm 12.18.

---

**Algorithm 12.19** Artin–Schreier root I

INPUT: Elements $\alpha, \beta, \gamma \in \mathbb{Z}_q$ with $\alpha$ a unit in $\mathbb{Z}_q$, $v_p(\beta) > 0$ and precision $N$.
OUTPUT: An element $x \in \mathbb{Z}_q$ such that $\alpha \Sigma(x) + \beta x + \gamma \equiv 0 \pmod{p^N}$.

1. $a_d, b_d \leftarrow$ Artin–Schreier root square multiply $(-\beta/\alpha, -\gamma/\alpha, n, N)$
2. **return** $b_d/(1 - a_d) \bmod p^N$

---

**Example 12.20** The ring $\mathbb{Z}_{2^{10}}$ admits a Gaussian normal basis of type 1 and can be represented as $\mathbb{Z}_2[X]/(M(X))$ with $M(X) = (X^{11} - 1)/(X - 1)$. An element $a = \sum_{i=0}^{9} a_i X^i$ is embedded in the ring $\mathbb{Z}_2[Y]/(Y^{11} - 1)$ as $\tilde{a} = \sum_{i=0}^{9} a_i Y^i$. Assume we want to find the integral solution $s$ to the equation $\Sigma(X) + \beta X + \gamma = 0$ where

$$\beta = 18X^9 + 804X^8 + 354X^7 + 56X^6 + 656X^5 + 892X^4 + 824X^3 + 578X^2 + 942X + 128,$$

$$\gamma = 248X^9 + 101X^8 + 64X^7 + 955X^6 + 399X^5 + 664X^4 + 313X^3 + 819X^2 + 1012X + 32,$$

then Algorithm 12.18 computes the following intermediate results:

| | |
|---|---|
| $a_1$ | $1006Y^9 + 220Y^8 + 670Y^7 + 968Y^6 + 368Y^5 + 132Y^4 + 200Y^3 + 446Y^2 + 82Y + 896$ |
| $b_1$ | $776Y^9 + 923Y^8 + 960Y^7 + 69Y^6 + 625Y^5 + 360Y^4 + 711Y^3 + 205Y^2 + 12Y + 992$ |
| $a_2$ | $420Y^{10} + 680Y^9 + 324Y^8 + 416Y^7 + 652Y^6 + 388Y^5 + 424Y^4 + 992Y^3 + 100Y^2 + 644Y + 96$ |
| $b_2$ | $287Y^{10} + 938Y^9 + 68Y^8 + 928Y^7 + 583Y^6 + 659Y^5 + 353Y^4 + 268Y^3 + 842Y^2 + 901Y + 698$ |
| $a_5$ | $768Y^{10} + 640Y^9 + 64Y^8 + 672Y^7 + 256Y^6 + 928Y^5 + 96Y^4 + 704Y^3 + 736Y^2 + 96Y + 160$ |
| $b_5$ | $969Y^{10} + 92Y^9 + 29Y^8 + 751Y^7 + 782Y^6 + 826Y^5 + 219Y^4 + 919Y^3 + 936Y^2 + 256Y + 954$ |
| $a_{10}$ | $0$ |
| $b_{10}$ | $992Y^{10} + 232Y^9 + 759Y^8 + 571Y^7 + 986Y^6 + 270Y^5 + 431Y^4 + 701Y^3 + 124Y^2 + 585Y + 58$ |

Since $a_{10} \equiv 0 \pmod{2^{10}}$, the unique solution $s$ is simply the reduction modulo $M$ of $b_{10}$, i.e.,

$$s = 264X^9 + 791X^8 + 603X^7 + 1018X^6 + 302X^5 + 463X^4 + 733X^3 + 156X^2 + 617X + 90.$$

### 12.6.2 Harley's algorithm

In an e-mail to the NMBRTHRY list [HAR 2002b], Harley sketched a doubly recursive algorithm to solve an Artin–Schreier equation of the form (12.12) assuming that $v_p(\beta) > 0$. Note that this implies that the solution is integral.

The main idea is as follows: assume we have an algorithm that returns a solution $x_{N'}$ to an equation of the form (12.12) to precision $N' = \lceil N/2 \rceil$. Write $x_N = x'_N + p^{N'} \Delta_N$ and substitute $x_N$ into (12.12), which leads to

$$\alpha \Sigma(\Delta_N) + \beta \Delta_N + \frac{\alpha \Sigma(x_{N'}) + \beta x_{N'} + \gamma}{p^{N'}} \equiv 0 \pmod{p^{N-N'}}.$$

Since $N - N' \leq N'$ we can use the same algorithm to determine $\Delta_N \bmod p^{N-N'}$ and therefore $x_N$. This immediately leads to a recursive algorithm if we can solve the base case, i.e., find a solution to (12.12) modulo $p$. If we assume that $v_p(\beta) > 0$, then the base case reduces to solving

$$\alpha \Sigma(x) + \gamma \equiv 0 \pmod{p}.$$

Since $\alpha$ is a unit, this uniquely determines $x \equiv (-\gamma/\alpha)^{1/p} \pmod{p}$.

## § 12.6 Artin–Schreier equations

---

**Algorithm 12.21** Artin–Schreier root II

INPUT: Elements $\alpha, \beta, \gamma \in \mathbb{Z}_q$, $\alpha$ a unit in $\mathbb{Z}_q$, $v_p(\beta) > 0$ and precision $N$.
OUTPUT: An element $x \in \mathbb{Z}_q$ such that $\alpha\Sigma(x) + \beta x + \gamma \equiv \pmod{p^N}$.

1. **if** $N = 1$ **then**
2. $\quad x \leftarrow (-\gamma/\alpha)^{1/p} \pmod{p}$
3. **else**
4. $\quad N' \leftarrow \lceil \frac{N}{2} \rceil$
5. $\quad x' \leftarrow$ Artin–Schreier root II $(\alpha, \beta, \gamma, N')$
6. $\quad \gamma' \leftarrow \dfrac{\alpha\Sigma(x') + \beta x' + \gamma}{p^{N'}} \bmod p^{N-N'}$
7. $\quad \Delta' \leftarrow$ Artin–Schreier root II $(\alpha, \beta, \gamma', N - N')$
8. $\quad x \leftarrow (x' + p^{N'}\Delta') \bmod p^N$
9. **return** $x$

---

The $p$-th root in Line 2 of Algorithm 12.21 should not be computed by naively taking the $p^{d-1}$-th power. Instead, let $\mathbb{F}_q = \mathbb{F}_p[\overline{\theta}]$, then

$$\left(\sum_{i=0}^{d-1} \overline{a}_i \overline{\theta}^i\right)^{1/p} = \sum_{j=0}^{p-1}\left(\sum_{0 \leqslant pk+j < d} \overline{a}_{pk+j}\overline{\theta}^k\right) C_j(\overline{\theta}), \quad \text{with } C_j(\overline{\theta}) = (\overline{\theta}^j)^{1/p} = \overline{\theta}^{jp^{d-1}}$$

This shows that for $\overline{z} \in \mathbb{F}_q$, we can compute $\overline{z}^{1/p}$ with $p - 1$ multiplications over $\mathbb{F}_q$.

The complexity of Algorithm 12.21 is determined by the recursive calls in Lines 5 and 7, the $O(1)$ multiplications in Line 6, and the Frobenius substitution in Line 6. If we assume that $\mathbb{Z}_q$ is represented using the Teichmüller modulus, the Frobenius substitution in $\mathbb{Z}_q/p^N\mathbb{Z}_q$ can be computed using $O(T_{d,N})$ bit-operations for $p$ fixed. If $T(N)$ is the running time of Algorithm 12.21 for precision $N$, then we have

$$T(N) \leqslant 2T(\lceil N/2 \rceil) + cT_{d,N},$$

for some constant $c$. The above relation implies by induction that the complexity of Algorithm 12.21 is $O(T_{d,N} \lg N)$.

**Example 12.22** Let $\mathbb{Z}_{2^8}$ be represented as $\mathbb{Z}_2[X]/(M(X))$ with $M(X) = X^8 + X^4 + X^3 + X^2 + 1$ and let

$$\beta = 186X^7 + 858X^6 + 810X^5 + 50X^4 + 208X^3 + 36X^2 + 2X + 652,$$
$$\gamma = 139X^7 + 911X^6 + 938X^5 + 970X^4 + 412X^3 + 1021X^2 + 99X + 667.$$

Then on input $(1, \beta, \gamma, 10)$ Algorithm 12.21 computes the following intermediate results given as a binary tree: the root of the tree is denoted T, a left child with L and a right child with R.

| Node | $N$ | $x$ |
|---|---|---|
| TLLLL | 1 | $X^7 + X^6 + X^4 + X$ |
| TLLLR | 1 | $X^7 + X^5 + X^3 + X^2$ |
| TLLL | 2 | $3X^7 + X^6 + 2X^5 + X^4 + 2X^3 + 2X^2 + X$ |
| TLLR | 1 | $X^6 + X^5 + X^2 + X + 1$ |
| TLL | 3 | $3X^7 + 5X^6 + 6X^5 + X^4 + 2X^3 + 6X^2 + 5X + 4$ |
| TLRL | 1 | $X^6 + X^5 + X^4 + X^3 + X + 1$ |
| TLRR | 1 | $X^7 + X^4 + X^3 + X + 1$ |
| TLR | 2 | $2X^7 + X^6 + X^5 + 3X^4 + 3X^3 + 3X + 3$ |
| TL | 5 | $19X^7 + 13X^6 + 14X^5 + 25X^4 + 26X^3 + 6X^2 + 29X + 28$ |
| TRLLL | 1 | $X^6 + X^5 + X^2 + X + 1$ |
| TRLLR | 1 | $X^7 + X^5 + X^3 + X^2 + X$ |
| TRLL | 2 | $2X^7 + X^6 + 3X^5 + 2X^3 + 3X^2 + 3X + 1$ |
| TRLR | 1 | $X^3 + X^2 + X$ |
| TRL | 3 | $2X^7 + X^6 + 3X^5 + 6X^3 + 7X^2 + 7X + 1$ |
| TRRL | 1 | $X^5 + X^4 + X + 1$ |
| TRRR | 1 | $X^6 + X^5 + X^4 + X^3 + 1$ |
| TRR | 2 | $2X^6 + 3X^5 + 3X^4 + 2X^3 + X + 3$ |
| TR | 5 | $2X^7 + 17X^6 + 27X^5 + 24X^4 + 22X^3 + 7X^2 + 15X + 25$ |
| T | 10 | $83X^7 + 557X^6 + 878X^5 + 793X^4 + 730X^3 + 230X^2 + 509X + 828$ |

## 12.7 Generalized Newton lifting

In this section we consider equations of the form $\phi(Y, \Sigma(Y)) = 0$ with $\phi(Y, Z) \in \mathbb{Z}_q[Y, Z]$. These equations arise naturally in the point counting algorithms described in Section 17.3. However, solving such an equation is also useful for more general applications, e.g., computing the Teichmüller lift of an element in $\mathbb{F}_q$. Indeed, let $\overline{a} \in \mathbb{F}_q$, then the Teichmüller lift $\omega(\overline{a})$ is the unique solution of the equation $Y^p - \Sigma(Y) = 0$ with $p_1(\omega(\overline{a})) = \overline{a}$.

Let $x \in \mathbb{Z}_q$ be a root of $\phi(Y, \Sigma(Y)) = 0$ and assume we know $x_N \equiv x \pmod{p^N}$. Define $\delta_N = (x - x_N)/p^N$, then the Taylor expansion around $x_N$ gives

$$0 = \phi(x, \Sigma(x)) = \phi(x_N + p^N \delta_N, \Sigma(x_N + p^N \delta_N)) \quad (12.13)$$
$$\equiv \phi(x_N, \Sigma(x_N)) + p^N(\delta_N \Delta_y + \Sigma(\delta_N)\Delta_z) \pmod{p^{2N}}, \quad (12.14)$$

with

$$\Delta_y \equiv \frac{\partial \phi}{\partial Y}(x_N, \Sigma(x_N)) \pmod{p^N} \text{ and } \Delta_z \equiv \frac{\partial \phi}{\partial Z}(x_N, \Sigma(x_N)) \pmod{p^N}.$$

This implies that $\delta_N$ has to be a solution of

$$\Delta_z \Sigma(X) + \Delta_y X + \frac{\phi(x_N, \Sigma(x_N))}{p^N} \equiv 0 \pmod{p^N}.$$

Let $k = v_p(\Delta_z)$, then if $v_p(\Delta_y) > k$ and $v_p(\phi(x_N, \Sigma(x_N))) \geqslant k + N$ and $N > k$, we recover the Artin–Schreier equation (12.12) with $\alpha = \Delta_z/p^k$, $\beta = \Delta_y/p^k$ and $\gamma = \phi(x_N, \Sigma(x_N))/p^{N+k}$ up to precision $N - k$. Note that any solution $\delta' \in \mathbb{Z}_q$ to the above equation satisfies

$$\delta' \equiv \delta_N \pmod{p^{N-k}}.$$

Let $x_{2N-k} = x_N + p^N \delta'$, then $x_{2N-k} \equiv x \pmod{p^{2N-k}}$ and (12.13) implies that

$$\phi(x_{2N-k}, \Sigma(x_{2N-k})) \equiv 0 \pmod{p^{2N}}.$$

Furthermore, since we assumed that $N > k$, we have

$$v_p\left(\frac{\partial \phi}{\partial Z}(x_{2N-k}, \Sigma(x_{2N-k}))\right) = k \text{ and } v_p\left(\frac{\partial \phi}{\partial Y}(x_{2N-k}, \Sigma(x_{2N-k}))\right) \geq k, \qquad (12.15)$$

so we can repeat the same procedure to find a solution up to arbitrary precision.

---

**Algorithm 12.23** Generalized Newton lift

INPUT: A polynomial $\phi(Y, Z) \in \mathbb{Z}_q$, an element $x_0 \in \mathbb{Z}_q$ satisfying the relation $\phi(x_0, \Sigma(x_0)) \equiv 0 \pmod{p^{2k+1}}$ with $k = v_p\left(\frac{\partial \phi}{\partial Z}(x_0, \Sigma(x_0))\right)$ and precision $N$.
OUTPUT: An element $x_N$ of $\mathbb{Z}_q$ such that $\phi(x_N, \Sigma(x_N)) \equiv 0 \pmod{p^{N+k}}$ and $x_N \equiv x_0 \pmod{p^{k+1}}$.

1. **if** $N \leq k+1$ **then**
2.     $x \leftarrow x_0$
3. **else**
4.     $N' \leftarrow \lceil \frac{N+k}{2} \rceil$
5.     $x' \leftarrow$ Generalized Newton lift $(\phi, x_0, N')$
6.     $y' \leftarrow \Sigma(x') \bmod p^{N+k}$
7.     $V \leftarrow \phi(x', y') \bmod p^{N+k}$
8.     $\Delta_y \leftarrow \frac{\partial \phi}{\partial Y}(x', y') \bmod p^{N'}$
9.     $\Delta_z \leftarrow \frac{\partial \phi}{\partial Z}(x', y') \bmod p^{N'}$
10.     $\delta \leftarrow$ Artin–Schreier root $(\Delta_z/p^k, \Delta_y/p^k, V/p^{N'+k}, N' - k)$
11.     $x \leftarrow (x' + p^{N'} \delta) \bmod p^N$
12. **return** $x$

---

Since the precision of the computations almost doubles in every step, the complexity of Algorithm 12.23 is the same as the complexity of the Artin–Schreier root subroutine in Line 10.

## 12.8 Applications

### 12.8.1 Teichmüller lift

Recall that the Teichmüller lift $\omega(\overline{a})$ of an element $\overline{a} \in \mathbb{F}_q$ is defined as follows: $\omega(0) = 0$ and for nonzero $\overline{a} \in \mathbb{F}_q$, $\omega(\overline{a})$ is the unique $(q-1)$-th root of unity with $p_1(\omega(\overline{a})) = \overline{a}$.

A trivial, but slow algorithm to compute $\omega(\overline{a}) \bmod p^N$ uses a simple Newton lifting on the polynomial $f(X) = X^{q-1} - 1$. Since evaluating $f$ requires $O(d)$ multiplications for $p$ fixed, the overall complexity of this approach is $O(dT_{d,N})$.

For $N \leqslant d$ there exists a faster algorithm based on repeated $p$-th powering: assume that $a$ satisfies $p_1(a) = \overline{a}$ and $a^{q-1} - 1 \equiv 0 \pmod{p^k}$, then we can write $a^{q-1} = 1 + p^k \Delta$ with $\Delta \in \mathbb{Z}_q$. Taking the $p$-th power of both sides then gives $(a^p)^{q-1} = 1 + p^{k+1}\Delta'$, with $\Delta' = \bigl((1+p^k\Delta)^p - 1\bigr)/p^{k+1}$. Reducing modulo $p^{k+1}$ shows that $a^p \equiv \omega(\overline{a}^p) \pmod{p^{k+1}}$ and thus is $a^p$, a better approximation of $\omega(a^p)$. This immediately leads to Algorithm 12.24, which has complexity $O(NT_{d,N})$, and thus is faster than the trivial algorithm for $N \leqslant d$.

---

**Algorithm 12.24** Teichmüller lift

---

INPUT: An element $\overline{a} \in \mathbb{F}_q$ and precision $N$.
OUTPUT: The Teichmüller lift $z \equiv \omega(\overline{a}) \pmod{p^N}$ of $\overline{a}$ to precision $N$.

1. $k \leftarrow N - 1$
2. $r \leftarrow \overline{a}^{\frac{1}{p^k}}$ [arbitrary lift]
3. $z \leftarrow \left(a^{p^k}\right) \bmod p^N$
4. **return** $z$

---

**Example 12.25** Let $\mathbb{F}_{2^8}$ be represented as $\mathbb{F}_2[X]/(M(X))$ with $M(X) = X^8 + X^4 + X^3 + X^2 + 1$ and let $\overline{a} = X^6 + X^2 + X + 1$. In Line 2 of Algorithm 12.24, we compute the $2^9$-th root $\overline{r}$ of $\overline{a}$ which is given by $\overline{r} = X^7 + X^3 + X^2 + X$. Let $r$ be an arbitrary lift of $\overline{r}$ to $\mathbb{Z}_{2^{10}} \simeq \mathbb{Z}_2[X]/(M(X))$, then
$$\omega(\overline{a}) \equiv r^{2^9}$$
$$\equiv 64X^7 + 871X^6 + 992X^5 + 784X^4 + 480X^3 + 615X^2 + 675X + 443 \pmod{2^{10}}.$$

The fastest algorithm, however, is based on the following observation: since the Teichmüller lift $\omega(\overline{a})$ is the unique $(q-1)$-th root of unity with $p_1\bigl(\omega(\overline{a})\bigr) = \overline{a}$, it also satisfies $\Sigma\bigl(\omega(\overline{a})\bigr) = \omega(\overline{a})^p$. Indeed, $\Sigma\bigl(\omega(\overline{a})\bigr)$ is a $(q-1)$-th root of unity and since $\omega(\cdot)$ is multiplicative and $\Sigma\bigl(\omega(\overline{a})\bigr) \equiv \overline{a}^p \pmod{p}$, the claim follows. The Teichmüller lift $\omega(\overline{a})$ can thus be computed as the solution of
$$\Sigma(X) - X^p = 0 \text{ and } X \equiv \overline{a} \pmod{p}.$$

Assuming that $\mathbb{Z}_q$ is represented using the Teichmüller modulus, Algorithm 12.23 then computes $\omega(\overline{a}) \pmod{p^N}$ using $O(T_{d,N} \lg N)$ bit-operations.

### 12.8.2 Logarithm

**Definition 12.26** Let $x \in \mathbb{Z}_q$ then the *p-adic logarithmic* function of $x$ is defined by
$$\log(x) = \sum_{i=1}^{\infty} (-1)^{i-1} \frac{(x-1)^i}{i}. \tag{12.16}$$

The function $\log(x)$ converges for $v_p(x-1) > 0$.

Assume that $a \in \mathbb{Z}_q$ satisfies $v_p(a-1) > 0$, then using Horner's rule, evaluating $\log(a)$ up to precision $N$ takes $O(N)$ multiplications over $\mathbb{Z}_q/p^N\mathbb{Z}_q$ or $O(NT_{d,N})$ bit-operations.

Satoh, Skjernaa, and Taguchi [SASK+ 2003] solve this problem by noting that $a^{p^k}$ for $k \in \mathbb{N}$ is very close to unity, i.e., $v_p(a^{p^k} - 1) > k$. If $a \in \mathbb{Z}_q/p^N\mathbb{Z}_q$, then $a^{p^k}$ is well defined in $\mathbb{Z}_q/p^{N+k}\mathbb{Z}_q$ and can be computed with $O(k)$ multiplications in $\mathbb{Z}_q/p^{N+k}\mathbb{Z}_q$. Furthermore, note that
$$\log(a) \equiv p^{-k}\bigl(\log(a^{p^k}) \pmod{p^{N+k}}\bigr) \pmod{p^N}$$

## § 12.8 Applications

and that $\log(a^{p^k})$ (mod $p^{N+k}$) can be computed with $O(N/k)$ multiplications over $\mathbb{Z}_q/p^{N+k}\mathbb{Z}_q$. So, if we take $k \simeq \sqrt{N}$, then $\log(a)$ (mod $p^N$) can be computed in $O(\sqrt{N}T_{d,N+\sqrt{N}})$ time.

In characteristic 2, Satoh, Skjernaa, and Taguchi suggested a further improvement. Without loss of generality, we can assume that $v_2(a-1) > 1$. Indeed, since $v_p(a-1) \geqslant 1$, we conclude that $v_p(a^2-1) > 1$ and $\log(a^2) = 2\log(a)$. Therefore, assume that $v_2(a-1) > 1$, then we have $a \equiv 1$ (mod $2^\nu$) for $\nu \geqslant 2$.

Let $z = a-1 \in 2^\nu \mathbb{Z}_q/2^N \mathbb{Z}_q$ and define $\gamma = \dfrac{z}{2+z} \in 2^{\nu-1}\mathbb{Z}_q/2^{N-1}\mathbb{Z}_q$, then $a = 1+z = \dfrac{1+\gamma}{1-\gamma}$ and thus

$$\log(a) = \log(1+z) = \log(1+\gamma) - \log(1-\gamma) = 2\sum_{j=1}^\infty \frac{\gamma^{2j-1}}{2j-1}.$$

Note that all the denominators in the above formula are odd. Reducing this equation modulo $2^N$ therefore leads to

$$\log(a) \equiv \log(1+z) \equiv 2 \sum_{1 \leqslant (\nu-1)(2j-1) < N-1} \frac{\gamma^{2j-1}}{2j-1} \pmod{2^N}.$$

**Example 12.27** Let $\mathbb{Z}_{2^8}$ be represented as $\mathbb{Z}_2[X]/(M(X))$ with $M(X) = X^8+X^4+X^3+X^2+1$ and let $a = 872X^7+376X^6+460X^5+476X^4+138X^3+462X^2+794X+381$. Since $a$ satisfies $v_2(a-1) = 1$, we can compute the logarithm of $a$ which is given by

$$\log(a) \equiv 540X^7 + 298X^6 + 944X^5 + 614X^4 + 390X^3 + 884X^2 + 586X + 244 \pmod{2^{10}}.$$

### 12.8.3 Exponential

**Definition 12.28** Let $x \in \mathbb{Z}_q$ then the *p-adic exponential* function of $x$ is defined by

$$\exp(x) = \sum_{i=0}^\infty \frac{x^i}{i!}. \tag{12.17}$$

The function $\exp(x)$ converges for $v_p(x) > 1/(p-1)$.

An easy calculation shows that

$$v_p(i!) = \sum_{k=1}^B \lfloor i/p^k \rfloor \text{ with } B = \lfloor \log_p i \rfloor.$$

The valuation $v_p(i!)$ can thus be bounded by $v_p(i!) \leqslant (i-1)/(p-1)$, which explains the radius of convergence. Let $a \in \mathbb{Z}_q$ then we have the following identities

$$\log\bigl(\exp(a)\bigr) = a, \text{ for } v_p(a) > 1/(p-1),$$
$$\exp\bigl(\log(a)\bigr) = a, \text{ for } v_p(a-1) > 1/(p-1).$$

First assume that $a \in \mathbb{Z}_p$, then since $v_p(a) > 1/(p-1)$, we have $v_p(a) \geqslant 1$ for $p \geqslant 3$ and $v_p(a) \geqslant 2$ for $p = 2$. So if we precompute $\exp(p)$ (mod $p^N$) for $p \geqslant 3$ or $\exp(4)$ (mod $2^N$) for $p = 2$, then

$$\exp(a) \equiv \exp(p)^{a/p} \pmod{p^N}, \text{ for } p \geqslant 3,$$
$$\exp(a) \equiv \exp(4)^{a/4} \pmod{2^N}, \text{ for } p = 2,$$

and we can use a simple square and multiply algorithm to perform the final exponentiation.

For $a \in \mathbb{Z}_q$ we simply evaluate the power series of the exponential function (12.17) modulo $p^N$. The bound $v_p(i!) \leq (i-1)/(p-1)$ implies that we have to compute

$$\exp(a) \equiv \sum_{1 \leq i < B} \frac{a^i}{i!} \pmod{p^N} \text{ with } B = \frac{(p-1)N - 1}{(p-1)v_p(a) - 1}.$$

Using the Paterson–Stockmeyer algorithm [PAST 1973], this requires $O(dN^{1+\mu} + \sqrt{N}T_{d,N})$ bit-operations and $O(dN^{1.5})$ space.

**Example 12.29** Let $\mathbb{Z}_{2^8}$ be represented as $\mathbb{Z}_2[X]/(M(X))$ with $M(X) = X^8 + X^4 + X^3 + X^2 + 1$ and let $a = 720X^7 + 752X^6 + 920X^5 + 952X^4 + 276X^3 + 924X^2 + 564X + 760$. Since $v_p(a) = 2$, we can compute the exponential function of $a$, which is given by

$$\exp(a) \equiv 496X^7 + 600X^6 + 552X^5 + 272X^4 + 388X^3 + 132X^2 + 308X + 57 \pmod{2^{10}}.$$

### 12.8.4 Trace

**Definition 12.30** Let $\Sigma$ denote the Frobenius substitution on $\mathbb{Q}_q$, then the trace of $x \in \mathbb{Q}_q$ is

$$\operatorname{Tr}_{\mathbb{Q}_q/\mathbb{Q}_p}(x) = x + \Sigma(x) + \cdots + \Sigma^{d-2}(x) + \Sigma^{d-1}(x). \tag{12.18}$$

Since $\Sigma$ generates $\operatorname{Gal}(\mathbb{Q}_q/\mathbb{Q}_p)$, the trace $\operatorname{Tr}_{\mathbb{Q}_q/\mathbb{Q}_p}(x)$ is an element of $\mathbb{Q}_p$.

Let $a \in \mathbb{Q}_q$, then $\operatorname{Tr}_{\mathbb{Q}_q/\mathbb{Q}_p}(p^k a) = p^k \operatorname{Tr}_{\mathbb{Q}_q/\mathbb{Q}_p}(a)$, so we can assume that $a$ is a unit in $\mathbb{Z}_q$. If $\mathbb{Z}_q$ is represented as $\mathbb{Z}_p[X]/(M(X))$ and

$$a = \sum_{i=0}^{d-1} a_i X^i \text{ with } a_i \in \mathbb{Z}_p,$$

then clearly

$$\operatorname{Tr}_{\mathbb{Q}_q/\mathbb{Q}_p}(a) = \sum_{i=0}^{d-1} a_i \operatorname{Tr}_{\mathbb{Q}_q/\mathbb{Q}_p}(X^i). \tag{12.19}$$

Each $\operatorname{Tr}_{\mathbb{Q}_q/\mathbb{Q}_p}(X^i)$ for $i = 0, \ldots, d-1$ can be precomputed using Newton's formula:

$$\operatorname{Tr}_{\mathbb{Q}_q/\mathbb{Q}_p}(X^i) + \sum_{j=1}^{i-1} \operatorname{Tr}_{\mathbb{Q}_q/\mathbb{Q}_p}(X^{i-j})M_{d-j} + iM_{d-i} \equiv 0 \pmod{p^N},$$

with $M(X) = \sum_{i=0}^{d} M_i X^i$. Assuming that the $\operatorname{Tr}_{\mathbb{Q}_q/\mathbb{Q}_p}(X^i)$ for $i = 0, \ldots, d-1$ are precomputed, the trace of an element $a \in \mathbb{Z}_q$ can be computed to precision $N$ in $O(dN^\mu)$ time.

**Example 12.31** Let $\mathbb{Z}_{2^8}$ be represented as $\mathbb{Z}_2[X]/(M(X))$ with $M(X) = X^8 + 644X^7 + 842X^6 + 134X^5 + 523X^4 + 21X^3 + 1019X^2 + 562X + 1$, then $M$ is a Teichmüller modulus to precision 10. Newton's formula gives

| $i$ | 1 | 2 | 3 | 4 | 5 | 6 | 7 |
|---|---|---|---|---|---|---|---|
| $\operatorname{Tr}_{\mathbb{Q}_q/\mathbb{Q}_p}(X^i)$ | 380 | 380 | 166 | 380 | 623 | 166 | 42 |

Note that the traces of $X^i$ and $X^{2^k i}$ are equal since $M$ is a Teichmüller modulus.

## 12.8.5 Norm

**Definition 12.32** Let $\Sigma$ denote the Frobenius substitution on $\mathbb{Q}_q$, then the norm of $x \in \mathbb{Q}_q$ is

$$N_{\mathbb{Q}_q/\mathbb{Q}_p}(x) = \prod_{i=0}^{d-1} \Sigma^i(x). \tag{12.20}$$

Since $\Sigma$ generates $\mathrm{Gal}(\mathbb{Q}_q/\mathbb{Q}_p)$, the norm $N_{\mathbb{Q}_q/\mathbb{Q}_p}(x)$ is an element of $\mathbb{Q}_p$.

In this section we give an overview of the existing algorithms to compute $N_{\mathbb{Q}_q/\mathbb{Q}_p}(a)$ for an element $a \in \mathbb{Q}_q$. Since $N_{\mathbb{Q}_q/\mathbb{Q}_p}(p^k a) = p^{dk} N_{\mathbb{Q}_q/\mathbb{Q}_p}(a)$, we can assume that $a$ is a unit in $\mathbb{Z}_q$.

### 12.8.5.a Norm computation I

Kedlaya [KED 2001] suggested a basic square and multiply approach by computing

$$\alpha_{i+1} = \Sigma^{2^i}(\alpha_i)\, \alpha_i, \ \text{ for } i = 0, \ldots, \lfloor \lg d \rfloor,$$

with $\alpha_0 = a$ and to combine these to recover $N_{\mathbb{Q}_q/\mathbb{Q}_p}(a) = \Sigma^{d-1}(a) \cdots \Sigma(a) a$. Let $d = \sum_{i=0}^{\ell-1} d_i 2^i$, with $d_i \in \{0, 1\}$ and $d_{\ell-1} = 1$, then we can write

$$N_{\mathbb{Q}_q/\mathbb{Q}_p}(a) = \prod_{i=0}^{\ell-1} \Sigma^{2^{i+1}+\cdots+2^{\ell-1}}(\alpha_i^{d_i}),$$

where the sum $2^{i+1} + \cdots + 2^{\ell-1}$ is defined to be zero for $i \geqslant \ell - 1$. This formula immediately leads to Algorithm 12.33. Note that this algorithm remains valid for matrices over $\mathbb{Z}_q$.

---

**Algorithm 12.33** Norm I

INPUT: An element $a \in \mathbb{Z}_q$ with $q = p^d$ and a precision $N$.
OUTPUT: The norm $N_{\mathbb{Q}_q/\mathbb{Q}_p}(a) \bmod p^N$.

1. $i \leftarrow d, j \leftarrow 0, r \leftarrow 1$ and $s \leftarrow a$
2. **while** $i > 0$ **do**
3.     **if** $i \equiv 1 \pmod 2$ **then** $r \leftarrow \Sigma^{2^j}(r)\, s \bmod p^N$
4.     **if** $i > 1$ **then** $s \leftarrow \Sigma^{2^j}(s)\, s \bmod p^N$
5.     $j \leftarrow j + 1$ and $i \leftarrow \lfloor i/2 \rfloor$
6. **return** $r$

---

This algorithm is particularly attractive for $p$-adic fields with Gaussian normal basis of small type, due to efficient repeated Frobenius substitutions. In this case the time complexity is determined by the $O(\lg d)$ multiplications in $\mathbb{Z}_q/p^N \mathbb{Z}_q$ or $O(T_{d,N} \lg d)$ bit-operations and the space complexity is $O(dN)$.

If the field does not admit a Gaussian normal basis, then Algorithm 12.33 should be adapted as follows: introduce a new variable $c$ that keeps track of $\Sigma^{2^j}(X)$, i.e., $c$ is initialized with $X$ and in Line 4, $c$ is evaluated at itself, since $\Sigma^{2^{j+1}}(X) = \Sigma^{2^j}(\Sigma^{2^j}(X))$. Computing $\Sigma^{2^j}(r)$ and $\Sigma^{2^j}(s)$ in Lines 3 and 4 then simply reduces to an evaluation at $c$. Using the Paterson–Stockmeyer algorithm [PAST 1973], the complexity of Algorithm 12.33 then becomes $O\big((d^2 N^\mu + \sqrt{d} T_{d,N}) \lg d\big)$.

**Example 12.34** The ring $\mathbb{Z}_{2^{10}}$ admits a Gaussian normal basis of type 1 and can be represented as $\mathbb{Z}_2[X]/\bigl(M(X)\bigr)$ with $M(X) = (X^{11} - 1)/(X - 1)$. An element $a = \sum_{i=0}^{9} a_i X^i$ is embedded in the ring $\mathbb{Z}_2[Y]/(Y^{11} - 1)$ as $\tilde{a} = \sum_{i=0}^{9} a_i Y^i$. On input $(a, 10)$ with

$$a = 93X^9 + 733X^8 + 164X^7 + 887X^6 + 106X^5 + 493X^4 + 348X^3 + 40X^2 + 609X + 603.$$

Algorithm 12.33 computes the following intermediate results at the end of the while loop:

| $r_{10}$ | 1 |
|---|---|
| $s_{10}$ | $93Y^9 + 733Y^8 + 164Y^7 + 887Y^6 + 106Y^5 + 493Y^4 + 348Y^3 + 40Y^2 + 609Y + 603$ |
| $r_5$ | 1 |
| $s_5$ | $112Y^{10} + 440Y^9 + 412Y^8 + 752Y^7 + 258Y^6 + 436Y^5 + 756Y^4 + 1007Y^3 + 384Y^2 + 439Y + 524$ |
| $r_2$ | $112Y^{10} + 440Y^9 + 412Y^8 + 752Y^7 + 258Y^6 + 436Y^5 + 756Y^4 + 1007Y^3 + 384Y^2 + 439Y + 524$ |
| $s_2$ | $522Y^{10} + 16Y^9 + 492Y^8 + 255Y^7 + 752Y^6 + 211Y^5 + 325Y^4 + 946Y^3 + 189Y^2 + 984Y + 684$ |
| $r_1$ | $112Y^{10} + 440Y^9 + 412Y^8 + 752Y^7 + 258Y^6 + 436Y^5 + 756Y^4 + 1007Y^3 + 384Y^2 + 439Y + 524$ |
| $s_1$ | $15Y^{10} + 173Y^9 + 743Y^8 + 648Y^7 + 90Y^6 + 712Y^5 + 646Y^4 + 996Y^3 + 87Y^2 + 349Y + 661$ |
| $r_0$ | $759Y^{10} + 759Y^9 + 759Y^8 + 759Y^7 + 759Y^6 + 759Y^5 + 759Y^4 + 759Y^3 + 759Y^2 + 759Y + 602$ |
| $s_0$ | $15Y^{10} + 173Y^9 + 743Y^8 + 648Y^7 + 90Y^6 + 712Y^5 + 646Y^4 + 996Y^3 + 87Y^2 + 349Y + 661$ |

Note that the element $r$ returned by the algorithm needs to be reduced modulo $M$, which finally gives $\mathrm{N}_{\mathbb{Q}_q/\mathbb{Q}_p}(a) \equiv 867 \pmod{2^{10}}$.

### 12.8.5.b  Norm computation II

Satoh, Skjernaa, and Taguchi [SASK+ 2003] also proposed a fast norm computation algorithm based on an analytic method. First assume that $a$ is close to unity, i.e., $v_p(a-1) > 1/(p-1)$, then

$$\mathrm{N}_{\mathbb{Q}_q/\mathbb{Q}_p}(a) = \exp\bigl(\mathrm{Tr}_{\mathbb{Q}_q/\mathbb{Q}_p}(\log(a))\bigr), \tag{12.21}$$

since $\Sigma$ is continuous and both series converge. Combining the algorithms described in Sections 12.8.2, 12.8.3 and 12.8.4, we conclude that if $a$ is close to unity then $\mathrm{N}_{\mathbb{Q}_q/\mathbb{Q}_p}(a) \pmod{p^N}$ can be computed in $O(\sqrt{N} T_{d,N+\sqrt{N}})$ bit-operations and $O(dN)$ space.

Algorithm 12.35 computes the norm of an element in $1 + 2^\nu \mathbb{Z}_q$, with $q = 2^d$, assuming that $\exp(4)$ and $\mathrm{Tr}_{\mathbb{Q}_q/\mathbb{Q}_p}(X^i)$ for $i = 0, \ldots, d-1$ are precomputed.

---

**Algorithm 12.35** Norm II

INPUT: An element $a \in 1 + 2^\nu \mathbb{Z}_q$ with $\nu \geq 2$ and a precision $N$.
OUTPUT: The norm $\mathrm{N}_{\mathbb{Q}_q/\mathbb{Q}_p}(a) \bmod 2^N$.

1. $s \leftarrow \lfloor \sqrt{N}/2 \rfloor$
2. $z \leftarrow \bigl(a^{2^s} - 1\bigr) \bmod 2^{N+s}$
3. $w \leftarrow \bigl(\log(1+z)\bigr) \bmod 2^{N+s}$
4. $w \leftarrow \bigl(2^{-s} w\bigr) \bmod 2^N$
5. $u \leftarrow \bigl(2^{-\nu} \mathrm{Tr}_{\mathbb{Q}_q/\mathbb{Q}_p}(w)\bigr) \bmod 2^{N-\nu}$
6. **return** $\bigl(\exp(4)^u\bigr) \bmod 2^N$

---

**Example 12.36** Let $\mathbb{Z}_{2^8}$ be represented as $\mathbb{Z}_2[X]/\bigl(M(X)\bigr)$ with $M(X) = X^8 + X^4 + X^3 + X^2 + 1$ and let $a = 572X^7 + 108X^6 + 660X^5 + 556X^4 + 456X^3 + 748X^2 + 36X + 569$. For $N = 10$, Algorithm 12.35 computes the following values: $s = 1$,

$$z = 936X^7 + 568X^6 + 760X^5 + 1880X^4 + 1840X^3 + 1176X^2 + 1656X + 1408,$$

$$w = 1864X^7 + 248X^6 + 1880X^5 + 728X^4 + 1648X^3 + 1304X^2 + 24X + 1632,$$

§ 12.8 Applications

$u = 163$, $\exp(4) = 333$ and finally $\mathrm{N}_{\mathbb{Q}_q/\mathbb{Q}_p}(a) \equiv 725 \pmod{2^{10}}$.

Now consider the more general situation where $a \in \mathbb{Z}_q$ is not close to unity. Let $\omega(p_1(a)) \in \mathbb{Z}_q$ denote the Teichmüller lift of $p_1(a)$, i.e., the unique $(q-1)$-th root of unity, which reduces to $p_1(a)$. Consider the equality

$$\mathrm{N}_{\mathbb{Q}_q/\mathbb{Q}_p}(a) = \mathrm{N}_{\mathbb{Q}_q/\mathbb{Q}_p}\big(\omega(p_1(a))\big) \mathrm{N}_{\mathbb{Q}_q/\mathbb{Q}_p}\big(\omega(p_1(a))^{-1} a\big),$$

then $v_p\big(\omega(p_1(a))^{-1} a - 1\big) \geqslant 1$. Furthermore, note that $\mathrm{N}_{\mathbb{Q}_q/\mathbb{Q}_p}\big(\omega(p_1(a))\big)$ is equal to the Teichmüller lift of $\mathrm{N}_{\mathbb{F}_q/\mathbb{F}_p}(p_1(a))$. For $p \geqslant 3$, (12.21) holds since

$$v_p\big(\omega(p_1(a))^{-1} a - 1\big) \geqslant 1 > 1/(p-1).$$

For $p = 2$ we need an extra trick: simply square $\omega(p_1(a))^{-1} a$ modulo $2^{N+1}$, compute the norm of the square using Algorithm 12.35 modulo $2^{N+1}$ and take the square root of the norm, which is determined modulo $2^N$. This shows that $\mathrm{N}_{\mathbb{Q}_q/\mathbb{Q}_p}(a) \bmod p^N$ for any $a \in \mathbb{Z}_q$ can be computed in $O(\sqrt{N} T_{d, N+\sqrt{N}})$ bit-operations and $O(dN)$ space.

**12.8.5.c  Norm computation III**

In an e-mail to the NMBRTHRY list [HAR 2002b], Harley suggested an asymptotically fast norm computation algorithm based on a formula from number theory that expresses the norm as a resultant. The resultant itself can be computed using an adaptation of Moenck's fast extended gcd algorithm [MOE 1973].

Let $\mathbb{Z}_q \simeq \mathbb{Z}_p[X]/(M(X))$ with $M \in \mathbb{Z}_p[X]$ a monic irreducible polynomial of degree $n$. Let $\theta \in \mathbb{Z}_q$ be a root of $M$, then $M$ splits completely over $\mathbb{Z}_q$ as

$$M(X) = \prod_{i=0}^{d-1} \big(X - \Sigma^i(\theta)\big).$$

For $a = \sum_{i=0}^{d-1} a_i X^i \in \mathbb{Z}_q$, define the polynomial $A(X) = \sum_{i=0}^{d-1} a_i X^i \in \mathbb{Z}_p[X]$. By definition of the norm and the resultant we have

$$\mathrm{N}_{\mathbb{Q}_q/\mathbb{Q}_p}(a) = \prod_{i=0}^{d-1} \Sigma^i(a) = \prod_{i=0}^{d-1} A\big(\Sigma^i(\theta)\big) = \mathrm{Res}\big(M(X), A(X)\big).$$

The resultant $\mathrm{Res}(M(X), A(X))$ can be computed in softly linear time using a variant of Moenck's fast extended gcd algorithm [MOE 1973]. The result is an algorithm to compute $\mathrm{N}_{\mathbb{Q}_q/\mathbb{Q}_p}(a) \bmod p^N$ in time $O\big((dN)^\mu \lg d\big)$.

**Example 12.37** Let $\mathbb{Z}_{2^8}$ be represented as $\mathbb{Z}_2[X]/(M(X))$ with $M(X) = X^8 + X^4 + X^3 + X^2 + 1$ and let $a = 572X^7 + 108X^6 + 660X^5 + 556X^4 + 456X^3 + 748X^2 + 36X + 569$. Computing the resultant of $M(X)$ and $A(X)$ as an integer gives

$$\mathrm{Res}\big(M(X), A(X)\big) = 110891016699366462823125 \equiv 725 \pmod{2^{10}},$$

which gives the same result as in Example 12.36. Of course, in practice we never compute the resultant as an integer, but always reduce modulo $2^{10}$.

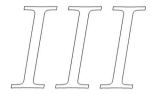

# Arithmetic of Curves

# Chapter 13

# Arithmetic of Elliptic Curves

*Christophe Doche and Tanja Lange*

**Contents in Brief**

| | | |
|---|---|---|
| **13.1** | **Summary of background on elliptic curves** | 268 |
| | First properties and group law • Scalar multiplication • Rational points • Torsion points • Isomorphisms • Isogenies • Endomorphisms • Cardinality | |
| **13.2** | **Arithmetic of elliptic curves defined over $\mathbb{F}_p$** | 280 |
| | Choice of the coordinates • Mixed coordinates • Montgomery scalar multiplication • Parallel implementations • Compression of points | |
| **13.3** | **Arithmetic of elliptic curves defined over $\mathbb{F}_{2^d}$** | 289 |
| | Choice of the coordinates • Faster doublings in affine coordinates • Mixed coordinates • Montgomery scalar multiplication • Point halving and applications • Parallel implementation • Compression of points | |

Elliptic curves constitute one of the main topics of this book. They have been proposed for applications in cryptography due to their fast group law and because so far no subexponential attack on their discrete logarithm problem (cf. Section 1.5) is known. We deal with security issues in later chapters and concentrate on the *group* arithmetic here. In an actual implementation this needs to be built on an efficient implementation of finite field arithmetic (cf. Chapter 11).

In the sequel we first review the background on elliptic curves to the extent needed here. For a more general presentation of elliptic curves, see Chapter 4. Then we address the question of efficient implementation in large odd and in even characteristics. We refer mainly to [HAME[+] 2003] for these sections.

Note that there are several softwares packages or libraries able to work on elliptic curves, for example PARI/GP [PARI] and apecs [APECS]. The former is a linkable library that also comes with an interactive shell, whereas the latter is a Maple package. Both come with full sources. The computer algebra systems Magma [MAGMA] and SIMATH [SIMATH] can deal with elliptic curves, too.

Elliptic curves have received a lot of attention throughout the past almost 20 years and many papers report experiments and timings for various field sizes and coordinates. We do not want to repeat the results but refer to [AVA 2004a, COMI[+] 1998] and Section 14.7 for odd characteristic and [HALÓ[+] 2000, LÓDA 1998, LÓDA 1999] for even characteristic. Another excellent and comprehensive reference comparing point multiplication costs and implementation results is [HAME[+] 2003, Tables 3.12, 3.13 and 3.14 and Chap. 5].

## 13.1 Summary of background on elliptic curves

### 13.1.1 First properties and group law

We start with a practical definition of the concept of an elliptic curve.

**Definition 13.1** An *elliptic curve* $E$ over a field $K$ denoted by $E/K$ is given by the *Weierstraß equation*

$$E : y^2 + a_1 xy + a_3 y = x^3 + a_2 x^2 + a_4 x + a_6 \tag{13.1}$$

where the coefficients $a_1, a_2, a_3, a_4, a_6 \in K$ are such that for each point $(x_1, y_1)$ with coordinates in $\overline{K}$ satisfying (13.1), the partial derivatives $2y_1 + a_1 x_1 + a_3$ and $3x_1^2 + 2a_2 x_1 + a_4 - a_1 y_1$ do not vanish simultaneously.

The last condition says that an elliptic curve is *nonsingular* or *smooth*. A point on a curve is called *singular* if both partial derivatives vanish (cf. the Jacobi criterion 4.94). For shorter reference we group the coefficients in (13.1) to the equation

$$E : y^2 + h(x)y = f(x), \quad h(x), f(x) \in K[x], \quad \deg(h) \leqslant 1, \deg(f) = 3 \text{ with } f \text{ monic.}$$

The smoothness condition can also be expressed more intrinsically. Indeed, let

$$b_2 = a_1^2 + 4a_2, \quad b_4 = a_1 a_3 + 2a_4,$$
$$b_6 = a_3^2 + 4a_6, \quad b_8 = a_1^2 a_6 - a_1 a_3 a_4 + 4a_2 a_6 + a_2 a_3^2 - a_4^2.$$

In odd characteristic, the transformation $y \mapsto y - (a_1 x + a_3)/2$ leads to an isomorphic curve given by

$$y^2 = x^3 + \frac{b_2}{4} x^2 + \frac{b_4}{2} x + \frac{b_6}{4}. \tag{13.2}$$

The cubic polynomial above has only simple roots over the algebraic closure $\overline{K}$ if and only if its discriminant is nonzero. The equation of the discriminant is therefore useful to determine if (13.2) is an elliptic curve or not. In addition, it is relevant for characteristic 2 fields as well.

**Definition 13.2** Let $E$ be a curve defined over $K$ by (13.1) and let $b_2, b_4, b_6$ and $b_8$ as above. The *discriminant of the curve* $E$ denoted by $\Delta$ satisfies

$$\Delta = -b_2^2 b_8 - 8b_4^3 - 27b_6^2 + 9b_2 b_4 b_6.$$

The curve $E$ is nonsingular, and thus is an elliptic curve, if and only if $\Delta$ is nonzero. In this case, we introduce the *j-invariant* of $E$, that is $j(E) = (b_2^2 - 24b_4)^3 / \Delta$.

**Example 13.3** In $\mathbb{F}_p$ with $p = 2003$, an elliptic curve is given by

$$E_1 : y^2 + 2xy + 8y = x^3 + 5x^2 + 1136x + 531. \tag{13.3}$$

Indeed, we have $b_2 = 24, b_4 = 285, b_6 = 185, \Delta = 1707 \neq 0$ and $j = 171$.

§ 13.1 Summary of background on elliptic curves

We now show how to turn the set of points of $E$ into a group with group operation denoted by $\oplus$. For this we visualize it over the reals as in Figure 13.1 and assume $h(x) = 0$.

**Figure 13.1** Group law on elliptic curve $y^2 = f(x)$ over $\mathbb{R}$.

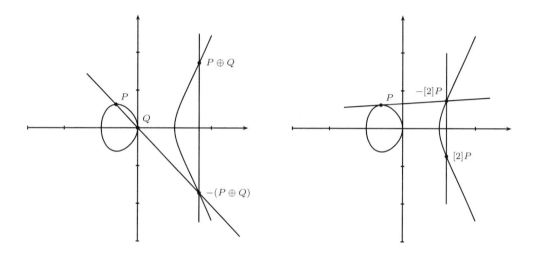

To add two points $P = (x_1, y_1)$ and $Q = (x_2, y_2)$ in general position one draws a line connecting them. There is a third point of intersection. Mirroring this point at the $x$-axis gives the sum $P \oplus Q$. The same construction can be applied to double a point where the connecting line is replaced by the tangent at $P$.

Furthermore, we need to define the sum of two points with the same $x$-coordinate since for them the group operation cannot be performed as stated. As $y^2 = f(x)$ there are at most 2 such points $(x_1, y_1)$ and $(x_1, -y_1)$. Furthermore, we have to find the neutral element of the group.

The way out is to include a further point $P_\infty$ called the *point at infinity*. It can be visualized as lying far out on the $y$-axis such that any line $x = c$, for some constant $c$, parallel to the $y$-axis passes through it. This point is the neutral element of the group. Hence, the line connecting $(x_1, y_1)$ and $(x_1, -y_1)$ passes through $P_\infty$. As it serves as the neutral element, the inflection process leaves it unchanged such that $(x_1, y_1) \oplus (x_1, -y_1) = P_\infty$, i.e., $(x_1, -y_1) = -P$.

This explanation might sound a little like hand-waving and only applicable to $\mathbb{R}$. We now derive the addition formulas for an arbitrary field $K$, which hold universally. For a proof we refer to Chapter 4.

Take $P \neq Q$ with $x_1 \neq x_2$ as above and let us compute the coordinates of $R = P \oplus Q = (x_3, y_3)$. The intersecting line has slope

$$\lambda = \frac{y_1 - y_2}{x_1 - x_2}$$

and passes through $P$. Its equation is thus given by

$$y = \lambda x + \frac{x_1 y_2 - x_2 y_1}{x_1 - x_2}.$$

We denote the constant term by $\mu$ and remark $\mu = y_1 - \lambda x_1$. The intersection points with the curve are obtained by equating the line and $E$

$$(\lambda x + \mu)^2 + (a_1 x + a_3)(\lambda x + \mu) = x^3 + a_2 x^2 + a_4 x + a_6.$$

This leads to the equation $r(x) = 0$ where

$$r(x) = x^3 + (a_2 - \lambda^2 - a_1\lambda)x^2 + (a_4 - 2\lambda\mu - a_3\lambda - a_1\mu)x + a_6 - \mu^2 - a_3\mu.$$

We already know two roots of $r(x)$, namely the $x$-coordinates of the other two points. Since

$$r(x) = (x - x_1)(x - x_2)(x - x_3)$$

one has $\lambda^2 + a_1\lambda - a_2 = x_1 + x_2 + x_3$. As $x_1, x_2$ are defined over $K$ so is $x_3$ and $\tilde{y}_3 = \lambda x_3 + \mu$. The inflection at the $x$-axis has to be translated to the condition that the second point has the same $x$-coordinate and also satisfies the curve equation. We observe that if $P = (x_1, y_1)$ is on the curve then so is $(x_1, -y_1 - a_1 x_1 - a_3)$, which corresponds to $-P$ since the point at infinity is the neutral element for this law. Accordingly, we find $y_3 = -\lambda x_3 - \mu - a_1 x_3 - a_3$.

Doubling $P = (x_1, y_1)$ works just the same with the slope obtained by implicit derivating. Thus we have $P \oplus Q = (x_3, y_3)$ and

$$\begin{aligned} -P &= (x_1, -y_1 - a_1 x_1 - a_3), \\ P \oplus Q &= (\lambda^2 + a_1\lambda - a_2 - x_1 - x_2, \lambda(x_1 - x_3) - y_1 - a_1 x_3 - a_3), \text{ where} \end{aligned}$$

$$\lambda = \begin{cases} \dfrac{y_1 - y_2}{x_1 - x_2} & \text{if } P \neq \pm Q, \\ \dfrac{3x_1^2 + 2a_2 x_1 + a_4 - a_1 y_1}{2y_1 + a_1 x_1 + a_3} & \text{if } P = Q. \end{cases}$$

It is immediate from the pictorial description that this law is commutative, has the point at infinity as neutral element, and that the inverse of $(x_1, y_1)$ is given by $(x_1, -y_1 - a_1 x_1 - a_3)$. The associativity can be shown to hold by simply applying the group law and comparing elements. We leave the lengthy computation to the reader. Note that Chapter 4 gives extensive background showing in an abstract way the group of points on $E$ to form a group. For a more geometrical proof, relying on Bezout's theorem, see e.g., [CAS 1991].

**Example 13.4** One can easily check that the points $P_1 = (1118, 269)$ and $Q_1 = (892, 529)$ lie on the curve $E_1/\mathbb{F}_p$ as defined by (13.3). Then

$$\begin{aligned} -P_1 &= (1118, 1493), \\ P_1 \oplus Q_1 &= (1681, 1706), \\ [2]P_1 &= (1465, 677) \end{aligned}$$

are also on $E_1$.

The point at infinity can be motivated by giving an alternative description of elliptic curves. Equation (13.1) expresses the curve in *affine coordinates*. The same elliptic curve $E$ in *projective coordinates* is then given by the equation

$$E : Y^2 Z + a_1 XYZ + a_3 YZ^2 = X^3 + a_2 X^2 Z + a_4 XZ^2 + a_6 Z^3.$$

Let us denote by $(X_1 : Y_1 : Z_1)$ an element of the *projective 2-space* $\mathbb{P}^2/K$, i.e., a class of $\overline{K}^3 \setminus \{(0,0,0)\}$ modulo the relation

$$(X_1 : Y_1 : Z_1) \sim (X_2 : Y_2 : Z_2) \iff \text{there is } \lambda \in \overline{K}^* \mid X_2 = \lambda X_1, Y_2 = \lambda Y_1 \text{ and } Z_2 = \lambda Z_1.$$

## § 13.1 Summary of background on elliptic curves

By abuse of notation, we identify a class with any of its representatives and call $(X_1 : Y_1 : Z_1)$ a *projective point*. We remark that only a single point of $E$ satisfies $Z_1 = 0$, namely the *point at infinity*, which is in this case $P_\infty = (0 : 1 : 0)$. When $Z_1 \neq 0$, there is a simple correspondence between the projective point $(X_1 : Y_1 : Z_1)$ and the affine point $(x_1, y_1)$ using the formula

$$(x_1, y_1) = (X_1/Z_1, Y_1/Z_1) \tag{13.4}$$

As the representation $(X_1 : Y_1 : Z_1)$ is not normalized, one can perform arithmetic in projective coordinates without any inversion. Note also that *generalized projective coordinates* involving suitable powers of $Z_1$ in (13.4) are commonly used, cf. Sections 13.2.1 and 13.3.1.

**Example 13.5** The point $P_1' = (917 : 527 : 687)$ lies on the curve $E_1$ of equation (13.3) expressed in projective coordinates, i.e.,

$$E_1 : Y^2 Z + 2XYZ + 8YZ^2 = X^3 + 5X^2 Z + 1136 XZ^2 + 531 Z^3.$$

In fact, $P_1'$ is in the same class as $(1118 : 269 : 1)$ and thus corresponds to the affine point $P_1 = (1118, 269)$.

### 13.1.2 Scalar multiplication

Take $n \in \mathbb{N} \setminus \{0\}$ and let us denote the *scalar multiplication by $n$ on $E$* by $[n]$, or $[n]_E$ to avoid confusion. Namely,

$$[n] : E \to E$$
$$P \mapsto [n]P = \underbrace{P \oplus P \oplus \cdots \oplus P}_{n \text{ times}}.$$

This definition extends trivially to all $n \in \mathbb{Z}$, setting $[0]P = P_\infty$ and $[n]P = [-n](-P)$ for $n < 0$. Chapter 9 deals with exponentiation, i.e., the computation of $x$ to some power $n$. In the context of elliptic curves, this corresponds to $[n]P$. Thus multiplications, squarings, and divisions are replaced by additions, doublings, and subtractions on $E$.

As an example, we give the analogue of Algorithm 9.10 with additive notation.

---

**Algorithm 13.6** Sliding window scalar multiplication on elliptic curves

INPUT: A point $P$ on an elliptic curve $E$, a nonnegative integer $n = (n_{l-1} \ldots n_0)_2$, a parameter $k \geqslant 1$ and the precomputed points $[3]P, [5]P, \ldots, [(2^k - 1)]P$.
OUTPUT: The point $[n]P$.

1. $Q \leftarrow P_\infty$ and $i \leftarrow l - 1$
2. **while** $i \geqslant 0$ **do**
3.     **if** $n_i = 0$ **then** $Q \leftarrow [2]Q$ and $i \leftarrow i - 1$
4.     **else**
5.         $s \leftarrow \max(i - k + 1, 0)$
6.         **while** $n_s = 0$ **do** $s \leftarrow s + 1$
7.         **for** $h = 1$ **to** $i - s + 1$ **do** $Q \leftarrow [2]Q$
8.         $u \leftarrow (n_i \ldots n_s)_2$         [$n_i = n_s = 1$ and $i - s + 1 \leqslant k$]
9.         $Q \leftarrow Q \oplus [u]P$         [$u$ is odd so that $[u]P$ is precomputed]

10.     $i \leftarrow s - 1$

11. **return** $Q$

---

**Remark 13.7** Since subtractions can be obtained in a straightforward way, signed-digit representations of $n$ are well suited to compute $[n]P$, cf. Section 9.1.4.

**Example 13.8** With the settings of Example 13.4, let us compute $[763]P_1$ with Algorithm 13.6 and a window of size 3. We precompute $[3]P_1 = (1081, 1674)$, $[5]P_1 = (851, 77)$, $[7]P_1 = (663, 1787)$ and since $763 = (\underset{5}{101}\ \underset{7}{111}\ \underset{5}{101}\ \underset{1}{1})_2$, the intermediate values of $Q$ are

$[5]P_1 = (851, 77)$,      $[10]P_1 = (4, 640)$,       $[20]P_1 = (836, 807)$,

$[40]P_1 = (1378, 1696)$,  $[47]P_1 = (1534, 747)$,    $[94]P_1 = (1998, 1094)$,

$[188]P_1 = (1602, 1812)$, $[376]P_1 = (478, 1356)$,   $[381]P_1 = (1454, 981)$,

$[762]P_1 = (1970, 823)$,  $[763]P_1 = (1453, 1428)$.

Using the NAF expansion of $763 = (\underset{3}{10\bar{1}}00000\underset{-5}{\overline{1}0\bar{1}})_s$ instead, one obtains

$[3]P_1 = (1081, 1674)$,    $[6]P_1 = (255, 1499)$,     $[12]P_1 = (459, 1270)$,

$[24]P_1 = (41, 1867)$,     $[48]P_1 = (1461, 904)$,    $[96]P_1 = (1966, 1808)$,

$[192]P_1 = (892, 529)$,    $[384]P_1 = (1928, 1803)$,  $[768]P_1 = (799, 1182)$,

$[763]P_1 = (1453, 1428)$.

The last step, namely $[763]P_1 = [768]P_1 \oplus [-5]P_1$, needs $[-5]P_1 = (851, 216)$ which is obtained directly from $[5]P_1$.

### 13.1.3 Rational points

When we consider a point $P$ on an elliptic curve $E/K$, it is implicit that $P$ has its coordinates in $\overline{K}$. To stress that $P$ has its coordinates in $K$, we introduce a new concept.

**Definition 13.9** Let $E$ be an elliptic curve defined over $K$. The points lying on $E$ with coordinates in $K$ form the set of $K$-*rational points* of $E$ denoted by $E(K)$. We have

$$E(K) = \{(x_1, y_1) \in K^2 \mid y_1^2 + a_1 x_1 y_1 + a_3 y_1 = x_1^3 + a_2 x_1^2 + a_4 x_1 + a_6\} \cup \{P_\infty\}.$$

The structure of the group of $\mathbb{F}_q$-rational points is easy to describe. Indeed, by Corollary 5.77, $E(\mathbb{F}_q)$ is either cyclic or isomorphic to a product of two cyclic groups, namely $E(\mathbb{F}_q) \simeq \mathbb{Z}/d_1\mathbb{Z} \times \mathbb{Z}/d_2\mathbb{Z}$ where $d_1 \mid d_2$ and $d_1 \mid q - 1$.

For cryptographic applications one usually works in a subgroup of prime order $\ell$. Hence, one is interested in curves and finite fields such that $|E(\mathbb{F}_q)| = c\ell$ for some small *cofactor* $c$. See [GAMC 2000] for conjectural probabilities that the number of points is a prime or has a small cofactor.

Finding a random $\mathbb{F}_q$-rational point $P$ on an elliptic curve $E/\mathbb{F}_q$ is quite easy. See Sections 13.2 and 13.3 for examples. If the curve has a cofactor $c > 1$ then this random point needs not lie inside the group of order $\ell$. However, the point $Q = [c]P$ either equals $P_\infty$, in which case one has to try with a different random point $P$, or is a point in the prime order subgroup.

**Example 13.10** Let us consider the curve $E_1$ as defined by (13.3). One can check that $|E_1(\mathbb{F}_p)| = 1956 = 12 \times 163$. So, there are 1955 affine points with coordinates in $\mathbb{F}_p$ and the point at infinity $P_\infty$ lying on $E_1$. The point $P_1 = (1118, 269)$ is of order 1956 which implies that the group $E_1(\mathbb{F}_p)$ is cyclic generated by $P_1$. The point $Q_1 = (892, 529)$ is of prime order 163.

## 13.1.4 Torsion points

**Definition 13.11** Let $E/K$ be an elliptic curve and $n \in \mathbb{Z}$. The *kernel of* $[n]$, denoted by $E[n]$, satisfies
$$E[n] = \{P \in E(\overline{K}) \mid [n]P = P_\infty\}.$$
An element $P \in E[n]$ is called a *$n$-torsion point*.

**Example 13.12** As $E_1(\mathbb{F}_p)$ is cyclic of order $1956 = 2^2 \times 3 \times 163$, there are $n$-torsion points in $E_1(\mathbb{F}_p)$ for every $n$ dividing 1956. For instance, $R_1 = (1700, 299)$ on $E_1$ satisfies $R_1 = -R_1$. Thus $R_1$ is a 2-torsion rational point. If $n$ is not a divisor of 1956, the corresponding $n$-torsion points have coordinates in some extension of $\mathbb{F}_p$. For example, there is a 9-torsion point with coordinates in the field $\mathbb{F}_{p^3} \simeq \mathbb{F}_p[\theta]$ with $\theta$ such that $\theta^3 + \theta^2 + 2 = 0$. Indeed, we can check that

$$\begin{aligned} S_1 &= (1239\theta^2 + 1872\theta + 112, 1263\theta^2 + 334\theta + 1752) \in E_1(\mathbb{F}_{p^3}), \\ [3]S_1 &= (520, 1568) \in E_1(\mathbb{F}_p), \\ [8]S_1 &= (1239\theta^2 + 1872\theta + 112, 265\theta^2 + 1931\theta + 19) = -S_1 \end{aligned}$$

so that $S_1$ is a 9-torsion point.
See also the related notion of division polynomial in Section 4.4.2.a.

**Theorem 13.13** Let $E$ be an elliptic curve defined over $K$. If the characteristic of $K$ is either zero or prime to $n$ then
$$E[n] \simeq \mathbb{Z}/n\mathbb{Z} \times \mathbb{Z}/n\mathbb{Z}.$$
Otherwise, when $\mathrm{char}(K) = p$ and $n = p^r$, then either
$$E[p^r] = \{P_\infty\}, \text{ for all } r \geqslant 1 \quad \text{or} \quad E[p^r] \simeq \mathbb{Z}/p^r\mathbb{Z}, \text{ for all } r \geqslant 1.$$

**Definition 13.14** Let $\mathrm{char}(K) = p$ and let $E$ be defined over $K$. If $E[p^r] = \{P_\infty\}$ for one and in fact for all positive integers $r$, then the curve is called *supersingular*. Otherwise the curve is called *ordinary*.

A curve defined over a prime field $\mathbb{F}_p, p > 3$ is supersingular if and only if $|E(\mathbb{F}_p)| = p + 1$, cf. Proposition 13.31. Note also that if $\mathrm{char}(\mathbb{F}_q) = 2$ or 3, $E$ is supersingular if and only if its $j$-invariant is zero.

**Example 13.15** The curve $E_1/\mathbb{F}_p$ is ordinary. This implies that $E_1[p]$ is a subgroup of $(E_1(\overline{\mathbb{F}}_p), \oplus)$ isomorphic to $(\mathbb{F}_p, +)$.

## 13.1.5 Isomorphisms

Some changes of variables do not fundamentally alter an elliptic curve. Let us first describe the transformations that keep the curve in Weierstraß form.

### 13.1.5.a Admissible change of variables and twists

Let $E/K$ be an elliptic curve
$$E: y^2 + a_1 xy + a_3 y = x^3 + a_2 x^2 + a_4 x + a_6.$$
The maps
$$x \mapsto u^2 x' + r \quad \text{and} \quad y \mapsto u^3 y' + u^2 s x' + t$$

with $(u, r, s, t) \in K^* \times K^3$ are invertible and transform the curve $E$ into

$$E' : y'^2 + a'_1 x'y' + a'_3 y' = x'^3 + a'_2 x'^2 + a'_4 x' + a'_6,$$

where the $a'_i$'s belong to $K$ and can be expressed in terms of the $a_i$'s and $u, r, s, t$. Via the inverse map, we associate to each point of $E$ a point of $E'$ showing that both curves are *isomorphic over $K$*. These changes of variables are the only ones leaving invariant the shape of the defining equation and, hence, they are the only *admissible change of variables*.

In case $(u, r, s, t)$ belongs to $\overline{K}^* \times \overline{K}^3$ whereas the curves $E$ and $E'$, as above, are still defined over $K$, then $E$ and $E'$ are *isomorphic over $\overline{K}$* or *twists* of each other.

**Corollary 13.16** Assume that the characteristic of $K$ is prime to 6 and let $E$ be given by a short Weierstraß equation

$$E : y^2 = x^3 + a_4 x + a_6.$$

- If $a_4 = 0$ then for every $a'_6 \in K^*$ the curve $E$ is isomorphic to $E' : y^2 = x^3 + a'_6$ over $K\big((a_6/a'_6)^{1/6}\big)$.
- If $a_6 = 0$ then for every $a'_4 \in K^*$ the curve $E$ is isomorphic to $E' : y^2 = x^3 + a'_4 x$ over $K\big((a_4/a'_4)^{1/4}\big)$.
- If $a_4 a_6 \neq 0$ then for every $v \in K^*$ the curve $E$ is isomorphic to $\widetilde{E}_v : y^2 = x^3 + a'_4 x + a'_6$ with $a'_4 = v^2 a_4$ and $a'_6 = v^3 a_6$ over the field $K(\sqrt{v})$.

The curves $\widetilde{E}_v$ are called *quadratic twists* of $E$. Note that $E$ is isomorphic to $\widetilde{E}_v$ over $K$ if and only if $v$ is a square in $K^*$. Therefore up to isomorphisms there is only one quadratic twist of a curve with $a_4 a_6 \neq 0$.

**Remark 13.17** Likewise one can define the quadratic twist of $E$ by a quadratic nonresidue $v$ as $\widetilde{E}_v : vy^2 = x^3 + a_4 x + a_6$, which is isomorphic to the above definition, as can be seen by dividing by $v^3$ and transforming $y \mapsto y/v, x \mapsto x/v$.

From this form one sees that $E$ and $\widetilde{E}_v$ together contain exactly two points $(x, y_i)$ for each field element $x \in \mathbb{F}_q$.

**Proposition 13.18** Let $E/K$ and $E'/K$ be two elliptic curves. If $E$ and $E'$ are isomorphic over $K$ then they have the same $j$-invariant. Conversely, if $j(E) = j(E')$ then $E$ and $E'$ are isomorphic over $\overline{K}$.

Using an adequate isomorphism over $K$, it is always possible to find a *short Weierstraß equation* that actually depends on the characteristic of the field and on the value of the $j$-invariant. All the cases and equations are summarized in Table 13.2.

**Table 13.2** Short Weierstraß equations.

| char $K$ | Equation | $\Delta$ | $j$ |
|---|---|---|---|
| $\neq 2, 3$ | $y^2 = x^3 + a_4 x + a_6$ | $-16(4a_4^3 + 27a_6^2)$ | $1728 a_4^3/4\Delta$ |
| 3 | $y^2 = x^3 + a_4 x + a_6$ | $-a_4^3$ | 0 |
| 3 | $y^2 = x^3 + a_2 x^2 + a_6$ | $-a_2^3 a_6$ | $-a_2^3/a_6$ |
| 2 | $y^2 + a_3 y = x^3 + a_4 x + a_6$ | $a_3^4$ | 0 |
| 2 | $y^2 + xy = x^3 + a_2 x^2 + a_6$ | $a_6$ | $1/a_6$ |

**Example 13.19** The change of variables $(x, y) \mapsto (x - 2, y - x - 2)$ transforms the curve $E_1$ given by (13.3) into
$$E_2 : y^2 = x^3 + 1132x + 278.$$
The point $P_1 = (1118, 269)$ is mapped to $P_2 = (1120, 1391) \in E_2(\mathbb{F}_p)$.

Let $v$ be a quadratic nonresidue modulo $p = 2003$ and let $u \in \mathbb{F}_{p^2}$ be a square root of $v$. Then the change of variables $(x, y) \mapsto (x/u^2, y/u^3)$ is an $\mathbb{F}_{p^2}$-isomorphism between
$$E_2 : y^2 = x^3 + 1132x + 278.$$
and its *quadratic twist by* $v$, namely
$$\widetilde{E}_{2,v} : y^2 = x^3 + 1132v^2 x + 278v^3.$$
We have $\Delta(\widetilde{E}_{2,v}) = v^6 \Delta(E_2)$ and $j(\widetilde{E}_{2,v}) = j(E_2) = 171$.

The curves $E_2$ and $\widetilde{E}_{2,v}$ are defined over $\mathbb{F}_p$ whereas the isomorphism has coefficients in $\mathbb{F}_{p^2}$.

**Remark 13.20** There are many other ways to represent an elliptic curve. For instance, we can cite the *Legendre form*
$$y^2 = x(x - 1)(x - \lambda)$$
or the *Jacobi model*
$$y^2 = x^4 + ax^2 + b.$$
Over a field of characteristic greater than 3, it is also possible to represent an elliptic curve as the intersection of two quadrics with a rational point [CAS 1991]. The resulting *Jacobi form* is used in [LISM 2001] to prevent SPA/DPA attacks, cf. Section 29.1.2.c. Quite recently, some attention has been given to another representation, namely the Hessian form, which presents some advantages from an algorithmic and cryptographic point of view [SMA 2001, FRI 2001, JOQU 2001].

### 13.1.5.b Hessian form

Let $\mathbb{F}_q$ be a finite field where $q$ is a prime power such that $q \equiv 2 \pmod{3}$ and consider an elliptic curve $E$ over $\mathbb{F}_q$ with a $\mathbb{F}_q$-rational point of order 3. These assumptions are not fundamentally necessary but they make the construction of the Hessian form easier and let the equation be defined over $\mathbb{F}_q$. In particular, one can assume that $E$ is given by the equation
$$E : y^2 + a_1 xy + a_3 y = x^3,$$
moving a point of order 3 to the origin, if necessary.

Let $\delta = (a_1^3 - 27a_3)$ so that $\Delta = a_3^3 \delta$. Now if $q \equiv 2 \pmod{3}$ every element $\alpha \in \mathbb{F}_q$ is a cube. Thus every $\alpha$ has a unique cube root, denoted by $\alpha^{1/3}$, which is equal to plus or minus the square root of $\alpha^{(q+1)/3}$. This implies that
$$\mu = \frac{1}{3}\left((-27a_3\delta^2 - \delta^3)^{1/3} + \delta\right) \in \mathbb{F}_q.$$
With these settings, to every point $(x_1, y_1)$ on $E$ corresponds $(X_1 : Y_1 : Z_1)$ with
$$X_1 = \frac{a_1(2\mu - \delta)}{3\mu - \delta} x_1 + y_1 + a_3, \quad Y_1 = \frac{-a_1\mu}{3\mu - \delta} x_1 - y_1, \quad Z_1 = \frac{-a_1\mu}{3\mu - \delta} x_1 - a_3$$
on the cubic
$$H : X^3 + Y^3 + Z^3 = cXYZ \quad \text{where} \quad c = 3\frac{\mu - \delta}{\mu}.$$

**Definition 13.21** The equation $H$ is called the *Hessian form* of $E$.

One of the main features of elliptic curves expressed in Hessian form is the simplicity of the group law, which is independent of the parameter $c$.

Namely, take $P = (X_1 : Y_1 : Z_1)$ and $Q = (X_2 : Y_2 : Z_2)$ on $H$ such that $P \neq Q$, then the point with coordinates $(X_3 : Y_3 : Z_3)$ such that

$$X_3 = Y_1^2 X_2 Z_2 - Y_2^2 X_1 Z_1, \quad Y_3 = X_1^2 Y_2 Z_2 - X_2^2 Y_1 Z_1, \quad Z_3 = Z_1^2 X_2 Y_2 - Z_2^2 X_1 Y_1$$

is on $H$ and corresponds to $P \oplus Q$.

One can check that the neutral element for that law is $(1 : -1 : 0)$ and that the opposite of $P_1$ is $-P_1 = (Y_1 : X_1 : Z_1)$.

The coordinates of $[2]P$ are

$$X_3 = Y_1(Z_1^3 - X_1^3), \qquad Y_3 = X_1(Y_1^3 - Z_1^3), \qquad Z_3 = Z_1(X_1^3 - Y_1^3).$$

An addition requires 12 field multiplication and 6 squarings, whereas a doubling needs 6 multiplications and 3 squarings and both operations can be implemented in a highly parallel way [SMA 2001]. It is also interesting to note that $[2]P$ is equal to $(Z_1 : X_1 : Y_1) \oplus (Y_1 : Z_1 : X_1)$. As a consequence the same formulas can be used to double, add, and subtract points, which makes Hessian curves interesting against side-channel attacks [JOQU 2001] (cf. Section 29.1.2.b).

To find the Hessian form of an elliptic curve $E/\mathbb{F}_q$ in the general case [FRI 2001], we remark that the $j$-invariant of $H$ is equal to

$$j = \frac{c^3(c^3 + 216)^3}{c^9 - 81c^6 + 2187c^3 - 19683}.$$

So the Hessian form of $E$ is defined over $\mathbb{F}_q$ if and only if there exists $c \in \mathbb{F}_q$ such that

$$c^3(c^3 + 216)^3 - j(c^9 - 81c^6 + 2187c^3 - 19683) = 0$$

where $j$ is the $j$-invariant of $E$.

**Example 13.22** Take
$$E_2 : y^2 = x^3 + 1132x + 278$$

defined over $\mathbb{F}_p$ with $p = 2003$. Moving the point $(522, 1914) \in E_2(\mathbb{F}_p)$ of order 3 to the origin by the transformation

$$(x, y) \mapsto (x + 522, y + 555x + 1914)$$

gives the curve
$$E_3 : y^2 + 1110xy + 1825y = x^3.$$

So, from above, $\delta = 1427$ and $\mu = 1322$ so that $E_3$, consequently $E_2$ and $E_1$, are all isomorphic to

$$H : X^3 + Y^3 + Z^3 = 274XYZ.$$

The point $(1118, 269)$ on $E_1$ is sent to $(1120, 1391)$ on $E_2$, from where it is in turn mapped to $(598, 85)$ on $E_3$, which is finally sent to $(1451 : 672 : 935)$ on $H$.

Note that all these transformations respect the group laws of the different curves. Indeed, a $K$-isomorphism between the curves $E$ and $E'$ always gives rise to a group homorphism between $E(K)$ and $E'(K)$. However, these notions are different. That is why we introduce a new concept in the following.

### 13.1.6 Isogenies

**Definition 13.23** Two curves $E/K$ and $E'/K$ are *isogenous over* $K$ if there exists a morphism $\psi : E \to E'$ with coefficients in $K$ mapping the neutral element of $E$ to the neutral element of $E'$. From this simple property, it is possible to show that $\psi$ is a group homomorphism from $E(K)$ to $E'(K)$.

One important property is that for every isogeny $\psi$, there exists a unique isogeny $\hat{\psi} : E' \to E$ called the *dual isogeny* such that

$$\hat{\psi} \circ \psi = [m]_E \quad \text{and} \quad \psi \circ \hat{\psi} = [m]_{E'}.$$

The *degree of the isogeny* $\psi$ is equal to this $m$. For more background on isogenies, we refer to Section 4.3.4

**Proposition 13.24** Two elliptic curves $E$ and $E'$ defined over $\mathbb{F}_q$ are isogenous over $\mathbb{F}_q$ if and only $|E(\mathbb{F}_q)| = |E'(\mathbb{F}_q)|$.

**Example 13.25** Take

$$E_2 : y^2 = x^3 + 1132x + 278$$

and

$$E_4 : y^2 = x^3 + 500x + 1005.$$

These two curves have the same cardinality, $|E_2(\mathbb{F}_p)| = |E_4(\mathbb{F}_p)| = 1956$. Then $E_2$ and $E_4$ must be isogenous over $\mathbb{F}_{2003}$. The isogeny of degree 2 is given by the formula [LER 1997]

$$\psi : (x, y) \longmapsto \left( \frac{x^2 + 301x + 527}{x + 301}, \frac{yx^2 + 602yx + 1942y}{x^2 + 602x + 466} \right).$$

For instance, the points, $P_2 = (1120, 1391)$ and $Q_2 = (894, 1425)$ in $E_2(\mathbb{F}_p)$ are respectively mapped by $\psi$ on $P_4 = (565, 302)$ and $Q_4 = (1818, 1002)$ which lie on $E_4$. Now

$$\begin{aligned}
P_2 \oplus Q_2 &= (1683, 1388), \\
P_4 \oplus Q_4 &= (1339, 821), \\
\psi(P_2 \oplus Q_2) &= (1339, 821), \\
&= \psi(P_2) \oplus \psi(Q_2).
\end{aligned}$$

Note that $E_2$ and $E_4$ are isogenous but not isomorphic since $j(E_2) = 171$ whereas $j(E_4) = 515$. Furthermore, the group structure is different as $E_2(\mathbb{F}_p)$ is cyclic while $E_4(\mathbb{F}_p)$ is the direct product of a group of order 2 generated by $(1829, 0)$ and a group of order 978 generated by $(915, 1071)$.

### 13.1.7 Endomorphisms

The multiplication by $n$ is an endomorphism of the curve $E$ for every $n \in \mathbb{Z}$. The set of all endomorphisms of $E$ defined over $K$ will be denoted by $\text{End}_K(E)$ or more simply by $\text{End}(E)$, and thus contains at least $\mathbb{Z}$.

**Definition 13.26** If $\text{End}(E)$ is strictly bigger than $\mathbb{Z}$ we say that $E$ has *complex multiplication*.

Let $E$ be a nonsupersingular elliptic curve over $\mathbb{F}_q$. Such an $E$ always has complex multiplication. Indeed, the Frobenius automorphism of $\mathbb{F}_q$ extends to the points of the curve by sending $P_\infty$ to itself and $P = (x_1, y_1)$ to $\phi_q(P) = (x_1^q, y_1^q)$. One can easily check that the point $\phi_q(P)$ is again a point on the curve irrespective of the field of definition of $P$. Hence, $\phi_q$ is an endomorphism of $E$, called the *Frobenius endomorphism* of $E/\mathbb{F}_q$. It is different from $[n]$ for all $n \in \mathbb{Z}$.

**Example 13.27** Take $P_1 = (1120, 391)$ on $E_1/\mathbb{F}_p$. Since $P_1$ has coordinates in $\mathbb{F}_p$, $\phi_p(P_1)$ is simply equal to $P_1$. At present, let us consider a point on $E_1$ with coordinates in an extension of $\mathbb{F}_p$. For instance, in Example 13.10, we give the point $S_1$ of order 9 in $E_1(\mathbb{F}_{p^3})$. We have

$$\begin{aligned} S_1 &= (1239\theta^2 + 1872\theta + 112, 1263\theta^2 + 334\theta + 1752), \\ \phi_p(S_1) &= (217\theta^2 + 399\theta + 1297, 681\theta^2 + 811\theta + 102), \\ \phi_p^2(S_1) &= (547\theta^2 + 1735\theta + 297, 59\theta^2 + 858\theta + 325), \\ \phi_p^3(S_1) &= (1239\theta^2 + 1872\theta + 112, 1263\theta^2 + 334\theta + 1752) = S_1. \end{aligned}$$

All of them are also 9-torsion points.

### 13.1.8 Cardinality

The cardinality of an elliptic curve $E$ over $\mathbb{F}_q$, i.e., the number of $\mathbb{F}_q$-rational points, is an important aspect for the security of cryptosystems built on $E(\mathbb{F}_q)$, cf. Section 19.3.

The theorem of Hasse–Weil relates the number of points to the field size.

**Theorem 13.28 (Hasse–Weil)** Let $E$ be an elliptic curve defined over $\mathbb{F}_q$. Then

$$|E(\mathbb{F}_q)| = q + 1 - t \text{ and } |t| \leqslant 2\sqrt{q}.$$

**Remarks 13.29**

(i) The integer $t$ is called the *trace of the Frobenius endomorphism*.
(ii) For any integer $t \in [-2\sqrt{p}, 2\sqrt{p}]$ there is at least one elliptic curve $E$ defined over $\mathbb{F}_p$ whose cardinality is $p + 1 - t$.

Concerning admissible cardinalities, the more general result is proved in [WAT 1969].

**Theorem 13.30** Let $q = p^d$. There exists an elliptic curve $E$ defined over $\mathbb{F}_q$ with $|E(\mathbb{F}_q)| = q + 1 - t$ if and only if one of the following conditions holds:

1. $t \not\equiv 0 \pmod{p}$ and $t^2 \leqslant 4q$.
2. $d$ is odd and either (i) $t = 0$ or (ii) $p = 2$ and $t^2 = 2q$ or (iii) $p = 3$ and $t^2 = 3q$.
3. $d$ is even and either (i) $t^2 = 4q$ or (ii) $p \not\equiv 1 \pmod 3$ and $t^2 = q$ or (iii) $p \not\equiv 1 \pmod 4$ and $t = 0$.

One associates to $\phi_q$ the polynomial

$$\chi_E(T) = T^2 - tT + q.$$

It is called the *characteristic polynomial of the Frobenius endomorphism*, since

$$\chi_E(\phi_q) = \phi_q^2 - [t]\phi_q + [q] = [0].$$

So, for each $P \in E(\overline{\mathbb{F}}_q)$, we have

$$\phi_q^2(P) \oplus [-t]\phi_q(P) \oplus [q]P = P_\infty.$$

As points in $E(\mathbb{F}_q)$ are fixed under $\phi_q$ they form the kernel of $(\mathrm{Id} - \phi_q)$ and $|E(\mathbb{F}_q)| = \chi_E(1)$.

## § 13.1 Summary of background on elliptic curves

From the complex roots $\tau$ and $\overline{\tau}$ of $\chi_E(\phi_q)$ one can compute the group order of $E(\mathbb{F}_{q^k})$, that is

$$|E(\mathbb{F}_{q^k})| = q^k + 1 - \tau^k - \overline{\tau}^k, \text{ for all } k \geq 1. \tag{13.5}$$

More explicitly, one has

$$|E(\mathbb{F}_{q^k})| = q^k + 1 - t_k$$

where the sequence $(t_k)_{k \in \mathbb{N}}$ satisfies $t_0 = 2$, $t_1 = t$ and $t_{k+1} = tt_k - qt_{k-1}$, for $k \geq 1$.

We also have the following properties.

**Proposition 13.31** Let $E$ be a curve defined over a field $\mathbb{F}_q$ of characteristic $p$. The curve $E$ is supersingular if and only if the trace $t$ of the Frobenius satisfies

$$t \equiv 0 \pmod{p}.$$

**Proposition 13.32** Let $E$ be a curve defined over $\mathbb{F}_q$ and let $\widetilde{E}$ be the quadratic twist of $E$. Then

$$|E(\mathbb{F}_q)| + |\widetilde{E}(\mathbb{F}_q)| = 2q + 2.$$

This can be easily seen to hold from Remark 13.17. One immediately gets $\chi_{\widetilde{E}}(T) = T^2 + tT + q$. When one tries to find a curve with a suitable cryptographic order, that is, an order with a large prime factor, Proposition 13.32 is especially useful since it gives two candidates for each computation, cf. Chapter 17.

**Example 13.33** The cardinality of $E_2(\mathbb{F}_p)$ is 1956. Therefore, $\phi_p$ satisfies

$$\chi_{E_2}(T) = T^2 - 48T + 2003.$$

Let $R_2 = (443\theta^2 + 1727\theta + 1809, 929\theta^2 + 280\theta + 946)$. Then

$$\phi_p(R_2) = (857\theta^2 + 1015\theta + 766, 1260\theta^2 + 1902\theta + 419),$$
$$\phi_p^2(R_2) = (7030\theta^2 + 1264\theta + 1568, 948\theta^2 + 1824\theta + 119)$$

and one can check that

$$\phi_p^2(R_2) - [48]\phi_{2003}(R_2) + [2003]R_2 = P_\infty.$$

Also, we deduce that $|E_2(\mathbb{F}_{p^2})| = 4013712$ and $|E_2(\mathbb{F}_{p^3})| = 8036231868$.

Finally the cardinality of the curve

$$\widetilde{E}_2 : y^2 = x^3 + 774x + 1867$$

which is the twist of $E_2$ by the quadratic nonresidue 78, satisfies $|\widetilde{E}_2| = 2052$, and the characteristic equation of the Frobenius of $\widetilde{E}_2/\mathbb{F}_p$ is

$$\chi_{\widetilde{E}_2}(T) = T^2 + 48T + 2003.$$

## 13.2 Arithmetic of elliptic curves defined over $\mathbb{F}_p$

In this section we consider curves defined over finite prime fields. As they should be used in cryptographic applications, we can assume $p$ to be large, hence, at least $p > 3$. We remark that all considerations in this section hold true for an elliptic curve defined over an arbitrary finite field $\mathbb{F}_q$ if $\mathrm{char}(\mathbb{F}_q) > 3$ and for supersingular curves over field of characteristic 3.

We already know that an elliptic curve $E$ can be represented with respect to several coordinate systems, e.g., affine or projective coordinates. In the following we deal with efficient addition and doubling in the group of points $E$. To this aim we introduce five different coordinate systems in which the speeds of addition and doubling differ. We measure the time by the number of field operations needed to perform the respective operation.

In characteristic $p > 3$, one can always take for $E$, cf. Table 13.2, an equation of the form

$$E : y^2 = x^3 + a_4 x + a_6,$$

where $a_4$ and $a_6$ are in $\mathbb{F}_p$. The points lying on the curve can have coordinates in $\mathbb{F}_p$ or in some extension $\mathbb{F}_q/\mathbb{F}_p$, for instance in an optimal extension field, cf. Section 11.3. This has two advantages. First, it is straightforward to obtain the cardinality of $E(\mathbb{F}_q)$ using (13.5) and one can use the Frobenius $\phi_p$ to speed up computations, cf. Section 15.1.

In the remainder of this section we deal with addition and doubling in different coordinate systems, give strategies for choosing optimal coordinates for scalar multiplication and introduce Montgomery coordinates and their arithmetic. Finally, we show how to compress the representation of a point.

An elementary multiplication in $\mathbb{F}_q$ (resp. a squaring and an inversion) will be abbreviated by M (resp. S and I).

### 13.2.1 Choice of the coordinates

This section is based on [COMI+ 1998].

In Section 13.1.1 we explained the group law in general. Here we shall give formulas for the coordinates of the result of the

- addition of two points $P$ and $Q \in E(\mathbb{F}_p)$ provided $P \neq \pm Q$,
- doubling of $P$.

#### 13.2.1.a Affine coordinates ($\mathcal{A}$)

We can assume that $E$ is given by

$$y^2 = x^3 + a_4 x + a_6.$$

By the arguments above, we know that the opposite of the point $(x_1, y_1)$ lying on $E$ is $(x_1, -y_1)$. Also we have:

**Addition**

Let $P = (x_1, y_1), Q = (x_2, y_2)$ such that $P \neq \pm Q$ and $P \oplus Q = (x_3, y_3)$. In this case, addition is given by

$$x_3 = \lambda^2 - x_1 - x_2, \qquad y_3 = \lambda(x_1 - x_3) - y_1, \qquad \lambda = \frac{y_1 - y_2}{x_1 - x_2}.$$

## Doubling

Let $[2]P = (x_3, y_3)$. Then

$$x_3 = \lambda^2 - 2x_1, \qquad y_3 = \lambda(x_1 - x_3) - y_1, \qquad \lambda = \frac{3x_1^2 + a_4}{2y_1}.$$

For these formulas one can easily read off that an addition and a doubling require $I + 2M + S$ and $I + 2M + 2S$, respectively.

## Doubling followed by an addition

Building on the ideas in [EILA+ 2003], the authors of [CIJO+ 2003] show how to speed up the computation of a doubling followed by an addition using $[2]P \oplus Q$ as $(P \oplus Q) \oplus P$. The basic idea, i.e., omitting the computation of the intermediate values $y_3$ and $x_3$, saves one multiplication and the new formulas are more efficient whenever a field inversion is more expensive than 6 multiplications. The formulas are as follows where we assume that $P \neq \pm Q$ and $[2]P \neq -Q$

$$A = (x_2 - x_1)^2, \qquad B = (y_2 - y_1)^2, \qquad C = A(2x_1 + x_2) - B,$$
$$D = C(x_2 - x_1), \qquad E = D^{-1}, \qquad \lambda = CE(y_2 - y_1),$$
$$\lambda_2 = 2y_1 A(x_2 - x_1)E - \lambda, \qquad x_4 = (\lambda_2 - \lambda)(\lambda + \lambda_2) + x_2, \qquad y_4 = (x_1 - x_4)\lambda_2 - y_1,$$

needing $I + 9M + 2S$.

### 13.2.1.b Projective coordinates ($\mathcal{P}$)

In *projective coordinates*, the equation of $E$ is

$$Y^2 Z = X^3 + a_4 X Z^2 + a_6 Z^3.$$

The point $(X_1 : Y_1 : Z_1)$ on $E$ corresponds to the affine point $(X_1/Z_1, Y_1/Z_1)$ when $Z_1 \neq 0$ and to the point at infinity $P_\infty = (0 : 1 : 0)$ otherwise. The opposite of $(X_1 : Y_1 : Z_1)$ is $(X_1 : -Y_1 : Z_1)$.

## Addition

Let $P = (X_1 : Y_1 : Z_1)$, $Q = (X_2 : Y_2 : Z_2)$ such that $P \neq \pm Q$ and $P \oplus Q = (X_3 : Y_3 : Z_3)$. Then set

$$A = Y_2 Z_1 - Y_1 Z_2, \qquad B = X_2 Z_1 - X_1 Z_2, \qquad C = A^2 Z_1 Z_2 - B^3 - 2B^2 X_1 Z_2$$

so that

$$X_3 = BC, \qquad Y_3 = A(B^2 X_1 Z_2 - C) - B^3 Y_1 Z_2, \qquad Z_3 = B^3 Z_1 Z_2.$$

## Doubling

Let $[2]P = (X_3 : Y_3 : Z_3)$ then put

$$A = a_4 Z_1^2 + 3X_1^2, \qquad B = Y_1 Z_1, \qquad C = X_1 Y_1 B, \qquad D = A^2 - 8C$$

and

$$X_3 = 2BD, \qquad Y_3 = A(4C - D) - 8Y_1^2 B^2, \qquad Z_3 = 8B^3.$$

No inversion is needed, and the computation times are $12M + 2S$ for a general addition and $7M + 5S$ for a doubling. If one of the input points to the addition is given by $(X_2 : Y_2 : 1)$, i.e., directly transformed from affine coordinates, then the requirements for an addition decrease to $9M + 2S$.

### 13.2.1.c  Jacobian and Chudnovsky Jacobian coordinates ($\mathcal{J}$ and $\mathcal{J}^c$)

With *Jacobian coordinates* the curve $E$ is given by

$$Y^2 = X^3 + a_4 X Z^4 + a_6 Z^6.$$

The point $(X_1 : Y_1 : Z_1)$ on $E$ corresponds to the affine point $(X_1/Z_1^2, Y_1/Z_1^3)$ when $Z_1 \neq 0$ and to the point at infinity $P_\infty = (1 : 1 : 0)$ otherwise. The opposite of $(X_1 : Y_1 : Z_1)$ is $(X_1 : -Y_1 : Z_1)$.

**Addition**

Let $P = (X_1 : Y_1 : Z_1)$, $Q = (X_2 : Y_2 : Z_2)$ such that $P \neq \pm Q$ and $P \oplus Q = (X_3 : Y_3 : Z_3)$. Then set

$$A = X_1 Z_2^2, \quad B = X_2 Z_1^2, \quad C = Y_1 Z_2^3, \quad D = Y_2 Z_1^3, \quad E = B - A, \quad F = D - C$$

and

$$X_3 = -E^3 - 2AE^2 + F^2, \quad Y_3 = -CE^3 + F(AE^2 - X_3), \quad Z_3 = Z_1 Z_2 E.$$

**Doubling**

Let $[2]P = (X_3 : Y_3 : Z_3)$. Then set

$$A = 4 X_1 Y_1^2, \qquad B = 3 X_1^2 + a_4 Z_1^4$$

and

$$X_3 = -2A + B^2, \qquad Y_3 = -8 Y_1^4 + B(A - X_3), \qquad Z_3 = 2 Y_1 Z_1.$$

The complexities are $12M + 4S$ for an addition and $4M + 6S$ for a doubling. If one of the points is given in the form $(X_1 : Y_1 : 1)$ the costs for addition reduce to $8M + 3S$.

The doubling involves one multiplication by the constant $a_4$. If it is small this multiplication can be performed by some additions and hence be neglected in the operation count. Especially if $a_4 = -3$ one can compute $T = 3X_1^2 - 3Z_1^4 = 3(X_1 - Z_1^2)(X_1 + Z_1^2)$ leading to only $4M + 4S$ for a doubling. Brier and Joye [BRJO 2003] study the use of isogenies to map a given curve to an isogenous one having this preferable parameter. Their conclusion is that for most randomly chosen curves there exists an isogeny of small degree mapping it to a curve with $a_4 = -3$, which justifies that the curves in the standards have this parameter.

The parameter $a_4 = 0$ is even more advantageous as the costs drop down to $3M + 4S$. However, this choice is far more special and the endomorphism ring $\text{End}(E)$ contains a third root of unity.

In Jacobian coordinates, doublings are faster and additions slower than for the projective coordinates. To improve additions, a point $P$ can be represented as a quintuple $(X_1, Y_1, Z_1, Z_1^2, Z_1^3)$. These coordinates are called *Chudnovsky Jacobian coordinates*. Additions and doublings are given by the same formulas as for $\mathcal{J}$ but the complexities are $11M + 3S$ and $5M + 6S$.

### 13.2.1.d  Modified Jacobian coordinates ($\mathcal{J}^m$)

*Modified Jacobian coordinates* were introduced by Cohen et al. [COMI+ 1998]. They are based on $\mathcal{J}$ but the internal representation of a point $P$ is the quadruple $(X_1, Y_1, Z_1, a_4 Z_1^4)$. The formulas are essentially the same as for $\mathcal{J}$. The main difference is the introduction of $C = 8 Y_1^4$ so that $Y_3 = B(A - X_3) - C$ and $a_4 Z_3^4 = 2C(a_4 Z_1^4)$ with the notation of Section 13.2.1.c. An addition takes $13M + 6S$ and a doubling $4M + 4S$. If one point is in affine coordinates, an addition takes $9M + 5S$. As I takes on average between 9 and 40M and S is about 0.8M, this system offers the fastest doubling procedure.

## § 13.2 Arithmetic of elliptic curves defined over $\mathbb{F}_p$

### 13.2.1.e Example

Take
$$E_2 : y^2 = x^3 + 1132x + 278$$
and let $P_2 = (1120, 1391)$ and $Q_2 = (894, 1425)$ be two affine points on $E_2$. We recall below the equation and the internal representation of $P_2$ and $Q_2$ for each coordinate system. Note that for projective like systems we put $Z$ to some random value and multiply $X$ and $Y$ by the respective powers.

| System | Equation | $P_2$ | $Q_2$ |
|---|---|---|---|
| $\mathcal{A}$ | $y^2 = x^3 + 1132x + 278$ | $(1120, 1391)$ | $(894, 1425)$ |
| $\mathcal{P}$ | $Y^2Z = X^3 + 1132XZ^2 + 278Z^3$ | $(450 : 541 : 1449)$ | $(1774 : 986 : 1530)$ |
| $\mathcal{J}$ | $Y^2 = X^3 + 1132XZ^4 + 278Z^6$ | $(1213 : 408 : 601)$ | $(1623 : 504 : 1559)$ |
| $\mathcal{J}^c$ | — | $(1213, 408, 601, 661, 667)$ | $(1623, 504, 1559, 842, 713)$ |
| $\mathcal{J}^m$ | — | $(1213, 408, 601, 1794)$ | $(1623, 504, 1559, 1232)$ |

With these particular values of $P_2$ and $Q_2$, let us compute $P_2 \oplus Q_2$, $[2]P_2$ and $[763]P_2$ within the different systems using the double and add method.

| System | $P_2 \oplus Q_2$ | $[2]P_2$ | $[763]P_2$ |
|---|---|---|---|
| $\mathcal{A}$ | $(1683, 1388)$ | $(1467, 143)$ | $(1455, 882)$ |
| $\mathcal{P}$ | $(185 : 825 : 1220)$ | $(352 : 504 : 956)$ | $(931 : 1316 : 1464)$ |
| $\mathcal{J}$ | $(763 : 440 : 1934)$ | $(1800 : 1083 : 1684)$ | $(752 : 1146 : 543)$ |
| $\mathcal{J}^c$ | $(763, 440, 1934, 755, 1986)$ | $(1800, 1083, 1684, 1611, 862)$ | $(752, 1146, 543, 408, 1214)$ |
| $\mathcal{J}^m$ | $(763, 440, 1934, 1850)$ | $(1800, 1083, 1684, 1119)$ | $(752, 1146, 543, 1017)$ |

For each computation, one can check that we obtain a result equivalent to the affine one.

### 13.2.2 Mixed coordinates

To compute scalar multiples of a point one can use all the methods introduced in Chapter 9, especially the signed-digit representations, which are useful, as the negative of $P$ is obtained by simply negating the $y$-coordinate.

The main idea here is to mix the different systems of coordinates defined above. This idea was already mentioned in adding an affine point to one in another system. In general, one can add points expressed in two different systems and give the result in a third one. For example $\mathcal{J} + \mathcal{J}^c = \mathcal{J}^m$ means that we add points in Jacobian and Chudnovsky Jacobian coordinates and express the result in the modified Jacobian system. So, we are going to choose the most efficient combination for each action we have to perform. See Table 13.3 on page 284 for a precise count of the required operations.

**Precomputations**

The following analysis is given in [COMI$^+$ 1998, Section 4]. Suppose that we want to compute $[n]P$. We shall use the NAF$_w$ representation of $n$; see Section 9.1.4. So, we need to precompute $[i]P$ for each odd $i$ such that $1 < i < 2^{w-1}$. For these precomputations, it is useful to choose either $\mathcal{A}$ if some inversions can be performed in the precomputation stage, or $\mathcal{J}^c$ otherwise, as these systems give rise to the most efficient (mixed) addition formulas. If $\mathcal{A}$ is selected, the Montgomery trick of simultaneous inversions in $\mathbb{F}_p$ should be used, cf. Algorithm 11.15. This leads to

$$(w-1)\mathrm{I} + \left(5 \times 2^{w-2} + 2w - 12\right)\mathrm{M} + \left(2^{w-2} + 2w - 5\right)\mathrm{S}$$

**Table 13.3** *Operations required for addition and doubling.*

| Doubling | | Addition | |
|---|---|---|---|
| Operation | Costs | Operation | Costs |
| $2\mathcal{P}$ | $7M + 5S$ | $\mathcal{J}^m + \mathcal{J}^m$ | $13M + 6S$ |
| $2\mathcal{J}^c$ | $5M + 6S$ | $\mathcal{J}^m + \mathcal{J}^c = \mathcal{J}^m$ | $12M + 5S$ |
| $2\mathcal{J}$ | $4M + 6S$ | $\mathcal{J} + \mathcal{J}^c = \mathcal{J}^m$ | $12M + 5S$ |
| $2\mathcal{J}^m = \mathcal{J}^c$ | $4M + 5S$ | $\mathcal{J} + \mathcal{J}$ | $12M + 4S$ |
| $2\mathcal{J}^m$ | $4M + 4S$ | $\mathcal{P} + \mathcal{P}$ | $12M + 2S$ |
| $2\mathcal{A} = \mathcal{J}^c$ | $3M + 5S$ | $\mathcal{J}^c + \mathcal{J}^c = \mathcal{J}^m$ | $11M + 4S$ |
| $2\mathcal{J}^m = \mathcal{J}$ | $3M + 4S$ | $\mathcal{J}^c + \mathcal{J}^c$ | $11M + 3S$ |
| $2\mathcal{A} = \mathcal{J}^m$ | $3M + 4S$ | $\mathcal{J}^c + \mathcal{J} = \mathcal{J}$ | $11M + 3S$ |
| $2\mathcal{A} = \mathcal{J}$ | $2M + 4S$ | $\mathcal{J}^c + \mathcal{J}^c = \mathcal{J}$ | $10M + 2S$ |
| — | — | $\mathcal{J} + \mathcal{A} = \mathcal{J}^m$ | $9M + 5S$ |
| — | — | $\mathcal{J}^m + \mathcal{A} = \mathcal{J}^m$ | $9M + 5S$ |
| — | — | $\mathcal{J}^c + \mathcal{A} = \mathcal{J}^m$ | $8M + 4S$ |
| — | — | $\mathcal{J}^c + \mathcal{A} = \mathcal{J}^c$ | $8M + 3S$ |
| — | — | $\mathcal{J} + \mathcal{A} = \mathcal{J}$ | $8M + 3S$ |
| — | — | $\mathcal{J}^m + \mathcal{A} = \mathcal{J}$ | $8M + 3S$ |
| — | — | $\mathcal{A} + \mathcal{A} = \mathcal{J}^m$ | $5M + 4S$ |
| — | — | $\mathcal{A} + \mathcal{A} = \mathcal{J}^c$ | $5M + 3S$ |
| $2\mathcal{A}$ | $I + 2M + 2S$ | $\mathcal{A} + \mathcal{A}$ | $I + 2M + S$ |

for the precomputations. Note also that it is possible to avoid some doublings as explained in Remark 9.11 (iii).

**Scalar multiplication**

A scalar multiplication $[n]P$ consists of a sequence of doublings and additions. If a signed windowing method is used with precomputations, there are often runs of doublings interfered with only a few additions. Thus it is worthwhile to distinguish between intermediate doublings, i.e., those followed by a further doubling, and final doublings, which are followed by an addition and choose different coordinate systems for them. Cohen et al. propose to perform the intermediate doublings within $\mathcal{J}^m$ and to express the result of the last doubling in $\mathcal{J}$ since the next step is an addition. More explicitly, for each nonzero coefficient in the expansion of $n$ the intermediate variable $Q$ is replaced in each step by some

$$[2^s]Q \pm [u]P,$$

where $[u]P$ is in the set of precomputed multiples. So, we actually perform $(s-1)$ doublings of the type $2\mathcal{J}^m = \mathcal{J}^m$, a doubling of the form $2\mathcal{J}^m = \mathcal{J}$, and then an addition $\mathcal{J} + \mathcal{A} = \mathcal{J}^m$ or $\mathcal{J} + \mathcal{J}^c = \mathcal{J}^m$ depending on the coordinates of the precomputed values.

Let the windowing work as

$$n = 2^{n_0}(2^{n_1}(\cdots 2^{n_{v-1}}(2^{n_v}W[v] + W[v-1])\cdots) + W[0]),$$

where $W[i]$ is an odd integer in the range $-2^{w-1}+1 \leqslant W[i] \leqslant 2^{w-1}-1$ for all $i$, $W[v] > 0$, $n_0 \geqslant 0$ and $n_i \geqslant w+1$ for $i \geqslant 1$.

§ 13.2 Arithmetic of elliptic curves defined over $\mathbb{F}_p$

In the main loop we perform $u = \sum_{i=0}^{v} n_i$ doublings and $v$ additions. Put $l_1 = l - (w-1)/2$ and $K = 1/2 - 1/(w+1)$. On average $l_1 + K$ doublings and $(l_1 - K)/(w+1)$ additions are used. Then we need approximately

$$\left(l_1 + K + \frac{l_1 - K}{w+1}\right)\mathrm{I} + \left(2(l_1 + K) + \frac{2}{w+1}(l_1 - K)\right)\mathrm{M} + \left(2(l_1 + K) + \frac{2}{w+1}(l_1 - K)\right)\mathrm{S}$$

to compute $[n]P$ excluding the costs for the precomputations if only affine coordinates are used,

$$\left(4(l_1 + K) + \frac{8}{w+1}(l_1 - K)\right)\mathrm{M} + \left(4(l_1 + K) + \frac{5}{w+1}(l_1 - K)\right)\mathrm{S}$$

if the precomputed points are in $\mathcal{A}$ and the computations are done without inversions using $\mathcal{J}$ and $\mathcal{J}^m$ for the intermediate points, and

$$\left(4(l_1 + K) + \frac{11}{w+1}(l_1 - K)\right)\mathrm{M} + \left(4(l_1 + K) + \frac{5}{w+1}(l_1 - K)\right)\mathrm{S}$$

if the precomputed points are in $\mathcal{J}^c$. Now depending on the ratio I/M, $\mathcal{A}$ or $\mathcal{J}^c$ should be chosen. For instance, for a 192-bits key length we choose $\mathcal{A}$ if $\mathrm{I} < 33.9\mathrm{M}$ and $\mathcal{J}^c$ otherwise, cf. [CoMi+ 1998].

### 13.2.3 Montgomery scalar multiplication

This technique was first described by Montgomery [Mon 1987] for a special type of curve in large characteristic and has been generalized to other curves and to even characteristic; see Section 13.3.4.

#### 13.2.3.a Montgomery form

Let $E_M$ be an elliptic curve expressed in *Montgomery form*, that is

$$E_M : By^2 = x^3 + Ax^2 + x. \tag{13.6}$$

The arithmetic on $E_M$ relies on an efficient $x$-coordinate only computation and can be easily implemented to resist side-channel attacks, cf. Chapter 29. Indeed, let $P = (x_1, y_1)$ be a point on $E_M$. In projective coordinates, we write $P = (X_1 : Y_1 : Z_1)$ and let $[n]P = (X_n : Y_n : Z_n)$. The sum $[n+m]P = [n]P \oplus [m]P$ is given by the following formulas where $Y_n$ never appears.

**Addition:** $n \neq m$

$$\begin{aligned} X_{m+n} &= Z_{m-n}\big((X_m - Z_m)(X_n + Z_n) + (X_m + Z_m)(X_n - Z_n)\big)^2, \\ Z_{m+n} &= X_{m-n}\big((X_m - Z_m)(X_n + Z_n) - (X_m + Z_m)(X_n - Z_n)\big)^2. \end{aligned}$$

**Doubling:** $n = m$

$$\begin{aligned} 4X_n Z_n &= (X_n + Z_n)^2 - (X_n - Z_n)^2, \\ X_{2n} &= (X_n + Z_n)^2 (X_n - Z_n)^2, \\ Z_{2n} &= 4X_n Z_n \big((X_n - Z_n)^2 + ((A+2)/4)(4X_n Z_n)\big). \end{aligned}$$

Thus an addition takes 4M and 2S whereas a doubling needs only 3M and 2S.

For some systems, the $x$-coordinate $x_n$ of $[n]P$ is sufficient but others, like some signature schemes, need the $y$-coordinate as well, cf. Chapter 1. To recover $y_n = Y_n/Z_n$, we use the following formula [OKSA 2001]

$$y_n = \frac{(x_1 x_n + 1)(x_1 + x_n + 2A) - 2A - (x_1 - x_n)^2 x_{n+1}}{2B y_1}, \quad (13.7)$$

where $P = (x_1, y_1)$ and $x_n$ and $x_{n+1}$ are the affine $x$-coordinates of $[n]P$ and $[n+1]P$.

### 13.2.3.b  General case

Brier et Joye [BRJO 2002] generalized Montgomery's idea to any curve in short Weierstraß equation

$$E : y^2 = x^3 + a_4 x + a_6.$$

Their formulas require more elementary operations.

**Addition:** $n \neq m$

$$\begin{aligned} X_{m+n} &= Z_{m-n}\bigl(-4a_6 Z_m Z_n (X_m Z_n + X_n Z_m) + (X_m X_n - a_4 Z_m Z_n)^2\bigr), \\ Z_{m+n} &= X_{m-n}(X_m Z_n - X_n Z_m)^2. \end{aligned}$$

**Doubling:** $n = m$

$$\begin{aligned} X_{2n} &= (X_n^2 - a_4 Z_n^2)^2 - 8 a_6 X_n Z_n^3, \\ Z_{2n} &= 4 Z_n \bigl(X_n(X_n^2 + a_4 Z_n^2) + a_6 Z_n^3\bigr). \end{aligned}$$

When $P$ is an affine point, an addition requires 9M and 2S whereas a doubling needs 6M and 3S. To recover $y_n$ in this case, we apply the formula

$$y_n = \frac{2a_6 + (x_1 x_n + a_4)(x_1 + x_n) - (x_1 - x_n)^2 x_{n+1}}{2 y_1}.$$

### 13.2.3.c  Transformation to Montgomery form

It is always possible to convert a curve in Montgomery form (13.6) into short Weierstraß equation, putting $a_4 = 1/B^2 - A^2/3B^2$ and $a_6 = -A^3/27B^3 - a_4 A/3B$. But the converse is false. Not all elliptic curves can be written in Montgomery form. However, this holds true as soon as $p \equiv 1 \pmod{4}$ and $x^3 + a_4 x + a_6$ has three roots in $\mathbb{F}_p$. More generally, a curve in short Weierstraß form can be converted to Montgomery form if and only if

- the polynomial $x^3 + a_4 x + a_6$ has at least one root $\alpha$ in $\mathbb{F}_p$,
- the number $3\alpha^2 + a_4$ is a quadratic residue in $\mathbb{F}_p$.

Put $A = 3\alpha s, B = s$ where $s$ is a square root of $(3\alpha^2 + a_4)^{-1}$ and the change of variables $(x, y) \mapsto (x/s + \alpha, y/s)$ is an isomorphism that transforms $E$ into $E_M$. For such curves $(0, 0)$ is a point of order 2 and $|E(\mathbb{F}_p)|$ is divisible by 4.

Note that recent standards [SEC, NIST] recommend that the cardinality of $E$ should be a prime number times a cofactor less than or equal to 4. One can state divisibility conditions in terms of the Legendre symbol $\left(\frac{\cdot}{p}\right)$. For a curve in Montgomery form $|E(\mathbb{F}_p)|$ is not divisible by 8 in the following cases:

## § 13.2 Arithmetic of elliptic curves defined over $\mathbb{F}_p$

| $p \equiv 1 \pmod 4$ | | | $p \equiv 3 \pmod 4$ | |
|---|---|---|---|---|
| $\left(\frac{A+2}{p}\right)$ | $\left(\frac{A-2}{p}\right)$ | $\left(\frac{B}{p}\right)$ | $\left(\frac{A+2}{p}\right)$ | $\left(\frac{A-2}{p}\right)$ |
| $-1$ | $+1$ | $-1$ | $-1$ | $+1$ |
| $+1$ | $-1$ | $-1$ | | |
| $+1$ | $+1$ | $-1$ | | |
| $-1$ | $-1$ | $+1$ | | |

Let $v$ be a quadratic nonresidue and let $\widetilde{E}_v$ be the quadratic twist of $E$ by $v$, cf. Example 13.19. Then either both $E$ and $\widetilde{E}_v$ are transformable to Montgomery form or none is. Together with Schoof's point counting algorithm (see Section 17.2) this gives an efficient method for generating a curve transformable to Montgomery form whose cofactor is equal to 4.

**Example 13.34** Let us show that $E_2/\mathbb{F}_p$

$$E_2 : y^2 = x^3 + 1132x + 278$$

can be expressed in Montgomery form.

First, $\alpha = 1702$ satisfies

$$\alpha^3 + 1132\alpha + 278 = 0$$

and $3\alpha^2 + a_4 = 527$ is a quadratic residue modulo $p = 2003$. Since $s = 899$ is an inverse square root of 527, we have $A = 1421$, $B = 899$ and the isomorphism $(x, y) \mapsto (899(x - 1702), 899y)$ maps the points of $E_2$ on the points of

$$E_{2,M} : 899y^2 = x^3 + 1421x^2 + x.$$

For instance, $P_2 = (1120, 1391)$ on $E_2$ is sent on $P_{2,M} = (1568, 637)$ on $E_{2,M}$.

### 13.2.3.d  Montgomery ladder

Whatever the form of the curve, we use a modified version of Algorithm 9.5 adapted to scalar multiplication to compute $[n]P$.

---

**Algorithm 13.35** Scalar multiplication using Montgomery's ladder

INPUT: A point $P$ on $E$ and a positive integer $n = (n_{\ell-1} \ldots n_0)_2$.
OUTPUT: The point $[n]P$.

1. $P_1 \leftarrow P$ and $P_2 \leftarrow [2]P$
2. **for** $i = \ell - 2$ **down to** $0$ **do**
3.     **if** $n_i = 0$ **then**
4.         $P_1 \leftarrow [2]P_1$ and $P_2 \leftarrow P_1 \oplus P_2$
5.     **else**
6.         $P_1 \leftarrow P_1 \oplus P_2$ and $P_2 \leftarrow [2]P_2$
7. **return** $P_1$

**Remarks 13.36**

(i) At each step, one performs one addition and one doubling, which makes this method interesting against side-channel attacks, cf. Chapter 29.

(ii) We can check that $P_2 \ominus P_1$ is equal to $P$ at each step so that $Z_{m-n} = Z_1$ in the formulas above. If $P$ is expressed in affine coordinates this saves an extra multiplication in the addition. So the total complexity to compute $[n]P$ is $(6M + 4S)(|n|_2 - 1)$ for elliptic curves in Montgomery form and $(14M + 5S)(|n|_2 - 1)$ in short Weierstraß form.

**Example 13.37** Let us compute $[763]P_{2,M}$ with Algorithm 13.35. We have $763 = (1011111011)_2$ and the different steps of the computation are given in the following table where $P$ stands for $P_{2,M}$ and the question mark indicates that the $y$-coordinate is unknown.

| $i$ | $n_i$ | $(P_1, P_2)$ | $P_1$ | $P_2$ |
|---|---|---|---|---|
| 9 | 1 | $(P, [2]P)$ | $(1568 : 637 : 1)$ | $(35 : ? : 1887)$ |
| 8 | 0 | $([2]P, [3]P)$ | $(35 : ? : 1887)$ | $(1887 : ? : 1248)$ |
| 7 | 1 | $([5]P, [6]P)$ | $(531 : ? : 162)$ | $(120 : ? : 1069)$ |
| 6 | 1 | $([11]P, [12]P)$ | $(402 : ? : 1041)$ | $(909 : ? : 1578)$ |
| 5 | 1 | $([23]P, [24]P)$ | $(1418 : ? : 1243)$ | $(1389 : ? : 1977)$ |
| 4 | 1 | $([47]P, [48]P)$ | $(613 : ? : 37)$ | $(1449 : ? : 231)$ |
| 3 | 1 | $([95]P, [96]P)$ | $(1685 : ? : 1191)$ | $(1256 : ? : 842)$ |
| 2 | 0 | $([190]P, [191]P)$ | $(119 : ? : 1871)$ | $(1501 : ? : 453)$ |
| 1 | 1 | $([381]P, [382]P)$ | $(1438 : ? : 956)$ | $(287 : ? : 868)$ |
| 0 | 1 | $([763]P, [764]P)$ | $(568 : ? : 746)$ | $(497 : ? : 822)$ |

To recover the $y$-coordinate of $[763]P_{2,M}$, we apply (13.7) with $x_1 = 1568$, $y_1 = 637$, $x_n$ and $x_{n+1}$ respectively equal to $568/746$ and $497/822$. Finally, $[763]P_{2,M} = (280, 1733)$.

### 13.2.4 Parallel implementations

For the addition formulas in affine coordinates only a few field operations are used and, hence, parallelization is not too useful. In the other coordinate systems two processors can be applied to reduce the time for a group operation.

For Montgomery coordinates a parallel implementation using two processors is immediate, namely one can take care of the addition while the other performs the doubling. This is possible as both operations need about the same amount of operations, reducing the idle time.

Smart [SMA 2001] investigates parallel implementations of Hessian coordinates.

For Jacobian coordinates on arbitrary curves, Izu and Takagi [IZTA 2002a] propose a parallel version that additionally proposes methods for $k$-fold doubling. It can be implemented together with precomputations and windowing methods for scalar multiplication. Also [FIGI$^+$ 2002] deals with parallel implementation. We come back to efficient parallel implementations in the chapter on side-channel attacks, cf. Chapter 29.

### 13.2.5 Compression of points

For some applications it might be desirable to store or transmit as few bits as possible and still keep the same amount of information.

The following technique works for elliptic curves $E/\mathbb{F}_q$ over arbitrary finite fields $\mathbb{F}_q = \mathbb{F}_{p^d} = \mathbb{F}_p(\theta)$ of odd characteristic $p$ (for details on the arithmetic of finite fields we refer to Chapter 11).

For an elliptic curve $E : y^2 = x^3 + a_2x^2 + a_4x + a_6$ there are at most two points with the same $x$-coordinate, namely $P = (x_1, y_1)$ and $-P = (x_1, -y_1)$. They are equal if and only if $y_1 = 0$, i.e., for the Weierstraß points.

**Compression**

To uniquely identify the point one saves $x_1$ and one bit $b(y_1)$. It is set to 0 if in the field representation $y_1 = \sum_{i=0}^{d-1} c_i\theta^i$ the value of $c_0$ taken as a nonnegative integer is even and set to 1 otherwise. This procedure works as $-y_1$ has $p-c_0$ as its least significant coefficient, which is of opposite parity as $p$ is odd. Hence, one simply needs to check for the least significant bit of the least significant coefficient of $y_1$.

**Decompression**

To recover the $y$-coordinate from $(x_1, b(y_1))$ some more work needs to be done. Namely, one evaluates $x_1^3 + a_2x_1^2 + a_4x_1 + a_6$, which has to be a square in $\mathbb{F}_q$ since $x_1$ is the $x$-coordinate of a point on $E$. Algorithms for square root computation, cf. Section 11.1.5, allow us to recover the two values $\pm y_1$ and the bit $b(y_1)$ determines the correct $y$-coordinate.

**Example 13.38** On the curve $E_2/\mathbb{F}_p$ the point $P_2 = (1120, 1391) \in E(\mathbb{F}_p)$ is coded by $(1120, 1)$ while $R_2 = (4430\theta^2 + 1727\theta + 1809, 9290\theta^2 + 280\theta + 946) \in E(\mathbb{F}_{p^3})$ is represented by $(4430\theta^2 + 1727\theta + 1809, 0)$.

## 13.3 Arithmetic of elliptic curves defined over $\mathbb{F}_{2^d}$

In this section we consider elliptic curves over $\mathbb{F}_{2^d}$. We first provide the transfer to short Weierstraß equations and state formulas for the arithmetic on supersingular and ordinary elliptic curves in affine coordinates. For the remainder of the section we concentrate on ordinary curves. The curves given in Weierstraß form

$$y^2 + a_1xy + a_3y = x^3 + a_2x^2 + a_4x + a_6$$

can be transformed depending on the value of $a_1$.

**Supersingular curves**

If $a_1 = 0$, we need to have $a_3 \neq 0$ as otherwise the curve is singular. The transformation $x \mapsto x + a_2$ leads to the equation

$$E : y^2 + a_3y = x^3 + a_4'x + a_6',$$

which is nonsingular as $a_3 \neq 0$. Such a curve $E$ has no point $P = (x_1, y_1)$ of order two over $\overline{\mathbb{F}}_{2^d}$, as these satisfy $P = -P$, i.e., $y_1 = y_1 + a_3$ and this would only be true for $a_3 = 0$. Therefore, $E[2] = \{P_\infty\}$ and $E$ is supersingular by Definition 13.14.

In Section 24.2.1, we extensively study supersingular curves as they come with an efficiently computable pairing. This has many consequences. For instance, the DLP is easier to solve for these curves. However, there also exist constructive aspects of pairings, e.g., see Chapter 24, and this justifies to investigate the arithmetic of these curves. Indeed, the arithmetic on the supersingular curve

$$E : y^2 + a_3y = x^3 + a_4x + a_6$$

is given by the following formulas where $P = (x_1, y_1)$ and $Q = (x_2, y_2)$ are two points in $E(\mathbb{F}_{2^d})$

- $-P = (x_1, y_1 + a_3)$.
- if $P \neq \pm Q$, we have $P \oplus Q = (x_3, y_3)$ where

$$x_3 = \lambda^2 + x_1 + x_2, \qquad y_3 = \lambda(x_1 + x_3) + y_1 + a_3, \qquad \lambda = \frac{y_1 + y_2}{x_1 + x_2}.$$

- if $P \neq -P$, we have $[2]P = (x_3, y_3)$ where

$$x_3 = \lambda^2, \qquad y_3 = \lambda(x_1 + x_3) + y_1 + a_3, \qquad \lambda = \frac{x_1^2 + a_4}{a_3}.$$

**Example 13.39** Let us consider $\mathbb{F}_{2^{11}}$, represented as $\mathbb{F}_2(\theta)$ with $\theta^{11} + \theta^2 + 1 = 0$. The elements of $\mathbb{F}_{2^{11}}$ will be represented using hexadecimal basis. For instance, 0x591 corresponds to the sequence of bits (0101 1001 0001) and therefore stands for the element $\theta^{10} + \theta^8 + \theta^7 + \theta^4 + 1$.

A supersingular elliptic curve over $\mathbb{F}_{2^{11}}$ is given by

$$E_5 : y^2 + a_3 y = x^3 + a_4 x + a_6$$

with $a_3 = $ 0x6EE, $a_4 = $ 0x1CC and $a_6 = $ 0x3F6. The discriminant of $E_5$ is $\Delta = $ 0x722 while its $j$-invariant is zero.

The points $P_5 = ($0x3DF$, $0x171$)$ and $Q_5 = ($0x732$, $0x27D$)$ belong to $E_5(\mathbb{F}_{2^{11}})$ and

$$\begin{aligned} -P_5 &= (0\text{x3DF}, 0\text{x79F}), \\ P_5 \oplus Q_5 &= (0\text{x314}, 0\text{x4BC}), \\ [2]P_5 &= (0\text{xEF}, 0\text{x6C3}). \end{aligned}$$

The cardinality of $E_5(\mathbb{F}_{2^{11}})$ is equal to $2^{11} + 2^6 + 1 = 2113$ which is prime. Thus the group $E_5(\mathbb{F}_{2^{11}})$ is cyclic and is generated by any one of its element.

**Ordinary curves**

If $a_1 \neq 0$, the transformations

$$y \mapsto a_1^3 y + \frac{a_3^2 + a_1^2 a_4}{a_1^3}, \qquad x \mapsto a_1^2 x + \frac{a_3}{a_1}$$

followed by a division by $a_1^6$ lead to an isomorphic curve given by

$$y^2 + xy = x^3 + a_2' x^2 + a_6',$$

which is nonsingular whenever $a_6' \neq 0$. In this case, the curve is ordinary.

**Remark 13.40** It is always possible to choose $a_2'$ small in the sense that multiplications by $a_2'$ can be carried out by a few additions only. Let $c$ be an element of absolute trace 0, i.e., $\mathrm{Tr}_{\mathbb{F}_{2^d}/\mathbb{F}_2}(c) = 0$, such that multiplications by $a_2' + c$ can be carried out efficiently. In practice, $d$ should be odd, (cf. Section 23.2.2.c) and in this case if $\mathrm{Tr}_{\mathbb{F}_{2^d}/\mathbb{F}_2}(a_2') = 1$ then $\mathrm{Tr}_{\mathbb{F}_{2^d}/\mathbb{F}_2}(a_2' + 1) = 0$. So in any case, $c$ can be taken equal to $a_2'$ or $a_2' + 1$ with the result that $a_2' + c$ is an element of $\mathbb{F}_2$. Let $\lambda$ be such that $\lambda^2 + \lambda + c = 0$. Indeed, (13.8) allows for a further transformation

$$x \mapsto x, \qquad y \mapsto y + \lambda x,$$

which leads to the curve

$$y^2 + xy = x^3 + (a_2' + c)x^2 + a_6'.$$

## § 13.3 Arithmetic of elliptic curves defined over $\mathbb{F}_{2^d}$

**Example 13.41** An ordinary elliptic curve over $\mathbb{F}_{2^{11}}$ is given by
$$E_6 : y^2 + xy = x^3 + a_2 x^2 + a_6$$
with $a_2 = $ 0x6EE and $a_6 = $ 0x1CC. As the trace of $a_2$ is 1 we can put $c = a_2 + 1$ which is of trace 0 and find $\lambda = $ 0x51E such that $\lambda^2 + \lambda = c$. Now the change of variables $x \mapsto x$, $y \mapsto y + \lambda x$, with $\lambda = $ 0x68B transforms the curve $E_6$ into
$$E_7 : y^2 + xy = x^3 + x^2 + a_6.$$
The discriminant of $E_7$ is $\Delta = a_6$ and its $j$-invariant is $1/a_6 = $ 0x37F. The points $P_7 = ($0x420, 0x5B3$)$ and $Q_7 = ($0x4B8, 0x167$)$ are on $E_7$. The curve $E_7$ has 2026 rational points in $\mathbb{F}_{2^{11}}$ and $E_7(\mathbb{F}_{2^{11}})$ is cyclic generated by $P_7$.

### 13.3.1 Choice of the coordinates

The remainder of this chapter is entirely devoted to ordinary curves, i.e., curves given by
$$E : y^2 + xy = x^3 + a_2 x^2 + a_6, \tag{13.8}$$
with $a_2, a_6 \in \mathbb{F}_{2^d}$ such that $a_6 \neq 0$. The coefficient $a_2$ can be chosen with a reduced number of terms and can even be taken in $\mathbb{F}_2$ when $d$ is odd, cf. Remark 13.40 for explanations.

We first give a study on the addition formulas in different coordinate systems and study mixed coordinate systems, then give a generalization of Montgomery coordinates and introduce a further endomorphism on the curve, the point halving. Finally we discuss compression techniques.

As in Section 13.2, an elementary multiplication in $\mathbb{F}_{2^d}$ (respectively a squaring and an inversion) will be represented by M (respectively S and I).

This section is mainly based on [HALÓ+ 2000]. As for curves over prime fields we study different systems of coordinates, namely affine, projective, Jacobian and López–Dahab. For these binary fields some extra tricks are applicable.

We shall give formulas for the

- addition of two points $P$ and $Q \in E(\mathbb{F}_{2^d})$ provided $P \neq \pm Q$,
- doubling of $P$.

#### 13.3.1.a Affine coordinates ($\mathcal{A}$)

Recall that we can choose an elliptic curve of the form
$$E : y^2 + xy = x^3 + a_2 x^2 + a_6.$$
The opposite of $P = (x_1, y_1)$ equals $-P = (x_1, x_1 + y_1)$.

**Addition**

Let $P = (x_1, y_1)$, $Q = (x_2, y_2)$ such that $P \neq \pm Q$ then $P \oplus Q = (x_3, y_3)$ is given by
$$x_3 = \lambda^2 + \lambda + x_1 + x_2 + a_2, \qquad y_3 = \lambda(x_1 + x_3) + x_3 + y_1, \qquad \lambda = \frac{y_1 + y_2}{x_1 + x_2}.$$

**Doubling**

Let $P = (x_1, y_1)$ then $[2]P = (x_3, y_3)$, where
$$x_3 = \lambda^2 + \lambda + a_2, \qquad y_3 = \lambda(x_1 + x_3) + x_3 + y_1, \qquad \lambda = x_1 + \frac{y_1}{x_1}.$$

Thus an addition and a doubling require exactly the same number of operations, that is, I + 2M + S.

**Doubling followed by an addition**

Extending an idea presented in Section 13.2.1.a (see also [EILA$^+$ 2003]), Ciet et al. [CIJO$^+$ 2003], propose a method to compute $[2]P \oplus Q$ as as single operation. The formulas are given below where we assume that $P \neq \pm Q$ and $[2]P \neq -Q$

$$A = x_2 + x_1, \qquad B = y_2 + y_1, \qquad C = A^2(x_2 + a_2) + B(B + A),$$
$$D = (AC)^{-1}, \qquad \lambda = BCD, \qquad \lambda_2 = A^3 D x_1 + \lambda + 1,$$
$$x_4 = (\lambda + \lambda_2)^2 + \lambda + \lambda_2 + x_2, \quad y_4 = (x_1 + x_4)\lambda_2 + y_1 + x_4,$$

requiring $I + 9M + 2S$.

### 13.3.1.b Projective coordinates ($\mathcal{P}$)

With *projective coordinates* the curve is parameterized by the equation

$$Y^2 Z + XYZ = X^3 + a_2 X^2 Z + a_6 Z^3.$$

Like in odd characteristic, we let $(X_1 : Y_1 : Z_1)$ represent the affine point $(X_1/Z_1, Y_1/Z_1)$ if $Z_1 \neq 0$ and $P_\infty = (0 : 1 : 0)$ otherwise. The opposite of $(X_1 : Y_1 : Z_1)$ is $(X_1 : X_1 + Y_1 : Z_1)$.

**Addition**

Let $P = (X_1 : Y_1 : Z_1)$, $Q = (X_2 : Y_2 : Z_2)$ such that $P \neq \pm Q$ then $P \oplus Q = (X_3 : Y_3 : Z_3)$ is given by

$$A = Y_1 Z_2 + Z_1 Y_2, \qquad B = X_1 Z_2 + Z_1 X_2, \qquad C = B^2,$$
$$D = Z_1 Z_2, \qquad E = (A^2 + AB + a_2 C)D + BC,$$
$$X_3 = BE, \qquad Y_3 = C(AX_1 + Y_1 B)Z_2 + (A + B)E, \qquad Z_3 = B^3 D.$$

**Doubling**

If $P = (X_1 : Y_1 : Z_1)$ then $[2]P = (X_3 : Y_3 : Z_3)$ is given by

$$A = X_1^2, \qquad B = A + Y_1 Z_1, \qquad C = X_1 Z_1,$$
$$D = C^2, \qquad E = (B^2 + BC + a_2 D),$$
$$X_3 = CE, \qquad Y_3 = (B + C)E + A^2 C, \qquad Z_3 = CD.$$

In projective coordinates, no inversion is needed. An addition needs $16M + 2S$ and a doubling requires $8M + 4S$.

If the addition receives one input point in affine coordinates, i.e., as $(X_2 : Y_2 : 1)$, the costs reduce to $12M + 2S$. Such an *addition in mixed coordinates* is studied in larger generality in the next section.

All operations profit from small $a_2$ as one multiplication is saved.

### 13.3.1.c Jacobian coordinates ($\mathcal{J}$)

In *Jacobian coordinates*, the curve is given by the equation

$$Y^2 + XYZ = X^3 + a_2 X^2 Z^2 + a_6 Z^6.$$

The point represented by $(X_1 : Y_1 : Z_1)$ corresponds to the affine point $(X_1/Z_1^2, Y_1/Z_1^3)$ when $Z_1 \neq 0$ and to $P_\infty = (1 : 1 : 0)$ otherwise. The opposite of $(X_1 : Y_1 : Z_1)$ is $(X_1 : X_1 Z_1 + Y_1 : Z_1)$.

## Addition

Let $P = (X_1 : Y_1 : Z_1), Q = (X_2 : Y_2 : Z_2)$ such that $P \neq \pm Q$ then $P \oplus Q = (X_3 : Y_3 : Z_3)$ is given by

$$A = X_1 Z_2^2, \qquad B = X_2 Z_1^2, \qquad C = Y_1 Z_2^3,$$
$$D = Y_2 Z_1^3, \qquad E = A + B, \qquad F = C + D,$$
$$G = E Z_1, \qquad H = F X_2 + G Y_2, \qquad Z_3 = G Z_2,$$
$$I = F + Z_3, \qquad X_3 = a_2 Z_3^2 + FI + E^3, \qquad Y_3 = I X_3 + G^2 H.$$

## Doubling

If $P = (X_1 : Y_1 : Z_1)$ then $[2]P = (X_3 : Y_3 : Z_3)$ is given by

$$A = X_1^2, \qquad B = A^2, \qquad C = Z_1^2,$$
$$X_3 = B + a_6 C^4, \qquad Z_3 = X_1 C, \qquad Y_3 = B Z_3 + (A + Y_1 Z_1 + Z_3) X_3.$$

In Jacobian coordinates an addition requires $16M + 3S$ in general and only $11M + 3S$ if one input is in affine coordinates. Also if $a_2 \in \{0, 1\}$ we need one multiplication less in the addition of points. A doubling needs $5M + 5S$ including one multiplication by $a_6$.

### 13.3.1.d  López–Dahab coordinates ($\mathcal{LD}$)

López and Dahab [LÓDA 1998] introduced a further set of coordinates in which the curve is given by the equation

$$Y^2 + XYZ = X^3 Z + a_2 X^2 Z^2 + a_6 Z^4.$$

The triple $(X_1 : Y_1 : Z_1)$ represents the affine point $(X_1/Z_1, Y_1/Z_1^2)$ when $Z_1 \neq 0$ and $P_\infty = (1 : 0 : 0)$ otherwise. The opposite of $(X_1 : Y_1 : Z_1)$ is $(X_1 : X_1 Z_1 + Y_1 : Z_1)$.

## Addition

Let $P = (X_1 : Y_1 : Z_1), Q = (X_2 : Y_2 : Z_2)$ such that $P \neq \pm Q$ then $P \oplus Q = (X_3 : Y_3 : Z_3)$ is given by

$$A = X_1 Z_2, \qquad B = X_2 Z_1, \qquad C = A^2,$$
$$D = B^2, \qquad E = A + B, \qquad F = C + D,$$
$$G = Y_1 Z_2^2, \qquad H = Y_2 Z_1^2, \qquad I = G + H,$$
$$J = IE, \qquad Z_3 = F Z_1 Z_2, \qquad X_3 = A(H + D) + B(C + G),$$
$$Y_3 = (AJ + FG)F + (J + Z_3) X_3.$$

A general addition $P \oplus Q$ in this coordinate system takes $13M + 4S$ as shown by Higuchi and Takagi [HITA 2000]. Note that the original formulas proposed in [LÓDA 1998] need $14M + 6S$.

## Mixed Addition

If $Q$ is in affine coordinates the costs drop to $10M + 3S$. In fact, it is possible to do a bit better, since Al–Daoud et al. [ALMA$^+$ 2002] proved that only $9M + 5S$ are sufficient in this case. The formulas are given below.

$$A = Y_1 + Y_2 Z_1^2, \qquad B = X_1 + X_2 Z_1, \qquad C = B Z_1,$$
$$Z_3 = C^2, \qquad D = X_2 Z_3, \qquad X_3 = A^2 + C(A + B^2 + a_2 C),$$
$$Y_3 = (D + X_3)(AC + Z_3) + (Y_2 + X_2) Z_3^2.$$

Note that when $a_2 \in \{0, 1\}$ one further multiplication is saved.

**Doubling**

If $P = (X_1 : Y_1 : Z_1)$ then $[2]P = (X_3 : Y_3 : Z_3)$ is given by [LÓDA 1998]

$$A = Z_1^2, \qquad B = a_6 A^2, \qquad C = X_1^2,$$
$$Z_3 = AC, \qquad X_3 = C^2 + B, \qquad Y_3 = (Y_1^2 + a_2 Z_3 + B)X_3 + Z_3 B.$$

To analyze the complexity, first note that in practice $a_2$ can be chosen in $\mathbb{F}_2$, cf. Remark 13.40, saving one product.

For fixed $a_2$ and $a_6$ it is also possible to use less additions if $\sqrt{a_6}$ can be precomputed. E.g., for $a_2 = 1$ one can use

$$A = X_1^2, \qquad B = \sqrt{a_6} Z_1^2, \qquad C = X_1 Z_1,$$
$$Z_3 = C^2, \qquad X_3 = (A + B)^2, \qquad Y_3 = \bigl(AC + (Y_1 + B)(A + B)\bigr)^2$$

requiring 4M + 5S including one multiplication by $\sqrt{a_6}$.

For fixed $a_2 = 0$, $X_3$ and $Z_3$ are given as above whereas $Y_3 = \bigl(BC + (Y_1 + B)(A + B)\bigr)^2$, which also requires 4M + 5S including one multiplication by $\sqrt{a_6}$.

It is also possible to trade this multiplication by a constant and a squaring for a general multiplication [LAN 2004b], which might be interesting if the curve varies or if $\sqrt{a_6}$ is big. The formulas are as follows

$$A = X_1 Z_1, \qquad B = X_1^2, \qquad C = B + Y_1, \qquad (13.9)$$
$$D = AC, \qquad Z_3 = A^2, \qquad X_3 = C^2 + D + a_2 Z_3,$$
$$Y_3 = (Z_3 + D)X_3 + B^2 Z_3$$

requiring 5M + 4S including one multiplication by $a_2$.

### 13.3.1.e Example

Take the curve

$$E_7 : y^2 + xy = x^3 + x^2 + a_6 \qquad (13.10)$$

with $a_6 = \text{0x1CC}$. We recall below the equation of $E_7$ as well as the coordinates of $P_7 = (\text{0x420}, \text{0x681})$ and $Q_7 = (\text{0x4B8}, \text{0x563})$ on $E_7$ for each coordinate system. Note that the third coordinate in projective, Jacobian and López–Dahab systems is chosen at random.

| System | Equation | $P_7$ | $Q_7$ |
|---|---|---|---|
| $\mathcal{A}$ | $y^2 + xy = x^3 + x^2 + a_6$ | (0x420, 0x5B3) | (0x4B8, 0x167) |
| $\mathcal{P}$ | $Y^2 Z + XYZ = X^3 + X^2 Z + a_6 Z^3$ | (0x64F : 0x5BA : 0x1C9) | (0x4DD : 0x1F0 : 0x3FA) |
| $\mathcal{J}$ | $Y^2 + XYZ = X^3 + X^2 Z^2 + a_6 Z^6$ | (0x4DA : 0x1F7 : 0x701) | (0x383 : 0x5BA : 0x1E1) |
| $\mathcal{LD}$ | $Y^2 + XYZ = X^3 Z + X^2 Z^2 + a_6 Z^4$ | (0x6BE : 0x15F : 0x7B3) | (0x757 : 0x3EF : 0xA1C) |

With these particular values of $P_7$ and $Q_7$, let us compute $P_7 \oplus Q_7$, $[2]P_7$ and $[763]P_7$ within the different systems using the double and add method.

### § 13.3 Arithmetic of elliptic curves defined over $\mathbb{F}_{2^d}$

| System | $P_7 \oplus Q_7$ | $[2]P_7$ | $[763]P_7$ |
|---|---|---|---|
| $\mathcal{A}$ | (0x724, 0x7B3) | (0x14D, 0x4CB) | (0x84, 0x475) |
| $\mathcal{P}$ | (0x675 : 0x6D5 : 0x4D5) | (0x4D5 : 0x21E : 0x705) | (0x582 : 0x14 : 0x543) |
| $\mathcal{J}$ | (0x12 : 0x46B : 0x5F) | (0x5B1 : 0x417 : 0x7D) | (0x2F7 : 0x572 : 0x3E2) |
| $\mathcal{LD}$ | (0x7C5 : 0x1D2 : 0x3D2) | (0x444 : 0x4A0 : 0x193) | (0x2F : 0x265 : 0x220) |

For each computation, the obtained result is equivalent to the affine one.

#### 13.3.2 Faster doublings in affine coordinates

Let $P = (x_1, y_1)$ be a point lying on the ordinary curve
$$E : y^2 + xy = x^3 + a_2 x^2 + a_6.$$
When the solution of a quadratic equation can be quickly found, e.g., if $\mathbb{F}_{2^d}$ is represented by a normal basis, the following method [SOL 1997] replaces one general multiplication by a multiplication by the fixed constant $a_6$.

Namely, compute $x_3 = x_1^2 + a_6/x_1^2$, which is also equal to $\lambda^2 + \lambda + a_2$. Then find $\mu$ such that $\mu^2 + \mu = x_3 + a_2$, see Section 11.2.6. So $\lambda = \mu + \varepsilon$ where $\varepsilon = 0$ or $1$. Therefore $\mu x_1 + x_1^2 + y_1 = \varepsilon x_1$ and we deduce $\varepsilon$ from this equation. Note that it is not necessary to perform $\mu x_1$ in full but rather to compute one well chosen coordinate in the product. Thus the computation of $\lambda$ is almost free and it remains to perform $y_3 = x_1^2 + (\lambda + 1)x_3$.

To perform several doublings in a row of $P = (x_1, y_1)$, it is faster to store the intermediate values by the $x$-coordinate and the slope of the tangent, i.e., to represent $[2^i]P$ as $(x_{2^i}, \lambda_{2^i})$. This is possible because

$$\begin{aligned}
x_2 &= \lambda_1^2 + \lambda_1 + a_2 \text{ and} \\
\lambda_2 &= \lambda_1^2 + \lambda_1 + a_2 + \frac{\lambda_1(x_1 + \lambda_1^2 + \lambda_1 + a_2) + \lambda_1^2 + \lambda_1 + a_2 + y_1}{\lambda_1^2 + \lambda_1 + a_2} \\
&= \lambda_1^2 + a_2 + \frac{a_6}{x_1^4 + a_6}.
\end{aligned}$$

This idea leads to the following algorithm described in [LÓDA 2000b].

**Algorithm 13.42** Repeated doublings

INPUT: A point $P = (x_1, y_1)$ on $E$ such that $[2^k]P \neq P_\infty$ and an integer $k \geq 2$.
OUTPUT: The point $[2^k]P$ of coordinates $(x_3, y_3)$.

1. $\lambda \leftarrow x_1 + y_1/x_1$ and $u \leftarrow x_1$
2. **for** $i = 1$ **to** $k-1$ **do**
3. $\quad x' \leftarrow \lambda^2 + \lambda + a_2$
4. $\quad \lambda' \leftarrow \lambda^2 + a_2 + \dfrac{a_6}{u^4 + a_6}$
5. $\quad u \leftarrow x'$ and $\lambda \leftarrow \lambda'$
6. $x_3 \leftarrow \lambda^2 + \lambda + a_2$ and $y_3 \leftarrow u^2 + (\lambda + 1)x_3$
7. **return** $(x_3, y_3)$

This algorithm needs $kI + (k+1)M + (3k-1)S$.

**Example 13.43** Take $P_7$ on $E_7$ as defined in Example 13.41 and let us compute $[2^5]P_7$. The values of $\lambda$ and $u$ along the execution of Algorithm 13.42 are given below.

| $i$ | — | 1 | 2 | 3 | 4 |
|---|---|---|---|---|---|
| $\lambda$ | 0x1C | 0x67 | 0x6F7 | 0x96 | 0x719 |
| $u$ | 0x420 | 0x14D | 0x479 | 0x344 | 0x1AB |

At the end, $x_3 = $ 0x67C and $y_3 = $ 0x71C.

Another strategy is to use a closed formula to get $[2^k]P$ directly rather than computing successive doublings. The interest is to perform only one inversion at the cost of extra multiplications [GUPA 1997]. We do not state these formulas here as the same number of operations can be obtained by using López–Dahab coordinates for the intermediate doublings and transforming the result to affine coordinates afterwards.

### 13.3.3 Mixed coordinates

In the previous part we introduced different representations for the point on $E$ together with the algorithms to perform addition and doubling. For the additions we also mentioned the number of operations needed if one of the input points is in affine coordinates. Like in odd characteristic we now study arbitrary mixes of coordinates to perform scalar multiplications where we use two (different) systems of coordinates as input and one as output. By $\mathcal{J} + \mathcal{A} = \mathcal{LD}$ we denote the addition taking as input one point in Jacobian coordinates and one in affine and giving the result in López–Dahab coordinates.

Additionally we use the abbreviations $\mathcal{A}'$ to denote the representation by $(x, \lambda)$ introduced in the previous section for multiple doublings. For $\mathcal{A}'$ coordinates the table entry refers to the asymptotic complexity of a doubling in a sequence of $k$ consecutive doublings, thus we neglect other marginal operations. Table 13.4 on page 297 gives the number of field operations needed depending on the coordinate systems. Compared to the case of odd characteristic, changes between the coordinate systems are not too interesting and are therefore not listed. We denote the costs for multiplication with $a_2$ by $M_2$ and concentrate on the most interesting cases. We do *not* take into account the effects of small $a_6$ as this cannot be achieved generically.

#### No precomputation

If the system offers no space to store precomputations one should use $\mathcal{A}$ if inversions are affordable, i.e., less than 8 times as expensive as a multiplication, and otherwise use $\mathcal{LD}$ for the doublings and $\mathcal{LD} + \mathcal{A} = \mathcal{LD}$ for the additions if the input is in affine coordinates and as $\mathcal{LD} + \mathcal{LD}$ otherwise.

#### Precomputations

Also in even characteristic, using the $\text{NAF}_w$ representation, cf. Section 9.1.4 is advantageous to compute scalar multiples $[n]P$. This requires precomputing all odd multiples $[i]P$ for $1 < i < 2^{w-1}$. They can be obtained as a sequence of additions and one doubling.

If inversions in $\mathbb{F}_{2^d}$ are not too expensive one should choose affine coordinates as a system for the precomputations as they offer the fastest mixed coordinates. Like in the case of odd characteristic this does not mean that one needs to perform $2^{w-2}$ inversions but one can follow [COMI+ 1998, section 4] and apply Montgomery's trick of simultaneous inversions. For details we refer to the study for odd characteristic, cf. Section 13.2.2. Then one needs:

$$(w-1)I + \left(5 \times 2^{w-2} + 2w - 12\right)M + \left(2^{w-2} + w - 3\right)S.$$

### § 13.3 Arithmetic of elliptic curves defined over $\mathbb{F}_{2^d}$

**Table 13.4** *Operations required for addition and doubling.*

| Doubling | | Addition | |
|---|---|---|---|
| Operation | Costs | Operation | Costs |
| $2\mathcal{P}$ | $7\text{M} + 4\text{S} + \text{M}_2$ | $\mathcal{J} + \mathcal{J}$ | $15\text{M} + 3\text{S} + \text{M}_2$ |
| $2\mathcal{J}$ | $5\text{M} + 5\text{S}$ | $\mathcal{P} + \mathcal{P}$ | $15\text{M} + 2\text{S} + \text{M}_2$ |
| $2\mathcal{LD}$ | $4\text{M} + 4\text{S} + \text{M}_2$ | $\mathcal{LD} + \mathcal{LD}$ | $13\text{M} + 4\text{S}$ |
| $2\mathcal{A} = \mathcal{P}$ | $5\text{M} + 2\text{S} + \text{M}_2$ | $\mathcal{P} + \mathcal{A} = \mathcal{P}$ | $11\text{M} + 2\text{S} + \text{M}_2$ |
| $2\mathcal{A} = \mathcal{LD}$ | $2\text{M} + 3\text{S} + \text{M}_2$ | $\mathcal{J} + \mathcal{A} = \mathcal{J}$ | $10\text{M} + 3\text{S} + \text{M}_2$ |
| $2\mathcal{A} = \mathcal{J}$ | $\text{M} + 2\text{S} + \text{M}_2$ | $\mathcal{LD} + \mathcal{A} = \mathcal{LD}$ | $8\text{M} + 5\text{S} + \text{M}_2$ |
| — | — | $\mathcal{A} + \mathcal{A} = \mathcal{LD}$ | $5\text{M} + 2\text{S} + \text{M}_2$ |
| $2\mathcal{A}$ | $\text{I} + 2\text{M} + \text{S}$ | $\mathcal{A} + \mathcal{A} = \mathcal{J}$ | $4\text{M} + \text{S} + \text{M}_2$ |
| $2\mathcal{A}'$ | $\text{I} + \text{M} + \text{S}$ | $\mathcal{A} + \mathcal{A} = \mathcal{A}'$ | $2\text{I} + 3\text{M} + \text{S}$ |
| $2\mathcal{A}' = \mathcal{A}$ | $\text{M} + 2\text{S}$ | $\mathcal{A} + \mathcal{A}$ | $\text{I} + 2\text{M} + \text{S}$ |

If inversions are prohibitively expensive one should choose $\mathcal{LD}$ coordinates as they are the most efficient inversion-free system, provided that one multiplication is at least as expensive as three squarings, which is usually the case in binary fields. This way

$$\left(13 \times 2^{w-2} - 8\right)\text{M} + 4 \times 2^{w-2}\text{S}$$

are needed for the precomputations.

Using one I and $3(2^{w-2} - 2)\text{M}$ the resulting precomputed points can be transformed to affine. Furthermore, the use of precomputations leads to long runs of doublings in the algorithms and they are much faster in $\mathcal{LD}$ than in $\mathcal{P}$, which otherwise would offer the lowest number of operations per addition.

**Scalar multiplication**

A scalar multiplication consists of a sequence of doublings and additions. If a signed windowing method is used with precomputations there are often runs of doublings interfered with only a few additions. Thus it is worthwhile to distinguish between intermediate doublings, i.e., those followed by a further doubling, and final doublings, which are followed by an addition, and to choose different coordinate systems for them.

We first assume that the precomputed points are in $\mathcal{A}$ as this leads to the most interesting mixes of coordinates. If one inversion per bit of the scalar is affordable one should use $\mathcal{A}'$ for the intermediate doublings.

More explicitly, the intermediate variable $Q$ is replaced each step by some

$$[2^s]Q \pm [u]P,$$

where $[u]P$ is in the set of precomputed multiples. So we actually perform $(s - 1)$ doublings of the type $2\mathcal{A}' = \mathcal{A}'$, a doubling of the form $2\mathcal{A}' = \mathcal{A}$ and then an addition $\mathcal{A} + \mathcal{A} = \mathcal{A}'$.

Let $l$ be the binary length of $n$, let $l_1 = l - (w-1)/2$, and $K = 1/2 - 1/(w+1)$.

In the main loop, we perform on average $l_1 + K - v$ doublings of the form $2\mathcal{A}' = \mathcal{A}'$, $v$ doublings of the form $2\mathcal{A}' = \mathcal{A}$, and $v$ additions, where $v = (l_1 - K)/(w + 1)$. Then we need approximately

$$\left(l_1 + K + \frac{l_1 - K}{w + 1}\right)\text{I} + \left(l_1 + K + 3\frac{l_1 - K}{w + 1}\right)\text{M} + \left(l_1 + K + 2\frac{l_1 - K}{w + 1}\right)\text{S}.$$

If the algorithm should not make use of inversions but the precomputed points are in $\mathcal{A}$, Table 13.4 shows that the doublings should be performed within $\mathcal{LD}$. This is followed by one addition of the type $\mathcal{LD} + \mathcal{A} = \mathcal{LD}$. This needs approximately

$$\left(4(l_1 + K) + 8\frac{l_1 - K}{w + 1}\right)\text{M} + \left(4(l_1 + K) + 5\frac{l_1 - K}{w + 1}\right)\text{S} + \left(l_1 + K + \frac{l_1 - K}{w + 1}\right)\text{M}_2.$$

If the precomputed points are in $\mathcal{LD}$ the most efficient way is to choose this coordinate system for all operations. In total this needs asymptotically

$$\left(4(l_1 + K) + 13\frac{l_1 - K}{w + 1}\right)\text{M} + \left(4(l_1 + K) + 4\frac{l_1 - K}{w + 1}\right)\text{S} + \left(l_1 + K\right)\text{M}_2.$$

### 13.3.4 Montgomery scalar multiplication

López and Dahab [LÓDA 1999] generalized Montgomery's idea, cf. Section 13.2.3, to binary curves. Let $P = (x_1, y_1)$ be a point on $E$. In projective coordinates, we write $P = (X_1 : Y_1 : Z_1)$ and let $[n]P = (X_n : Y_n : Z_n)$. The sum $[n+m]P = [n]P \oplus [m]P$ is given by the following formulas where $Y_n$ does not occur.

**Addition:** $n \neq m$

$$\begin{aligned} Z_{m+n} &= (X_m Z_n)^2 + (X_n Z_m)^2, \\ X_{m+n} &= Z_{m+n} X_{m-n} + X_m Z_n X_n Z_m. \end{aligned}$$

**Doubling:** $n = m$

$$\begin{aligned} X_{2n} &= X_n^4 + a_6 Z_n^4 = \left(X_n^2 + \sqrt{a_6} Z_n^2\right)^2, \\ Z_{2n} &= X_n^2 Z_n^2. \end{aligned}$$

An addition takes 4M and 1S whereas a doubling needs only 2M and 3S, if $\sqrt{a_6}$ is precomputed. For the full scalar multiplication $[n]P$, we use *Montgomery's ladder*, cf. Algorithm 13.35, which requires $(6\text{M} + 4\text{S})(|n|_2 - 1)$ in total.

To recover the $y$-coordinate of $[n]P = (X_n : Y_n : Z_n)$ we first compute the affine $x$-coordinates of $[n]P$ and $[n+1]P$, that is $x_n = X_n/Z_n$ and $x_{n+1} = X_{n+1}/Z_{n+1}$ and then use the formula [LÓDA 1999, OKSA 2001]

$$y_n = \frac{(x_n + x_1)\big((x_n + x_1)(x_{n+1} + x_1) + x_1^2 + y_1\big)}{x_1} + y_1. \tag{13.11}$$

**Example 13.44** Let us compute $[763]P_7$ with Algorithm 13.35. The different steps of the compu-

*§ 13.3 Arithmetic of elliptic curves defined over $\mathbb{F}_{2^d}$*

tation are given in the following table where $P$ stands for $P_7 \in E_7$, given by (13.10).

| $i$ | $n_i$ | $(P_1, P_2)$ | $P_1$ | $P_2$ |
|---|---|---|---|---|
| 9 | 1 | $(P, [2]P)$ | (0x420 : ? : 0x1) | (0x158 : ? : 0x605) |
| 8 | 0 | $([2]P, [3]P)$ | (0x158 : ? : 0x605) | (0x7E9 : ? : 0x2FD) |
| 7 | 1 | $([5]P, [6]P)$ | (0x295 : ? : 0x56B) | (0x620 : ? : 0x43B) |
| 6 | 1 | $([11]P, [12]P)$ | (0x5D0 : ? : 0x247) | (0xA6 : ? : 0x6CE) |
| 5 | 1 | $([23]P, [24]P)$ | (0x755 : ? : 0x21B) | (0x409 : ? : 0x93) |
| 4 | 1 | $([47]P, [48]P)$ | (0xBD : ? : 0x25E) | (0x26 : ? : 0x4BE) |
| 3 | 1 | $([95]P, [96]P)$ | (0x4EE : ? : 0x51D) | (0x4D6 : ? : 0x304) |
| 2 | 0 | $([190]P, [191]P)$ | (0x4C1 : ? : 0x58C) | (0x553 : ? : 0x386) |
| 1 | 1 | $([381]P, [382]P)$ | (0x613 : ? : 0x7E4) | (0x2BB : ? : 0x60B) |
| 0 | 1 | $([763]P, [764]P)$ | (0x6C4 : ? : 0x105) | (0x655 : ? : 0x485) |

To end the computation, we apply (13.11) to obtain that $[763]P_7 = (0\text{x}84, 0\text{x}475)$.

### 13.3.5 Point halving and applications

In this section we introduce a further map on the group of points of an elliptic curve.

Let $|E(\mathbb{F}_{2^d})| = 2^k \ell$, where $\ell$ is odd. If $k = 1$ then $E$ is said to have *minimal 2-torsion* as curves of the form

$$E : y^2 + xy = x^3 + a_2 x^2 + a_6 \qquad (13.12)$$

considered here always have one point of order 2, namely the point $T = (0, \sqrt{a_6})$. Hence, the doubling map $[2]$ is not injective. Now assume $E$ to have minimal 2-torsion and let $G$ be a subgroup of odd order. If $P$ belongs to $G$ then there is a unique point $Q \in G$ such that $P = [2]Q$. Then denote $Q = \left[\frac{1}{2}\right]P$ and define the one-to-one *halving map* by

$$\begin{aligned} \left[\tfrac{1}{2}\right] : G &\to G \\ P &\mapsto Q \text{ such that } [2]Q = P. \end{aligned}$$

In the following, we shall represent a point $P = (x_1, y_1)$ as $(x_1, \lambda_1)$ where $\lambda_1 = x_1 + y_1/x_1$. In the context of a scalar multiplication based on halvings, this representation leads to a faster implementation as for repeated doublings.

In [KNU 1999] Knudsen develops an efficient technique to halve a point in affine coordinates lying on an elliptic curve with minimal 2-torsion. Independently, Schroeppel proposed the same method [SCH 2000c].

Note that half the curves of the form (13.12) defined over $\mathbb{F}_{2^d}$ have minimal 2-torsion, since this property is equivalent to $\text{Tr}(a_2) = 1$.

Let $P = (x_1, \lambda_1) \in G$ and $Q = \left[\frac{1}{2}\right]P = (x_2, \lambda_2)$. Inverting doubling formulas, one has

$$\begin{aligned} \lambda_2^2 + \lambda_2 &= a_2 + x_1, \\ x_2^2 &= x_1(\lambda_2 + 1) + y_1 = x_1(\lambda_2 + \lambda_1 + x_1 + 1), \\ y_2 &= x_2(x_2 + \lambda_2). \end{aligned}$$

The algorithm is as follows. First find $\gamma$ such that $\gamma^2 + \gamma = a_2 + x_1$. The other solution of the equation is then $\gamma + 1$, cf. Lemma 11.56. One corresponds to $\lambda_2$ and $Q$ and the other one to $\lambda_2 + 1$ and $Q \oplus T$. If $E$ has minimal 2-torsion it is possible to determine if $\gamma$ is equal to $\lambda_2$ or not. Indeed only $Q$ can be halved but not $Q \oplus T$. So $(x_2, \lambda_2)$ is equal to $\left[\frac{1}{2}\right]P$ if and only if the equation $X^2 + X = a_2 + x_2$ has a solution in $\mathbb{F}_{2^d}$. This holds true if and only if $\mathrm{Tr}(a_2 + x_2) = 0$. Clearly $\mathrm{Tr}(a_2 + x_2) = \mathrm{Tr}(a_2^2 + x_2^2)$ and this remark saves a square root computation.

So one first obtains $w = x_1(\gamma + \lambda_1 + x_1 + 1)$, which is a candidate for $x_2^2$. If $\mathrm{Tr}(a_2^2 + w) = 0$ then $\lambda_2 = \gamma$ and $x_2 = \sqrt{w}$. Otherwise $\lambda_2 = \gamma + 1$ and $x_2 = \sqrt{w + x_1}$.

All these steps are summarized in the following algorithm.

---

**Algorithm 13.45** Point halving

INPUT: The point $P = (x_1, y_1) \in G$ represented as $(x_1, \lambda_1)$.
OUTPUT: The point $\left[\frac{1}{2}\right]P = (x_2, y_2)$ represented as $(x_2, \lambda_2)$.

1. compute $\gamma$ such that $\gamma^2 + \gamma = a_2 + x_1$
2. $w \leftarrow x_1(\gamma + \lambda_1 + x_1 + 1)$
3. **if** $\mathrm{Tr}(a_2^2 + w) = 1$ **then** $\gamma \leftarrow \gamma + 1$ and $w \leftarrow w + x_1$
4. $\lambda_2 \leftarrow \gamma$ and $x_2 \leftarrow \sqrt{w}$
5. **return** $Q = (x_2, \lambda_2)$

---

**Remarks 13.46**

(i) To determine $(x_2, \lambda_2)$ Algorithm 13.45 requires us to compute the solution of a quadratic equation, one square root, one multiplication, and one absolute trace. A further multiplication is necessary to obtain $y_2$.

(ii) See Section 11.2.6 for a description of algorithms to compute $\gamma$ and $\sqrt{w}$.

(iii) The computation of the trace in Line 3 is straightforward, cf. Remarks 11.57.

(iv) Algorithm 13.45 can be easily generalized when $E(\mathbb{F}_{2^d})$ has a subgroup isomorphic to $\mathbb{Z}/2^k\mathbb{Z}$ with $k > 1$ [KNU 1999]. Nevertheless, it is necessary, in this case, to solve $k$ equations, perform $k + 1$ multiplications, one test, and $k$ or $k + 1$ square root computations to find $(x_2, y_2)$, so that in practice the technique is usually not interesting for $k > 1$.

**Example 13.47** The point $P_7 = (\mathtt{0x420}, \mathtt{0x5B3})$ on $E_7$ is a point of order 2026. This implies that $R_7 = [2]P_7 = (\mathtt{0x14D}, \mathtt{0x4CB})$ is a point of odd order $\ell = 1013$. Thus in the group $G = \langle R_7 \rangle$, the halving map is well defined. Let us compute $\left[\frac{1}{2}\right]R_7$ with Algorithm 13.45. First, we have $\lambda_1 = \mathtt{0x67}$. We deduce that $\gamma = \mathtt{0x1C}$ and $w = \mathtt{0x605}$. Since the trace of $a_2^2 + w$ is equal to one, the values of $\gamma$ and $w$ are changed to $\mathtt{0x1D}$ and $\mathtt{0x748}$. Finally, $\lambda_2 = \mathtt{0x1D}$ and $x_2 = \mathtt{0x3B8}$. It follows that the unique point $S_7 \in G$ such that $[2]S_7 = R_7$ is $(\mathtt{0x3B8}, \mathtt{0x441})$.

We also have $P_7 = S_7 \oplus T_7$ where $T_7 = (\mathtt{0x0}, \mathtt{0x19A})$ is the 2-torsion point of $E_7$.

Now, let us explain how to compute the scalar multiplication $[n]P$ of a point $P$ of odd order $\ell_1 \mid \ell$. Let $m = \lceil \lg \ell_1 \rceil$. Then if

$$2^{m-1}n = \sum_{i=0}^{m-1} \widehat{n}_i 2^i \bmod \ell_1, \text{ with } n_i \in \{0, 1\}$$

§ 13.3 Arithmetic of elliptic curves defined over $\mathbb{F}_{2^d}$

one has
$$n \equiv \sum_{i=0}^{m-1} \frac{\widehat{n}_{m-1-i}}{2^i} \pmod{\ell_1}$$

and $[n]P$ can be obtained by the following algorithm. We additionally put $\left[\frac{1}{2}\right]P_\infty = P_\infty$.

---
**Algorithm 13.48** Halve and add scalar multiplication

INPUT: A point $P \in E(\mathbb{F}_{2^d})$ of odd order $\ell_1$ and a positive integer $n$.
OUTPUT: The point $[n]P$.

1. $m \leftarrow \lceil \lg \ell_1 \rceil$
2. $\widehat{n} \leftarrow (2^{m-1}n) \bmod \ell_1$      $[\widehat{n} = (\widehat{n}_{m-1} \ldots \widehat{n}_0)_2]$
3. $Q \leftarrow P_\infty$
4. **for** $i = 0$ **to** $m-1$ **do**
5.      $Q \leftarrow \left[\frac{1}{2}\right]Q$
6.      **if** $\widehat{n}_i = 1$ **then** $Q \leftarrow Q \oplus P$
7. **return** $Q$

---

**Remarks 13.49**

(i) All the window and recoding techniques seen in Chapter 9 apply as well. In particular, if $\sum_{i=0}^{m} \widehat{n}_i 2^i$ is the $\text{NAF}_w$ representation of $2^m n$ modulo $\ell_1$, then
$$n \equiv \sum_{i=0}^{m} \frac{\widehat{n}_{m-i}}{2^i} \pmod{\ell_1}.$$

(ii) No method is currently known to halve a point in projective coordinates. In [HAME$^+$ 2003] two halve-and-add algorithms for the $\text{NAF}_w$ representation are given. The one operating from the right to the left halves the input $P$ rather than the accumulators, which can therefore be represented in projective coordinates. In this case mixed addition formulas can be used for a better efficiency.

(iii) Point halving can be used to achieve faster scalar multiplication on Koblitz curves; see Chapter 15 and [AvCI$^+$ 2004].

**Example 13.50** Let us compute $[763]R_7$ with Algorithm 13.48. As $R_7$ is of odd order 1013, we have $m = 10$ and $2^9 \times 763 \equiv 651 \pmod{1013}$. Now $651 = (1010001011)_2$ which implies that
$$763 \equiv \frac{1}{2^9} + \frac{1}{2^8} + \frac{1}{2^6} + \frac{1}{2^2} + 1 \pmod{1013}.$$

Thus, the main steps of the computation, expressed in the form $(x, \lambda)$, are

$$\left[\tfrac{1}{2}\right]R_7 \oplus R_7 = (\mathtt{0x1}, \mathtt{0x21D}),$$
$$\left[\tfrac{1}{2}\right]^3 R_7 \oplus \left[\tfrac{1}{2}\right]^2 R_7 \oplus R_7 = (\mathtt{0x644}, \mathtt{0x184}),$$
$$\left[\tfrac{1}{2}\right]^7 R_7 \oplus \left[\tfrac{1}{2}\right]^6 R_7 \oplus \left[\tfrac{1}{2}\right]^4 R_7 \oplus R_7 = (\mathtt{0x77C}, \mathtt{0x3EC}),$$
$$\left[\tfrac{1}{2}\right]^9 R_7 \oplus \left[\tfrac{1}{2}\right]^8 R_7 \oplus \left[\tfrac{1}{2}\right]^6 R_7 \oplus \left[\tfrac{1}{2}\right]^2 R_7 \oplus R_7 = (\mathtt{0x2EA}, \mathtt{0x281}).$$

We deduce that $[763]R_7 = (\text{0x2EA}, \text{0x7C8})$ in affine coordinates.

It is also easy to obtain the multiple of a point that does not belong to $G$. For instance, let us compute $[763]P_7$. We have $P_7 = \left[\frac{1}{2}\right]R_7 \oplus T_7$ and since the maps $\left[\frac{1}{2}\right]$ and $[n]$ commute, it follows that

$$\begin{aligned}
[763]P_7 &= \left[\tfrac{1}{2}\right][763]R_7 \oplus T_7, \\
&= (\text{0x5CF}, \text{0x485}) \oplus (\text{0x0}, \text{0x19A}), \\
&= (\text{0x84}, \text{0x475}).
\end{aligned}$$

### 13.3.6 Parallel implementation

Also, for fields of even characteristic, parallel implementations have gained some interest, one of the first works being [KOTS 1993]. However, applications using affine coordinates usually try to achieve parallelism on the lower level of field arithmetic.

In the chapter on side-channel attacks, cf. Chapter 29, we discuss several parallel implementations as this is mainly of interest for small devices like smart cards. There, the additional restriction is that the implementation should be secured against some particular attacks. Common choices are Montgomery coordinates distributed on two processors. We refer the reader to that chapter for details and mention here only the work of Mishra [MIS 2004a], who derives a pipelined computation such that in a scalar multiplication the average number of clock cycles needed per group operation is only 6 when using two processors.

### 13.3.7 Compression of points

Let $P = (x_1, y_1)$ be a point on $E/\mathbb{F}_{2^d} : y^2 + xy = f(x)$. As for odd characteristic, we show how to represent $P$ by $(x_1, b(y_1))$, where $b(y_1)$ is a bit distinguishing $P$ from $-P = (x_1, y_1 + x_1)$.

There exists exactly one Weierstraß point having $x_1 = 0$. For the other points we follow the steps in the next paragraphs.

#### Decompression

In even characteristic it is easier to explain decompression first. Thus, assume that $P$ is given by $(x_1, b(y_1))$, and $b(y_1) \in \{0, 1\}$. As $x_1$ is the $x$-coordinate of a point, the quadratic equation $y^2 + x_1 y + x_1^3 + a_2 x_1^2 + a_6$ has two solutions. It is clear that such a solution exists if $Y^2 + Y + (x_1^3 + a_2 x_1^2 + a_6)/x_1^2$ has a solution, i.e., if $\text{Tr}\big((x_1^3 + a_2 x_1^2 + a_6)/x_1^2\big) = 0$, cf. Section 11.2.6. If $y_1'$ is one solution then $y_1' + 1$ is the other. Hence, for the roots $y_1'$ the least significant bit allows us to distinguish between the solutions and we need to resort to the equation in $Y$ to compute the roots. To find the solutions of the original equation we put $y_1 = y_1' x_1$ for the $y_1'$ determined by $b(y_1)$.

#### Compression

We have just seen that the least significant bit of $y_1' = y_1/x_1$ should be used as $b(y_1)$. Unfortunately, this requires one inversion, hence, some work is also needed to compress a point. This is in contrast to the case of odd characteristic.

# Chapter 14

# Arithmetic of Hyperelliptic Curves

*Sylvain Duquesne and Tanja Lange*

## Contents in Brief

| | | |
|---|---|---|
| **14.1** | **Summary of background on hyperelliptic curves** | 304 |
| | Group law for hyperelliptic curves • Divisor class group and ideal class group • Isomorphisms and isogenies • Torsion elements • Endomorphisms • Cardinality | |
| **14.2** | **Compression techniques** | 311 |
| | Compression in odd characteristic • Compression in even characteristic | |
| **14.3** | **Arithmetic on genus 2 curves over arbitrary characteristic** | 313 |
| | Different cases • Addition and doubling in affine coordinates | |
| **14.4** | **Arithmetic on genus 2 curves in odd characteristic** | 320 |
| | Projective coordinates • New coordinates in odd characteristic • Different sets of coordinates in odd characteristic • Montgomery arithmetic for genus 2 curves in odd characteristic | |
| **14.5** | **Arithmetic on genus 2 curves in even characteristic** | 334 |
| | Classification of genus 2 curves in even characteristic • Explicit formulas in even characteristic in affine coordinates • Inversion-free systems for even characteristic when $h_2 \neq 0$ • Projective coordinates • Inversion-free systems for even characteristic when $h_2 = 0$ | |
| **14.6** | **Arithmetic on genus 3 curves** | 348 |
| | Addition in most common case • Doubling in most common case • Doubling on genus 3 curves for even characteristic when $h(x) = 1$ | |
| **14.7** | **Other curves and comparison** | 352 |

In Chapter 1 we introduced the *discrete logarithm problem* and showed that the main operation in a public-key cryptosystem is the computation of scalar multiples in a cyclic group. Chapter 9 showed how the computation of scalar multiples can be reduced to a sequence of additions and doublings in the group. Hence, for an efficient system we need to have groups with efficient group laws.

In Chapter 13 we detailed the arithmetic on elliptic curves. This chapter deals with hyperelliptic curves, which can be seen as a generalization of elliptic curves. We first give a brief overview of the main properties of hyperelliptic curves repeating the definitions for the convenience of the reader. The details can be found in Chapter 4. In the applications, group elements must be stored and transmitted. For restricted environments or restricted bandwidth it might be useful to use compression even though recovering the original coordinates needs some efforts. Accordingly, we consider compression techniques.

The main emphasis of this chapter is put on the arithmetic properties, i.e., on algorithms to perform the group operation. We state Cantor's algorithm, which works for arbitrary ground field and genus of the curve. To obtain better performance one needs to fix the genus and develop explicit formulas as for elliptic curves (cf. Chapters 13.2 and 13.3). We first specialize to considering curves of genus 2, separately over finite fields of odd and then of even characteristic. For both cases we give formulas for different coordinate systems, namely affine, projective, and new coordinates. The latter two systems allow us to avoid inversions in the group operation. For odd characteristic we also state two possible generalizations of Montgomery coordinates (cf. Section 13.2.3); for even characteristic there is no such generalization yet.

Also for genus 3 hyperelliptic curves, explicit formulas have been proposed. We give explicit formulas in affine coordinates in Section 14.6. Also nonhyperelliptic curves of genus 3 have been proposed for cryptographic applications. The final section gives references to these publications and also for genus 4 hyperelliptic curves before we conclude with a comparison and timings.

## 14.1 Summary of background on hyperelliptic curves

For cryptographic purposes we concentrate on *imaginary quadratic hyperelliptic curves* given by an equation, as below. Only in Chapter 18 we need to deal with the most general definition as given in Definition 4.121.

**Definition 14.1** A curve given by an equation of the form

$$C : y^2 + h(x)y = f(x), \ h, f \in K[x], \ \deg(f) = 2g+1, \deg(h) \leqslant g, f \text{ monic} \qquad (14.1)$$

is called a *hyperelliptic curve of genus g over* $K$ if no point on the curve over the algebraic closure $\overline{K}$ of $K$ satisfies both partial derivatives $2y + h = 0$ and $f' - h'y = 0$.

The last condition ensures that the curve is *nonsingular*. The negative of a point $P = (x, y)$ is given by $-P = (x, -y - h(x))$. The points fixed under this *hyperelliptic involution* are called *Weierstraß points*. Elliptic curves are subsumed under this definition as curves of genus one (cf. Definition 13.1). Even though this is not entirely standard from the classical viewpoint, the algorithmic properties that constitute our focus are equal. For a discussion see the remark after Definition 4.121.

**Example 14.2** Let $p = 2003$. Over the finite field $\mathbb{F}_p$, the equation $y^2 = x^5 + 1184x^3 + 1846x^2 + 956x + 560$ gives a hyperelliptic curve of genus 2.

We can check explicitly that the partial derivative $2y$ equals $0$ only if $y = 0$ and this leads to a point $P = (x_1, 0)$ only if $f(x_1) = 0$. The partial derivative with respect to $x$ gives $f'(x) = 5x^4 + 1549x^2 + 1689x + 956$. One can check by direct calculation that no root of $f$ simultaneously satisfies $f'$.

In general, $f'$ evaluates to $0$ at a root $x_1$ of $f$ if and only if $x_1$ is a multiple root of $f$. For odd characteristic, the equation $y^2 = f(x)$ defines a nonsingular and hence hyperelliptic curve if and only if $f$ has no multiple roots.

### 14.1.1 Group law for hyperelliptic curves

For elliptic curves one can take the set of points together with a point at infinity as a group. For curves of genus larger than one this is no longer possible. The way out is to take finite sums of points as group elements and perform the addition coefficient-wise like $(P+Q) \oplus (R+Q) = P + 2Q + R$. This would lead to an infinite group and longer and longer representations of the group elements. The group one actually uses is the quotient group of this group by all sums of points that lie on a function.

§ 14.1 Summary of background on hyperelliptic curves

Before stating this as a formal definition we give a pictorial description for a genus 2 curve over the reals given by an equation $y^2 = f(x)$ with $f$ monic of degree 5. As for elliptic curves, $f$ is not allowed to have multiple roots over the algebraic closure to satisfy the condition of the definition.

**Figure 14.1** Group law on genus 2 curve over the reals $\mathbb{R}$, $y^2 = f(x)$, $\deg f = 5$ for $(P_1 + P_2) \oplus (Q_1 + Q_2) = R_1 + R_2$.

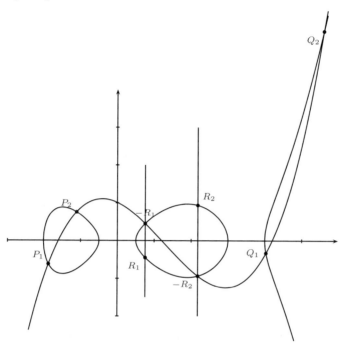

Figure 14.1 demonstrates again that one cannot continue using the chord-and-tangent method from elliptic curves as a line intersects in 5 instead of 3 points. To build a group we take the quotient of the group of sums of points on the curve by the subset of those sums where the points lie on a function, e.g., $R_1 = (x_{R_1}, y_{R_1})$ and $-R_1 = (x_{R_1}, -y_{R_1})$ lie on the curve given by $x = x_{R_1}$ and, hence, $R_1 \oplus (-R_1) = 0$. Likewise the six points $P_1, P_2, Q_1, Q_2, -R_1, -R_2$ on the cubic add up to zero in the quotient group we consider.

This way one sees that each element can be represented by at most two points that do not have the same $x$-coordinate and inverse $y$-coordinate. Namely, any $n > 1$ points give rise to a polynomial of degree $n - 1$. There are $\max\{5, 2(n-1)\} - n$ other points of intersection. As soon as $n > 2$ the inverse of this sum of points, obtained by inflecting all points at the $x$-axis, contains fewer points. Repeating this process gives a reduced group element with at most 2 points. The second condition can be seen to hold as points $(x_1, y_1)$ and $(x_1, -y_1)$ both lie on the function $x = x_1$.

Adding two elements is done in two steps. First the formal sum is formed and then it is reduced. In the general case both group elements consist of 2 points given by $P_1 + P_2$ and $Q_1 + Q_2$ and the 4 points are all different. A function $y = s(x)$ of degree 3 in $x$ passes through all of them having 2 more points of intersection with $C$. The two new points $-R_1$ and $-R_2$ are inflected and give the result of the addition $(P_1 + P_2) \oplus (Q_1 + Q_2) = R_1 + R_2$.

As in the case of elliptic curves, one can derive the group law from this description by making all steps explicit. For genus 2 curves over finite fields this is done in Sections 14.3, 14.4, and 14.5. If $h \neq 0$ there are still two points with equal $x$-coordinate but the opposite of $P = (x_1, y_1)$ is given by $-P = (x_1, -y_1 - h(x_1))$.

For hyperelliptic curves of arbitrary genus $g$ one obtains that each group element is represented by at most $g$ points and that in the reduction step one might need several rounds to find a minimal representative. Note that so far we did not touch the problem over which fields the points should be defined. In the following an isomorphic representation is introduced, which is advantageous for implementation purposes.

### 14.1.2 Divisor class group and ideal class group

The group we described so far is called the *divisor class group* $\mathrm{Pic}_C^0$ of $C$. To formally define the group law we need to take into account a further point $P_\infty$ called the point at infinity. (For curves of the form (14.1) there is only a single point at infinity and this is crucial for the way we implement the arithmetic.) In the picture it can be visualized as lying far out on the $y$-axis such that any line parallel to it passes through $P_\infty$.

We repeat the main definitions from Chapter 4.

**Definition 14.3** Let $C$ be a hyperelliptic curve of genus $g$ over $K$ given by an equation of the form (14.1). The *group of divisors of $C$ of degree* 0 is given by

$$\mathrm{Div}_C^0 = \Big\{ D = \sum_{P \in C} n_P P \mid n_P \in \mathbb{Z}, n_P = 0 \text{ for almost all } P \in C,$$
$$\sum_{P \in C} n_P = 0, \text{ and such that } \sigma(D) = D \text{ for all } \sigma \in G_K \Big\}.$$

This latter condition means that the divisor is defined over $K$. This is equivalent to $n_{\sigma(P)} = n_P$ for all $\sigma \in G_K$, the Galois group of $K$.

**Definition 14.4** The *divisor class group* $\mathrm{Pic}_C^0$ of $C$ is the quotient group of $\mathrm{Div}_C^0$ by the group of principal divisors, that are divisors of degree zero resulting from functions.

Each divisor class can be uniquely represented by a finite sum

$$\sum_{i=1}^r P_i - rP_\infty, \quad P_i \in C \smallsetminus \{P_\infty\}, r \leqslant g,$$

where for $i \neq j$ we have $P_i = (x_i, y_i) \neq (x_j, -y_j - h(x_j)) = -P_j$.

The following theorem introduces a different representation that is more useful for implementations, and for which one can simply read off the field of definition of the group elements. The theoretical background for this alternative representation is the fact that for curves of the form (14.1) the divisor class group is isomorphic to the *ideal class group* of the function field $K(C)$. Furthermore, the divisor class group is isomorphic to the group of $K$-rational points of the Jacobian $J_C$ of $C$. (For details see Section 4.4.6.a.) Mumford representation makes explicit this isomorphism and we will use the representation as an ideal class group for the arithmetic. However, to develop the formulas we come back to the representation as a finite sum of points. To fix names we keep speaking of the divisor class group and call the group elements divisor classes even when using the notation as ideal classes.

**Theorem 14.5 (Mumford representation)**
Let $C$ be a genus $g$ hyperelliptic curve as in (14.1) given by $C : y^2 + h(x)y = f(x)$, where $h, f \in K[x]$, $\deg f = 2g + 1$, $\deg h \leqslant g$. Each nontrivial divisor class over $K$ can be represented via a unique pair of polynomials $u(x)$ and $v(x)$, $u, v \in K[x]$, where

## § 14.1 Summary of background on hyperelliptic curves

1. $u$ is monic,
2. $\deg v < \deg u \leqslant g$,
3. $u \mid v^2 + vh - f$.

Let $D = \sum_{i=1}^{r} P_i - rP_\infty$, where $P_i \neq P_\infty, P_i \neq -P_j$ for $i \neq j$ and $r \leqslant g$. Put $P_i = (x_i, y_i)$. Then the divisor class of $D$ is represented by

$$u(x) = \prod_{i=1}^{r}(x - x_i)$$

and if $P_i$ occurs $n_i$ times then

$$\left(\frac{d}{dx}\right)^j \left[v(x)^2 + v(x)h(x) - f(x)\right]_{|x=x_i} = 0, \text{ for } 0 \leqslant j \leqslant n_i - 1.$$

A divisor with at most $g$ points in the support satisfying $P_i \neq P_\infty, P_i \neq -P_j$ for $i \neq j$ is called a *reduced divisor*. The first part states that each class can be represented by a reduced divisor. The second part of the theorem means that for all points $P_i = (x_i, y_i)$ occurring in $D$ we have $u(x_i) = 0$ and the third condition guarantees that $v(x_i) = y_i$ with appropriate multiplicity.

We denote the class represented by $u(x)$ and $v(x)$ by $[u(x), v(x)]$. To unify notation we denote the neutral element of the group by $[1, 0]$.

There are basically two ways for finding a $K$-rational divisor class. This can be done by building it from $K$-rational points on the curve. For instance, choose a random $x_1 \in K$, and try to find $y_1 \in K$ such that

$$y_1^2 + h(x_1)y_1 - f(x_1) = 0. \tag{14.2}$$

In odd characteristic or characteristic zero when $h = 0$, the problem reduces to computing a square root when there exists one. This can be checked with the Legendre symbol, see Section 2.3.4. If the result of this test, performed with Algorithm 11.19 in a prime field and Algorithm 11.69 in an extension field, is negative, then one chooses another $x_1$ and repeats the process. Otherwise, one deduces $y_1$ with one of the Algorithms 11.23 and 11.26.

For a field of characteristic 2, see Section 11.2.6 and Lemma 11.56 for a criterion of existence of a solution to (14.2) and a method to compute $y_1$.

Already a single point gives rise to a divisor and $[x - x_1, y_1]$ is a valid representative of a divisor class $\overline{D}$. Unless $y_1 = 0$ (or in general $\text{ord}(\overline{D})$ small) the multiples $[n]\overline{D}$ will have the first polynomial of full degree $g$ for $n \geqslant g$.

Later in this chapter we consider explicit formulas to perform the group operations. For applications we suggest implementing only the most frequent cases of inputs — which implies that the first polynomial $u$ has to have degree $g$ before one can start computing scalar multiples. To build such a class of full degree one takes $g$ random distinct points as above and combines them using Lagrange interpolation: to the points $P_1 = (x_1, y_1), \ldots, P_g = (x_g, y_g)$ correspond the polynomials

$$u(x) = \prod_{i=1}^{g}(x - x_i) \text{ and } v(x) = \sum_{i=g}^{t} \frac{\prod_{j \neq i}(x - x_j)}{\prod_{j \neq i}(x_i - x_j)} y_i.$$

The resulting classes are not completely random as they are built from points defined over $K$ while a $K$-rational class may also contain points defined over an extension field $L/K$ of degree $[L:K] \leqslant g$. The second strategy avoids this possible drawback. One chooses a random monic polynomial $u(x) \in K[x]$ of degree $g$ by randomly choosing its $g$ free coefficients. Using the decompression techniques as explained in Sections 14.2.1 and 14.2.2 one tries to recover a polynomial $v$ satisfying $u \mid v^2 + vh - f$. If this fails one starts anew with a different choice of $u$. By Theorem 14.5 the tuple $[u, v]$ represents a divisor class.

The amount of work to find $v$ is equal to solving the $g$ quadratic equations in the first approach. However, checking if a $u$ belongs to a class requires more effort. Hence, for an implementation one can trade off the generality of the class for less work.

**Example 14.6** On the curve from Example 14.2, a divisor class is given by $\overline{D} = [x^2 + 376x + 245, 1015x + 1368]$. We have $[2]\overline{D} = [x^2 + 1226x + 335, 645 + 1117]$ and $[3874361]\overline{D} = [1, 0]$.

Using this compact description of the elements, one can transfer the group law that was derived above as a sequence of composition and reduction to an algorithm operating on the representing polynomials and using only polynomial arithmetic over the field of definition $K$. This algorithm was described by Cantor [CAN 1987] for odd characteristic and by Koblitz [KOB 1989] for arbitrary fields.

---

**Algorithm 14.7** Cantor's algorithm

INPUT: Two divisor classes $\overline{D}_1 = [u_1, v_1]$ and $\overline{D}_2 = [u_2, v_2]$ on the curve $C : y^2 + h(x)y = f(x)$.
OUTPUT: The unique reduced divisor $D$ such that $\overline{D} = \overline{D}_1 \oplus \overline{D}_2$.

1. $d_1 \leftarrow \gcd(u_1, u_2)$      $[d_1 = e_1 u_1 + e_2 u_2]$
2. $d \leftarrow \gcd(d_1, v_1 + v_2 + h)$      $[d = c_1 d_1 + c_2(v_1 + v_2 + h)]$
3. $s_1 \leftarrow c_1 e_1, s_2 \leftarrow c_1 e_2$ and $s_3 \leftarrow c_2$
4. $u \leftarrow \dfrac{u_1 u_2}{d^2}$ and $v \leftarrow \dfrac{s_1 u_1 v_2 + s_2 u_2 v_1 + s_3(v_1 v_2 + f)}{d} \bmod u$
5. **repeat**
6.      $u' \leftarrow \dfrac{f - vh - v^2}{u}$ and $v' \leftarrow (-h - v) \bmod u'$
7.      $u \leftarrow u'$ and $v \leftarrow v'$
8. **until** $\deg u \leqslant g$
9. **make** $u$ monic
10. **return** $[u, v]$

---

We remark that Cantor's algorithm is completely general and holds for any field and genus. It is a nice exercise to check that the addition formulas derived in Section 13.1.1 for elliptic curves can be obtained as a special case of Cantor's algorithm making all steps explicit for $g = 1$.

### 14.1.3 Isomorphisms and isogenies

Some changes of variables do not fundamentally alter the hyperelliptic curve. More precisely, let the hyperelliptic curve $C/K$ of genus $g$ be given by $C : y^2 + h(x)y = f(x)$. The maps

$$y \mapsto u^{2g+1} y' + a_g x'^g + \cdots + a_1 x' + a_0 \text{ and } x \mapsto u^2 x' + b \text{ with } (a_g, \ldots, a_1, a_0, b, u) \in K^{g+2} \times K^*$$

are invertible and map each point of $C$ to a point of $C' : y'^2 + \tilde{h}(x')y' = \tilde{f}(x')$, where $\tilde{h}, \tilde{f}$ are defined over $K$ and can be expressed in terms of $h, f, a, b, c, d$ and $u$. Via the inverse map we associate to each point of $C'$ a point of $C$ showing that both curves are *isomorphic*. These changes of variables are the only ones leaving invariant the shape of the defining equation and, hence, they are the only admissible *isomorphisms*.

The following examples study even and odd characteristic separately as they allow us different isomorphic transformations of the curve equation. These transformations will be useful for the arithmetic of the curves.

**Example 14.8** In the case of *odd characteristic* the transformation $y \mapsto y' - h(x)/2$ allows to consider an isomorphic curve of the form

$$y'^2 = \tilde{f}(x) = x^{2g+1} + \tilde{f}_{2g}x^{2g} + \cdots + \tilde{f}_3 x^3 + \tilde{f}_2 x^2 + \tilde{f}_1 x + \tilde{f}_0, \text{ with } \tilde{f}_i \in K, \qquad (14.3)$$

where $\tilde{f}$ has no multiple roots over the algebraic closure $\overline{K}$.

Furthermore, if $\text{char}(K) \neq 2g+1$ we can obtain a further coefficient being zero. Namely, putting $x \mapsto x' - \tilde{f}_{2g}/(2g+1)$ leads to

$$y'^2 = \hat{f}(x') = x'^{2g+1} + \hat{f}_{2g-1} x'^{2g-1} + \cdots + \hat{f}_1 x' + \hat{f}_0, \text{ with } \hat{f}_i \in K. \qquad (14.4)$$

**Example 14.9** In the case of even characteristic a nonsingular curve must have $h(x) \neq 0$ as will be shown now. The partial derivative equations are given by $2y - h(x) = h(x)$ and $h'(x)y - f'(x)$.

If $h(x) = 0$ identically then the partial derivative for $y$ vanishes completely and the one for $x$ simplifies to $f'(x) = 0$. Over the algebraic closure $\overline{K}$ this equation has $2g$ roots. Let $x_1$ be one of them. Over $\overline{K}$, there exists an $y_1$ with $f(x_1) = y_1^2$, and, hence, a singular point $P = (x_1, y_1)$ satisfying the curve equation and both partial derivative equations.

Hence, we can always assume $h(x) \neq 0$ and the transformation $x \mapsto x' + f_{2g}$ leads to an equation

$$y^2 + \tilde{h}(x')y = x'^{2g+1} + \tilde{f}_{2g-1} x'^{2g-1} + \cdots + \tilde{f}_1 x' + \tilde{f}_0, \text{ with } \tilde{f}_i \in K.$$

Some more transformations are possible and useful to obtain efficient doublings. We come back to this in Section 14.5 and Section 14.6.3.

Even if the curves are not isomorphic, the Jacobians of $C$ and $C'$ might share some common properties. One calls $J_C$ and $J_{C'}$ *isogenous* if there exists a morphism $\psi : J_C \to J_{C'}$ mapping $[1, 0]$ to the neutral element of $J_{C'}$. One important property of isogenies is that for every isogeny there exists a unique isogeny $\hat{\psi} : J_{C'} \to J_C$ called the *dual isogeny* such that

$$\hat{\psi} \circ \psi = [n] \text{ and } \psi \circ \hat{\psi} = [n]',$$

where $[n]'$ denotes the multiplication by $n$ map on $J_{C'}$. The *degree of the isogeny* $\psi$ is equal to this $n$. For more background on isogenies we refer to Section 4.3.4.

By abuse of notation, two curves $C/K$ and $C'/K$ are called *isogenous* if the corresponding Jacobian varieties $J_C$ and $J_{C'}$ are isogenous.

### 14.1.4 Torsion elements

**Definition 14.10** The kernel of $[n]$ on $J_C$ is denoted by $J_C[n]$. An element $\overline{D} \in J_C[n]$ is called an element of *n-torsion*.

**Theorem 14.11** Let $C$ be a hyperelliptic curve defined over $K$. If the characteristic of $K$ is either zero or prime to $n$ then
$$J_C[n] \simeq (\mathbb{Z}/n\mathbb{Z})^{2g}.$$
Otherwise, when $\text{char}(K) = p$ and $n = p^e$ then
$$J_C[p^e] \simeq (\mathbb{Z}/p^e\mathbb{Z})^r,$$
with $0 \leqslant r \leqslant g$, for all $e \geqslant 1$.

The following definition was given for general abelian varieties in Definition 4.74. We repeat it here for easy reference in the special case of Jacobians of curves.

**Definition 14.12** Let $\operatorname{char}(K) = p$. The *p-rank of $C/K$* is defined to be the integer $r$ in Theorem 14.11.

An elliptic curve $E$ is called *supersingular* if it has $E[p^e] \simeq \{P_\infty\}$, i.e., if it has $p$-rank 0. A Jacobian of a curve is called *supersingular* if it is the product of supersingular elliptic curves. Thus, especially the $p$-rank of a supersingular Jacobian variety is 0 but the converse does not have to hold.

One also uses the term *supersingular curve* to denote that the Jacobian of this curve is supersingular.

### 14.1.5 Endomorphisms

The multiplication by $n$ is an endomorphism of $J_C$. The set of all endomorphisms of $J_C$ defined over $K$ will be denoted by $\operatorname{End}_K(J_C)$ or more simply by $\operatorname{End}(J_C)$, and contains at least $\mathbb{Z}$.

**Definition 14.13** If $\operatorname{End}(J_C) \otimes \mathbb{Q}$ contains a number field $F$ of degree $2g$ over $\mathbb{Q}$ we say that $J_C$ has *complex multiplication*.

Thus, one has complex multiplication if the endomorphism ring is strictly bigger than $\mathbb{Z}$.

**Remark 14.14** Let $C$ be a nonsupersingular hyperelliptic curve over $\mathbb{F}_q$. Then $J_C$ always has complex multiplication. Indeed, the Frobenius automorphism $\phi_q$ of $\mathbb{F}_q$ extends to the points of the curve by sending $P_\infty$ to itself and $P = (x_1, y_1)$ to $(x_1^q, y_1^q)$. One can easily check that the point $\phi_q(P) = (x_1^q, y_1^q)$ is again a point on the curve irrespective of the field of definition of $P$. The map also extends to the divisors and divisor classes. Hence, $\phi_q$ is an endomorphism of $J_C$, called the *Frobenius endomorphism*. It is different from $[n]$ for all $n \in \mathbb{Z}$.

### 14.1.6 Cardinality

Let the hyperelliptic curve $C$ be defined over a finite field $\mathbb{F}_q$. As for elliptic curves we have bounds on the number of points over a finite field and on the group order of the divisor class group. The following bounds depend only on the finite field and the genus of the curve:

**Theorem 14.15 (Hasse–Weil)**

$$(q^{1/2} - 1)^{2g} \leqslant |\operatorname{Pic}_C^0| \leqslant (q^{1/2} + 1)^{2g}.$$

The Frobenius endomorphism $\phi_q$ operates on the points by raising the coordinates to the power of $q$. This action is inherited by the divisor class group and there $\phi_q$ maps the divisor class $[u(x), v(x)]$ to $[\phi_q(u(x)), \phi_q(v(x))]$, where it is applied to the coefficients of the polynomials. The Frobenius endomorphism satisfies a *characteristic polynomial* defined over the integers.

**Theorem 14.16** Let $C$ by a hyperelliptic curve of genus $g$ defined over $\mathbb{F}_q$.
The Frobenius endomorphism satisfies a characteristic polynomial of degree $2g$ given by

$$\chi(\phi_q)_C(T) = T^{2g} + a_1 T^{2g-1} + \cdots + a_g T^g + \cdots + a_1 q^{g-1} T + q^g,$$

where $a_i \in \mathbb{Z}, 1 \leqslant i \leqslant g$.

Denote by $M_r$ the number of points of $C(\overline{\mathbb{F}}_q)$ that are defined over $\mathbb{F}_{q^r}$ or a subfield $\mathbb{F}_{q^s}$ with $s \mid r$, and put $N_k = |\operatorname{Pic}_{C \cdot \mathbb{F}_{q^k}}^0|$. There is a relationship between the $N_i$ and the numbers $M_r$ for $1 \leqslant r \leqslant g$ given by the following theorem, which also provides the characteristic polynomial $\chi(\phi_q)_C(T)$ of the Frobenius endomorphism.

**Theorem 14.17** Let $C$ be a hyperelliptic curve of genus $g$ defined over $\mathbb{F}_q$ and let the factorization of the characteristic polynomial $\chi(\phi_q)_C(T)$ of the Frobenius endomorphism over $\mathbb{C}$ be

$$\chi(\phi_q)_C(T) = \prod_{i=1}^{2g}(T - \tau_i) \text{ with } \tau_i \in \mathbb{C}.$$

Then

(i) The roots of $\chi(\phi_q)_C$ satisfy $|\tau_i| = \sqrt{q}$ for $1 \leqslant i \leqslant 2g$.
(ii) There exists an ordering with $\tau_{i+g} = \bar{\tau}_i$, hence, $\tau_i \tau_{i+g} = q$ for $1 \leqslant i \leqslant g$.
(iii) For any integer $k$, we have

$$N_k = \prod_{i=1}^{2g}(1 - \tau_i^k),$$

$$M_k = q^k + 1 - \sum_{i=1}^{2g}\tau_i^k,$$

$$|M_k - (q^k + 1)| \leqslant g \lfloor 2q^{k/2} \rfloor.$$

(iv) Put $a_0 = 1$, then for $1 \leqslant i \leqslant g$ we have

$$ia_i = (M_i - (q^i + 1))a_0 + (M_{i-1} - (q^{i-1} + 1))a_1 + \cdots + (M_1 - (q+1))a_{i-1}.$$

By the last property one sees that from the first $g$ numbers $M_i$ of points on the curve one can obtain the whole polynomial $\chi(\phi_q)_C(T)$ and thus the cardinality of $\text{Pic}_C^0$ as $\chi(\phi_q)_C(1)$. To illustrate this relation: for a genus 2 curve we have to count the number of points defined over $\mathbb{F}_q$ and $\mathbb{F}_{q^2}$ to obtain $a_1 = M_1 - q - 1$ and $a_2 = (M_2 - q^2 - 1 + a_1^2)/2$. For curves over large prime fields this poses a nontrivial problem. For details on point counting we refer to Chapter 17. For curves defined over small finite fields this way of determining the group order over extension fields is advantageous. We deal with such curves, called *Koblitz curves*, in Section 15.1.

## 14.2 Compression techniques

We now study the question of how to compress divisor classes $[u, v]$ for a genus $g$ curve given by (14.1). To this aim we fist consider compression of points as done for elliptic curves.

We observe that for a point $P = (x_1, y_1)$ there is at most one more point with the same $x$-coordinate, namely $-P = (x_1, -y_1 - h(x_1))$. There is no further point exactly if $y_1 = -y_1 - h(x_1)$. Therefore, the $x$-coordinate determines a point up to the choice of two $y$-coordinates. This choice can be given by a further bit, hence, a point can be represented by its $x$-coordinate and a further bit.

The compression technique by Hess, Seroussi, and Smart [HESE+ 2001] uses Mumford representation (cf. Theorem 14.5) and recovers the points representing the divisor class $[u, v]$.

For our exposition we first follow the approach of Stahlke [STA 2004]. To obtain efficient compression and decompression techniques we need to study odd and even characteristic independently and use the isomorphic transformations from Examples 14.8 and 14.9.

### 14.2.1 Compression in odd characteristic

In this paragraph we show how to compress a representative $[u, v]$ of a divisor class by storing $u$ and some more bits, such that we can reconstruct $v$. Since the characteristic is odd we can assume $h(x) = 0$. The same considerations hold in characteristic 0. Assume $v$ to be unknown.

Since $u \mid v^2 - f$, there is a $s \in K[x]$ such that $us = v^2 - f$. This is an equation between two polynomials of degree $2g+1$ as $\deg f = 2g+1$. The unknowns are the $(2g+2-\deg u)$ coefficients of $s$ and at most $\deg u$ coefficients of $v$. Therefore, by comparison of coefficients we have $2g + 2$ equations with at most $2g + 2$ unknowns. We expect at most $2^g$ solutions for $v$ in which case the choice of one solution can be encoded in $g$ bits.

This expectation has to be verified in each case, of course. For elliptic curves we get two solutions for $v$ and with just one bit we can reconstruct $v$, which corresponds to calculating the $y$-coordinate from the $x$-coordinate of a point (cf. Section 13.2.5). By way of illustration, from now on we restrict ourselves to genus $g = 2$.

Then a curve can be defined by

$$y^2 = x^5 + f_4 x^4 + f_3 x^3 + f_2 x^2 + f_1 x + f_0,$$

where $f$ has only simple roots in the algebraic closure. Let $f_i$ be the coefficients of $f$, (respectively $u_i$ of $u$, $v_i$ of $v$ and $s_i$ of $s$). From $us + f = v^2$ by comparison of coefficients we get

$$s(x) = -x^3 + (u_1 - f_4) x^2 + (u_0 - u_1^2 + f_4 u_1 - f_3) x + s_0.$$

The discriminant of

$$\begin{aligned} u(x)s(x) + f(x) &= \left(s_0 + f_2 - f_3 u_1 - f_4(u_0 - u_1^2) + u_1(2u_0 - u_1^2)\right) x^2 \quad (14.5) \\ &+ \left(u_1 s_0 + f_1 - f_3 u_0 + f_4 u_0 u_1 + u_0(u_0 - u_1^2)\right) x \\ &+ u_0 s_0 + f_0 \end{aligned}$$

is of degree at most 2 in $s_0$ and all coefficients are known. It is zero because $u(x)s(x) + f(x)$ is a square (namely $v(x)^2$). From $v(x)^2 = u(x)s(x) + f(x)$ we get relations for $v_0$ and $v_1$:

$$v_0^2 = u_0 s_0 + f_0, \quad (14.6)$$
$$2 v_0 v_1 = u_1 s_0 + f_1 - f_3 u_0 + f_4 u_0 u_1 + u_0(u_0 - u_1^2), \quad (14.7)$$
$$v_1^2 = s_0 + f_2 - f_3 u_1 - f_4(u_0 - u_1^2) + u_1(2u_0 - u_1^2). \quad (14.8)$$

**Compression**

The $f_i$, $u_i$ and $v_i$ are known. Calculate $s_0$ from (14.6) or (if $u_0 = 0$) from (14.8). Consider the right hand side of (14.5) as polynomial in $x$ and let $d(s_0)$ be its discriminant, which we consider as polynomial in $s_0$. If $u_1^2 - 4u_0 \neq 0$, consider the discriminant of $d$ and decide which root gives the correct value for $s_0$. Store this choice in $\text{Bit}_1$. As $q$ is odd, the most convenient choice might be to take as $\text{Bit}_1$ the least significant bit of the root (i.e., the parity of a coordinate of the root considered as number in $[0, p-1]$).

*Exception:* If $u_1^2 - 4u_0 = 0$ then $d(s_0)$ is of degree 1 and $\text{Bit}_1$ can be chosen arbitrarily. In fact $d(s_0)$ is never of degree 0, otherwise a short calculation shows that $f(-u_1/2) = 0$ and $f'(-u_1/2) = 0$, so $-u_1/2$ would be a singular point of the curve.

Now store in $\text{Bit}_2$ the correct choice of $v_0$ as root of $u_0 s_0 + f_0$ (cf. (14.6)). (Again the most convenient choice might be to take as $\text{Bit}_2$ the least significant bit of $v_0$.) But if $v_0 = 0$ then instead store in $\text{Bit}_2$ the correct choice of $v_1$ as root of the right hand side of (14.8).

The compressed point is the tuple $(u_0, u_1, \text{Bit}_1, \text{Bit}_2)$.

**Decompression**

The $f_i$ and $u_i$ are known. Also known are $\text{Bit}_1$ and $\text{Bit}_2$. We need to recover $v_0$ and $v_1$. Take the discriminant of $u(x)s(x) + f(x)$ (see (14.5)) and consider this discriminant $d(s_0)$ as polynomial in $s_0$. Calculate $s_0$ from $d(s_0) = 0$ according to $\text{Bit}_1$. Calculate $v_0$ from (14.6) according to $\text{Bit}_2$. If $v_0 \neq 0$ then calculate $v_1$ from (14.7). If $v_0 = 0$ then calculate $v_1$ from (14.8) according to $\text{Bit}_2$.

## 14.2.2 Compression in even characteristic

Here, we focus on finite fields of characteristic 2. Thus, we now assume that the genus $g$ curve $C$ is given by $C : y^2 + h(x)y = f(x)$, with $h(x) \neq 0$ and $h, f \in \mathbb{F}_{2^d}[x]$ and that the divisor class $[u(x), v(x)]$ has $u, v \in \mathbb{F}_{2^d}[x]$. The following algorithm is a special case of [HESE+ 2001] and we provide the version for arbitrary genus. Hence, one needs to factor the first polynomial $u(x)$. Note that for fixed genus and type of the curve it is again possible to take advantage of ideas as in the previous section and work without factoring $u(x)$, as shown for genus $g = 2$ in [LAN 2005a].

### Compression

Let $m_1(x), m_2(x), \ldots, m_l(x)$ be the factorization of $u(x)$ over $\mathbb{F}_{2^d}$ ordered in some prescribed way. For each factor $m_i(x)$ compute $r_i(x) = v(x) \bmod m_i(x)$ and $\tilde{r}_i(x) = v(x) + h(x) \bmod m_i(x)$, and determine a bit to carry the correct choice as follows: if $r_i(x)$ is smaller than $\tilde{r}_i(x)$ according to the ordering, then put $\text{Bit}_i$ to 0, otherwise assign $\text{Bit}_i = 1$.

### Remarks 14.18

(i) For fixed $h(x)$ it is possible to omit the computation of $\tilde{r}_i(x)$ and derive a condition depending on $r_i(x)$ only to distinguish the two choices.

(ii) If two factors $m_i(x)$ and $m_{i+1}(x)$ are equal, the computation of $r_i$ needs to be done only once and one can put $\text{Bit}_{i+1} = \text{Bit}_i$.

The class is then represented by $[u(x), \text{Bit}_1, \ldots, \text{Bit}_l]$ needing at most $gd + g$ bits.

### Decompression

To recover the full representation one needs to invert the above process. Again one starts with factoring $u(x)$ into $m_1(x), m_2(x), \ldots, m_l(x)$.

For each polynomial $m_i(x)$ one performs the following procedure. Note that these factors are irreducible and hence define a field extension $L = \mathbb{F}_{2^d}[x]/(m_i(x))$ of degree $\deg(m_i)$. In particular this means that one can compute in $L$ and solve quadratic equations in the polynomial ring $L[Y]$ as explained in Section 11.2.6.

Compute the two solutions $r_{i1}(x)$ and $r_{i2}(x)$ of $Y^2 + h(x)Y + f(x) \pmod{m_i(x)}$. The results are interpreted as polynomials of degree less than $\deg(m_i)$ over $\mathbb{F}_{2^d}$.

The bit $\text{Bit}_i$ determines which one to choose as the solutions correspond exactly to the polynomials $r_i(x)$ and $\tilde{r}_i(x)$ computed above.

As above, these computations need not be performed if $m_i = m_{i+1}$.

This way one recovers $[u(x), v(x)] = [m_1(x), r_1(x)] \oplus [m_2(x), r_2(x)] \oplus \cdots \oplus [m_l(x), r_l(x)]$. Note that these additions do not involve reductions as the combined divisor class is reduced. This means that $v(x)$ can be obtained by interpolation, while $u$ is known directly from the transmitted values.

## 14.3 Arithmetic on genus 2 curves over arbitrary characteristic

For elliptic curves the group law can be stated as a sequence of multiplications, squarings, inversions, and additions over the field. For hyperelliptic curves we have Cantor's algorithm to perform addition and doubling. Here we aim at deriving explicit formulas for genus two curves.

The first attempt to find such explicit formulas was done by Spallek [SPA 1994] and by Krieger [KRI 1997]. The first practical formulas were obtained by Harley [HAR 2000], which were generalized to even characteristic by Lange [LAN 2001a]; an improvement of the former paper can be found in Matsuo, Chao, and Tsujii [MACH⁺ 2001]. A significant improvement was obtained independently by Takahashi [TAK 2002] and Miyamoto, Doi, Matsuo, Chao, and Tsujii [MIDO⁺ 2002]; this was generalized to even characteristic in [SUMA⁺ 2002] and independently in [LAN 2002b]. The second reference allows more general curves and manages to trade more multiplications for squarings, which is desirable for characteristic 2 implementations.

In this section we give the main ingredients to study explicit formulas for the group law for genus two curves. To this end we first give the case study of [LAN 2001a] investigating what can be the input of Cantor's Algorithm 14.7 and proceed in considering these different cases. We state algorithms for the addition and doubling in the most frequent cases for even and odd characteristic.

These formulas involve (at least) 1 inversion per addition or doubling respectively. In some environments inversions are extremely time or space critical. In the following two Sections 14.4 and 14.5 we consider in more detail the case of odd and even characteristic and also describe other coordinate systems that avoid inversions on the cost of more multiplications.

Unless stated otherwise, the following formulas hold independently of the characteristic, therefore we take care of the signs; in characteristic 2, $2y$ is understood as zero.

For all following sections we fix the notation to refer to the coefficient of $x^i$ in a polynomial $l(x)$ as $l_i$.

### 14.3.1 Different cases

Consider the composition step of Cantor's Algorithm 14.7. The input are two classes represented by two polynomials each, namely $[u_1, v_1]$ and $[u_2, v_2]$. Recall that Mumford representation 14.5 links each such class to at most two finite points of the curve.

Without loss of generality, let $\deg u_1 \leqslant \deg u_2$.

1. If $u_1$ is of degree 0 then $\overline{D}_1$ must be the neutral element $[u_1, v_1] = [1, 0]$. The result of the combination and reduction is the second class $[u_2, v_2]$.

2. If $u_1$ is of degree 1, then either $u_2$ is of degree 1 as well or it has full degree.

   A. Assume $\deg u_2 = 1$, i.e., $u_i = x + u_{i0}$ and the $v_i$ are constant. Then if $u_1 = u_2$ we obtain for $v_1 = -v_2 - h(-u_{10})$ the zero element $[1, 0]$ and for $v_1 = v_2$ we double the divisor to obtain

$$u = u_1^2, \tag{14.9}$$
$$v = \frac{\bigl(f'(-u_{10}) - v_1 h'(-u_{10})\bigr)x + \bigl(f'(-u_{10}) - v_1 h'(-u_{10})\bigr)u_{10}}{2v_1 + h(-u_{10})} + v_1.$$

   Otherwise the composition leads to

$$u = u_1 u_2 \text{ and } v = \bigl((v_2 - v_1)x + v_2 u_{10} - v_1 u_{20}\bigr)/(u_{10} - u_{20}).$$

   In all cases the results are already reduced.

   B. Now let the second polynomial be of degree 2, i.e., $u_2 = x^2 + u_{21}x + u_{20}$. Then the corresponding divisors are given by

$$D_1 = P_1 - P_\infty \text{ and } D_2 = P_2 + P_3 - 2P_\infty, \text{ with } P_i \neq P_\infty.$$

   i. If $u_2(-u_{10}) \neq 0$ then $P_1$ and $-P_1$ do not occur in $D_2$. This case will be dealt with below in Section 14.3.2.b.

   ii. Otherwise if $D_2 = 2P_1 - 2P_\infty$, which holds if $u_{21} = 2u_{10}$ and $u_{20} = u_{10}^2$ one first doubles $D_2$ as in 3.A.ii and then subtracts $D_1$ using 2.B.i. If $v_2(-u_{10}) =$

### § 14.3 Arithmetic on genus 2 curves over arbitrary characteristic

$v_1 + h(-u_{10})$ then $-P_1$ occurs in $D_2$ and the resulting class is given by $u = x + u_{21} - u_{10}$ and $v = v_2(-u_{21} + u_{10})$. Otherwise one first doubles $[u_1, v_1]$ by (14.9) and then adds

$$[x + u_{21} - u_{10}, v_2(-u_{21} + u_{10})],$$

hence, reduces the problem to the case 2.B.i.

3. Let $\deg u_1 = \deg u_2 = 2$.

   A. Let first $u_1 = u_2$. This means that for an appropriate ordering

   $$D_1 = P_1 + P_2 - 2P_\infty \text{ and } D_2 = P_3 + P_4 - 2P_\infty,$$

   where the $x$-coordinates of $P_i$ and $P_{i+2}$ are equal.

   i. If $v_1 \equiv -v_2 - h \pmod{u_1}$ then the result is $[1, 0]$.

   ii. If $v_1 = v_2$ then we are in the case in which we double a class of order different from 2 and with first polynomial of full degree. Again we need to consider two subcases depending on whether a point in the support has order 2.
   If $D_1 = P_1 + P_2 - 2P_\infty$ where $P_1$ is equal to its opposite, then the result is $2P_2$ and can be computed as above. The point $P_1 = (x_1, y_1)$ is equal to its opposite, if and only if $h(x_1) = -2y_1$. To check for this case we compute the resultant of $h + 2v_1$ and $u_1$.

      a. If $\text{Res}(h + 2v_1, u_1) \neq 0$ then we are in the usual case where both points are not equal to their opposite. This will be considered in Section 14.3.2.c.

      b. Otherwise we compute the $\gcd(h + 2v_1, u_1) = (x - x_1)$ to get the coordinate of $P_1$ and double $[x + u_{11} + x_1, v_1(-u_{11} - x_1)]$.

   iii. Now we know that without loss of generality $P_1 = P_3$ and $P_2 \neq P_4$ is the opposite of $P_4$. Let $v_i = v_{i1}x + v_{i0}$, then the result $2P_1$ is obtained by doubling

   $$\left[x - (v_{10} - v_{20})/(v_{21} - v_{11}), v_1\big((v_{10} - v_{20})/(v_{21} - v_{11})\big)\right]$$

   using (14.9).

   B. For the remaining case $u_1 \neq u_2$, we need to consider the following possibilities:

   i. If $\text{Res}(u_1, u_2) \neq 0$ then no point of $D_1$ is equal to a point or its opposite in $D_2$. This is the most frequent case. We deal with it in Section 14.3.2.a.

   ii. If the above resultant is zero then $\gcd(u_1, u_2) = x - x_1$ and we know that either $D_1 = P_1 + P_2 - 2P_\infty$, $D_2 = P_1 + P_3 - 2P_\infty$ or $D_2$ contains the opposite of $P_1$ instead. This can be checked by inserting $x_1$ in both $v_1$ and $v_2$.

      a. If the results are equal then we are in the first case and proceed by computing $D' = 2(P_1 - P_\infty)$, then $D'' = D' + P_2 - P_\infty$ and finally $D = D'' + P_3 - P_\infty$ by the formulas in 2. We extract the coordinates of $P_2$ and $P_3$ by

      $$P_2 = \big(-u_{11} - x_1, v_1(-u_{11} - x_1)\big)$$
      $$P_3 = \big(-u_{21} - x_1, v_2(-u_{21} - x_1)\big).$$

      b. In case $v_1(x_1) \neq v_2(x_1)$ the result is $P_2 + P_3 - 2P_\infty$.

If one uses the resultant as recommended in 3.A.ii and 3.B.i then one needs to compute a greatest common divisor as well, to extract the coordinates of $P_1$ when needed. However, most frequently we are in the case of nonzero resultant and thus we save on average.

### 14.3.2 Addition and doubling in affine coordinates ($\mathcal{A}$)

We now present in detail the algorithms for the cases left out above. These are the most common cases. For the complexity estimates we always assume $h_2 \in \{0, 1\}$ and $f_4 = 0$ as this can always be achieved by isomorphic transformations unless the characteristic is 5. If the curve is not brought to this form some computations should be performed differently (e.g., $s_0(s_0+h_2)$ instead of $s_0^2+s_0h_2$). We would like to stress that the formulas remain correct for other values of $h_2$ and $f_4$, only the operation count changes. Depending on the equation of the curve and the characteristic some further transformations can save operations. Later sections specialize the characteristic and thus obtain a lower operation count.

Finally, we mention that we only count multiplications, squarings, and inversions because additions and subtractions are comparably cheap.

#### 14.3.2.a Addition in most common case

In this case the two divisor classes to be combined consist of four points different from each other and from each other's negative. The results of the composition in Cantor's Algorithm 14.7 are $u = u_1 u_2$ and a polynomial $v$ of degree $\leqslant 3$ satisfying $u \mid v^2 + vh - f$ (see Theorem 14.5). As we started with $u_i \mid v_i^2 + v_i h - f$ we can obtain $v$ using the Chinese remainder theorem (cf. Section 10.6.4).

$$v \equiv v_1 \pmod{u_1}, \qquad (14.10)$$
$$v \equiv v_2 \pmod{u_2}.$$

Then we compute the resulting first polynomial $u'$ by making $(f - vh - v^2)/(u_1 u_2)$ monic and taking $v' = (-h - v) \bmod u'$.

To optimize the computations we do not follow this literally. We now list the needed subexpressions and then show that in fact we obtain the desired result.

$$
\begin{aligned}
t &\leftarrow (f - v_2 h - v_2^2)/u_2 \\
s &\leftarrow \bigl((v_1 - v_2)/u_2\bigr) \bmod u_1 \\
l &\leftarrow s u_2 \\
u &\leftarrow \bigl(t - s(l + h + 2v_2)\bigr)/u_1 \\
u' &\leftarrow u \text{ made monic} \\
v' &\leftarrow \bigl(-h - (l + v_2)\bigr) \bmod u'
\end{aligned}
$$

The divisions made to get $t$ and $u$ are exact divisions due to the definition of the polynomials. Let us first verify that $v = l + v_2 = su_2 + v_2$ satisfies the system of equations (14.10). This is obvious for the second equation. For the first one we consider

$$v \equiv su_2 + v_2 \equiv \bigl((v_1 - v_2)/u_2\bigr)u_2 + v_2 \equiv v_1 \pmod{u_1}.$$

Now we check that $u = (f - vh - v^2)/(u_1 u_2)$ by expanding out

$$u_1 u_2 u = u_2\bigl(t - s(l + h + 2v_2)\bigr) = f - v_2 h - v_2^2 - l(l+h) - 2lv_2 = f - vh - v^2.$$

In the course of computing we do not need all coefficients of the polynomials defined above. As $f = x^5 + \sum_{i=0}^{4} f_i x^i$ is monic and of degree 5, $u_2$ is monic of degree 2, $\deg h \leqslant 2$, and $\deg v_2 = 1$ we have that $t(x) = x^3 + (f_4 - u_{21})x^2 + cx + c'$, where $c$, $c'$ are some constants. In the computation of $u$ we divide an expression involving $t$ by a polynomial of degree 2, thus we only need the above

§ 14.3 Arithmetic on genus 2 curves over arbitrary characteristic

known part of $t$. In the computation of a product of polynomials we use the following Karatsuba style formula (cf. Section 10.3.2) to save one multiplication:

$$(ax+b)(cx+d) = acx^2 + \big((a+b)(c+d) - ac - bd\big)x + bd.$$

To reduce a polynomial of degree 3 modulo a monic one of degree 2 we use

$$ax^3 + bx^2 + cx + d \equiv \big(c - (i+j)\big(a + (b - ia)\big) + ia + j(b - ia)\big)x + d - j(b - ia) \pmod{x^2 + ix + j}$$

using only 3 multiplications instead of 4. Furthermore, we use an almost-inverse in the computation of $s$ and compute $rs$ instead, where $r$ is the resultant of $u_1$ and $u_2$, postponing and combining the inversion of $r$ with that of $s$. This leads us to consider a further subcase, namely $\deg s = 1$.

In the following table we list the intermediate steps together with the number of multiplications, squarings and inversions needed (respectively denoted by M, S and I). The names of the intermediate variables refer to the above computations. A prime $'$ indicates relative changes. For an actual implementation less variables are needed. However, we favored readability by humans in this exposition.

In the case study we have already computed the resultant of $u_1$ and $u_2$ when we arrive at this algorithm. Hence, we can assume that $\tilde{u}_2 = u_2 \pmod{u_1}$ and $\text{Res}(\tilde{u}_2, u_1)$ are known. However, we include the costs in the table, as we use these expressions to compute $1/\tilde{u}_2 \bmod u_1$.

The following Algorithm 14.19 presents the complete addition formula. We apply the trick introduced by Takahashi [TAK 2002] to use a monic $s''$. Note that the following algorithm needs the same number of operations (assuming M = S) as his but manages to trade one more multiplication for a squaring, which might be advantageous for implementations.

For even characteristic the independent work [SUMA+ 2002] needs the same number of operations but considers only the case of $\deg h = 2$. Furthermore, in even characteristic squarings are much cheaper than multiplications and therefore our algorithm is faster. If $h_1 = 1$ the algorithm presented now saves one multiplication in Line 6. The complexity for each step is given in brackets.

---

**Algorithm 14.19** Addition ($g = 2$ and $\deg u_1 = \deg u_2 = 2$)

INPUT: Two divisor classes $[u_1, v_1], [u_2, v_2]$ with $u_i = x^2 + u_{i1}x + u_{i0}$ and $v_i = v_{i1}x + v_{i0}$.
OUTPUT: The divisor class $[u', v'] = [u_1, v_1] \oplus [u_2, v_2]$.

1. **compute $r = \text{Res}(u_1, u_2)$** [3M + S]
   $z_1 \leftarrow u_{11} - u_{21}$, $z_2 \leftarrow u_{20} - u_{10}$, $z_3 \leftarrow u_{11}z_1 + z_2$ and $r \leftarrow z_2 z_3 + z_1^2 u_{10}$

2. **compute almost inverse of $u_2$ modulo $u_1$, i.e., $\text{inv} = (r/u_2) \bmod u_1$**
   $\text{inv}_1 \leftarrow z_1$ and $\text{inv}_0 \leftarrow z_3$

3. **compute $s' = rs = \big((v_1 - v_2)\text{inv}\big) \bmod u_1$** [5M]
   $w_0 \leftarrow v_{10} - v_{20}$, $w_1 \leftarrow v_{11} - v_{21}$, $w_2 \leftarrow \text{inv}_0 w_0$ and $w_3 \leftarrow \text{inv}_1 w_1$
   $s'_1 \leftarrow (\text{inv}_0 + \text{inv}_1)(w_0 + w_1) - w_2 - w_3(1 + u_{11})$ and $s'_0 \leftarrow w_2 - u_{10}w_3$
   if $s'_1 = 0$ see below

4. **compute $s'' = x + s_0/s_1 = x + s'_0/s'_1$ and $s_1$** [I + 5M + 2S]
   $w_1 \leftarrow (rs'_1)^{-1}$, $w_2 \leftarrow rw_1$ and $w_3 \leftarrow {s'_1}^2 w_1$
   $w_4 \leftarrow rw_2$, $w_5 \leftarrow w_4^2$ and $s''_0 \leftarrow s'_0 w_2$
   we have $w_1 = 1/r^2 s_1$, $w_2 = 1/s'_1$, $w_3 = s_1$ and $w_4 = 1/s_1$

5. **compute $l' = s''u_2 = x^3 + l'_2 x^2 + l'_1 x + l'_0$** [2M]
   $l'_2 \leftarrow u_{21} + s''_0$, $l'_1 \leftarrow u_{21}s''_0 + u_{20}$ and $l'_0 \leftarrow u_{20}s''_0$

6. **compute $u' = (s(l + h + 2v_2) - t)/u_1 = x^2 + u'_1 x + u'_0$** [3M]
   $u'_0 \leftarrow (s''_0 - u_{11})(s''_0 - z_1 + h_2 w_4) - u_{10}$
   $u'_0 \leftarrow u'_0 + l'_1 + (h_1 + 2v_{21})w_4 + (2u_{21} + z_1 - f_4)w_5$
   $u'_1 \leftarrow 2s''_0 - z_1 + h_2 w_4 - w_5$

| 7. | **compute** $v' = \bigl(-h - (l + v_2)\bigr) \bmod u' = v'_1 x + v'_0$ | [4M] |
|---|---|---|
| | $w_1 \leftarrow l'_2 - u'_1$, $w_2 \leftarrow u'_1 w_1 + u'_0 - l'_1$ and $v'_1 \leftarrow w_2 w_3 - v_{21} - h_1 + h_2 u'_1$ | |
| | $w_2 \leftarrow u'_0 w_1 - l'_0$ and $v'_0 \leftarrow w_2 w_3 - v_{20} - h_0 + h_2 u'_0$ | |
| 8. | **return** $[u', v']$ | [total complexity: I + 22M + 3S] |

In case $s = s_0$, one needs to replace Lines 4–6 with the following.

| 4'. | **compute** $s$ | [I + M] |
|---|---|---|
| | inv $\leftarrow 1/r$ and $s_0 \leftarrow s'_0$inv | |
| 5'. | **compute** $u' = \bigl(t - s(l + h + 2v_2)\bigr)/u_1 = x + u'_0$ | [S] |
| | $u'_0 \leftarrow f_4 - u_{21} - u_{11} - s_0^2 - s_0 h_2$ | |
| 6'. | **compute** $v' = \bigl(-h - (l + v_2)\bigr) \bmod u' = v'_0$ | [3M] |
| | $w_1 \leftarrow s_0(u_{21} + u'_0) + h_1 + v_{21} - h_2 u'_0$ and $w_2 \leftarrow u_{20} s_0 + v_{20} + h_0$ | |
| | $v'_0 \leftarrow u'_0 w_1 - w_2$ | |

In this case the total complexity drops to I + 12M + 2S.

### 14.3.2.b Addition in case $\deg u_1 = 1$ and $\deg u_2 = 2$

We now treat case 2.B.i of Section 14.3.1 in which for $u_1 = x + u_{10}$ we have $u_2(-u_{10}) \neq 0$.

In principle we follow the same algorithm as stated in the previous section. But to obtain $u$ we divide by a polynomial of degree one, therefore we need an additional coefficient of $t$ and save a lot in the other operations. Algorithm 14.20 shows that this case is much cheaper than the general one; however, it is not too likely to happen as with all special cases.

**Algorithm 14.20** Addition ($g = 2$, $\deg u_1 = 1$, and $\deg u_2 = 2$)

INPUT: Two divisor classes $[u_1, v_1]$, $[u_2, v_2]$ with $u_1 = x + u_{10}$, $u_2 = x^2 + u_{21}x + u_{20}$, $v_1 = v_{10}$ and $v_2 = v_{21} x + v_{20}$.
OUTPUT: The divisor class $[u', v'] = [u_1, v_1] \oplus [u_2, v_2]$.

| 1. | **compute** $r = u_2 \bmod u_1$ | [M] |
|---|---|---|
| | $r \leftarrow u_{20} - (u_{21} - u_{10})u_{10}$ | |
| 2. | **compute inverse of** $u_2$ **modulo** $u_1$ | [I] |
| | inv $\leftarrow 1/r$ | |
| 3. | **compute** $s = \bigl((v_1 - v_2)\text{inv}\bigr) \bmod u_1$ | [2M] |
| | $s_0 \leftarrow \text{inv}(v_{10} - v_{20} - v_{21} u_{10})$ | |
| 4. | **compute** $l = s u_2 = s_0 x^2 + l_1 x + l_0$ | [2M] |
| | $l_1 \leftarrow s_0 u_{21}$ and $l_0 = s_0 u_{20}$ | |
| 5. | **compute** $t = (f - v_2 h - v_2^2)/u_2 = x^3 + t_2 x^2 + t_1 x + t_0$ | [M] |
| | $t_2 \leftarrow f_4 - u_{21}$ and $t_1 \leftarrow f_3 - (f_4 - u_{21})u_{21} - v_{21} h_2 - u_{20}$ | |
| 6. | **compute** $u' = \bigl(t - s(l + h + 2v_2)\bigr)/u_1 = x^2 + u'_1 x + u'_0$ | [2M + S] |
| | $u'_1 \leftarrow t_2 - s_0^2 - s_0 h_2 - u_{10}$ | |
| | $u'_0 \leftarrow t_1 - s_0(l_1 + h_1 + 2v_{21}) - u_{10} u'_1$ | |
| 7. | **compute** $v' = \bigl(-h - (l + v_2)\bigr) \bmod u' = v'_1 x + v'_0$ | [2M] |
| | $v'_1 \leftarrow (h_2 + s_0)u'_1 - (h_1 + l_1 + v_{21})$ | |
| | $v'_0 \leftarrow (h_2 + s_0)u'_0 - (h_0 + l_0 + v_{20})$ | |
| 8. | **return** $[u', v']$ | [total complexity: I + 10M + S] |

### 14.3.2.c Doubling

The above case study left open how one computes the double of a class where the first polynomial has degree 2 and both points of the representing divisor are not equal to their opposites. Put $u = x^2 + u_1 x + u_0$, $v = v_1 x + v_0$. Composing $[u, v]$ with itself should result in a class $[u_{\text{new}}, v_{\text{new}}]$, where

$$u_{\text{new}} = u^2,$$
$$v_{\text{new}} \equiv v \pmod{u}, \qquad (14.11)$$
$$u_{\text{new}} \mid v_{\text{new}}^2 + v_{\text{new}} h - f. \qquad (14.12)$$

Then this class is reduced to obtain $[u', v']$. We use the following subexpressions:

$$
\begin{aligned}
t &\leftarrow (f - hv - v^2)/u \\
s &\leftarrow t/(h + 2v) \bmod u \\
l &\leftarrow su \\
\tilde{u} &\leftarrow s^2 - \bigl((h + 2v)s - t\bigr)/u \\
u' &\leftarrow \tilde{u} \text{ made monic} \\
v' &\leftarrow \bigl(-h - (l + v)\bigr) \bmod u'
\end{aligned}
$$

Note that as above we do not compute the semi-reduced divisor explicitly, here

$$v_{\text{new}} = l + v = su + v.$$

Hence, we see that (14.11) holds. To prove (14.12) we consider

$$v_{\text{new}}^2 + v_{\text{new}} h - f = l^2 + 2lv + v^2 + hl + hv - f = s^2 u^2 + u(s(h + 2v) - t)$$

and

$$(h + 2v)s - t \equiv (h + 2v)t/(h + 2v) - t \equiv 0 \pmod{u}.$$

Finally, one finds by

$$(v_{\text{new}}^2 + v_{\text{new}} h - f)/u_{\text{new}} = (s^2 u^2 + (h + 2v)su - tu)/u^2$$

that $\tilde{u}$ is in fact obtained as described in the reduction algorithm.

Unlike in the addition case we now need the exact polynomial $t$ to compute the result. For the doublings it is necessary to separately count the operations for odd and even characteristic. The given formulas are most general, but for an actual implementation they should be modified, depending on the characteristic. For odd characteristic we assume $h = 0$, as this can be achieved by replacing $y$ by $y - h(x)/2$ in the defining equation, which transforms the curve to an isomorphic one.

For even characteristic the different types of defining equation lead to many ways of improvement. They are studied in Section 14.5 in more detail. Therefore, we skip counting them in the following Algorithm 14.21.

---

**Algorithm 14.21** Doubling ($g = 2$ and $\deg u = 2$)

INPUT: A divisor class $[u, v]$ with $u = x^2 + u_1 x + u_0$ and $v = v_1 x + v_0$.
OUTPUT: The divisor class $[u', v'] = [2][u, v]$.

---

1. **compute $\tilde{v} = (h + 2v) \bmod u = \tilde{v}_1 x + \tilde{v}_0$**
   $\tilde{v}_1 \leftarrow h_1 + 2v_1 - h_2 u_1$ and $\tilde{v}_0 = h_0 + 2v_0 - h_2 u_0$

2. **compute resultant** $r = \text{Res}(\tilde{v}, u)$  [3M + 2S]
   $w_0 \leftarrow v_1^2$, $w_1 \leftarrow u_1^2$ and $w_2 \leftarrow \tilde{v}_1^2$ (note that $w_2 = 4w_0$)
   $w_3 \leftarrow u_1 \tilde{v}_1$ and $r \leftarrow u_0 w_2 + \tilde{v}_0(\tilde{v}_0 - w_3)$

3. **compute almost inverse** $\text{inv}' = r\,\text{inv}$
   $\text{inv}_1' \leftarrow -\tilde{v}_1$ and $\text{inv}_0' \leftarrow \tilde{v}_0 - w_3$

4. **compute** $t' = ((f - hv - v^2)/u) \bmod u = t_1' x + t_0'$  [M]
   $w_3 \leftarrow f_3 + w_1$, $w_4 \leftarrow 2u_0$ and $t_1' \leftarrow 2(w_1 - f_4 u_1) + w_3 - w_4 - h_2 v_1$
   $t_0' \leftarrow u_1(2w_4 - w_3 + f_4 u_1 + h_2 v_1) + f_2 - w_0 - 2f_4 u_0 - h_1 v_1 - h_2 v_0$

5. **compute** $s' = (t'\text{inv}') \bmod u$  [5M]
   $w_0 \leftarrow t_0' \text{inv}_0'$ and $w_1 \leftarrow t_1' \text{inv}_1'$
   $s_1' \leftarrow (\text{inv}_0' + \text{inv}_1')(t_0' + t_1') - w_0 - w_1(1 + u_1)$ and $s_0' \leftarrow w_0 - u_0 w_1$
   if $s_1' = 0$ see below

6. **compute** $s'' = x + s_0/s_1$ **and** $s_1$  [I + 5M + 2S]
   $w_1 \leftarrow 1/(rs_1')$, $w_2 \leftarrow rw_1$ and $w_3 \leftarrow s_1'^2 w_1$
   $w_4 \leftarrow rw_2$, $w_5 \leftarrow w_4^2$ and $s_0'' \leftarrow s_0' w_2$
   we have $w_1 = 1/r^2 s_1$, $w_2 = 1/s_1'$, $w_3 = s_1$ and $w_4 = 1/s_1$

7. **compute** $l' = s''u = x^3 + l_2' x^2 + l_1' x + l_0'$  [2M]
   $l_2' \leftarrow u_1 + s_0''$, $l_1' \leftarrow u_1 s_0'' + u_0$ and $l_0' \leftarrow u_0 s_0''$

8. **compute** $u' = s^2 + (h + 2v)s/u + (v^2 + hv - f)/u^2$  [2M + S]
   $u_0' \leftarrow s_0''^2 + w_4\bigl(h_2(s_0'' - u_1) + 2v_1 + h_1\bigr) + w_5(2u_1 - f_4)$
   $u_1' \leftarrow 2s_0'' + h_2 w_4 - w_5$

9. **compute** $v' = \bigl(-h - (l + v)\bigr) \bmod u' = v_1' x + v_0'$  [4M]
   $w_1 \leftarrow l_2' - u_1'$, $w_2 \leftarrow u_1' w_1 + u_0' - l_1'$ and $v_1' \leftarrow w_2 w_3 - v_1 - h_1 + h_2 u_1'$
   $w_2 \leftarrow u_0' w_1 - l_0'$ and $v_0' \leftarrow w_2 w_3 - v_0 - h_0 + h_2 u_0'$

10. **return** $[u', v']$  [total complexity: I + 22M + 5S]

In case $s = s_0$, one replaces Lines 6–9 by the following.

6'. **compute** $s$ **and precomputations**  [I + 2M]
   $w_1 \leftarrow 1/r$, $s_0 \leftarrow s_0' w_1$ and $w_2 \leftarrow u_0 s_0 + v_0 + h_0$

7'. **compute** $u' = (f - hv - v^2)/u^2 - (h + 2v)s/u - s^2$  [S]
   $u_0' \leftarrow f_4 - s_0^2 - s_0 h_2 - 2u_1$

8'. **compute** $v' = \bigl(-h - (su + v)\bigr) \bmod u'$  [2M]
   $w_1 \leftarrow s_0(u_1 - u_0') - h_2 u_0' + v_1 + h_1$ and $v_0' \leftarrow u_0' w_1 - w_2$

In this case the total complexity drops to I + 13M + 3S.

## 14.4 Arithmetic on genus 2 curves in odd characteristic

In the previous section we derived explicit formulas for addition and doubling on genus 2 curves over finite fields of arbitrary characteristic. Already for the operation count of the doubling we needed to separate even characteristic fields from those of odd characteristic. In this section we consider further improvements for different coordinate systems. Here, we concentrate on fields of odd characteristic.

We introduce different sets of coordinates and consider the most frequent cases of inputs, namely those given in detail in the previous section, and assume the most general output cases. Thus the special case of $s = s_0$ is no longer considered. Then we consider the task of computing scalar mul-

§ 14.4 Arithmetic on genus 2 curves in odd characteristic

tiplication in the divisor class group and investigate the advantages of the different representations with and without precomputations.

### 14.4.1 Projective coordinates ($\mathcal{P}$)

So far, in the process of computation one inversion is required for each addition or doubling. Now, instead of following this line, we introduce a further coordinate called $Z$ as with elliptic curves and let the quintuple $[U_1, U_0, V_1, V_0, Z]$ stand for $[x^2 + U_1/Z\, x + U_0/Z, V_1/Z\, x + V_0/Z]$. Due to the obvious resemblance with the case of elliptic curves we refer to this representation as *projective coordinates* and to the original one as *affine coordinates*. If the output of a scalar multiplication should be in the usual affine representation we need one inversion and four multiplications at the end of the computations.

This idea was first proposed for genus 2 curves in [MIDO$^+$ 2002] and then largely improved and generalized by Lange in [LAN 2002c]. The idea of using affine and projective inputs together for the addition algorithm was proposed for elliptic curves in [COMI$^+$ 1998] and generalized to genus 2 curves in [LAN 2002c].

We now proceed to investigate the arithmetic in the main cases. For applications one usually chooses prime fields $\mathbb{F}_q = \mathbb{F}_p$, where $p$ is large, or optimal extension fields $\mathbb{F}_{p^d}$ (cf. Chapter 2 for the definitions) for $p \sim 2^{32}$. In particular one can assume that $p \neq 5$ and thus choose $f_4 = 0$ by Example 14.4.

#### 14.4.1.a Addition in projective coordinates for odd characteristic

Here we consider the case that we add two classes, both in projective representation. This is needed if the whole system avoids inversion and classes are transmitted using this quintuple representation, if during the verification of a signature intermediate results should be added, or when using precomputations given in projective representation. Obviously this algorithm also works for affine inputs if one writes $[u_1, v_1]$ as $[u_{11}, u_{10}, v_{11}, v_{10}, 1]$. The numbers in brackets refer to the case, that the first input is affine, i.e., has $Z_1 = 1$. A more careful implementation of this case allows us to save some further multiplications (see [LAN 2002c]), namely then one addition needs $40M + 3S$.

---

**Algorithm 14.22** Addition in projective coordinates ($g = 2$ and $q$ odd)

INPUT: Two divisor classes $\overline{D}_1$ and $\overline{D}_2$ represented by $\overline{D}_1 = [U_{11}, U_{10}, V_{11}, V_{10}, Z_1]$ and $\overline{D}_2 = [U_{21}, U_{20}, V_{21}, V_{20}, Z_2]$.
OUTPUT: The divisor class $[U'_1, U'_0, V'_1, V'_0, Z'] = \overline{D}_1 \oplus \overline{D}_2$.

1. ***precomputations***                                                                               [5M (none)]
   $Z \leftarrow Z_1 Z_2,\ \widetilde{U}_{21} \leftarrow Z_1 U_{21},\ \widetilde{U}_{20} \leftarrow Z_1 U_{20},\ \widetilde{V}_{21} \leftarrow Z_1 V_{21}$ and $\widetilde{V}_{20} \leftarrow Z_1 V_{20}$

2. ***compute resultant*** $r = \mathrm{Res}(U_1, U_2)$                                       [6M + S (5M + S)]
   $z_1 \leftarrow U_{11} Z_2 - \widetilde{U}_{21}$ and $z_2 \leftarrow \widetilde{U}_{20} - U_{10} Z_2$
   $z_3 \leftarrow U_{11} z_1 + z_2 Z_1$ and $r \leftarrow z_2 z_3 + z_1^2 U_{10}$

3. ***compute almost inverse of*** $u_2$ ***modulo*** $u_1$
   $\mathrm{inv}_1 \leftarrow z_1$ and $\mathrm{inv}_0 \leftarrow z_3$

4. ***compute*** $s$                                                                                                      [8M (7M)]
   $w_0 \leftarrow V_{10} Z_2 - \widetilde{V}_{20}$ and $w_1 \leftarrow V_{11} Z_2 - \widetilde{V}_{21}$
   $w_2 \leftarrow \mathrm{inv}_0 w_0$ and $w_3 \leftarrow \mathrm{inv}_1 w_1$
   $s_1 \leftarrow (\mathrm{inv}_0 + Z_1 \mathrm{inv}_1)(w_0 + w_1) - w_2 - w_3(Z_1 + U_{11})$
   $s_0 \leftarrow w_2 - U_{10} w_3$

5. **precomputations** [8M + S]
   $R \leftarrow Zr, s_0 \leftarrow s_0 Z, s_3 \leftarrow s_1 Z$ and $\widetilde{R} \leftarrow Rs_3$
   $S_3 \leftarrow s_3^2, S \leftarrow s_0 s_1, \widetilde{S} \leftarrow s_3 s_1, \widehat{S} \leftarrow s_0 s_3$ and $\widehat{R} \leftarrow \widetilde{R}\widetilde{S}$

6. **compute** $l$ [3M]
   $l_2 \leftarrow \widetilde{S}\widetilde{U}_{21}, l_0 \leftarrow S\widetilde{U}_{20}$ and $l_1 \leftarrow (\widetilde{S} + S)(\widetilde{U}_{21} + \widetilde{U}_{20}) - l_2 - l_0$
   $l_2 \leftarrow l_2 + \widehat{S}$

7. **compute** $U'$ [8M + 2S]
   $U_0' \leftarrow s_0^2 + s_1 z_1 (s_1(z_1 + \widetilde{U}_{21}) - 2s_0) + z_2 \widetilde{S} + R(2s_1 \widetilde{V}_{21} + r(z_1 + 2\widetilde{U}_{21} - f_4 Z))$
   $U_1' \leftarrow 2\widetilde{S} - \widetilde{S}z_1 - R^2$

8. **precomputations** [4M]
   $l_2 \leftarrow l_2 - U_1', w_0 \leftarrow U_0' l_2 - S_3 l_0$ and $w_1 \leftarrow U_1' l_2 + S_3(U_0' - l_1)$

9. **adjust** [3M]
   $Z' \leftarrow \widetilde{R}S_3, U_1' \leftarrow \widetilde{R}U_1'$ and $U_0' \leftarrow \widetilde{R}U_0'$

10. **compute** $V'$ [2M]
    $V_0' \leftarrow w_0 - \widehat{R}\widetilde{V}_{20}$ and $V_1' \leftarrow w_1 - \widehat{R}\widetilde{V}_{21}$

11. **return** $[U_1', U_0', V_1', V_0', Z']$ [total complexity: 47M + 4S (40M + 4S)]

### 14.4.1.b  Doubling in projective coordinates for odd characteristic

For the doubling algorithm the input is almost always in projective representation. Multiplications by $f_4$ are not counted.

**Algorithm 14.23** Doubling in projective coordinates ($g = 2$ and $q$ odd)

INPUT: A divisor class represented by $[U_1, U_0, V_1, V_0, Z]$.
OUTPUT: The divisor class $[U_1', U_0', V_1', V_0', Z'] = [2][U_1, U_0, V_1, V_0, Z]$.

1. **compute resultant and precomputations** [4M + 3S]
   $Z_2 \leftarrow Z^2, \widetilde{V}_1 \leftarrow 2V_1, \widetilde{V}_0 \leftarrow 2V_0, w_0 \leftarrow V_1^2, w_1 \leftarrow U_1^2$ and $w_2 \leftarrow 4w_0$
   $w_3 \leftarrow \widetilde{V}_0 Z - U_1 \widetilde{V}_1$ and $r \leftarrow \widetilde{V}_0 w_3 + w_2 U_0$

2. **compute almost inverse**
   $\text{inv}_1 \leftarrow -\widetilde{V}_1$ and $\text{inv}_0 \leftarrow w_3$

3. **compute** $t$ [5M]
   $w_3 \leftarrow f_3 Z_2 + w_1, w_4 \leftarrow 2U_0, t_1 \leftarrow 2w_1 + w_3 - Z(w_4 + 2f_4 U_1)$
   $t_0 \leftarrow U_1 \big(Z(2w_4 + f_4 U_1) - w_3\big) + Z\big(Z(f_2 Z - 2f_4 U_0) - w_0\big)$
   see Remark 14.24 (ii)

4. **compute** $s$ [7M]
   $w_0 \leftarrow t_0 \text{inv}_0, w_1 \leftarrow t_1 \text{inv}_1$ and $s_3 \leftarrow (\text{inv}_0 + \text{inv}_1)(t_0 + t_1) - w_0 - (1 + U_1)w_1$
   $s_1 \leftarrow s_3 Z$ and $s_0 \leftarrow w_0 - ZU_0 w_1$

5. **precomputations** [6M + 2S]
   $R \leftarrow Z_2 r, \widetilde{R} \leftarrow Rs_1, S_1 \leftarrow s_1^2$ and $S_0 \leftarrow s_0^2$
   $s_1 \leftarrow s_1 s_3, s_0 \leftarrow s_0 s_3, S \leftarrow s_0 Z$ and $\widehat{R} \leftarrow \widetilde{R}s_1$

6. **compute** $l$ [3M]
   $l_2 \leftarrow U_1 s_1, l_0 \leftarrow U_0 s_0$ and $l_1 \leftarrow (s_1 + s_0)(U_1 + U_0) - l_2 - l_0$

7. **compute** $U'$ [4M + S]
   $U_0' \leftarrow S_0 + R\big(2s_3 V_1 + Zr(2U_1 - f_4 Z)\big)$ and $U_1' \leftarrow 2S - R^2$

8. **precomputations** [4M]
$l_2 \leftarrow l_2 + S - U_1'$, $w_0 \leftarrow U_0' l_2 - S_1 l_0$ and $w_1 \leftarrow U_1' l_2 + S_1(U_0' - l_1)$

9. **adjust** [3M]
$Z' \leftarrow S_1 \widetilde{R}$, $U_1' \leftarrow \widetilde{R} U_1'$ and $U_0' \leftarrow \widetilde{R} U_0'$

10. **compute** $V'$ [2M]
$V_0' \leftarrow w_0 - \widehat{R} V_0$ and $V_1' \leftarrow w_1 - \widehat{R} V_1$

11. **return** $[U_1', U_0', V_1', V_0', Z']$ [total complexity: 38M + 6S]

**Remarks 14.24**

(i) First of all one notices that doublings are much faster than general additions; this is especially interesting as doublings occur much more frequently than additions in any algorithm to compute scalar multiples, most striking in windowing methods.

(ii) If $f_4 = 0$, then $t_0$ in Line 3 is computed as $t_0 \leftarrow U_1(2Zw_4 - w_3) + Z(f_2 Z_2 - w_0)$. Here, we use that $Zw_4$ is already obtained during the computation of $t_1$.

It is interesting to note that the value of the additional coordinate $Z'$ was not kept minimal. One could have avoided (at least) a factor of $Z_1^2 Z_2$ of the denominator in the addition and of $Z$ in the doubling. However, as we tried to minimize the number of operations, we allowed the larger value of $Z^4 r s_1^3$ in both cases as this proved to be more efficient. Besides one sees that $U_1'$ and $U_0'$ have to be adjusted to have the same (larger) denominator $Z'$ as $V_1', V_0'$.

### 14.4.2 New coordinates in odd characteristic ($\mathcal{N}$)

As we just noticed, the necessary denominator of the $V_i$'s differs from that of the $U_i$'s. This leads us to consider a further set of coordinates. Here, we suggest letting $[U_1, U_0, V_1, V_0, Z_1, Z_2]$ correspond to the affine class $[x^2 + U_1/Z_1^2 x + U_0/Z_1^2, V_1/(Z_1^3 Z_2) x + V_0/(Z_1^3 Z_2)]$. This means that now a point corresponds to a sextuple; thus one needs one more entry than for projective coordinates. Coordinate systems in which not all entries have the same denominator are called *weighted coordinates*. For elliptic curves this corresponds to Jacobian coordinates. Compared to the case of elliptic curves, for equal security the entries for $g = 2$ are of only half the size, thus the space requirements are similar.

As usual we assume $f_4 = 0$. This time we do not include this coefficient in the formulas. To increase the performance we enlarge the set of coordinates to $[U_1, U_0, V_1, V_0, Z_1, Z_2, z_1, z_2]$, where $z_1 = Z_1^2$ and $z_2 = Z_2^2$. These additional entries are computed anyway during each addition or doubling and keeping them saves in the following operation. Both addition and doubling profit from $z_1$ whereas $z_2$ is only used for the doublings. We do not include $Z_1 Z_2$ because it is not useful in doublings and is not automatically computed.

If space is more restricted such that we can only use the sextuple of coordinates, we need two extra squarings in the first step of the doubling or addition.

These *new coordinates* were first proposed by the Lange [LAN 2002d]; they generalize the concepts of Jacobian, Chudnovsky Jacobian and modified Jacobian coordinates from elliptic to hyperelliptic curves. The respective counterparts can be seen if one varies the additional coordinates — the original Jacobian coordinates correspond to allowing only $Z_1, Z_2$. In the following we state the most efficient algorithms, the additional coordinates become more and more useful with the increase of the window size in scalar multiplications, and hence they form the counterpart to modified Jacobian coordinates. The topic of which coordinates to choose in which system is treated further in Section 14.4.3.

### 14.4.2.a Addition in new coordinates in odd characteristic

If one computes a scalar multiple of a point given in affine coordinates and has the intermediate results non-normalized, then in the addition, the intermediate result enters in new coordinates whereas the other class enters always as $[U_{11}, U_{10}, V_{11}, V_{10}, 1, 1, 1, 1]$. The numbers in brackets refer to this (cheaper) case. For an algorithm devoted to this case see [LAN 2002d]. It needs only $36M + 5S$.

---

**Algorithm 14.25** Addition in new coordinates ($g = 2$ and $q$ odd)

INPUT: Two divisor classes $\overline{D}_1$ and $\overline{D}_2$ represented by $\overline{D}_1 = [U_{11}, U_{10}, V_{11}, V_{10}, Z_{11}, Z_{12}, z_{11}, z_{12}]$ and $\overline{D}_2 = [U_{21}, U_{20}, V_{21}, V_{20}, Z_{21}, Z_{22}, z_{21}, z_{22}]$.
OUTPUT: The divisor class $\overline{D}' = [U'_1, U'_0, V'_1, V'_0, Z'_1, Z'_2, z'_1, z'_2] = \overline{D}_1 \oplus \overline{D}_2$.

1. **precomputations** [8M (2M)]
   $z_{13} \leftarrow Z_{11}Z_{12}, z_{23} \leftarrow Z_{21}Z_{22}, z_{14} \leftarrow z_{11}z_{13}$ and $z_{24} \leftarrow z_{21}z_{23}$
   $\widetilde{U}_{21} \leftarrow U_{21}z_{11}, \widetilde{U}_{20} \leftarrow U_{20}z_{11}, \widetilde{V}_{21} \leftarrow V_{21}z_{14}$ and $\widetilde{V}_{20} \leftarrow V_{20}z_{14}$

2. **compute resultant** $r = \mathrm{Res}(U_1, U_2)$ [11M + 4S (8M + 3S)]
   $y_1 \leftarrow U_{11}z_{21} - \widetilde{U}_{21}, y_2 \leftarrow \widetilde{U}_{20} - U_{10}z_{21}$ and $y_3 \leftarrow U_{11}y_1 + y_2z_{11}$
   $r \leftarrow y_2y_3 + y_1^2 U_{10}, Z'_2 \leftarrow Z_{11}Z_{21}, \widetilde{Z}_2 \leftarrow Z_{12}Z_{22}$ and $Z'_1 \leftarrow Z'^2_2$
   $\widetilde{Z}_2 \leftarrow \widetilde{Z}_2 Z'_1 r, Z'_2 \leftarrow Z'_2 \widetilde{Z}_2, \widetilde{Z}_2 \leftarrow \widetilde{Z}_2^2$ and $z'_2 \leftarrow Z'^2_2$

3. **compute almost inverse of $u_2$ modulo $u_1$**
   $\mathrm{inv}_1 \leftarrow y_1$ and $\mathrm{inv}_0 \leftarrow y_3$

4. **compute $s$** [8M (7M)]
   $w_0 \leftarrow V_{10}z_{24} - \widetilde{V}_{20}, w_1 \leftarrow V_{11}z_{24} - \widetilde{V}_{21}, w_2 \leftarrow \mathrm{inv}_0 w_0$ and $w_3 \leftarrow \mathrm{inv}_1 w_1$
   $s_1 \leftarrow (\mathrm{inv}_0 + z_{11}\mathrm{inv}_1)(w_0 + w_1) - w_2 - w_3(z_{11} + U_{11})$
   $s_0 \leftarrow w_2 - U_{10}w_3$

5. **precomputations** [6M + 3S]
   $S_1 \leftarrow s_1^2, S_0 \leftarrow s_0 Z'_1, Z'_1 \leftarrow s_1 Z'_1, S \leftarrow Z'_1 S_0$ and $S_0 \leftarrow S_0^2$
   $R \leftarrow r Z'_1, s_0 \leftarrow s_0 Z'_1, s_1 \leftarrow s_1 Z'_1$ and $z'_1 \leftarrow Z'^2_1$

6. **compute $l$** [3M]
   $l_2 \leftarrow s_1 \widetilde{U}_{21}, l_0 \leftarrow s_0 \widetilde{U}_{20}$ and $l_1 \leftarrow (s_0 + s_1)(\widetilde{U}_{20} + \widetilde{U}_{21}) - l_0 - l_2$
   $l_2 \leftarrow l_2 + S$

7. **compute $U'$** [6M]
   $V'_1 \leftarrow R\widetilde{V}_{21}$
   $U'_0 \leftarrow S_0 + y_1(S_1(y_1 + \widetilde{U}_{21}) - 2s_0) + y_2 s_1 + 2V'_1 + (2\widetilde{U}_{21} + y_1)\widetilde{Z}_2$
   $U'_1 \leftarrow 2S - y_1 s_1 - z'_2$

8. **precomputations** [2M]
   $l_2 \leftarrow l_2 - U'_1, w_0 \leftarrow l_2 U'_0$ and $w_1 \leftarrow l_2 U'_1$

9. **compute $V'$** [3M]
   $V'_1 \leftarrow w_1 - z'_1(l_1 + V'_1 - U'_0)$ and $V'_0 \leftarrow w_0 - z'_1(l_0 + R\widetilde{V}_{20})$

10. **return** $[U'_1, U'_0, V'_1, V'_0, Z'_1, Z'_2, z'_1, z'_2]$ [total complexity: $47M + 7S$ ($37M + 6S$)]

---

### 14.4.2.b Doubling

The formulas for doubling make obvious why we include $z_2 = Z_2^2$ as well. If space is very limited such that one cannot apply windowing methods at all, this is the first coordinate to drop if one needs to restrict the storage. Still, any binary method is fastest when including $z_2$.

### §14.4 Arithmetic on genus 2 curves in odd characteristic

---

**Algorithm 14.26** Doubling in new coordinates ($g = 2$ and $q$ odd)

INPUT: A divisor class represented by $\overline{D} = [U_1, U_0, V_1, V_0, Z_1, Z_2, z_1, z_2]$.
OUTPUT: The divisor class $[U'_1, U'_0, V'_1, V'_0, Z'_1, Z'_2, z'_1, z'_2] = [2]\overline{D}$.

1. **compute resultant and precomputations** [8M + 3S]
   $\widetilde{U}_0 \leftarrow U_0 z_1,\ w_0 \leftarrow V_1^2,\ w_1 \leftarrow U_1^2$ and $w_3 \leftarrow V_0 z_1 - U_1 V_1$
   $r \leftarrow w_0 U_0 + V_0 w_3,\ \widetilde{Z}_2 \leftarrow Z_2 r z_1,\ Z'_2 \leftarrow 2\widetilde{Z}_2 Z_1$ and $\widetilde{Z}_2 \leftarrow \widetilde{Z}_2^2$

2. **compute almost inverse**
   $\mathrm{inv}_1 \leftarrow -V_1$ and $\mathrm{inv}_0 \leftarrow w_3$

3. **compute $t$** [6M + S]
   $z_3 \leftarrow z_1^2,\ w_3 \leftarrow f_3 z_3 + w_1$ and $t_1 \leftarrow z_2\bigl(2(w_1 - \widetilde{U}_0) + w_3\bigr)$
   $z_3 \leftarrow z_3 z_1$ and $t_0 \leftarrow z_2\bigl(U_1(4\widetilde{U}_0 - w_3) + z_3 f_2\bigr) - w_0$

4. **compute $s$** [5M]
   $w_0 \leftarrow t_0 \mathrm{inv}_0$ and $w_1 \leftarrow t_1 \mathrm{inv}_1$
   $s_1 \leftarrow (\mathrm{inv}_0 + \mathrm{inv}_1)(t_0 + t_1) - w_0 - w_1(1 + U_1)$ and $s_0 \leftarrow w_0 - w_1 \widetilde{U}_0$

5. **precomputations** [5M + 3S]
   $S_0 \leftarrow s_0^2,\ Z'_1 \leftarrow s_1 z_1,\ z'_1 \leftarrow Z'^2_1$ and $S \leftarrow s_0 Z'_1$
   $R \leftarrow r Z'_1,\ z'_2 \leftarrow Z'^2_2,\ s_0 \leftarrow s_0 s_1$ and $s_1 \leftarrow Z'_1 s_1$

6. **compute $l$** [3M]
   $l_2 \leftarrow s_1 U_1,\ l_0 \leftarrow s_0 U_0$ and $l_1 \leftarrow (s_0 + s_1)(U_0 + U_1) - l_0 - l_2$
   $l_2 \leftarrow l_2 + S$

7. **compute $U'$** [2M]
   $V'_1 \leftarrow R V_1,\ U'_0 \leftarrow S_0 + 4(V'_1 + 2\widetilde{Z}_2 U_1)$ and $U'_1 \leftarrow 2S - z'_2$

8. **precomputations** [2M]
   $l_2 \leftarrow l_2 - U'_1,\ w_0 \leftarrow l_2 U'_0$ and $w_1 \leftarrow l_2 U'_1$

9. **compute $V'$** [3M]
   $V'_1 \leftarrow w_1 - z'_1(l_1 + 2V'_1 - U'_0)$ and $V'_0 \leftarrow w_0 - z'_1(l_0 + 2R V_0)$

10. **return** $[U'_1, U'_0, V'_1, V'_0, Z'_1, Z'_2, z'_1, z'_2]$  [total complexity: 34M + 7S]

---

### 14.4.3 Different sets of coordinates in odd characteristic

So far we have given algorithms to perform the computations within one system and briefly mentioned additions involving one input in affine coordinates. Now we are concerned with mixes of coordinate systems. To have suitable abbreviations, we denote by $\mathcal{C}_1 + \mathcal{C}_2 = \mathcal{C}_3$ the computation of an addition, where the first input is in coordinate system $\mathcal{C}_1$, the second in $\mathcal{C}_2$ and the output is in $\mathcal{C}_3$. Similarly, $2\mathcal{C}_1 = \mathcal{C}_2$ denotes a doubling with input in system $\mathcal{C}_1$ and output in $\mathcal{C}_2$. We denote the affine system by $\mathcal{A}$, the projective by $\mathcal{P}$ and the new by $\mathcal{N}$. In the following we estimate the costs of computing scalar multiples using various systems of coordinates. To have the figures in mind, the following Table 14.2 lists the costs for the most useful additions and doublings.

#### 14.4.3.a Scalar multiples in odd characteristic

In this section we concentrate on the computation of $n$-folds $[n]\overline{D}$, where $n$ is an integer and $\overline{D}$ is a divisor class. For references on how to compute the respective expansions of $n$, see Chapter 9 and the references given therein.

**Table 14.2** Addition and doubling in different systems and in odd characteristic.

| Doubling | | Addition | |
|---|---|---|---|
| Operation | Costs | Operation | Costs |
| $2\mathcal{N} = \mathcal{P}$ | $38\text{M} + 7\text{S}$ | $\mathcal{N} + \mathcal{N} = \mathcal{P}$ | $51\text{M} + 7\text{S}$ |
| $2\mathcal{P} = \mathcal{P}$ | $38\text{M} + 6\text{S}$ | $\mathcal{N} + \mathcal{P} = \mathcal{P}$ | $51\text{M} + 4\text{S}$ |
| $2\mathcal{P} = \mathcal{N}$ | $35\text{M} + 6\text{S}$ | $\mathcal{N} + \mathcal{N} = \mathcal{N}$ | $47\text{M} + 7\text{S}$ |
| $2\mathcal{N} = \mathcal{N}$ | $34\text{M} + 7\text{S}$ | $\mathcal{N} + \mathcal{P} = \mathcal{N}$ | $48\text{M} + 4\text{S}$ |
| — | — | $\mathcal{P} + \mathcal{P} = \mathcal{P}$ | $47\text{M} + 4\text{S}$ |
| — | — | $\mathcal{P} + \mathcal{P} = \mathcal{N}$ | $44\text{M} + 4\text{S}$ |
| — | — | $\mathcal{A} + \mathcal{N} = \mathcal{P}$ | $40\text{M} + 5\text{S}$ |
| — | — | $\mathcal{A} + \mathcal{P} = \mathcal{P}$ | $40\text{M} + 3\text{S}$ |
| — | — | $\mathcal{A} + \mathcal{N} = \mathcal{N}$ | $36\text{M} + 5\text{S}$ |
| — | — | $\mathcal{A} + \mathcal{P} = \mathcal{N}$ | $37\text{M} + 3\text{S}$ |
| $2\mathcal{A} = \mathcal{A}$ | $\text{I} + 22\text{M} + 5\text{S}$ | $\mathcal{A} + \mathcal{A} = \mathcal{A}$ | $\text{I} + 22\text{M} + 3\text{S}$ |

Let $n = \sum_{i=0}^{l-1} n_i 2^i$. The direct approach to compute $[n]\overline{D}$ for a given class $\overline{D}$ is to use the binary double and add method, starting with the most significant bit of $n$. For every $n_i = 1$ we need to perform an addition as well as a doubling, while for a 0 one only doubles. The density is asymptotically $1/2$.

Here we deal with the divisor class group of hyperelliptic curves and the negative of a class is obtained by negating the coordinates $V_i$ or $v_i$ respectively. Hence, signed binary expansions are useful. They have the lower density of $1/3$ if one uses an NAF of the multiplier, and are approximately of the same length.

If we can afford some precomputations, windowing methods (cf. Section 9.1.3) provide better performance; we consider signed expansions here. The $\text{NAF}_w$ representation, cf. Section 9.1.4 is advantageous to compute scalar multiples $[n]P$. This requires us to precompute all odd multiples $[i]P$ for $1 < i < 2^{w-1}$. They can be obtained as a sequence of additions and one doubling.

We will first consider systems without precomputations and then investigate good matches of coordinate systems for windowing methods. The reason for treating these cases separately is that for precomputations, the addition will involve the set of coordinates that is advantageous for the precomputations, whereas in the system without precomputations this choice depends on the efficiency of the mixed addition only.

**No precomputation**

In this approach we perform approximately $l$ doublings and $\frac{l}{3}$ additions per scalar multiple of length $l$. Table 14.3 lists the number of operations depending on the coordinate system; details are given below. We assume $l$ to be large and therefore leave out the costs for the initial moving from one system to the other as they occur only once. However, note that except for the first line, where inversions are assumed to be cheap, this conversion involves no divisions.

If inversions are relatively cheap, affine coordinates provide the best performance; thus, if the class is given in a non-normalized system we first normalize it. This takes $\text{I} + 4\text{M}$ for $\mathcal{P} \rightarrow \mathcal{A}$ and

§ 14.4 Arithmetic on genus 2 curves in odd characteristic

I+7M for $\mathcal{N} \to \mathcal{A}$. Then we double $l$ times and add on average $\frac{l}{3}$ times using $\frac{l}{3}(4\mathrm{I} + 88\mathrm{M} + 18\mathrm{S})$.

**Table 14.3** Without precomputations, odd characteristic.

| Systems | Cost |
|---|---|
| $2\mathcal{A} = \mathcal{A}$ and $\mathcal{A} + \mathcal{A} = \mathcal{A}$ | $\frac{l}{3}(4\mathrm{I} + 88\mathrm{M} + 18\mathrm{S})$ |
| $2\mathcal{N} = \mathcal{N}$ and $\mathcal{A} + \mathcal{N} = \mathcal{N}$ | $\frac{l}{3}(138\mathrm{M} + 26\mathrm{S})$ |
| $2\mathcal{N} = \mathcal{N}$ and $\mathcal{N} + \mathcal{P} = \mathcal{N}$ | $\frac{l}{3}(150\mathrm{M} + 25\mathrm{S})$ |

Otherwise, for affine input the new system is best, as the most common operation (doubling) is cheaper than in any other fixed system and the mixed addition is also fast.

If the input is in $\mathcal{P}$ and inversions are very expensive, we need to find two systems $\mathcal{C}_1$ and $\mathcal{C}_2$ such that $2\mathcal{C}_1 = \mathcal{C}_1$, $2\mathcal{C}_1 = \mathcal{C}_2$ and $\mathcal{C}_2 + \mathcal{P} = \mathcal{C}_1$ are as cheap as possible, the first being the most frequent operation. By Table 14.2 we choose $\mathcal{C}_1 = \mathcal{N}$. For $\mathcal{C}_2$ it is equal to choose $\mathcal{N}$ or $\mathcal{P}$, therefore we use $\mathcal{N}$ to save some bookkeeping. Thus the first doubling is done as $2\mathcal{P} = \mathcal{N}$ and all further as $2\mathcal{N} = \mathcal{N}$. There are approximately $l$ doublings $2\mathcal{N} = \mathcal{N}$ and $\frac{l}{3}$ additions $\mathcal{N} + \mathcal{P} = \mathcal{N}$ leading to $\frac{l}{3}(150\mathrm{M} + 25\mathrm{S})$.

If the input is in new coordinates, we do the same, except that the first doubling is $2\mathcal{N} = \mathcal{N}$ needing 1 more S, and we use 4M for $\mathcal{N} \to \mathcal{P}$ of the initial point.

To compare, using only projective coordinates we would need $\frac{l}{3}(161\mathrm{M} + 22\mathrm{S})$ and using only new coordinates results in $\frac{l}{3}(149\mathrm{M} + 28\mathrm{S})$; thus mixing the coordinate systems is advantageous. We state the overview in Table 14.3.

**Windowing methods**

To obtain the table of precomputed values, i.e., all odd multiples $[i]D$ for $1 < i < 2^{w-1}$, we need one doubling and $2^{w-2} - 1$ additions.

As before, we distinguish cases depending on the relative cost of inversions. If inversions are not too expensive, the precomputations are performed in affine coordinates. To still trade off some inversions for multiplications, we make use of Montgomery's trick of simultaneous inversions. As in [CoMi+ 1998] (cf. Chapter 13) we first compute $[2]\overline{D}$, then $([3]\overline{D}, [4]\overline{D})$, then $([5]\overline{D}, [7]\overline{D}, [8]\overline{D})$, ..., $([2^{w-3}+1]\overline{D}, \ldots, [2^{w-2}-1]\overline{D}, [2^{w-2}]\overline{D})$, and finally $([2^{w-2}+1]\overline{D}, \ldots, [2^{w-1}-1]\overline{D})$, where each sequence involves only one I. Computing $m$ inversions simultaneously is done by $\mathrm{I} + 3(m-1)\mathrm{M}$. Thus we need

$(w-1)\mathrm{I}, (w-2)$ class-doublings, $2^{w-2} - 1$ class-additions, and $3(2^{w-2} - 2)$ extra M.

As most of the operations for the precomputations are additions, we choose projective coordinates $\mathcal{P}$ if we want to perform the precomputations avoiding inversions. Table 14.2 shows that additions involving at least one point in affine coordinates and leading to inversion-free coordinates are much faster than those involving two non-affine points. Therefore, it can be useful to allow some more multiplications and 1 inversion to transform the precomputed points to affine coordinates. The costs for these three approaches leading to $\mathcal{A}$ and also for precomputations in $\mathcal{P}$ are listed in Table 14.4; in the last row we assume that the class was given in projective coordinates.

If inversions are cheap we stick to the affine system to compute the scalar multiplication. If we can afford the $w$ inversions (or one more for non-affine input) to do the precomputations in affine, we use the new system for doublings, and perform the additions using the new mixed system $\mathcal{A} + \mathcal{N} = \mathcal{N}$. Finally, if inversions are very expensive, the best match is obtained if one uses projective coordinates for the precomputations, and the doublings are performed as $2\mathcal{N} = \mathcal{N}$. Then the addition is done as $\mathcal{N} + \mathcal{P} = \mathcal{N}$.

**Table 14.4** Precomputations in odd characteristic.

| System | I | M | S |
|---|---|---|---|
| $\mathcal{A}$ | $2^{w-2}$ | $22 \times 2^{w-2}$ | $3 \times 2^{w-2} + 2$ |
| $\mathcal{A}$ | $w - 1$ | $25 \times 2^{w-2} + 22w - 72$ | $3 \times 2^{w-2} + 5w - 3$ |
| $\mathcal{A}$ | 1 | $50 \times^{w-2} - 15$ | $4 \times 2^{w-2} + 2$ |
| $\mathcal{P}$ | | $47 \times 2^{w-2} - 9$ | $4 \times 2^{w-2} + 2$ |

As in the case of elliptic curves, we put $l_1 = l - (w-1)/2$ and $K = 1/2 - 1/(w+1)$. In a scalar multiplication using an $\text{NAF}_w$ representation, on average $l_1 + K$ doublings and $(l_1 - K)/(w+1)$ additions are used. The costs are listed in Table 14.5. Some more computations can be saved using the tricks of [COMI$^+$ 1998]. Again, we leave out the initial costs for conversions and state the precomputations separately in Table 14.4.

**Table 14.5** Windowing method in odd characteristic.

| Systems | I | M | S |
|---|---|---|---|
| $2\mathcal{A} = \mathcal{A}, \mathcal{A} + \mathcal{A} = \mathcal{A}$ | $l_1 + K + \frac{l_1 - K}{w+1}$ | $22\left(l_1 + K + \frac{l_1 - K}{w+1}\right)$ | $5(l_1 + K) + 3\frac{l_1 - K}{w+1}$ |
| $2\mathcal{N} = \mathcal{N}, \mathcal{A} + \mathcal{N} = \mathcal{N}$ | | $34(l_1 + K) + 36\frac{l_1 - K}{w+1}$ | $7(l_1 + K) + 5\frac{l_1 - K}{w+1}$ |
| $2\mathcal{N} = \mathcal{N}, \mathcal{N} + \mathcal{P} = \mathcal{N}$ | | $34(l_1 + K) + 48\frac{l_1 - K}{w+1}$ | $7(l_1 + K) + 4\frac{l_1 - K}{w+1}$ |

### 14.4.4 Montgomery arithmetic for genus 2 curves in odd characteristic

For elliptic curves we have shown in Section 13.2.3 that for certain curves a more efficient arithmetic is possible, one that also avoids inversions and uses the Montgomery ladder (cf. Algorithm 9.5). This allows us to have the uniform sequence of operations in the scalar multiplication of performing an addition and a doubling for each bit of the scalar. This is advantageous if one tries to find countermeasures against side-channel attacks as considered in Chapter 29. Furthermore, compared to the direct application of ladder techniques, one can save some space by neglecting the $y$-coordinate.

In this section we provide an analogue for genus 2 curves over fields of odd characteristic and show how one can also neglect the second part of the representations here, i.e., the $v$ in $[u, v]$. Note that this is still an active area of research and that so far explicit formulas have only been derived for odd characteristic fields. Furthermore, both publications [DUQ 2004, LAN 2004a] take their main motivation from achieving countermeasures against side-channel attacks. Finding an analogue of Montgomery coordinates in the sense of obtaining a very efficient projective coordinate system is still an open problem.

To describe Duquesne's generalization [DUQ 2004] we need to consider a new object related to genus 2 curves, the *Kummer surface*. His approach works only for curves that have a second $\mathbb{F}_q$-rational Weierstraß point, hence they have at least a cofactor of 2. Lange's proposal [LAN 2004a] can be applied to arbitrary curves, but is less efficient. Therefore, we decided not to put it in this chapter on *efficient* implementations.

### 14.4.4.a   A Montgomery-like form for genus 2 curves

**Definition**

In the following, we will say that a curve $C$ is transformable into Montgomery-like form if it is isomorphic to a curve given by an equation of the type

$$By^2 = x^5 + f_4 x^4 + f_3 x^3 + f_2 x^2 + x. \tag{14.13}$$

It is easy to prove that a curve $C$ as defined in (14.1) is transformable into Montgomery-like form if and only if

- the polynomial $f(x)$ has at least one root $\alpha$ in $\mathbb{F}_q$.
- for this root, $f'(\alpha)$ is a fourth power in $\mathbb{F}_q$.

Thus, as in the case of elliptic curves, not all the curves are transformable into the Montgomery-like form. Nevertheless, there are $O(q^3)$ genus 2 curves over $\mathbb{F}_q$ showing that the choice of curves in Montgomery-like form is not too special. Note that different choices of $f_4$, $f_3$ and $f_2$ give rise to different isomorphism classes while $B$ distinguishes only between the two quadratic twists.

**The Kummer surface**

Montgomery coordinates for elliptic curves allow us to avoid storing the $y$-coordinate (cf. Section 13.2.3). This means that for the representation $P$ and $-P$ are identified. The analog for genus 2 curves is called the *Kummer surface*, where a divisor and its opposite are identified. The Kummer surface is a quartic surface in $\mathbb{P}^3$. We give here the definition of the Kummer surface and its properties without proofs for curves in Montgomery-like form. They were obtained using the same method as in the book of Cassels and Flynn on genus 2 curves (cf. [CAFL 1996] or [FLY 1993]). We use the fact that each divisor class $\overline{D}$ of degree 0 in $\mathrm{Pic}_C^0$ can be uniquely represented by at most 2 points $P = (x_1, y_1)$ and $Q = (x_2, y_2)$ and identify the class $\overline{D}$ with this representation. Let $S$ be the subset of $\mathrm{Pic}_C^0$ such that each divisor class is represented by a reduced divisor with exactly two different points in the support.

The Kummer surface is the image of the map

$$\kappa : S \longrightarrow \mathbb{P}^3(\mathbb{F}_q)$$
$$\{(x_1, y_1), (x_2, y_2)\} \longmapsto \left(1 : x_1 + x_2 : x_1 x_2 : \frac{F_0(x_1, x_2) - 2By_1 y_2}{(x_1 - x_2)^2}\right),$$

with

$$F_0(x_1, x_2) = (x_1 + x_2) + 2 f_2 x_1 x_2 + f_3 (x_1 + x_2) x_1 x_2 + 2 f_4 x_1^2 x_2^2 + (x_1 + x_2) x_1^2 x_2^2,$$

together with the neutral element, which is represented by $(0 : 0 : 0 : 1)$.

In the following, for any divisor class $\overline{D}$, we will denote the components of the map $\kappa$ as

$$\kappa(\overline{D}) = \bigl(k_1(\overline{D}) : k_2(\overline{D}) : k_3(\overline{D}) : k_4(\overline{D})\bigr).$$

More precisely, the Kummer surface is the projective locus given by an equation $\mathcal{K}$ of degree four in the first three variables and of degree two in the last one. The exact equation can be found online [FLY]. In passing from the Jacobian to the Kummer surface, we have lost the group structure (as was already the case with elliptic curves) but traces of it remain. For example, it is possible to double on the Kummer surface.

Nevertheless, for general divisor classes $\overline{D}_1$ and $\overline{D}_2$, we cannot determine the values of the $k_i(\overline{D}_1 \oplus \overline{D}_2)$ from the values of the $k_i(\overline{D}_1)$ and $k_i(\overline{D}_2)$ since the latter do not distinguish between $\pm \overline{D}_1$ and $\pm \overline{D}_2$, and so not between $\overline{D}_1 \oplus \overline{D}_2$ and $\overline{D}_1 \ominus \overline{D}_2$. However the values of the $k_i(\overline{D}_1 \oplus \overline{D}_2)k_j(\overline{D}_1 \ominus \overline{D}_2) + k_i(\overline{D}_1 \ominus \overline{D}_2)k_j(\overline{D}_1 \oplus \overline{D}_2)$ are well defined.

**Theorem 14.27** There are explicit polynomials $\varphi_{ij}$ biquadratic in the $k_i(\overline{D}_1), k_i(\overline{D}_2)$ such that projectively

$$k_i(\overline{D}_1 \oplus \overline{D}_2)k_j(\overline{D}_1 \ominus \overline{D}_2) + k_i(\overline{D}_1 \ominus \overline{D}_2)k_j(\overline{D}_1 \oplus \overline{D}_2) = \varphi_{ij}\big(\kappa(\overline{D}_1), \kappa(\overline{D}_2)\big). \qquad (14.14)$$

Using these biquadratic forms, we can easily compute the $k_i(\overline{D}_1 \oplus \overline{D}_2)$ if the $k_j(\overline{D}_1 \ominus \overline{D}_2)$ are known. We can also compute the $k_i([2]\overline{D}_1)$ by putting $\overline{D}_1 = \overline{D}_2$.

By abuse of notation we put $[2]\kappa(\overline{D}) = \kappa([2]\overline{D})$ and $\kappa(\overline{D}_1) \oplus \kappa(\overline{D}_2) = \kappa(\overline{D}_1 \oplus \overline{D}_2)$ defining a group law on the Kummer surface, provided that this is possible, i.e., that $\kappa(\overline{D}_1) \ominus \kappa(\overline{D}_2) = \kappa(\overline{D}_1 \ominus \overline{D}_2)$ is known.

**Addition**

**Proposition 14.28** Let $\mathbb{F}_q$ be a field of characteristic $p \neq 2, 3$ and let $C/\mathbb{F}_q$ be a curve of genus 2 in Montgomery form (14.13). Let $K_C$ denote the Kummer surface of $C$. Let $\overline{D}_1$ and $\overline{D}_2$ be two divisor classes of $C$, and assume that the difference $\overline{D}_1 \ominus \overline{D}_2$ is known and that the third coordinate of its image in the Kummer surface is 1 (remember we are in $\mathbb{P}^3(\mathbb{F}_q)$ and thus can normalize as long as $x_1 x_2 \neq 0$).

Then we obtain the Kummer coordinates for $\overline{D}_1 \oplus \overline{D}_2$ by the following formulas :

$$\begin{aligned}
k_1(\overline{D}_1 \oplus \overline{D}_2) &= \varphi_{11}\big(\kappa(\overline{D}_1), \kappa(\overline{D}_2)\big), \\
k_2(\overline{D}_1 \oplus \overline{D}_2) &= 2\big(\varphi_{12}\big(\kappa(\overline{D}_1), \kappa(\overline{D}_2)\big) - k_1(\overline{D}_1 \oplus \overline{D}_2)k_2(\overline{D}_1 \ominus \overline{D}_2)\big), \\
k_3(\overline{D}_1 \oplus \overline{D}_2) &= k_1(\overline{D}_1 \ominus \overline{D}_2)\varphi_{33}\big(\kappa(\overline{D}_1), \kappa(\overline{D}_2)\big), \\
k_4(\overline{D}_1 \oplus \overline{D}_2) &= 2\big(\varphi_{14}\big(\kappa(\overline{D}_1), \kappa(\overline{D}_2)\big) - k_1(\overline{D}_1 \oplus \overline{D}_2)k_4(\overline{D}_1 \ominus \overline{D}_2)\big),
\end{aligned} \qquad (14.15)$$

where the $\varphi_{ij}$ are the biquadratic forms described in Theorem 14.27.

The expressions of the $\varphi_{ij}\big(\kappa(\overline{D}_1 \oplus \overline{D}_2)\big)$, are available by anonymous ftp [FLY] but require a large number of operations in the base field to be computed. The main difficulty is to find expressions that require the least possible number of multiplications in $\mathbb{F}_q$. We now give more precisely these expressions for those $\varphi_{ij}$ we are interested in. We use the notation $\kappa(\overline{D}_1) = (k_1 : k_2 : k_3 : k_4)$ and $\kappa(\overline{D}_2) = (l_1 : l_2 : l_3 : l_4)$.

$$\begin{aligned}
\varphi_{11}\big(\kappa(\overline{D}_1), \kappa(\overline{D}_2)\big) &= \big((k_4 l_1 - k_1 l_4) + (k_2 l_3 - k_3 l_2)\big)^2, \\
\varphi_{12}\big(\kappa(\overline{D}_1), \kappa(\overline{D}_2)\big) &= \big((k_2 l_3 + k_3 l_2) + (k_1 l_4 + k_4 l_1)\big)\big(f_3(k_1 l_3 + k_3 l_1) + (k_2 l_4 + k_4 l_2)\big) \\
&\quad + 2(k_1 l_3 + k_3 l_1)\big(f_2(k_1 l_3 + k_3 l_1) + (k_1 l_2 + k_2 l_1) - (k_3 l_4 + k_4 l_3)\big) \\
&\quad + 2 f_4(k_1 l_4 + k_4 l_1)(k_2 l_3 + k_3 l_2), \\
\varphi_{33}\big(\kappa(\overline{D}_1), \kappa(\overline{D}_2)\big) &= \big((k_3 l_4 - k_4 l_3) + (k_1 l_2 - k_2 l_1)\big)^2, \\
\varphi_{14}\big(\kappa(\overline{D}_1), \kappa(\overline{D}_2)\big) &= (k_1 l_1 - k_3 l_3)\big[f_3\big((k_1 l_4 + k_4 l_1) - (k_2 l_3 + k_3 l_2)\big) \\
&\quad + 2\big((k_1 l_2 + k_2 l_1) - (k_3 l_4 + k_4 l_3)\big) \\
&\quad + f_2(k_4 l_4 + k_2 l_2) + 2 f_4(k_1 l_1 - k_3 l_3)\big] \\
&\quad + (k_2 l_2 - k_4 l_4)\big((k_2 l_3 + k_3 l_2) - (k_1 l_4 + k_4 l_1) - f_2(k_1 l_1 + k_3 l_3)\big).
\end{aligned}$$

## Doubling

**Proposition 14.29** Let $\mathbb{F}_q$ be a field of characteristic $p \neq 2, 3$ and let $C/\mathbb{F}_q$ be a curve of genus 2 in Montgomery form (14.13). Let $\mathcal{K}_C$ denote the Kummer surface of $C$. Let also $\overline{D}_1$ be a divisor class of $C$ and $\kappa(\overline{D}_1) = (k_1 : k_2 : k_3 : k_4)$, its image in the Kummer surface. Then we obtain the Kummer coordinates for $[2]\overline{D}_1$ given by $\kappa([2]\overline{D}_1) = (\delta_1 : \delta_2 : \delta_3 : \delta_4)$ by the following formulas:

$$\begin{aligned}
\delta_1 &= 2\varphi_{14}\big(\kappa(\overline{D}_1), \kappa(\overline{D}_1)\big), \\
\delta_2 &= 2\varphi_{24}\big(\kappa(\overline{D}_1), \kappa(\overline{D}_1)\big) + 2f_3 \mathcal{K}(\overline{D}_1), \\
\delta_3 &= 2\varphi_{34}\big(\kappa(\overline{D}_1), \kappa(\overline{D}_1)\big), \\
\delta_4 &= \varphi_{44}\big(\kappa(\overline{D}_1), \kappa(\overline{D}_1)\big) + 2\mathcal{K}(\overline{D}_1),
\end{aligned} \qquad (14.16)$$

where the $\varphi_{ij}$ are the biquadratic forms described in Theorem 14.27 and $\mathcal{K}$ is the equation of the Kummer surface such that $\mathcal{K}(\overline{D}_1) = 0$.

This is just a consequence of Theorem 14.27. Let us note that in $\delta_2$ and $\delta_4$ we added a multiple of the equation of the Kummer surface in order to simplify the expressions as much as possible. We give now more precisely these expressions for the $\delta_i$.

$$\begin{aligned}
\delta_1 &= 8(k_1^2 - k_3^2)\big(f_4(k_1^2 - k_3^2) + 2(k_1 k_2 - k_3 k_4)\big) \\
&\quad + 8(k_1 k_4 - k_2 k_3)\big(k_4^2 - k_2^2 + f_2(k_1 k_4 - k_2 k_3) + f_3(k_1^2 - k_3^2)\big), \\
\delta_2 &= 8(k_1^2 + k_3^2 - f_3 k_1 k_3 - 3 k_2 k_4)(k_2^2 + k_4^2 - f_3(k_1^2 + k_3^2) + 4 k_1 k_3) \\
&\quad + 16(k_2 k_4 + f_3 k_1 k_3)\big(f_4(k_1 k_2 + k_3 k_4) + 2(k_2^2 + k_4^2) + f_2(k_1 k_4 + k_2 k_3)\big) \\
&\quad + 32 k_1 k_3 \big(4 k_2 k_4 + f_2(k_1 k_2 + k_3 k_4) + (f_2^2 + f_4^2) k_1 k_3 + 8 f_4(k_1 k_4 + k_2 k_3)\big), \\
\delta_3 &= 8(k_1^2 - k_3^2)(f_2(k_1^2 - k_3^2) + 2(k_1 k_4 - k_2 k_3) + f_3(k_1 k_2 - k_3 k_4)) + \\
&\quad 8(k_3 k_4 - k_1 k_2)\big(k_4^2 - k_2^2 + f_4(k_3 k_4 - k_1 k_2)\big), \\
\delta_4 &= (k_2^2 + k_4^2)\big((k_2^2 + k_4^2) - 2 f_3(k_1^2 + k_3^2) - 8 k_1 k_3\big) \\
&\quad + (k_1^2 + k_3^2)\big[f_3 k_1 k_3 + f_4(k_1 k_4 + k_2 k_3) + 2 k_2 k_4 + f_2(k_1 k_2 + k_3 k_4) \\
&\quad + (f_3^2 - 4 f_2 f_4)(k_1^2 + k_3^2)\big] - 8 f_2 f_4 (k_1 k_3)^2.
\end{aligned}$$

### 14.4.4.b Montgomery scalar multiplication on genus 2 curves in Montgomery-like form

We now show how to compute scalar multiplications $[n]\overline{D}$ for some integer $n$ and divisor class $\overline{D}$. To add two classes in the representation on the Kummer surface we need to know their difference. Therefore, we can apply the Montgomery ladder (cf. Algorithm 9.5).

---

**Algorithm 14.30** Montgomery scalar multiplication for genus 2 curves

INPUT: A divisor class $\overline{D} \in \mathrm{Pic}_C^0$ and a positive integer $n = (n_{l-1} \ldots n_0)_2$.
OUTPUT: The image $\kappa([n]\overline{D})$ of $[n]\overline{D}$ in the Kummer surface.

1. $(D_1, D_2) \leftarrow \big((0:0:0:1), \kappa(\overline{D})\big)$
2. **for** $i = l - 1$ **down to** $0$ **do**
3.     **if** $n_i = 0$ **then** $(D_1, D_2) \leftarrow ([2]D_1, D_1 \oplus D_2)$
4.     **else** $(D_1, D_2) \leftarrow (D_1 \oplus D_2, [2]D_2)$
5. **return** $D_1$

Note that, at each step, we always have $D_2 \ominus D_1 = \kappa(\overline{D})$ so that the addition of $D_1$ and $D_2$ is possible on the Kummer surface.

**Number of operations**

At each step of the algorithm, we perform both an addition and a doubling, hence we just have to count the number of operations required for each of them. In the following, M will denote a multiplication in $\mathbb{F}_q$ and S a squaring.

**Table 14.6** Doubling of $\overline{D}_1$ in $K_C$ and addition of $\overline{D}_1$ and $\overline{D}_2$ in $K_C$ if $\overline{D}_1 \ominus \overline{D}_2$ is known.

| Doubling | | Addition | |
|---|---|---|---|
| Operation | Costs | Operation | Costs |
| Precomputations $\{k_i k_j\}_{i,j=1,\ldots,4}$ | 6M + 4S | Precomputations $\{k_i l_j\}_{i,j=1,\ldots,4}$ | 16M |
| $\delta_1$ | 5M | $\varphi_{11}(\kappa(\overline{D}_1), \kappa(\overline{D}_2))$ | S |
| $\delta_2$ | 11M | $\varphi_{12}(\kappa(\overline{D}_1), \kappa(\overline{D}_2))$ | 6M |
| $\delta_3$ | 5M | $\varphi_{33}(\kappa(\overline{D}_1), \kappa(\overline{D}_2))$ | S |
| $\delta_4$ | 5M | $\varphi_{14}(\kappa(\overline{D}_1), \kappa(\overline{D}_2))$ | 6M |
| — | — | $\kappa(\overline{D}_1 \oplus \overline{D}_2)$ | 3M |
| Total | 31M + 5S | Total | 31M + 2S |

**Remarks 14.31**

(i) To double one uses 31 multiplications including 16 multiplications by coefficients of the curve. The multiplications $f_3 k_1 k_3$, $f_3(k_1^2 + k_3^2)$, $f_4(k_1 k_4 + k_2 k_3)$ and $f_2(k_1 k_2 + k_3 k_4)$ are not counted in $\delta_4$ since they were already computed in $\delta_2$. Furthermore, we assumed that $f_2 f_4$, $f_3^2 - 4 f_2 f_4$ and $f_2^2 + f_4^2$ were precomputed.

(ii) The 31 multiplications needed for an addition include 7 multiplications by coefficients of the curve.

Hence, on a curve in the Montgomery form as in (14.13), the scalar multiplication $[n]\overline{D}$ by $n = \sum_{i=0}^{l-1} n_i 2^i$ using the Montgomery method requires $(62M + 7S)l$.

### 14.4.4.c  Comparison with usual algorithms for scalar multiplication

To date, the best algorithms for scalar multiplication on genus 2 curves defined over a field of odd characteristic are obtained by using mixed new coordinates (cf. Section 14.4.2). In this case, one needs 34M + 7S for a doubling and 36M + 5S for an addition. Hence, a single operation is more expensive but the Montgomery ladder requires us to compute both addition and doubling for each bit of the scalar while the usual arithmetic can be used with $\mathcal{N}$. Assuming an average density of $1/3$ of the scalar, new coordinates need only about 46M + 9S per bit. The use of windowing methods with precomputations allows us to reduce this even further.

Nevertheless, this algorithm is still interesting for many reasons.

**Remarks 14.32**

(i) As for elliptic curves, the Montgomery algorithm is automatically resistant against simple side-channel attacks (cf. Section 29.1). Therefore, it is of interest for implementations on restricted devices including smart cards.

Compared to the easiest countermeasure of performing a *double and always add* Algorithm 28.3, the use of Montgomery arithmetic is much more efficient as 62M+7S instead of 70M + 12S are needed per bit. Furthermore, one does not need dummy operations that could weaken the system against fault attacks. For more details on side-channel attacks see Chapters 28 and 29.

(ii) The algorithm is very dependent on the coefficients of the curve. Indeed, there are 23 multiplications by these coefficients but only 2 in Lange's formulas. Hence, a good choice of the coefficients of the curve allows better timings. We consider this issue now.

### 14.4.4.d  Some special cases

In order to decrease the number of base field operations for the Montgomery algorithm, certain choices of coefficients of the curve are better to use. For example, there are 6 multiplications by $f_3$ in the formulas in (14.15) and (14.16) so that, if one chooses $f_3 = 0$ or 1, the total amount of multiplications necessary for each bit of the scalar is 63 instead of 69. In Table 14.7, we summarize the gain obtained in each operation. Let us note that there is no gain for the calculation of $\varphi_{11}$, $\varphi_{33}$ and precomputations.

**Table 14.7** Gain obtained in different cases.

|  | $f_2 = 0$ | $f_2$ small | $f_3 = 0$ or small | $f_4 = 0$ | $f_4$ small |
|---|---|---|---|---|---|
| $\varphi_{12}$ | M | M | M | 2M | M |
| $\varphi_{14}$ | 2M | 2M | M | M | M |
| $\delta_1$ | M | M | M | M | M |
| $\delta_2$ | 2M | 2M | 2M | 2M | 2M |
| $\delta_3$ | M | M | M | M | M |
| $\delta_4$ | M + S | M | 0 | M + S | M |
| Total | 8M + S | 8M | 6M | 8M + S | 7M |

**Remark 14.33** If two of these conditions on the coefficients are satisfied, the gain obtained is at least the sum of the gains. If $f_2$ and $f_4$ are both small a further M is saved in $(f_2^2 + f_4^2)k_1k_3$.

Of course, this kind of restriction implies that fewer curves are taken into account. For example, if $f_3 = 0$, there are basically two free coefficients (namely $f_2$ and $f_4$) so that the number of such curves is $O(q^2)$. Thus, we lose in generality. However, the cryptographic applications require only one curve such that the divisor class group has a large prime order subgroup.

Let us now examine more precisely a particular case and compare our algorithm to the usual ones. Let $C$ be a genus 2 curve defined over $\mathbb{F}_q$ by an equation of the form

$$By^2 = x^5 + f_3 x^3 + \varepsilon x^2 + x \text{ with } \varepsilon = \pm 1 \text{ and } B \text{ and } f_3 \in \mathbb{F}_p. \qquad (14.17)$$

Here, Montgomery algorithm of scalar multiplication requires 46M+6S for each bit of the exponent, whereas with mixed new coordinates,

- a sliding window method with window size equal to 4 requires on average 40M + 8S per bit,

- using the NAF representation requires $(46M + 9S)l$ on average.

Thus, this algorithm is 7 percent faster than a double and add and not too far from the sliding window method (less than 6 percent).

The gain compared to the double-and-always-add algorithm is particularly striking, as there only $70M + 12S$ are needed per bit. In fact, this is a gain of 37 percent.

Of course, one can even be faster than the sliding window method by choosing a small coefficient $f_3$, but the number of such curves becomes very small.

**Remark 14.34** Another means to accelerate this algorithm would be to choose $f_2$, $f_3$ and $f_4$ to fit in one word. If the field needs $b$ words then each multiplication is about $b$ times faster, as for such small field sizes the schoolbook method (Algorithm 10.8) is applied. E.g., if $b = 3$ then the number of operations per bit reduces to about $47M + 7S$.

#### 14.4.4.e Examples

Here we give examples of genus 2 curves in Montgomery form. The base field is the prime field $\mathbb{F}_p$ with $p = 2^{80} + 13$ (so that the group size is the same as for elliptic curves over fields of 160 bits). Let $C_1$, $C_2$ and $C_3$ be the genus 2 curves defined by the equations

$$
\begin{aligned}
C_1: \; & 44294637780422381596577\, y^2 = x^5 + 27193747657561668783534\, x^4 \\
& \qquad\qquad\qquad\qquad\qquad\qquad\quad + 29755922098575219239037\, x^3 \\
& \qquad\qquad\qquad\qquad\qquad\qquad\quad + 76047862040141126737826\, x^2 + x, \\
C_2: \; & 10377931047456722522292\, y^2 = x^5 + 77304198867988157865677\, x^3 + x^2 + x, \\
C_3: \; & 69418837068442493864220\, y^2 = x^5 + x^3 + x.
\end{aligned}
$$

Timings for these curves are provided in [DUQ 2004]. Montgomery arithmetic on $C_3$ outperforms new coordinates with window width $w = 4$ and gets close to this on $C_2$. The general type curve $C_1$ is slower than new coordinates except for the side-channel resistant implementation.

## 14.5 Arithmetic on genus 2 curves in even characteristic

In this section we consider hyperelliptic curves of genus 2 over binary fields given by an equation of the form (14.1) $y^2 + h(x)y = f(x)$, where $h(x) = h_2 x^2 + h_1 x + h_0$ and $f(x) = x^5 + f_4 x^4 + f_3 x^3 + f_2 x^2 + f_1 x + f_0$ are polynomials defined over $\mathbb{F}_{2^d}$. By Example 14.9 we have $h(x) \neq 0$.

In order to get the best possible arithmetic, we first classify genus 2 curves and then give formulas for addition and doubling for each type of curves following [BYDU 2004, LAST 2005, PEWO+ 2004]. Then we deal with inversion-free systems as in odd characteristic.

In this section, we will call "small" an element in $\mathbb{F}_{2^d}$ that is represented by a polynomial with almost all its coefficients equal to zero, so that multiplications by such an element can be performed via few additions and are almost for free.

### 14.5.1 Classification of genus 2 curves in even characteristic

Classification of genus 2 curves in characteristic 2 is considered in, for example, [CHYU 2002, BYDU 2004, LAST 2005]. We divide them into into three types depending on the polynomial $h$:

- Type I if $\deg h = 2$. This type can be split in a Type Ia where $h$ has no root in $\mathbb{F}_{2^d}$ and a Type Ib where such a root exists.
- Type II if $\deg h = 1$.

## § 14.5 Arithmetic on genus 2 curves in even characteristic

- Type III if $\deg h = 0$.

We will first find equations for these types of curves with the minimal number of nonzero coefficients. In other words, we give an analogue of the short Weierstraß equations for elliptic curves.

Thanks to a change of variables of the form

$$x \mapsto \mu^2 x' + \lambda \quad \text{and} \quad y \mapsto \mu^5 y' + \mu^4 \alpha x'^2 + \mu^2 \beta x' + \gamma \tag{14.18}$$

after dividing the new equation by $\mu^{10}$, we can eliminate some coefficients from the equation (14.1).

**Proposition 14.35** A genus 2 curve of Type I defined over $\mathbb{F}_{2^d}$ by an equation of the form (14.1) is isomorphic to a curve defined by an equation of one of the following forms:

$$\text{Type Ia} : \quad y^2 + (x^2 + h_1 x + t h_1^2) y = x^5 + \varepsilon t x^4 + f_1 x + f_0,$$
$$\text{Type Ib} : \quad y^2 + x(x + h_1) y = x^5 + \varepsilon t x^4 + f_1 x + f_0,$$

where $\varepsilon \in \mathbb{F}_2$ and $t$ denotes an element of trace 1 ($t = 1$ if $d$ is odd). The isomorphism is explicit and uses the solution of quadratic equations in characteristic 2 explained in detail in Section 11.2.6. It is obtained using the change of variables (14.18) with

- $\mu = h_2$,
- $\lambda$ a root of $h$, if the Type is Ib (i.e., if $\operatorname{Tr}(h_0 h_2 h_1^{-2}) = 0$) and a solution of $h(x) = t h_1^2 h_2^{-1}$, if the Type is Ia (i.e., if $\operatorname{Tr}(h_0 h_2 h_1^{-2}) = 1$),
- $\alpha$ a root of $x^2 + h_2 x + f_4 + \lambda + \varepsilon t h_2^{-2}$ with $\varepsilon = \operatorname{Tr}((f_4 + \lambda) h_2^2)$,
- $\beta = (f_3 + h_1 \alpha) h_2^{-1}$,
- $\gamma = (\beta^2 + h_1 \beta + \alpha h(\lambda) + f_3 \lambda + f_2) h_2^{-1}$.

**Remark 14.36** For curves of Type I, the three parameters $h_1$, $f_1$ and $f_0$ are free so that this change of variables provides $4q^3$ isomorphism classes (with $q = 2^d$).

**Proposition 14.37** A genus 2 curve of Type II defined over $\mathbb{F}_{2^d}$ by an equation of the form (14.1) is isomorphic to a curve defined by an equation of the form

$$y^2 + xy = x^5 + f_3 x^3 + \varepsilon x^2 + f_0, \quad \text{if } d \text{ is odd}$$
$$y^2 + h_1 xy = x^5 + \varepsilon' x^3 + \varepsilon t h_1^2 x^2 + f_0, \quad \text{if } d \text{ is even}$$

where $\varepsilon$ and $\varepsilon'$ are in $\mathbb{F}_2$ and $t$ denotes an element of trace 1.

The isomorphism is explicit and it is obtained using the change of variables (14.18) with

- $\mu$ such that $\mu^3 = h_1$ if $d$ is odd and $\mu^4 = f_3 + h_1 \alpha$ if $d$ is even,
- $\lambda = h_0 h_1^{-1}$,
- $\alpha = \sqrt{\lambda + f_4}$,
- $\beta$ a root of $x^2 + h_1 x + f_2 + f_3 \lambda + \varepsilon t h_1^2$ with $\varepsilon = \operatorname{Tr}((f_2 + f_3 \lambda) h_1^{-2})$ and $t = 1$ if $d$ is odd,
- $\gamma = (\lambda^2 f_3 + \lambda^4 + f_1) h_1^{-1}$.

**Remarks 14.38**

(i) Finding an element $\mu$ such that $\mu^3 = h_1$ is always possible if $d$ is odd, as then $3 \nmid 2^d - 1$ (cf. Remark 2.94). Even though $d$ odd is the most common case in cryptographic applications (because of the Weil descent attack, Section 22.3), we also consider the case where $d$ is even. In this case, if $h_1$ is not a cube, an element $b$ can be chosen in $\mathbb{F}_{2^d}$ such that $h_1 b$ is a cube. Moreover, the probability that this element can be chosen "small" is very high so that multiplications by $b$ are almost for free. In this case one can obtain an isomorphic curve given by an equation that is very similar to the one obtained if $d$ is odd [LaSt 2005]:

$$y^2 + b^{-1}xy = x^5 + f_3 x^3 + \varepsilon t s^{-2} x^2 + f_0. \tag{14.19}$$

(ii) There are only two free parameters, $f_3$ and $f_0$ in the first form and $h_1$ and $f_0$ in the second one, as opposed to three in the general case showing that this type is indeed special. Thus, we obtain at most $2q^2$ isomorphism classes of curves of Type II defined over $\mathbb{F}_q$ if $d$ is odd and $4q^2$ if not.

(iii) It is also possible to have $f_3$ zero at the cost of a nonzero $h_0$, but we will see later that it is much more useful to have $h_0$ zero.

For curves of Type Ib the group order is always divisible by 4 since there exist three divisor classes of order 2 resulting from the 2 points with $x_1 = 0$ and $x_1 = h_1$. Over $\mathbb{F}_{2^d}$ the group order of Type Ia curves is divisible by 2 but not by 4, which needs a quadratic extension. Both types have full 2-rank, i.e., $J_C[2] \simeq \mathbb{Z}/2\mathbb{Z} \times \mathbb{Z}/2\mathbb{Z}$.

Type II curves have group order divisible by 2 as $h$ has a single root. These curves have 2-rank 1.

Type III curves have 2-rank 0 as $h$ is constant and are thus supersingular (cf. Definition 4.74 and the remark thereafter). This makes them weak under the Frey–Rück attack [FRRÜ 1994] as they always have a small embedding degree (cf. Section 24.2.2 and Galbraith [GAL 2001a]). Hence, they should be avoided for discrete logarithm systems. However, such curves have found an application in pairing based cryptography so that they must be considered.

**Proposition 14.39** *A genus 2 curve of Type III defined over $\mathbb{F}_{2^d}$ by an equation of the form (14.1) is isomorphic to a curve defined by an equation of the form*

$$y^2 + b^{-1}y = x^5 + f_3 x^3 + f_1 x + \varepsilon t s^{-2},$$

*where we may assume that $b$ is a "small" element of $\mathbb{F}_{2^d}$ such that $b h_0$ is a fifth power and $\varepsilon$ is in $\mathbb{F}_2$. For odd $d$ we can again achieve $b = 1$.*

**Remark 14.40** With this form we do not get a unique representative equation for each isomorphism class, because it is proven in [CHYU 2002] that there are between $2q$ and $32q$ isomorphism classes for curves of Type III and the form presented here has two free parameters ($f_3$ and $f_1$). In fact, a further change of variables leads to restrictions on $f_1$ but this involves equations of degree 16.

### 14.5.2 Explicit formulas in even characteristic in affine coordinates ($\mathcal{A}$)

The classification of the previous section allows some improvements in the formulas for the doubling. Addition works the same as in the general case given in Section 14.3.2.a. Of course, as some coefficients of the curve become zero or "small", some multiplications can be easily saved in the formulas. We now show how the doubling Algorithm 14.21 can be sped up for the individual types.

In the following, the element $t$ of trace 1 will always be chosen "small" and multiplications by $t$ are not taken into account when listing the costs per step.

§ 14.5 Arithmetic on genus 2 curves in even characteristic

The major speedup is obtained when $h_0 = 0$ which holds for curves of Type Ib and II. Lange and Stevens noticed in [LaSt 2005] that $r$, the resultant computed in the general formulas (see Section 14.3.2.c for more details) will simplify to the form $r = u_0\tilde{r}$ for some $\tilde{r}$. This allows us to cancel $r$ in the expressions, so its inverse is no longer needed.

Moreover, they use the equation

$$f + hv + v^2 = ut' + u^2(x + f_4), \tag{14.20}$$

to avoid the computation of $t'_0$ and also the exact computation of $s'_0$ is not necessary.

We will now give explicit formulas for doubling an element $[u, v]$ with $\deg u = 2$ (this is the general case) on each type of curve in affine coordinates.

In Algorithm 14.41 we give doubling formulas for a curve of Type Ia given by an equation of the form as in Proposition 14.35. In this case, $h_0 \neq 0$ so that the improvement of [LaSt 2005] cannot be used. However, it is possible to use the equation (14.20) to trade off a multiplication for a squaring (which is usually more efficient in characteristic 2) as explained in [LaSt 2005]. Finally, we use the fact that $h_0 = th_1^2$ to save a multiplication compared to the doubling formulas in [LaSt 2005].

Formulas for such curves contain a lot of multiplications by $h_1$ so that it is interesting for efficiency's sake to choose $h_1$ "small". For $h_1 = 1$ and thus $h_0 = t$, only 15 multiplications, 7 squarings and 1 inversion are required for doubling.

---

**Algorithm 14.41** Doubling on Type Ia curves ($g = 2$ and $q$ even)

INPUT: A divisor class $[u, v]$ with $u = x^2 + u_1 x + u_0$, $v = v_1 x + v_0$, $h_1^2$ and $t$ with $\text{Tr}(t) = 1$.
OUTPUT: The divisor class $[u', v'] = [2][u, v]$.

1. **compute $t'_1$ and precomputations**     [3M + 2S]
   $z_0 \leftarrow u_0^2$, $z_1 \leftarrow u_1^2$ and $w_0 \leftarrow h_1 v_1(h_1 + v_1)$
   $t'_1 \leftarrow z_1 + v_1$ and $w_1 \leftarrow h_1 u_1$

2. **compute resultant $r = \text{Res}(\tilde{v}, u)$**     [2M]
   $r \leftarrow (u_0 + th_1^2)(u_0 + th_1^2 + w_1) + h_1^2(u_0 + tz_1)$

3. **compute $s'_1$ and almost $s'_0$**     [3M + S]
   $s'_1 \leftarrow f_1 + z_0 + w_1(t'_1 + t(w_1 + \varepsilon u_1)) + w_0$
   $y \leftarrow f_0 + \varepsilon t z_0 + (v_0 + \varepsilon t w_1)^2 + h_1(u_0 t'_1 + t w_0)$
   If $s'_1 = 0$ see below

4. **compute $s'' \leftarrow x + s_0/s_1$ and $s_1$**     [I + 5M + 2S]
   $w_1 \leftarrow 1/(rs'_1)$, $w_2 \leftarrow rw_1$ and $w_3 \leftarrow s'^2_1 w_1$
   $w_4 \leftarrow rw_2$, $w_5 \leftarrow w_4^2$ and $s''_0 \leftarrow u_1 + yw_2$
   note that $w_1 = 1/r^2 s_1$, $w_2 = 1/s'_1$, $w_3 = s_1$ and $w_4 = 1/s_1$

5. **compute $l'$**     [2M]
   $l'_2 \leftarrow u_1 + s''_0$, $l'_1 \leftarrow u_1 s''_0 + u_0$ and $l'_0 \leftarrow u_0 s''_0$

6. **compute $u'$**     [M + S]
   $u'_0 \leftarrow s''^2_0 + w_4(s''_0 + u_1 + h_1) + \varepsilon t w_5$ and $u'_1 \leftarrow w_4 + w_5$

7. **compute $v'$**     [4M]
   $w_1 \leftarrow l'_2 + u'_1$ and $w_2 \leftarrow u'_1 w_1 + u'_0 + l'_1$
   $v'_1 \leftarrow w_2 w_3 + v_1 + h_1 + u'_1$
   $w_2 \leftarrow u'_0 w_1 + l'_0$ and $v'_0 \leftarrow w_2 w_3 + v_0 + h_0 + u'_0$

8. **return** $[u', v']$     [total complexity: I + 20M + 6S]

In case $s = s_0$, one replaces Lines 4–7 by the following lines.

| | | |
|---|---|---|
| 4'. | **compute $s$ and precomputations** | $[I + 2M]$ |
| | $w_1 \leftarrow 1/r$, $s_0 \leftarrow yw_1$ and $w_2 \leftarrow u_0 s_0 + v_0 + h_0$ | |
| 5'. | **compute $u'$** | $[S]$ |
| | $u'_0 \leftarrow s_0^2 + s_0 + \varepsilon t$ | |
| 6'. | **compute $v'$** | $[2M]$ |
| | $w_1 \leftarrow s_0(u_1 + u'_0) + u'_0 + v_1 + h_1$ and $v'_0 \leftarrow u'_0 w_1 + w_2$ | |

In this case the total complexity drops to $I + 12M + 4S$.

In Algorithm 14.42 we give doubling formulas for a curve of Type Ib given by an equation of the form given in Proposition 14.35, assuming that $h_1^2$ is precomputed. There are, again, a lot of multiplications by $h_1$ so that we can get many more operations for free if we are willing to choose $h_1$ "small". Therefore we also include this possibility in parentheses.

---

**Algorithm 14.42** Doubling on Type Ib curves ($g = 2$ and $q$ even)

INPUT: A divisor class $[u, v]$ with $u = x^2 + u_1 x + u_0$, $v = v_1 x + v_0$ and $h_1^2$.
OUTPUT: The divisor class $[u', v'] = [2][u, v]$.

| | | |
|---|---|---|
| 1. | **compute $t'_1$ and precomputations** | $[2M + 2S\,(3S)]$ |
| | $z_0 \leftarrow u_0^2$, $z_1 \leftarrow u_1^2$ and $w_0 \leftarrow v_1(h_1 + v_1)$ | |
| | $t'_1 \leftarrow z_1 + v_1$, $z_2 \leftarrow h_1 u_1$ and $z_3 \leftarrow \varepsilon t u_1$ | |
| 2. | **compute resultant $r = \mathrm{Res}(\tilde{v}, u)$** | |
| | $\tilde{r} \leftarrow u_0 + h_1^2 + z_2$ note that $\tilde{r} = r/u_0$ | |
| 3. | **compute $s'_1$ and almost $s'_0$** | $[3M\,(M)]$ |
| | $w_2 \leftarrow u_1(t'_1 + z_3) + w_0$ and $w_3 \leftarrow v_0 + h_1 t'_1$ | |
| | $s'_1 \leftarrow f_1 + z_0 + h_1 w_2$ | |
| | $m_0 \leftarrow w_2 + w_3$ note that $m_0 = (s'_0 - u_1 s'_1)/u_0$ | |
| | If $s'_1 = 0$ see below | |
| 4. | **compute $s'' = x + s_0/s_1$ and $s_1$** | $[I + 3M + S]$ |
| | $w_2 \leftarrow 1/(s'_1)$ and $w_3 \leftarrow u_0 w_2$ | |
| | $w_4 \leftarrow \tilde{r} w_3$ and $w_5 \leftarrow w_4^2$ | |
| | $s''_0 \leftarrow u_1 + m_0 w_3$ | |
| | note that $w_2 = 1/rs_1$ and $w_4 = 1/s_1$ | |
| 5. | **compute $u'$** | $[M + S]$ |
| | $z_4 \leftarrow \varepsilon t w_4$ and $u'_1 \leftarrow w_4 + w_5$ | |
| | $u'_0 \leftarrow s''^2_0 + w_4(s''_0 + h_1 + u_1 + z_4)$ | |
| 6. | **compute $v'$** | $[6M + S\,(5M + S)]$ |
| | $z_5 \leftarrow w_2\bigl(m_0^2 + t'_1(s'_1 + h_1 m_0)\bigr)$ | |
| | $z_6 \leftarrow s''_0 + h_1 + z_4 + z_5$ | |
| | $v'_0 \leftarrow v_0 + z_2 + z_1 + w_4(u'_0 + z_3) + s''_0 z_6$ | |
| | $v'_1 \leftarrow v_1 + w_4(u'_1 + s''_0 + \varepsilon t + u_1) + z_5$ | |
| 7. | **return** $[u', v']$ | [total complexity: $I + 15M + 5S\,(I + 10M + 6S)$] |

---

In case $s = s_0$, one replaces Lines 4–6 by the following lines.

| | | |
|---|---|---|
| 4'. | **compute $s$ and precomputations** | $[I + 2M]$ |
| | $w_1 \leftarrow 1/\tilde{r}$, $s_0 \leftarrow m_0 w_1$ and $w_2 \leftarrow u_0 s_0 + v_0$ | |
| 5'. | **compute $u'$** | $[S]$ |
| | $u'_0 \leftarrow s_0^2 + s_0$ | |

*§ 14.5 Arithmetic on genus 2 curves in even characteristic*

6'. **compute $v'$** [2M]
$w_1 \leftarrow s_0(u_1 + u'_0) + u'_0 + v_1 + h_1$ and $v'_0 \leftarrow u'_0 w_1 + w_2$

In this case the total complexity drops to $I + 9M + 3S$, resp. $I + 5M + 4S$.

In Algorithm 14.43 we give doubling formulas for a curve of Type II given by an equation of the form
$$y^2 + h_1 xy = x^5 + f_3 x^3 + f_2 x^2 + f_0.$$

The number of operations required for each step is given for $d$ odd (since it is the most frequently used case) and in parentheses for $d$ even. If $d$ is even, $h_1^2$ is precomputed.

---

**Algorithm 14.43** Doubling on Type II curves ($g = 2$ and $q$ even)

INPUT: A divisor class $[u, v]$ with $u = x^2 + u_1 x + u_0$, $v = v_1 x + v_0$, $h_1^{-1}$, and $h_1^2$.
OUTPUT: The divisor class $[u', v'] = [2][u, v]$.

1. **compute $rs_1$** [3S]
   $z_0 \leftarrow u_0^2$ and $t'_1 \leftarrow u_1^2 + f_3$
   $w_0 \leftarrow f_0 + v_0^2$ note that $w_0 = rs_1'/h_1^3$
   If $w_0 = 0$ see below

2. **compute $1/s_1$ and $s''_0$** [I + 2M]
   $w_1 \leftarrow (1/w_0)z_0$ note that $w_1 = h_1/s_1$
   $z_1 \leftarrow t'_1 w_1$ and $s''_0 \leftarrow z_1 + u_1$

3. **compute $u'$** [2S (2M + S)]
   $w_2 \leftarrow h_1^2 w_1$, $u'_1 \leftarrow w_2 w_1$ and $u'_0 \leftarrow s''^2_0 + w_2$

4. **compute $v'$** [3M + S (5M + S)]
   $w_3 \leftarrow w_2 + t'_1$
   $v'_1 \leftarrow h_1^{-1}(w_3 z_1 + w_2 u'_1 + f_2 + v_1^2)$ and $v'_0 \leftarrow h_1^{-1}(w_3 u'_0 + z_0)$

5. **return** $[u', v']$ [total complexity: $I + 5M + 6S$ ($I + 9M + 5S$)]

---

If $h_1^{-1}$ is small then the complexity drops down by 2M in Line 4.

In case $w_0 = 0$, one replaces Lines 2–4 by the following lines.

2'. **compute $s$ and precomputations** [M (2M)]
    $s_0 \leftarrow h_1^{-1} t'_1$ and $w_1 \leftarrow u_0 s_0 + v_0$

3'. **compute $u'$** [S]
    $u'_0 \leftarrow s_0^2$

4'. **compute $v'$** [2M]
    $w_2 \leftarrow s_0(u_1 + u'_0) + v_1 + h_1$ and $v'_0 \leftarrow u'_0 w_2 + w_1$

In this case the total complexity is $3M + 4S$ for $h_1^{-1} = 1$ or small and $4M + 4S$ otherwise.

## Summary

The classification of the different types of genus 2 curves in characteristic 2 allows significant speedups in the formulas for doublings given in Section 14.3.2.c for general curves. Indeed, the formulas for general curves in the general case require $I + 22M + 5S$ ([LAN 2005b]) (in affine coordinates) whereas only $I + 5M + 6S$ are needed for $h(x) = x$.

We summarize the results in Table 14.8 listing only the general cases; for $h$ of degree 1 and general $h$ the case "$f_4$ not small" does not apply, since then $f_4 = 0$.

**Table 14.8** Overview.

|  | $f_4$ small | $f_4$ not small |
|---|---|---|
| $h = x$ | I + 5M + 6S | n. a. |
| $h = h_1 x$ with $h_1^{-1}$ small | I + 7M + 5S | n. a. |
| $h = h_1 x$ | I + 9M + 5S | n. a. |
| $h = x^2 + h_1 x$ with $h_1^{-1}$ small | I + 10M + 6S | I + 12M + 6S |
| Type Ib | I + 15M + 5S | I + 17M + 5S |
| Type Ia, $h_1 = 1$ | I + 15M + 7S | n. a. |
| Type Ia | I + 20M + 6S | n. a. |

## Supersingular curves

Finally, Algorithm 14.44 presents the doubling formulas for supersingular curves, i.e., curves of Type III. This case is included due to the recent work done with supersingular curves and identity based encryption (cf. Chapter 24). The Tate–Lichtenbaum pairing on these curves (cf. Chapter 16) builds on addition and doubling. The costs of the doubling are given for $h_0^{-1} = 1$ as for $d$ odd this case can always be achieved. Otherwise we include $h_0^{-1}$ in the curve parameters. We state the costs for arbitrary $h_0$ in parentheses and comment on small $h_0^{-1}$ below.

---

**Algorithm 14.44** Doubling on Type III curves ($g = 2$ and $q$ even)

INPUT: A divisor class $[u, v]$ with $u = x^2 + u_1 x + u_0$, $v = v_1 x + v_0$ and $h_0^{-1}$.
OUTPUT: The divisor class $[u', v'] = [2][u, v]$.

1. **compute $s_1$**    [2S]
   $z_0 \leftarrow u_1^2$ and $z_1 \leftarrow v_1^2$
   $w_0 \leftarrow f_3 + z_0$ note that $w_0 \leftarrow s_1'/h_0$
   If $w_0 = 0$ see below

2. **compute $1/s_1$ and $s_0''$**    [I + M + S]
   $w_1 \leftarrow 1/w_0$ note that $w_1 \leftarrow h_0/s_1'$
   $s_0'' \leftarrow (f_2 + z_1)w_1 + u_1$ and $z_3 \leftarrow s_0''^2$

3. **compute $u'$**    [S (2M)]
   $w_2 \leftarrow h_0^2 w_1$, $u_1' \leftarrow w_2 w_1$ and $u_0' \leftarrow z_3$

4. **compute $v'$**    [3M + 2S (5M + 2S)]
   $v_1' \leftarrow h_0^{-1}\bigl(f_1 + u_0^2 + u_0' w_0 + w_2(u_1' + u_1 + s_0'')\bigr)$
   $v_0' \leftarrow h_0^{-1}\bigl(f_0 + v_0^2 + u_0'(f_2 + z_1 + w_2)\bigr) + h_0$

5. **return** $[u', v']$    [total complexity: I + 4M + 6S (I + 8M + 5S)]

---

For small $h_0^{-1}$ we save 2M in Line 4. In case $w_0 = 0$, one replaces Lines 2–4 by the following lines.

2'. **compute $s$ and precomputations**    [M (2M)]
   $s_0 \leftarrow h_0^{-1}(f_2 + z_1)$ and $w_1 \leftarrow u_0 s_0 + v_0 + h_0$

3'. **compute $u'$**    [S]
   $u_0' \leftarrow s_0^2$

4'. **compute $v'$**    [2M]
   $w_2 \leftarrow s_0(u_1 + u_0') + v_1$ and $v_0' \leftarrow u_0' w_2 + w_1$

In this case the total complexity drops to $4M + S$ for arbitrary $h_0$ and $3M + 3S$ for $h_0^{-1}$ small including the case $h_0 = 1$.

### 14.5.3 Inversion-free systems for even characteristic when $h_2 \neq 0$

In this section we discuss inversion-free coordinate systems starting with some comments on projective coordinates. It is also possible to design formulas for the Types Ia and Ib separately resulting in slightly better results.

All considerations for $\deg(h) = 1$ are postponed until the next section.

### 14.5.4 Projective coordinates ($\mathcal{P}$)

Addition in projective coordinates works almost the same as in Algorithm 14.22. In Line 5 one additionally computes $w_4 = s_1(w_1 + \widetilde{U}_{21})$ and $\tilde{h}_1 = h_1 Z$. The expressions for the output change to

$$\begin{aligned}
U_0' &\leftarrow s_0^2 + s_1 z_1 w_4 + z_2 \widetilde{S} + R\big(h_2(s_0 + w_4) + s_1 \tilde{h}_1 + r(z_1 + f_4 Z)\big) \\
U_1' &\leftarrow \widetilde{S} z_1 + h_2 R + R^2 \\
V_0' &\leftarrow w_0 + h_2 u_0' + \widehat{R} \widetilde{V}_{20} + h_0 Z \\
V_1' &\leftarrow w_1 + h_2 U_1' + \widehat{R}(\widetilde{V}_{21} + \tilde{h}_1)
\end{aligned}$$

The doubling algorithm differs so much that we give the general formulas for even characteristic only. For the counting we assume $h_2 = 1$ and $f_4 = f_3 = f_2 = 0$ as this can be reached for each curve of Type I by a slightly different change of variables for $x$ and allowing an arbitrary $h_0$.

---

**Algorithm 14.45** Doubling in projective coordinates ($g = 2$, $h_2 \neq 0$, and $q$ even)

INPUT: A divisor class represented by $[U_1, U_0, V_1, V_0, Z]$.
OUTPUT: The divisor class $[U_1', U_0', V_1', V_0', Z'] = [2][U_1, U_0, V_1, V_0, Z]$.

1. ***compute resultant and precomputations***   [6M + 4S]
   $\tilde{h}_1 \leftarrow h_1 Z,\ \tilde{h}_0 \leftarrow h_0 Z,\ Z_2 \leftarrow Z^2$ and $\widetilde{V}_1 \leftarrow \tilde{h}_1 + h_2 U_1$
   $\widetilde{V}_0 \leftarrow \tilde{h}_0 + h_2 U_0,\ w_0 \leftarrow V_1^2,\ w_1 \leftarrow U_1^2$ and $w_2 \leftarrow \widetilde{V}_1^2$
   $w_3 \leftarrow \widetilde{V}_0 Z + U_1 \widetilde{V}_1$ and $r \leftarrow \widetilde{V}_0 w_3 + w_2 U_0$

2. ***compute almost inverse***
   $\text{inv}_1 \leftarrow \widetilde{V}_1$ and $\text{inv}_0 \leftarrow w_3$

3. ***compute t***   [5M]
   $t_1 \leftarrow w_1 + f_3 Z_2 + Z h_2 V_1$ see Remark 14.46
   $t_0 \leftarrow U_1\big(Z(f_4 U_1 + h_2 V_1) + w_1 + f_3 Z_2\big) + Z\big(Z(f_2 Z + V_1 h_1 + V_0 h_2) + w_0\big)$

4. ***compute s***   [7M]
   $w_0 \leftarrow t_0 \text{inv}_0$ and $w_1 \leftarrow t_1 \text{inv}_1$
   $s_3 \leftarrow (\text{inv}_0 + \text{inv}_1)(t_0 + t_1) + w_0 + (1 + U_1) w_1$
   $s_1 \leftarrow s_3 Z$ and $s_0 \leftarrow w_0 + Z U_0 w_1$

5. ***precomputations***   [6M + 2S]
   $R \leftarrow Z_2 r,\ \widetilde{R} \leftarrow R s_1,\ S_1 \leftarrow s_1^2,\ S_0 \leftarrow s_0^2$ and $w_2 \leftarrow h_2 s_0$
   $s_1 \leftarrow s_1 s_3,\ s_0 \leftarrow s_0 s_3,\ S \leftarrow s_0 Z$ and $\widehat{R} \leftarrow \widetilde{R} s_1$

6. ***compute l***   [3M]
   $l_2 \leftarrow U_1 s_1,\ l_0 \leftarrow U_0 s_0$ and $l_1 \leftarrow (s_1 + s_0)(U_1 + U_0) + l_2 + l_0$

| | | |
|---|---|---|
| 7. | **compute $U'$** | [2M + S] |
| | $U_0' \leftarrow S_0 + R\big(s_3(h_2 U_1 + \tilde{h}_1)\big) + w_2 + f_4 R\big)$ and $U_1' \leftarrow h_2 \widetilde{R} + R^2$ | |
| 8. | **precomputations** | [4M] |
| | $l_2 \leftarrow l_2 + S + U_1'$, $w_0 \leftarrow U_0' l_2 + S_1 l_0$ and $w_1 \leftarrow U_1' l_2 + S_1(U_0' + l_1)$ | |
| 9. | **adjust** | [3M] |
| | $Z' \leftarrow S_1 \widetilde{R}$, $U_1' \leftarrow \widetilde{R} U_1'$ and $U_0' \leftarrow \widetilde{R} U_0'$ | |
| 10. | **compute $V'$** | [2M] |
| | $V_0' \leftarrow w_0 + h_2 U_0' + \widehat{R}(V_0 + \tilde{h}_0)$ and $V_1' \leftarrow w_1 + h_2 U_1' + \widehat{R}(V_1 + \tilde{h}_1)$ | |
| 11. | **return** $[U_1', U_0', V_1', V_0', Z']$ | [total complexity: 38M + 7S] |

**Remark 14.46** In fact if $f_4 = f_3 = f_2 = 0$ one computes $t_0$ differently as $t_0 = U_1 t_1 + Z_2(h_2 V_0 + h_1 V_1) + Z w_0$ using $t_1 = w_1 + h_2 Z V_1$ as precomputation.

#### 14.5.4.a New coordinates in even characteristic ($\mathcal{N}$)

To achieve better performance in inversion-free coordinates one can introduce weighted coordinates. In the following we present Lange's [LAN 2002d, LAN 2005b] *new coordinates*. For the general case in even characteristic it is most useful to use the set of coordinates extended by some precomputations and let $\mathcal{N}$ denote $[U_1, U_0, V_1, V_0, Z_1, Z_2, z_1, z_2, z_3, z_4]$ with $u_i = U_i/Z_1^2$, $v_i = V_i/(Z_1^3 Z_2)$ and the precomputations $z_1 = Z_1^2$, $z_2 = Z_2^2$, $z_3 = Z_1 Z_2$ and $z_4 = z_1 z_3$. The latter is useful for additions only and leaves the costs for doublings unchanged. The formulas show that $Z_1$ and $Z_2$ are no longer used separately. Therefore they can be left out leading again to 6 coordinates only.

As $p = 2$ and $h_2 \neq 0$, we assume $f_3 = f_2 = 0$, $h_2 = 1$ and include them in the algorithm (but not in the counting) only for the sake of completeness; $f_4$ is left out completely.

For the addition we assume that both classes are in $\mathcal{N}$. If one is in $\mathcal{A}$ the costs are given in brackets. A dedicated algorithm for $\mathcal{N} + \mathcal{A} = \mathcal{N}$ needs 37M + 5S (see [LAN 2002d]).

---

**Algorithm 14.47** Addition in new coordinates ($g = 2$, $h_2 \neq 0$, and $q$ even)

INPUT: Two divisor classes $\overline{D}_1$ and $\overline{D}_2$ represented by $\overline{D}_1 = [U_{11}, U_{10}, V_{11}, V_{10}, Z_{11}, Z_{12}, z_{11}, z_{12}, z_{13}, z_{14}]$ and $\overline{D}_2 = [U_{21}, U_{20}, V_{21}, V_{20}, Z_{21}, Z_{22}, z_{21}, z_{22}, z_{23}, z_{24}]$.
OUTPUT: The divisor class $\overline{D}' = [U_1', U_0', V_1', V_0', Z_1', Z_2', z_1', z_2', z_3', z_4'] = \overline{D}_1 \oplus \overline{D}_2$.

| | | |
|---|---|---|
| 1. | **precomputations** | [6M (none)] |
| | $\widetilde{U}_{21} \leftarrow U_{21} z_{11}$, $\widetilde{U}_{20} \leftarrow U_{20} z_{11}$, $\widetilde{V}_{21} \leftarrow V_{21} z_{14}$ and $\widetilde{V}_{20} \leftarrow V_{20} z_{14}$ | |
| | $Z_1 \leftarrow z_{11} z_{21}$ and $Z_3 \leftarrow z_{13} z_{23}$ | |
| 2. | **compute resultant $r = \text{Res}(U_1, U_2)$** | [8M + S (7M + S)] |
| | $y_1 \leftarrow U_{11} z_{21} + \widetilde{U}_{21}$, $y_2 \leftarrow U_{10} z_{21} + \widetilde{U}_{20}$ and $y_3 \leftarrow U_{11} y_1 + y_2 z_{11}$ | |
| | $r \leftarrow y_2 y_3 + y_1^2 U_{10}$, $\widetilde{Z}_2 \leftarrow r Z_3$ and $Z_2' \leftarrow \widetilde{Z}_2 Z_1$ | |
| 3. | **compute almost inverse of $u_2$ modulo $u_1$** | |
| | $\text{inv}_1 \leftarrow y_1$ and $\text{inv}_0 \leftarrow y_3$ | |
| 4. | **compute $s$** | [8M (7M)] |
| | $w_0 \leftarrow V_{10} z_{24} + \widetilde{V}_{20}$, $w_1 \leftarrow V_{11} z_{24} + \widetilde{V}_{21}$, $w_2 \leftarrow \text{inv}_0 w_0$ and $w_3 \leftarrow \text{inv}_1 w_1$ | |
| | $s_1 \leftarrow (\text{inv}_0 + z_{11} \text{inv}_1)(w_0 + w_1) + w_2 + w_3(z_{11} + U_{11})$ | |
| | $s_0 \leftarrow w_2 + U_{10} w_3$ | |
| 5. | **precomputations** | [10M + 3S] |
| | $\tilde{s}_0 \leftarrow s_0 Z_1$, $S_0 \leftarrow \tilde{s}_0^2$, $Z_1' \leftarrow s_1 Z_1$ and $R \leftarrow r Z_1'$ | |

*§ 14.5 Arithmetic on genus 2 curves in even characteristic*  343

$y_4 \leftarrow s_1(y_1 + \widetilde{U}_{21}), U_1' \leftarrow y_1 s_1, s_1 \leftarrow s_1 Z_1'$ and $s_0 \leftarrow s_0 Z_1'$
$z_1' \leftarrow Z_1'^2, z_2' \leftarrow Z_2'^2, z_3' \leftarrow Z_1' Z_2', z_4' \leftarrow z_1' z_3'$ and $\tilde{h}_1 \leftarrow h_1 z_3'$

6. **compute $l$**  [3M]
   $l_2 \leftarrow s_1 \widetilde{U}_{21}, l_0 \leftarrow s_0 \widetilde{U}_{20}$ and $l_1 \leftarrow (s_0 + s_1)(\widetilde{U}_{20} + \widetilde{U}_{21}) + l_0 + l_2$

7. **compute $U'$**  [5M]
   $U_0' \leftarrow S_0 + y_4 U_1' + y_2 s_1 + Z_2'\big(h_2(\tilde{s}_0 + y_4) + y_1 \widetilde{Z}_2\big) + \tilde{h}_1$
   $U_1' \leftarrow U_1' Z_1' + h_2 z_3' + z_2'$

8. **precomputations**  [3M]
   $l_2 \leftarrow l_2 + Z_1' \tilde{s}_0 + h_2 z_3' + U_1', w_0 \leftarrow l_2 U_0'$ and $w_1 \leftarrow l_2 U_1'$

9. **compute $V'$**  [5M]
   $V_1' \leftarrow w_1 + z_1'(l_1 + R\widetilde{V}_{21} + U_0' + \tilde{h}_1)$ and $V_0' \leftarrow w_0 + z_1'(l_0 + R\widetilde{V}_{20}) + h_0 z_4'$
   $\overline{D}' \leftarrow [U_1', U_0', V_1', V_0', Z_1', Z_2', z_1', z_2', z_3', z_4']$

10. **return** $\overline{D}'$  [total complexity: 48M + 4S (40M + 4S)]

Now we consider doublings.

---

**Algorithm 14.48** Doubling in new coordinates ($g = 2$, $h_2 \neq 0$, and $q$ even)

INPUT: A divisor class represented by $\overline{D} = [U_1, U_0, V_1, V_0, Z_1, Z_2, z_1, z_2, z_3, z_4]$.
OUTPUT: The divisor class $[U_1', U_0', V_1', V_0', Z_1', Z_2', z_1', z_2', z_3', z_4'] = [2]\overline{D}$.

1. **compute resultant and precomputations**  [8M + 3S]
   $\tilde{h}_1 \leftarrow z_1 h_1$ and $\tilde{h}_0 \leftarrow z_1 h_0$
   $\widetilde{V}_1 \leftarrow \tilde{h}_1 + h_2 U_1$ and $\widetilde{V}_0 \leftarrow \tilde{h}_0 + h_2 U_0$
   $w_0 \leftarrow V_1^2, w_1 \leftarrow U_1^2$ and $w_2 \leftarrow \tilde{h}_1^2 + h_2^2 w_1$
   $w_3 \leftarrow z_1(h_1 U_1 + h_2 U_0 + \tilde{h}_0) + h_2 w_1$
   $r \leftarrow w_2 U_0 + \widetilde{V}_0 w_3, \widetilde{Z}_2 \leftarrow z_3 r$ and $Z_2' \leftarrow \widetilde{Z}_2 z_4$

2. **compute almost inverse**
   $\text{inv}_1 \leftarrow \widetilde{V}_1$ and $\text{inv}_0 \leftarrow w_3$

3. **compute $t$**  [5M]
   $w_3 \leftarrow f_3 z_1^2 + w_1$ and $t_1 \leftarrow w_3 z_2 + V_1 h_2 z_3$
   $t_0 \leftarrow U_1 t_1 + w_0 + z_4(V_1 h_1 + V_0 h_2 + f_2 z_4)$

4. **compute $s = (t\,\text{inv}) \bmod u$**  [6M]
   $w_0 \leftarrow t_0 \text{inv}_0$ and $w_1 \leftarrow t_1 \text{inv}_1$
   $s_1 \leftarrow (\text{inv}_0 + \text{inv}_1)(t_0 + t_1) + w_0 + w_1(1 + U_1)$
   $s_0 \leftarrow w_0 + U_0 w_1 z_1$

5. **precomputations**  [8M + 3S]
   $y \leftarrow h_2 s_0 + s_1(h_2 U_1 + \tilde{h}_1), Z_1' \leftarrow s_1 z_1, S_0 \leftarrow s_0^2$ and $z_1' \leftarrow Z_1'^2$
   $S \leftarrow s_0 Z_1', R \leftarrow \widetilde{Z}_2 Z_1', s_0 \leftarrow s_0 s_1$ and $s_1 \leftarrow Z_1' s_1$
   $z_2' \leftarrow Z_2'^2, z_3' \leftarrow Z_1' Z_2'$ and $z_4' \leftarrow z_1' z_3'$

6. **compute $l$**  [3M]
   $l_2 \leftarrow s_1 U_1, l_0 \leftarrow s_0 U_0$ and $l_1 \leftarrow (s_1 + s_0)(U_1 + U_0) + l_0 + l_2$
   $l_2 \leftarrow l_2 + S + h_2 z_3'$

7. **compute $U'$**  [M]
   $U_0' \leftarrow S_0 + Z_2' y$ and $U_1' \leftarrow z_2' + h_2 z_3'$

8. ***precomputations***  [2M]
   $l_2 \leftarrow l_2 + U_1'$, $w_0 \leftarrow l_2 U_0'$ and $w_1 \leftarrow l_2 U_1'$

9. ***compute V'***  [6M]
   $V_1' \leftarrow w_1 + z_1'(l_1 + RV_1 + U_0') + z_4' h_1$
   $V_0' \leftarrow w_0 + z_1'(l_0 + RV_0) + z_4' h_0$

10. **return** $[U_1', U_0', V_1', V_0', Z_1', Z_2', z_1', z_2', z_3', z_4']$  [total complexity: $39M + 6S$]

#### 14.5.4.b  Different sets of coordinates

Using the same abbreviations as in odd characteristic, we state the costs for the operations in different coordinate systems in Table 14.9. Note that contrary to the odd characteristic case the advantage of using the new coordinates is smaller. We state the operation count for curves of Type Ia. If in fact $h$ has a root it is possible to design faster algorithms.

**Table 14.9** Addition and doubling in different systems and in even characteristic with $h_2 \neq 0$.

| Doubling | | Addition | |
|---|---|---|---|
| Operation | Costs | Operation | Costs |
| $2\mathcal{N} = \mathcal{P}$ | $39M + 6S$ | $\mathcal{N} + \mathcal{P} = \mathcal{P}$ | $51M + 4S$ |
| $2\mathcal{P} = \mathcal{P}$ | $38M + 7S$ | $\mathcal{N} + \mathcal{N} = \mathcal{P}$ | $50M + 4S$ |
| $2\mathcal{N} = \mathcal{N}$ | $37M + 6S$ | $\mathcal{N} + \mathcal{P} = \mathcal{N}$ | $49M + 4S$ |
| $2\mathcal{P} = \mathcal{N}$ | $36M + 7S$ | $\mathcal{P} + \mathcal{P} = \mathcal{P}$ | $49M + 4S$ |
| — | — | $\mathcal{N} + \mathcal{N} = \mathcal{N}$ | $48M + 4S$ |
| — | — | $\mathcal{P} + \mathcal{P} = \mathcal{N}$ | $47M + 4S$ |
| — | — | $\mathcal{A} + \mathcal{N} = \mathcal{P}$ | $39M + 5S$ |
| — | — | $\mathcal{A} + \mathcal{P} = \mathcal{P}$ | $39M + 4S$ |
| — | — | $\mathcal{A} + \mathcal{P} = \mathcal{N}$ | $37M + 4S$ |
| — | — | $\mathcal{A} + \mathcal{N} = \mathcal{N}$ | $37M + 5S$ |
| $2\mathcal{A} = \mathcal{A}$ | $I + 20M + 6S$ | $\mathcal{A} + \mathcal{A} = \mathcal{A}$ | $I + 22M + 3S$ |

#### 14.5.4.c  Computation of scalar multiples

We follow the same lines as in the odd characteristic and distinguish between precomputations and no precomputations.

**No precomputation**

For cheap inversions one again uses the affine system alone. If one wants to avoid inversions and has an affine input (or can allow one I to achieve this) we perform the doublings as $2\mathcal{N} = \mathcal{N}$ and the addition as $\mathcal{A} + \mathcal{N} = \mathcal{N}$. For non-normalized input we use the new coordinates for doublings and as non-normalized input system if necessary.

## § 14.5 Arithmetic on genus 2 curves in even characteristic

**Table 14.10** Without precomputations in even characteristic with $h_2 \neq 0$.

| Systems | Cost |
|---|---|
| $2\mathcal{A} = \mathcal{A}, \mathcal{A} + \mathcal{A} = \mathcal{A}$ | $\frac{1}{3}(4\text{I} + 82\text{M} + 21\text{S})$ |
| $2\mathcal{N} = \mathcal{N}, \mathcal{A} + \mathcal{N} = \mathcal{N}$ | $\frac{1}{3}(148\text{M} + 23\text{S})$ |
| $2\mathcal{N} = \mathcal{N}, \mathcal{N} + \mathcal{N} = \mathcal{N}$ | $\frac{1}{3}(159\text{M} + 22\text{S})$ |

**Windowing methods**

To obtain the table of precomputed values we need one doubling and $2^{w-2} - 1$ additions. Here it is advantageous to choose either $\mathcal{C}_3 = \mathcal{A}$ or $\mathcal{C}_3 = \mathcal{N}$.

**Table 14.11** Precomputations in even characteristic with $h_2 \neq 0$.

| System | I | M | S |
|---|---|---|---|
| $\mathcal{A}$ | $2^{w-2}$ | $22 \times 2^{w-2} - 2$ | $3 \times 2^{w-2} + 3$ |
| $\mathcal{A}$ | $w - 1$ | $25 \times 2^{w-2} + 20w - 68$ | $3 \times 2^{w-2} + 6w - 15$ |
| $\mathcal{A}$ | 1 | $51 \times 2^{w-2} - 17$ | $4 \times 2^{w-2} + 2$ |
| $\mathcal{P}$ | | $48 \times 2^{w-2} - 11$ | $4 \times 2^{w-2} + 2$ |

The costs of computing scalar multiples are listed in Table 14.12 for the most useful matches of sets of coordinates. We use the same abbreviations as in the odd characteristic case. Again we leave out the costs for the initial conversions and mention that some constant number of operations can be saved if one considers in more detail the first doubling and the final addition/doubling.

**Table 14.12** Windowing method in even characteristic with $h_2 \neq 0$.

| Systems | I | M | S |
|---|---|---|---|
| $2\mathcal{A} = \mathcal{A}, \mathcal{A} + \mathcal{A} = \mathcal{A}$ | $l_1 + K + \frac{l_1 - K}{w+1}$ | $20(l_1 + K) + 22\frac{l_1 - K}{w+1}$ | $6(l_1 + K) + 3\frac{l_1 - K}{w+1}$ |
| $2\mathcal{N} = \mathcal{N}, \mathcal{A} + \mathcal{N} = \mathcal{N}$ | | $37\left((l_1 + K) + \frac{l_1 - K}{w+1}\right)$ | $6(l_1 + K) + 5\frac{l_1 - K}{w+1}$ |
| $2\mathcal{N} = \mathcal{N}, \mathcal{N} + \mathcal{N} = \mathcal{N}$ | | $37(l_1 + K) + 48\frac{l_1 - K}{w+1}$ | $6(l_1 + K) + 4\frac{l_1 - K}{w+1}$ |

Compared to the results in odd characteristic this case is a bit more expensive. On the other hand the arithmetic in binary fields is easier to implement and usually faster and there is space for improvements taking into account the different types of curves.

### 14.5.5 Inversion-free systems for even characteristic when $h_2 = 0$

Obviously this case can be considered as a special case of the previous section, but as in affine coordinates specialized doubling algorithms are much faster. For the whole section we assume that the curve is given by an affine equation of the form (14.19).

### 14.5.5.a Doubling in projective coordinates

For the additions the changes are quite obvious and are simply obtained by fixing the respective curve parameters to be zero. Hence, we only treat doublings in the following.

---

**Algorithm 14.49** Doubling in projective coordinates ($g = 2$, $h_2 = 0$, and $q$ even)

INPUT: A divisor class represented by $\overline{D} = [U_1, U_0, V_1, V_0, Z]$ and the precomputed values $h_1^2$ and $h_1^{-1}$.
OUTPUT: The divisor class $[U_1', U_0', V_1', V_0', Z'] = [2][U_1, U_0, V_1, V_0, Z]$.

1. ***precomputations***                                                                                                 [9M + 4S]
   $Z_2 \leftarrow Z^2$, $z_0 \leftarrow U_0^2$, $t_1 \leftarrow U_1^2 + f_3 Z_2$, $w_0 \leftarrow f_0 Z_2 + V_0^2$ and $w_1 \leftarrow z_0 Z_2$
   $z_1 \leftarrow t_1 z_0$, $w_2 \leftarrow h_1^2 w_1$, $w_3 \leftarrow w_2 + t_1 w_0$ and $w_4 \leftarrow w_0 Z$
   $s_0 \leftarrow z_1 + U_1 w_4$ and $w_4 \leftarrow w_4 Z$

2. ***compute $U'$***                                                                                                          [2M + S]
   $U_1' \leftarrow w_1 w_2$ and $U_0' \leftarrow s_0^2 + w_2 w_4$

3. ***compute $V'$***                                                                                             [11M + S]
   $w_5 \leftarrow w_0 w_4$ and $V_1' \leftarrow h_1^{-1}\bigl(w_2 U_1' + (w_3 z_1 + (f_2 Z_2 + V_1^2)w_5)w_4\bigr)$
   $w_5 \leftarrow w_5 w_4$ and $V_0' \leftarrow h_1^{-1}(w_3 U_0' + z_0 w_5)$

4. ***adjust***                                                                                                                [3M]
   $Z' \leftarrow w_5 Z_2$, $U_1' \leftarrow U_1' w_4$ and $U_0' \leftarrow U_0' w_4$

5. **return** $[U_1', U_0', V_1', V_0', Z']$                                            [total complexity: 25M + 6S]

---

If $h_1^{-1}$ is small one saves 2M, and if $h_1 = 1$ — as one can always achieve for odd extension degrees — 22M + 6S are used in total.

### 14.5.5.b Recent coordinates in even characteristic ($\mathcal{R}$)

For $h_2 = 0$ we follow [LAN 2005a] and use $[U_1, U_0, V_1, V_0, Z, z]$ with $u_i = U_i/Z$, $v_i = V_i/Z^2$ and the precomputation $z = Z^2$. These coordinates have the advantage of allowing faster doublings while the additions are more expensive. However, usually mixed additions are chosen for implementations. They are not too much slower, and furthermore, in windowing methods the number of additions is reduced considerably.

The formulas for new coordinates (in the sense of section 14.5.4.a) can be found in [LAN 2005b]. An addition $\mathcal{N} + \mathcal{N}$ takes 44M + 6S and in mixed coordinates $\mathcal{A} + \mathcal{N} = \mathcal{N}$ one needs 36M + 4S. Using the conditions on the curve parameters given in (14.19) for extension of odd degrees the costs for a doubling reduce to 28M + 7S.

The results in brackets refer to the case in which the second input is in affine coordinates.

---

**Algorithm 14.50** Addition in recent coordinates ($g = 2$, $h_2 = 0$, and $q$ even)

INPUT: Two divisor classes $\overline{D}_1$ and $\overline{D}_2$ represented by $\overline{D}_1 = [U_{11}, U_{10}, V_{11}, V_{10}, Z_1, z_1]$ and $\overline{D}_2 = [U_{21}, U_{20}, V_{21}, V_{20}, Z_2, z_2]$.
OUTPUT: The divisor class $[U_1', U_0', V_1', V_0', Z', z'] = \overline{D}_1 \oplus \overline{D}_2$.

1. ***precomputations***                                                                             [5M + S (none)]
   $Z \leftarrow Z_1 Z_2$, $z \leftarrow Z^2$, $\widetilde{U}_{21} \leftarrow U_{21} Z_1$ and $\widetilde{U}_{20} \leftarrow U_{20} Z_1$
   $\widetilde{V}_{21} \leftarrow V_{21} z_1$ and $\widetilde{V}_{20} \leftarrow V_{20} z_1$

2. ***compute resultant $r = \mathrm{Res}(U_1, U_2)$***                                 [6M + S (5M + S)]
   $y_1 \leftarrow U_{11} Z_2 + \widetilde{U}_{21}$ and $y_2 \leftarrow U_{10} Z_2 + \widetilde{U}_{20}$
   $y_3 \leftarrow U_{11} y_1 + y_2 Z_1$ and $r \leftarrow y_2 y_3 + y_1^2 U_{10}$

§ 14.5 Arithmetic on genus 2 curves in even characteristic                                                                                      347

3. **compute almost inverse of $u_2$ modulo $u_1$**
   $\text{inv}_1 \leftarrow y_1$ and $\text{inv}_0 \leftarrow y_3$

4. **compute $s$**                                                                                              [8M (7M)]
   $w_0 \leftarrow V_{10}z_2 + \widetilde{V}_{20}$ and $w_1 \leftarrow V_{11}z_2 + \widetilde{V}_{21}$
   $w_2 \leftarrow \text{inv}_0 w_0$ and $w_3 \leftarrow \text{inv}_1 w_1$
   $s_1 \leftarrow (\text{inv}_0 + \text{inv}_1 Z_1)(w_0 + w_1) + w_2 + w_3(Z_1 + U_{11})$
   $s_0 \leftarrow w_2 + U_{10}w_3$

5. **precomputations**                                                                                          [7M + S]
   $\overline{Z} \leftarrow s_1 r$, $w_4 \leftarrow rZ$, $w_5 \leftarrow w_4^2$, $S \leftarrow s_0 Z$ and $Z' \leftarrow Z\overline{Z}$
   $\tilde{s}_0 \leftarrow s_0 Z'$, $\bar{s}_1 \leftarrow s_1 \overline{Z}$ and $\tilde{s}_1 \leftarrow \bar{s}_1 Z$

6. **compute $l$**                                                                                              [5M]
   $L_2 \leftarrow \bar{s}_1 \widetilde{U}_{21}$, $l_2 \leftarrow L_2 Z$ and $l_0 \leftarrow \tilde{s}_0 \widetilde{U}_{20}$
   $l_1 \leftarrow (\widetilde{U}_{21} + \widetilde{U}_{20})(\tilde{s}_0 + \tilde{s}_1) + l_2 + l_0$, $l_2 \leftarrow L_2 + \tilde{s}_0$ and $\tilde{h}_1 \leftarrow h_1 z$

7. **compute $U'$**                                                                                             [8M + 2S]
   $U'_0 \leftarrow r(S^2 + y_1(s_1^2(y_1 + \widetilde{U}_{21}) + Zw_5) + \tilde{h}_1 Z') + y_2 \tilde{s}_1$
   $U'_1 \leftarrow y_1 \bar{s}_1 + w_4 w_5$

8. **precomputations**                                                                                          [5M + S]
   $w_1 \leftarrow l_2 + U'_1$, $U'_1 \leftarrow U'_1 w_4$, $\overline{Z} \leftarrow Z'\overline{Z}$ and $l_0 \leftarrow l_0 \overline{Z}$
   $w_2 \leftarrow U'_1 w_1 + (U'_0 + l_1)\overline{Z}$ and $\overline{Z} \leftarrow \overline{Z}^2$

9. **compute $V'$**                                                                                             [6M + 2S]
   $V'_1 \leftarrow w_2 s_1 + (\widetilde{V}_{21} + \tilde{h}_1)\overline{Z}$, $U'_0 \leftarrow U'_0 r$ and $w_2 \leftarrow U'_0 w_1 + l_0$
   $V'_0 \leftarrow w_2 s_1 + \widetilde{V}_{20} \overline{Z}$, $Z' \leftarrow Z'^2$ and $z' \leftarrow Z'^2$

10. **return** $[U'_1, U'_0, V'_1, V'_0, Z', z']$                                       [total complexity: 50M + 8S (43M + 7S)]

If $h_1 = 1$ as we can always assume for $d$ odd one more multiplication is saved in Line 6.

---

**Algorithm 14.51** Doubling in recent coordinates ($g = 2$, $h_2 = 0$, and $q$ even)

INPUT: A divisor class $[U_1, U_0, V_1, V_0, Z, z]$ and the precomputed values $h_1^2$ and $h_1^{-1}$.
OUTPUT: The divisor class $[U'_1, U'_0, V'_1, V'_0, Z', z'] = [2][U_1, U_0, V_1, V_0, Z, z]$.

1. **precomputations**                                                                                          [10M + 4S]
   $Z_4 \leftarrow z^2$, $y_0 \leftarrow U_0^2$, $t_1 \leftarrow U_1^2 + f_3 z$ and $w_0 \leftarrow Z_4 f_0 + V_0^2$
   $\overline{Z} \leftarrow z w_0$, $w_1 \leftarrow y_0 Z_4$, $y_1 \leftarrow t_1 y_0 z$ and $s_0 \leftarrow y_1 + U_1 w_0 Z$
   $w_2 \leftarrow h_1^2 w_1$ and $w_3 \leftarrow w_2 + t_1 w_0$

2. **compute $U'$**                                                                                             [2M + S]
   $U'_1 \leftarrow w_2 w_1$, $w_2 \leftarrow w_2 \overline{Z}$ and $U'_0 \leftarrow s_0^2 + w_2$

3. **compute $V'$**                                                                                             [11M + 3S]
   $Z' \leftarrow \overline{Z}^2$ and $V'_1 \leftarrow h_1^{-1}\big(w_2 U'_1 + (w_3 y_1 + f_2 Z' + (V_1 w_0)^2)Z'\big)$
   $V'_0 \leftarrow h_1^{-1}\big(\overline{Z}(w_3 U'_0 + y_0 w_0 Z')\big)$, $z' \leftarrow Z'^2$

4. **return** $[U'_1, U'_0, V'_1, V'_0, Z', z']$                                                    [total complexity: 23M + 8S]

---

For small $h_1^{-1}$ we save 2M, if even $h_1 = 1$ a total of only 20M + 8S is needed.

A comparison of different sets of coordinates is given in [LAN 2005a]. It also contains formulas for operations in $[U_1, U_0, V_1, V_0, Z, Z^2 Z^3]$ in which the doublings are less efficient than in $\mathcal{R}$ but the additions do not introduce such a big overhead. In general, for curves of form (14.19) inversion-free systems will be useful only for very expensive inversions and when combined with windowing methods.

## 14.6 Arithmetic on genus 3 curves

Cantor's algorithm applies to hyperelliptic curves of arbitrary genus. In this section we study arithmetic on curves of genus 3. Again the most frequent input for addition consists of two divisor classes represented by $[u_1, v_1], [u_2, v_2]$, where $\deg(u_1) = \deg(u_2) = 3$ and $\gcd(u_1, u_2) = 1$. These conditions guarantee that the associated reduced divisors $D_1, D_2$ do not have any point or its opposite in common and both divisors have 3 affine points in the support. For the doubling we may assume that the class is represented by $[u_1, v_1]$ with $\deg(u_1) = 3$ and that $\gcd(h + 2v_1, u_1) = 1$. This means that the support of $D_1$ contains no Weierstraß point and there are 3 affine points.

We omit the complete case study here. It can be found in [PEL 2002, WOL 2004]. There exists a generalization of projective coordinates to genus 3 curves [FAWA 2004] such that one does not need inversions for the group operations, but we only state arithmetic in affine coordinates. For smaller fields an inversion is less expensive in terms of multiplications and on the other hand more multiplications are needed to save the one remaining inversion.

The following sections give algorithms for addition of general divisor classes and for doubling of a general class. For these we allow arbitrary equations of the curve and arbitrary finite ground fields. Note that the number of operations might still depend on this. For even characteristic we additionally state doubling formulas for one special curve. As for genus 2, the addition formulas do barely change with the equation of the curve but the doubling needs far fewer field operations for special equations.

These formulas were taken from [WOL 2004]. Formulas for genus 3 curves can also be found in [GOMA+ 2005, KUGO+ 2002] and [GUKA+ 2004].

### 14.6.1 Addition in most common case

This section treats the addition of two different divisor classes. In odd characteristic we can transform to an isomorphic curve $y^2 = f(x)$. In even characteristic we assume for simplicity that $h(x) \in \mathbb{F}_2[x]$. For other values of the $h_i$ some operations should be performed differently.

---

**Algorithm 14.52** Addition on curves of genus 3 in the general case

INPUT: Two divisor classes $[u_1, v_1]$ and $[u_2, v_2]$ with $u_i = x^3 + u_{i2}x^2 + u_{i1}x + u_{i0}, v_i = v_{i2}x^2 + v_{i1}x + v_{i0}$.
OUTPUT: The divisor class $[u'', v''] = [u_1, v_1] \oplus [u_2, v_2]$ with $u'' = x^3 + u_2''x^2 + u_1''x + u_0'', v'' = v_2''x^2 + v_1''x + v_0''$.

1. **compute resultant** $r = \text{Res}(u_1, u_2)$ **(Bezout)** [12M + 2S]

   $t_1 \leftarrow u_{12}u_{21}, t_2 \leftarrow u_{11}u_{22}, t_3 \leftarrow u_{11}u_{20}, t_4 \leftarrow u_{10}u_{21}$ and $t_5 \leftarrow u_{12}u_{20}$
   $t_6 \leftarrow u_{10}u_{22}, t_7 \leftarrow (u_{20} - u_{10})^2, t_8 \leftarrow (u_{21} - u_{11})^2$ and $t_9 \leftarrow (u_{22} - u_{12})(t_3 - t_4)$
   $t_{10} \leftarrow (u_{22} - u_{12})(t_5 - t_6)$ and $t_{11} \leftarrow (u_{21} - u_{11})(u_{20} - u_{10})$
   $r \leftarrow (u_{20} - u_{10} + t_1 - t_2)(t_7 - t_9) + (t_5 - t_6)(t_{10} - 2t_{11}) + t_8(t_3 - t_4)$
   If $r \leftarrow 0$ perform Cantor's Algorithm 14.7

2. **compute almost inverse** $\text{inv} = r/u_1 \bmod u_2$ [4M]

   $\text{inv}_2 \leftarrow (t_1 - t_2 - u_{10} + u_{20})(u_{22} - u_{12}) - t_8$ and $\text{inv}_1 \leftarrow \text{inv}_2 u_{22} - t_{10} + t_{11}$
   $\text{inv}_0 \leftarrow \text{inv}_2 u_{21} - u_{22}(t_{10} - t_{11}) + t_9 - t_7$

3. **compute** $s' = rs \equiv (v_2 - v_1)\text{inv} \pmod{u_2}$ **(Karatsuba)** [11M]

   $t_{12} \leftarrow (\text{inv}_1 + \text{inv}_2)(v_{22} - v_{12} + v_{21} - v_{11})$ and $t_{13} \leftarrow (v_{21} - v_{11})\text{inv}_1$

§ 14.6 Arithmetic on genus 3 curves

$t_{14} \leftarrow (\mathrm{inv}_0 + \mathrm{inv}_2)(v_{22} - v_{12} + v_{20} - v_{10})$ and $t_{15} \leftarrow (v_{20} - v_{10})\mathrm{inv}_0$
$t_{16} \leftarrow (\mathrm{inv}_0 + \mathrm{inv}_1)(v_{21} - v_{11} + v_{20} - v_{10})$ and $t_{17} \leftarrow (v_{22} - v_{12})\mathrm{inv}_2$
$r'_0 \leftarrow t_{15}, r'_1 \leftarrow t_{16} - t_{13} - t_{15}$ and $r'_2 \leftarrow t_{13} + t_{14} - t_{15} - t_{17}$
$r'_3 \leftarrow t_{12} - t_{13} - t_{17}, r'_4 \leftarrow t_{17}$ and $t_{18} \leftarrow u_{22}r'_4 - r'_3$
$t_{15} \leftarrow u_{20}t_{18}, t_{16} \leftarrow u_{21}r'_4$ and $s'_0 \leftarrow r'_0 + t_{15}$
$s'_1 \leftarrow r'_1 - (u_{21} + u_{20})(r'_4 - t_{18}) + t_{16} - t_{15}$
$s'_2 \leftarrow r'_2 - t_{16} + u_{22}t_{18}$
If $s'_2 = 0$ perform Cantor's Algorithm 14.7

4. **compute $s = (s'/r)$ and make $s$ monic**    [I + 6M + 2S]
$w_1 \leftarrow (rs'_2)^{-1}, w_2 \leftarrow rw_1, w_3 \leftarrow w_1{s'_2}^2, w_4 \leftarrow rw_2$ and $w_5 \leftarrow w_4^2$
$s_0 \leftarrow w_2 s'_0$ and $s_1 \leftarrow w_2 s'_1$

5. **compute $z = su_1$**    [6M]
$z_0 \leftarrow s_0 u_{10}, z_1 \leftarrow s_1 u_{10} + s_0 u_{11}$ and $z_2 \leftarrow s_0 u_{12} + s_1 u_{11} + u_{10}$
$z_3 \leftarrow s_1 u_{12} + s_0 + u_{11}$ and $z_4 \leftarrow u_{12} + s_1$

6. **compute $u' = [s(z + w_4(h + 2v_1)) - w_5((u_{20} - v_1 h - v_1^2)/u_1)]/u_2$**    [15M]
$u'_3 \leftarrow z_4 + s_1 - u_{22}$ and $u'_2 \leftarrow -u_{22}u'_3 - u_{21} + z_3 + s_0 + w_4 h_3 + s_1 z_4$
$u'_1 \leftarrow w_4(h_2 + 2v_{12} + s_1 h_3) + s_1 z_3 + s_0 z_4 + z_2 - w_5 - u_{22}u'_2 - u_{21}u'_3 - u_{20}$
$u'_0 \leftarrow w_4(s_1 h_2 + h_1 + 2v_{11} + 2s_1 v_{12} + s_0 h_3) + s_1 z_2 + z_1 + s_0 z_3 + w_5(u_{12} - f_6) - u_{22}u'_1 - u_{21}u'_2 - u_{20}u'_3$

7. **compute $v' = -(w_3 z + h + v_1) \bmod u'$**    [8M]
$t_1 \leftarrow u'_3 - z_4$ and $v'_0 \leftarrow -w_3(u'_0 t_1 + z_0) - h_0 - v_0$
$v'_1 \leftarrow -w_3(u'_1 t_1 - u'_0 + z_1) - h_1 - v_{11}$
$v'_2 \leftarrow -w_3(u'_2 t_1 - u'_1 + z_2) - h_2 - v_{12}$
$v'_3 \leftarrow -w_3(u'_3 t_1 - u'_2 + z_3) - h_3$

8. **reduce $u'$, i.e., $u'' = (f - v'h - v'^2)/u'$**    [5M + 2S]
$u''_2 \leftarrow f_6 - u'_3 - {v'_3}^2 - v'_3 h_3$
$u''_1 \leftarrow -u'_2 - u''_2 u'_3 + f_5 - 2v'_2 v'_3 - v'_3 h_2 - v'_2 h_3$
$u''_0 \leftarrow -u'_1 - u''_2 u'_2 - u''_1 u'_3 + f_4 - 2v'_1 v'_3 - {v'_2}^2 - v'_2 h_2 - v'_3 h_1 - v'_1 h_3$

9. **compute $v'' = -(v' + h) \bmod u_3$**    [3M]
$v''_2 \leftarrow -v'_2 + (v'_3 + h_3)u''_2 - h_2$
$v''_1 \leftarrow -v'_1 + (v'_3 + h_3)u''_1 - h_1$
$v''_0 \leftarrow -v'_0 + (v'_3 + h_3)u''_0 - h_0$

10. **return** $[u'', v'']$    [total complexity: I + 70M + 6S]

If $\mathrm{char}(\mathbb{F}_q)$ is even, $h(x) \in \mathbb{F}_2[x]$, and $f_6 = 0$ then the total complexity reduces to I + 65M + 6S.

### 14.6.2 Doubling in most common case

We now state the formulas to double on general curves. Compared to the addition the formulas depend much more on the equation of the curve. We give the number of operations for arbitrary characteristic including characteristic 2. For the counting we assume for simplicity that $h(x) \in \mathbb{F}_2[x]$ and $f_6 = 0$.

For other values of the $h_i$ some operations should be performed differently. The special case of $h(x) = 1$ will be discussed in more detail in the next section.

**Algorithm 14.53** Doubling on curves of genus $3$ in the general case

INPUT: A divisor class $[u, v]$ with $u = x^3 + u_2x^2 + u_1x + u_0$ and $v = v_2x^2 + v_1x + v_0$.
OUTPUT: The divisor class $[u'', v''] = [2][u, v]$.

1. **compute resultant** $r = \text{Res}(u, \tilde{h})$ **where** $\tilde{h} = h + 2v$ **(Bezout)** $\hspace{1cm}$ [12M + 2S]
   $t_1 \leftarrow u_2\tilde{h}_1, t_2 \leftarrow u_1\tilde{h}_2, t_3 \leftarrow u_1\tilde{h}_0, t_4 \leftarrow u_0\tilde{h}_1, t_5 \leftarrow u_2\tilde{h}_0$ and $t_6 \leftarrow u_0\tilde{h}_2$
   $t_7 \leftarrow (\tilde{h}_0 - h_3u_0)^2, t_8 \leftarrow (\tilde{h}_1 - h_3u_1)^2$ and $t_9 \leftarrow (\tilde{h}_2 - h_3u_2)(t_3 - t_4)$
   $t_{10} \leftarrow (\tilde{h}_2 - h_3u_2)(t_5 - t_6)$ and $t_{11} \leftarrow (\tilde{h}_1 - h_3u_1)(\tilde{h}_0 - h_3u_0)$
   $r \leftarrow (\tilde{h}_0 - h_3u_0 + t_1 - t_2)(t_7 - t_9) + (t_5 - t_6)(t_{10} - 2t_{11}) + t_8(t_3 - t_4)$
   If $r = 0$ perform Cantor's Algorithm 14.7

2. **compute almost inverse** $\text{inv} = r/(h + 2v) \bmod u$ $\hspace{1cm}$ [4M]
   $\text{inv}_2 \leftarrow -(t_1 - t_2 - h_3u_0 + \tilde{h}_0)(\tilde{h}_2 - h_3u_2) + t_8$
   $\text{inv}_1 \leftarrow \text{inv}_2 u_2 + t_{10} - t_{11}$
   $\text{inv}_0 \leftarrow \text{inv}_2 u_1 + u_2(t_{10} - t_{11}) - t_9 + t_7$

3. **compute** $z = ((f - hv - v^2)/u) \bmod u$ $\hspace{1cm}$ [8M + 2S]
   $t_{12} \leftarrow v_2^2, z_3' \leftarrow f_6 - u_2, t_{13} \leftarrow z_3'u_1$ and $z_2' \leftarrow f_5 - h_3v_2 - u_1 - u_2f_6 + u_2^2$
   $z_1' \leftarrow f_4 - h_2v_2 - h_3v_1 - t_{12} - u_0 - t_{13} - z_2'u_2$
   $z_2 \leftarrow f_5 - h_3v_2 - 2u_1 + u_2(u_2 - 2z_3')$ and $z_1 \leftarrow z_1' - t_{13} + u_2u_1 - u_0$
   $z_0 \leftarrow f_3 - h_2v_1 - h_1v_2 - 2v_2v_1 - h_3v_0 + u_0(u_2 - 2z_3') - z_2'u_1 - z_1'u_2$

4. **compute** $s' = (z \, \text{inv}) \bmod u$ **(Karatsuba)** $\hspace{1cm}$ [11M]
   $t_{12} \leftarrow (\text{inv}_1 + \text{inv}_2)(z_1 + z_2)$ and $t_{13} \leftarrow z_1\text{inv}_1$
   $t_{14} \leftarrow (\text{inv}_0 + \text{inv}_2)(z_0 + z_2)$ and $t_{15} \leftarrow z_0\text{inv}_0$
   $t_{16} \leftarrow (\text{inv}_0 + \text{inv}_1)(z_0 + z_1)$ and $t_{17} \leftarrow z_2\text{inv}_2$
   $r_0' \leftarrow t_{15}, r_1' \leftarrow t_{16} - t_{13} - t_{15}$ and $r_2' \leftarrow t_{13} + t_{14} - t_{15} - t_{17}$
   $r_3' \leftarrow t_{12} - t_{13} - t_{17}, r_4' \leftarrow t_{17}$ and $t_{18} \leftarrow u_2r_4' - r_3'$
   $t_{15} \leftarrow u_0t_{18}, t_{16} \leftarrow u_1r_4', s_0' \leftarrow r_0' + t_{15}$ and $s_1' \leftarrow r_1' - (u_1+u_0)(r_4' - t_{18}) + t_{16} - t_{15}$
   $s_2' \leftarrow r_2' - t_{16} + u_2t_{18}$
   If $s_2' = 0$ perform Cantor's Algorithm 14.7

5. **compute** $s = (s'/r)$ **and make** $s$ **monic** $\hspace{1cm}$ [I + 6M + 2S]
   $w_1 \leftarrow (rs_2')^{-1}, w_2 \leftarrow w_1r, w_3 \leftarrow w_1(s_2')^2$, and $w_4 \leftarrow w_2r$ note that $w_4 = r/s_2'$
   $w_5 \leftarrow w_4^2, s_0 \leftarrow w_2s_0'$ and $s_1 \leftarrow w_2s_1'$

6. **compute** $G = su$ $\hspace{1cm}$ [6M]
   $g_0 \leftarrow s_0u_0, g_1 \leftarrow s_1u_0 + s_0u_1$ and $g_2 \leftarrow s_0u_2 + s_1u_1 + u_0$
   $g_3 \leftarrow s_1u_2 + s_0 + u_1$ and $g_4 \leftarrow u_2 + s_1$

7. **compute** $u' = u^{-2}[(G + w_4v)^2 + w_4hG + w_5(hv - f)]$ $\hspace{1cm}$ [6M + 2S]
   $u_3' \leftarrow 2s_1, u_2' \leftarrow s_1^2 + 2s_0 + w_4h_3$
   $u_1' \leftarrow 2s_0s_1 + w_4(2v_2 + h_3s_1 + h_2 - h_3u_2) - w_5$
   $u_0' \leftarrow w_4(2v_1 + h_1 + h_3s_0 - h_3u_1 + 2v_2s_1 + u_2(u_2h_3 - 2v_2 - h_2 - s_1h_3) + h_2s_1)$
   $u_0' \leftarrow u_0' + w_5(-f_6 + 2u_2) + s_0^2$

8. **compute** $v' = -(Gw_3 + h + v) \bmod u'$ $\hspace{1cm}$ [8M]
   $t_1 \leftarrow u_3' - g_4$
   $v_3' \leftarrow -(t_1u_3' - u_2' + g_3)w_3 - h_3$ and $v_2' \leftarrow -(t_1u_2' - u_1' + g_2)w_3 - h_2 - v_2$
   $v_1' \leftarrow -(t_1u_1' - u_0' + g_1)w_3 - h_1 - v_1$ and $v_0' \leftarrow -(t_1u_0' + g_0)w_3 - h_0 - v_0$

9. **reduce** $u'$ **i.e.,** $u'' = (f - v'h - v'^2)/u'$ $\hspace{1cm}$ [5M + 2S]

$$u_2'' \leftarrow f_6 - u_3' - v_3'^2 - v_3' h_3$$
$$u_1'' \leftarrow -u_2' - u_2'' u_3' + f_5 - 2v_2' v_3' - v_3' h_2 - v_2' h_3$$
$$u_0'' \leftarrow -u_1' - u_2'' u_2' - u_1'' u_3' + f_4 - 2v_1' v_3' - v_2'^2 - v_2' h_2 - v_3' h_1 - v_1' h_3$$

10. **compute $v_2 = -(v' + h) \bmod u_2$** [3M]
$$v_2'' \leftarrow -v_2' + (v_3' + h_3) u_2'' - h_2$$
$$v_1'' \leftarrow -v_1' + (v_3' + h_3) u_1'' - h_1$$
$$v_0'' \leftarrow -v_0' + (v_3' + h_3) u_0'' - h_0$$

11. **return** $[u'', v'']$ [total complexity: I + 69M + 10S]

For characteristic 2 fields, the costs drop down to I + 53M + 10S if we assume in addition that $h_i \in \mathbb{F}_2$ and to I + 22M + 7S when $h(x) = 1$. Note that in the latter case the following section gives a much better result.

### 14.6.3 Doubling on genus 3 curves for even characteristic when $h(x) = 1$

For genus 2 curves we gave a complete characterization of all types of curves (cf. Section 14.5). There the fastest doubling occurred for constant $h \in \mathbb{F}_{2^d}^*$ but we did not further investigate these curves as they are supersingular.

For genus 3 the situation is different: constant polynomials $h$ again lead to a minimal number of field operations, but these curves are not supersingular. They have 2-rank zero (cf. Definition 14.12) but so far there is no method known making the computation of discrete logarithms on such curves easier than on arbitrary curves.

Here, we detail the doubling on
$$C : y^2 + y = x^7 + f_5 x^5 + f_4 x^4 + f_3 x^3 + f_2 x^2 + f_1 x + f_0, \ f_i \in \mathbb{F}_{2^d}.$$

**Algorithm 14.54** Doubling on curves of genus 3 with $h(x) = 1$

INPUT: A divisor class $[u, v]$ with $u = x^3 + u_2 x^2 + u_1 x + u_0$ and $v = v_2 x^2 + v_1 x + v_0$.
OUTPUT: The divisor class $[u', v'] = [2][u, v]$.

1. **compute $d = \gcd(u_1, 1) = 1 = s_1 a + s_3 h$**
   $s_3 \leftarrow 1$ and $s_1 \leftarrow 0$

2. **compute $u^2$** [3S]
   $t_1 \leftarrow u_2^2, t_2 \leftarrow u_1^2$ and $t_3 \leftarrow u_0^2$

3. **compute $w = v^2 + f \bmod u'$** [3S]
   $t_4 \leftarrow v_2^2, t_5 \leftarrow v_1^2, t_6 \leftarrow v_0^2, w_5 \leftarrow f_5 + t_1$ and $w_4 \leftarrow f_4 + t_4$
   $w_3 \leftarrow f_3 + t_2, w_2 \leftarrow f_2 + t_5, w_1 \leftarrow f_1 + t_3$ and $w_0 \leftarrow f_0 + t_6$

4. **compute $u' = ((f - hw - w^2)/u^2)$** [3M + 3S]
   $u_4' \leftarrow w_5^2, u_3' \leftarrow 0, u_2' \leftarrow w_4^2 + t_1 u_4', u_1' \leftarrow 1$ and $u_0' \leftarrow w_3 + t_2 u_4' + t_1 u_2'$

5. **compute $u_2 = u''$ made monic, i.e., $u' \leftarrow u'$ made monic** [I + 2M]
   $u_1' \leftarrow u_4'^{-1}, u_2' \leftarrow u_2' u_1'$ and $u_0' \leftarrow u_0' u_1'$

6. **compute $v' = -(w + h) \bmod u'$** [5M]
   $t_1 \leftarrow w_5 u_2', t_2 \leftarrow w_4 u_1', t_3 \leftarrow (w_5 + w_4)(u_2' + u_1')$ and $v_3' \leftarrow w_3 + t_1$
   $v_2' \leftarrow t_3 + t_1 + t_2 + w_2, v_1' \leftarrow t_2 + w_5 u_0' + w_1$ and $v_0' \leftarrow w_4 u_0' + w_0 + 1$

7. **compute $u'' = (f - v'h - v'^2)/u'$** [M + 2S]

$u_2'' \leftarrow v_3'^2, u_1'' \leftarrow u_2' + f_5$ and $u_0'' \leftarrow u_2'u_2'' + f_4 + v_2'^2 + u_1'$

8. **compute $v''$** $:\leftarrow -(v'+h) \bmod u''$  [3M]
   $v_2'' \leftarrow v_2' + v_3'u_2'', v_1'' \leftarrow v_1' + v_3'u_1''$ and $v_0'' \leftarrow v_0' + v_3'u_0'' + 1$

9. **return** $[u'', v'']$  [total complexity: $I + 14M + 11S$]

## 14.7 Other curves and comparison

In principle such explicit formulas can be derived for arbitrarily large genus and also for nonhyperelliptic curves. Picard curves and more generally $C_{3,4}$ curves are nonhyperelliptic curves of genus 3 (cf. Section 4.4.6.a). Starting from a generalization of Cantor's algorithm or a geometrical description of the steps, respectively, [BAEN[+] 2002, FLOY 2004] deal with the arithmetic of Picard curves. These curves are a special class of $C_{3,4}$; the general curves are studied in [BAEN[+] 2004, FLOY[+] 2004].

Also, hyperelliptic curves of genus 4 received some attention and explicit formulas are given for the most frequent cases. For this we refer to [WOL 2004] and the references given therein. His thesis also provides timings for hyperelliptic curves over binary fields for group sizes of use for cryptographic applications. Depending on the processor, genus 2 curves can outperform elliptic curves for the same group size. We remark that the genus 2 curves were chosen of the special Type II (cf. Section 14.5) but more field operations than proposed here were used. Thus, the results have to be taken with a grain of salt.

In [LAST 2005] implementations of binary elliptic curves and curves of genus 2 are reported. Their study considers all different choices detailed above. Curves of Type II lead to faster scalar multiplication than elliptic curves with the same number of precomputations. Scalar multiplication on general curves of Type I is a bit slower than on elliptic curves.

Avanzi [AVA 2004a] gives a thorough comparison in implementing the explicit formulas in prime fields of cryptographically relevant size. His results show that computing scalar multiples in the Jacobian of hyperelliptic curves of genus 2 is only 10% slower than on elliptic curves of the same group size. The chosen curves were general using no special properties. For all field sizes the basic arithmetic was optimized to the same extent.

To conclude, we present the timings from [AVA 2004a] for all coordinate systems presented above and in Chapter 13 for odd characteristic. To compute scalar multiples on the curves signed windowing methods were used (cf. Chapter 9). The parameter $w$ states the width of the window.

§ 14.7 Other curves and comparison 353

**Table 14.13** Comparison of running times, in msec (1 GHz AMD Athlon PC).

| curve | coord. | scalar mult. | Bit length of group order (approximate) | | | | | | | |
|---|---|---|---|---|---|---|---|---|---|---|
| | | | 128 | 144 | 160 | 192 | 224 | 256 | 320 | 512 |
| ec | $\mathcal{A}$ | binary | 1.671 | 2.521 | 3.074 | 5.385 | 8.536 | 12.619 | | |
| | | NAF | 1.488 | 2.252 | 2.701 | 4.809 | 7.596 | 11.315 | | |
| | | $\text{NAF}_w$ | 1.363 | 2.205 | 2.489 | 4.335 | 6.841 | 10.099 | | |
| | | | ($w=4$) | ($w=3$) | ($w=4$) | ($w=4$) | ($w=4$) | ($w=4$) | | |
| | $\mathcal{P}$ | binary | 0.643 | 0.94 | 1.152 | 1.879 | 3.22 | 4.243 | | |
| | | NAF | 0.575 | 0.841 | 1.017 | 1.685 | 2.881 | 3.747 | | |
| | | $\text{NAF}_w$ | 0.551 | 0.808 | 0.982 | 1.591 | 2.711 | 3.523 | | |
| | | | ($w=3$) | ($w=3$) | ($w=3$) | ($w=3$) | ($w=4$) | ($w=4$) | | |
| | $\mathcal{J}$ | binary | 0.584 | 0.856 | 1.05 | 1.702 | 2.912 | 3.876 | | |
| | | NAF | 0.517 | 0.776 | 0.907 | 1.499 | 2.558 | 3.325 | | |
| | | $\text{NAF}_w$ | 0.492 | 0.713 | 0.864 | 1.397 | 2.357 | 3.086 | | |
| | | | ($w=3$) | ($w=3$) | ($w=3$) | ($w=3$) | ($w=3$) | ($w=4$) | | |
| | $\mathcal{J}^c$ | binary | 0.614 | 0.901 | 1.109 | 1.812 | 3.081 | 3.995 | | |
| | | NAF | 0.546 | 0.802 | 0.965 | 1.6 | 2.727 | 3.583 | | |
| | | $\text{NAF}_w$ | 0.517 | 0.756 | 0.922 | 1.499 | 2.527 | 3.275 | | |
| | | | ($w=3$) | ($w=3$) | ($w=3$) | ($w=3$) | ($w=3$) | ($w=3$) | | |
| | $\mathcal{J}^m$ | binary | 0.607 | 0.872 | 1.076 | 1.782 | 3.005 | 3.945 | | |
| | | NAF | 0.512 | 0.748 | 0.906 | 1.515 | 2.592 | 3.35 | | |
| | | $\text{NAF}_w$ | 0.474 | 0.684 | 0.838 | 1.395 | 2.296 | 3.048 | | |
| | | | ($w=3$) | ($w=3$) | ($w=3$) | ($w=3$) | ($w=3$) | ($w=3$) | | |
| hec g=2 | $\mathcal{A}$ | binary | 0.888 | 1.614 | 1.899 | 2.546 | 4.612 | 5.514 | 10.409 | 39.673 |
| | | NAF | 0.797 | 1.449 | 1.706 | 2.265 | 4.139 | 4.952 | 9.298 | 35.430 |
| | | $\text{NAF}_w$ | 0.73 | 1.421 | 1.558 | 2.053 | 3.73 | 4.464 | 8.343 | 31.246 |
| | | | ($w=4$) | ($w=4$) | ($w=4$) | ($w=4$) | ($w=4$) | ($w=4$) | ($w=4$) | ($w=5$) |
| | $\mathcal{P}$ | binary | 0.839 | 1.473 | 1.642 | 2.102 | 3.996 | 4.712 | 8.653 | 30.564 |
| | | NAF | 0.755 | 1.325 | 1.48 | 1.901 | 3.588 | 4.203 | 7.758 | 27.359 |
| | | $\text{NAF}_w$ | 0.703 | 1.211 | 1.352 | 1.742 | 3.256 | 3.842 | 6.998 | 24.451 |
| | | | ($w=4$) | ($w=4$) | ($w=4$) | ($w=4$) | ($w=4$) | ($w=4$) | ($w=4$) | ($w=5$) |
| | $\mathcal{N}$ | binary | 0.844 | 1.395 | 1.564 | 2.038 | 3.777 | 4.413 | 8.265 | 29.11 |
| | | NAF | 0.746 | 1.247 | 1.391 | 1.778 | 3.357 | 4.002 | 7.329 | 25.816 |
| | | $\text{NAF}_w$ | 0.675 | 1.14 | 1.262 | 1.623 | 3.02 | 3.575 | 6.53 | 22.73 |
| | | | ($w=4$) | ($w=4$) | ($w=4$) | ($w=3$) | ($w=4$) | ($w=4$) | ($w=4$) | ($w=5$) |
| hec g=3 | $\mathcal{A}$ | binary | 1.896 | 1.984 | 2.992 | 3.597 | 5.39 | 6.001 | 12.66 | 42.907 |
| | | NAF | 1.64 | 1.744 | 2.538 | 3.085 | 4.82 | 5.39 | 11.24 | 38.326 |
| | | $\text{NAF}_w$ | 1.424 | 1.528 | 2.077 | 2.584 | 4.33 | 4.86 | 9.92 | 34.117 |
| | | | ($w=4$) | ($w=4$) | ($w=5$) | ($w=5$) | ($w=4$) | ($w=5$) | ($w=4$) | ($w=4$) |

# Chapter 15

# Arithmetic of Special Curves

*Christophe Doche and Tanja Lange*

### Contents in Brief

| | | |
|---|---|---|
| **15.1 Koblitz curves** | | 355 |
| Elliptic binary Koblitz curves • Generalized Koblitz curves • Alternative setup | | |
| **15.2 Scalar multiplication using endomorphisms** | | 376 |
| GLV method • Generalizations • Combination of GLV and Koblitz curve strategies • Curves with endomorphisms for identity-based parameters | | |
| **15.3 Trace zero varieties** | | 383 |
| Background on trace zero varieties • Arithmetic in $G$ | | |

In this chapter we present the arithmetic for special choices of curves that offer advantages in the computation of scalar multiples.

The first example is given by Koblitz curves. There the improvement results from a clever use of the Frobenius endomorphism. The use of endomorphisms on special curves over prime fields is the topic of the following section. An approach similar to Koblitz curves but for large characteristic is the key to trace zero varieties, which we present in the last section.

For a curve $C$ defined over $\mathbb{F}_q$, a common strategy in this chapter is to consider $C$ over an extension field $\mathbb{F}_{q^k}$, i.e., $C \cdot \mathbb{F}_{q^k}$, and to study the divisor class group there. For $C = E$, an elliptic curve, we use the fact that $E$ is equal to its divisor class group and write $E(\mathbb{F}_{q^k})$ to denote the set of points defined over $\mathbb{F}_{q^k}$. For a curve $C$ of higher genus, we put $\operatorname{Pic}_C^0(\mathbb{F}_{q^d})$ as a shorthand for $\operatorname{Pic}_{C \cdot \mathbb{F}_{q^k}}^0$ to abbreviate notation and to highlight the field in which we are working. Finally we assume an embedding $\operatorname{Pic}_C^0(\mathbb{F}_q) \subset \operatorname{Pic}_C^0(\mathbb{F}_{q^k})$.

## 15.1 Koblitz curves

The notion of Koblitz curves is not clearly defined. In the scope of this book we associate a broad meaning with this name and use it to denote subfield curves for which the Frobenius endomorphism is used in the computation of scalar multiples.

We introduce Koblitz curves first in the special case initially proposed by Koblitz and then continue with the general case of arbitrary genus and arbitrary characteristic. Finally we present an alternative approach that is better suited for applications.

### 15.1.1 Elliptic binary Koblitz curves

We refer to [KOB 1992] and [SOL 1999b, SOL 2000] for this part and treat in detail elliptic curves defined over $\mathbb{F}_2$ and considered over the extension field $\mathbb{F}_{2^d}$.

The first attempt to use the Frobenius endomorphism in the computation of scalar multiples was made by Menezes and Vanstone [MEVA 1990] using the curve

$$E : y^2 + y = x^3.$$

In this case, the characteristic polynomial of the Frobenius endomorphism denoted by $\phi_2$ (cf. Example 4.87 and Section 13.1.8), which sends $P_\infty$ to itself and $(x_1, y_1)$ to $(x_1^2, y_1^2)$, is

$$\chi_E(T) = T^2 + 2.$$

Thus doubling is replaced by a twofold application of the Frobenius endomorphism and taking the negative as for all points $P \in E(\mathbb{F}_{2^d})$, we have $\phi_2^2(P) = -[2]P$. Since the computation of the Frobenius map is almost free in normal basis representation and requires at most three squarings using a polynomial basis and projective coordinates, this idea led to very efficient implementations. However, the curve $E$ is supersingular and should not be used in applications if one only needs a DL system, (cf. Section 22.2).

As "the next best thing," Koblitz [KOB 1992] suggested using the remaining two nonsupersingular curves defined over $\mathbb{F}_2$.

**Definition 15.1** In the context of an elliptic curve defined over $\mathbb{F}_2$, a *Koblitz curve* is given by the equation

$$E_{a_2} : y^2 + xy = x^3 + a_2 x^2 + 1, \text{ with } a_2 = 0 \text{ or } 1. \tag{15.1}$$

These curves are sometimes referred to as *anomalous binary curves*, ABC for short.

In this case, the characteristic polynomial of the Frobenius endomorphism is

$$\chi_{a_2}(T) = T^2 - \mu T + 2 \tag{15.2}$$

where $\mu = (-1)^{1-a_2}$. It follows that doublings can be replaced by computations involving the Frobenius endomorphism as

$$[2]P = [\mu]\phi_2(P) \ominus \phi_2^2(P). \tag{15.3}$$

The following shows that scalar multiplications by powers of 2 can be easily obtained using the Frobenius endomorphism. For instance when $a_2 = 1$, we have

$$[2]P = \phi_2(P) \ominus \phi_2^2(P), \qquad [4]P = -\phi_2^2(P) \ominus \phi_2^3(P),$$
$$[8]P = -\phi_2^3(P) \oplus \phi_2^5(P), \qquad [16]P = \phi_2^4(P) \ominus \phi_2^8(P),$$

as one can easily check from using (15.2). This idea gives an efficient general scalar multiplication when combined with the $2^k$-ary or the sliding window method; see Section 9.1. But one can use it even more efficiently in the computation of arbitrary scalar multiples, as we shall see in the sequel. Let $\tau$ be a complex root of $\chi_{a_2}(T)$. As the Frobenius endomorphism operates on $E_{a_2}$, the curve has complex multiplication by the ring $\mathbb{Z}[\tau]$.

### 15.1.1.a Properties of the ring $\mathbb{Z}[\tau]$

First, fix $\tau = \frac{\mu+\sqrt{-7}}{2}$ as a root of $\chi_{a_2}$. The other root is the complex conjugate of $\tau$, namely $\overline{\tau} = \frac{\mu-\sqrt{-7}}{2}$. Let us review some properties of the ring $\mathbb{Z}[\tau]$.

The field $\mathbb{Q}(\tau)$ is an imaginary quadratic field with ring of integers $\mathbb{Z}[\tau]$. Every element $\eta \in \mathbb{Z}[\tau]$ can be written in the form $\eta = n_0 + n_1\tau$ with $n_0, n_1 \in \mathbb{Z}$. To every element $\eta = n_0 + n_1\tau \in \mathbb{Z}[\tau]$ one associates the norm of $\eta$ (cf. Proposition 2.77), given by

$$N(\eta) = (n_0 + n_1\tau)(n_0 + n_1\overline{\tau}) = n_0^2 + \mu n_0 n_1 + 2n_1^2.$$

As a consequence, $N(\tau) = 2$. As $\phi_2^d$ operates on $E_{a_2}(\mathbb{F}_{2^d})$ as identity, one gets

$$N(1 - \tau^d) = |E_{a_2}(\mathbb{F}_{2^d})|.$$

In classical number theory one associates *Lucas sequences* to a quadratic polynomial using the equation to define a recurrence relation. In our case, (15.2) gives rise to

$$L_{k+1} = \mu L_k - 2L_{k-1}, \text{ for } k \geqslant 1. \tag{15.4}$$

The first and second Lucas sequence differ by the initialization of the first two sequence elements and use the same recurrence relation (15.4). Define $(U_k)_{k \geqslant 0}$ by

$$U_0 = 0, U_1 = 1, \tag{15.5}$$

and $(V_k)_{k \geqslant 0}$ by

$$V_0 = 2, V_1 = \mu. \tag{15.6}$$

By using (15.4) one immediately sees

$$\tau^k = -2U_{k-1} + U_k \tau \tag{15.7}$$

and

$$\tau^k + \overline{\tau}^k = -2V_{k-1} + V_k \tau. \tag{15.8}$$

This second sequence gives a recurrence formula to compute the cardinality of $E_{a_2}(\mathbb{F}_{2^d})$. Indeed, we have

$$\begin{aligned} |E_{a_2}(\mathbb{F}_{2^d})| &= (1 - \tau^d)(1 - \overline{\tau}^d) \\ &= 2^d + 1 - (\tau^d + \overline{\tau}^d) = 2^d + 1 - V_d. \end{aligned} \tag{15.9}$$

As the number of points over the ground field always divides the group order over extension fields, we have an inevitable factor $c$ of the group order given by

$$|E_{a_2}(\mathbb{F}_{2^d})| = cN \text{ with } \begin{cases} c = 4, & \text{if } a_2 = 0 \\ c = 2, & \text{if } a_2 = 1. \end{cases} \tag{15.10}$$

Here, $c = |E_{a_2}(\mathbb{F}_2)| = N(\tau - 1)$ and $N = N(\delta)$ where $\delta = \frac{\tau^d - 1}{\tau - 1}$.

**Example 15.2** Take

$$E_1 : y^2 + xy = x^3 + x^2 + 1$$

over $\mathbb{F}_{2^{11}}$. Then we have $E_1(\mathbb{F}_{2^{11}}) = 2 \times 991$ where 991 is prime.

Further developments would be irrelevant if we were not able to find extension fields $\mathbb{F}_{2^d}$ with a large prime order subgroup of $E(\mathbb{F}_{2^d})$. Fortunately, the integer $N$ in (15.10) is prime for many extension degrees and in particular for the following ones of cryptographic interest

| $a_2$ | Degree $d$ |
|---|---|
| 0 | 233, 239, 277, 283, 349, 409, 571 |
| 1 | 163, 283, 311, 331, 347, 359 |

For more information on point counting we refer to Chapter 17.

### 15.1.1.b  Computation of $\tau$-adic expansions

We have seen how to use (15.3) to replace computations of $[2^k]P$ by operations involving the Frobenius endomorphism. In the endomorphism ring of $E_{a_2}$ we have the relation $2 = \mu\tau - \tau^2$ and we can try to find a similar expansion for every integer.

**Definition 15.3** In analogy to binary expansions, we define the $\tau$-adic expansion of $\eta \in \mathbb{Z}[\tau]$ as

$$\eta = \sum_{i=0}^{l-1} r_i \tau^i, \qquad (15.11)$$

where $r_i \in \{0, 1\}$, for all $i$. Such an expansion is denoted by $(r_{l-1} \ldots r_0)_\tau$.

The key observation to compute the expansion (15.11) of $\eta = n_0 + n_1\tau \in \mathbb{Z}[\tau]$ is that the coefficients can be obtained from the least significant one by repeatedly dividing $n_0$ with remainder by 2 as in the binary expansion and replacing 2 by $\mu\tau - \tau^2$. The remainders constitute the sequence of coefficients and the norm decreases. This shows that every $\eta \in \mathbb{Z}[\tau]$ has a $\tau$-adic expansion.

**Example 15.4** For instance, put $\mu = 1$ and let us compute the $\tau$-adic expansion of 7. We have the following equalities

$$\begin{aligned} 7 &= 1 + 2 \times 3 \\ &= 1 + \tau(3 - 3\tau) \\ &= 1 + \tau + \tau^2(-2 - \tau) \\ &= 1 + \tau + \tau^3(-2 + \tau) \\ &= 1 + \tau + \tau^4(\tau) \\ &= 1 + \tau + \tau^5. \end{aligned}$$

Once it is determined, the $\tau$-adic expansion of an integer $n$ gives a way to compute the scalar multiplication by $n$ using only the Frobenius endomorphism and additions. For example, if $P$ is a point of $E_1(\mathbb{F}_{2^d})$, we have

$$[7]P = P \oplus \phi_2(P) \oplus \phi_2^5(P).$$

The number of nonzero coefficients in the $\tau$-adic expansion of $n$ determines the number of additions to perform in order to get $[n]P$ and thus rules the complexity of the computation. To obtain a sparser representation, Koblitz [KOB 1992] suggests using a signed $\tau$-adic expansion. Solinas [SOL 1997] generalizes the notion of binary non-adjacent forms NAFs.

**Definition 15.5** An element $\eta = n_0 + n_1\tau \in \mathbb{Z}[\tau]$ is written in $\tau$-adic non-adjacent form, $\tau$NAF for short, if

$$\eta = \sum_{i=0}^{l-1} r_i \tau^i, \qquad (15.12)$$

## § 15.1 Koblitz curves

with the extra conditions $r_i \in \{0, \pm 1\}$ and $r_i r_{i+1} = 0$ for all $i$. Such an expansion is denoted by $(r_{l-1} \ldots r_0)_{\tau\text{NAF}}$.

In [SOL 1999b], an algorithm to compute the $\tau$NAF of every $\eta \in \mathbb{Z}[\tau]$ is proposed. In addition, it is shown that the $\tau$NAF expansion is unique and Avanzi et al. show in [AVHE$^+$ 2004] that its number of nonzero coefficients is minimal among all the representations with coefficients in $\{0, \pm 1\}$.

---

**Algorithm 15.6** $\tau$NAF representation

INPUT: An element $\eta = n_0 + n_1\tau \in \mathbb{Z}[\tau]$.
OUTPUT: The $\tau$-adic expansion $(r_{l-1} \ldots r_0)_{\tau\text{NAF}}$ of $\eta$ in non-adjacent form.

1. $S \leftarrow ()$
2. **while** $|n_0| + |n_1| \neq 0$ **do**
3.     **if** $n_0 \equiv 1 \pmod{2}$ **then**
4.         $r \leftarrow 2 - ((n_0 - 2n_1) \bmod 4)$
5.         $n_0 \leftarrow n_0 - r$
6.     **else** $r \leftarrow 0$
7.     $S \leftarrow r \| S$                                                          [$r$ prepended to $S$]
8.     $(n_0, n_1) \leftarrow (n_1 + \mu n_0/2, -n_0/2)$
9. **return** $S$

---

**Remark 15.7** The length $l$ of the $\tau$-adic NAF expansion of $n \in \mathbb{Z}$ obtained by Algorithm 15.6 is approximately equal to $2 \lg n$ while its density is $1/3$ as with the ordinary NAF, cf. Section 9.1.4. In order to obtain an expansion of size $\lg n$, we shall introduce the notion of reduction (see Section 15.1.1.c) and this will allow us to get expansions of the same density and length $\lg n$.

**Example 15.8** Take $\mu = 1$ and let us compute the $\tau$NAF representation of $n = 409$ in $\mathbb{Z}[\tau]$. Using Algorithm 15.6, we obtain

$$409 = (\bar{1}00\bar{1}000010\bar{1}01001001)_{\tau\text{NAF}}$$

whereas it is easily checked that the $\tau$-adic expansion of 409 is $(11000111101001001)_\tau$. The length is the same in both cases but the density of the $\tau$NAF expansion is less.

From the $\tau$-adic or the $\tau$NAF expansion, it is possible to derive a left-to-right scalar multiplication algorithm similar to the double and add method, where every doubling is replaced by the Frobenius action. In Chapter 9, we have seen that some precomputations can reduce significantly the number of additions involved in a scalar multiplication, and the same considerations hold here as well.

### 15.1.1.c Reducing the length

Let $P$ be a point in $E_{a_2}(\mathbb{F}_{2^d})$. Since $\phi_2^d$ is the identity map in the finite field $\mathbb{F}_{2^d}$, we have, as pointed out in [MEST 1993], that

$$[n]P = [\eta]P \quad \text{whenever} \quad n \equiv \eta \pmod{\tau^d - 1}.$$

Now recall that $|E_{a_2}(\mathbb{F}_{2^d})| = cN$ with $c = 2$ or $4$. When there exists a point $Q$ such that $P = [c]Q$, then it is even possible to reduce $n$ modulo $\delta = (\tau^d - 1)/(\tau - 1)$ to compute $[n]P$. Note that this condition means no restriction for cryptographic applications since the base point $P$ is usually chosen to have prime order. Furthermore, for an arbitrary $P$ this property is also very easy to check for. Indeed, if $P = (x_1, y_1)$ there exists $Q \in E_{a_2}(\mathbb{F}_{2^d})$ such that $P = [c]Q$ if and only if

- $\mathrm{Tr}(x_1) = \mathrm{Tr}(a_2)$, when $c = 2$
- $\mathrm{Tr}(x_1) = 0$ and $\mathrm{Tr}(y_1) = \mathrm{Tr}(\lambda x_1)$ where $\lambda$ satisfies $\lambda^2 + \lambda = x_1$, when $c = 4$.

Taking such a $Q$, we see that

$$\begin{aligned}
[\delta]P &= [\delta][c]Q \\
&= [\delta][\tau - 1][\bar{\tau} - 1]Q \\
&= [\bar{\tau} - 1]([\tau^d - 1]Q) \\
&= P_\infty,
\end{aligned}$$

where the second equality used $\mathrm{N}(\tau - 1) = c$. As a consequence, if $n \equiv \rho \pmod{\delta}$ we have $n = \kappa\delta + \rho$ and

$$\begin{aligned}
[n]P &= [\kappa][\delta]P \oplus [\rho]P \\
&= [\rho]P.
\end{aligned}$$

This reduction is of great interest, since the length of a $\tau$-adic expansion of $\rho$, computed with Algorithm 15.6 or 15.17, will be shown to be approximately half the one of $n$ and is now of the same size as the binary expansion of $n$.

To actually compute the reduction, we first need to determine $\delta$. For that, we use the Lucas sequence $(U_k)_{k \geqslant 0}$ as defined in (15.5) and get $\tau^d = -2U_{d-1} + U_d\tau$. Then

$$\begin{aligned}
\delta &= (\tau^d - 1)/(\tau - 1) = (-2U_{d-1} + U_d\tau - 1)(\bar{\tau} - 1)/c \quad (15.13) \\
&= \bigl((2 - 2\mu)U_{d-1} + 2U_d - \mu + 1 + (2U_{d-1} - U_d + 1)\tau\bigr)/c \\
&=: \delta_0 + \delta_1\tau
\end{aligned}$$

using $c = (\tau - 1)(\bar{\tau} - 1)$ and $\mu = \tau + \bar{\tau}$.

To be able to compute a division with remainder as required in $n = \kappa\delta + \rho$ we need a *rounding* notion. For $\lambda \in \mathbb{Q}$, the rounding is ambiguous for half integers. In the following, we use two different rounding notions for $\lambda \in \mathbb{Q}$, namely

$$\lfloor \lambda \rceil = \lfloor \lambda + 1/2 \rfloor \quad \text{and} \quad \lceil \lambda \rfloor = \begin{cases} \lceil \lambda - 1/2 \rceil & \text{if } \lambda > 0 \\ \lfloor \lambda + 1/2 \rfloor & \text{else.} \end{cases}$$

The second definition ensures that in the ambiguous cases the integer with least absolute value is chosen.

Similarly, for $\lambda \in \mathbb{Q}(\tau)$, we need to find a closest neighbor of $\lambda$ in $\mathbb{Z}[\tau]$. As $\mathbb{Q}(\tau) \subset \mathbb{C}$ and $\mathbb{Z}[\tau]$ forms a two-dimensional lattice, we use the complex absolute value as it captures the distance in the complex plane. The complex absolute value equals the positive square root of the norm. The element $q_0 + q_1\tau \in \mathbb{Z}[\tau]$ is a closest lattice element of $\lambda$ if

$$\mathrm{N}(\lambda - q_0 - q_1\tau) \leqslant \mathrm{N}(\lambda - \alpha) \quad \text{for all } \alpha \in \mathbb{Z}[\tau].$$

Algorithm 15.9 provided in [MEST 1993, SOL 1999b] computes such an element denoted by $\lfloor \lambda \rceil_\tau$.

## Algorithm 15.9 Rounding-off of an element of $\mathbb{Q}[\tau]$

INPUT: Two rational numbers $\lambda_0$ and $\lambda_1$ specifying $\lambda = \lambda_0 + \lambda_1 \tau \in \mathbb{Q}(\tau)$.
OUTPUT: Two integers $q_0$ and $q_1$ such that $q_0 + q_1 \tau = \lfloor \lambda \rceil_\tau \in \mathbb{Z}[\tau]$.

1. $f_0 \leftarrow \lfloor \lambda_0 \rceil$ and $f_1 \leftarrow \lfloor \lambda_1 \rceil$
2. $\eta_0 \leftarrow \lambda_0 - f_0$ and $\eta_1 \leftarrow \lambda_1 - f_1$
3. $h_0 \leftarrow 0$ and $h_1 \leftarrow 0$
4. $\eta \leftarrow 2\eta_0 + \mu\eta_1$
5. **if** $\eta \geqslant 1$ **then**
6.      **if** $\eta_0 - 3\mu\eta_1 < -1$ **then** $h_1 \leftarrow \mu$ **else** $h_0 \leftarrow 1$
7. **else**
8.      **if** $\eta_0 + 4\mu\eta_1 \geqslant 2$ **then** $h_1 \leftarrow \mu$
9. **if** $\eta < -1$ **then**
10.     **if** $\eta_0 - 3\mu\eta_1 \geqslant 1$ **then** $h_1 \leftarrow -\mu$ **else** $h_0 \leftarrow -1$
11. **else**
12.     **if** $\eta_0 + 4\mu\eta_1 < -2$ **then** $h_1 \leftarrow -\mu$
13. $q_0 \leftarrow f_0 + h_0$ and $q_1 \leftarrow f_1 + h_1$
14. **return** $(q_0, q_1)$

**Example 15.10** Let $\lambda = 1.6 + 2.4\tau$ then $\lfloor \lambda \rceil_\tau = 1 + 2\tau$ and not $2 + 2\tau$ as we might expect. We can check that $N(\lambda - (1 + 2\tau)) = 0.44$ whereas $N(\lambda - (2 + 2\tau)) = 0.64$.

Building on this notion, one can now define a division with remainder in $\mathbb{Z}[\tau]$. The idea behind the following algorithm is to first compute $\eta/\delta \in \mathbb{Q}(\tau)$, which is done via $\eta/\delta = \eta\bar{\delta}/N(\delta) = ((\delta_0 + \mu\delta_1)n_0 + 2d_1n_2 + (d_0n_1 - \delta_1n_0)\tau)/N(\delta)$. After rounding the result, one determines the remainder as
$$\eta - \lfloor \eta/\delta \rceil_\tau \delta,$$
where $\tau^2$ is replaced by $\mu\tau - 2$.

## Algorithm 15.11 Division with remainder in $\mathbb{Z}[\tau]$

INPUT: Two elements $\eta = n_0 + n_1\tau$ and $\delta = d_0 + d_1\tau$ in $\mathbb{Z}[\tau]$.
OUTPUT: Two elements $\kappa = q_0 + q_1\tau$ and $\rho = r_0 + r_1\tau$ in $\mathbb{Z}[\tau]$ with $\eta = \kappa\delta + \rho$, $N(\rho) < N(\delta)$.

1. $g_0 \leftarrow n_0 d_0 + \mu n_0 d_1 + 2 n_1 d_1$
2. $g_1 \leftarrow n_1 d_0 - n_0 d_1$
3. $N \leftarrow N(\delta) = d_0^2 + \mu d_0 d_1 + 2 d_1^2$
4. $q_0 + q_1 \tau \leftarrow \left\lfloor \dfrac{g_0}{N} + \dfrac{g_1}{N}\tau \right\rceil_\tau$      [use Algorithm 15.9]
5. $r_0 \leftarrow n_0 - d_0 q_0 + 2 d_1 q_1$
6. $r_1 \leftarrow n_1 - d_1 q_0 - d_0 q_1 - \mu d_1 q_1$
7. $\kappa \leftarrow q_0 + q_1 \tau$ and $\rho \leftarrow r_0 + r_1 \tau$
8. **return** $\kappa$ and $\rho$

**Remark 15.12** The remainder $\rho$ given by Algorithm 15.11 does not only satisfy $\mathrm{N}(\rho) < \mathrm{N}(\delta)$ but due to the lattice structure one even has $\mathrm{N}(\rho) \leqslant \frac{4}{7}\mathrm{N}(\delta)$, cf. [MEST 1993].

This algorithm can be used to compute reductions of integers $n$ modulo

$$\delta = \delta_0 + \delta_1 \tau = (\tau^d - 1)/(\tau - 1).$$

---

**Algorithm 15.13** Reduction of $n$ modulo $\delta$

INPUT: An integer $n \in [1, N-1]$ where $N = \mathrm{N}(\delta)$ and $\delta_0, \delta_1$ as in (15.13).
OUTPUT: The element $\rho = r_0 + r_1\tau \equiv n \pmod{\delta}$.

1. $d_0 \leftarrow \delta_0 + \mu\delta_1$
2. $\lambda_0 \leftarrow d_0 n/N$ and $\lambda_1 \leftarrow -\delta_1 n/N$
3. $q_0 + q_1\tau \leftarrow \lfloor \lambda_0 + \lambda_1\tau \rceil_\tau$      [use Algorithm 15.9]
4. $r_0 \leftarrow n - \delta_0 q_0 + 2\delta_1 q_1$
5. $r_1 \leftarrow -\delta_1 q_0 - d_0 q_1$
6. **return** $r_0 + r_1\tau$

---

**Remarks 15.14**

(i) The length of the $\tau$-adic expansion of $\rho$ is at most $d + a_2$ and grows with $\mathrm{N}(\rho)$.

(ii) To avoid the multiprecision division by $N$ in Line 2, a *partial reduction* algorithm has been proposed [SOL 1999b]. Let $C$ be a constant greater than 1, $K = (d+5)/2 + C$, $n' = \lfloor n/(2^{d-K-2+a_2}) \rfloor$ and $V_d$ as in (15.6). Then perform the following operations to obtain an approximation $\lambda'_i$ of $\lambda_i$, for $i = 0$ and $1$.

2'.    $g'_i \leftarrow s_i n'$, $h'_i \leftarrow \lfloor g'_i/2^d \rfloor$, $j'_i \leftarrow V_d h'_i$ and $\lambda'_i \leftarrow \lfloor (g'_i + j'_i)/2^{K-C} + 1/2 \rfloor/2^C$

In any case, $\rho' = r'_0 + r'_1\tau$ obtained in that way is equivalent to $n$ modulo $\delta$. However, $\rho'$ can be different from $\rho$ computed exactly. This occurs with probability less than $1/2^{C-5}$ and the length of the $\tau$-adic expansion of $\rho'$ is always at most $d + a_2 + 3$.

(iii) After the reduced element $\rho$ has been computed, we can compute its $\tau$NAF. To obtain sparser representations, windowing methods can be used (cf. Section 15.1.1.d).

**Example 15.15** Take
$$E_1 : y^2 + xy = x^3 + x^2 + 1$$

over $\mathbb{F}_{2^{11}}$, realized as $\mathbb{F}_2[\theta]$ with $\theta$ such that $\theta^{11} + \theta^2 + 1 = 0$. We know from Example 15.2 that $E_1(\mathbb{F}_{2^{11}}) = 2 \times 991$. Let us take $P = (\texttt{0x34A}, \texttt{0x69B})$ a random point in $E_1(\mathbb{F}_{2^{11}})$, expressed in hexadecimal notation, and let us compute $[409]P$. We remark that $\mathrm{Tr}(\texttt{0x34A}) = 1$. So there is $Q$ such that $[2]Q = P$ and we can reduce 409 modulo $\delta$, where $\delta = (\tau^{11} - 1)/(\tau - 1)$. Algorithm 15.13 shows that $409 \equiv 13 - 9\tau \pmod{\delta}$ and Algorithm 15.6 gives

$$13 - 9\tau = (\bar{1}010100\bar{1})_{\tau\mathrm{NAF}}$$

having only 8 digits instead of 17.

### 15.1.1.d  Windowing methods

It is possible to generalize windowing methods to $\tau$-adic windowing methods. We present a generalization of the notion of width-$w$ NAF expansion, cf. Definition 9.19.

**Definition 15.16** Let $w$ be a parameter greater than 1. Then every element $\eta \in \mathbb{Z}[\tau]$ can be written as

$$\eta = \sum_{i=0}^{l-1} r_i \tau^i$$

where

- each $r_i$ is zero or $r_i = \pm \alpha_u$ where $\alpha_u \equiv u \pmod{\tau^w}$ for some odd $u \in [1, 2^{w-1} - 1]$
- $r_{l-1} \neq 0$
- among any $w$ consecutive coefficients, at most one is nonzero.

Such a representation is called a *width-$w$ $\tau$-adic expansion in non-adjacent form*, $\tau\mathrm{NAF}_w$ for short, and is denoted by $(r_{l-1} \ldots r_0)_{\tau\mathrm{NAF}_w}$.

Mimicking Algorithm 9.20, we derive a method to compute the $\tau\mathrm{NAF}_w$, which we shall describe now. More explanations can be found in [SOL 2000].

Let $(h_k)_{k \geqslant 1}$ be the sequence of integers defined by

$$h_k = 2U_{k-1}U_k^{-1} \bmod 2^k, \quad \text{for all } k \geqslant 1,$$

where $(U_k)_{k \geqslant 0}$ is the Lucas sequence given by (15.5). We can check that

$$h_k^2 - \mu h_k + 2 \equiv 0 \pmod{2^k}, \quad \text{for all } k \geqslant 1.$$

The next step is to precompute the elements $\alpha_u \in \mathbb{Z}[\tau]$ such that $\alpha_u \equiv u \pmod{\tau^w}$ for $u$ odd in $[1, 2^{w-1} - 1]$. With Algorithm 15.9, we obtain the remainder $\beta_u + \gamma_u \tau$ of $u$ divided by $\tau^w$ expressed as a linear term in $\tau$ using (15.7). Then, when it is possible, a simpler expression for $\alpha_u$ is computed using, for instance, the $\tau\mathrm{NAF}$ or previous precomputations. As a result, at most two additions are needed to determine $\alpha_u P$ in each case.

| $w$ | $u$ | $\beta_u + \gamma_u \tau$ | $\alpha_u$ |
|---|---|---|---|
| 3 | 3 | $1 - \mu\tau$ | — |
| 4 | 3 | $-3 + \mu\tau$ | $-1 + \tau^2$ |
|   | 5 | $-1 + \mu\tau$ | — |
|   | 7 | $1 + \mu\tau$ | — |
| 5 | 3 | $-3 + \mu\tau$ | $-1 + \tau^2$ |
|   | 5 | $-1 + \mu\tau$ | $1 + \tau^2$ |
|   | 7 | $1 + \mu\tau$ | — |
|   | 9 | $-3 + 2\mu\tau$ | $1 - \mu\tau^3 \alpha_5$ |
|   | 11 | $-1 + 2\mu\tau$ | $\alpha_5 + \mu\tau$ |
|   | 13 | $1 + 2\mu\tau$ | $\alpha_7 + \mu\tau$ |
|   | 15 | $1 - 3\mu\tau$ | $-\alpha_{11} - \mu\tau$ |

Now we can give the algorithm to compute the $\tau\mathrm{NAF}_w$. It was first proposed in [SOL 1999b] but the given routine fails in some situations. We present here the corrected version [SOL 2000].

**Algorithm 15.17** $\tau\text{NAF}_w$ representation

INPUT: An element $\eta = n_0 + n_1\tau \in \mathbb{Z}[\tau]$, a parameter $w > 1$ and $h_w, \beta_u, \gamma_u$ and $\alpha_u$ as above.
OUTPUT: The width-$w$ $\tau$-adic expansion $(r_{l-1}\ldots r_0)_{\tau\text{NAF}_w}$ of $\eta$ in non-adjacent form.

1. $S \leftarrow ()$
2. **while** $|n_0| + |n_1| \neq 0$ **do**
3.     **if** $n_0$ is odd **then**
4.         $u \leftarrow (n_0 + n_1 h_w) \text{ mods } 2^w$         [see Remark (i)]
5.         **if** $u > 0$ **then** $\xi \leftarrow 1$
6.         **else** $\xi \leftarrow -1$ and $u \leftarrow -u$
7.         $n_0 \leftarrow n_0 - \xi\beta_u$, $n_1 \leftarrow n_1 - \xi\gamma_u$ and $r \leftarrow \xi\alpha_u$
8.     **else** $r \leftarrow 0$
9.     $S \leftarrow r \parallel S$
10.     $(n_0, n_1) \leftarrow (n_1 + \mu n_0/2, -n_0/2)$
11. **return** $S$

**Remarks 15.18**

(i) The integer $(n_0 + n_1 h_w) \text{ mods } 2^w$ is the unique integer in $[-2^{w-1}, 2^{w-1}]$ which is congruent to $n_0 + n_1 h_w$ modulo $2^w$.

(ii) When $w = 2$ the width-$w$ $\tau$-adic NAF coincides with the classical $\tau$-adic NAF, in other words $\tau\text{NAF}_2 = \tau\text{NAF}$.

(iii) The length $l$ of the $\tau\text{NAF}_w$ expansion of $\eta = n_0 + n_1\tau$ is approximately equal to the binary length of the norm of $\eta$. If $\eta = n \in \mathbb{Z}$ then $l \approx 2\lg n$.

(iv) Since $P_{\alpha_1}$ simply equals $P$, we need to precompute only $2^{w-2} - 1$ points.

(v) The average density of the $\tau\text{NAF}_w$ expansion of $\eta$ is equal to $1/(w+1)$. As for the ordinary windowing methods, the expansions do not get longer. Hence, on average $2^{w-2} - 1 + \frac{d}{w+1}$ elliptic curve additions are sufficient to compute the scalar multiplication $[n]P$, including the cost of the precomputations and assuming a reduced representation of $n$ as input of Algorithm 15.17.

**Example 15.19** We continue with the setting of Example 15.15, where we have already obtained that $[409]P = [13]P \ominus [9]\phi_2(P)$. Now it is possible to start with the $\tau\text{NAF}_4$ of $13 - 9\tau$ instead, and Algorithm 15.17 returns

$$13 - 9\tau = (\alpha_5\ 000\ \overline{\alpha}_7\ 000\ \alpha_7)_{\tau\text{NAF}_w}.$$

For instance, using this last expansion, we see that

$$[409]P = \phi_2^8(P_{\alpha_5}) \ominus \phi_2^4(P_{\alpha_7}) \oplus P_{\alpha_7},$$

where

$$\begin{aligned} P_{\alpha_5} &= [\alpha_5]P = \phi_2(P) \ominus P = (0\text{x}256, 0\text{x}61\text{B}) \\ P_{\alpha_7} &= [\alpha_7]P = \phi_2(P) \oplus P = (0\text{x}2\text{F}2, 0\text{x}11\text{A}) \end{aligned}$$

## § 15.1 Koblitz curves

have been precomputed, following the table given on page 363. We get $[409]P = (\text{0x606}, \text{0x55A})$ with only two extra additions.

To obtain expansions of lower density, it is possible to mix point halving (cf. Section 13.3.5) and $\tau$-adic NAF recoding to form a double expansion (cf. Section 9.1.5) with joint weight $1/4$ [AVCI+ 2004, AVHE+ 2004], which allows faster scalar multiplication if a normal basis is chosen and some precomputations can be made.

### 15.1.1.e Computation of the $\tau$-adic joint sparse form

By analogy with the joint sparse form (see Section 9.1.5) a $\tau$-adic joint sparse form is developed in [CILA+ 2003]. It can be applied to Koblitz curves for faster signature verification, but also for faster scalar multiplication if one long-term precomputation, namely that of $\phi_2^{\lfloor d/2 \rfloor}$, is possible.

**Definition 15.20** Let

$$\eta_0 = \sum_{j=0}^{l-1} n_{0,j} \tau^j \quad \text{and} \quad \eta_1 = \sum_{j=0}^{l-1} n_{1,j} \tau^j$$

be two elements of $\mathbb{Z}[\tau]$ with $n_{i,j} \in \{0, \pm 1\}$. The $\tau$-*adic joint sparse form*, $\tau$JSF for short, of $\eta_0$ and $\eta_1$ is a signed representation of the form

$$\begin{pmatrix} \eta_0 \\ \eta_1 \end{pmatrix} = \begin{pmatrix} r_{0,l+2} \cdots r_{0,0} \\ r_{1,l+2} \cdots r_{1,0} \end{pmatrix}_{\tau\text{JSF}}$$

where $r_{i,j} \in \{0, \pm 1\}$, and such that

- of any three consecutive positions, at least one is a zero column, that is for all $i$ and all positive $j$ one has $r_{i,j+k} = r_{1-i,j+k} = 0$, for at least one $k$ in $\{0, \pm 1\}$
- it is never the case that $r_{i,j} r_{i,j+1} = \mu$
- if $r_{i,j+1} r_{i,j} \neq 0$ then one has $r_{1-i,j+1} = \pm 1$ and $r_{1-i,j} = 0$.

Next, we give an algorithm to actually compute such a form.

---

**Algorithm 15.21** Recoding in $\tau$-adic joint sparse form

INPUT: Two $\tau$-adic expansions $\eta_i = \sum_{j=0}^{l-1} n_{i,j} \tau^j$, with $n_{i,j} \in \{0, \pm 1\}$, for $i = 0$ and 1.
OUTPUT: The $\tau$-adic joint sparse form of $\eta_0$ and $\eta_1$.

---

1. $j \leftarrow 0$, $S_0 = ()$ and $S_1 = ()$
2. **for** $i = 0$ **to** 1 **do**
3.     $d_{i,0} \leftarrow 0$ and $d_{i,1} \leftarrow 0$
4.     $a_i \leftarrow n_{i,0}$, $b_i \leftarrow n_{i,1}$ and $c_i \leftarrow n_{i,2}$
5. **while** $l - j > 0$ **or** $|d_{0,0}| + |d_{0,1}| + |d_{1,0}| + |d_{1,1}| > 0$ **do**
6.     **for** $i = 0$ **to** 1 **do**
7.         **if** $d_{i,0} \equiv a_i \pmod{2}$ **then** $r_i \leftarrow 0$
8.         **else** $r_i \leftarrow \big(d_{i,0} + a_i + 2\mu(d_{i,1} + b_i)\big) \bmod 4$
9.         $t_{i,0} \leftarrow d_{i,0} + a_i - 2\mu(d_{i,1} + b_i) - 4c_i$
10.         **if** $t_{i,0} \equiv \pm 3 \pmod{8}$ **then** $t_{i,1} \leftarrow d_{1-i,0} + a_{1-i} + 2(d_{1-i,1} + b_{1-i})$
11.             **if** $t_{i,1} \equiv 2 \pmod{4}$ **then** $r_i \leftarrow -r_i$

12. $\quad\quad\quad\quad S_i \leftarrow r_i \| S_i$
13. $\quad\quad$ **for** $i = 0$ **to** $1$ **do**
14. $\quad\quad\quad\quad d_{i,0} \leftarrow \mu(d_{i,0} + a_i - r_i)/2 + d_{i,1}$ and $d_{i,1} \leftarrow \mu(d_{i,1} - d_{i,0})$
15. $\quad\quad\quad\quad a_i \leftarrow b_i, b_i \leftarrow c_i$ and $c_i \leftarrow \eta_{i,j+3}$
16. $\quad\quad j \leftarrow j + 1$
17. **return** $S_0$ and $S_1$

---

**Remarks 15.22**

(i) Algorithm 15.21 works for every signed $\tau$-adic expansions; however, in signature verification it is usually applied to two integers $\eta_1, \eta_2$ after reducing them modulo $\delta$. In this case one saves the time to compute the $\tau$-adic expansion and just computes the reduction. To use it for single scalar multiplication, one starts with a $\tau$-adic expansion of length $d + a_2$ and splits it as

$$\eta_0 = \sum_{i=0}^{\lfloor d/2 \rfloor - 1} r_i \tau^i \text{ and } \eta_1 = \sum_{i=\lfloor d/2 \rfloor}^{d-1+a_2} r_i \tau^i$$

such that $n = \eta_0 + \tau^{\lfloor d/2 \rfloor} \eta_1$.

(ii) Although Algorithm 15.21 is inspired from Algorithm 9.27, there are two main differences. First, two carries $d_{i,0}$ and $d_{i,1}$ are used for each $i$. Second, the conditions modulo 2, 4, and 8 are translated to conditions modulo $\tau$, $\tau^2$ and $\tau^3$.

(iii) Like the JSF, the $\tau$JSF exists for any two elements $\eta_0, \eta_1 \in \mathbb{Z}[\tau]$ and is unique. However, the $\tau$JSF of $-3 + \mu\tau = (10\bar{1})_{\tau\text{NAF}}$ and of $\mu\tau = (0\mu 0)_{\tau\text{NAF}}$ is

$$\begin{pmatrix} -3 + \mu\tau \\ \mu\tau \end{pmatrix} = \begin{pmatrix} \bar{\mu}0\bar{\mu}0\bar{\mu}1 \\ 0\,0\,0\,0\,\mu 0 \end{pmatrix}_{\tau\text{JSF}}$$

showing that the optimality of the JSF, cf. Remark 9.28 (i), does not carry over to the $\tau$JSF.

**Example 15.23** Let $P$ and $Q$ be two elements of $E_1(\mathbb{F}_{2^{11}})$ and let us compute $[409]P \oplus [457]Q$. We have already seen that $409 \equiv 13 - 9\tau \pmod{\delta}$. In the same way, $457 \equiv 17 - 10\tau \pmod{\delta}$. Their $\tau$NAF expansions are

$$13 - 9\tau = (000\bar{1}0101001\bar{1})_{\tau\text{NAF}}$$
$$17 - 10\tau = (10100 0\bar{1}0101)_{\tau\text{NAF}}$$

which allows to compute $[409]P \oplus [457]Q$ with essentially 7 additions, if $P \oplus Q$ and $P \ominus Q$ are precomputed. In contrast, Algorithm 15.21 returns

$$\begin{pmatrix} 13 - 9\tau \\ 17 - 10\tau \end{pmatrix} = \begin{pmatrix} \bar{1}010100\bar{1} \\ \bar{1}0101\bar{1}01 \end{pmatrix}_{\tau\text{JSF}}$$

giving the same result with only 4 additions.

**Remark 15.24** For larger ground fields, similar subfield curves have been studied by Müller in characteristic 2 [MÜL 1998] and by Smart in characteristic $p$ [SMA 1999b]. In both cases, the field of definition is small so that $\chi(T)$ can be computed easily. The process of expanding is as described above; however, the study is not as detailed as Solinas'.

In [GÜLA+ 2000], Günther, Lange, and Stein generalized the concept of Koblitz curves to larger genus curves and studied two curves of genus two over $\mathbb{F}_2$. In [LAN 2001b] it has been shown that this approach works for any genus and characteristic and this study has been detailed in [LAN 2001a]. The following section deals with Koblitz curves of arbitrary characteristic and curves of larger genera.

### 15.1.2 Generalized Koblitz curves

In the previous section, we introduced the main strategies one applies to use the Frobenius endomorphism for scalar multiplication. Here we consider generalized Koblitz curves of arbitrary genus over finite fields $\mathbb{F}_q$ of arbitrary characteristic $p$. In spite of the generality, the reader should keep in mind that for use as DL systems only the Jacobians of small genus $g \leqslant 3$ curves are useful and that this approach should be applied only to small characteristic as the number of precomputations grows exponentially with the characteristic. Usually we assume prime fields as ground fields to reduce the risks of Weil descent attacks (cf. Section 22.3).

We sometimes require that the characteristic polynomial of the Frobenius endomorphism is irreducible. In our case, this means no restriction as one always chooses curves with irreducible $\chi$ to avoid cofactors in the group order. We additionally require that the degree $k$ of extension should be prime to get an almost prime group order. Let $\tau$ be a complex root of the characteristic polynomial of the Frobenius endomorphism.

Then

$$\begin{aligned}|\mathrm{Pic}_C^0(\mathbb{F}_{q^k})| &= \prod_{i=1}^{2g}(1-\tau_i^k) \\ &= \prod_{i=1}^{2g}(1-\tau_i)(1+\tau_i+\cdots+\tau_i^{k-1}) = |\mathrm{Pic}_C^0(\mathbb{F}_q)|\prod_{i=1}^{2g}(1+\tau_i+\cdots+\tau_i^{k-1})\end{aligned}$$

where the $\tau_i$'s are the conjugates of $\tau$. Thus we cannot avoid having a cofactor of size about $q^g$, and any divisor of $k$ will lead to additional factors. Likewise a composite $\chi$ gives rise to cofactors for any degree of extension. Hence, the condition that $\chi$ should be irreducible means no restriction. Furthermore, for composite or medium degree extensions, Weil descent attacks have to be taken seriously.

Formally, we have
$$|\mathrm{Pic}_C^0(\mathbb{F}_{q^k})| = cN \quad \text{where} \quad c = |\mathrm{Pic}_C^0(\mathbb{F}_q)|.$$

For cryptographic applications, we work in the cyclic subgroup of prime order $\ell \mid N$ of $\mathrm{Pic}_C^0(\mathbb{F}_{q^k})$ generated by some divisor class $\overline{D}$. As our aim is to find groups with fast scalar multiplication and hard DLP we avoid supersingular curves here.

We follow the same approach as in Section 15.1.1 in showing how to expand integers to the base of $\tau$. As before, we need to show how to reduce the length of the expansions by using a fixed degree of the extension field.

Throughout this section we let $C/\mathbb{F}_q$ be a hyperelliptic curve of genus $g$ given by an equation

$$C: y^2 + h(x)y = f(x), \; f,h \in \mathbb{F}_q[x], \; \deg f = 2g+1, \deg h \leqslant g, f \text{ monic}, \qquad (15.14)$$

and we consider the curve over the extension field $\mathbb{F}_{q^k}$. Our aim is to compute the scalar multiplication $[n]\overline{D}$ for a divisor class $\overline{D} \in \mathrm{Pic}_C^0(\mathbb{F}_{q^k})$. For full details we refer to Lange [LAN 2005c].

### 15.1.2.a Computation of $\chi_C(T)$

To make use of the Frobenius endomorphism we need to know the way it operates. By Theorem 14.17, it is enough to count the number of points on $C$ over $\mathbb{F}_q, \mathbb{F}_{q^2}, \ldots, \mathbb{F}_{q^g}$ to compute the characteristic polynomial. This is very feasible as the ground field and the genus are small enough to perform this by brute force computation.

We repeat that for a curve of genus $g$ defined over $\mathbb{F}_q$, the characteristic polynomial $\chi_C(T)$ is of the following form:

$$\chi_C(T) = T^{2g} + a_1 T^{2g-1} + \cdots + a_g T^g + \cdots + a_1 q^{g-1} T + q^g,$$

where $a_i \in \mathbb{Z}$.

**Example 15.25** Over $\mathbb{F}_2$, we can classify up to isogeny the nine classes of hyperelliptic curves of genus 2 given by an equation of the form (15.14) with irreducible $\chi_C(T)$, which are given in Table 15.1.

**Table 15.1** Binary curves of genus 2.

| Equation of $C$ | Characteristic polynomial $\chi_C(T)$ |
|---|---|
| $y^2 + y = x^5 + x^3$ | $T^4 + 2T^3 + 2T^2 + 4T + 4$ |
| $y^2 + y = x^5 + x^3 + 1$ | $T^4 - 2T^3 + 2T^2 - 4T + 4$ |
| $y^2 + y = x^5 + x^3 + x$ | $T^4 + 2T^2 + 4$ |
| $y^2 + xy = x^5 + 1$ | $T^4 + T^3 + 2T + 4$ |
| $y^2 + xy = x^5 + x^2 + 1$ | $T^4 - T^3 - 2T + 4$ |
| $y^2 + (x^2 + x)y = x^5 + 1$ | $T^4 - T^2 + 4$ |
| $y^2 + (x^2 + x + 1)y = x^5 + 1$ | $T^4 + T^2 + 4$ |
| $y^2 + (x^2 + x + 1)y = x^5 + x$ | $T^4 + 2T^3 + 3T^2 + 4T + 4$ |
| $y^2 + (x^2 + x + 1)y = x^5 + x + 1$ | $T^4 - 2T^3 + 3T^2 - 4T + 4$ |

The first five examples were given in Koblitz [KOB 1989]. Besides the first three classes these curves are nonsupersingular. The fourth and fifth case were studied by Günter, Lange, and Stein in [GÜLA$^+$ 2000].

Group orders and characteristic polynomials $\chi_C(T)$ for all Koblitz curves of genus $g \leqslant 4$ over $\mathbb{F}_q$ with $q \leqslant 7$ can be found at [LAN 2001C]. More details on how to obtain the characteristic polynomial and the number of points over extension fields can be found in Section 17.1.1.

### 15.1.2.b Computation of $\tau$-adic expansions

As before, let $\chi_C(T)$ denote the characteristic polynomial of the Frobenius endomorphism and let $\tau$ be one of its complex roots. To make use of the Frobenius endomorphism, we need to be able to represent $[n]\overline{D}$ as a linear combination of the $\phi_q^i(\overline{D})$ with bounded coefficients. This is equivalent to expanding $n$ to the base of $\tau$ as

$$n = \sum_{i=0}^{l-1} r_i \tau^i$$

where the $r_i \in R$ are elements of a set of coefficients $R$ to be defined later. If one precomputes $[r]\overline{D}$ for all occurring coefficients $r \in R$, then the computation of $[n]\overline{D}$ is realized by applications of the Frobenius endomorphism, table lookups, and additions of divisor classes whenever the coefficient is nonzero. The elements of $\mathbb{Z}[\tau]$ are of the form $\eta = n_0 + n_1\tau + \cdots + n_{2g-1}\tau^{2g-1}$ with $n_i \in \mathbb{Z}$. By definition, $\tau$ satisfies a polynomial of degree $2g$ with constant term $q^g$. Thus one can replace the computation of $[q^g]\overline{D}$ with

$$-\left([q^{g-1}a_1]\phi_q(\overline{D}) \oplus [q^{g-2}a_2]\phi_q^2(\overline{D}) \oplus \cdots \oplus [a_g]\phi_q^g(\overline{D}) \oplus \cdots \oplus [a_1]\phi_q^{2g-1}(\overline{D}) \oplus \phi_q^{2g}(\overline{D})\right).$$

But this need not be faster than computing $[q^g]\overline{D}$ by the usual method of double and add. Still it is the key observation used in expanding an integer. To compute the expansion we need a division by $\tau$ with remainder. We give the following result with proof to explain the details behind the expansion mechanism.

**Lemma 15.26** The element $\eta = n_0 + n_1\tau + \cdots + n_{2g-1}\tau^{2g-1} \in \mathbb{Z}[\tau]$ is divisible by $\tau$ if and only if $q^g \mid n_0$.

Indeed, let us suppose that $\tau \mid \eta$. This is equivalent to

$$\begin{aligned}
\eta &= \tau\eta' = \tau(n_0' + n_1'\tau + \cdots + n_{2g-1}'\tau^{2g-1}) \\
&= n_0'\tau + n_1'\tau^2 + \cdots + n_{2g-2}'\tau^{2g-2} - n_{2g-1}'(q^g + a_1q^{g-1}\tau + \cdots + a_1\tau^{2g-1}) \\
&= -n_{2g-1}'q^g + n_1\tau + \cdots + n_{2g-1}\tau^{2g-1}
\end{aligned}$$

which is in turn equivalent to $q^g \mid n_0$.

To allow computing an expansion by dividing with remainder by $\tau$, the set $R$ needs to contain at least a complete set of remainders modulo $q^g$. Since taking the negative of a class is essentially for free, we will use

$$R = \left\{0, \pm 1, \pm 2, \ldots, \pm \left\lceil \frac{q^g - 1}{2} \right\rceil\right\}$$

as a minimal set of remainders. Note that we would not need to include $-q^g/2$ in the case of even characteristic. But as we get it for free we will make use of it.

To derive a $\tau$-adic expansion of $n \in \mathbb{Z}$, we apply Lemma 15.26 repeatedly. Put $r_0 = n \bmod q^g$ for $r_0 \in R$, $s_1 = (n-r_0)/q^g$, $r_1 = -s_1a_1q^{g-1} \bmod q^g$ for $r_1 \in R$ and $s_2 = (-s_1a_1q^{g-1} - r_1)/q^g$. Then

$$\begin{aligned}
n &= r_0 + n - r_0 = r_0 + s_1q^g \\
&= r_0 - s_1(a_1q^{g-1}\tau + a_2q^{g-2}\tau^2 + \cdots + a_g\tau^g + \cdots + a_1\tau^{2g-1} + \tau^{2g}) \\
&= r_0 + \tau(-s_1a_1q^{g-1} - s_1a_2q^{g-2}\tau - \cdots - s_1a_g\tau^{g-1} - \cdots - s_1a_1\tau^{2g-2} - s_1\tau^{2g-1}) \\
&= r_0 + r_1\tau + \tau(s_2q^g - s_1a_2q^{g-2}\tau - \cdots - s_1a_g\tau^{g-1} - \cdots - s_1a_1\tau^{2g-2} - s_1\tau^{2g-1}) \\
&= r_0 + r_1\tau + \tau^2(\ldots).
\end{aligned}$$

For a concrete application of this idea in the context of elliptic curves, see Example 15.4.

The expansions derived by repeatedly applying this process with minimal remainders $|r_i| \leqslant \left\lceil \frac{q^g-1}{2} \right\rceil$ might become periodic in some rare cases.

**Remarks 15.27**

(i) For elliptic curves in even characteristic, these expansions are finite for every input (cf. Müller [MÜL 1998]), whereas in odd characteristic, Smart [SMA 1999b] shows that only for small field sizes they can turn out to be periodic and that this can only happen for $q = 5, a_1 = \pm 4$ and $q = 7, a_1 = \pm 5$. To surround this problem these numbers are included in the set of allowed coefficients. Therefore, these "bad" values are no longer expanded and each integer has a finite expansion.

(ii) In general, let $C$ be a hyperelliptic curve over $\mathbb{F}_q$ of genus $g$. An expansion using the set of remainders $R$ with maximal coefficient $r_{\max}$ can be periodic of period length 1 up to change of sign only if
$$r_{\max} \geqslant |\operatorname{Pic}^0_{\widetilde{C}}(\mathbb{F}_q)|,$$
where $\widetilde{C}$ is either the curve or its quadratic twist. If in this case the period starts at some $\eta \in \mathbb{Z}[\tau]$, then $\eta$ is small and one can adjoin $\pm r(q^g - |\operatorname{Pic}^0_{\widetilde{C}}(\mathbb{F}_q)|)$ to $R$ for all integers $1 \leqslant r \leqslant r_{\max}/|\operatorname{Pic}^0_{\widetilde{C}}(\mathbb{F}_q)|$ to guarantee finite expansions, cf. [LAN 2005c] for full details and for considerations of larger periods.

In the following, we assume that $R$ has been chosen to contain a complete set of remainders and some further coefficients if necessary. Later in the text we shall impose conditions to achieve a sparse representation, and therefore we will use different choices of the set of coefficients $R$ depending on the structure of $\chi_C(T)$.

Now we state the algorithm for expanding an element of $\mathbb{Z}[\tau]$ to the base of $\tau$. Note that at the moment we would only need to represent integers, but in the further sections we will reduce the length of the representation. Thereby we stumble over this more general problem.

---

**Algorithm 15.28** Expansion in $\tau$-adic form

INPUT: The element $\eta = n_0 + n_1 \tau + \cdots + n_{2g-1} \tau^{2g-1}$, $\chi_C(T)$ given by the coefficients $a_i$ and a suitable set $R$.
OUTPUT: The $\tau$-adic expansion $(r_{l-1} \ldots r_0)_\tau$ of $\eta$.

---

1. $S \leftarrow ()$
2. **while** $\sum_{j=0}^{2g-1} |n_j| > 0$ **do**
3.     **if** $q^g \mid n_0$ **then** $r \leftarrow 0$
4.     **else** $r \leftarrow n_0 \bmod q^g$ with $r \in R$     [see Remark 15.29]
5.     $S \leftarrow r \| S$
6.     $s \leftarrow (n_0 - r)/q^g$
7.     **for** $i = 0$ **to** $g - 1$ **do**
8.         $n_i \leftarrow n_{i+1} - a_{i+1} q^{g-i-1} s$
9.     **for** $i = 0$ **to** $g - 2$ **do**
10.         $n_{g+i} \leftarrow n_{g+i+1} - a_{g-i-1} s$
11.     $n_{2g-1} \leftarrow -s$
12. **return** $S$

---

**Remark 15.29** Further requirements may be taken into account in Line 4, for instance, in even characteristic set $r \leftarrow n_0$ if $|n_0| = q^g/2$.

## § 15.1 Koblitz curves

As in the binary elliptic case, the expansions are longer than desired but as there, we can use a reduction technique.

### 15.1.2.c Reducing the length

The strategy explained so far would lead to expansions of length $2\log_q n \approx 2gk$ — thus expansions that are $2g$ times as long as a $q^g$-adic expansion, which mitigates the advantage of using the Frobenius endomorphism. Thus, actually one does not expand $n$ itself but looks for an element $\eta \in \mathbb{Z}[\tau]$ having a short expansion and satisfying $[n]\overline{D} = [\eta]\overline{D}$ for all $\overline{D} \in \operatorname{Pic}^0_C(\mathbb{F}_{q^k})$. Once we decide to use such a curve we need to fix the field $\mathbb{F}_{q^k}$, i.e., the degree of extension. This gives us the additional equation $\phi_q^k = \operatorname{Id}$. Therefore, if $n \equiv \eta \bmod (\tau^k - 1)$ then $[n]\overline{D} = [\eta]\overline{D}$ for $\overline{D} \in \operatorname{Pic}^0_C(\mathbb{F}_{q^k})$ and we can choose an equivalent $\eta$ with a short expansion.

**Remark 15.30** Note that for a fixed extension field, $\tau$ satisfies two equations. Since we consider only irreducible polynomials $\chi_C$ and since the constant term of $\chi_C$ is $q^g \neq \pm 1$, the polynomials $\chi_C(T)$ and $T^k - 1$ are coprime. Thus their gcd over $\mathbb{Q}[T]$ is one. But we are working in $\mathbb{Z}[T]$. The ideal generated by these polynomials is a principal ideal generated by an integer (since the gcd over $\mathbb{Q}[T]$ is one). In fact this number is equal to the cardinality of the divisor class group over $\mathbb{F}_{q^k}$. To rephrase this, modulo $|\operatorname{Pic}^0_C(\mathbb{F}_{q^k})|$ these polynomials have a common linear factor. Now recall that we work in the subgroup of prime order $\ell \mid N$ of $|\operatorname{Pic}^0_C(\mathbb{F}_{q^k})|$. Hence, if we consider only this cyclic group, the polynomials have a common factor $T - s$ in $\mathbb{F}_\ell[T]$. This means that the operation of the Frobenius endomorphism on a divisor class corresponds to the multiplication of the ideal class by the integer $s$ modulo $\ell$, i.e., $\phi_q(\overline{D}) = [s]\overline{D}$ for $\overline{D}$ in the subgroup of order $\ell$.

**Example 15.31** Here, we present one example; however, further good instances are easy to get [LAN 2001c]. Consider the binary curve of genus 2 given by

$$C: y^2 + (x^2 + x + 1)y = x^5 + x + 1$$

with characteristic polynomial of the Frobenius endomorphism $\chi_C(T) = T^4 - 2T^3 + 3T^2 - 4T + 4$. For the extension of degree 89 the class number is almost prime, namely

$$|\operatorname{Pic}^0_C(\mathbb{F}_{2^{89}})| = 2 \times 191561942608242456073498418252108663615312031512914969.$$

Let $\ell$ be the large prime number. The operation of $\phi_q$ on the group of order $\ell$ corresponds to the multiplication by

$$s \equiv 82467179009623045188999864044344866954789403836113928 \pmod{\ell}.$$

Now let us suppose that $\ell^2 \nmid |\operatorname{Pic}^0_C(\mathbb{F}_{q^k})|$, which is not a restriction in practice. Then we can even reduce $n$ modulo $\delta = (\tau^k - 1)/(\tau - 1) = \tau^{k-1} + \tau^{k-2} + \cdots + \tau + 1$ to compute $[n]\overline{D}$, as the Frobenius endomorphism cannot correspond to the identity.

In summary, the following theorem holds.

**Theorem 15.32** Let $\tau$ be a complex root of the characteristic polynomial $\chi_C(T)$ of the Frobenius endomorphism $\phi_q$ of the hyperelliptic curve $C$ of genus $g$ defined over $\mathbb{F}_q$. Consider the curve over $\mathbb{F}_{q^k}$ and let $n \in \mathbb{Z}$. Using a set of remainders $R$ for the expansion such that no periodic expansions occur, there is an element $\eta \in \mathbb{Z}[\tau]$ such that

- $\eta \equiv n \pmod{\delta}$
- $\eta$ has a $\tau$-adic expansion with coefficients in $R$ of length at most $k + 4g$.

For an element $\lambda \in \mathbb{Q}$, let us recall that $z = \lfloor \lambda \rceil$ is the nearest integer to $\lambda$ with the least absolute value; see p. 360 for a computational realization. We will also use $\lceil \cdot \rfloor$ for elements of $\mathbb{Q}(\tau)$ represented as $\mathbb{Q}[T]/(\chi_C(T))$, where it is understood coefficient-wise.

The idea for reduction uses the fact that one can invert elements in the field $\mathbb{Q}(\tau)$. Thus, put

$$\lambda = n/\delta \in \mathbb{Q}(\tau),$$

so $\lambda = \sum_{i=0}^{2g-1} \lambda_i \tau^i$, where $\lambda_i \in \mathbb{Q}$. For $0 \leqslant i \leqslant 2g-1$, put $z_i = \lceil \lambda_i \rfloor$ and put

$$z := \sum_{i=0}^{2g-1} z_i \tau^i \quad \text{and} \quad \eta := n - z\delta.$$

Thus it is easy to see that $\eta \equiv n \pmod{\delta}$ and we use this equivalent multiplier.

**Remark 15.33** The usage of $\lceil \cdot \rfloor$ might not be the best choice, but nevertheless it provides an efficient way to compute a length-reduced representation that works for every genus $g$, ground field $\mathbb{F}_q$, and degree of extension $k$. For example, for the two binary elliptic curves, Algorithm 15.9 investigates in more detail an optimal way of reduction. Considering the lattice spanned by $\{1, \tau\}$, the algorithm uses that for each element of $\mathbb{Q}(\tau)$ there is a unique lattice point within distance less than $4/7$. For larger genera the computation of the nearest point is computationally hard to realize and we do not lose much choosing the "rounded" elements the way it is presented here.

The remainder of this section is devoted to computational aspects. One first needs to compute $\delta$ and its inverse in $\mathbb{Q}(\tau)$, which is done *only once* for $C$ and $k$. The computations are performed using recurrence sequences. This corresponds to the use of Lucas sequences in the case of elliptic curves (15.13). To derive the inverse in $\mathbb{Q}(\tau)$ one uses the extended Euclidean algorithm. If $C$ and $k$ are system parameters, these elements can be computed externally and stored on the device, as they are independent of the chosen divisor classes. In Section 15.1.3 we state a way to circumvent this; see also Remark 15.36 (iii).

First of all one needs the representation of $\delta$ in $\mathbb{Z}[\tau]$.

---

**Algorithm 15.34** Representation of $\delta = (\tau^k - 1)/(\tau - 1)$ in $\mathbb{Z}[\tau]$

INPUT: An extension degree $k \geqslant 1$ and $\chi_C(T)$.
OUTPUT: The $\tau$-adic expansion $(d_{2g-1}d_{2g-2}\ldots d_0)_\tau$ of $\delta$.

---

1.    $c_0 \leftarrow 1$ and $d_0 \leftarrow 1$
2.    **for** $i = 1$ **to** $2g - 1$ **do**
3.        $c_i \leftarrow 0$ and $d_i \leftarrow 0$
4.    **for** $j = 1$ **to** $k - 1$ **do**
5.        $c' \leftarrow c_{2g-1}$
6.        **for** $i = 2g - 1$ **down to** $g$ **do**
7.            $c_i \leftarrow c_{i-1} - a_{2g-i}c'$ and $d_i \leftarrow d_i + c_i$
8.        **for** $i = g - 1$ **down to** $1$ **do**
9.            $c_i \leftarrow c_{i-1} - a_i q^{g-i} c'$ and $d_i \leftarrow d_i + c_i$
10.       $c_0 \leftarrow -q^g c'$

## § 15.1 Koblitz curves

11.      $d_0 \leftarrow d_0 + c_0$
12. **return** $(d_{2g-1} d_{2g-2} \ldots d_0)_\tau$

---

To find the inverse $\delta_1$ of $\delta$ in $\mathbb{Q}(\tau)$, we consider $\delta(T) = \sum d_i T^i$ and compute the inverse of $\delta(T)$ modulo $\chi_C(T)$ in $\mathbb{Q}[T]$ by the extended Euclidean algorithm; as for

$$\gcd\big(\delta(T), \chi_C(T)\big) = \delta(T)\delta_1(T) + \chi_C(T)V(T)$$

one has $\delta^{-1}(T) \equiv \delta_1(T) \pmod{\chi_C(T)}$, which gives $\delta^{-1} = \delta_1$, by abuse of notation. For fixed genus, and hence degree of the involved polynomials, this can be made explicit.

We now present the algorithm for computing scalar multiples as a whole.

---

**Algorithm 15.35** Computation of $n$-folds using $\tau$-adic expansions

INPUT: An integer $n$, $\overline{D} = [u, v]$, $u, v \in \mathbb{F}_{q^k}[x]$, $\chi_C(T)$, the set $R$, $\delta$ and $\delta_1$ as above.
OUTPUT: The divisor class $[n]\overline{D}$ represented by $H = [s, t]$ with $s, t \in \mathbb{F}_{q^k}[x]$.

1. **for** $i \in R, i > 0$ **do**     [precomputation]
2.      $\overline{D}(i) \leftarrow [i]\overline{D}$ and $\overline{D}(-i) \leftarrow -\overline{D}(i)$
3. $z(T) \leftarrow \lceil n\delta_1(T) \rfloor$
4. $\sum_{i=0}^{2g-1} n_i T^i \leftarrow n - \delta(T)z(T) \bmod \chi_C(T)$
5. $\eta \leftarrow \sum_{i=0}^{2g-1} n_i \tau^i$
6. compute the $\tau$-adic expansion $(r_{l-1} \ldots r_0)_\tau$ of $\eta$     [use Algorithm 15.28]
7. $\overline{H} \leftarrow \overline{D}(r_{l-1})$
8. **for** $i = l - 2$ **down to** $0$ **do**
9.      $\overline{H} \leftarrow \phi_q(\overline{H})$
10.      **if** $r_i \neq 0$ **then**
11.          $\overline{H} \leftarrow \overline{H} \oplus \overline{D}(r_i)$
12. **return** $\overline{H}$

---

**Remarks 15.36**

(i) The first two lines are precomputations and need to be performed only once per curve and base $\overline{D}$, so in some applications one saves the precomputed points on the device and skips this step. In this case or if space is more restricted it is probably better to precompute only the $\overline{D}(i)$'s and compute $-\overline{D}(i)$ on the fly when necessary.

(ii) Lines 3 to 6 compute the $\tau$-adic expansion of $\eta \equiv n \pmod{\delta}$. The computations are in fact performed in $\mathbb{Q}[T]$ modulo $\chi_C(T)$.

(iii) Arithmetic in $\mathbb{Q}$ has high system requirements. Therefore, for binary elliptic curves, Solinas [SOL 2000] proposes a partial modular reduction, cf. Remark 15.14 (ii). Instead of computing $\eta \in \mathbb{Z}[\tau]$ of minimal norm he obtains an element $\eta' \equiv n \pmod{\delta}$ which might have a slightly longer expansion, but the computations involve only truncated divisions by powers of 2, which can easily be realized in soft- and hardware. For the particular curves he considers one can explicitly state the group order as an expression of the degree of extension $k$ and therefore find appropriate denominators giving a good approximation. In our general case this is not possible, but one can work with an arbitrary good approximation $\delta_1'(T)$ of $\delta_1(T)$ in which all denominators are powers of 2. The idea of Solinas of using the number theoretic norm can be generalized to computing

$$\frac{(\tau_1-1)}{(\tau_1^k-1)} = \prod_{i=1}^{2g} \frac{(\tau_i-1)}{(\tau_i^k-1)} \times \prod_{i=2}^{2g} \frac{(\tau_i^k-1)}{(\tau_i-1)} = N^{-1} \times \prod_{i=2}^{2g} \frac{(\tau_i^k-1)}{(\tau_i-1)}.$$

Thus, one can also precompute a Barrett inversion of $N$ and get the inversion by multiplications and modular reductions.

Note that in any case the resulting $\eta'$ will still be in the same class as $n$ since

$$\eta' = n - \delta \lceil n\delta_1' \rfloor \equiv n \pmod{\delta}.$$

### 15.1.2.d Complexity and comparison

The estimates for the complexity are given as number of group operations. Using precomputations as suggested, one only needs to use group additions. If the elements are represented with respect to a normal basis, then $\phi_q(\overline{D})$ can be computed for free. In any case, the expansions are all of approximate length $k$ such that $k$ applications of $\phi_q$ are always needed. Thus we ignore these operations in the following. Besides the length, the second important characteristic for the complexity is the density of the expansion.

We first consider the minimal set $R = \{0, \pm 1, \ldots, \pm \lfloor q^g/2 \rfloor\}$ of coefficients. A zero coefficient occurs with probability $1/q^g$, therefore the asymptotic density is $(q^g - 1)/q^g < 1$. Certainly all usual (signed) windowing methods carry through to $\tau$-adic windows; thus if one precomputes all multiples

$$[r_0]\overline{D} \oplus [r_1]\phi_q(\overline{D}) \oplus \cdots \oplus [r_{w-1}]\phi_q^{w-1}(\overline{D}), \quad r_i \in R, \ r_0 \neq 0$$

the density reduces. Thus one can trade off storage for larger speedup. Depending on the curve one can also use other sets of coefficients, cf. Chapter 9 and [GÜLA+ 2000, LAN 2001a]. This includes precomputing all multiples $[r]\overline{D}$ for the set $R' = \{r \in [2, \lfloor q^{2g}/2 \rfloor] : q^g \nmid r\}$ and leads to a density of $(q^g - 1)/(2q^g - 1)$. We also like to mention that the comb methods can be applied as well, which is interesting if the Frobenius operation is not for free and the precomputations are done for a fixed base point as this method reduces not only the number of additions but also the number of Frobenius operations. As a drawback, more additions than with the windowing method are used for the same number of precomputations.

To estimate the advantage of using Koblitz curves these numbers need to be compared to the usual arithmetic. Using binary double and add, the number of operations is $\frac{3}{2} \lg n \approx \frac{3}{2} kg \lg q$ and using an NAF of $n$ it still is $\approx \frac{4}{3} kg \lg q$.

To sum up we have obtained the following result:

**Fact 15.37** If we disregard storage and time for precomputations and assume a $\tau$-adic expansion of length $\approx k + 2g + 1$, the speedup factor is approximately

$$\frac{3gq^g \lg q}{2(q^g - 1)} > 1.5g$$

compared to the binary expansion and

$$\frac{4gq^g \lg q}{3(q^g - 1)} > 1.3g$$

compared to the NAF expansion, for the minimal set of coefficients and for $k$ large compared to $g$ and $q$.

Precomputations and signed digit expansions cannot be taken into account in a general formula, as it is a bit tricky to allow the same number of precomputations for comparison. We state $q = 2$ and $q = 5$ as examples, allowing windows of length at most 2, and using both, the minimal set of remainders and $R'$. The tables list the average number of group operations to compute a scalar multiple using a signed digit windowing method and using the Frobenius endomorphism. For $q = 2$ and $w = 1$ the corresponding binary system is allowed to use a window of width $g + 1$; for $w = 2$ a width of $2g + 1$ is more than fair.

Entries may be rounded to the nearest integer. We also include the case of elliptic curves. For them $w = 2$ does not require any storage and hence, we consider only that case.

| $g$ | binary window | $\tau$-adic $w=1$ | speedup factor | binary window | $\tau$-adic $w=2$ | speedup factor |
|---|---|---|---|---|---|---|
| 1 | | | | $4k/3$ | $k/3$ | 4 |
| 2 | $5k/2$ | $3k/4$ | 3 | $7k/3$ | $3k/7$ | 5.5 |
| 3 | $18k/5$ | $7k/8$ | 4 | $27k/8$ | $7k/15$ | 7 |

For $q = 5$, we cannot directly express the width $w_{\text{bin}}$ for the binary method as a function in $g$, thus we include this parameter in the table. In all cases, $w_{\text{bin}}$ was chosen as the *ceiling* of $\lg(q^g - 1/2) + 2$ respectively $\lg(q^g q^g - 1/2) + 2$, hence favoring the binary method.

| $g$ | $w_{\text{bin}}$ | binary window | $\tau$-adic $w=1$ | speedup factor | $w_{\text{bin}}$ | binary window | $\tau$-adic $w=2$ | speedup factor |
|---|---|---|---|---|---|---|---|---|
| 1 | 3 | $4\lg(5)k/5$ | $4k/5$ | 3.5 | 6 | $8\lg(5)k/7$ | $4k/9$ | 6 |
| 2 | 6 | $16\lg(5)k/7$ | $24k/25$ | 5.5 | 11 | $13\lg(5)k/6$ | $24k/49$ | 10 |
| 3 | 8 | $10\lg(5)k/3$ | $124k/125$ | 8 | 15 | $51\lg(5)k/16$ | $124k/249$ | 15 |

### 15.1.3 Alternative setup

Like before, we concentrate on the fast computation of scalar multiples. If this is used in a DL system (cf. Chapter 23), the scalars are often randomly chosen. Hence, one can as well start with an expansion of fixed length and use this as the secret scalar — not caring which integer it corresponds to if at all. This idea, suggested by H. Lenstra, as mentioned by Koblitz [KOB 1992], implies that the device need not be able to perform polynomial arithmetic and to deal with arithmetic, both in finite fields and in $\mathbb{Q}$.

If, as usual, we restrict ourselves to divisor classes $\overline{D}$ of prime order $\ell$, we work in a cyclic group and $\phi_q$ operates as a group automorphism. Then for the action of the Frobenius, we have $\phi_q(\overline{D}) = [s]\overline{D}$, where $s$ is an integer modulo $\ell$; see Remark 15.30. Hence, *any sum*

$$\sum_{i=0}^{l-1} r_i \tau^i \quad \text{corresponds to an integer mod } \ell, \text{ namely to} \quad \sum_{i=0}^{l-1} r_i s^i \bmod \ell,$$

and one can replace the whole procedure of choosing random integers and computing a reduced expansion described above by choosing a random $l$-tuple of coefficients $r_i \in R$. The integer $s$ is obtained via

$$\gcd\bigl(\chi_C(T), T^{d-1} + \cdots + T + 1\bigr) = (T - s) \quad \text{in} \quad \mathbb{F}_\ell[T].$$

This computation is done only once and $s$ is included in the curve parameters. Here, we use the minimal set

$$R = \left\{0, \pm 1, \ldots, \pm \frac{q^g - 1}{2}\right\}$$

in odd characteristic and

$$R = \left\{0, \pm 1, \ldots, \pm \left(\frac{q^g}{2} - 1\right), \frac{q^g}{2}\right\}$$

in even characteristic to avoid ambiguity.

**Remark 15.38** Likewise we can use the enlarged sets $R'$ of size $q^g(q^g - 1)$ and impose conditions on the density to obtain sequences $(r_{l-1} \ldots r_0)_\tau$ resembling outputs of the reduction and expansion procedure. Obviously, this leads to faster computations but it requires more precomputed points. We skip the details as most considerations are very similar.

By a random expansion of length $l$ we mean a tuple $(r_{l-1}, \ldots, r_0)$ with $r_i$ chosen randomly in $R$ along with the interpretation as $\sum_{i=0}^{l-1} r_i \tau^i$. In [LAN 2005c, LASH 2005b] it is shown that a reasonable choice is $l = k - 1$. The main reason against longer expansions is that collisions are then more likely, i.e., there are several expansions corresponding to the same integer modulo $\ell$ and the probability of multiple occurrence is not equally distributed.

As a possible drawback, the most significant $\tau$-adic digits are always zero. One can design a $\tau$-adic baby-step giant-step algorithm to exploit this and it slightly reduces the security, but usually the storage requirements are prohibitively large. To play it safe one should choose a slightly bigger value for the extension degree $k$.

This is also suggested as the Frobenius endomorphism speeds up the ordinary baby-step giant-step algorithm by allowing us to consider equivalence classes under $\phi_q$, cf. Section 19.5.5.

For signature schemes this setup might sound suspicious to attacks as described in [BOVE 1996, HOSM 2001, NGSH 2003]. Lange [LAN 2001a, LAN 2005c] generalizes the attack to hyperelliptic curves and shows that the $\tau$-adic variant is much harder to break and that up to current knowledge no extra weakness is implied. An extremely careful user might feel better using this approach only for ElGamal and Diffie-Hellman.

**Remarks 15.39**

(i) One can restrict the key size even more by choosing a smaller set of coefficients for the $\tau$-adic expansion. This reduces the storage requirements and the probability of collisions, but for extreme choices, e.g., if $R' = \{0, \pm 1\}$ chosen also for $g, q > 2$ to avoid precomputations, one has to be aware of lattice based attacks on the subset sum problem [COJO+ 1992, NGST 1999]. If one tries to get around these by using longer keys of length $k + \epsilon$, collisions get more likely since one has to deal with $1 + s + \cdots + s^{k-1} \equiv 0 \pmod{\ell}$. Then the zero element occurs at least

$$2\binom{\epsilon + r'_{\max} - 1}{r'_{\max}} + 1 \text{ times}$$

where $r'_{\max}$ is the maximal coefficient of $R'$.

Another idea is to consider only sparse representations to reduce the complexity. Although this reduces the size of the key-space as well, the implications are less dramatic.

## 15.2 Scalar multiplication using endomorphisms

In the previous section we detailed how to use the Frobenius endomorphism to speed up the computation of scalar multiples. Gallant, Lambert, and Vanstone [GALA+ 2001] observe that for curves that have an efficiently computable endomorphism of small norm, a speedup can be obtained. Such specially designed curves are often called *GLV curves*. Every curve over a finite field comes with a nontrivial endomorphism ring, cf. Remark 14.14, but for a random curve over a prime field one cannot expect to find such a special endomorphism. We refer to [SICI+ 2002] for arguments as to why the class of elliptic curves with an endomorphism of small norm is small.

## 15.2.1  GLV method

Here we describe how to compute scalar multiples using efficient endomorphisms. To ease the exposition we start with elliptic curves only. Let $E/\mathbb{F}_q$ be an elliptic curve and let $|E(\mathbb{F}_q)| = c\ell$, where the cofactor $c$ is small, in particular $\ell \nmid c$. Hence, there is only one cyclic group $G$ of order $\ell$ contained in $E(\mathbb{F}_q)$. Let $\psi$ be a nontrivial endomorphism defined over $\mathbb{F}_q$ with characteristic polynomial (cf. Definition 4.84)

$$\chi_\psi(T) = T^2 + t_\psi T + n_\psi,$$

where the integers $t_\psi$ and $n_\psi$ are respectively the *trace* and the *norm* of $\psi$. In the sequel, we assume that both the trace and the norm of $\psi$ are small. In $G$, the endomorphism corresponds to the multiplication with some integer $\psi(P) = [s_\psi]P$ for $s_\psi \in [0, \ell-1]$ and $s_\psi$ can be obtained as one of the roots of $\chi_\psi(T)$ modulo $\ell$. This is a complete analogy with the case of Koblitz curves considered above where $\psi$ is the Frobenius endomorphism and $s_\psi = s$, cf. Section 15.1.

To every pair $(n_0, n_1)$ there corresponds the endomorphism $n_0 + n_1\psi$ operating as

$$[n_0]P \oplus [n_1]\psi(P), \text{ for } P \in G.$$

So we can associate the integer $n = n_0 + n_1 s_\psi \bmod \ell$ to $(n_0, n_1)$. Conversely, given $n$, the GLV method aims at finding $n_0$ and $n_1$ with the extra condition that $n_0$ and $n_1$ should be sufficiently small, i.e., of half the binary length of $n$. Then one can use multi-exponentiation, cf. Section 9.1.5, to compute $[n]P$ faster than with ordinary scalar multiplication.

We first give some examples of curves with efficiently computable $\psi$ and then state the algorithms to compute the $\psi$-adic expansion.

### 15.2.1.a  Examples

The following examples can be found in [COH 2000, GALA+ 2001].

1. Let $p \equiv 1 \pmod 4$ be a prime. Let $E_1/\mathbb{F}_p$ be given by

    $$E_1 : y^2 = x^3 + a_4 x.$$

    Let $\alpha$ be an element of order 4 in $\mathbb{F}_p$ and let $P = (x_1, y_1) \in E_1$. Then $\psi_1(P) = (-x_1, \alpha y_1)$ is also on $E_1$ and the map $\psi_1$ is an endomorphism of $E_1$ defined over $\mathbb{F}_p$ with endomorphism ring $\mathbb{Z}[\psi_1] = \mathbb{Z}[\sqrt{-1}]$. One can check that $\psi_1$ satisfies $\psi_1^2 + 1 = 0$.

2. Let $p \equiv 1 \pmod 3$ be a prime. Define an elliptic curve $E_2/\mathbb{F}_p$ by

    $$E_2 : y^2 = x^3 + a_6.$$

    If $\beta$ is a third root of unity in $\mathbb{F}_p$, then $\psi_2$ defined in the affine plane by $\psi_2(x_1, y_1) = (\beta x_1, y_1)$ is an endomorphism of $E_2$ defined over $\mathbb{F}_p$ with $\mathbb{Z}[\psi_2] = \mathbb{Z}\left[\frac{1+\sqrt{-3}}{2}\right]$. As $\beta$ is a third root of unity, $\psi_2$ satisfies the equation $\psi_2^2 + \psi_2 + 1 = 0$.

3. Let $p > 3$ be a prime such that $-7$ is a quadratic residue modulo $p$. Define an elliptic curve $E_3/\mathbb{F}_p$ by

    $$E_3 : y^2 = x^3 - \frac{3}{4}x^2 - 2x - 1.$$

If $\gamma = \frac{1+\sqrt{-7}}{2}$ and $a = (\gamma - 3)/4$, then the map $\psi_3$ defined in the affine plane by

$$\psi_3(x_1, y_1) = \left( \frac{x_1^2 - \gamma}{\gamma^2(x_1 - a)}, \frac{y_1(x_1^2 - 2ax_1 + \gamma)}{\gamma^3(x_1 - a)^2} \right)$$

is an endomorphism of $E_3$ defined over $\mathbb{F}_p$ with $\mathbb{Z}[\psi_3] = \mathbb{Z}\left[\frac{1+\sqrt{-7}}{2}\right]$. Moreover $\psi_3$ satisfies the equation $\psi_3^2 - \psi_3 + 2 = 0$.

4. Let $p > 3$ be a prime such that $-2$ is a quadratic residue modulo $p$. Let $E_4/\mathbb{F}_p$ be defined by

$$E_4 : y^2 = 4x^3 - 30x - 28.$$

The map $\psi_4$ defined in the affine plane by

$$\psi_4(x_1, y_1) = \left( -\frac{2x_1^2 + 4x_1 + 9}{4(x_1 + 2)}, -\frac{2x_1^2 + 8x_1 - 1}{4\sqrt{-2}(x_1 + 2)^2} \right)$$

is an endomorphism of $E_4$ defined over $\mathbb{F}_p$ with $\mathbb{Z}[\psi_4] = \mathbb{Z}[\sqrt{-2}]$. Moreover, $\psi_4$ satisfies the equation $\psi_4^2 + 2 = 0$.

5. Iijima et al. [IIMA+ 2002] suggest using a quadratic twist $\widetilde{E}_v/\mathbb{F}_{q^k}$ of a Koblitz curve $E/\mathbb{F}_q$, with $\mathrm{char}(\mathbb{F}_q) \neq 2, 3$, to get an endomorphism derived from the Frobenius endomorphism without the drawback that the group order needs to contain a factor $|E(\mathbb{F}_q)|$ from the ground field. Therefore, $k$ needs to be even and $v$ is a nonsquare in $\mathbb{F}_{q^k}$. More precisely, let $E$ be given by

$$E : y^2 = x^3 + a_4 x + a_6,$$

and let

$$\widetilde{E}_v : y^2 = x^3 + a_4 v^2 x + a_6 v^3$$

be the quadratic twist of $E$ by $v$. Let $P = (x_1, y_1) \in \widetilde{E}_v(\mathbb{F}_{q^k})$. Then one can check that $\widetilde{\phi}_q$ defined by

$$\widetilde{\phi}_q(x_1, y_1) = \left( v^{1-q} x_1^q, v^{3(1-q)/2} y_1^q \right)$$

is an endomorphism of $\widetilde{E}_v$ over $\mathbb{F}_{q^k}$, satisfying also the characteristic polynomial of $\phi_q$.

### 15.2.1.b Computing a basis of the endomorphism ring

In each case the endomorphism ring can be seen as a two-dimensional lattice spanned by 1 and $s_\psi$. Finding integers $n_0$ and $n_1$ such that $n = n_0 + n_1 s_\psi$ with $n_i \approx \ell^{1/2}$ corresponds to finding the closest vector to $n$. To avoid this heavy machinery, [GALA+ 2001] suggest starting with applying several steps of the extended Euclidean algorithm 10.42 to first find short vectors $v_0, v_1$ generating the lattice and then to compute the splitting of $n$.

Note that the initial part of finding $v_0, v_1$ does not depend on the integer; hence, the vectors can be precomputed and kept with the curve parameters. Let us detail how to compute $v_0$ and $v_1$. By repeated division with remainder, it is possible to obtain a sequence of relations

$$s_i \ell + t_i s_\psi = r_i, \quad \text{for } i = 0, 1, 2, \ldots,$$

where

- $|s_i| < |s_{i+1}|$, for $i \geq 1$
- $|t_i| < |t_{i+1}|$ and $r_i > r_{i+1} \geq 0$, for $i \geq 0$.

## § 15.2 Scalar multiplication using endomorphisms

This can be explicitly done executing a modified version of Algorithm 10.42 with arguments $s_\psi$ and $\ell$. Namely, start with $i = 0$, put the statements $t_i \leftarrow U_A$ and $r_i \leftarrow A$ at the beginning of the while loop in Algorithm 10.42, and increment the index after each loop.

This procedure would terminate with $r_{l-1} = 1$ for some $l$ as $\ell$ is prime and $s_\psi < \ell$, but we do not need to compute all these steps. We use the additional property that at each step

$$r_i|t_{i+1}| + r_{i+1}|t_i| = \ell.$$

As one aims to find $t_i$ of size $\ell^{1/2}$ the GLV algorithm defines the index $m$ as the largest integer for which $r_m > \ell^{1/2}$. The coefficients $r_{m+1}$ and $t_{m+1}$ satisfy $r_{m+1} - s_\psi t_{m+1} \equiv 0 \pmod{\ell}$ and

$$(r_{m+1}^2 + t_{m+1}^2)^{1/2} \leqslant (2\ell)^{1/2}.$$

So $v_0 = (r_{m+1}, -t_{m+1})$ is a short vector. The GLV algorithm then sets $v_1$ to be the shorter of $(r_m, -t_m)$ and $(r_{m+2}, -t_{m+2})$.

**Remark 15.40** In [SICI$^+$ 2002], it is shown that

$$\min\{\max\{r_m, |t_m|\}, \max\{r_{m+2}, |t_{m+2}|\}\} \leqslant K\ell^{1/2}$$

with an explicitly computable constant $K$. The size of $K$ depends heavily on the norm of $\psi$ and, hence, only endomorphisms with small norm lead to representations with small integers. Furthermore, this paper gives an optimal strategy to find a short vector such that the entries have a small size. This is done by using a different norm.

Given the short vectors $v_0$ and $v_1$ from the previous step we now show how to find an expansion of an integer $n$. Solve the linear equation $n = n_0'' v_0 + n_1'' v_1$ in the two-dimensional vector space over the rationals $\mathbb{Q}$ and then choose $n_i' = \lfloor n_i'' \rceil = \lfloor n_i'' + 1/2 \rfloor$, a closest integer to $n_i''$. As for Koblitz curves this approximation is sufficiently good. Then we deduce

$$n \equiv n - n_0' v_0 - n_1' v_1 s_\psi \pmod{\ell}$$

and we assign the result to $n_0 + n_1 s_\psi$, with small $n_i$ as the entries of $v_i$ are small. The algorithm is as follows.

---

**Algorithm 15.41** GLV representation

INPUT: The integers $n \in [0, \ell - 1]$, $s_\psi$ and the vectors $v_0 = (a_0, b_0)$, $v_1 = (a_1, b_1)$ as above.
OUTPUT: The integers $(n_0, n_1)$ such that $n \equiv n_0 + n_1 s_\psi \pmod{\ell}$.

1. $n_0'' \leftarrow b_1 n/\ell$ and $n_1'' \leftarrow -b_0 n/\ell$
2. $n_0' \leftarrow \lfloor n_0'' \rceil$ and $n_1' \leftarrow \lfloor n_1'' \rceil$
3. $n_0 \leftarrow n - n_0' a_0 - n_1' a_1$ and $n_1 \leftarrow -n_0' b_0 - n_1' b_1$
4. **return** $(n_0, n_1)$

---

**Example 15.42** Let $p = 2029$ and let us consider a curve of the form $E_2$

$$E : y^2 = x^3 + 10$$

defined over $\mathbb{F}_p$. The group $E(\mathbb{F}_p)$ is generated by $P = (1620, 334)$ and has prime cardinality $\ell = 1951$. In $\mathbb{F}_p$, $\beta = 975$ is a primitive third root of unity and $\psi_2(x, y) = (\beta x, y)$ is an endomorphism of $E$. Also, the action of $\psi_2$ corresponds to the scalar multiplication by $s_\psi = 1874$.

To compute $[n]P$ with the GLV method, one must first compute short vectors $v_0 = (a_0, b_0)$ and $v_1 = (a_1, b_1)$ such that $a_i + s_\psi b_i \equiv 0 \pmod{\ell}$, for $i = 0, 1$. One applies the modified version of Algorithm 10.42 explained above, which gives the relations

$$
\begin{aligned}
1 \times 1951 + 0 \times 1874 &= 1951 \\
0 \times 1951 + 1 \times 1874 &= 1874 \\
1 \times 1951 - 1 \times 1874 &= 77 \\
-24 \times 1951 + 25 \times 1874 &= 26 \\
49 \times 1951 - 51 \times 1874 &= 25.
\end{aligned}
$$

It follows that $v_0 = (26, -25)$ and $v_1 = (25, 51)$. Now let us compute $[1271]P$ using Algorithm 15.41. One has $1271 \equiv 13 + 9s_\psi \pmod{1951}$. The joint sparse form of 13 and 9 returned by Algorithm 9.27 is

$$\begin{pmatrix} 13 \\ 9 \end{pmatrix} = \begin{pmatrix} 1101 \\ 1001 \end{pmatrix}_{\text{JSF}}.$$

Now precomputing $\psi_2(P) = Q = (938, 334)$ and $R = P \oplus Q = (1500, 1695)$, one has

$$
\begin{aligned}
[1271]P &= [13]P \oplus [9]Q \\
&= [4]([2]R \oplus P) \oplus R \\
&= (370, 359).
\end{aligned}
$$

with only 3 additions and 3 doublings. A direct computation using $1271 = (10100001\bar{0}0\bar{1})_{\text{NAF}}$ requires 10 doublings and 4 additions.

### 15.2.2 Generalizations

The approach obviously generalizes to hyperelliptic curves as soon as one has an efficiently computable endomorphism of small norm. In [PAJE$^+$ 2002b] this idea is outlined and studied in more detail, also giving bounds for the constants in [SICI$^+$ 2002, PAJE$^+$ 2002a].

So far, only very few examples were stated in the literature. The following curves are generalizations of the above examples.

1. Let $p \equiv 1 \pmod 4$ and consider the hyperelliptic curve $C_1/\mathbb{F}_p$ of genus $g$ given by

$$C_1 : y^2 = x^{2g+1} + f_{2g-1}x^{2g-1} + \cdots + f_3 x^3 + f_1 x.$$

For $P = (x_1, y_1) \in C_1$ also $\psi_1(P) = (-x_1, \alpha y_1)$ is on $C_1$ where as before $\alpha^4 = 1$. This endomorphism operates on the divisor classes in a similar manner by

$$\psi_1\left(\left[x^r + \sum_{i=0}^{r-1} u_i x^i, \sum_{i=0}^{r-1} v_i x^i\right]\right) = \left[x^r + \sum_{i=0}^{r-1}(-1)^i u_i x^i, \alpha \sum_{i=0}^{r-1}(-1)^i v_i x^i\right]$$

and has the characteristic polynomial $T^2 + 1$.

2. Let $p \equiv 1 \pmod 8$. Let the genus 2 curve $C_2/\mathbb{F}_p$ be given by

$$C_2 : y^2 = x^5 + ax,$$

for an arbitrary $a \in \mathbb{F}_p^*$. For $\xi$ an eighth root of unity the map $\psi_5(x_1, y_1) = (\xi^2 x_1, \xi y_1)$ is an endomorphism of the curve with characteristic polynomial $T^4 + 1$.

## § 15.2 Scalar multiplication using endomorphisms

Likewise, one can construct a genus 3 curve with complex multiplication by a twelfth root of unity $\zeta$

$$C_3 : y^2 = x^7 + ax,$$

satisfying $\psi_6(x_1, y_1) = (\zeta^2 x_1, \zeta y_1)$ and $\chi_{C_3}(T) = T^6 + 1$.

3. Let $m = 2g + 1$ be an odd prime and let $p \equiv 1 \pmod{m}$. Then the curve $C_4/\mathbb{F}_p$ given by

$$C_4 : y^2 = x^m + a$$

has complex multiplication by an $m$-th root of unity and characteristic polynomial

$$\frac{T^{2g+1} - 1}{T - 1} = T^{2g} + T^{2g-1} + \cdots + T + 1.$$

All these examples have complex multiplication by a root of unity. There is however no technical reason against constructing curves with an endomorphism with small norm, e.g., using the CM method 18.

Takashima [TAK 2004] proposes to combine the GLV method of scalar splitting with real multiplication by some $\psi_{\text{RM}}$ on genus 2 curves. He states how to apply $\psi_{\text{RM}}$ on divisor classes in Mumford representation and presents two types of curves on which the endomorphism can be applied particularly efficiently. These curves are far more general than the examples above and the splitting is efficient.

On the *BH curves*

$$C_{\beta,\gamma} : y^2 = \beta x^5 - (\beta + \gamma - 3)x^4 + (\beta^2 - 3\beta + 5 - 2\gamma)x^3 - \gamma x^2 + (\beta - 3)x - 1, \text{ with } \beta \in \mathbb{F}_q^*, \gamma \in \mathbb{F}_q$$

the application of $\psi_{\text{RM}}$ can be computed with only slightly more field operations than an addition of divisor classes, and the characteristic polynomial of $\psi_{\text{RM}}$ is

$$\chi(\psi_{\text{RM}})(T) = T^2 + T - 1,$$

such that the splitting leads to small coefficients.

The idea of using real multiplication can be extended to other curves and Takashima provides one more example.

### 15.2.3 Combination of GLV and Koblitz curve strategies

In [CILA⁺ 2003], it is shown how to combine the methods presented so far. Let $\psi$ be an endomorphism of a GLV curve of cardinality $\ell$ with endomorphism $\psi$ and let

$$\chi_\psi(T) = T^2 + t_\psi T + n_\psi.$$

We denote by $\nu$ a complex root of $\chi_\psi(T)$. In Section 15.2.1, we have given an algorithm to compute an expansion of the form $n = n_0 + n_1 \nu \in \mathbb{Z}[\nu]$ where $n_i \approx \ell^{1/2}$ while for Koblitz curves we have shown how to obtain a long expansion with small coefficients, cf. Sections 15.1.1.b and 15.1.2.b.

The ideas used to achieve the long expansion can be applied successfully to every other endomorphism as long as the norm $n_\psi$ is larger than 1. Namely, set

$$R = \left\{ 0, \pm 1, \ldots, \pm \left\lfloor \frac{n_\psi}{2} \right\rfloor \right\}$$

and precompute $[r_i]P$ for every $r_i \in R$. If $(t_\psi, n_\psi) \neq (\pm 2, 2), (\pm 3, 3), (\pm 4, 5)$ or $(\pm 5, 7)$, the relation $n_\psi = -t_\psi \nu - \nu^2$ used inductively on $n_0 + n_1 \nu$ gives rise to a $\nu$-adic expansion of the form

$$n = \sum_{i=0}^{l-1} r_i \nu^i$$

with $r_i \in R$ and $l \approx \log_{n_\psi} \ell$.

When the pair $(t_\psi, n_\psi)$ belongs to the set of exceptions listed above, the coefficients $r_i$ have to be taken in the larger set

$$R \cup \left\{ \pm \left\lceil \frac{n_\psi + 1}{2} \right\rceil \right\}.$$

This longer expansion with small coefficients is useful only if applying $\psi$ to $P$ is less expensive than two additions. In [CILA$^+$ 2003] some examples are given where this can be achieved in projective coordinates, namely for the examples $E_3$ and $E_4$.

For these two cases one can obtain better performance for either simple scalar multiplications, if one can precompute $Q = \psi^{\lceil l/2 \rceil}(P)$, or double scalar multiplications. In the former case, once a $\nu$-adic expansion of length $l$ is found, we can split the expansion in two parts,

$$\sum_{i=0}^{\lceil l/2 \rceil - 1} r_i \psi^i(P) \quad \text{and} \quad \sum_{i=0}^{\lfloor l/2 \rfloor - 1} r_{\lceil l/2 \rceil + i} \psi^i(Q).$$

such that in both cases one deals with a double scalar multiplication and can try to reduce the *joint weight* of them.

The characteristic polynomial of $\psi_4$ for $E_4$ is given by $T^2 + 2$ and thus a slightly modified version of the original JSF for integers can be used leading to a joint density of $1/2$.

We observe that the characteristic polynomial of $\psi_3$ is the same as that of the binary elliptic Koblitz curves considered in Section 15.1.1 and thus exactly the same Algorithm 15.21 can be used to derive a joint $\psi_3$-adic expansion of joint density $1/2$.

Hence, in the two cases that one either does a double expansion as for signature verification (cf. Algorithm 1.20) or one has a long-term precomputation of $Q$, the longer expansions lead to faster computation of scalar multiples.

### 15.2.4 Curves with endomorphisms for identity-based parameters

Lenstra [LEN 1999] and Brown, Myers, and Solinas [BRMY$^+$ 2001] suggest curves with endomorphisms not only because they offer the fast scalar multiplication detailed in the previous sections but also because their numbers of points are easy to determine due to their simple endomorphism rings.

All examples given in Section 15.2.1.a have small discriminant and Lenstra lists some more curves with discriminant up to 163 together with divisibility conditions on $p$. Using the CM-theory as shown in Chapter 18 in an explicit way, one can easily state the number of points depending on the representation of $p = a^2 + db^2$ with integers $a, b$, where $d$ divides the discriminant. He also shows that there are several primes $p$ such that the curves have an almost prime group order over $\mathbb{F}_p$.

The advantage of this idea lies in the cryptographic applications (cf. Chapters 1 and 23). The tables can be used to associate a curve with known group order and base point of almost prime order in a deterministic manner to any bit-string, e.g., the name and properties of an entity. This is done by using the bit-string to define $a$ and $x$, where first $a$ is modified in a prescribed way till there exists a $b$ such that $a^2 + db^2$ is a prime of the desired size. Then the base point is constructed using $x$. The resulting curve and base point can be used as parameters for a DL system assigned to the entity given by $a, x$. The advantage is that the public parameters can be computed by everybody, reducing the requirements for certificates. Furthermore, the scalar multiplication on these curves is fast.

In [BRMY$^+$ 2001], Brown et al. restrict to fewer curves that correspond to the first two in [LEN 1999], which they call *compact curves*, and give the group order more explicitly. This implies that the parameters can be constructed much faster even by restricted devices. These curves are special cases of the curves $E_1$ and $E_2$ given in Section 15.2.1.a and thus the GLV method leads to a fast scalar multiplication. Moreover, they provide the short vectors $v_0, v_1$, as defined in Section 15.2.1.b, which allow us to obtain the decomposition in the GLV method efficiently and then apply the JSF. The following two examples are provided in [BRMY[+] 2001].

1. Let $f \equiv 3 \pmod 4$ and $g \equiv 2 \pmod 8$ be integers such that $p = f^2 + g^2$ and $r = p + 1/2 + g$ are both prime. The curve $C_{f,g}^\alpha$, having complex multiplication with a fourth root of unity $\alpha$ as $\psi_1(x_1, y_1) = (-x_1, \alpha y_1)$, is given by

$$C_{f,g}^\alpha : y^2 = x^3 - 2x.$$

Obviously the point $P = (-1, 1) \in C_{f,g}^\alpha$ for all choices of $f, g$. The curve $C_{f,g}^\alpha$ has order $2r$.

2. Let $f \equiv 2 \pmod 3$ and $g \equiv 3 \pmod 6$ be integers such that $p = f^2 - fg + g^2$ and $r = p + 1 - (2f - g)$ are both prime. For $p \equiv 2, 5 \pmod 8$ define

$$C_{f,g}^\beta : y^2 = x^3 + 2.$$

Obviously the point $P = (-1, 1) \in C_{f,g}^\beta$ for all choices of $f, g$.
If $p \equiv 7 \pmod 8$ let

$$C_{f,g}^\beta : y^2 = x^3 - 2,$$

with point $P = (3, 5)$.
For these cases, $C_{f,g}^\beta$ has order $r$ and the curves have an automorphism $\psi_2$ given by $\psi_2(x_1, y_1) = (\beta x_1, y_1)$ with $\beta^3 = 1$.

These conditions can be verified very easily such that they can be used to implement Lenstra's idea even for restricted environments. We like to point out that in spite of these positive applications the curves are very special and thus susceptible to specialized attacks.

The same authors also proposed such a system based on genus two GLV curves of type $C_4$.

## 15.3 Trace zero varieties

Trace zero varieties were suggested for cryptographic applications by Frey [FRE 1998, FRE 2001]. The construction is based on the Weil restriction of a curve over $\mathbb{F}_{p^d}$ to $\mathbb{F}_p$, cf. Section 7.4.2. To obtain fast arithmetic in the group, one makes use of efficient arithmetic in the finite field $\mathbb{F}_{p^d}$ and of the Frobenius endomorphism. The strategy can be seen as a Koblitz curve method applied to small extension degrees. This way, scalar multiplications in the group can be performed faster than on a Jacobian of the same size.

In the genus 1 case, these varieties were studied by Naumann [NAU 1999] and Blady [BLA 2002] for $d = 3$ and by Weimerskirch [WEI 2001] for $d = 5$. Diem [DIE 2001] studies the background of the general case and [LAN 2001a, LAN 2004c] studies in detail the case of genus 2 curves over $\mathbb{F}_{p^3}$. The implementation details are considered in [AVLA 2005]. (See also [CES 2005, AVCE 2005] for the case of trace zero varieties over fields of even characteristic.) Following [DISC 2003] we argue in Section 22.3.4.b that these are the only cases of relevance for cryptographic applications as the others achieve lower security for the same group size and thus are not competitive. Clearly, the theoretical results can easily be generalized to larger genera and extension fields $\mathbb{F}_{p^d}$. The complete mathematical background is detailed in Section 7.4.2. Therefore, we concentrate on the algorithmic details here.

### 15.3.1 Background on trace zero varieties

The starting point for our construction is a hyperelliptic curve of genus $g$ including elliptic curves defined over a prime field $\mathbb{F}_p$, where $p$ is chosen such that $p^{d-1}$ is of the desired group size. In particular, we assume $p > 5$. We consider the divisor class group over the finite field extension $\mathbb{F}_{p^d}$ and restrict the computations to the subgroup $G$ defined by the property that its elements $\overline{D}$ are of trace zero, i.e.,

$$G := \left\{ \overline{D} \in \mathrm{Pic}_C^0(\mathbb{F}_{p^d}) \mid \overline{D} \oplus \phi_p(\overline{D}) \oplus \cdots \oplus \phi_p^{d-1}(\overline{D}) = 0 \right\}.$$

The group $G$ is a subgroup of $\mathrm{Pic}_C^0(\mathbb{F}_{p^d})$ as it is the kernel of the trace map. Obviously, $\phi_p$ is a group automorphism of $G$.

Since in the case considered here, $p$ is not too large, we may assume that the characteristic polynomial of the Frobenius endomorphism is known. Therefore, we can compute the order of $G$ as

$$|G| = \frac{|\mathrm{Pic}_C^0(\mathbb{F}_{p^d})|}{|\mathrm{Pic}_C^0(\mathbb{F}_p)|} = \frac{\prod_{i=1}^{2g}(1 - \tau_i^d)}{\prod_{i=1}^{2g}(1 - \tau_i)}, \tag{15.15}$$

where $\tau_i$ are the roots of $\chi_C(T)$, the characteristic polynomial of the Frobenius endomorphism.

Explicitly, if $a_i$ are the coefficients of $\chi_C(T)$, one has

- for $g = 1$ and $d = 3$

$$|G| = p^2 - p(1 + a_1) + a_1^2 - a_1 + 1$$

- for $g = 1$ and $d = 5$

$$|G| = p^4 - (a_1 + 1)p^3 + (a_1 + 1)^2 p^2 + \\ \left(5a_1 - (a_1 + 1)^3\right)p - \left(5a_1(a_1^2 + a_1 + 1) - (a_1 + 1)^4\right)$$

- for $g = 2$ and $d = 3$

$$|G| = p^4 - a_1 p^3 + (a_1^2 + 2a_1 - a_2 - 1)p^2 + \\ (-a_1^2 - a_1 a_2 + 2a_1)p + a_1^2 + a_2^2 - a_1 a_2 - a_1 - a_2 + 1.$$

The equations above with $|\tau_i| \leqslant p^{1/2}$ allow to obtain upper bounds on $|G|$. Namely

- for $g = 1$ and $d = 3$
$$|G| \leqslant p^2 + 2p^{3/2} + 3p + 2p^{1/2} + 1$$

- for $g = 1$ and $d = 5$
$$|G| \leqslant p^4 + 2p^{7/2} + 3p^3 + 4p^{5/2} + 5p^2 + 4p^{3/2} + 3p + 2p^{1/2} + 1$$

- for $g = 2$ and $d = 3$
$$|G| \leqslant p^4 + 4p^{7/2} + 10p^3 + 16p^{5/2} + 19p^2 + 16p^{3/2} + 10p + 4p^{1/2} + 1.$$

In the cryptographic applications which we envision, a cyclic group of prime order is used. Therefore we shall assume that $G$ has a subgroup $G_1$ of large prime order $\ell$ with a small cofactor. In particular, we may assume that $G_1$ is the only subgroup of order $\ell$ of $G$, hence the Frobenius operation maps $G_1$ onto itself. Like in the case of Koblitz curves (cf. Remark 15.30), there is an integer $s$ (modulo $\ell$) such that $\phi_p(\overline{D}) = [s]\overline{D}$ for all $\overline{D} \in G_1$ (this must hold for a generator of $G_1$, hence for all elements). The integer $s$ is computed as follows

## § 15.3 Trace zero varieties

- for $g = 1$ and $d = 3$

$$s = \frac{p-1}{1-a_1} \mod \ell$$

- for $g = 1$ and $d = 5$

$$s = \frac{p^2 - p - a_1^2 p + a_1 p + 1}{p - 2a_1 p + a_1^3 - a_1^2 + a_1 - 1} \mod \ell$$

- for $g = 2$ and $d = 3$

$$s = -\frac{p^2 - a_2 + a_1}{a_1 p - a_2 + 1} \mod \ell.$$

In other words, it can be computed explicitly in terms of $a_1$ (and $a_2$) and, hence, it depends only on the curve parameters; if all users of a cryptographic system use the same curve, or may choose from a small set, $s$ can as well be hard-coded.

### 15.3.2  Arithmetic in $G$

To perform the arithmetic in the trace zero subvariety, one can use the formulas and algorithms for the whole divisor class group, cf. Chapters 13 and 14. So far no formulas are known that succeed in making use of the subgroup properties.

**Remark 15.43**  In [LAN 2004c] it is shown that for $g = 2$ only the main cases of representation need to be considered in the case distinction of Section 14.3.1. This has advantages for implementations as the code size can be reduced and conditional branches can be avoided to a larger extent.

The fast arithmetic of trace zero varieties results from a fast arithmetic of the finite field $\mathbb{F}_{p^d}$ with small extension degree (cf. Section 11.3.6) and from using the Frobenius endomorphism $\phi_p$ of the Jacobian which can be restricted to $G$. This approach is similar to the one for Koblitz curves presented in Section 15.1 but due to the small extension degree, some different choices have to be made.

**Remark 15.44**  It is still an open problem to find explicit formulas for the arithmetic in $G$ or $G_1$ only, that are faster than the general formulas. For $g = 1$, see Section 7.4.2 for a geometrical description of the trace zero variety.

On the one hand, $\phi_p$ satisfies its characteristic polynomial inherited from $\mathrm{Pic}_C^0$ and by construction it also satisfies $T^{d-1} + \cdots + T + 1 = 0$. Using it, we want to replace the scalar multiplication $[n]\overline{D}$ with computations of the type

$$[n_0]\overline{D} \oplus \cdots \oplus [n_{d-1}]\phi_p^{d-1}(\overline{D}),$$

where the $n_i$ are of size $O\bigl(\ell^{1/(d-1)}\bigr)$. The multiple scalar product can be carried out simultaneously reducing the number of doublings to about $1/(d-1)$-th of those necessary for $[n]D$, if the implied constants are small enough. For more details on multiple scalar multiplications see Section 9.1.5.

There are two approaches to get such a representation, both being generalizations of the different approaches for Koblitz curves.

- One is, given $n$, to *split* it into suitable $n_i$'s like in the expansion method of Koblitz curves Section 15.1 or in the GLV method Section 15.2. As we started with random $n < \ell$ we need to ensure that the $n_i$'s are small, i.e., $n_i = O\bigl(\ell^{1/(d-1)}\bigr)$ with small constants, but need not worry about the distribution of the scalars.

- The second approach is to *start with the $n_i$'s* and with suitable bounds on their size. We give upper bounds on the $n_i$'s ensuring that collisions cannot occur, i.e., any two different sets of $n_i$'s lead to different elements of $G$. One can use this approach if one can obtain close to $\ell$ different elements. It has the advantage that one saves the time for the splitting of $n$; for a restricted environment it is even more important that one saves space for the code, cf. Section 15.1.3 for similar discussions of Koblitz curves.

Starting with the expansion is preferable if one can randomly choose the integer, however, in applications like electronic signatures, cf. Section 1.6.3, one needs to multiply with a given integer.

To obtain the splitting, one can use the same method as developed for the GLV curves because the Frobenius endomorphism satisfies $T^{d-1} + \cdots + T + 1$ for all points in $G$ and this polynomial plays the role of the characteristic polynomial of the endomorphism in that method. As the constant term is 1, we can expect the expansions to be short. A different version working directly with the *two* equations $T^{d-1} + \cdots + T + 1$ and $\chi_{\phi_p}(T)$ is provided in [AVLA 2005]. This method has lower requirements than needed for implementing the GLV method but leads to slightly longer expansions. For example, the splitting of the initial $n$ can be done with 10M for $g = 1, d = 3$ and some more multiplications in the other two recommended cases.

We now state bounds on the $n_i$ such that the resulting $n = n_0 + \cdots + n_{d-2}s^{d-2}$ are all distinct modulo $\ell$. Thus, using $n_i$ within the given bounds avoids collisions. The proofs can be found in [NAU 1999, AVLA 2005, LAN 2004c] in the given order.

**Theorem 15.45** Let $C$ be a curve of genus $g$ defined over $\mathbb{F}_p$ and consider a base field extension of degree $d$. Let $\overline{D}$ be a generator of a subgroup of prime order $\ell$ of the corresponding trace zero variety $G$.

1. In case $g = 1$, let $T^2 + a_1 T + p$ be the characteristic polynomial of the Frobenius endomorphism.

    (i) If $d = 3$, let
    $$r := \min\left\{\frac{\ell}{p - a_1}, \frac{p - 1}{\gcd(p - 1, a_1 - 1)}\right\}.$$
    Then the $r^2$ classes $[n_0]\overline{D} \oplus [n_1]\phi_p(\overline{D})$, $0 \leqslant n_i < r$ are distinct.

    (ii) If $d = 5$, let
    $$r := \min\left\{\frac{\ell}{(1 + p + |a_1|p)M}, \frac{|p^2 - a_1^2 p + a_1 p - p + 1|}{\gamma}\right\}.$$
    Then the $r^4$ classes $[n_0]\overline{D} \oplus [n_1]\phi_p(\overline{D}) \oplus [n_2]\phi_p^2(\overline{D}) \oplus [n_3]\phi_p^3(\overline{D})$, $0 \leqslant n_i < r$ are pairwise distinct.

2. In case $g = 2$, let $T^4 + a_1 T^3 + a_2 T^2 + a_1 p T + p^2$ be the characteristic polynomial of the Frobenius endomorphism. If $d = 3$, let
    $$r := \min\left\{\frac{\ell}{M}, \frac{p^2 - a_2 + a_1}{\gcd(p^2 - a_2 + a_1, a_1 p - a_2 + 1)}\right\},$$
    where $M = \max\left\{|p^2 + a_1 p - 2a_2 + a_1 + 1|, |p^2 + a_1 - a_1 p - 1|\right\}$.
    Then the $r^2$ classes $[n_0]\overline{D} \oplus [n_1]\phi_p(\overline{D})$, $0 \leqslant n_i < r$ are distinct.

These bounds show how to avoid collisions. If we assume $\ell$ to be of size $p^{g(d-1)}$ and if the involved greatest common divisor is not too large, we can hope for $r \sim p^g$. Then there are $\sim p^{g(d-1)}$ distinct multiples of $\overline{D}$ obtainable using this approach. Since $|G| \sim p^{g(d-1)}$ these multiples represent almost all of $G$. As the size of $r$ can easily be computed, one can check if a trace zero variety is good for this approach.

For $d = 3$ one uses the JSF of the two integers applied to $\overline{D}$ and $\phi_p(\overline{D})$ while for $d = 5$ it is more useful to work with all four integers in NAF representation and use the plain Straus–Shamir trick for joint doublings (cf. Chapter 9 for details).

# Chapter 16

# Implementation of Pairings

*Sylvain Duquesne and Gerhard Frey*

### Contents in Brief

| | | |
|---|---|---|
| **16.1** | **The basic algorithm** | **389** |
| | The setting • Preparation • The pairing computation algorithm • The case of nontrivial embedding degree $k$ • Comparison with the Weil pairing | |
| **16.2** | **Elliptic curves** | **396** |
| | The basic step • The representation • The pairing algorithm • Example | |
| **16.3** | **Hyperelliptic curves of genus 2** | **398** |
| | The basic step • Representation for $k > 2$ | |
| **16.4** | **Improving the pairing algorithm** | **400** |
| | Elimination of divisions • Choice of the representation • Precomputations | |
| **16.5** | **Specific improvements for elliptic curves** | **400** |
| | Systems of coordinates • Subfield computations • Even embedding degree • Example | |

## 16.1 The basic algorithm

In this chapter, we will be dealing with the computation of the Tate–Lichtenbaum pairing on Jacobians $J_C$ of hyperelliptic curves $C$ of genus $g$ defined over some finite field $\mathbb{F}_q$ of characteristic $p$. It is defined in Definition 6.16. The mathematical background can be found in Chapter 6. The idea to use pairings for cryptographic purposes was introduced during the 1990s, and because of the importance for applications a lot of valuable work was done to make the computation of these pairings as efficient as possible. There is a vast and rapidly growing literature; here we restrict ourselves to mentioning [GAHA+ 2002, BAKI+ 2002, BALY+ 2004b] and recommending a visit to the most interesting crypto lounge of Barreto [BAR] which provides a very complete source of information.

For simplicity, we assume that $C$ has an $\mathbb{F}_q$-rational Weierstraß point, which we choose as the point at infinity $P_\infty$. That implies that we can describe the affine part of $C$ by an equation $y^2 + h(x)y = f(x)$ with a polynomial $f$ of odd degree.

As explained in Section 4.4.6.a, we can represent elements in $J_C(\mathbb{F}_{q^k})$, or alternatively divisor classes $\overline{D}$ of degree 0 in $\text{Pic}^0_{C \cdot \mathbb{F}_{q^k}}$, by divisors of the form

$$D - gP_\infty$$

with $D$ an effective divisor of $C$ of degree $g$ rational over $\mathbb{F}_{q^k}$. If we assume that $D$ is reduced (see Section 4.4.6.a) the representation of $\overline{D}$ depends only on the choice of the Weierstraß point.

We will first specify the context we are interested in and recall the definition of the Tate–Lichtenbaum pairing in this context.

### 16.1.1 The setting

As usual in cryptographic applications, we shall work in a cyclic subgroup of prime order $\ell$. We shall assume that $\ell$ is large. The most interesting case is that $\ell^2$ does not divides the order of $J_C(\mathbb{F}_q)$. We shall assume this in the future.

Let $k$ be the smallest integer such that $\ell$ divides $q^k - 1$. Thus, $\mathbb{F}_{q^k}$ is obtained by adjoining the $\ell$-th roots of unity to $\mathbb{F}_q$. The number $k$ is called the *embedding degree* (with respect to $\ell$). In most of the constructive applications of the Tate pairing we have $k > 1$.

The following remarks are easy but important consequences of our definitions and assumptions.

**Remarks 16.1**

(i) For $e < k$ there are no $\ell$-th roots of unity in the field $\mathbb{F}_{q^e}$ and so every element in $\mathbb{F}_{q^e}$ is an $\ell$-th power.

(ii) The group $J_C(\mathbb{F}_{q^k})$ has no element of order $\ell^2$.

More generally, it may happen that we have to deal with the whole group $J_C(\mathbb{F}_q)$. In this case, $k$ is chosen such that the exponent of $J_C(\mathbb{F}_q)$ divides $q^k - 1$.

Using Remarks 16.1, we can identify $J_C(\mathbb{F}_{q^k})/\ell J_C(\mathbb{F}_{q^k})$ with $J_C(\mathbb{F}_{q^k})[\ell]$. But on the other side we have great freedom to choose a convenient representative in the class $Q + \ell J_C(\mathbb{F}_{q^k})$, which can be used to simplify computations, and we shall do this rather often in the following.

So, even when we are interested in computing the Tate–Lichtenbaum pairing $T_\ell$ over the field $\mathbb{F}_{q^k}$ between elements $P$ of order $\ell$ in $J_C(\mathbb{F}_q)$ and elements $Q$ of order $\ell$ in $J_C(\mathbb{F}_{q^k})$, we shall identify $Q$ with a class in $J_C(\mathbb{F}_{q^k})/\ell J_C(\mathbb{F}_{q^k})$. Since the value of the pairing does not depend on the choice of the representative of the class, we can give up the condition that $Q$ has order $\ell$ and define the Tate–Lichtenbaum pairing for arbitrary $Q \in J_C(\mathbb{F}_{q^k})$ without changing notation.

Next we remark that by definition $q^k$ is congruent to 1 modulo $\ell$. This implies that the subgroup $J_0 \subset J_C(\mathbb{F}_{q^k})$ (defined as in Theorem 6.15 but with respect to the ground field $\mathbb{F}_{q^k}$) is equal to $J_C(\mathbb{F}_{q^k})[\ell]$), and so contains $J_C(\mathbb{F}_q)[\ell]$.

Hence, we interpret (without changing the notation) the Tate–Lichtenbaum pairing for our purposes as pairing

$$T_\ell : J_C(\mathbb{F}_q)[\ell] \times J_C(\mathbb{F}_{q^k}) \rightarrow \mathbb{F}_{q^k}^*/(\mathbb{F}_{q^k}^*)^\ell.$$

Here is its explicit description. Take $P \in J_C(\mathbb{F}_q)[\ell]$ and $Q \in J_C(\mathbb{F}_{q^k})$.

1. Represent $P$ by an $\mathbb{F}_q$-rational divisor $D_P$ of degree 0.
2. Let $f_P$ be a function on $C$ with $\text{div}(f_P) = \ell(D_P)$.
3. Represent $Q$ by a divisor $D_Q$ of degree 0 coprime to $D_P$ and evaluate $f_P(D_Q)$. Recall that for $D = \sum_{R \in C} n_R R$ the value $f_P(D)$ is defined by $\prod_{R \in C} f_P(R)^{n_R}$.

**Remark 16.2** In this form, the value set of the pairing consists of classes modulo $\ell$-th powers. To get as value set the elements of order $\ell$ in $\mathbb{F}_{q^k}$ we compose the pairing with the exponentiation map with exponent $\frac{q^k-1}{\ell}$. In the following, we shall always assume that we use this slight modification of the Tate–Lichtenbaum pairing without changing the notation.

## 16.1.2 Preparation

To implement this pairing we shall need a result following from the Riemann–Roch theorem.

**Lemma 16.3** Let $C$ be as above and $k \in \mathbb{N}$.

Let $s \in \mathbb{N}$ with $s \leqslant O(\lg q)$ and take effective divisors $D^1, \ldots, D^s$ of degree $\leqslant g$ rational over $\mathbb{F}_q$. Let $D$ be an effective divisor of degree $g$ rational over $\mathbb{F}_{q^k}$ and $D_2$ a randomly chosen effective $\mathbb{F}_q$-rational divisor of $C$.

Then, with a high probability (depending on $q$) the divisor $D_1 := D \oplus D_2$ is relatively prime to $\sum_{j=1}^{s} D^j + P_\infty$.

**Remark 16.4** This lemma is of no real practical importance. In the cases that are important for cryptographic applications we shall see directly how to choose $D_2$.

Using a variant of Lemma 16.3, we give a heuristic algorithm to represent a divisor class $\overline{(D - gP_\infty)}$ in $J_C(\mathbb{F}_{q^k})$ by a difference $D_1 - D_2$ of effective divisors (with $D_2$ being $\mathbb{F}_q$-rational) and $D_1 + D_2$ prime to a finite set of $\mathbb{F}_q$-rational divisors $D^1, \ldots, D^s$ of degree $\leqslant g$ with $s = O(\lg q)$.

---

**Algorithm 16.5** Relative prime representation

INPUT: Effective divisors $D, D^1, \ldots, D^s$ of degree $g$.
OUTPUT: Divisors $D_1, D_2$ with $\overline{D_1 - D_2} = \overline{D - gP_\infty}$ and $D_1 + D_2$ prime to $P_\infty + \sum_{j=1}^{s} D^j$.

1. **repeat**
2.     choose $P \in C(\mathbb{F}_q)$ and $m \in \mathbb{N}$ such that $m \leqslant |J_C(\mathbb{F}_q)|$
3.     compute $D_2$ effective with $\overline{D_2 - gP_\infty} = \overline{m(P - P_\infty)}$
4.     compute $D_1$ such that $\overline{(D - gP_\infty) - (D_2 - D_1)} = 0$.
5. **until** $D_1 + D_2$ is prime to $\sum_{j=1}^{s} D^j + P_\infty$
6. **return** $(D_1, D_2)$

---

**Remarks 16.6**

(i) By Lines 2 and 3, we get a "nearly random" element in the set of effective divisors of degree $g$ on $C$. Note that for many instances these steps can be done by a precomputation.

(ii) In very rare cases $(D_1, D_2)$ will not satisfy the relative primeness condition. So the choice of another random pair $(P, m)$ will never be necessary in practice.

## 16.1.3 The pairing computation algorithm

We shall give a procedure to compute the Tate–Lichtenbaum pairing that works in the general case.

### 16.1.3.a The basic step

To compute the Tate–Lichtenbaum pairing we first have to compute the function $f_P$ and then evaluate it at $D_Q$. The basic step for the computation of $f_P$ consists of solving the following task, which is also the key ingredient for the addition law in the Jacobian.

For given effective divisors $A, A'$ of degree $g$ and rational over $\mathbb{F}_q$, find an effective divisor $B$ of degree $g$ and a function $G$ on $C$ such that $A + A' - B - gP_\infty = \text{div}(G)$.

We remark that $G$ is a function on $C$ defined over $\mathbb{F}_q$, whose zero divisor and pole divisor have degree $\leqslant 2g$.

We shall always assume that the divisor $B$ is reduced. By adding a suitable multiple of $P_\infty$ we can and will assume that $B$ has degree $g$.

#### 16.1.3.b  Representation of elements

We want to compute $T_\ell(P,Q)$ with $P \in J_C(\mathbb{F}_q)[\ell]$ and $Q \in J_C(\mathbb{F}_{q^k})$.

We give $P$ by a representative $D_P - gP_\infty$ with $D_P$ reduced of degree $g$.

For every multiple $[i]P$, $1 \leqslant i \leqslant \ell$ we choose the same type of representation: $[i]P$ is given by $D_{P_i} - gP_\infty$ with $D_{P_i}$ an $\mathbb{F}_q$-rational effective divisor of degree $g$. So we have an identity of divisors

$$D_{P_i} + D_{P_j} - D_{P_{i+j}} - gP_\infty = \operatorname{div}(h_{i,j}) \text{ for } i+j \leqslant \ell$$

and $D_{P_i} \oplus D_{P_j} = D_{P_{i+j}}$.

By using Algorithm 16.5 we choose a representative $D_Q = D_1 - D_2$ for $Q$ with $D_2$ an $\mathbb{F}_q$-rational effective divisor on $C$ such that $D_1 + D_2$ is prime to $D_P + P_\infty$.

**Remark 16.7** The reader should not be confused: It can happen that $P = Q$. Nevertheless, we choose different representations according to the different roles the elements play in the pairing.

#### 16.1.3.c  The pairing algorithm

We get the following algorithm for the computation of the Tate–Lichtenbaum pairing. For elliptic curves this algorithm was proposed by Miller to compute the Weil pairing [MIL 1986, MIL 2004].

---

**Algorithm 16.8** Tate–Lichtenbaum pairing

INPUT: The integer $\ell = (\ell_{l-1} \ldots \ell_0)_2$ with $\ell_{l-1} = 1$, a point $P \in J_C(\mathbb{F}_q)[\ell]$, the divisor $D_P$ with $P = \overline{D_P - gP_\infty}$, and $Q = \overline{D_Q - gP_\infty} \in J_C(\mathbb{F}_{q^k})$.
OUTPUT: The Tate–Lichtenbaum pairing $T_\ell(P,Q)$.

1. compute $D_1, D_2$                                       [using Algorithm 16.5 on $D_Q$]
2. $D_Q \leftarrow D_1 - D_2$
3. $T \leftarrow D_P$ and $f \leftarrow 1$
4. **for** $i = l-2$ **down to** $0$ **do**
5.     $T \leftarrow [2]T$
6.     $f \leftarrow f^2 G(D_Q)$                                      [$\operatorname{div}(G) = 2T - [2]T - gP_\infty$]
7.     **if** $\ell_i = 1$ **then**
8.         $T \leftarrow T \oplus D_P$
9.         $f \leftarrow f G(D_Q)$                           [$\operatorname{div}(G) = T + D_P - (T \oplus D_P) - gP_\infty$]
10. **return** $(f)^{\frac{q^k-1}{\ell}}$

---

### Remarks 16.9

(i) In Algorithm 16.8, we have to evaluate several functions $G$ at $D_Q$. They have zeroes and poles at $P_\infty$ and $D_{P_i}$ for indices $i$ occurring in the addition chain, depending only on the binary expansion of $\ell$. We need the divisor $D_Q$ to be prime to the divisors of these functions. By Lemma 16.3 we know that we have a very good chance that this is satisfied; or else we have to choose a new random representation for $Q$.

(ii) The algorithm is presented here in a double and add form and requires $O(\lg \ell)$ basic steps to evaluate $f_P$ at $D_Q$ (i.e., to compute the Tate–Lichtenbaum pairing). In the following, for clarity, we will always use this form, but the reader must keep in mind that better algorithms are available and must be used in practice. These algorithms (window methods, recoding of the exponent, use of endomorphisms of special curves if available) are described in Chapters 9 and 15 and can be applied to our situation for pairings without difficulties.

(iii) While executing Algorithm 16.8 we have to evaluate quotients of polynomials and so inversions occur. But, as is easily seen, we can postpone these inversions by multiplying and squaring denominators, and then we have to execute just one inversion in $\mathbb{F}_{q^k}$ at the end of the algorithm.

(iv) The algorithm depends heavily on the Hamming weight of $\ell$, and if we have the opportunity to have an $\ell$ with small Hamming weight, such as a Solinas prime [SOL 1999a], for instance, it should be taken.

In fact, the same remark applies if the order $N$ of $J_C(\mathbb{F}_q)$ has low Hamming weight [GAHA+ 2002]. We replace $\ell$-elements by the whole Mordell–Weil group $J_C(\mathbb{F}_q)$. If $N/\ell$ is sufficiently small such a choice provides computational savings. But be careful: we need more roots of unity and so $k$ would be much larger in general (without any positive effect for security).

In constructive applications we are often in a context that $k > g$. For $g = 1$ this just means that $k$ is larger than 1. In applications to $g \geqslant 2$ the assumption on $k$ is reasonable, too. In the following section we explain which accelerations this implies.

### 16.1.4 The case of nontrivial embedding degree $k$

In this section we shall assume that $k > g$.

As always we assume that the element $P$ is rational over $\mathbb{F}_q$ and has order $\ell$.

It is useful to recall that for any element $P'$ in $J_C(\mathbb{F}_{q^e})$ with $e < k$ we get that

$$T_\ell(P, P') = 1.$$

This can be seen by either using Example 6.10 or directly in the following way.

Remember we are computing the Tate–Lichtenbaum pairing over $\mathbb{F}_{q^k}$ by using Algorithm 16.8. So, we have to take a representation of $P'$ by divisors prime to (the representation of multiples of) $D_P - gP_\infty$. This can be done over $\mathbb{F}_{q^e}$. Then the result of the evaluation lies in $\mathbb{F}_{q^e}$ and hence (Remarks 16.1) is an $\ell$-th power.

Hence, we get for all elements $Q \in J_C(\mathbb{F}_{q^k})$

$$T_\ell(P, Q) = T_\ell(P, Q + P').$$

Now assume that $Q \in J_C(\mathbb{F}_{q^k}) \smallsetminus J_C(\mathbb{F}_{q^e})$ for all $e < k$.

Let $P$ be represented by $D_P - gP_\infty$ and $Q$ by $D_Q - gP_\infty$.

We choose a random point $P_0 \in C(\mathbb{F}_q)$. Since $q$ is assumed to be large we can assume that $P_0$ is different from $P_\infty$ and prime to the divisors $D_{P_j}$, which occur in the addition chain during the execution of Algorithm 16.8.

Let $Q'$ be the class of $D_Q - gP_0$. Since $D_Q - gP_0 = (D_Q - gP_\infty) + (gP_\infty - gP_0)$ and $gP_\infty - gP_0$ defines an $\mathbb{F}_q$-rational element in $J_C(\mathbb{F}_q)$ we get

$$T_\ell(P, Q) = T_\ell(P, Q').$$

Let $P'$ be a point on $C(\overline{\mathbb{F}}_q)$, which appears in $D_{P_j} =: D_j$ (for some $j$) as well as in $D_Q$. Then all conjugates of $P'$ appear in $D_j$ and so $P'$ is rational over a field $\mathbb{F}_{q^e}$ with $e < k$. First assume that $e \mid k$. Then the divisor class of $P' - P_0$ lies in $J_C(\mathbb{F}_{q^e})$ and so we can subtract $P' - P_0$ from $D_Q - gP_0$ without changing the Tate–Lichtenbaum pairing (see discussion above).

Assume now that $e$ does not divide $k$ and define

$$e_1 = \frac{k}{\gcd(k,e)}, \quad k_1 = \frac{e}{\gcd(k,e)}.$$

So $\mathbb{F}_{q^{e_1}}$ is a proper subfield of $\mathbb{F}_{q^k}$ and the composite field of $\mathbb{F}_{q^k}$ with $\mathbb{F}_{q^e}$ is equal to $\mathbb{F}_{q^{kk_1}}$. It follows that the conjugates of $P'$ over $\mathbb{F}_{q^{e_1}}$ are the same as the conjugates of $P'$ over $\mathbb{F}_{q^k}$ and so the divisor

$$\sum_{j=1}^{k_1} \phi_{q^{e_1}}^j P' - k_1 P_0$$

is an $\mathbb{F}_{q^{e_1}}$-rational sub-summand of $D_Q - gP_0$. Hence, it can be subtracted from $D_Q - gP_0$ without changing the Tate–Lichtenbaum pairing.

We summarize and get the following proposition.

**Proposition 16.10** Let

- $C$ be a hyperelliptic curve of genus $g$ defined over $\mathbb{F}_q$ such that $\ell \mid |J_C(\mathbb{F}_q)|$ and $\ell^2$ does not divide $|J_C(\mathbb{F}_q)|$
- $k$ be the smallest integer such that $\ell \mid (q^k - 1)$ and assume that $k > g$
- $P \in J_C(\mathbb{F}_q)[\ell]$, $Q \in J_C(\mathbb{F}_{q^k})$, $P$ and $Q$ represented by $D_P - gP_\infty$ resp. $D'_Q - gP_\infty$
- $D_Q$ be the divisor obtained from $D'_Q$ by removing all sub-summands that are rational over $\mathbb{F}_{q^e}$ for some $e < k$.

Then

1. the divisor $D_Q$ is prime to all divisors $D_{P_j} + P_\infty$ where $D_{P_j}$ occurs as positive part in the standard representation of $[j]P$
2. the Tate–Lichtenbaum pairing satisfies

$$T_\ell(P,Q) = \bigl(f_P(D_Q)\bigr)^{\frac{q^k-1}{\ell}}.$$

To prove Proposition 16.10, we use the discussion from above to replace $D'_Q$ by $D_Q$ in Algorithm 16.8. To get rid of $P_0$ we remark that any factor in the evaluation of the functions $G$ in the pairing algorithm that arises by evaluating $\mathbb{F}_q$-rational divisors can be omitted since we are only interested in values modulo $\ell$-th powers.

So we get a much simpler and faster algorithm.

---

**Algorithm 16.11** Tate–Lichtenbaum pairing if $k > g$

INPUT: The integer $\ell = (\ell_{l-1} \ldots \ell_0)_2$ with $\ell_{l-1} = 1$, $P = \overline{D_P - gP_\infty} \in J_C(\mathbb{F}_q)[\ell]$, $Q = \overline{D'_Q - gP_\infty} \in J_C(\mathbb{F}_{q^k})$, $D'_Q$ effective of degree $g$.
OUTPUT: The Tate–Lichtenbaum pairing $T_\ell(P,Q)$.

1. compute $D_Q$ from $D'_Q$ by removing all sub-summands rational over $\mathbb{F}_{q^e}$ with $e < k$
2. $T \leftarrow D_P$ and $f \leftarrow 1$
3. **for** $i = l-2$ **down to** $0$ **do**
4. $\quad T \leftarrow [2]T$

## § 16.1 The basic algorithm

5.   $\quad f \leftarrow f^2 G(D_Q)$ $\hspace{4cm}$ $[\operatorname{div}(G) = 2T - [2]T - gP_\infty\,]$
6.   **if** $\ell_i = 1$ **then**
7.   $\quad\quad T \leftarrow T \oplus D_P$
8.   $\quad\quad f \leftarrow f G(D_Q)$ $\hspace{3.5cm}$ $[\operatorname{div}(G) = T + D_P - (T \oplus D_P) - gP_\infty]$
9.   **return** $(f)^{\frac{q^k - 1}{\ell}}$

The remaining computations in the pairing algorithm are performed in $\mathbb{F}_q$ with the exception of the evaluations of the functions $G$ at $D_Q$, which are executed by multiplications of elements of $\mathbb{F}_q$ with elements of $\mathbb{F}_{q^k}$, since the coefficients of the functions $G$ are in $\mathbb{F}_q$. For implementation of extension field arithmetic, we refer to Chapter 11.

### 16.1.5 Comparison with the Weil pairing

Both in a destructive manner [MEOK+ 1993] and in a constructive manner [BOFR 2001], the use of pairings in cryptography first appeared through the use of the *Weil pairing*. It is defined by

$$W_\ell : J_C(\overline{\mathbb{F}}_q)[\ell] \times J_C(\overline{\mathbb{F}}_q)[\ell] \;\to\; \mu_\ell$$
$$(P, Q) \;\mapsto\; \frac{f_P(D_Q)}{f_Q(D_P)}$$

where $\mu_\ell$ is the multiplicative groups of the $\ell$-th roots of unity in the algebraic closure $\overline{\mathbb{F}}_q$ of $\mathbb{F}_q$. Note that no final powering is required for the Weil pairing.

To evaluate the pairing, we have to work in a finite field $\mathbb{F}_{q^{k'}}$. We can take $k'$ as the degree of the smallest extension of $\mathbb{F}_q$ over which the rank of $\ell$-torsion elements is larger than $g$.

It follows that $k \leqslant k'$. In most cases, $k$ is equal to $k'$ (for instance, for elliptic curves such that $\ell$ does not divide $q - 1$ [BAKO 1998]) but there are cases in which we have inequality even for elliptic curves and so the underlying field used for the Tate–Lichtenbaum pairing is smaller. Related to this is the following observation: by definition $W_\ell(P, P)$ is always trivial. This is not always the case for the Tate–Lichtenbaum pairing [FRMÜ+ 1999]. This point will be discussed in more detail in Section 24.2.1.b.

This first advantage of the Tate–Lichtenbaum pairing concerns mostly its destructive role. On the constructive side, we want $k$ to be different from 1.

The computation of the Weil pairing evidently requires two evaluations of functions, whereas only one is required for the Tate pairing so that it is usually assumed that the computation of the Weil pairing takes roughly twice as long as the computation of the Tate–Lichtenbaum pairing. The situation is in fact worse if we take the results in Section 16.1.4 into account. The accelerations obtained there crucially depend on the fact that $P$ is defined over $\mathbb{F}_q$. It shows that the evaluation of $f_P(Q)$ modulo $\ell$-th powers is faster than the evaluation of $f_Q(P)$ if $k > 1$ or even $k > g$.

Moreover, in the Weil pairing we must take $Q$ as an element of order $\ell$ not lying in the cyclic group generated by $P$. To find such an element it is often necessary to take a random element in $J_C(\mathbb{F}_{q^k})$ and then to multiply it by $|J_C(\mathbb{F}_{q^k})|/\ell$, which corresponds to the final exponentiation in the Tate–Lichtenbaum pairing. In any case we cannot use the freedom to choose a suitable representative in the class of $Q$ modulo $\ell J_C(\mathbb{F}_{q^k})$, which enabled us to considerably simplify the pairing algorithm.

There is one additional step when we compute the Tate–Lichtenbaum pairing: the final exponentiation. Since it is costly it should be postponed whenever possible, and in fact this can be done in many protocols (cf. Chapter 24).

So, there are reasons to prefer the Tate–Lichtenbaum pairing for cryptographic use. We shall now

## 16.2 Elliptic curves

In this section, $E$ will denote an elliptic curve defined over $\mathbb{F}_q$. In this case the Tate–Lichtenbaum pairing is described in full detail in [FRMÜ+ 1999]. The special situation is that $E$ can be identified with its Jacobian (after the choice of $P_\infty$ as zero point).

### 16.2.1 The basic step

As we know, every divisor class of degree 0 in $E(\mathbb{F}_q)$ can be uniquely represented by a divisor $P - P_\infty$ with $P \in E(\mathbb{F}_q)$. We will describe how to find the function $G$ that occurs in the basic step of the computation.

Take $P$ and $P'$ in $E(\mathbb{F}_q)$. We have to find a point $B$ on $E$ and a function $G$ such that we have the divisor identity
$$P + P' - B - P_\infty = \mathrm{div}(G).$$

We observe that $B = P \oplus P'$ is in fact the usual sum of $P$ with $P'$ on $E$, and $G$ is given by the equations of the lines used to compute $B$.

Let $L_1$ be the line through $P$ and $P'$ (which is the tangent to the curve at $P$ if $P = P'$). This line intersects $E$ at a third point $C$. Let $L_2$ be the (vertical) line through $C$ and $P_\infty$.

The equations of these two lines induce two functions on the curve, the function $G$ is nothing but $L_1/L_2$. In fact
$$\begin{aligned}
\mathrm{div}(L_1) &= P + P' + C - 3P_\infty \\
\mathrm{div}(L_2) &= C + B - 2P_\infty \\
\mathrm{div}(L_1/L_2) &= P + P' - B - P_\infty.
\end{aligned}$$

To clarify, we choose the usual affine coordinates and assume that $P = (x_1, y_1)$ and $Q = (x_2, y_2)$ and that $x_1 \neq x_2$. We get
$$G = \frac{Y - \lambda(X - x_1) - y_1}{X + (x_1 + x_2) - \lambda^2} \quad \text{where} \quad \lambda = \frac{y_1 - y_2}{x_1 - x_2}.$$

If $P = Q$ we have to replace $\lambda$ by the slope of the tangent to $E$ at $P$ to get $G$.

We remark that we can avoid the inversion needed to compute $\lambda$ by using homogeneous coordinates. We cannot avoid that $G$ is a rational function, so divisions occur when we evaluate the functions $G$ at a point $R$. But we can reduce them easily to one inversion at the end of the addition chain (for the cost of one multiplication and, in some instances, one squaring in addition, at each step of the pairing algorithm).

### 16.2.2 The representation

We restrict ourselves to the most interesting case and assume that $k > 1$.

As usual, we represent points $R$ on $E$ by the divisor $R - P_\infty$. So, points on $E$ play an ambiguous role, being interpreted as prime divisors or as elements on the Jacobian of $E$ associated to the class $R - P_\infty$.

## § 16.2 Elliptic curves

As $\ell$ is given, the pairing algorithm involves a fixed set of prime divisors $[d_j]P$ (denoted by $D_{P_j}$ in the general case) whose number $s$ is of size $O(\lg \ell)$. Let $Q$ be given in the standard form $Q - P_\infty$ with $Q$ an $\mathbb{F}_{q^k}$-rational point on $E$ which is not rational over $\mathbb{F}_{q^e}$ for $e < k$. In particular, $Q$ is prime to $P_\infty$ and $[d_j]P$.

To justify the following algorithm we choose a number $m \leqslant \ell - 1$ different from all of the numbers $d_j$ and take $P_0 = [m]P$. So $P_0$ is prime to all divisors occurring in the pairing algorithm. In the evaluation *only the existence* of the point $P_0$ is needed.

### 16.2.3 The pairing algorithm

We now give the algorithm for computing the Tate–Lichtenbaum pairing. We refer to the remark that the inversion operation in each step of the algorithm can be postponed till the end and we write down the algorithm in this version. Afterwards we give a baby example.

**Algorithm 16.12** Tate–Lichtenbaum pairing for $g = 1$ if $k > 1$

INPUT: The integer $\ell = (\ell_{l-1} \ldots \ell_0)_2$ with $\ell_{l-1} = 1$, the points $P = (x_1, y_1) \in E(\mathbb{F}_q)[\ell]$ and $Q = (x_2, y_2) \in E(\mathbb{F}_{q^k})$.
OUTPUT: The Tate–Lichtenbaum pairing $T_\ell(P, Q)$.

1. $T \leftarrow P, f_1 \leftarrow 1$ and $f_2 \leftarrow 1$
2. **for** $i = l - 2$ **down to** $0$ **do**
3.      $T \leftarrow [2]T$      $[T = (x_3, y_3)]$
4.      $\lambda \leftarrow$ the slope of the tangent of $E$ at $T$
5.      $f_1 \leftarrow f_1^2(y_2 - \lambda(x_2 - x_3) - y_3)$
6.      $f_2 \leftarrow f_2^2(x_2 + (x_1 + x_3) - \lambda^2)$
7.      **if** $\ell_i = 1$ **then**
8.          $T \leftarrow T \oplus P$
9.          $\lambda \leftarrow$ the slope of the line through $T$ and $P$
10.         $f_1 \leftarrow f_1(y_2 - \lambda(x_2 - x_3) - y_3)$
11.         $f_2 \leftarrow f_2(x_2 + (x_3 + x_1) - \lambda^2)$
12. **return** $\left(\frac{f_1}{f_2}\right)^{\frac{q^k - 1}{\ell}}$

### 16.2.4 Example

Consider the elliptic curve $E$ defined over $\mathbb{F}_{13}$ by

$$y^2 = x^3 + 6.$$

Its order is $\ell = 7 = (111)_2$ and $E(\mathbb{F}_{13})$ is generated by the point $P = (2, 1)$. In the addition chain, there occur $T = P$, $T = [2]P$, $T = [3]P$, $T = [6]P$. We can choose $P_0 = [5]P$. The embedding degree is 2 because 7 divides $13^2 - 1$ but not $13 - 1$. Since 2 is not a square in $\mathbb{F}_{13}$, $\mathbb{F}_{13^2} \simeq \mathbb{F}_{13}[\alpha]$ where $\alpha^2 = 2$. We want to compute the Tate–Lichtenbaum pairing of $P$ and $Q = (10 + 3\alpha, 11 + 2\alpha)$. Let us apply Algorithm 16.12.

First, initialize $f_1 = f_2 = 1$ and $T = (2, 1)$.

Then $i = 1$. We compute the lines $L_1$ and $L_2$ arising in the doubling of $T = (x_3, y_3)$:

$$\begin{aligned} \lambda &= 3x_3^2/2y_3 &&= 6, \\ L_1 &= y - y_3 - \lambda(x - x_3) &&= y + 7x + 11, \\ L_2 &= x + 2x_3 - \lambda^2 &&= x + 7. \end{aligned}$$

Then, we evaluate these functions at $Q$:

$$\begin{aligned} L_1(Q) &= \bigl(11 + 2\alpha + 7(10 + 3\alpha) + 11\bigr) = 1 + 10\alpha, \\ L_2(Q) &= (10 + 3\alpha) + 7 = 4 + 3\alpha, \end{aligned}$$

so that $T = [2]P = (6, 1)$ and

$$\begin{aligned} f_1 &= L_1(Q) = 1 + 10\alpha, \\ f_2 &= L_2(Q) = 4 + 3\alpha. \end{aligned}$$

Since $\ell_1 = 1$, we compute now the lines arising in the addition of $T$ and $P$:

$$L_1 = y + 12 \text{ and } L_2 = x + 8,$$

so that $T = [3]P = (5, 12)$ and

$$\begin{aligned} f_1 &= f_1 L_1(Q) = (1 + 10\alpha)(10 + 2\alpha) = 11 + 11\alpha, \\ f_2 &= f_2 L_2(Q) = (4 + 3\alpha)(5 + 3\alpha) = 12 + \alpha. \end{aligned}$$

The next value of $i$ is 0. We compute the lines $L_1$ and $L_2$ arising in the doubling of $T$:

$$L_1 = y + 5x + 2 \text{ and } L_2 = x + 11,$$

so that $T = [6]P = (2, 12)$ and

$$\begin{aligned} f_1 &= f_1^2 L_1(Q) = (11 + 11\alpha)^2(11 + 4\alpha) = 1 + 6\alpha, \\ f_2 &= f_2^2 L_2(Q) = (12 + \alpha)^2(8 + 3\alpha) = 12 + 6\alpha. \end{aligned}$$

Since $\ell_0 = 1$, we now compute the lines arising in the addition of $T$ and $P$:

$$L_1 = x - 2 \text{ and } L_2 = 1,$$

so that $T = 7P = P_\infty$, $f_1 = (1 + 6\alpha)(8 + 3\alpha) = 5 + 12\alpha$ and $f_2 = 12 + 6\alpha$.

Thus the Tate–Lichtenbaum pairing of $P$ and $Q$ is

$$\left( \frac{5 + 12\alpha}{12 + 6\alpha} \right)^{24} = 4 + \alpha.$$

## 16.3 Hyperelliptic curves of genus 2

In Section 16.1.3.c, it is shown how to evaluate the Tate–Lichtenbaum pairing on the Jacobian of a curve, assuming an explicit reduction algorithm for divisors on the curve. For hyperelliptic curves, such an algorithm can be given by Cantor's algorithm or by explicit formulas (see Chapter 14). We shall make this a bit more explicit in the case that $g = 2$.

### 16.3.1 The basic step

Let $A = [u, v]$ and $A' = [u', v']$ be divisors in Mumford representation. Cantor's Algorithm 14.7 returns the divisor $B$ in Mumford representation such that there exists a function $G$ on $C$ and

$$A + A' - B - 2P_\infty = \text{div}(G).$$

We will now explain how to compute $G$.

In the composition step of Cantor's algorithm, we compute the polynomial $\delta$ which is the greatest common divisor of $u$, $u'$ and $v + v' + h$ and a divisor $[U, V]$, which is in the same divisor class as $B$. At this point, two cases may occur:

- $\deg(U) \leqslant 2$: this means that $[U, V]$ is already reduced, so we set $G(x, y)$ equal to $\delta(x)$,
- $\deg(U) > 2$: the divisor $[U, V]$ must be reduced to obtain $B$ and $G(x, y) = \dfrac{U(x)}{V(x) + y} \delta(x)$.

### 16.3.2 Representation for $k > 2$

Assume that $P \in J_C(\mathbb{F}_q)$ and $Q \in J_C(\mathbb{F}_{q^k}) \smallsetminus J_C(\mathbb{F}_{q^e})$ for all $e < k$ are given in the standard representation by divisors $D_P - 2P_\infty$ respectively $D'_Q - 2P_\infty$.

For cryptographic applications we can assume that $q$ is so big that for fixed $\ell$ there is a point $P_0 \in C(\mathbb{F}_q)$, which is (as divisor) prime to the divisors $D_{P_j}$ occurring in the addition chain of the pairing algorithm. Again we only need the existence of this point to justify Algorithm 16.11.

In Proposition 16.10 we have proved that we can remove from $D'_Q$ sub-summands that are rational over proper subfields. We shall make this more explicit now.

**Proposition 16.13** Let

- $C$ be a hyperelliptic curve of genus 2 defined over $\mathbb{F}_q$ such that $\ell \mid |J_C(\mathbb{F}_q)|$ and $\ell^2$ does not divide $|J_C(\mathbb{F}_q)|$
- $k$ be the smallest integer such that $\ell \mid (q^k - 1)$ and assume that $k > 2$
- $P \in J_C(\mathbb{F}_q)[\ell]$ and $Q \in J_C(\mathbb{F}_{q^k})$, $P$ and $Q$ represented by $D_P - 2P_\infty$ respectively $D'_Q - 2P_\infty$ with $D'_Q = Q_1 + Q_2$ and assume that $Q$ is not rational over any field $\mathbb{F}_{q^e}$ with $e < k$
- $D_Q$ be the divisor obtained from $D'_Q$ by removing $Q_i$ if it is rational over $\mathbb{F}_{q^e}$ with $e < k$.

Then

1. the divisor $D_Q$ is prime to all divisors $D_{P_j} + P_\infty$ where $D_{P_j}$ occurs as positive part in the standard representation of $[j]P$
2. the Tate–Lichtenbaum pairing satisfies

$$T_\ell(P, Q) = \bigl(f_P(D_Q)\bigr)^{\frac{q^k - 1}{\ell}}.$$

**Proof.** Let $D'_Q = Q_1 + Q_2$ with $Q_i \in C$. If $Q_1$ is rational over $\mathbb{F}_{q^e}$ with $e < k$ we can remove it from $D'_Q$ without changing the result of the Tate–Lichtenbaum pairing.

If $Q_1$ is rational over $\mathbb{F}_{q^k}$ but not over a proper subfield containing $\mathbb{F}_q$, it has $k$ conjugates over $\mathbb{F}_q$ and hence it cannot divide any effective divisor of degree $\leqslant 2$ rational over $\mathbb{F}_q$.

If $Q_1$ is not rational over $\mathbb{F}_{q^k}$ then it is Galois conjugate to $Q_2$ over $\mathbb{F}_{q^k}$ and so over $\mathbb{F}_q$. If it divided an $\mathbb{F}_q$-rational effective divisor $D$ of degree $\leqslant 2$, $Q_2$ would also divide $D$ and so $D = Q_1 + Q_2 = D'_Q$, which contradicts our assumption on $Q$. □

So, we can remove in $D'_Q$ exactly the points that are defined over $\mathbb{F}_{q^e}$ with $e < k$ to get $D_Q$.

Having an explicit description of the basic step and of the relevant divisor $D_Q$, we can now use Algorithm 16.11 with a simple first step.

## 16.4 Improving the pairing algorithm

For the convenience of the reader we summarize some possibilities to improve the performance of the pairing algorithm.

### 16.4.1 Elimination of divisions

One of the main improvements to the evaluation algorithm consists of avoiding inversions, since these are time-consuming operations, especially in $\mathbb{F}_{q^k}$. Inversions appear twice at each step of the algorithm: once in the computation of the functions $G$, and once in the function evaluations.

Inversions in the first case can be eliminated by using alternative systems of coordinates, such as projective coordinates, as explained in Chapter 13 for elliptic curves and in Chapter 14 for genus 2 curves. More details for elliptic curves are given in Section 16.5.1.

Concerning inversions occurring in function evaluations, we have already seen that they can be postponed at the end of the algorithm.

We will even see in Section 16.5.2 that these denominators can be removed for interesting instances.

### 16.4.2 Choice of the representation

To make the pairing algorithm explicit, we have to represent $P$ and $Q$ by relatively prime divisors. Because of the optimization of addition formulas, it is wise to choose for $P$ a standard representation.

The choice for the representing divisors for $Q$ is especially important if $k > g$. We have explained in Proposition 16.10 and in Example 16.2.4 how to do this.

Also, the extension field arithmetic has to be implemented with care [GAHA+ 2002].

### 16.4.3 Precomputations

If $P$ is either fixed or used several times, the computation of the pairing can easily been sped up by precomputing the functions $G$. Indeed, these functions depend only on $P$. Then, if $k > g$, we simply evaluate these functions at $D_Q$. Note that actual cryptosystems based on pairings often need such a $P$ (base divisor on the Jacobian, public key). On the other side the number of needed precomputations is of size $O(\lg \ell)$, which is quite large.

## 16.5 Specific improvements for elliptic curves

We list now some possibilities for improving the evaluation algorithm in the special case that $C$ is an elliptic curve $E$.

## 16.5.1 Systems of coordinates

We have already seen that other systems of coordinates should be used to avoid inversions, as it is the case for scalar multiplication (see Section 16.4.1). Moreover, Izu and Takagi observe [IZTA 2003a] that in using Jacobian coordinates, many squarings of the $Z$-coordinate are needed in the computation of the lines $L_1$ and $L_2$. Therefore, they introduce a new system of coordinates $(X, Y, Z, Z^2)$ in which $(X, Y, Z)$ represent a point in the Jacobian coordinates. This representation is called the simplified Chudnovsky Jacobian coordinates and denoted $\mathcal{J}^s$. This system of coordinates requires the same number of operations as Jacobian coordinates for doubling and addition, but a square can be removed both in the computation of the line for addition and for doubling.

The authors also provide iterated doubling formulas with this new system of coordinates. It consists in directly computing $2^j P$. This method avoids $j - 1$ squarings compared to $j$ standard doublings. We can take advantage of this if a large window size is used in the scalar multiplication or if the binary expansion of $\ell$ has many zeroes. The same kind of idea is used in [BLMU$^+$ 2004] where some operations are saved using a 4-ary algorithm with direct formulas.

Nevertheless, we have to keep in mind that, since $P$ is defined over $\mathbb{F}_q$, the inversions saved by these changes of coordinate are inversions in $\mathbb{F}_q$. Also, extra multiplications are introduced (as usual, when one uses projective coordinates), but with one of the factors in $\mathbb{F}_{q^k}$. So projective-like coordinates must be used very carefully for the computation of the Tate–Lichtenbaum pairing [GAHA$^+$ 2002]. By coupling the pairing algorithm with Montgomery's scalar multiplication ladder, Scott and Barreto obtained a laddering version of Algorithm 16.12 [SCBA 2004]. This can of course be combined with efficient formulas for doubling and addition if the curve can be transformed into Montgomery form as described in Section 13.2.3. Therefore, the advantages of this scalar multiplication method (little memory required, parallel computing eased and resistance to side-channel attacks enabled [JOYE 2000]) can be exploited in pairing-based cryptosystems.

## 16.5.2 Subfield computations

Let $e$ be a nontrivial divisor of $k$ and let $k_1 = k/e$. We shall use the fact that elements in $\mathbb{F}_{q^e}$ are $\ell$-th powers. This makes the inversion needed in the evaluation step faster.

**Lemma 16.14** Let $z$ be an element of $\mathbb{F}_{q^k}^*$. Then

$$\prod_{j=1}^{k_1-1} \phi_{q^e}^j(z) \equiv z^{-1} \bmod \left( (\mathbb{F}_{q^k}^*)^\ell \right). \tag{16.1}$$

The proof follows from the fact that $z$ times the left-hand term of (16.1) is in $\mathbb{F}_{q^e}$ and hence is an $\ell$-th power.

**Remark 16.15** In Chapter 11, it is explained how to evaluate $\prod_{j=1}^{k_1-1} \phi_{q^e}^j(z)$ quickly. There, the norm map is used to compute inverse elements in extension fields exactly and not only up to $\ell$-th powers.

So, we can rewrite our pairing algorithm for the Tate–Lichtenbaum pairing of elliptic curves.

We assume that $k > 1$ and that $k_1$ is the smallest prime divisor of $k$.

As usual we take $P \in E(\mathbb{F}_q)$ and $Q \in E(\mathbb{F}_{q^k}) \smallsetminus E(\mathbb{F}_q)$.

**Algorithm 16.16** Tate–Lichtenbaum pairing for $g = 1$ if $k > 1$ and $k = k_1 e$

INPUT: The integer $\ell = (\ell_{l-1} \ldots \ell_0)_2$ with $\ell_{l-1} = 1$, the points $P = (x_1, y_1) \in E(\mathbb{F}_q)[\ell]$ and $Q = (x_2, y_2) \in E(\mathbb{F}_{q^k})$.
OUTPUT: The Tate–Lichtenbaum pairing $T_\ell(P, Q)$.

1. $T \leftarrow P$ and $f_1 \leftarrow 1$ and $f_2 \leftarrow 1$
2. **for** $i = l - 2$ **down to** $0$ **do**
3.      $T \leftarrow [2]T$             $[T = (x_3, y_3)]$
4.      $\lambda \leftarrow$ the slope of the tangent of $E$ at $T$
5.      $f_1 \leftarrow f_1^2 \left(y_2 - \lambda(x_2 - x_3) - y_3\right)$
6.      $f_2 \leftarrow f_2^2 \left(x_2 + (x_3 + x_1) - \lambda^2\right)$
7.      **if** $\ell_i = 1$ **then**
8.          $T \leftarrow T \oplus P$
9.          $\lambda \leftarrow$ the slope of the line through $T$ and $P$
10.         $f_1 \leftarrow f_1(y_2 - \lambda(x_2 - x_3) - y_3)$
11.         $f_2 \leftarrow f_2(x_2 + (x_3 + x_1) - \lambda^2)$
12. **return** $\left(f_1 \prod_{j=1}^{k_1-1} \phi_{q^e}^j(f_2)\right)^{\frac{q^k-1}{\ell}}$

For small number $k_1$ the inversion would be computed using the norm map and an inversion in $\mathbb{F}_{q^k}$. So, our observation saves one inversion and also $2k_1$ multiplications in $\mathbb{F}_{q^e}$.

The nicest case is $k_1 = 2$ and so $k$ even. We have to apply $\phi_{q^e}$ just once. But we shall see in the next section that with the cost of at most one addition on elliptic curves (which is very often not necessary) we shall get a further simplification.

### 16.5.3 Even embedding degree

We assume that $k = 2e$ and choose $k_1 = 2$. The Frobenius automorphism $\phi_{q^e}$ operates as an involution on $E(\mathbb{F}_{q^k})[\ell]$ and has one eigenvalue $1$ with $P$ as eigenvector. By elementary linear algebra it follows that it has $-1$ as eigenvalue with eigenspace $V^-$ of rank $1$. If $Q' \in V^-$ we get that $\phi_{q^e}(Q')$ has the same $x$-coordinate as $Q'$ and so the $x$-coordinate of $Q'$ lies in $\mathbb{F}_{q^e}$.

If $Q$ is any point of $E[\ell]$ the point $Q' := Q - \phi_{q^e} Q$ lies in $V^-$.

Since
$$T_\ell(P, Q') = T_\ell(P, Q) \cdot T_\ell(P, -\phi_{q^e} Q) = T_\ell(P, Q)^2$$

it follows that we can compute $T_\ell(P, Q)^2$ by using $Q'$, a point whose $x$-coordinate lies in $\mathbb{F}_{q^e}$. Again, given that elements in this field are $\ell$-th powers, we can simplify the evaluation algorithm for the Tate–Lichtenbaum pairing.

**Algorithm 16.17** Tate–Lichtenbaum pairing for $g = 1$ and $k = 2e$

INPUT: The integer $\ell = (\ell_{l-1} \ldots \ell_0)_2$ with $\ell_{l-1} = 1$, points $P \in E(\mathbb{F}_q)[\ell]$ and $Q \in E(\mathbb{F}_{q^k})$.
OUTPUT: The quantity $T_\ell(P, Q)^2$

1. $Q \leftarrow Q - \phi_{q^e} Q$
2. $T \leftarrow P$ and $f \leftarrow 1$

## § 16.5 Specific improvements for elliptic curves

3.  **for** $i = l - 2$ **down to** $0$ **do**
4.     $T \leftarrow [2]T$
5.     $f \leftarrow f^2\, L(Q)$                                  [$L$ is the tangent line of $E$ at $T$]
6.     **if** $\ell_i = 1$ **then**
7.         $T \leftarrow T \oplus P$
8.         $f \leftarrow f\, L(Q)$                                   [$L$ is the line through $T$ and $P$]
9.  **return** $f^{\frac{q^k-1}{\ell}}$

**Remark 16.18** In most applications, the knowledge of $T_\ell(P, Q)^2$ will be as good as that of $T_\ell(P, Q)$. If not, we can proceed as follows.

- If $q$ is even, compute the unique square root of $f^{\frac{q^k-1}{\ell}}$ in $\mathbb{F}_{q^k}$ which is equal to $T_\ell(P, Q)$.
- Otherwise, set

$$\chi(f) = \begin{cases} -1 & \text{if } f \text{ is not a square in } \mathbb{F}_{q^k}, \\ 1 & \text{else.} \end{cases}$$

and $T_\ell(P, Q) = f^{\frac{q^k-1}{2\ell}} \chi(f)$.

### 16.5.4 Example

We consider again the elliptic curve $E$ defined over $\mathbb{F}_{13}$ by

$$y^2 = x^3 + 6.$$

We take $P = (2, 1)$.

The embedding degree is 2 because 7 divides $13^2 - 1$ but not $13 - 1$. Since 2 is not a square in $\mathbb{F}_{13}$, $\mathbb{F}_{13^2} \simeq \mathbb{F}_{13}[\alpha]$ where $\alpha^2 = 2$. We want to compute the Tate–Lichtenbaum pairing of $P$ and $Q = (10 + 3\alpha, 11 + 2\alpha)$.

We apply Algorithm 16.17. So, we have to compute $Q' = Q - \phi_{13^2}(Q)$. Since

$$-\phi_{13^2}(Q) = (10 + 10\alpha, 2 + 2\alpha)$$

we get

$$Q' = (12, 3\alpha).$$

First, let us initialize $f = 1$ and $T = (2, 1)$. Then $i = 1$. We compute the line $L$ arising in the doubling of $T = (x_3, y_3)$:

$$\lambda = 3x_3^2/2y_3 = 6,$$
$$L = y - y_3 - \lambda(x - x_3) = y + 7x + 11.$$

We evaluate $L$ at $Q'$:

$$L(Q') = (3\alpha + 7 \times 12 + 11) = 4 + 3\alpha.$$

Now, we have $T = [2]P = (6, 1)$ and $f = 4 + 3\alpha$.

Since $\ell_1 = 1$, we compute the line arising in the addition of $T$ and $P$:

$$L = y + 12$$

and compute $T = [3]P = (5, 12)$. So

$$f = (4 + 3\alpha)(12 + 3\alpha) = 1 + 9\alpha.$$

Finally, $i = 0$. We compute the line $L$ arising in the doubling of $T$:

$$L = y + 5x + 2$$

and compute $T = [6]P = (2, 12)$ and

$$f = f^2 L(Q') = (1 + 9\alpha)^2 (10 + 3\alpha = 9 + 6\alpha = 3(3 + 2\alpha).$$

Since $\ell_0 = 1$, we compute the line arising in the addition of $T$ and $P$:

$$L = x - 2$$

We get, as expected $T = 7P = P_\infty$.

Since the $x$-coordinate of $Q'$ is in $\mathbb{F}_{13}$, we are done:

$$T_\ell(P, Q') = (3 + 2\alpha)^{24} = 5 + 8\alpha.$$

We note that $T_\ell(P, Q) = (3 + 2\alpha)^{12} = 4 + \alpha$, and we get the result from Example 16.2.4 back.

# IV

# Point Counting

# Chapter 17

# Point Counting on Elliptic and Hyperelliptic Curves

*Reynald Lercier, David Lubicz, and Frederik Vercauteren*

**Contents in Brief**

| | | |
|---|---|---|
| **17.1** | **Elementary methods** | **407** |
| | Enumeration • Subfield curves • Square root algorithms • Cartier–Manin operator | |
| **17.2** | **Overview of $\ell$-adic methods** | **413** |
| | Schoof's algorithm • Schoof–Elkies–Atkin's algorithm • Modular polynomials • Computing separable isogenies in finite fields of large characteristic • Complete SEA algorithm | |
| **17.3** | **Overview of $p$-adic methods** | **422** |
| | Satoh's algorithm • Arithmetic–Geometric–Mean algorithm • Kedlaya's algorithm | |

## 17.1 Elementary methods

This section describes various elementary methods to compute the characteristic polynomial $\chi(\phi_q)_C$ of a nonsingular curve $C$ over a finite field $\mathbb{F}_q$. Recall that the $L$-polynomial of $C$ is defined as $L_C(T) = T^{2g}\chi(\phi_q)_C(1/T)$, which shows that computing either polynomial is sufficient.

### 17.1.1 Enumeration

Let $C$ be a projective, nonsingular curve over a finite field $\mathbb{F}_q$ with $q = p^d$ and let

$$L_C(T) = \sum_{i=0}^{2g} a_i T^i = \prod_{i=1}^{2g} (1 - \alpha_i T)$$

denote the $L$-polynomial of $C$. By the Weil conjectures we have

$$|C(\mathbb{F}_{q^k})| = q^k + 1 - \sum_{i=0}^{2g} \alpha_i^k, \tag{17.1}$$

and if $S_k = -\sum_{i=0}^{2g} \alpha_i^k$, then the Newton–Girard formula gives the following recurrence formula:

$$ka_k = S_k a_0 + S_{k-1} a_1 + \cdots + S_1 a_{k-1}, \qquad (17.2)$$

with $a_i = 0$, for $i > 2g$. The functional equation of the zeta function implies that $a_{i+g} = q^{g-i} a_i$ for $i = 0, \ldots, g$, so it suffices to compute $a_0, \ldots, a_g$ or $S_1, \ldots, S_g$ by (17.2).

### 17.1.1.a Hyperelliptic curves over finite fields of characteristic $\geqslant 3$

In this section, we will assume that $p$ is an odd prime and $q = p^d$. As shown in Section 4.4.2.b, a genus $g$ hyperelliptic curve $C$ with $\mathbb{F}_q$-rational Weierstraß point can be defined by an affine equation of the form $y^2 = f(x)$ with $f \in \mathbb{F}_q[x]$ a squarefree polynomial of degree $2g + 1$.

To compute $|C(\mathbb{F}_{q^k})|$ for $k = 1, \ldots, g$, we simply test which $\alpha \in \mathbb{F}_{q^k}$ give rise to points on $C$ and add the unique place at infinity. Clearly, to every zero $\alpha \in \mathbb{F}_{q^k}$ of $f$ only one point $(\alpha, 0)$ corresponds and to every $\alpha \in \mathbb{F}_{q^k}$ with $f(\alpha)$ a nonzero square, two points $(\alpha, \pm\sqrt{f(\alpha)})$ correspond. Let $\mathbb{F}_{q^k}$ be represented by $\mathbb{F}_p[X]/(g(X))$ with $g(X)$ an irreducible polynomial of degree $dk$, then by the definition of the (generalized) Legendre symbol (see Section 2.3.4), we have

$$|C(\mathbb{F}_{q^k})| = 1 + \sum_{\alpha \in \mathbb{F}_{q^k}} \left(1 + \left(\frac{f(\alpha)}{g(X)}\right)\right) = q^k + 1 + \sum_{\alpha \in \mathbb{F}_{q^k}} \left(\frac{f(\alpha)}{g(X)}\right), \qquad (17.3)$$

where the 1 corresponds to the place at infinity. Comparing with (17.1) shows that

$$S_k = \sum_{\alpha \in \mathbb{F}_{q^k}} \left(\frac{f(\alpha)}{g(X)}\right).$$

To compute the generalized Legendre symbol, we can use Algorithm 11.69. However, for small $q$ it is often much faster to precompute a hash table of squares in $\mathbb{F}_{q^k}$ and test if $f(\alpha)$ is in the table.

**Remark 17.1** Precomputing a hash table of squares in a finite field $\mathbb{F}_{q^k}$ can be done quite efficiently as follows: first assume that $q = p$ and $k = 1$, then clearly $1 \pmod{p}$ is a square and the other squares can be computed using the recursion

$$(\alpha + 1)^2 \equiv \alpha^2 + 2\alpha + 1 \pmod{p},$$

for $\alpha = 1, \ldots, (p-1)/2$. A similar, though slightly more complicated trick can be used to compute the table for $k > 1$ based on the recursion

$$\left(\alpha(X) + X^i\right)^2 = \alpha(X)^2 + 2\alpha(X)X^i + X^{2i}.$$

The product $\alpha(X) X^i$ is in fact a simple shift, which leaves the term $X^{2i}$. Here we make the distinction between $i < \lfloor dk/2 \rfloor$ and $i \geqslant \lceil dk/2 \rceil$, since in this case we also need $X^{2i} \pmod{g(X)}$. Note that a similar trick should be used to compute $f(\alpha)$ for all $\alpha \in \mathbb{F}_{q^k}$.

**Example 17.2** Let $p = 5$, $d = 2$ and $\mathbb{F}_q \simeq \mathbb{F}_p(\theta)$ with $\theta^2 + 4\theta + 2 = 0$. Let the genus 3 hyperelliptic curve be defined by

$$y^2 = x^7 + (4\theta + 4)x^6 + (3\theta + 1)x^5 + (3\theta + 1)x^4 + (3\theta + 3)x^3 + (3\theta + 2)x + 3\theta + 3,$$

then $S_1 = 3$, $S_2 = 7$ and $S_3 = -351$. Formula (17.2) gives $a_0 = 1, a_1 = 3, a_2 = 8, a_3 = -102$, so the $L$-polynomial of $C$ is

$$15625T^6 + 1875T^5 + 200T^4 - 102T^3 + 8T^2 + 3T + 1.$$

## 17.1.1.b Hyperelliptic curves over finite fields of characteristic 2

Recall from Section 4.4.2.b that a genus $g$ hyperelliptic curve $C$ with $\mathbb{F}_q$-rational Weierstraß point for $q = 2^d$ can be defined by an affine equation $y^2 + h(x)y = f(x)$, with $\deg h \leqslant g$ and $\deg f = 2g + 1$. To compute $|C(\mathbb{F}_{q^k})|$, we again count the number of points corresponding to every element $\alpha \in \mathbb{F}_{q^k}$ and add the unique place at infinity. Every zero $\alpha \in \mathbb{F}_{q^k}$ of $h$ is the $x$-coordinate of only one point $(\alpha, \sqrt{f(\alpha)})$; every $\alpha \in \mathbb{F}_{q^k}$ with $h(\alpha) \neq 0$ is the $x$-coordinate of either 0 or 2 points depending on the trace of $f(\alpha)/h(\alpha)^2$. Indeed, define $z = y/h(\alpha)$, then the equation of the curve becomes $z^2 + z = f(\alpha)/h(\alpha)^2$ and according to Hilbert Satz 90 this equation has two distinct solutions if and only if $\mathrm{Tr}_{\mathbb{F}_{q^k}/\mathbb{F}_2}(f(\alpha)/h(\alpha)^2) = 0$. Let $V_h$ be the affine variety defined by $h(x) = 0$, then we conclude that

$$|C(\mathbb{F}_{q^k})| = 1 + |V_h(\mathbb{F}_{q^k})| + 2 \sum_{\alpha \in \mathbb{F}_{q^k} \setminus V_h} \left(1 - \mathrm{Tr}_{\mathbb{F}_{q^k}/\mathbb{F}_2}\left(\frac{f(\alpha)}{h(\alpha)^2}\right)\right),$$

where again the 1 corresponds to the unique place at infinity.

**Example 17.3** Let $d = 5$ and $\mathbb{F}_{2^5} \simeq \mathbb{F}_2(\theta)$ with $\theta^5 + \theta^2 + 1 = 0$ and let $C$ be defined by $y^2 + h(x)y = f(x)$ with

$$h(x) = (\theta^4 + \theta^3 + \theta + 1)x^3 + (\theta^4 + \theta^2 + \theta + 1)x^2 + (\theta^3 + \theta)x + \theta^2 + \theta + 1,$$
$$f(x) = x^7 + (\theta^4 + \theta^2)x^6 + (\theta^4 + \theta^2 + \theta)x^5 + (\theta^4 + \theta^3 + \theta^2 + \theta)x^4 + (\theta^3 + \theta^2)x^3$$
$$+ (\theta^3 + \theta)x^2 + (\theta^4 + \theta^3)x + \theta^3 + \theta$$

then $S_1 = 3, S_2 = 13, S_3 = 423$ and formula (17.2) leads to $a_0 = 1, a_1 = 3, a_2 = 11, a_3 = 165$, so the $L$-polynomial of $C$ is given by

$$L_C(T) = 32768T^6 + 3072T^5 + 352T^4 + 165T^3 + 11T^2 + 3T + 1.$$

### 17.1.2 Subfield curves

Curves defined over a small finite field admit faster Jacobian arithmetic using the Frobenius endomorphism as shown in Section 15.1. Since the curve is defined over a small finite field, it is easy to compute its $L$-polynomial using an elementary method such as enumeration.

Let $C$ be a nonsingular curve over $\mathbb{F}_q$ with Jacobian variety $J_C$ and let $L_k \in \mathbb{Z}[T]$ denote the $L$-polynomial of $C/\mathbb{F}_{q^k}$, then the order $|J_C(\mathbb{F}_{q^k})|$ is given by $L_k(1)$. Assume we have computed $L_1(T) = \prod_{i=1}^{2g}(1 - \alpha_i T)$, then by the Weil conjectures (see Section 8.3) we can easily recover $L_k(T)$ as

$$L_k(T) = \prod_{i=1}^{2g}(1 - \alpha_i^k T).$$

A first approach therefore is to factor $L_1(T)$ over $\mathbb{C}$ with suitable precision, compute the $k$-th powers of the roots and recover $|J_C(\mathbb{F}_{q^k})| = \prod_{i=1}^{2g}(1 - \alpha_i^k)$. Since $J_C(\mathbb{F}_q)$ is a subgroup of $J_C(\mathbb{F}_{q^k})$ we conclude that $L_1(1) \mid L_k(1)$. This observation can be used to lower the precision with which the $\alpha_i$ have to be computed, since it suffices to round $\prod_{i=1}^{2g}(1 - \alpha_i^k)$ to the nearest multiple of $L_1(1)$.

The second method solely relies on integer arithmetic and is thus far easier to implement. Let $L_1(T) = \sum_{i=0}^{2g} a_i T^i$ and define $S_k = -\sum_{i=1}^{2g} \alpha_i^k$, then again we can use the recurrence formula due to Newton–Girard

$$ka_k = S_k a_0 + S_{k-1} a_1 + \cdots + S_1 a_{k-1} \tag{17.4}$$

where $a_i = 0$, for $i > 2g$.

The above formula can be used in two ways: firstly, given $a_i$ for $i = 0, \ldots, 2g$ compute $S_k$ for any $k$ or secondly, given $S_k$ for $k = 1, \ldots, g$ compute $a_i$ for $i = 0, \ldots, g$ and thus also for $i = g+1, \ldots, 2g$ since $a_{i+g} = q^{g-i}a_i$. Write

$$L_k(T) = \prod_{i=1}^{2g}(1 - \alpha_i^k T) = \sum_{i=0}^{2g} a_{k,i}T^i,$$

then we can easily compute the $a_{k,i}$ given $S_{jk}$ for $j = 1, \ldots, g$ by using the Newton–Girard formula for $L_k$:

$$ia_{k,i} = S_{ik}a_{k,0} + S_{(i-1)k}a_{k,1} + \cdots + S_k a_{k,i-1}. \tag{17.5}$$

The algorithm thus proceeds in two phases: in the first phase compute $S_{jk}$ for $j = 1, \ldots, g$ from the coefficients $a_i$ using (17.4); in the second phase, recover the $a_{k,i}$ using (17.5). A similar approach can be found in the thesis of Lange [LAN 2001a].

**Example 17.4** Let $C$ be an elliptic curve over $\mathbb{F}_q$, then $L_1(T) = qT^2 - tT + 1$, with $t = \alpha_1 + \alpha_2$ the trace of the $q$-th power Frobenius. Let $t_k = \alpha_1^k + \alpha_2^k$ be the trace of the $q^k$-th power Frobenius, then the $t_k$ are given by the recursion $t_1 = t$, $t_2 = t^2 - 2q$ and

$$t_k = t\,t_{k-1} - qt_{k-2}.$$

The number of points on $C(\mathbb{F}_{q^k})$ then follows easily from $|C(\mathbb{F}_{q^k})| = q^k + 1 - t_k$.

### 17.1.3 Square root algorithms

Let $C$ be a projective nonsingular curve over $\mathbb{F}_q$ of genus $g$ and let $J_C$ denote its Jacobian variety. This section shows how the structure of the abelian group $J_C(\mathbb{F}_q)$ helps to compute its order.

Recall that for an arbitrary abelian group $G$ the exponent $e$ is defined as the smallest nonzero integer such that $[e]\alpha = 0$ for all $\alpha \in G$. Denote with $\mathrm{ord}(\alpha)$ the order of $\alpha$. The square root algorithms described in Chapter 19 compute $\mathrm{ord}(\alpha)$ as the discrete logarithm of 0 to the base $\alpha$ in time $O\bigl(\mathrm{ord}(\alpha)^{1/2}\bigr)$.

Since clearly $e = \mathrm{lcm}_{\alpha \in G}\bigl(\mathrm{ord}(\alpha)\bigr)$, in many cases $e$ can be computed very efficiently by picking a few random $\alpha_i \in G$ for $i = 1, \ldots, n$ and computing $e_n = \mathrm{lcm}\bigl(\mathrm{ord}(\alpha_1), \ldots, \mathrm{ord}(\alpha_n)\bigr)$. Note that $e_i$ should be used to speed up the computation of the order of $\alpha_{i+1}$. If the group order is bounded from above by $C$ and if $e_n > C/2$, then $e_n$ is the group order of $G$. Furthermore, if the group order is also bounded from below by $C/2 < B \leqslant |G| \leqslant C$ and $e_n > C - B$, then the unique multiple of $e_n$ in the interval $[B, C]$ is the group order of $G$. In the unlikely event that the above cases do not apply, one can always resort to a probabilistic algorithm that computes the group order of any abelian group [COH 2000, Algorithm 5.4.1] with expected running time of $O(\sqrt{C - B})$ group operations. Recall from Proposition 5.78 that $J_C(\mathbb{F}_q)$ is the direct sum of at most $2g$ cyclic groups

$$J_C(\mathbb{F}_q) \simeq \bigoplus_{1 \leqslant i \leqslant 2g} \mathbb{Z}/d_i\mathbb{Z},$$

with $d_i \mid d_{i+1}$ for $1 \leqslant i < 2g$ and for all $1 \leqslant i \leqslant g$ one has $d_i \mid q - 1$.

Since the curve $C$ is of genus $g$, the Weil conjectures imply

$$(\sqrt{q} - 1)^{2g} \leqslant |J_C(\mathbb{F}_q)| \leqslant (\sqrt{q} + 1)^{2g}.$$

In particular, the width of the Hasse–Weil interval is $w = 4gq^{g-1/2} + O(q^{g-1})$ and the square root algorithms compute $|J_C(\mathbb{F}_q)|$ in time $O\bigl(\sqrt{g}\,q^{\frac{2g-1}{4}}\bigr)$. Note that since $|J_C(\mathbb{F}_q)| \geqslant (\sqrt{q} - 1)^{2g}$, the

§ 17.1 Elementary methods

exponent $e = d_{2g}$ satisfies $e \geqslant (\sqrt{q} - 1)$. However, in most cases $e > w$ so there is only one multiple of $e$ in the Hasse–Weil interval.

The above approach is especially useful in combination with other algorithms that provide partial information, i.e., $J_C(\mathbb{F}_q) \pmod{N}$ for some modulus $N \in \mathbb{N}$. Clearly, if $N > w$ then there is only one candidate for $|J_C(\mathbb{F}_q)|$ in the Hasse–Weil interval. When $N < w$, the square root algorithms can be easily adapted to work in time $O(\sqrt{w/N})$.

If not only $J_C(\mathbb{F}_q) \pmod{N}$ is known, but also $\chi(\phi_q)_C(T) \pmod{N}$, then it is possible to obtain an $N$-fold speedup instead of a $\sqrt{N}$-fold speedup. This was first noticed by Matsuo, Chao and Tsujii [MACH$^+$ 2002] who adapted the classical baby-step giant-step algorithm to take this extra information into account. Gaudry and Schost [GASC 2004b] applied the same ideas to a parallelized lambda method (see Section 19.6) with distinguished points. The latter version only requires negligible space, whereas the former needs $O\bigl(\sqrt{g} q^{\frac{2g-1}{4}}/N\bigr)$ space; a slight disadvantage is that it runs about 3 times slower than the original algorithm.

### 17.1.4 Cartier–Manin operator

This section introduces the Cartier–Manin operator and shows how it can be used to recover the characteristic polynomial of Frobenius modulo the characteristic of the base field.

Let $\mathbb{F}_q$ be a finite field of characteristic $p > 2$ and let $C$ be a projective nonsingular curve over $\mathbb{F}_q$, with function field $K = \mathbb{F}_q(C)$. Let $K^p$ denote the subfield of $p$-th powers and choose a separable generating transcendental element $x \in K \smallsetminus K^p$.

Denote with $\Omega(K)$ the $K$-vector space of differentials as introduced in Section 4.4.2.c, then every differential $\omega \in \Omega(K)$ can be written uniquely as

$$\omega = d\lambda + \alpha^p x^{p-1} dx \quad \text{with } \lambda, \alpha \in K.$$

**Definition 17.5** The *Cartier operator* $\mathfrak{C} : \Omega(K) \to \Omega(K)$ is defined as

$$\mathfrak{C}(\omega) = \alpha dx.$$

Recall that the set of holomorphic differentials $\Omega^0(K)$ (see Section 4.4.2.c) forms an $\mathbb{F}_q$-vector space of dimension $g$.

**Proposition 17.6** The $\mathbb{F}_q$-vector space $\Omega^0(K)$ is closed under the Cartier operator. Let $(\omega_1, \ldots, \omega_g)$ be a basis of $\Omega^0(K)$, then there exists a $g \times g$ matrix $A = (a_{i,j})$ with coefficients in $\mathbb{F}_q$ such that

$$\mathfrak{C} \begin{bmatrix} \omega_1 \\ \vdots \\ \omega_g \end{bmatrix} = A^{(1/p)} \begin{bmatrix} \omega_1 \\ \vdots \\ \omega_g \end{bmatrix},$$

where $A^{(1/p)}$ denotes the matrix $\bigl(a_{i,j}^{1/p}\bigr)$.

The matrix $A$ in the above proposition is called the Cartier–Manin matrix of $C$ and is determined up to a transformation of the form $S^{-1}AS^{(p)}$, with $S$ a nonsingular $g \times g$ matrix and $S^{(p)}$ is obtained from $S$ by raising each of its elements to the $p$-th power. Manin [MAN 1961] proved that the matrix $A$ can be linked to the characteristic polynomial of Frobenius $\chi(\phi_q)_C$.

**Theorem 17.7** Let $C$ be a projective nonsingular curve of genus $g$ over a finite field $\mathbb{F}_q$ with $q = p^d$. Since the Cartier–Manin matrix $A$ is determined up to a transformation of the form $S^{-1}AS^{(p)}$, the matrix $M = AA^{(p)} \cdots A^{(p^{d-1})}$ is an invariant of the curve. Let $\kappa(T) = \det(TI_g - M)$ be the characteristic polynomial of $M$ and $\chi(\phi_q)_C$ the characteristic polynomial of Frobenius, then

$$\chi(\phi_q)_C(T) \equiv T^g \kappa(T) \pmod{p}.$$

To apply the above theorem, we need to compute the Cartier–Manin matrix $A$. For a hyperelliptic curve of genus $g$ given by an equation $y^2 = f(x)$ with $f \in \mathbb{F}_q[x]$ a monic squarefree polynomial of degree $2g+1$, Manin [MAN 1962] proved the following proposition.

**Proposition 17.8** Let $c_i$ denote the coefficient of $x^i$ in the polynomial $f(x)^{(p-1)/2}$, then the Cartier–Manin matrix with respect to the basis $\omega_i = x^{i-1} dx/y$ for $i = 1, \ldots, g$ is given by

$$A = \begin{bmatrix} c_{p-1} & c_{p-2} & \cdots & c_{p-g} \\ c_{2p-1} & c_{2p-2} & \cdots & c_{2p-g} \\ \vdots & \vdots & \ddots & \vdots \\ c_{gp-1} & c_{gp-2} & \cdots & c_{gp-g} \end{bmatrix}. \tag{17.6}$$

The trivial method to apply the above proposition is to expand $h = f(x)^{(p-1)/2}$ and to select the desired coefficients. As noted by Flajolet and Salvy [FLSA 1997], the coefficients of $h$ satisfy a linear recurrence based on the identity:

$$f(x) h'(x) - \frac{p-1}{2} f'(x) h(x) = 0.$$

Let $f(x) = \sum_{n \in \mathbb{Z}} f_n x^n$ and $h(x) = \sum_{n \in \mathbb{Z}} h_n x^n$, then from the above equation we deduce that:

$$(n+1) f_0 h_{n+1} + \left(n - \frac{p-1}{2}\right) f_1 h_n + \cdots + \left(n - 2g - \frac{(2g+1)(p-1)}{2}\right) f_{2g+1} h_{n-2g} = 0.$$

Bostan, Gaudry, and Schost [BOGA$^+$ 2004] describe an optimized algorithm to compute terms in linear recurrences and prove that the Cartier–Manin matrix can be computed in time

$$O\big((R_M(g) g \sqrt{p} + g^3 R_P(\sqrt{p}))(d \lg p)^\mu\big),$$

with $R_M(s)$ the number of ring operations to multiply two $s \times s$ matrices over a ring, $R_P(s)$ the number of ring operations to multiply two degree $s$ polynomials over a ring, and $\mu$ a constant such that two $B$-bit integers can be multiplied in time $B^\mu$.

**Example 17.9** Let $p = 101$ and consider the genus 3 hyperelliptic curve $C$ defined by

$$y^2 = x^7 + 93x^6 + 64x^5 + 2x^4 + 65x^3 + 90x^2 + 18x + 10.$$

The Cartier–Manin matrix $A$ of $C$ can be easily computed and is given by

$$A = \begin{bmatrix} 4 & 54 & 49 \\ 13 & 67 & 10 \\ 78 & 40 & 44 \end{bmatrix}.$$

Computing the characteristic polynomial of $A$ then shows that

$$\chi(\phi_q)_C(T) \equiv T^6 + 87T^5 + 84T^4 + 100T^3 \pmod{p}.$$

which is indeed the reduction of the characteristic polynomial of Frobenius

$$\chi(\phi_q)_C(T) = T^6 - 14T^5 - 17T^4 + 1716T^3 - 1717T^2 - 142814T + 1030301.$$

## 17.2 Overview of $\ell$-adic methods

This section gives an overview of the $\ell$-adic methods to compute the number of points on an elliptic curve over a finite field. Although the $\ell$-adic approach can be made to work in all characteristics, we will mainly focus on elliptic curves over large prime fields $\mathbb{F}_p$. For curves over finite fields of small characteristic, the $p$-adic algorithms described in Section 17.3 are much faster and easier to implement.

### 17.2.1 Schoof's algorithm

In 1985 Schoof [SCH 1985] was the first to describe a polynomial time algorithm to count the number of points on an elliptic curve $E$ over a large prime field $\mathbb{F}_p$. In the remainder of this section, we will assume that $p > 3$. As shown in Corollary 4.118, this means that $E$ can be given by an equation of the form
$$E : y^2 = x^3 + a_4 x + a_6, \text{ with } a_4, a_6 \in \mathbb{F}_p. \tag{17.7}$$

Recall that $|E(\mathbb{F}_p)| = p + 1 - t$ with $t$ the trace of the Frobenius endomorphism $\phi_p$ and by Hasse's Theorem 5.83, we have $|t| \leq 2\sqrt{p}$. The main idea of Schoof's algorithm is to compute $t$ modulo various small primes $\ell_1, \ldots, \ell_r$ such that $\prod_{i=1}^{r} \ell_i > 4\sqrt{p}$. The trace $t$ can then be determined using the Chinese remainder theorem and the group order follows. From the prime number theorem, it follows that $r$ is $O(\lg p / \lg \lg p)$ and that the largest prime $\ell_r$ is of order $O(\lg p)$.

To illustrate the idea, we show how to compute $t \pmod{2}$. Since $p$ is an odd prime, we have $|E(\mathbb{F}_p)| \equiv t \pmod{2}$, so $t \equiv 0 \pmod{2}$ if and only if $E(\mathbb{F}_p)$ has a nontrivial $\mathbb{F}_p$-rational point of order two. The nontrivial points of order two are given by $(\xi_i, 0)$ with $\xi_i$ a root of $X^3 + a_4 X + a_6$. Therefore, if $X^3 + a_4 X + a_6$ is irreducible over $\mathbb{F}_p$ we have $t \equiv 1 \pmod{2}$; otherwise, $t \equiv 0 \pmod{2}$. Note that the polynomial $X^3 + a_4 X + a_6$ is irreducible over $\mathbb{F}_p$ if and only if $\gcd(X^3 + a_4 X + a_6, X^p - X) = 1$. The computation of $t \pmod{2}$ thus boils down to polynomial arithmetic modulo $X^3 + a_4 X + a_6$.

More generally, we obtain the trace $t$ modulo a prime $\ell$ by computing with the $\ell$-torsion points. Recall that the Frobenius endomorphism $\phi_p$ is defined by $\phi_p : E(\overline{\mathbb{F}}_p) \to E(\overline{\mathbb{F}}_p) : (x, y) \mapsto (x^p, y^p)$ and that it satisfies the equation $\chi(\phi_q)_E(T) = T^2 - tT + p$, i.e., for all points $P \in E(\overline{\mathbb{F}}_p)$ we have
$$\phi_p^2(P) - [t]\phi_p(P) + [p](P) = P_\infty.$$

By restricting to nontrivial $\ell$-torsion points $P \in E[\ell] \smallsetminus \{P_\infty\}$, we obtain the reduced equation
$$\phi_p^2(P) - [t_\ell]\phi_p(P) + [p_\ell]P = P_\infty$$

with $t_\ell \equiv t \pmod{\ell}$ and $p_\ell \equiv p \pmod{\ell}$ and $0 \leq t_\ell, p_\ell < \ell$.

As shown in Section 4.4.5.a, $P = (x_1, y_1) \in E(\overline{\mathbb{F}}_p)$ is a nontrivial $\ell$-torsion point if and only if $x_1$ is a root of the $\ell$-th division polynomial $f_\ell$. The nontrivial $\ell$-torsion points can therefore be described as the solutions of the system of equations
$$Y^2 - X^3 - a_4 X - a_6 = 0 \text{ and } f_\ell(X) = 0.$$

This implies that the equation
$$\left( X^{p^2}, Y^{p^2} \right) \oplus [p_\ell](X, Y) = [t_\ell](X^p, Y^p) \tag{17.8}$$

holds modulo the polynomials $f_\ell(X)$ and $E(X, Y) = Y^2 - X^3 - a_4 X - a_6$. Note that the addition $\oplus$ in (17.8) is the elliptic curve addition and that $[p_\ell](X, Y)$ and $[t_\ell](X^p, Y^p)$ can be easily computed using the division polynomials (see Section 4.4.5.a). To compute $t_\ell$, we simply try all $\tau \in \{0, \ldots, \ell - 1\}$ until we find the unique value $\tau$ for which $(X^{p^2}, Y^{p^2}) \oplus [p_\ell](X, Y) = [\tau](X^p, Y^p)$ holds modulo $f_\ell(X)$ and $E(X, Y)$.

**Example 17.10** Let $p = 1009$ and consider the elliptic curve $E/\mathbb{F}_p$ defined by the equation

$$E : y^2 = x^3 + 184x + 896.$$

For $\ell = 3$, the $\ell$-th division polynomial is given by $f_\ell(X) = 3X^4 + 95X^2 + 338X + 450$. An easy calculation shows that $q_\ell \equiv 1 \pmod{\ell}$ and

$$X^p \equiv 278X^3 + 467X^2 + 479X + 447 \pmod{f_\ell(X)}$$
$$Y^p \equiv (357X^3 + 734X^2 + 822X + 967)Y \pmod{(f_\ell(X), E(X,Y))}$$

and $X^{p^2} \equiv X \pmod{f_\ell(X)}$ and $Y^{p^2} \equiv -Y \pmod{(f_\ell(X), E(X, Y))}$. This clearly shows that $\phi_p^2(X, Y) \oplus (X, Y) = P_\infty$ and thus $t \equiv 0 \pmod{3}$. Computing $t$ modulo 2, 3, 5, and 7 shows that $\chi(\phi_q)_E(T) = T^2 + 12T + p$.

Recall that for $\ell \neq 2$ and $\gcd(\ell, p) = 1$, we have $E[\ell] \simeq \mathbb{Z}/\ell\mathbb{Z} \times \mathbb{Z}/\ell\mathbb{Z}$ and thus $\deg f_\ell = (\ell^2 - 1)/2$. The computation of $(X^{p^2}, Y^{p^2})$ and $(X^p, Y^p)$ modulo $f_\ell$ and $E(X, Y)$ clearly takes $O(\lg p)$ multiplications in the ring $\mathbb{F}_p[X, Y]/(E(X, Y), f_\ell(X))$. Since $\deg f_\ell$ is of the order $O(\ell^2)$, each of these multiplications takes $O(\ell^{2\mu} \lg^\mu p)$ bit-operations, so computing $t \pmod{\ell}$ requires $O(\ell^{2\mu} \lg^{1+\mu} p)$ bit-operations. Summing over all primes $\ell_1, \ldots, \ell_r$ and using the prime number theorem gives an overall complexity of $O(\lg^{2+3\mu} p)$ bit-operations.

Note that if we could replace the division polynomials $f_\ell$ by alternative polynomials of lower degree, the complexity of the algorithm would drop considerably. In the next section, we show that for ordinary elliptic curves it is possible to use alternative polynomials for about half the primes $\ell$ and show how to deal with the other primes.

### 17.2.2 Schoof–Elkies–Atkin's algorithm

The improvements of Atkin and Elkies rely on a detailed analysis of the action of the Frobenius endomorphism $\phi_p$ on the group of $\ell$-torsion points $E[\ell]$ with $\ell \neq p$. Since $E[\ell] \simeq \mathbb{F}_\ell \times \mathbb{F}_\ell$, the Frobenius endomorphism can be represented by an element of $\mathrm{PGL}_2(\mathbb{F}_\ell)$. Atkin [ATK 1988, ATK 1991] devised an algorithm to compute the order of $\phi_p$ in $\mathrm{PGL}_2(\mathbb{F}_\ell)$ and showed how this can be used to count the number of points on $E(\mathbb{F}_p)$. Elkies [ELK 1991] on the other hand devised a method to replace the division polynomials of degree $(\ell^2 - 1)/2$ by a factor of degree $(\ell - 1)/2$ for about half the primes $\ell$.

Recall that the restriction of $\phi_q$ to $E[\ell]$ satisfies the equation

$$T^2 - t_\ell T + q_\ell = 0.$$

Depending on whether the discriminant $\Delta = t^2 - 4q$ is a square or a nonsquare in $\mathbb{F}_\ell$, the roots of this polynomial are defined over $\mathbb{F}_\ell$ or $\mathbb{F}_{\ell^2}$. In the former case, the prime $\ell$ is called an *Elkies prime* and in the latter case, $\ell$ is called an *Atkin prime*. Of course, since we do not know $t$, we cannot use the above definition to decide if $\ell$ is an Elkies or Atkin prime. Atkin proved that the $\ell$-th modular polynomial $\Phi_\ell(X, Y) \in \mathbb{Z}[X, Y]$ can be used to distinguish both cases.

The $\ell$-th modular polynomial $\Phi_\ell(X, Y) \in \mathbb{Z}[X, Y]$ (see Section 17.2.3) is a symmetric polynomial of degree $\ell + 1$ and has the following crucial property: let $\mathbb{F}_p$ be a finite field with $p \neq \ell$ and $E$ an elliptic curve over $\mathbb{F}_p$, then the $\ell + 1$ zeroes $\tilde{j} \in \overline{\mathbb{F}}_p$ of the polynomial $\Phi_\ell(X, j(E)) = 0$

## § 17.2 Overview of ℓ-adic methods

are precisely the $j$-invariants of the isogenous curves $\widetilde{E} = E/C$ with $C$ one of the $\ell+1$ cyclic subgroups of $E[\ell]$. Vélu's formulas [VÉL 1971] can be used to write down a Weierstraß equation for $\widetilde{E}$ and the isogeny $E \to \widetilde{E}$. The following proposition shows that $C$ can be defined over $\mathbb{F}_p[\tilde{j}]$ with $\tilde{j} \in \overline{\mathbb{F}}_p$ a zero of $\Phi_\ell(X, j(E)) = 0$.

**Proposition 17.11** Let $E$ be an ordinary elliptic curve over $\mathbb{F}_p$ with $j$-invariant $j \neq 0$ or 1728. Then

- the polynomial $\Phi_\ell(X, j)$ has a zero $\tilde{j} \in \mathbb{F}_{p^r}$ if and only if the kernel $C$ of the corresponding isogeny $E \to E/C$ is a one-dimensional eigenspace of $\phi_p^r$ in $E[\ell]$, with $\phi_p$ the Frobenius endomorphism of $E$
- the polynomial $\Phi_\ell(X, j)$ splits completely in $\mathbb{F}_{p^r}[T]$ if and only if $\phi_p^r$ acts as a scalar matrix on $E[\ell]$.

The above proposition already shows that $\phi_p$ has a one-dimensional eigenspace defined over $\mathbb{F}_p$ precisely when $\Phi_\ell(X, j)$ has a root in $\mathbb{F}_p$. However, Atkin [ATK 1991] proved that only certain factorizations can occur.

**Theorem 17.12 (Atkin)** Let $E$ be an ordinary elliptic curve defined over $\mathbb{F}_p$ with $j$-invariant $j \neq 0$ or 1728. Let $\Phi_\ell(X, j) = f_1 f_2 \cdots f_s$ be the factorization of $\Phi_\ell(X, j) \in \mathbb{F}_p[X]$ as a product of irreducible polynomials. Then there are the following possibilities for the degrees of $f_1, \ldots, f_s$:

1. $(1, \ell)$ or $(1, 1, \ldots, 1)$. In either case we have $t^2 - 4p \equiv 0 \pmod{\ell}$. In the former case we set $r = \ell$ and in the latter case $r = 1$.
2. $(1, 1, r, r, \ldots, r)$. In this case $t^2 - 4p$ is a square modulo $\ell$, $r$ divides $\ell - 1$ and $\phi_p$ acts on $E[\ell]$ as a diagonal matrix $\begin{bmatrix} \lambda & 0 \\ 0 & \mu \end{bmatrix}$ with $\lambda, \mu \in \mathbb{F}_\ell^*$.
3. $(r, r, \ldots, r)$ for some $r > 1$. In this case $t^2 - 4p$ is a nonsquare modulo $\ell$, $r$ divides $\ell + 1$ and the restriction of $\phi_p$ to $E[\ell]$ has an irreducible characteristic polynomial over $\mathbb{F}_\ell$.

In all cases, $r$ is the order of $\phi_p$ in the projective general linear group $\mathrm{PGL}_2(\mathbb{F}_\ell)$ and the trace $t$ of the Frobenius satisfies

$$t^2 \equiv p(\xi + \xi^{-1})^2 \pmod{\ell}, \tag{17.9}$$

for some primitive $r$-th root of unity $\xi \in \overline{\mathbb{F}}_\ell$. Furthermore, the number of irreducible factors $s$ satisfies

$$(-1)^s = \left(\frac{p}{\ell}\right).$$

To decide if a prime $\ell$ is an Atkin or Elkies prime, the above theorem shows that it suffices to compute $g(X) = \gcd(\Phi_\ell(X, j), X^p - X)$. Indeed, for $g(X) = 1$, $\ell$ is an Atkin prime, else $\ell$ is an Elkies prime. Since $\Phi_\ell(X, j)$ has degree $\ell + 1$, computing $g(X)$ only requires $O(\ell^\mu \lg p^{1+\mu})$ bit-operations.

**Atkin primes**

To limit the number of possibilities for $t \pmod \ell$ in case of an Atkin prime $\ell$, we need to compute the exact order $r$ of the Frobenius endomorphism in $\mathrm{PGL}_2(\mathbb{F}_\ell)$ by computing

$$g_i(X) = \gcd(\Phi_\ell(X, j), X^{p^i} - X),$$

for $i = 2, 3, \ldots$ until $g_i(X) = \Phi_\ell(X, j)$. To speed up this computation, $i$ should be limited to the divisors of $\ell + 1$ that furthermore satisfy $(-1)^{(\ell+1)/i} = \left(\frac{p}{\ell}\right)$. Once $r$ is determined, there are only $\varphi(r) \leq (\ell+1)/2$ choices for the $r$-th root of unity $\xi$, where $\varphi$ denotes Euler's totient function. By symmetry there are $\varphi(r)/2$ possible values for $t_\ell^2$ and accordingly $\varphi(r)$ values for $t_\ell$.

**Atkin** repeats this computation for various small primes $\ell$ and then uses a baby-step/giant-step-like algorithm to determine the correct value of the trace of Frobenius. Note that only the computations with the modular polynomials are polynomial time; the baby-step giant-step algorithm on the other hand is an exponential time algorithm.

### Elkies primes

If $\ell$ is an Elkies prime, then Theorem 17.12 implies that there exists a subgroup $C_\ell$ of order $\ell$ that is stable under $\phi_p$, i.e., $\phi_p(P) \in C_\ell$ for $P \in C_\ell$. This subgroup $C_\ell$ is an eigenspace of $\phi_p$ corresponding to one of the eigenvalues $\lambda$ or $\mu$ respectively, i.e., $\phi_p$ acts as multiplication by $\lambda$ or $\mu$ on $C_\ell$. Let

$$g_\ell(X) = \prod_{\pm P \in C_\ell \setminus \{P_\infty\}} (X - x(P)),$$

where $x(P)$ denotes the $x$-coordinate of $P$, then $\deg g_\ell = (\ell - 1)/2$ and $g_\ell$ could replace the division polynomial in the above explanation if we knew how to compute it (see Section 17.2.4). Furthermore as we can obtain $t_\ell$ as $t_\ell = \lambda + p_\ell/\lambda \in \mathbb{F}_\ell$ it suffices to find $0 \leqslant \lambda < \ell$ with $\phi(P) = [\lambda]P$, for all $P \in C_\ell$. The complexity of this calculation is determined by the computation of $X^p$ and $Y^p$ modulo $g_\ell(X)$; since $g_\ell$ has degree $O(\ell)$ instead of $O(\ell^2)$, this computation only takes $O(\ell^\mu \lg p^{1+\mu})$ bit-operations.

The main idea to obtain $g_\ell$ is to use an isogeny whose kernel is precisely $C_\ell$. This is possible as for every finite subgroup $C_\ell$ of $E(\overline{\mathbb{F}}_p)$ which is stable under the Frobenius endomorphism $\phi_p$, there exists an isogenous curve $E^{(\ell)}$ defined over $\mathbb{F}_p$ and a separable isogeny $\psi_\ell : E \to E^{(\ell)}$ with kernel equal to $C_\ell$. A detailed description of how to compute $g_\ell$ is given in Section 17.2.4.

## 17.2.3 Modular polynomials

In this section, we define the $\ell$-th modular polynomials $\Phi_\ell(X, Y)$ over $\mathbb{C}$ as minimum polynomials for $j$-invariants of curves isogenous to $E$ considered as modular functions for certain subgroups $\Gamma_0(\ell)$ of $\mathrm{SL}_2(\mathbb{Z})$. However, the classical modular polynomials $\Phi_\ell(X, Y)$ have two main drawbacks: the size of their coefficients increases badly as $\ell$ increases and their degree in $Y$ is too high. Polynomials with fewer and smaller coefficients can be obtained as minimum polynomials of different modular functions. In this section we only review one such alternative, called the canonical modular polynomials, and refer the interested reader to [MOR 1995, MÜL 1995] for more complete details. The mathematical background can be found in [SER 1970, SCH 1974, SIL 1994].

### 17.2.3.a Modular functions

As shown in Corollary 5.18, an elliptic curve $E$ defined over $\mathbb{C}$ is analytically isomorphic to the quotient of $\mathbb{C}$ by a lattice $\Lambda_\tau = \mathbb{Z} + \tau\mathbb{Z}$ where $\tau$ belongs to the *Poincaré upper half plane* $\mathcal{H} = \{z \in \mathbb{C} \mid \Im m\, z > 0\}$. The $j$-invariant of the elliptic curve defined by $\Lambda_\tau$ can be computed by the series

$$\forall \tau \in \mathcal{H}, \quad j(\tau) = \frac{1}{q} + 744 + \sum_{n \geqslant 1} c_n q^n$$

where the coefficients $c_n$ are integers and $q = e^{2i\pi\tau}$. For a matrix $M = \begin{bmatrix} a & b \\ c & d \end{bmatrix} \in \mathrm{SL}_2(\mathbb{Z})$, the lattice $(a\tau + b)\mathbb{Z} + (c\tau + d)\mathbb{Z}$ is isomorphic to $\Lambda_\tau$. The *fundamental domain* $\mathcal{F}$ of a subgroup $\Gamma$ of $\mathrm{SL}_2(\mathbb{Z})$ is defined as the connected open subset of $\mathcal{H}$ such that $\tau$ and $M(\tau) = (a\tau + b)/(c\tau + d)$ for $M \in \Gamma$ do not simultaneously belong to this subset. The fundamental domain $\mathcal{F}$ for $\mathrm{SL}_2(\mathbb{Z})$ is shown in Figure 17.1

**Figure 17.1** Fundamental domain for $SL_2(\mathbb{Z})$.

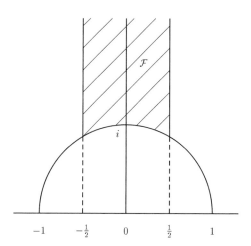

Since $E_\tau$ and $E_{M(\tau)}$ are isomorphic, we have $\forall \tau \in \mathcal{H}, \forall M \in SL_2(\mathbb{Z}), j(M(\tau)) = j(\tau)$, which shows that $j(\tau)$ is a *modular function*. More precisely, denote with

$$\mathcal{H}^* = \{z \in \mathbb{C} \mid \Im m\, z > 0\} \cup \mathbb{Q} \cup \{\infty\}$$

the *compactification of* $\mathcal{H}$, then a modular function for a discrete subgroup $\Gamma$ of $SL_2(\mathbb{Z})$ is a meromorphic function on the compact Riemann surface $\Gamma \setminus \mathcal{H}^*$.

**Definition 17.13** Let $\Gamma$ be a subgroup of finite index in $SL_2(\mathbb{Z})$; a modular function for $\Gamma$ is a complex function $f$ of $\mathcal{H}^*$ such that

1. $f$ is invariant under $\Gamma$, i.e., $\forall M \in \Gamma, f(M(\tau)) = f(\tau)$,
2. $f$ is meromorphic in $\mathcal{H}^*$ (see [SCH 1974] for a precise definition of this condition).

It is not difficult to see that modular functions for a subgroup $\Gamma$ form a field, denoted $K_\Gamma$. They satisfy the following two additional properties.

**Proposition 17.14** If $f$ is a modular function for $\Gamma$ and if $\Gamma'$ is a subgroup of finite index of $\Gamma$, then $f$ is a modular function for $\Gamma'$.

**Proposition 17.15** If $f$ is a modular function for $\Gamma$ and $M \in SL_2(\mathbb{Z})$, then $f_{|M} : \mathcal{H}^* \to \mathbb{C}$, $\tau \mapsto f(M(\tau))$, is a modular function for $M^{-1}\Gamma M$.

Modular functions are worth studying in our context mainly because a modular function for a discrete subgroup $\Gamma$ of $SL_2(\mathbb{Z})$ is a root of an algebraic equation of degree $d = [SL_2(\mathbb{Z}) : \Gamma]$ whose coefficients are rational functions of $j$. In other words, $K_\Gamma$ is an algebraic extension of $K_{SL_2(\mathbb{Z})} = \mathbb{C}(j)$ of degree $d$.

**Theorem 17.16** Any modular function $f$ for a subgroup $\Gamma$ of $SL_2(\mathbb{Z})$ with finite index $d$ is the root of an equation of degree $d$,

$$G(f, j) = \sum_{i=0}^{d} R_i(j) f^i = 0$$

where $R_i(j)$ are rational functions in $j(\tau)$ over $\mathbb{C}$. This equation is $G(X,j) = \prod_{i=1}^{d}(X - f_{|S_i})$ where $S_i$ are representatives of $\mathrm{SL}_2(\mathbb{Z})/\Gamma$.

The modular equations used in the SEA algorithm are built using the above theorem and depend on the specific discrete subgroups of $\mathrm{SL}_2(\mathbb{Z})$ and functions $f(\tau)$ chosen.

### 17.2.3.b Classical modular polynomials

For a prime $\ell$, the modular polynomial $\Phi_\ell(X,Y)$ is the minimum polynomial of the modular function $j_\ell : \tau \mapsto j(\ell\tau)$ for the subgroup $\Gamma_0(\ell)$ of $\mathrm{SL}_2(\mathbb{Z})$ defined by

$$\Gamma_0(\ell) = \left\{ \begin{bmatrix} a & b \\ c & d \end{bmatrix} \text{ with } ad - bc = 1 \text{ and } c \equiv 0 \pmod{\ell} \right\}.$$

In order to apply Theorem 17.16 to $j_\ell$, we need to know the structure of $\mathrm{SL}_2(\mathbb{Z})/\Gamma_0(\ell)$.

**Proposition 17.17** $\Gamma_0(\ell)$ is a subgroup of $\mathrm{SL}_2(\mathbb{Z})$ of finite index $\ell+1$. A set of representatives for the cosets of $\mathrm{SL}_2(\mathbb{Z})/\Gamma_0(\ell)$ is given by

$$\begin{bmatrix} 1 & 0 \\ 0 & 1 \end{bmatrix} \text{ and } \begin{bmatrix} 0 & -1 \\ 1 & v \end{bmatrix} \text{ for } 0 \leqslant v < \ell.$$

Therefore, the minimum polynomial of $j_\ell$ can be computed as

$$\Phi_\ell(X, j(\tau)) = (X - j(\ell\tau)) \prod_{i=0}^{\ell-1} \left( X - j\left(\frac{-\ell}{\tau+i}\right) \right).$$

The main disadvantage of this polynomial is its huge coefficients, caused by the fact that $j_\ell$ has a pole of large order at infinity.

**Example 17.18** To illustrate this disadvantage we give the classical modular polynomial for $\ell = 3$

$$\begin{aligned}
\Phi_3(X,Y) = {} & X^4 - X^3Y^3 + Y^4 + 2232(X^3Y^2 + X^2Y^3) - 1069956(X^3Y + XY^3) \\
& + 36864000(X^3 + Y^3) + 2587918086X^2Y^2 + 8900222976000(X^2Y + XY^2) \\
& + 452984832000000(X^2 + Y^2) - 770845966336000000XY \\
& + 1855425871872000000000(X+Y).
\end{aligned}$$

### 17.2.3.c Canonical modular polynomials

Instead of using $\Phi_\ell(X,Y)$, Atkin proposed to replace the function $j_\ell$ by the function $f_\ell$ defined by

$$f_\ell(\tau) = \ell^s \left( \frac{\eta(\ell\tau)}{\eta(\tau)} \right)^{2s}$$

with $s = 12/\gcd(12, \ell - 1)$ and Dedekind's $\eta$-function

$$\eta(\tau) = q^{1/24} \prod_{n=1}^{\infty}(1 - q^n),$$

where $q = e^{2i\pi\tau}$. One can show that $f_\ell(\tau)$ is given by a power series expansion of the form

$$f_\ell(\tau) = \ell^s q^v + \sum_{n \geqslant v+1} a_n q^n,$$

## § 17.2 Overview of $\ell$-adic methods

with $v = s(\ell - 1)/12$ and that $f_\ell$ is a modular function for $\Gamma_0(\ell)$.

By definition, the *canonical modular polynomial* is the minimum polynomial of $f_\ell$ given by

$$\Phi_\ell^c(X, j(\tau)) = (X - f_\ell(\tau)) \prod_{i=0}^{\ell-1} \left( X - f_\ell\left(\frac{-1}{\tau + i}\right) \right).$$

One can prove that the splitting behavior of $\Phi_\ell^c(X, j)$ is the same as the splitting behavior of $\Phi_\ell(X, j)$. However, to determine the $j$-invariant of the isogenous curve, we cannot simply compute a zero of $\Phi_\ell^c(X, j)$, but have to work in two stages. First, compute a zero $f$ of $\Phi_\ell^c(X, j)$ and then compute a zero of $\Phi_\ell^c(\ell^s/f, Y)$ which, gives the $j$-invariant of the isogenous curve. This follows immediately from the definition of $\Phi_\ell^c(X, Y)$, since $\Phi_\ell^c(f_\ell, j) = 0$ implies $\Phi_\ell^c(\ell^s/f_\ell, j_\ell) = 0$.

**Example 17.19** To illustrate the fact that the canonical modular polynomials have smaller coefficients and lower degree in $Y$ compared to the classical ones, we give $\Phi_5^c(X, Y)$:

$$\Phi_5^c(X, Y) = X^6 + 30X^5 + 315X^4 + 1300X^3 + 1575X^2 + (-Y + 750)X + 125.$$

### 17.2.4 Computing separable isogenies in finite fields of large characteristic

Assuming $\ell$ is an Elkies prime, Theorem 17.12 implies that there exists a subgroup $C_\ell \subset E[\ell]$ of order $\ell$ which is an eigenspace of the Frobenius endomorphism $\phi_p$. Furthermore, there exists an isogenous elliptic curve $\widetilde{E} = E/C_\ell$ and a separable isogeny of degree $\ell$ between $E$ and $\widetilde{E}$. This section not only shows how to compute a Weierstraß equation for the elliptic curve $\widetilde{E}$, but also the polynomial

$$g_\ell(X) = \prod_{\pm P \in C_\ell \smallsetminus \{P_\infty\}} (X - x(P)).$$

Both algorithms are due to Elkies [ELK 1991].

#### 17.2.4.a Isogenous curve

Clearly, the $j$-invariant $\tilde{j}$ of $\widetilde{E}$ can simply be obtained as a root of the polynomial $\Phi_\ell(X, j(E)) = 0$. However, in finite fields of large characteristic, the equation of $\widetilde{E}$ can not be recovered directly from $\tilde{j}$, since a twisted curve has the same $j$-invariant.

Elkies [ELK 1991] proved the following theorem, which provides a Weierstraß equation for $\widetilde{E}$, given the Weierstraß equation for $E$ and the $j$-invariant $\tilde{j}$. A detailed proof of this theorem can be found in an article by Schoof [SCH 1995].

**Theorem 17.20** Let $E$ be an ordinary elliptic curve over a large prime finite field $\mathbb{F}_p$ with $j$-invariant $j$ not $0$ or $1728$. Assume that $E$ is given by the Weierstraß equation $E : y^2 = x^3 + a_4 x + a_6$ and that $\widetilde{E}$ is $\ell$-isogenous over $\mathbb{F}_p$ to $E$. Let $\tilde{j}$ be the $j$-invariant of $\widetilde{E}$ then a Weierstraß equation for $\widetilde{E}$ is given by

$$\widetilde{E} : y^2 = x^3 + \tilde{a}_4 x + \tilde{a}_6$$

with

$$\tilde{a}_4 = -\frac{1}{48}\frac{\tilde{j}'^2}{\tilde{j}(\tilde{j} - 1728)} \quad \text{and} \quad \tilde{a}_6 = -\frac{1}{864}\frac{\tilde{j}'^3}{\tilde{j}^2(\tilde{j} - 1728)},$$

where $\tilde{j}' \in \mathbb{F}_p$ is given by

$$\tilde{j}' = -\frac{18}{\ell}\frac{a_6}{a_4}\frac{\Phi_{\ell,X}(j, \tilde{j})}{\Phi_{\ell,Y}(j, \tilde{j})} j$$

and $\Phi_{\ell,X}$ (resp. $\Phi_{\ell,Y}$) denotes the partial derivative of $\Phi_\ell(X, Y)$ with respect to $X$ (resp. $Y$).

**Remark 17.21** A similar theorem can be obtained for the canonical modular polynomial $\Phi_\ell^c(X,Y)$; the explicit formulas in this case can be found in [MOR 1995, MÜL 1995].

### 17.2.4.b Kernel of a separable isogeny

Let $E$ be defined by the Weierstraß equation $E : y^2 = x^3 + a_4 x + a_6$ and the isogenous elliptic curve $\widetilde{E} = E/C_\ell$ by $\widetilde{E} : y^2 = x^3 + \tilde{a}_4 x + \tilde{a}_6$. To compute the polynomial

$$g_\ell(X) = \prod_{\pm P \in C_\ell \smallsetminus \{P_\infty\}} (X - x(P)),$$

Elkies first computes the sum $p_1$ of the $x$-coordinates of the nonzero points in $E[\ell]$. A proof of the following theorem can again be found in [SCH 1995].

**Theorem 17.22** Let $E_4 = -48 a_4$ and $E_6 = 864 a_6$ and similarly, $\widetilde{E}_4 = -48 \tilde{a}_4$ and $\widetilde{E}_6 = 864 \tilde{a}_6$, then

$$p_1 = \sum_{P \in C_\ell \smallsetminus \{P_\infty\}} x(P)$$

is given by

$$p_1 = \frac{\ell}{2} J + \frac{\ell}{4}\left( \frac{E_4^2}{E_6} - \ell \frac{\widetilde{E}_4^2}{\widetilde{E}_6} \right) + \frac{\ell}{3}\left( \frac{E_6}{E_4} - \ell \frac{\widetilde{E}_6}{\widetilde{E}_4} \right),$$

where $J$ is defined as

$$J = -\frac{j'^2 \Phi_{\ell,XX}(j,\tilde{j}) + 2\ell j' \tilde{j}' \Phi_{\ell,XY}(j,\tilde{j}) + \ell^2 \tilde{j}'^2 \Phi_{\ell,YY}(j,\tilde{j})}{j' \Phi_{\ell,X}(j,\tilde{j})}$$

and $j' = -j E_6 / E_4$ and $\tilde{j}' = -\tilde{j} \widetilde{E}_6 / \widetilde{E}_4$. The notation for the partial derivatives is as above, for instance $\Phi_{\ell,XX}$ is shorthand for $\partial^2 \Phi_\ell / \partial X^2$.

Given the equation for $E$ and $\widetilde{E}$ and the value of $p_1$, there are various algorithms to compute the polynomial $g_\ell(X)$. In this section we follow the description in [SCH 1995] and refer to [MOR 1995, MÜL 1995] for alternative algorithms.

Instead of working with $\widetilde{E}$, Schoof works with the isomorphic curve $\widehat{E} : y^2 = x^3 + \hat{a}_4 x + \hat{a}_6$ with $\hat{a}_4 = \ell^4 \tilde{a}_4$ and $\hat{a}_6 = \ell^6 \tilde{a}_6$. To $E$ we can associate the reduced Weierstraß $\wp$-function by

$$\wp(z) = \frac{1}{z^2} + \sum_{k=1}^\infty c_k z^{2k}$$

where the coefficients are given by

$$c_1 = -\frac{a_4}{5}, \quad c_2 = -\frac{a_6}{7}, \quad c_k = \frac{3}{(k-2)(2k+3)} \sum_{j=1}^{k-2} c_j c_{k-1-j}, \quad \text{for } k \geqslant 3.$$

The function $\hat{\wp}$ for $\widehat{E}$ is defined similarly using the coefficients $\hat{a}_4$ and $\hat{a}_6$. The following theorem provides an efficient method to compute $g_\ell(X)$ given $E$, $\widetilde{E} = E/C_\ell$ and $p_1$.

**Theorem 17.23** Let $g_\ell(X)$ be the polynomial that vanishes on the $x$-coordinates of the points in the kernel of the isogeny $\psi_\ell : E \to \widetilde{E}$, then

$$z^{\ell-1} g_\ell(\wp(z)) = \exp\left( -\frac{1}{2} p_1 z^2 - \sum_{k=1}^\infty \frac{\hat{c}_k - \ell c_k}{(2k+1)(2k+2)} z^{2k+2} \right).$$

§ 17.2 Overview of $\ell$-adic methods

Let $g_\ell(X) = X^d + \sum_{k=0}^{d-1} g_{\ell,k} X^k$, then by the above theorem, the first few coefficients of $g_\ell(X)$ are

$$g_{\ell,d-1} = -\frac{p_1}{2},$$

$$g_{\ell,d-2} = \frac{p_1^2}{8} - \frac{\hat{c}_1 - \ell c_1}{12} - \frac{\ell - 1}{2} c_1,$$

$$g_{\ell,d-3} = -\frac{p_1^3}{48} - \frac{\hat{c}_2 - \ell c_2}{30} + p_1 \frac{\hat{c}_1 - \ell c_1}{24} - \frac{\ell - 1}{2} c_2 + \frac{\ell - 3}{4} c_1 p_1.$$

**Remark 17.24** The above algebraic relations are in fact reductions of relations over $\mathbb{C}$; for these relations to hold over a finite field, the characteristic $p$ must be large enough. In particular, the coefficients $c_k$ and $\hat{c}_k$ need inversions of integers of the form $(k-2)(2k+3)$, which can be equal to zero when $k \geqslant (p-3)/2$. For finite fields of small characteristic, alternative algorithms to compute $g_\ell(X)$ have been developed; the most important references are [LER 1996, COU 1996].

### 17.2.5 Complete SEA algorithm

To conclude we give an overview of the SEA algorithm in pseudo-code. The time complexity of Algorithm 17.25 amounts to $O\big((\lg p)^{2\mu+2}\big)$ bit-operations and the space complexity is $O\big((\lg p)^2\big)$.

---

**Algorithm 17.25** SEA

INPUT: An ordinary elliptic curve $E$ over $\mathbb{F}_p$, with $j$-invariant not $0$ or $1728$.
OUTPUT: The number of points $|E(\mathbb{F}_p)|$.

1. $\ell \leftarrow 3, M_A \leftarrow 1, A \leftarrow \{\}, M_E \leftarrow 2$ and $E \leftarrow \{(t_2, 2)\}$      [$t_2$ like in the introduction]
2. **while** $M_E \times M_A < 4\sqrt{p}$ **do**
3.      compute a modular equation $\Phi_\ell(X, Y) \pmod{p}$
4.      find the splitting of $\Phi_\ell(X, j) \pmod{p}$
5.      **if** $\ell$ is an Elkies prime **then**
6.          compute $j^{(\ell)}$ from $\Phi_\ell(X, j)$ in $\mathbb{F}_p$
7.          determine an isogeny $\psi_\ell : E \to E^{(\ell)}$ and $g_\ell$
8.          find an eigenvalue $\lambda$ of $\phi$ in $\mathbb{F}_\ell$
9.          $t_\ell \leftarrow \lambda + p/\lambda \bmod \ell$
10.         $E \leftarrow E \cup \{(t_\ell, \ell)\}$ and $M_E \leftarrow M_E \times \ell$
11.      **else**                                                                                                            [$\ell$ is an Atkin prime]
12.          determine the set $T_\ell$ of possible values for $t \pmod \ell$
13.          $A \leftarrow A \cup \{(T_\ell, \ell)\}$ and $M_A \leftarrow M_A \times \ell$
14.      $\ell \leftarrow \texttt{nextprime}(\ell)$
15. determine $t$ from $E \cup A$ using match and sort algorithm [ATK 1988, LER 1997]
16. **return** $p + 1 - t$

---

**Example 17.26** Let $E$ be the curve defined over $\mathbb{F}_{257}$ by $y^2 = x^3 + x + 2$. Assume we want to compute the trace $t_5$ of the Frobenius endomorphism modulo 5. Evaluating $\Phi_5^c(X, j) \pmod{p}$ with

$j = 25$ shows that 5 is an Elkies prime for $E$ since

$$\Phi_5^c(X, 25) \equiv (X+74)(X+137)(X^4 + 76\,X^3 + 98\,X^2 + 152\,X + 124) \pmod{p}.$$

This polynomial has two roots, $f_1 = 120$ and $f_2 = 183$. Assume we choose the root $f_2 = 183$, then the $j$-invariant of the isogenous curve can be determined as the root of $\Phi_5^c(\ell^s/f_2, Y) \equiv \Phi_5^c(231, Y) \equiv 0 \pmod{p}$. The root of this polynomial is $j^{(5)} = 166$. Using Elkies' formulas, we know that the curve $E$ is therefore isogenous to

$$E^{(5)} : y^2 = x^3 + 86x + 204,$$

and that the sum of the $x$-coordinates of the nonzero points in the kernel of the isogeny is equal to $p_1 = 225$. Finally, we recover

$$g_5(X) = X^2 + 16X + 21.$$

An easy calculation then shows that

$$X^{257} \equiv 256X + 241 \pmod{g_\ell(X)}$$
$$Y^{257} \equiv (144X + 61)Y \pmod{g_\ell(X), Y^2 - X^3 - X - 2}.$$

Determining the eigenvalue gives $\lambda = 3$ and finally we conclude that $t_5 = \lambda + 257/\lambda \equiv 2 \pmod{5}$.

**Remarks 17.27**

(i) In 1990, Pila [PIL 1990] described a generalization of Schoof's algorithm that computes the characteristic polynomial of Frobenius on an arbitrary abelian variety over a finite field. Although the algorithm has a polynomial time complexity in the size of the finite field, it is rather impractical since it requires as input a system of polynomial equations describing the abelian variety and a set of rational functions that express the group law. Even for the Jacobian of a genus 2 curve, this requires 72 quadratic equations in 15 variables.

(ii) Kampkötter [KAM 1991] developed an algorithm specifically for hyperelliptic curves including recurrence formulas describing the $\ell$-torsion. Although the algorithm is much faster than Pila's, it is still too slow to be of cryptographic interest.

(iii) Adleman and Huang [ADHU 1996, ADHU 2001] and Huang and Ierardi [HUIE 1998] improved Pila's work by restricting to curves, but again the algorithms are more of theoretical interest and far too slow for cryptographic purposes.

(iv) For curves over finite fields of large characteristic, only the genus 2 case is more or less practical thanks to the work of Gaudry and Harley [GAHA 2000] and the improvements by Gaudry and Schost [GASC 2004a]. The current situation is best compared to Schoof's algorithm for elliptic curves and as such, the algorithm is just fast enough to reach the cryptographic range. The corresponding improvements by Elkies and Atkin are still lacking for higher genus curves, although Gaudry and Schost [GASC 2005] have introduced modular polynomials using a purely algebraic construction based on ideas by Charlap, Coley and Robbins [CHCO$^+$ 1991].

## 17.3 Overview of $p$-adic methods

The history of $p$-adic algorithms to compute the zeta function of curves or even algebraic varieties, is a rather odd one. Indeed, in 1960 Dwork [DWO 1960] already illustrated the power of the $p$-adic

## § 17.3 Overview of $p$-adic methods

approach in his proof of the rationality of the zeta function of an algebraic variety. Furthermore, at the end of the 1970's, Kato and Lubkin [KALU 1982] designed a $p$-adic algorithm to compute the number of points on an elliptic curve that runs in polynomial time for fixed $p$, but nobody actually implemented their algorithm.

Finally, at the end of 1999, Satoh [SAT 2000] realized that the $p$-adic approach is, at least for small $p$, much more powerful than the existing $\ell$-adic methods. Not only did Satoh describe a $p$-adic algorithm to compute the number of points on an ordinary elliptic curve over a finite field, but also illustrated its efficiency with an implementation.

Since then, many $p$-adic algorithms have been designed and in this section we give an overview of the three most practical ones: Satoh's algorithm, the Arithmetic-Geometric-Mean algorithm and finally, an algorithm based on Monsky–Washnitzer cohomology.

### 17.3.1 Satoh's algorithm

In [SAT 2000], Satoh shows how to efficiently compute a $p$-adic approximation of the canonical lift of an ordinary elliptic curve $E$ over a finite field and how this can be used to count the number of points on $E$.

#### 17.3.1.a The canonical lift of an ordinary elliptic curve

In this section, we specialize the theory of the canonical lift of ordinary abelian varieties to ordinary elliptic curves over a finite field $\mathbb{F}_q$, with $q = p^d$ and $p$ prime.

**Definition 17.28** The canonical lift $\mathcal{E}$ of an ordinary elliptic curve $E$ over $\mathbb{F}_q$ is an elliptic curve over $\mathbb{Q}_q$ which satisfies:

- the reduction of $\mathcal{E}$ modulo $p$ equals $E$,
- the ring homomorphism $\mathrm{End}(\mathcal{E}) \to \mathrm{End}(E)$ induced by reduction modulo $p$ is an isomorphism.

Deuring [DEU 1941] showed that the canonical lift $\mathcal{E}$ always exists and is unique up to isomorphism. The above definition is actually one of many ways in which the canonical lift can be characterized.

**Theorem 17.29** Let $E/\mathbb{F}_q$ denote an ordinary elliptic curve and let $\mathcal{E}/\mathbb{Q}_q$ be a lift of $E$, then the following statements are equivalent:

- $\mathcal{E}$ is the canonical lift of $E$.
- Reduction modulo $p$ induces an isomorphism $\mathrm{End}(\mathcal{E}) \simeq \mathrm{End}(E)$.
- The $q$-th power Frobenius $\phi_q \in \mathrm{End}(E)$ lifts to an endomorphism $\mathcal{F}_q \in \mathrm{End}(\mathcal{E})$.
- The $p$-th power Frobenius isogeny $\phi_p : E \to E^\sigma$ lifts to an isogeny $\mathcal{F}_p : \mathcal{E} \to \mathcal{E}^\Sigma$, with $\Sigma$ the Frobenius substitution on $\mathbb{Q}_q$.

The last property implies that there exists an isogeny of degree $p$ between the canonical lift $\mathcal{E}$ and its conjugate $\mathcal{E}^\Sigma$. By the properties of the $p$-th modular polynomial $\Phi_p(X, Y) \in \mathbb{Z}[X, Y]$, this implies that $\Phi_p\bigl(j(\mathcal{E}), \Sigma(j(\mathcal{E}))\bigr) = 0$. The next theorem by Lubin, Serre and Tate [LUSE+ 1964] shows that this is in fact a sufficient condition.

**Theorem 17.30 (Lubin–Serre–Tate)** Let $E$ be an ordinary elliptic curve over $\mathbb{F}_q$ with $j$-invariant $j(E) \in \mathbb{F}_q \smallsetminus \mathbb{F}_{p^2}$. Denote with $\Sigma$ the Frobenius substitution on $\mathbb{Q}_q$ and with $\Phi_p(X, Y)$ the $p$-th modular polynomial. Then the system of equations

$$\Phi_p\bigl(X, \Sigma(X)\bigr) = 0 \text{ and } X \equiv j(E) \pmod{p}, \tag{17.10}$$

has a unique solution $J \in \mathbb{Z}_q$, which is the $j$-invariant of the canonical lift $\mathcal{E}$ of $E$ (defined up to isomorphism).

The hypothesis $j(E) \notin \mathbb{F}_{p^2}$ in Theorem 17.30 is necessary to ensure that a certain partial derivative of $\Phi_p$ does not vanish modulo $p$ and guarantees the uniqueness of the solution of (17.10). The case $j(E) \in \mathbb{F}_{p^2}$ can be handled very easily as shown in Section 17.1.2, since the curve is then isomorphic to a curve defined over $\mathbb{F}_p$ or $\mathbb{F}_{p^2}$. In the remainder of this section we will therefore assume that $j(E) \notin \mathbb{F}_{p^2}$ and in particular that $E$ is ordinary [SIL 1986, Theorem V.3.1].

### 17.3.1.b Isogeny cycles

To compute $j(\mathcal{E}) \bmod p^N$, Satoh considered $E$ together with all its conjugates $E_i = E^{\sigma^i}$ with $0 \leqslant i < d$ and $\sigma$ the $p$-th power Frobenius automorphism of $\mathbb{F}_q$. Let $\phi_{p,i}$ denote the $p$-th power Frobenius isogeny $\phi_{p,i} : E_i \to E_{i+1} : (x, y) \mapsto (x^p, y^p)$, then we obtain the following cycle

$$E_0 \xrightarrow{\phi_{p,0}} E_1 \xrightarrow{\phi_{p,1}} \cdots \xrightarrow{\phi_{p,d-2}} E_{d-1} \xrightarrow{\phi_{p,d-1}} E_0.$$

Composing these isogenies, we can express the Frobenius endomorphism as

$$\phi_q = \phi_{p,d-1} \circ \phi_{p,d-2} \circ \cdots \circ \phi_{p,0}.$$

Instead of lifting $E$ separately, Satoh lifts the whole cycle $(E_0, E_1, \ldots, E_{d-1})$ simultaneously leading to the diagram

$$\begin{array}{ccccccccc}
\mathcal{E}_0 & \xrightarrow{\mathcal{F}_{p,0}} & \mathcal{E}_1 & \xrightarrow{\mathcal{F}_{p,1}} & \cdots & \xrightarrow{\mathcal{F}_{p,d-2}} & \mathcal{E}_{d-1} & \xrightarrow{\mathcal{F}_{p,d-1}} & \mathcal{E}_0 \\
\pi_1 \downarrow & & \pi_1 \downarrow & & & & \pi_1 \downarrow & & \pi_1 \downarrow \\
E_0 & \xrightarrow{\phi_{p,0}} & E_1 & \xrightarrow{\phi_{p,1}} & \cdots & \xrightarrow{\phi_{p,d-2}} & E_{d-1} & \xrightarrow{\phi_{p,d-1}} & E_0,
\end{array} \quad (17.11)$$

with $\mathcal{E}_i$ the canonical lift of $E_i$ and $\mathcal{F}_{p,i}$ the corresponding lift of $\phi_{p,i}$, which exists by Theorem 17.29.

### 17.3.1.c Computing the canonical lift

The theorem of Lubin, Serre, and Tate implies that the $j$-invariant of $\mathcal{E}$ is uniquely determined by $\Phi_p\big(j(\mathcal{E}), \Sigma(j(\mathcal{E}))\big) = 0$ and $j(\mathcal{E}) \equiv j(E) \pmod{p}$. Note that this system of equations can be solved efficiently using the algorithms described in Section 12.7, which were designed specifically with this application in mind.

In Satoh's original algorithm however, this lifting problem is solved as follows: clearly, the $j$-invariants of the $\mathcal{E}_i$ satisfy

$$\Phi_p\big(j(\mathcal{E}_{i+1}), j(\mathcal{E}_i)\big) = 0 \quad \text{and} \quad j(\mathcal{E}_i) \equiv j(E_i) \pmod{p},$$

for $i = 0, \ldots, d-1$. Define $\Theta : \mathbb{Z}_q^d \longrightarrow \mathbb{Z}_q^d$ by

$$\Theta(x_0, x_1, \ldots, x_{d-1}) = \big(\Phi_p(x_0, x_1), \Phi_p(x_1, x_2), \ldots, \Phi_p(x_{d-1}, x_0)\big),$$

then we have $\Theta\big(j(\mathcal{E}_{d-1}), j(\mathcal{E}_{d-2}), \ldots, j(\mathcal{E}_0)\big) = (0, 0, \ldots, 0)$. Using a multivariate Newton iteration on $\Theta$, we can lift the cycle of $j$-invariants $\big(j(E_{d-1}), j(E_{d-2}), \ldots, j(E_0)\big)$ to $\mathbb{Z}_q^d$ with arbitrary precision. The iteration is given by

$$(x_0, x_1, \ldots, x_{d-1}) \leftarrow (x_0, x_1, \ldots, x_{d-1}) - \big((D\Theta)^{-1}\Theta\big)(x_0, x_1, \ldots, x_{d-1}),$$

## § 17.3 Overview of $p$-adic methods

with $(D\Theta)(x_0, \ldots, x_{d-1})$ the Jacobian matrix

$$\begin{bmatrix} \Phi_p'(x_0, x_1) & \Phi_p'(x_1, x_0) & 0 & \cdots & 0 \\ 0 & \Phi_p'(x_1, x_2) & \Phi_p'(x_2, x_1) & \cdots & 0 \\ \vdots & \vdots & \vdots & \ddots & \vdots \\ 0 & 0 & 0 & \cdots & \Phi_p'(x_{d-1}, x_{d-2}) \\ \Phi_p'(x_0, x_{d-1}) & 0 & 0 & \cdots & \Phi_p'(x_{d-1}, x_0) \end{bmatrix},$$

where $\Phi_p'(X, Y)$ denotes the partial derivative with respect to $X$. Note that $\Phi_p'(Y, X)$ is the partial derivative of $\Phi_p(X, Y)$ with respect to $Y$ since $\Phi_p(X, Y)$ is symmetric.

The $p$-th modular polynomial satisfies the Kronecker relation

$$\Phi_p(X, Y) \equiv (X^p - Y)(X - Y^p) \pmod{p}$$

and since $j(E_i) \notin \mathbb{F}_{p^2}$ and $j(E_{i+1}) = j(E_i)^p$, we have

$$\begin{cases} \Phi_p'\big(j(E_{i+1}), j(E_i)\big) \equiv j(E_i)^{p^2} - j(E_i) \not\equiv 0 \pmod{p}, \\ \Phi_p'\big(j(E_i), j(E_{i+1})\big) \equiv j(E_i)^p - j(E_i)^p \equiv 0 \pmod{p}. \end{cases}$$

The above equations imply that $(D\Theta)(x_0, \ldots, x_{d-1}) \pmod{p}$ is a diagonal matrix with nonzero diagonal elements. Therefore, the Jacobian matrix is invertible over $\mathbb{Z}_q$ and we conclude that $\delta = \big((D\Theta)^{-1}\Theta\big)(x_0, x_1, \ldots, x_{d-1}) \in \mathbb{Z}_q^d$. Note that we can simply apply Gaussian elimination to solve

$$(D\Theta)(x_0, \ldots, x_{d-1})\delta = \Theta(x_0, \ldots, x_{d-1})$$

since the diagonal elements are all invertible. Using row operations we move the bottom left element $\Phi_p'(x_0, x_{d-1})$ towards the right. After $k$ row operations this element becomes

$$(-1)^k \Phi_p'(x_0, x_{d-1}) \prod_{i=0}^{k-1} \frac{\Phi_p'(x_{i+1}, x_i)}{\Phi_p'(x_i, x_{i+1})}$$

which clearly is divisible by $p^k$ since $\Phi_p'(x_{i+1}, x_i) \equiv 0 \pmod{p}$. This procedure is summarized in Algorithm 17.31.

---

**Algorithm 17.31** Lift $j$-invariants

INPUT: A cycle $j_i \in \mathbb{F}_q \setminus \mathbb{F}_{p^2}$ with $\Phi_p(j_{i+1}, j_i) \equiv 0 \pmod{p}$ for $0 \leqslant i < d$ and precision $N$.
OUTPUT: A cycle $J_i \in \mathbb{Z}_q$ with $\Phi_p(J_{i+1}, J_i) \equiv 0 \pmod{p^N}$ and $J_i \equiv j_i \pmod{p}$ for all $0 \leqslant i < d$.

1. **if** $N = 1$ **then**
2.     **for** $i = 0$ **to** $d - 1$ **do** $J_i \leftarrow j_i$
3. **else**
4.     $N' \leftarrow \lceil \frac{N}{2} \rceil$ and $M \leftarrow N - N'$
5.     $(J_0, \ldots, J_{d-1}) \leftarrow$ Lift $j$-invariants$\big((j_0, \ldots, j_{d-1}), N'\big)$
6.     **for** $i = 0$ **to** $d - 2$ **do**
7.         $t \leftarrow \Phi_p'(J_i, J_{i+1})^{-1} \bmod p^M$
8.         $D_i \leftarrow t\Phi_p'(J_{i+1}, J_i) \bmod p^M$

9. $\quad P_i \leftarrow t\big((\Phi_p(J_{i+1}, J_i) \bmod p^N)/p^{N'}\big) \bmod p^M$
10. $\quad R \leftarrow \Phi'_p(J_0, J_{d-1}) \bmod p^M$
11. $\quad S \leftarrow \Big(\big((\Phi_p(J_0, J_{d-1}) \bmod p^N)\big)/p^{N'}\Big) \bmod p^M$
12. $\quad$ **for** $i = 0$ **to** $\min\{M, d-2\}$ **do**
13. $\quad\quad S \leftarrow (S - RP_i) \bmod p^M$
14. $\quad\quad R \leftarrow (-RD_i) \bmod p^M$
15. $\quad R \leftarrow \big(R + \Phi'_p(J_{d-1}, J_0)\big) \bmod p^M$
16. $\quad P_{d-1} \leftarrow (SR^{-1}) \bmod p^M$
17. $\quad$ **for** $i = d-2$ **down to** $0$ **do** $P_i \leftarrow (P_i - D_i P_{i+1}) \bmod p^M$
18. $\quad$ **for** $i = 0$ **to** $d-1$ **do** $J_i \leftarrow (J_i - p^{N'} P_i) \bmod p^N$
19. **return** $(J_0, \ldots, J_{d-1})$

**Example 17.32** Let $p = 7$, $d = 5$ and $\mathbb{F}_{p^d} \simeq \mathbb{F}_p(\theta)$ with $\theta^5 + \theta + 4 = 0$. Consider the $j$-invariant $j_0 = 3\theta^4 + 6\theta^3 + 2\theta$, then Algorithm 17.31 computes the $j$-invariant of the canonical lift to precision $N = 10$ with $\mathbb{Z}_q \simeq \mathbb{Z}_p[T]/\big(G(T)\big)$ and $G(T) = T^5 + T + 4$ as

$$J_0 \equiv 249888299T^4 + 236778044T^3 + 9871351T^2 + 169542361T + 26531974 \pmod{p^N}.$$

As shown in Section 4.4.2.a, we can assume that either $E$ or its quadratic twist is given by an equation of the form

$$\begin{aligned} p = 2 &: \quad y^2 + xy = x^3 + a_6, & j(E) &= 1/a_6, \\ p = 3 &: \quad y^2 = x^3 + x^2 + a_6, & j(E) &= -1/a_6, \\ p > 5 &: \quad y^2 = x^3 + 3ax + 2a, & j(E) &= 1728a/(1+a). \end{aligned}$$

Once the $j$-invariant $j(\mathcal{E})$ of the canonical lift of $E$ is computed, a Weierstraß model for $\mathcal{E}$ is given by

$$\begin{aligned} p = 2 &: \quad y^2 + xy = x^3 + 36\alpha x + \alpha, & \alpha &= 1/\big(1728 - j(\mathcal{E})\big), \\ p = 3 &: \quad y^2 = x^3 + x^2/4 + 36\alpha x + \alpha, & \alpha &= 1/\big(1728 - j(\mathcal{E})\big), \\ p > 5 &: \quad y^2 = x^3 + 3\alpha x + 2\alpha, & \alpha &= j(\mathcal{E})/\big(1728 - j(\mathcal{E})\big). \end{aligned}$$

Note that the above models have the correct $j$-invariant $j(\mathcal{E})$ and reduce to $E$ modulo $p$.

### 17.3.1.d The trace of Frobenius

By analyzing the action of $\mathcal{F}_q$ on a holomorphic differential on $\mathcal{E}$ (see Section 4.4.2.c), we can easily recover the trace $\text{Tr}(\phi_q)$ of the Frobenius endomorphism $\phi_q$.

**Proposition 17.33 (Satoh)** Let $E$ be an elliptic curve over $\mathbb{Q}_q$ having good reduction modulo $p$ and let $f \in \text{End}(E)$ of degree $n$. Let $\omega$ be a holomorphic differential on $E$ and let $f^*(\omega) = c\omega$ be the action of $f$ on $\omega$, then

$$\text{Tr}(f) = c + \frac{n}{c}.$$

Note that if we apply the above proposition to the lifted Frobenius endomorphism $\mathcal{F}_q$, we get $\text{Tr}(\mathcal{F}_q) = b + q/b$ with $\mathcal{F}_q^*(\omega) = b\omega$. Since $E$ is ordinary it follows that either $b$ or $q/b$ is a unit in $\mathbb{Z}_q$. However, $\phi_q$ is inseparable and thus $b \equiv 0 \pmod{p}$, which implies that $b \equiv 0 \pmod{q}$, since

§ 17.3 Overview of $p$-adic methods

$q/b$ has to be a unit in $\mathbb{Z}_q$. So if we want to compute $\mathrm{Tr}(\mathcal{F}_q)$ mod $p^N$, we would need to determine $b$ mod $p^{d+N}$. Furthermore, it turns out to be quite difficult to compute $b$ directly. As will become clear later, we would need to know $\ker(\mathcal{F}_{p,i})$, which is subgroup of $\mathcal{E}_i[p]$ of order $p$. However, since $\ker(\phi_{p,i})$ is trivial we cannot use a simple lift of $\ker(\phi_{p,i})$ to $\ker(\mathcal{F}_{p,i})$, but would need to factor the $p$-division polynomial of $\mathcal{E}_i$.

To avoid these problems, Satoh works with the Verschiebung $\overline{V}_q$, i.e., the dual isogeny of $\phi_q$, which is separable since $E$ is ordinary and thus $\ker(\overline{V}_q)$ can be easily lifted to $\ker(\mathcal{V}_q)$, with $\mathcal{V}_q$ the image of $\overline{V}_q$ under the ring isomorphism $\mathrm{End}(E) \simeq \mathrm{End}(\mathcal{E})$. Furthermore, the trace of an endomorphism equals the trace of its dual, so we have $\mathrm{Tr}(\mathcal{F}_q) = \mathrm{Tr}(\mathcal{V}_q) = c + q/c$ with $\mathcal{V}_q^*(\omega) = c\,\omega$ and $c$ a unit in $\mathbb{Z}_q$. Diagram (17.11) shows that $\mathcal{V}_q = \widehat{\mathcal{F}}_{p,0} \circ \widehat{\mathcal{F}}_{p,1} \circ \cdots \circ \widehat{\mathcal{F}}_{p,d-1}$ and therefore we can compute $c$ from the action of $\widehat{\mathcal{F}}_{p,i}$ on a holomorphic differential $\omega_i$ of $\mathcal{E}_i$ for $i = 0, \ldots, d-1$. More precisely, take $\omega_i = \omega^{\Sigma^i}$ for $0 \leqslant i < d$ and let $c_i$ be defined by

$$\widehat{\mathcal{F}}_{p,i}^*(\omega_i) = c_i\,\omega_{i+1},$$

then $c = \prod_{0 \leqslant i < d} c_i$. Since $\overline{V}_q$ is separable, $c$ will be nonzero modulo $p$ and we conclude

$$\mathrm{Tr}(\phi_q) \equiv \prod_{0 \leqslant i < d} c_i \pmod{q}.$$

Since all commutative squares in diagram (17.11) are conjugates of each other, we can also recover the trace of Frobenius as the norm of $c_0$, i.e.,

$$\mathrm{Tr}(\phi_q) \equiv \mathrm{N}_{\mathbb{Q}_q/\mathbb{Q}_p}(c_0) \pmod{q}.$$

#### 17.3.1.e Computing the $c_i$

The final step in Satoh's algorithm is to compute the coefficients $c_i$ using the equations for $\mathcal{E}_i$ and $\mathcal{E}_{i+1}$ and the kernel of $\widehat{\mathcal{F}}_{p,i}$. Consider the following diagram:

$$\begin{array}{ccc} \mathcal{E}_{i+1} & \xrightarrow{\widehat{\mathcal{F}}_{p,i}} & \mathcal{E}_i \\ & \searrow\nu_i \quad \nearrow\lambda_i & \\ & \mathcal{E}_{i+1}/\ker(\widehat{\mathcal{F}}_{p,i}) & \end{array} \qquad (17.12)$$

Given $\ker(\widehat{\mathcal{F}}_{p,i})$, Satoh uses Vélu's formulas [VÉL 1971] to compute an equation for the curve $\mathcal{E}_{i+1}/\ker(\widehat{\mathcal{F}}_{p,i})$ and the isogeny $\nu_i$. Since $\nu_i$ and $\widehat{\mathcal{F}}_{p,i}$ are both separable and $\ker(\nu_i) = \ker(\widehat{\mathcal{F}}_{p,i})$, there exists an isomorphism $\lambda_i : \mathcal{E}_{i+1}/\ker(\widehat{\mathcal{F}}_{p,i}) \longrightarrow \mathcal{E}_i$ that makes the above diagram commutative. Due to Vélu's construction, the action of $\nu_i$ on the chosen holomorphic differentials is trivial, i.e., $\nu_i^*(\omega_{i+1,K}) = \omega_{i+1}$ with $\omega_{i+1,K}$ the holomorphic differential on $\mathcal{E}_{i+1}/\ker(\widehat{\Sigma}_i)$. Therefore, it is sufficient to compute the action of $\lambda_i$ on $\omega_i$.

Note that $\ker(\widehat{\mathcal{F}}_{p,i})$ is a subgroup of order $p$ of $\mathcal{E}_{i+1}[p]$. Let $H_i(x)$ be

$$H_i(x) = \prod_{P \in (\ker(\widehat{\mathcal{F}}_{p,i}) \smallsetminus \{P_\infty\})/\pm} (x - x(P)),$$

then $H_i(x)$ divides the $p$-division polynomial $\Psi_{p,i+1}(x)$ of $\mathcal{E}_{i+1}$. To find the correct factor of $\Psi_{p,i+1}(x)$ Satoh proves the following lemma.

**Lemma 17.34 (Satoh)** Let $p \geq 3$, then $\ker \widehat{\Sigma}_i = \mathcal{E}_{i+1}[p] \cap \mathcal{E}_{i+1}(\mathbb{Z}_q^{\mathrm{ur}})$, with $\mathbb{Z}_q^{\mathrm{ur}}$ the valuation ring of the maximal unramified extension $\mathbb{Q}_q^{\mathrm{ur}}$ of $\mathbb{Q}_q$.

The above lemma implies that $H_i(x) \in \mathbb{Z}_q[x]$ is the unique monic polynomial that divides $\Psi_{p,i+1}(x)$ and such that $H_i(x)$ is squarefree modulo $p$ of degree $(p-1)/2$. Since $E_i$ is ordinary, $\ker(\phi_{p,i}) = E_{i+1}[p]$ and $\Psi_{p,i+1}(x) \pmod{p}$ has inseparable degree $p$. Therefore, $\delta H_i(x)^p \equiv \Psi_{p,i+1}(x) \pmod{p}$, with $\delta$ the leading coefficient of $\Psi_{p,i+1}$. This implies that we cannot apply Hensel's lemma 3.17, since the polynomials $H_i(x) \bmod p$ and $\Psi_{p,i+1}(x)/H_i(x) \bmod p$ are not coprime. To solve this problem, Satoh devised a modified Hensel lifting [SAT 2000, Lemma 2.1], which also has quadratic convergence.

**Lemma 17.35 (Satoh)** Let $p \geq 3$ be a prime and $\Psi(x) \in \mathbb{Z}_q[x]$ satisfying $\Psi'(x) \equiv 0 \pmod{p}$ and $\Psi'(x) \not\equiv 0 \pmod{p^2}$. Let $h(x) \in \mathbb{Z}_q[x]$ be a monic polynomial such that

1. $h(x) \bmod p$ is squarefree and coprime to $(\Psi'(x)/p) \bmod p$,
2. $\Psi(x) \equiv f(x)h(x) \pmod{p^{n+1}}$,

then the polynomial

$$H(x) = h(x) + \left( \left( \frac{\Psi(x)}{\Psi'(x)} h'(x) \right) \bmod h(x) \right)$$

satisfies $H(x) \equiv h(x) \pmod{p^n}$ and $\Psi(x) \equiv F(x)H(x) \pmod{p^{2n+1}}$.

The following algorithm computes

$$H(x) = \prod_{P \in (\ker \widehat{\Sigma}_i \setminus \{P_\infty\})/\pm} (x - x(P)) \bmod p^{N-1} \tag{17.13}$$

---

**Algorithm 17.36** Lift kernel

---

INPUT: The $p$-division polynomial $\Psi_p(x)$ of an elliptic curve $\mathcal{E}$ over $\mathbb{Z}_q/p^N \mathbb{Z}_q$ and precision $N$.
OUTPUT: The polynomial $H(x)$ as in (17.13).

1. **if** $N = 1$ **then**
2. $\quad H(x) \leftarrow h(x)$ such that $\Psi_p(x) \equiv \delta h(x)^p \pmod{p}$ $\qquad$ [$h(x)$ monic]
3. **else**
4. $\quad N' \leftarrow \lceil \frac{N-1}{2} \rceil$
5. $\quad H(x) \leftarrow$ Lift kernel($\Psi_p(x), N'$)
6. $\quad H(x) \leftarrow H(x) + \left( \frac{H'(x)\Psi_p(x)}{\Psi'_p(x)} \bmod H(x) \right) \bmod p^N$
7. **return** $H(x)$

---

**Example 17.37** Let $p = 7$, $d = 3$, $\mathbb{Z}_q \simeq \mathbb{Z}_p[T]/(G(T))$ with $G(T) = T^3 + 6T^2 + 4$ and consider the elliptic curve $\mathcal{E} : y^2 = x^3 + a_4 x + a_6$ with

$$a_4 \equiv 1409T^2 + 2308T + 2293 \pmod{p^4} \text{ and } a_6 \equiv 139T^2 + 2339T + 2329 \pmod{p^4}.$$

## § 17.3 Overview of $p$-adic methods

The $p$-division polynomial $\Psi_p(x)$ of $\mathcal{E}$ is then equivalent to

$7x^{24} + (1792T^2 + 168T + 350)x^{22} + (788T^2 + 374T + 1751)x^{21} + (364T^2 + 1330T + 2051)x^{20}$
$+ (1974T^2 + 2226T + 1057)x^{19} + (98T^2 + 1526T + 1995)x^{18} + (1673T^2 + 546T + 70)x^{17}$
$+ (77T^2 + 910T + 378)x^{16} + (1302T^2 + 2289T + 2058)x^{15} + (631T^2 + 2008T + 1189)x^{14}$
$+ (791T^2 + 504T + 2268)x^{13} + (21T^2 + 2247T + 1953)x^{12} + (1519T^2 + 1106T + 945)x^{11}$
$+ (490T^2 + 525T + 1526)x^{10} + (532T^2 + 1792T + 1575)x^9 + (434T^2 + 735T + 147)x^8$
$+ (843T^2 + 879T + 925)x^7 + (1274T^2 + 2212T + 2009)x^6 + (1764T^2 + 903T + 882)x^5$
$+ (2107T^2 + 1946T + 2324)x^4 + (784T^2 + 1197T + 1673)x^3 + (245T^2 + 560T + 2261)x^2$
$+ (245T^2 + 679T + 1582)x + 1014T^2 + 282T + 1562$

modulo $p^4$ and Algorithm 17.36 computes the following factor of $\Psi_p$:

$$H(x) \equiv x^3 + (502T^2 + 1965T + 742)x^2 + (1553T^2 + 2106T + 474)x + 2329T^2 + 1521T + 2058 \pmod{p^4}.$$

For $p > 3$, $\mathcal{E}_{i+1}$ can be defined by the equation $y^2 = x^3 + a_{i+1}x + b_{i+1}$. Using Vélu's formulas, Satoh [SAT 2000, Proposition 4.3] shows that $\mathcal{E}_{i+1}/\ker(\widehat{\mathcal{F}}_{p,i})$ is given by the equation $y^2 = x^3 + \alpha_{i+1}x + \beta_{i+1}$ with

$$\alpha_{i+1} = (6 - 5p)a_{i+1} - 30(h_{i,1}^2 - 2h_{i,2})$$
$$\beta_{i+1} = (15 - 14p)b_{i+1} - 70(-h_{i,1}^3 + 3h_{i,1}h_{i,2} - 3h_{i,3}) + 42a_{i+1}h_{i,1}$$

where $h_{i,k}$ denotes the coefficient of $x^{(p-1)/2-k}$ in $H_i(x)$ and we define $h_{i,k} = 0$ for $(p-1)/2 < k$.

Given the above Weierstraß model for $\mathcal{E}_{i+1}/\ker(\widehat{\mathcal{F}}_{p,i})$ we can now compute the isomorphism $\lambda_i$ to $\mathcal{E}_i : y^2 = x^3 + a_i x + b_i$. The only change of variables preserving the form of these equations is $\lambda_i : (x, y) \longrightarrow (u_i^2 x, u_i^3 y)$ with

$$u_i^2 = \frac{\alpha_{i+1}}{\beta_{i+1}} \frac{b_i}{a_i}.$$

The action of $\lambda_i$ on $\omega_i = dx/y$ is given by $\lambda_i^*(\omega_i) = u_i^{-1} \omega_{i+1,K}$ with $\omega_{i+1,K} = dx/y$ and therefore

$$c_i^2 = \frac{\beta_{i+1}}{\alpha_{i+1}} \frac{a_i}{b_i}. \tag{17.14}$$

Computing

$$c^2 = \prod_{i=0}^{d-1} c_i^2 = N_{\mathbb{Q}_q/\mathbb{Q}_p}(c_0^2)$$

and taking the square root gives the trace of Frobenius up to sign. This ambiguity can be resolved using the Cartier–Manin operator of Section 17.1.4:

$$t \equiv \gamma \gamma^\sigma \cdots \gamma^{\sigma^{d-1}} \pmod{p}$$

where $\gamma$ is the coefficient of $x^{p-1}$ in the polynomial $(x^3 + a_4 x + a_6)^{(p-1)/2}$.

This finally leads to Algorithm 17.38.

**Algorithm 17.38** Satoh's point counting method

INPUT: The elliptic curve $E: y^2 = x^3 + a_4 x + a_6$ over $\mathbb{F}_{p^d}$, $j(E) \notin \mathbb{F}_{p^2}$.
OUTPUT: The number of points on $E(\mathbb{F}_{p^d})$

1. $N \leftarrow \lceil \log_p 4 + d/2 \rceil$
2. $S \leftarrow 1$ and $T \leftarrow 1$
3. $a_0 \leftarrow a_4, a_d \leftarrow a_4, b_0 \leftarrow a_6, b_d \leftarrow a_6, j_0 \leftarrow j(E)$ and $j_d \leftarrow j(E)$
4. **for** $i = 0$ **to** $d-2$ **do**
5. $\quad j_{i+1} \leftarrow j_i^p$
6. $(J_{d-1}, \ldots, J_0) \leftarrow$ Lift j-invariants$\big((j_{d-1}, \ldots, j_0), N\big)$
7. **for** $i = 0$ **to** $d-1$ **do**
8. $\quad \gamma \leftarrow J_i/(1728 - J_i) \bmod p^N$
9. $\quad a \leftarrow 3\gamma \bmod p^N$ and $b \leftarrow 2\gamma \bmod p^N$
10. $\quad \Psi_p(x) \leftarrow$ p-division polynomial of $y^2 = x^3 + ax + b$
11. $\quad H(x) \leftarrow$ Lift kernel $(\Psi_p(x), N+1)$
12. $\quad$ **for** $j = 1$ **to** $3$ **do** $h_j \leftarrow$ coefficient of $H(x)$ of degree $(p-1)/2 - j$
13. $\quad \alpha \leftarrow (6 - 5p)a - 30(h_1^2 - 2h_2)$
14. $\quad \beta \leftarrow (15 - 14p)b - 70(-h_1^3 + 3h_1 h_2 - 3h_3) + 42ah_1$
15. $\quad S \leftarrow \beta a S$ and $T \leftarrow \alpha b T$
16. $t \leftarrow \text{Sqrt}(S/T, N)$
17. $\gamma \leftarrow$ coefficient of $(x^3 + a_4 x + a_6)^{(p-1)/2}$ of degree $p - 1$
18. **if** $t \not\equiv \gamma \gamma^\sigma \cdots \gamma^{\sigma^{n-1}} \pmod{p}$ **then** $t \leftarrow -t \pmod{p^N}$
19. **if** $t^2 > 4p^d$ **then** $t \leftarrow t - p^N$
20. **return** $p^d + 1 - t$

**Example 17.39** Let $p = 5$, $d = 7$, $\mathbb{F}_{p^d} \simeq \mathbb{F}_p(\theta)$ with $\theta^7 + 3\theta + 3 = 0$ and consider the elliptic curve defined by $y^2 = x^3 + x + a_6$ with

$$a_6 = 4\theta^6 + 3\theta^5 + 3\theta^4 + 3\theta^3 + 3\theta^2 + 3.$$

Algorithm 17.38 then computes the following intermediate results: $N = 6$, $j_0 = 4T^6 + T^5 + 2T^4 + 2T^2$ and

$$J_0 \equiv 6949T^6 + 6806T^5 + 14297T^4 + 2260T^3 + 13542T^2 + 13130T + 15215 \pmod{p^N},$$

with $\mathbb{Z}_q \simeq \mathbb{Z}_p[T]/\big(G(T)\big)$ and $G(T) = T^7 + 3T + 3$. This gives the following values for $a$, $b$

$$a \equiv 6981T^6 + 8408T^5 + 1033T^4 + 8867T^3 + 15614T^2 + 3514T + 675 \pmod{p^N}$$
$$b \equiv 4654T^6 + 397T^5 + 5897T^4 + 703T^3 + 5201T^2 + 7551T + 450 \pmod{p^N}$$

and the polynomial $H$ describing the kernel of $\mathcal{F}_p$

$$H(x) \equiv x^2 + (1395T^6 + 7906T^5 + 3737T^4 + 9221T^3 + 9207T^2 + 5403T + 7401)x$$
$$+ 6090T^6 + 206T^5 + 5259T^4 + 7576T^3 + 3863T^2 + 8903T + 7926 \pmod{p^N}.$$

## § 17.3 Overview of $p$-adic methods

Finally, we recover $\alpha$ and $\beta$ as

$$\alpha \equiv 11086T^6 + 2618T^5 + 6983T^4 + 13192T^3 + 15324T^2 + 13544T + 10550 \pmod{p^N}$$
$$\beta \equiv 4940T^6 + 3060T^5 + 14966T^4 + 6589T^3 + 7934T^2 + 6060T + 12470 \pmod{p^N}.$$

By computing the norm of $(\alpha b)/(\beta a)$, taking the square root and determining the correct sign, Algorithm 17.38 computes $\text{Tr}(\phi_q) = 433$ and $|E(\mathbb{F}_{p^d})| = 77693$.

The case $p = 3$ is very similar to the case $p \geqslant 5$. There are only two minor adaptations: firstly, note that $\ker(\mathcal{F}_{p,i}) = \{Q, -Q, P_\infty\}$ with $Q$ a 3-torsion point on $\mathcal{E}_{i+1}$ with integral coordinates, so Algorithm 17.36 reduces to a simple Newton iteration on the 3-division polynomial of $\mathcal{E}_{i+1}$; secondly, the Weierstraß equation for $\mathcal{E}_i$ is different from the one for $p \geqslant 5$, which slightly changes Vélu's formulas. Let $x_2$ denote the $x$-coordinate of $Q \in \ker(\widehat{\mathcal{F}}_{p,i})$ and let $\mathcal{E}_{i+1}$ be defined by $y^2 = x^3 + x^2/4 + a_{i+1}x + b_{i+1}$, then $\mathcal{E}_{i+1}/\ker(\widehat{\mathcal{F}}_{p,i})$ is given by the equation $y^2 = x^3 + x^2/4 + \alpha_{i+1}x + \beta_{i+1}$, with

$$\alpha_{i+1} = -30x_2^2 - 5x_2 - 9a_{i+1},$$
$$\beta_{i+1} = -70x_2^3 - 20x_2^2 - (42a_{i+1} + 1)x_2 - 2a_{i+1} - 27b_{i+1}.$$

To compute $u_i^2$ and thus $c_i^2$, we first translate the $x$-axis to get rid of the quadratic term and then apply (17.14) which then gives

$$c_i^2 = \frac{(48a_i - 1)(864\beta_{i+1} - 72\alpha_{i+1} + 1)}{(864b_i - 72a_i + 1)(48\alpha_{i+1} - 1)}.$$

Taking the square root of

$$c^2 = \prod_{i=0}^{d-1} c_i^2$$

determines the trace of Frobenius $t$ up to sign. Furthermore, since the curve $E$ is defined by an equation of the form $y^2 = x^3 + x^2 + a_6$, the correct sign follows from $t \equiv 1 \pmod{3}$.

**Example 17.40** Let $d = 7$ and $\mathbb{F}_{3^d} \simeq \mathbb{F}_3(\theta)$ with $\theta^7 + 2\theta^2 + 1 = 0$ and consider the elliptic curve $E$ given by

$$E : y^2 = x^3 + x^2 + \theta^6 + \theta^4 + 2\theta^3 + 2\theta^2.$$

The $j$-invariant of $E$ is $2\theta^6 + \theta^4 + \theta^2$ and if $\mathbb{Z}_q \simeq \mathbb{Z}_3[T]/(G(T))$ with $G(T) = T^7 + 2T^2 + 1$, then

$$j(\mathcal{E}) \equiv 29T^6 + 378T^5 + 310T^4 + 528T^3 + 337T^2 + 621T + 474 \pmod{p^6}.$$

Computing $a, b, \alpha, \beta$ then gives

$$a \equiv 387T^6 + 117T^4 + 315T^3 + 531T^2 + 54T \pmod{p^6}$$
$$b \equiv 31T^6 + 81T^5 + 145T^4 + 191T^3 + 521T^2 + 123T + 81 \pmod{p^6}$$
$$\alpha \equiv 150T^6 + 204T^5 + 531T^4 + 218T^3 + 329T^2 + 178T + 39 \pmod{p^6}$$
$$\beta \equiv 488T^6 + 53T^5 + 675T^4 + 151T^3 + 97T^2 + 320T + 510 \pmod{p^6}.$$

Computing the norm of the formula for $c_i^2$ and taking the square root finally leads to $\text{Tr}(\phi_q) = 73$ and thus $|E(\mathbb{F}_{3^d})| = 2115$.

For $p = 2$, Lemma 17.34 no longer holds. Indeed, the Newton polygon of the 2-division polynomial shows that there are two nontrivial points in $\mathcal{E}_{i+1}[p] \cap \mathcal{E}_{i+1}(\mathbb{Z}_q^{\text{ur}})$, whereas $\ker(\widehat{\mathcal{F}}_{p,i})$ has only one nontrivial point. The main problem in extending Satoh's algorithm to characteristic 2 therefore lies in choosing the correct 2-torsion point. There are two algorithms that are both based on diagram (17.12). Let $\ker(\widehat{\mathcal{F}}_{p,i}) = \langle Q \rangle$, then since $\lambda$ is an isomorphism we conclude $j(\mathcal{E}_{i+1}/\langle Q \rangle) = j(\mathcal{E}_i)$.

The first algorithm to compute $Q = (x_2, y_2)$ is due to Skjernaa [SKJ 2003], who gives an explicit formula for $x_2$ as a function of $j(\mathcal{E}_i)$ and $j(\mathcal{E}_{i+1})$. Since $Q$ is a 2-torsion point, it follows that $2y_2 + x_2 = 0$. Substituting $y_2$ in the equation of the curve and using the equality $j(\mathcal{E}_{i+1}/\langle Q \rangle) = j(\mathcal{E}_i)$, Skjernaa deduces an explicit expression for $x_2$. A proof of the following proposition can be found in [SKJ 2003, Lemma 4.1].

**Proposition 17.41** Let $Q = (x_2, y_2)$ be the nontrivial point in $\ker(\widehat{\mathcal{F}}_{p,i+1})$ and let $z_2 = x_2/2$, then

$$z_2 = -\frac{j(\mathcal{E}_i)^2 + 195120 j(\mathcal{E}_i) + 4095 j(\mathcal{E}_{i+1}) + 660960000}{8\big(j(\mathcal{E}_i)^2 + j(\mathcal{E}_i)(563760 - 512 j(\mathcal{E}_{i+1})) + 372735 j(\mathcal{E}_{i+1}) + 8981280000\big)}.$$

Skjernaa shows that the 2-adic valuation of both the numerator and denominator is 12, so we have to compute $j(\mathcal{E}_i) \pmod{2^{N+12}}$ to recover $z_2 \pmod{2^N}$.

The second algorithm is due to Fouquet, Gaudry, and Harley [FOGA+ 2000] and is based on the fact that $\ker(\widehat{\mathcal{F}}_{p,i}) = \langle Q \rangle \subset \mathcal{E}_{i+1}[2]$. Let $\mathcal{E}_{i+1}$ be given by the equation $y^2 + xy = x^3 + 36 a_{i+1} x + a_{i+1}$ with $a_{i+1} = 1/(1728 - j(\mathcal{E}_{i+1}))$. Since $Q$ is a 2-torsion point, we have $2y_2 + x_2 = 0$ and $x_2$ is a zero of the 2-division polynomial $4x^3 + x^2 + 144 a_{i+1} x + 4 a_{i+1}$. Clearly we have $x_2 \equiv 0 \pmod{2}$, so Fouquet, Gaudry, and Harley compute $z_2 = x_2/2$ as a zero of the modified 2-division polynomial $8z^3 + z^2 + 72 a_{i+1} z + a_{i+1}$. The main problem is choosing the correct starting value when considering this equation modulo 8. Using $j(\mathcal{E}_{i+1}/\langle Q \rangle) = j(\mathcal{E}_i)$ they proved that $z \equiv 1/j(\mathcal{E}_i) \pmod{8}$ is the correct starting value giving $x_2$.

Vélu's formulas show that $\mathcal{E}_{i+1}/\ker(\widehat{\mathcal{F}}_{p,i})$ is given by the Weierstraß equation $y^2 + xy = x^3 + \alpha_{i+1} x + \beta_{i+1}$ with

$$\alpha_{i+1} = -\frac{36}{j(\mathcal{E}_{i+1}) - 1728} - 5\gamma_{i+1},$$

$$\beta_{i+1} = -\frac{1}{j(\mathcal{E}_{i+1}) - 1728} - (1 + 7x_2)\gamma_{i+1},$$

where $\gamma_{i+1} = 3x_2^2 - 36/(j(\mathcal{E}_{i+1}) - 1728) + x_2/2$. The isomorphism $\lambda_i$ now has the general form

$$(x, y) \longrightarrow (u_i^2 x + r_i, u_i^3 y + u_i^2 s_i x + t_i), \quad (u_i, r_i, s_i, t_i) \in \mathbb{Q}_q^* \times \mathbb{Q}_q^3,$$

but an easy calculation shows that $c_i^2 = u_i^{-2}$. Solving the equations satisfied by $(u_i, r_i, s_i, t_i)$ given in [SIL 1986, Table 1.2] finally leads to

$$c_i^2 = -\frac{864 \beta_i - 72 \alpha_i + 1}{48 \alpha_i - 1}. \tag{17.15}$$

The complexity of Algorithm 17.38 directly follows from Hasse's theorem, i.e., $|t| \leqslant 2\sqrt{q}$. Therefore it suffices to lift all data with precision $N \simeq d/2$. Since elements of $\mathbb{Z}_q/p^N \mathbb{Z}_q$ are represented as polynomials of degree less than $d$ with coefficients in $\mathbb{Z}/p^N \mathbb{Z}$, every element takes $O(dN)$ memory for fixed $p$. As shown in Section 12.2.1, multiplication and division in $\mathbb{Z}_q/p^N \mathbb{Z}_q$ takes $O(d^\mu N^\mu)$ time.

For each curve $E_i$ with $0 \leqslant i < d$ we need $O(1)$ elements of $\mathbb{Z}_q/p^N \mathbb{Z}_q$, so the total memory needed is $O(d^2 N)$ bits. Lifting the cycle of $j$-invariants to precision $N$ requires $O(\lg N)$ iterations.

## § 17.3 Overview of $p$-adic methods

In every iteration the precision of the computations almost doubles, so the complexity is determined by the last iteration that takes $O(d^{\mu+1}N^{\mu})$ bit-operations. Computing one coefficient $c_i^2$ requires $O(1)$ multiplications, so to compute all $c_i$ we also need $O(d^{\mu+1}N^{\mu})$ bit-operations.

**Theorem 17.42 (Satoh)** There exists a deterministic algorithm to compute the number of points on an elliptic curve $E$ over a finite field $\mathbb{F}_q$ with $q = p^d$ and $j(E) \notin \mathbb{F}_{p^2}$, which requires $O(d^{2\mu+1})$ bit-operations and $O(d^3)$ space for fixed $p$.

### 17.3.1.f Optimizations of Satoh's algorithm

Satoh's algorithm basically consists of two steps: in the first step, a sufficiently precise approximation of the canonical lift of an ordinary elliptic curve is computed; in the second step, the trace of Frobenius is recovered as the norm of an element in $\mathbb{Z}_q$. As such, the optimizations of Satoh's algorithm can be categorized accordingly. Since computing the norm of an element in $\mathbb{Z}_q$ is not specific to point counting algorithms, we simply refer the interested reader to Section 12.8.5.

Recall that if $E$ is an ordinary elliptic curve over $\mathbb{F}_q$ with $q = p^d$ and $p$ a prime, then by the theorem of Lubin, Serre, and Tate (Theorem 17.30), the $j$-invariant of the canonical lift $\mathcal{E}$ satisfies

$$\Phi_p\big(j(\mathcal{E}), \Sigma\big(j(\mathcal{E})\big)\big) = 0 \text{ and } j(\mathcal{E}) \equiv j(E) \pmod{p}.$$

The optimizations of the lifting step are essentially algorithms to compute the root of an equation of the form $\phi\big(X, \Sigma(X)\big) = 0$ with $\phi(X,Y) \in \mathbb{Z}_q[X,Y]$, starting from an initial approximate root $x_0 \in \mathbb{Z}_q$. The algorithm used by Satoh is a simple multivariate Newton iteration that lifts the whole cycle of conjugates of $x_0$ simultaneously. Computing the solution to precision $p^N$ then takes $O(d^{1+\mu}N^{\mu})$ bit-operations and $O(d^2N)$ space, for fixed $p$ and fixed degree of $\phi$.

Vercauteren [VEPR+ 2001] devised a lifting algorithm that requires $O(d^{\mu}N^{1+\mu})$ bit-operations, but only $O(dN)$ space. This algorithm is based on a repeated application of Proposition 17.51, since this allows us to avoid costly Frobenius substitutions. Further details can be found in the paper [VEPR+ 2001].

Mestre [MES 2000b] devised an elliptic curve point counting algorithm in characteristic 2, based on a 2-adic version of the Arithmetic-Geometric-Mean (AGM). The AGM algorithm has the same time and space complexity as Vercauteren's algorithm, but is far easier to implement and also runs faster in practice. The main difference with the other optimizations is that the AGM cannot be used to solve a general equation of the form $\phi\big(X, \Sigma(X)\big) = 0$, although implicitly it computes a root of $\Phi_2\big(X, \Sigma(X)\big) = 0$ with $\Phi_2(X,Y)$ the 2-nd modular polynomial. A detailed description of the AGM algorithm and its generalization to hyperelliptic curves is given in Section 17.3.2.

Satoh, Skjernaa, and Taguchi [SASK+ 2003] proposed to use a Teichmüller modulus (see Section 12.1) to represent $\mathbb{Z}_q$, i.e., if $\mathbb{F}_q$ is represented as $\mathbb{F}_p[T]/\big(\overline{f}(T)\big)$ with $\overline{f}(T)$ a monic irreducible polynomial of degree $d$, then $\mathbb{Z}_q$ is represented as $\mathbb{Z}_p[T]/\big(f(T)\big)$ with $f(T)$ the unique monic polynomial with $f(T) \mid T^q - T$ and $f(T) \equiv \overline{f}(T) \pmod{p}$. The advantage of this representation is that it allows efficient Frobenius substitutions. Combining this with a Taylor expansion leads to an algorithm to compute the root of an equation of the form $\phi\big(X, \Sigma(X)\big) = 0$ to precision $p^N$ in time $O\big(d^{\mu}N^{\mu+1/(\mu+1)}\big)$ and $O(dN)$ space, for fixed $p$ and fixed degree of $\phi$. Furthermore, they also proposed a fast norm computation algorithm, which is described in detail in Section 12.8.5.

Kim, Park, Cheon, Park, Kim, and Hahn [KIPA+ 2002] showed that a Gaussian normal basis can be lifted trivially to $\mathbb{Z}_q$ and allows efficient computation of arbitrary iterates of the Frobenius substitution. Their point counting algorithm is a combination of the SST algorithm and a norm computation algorithm due to Kedlaya [KED 2001], which can be found in Section 12.8.5.

Gaudry [GAU 2002] used the AGM iteration to devise a modified modular polynomial in characteristic 2 (see Section 17.3.2.b), which has lower degree than $\Phi_2(X,Y)$ and thus leads to a faster algorithm, but with the same time and space complexity as the SST algorithm.

Lercier and Lubicz [LELU 2003] devised an algorithm to solve Artin–Schreier equations over $\mathbb{Z}_q$ (see Section 12.6) and used this to compute a root of an equation of the form $\phi(X, \Sigma(X)) = 0$ to any precision $p^N$. For finite fields with Gaussian normal basis, this algorithm runs in $O(d^\mu N^\mu \lg d)$ time and $O(dN)$ space, for $p$ fixed and fixed degree of $\phi$. A detailed description of this algorithm can be found in Section 12.7.

Finally, Harley [HAR 2002b] presented three very efficient algorithms: an algorithm to compute the Teichmüller modulus (see Section 12.1), an algorithm to solve Artin–Schreier equations modulo $p^N$ in time $O(d^\mu N^\mu \lg N)$ (see Section 12.7) and a fast norm computation algorithm (see Section 12.8.5). Combining these algorithms leads to an asymptotically optimal elliptic curve point counting algorithm that runs in time $O(d^{2\mu} \lg d)$ and requires $O(d^2)$ space, for $p$ fixed.

### 17.3.2 Arithmetic–Geometric–Mean algorithm

Classically, the *Arithmetic-Geometric-Mean* (AGM) was introduced by Lagrange [LAG 1973] and Gauß [GAU 1973] to compute elliptic integrals or equivalently the period matrix of an elliptic curve over $\mathbb{C}$. Mestre [MES 2000b] showed how a 2-adic version of the AGM can be used to count the number of points on an ordinary elliptic curve over a finite field of characteristic 2. Later, Mestre [MES 2002] reinterpreted this algorithm as a special case of Riemann's duplication formula for theta functions and generalized it to ordinary hyperelliptic curves. This section first reviews the AGM over $\mathbb{C}$ and then describes both approaches to the 2-adic AGM.

#### 17.3.2.a AGM in $\mathbb{C}$

**Definition 17.43** Let $a_0, b_0 \in \mathbb{R}$ with $a_0 \geqslant b_0 > 0$, then the *AGM iteration* for $k \in \mathbb{N}$ is defined as

$$(a_{k+1}, b_{k+1}) = \left( \frac{a_k + b_k}{2}, \sqrt{a_k b_k} \right).$$

Since $b_k \leqslant b_{k+1} \leqslant a_{k+1} \leqslant a_k$ and $0 \leqslant a_{k+1} - b_{k+1} \leqslant (a_k - b_k)/2$, we conclude that the limits of $a_k$ and $b_k$ when $k$ tends to infinity exist and are equal. This common value is called the AGM of $a_0$ and $b_0$ and is denoted as $\mathrm{AGM}(a_0, b_0)$. An easy calculation shows that

$$\frac{a_k}{b_k} - 1 \leqslant \frac{a_0 - b_0}{2^k b_k} \leqslant \frac{1}{2^k}\left(\frac{a_0}{b_0} - 1\right),$$

so after a logarithmic number of steps we have $a_k/b_k = 1 + \varepsilon_k$ with $\varepsilon_k < 1$. The Taylor series expansion of $1/\sqrt{1+\varepsilon_k}$ shows that convergence is quadratic when $\varepsilon_k < 1$:

$$\begin{aligned}
\frac{a_{k+1}}{b_{k+1}} &= \frac{a_k + b_k}{2\sqrt{a_k b_k}} = \frac{2 + \varepsilon_k}{2\sqrt{1 + \varepsilon_k}} \\
&= \left(1 + \frac{\varepsilon_k}{2}\right) \left( \sum_{n=0}^{\infty} \binom{-1/2}{n} \varepsilon_k^n \right) \\
&= 1 + \frac{\varepsilon_k^2}{8} - \frac{\varepsilon_k^3}{8} + O(\varepsilon_k^4).
\end{aligned} \qquad (17.16)$$

**Remark 17.44** The coefficient of $\varepsilon_k^n$ in the above Taylor expansion is easily seen to be

$$\binom{-1/2}{n} \frac{1-n}{1-2n}.$$

Since $1 - 2n$ is odd, the 2-adic valuation of the denominator of the above coefficient is at most $n + v_2(n!) \leqslant n + \lfloor \lg n \rfloor (1 + \lfloor \lg n \rfloor)/2$. For the 2-adic AGM to converge, it is necessary that $v_2(\varepsilon_{k+1}) < v_2(\varepsilon_k)$, which implies $v_2(\varepsilon_k) > 3$ due to the quadratic term.

## § 17.3 Overview of $p$-adic methods

The main reason for Legendre and Gauß to introduce the AGM is that it allows us to compute elliptic integrals and periods of elliptic curves.

**Theorem 17.45** Let $a, b \in \mathbb{R}$ with $0 < b \leqslant a$, then

$$\int_0^{\pi/2} \frac{dt}{\sqrt{a^2 \cos^2 t + b^2 \sin^2 t}} = \frac{\pi}{2\mathrm{AGM}(a,b)}.$$

The proof of this theorem is given later in this section, but the main idea is that the elliptic integral is invariant under the transformation $(a, b) \to (a_k, b_k)$. By a limit argument, we obtain the integral $\int_0^{\pi/2} \mathrm{AGM}(a,b)^{-1} dt$, which proves the result.

The link with elliptic curves is the following classical result: consider the elliptic curve

$$E_\tau : y_0^2 = x_0(x_0 - a_0^2)(x_0 - b_0^2),$$

with $a_0 = a$ and $b_0 = b$, then after a suitable change of variables, the above integral becomes

$$-\frac{i}{2} \int_0^{-\infty} \frac{dx_0}{y_0} = \frac{\pi}{2\mathrm{AGM}(a,b)}, \tag{17.17}$$

which also equals half of one of the periods of the elliptic curve $E_\tau$. To prove the above equation, recall that an elliptic curve $E_\tau$ over $\mathbb{C}$ is isomorphic as an analytic variety to a torus $\mathbb{C}/(\mathbb{Z} + \tau \mathbb{Z})$ with $\Im m\, \tau > 0$. Consider the following diagram

$$\begin{array}{ccc} \mathbb{C}/(\mathbb{Z} + \tau\mathbb{Z}) & \xrightarrow{\simeq}_{\mu_0} & E_\tau(\mathbb{C}) \\ F \uparrow & & \uparrow \phi \\ \mathbb{C}/(\mathbb{Z} + 2\tau\mathbb{Z}) & \xrightarrow{\simeq}_{\mu_1} & E_{2\tau}(\mathbb{C}) \end{array}$$

with $F : z \mapsto z$ and $\phi$ the isogeny of degree 2 such that the above diagram commutes. The elliptic curve $E_{2\tau}$ has as equation $E_{2\tau} : y_1^2 = x_1(x_1 - a_1^2)(x_1 - b_1^2)$ with $a_1 = (a_0+b_0)/2$ and $b_1 = \sqrt{a_0 b_0}$. Furthermore, the isogeny $\phi$ is given by

$$\phi : (x_1, y_1) \mapsto \left( \frac{x_1(x_1 - b_1^2)}{x_1 - a_1^2}, y \frac{(x^2 - 2x_1 a_1^2 + a_1^2 b_1^2)}{(x_1 - a_1^2)^2} \right). \tag{17.18}$$

Since $\phi = \mu_0 \circ F \circ \mu_1^{-1}$, the action of $\phi$ on the holomorphic differential form $dx_0/y_0$ is given by

$$\phi^* \left( \frac{dx_0}{y_0} \right) = (\mu_0 \circ F \circ \mu_1^{-1})^* \left( \frac{dx_0}{y_0} \right) = (F \circ \mu_1^{-1})^* dz = (\mu_1^{-1})^* dz = \left( \frac{dx_1}{y_1} \right),$$

since clearly $F^*(dz) = dz$. This shows that the integral (17.17) remains unchanged under the transformation $\phi$ and thus

$$-\frac{i}{2} \int_0^{-\infty} \frac{dx}{\sqrt{x(x - a_0^2)(x - b_0^2)}} = -\frac{i}{2} \int_0^{-\infty} \frac{dx}{\sqrt{x(x - a_1^2)(x - b_1^2)}}.$$

Repeating the same argument and by taking limits, we obtain $E_\infty : y^2 = x(x - \mathrm{AGM}(a,b)^2)^2$. Note that $E_\infty$ is of genus 0, which implies that the corresponding integral can be evaluated using elementary functions. This finally proves

$$-\frac{i}{2} \int_0^{-\infty} \frac{dx}{\sqrt{x(x - a^2)(x - b^2)}} = -\frac{i}{2} \int_0^{-\infty} \frac{dx}{\sqrt{x}(x - \mathrm{AGM}(a,b)^2)} = \frac{\pi}{2\mathrm{AGM}(a,b)}.$$

### 17.3.2.b  Elliptic curve AGM

Let $\mathbb{F}_q$ denote the finite field with $2^d$ elements and let $\mathbb{Q}_q$ be the unramified extension of $\mathbb{Q}_2$ of degree $d$ with valuation ring $\mathbb{Z}_q$. Let $\Sigma \in \text{Gal}(\mathbb{Q}_q/\mathbb{Q}_2)$ be the Frobenius substitution, i.e., the unique automorphism of $\mathbb{Q}_q$ with $\Sigma(x) \equiv x^2 \pmod{2}$ for $x \in \mathbb{Z}_q$.

For $c \in 1 + 8\mathbb{Z}_q$, denote by $\sqrt{c}$ the unique element $e \in 1 + 4\mathbb{Z}_q$ with $e^2 = c$. Given $a, b \in \mathbb{Z}_q$ with $a/b \in 1 + 8\mathbb{Z}_q$, then $a' = (a+b)/2$ and $b' = b\sqrt{a/b}$ also belong to $\mathbb{Z}_q$ and $a'/b' \in 1 + 8\mathbb{Z}_q$. Furthermore, if $a, b \in 1 + 4\mathbb{Z}_q$, then also $a', b' \in 1 + 4\mathbb{Z}_q$.

**Remark 17.46** The analysis in Remark 17.44, shows that the 2-adic AGM sequence will converge if and only if $a/b \in 1 + 16\mathbb{Z}_q$. For $a/b \in 1 + 8\mathbb{Z}_q$ the AGM sequence will not converge at all, so $\text{AGM}(a, b)$ is not defined. Nonetheless, the AGM sequence $(a_k, b_k)_{k=0}^{\infty}$ can be used to compute the number of points on an ordinary elliptic curve.

Let $a, b \in 1 + 4\mathbb{Z}_q$ with $a/b \in 1 + 8\mathbb{Z}_q$ and $E_{a,b}$ the elliptic curve defined by

$$E_{a,b} : y^2 = x(x - a^2)(x - b^2).$$

Similar to the AGM over $\mathbb{C}$ of Section 17.3.2.a, the 2-adic AGM iteration constructs a sequence of elliptic curves all of which are 2-isogenous.

**Proposition 17.47** Let $a, b \in 1 + 4\mathbb{Z}_q$ with $a/b \in 1 + 8\mathbb{Z}_q$ and $E_{a,b}$ the elliptic curve defined by the equation $y_0^2 = x_0(x_0 - a^2)(x_0 - b^2)$. Let $a' = (a+b)/2$, $b' = \sqrt{ab}$ and $E_{a',b'} : y_1^2 = x_1(x_1 - a'^2)(x_1 - b'^2)$, then $E_{a,b}$ and $E_{a',b'}$ are 2-isogenous. The isogeny is given by

$$\psi : E_{a,b} \longrightarrow E_{a',b'}$$
$$(x, y) \longmapsto \left( \frac{(x+ab)^2}{4x}, y\frac{(x-ab)(x+ab)}{8x^2} \right),$$

and the kernel of $\psi$ is $\langle (0, 0) \rangle$. Furthermore, the action of $\psi$ on the holomorphic differential is

$$\psi^* \left( \frac{dx_1}{y_1} \right) = 2 \frac{dx_0}{y_0}.$$

The isogeny $\phi$ defined in (17.18) is in fact the dual of the isogeny $\psi$ and the same formula clearly remains valid over $\mathbb{Q}_q$.

**Remark 17.48** Note that the above proposition is in fact a special case of the general construction described in Section 8.3.1. Indeed, the kernel of $\psi$ is precisely $E_{a,b}[2]^{\text{loc}}$, i.e., the 2-torsion points in the kernel of reduction. This implies that $E_{a',b'} \simeq E_{a,b}/E_{a,b}[2]^{\text{loc}}$.

Let $(a_k, b_k)_{k=0}^{\infty}$ be the AGM sequence and consider the elliptic curves $E_{a_k,b_k}$. To relate this back to a sequence of elliptic curves over $\mathbb{F}_q$, we cannot simply reduce the equations defining $E_{a_k,b_k}$ since the result would be singular. The following lemma construct an isomorphic curve which can then be reduced modulo 2.

**Lemma 17.49** Let $a, b \in 1 + 4\mathbb{Z}_q$ with $a/b \in 1 + 8\mathbb{Z}_q$ and $E_{a,b} : y^2 = x(x - a^2)(x - b^2)$. The isomorphism

$$(x, y) \mapsto \left( \frac{x - ab}{4}, \frac{y - x + ab}{8} \right)$$

## § 17.3 Overview of p-adic methods

transforms $E_{a,b}$ in $y^2 + xy = x^3 + rx^2 + sx + t$ with

$$r = \frac{-a^2 + 3ab - b^2 - 1}{4},$$
$$s = \frac{-a^3 b + 2a^2 b^2 - ab^3}{8},$$
$$t = \frac{-a^4 b^2 + 2a^3 b^3 - a^2 b^4}{64}.$$

Furthermore, $r \in 2\mathbb{Z}_q$, $s \in 8\mathbb{Z}_q$ and $t \equiv -\left(\frac{a-b}{8}\right)^2 \pmod{16}$, which shows that the reduction is nonsingular.

Let $E$ be an ordinary elliptic curve defined by $y^2 + xy = x^3 + \bar{c}$ with $\bar{c} \in \mathbb{F}_q^*$ and let $\mathcal{E}$ be its canonical lift. Take any $e \in \mathbb{Z}_q$ such that $e^2 \equiv c \pmod 2$ and let $a_0 = 1 + 4e$ and $b_0 = 1 - 4e$, then Lemma 17.49 shows that $E_{a_0,b_0}$ is isomorphic to a lift of $E$ to $\mathbb{Z}_q$ and thus $j(E_{a_0,b_0}) \equiv j(E) \pmod 2$.

As indicated in Remark 17.48, the AGM sequence is a special case of the the general construction described in Section 8.3.1 and thus must converge linearly to the canonical lift $\mathcal{E}$ of $E$.

**Theorem 17.50** The sequence of elliptic curves $E_{a_k,b_k}$ converges linearly towards the canonical lift $\mathcal{E}$ of $E$ in the following sense:

$$j(E_{a_k,b_k}) \equiv \Sigma^k(j(\mathcal{E})) \pmod{2^{k+1}}.$$

The proof of this theorem is not difficult and is based on the following proposition due to Vercauteren [VEPR+ 2001].

**Proposition 17.51** Let $\mathbb{Q}_q$ be the unramified extension of $\mathbb{Q}_p$ of degree $d$ and denote with $\mathbb{Z}_q$ its valuation ring. Let $g \in \mathbb{Z}_q[X,Y]$ and assume that $x_0, y_0 \in \mathbb{Z}_q$ satisfy

$$g(x_0,y_0) \equiv 0 \pmod p, \quad \frac{\partial g}{\partial X}(x_0,y_0) \not\equiv 0 \pmod p, \quad \frac{\partial g}{\partial Y}(x_0,y_0) \equiv 0 \pmod p.$$

Then the following properties hold:

1. For every $y \in \mathbb{Z}_q$ with $y \equiv y_0 \pmod p$ there exists a unique $x \in \mathbb{Z}_q$ such that $x \equiv x_0 \pmod p$ and $g(x,y) = 0$.
2. Let $y' \in \mathbb{Z}_q$ with $y \equiv y' \pmod{p^n}$, $n \geqslant 1$ and let $x' \in \mathbb{Z}_q$ be the unique element with $x' \equiv x_0 \pmod p$ and $g(x',y') = 0$. Then $x'$ satisfies $x' \equiv x \pmod{p^{n+1}}$.

By Proposition 17.47 the curves $E_{a_k,b_k}$ and $E_{a_{k+1},b_{k+1}}$ are 2-isogenous, so

$$\Phi_2\left(j(E_{a_k,b_k}), j(E_{a_{k+1},b_{k+1}})\right) = 0,$$

with $\Phi_2(X,Y)$ the second modular polynomial. Furthermore, an easy computation shows that $j(E_{a_{k+1},b_{k+1}}) \equiv j(E_{a_k,b_k})^2 \pmod 2$, so we can apply Proposition 17.51 which proves the result.

As noted in Remark 17.46, the AGM sequence itself does not converge, but by the above theorem the sequence of elliptic curves $E_{a_{dk},b_{dk}}$ does converge to the canonical lift $\mathcal{E}$. The first approach would thus be to use the AGM iteration to compute $j(\mathcal{E}) \bmod 2^N$ and then to apply the second stage of Satoh's algorithm. However, the AGM also provides an elegant formula for the trace of Frobenius.

Assume we have computed $a, b \in 1 + 4\mathbb{Z}_q$ with $a/b \in 1 + 8\mathbb{Z}_q$ such that $j(\mathcal{E}_{a,b}) = j(\mathcal{E})$. Let $(a', b') = ((a+b)/2, \sqrt{ab})$, $\psi : \mathcal{E}_{a,b} \longrightarrow \mathcal{E}_{a',b'}$ the AGM isogeny and $\mathcal{F}_2 : \mathcal{E}_{a,b} \longrightarrow \mathcal{E}_{\Sigma(a),\Sigma(b)}$ the lift of the 2-nd power Frobenius, then we have the following diagram:

$$\begin{array}{ccc} \mathcal{E}_{a,b} & \xrightarrow{\mathcal{F}_2} & \mathcal{E}_{\Sigma(a),\Sigma(b)} \\ & \psi \searrow \quad \nearrow \lambda & \\ & \mathcal{E}_{a',b'} & \end{array} \qquad (17.19)$$

The kernel of the Frobenius isogeny $\mathcal{F}_2 : \mathcal{E}_{a,b} \longrightarrow \mathcal{E}_{\Sigma(a),\Sigma(b)}$ is a subgroup of order 2 of

$$\mathcal{E}_{a,b}[2] = \{P_\infty, (0,0), (a^2, 0), (b^2, 0)\}.$$

Using the isomorphism given in Lemma 17.49, we now analyze the reduction of the 2-torsion points. An easy calculation shows that $(0,0)$ is mapped onto $P_\infty$ whereas $(a^2, 0)$ and $(b^2, 0)$ are mapped to $(0, (a-b)/8 \bmod 2)$. Therefore we conclude that $\ker(\mathcal{F}_2) = \{P_\infty, (0,0)\}$. Proposition 17.47 shows that $\ker(\mathcal{F}_2) = \ker(\psi)$ and since both isogenies are separable, there exists an isomorphism $\lambda : \mathcal{E}_{a',b'} \longrightarrow \mathcal{E}_{\Sigma(a),\Sigma(b)}$, such that $\mathcal{F}_2 = \lambda \circ \psi$. The following proposition shows that this isomorphism has a very simple form.

**Proposition 17.52** Given two elliptic curves $\mathcal{E}_{a,b} : y^2 = x(x-a^2)(x-b^2)$ and $\mathcal{E}_{a',b'} : y'^2 = x'(x'-a'^2)(x'-b'^2)$ over $\mathbb{Q}_q$ with $a, b, a', b' \in 1 + 4\mathbb{Z}_q$ and $a/b, a'/b' \in 1 + 8\mathbb{Z}_q$, then $\mathcal{E}_{a,b}$ and $\mathcal{E}_{a',b'}$ are isomorphic if and only if $x' = u^2 x$ and $y' = u^3 y$ with

$$u^2 = \frac{a'^2 + b'^2}{a^2 + b^2}.$$

Furthermore, we have $\left(\frac{a}{b}\right)^2 = \left(\frac{a'}{b'}\right)^2$ or $\left(\frac{a}{b}\right)^2 = \left(\frac{b'}{a'}\right)^2$.

Let $\omega = dx/y$ and $\omega' = dx'/y'$ be the holomorphic differentials on $\mathcal{E}_{a,b}$ and $\mathcal{E}_{\Sigma(a),\Sigma(b)}$ respectively, then

$$\mathcal{F}_2^*(\omega') = (\lambda \circ \psi)^*(\omega') = 2u^{-1}\omega \text{ with } u^2 = \frac{\Sigma(a)^2 + \Sigma(b)^2}{a'^2 + b'^2}.$$

Define $\zeta = a/b = 1 + 8c$ and $\zeta' = a'/b' = 1 + 8c'$, then Lemma 17.52 also implies that

$$\zeta'^2 = \Sigma(\zeta)^2 \text{ or } \zeta'^2 = \frac{1}{\Sigma(\zeta)^2}.$$

Substituting $\zeta = 1 + 8c$ and $\zeta' = 1 + 8c'$ in the above equation and dividing by 16, we conclude that $c' \equiv \Sigma(c) \pmod 4$ or $c' \equiv -\Sigma(c) \pmod 4$. The Taylor expansion of $1 + 8c' = (1 + 4c)/\sqrt{1 + 8c}$ modulo 32 shows that $c' \equiv c^2 \pmod 4$. Since after the first iteration, $c$ itself is a square $\alpha^2$ modulo 4 and $\Sigma(\alpha^2) \equiv \alpha^4 \pmod 4$, we conclude that $\zeta'^2 = \Sigma(\zeta)^2$. Note that since $\zeta \equiv \zeta' \equiv 1 \pmod 8$, we conclude that

$$\zeta' = \Sigma(\zeta). \qquad (17.20)$$

Substituting $b'^2 = a'^2 \Sigma(b)/\Sigma(a)^2$ in the expression for $u^2$ and taking square roots leads to

$$u = \pm \frac{\Sigma(a)}{a'}. \qquad (17.21)$$

## § 17.3 Overview of $p$-adic methods

Let $(a_k, b_k)_{k=0}^{\infty}$ be the AGM sequence with $a_0 = a$ and $b_0 = b$ and consider the following diagram where $\mathcal{E}_k = \mathcal{E}_{\Sigma^k(a), \Sigma^k(b)}$ and $\mathcal{F}_{2,k} : \mathcal{E}_k \longrightarrow \mathcal{E}_{k+1}$ the lift of the 2-nd power Frobenius isogeny.

$$\begin{array}{ccccccccc}
\mathcal{E}_{a,b} & \xrightarrow{\psi_0} & \mathcal{E}_{a_1, b_1} & \xrightarrow{\psi_1} & \mathcal{E}_{a_2, b_2} & \xrightarrow{\psi_2} & \cdots & \xrightarrow{\psi_{d-1}} & \mathcal{E}_{a_d, b_d} \\
{\scriptstyle \mathrm{Id}} \downarrow & & {\scriptstyle \lambda_1} \downarrow & & {\scriptstyle \lambda_2} \downarrow & & & & {\scriptstyle \lambda_d} \downarrow \\
& {\scriptstyle \mathcal{F}_{2,0}} & & {\scriptstyle \mathcal{F}_{2,1}} & & {\scriptstyle \mathcal{F}_{2,2}} & & {\scriptstyle \mathcal{F}_{2,d-1}} & \\
\mathcal{E}_0 & \longrightarrow & \mathcal{E}_1 & \longrightarrow & \mathcal{E}_2 & \longrightarrow & \cdots & \longrightarrow & \mathcal{E}_d = \mathcal{E}_0
\end{array} \quad (17.22)$$

Since $\ker(\mathcal{F}_{2,k} \circ \lambda_k) = \ker(\psi_k)$ for $k \in \mathbb{N}$, we can repeat the same argument as for diagram (17.19) and find an isomorphism $\lambda_{k+1}$ such that $\mathcal{F}_{2,k} = \lambda_{k+1} \circ \psi_k \circ \lambda_k^{-1}$. Since $\mathcal{E}_{a,b}$ is isomorphic to the canonical lift $\mathcal{E}$ of $E$, we conclude that

$$\mathrm{Tr}(\mathcal{F}_{2,d-1} \circ \cdots \circ \mathcal{F}_{2,0}) = \mathrm{Tr}(\mathcal{F}_q) = \mathrm{Tr}(\phi_q).$$

The above diagram shows that $\mathcal{F}_{2,d-1} \circ \cdots \circ \mathcal{F}_{2,0} = \lambda_d \circ \psi_{d-1} \circ \cdots \circ \psi_0$ and since $\psi_k$ acts on the holomorphic differential $\omega$ as multiplication by 2 and $\lambda_d$ as multiplication by $\pm a_d/a_0$, we conclude that

$$\mathcal{F}_q^*(\omega) = \pm 2^d \frac{a_d}{a_0}(\omega).$$

The Weil conjectures imply that the product of the roots of the characteristic polynomial of Frobenius is $2^d$ and thus

$$\mathrm{Tr}(\mathcal{F}_q) = \mathrm{Tr}(\phi_q) = \pm \frac{a_0}{a_d} \pm 2^d \frac{a_d}{a_0}. \quad (17.23)$$

**Remark 17.53** If the curve $E$ is defined by the equation $y^2 + xy = x^3 + \overline{c}$ with $\overline{c} \in \mathbb{F}_q^*$, then $(\sqrt[4]{\overline{c}}, \sqrt{\overline{c}})$ is a point of order 4, which implies that $\mathrm{Tr}(\phi_q) \equiv 1 \pmod{4}$. Since $a_0/a_d \in 1 + 4\mathbb{Z}_q$, we can choose the correct sign in (17.23) and conclude that $\mathrm{Tr}(\phi_q) \equiv a_0/a_d \pmod{q}$.

The only remaining problem is that we are working with approximations to $a$ and $b$ and not the exact values as assumed in the previous section, so the question of precision remains.

Let $E$ be an ordinary elliptic curve defined by $y^2 + xy = x^3 + \overline{c}$ with $c \in \mathbb{F}_q^*$. Take any $r \in \mathbb{Z}_q$ such that $r^2 \equiv \overline{c} \pmod 2$ and let $a_0 = 1 + 4r$ and $b_0 = 1 - 4r$. Theorem 17.50 shows that if $(a_k, b_k)_{k=0}^{\infty}$ is the AGM sequence, then

$$j(\mathcal{E}_{a_k, b_k}) \equiv j(\mathcal{E}_k) \pmod{2^{k+1}},$$

where $\mathcal{E}_k$ is the canonical lift of $\sigma^k(E)$. Expressing $j(a_k, b_k)$ as a function of $a_k$ and $b_k$ shows that $a_k$ and $b_k$ must be correct modulo $2^{k+3}$. Therefore, we conclude that

$$\mathrm{Tr}(\phi_q) \equiv \frac{a_{N-3}}{a_{N-3+d}} + 2^d \frac{a_{N-3+d}}{a_{N-3}} \pmod{2^N}.$$

---

**Algorithm 17.54** Elliptic curve AGM

INPUT: An elliptic curve $E : y^2 + xy = x^3 + \overline{c}$ over $\mathbb{F}_{2^d}$ with $j(E) \neq 0$.
OUTPUT: The number of points on $E(\mathbb{F}_{2^d})$.

1. $N \leftarrow \lceil \frac{d}{2} \rceil + 3$
2. $a \leftarrow 1$ and $b \leftarrow (1 + 8c) \bmod 2^4$            [$c$ arbitrary lift of $\overline{c}$]
3. **for** $i = 5$ **to** $N$ **do**
4.      $(a, b) \leftarrow \big((a+b)/2, \sqrt{ab}\big) \bmod 2^i$

5.　$a_0 \leftarrow a$
6.　**for** $i = 0$ **to** $d - 1$ **do**
7.　　　$(a, b) \leftarrow ((a+b)/2, \sqrt{ab}) \bmod 2^N$
8.　　　$t \leftarrow \dfrac{a_0}{a} \bmod 2^{N-1}$
9.　　　**if** $t^2 > 2^{d+2}$ **then** $t \leftarrow t - 2^{N-1}$
10.　**return** $2^d + 1 - t$

---

The complexity of Algorithm 17.54 is determined by the AGM iterations in the for loops, which require $O(d)$ square root computations to precision $O(d)$. Since each square root computation takes $O(1)$ multiplications at full precision, the complexity of Algorithm 17.54 is $O(d^{2\mu+1})$ bit-operations. The space complexity clearly is $O(d^2)$, since only $O(1)$ elements of $\mathbb{Z}_q/2^N\mathbb{Z}_q$ are required.

**Remark 17.55** Note that it is possible to replace the loop starting in Line 6 of Algorithm 17.54 by one AGM iteration and a norm computation. Indeed, equation (17.21) shows that

$$\mathcal{F}_2^*\left(\frac{dx}{y}\right) = \pm 2 \frac{a_1}{\Sigma(a_0)} \left(\frac{dx}{y}\right),$$

and since all curves $\mathcal{E}_k$ are conjugates, we have

$$\mathcal{F}_q^*\left(\frac{dx}{y}\right) = \pm 2^d \, \mathrm{N}_{\mathbb{Q}_q/\mathbb{Q}_p}\left(\frac{a_1}{\Sigma(a_0)}\right)\left(\frac{dx}{y}\right) = \pm 2^d \, \mathrm{N}_{\mathbb{Q}_q/\mathbb{Q}_p}\left(\frac{a_1}{a_0}\right)\left(\frac{dx}{y}\right).$$

Therefore, it suffices to compute $a_1$ with one AGM iteration and to set

$$t \equiv \mathrm{N}_{\mathbb{Q}_q/\mathbb{Q}_p}(a_0/a_1) \pmod{2^N}.$$

We refer to Section 12.8.5 for efficient norm computation algorithms.

**Remark 17.56** The AGM sequence has been generalized, at least in theory, to ordinary abelian varieties in the thesis of Carls [CAR 2003].

**Example 17.57** Let $p = 2$, $d = 7$ and $\mathbb{F}_{p^d} \simeq \mathbb{F}_p(\theta)$ with $\theta^7 + \theta + 1 = 0$. Consider the elliptic curve $E$ given by the affine equation $y^2 + xy = x^3 + c$ with $c = 1 + \theta^4$. It is easy to check that $j(E) \neq 0$. Let $\mathbb{Z}_q \simeq \mathbb{Z}_2[T]/(T^7 + T + 1)$, then after the initialization step

$$a \equiv 1 \pmod{2^4} \text{ and } b \equiv 8T^4 + 9 \pmod{2^4}.$$

The values of $a_0$ and $a$ in Line 8 are then given by

$$a_0 \equiv 16T^6 + 32T^4 + 8T^3 + 16T^2 + 36T + 5 \pmod{2^6}$$
$$a \equiv 48T^6 + 32T^5 + 32T^4 + 40T^3 + 48T^2 + 52T + 57 \pmod{2^6}$$

and finally $t = 13$ and $|E(\mathbb{F}_{p^d})| = 116$.

A first improvement of Algorithm 17.54 is to consider a univariate AGM sequence. Let $(a_k, b_k)_{k=0}^\infty$ with $a_k \equiv b_k \equiv 1 \pmod 4$ and $a_k \equiv b_k \pmod 8$ be the bivariate AGM sequence, then we define the *univariate AGM sequence* $(\xi_k)_{k=0}^\infty$ by $\xi_k = a_k/b_k$, which corresponds to the curves

$$E_{\xi_k} : y^2 = x(x-1)(x - \xi_k^2).$$

§ 17.3 Overview of p-adic methods

Since $(a_{k+1}, b_{k+1}) = ((a_k + b_k)/2, \sqrt{a_k b_k})$, we can compute $\xi_{k+1}$ from $\xi_k$ by

$$\xi_{k+1} = \frac{1 + \xi_k}{2\sqrt{\xi_k}}. \tag{17.24}$$

Note that this only requires 1 *inverse* square root computation; as shown in Section 12.3.3, this saves one multiplication over the square root computation $\sqrt{a_k b_k}$. Given an ordinary elliptic curve $E : y^2 + xy = x^3 + \bar{c}$ with $\bar{c} \in \mathbb{F}_q^*$, the univariate AGM sequence should be initialized by $\xi_1 \equiv 1 + 8c$ (mod 16) with $c$ an arbitrary lift of $\bar{c}$. Unlike the bivariate AGM sequence, the univariate AGM sequence does converge, in the sense that $\xi_k \equiv \xi_{k+d} \pmod{2^{k+3}}$.

Remark 17.55 shows that given the bivariate AGM sequence $(a_k, b_k)_{k=0}^{\infty}$ we can compute the trace of Frobenius as

$$\mathrm{Tr}(\phi_q) \equiv t_k + \frac{q}{t_k} \pmod{2^{k+3}},$$

with $t_k = \mathrm{N}_{\mathbb{Q}_q/\mathbb{Q}_p}(a_k/a_{k+1})$. Substituting $a_{k+1} = (a_k + b_k)/2$, $\xi_k = a_k/b_k$ finally gives

$$t_k = \mathrm{N}_{\mathbb{Q}_q/\mathbb{Q}_p}\left(\frac{2\xi_k}{1 + \xi_k}\right).$$

---

**Algorithm 17.58** Univariate elliptic curve AGM

INPUT: An elliptic curve $E : y^2 + xy = x^3 + \bar{c}$ over $\mathbb{F}_{2^d}$ with $j(E) \neq 0$.
OUTPUT: The number of points on $E(\mathbb{F}_{2^d})$.

1. $N \leftarrow \lceil \frac{d}{2} \rceil + 3$
2. $c \leftarrow \bar{c} \bmod 2$                                                            [$c$ arbitrary lift of $\bar{c}$]
3. $\xi \leftarrow (1 + 8c) \bmod 2^4$
4. **for** $i = 5$ **to** $N$ **do**
5.      $\xi \leftarrow (1 + \xi)/(2\sqrt{\xi}) \bmod 2^i$
6. $t \leftarrow \mathrm{N}_{\mathbb{Q}_q/\mathbb{Q}_p}(2\xi/(1 + \xi)) \bmod 2^{N-1}$
7. **if** $t^2 > 2^{d+2}$ **then** $t \leftarrow t - 2^{N-1}$
8. **return** $2^d + 1 - t$

---

**Example 17.59** We keep the settings of Example 17.57, i.e., $p = 2$, $d = 7$ and $\mathbb{F}_{p^d} \simeq \mathbb{F}_p(\theta)$ with $\theta^7 + \theta + 1 = 0$ and let $E$ be the elliptic curve given by the affine equation $y^2 + xy = x^3 + c$ with $c = 1 + \theta^4$. Let $\mathbb{Z}_q \simeq \mathbb{Z}_2[T]/(T^7 + T + 1)$, then after the initialization step we get $\xi \equiv 8T^4 + 9$ (mod $2^4$) and finally in Line 6, $\xi \equiv 16T^6 + 56T^4 + 8T^2 + 24T + 9 \pmod{2^6}$ and $t = 13$.

A second improvement of Algorithm 17.54 is due to Gaudry [GAU 2002], who devised a modified modular polynomial based on the univariate AGM sequence. In this case, define $\lambda_k = b_k/a_k$ (i.e., $\lambda_k = 1/\xi_k$) which gives the iteration

$$\lambda_{k+1} = \frac{2\sqrt{\lambda_k}}{1 + \lambda_k}. \tag{17.25}$$

Note that this iteration is less efficient than the one for $\xi_k$, since it requires a square root computation (not an inverse square root) and an extra inverse. The sequence can again be initialized by $\lambda_1 = 1 + 8c$ with $c$ an arbitrary lift of $\bar{c}$. From 17.20 follows that

$$\lambda_{k+1} \equiv \Sigma(\lambda_k) \pmod{2^{k+3}}.$$

Substituting this in equation (17.25) shows that $\lambda_k$ satisfies

$$\Sigma(Z)^2(1+Z)^2 - 4Z \equiv 0 \pmod{2^{k+3}} \text{ and } Z \equiv 1 + 8\Sigma^{k-1}(c) \pmod{16}.$$

Let $\Lambda_2(X,Y) = Y^2(1+X)^2 - 4X$, then $\lambda_k$ satisfies $\Lambda_2(X, \Sigma(X)) \equiv 0 \pmod{2^{k+3}}$. This equation is almost of the form considered in Section 12.7, except that both partial derivatives vanish modulo 2. Therefore, we make the change of variables $X \leftarrow 1 + 8X$ and $Y \leftarrow 1 + 8Y$ to obtain the modified modular polynomial

$$\Upsilon_2(X,Y) = (X + 2Y + 8XY)^2 + Y + 4XY,$$

Let $\gamma_k$ be defined by $\lambda_k = 1 + 8\gamma_k$, then $\gamma_k$ satisfies $\Upsilon_2(X, \Sigma(X)) \equiv 0 \pmod{2^k}$ and $\gamma_k \equiv \sigma^{k-1}(c) \pmod 2$. The partial derivatives of $\Upsilon_2$ are

$$\frac{\partial \Upsilon_2}{\partial X}(X,Y) = 2(X + 2Y + 8XY)(1 + 8Y) + 4Y,$$
$$\frac{\partial \Upsilon_2}{\partial Y}(X,Y) = (4(X + 2Y + 8XY) + 1)(1 + 4X),$$

which shows that $\frac{\partial \Upsilon_2}{\partial X} \equiv 0 \pmod 2$, but $\frac{\partial \Upsilon_2}{\partial Y} \equiv 1 \pmod 2$. Using any of the algorithms described in Section 12.7, we can compute $\gamma_k$ for any $k \in \mathbb{N}$. The trace of Frobenius can be computed as

$$\mathrm{Tr}(\phi_q) \equiv t_k + \frac{q}{t_k} \pmod{2^{k+3}},$$

with $t_k = \mathrm{N}_{\mathbb{Q}_q/\mathbb{Q}_p}(2/(1+\lambda_k)) = \mathrm{N}_{\mathbb{Q}_q/\mathbb{Q}_p}(1/(1+4\gamma_k))$.

**Remark 17.60** This optimization by Gaudry has been extended to other characteristics by Madsen in [MAD 2003] using $\lambda$-modular polynomials. A related approach based on lifting of Heegner points on modular curves $X_0(N)$ was described by Kohel in [KOH 2003].

### 17.3.2.c  Hyperelliptic curve AGM

To extend the AGM to ordinary curves of genus $g \geqslant 2$, Mestre [MES 2002] showed how to interpret the AGM as a special case of the Riemann duplication formula for theta functions.

Let $\mathbb{F}_q$ a finite field with $q = 2^d$ elements. Let $\overline{C}$ be a curve defined over $\mathbb{F}_q$ such that its Jacobian variety $\overline{J} = J_{\overline{C}}$ is ordinary and assume that $J_{\overline{C}}[2]$ is $\mathbb{F}_q$-rational. Denote with $\overline{J}_i$ the $i$-th conjugate of $\overline{J}$ and let $\phi_{2,i} : \overline{J}_i \longrightarrow \overline{J}_{i+1}$ denote the corresponding 2-nd power Frobenius isogeny.

Let $\mathbb{Q}_q$ be the degree $d$ unramified extension of $\mathbb{Q}_2$, with valuation ring $\mathbb{Z}_q$ such that $\mathbb{Z}_q/2\mathbb{Z}_q \simeq \mathbb{F}_q$ and denote with $\mathcal{J}_c$ the canonical lift of $J_{\overline{C}}$ and $\mathcal{J}_{i,c} = \mathcal{J}_c^{\Sigma^i}$ for $i$ positive. Since $\Sigma^d = \mathrm{Id}$, we clearly have $\mathcal{J}_{i,c} = \mathcal{J}_{i+d,c}$. Let $\mathcal{F}_{2,i}$ be the image of $\phi_{2,i}$ under the isomorphism $\mathrm{Hom}(\mathcal{J}_{i,c}, \mathcal{J}_{i+1,c}) \simeq \mathrm{Hom}(\overline{J}_i, \overline{J}_{i+1})$, then $\mathcal{F}_q = \mathcal{F}_{2,d-1} \circ \cdots \circ \mathcal{F}_{2,0}$ is the lift of the Frobenius endomorphism $\phi_q$ under the ring isomorphism $\mathrm{End}(\mathcal{J}_c) \simeq \mathrm{End}(J_{\overline{C}})$.

Consider the following sequence of abelian varieties

$$\mathcal{J}_0 \xrightarrow{\mathcal{I}_0} \mathcal{J}_1 \xrightarrow{\mathcal{I}_1} \mathcal{J}_2 \xrightarrow{\mathcal{I}_2} \mathcal{J}_3 \xrightarrow{\mathcal{I}_3} \ldots, \qquad (17.26)$$

with $\mathcal{J}_0 = \mathcal{J}_c$ and for $i \in \mathbb{N}$, $\mathcal{J}_{i+1} = \mathcal{J}_i/\mathcal{J}_i[2]^{\mathrm{loc}}$ with $\mathcal{J}_i[2]^{\mathrm{loc}} = \mathcal{J}_i[2] \cap \ker(\pi)$, i.e., the 2-torsion points in the kernel of reduction. Since the isogenies $\mathcal{I}_i : \mathcal{J}_i \longrightarrow \mathcal{J}_{i+1}$ reduce to the 2-nd power Frobenius isogeny $\phi_{2,i} : \overline{J}_i \longrightarrow \overline{J}_{i+1}$, we conclude that $\mathcal{J}_i$ must also be a canonical lift of $\overline{J}_i$. Due to the uniqueness of the canonical lift, there must exists an isomorphism $\lambda_i : \mathcal{J}_{i,c} \longrightarrow \mathcal{J}_i$. Consider

### § 17.3 Overview of p-adic methods

the following diagram

$$
\begin{array}{ccccccccc}
& \mathcal{I}_0 & & \mathcal{I}_1 & & \mathcal{I}_2 & & \mathcal{I}_{d-1} & \\
\mathcal{J}_0 & \longrightarrow & \mathcal{J}_1 & \longrightarrow & \mathcal{J}_2 & \longrightarrow & \cdots & \longrightarrow & \mathcal{J}_d \\
\text{Id} \uparrow & & \lambda_1 \uparrow & & \lambda_2 \uparrow & & \vdots & & \lambda_d \uparrow \\
& \mathcal{F}_{2,0} & & \mathcal{F}_{2,1} & & \mathcal{F}_{2,2} & & \mathcal{F}_{2,d-1} & \\
\mathcal{J}_{0,c} & \longrightarrow & \mathcal{J}_{1,c} & \longrightarrow & \mathcal{J}_{2,c} & \longrightarrow & \cdots & \longrightarrow & \mathcal{J}_{d,c} = \mathcal{J}_{0,c}
\end{array} \quad (17.27)
$$

then we can express the lift $\mathcal{V}_q$ of the dual of the $q$-th power Frobenius $\phi_q$ as

$$\mathcal{V}_q = \widehat{\mathcal{F}}_{2,0} \circ \cdots \circ \widehat{\mathcal{F}}_{2,d-1} = \widehat{\mathcal{I}}_0 \circ \cdots \circ \widehat{\mathcal{I}}_{d-1} \circ \lambda_d.$$

As in the elliptic curve AGM, we can choose a basis $\mathfrak{B}$ of the vector space $\mathfrak{D}_0(\mathcal{J}_c, \mathbb{Q}_q)$ of holomorphic differential forms of degree 1 on $\mathcal{J}_c$ such that the action of $\widehat{\mathcal{I}}_i$, i.e., the dual of $\mathcal{I}_i$, is the identity. The action of the lift of Verschiebung $\mathcal{V}_q$ on $\mathfrak{B}$ is thus simply given by the action of $\lambda_d$.

To turn the above approach into an algorithm, Mestre considers the diagram (17.27) over $\mathbb{C}$ and computes on the corresponding analytic varieties. Let $\mathbb{Q}_2^{\mathrm{ur}}$ denote the maximal unramified extension of $\mathbb{Q}_2$ and fix an embedding $\sigma : \mathbb{Q}_2^{\mathrm{ur}} \hookrightarrow \mathbb{C}$. Since $\mathbb{Q}_q \subset \mathbb{Q}_2^{\mathrm{ur}}$, we can associate to the $\mathcal{J}_i$ in the above diagram, an abelian variety over $\mathbb{C}$ given by $\mathcal{J}_{\mathbb{C},i} = \mathcal{J}_i \otimes_\sigma \mathbb{C}$.

As shown in Section 5.1.3, each abelian variety $\mathcal{J}_{\mathbb{C},i}$ is isomorphic as an analytic variety to a complex torus $T_i = \mathbb{C}^g/(\mathbb{Z}^g + \Omega_i \mathbb{Z}^g)$ with $\Omega_i$ a symmetric matrix such that $\Im m\, \Omega_i > 0$, i.e., the period matrix of $\mathcal{J}_{\mathbb{C},i}$. Let $\mu_i : \mathcal{J}_{\mathbb{C},i} \longrightarrow T_i$ denote this isomorphism, then the following proposition shows that the isogenies $\mathcal{I}_i$ have a particular simple expression in the analytic setting.

**Proposition 17.61** The complex torus $T_i$ is given by $\mathbb{C}^g/(\mathbb{Z}^g + 2^i \Omega \mathbb{Z}^g)$ and the isogeny $\mathcal{I}_i$ corresponds to the isogeny $\mathcal{I}_{i,\mathrm{an}}$ defined by the inclusion of the lattices

$$2(\mathbb{Z}_g + 2^i \Omega \mathbb{Z}^g) \subset (\mathbb{Z}^g + 2^{i+1} \Omega \mathbb{Z}^g),$$

i.e., $\mathcal{I}_{i,\mathrm{an}} \circ \mu_i = \mu_{i+1} \circ \mathcal{I}_i$.

Since $\lambda_d : \mathcal{J}_0 \longrightarrow \mathcal{J}_d$ is an isomorphism, there exists an isomorphism $\lambda_{\mathbb{C},d} : T_0 \longrightarrow T_d$ over $\mathbb{C}$.

**Definition 17.62** Let $\mathrm{Sp}(2g, \mathbb{Z})$ denote the symplectic group over $\mathbb{Z}$, i.e., the set of $2g \times 2g$ matrices $\begin{bmatrix} A & B \\ C & D \end{bmatrix}$ with $A, B, C, D \in \mathbb{Z}^{(g \times g)}$, such that

$$\begin{bmatrix} A & B \\ C & D \end{bmatrix}^t \begin{bmatrix} 0 & I_g \\ -I_g & 0 \end{bmatrix} \begin{bmatrix} A & B \\ C & D \end{bmatrix} = \begin{bmatrix} 0 & I_g \\ -I_g & 0 \end{bmatrix}.$$

Denote with $\Gamma_N$ the subgroup of elements $M \in \mathrm{Sp}(2g, \mathbb{Z})$ with $M \equiv I_{2g} \pmod{N}$.

Then there exists a group action of $\mathrm{Sp}(2g, \mathbb{Z})$ on $\mathbb{C}^g \times \mathbb{H}_g$ given by

$$\begin{bmatrix} A & B \\ C & D \end{bmatrix} (z, \Omega) = \left( (C\Omega + D)^{-1} z, (A\Omega + B)(C\Omega + D)^{-1} \right).$$

By Theorem 5.50, the isomorphism $\lambda_{\mathbb{C},d} : T_0 \to T_d$ is defined by a matrix $\begin{bmatrix} A & B \\ C & D \end{bmatrix} \in \mathrm{Sp}(2g, \mathbb{Z})$ and the action of $\lambda_{\mathbb{C},d}$ on $\mathbb{C}^g$ is given by $z \mapsto (C\Omega + D)^{-1} z$. For every $i \in \mathbb{N}$, we have that $\mathfrak{B}_{i,\mathrm{an}} = \left( dz_1^{(i)}, \ldots, dz_g^{(i)} \right)$ forms a basis of the holomorphic differentials on $T_i$. By the description of the $\mathcal{I}_{i,\mathrm{an}}$ in Proposition 17.61, we conclude that

$$\widehat{\mathcal{I}}_{i,\mathrm{an}}^* \mathfrak{B}_{i,\mathrm{an}} = \mathfrak{B}_{i+1,\mathrm{an}}.$$

The action of $\mathcal{V}_{q,\mathrm{an}}$ on $\mathfrak{B}_{0,\mathrm{an}}$ is thus simply given by $M_{\mathcal{V}_q} = (C\Omega + D)^{-1}$. If we would be able to compute $M_{\mathcal{V}_q}$ we could easily recover the characteristic polynomial of $\phi_q$ from the minimal polynomial of $M_{\mathcal{V}_q} + qM_{\mathcal{V}_q}^{-1}$. Unfortunately, we will only be able to determine $\pm \det(M_{\mathcal{V}_q})$ and not $M_{\mathcal{V}_q}$ itself. Since $J_{\overline{C}}$ is ordinary, the characteristic polynomial $\chi(\phi_q)_{\overline{C}}$ has exactly $g$ roots $\pi_1, \ldots, \pi_g$ that are 2-adic units and since $\mathcal{V}_q$ is separable, we conclude that $\det(M_{\mathcal{V}_q}) = \pi_1 \cdots \pi_g$. Although this only determines $\chi(\phi_q)_{\overline{C}}$ uniquely in genus 1 and 2, we can use the LLL algorithm to determine $\chi(\phi_q)_{\overline{C}}(\pm T)$. To decide the correct sign, test which choice $\chi(\phi_q)_{\overline{C}}(\pm 1)$ gives the correct order.

The determinant $\pm \det(C\Omega + D)^{-1}$ can be recovered from the theta constants associated to $T_n$ and $T_{n+d}$ for any $n \in \mathbb{N}$. As shown in Section 5.1.6.a, we can associate to a principally polarized abelian variety $\mathcal{A}_{\mathbb{C}}$ over $\mathbb{C}$ with period matrix $\Omega$, the Riemann theta function, which for $z \in \mathbb{C}^g$ is given by

$$\theta(z, \Omega) = \sum_{\mathbf{n} \in \mathbb{Z}^g} \exp\bigl(\pi i (\mathbf{n}^t \Omega \mathbf{n} + 2\mathbf{n}^t z)\bigr).$$

More generally, we consider translations of $\theta(z, \Omega)$ by an element of $\frac{1}{2}\mathbb{Z}^g + \Omega\left(\frac{1}{2}\mathbb{Z}^g\right)$, leading to the theta functions with characteristics

$$\theta\begin{bmatrix}\delta\\\varepsilon\end{bmatrix}(z, \Omega) = \sum_{\mathbf{n} \in \mathbb{Z}^g} \exp\left(\pi i \left(\mathbf{n} + \frac{1}{2}\delta\right)^t \Omega\left(\mathbf{n} + \frac{1}{2}\delta\right) + 2\left(\mathbf{n} + \frac{1}{2}\delta\right)^t\left(z + \frac{1}{2}\varepsilon\right)\right),$$

with $\delta, \varepsilon \in \mathbb{R}^g$. When $\delta, \varepsilon \in \mathbb{Z}^g$ and $\delta^t \varepsilon \equiv 0 \pmod 2$ the *theta characteristic* is called *even*; otherwise, it is called *odd*.

**Definition 17.63** For integral theta characteristic $\begin{bmatrix}\delta\\\varepsilon\end{bmatrix}$, the value $\theta\begin{bmatrix}\delta\\\varepsilon\end{bmatrix}(0, \Omega)$ is called the *theta constant of characteristic* $\begin{bmatrix}\delta\\\varepsilon\end{bmatrix}$.

The relation with the isomorphism $\lambda_{\mathbb{C},d} : T_0 \to T_d$ given by the matrix $\begin{bmatrix}A & B\\C & D\end{bmatrix} \in \mathrm{Sp}(2g, \mathbb{Z})$ is the following proposition.

**Proposition 17.64** For all $\delta, \varepsilon \in \mathbb{Z}^g$ and $\begin{bmatrix}A & B\\C & D\end{bmatrix} \in \Gamma_2$ we have

$$\theta\begin{bmatrix}\delta\\\varepsilon\end{bmatrix}^2\bigl(0, (A\Omega + B)(C\Omega + D)^{-1}\bigr) = \pm \det(C\Omega + D)\theta\begin{bmatrix}\delta\\\varepsilon\end{bmatrix}^2(0, \Omega).$$

Since $T_0$ has period matrix $\Omega$ and $T_d$ has period matrix $2^d\Omega$, the above proposition leads to the main theorem.

**Theorem 17.65** Let $\pi_1, \ldots, \pi_g$ be the unit roots of $\chi(\phi_q)_{\overline{C}}$, then if there exists $\delta, \varepsilon \in \mathbb{Z}^g$ such that $\theta\begin{bmatrix}\delta\\\varepsilon\end{bmatrix}^2(0, \Omega) \neq 0$, then $\theta\begin{bmatrix}\delta\\\varepsilon\end{bmatrix}^2(0, 2^d\Omega) \neq 0$ is nonzero and

$$\frac{\theta\begin{bmatrix}\delta\\\varepsilon\end{bmatrix}^2(0, \Omega)}{\theta\begin{bmatrix}\delta\\\varepsilon\end{bmatrix}^2(0, 2^d\Omega)} = \pm(\pi_1 \cdots \pi_g).$$

To apply the above theorem, we need to compute $\theta\begin{bmatrix}\delta\\\varepsilon\end{bmatrix}^2(0, \Omega)$ and $\theta\begin{bmatrix}\delta\\\varepsilon\end{bmatrix}^2(0, 2^d\Omega)$ for a given $\delta, \varepsilon \in \mathbb{Z}^g$. Unfortunately, there is currently no algorithm that computes these values for one fixed pair $\delta, \varepsilon \in \mathbb{Z}^g$. Luckily it is possible to compute $\theta\begin{bmatrix}0\\\varepsilon\end{bmatrix}^2(0, 2^d\Omega)$ for all vectors $\varepsilon \in (\mathbb{Z}/2\mathbb{Z})^g$ simultaneously. The following duplication formula, which is a special case of Riemann's duplication formula, is at the heart of this algorithm.

**Proposition 17.66** For $\varepsilon \in (\mathbb{Z}/2\mathbb{Z})^g$ we have

$$\theta\begin{bmatrix}0\\\varepsilon\end{bmatrix}^2(0, 2\Omega) = \frac{1}{2^g} \sum_{\mathbf{e} \in (\mathbb{Z}/2\mathbb{Z})^g} \theta\begin{bmatrix}0\\\varepsilon+\mathbf{e}\end{bmatrix}(0, \Omega) \times \theta\begin{bmatrix}0\\\mathbf{e}\end{bmatrix}(0, \Omega). \qquad (17.28)$$

**Example 17.67** Let $g = 1$ and define $\theta_{i,k} = \theta \left[\begin{smallmatrix} 0 \\ \varepsilon_i \end{smallmatrix}\right]^2 (0, 2^k \Omega)$ with $\varepsilon_i$ the vector containing the bits in the binary representation of $i$. The above proposition then gives:

$$\theta_{0,k+1}^2 = \frac{1}{2}\left(\theta_{0,k}^2 + \theta_{1,k}^2\right)$$
$$\theta_{1,k+1}^2 = \theta_{0,k}\theta_{1,k},$$

which is exactly the AGM iteration if we set $a_k = \theta_{0,k}^2$ and $b_k = \theta_{1,k}^2$.

In the elliptic curve case, we considered the univariate AGM instead of the bivariate one since it was more efficient. Similarly, we can divide out the theta constant $\theta \left[\begin{smallmatrix} 0 \\ 0 \end{smallmatrix}\right]^2 (0, 2^k \Omega)$ to obtain a nonhomogeneous variant of the above proposition.

**Proposition 17.68** Define $\tau_{i,k} = \theta_{i,k}/\theta_{0,k}$ for $i = 1, \ldots, 2^g - 1$ with $\theta_{i,k} = \theta \left[\begin{smallmatrix} 0 \\ \varepsilon_i \end{smallmatrix}\right]^2 (0, 2^k \Omega)$ and $\varepsilon_i$ the vector containing the bits in the binary representation of $i$. Let $G : \mathbb{Z}_q^{2^g - 1} \longrightarrow \mathbb{Z}_q$ be defined by $G(t_1) = \frac{2\sqrt{t_1}}{1 + t_1}$ if $g = 1$ and more generally for $g > 1$ by

$$G(t_1, \ldots, t_{2^g - 1}) = 2 \frac{\sqrt{t_1} + \sqrt{t_2}\sqrt{t_3} + \cdots + \sqrt{t_{2^g - 2}}\sqrt{t_{2^g - 1}}}{1 + t_1 + \cdots + t_{2^g - 1}}. \qquad (17.29)$$

Then
$$\forall i \in \{1, \ldots, 2^g - 1\}, \ \tau_{i,k+1} = G(\tau_{i,k}, \tau_{i_2,k}, \tau_{i_3,k}, \ldots, \tau_{i_{2^g - 2},k}, \tau_{i_{2^g - 1},k}),$$

where, for each $i$, the indices $i_2, \ldots, i_{2^g - 1}$ are such that

$$\{\{0, i\}, \{i_2, i_3\}, \ldots, \{i_{2^g - 2}, i_{2^g - 1}\}\} = \{\{j, j \text{ XOR } i\} \mid j \in \{1, \ldots, 2^g - 1\}\}. \qquad (17.30)$$

The iteration in the above proposition satisfies similar properties as the univariate AGM iteration described in Section 17.3.2.b.

**Lemma 17.69** The generalized AGM sequence defined in Proposition 17.68 satisfies:

- Linear convergence: let $\eta_{i,k} \equiv \tau_{i,k} \pmod{p^n}$ for $i = 1, \ldots, 2^g - 1$ and let $\eta_{i,k+1}$ be obtained by applying $G$, then

$$\eta_{i,k+1} \equiv \tau_{i,k+1} \pmod{p^{n+1}}.$$

- Conjugacy property:
$$\tau_{i,k+1} = \Sigma(\tau_{i,k}).$$

The linear convergence property shows that if we start from an approximation of $\tau_{i,0} \pmod{2^s}$ for $i = 1, \ldots, 2^g - 1$, then after $n$ iterations we have obtained $\tau_{i,n} \pmod{2^{s+n}}$. To recover $\pm (\pi_1 \cdots \pi_g)$ to precision $N$, we therefore simply compute

$$\frac{\tau_{i,N-s}}{\tau_{i,N-s+d}} \equiv \pm(\pi_1 \cdots \pi_g) \pmod{2^N}.$$

Like in the univariate AGM, there are two further optimizations possible based on the conjugacy property. Instead of using the generalized AGM iteration to compute $\tau_{i,k}$ for $i = 1, \ldots, 2^g - 1$, we compute $\tau_i \pmod{2^N}$ for $i = 1, \ldots, 2^g - 1$ as the solution of the system of equations

$$\Sigma(\tau_i) = G(\tau_i, \tau_{i_2}, \tau_{i_3}, \ldots, \tau_{i_{2^g - 2}}, \tau_{i_{2^g - 1}}), \text{ for } i = 1, \ldots, 2^g - 1. \qquad (17.31)$$

This system of equations can be solved efficiently using a vectorial version of the generalized Newton lift described in Section 12.7.

The second optimization replaces the extra $d$ AGM iterations by a fast norm computation. Indeed, let $\tau_i$ with $i = 1, \ldots, 2^g - 1$ be a solution of (17.31) to precision $N$, then

$$\pm (\pi_1 \cdots \pi_g) \equiv N_{\mathbb{Q}_q/\mathbb{Q}_p} \left( \frac{2^g}{1 + \tau_1 + \cdots + \tau_{2^g-1}} \right) \pmod{2^N}. \tag{17.32}$$

The only remaining problem is to devise an initial approximation to the $\tau_i$ for $i = 1, \ldots, 2^g - 1$. This is in fact the only step that really depends on the curve $\overline{C}$ itself; all the steps described so far only require that $J_{\overline{C}}$ is ordinary and that $J_{\overline{C}}[2]$ is defined over $\mathbb{F}_q$.

For $\overline{C}$ a hyperelliptic curve, the initial approximations can be computed using formulas due to Thomae as follows: assume that $\overline{C}$ is defined by an equation of the form

$$\overline{C} : y^2 + \overline{h}(x)y = \overline{h}(x)\overline{f}(x),$$

with $\overline{f}, \overline{h} \in \mathbb{F}_q[x]$, $\deg \overline{f} = \deg \overline{h} = g + 1$ and such that $\overline{h}$ has $g + 1$ roots in $\mathbb{F}_q$ with multiplicity one. This latter condition ensures that $J_{\overline{C}}[2]$ is defined over $\mathbb{F}_q$. Note that all ordinary hyperelliptic curves can be written in the above form after a suitable extension of the base field and a change of variables.

Take arbitrary lifts $f, h \in \mathbb{Z}_q[x]$ of $\overline{f}$ and $\overline{h}$ with $\deg f = \deg h = g + 1$, and consider the curve $C : y^2 + h(x)y = h(x)f(x)$. Multiplying both sides by 4 and completing the square gives $\bigl(2y + h(x)\bigr)^2 = h(x)\bigl(h(x) + 4f(x)\bigr)$. Since $\overline{h}(x)$ splits completely over $\mathbb{F}_q$, Hensel's lemma implies that $h(x)$ and also $h(x) + 4f(x)$ have $g + 1$ roots over $\mathbb{Z}_q$ with multiplicity 1. We can thus consider the isomorphic curve defined by the equation

$$y^2 = \prod_{i=0}^{g}(x - \alpha_i)\bigl(x - (\alpha_i + 4\beta_i)\bigr).$$

Let $a_{2i} = \alpha_i$ and $a_{2i+1} = \alpha_i + 4\beta_i$ for $i = 0, \ldots, g$, then we clearly have $a_{2i} \equiv a_{2i+1} \pmod{4}$. The following lemma finally provides the last missing step in the algorithm.

**Lemma 17.70 (Thomae formula)** Let $C$ be a hyperelliptic curve over $\mathbb{Z}_q$ given by an equation of the form $C : y^2 = \prod_{i=0}^{2g+1}(x - a_i)$ with $a_i \in \mathbb{Z}_q$ and $a_{2i} \equiv a_{2i+1} \pmod 4$, then for $k = 0, \ldots, 2^g - 1$

$$\theta_{1,k} = \sqrt{\prod_{0 \leq i < j \leq g} (a_{2i+k_i} - a_{2j+k_j})(a_{2i+1-k_i} - a_{2j+1-k_j})},$$

with $k_0 = 0$, $k = \sum_{i=0}^{g-1} k_{i+1} 2^i$ and $k_i \in \{0, 1\}$. The square roots should be chosen such that $\theta_{1,k} \equiv 1 \pmod 4$.

This finally leads to Algorithm 17.71 which computes $\pm (\pi_1 \cdots \pi_g) \pmod{2^N}$ for a hyperelliptic curve.

---

**Algorithm 17.71** Hyperelliptic curve AGM

INPUT: Polynomials $\overline{f}, \overline{h} \in \mathbb{F}_q[x]$, with $\deg \overline{f} = \deg \overline{h} = g + 1$, $\overline{h}$ separable and precision $N$.
OUTPUT: The product of unit roots (up to sign) $\lambda \equiv \pm (\pi_1 \cdots \pi_g) \pmod{2^N}$.

1.  $f \leftarrow \overline{f}$ and $h \leftarrow \overline{h}$            [arbitrary lift to $\mathbb{Z}_q$ with $\deg f = \deg h = g + 1$]
2.  $(a_0, \ldots, a_{2g+2}) \leftarrow $ Roots $(h) \bmod 8$
3.  $(a_1, \ldots, a_{g+1}) \leftarrow $ Roots $(h + 4f) \bmod 8$     [with $a_{2i} \equiv a_{2i+1} \pmod 4$]
4.  **for** $k = 0$ **to** $2^g - 1$ **do**

## § 17.3 Overview of $p$-adic methods

5.      $\theta_k \leftarrow \sqrt{\prod_{0 \leqslant i < j \leqslant g}(a_{2i+k_i} - a_{2j+k_j})(a_{2i+1-k_i} - a_{2j+1-k_j})} \mod 2^{10g+10}$

6.      **for** $k = 1$ **to** $2^g - 1$ **do**

7.          $\tau_k \leftarrow (\theta_k/\theta_0) \mod 2^{10g+10}$

8.      $\phi(X, Y) \leftarrow \left\{ y_i^2(1 + x_1^2 + \cdots + x_{2^g-1}^2) - \sum_{0 \leqslant j \leqslant 2^g-1} x_j x_{i \text{ XOR } j} \right\}_{i=1,\ldots,2^g-1}$

9.      $(\tau_1, \ldots, \tau_{2^g-1}) \leftarrow$ Generalized Newton lift $\big(\phi(X, Y), (\tau_1, \ldots, \tau_{2^g-1}), N\big)$

10.      $\lambda \leftarrow \mathrm{N}_{\mathbb{Q}_q/\mathbb{Q}_p}\left(\frac{2^g}{1+\tau_1+\cdots \tau_{2^g-1}}\right) \mod 2^N$

11.      **return** $\lambda$

**Example 17.72** Consider the hyperelliptic curve of genus 3 defined over $\mathbb{F}_{2^4} \simeq \mathbb{F}_2(\alpha)$ with $\alpha^4 + \alpha^3 + \alpha^2 + \alpha + 1 = 0$ an affine model of which is given by the equation $y^2 + h(x)y + h(x)f(x) = 0$ where

$$f(x) = x^4 + (\alpha + 1)x^3 + (\alpha^3 + \alpha)x^2 + \alpha^3 x + \alpha^3 + \alpha + 1,$$
$$h(x) = (x + \alpha^2 + 1)(x + \alpha^3)(x + \alpha^3 + 1)(x + \alpha^2).$$

The $x$-coordinates of the eight 2-torsion points of a lift of the curve in the unramified 2-adic extension defined by $T^4 + T^3 + T^2 + T + 1$ are, at precision 6,

$$a = [-T^2, 24T^3 + 7T^2 + 20, -T^2 - 1, -8T^3 + 15T^2 - 28T + 3,$$
$$- 13T^3 - 24T^2 - 8T + 32, -T^3, 19T^3 + 24T^2 - 8T - 13, -T^3 - 1].$$

The Thomae–Fay formulas yield seven constants $\tau_e = \theta_e/\theta_0$ which, at precision 7, are equal to

$$\tau = [-48T^3 + 40T^2 - 8T - 31, -8T^3 + 24T^2 + 56T + 25,$$
$$-24T^3 + 64T^2 - 55, -56T^3 + 16T^2 - 56T + 57,$$
$$-8T^3 + 56T^2 + 16T + 41, -24T^2 + 16T - 47,$$
$$-48T^3 + 16T^2 + 40T + 17].$$

A call to `NewtonLift` successively lifts $\tau$ at precision 82. This yields,

$$\tau = [-312726215120141988400432\,T^3 - 1933727213108832572762328\,T^2$$
$$- 1494080419622419245495432\,T + 3693732824333787018444449,$$
$$1976425370164879348289528\,T^3 + 1563232078799037272755224\,T^2$$
$$- 1149644917000765525620040\,T - 2354501050379432611581927,$$
$$- 936875050472473454654744\,T^3 - 1319227091748533942659264\,T^2$$
$$+ 2656963514065467867444448\,T + 8856933584329453881851 61,$$
$$- 1794275179996455307177912\,T^3 - 1319924955112374713058 40\,T^2$$
$$+ 8364955831783298344940 24\,T - 2038182564118844375928135,$$
$$8394583005213698031766 96\,T^3 - 59109618385840075438535 2\,T^2$$
$$- 7043542755196500818211 68\,T + 595721996026176100397993,$$
$$11921687141564846260620 8\,T^3 - 27933962842873584900674 4\,T^2$$
$$- 1478145604901852055715 568\,T + 1355279383057163288500689,$$
$$20922924694676668879726 4\,T^3 + 947564514383409075790 352\,T^2$$
$$+ 916123454126249969343784\,T + 3602780223856537190376 1]$$

and
$$\lambda \equiv 2876672816405499004481849 \pmod{2^{82}}.$$

The only remaining problem is to devise $\chi(\phi_q)_C$ from the product of the unit roots $\pm(\pi_1 \cdots \pi_g)$ $\pmod{2^N}$. Factoring $\chi(\phi_q)_C$ gives

$$\chi(\phi_q)_C(T) = \prod_{i=1}^{g}(T - \pi_i)(T - \overline{\pi}_i).$$

As an intermediate step, we will compute a polynomial of degree $2^{g-1}$ with coefficients in $\mathbb{Z}$.

**Definition 17.73** The symmetric polynomial $P_{\text{sym}}$ of a projective nonsingular curve $C$ over $\mathbb{F}_q$ is the unique monic polynomial of degree $2^{g-1}$ whose roots are $\alpha + q^g/\alpha$ where $\alpha$ is the product of $g$ successive terms in $\{\pi_1, \overline{\pi}_1\}, \ldots, \{\pi_g, \overline{\pi}_g\}$.

It is not difficult to see that $P_{\text{sym}}$ has integer coefficients. Recall that since $C$ is assumed to be ordinary, the roots $\pi_1, \ldots, \pi_g$ are 2-adic units. The following lemma shows that in this case $P_{\text{sym}}$ almost determines $\chi(\phi_q)_C$.

**Lemma 17.74** Let $C$ be an ordinary, projective nonsingular curve of genus $g > 1$, then $P_{\text{sym}}$ determines the set $\{\pi_i^2\}_{i=1,\ldots,g}$. Furthermore, if $\chi(\phi_q)_C$ is irreducible, then $P_{\text{sym}}$ determines $\chi(\phi_q)_C(\pm T)$.

According to a theorem by Tate [TAT 1966], if $J_C$ is $\mathbb{F}_q$-simple then $\chi(\phi_q)_C$ is irreducible, so in this case $P_{\text{sym}}$ determines $\chi(\phi_q)_C$.

Given $\lambda \equiv \pm(\pi_1 \cdots \pi_g) \pmod{p^N}$ for some sufficiently large precision $N$, $P_{\text{sym}}$ can be computed using the LLL algorithm [LELE$^+$ 1982] as follows: let $\eta = \lambda + 2^{dg}/\lambda$ and consider the lattice $\mathcal{L}$ over $\mathbb{Z}$ given by

$$\begin{bmatrix} SM_0 & SM_1 & \cdots & SM_m & S2^N \\ 0 & 0 & \cdots & q^{\lfloor S_m \rfloor} & 0 \\ 0 & 0 & \cdots & 0 & 0 \\ \vdots & \vdots & \ddots & \vdots & \vdots \\ 0 & q^{\lfloor S_1 \rfloor} & \cdots & 0 & 0 \\ q^{\lfloor S_0 \rfloor} & 0 & \cdots & 0 & 0 \end{bmatrix}$$

where the columns are the basis vectors, $m = 2^{g-1}$, $S$ an arbitrarily large constant and

$$M_i \equiv 2^{d(m-1-i)}\eta^i \pmod{2^N}, \quad i = 0, \ldots, m-1, \quad M_m \equiv \eta^m \pmod{p^N}$$
$$S_i = \frac{i(g-2)}{2}, \quad i = 0, \ldots, m-1, \quad S_m = \frac{m(g-2)}{2} + 1.$$

To see where the above lattice comes from, note that $P_{\text{sym}}(T) = T^m + \sum_{i=0}^{m-1} r_i q^{m-1-i} T^i$ and that the $r_i$ can be bounded by

$$|r_i| \leq \binom{m}{i} 2^{(m-i)} q^{\frac{(g-2)(m-i)}{2}} + 1.$$

The $S_i$ are thus weight factors such that $S_i r_i \simeq S_j r_j$ for all $0 \leq i, j \leq m$. Multiplying the above matrix by the column vector $[r_0, \ldots, r_{m-1}, 1, R]^t$ shows that we can recover the coefficients of $P_{\text{sym}}$ from a short vector $\Pi$ in the lattice. The LLL algorithm can compute this vector $\Pi$ if its

*§ 17.3 Overview of p-adic methods*

Euclidean norm $||\cdot||_2$ (or sup-norm $||\ ||_1$) satisfy $||\Pi||_1 \leqslant ||\Pi||_2 \leqslant \det(\mathcal{L})^{1/\dim \mathcal{L}}$. This leads to the following asymptotic estimates for the necessary precision $N$:

$$N > \frac{2^{2g}(g-2) + 2^{g+1}(g+2)}{16} d.$$

**Example 17.75** Continuing Example 17.72, after the lattice reduction step, one obtains,

$$P_{\text{sym}}(T) = T^4 + 23\,T^3 - 18384\,T^2 - 264960\,T + 58593280,$$

and finally,

$$\chi(\phi_q)_C(T) = T^6 - 3\,T^5 + 11\,T^4 - 73\,T^3 + 176\,T^2 - 768\,T + 4096.$$

### 17.3.3 Kedlaya's algorithm

Let $p \geqslant 3$ be a prime number and $\mathbb{F}_q$ a finite field with $q = p^d$ elements and algebraic closure $\overline{\mathbb{F}}_q$. Let $\overline{\mathcal{C}}$ be a hyperelliptic curve of genus $g$ defined by the equation

$$y^2 = \overline{f}(x),$$

with $\overline{f} \in \mathbb{F}_q[x]$ a monic polynomial of degree $2g+1$ without repeated roots. In particular, $\overline{\mathcal{C}}$ is nonsingular in its affine part and has one rational Weierstraß place at infinity. Kedlaya [KED 2001] does not work with the curve $\overline{\mathcal{C}}$ itself, but with the affine curve $\overline{C}$, which is obtained from $\overline{\mathcal{C}}$ by removing the point at infinity and the locus of $y = 0$, i.e., the points $(\overline{\xi}_i, 0) \in \overline{\mathbb{F}}_q \times \overline{\mathbb{F}}_q$ where $\overline{\xi}_i$ is a zero of $\overline{f}$. The reason for working with $\overline{C}$ instead of $\overline{\mathcal{C}}$ will become clear during the exposition of the algorithm.

#### 17.3.3.a Dagger algebra and Frobenius lift

Let $\mathbb{Q}_q$ be the degree $d$ unramified extension of $\mathbb{Q}_p$, with valuation ring $\mathbb{Z}_q$, such that $\mathbb{Z}_q/p\mathbb{Z}_q = \mathbb{F}_q$. Take any monic lift $f \in \mathbb{Z}_q[x]$ of $\overline{f}$ and let $\mathcal{C}$ be the hyperelliptic curve defined by $y^2 = f(x)$. Let $C$ be the curve obtained from $\mathcal{C}$ by removing the point at infinity and the locus of $y = 0$ and let $A = \mathbb{Q}_q[x, y, y^{-1}]/(y^2 - f(x))$ be its coordinate ring.

The dagger algebra corresponding to $\overline{C}$ is given by

$$A^\dagger = \mathbb{Q}_q\langle x, y, y^{-1}\rangle^\dagger / (y^2 - f(x)),$$

with $\mathbb{Q}_q\langle x, y, y^{-1}\rangle^\dagger$ the dagger algebra of $\mathbb{Q}_q\langle x, y, y^{-1}\rangle$ as described in Section 8.3.2. Every element of $A^\dagger$ can be written as a power series $\sum_{i \in \mathbb{Z}} A_i(x) y^i$ with $A_i \in \mathbb{Q}_q[x]$, $\deg(A_i) \leqslant 2g$ and $v_p(A_i) > \alpha|i| + \beta$ for constants $\alpha, \beta$ with $\alpha > 0$. Note that these constants are not fixed for all of $A^\dagger$, but depend on the particular element chosen.

Since $\phi_q = \phi_p^d$, with $\phi_p$ the $p$-power Frobenius, it suffices to lift $\phi_p$ to an endomorphism $\mathcal{F}_p$ of $A^\dagger$. It is natural to define $\mathcal{F}_p$ as the Frobenius substitution on $\mathbb{Z}_q$ and to extend it to $A^\dagger$ by mapping $x$ to $\mathcal{F}_p(x) = x^p$ and $y$ to $\mathcal{F}_p(y)$ with

$$\mathcal{F}_p(y) = y^p \left(1 + \frac{\mathcal{F}_p(f(x)) - f(x)^p}{y^{2p}}\right)^{1/2} = y^p \sum_{i=0}^{\infty} \binom{1/2}{i} \frac{(\mathcal{F}_p(f(x)) - f(x)^p)^i}{y^{2pi}}. \qquad (17.33)$$

An easy calculation shows that $\text{ord}_p\binom{1/2}{i} \geqslant 0$ which implies that $\mathcal{F}_p(y)$ is an element of $A^\dagger$, since $p$ divides $\mathcal{F}_p(f(x)) - f(x)^p$. Note that it is essential that $y^{-1}$ is an element of $A^\dagger$, which explains why we compute with $\overline{C}$ instead of $\overline{\mathcal{C}}$.

In the final algorithm, we actually need $\mathcal{F}_p(y)^{-1}$, which can be computed as $\mathcal{F}_p(y)^{-1} = y^{-p} R$ where $R$ is a root of the equation $F(Z) = SZ^2 - 1$ with $S = \bigl(1 + (\mathcal{F}_p(f(x)) - f(x)^p)/y^{2p}\bigr)$. The Newton iteration to compute $R$ is given by

$$Z \leftarrow \frac{Z(3 - SZ^2)}{2} \tag{17.34}$$

starting from $Z \equiv 1 \pmod{p}$. In each step, the truncated power series should be reduced modulo the equation of the curve to keep the degree of the coefficients less than or equal to $2g$, i.e., for each term $B_i(x) y^i$ with $\deg B_i > 2g$, write

$$B_i(x) y^i = \bigl(Q_i(x) f(x) + R_i(x)\bigr) y^i = R_i(x) y^i + Q_i(x) y^{i+2}$$

with $Q_i$ and $R_i$ the quotient and remainder in the division of $B_i$ by $f$.

### 17.3.3.b Reduction in cohomology

Analogous to Section 4.4.2.c, we associate to each element $h$ of $A^\dagger$ a differential $dh$ such that the Leibniz rule holds $d(hg) = h\,dg + g\,dh$ and such that $da = 0$ for $a \in \mathbb{Q}_q$. Let $\Omega$ be the module of these differentials, then the operator $d$ defines a $\mathbb{Q}_q$-derivation from $A^\dagger$ to $\Omega$. Since $y^2 - f(x) = 0$, we conclude that

$$dy = \frac{f'(x)}{2y} dx \quad \text{and thus} \quad \Omega = A^\dagger \frac{dx}{y}.$$

By definition we have

$$H^0(\overline{C}/\mathbb{Q}_q) = \ker(d) = \{h \in A^\dagger \mid dh = 0\}, \qquad H^1(\overline{C}/\mathbb{Q}_q) = \mathrm{coker}(d) = \left(A^\dagger \frac{dx}{y}\right)/(dA^\dagger),$$

thus elements of $H^1(\overline{C}/\mathbb{Q}_q)$ are differentials modulo exact differentials $dh$ for some $h \in A^\dagger$. The next lemma gives bases for both $H^0(\overline{C}/\mathbb{Q}_q)$ and $H^1(\overline{C}/\mathbb{Q}_q)$.

**Lemma 17.76** $H^0(\overline{C}/\mathbb{Q}_q) = \mathbb{Q}_q$ and $H^1(\overline{C}/\mathbb{Q}_q)$ splits into eigenspaces under the hyperelliptic involution $\iota$:

- a positive eigenspace $H^1(\overline{C}/\mathbb{Q}_q)^+$ with basis $x^i/y^2\, dx$ for $i = 0, \ldots, 2g$,
- a negative eigenspace $H^1(\overline{C}/\mathbb{Q}_q)^-$ with basis $x^i/y\, dx$ for $i = 0, \ldots, 2g-1$.

To compute in $H^1(\overline{C}/\mathbb{Q}_q)$ we need to express an arbitrary differential form as the sum of a $\mathbb{Q}_q$-linear combination of the basis in Lemma 17.76 and an exact differential. This process is called reduction in cohomology and works as follows. It is clear that any differential form can be written as

$$\sum_{k=-\infty}^{+\infty} \sum_{i=0}^{2g} a_{i,k} x^i / y^k \, dx$$

with $a_{i,k} \in \mathbb{Q}_q$. Indeed, using the equation of the curve we can repeatedly replace $h(x) f(x)$ by $h(x) y^2$. A differential $P(x)/y^s\, dx$ with $P(x) \in \mathbb{Q}_q[x]$ and $s \in \mathbb{N}$ can be reduced as follows. Since $f(x)$ has no repeated roots, we can always write $P(x) = U(x) f(x) + V(x) f'(x)$. Using the fact that $d(V(x)/y^{s-2})$ is exact, we obtain

$$\frac{P(x)}{y^s} dx \equiv \left(U(x) + \frac{2V'(x)}{s-2}\right) \frac{dx}{y^{s-2}}, \tag{17.35}$$

where $\equiv$ means equality modulo exact differentials. This congruence can be used to reduce a differential form involving negative powers of $y$ to the case $s = 1$ and $s = 2$. A differential

§ 17.3 Overview of p-adic methods

$P(x)/y\,dx$ with $\deg P = n \geqslant 2g$ can be reduced by repeatedly subtracting suitable multiples of the exact differential $d(x^{i-2g}y)$ for $i = n, \ldots, 2g$. Finally, it is clear that the differential $P(x)/y^2\,dx$ is congruent to $(P(x) \bmod f(x))/y^2\,dx$ modulo exact differentials. A differential of the form $P(x)y^s\,dx$ with $P(x) \in \mathbb{Q}_q[x]$ and $s \in \mathbb{N}$ is exact if $s$ is even and equal to $P(x)f(x)^{\lceil s/2 \rceil}/y\,dx$ if $s$ is odd and thus can be reduced using the above reduction formula.

### 17.3.3.c Recovering the zeta function

The first Monsky–Washnitzer cohomology group $H^1(\overline{C}/\mathbb{Q}_q)$ decomposes as the direct sum of the $\iota$-invariant part $H^1(\overline{C}/\mathbb{Q}_q)^+$ on which $\iota$ acts trivially and the $\iota$-anti-invariant part $H^1(\overline{C}/\mathbb{Q}_q)^-$ on which $\iota$ acts as multiplication by $-1$. The $\iota$-invariant part of $A^\dagger$ is given by

$$\mathbb{Z}_q\langle x, y^2, y^{-2}\rangle^\dagger/(y^2 - f(x)),$$

which clearly is isomorphic to $\mathbb{Z}_q\langle x, (f(x))^{-1}\rangle^\dagger$. The latter ring is the dagger ring of the curve $\mathbb{A}^1 \smallsetminus V_{\overline{f}}$ with $V_{\overline{f}}$ the set of zeroes of $\overline{f}$. The Lefschetz fixed point formula applied to $\overline{C}$ and $\mathbb{A}^1 \smallsetminus V_{\overline{f}}$ then gives

$$\left|\overline{C}(\mathbb{F}_{q^k})\right| = q^k - r_k - \mathrm{Tr}\left(q^k \mathcal{F}_q^{-k}; H^1(\overline{C}/\mathbb{Q}_q)^-\right)$$

with $r_k$ the number of zeroes of $\overline{f}$ defined over $\mathbb{F}_{q^k}$. Let $\widetilde{C}$ be the unique smooth projective curve birational to $\overline{C}$, then

$$\left|\widetilde{C}(\mathbb{F}_{q^k})\right| = \left|\overline{C}(\mathbb{F}_{q^k})\right| + r_k + 1 = q^k + 1 - \sum_{i=1}^{2g} \alpha_i^k, \tag{17.36}$$

with $\alpha_i$ the eigenvalues of $q\mathcal{F}_q^{-1}$ on $H^1(\overline{C}/\mathbb{Q}_q)^-$.

The Weil conjectures from Section 8.1.1 imply that the $\alpha_i$ can be labeled such that $\alpha_i \alpha_{g+i} = q$ for $i = 1, \ldots, 2g$ where the indices are taken modulo $2g$. Since $\alpha_i$ is an eigenvalue of $q\mathcal{F}_q^{-1}$ on $H^1(\overline{C}/\mathbb{Q}_q)^-$, it is clear that $q/\alpha_i = \alpha_{i+g}$ is an eigenvalue of $\mathcal{F}_q$ on $H^1(\overline{C}/\mathbb{Q}_q)^-$. This proves the following proposition.

**Proposition 17.77** Let $\chi(T)$ be the characteristic polynomial of $\mathcal{F}_q$ on $H^1(\overline{C}/\mathbb{Q}_q)^-$, then the zeta function of $\widetilde{C}$ is given by

$$Z(\widetilde{C}/\mathbb{F}_q; T) = \frac{T^{2g}\chi(1/T)}{(1-T)(1-qT)}.$$

Since $\mathcal{F}_q = \mathcal{F}_p^d$, it suffices to compute the matrix $M$ through which the $p$-power Frobenius $\mathcal{F}_p$ acts on the anti-invariant part $H^1(\overline{C}/\mathbb{Q}_q)^-$ of $H^1(\overline{C}/\mathbb{Q}_q)$; the matrix of the $q$-power Frobenius can then be easily obtained as $M_{\mathcal{F}_q} = \Sigma^{d-1}(M) \cdots \Sigma(M) M$.

The action of $\mathcal{F}_p$ on the basis of $H^1(\overline{C}/\mathbb{Q}_q)^-$ can be computed as

$$\mathcal{F}_p\left(\frac{x^i}{y}\,dx\right) = \frac{px^{p(i+1)-1}}{\mathcal{F}_p(y)}\,dx,$$

for $i = 0, \ldots, 2g-1$. Given a sufficiently precise approximation to $1/\mathcal{F}_p(y)$, we can use reduction in cohomology to express $\mathcal{F}_p(x^i/y\,dx)$ on the basis of $H^1(\overline{C}/\mathbb{Q}_q)^-$ and compute the matrix $M$.

**Remark 17.78** In general, the matrix $M$ has coefficients in $p^{-s}\mathbb{Z}_q$ where $s$ is small and only depends on $p$ and $g$. However, by comparison with crystalline cohomology there always exists a basis such that $M$ has integral coefficients, e.g., the basis $x^i/y^3\,dx$ with $0 \leqslant i < 2g$ leads to integral $M$ for all $p$ and $g$.

Write $T^{2g}\chi(1/T) = \sum_{i=0}^{2g} a_i T^i$; then the Weil conjectures imply that for $1 \leqslant i \leqslant g$, $a_{g+i} = q^i a_{g-i}$ and
$$|a_i| \leqslant \binom{2g}{i} q^{i/2} \leqslant \binom{2g}{g} q^{g/2}.$$

In particular, if we compute $\chi \pmod{p^B}$ with $p^B$ greater than twice the above bound, we can recover $\chi$ uniquely. Note that this does not imply that all computations should be performed modulo $p^B$ since the reduction process introduces denominators and thus causes a loss of precision. The following lemma quantifies this loss of precision and can be found in [KED 2001, Lemma 2-3].

**Lemma 17.79 (Kedlaya)** Let $h \in \mathbb{Z}_q[x]$ be a polynomial of degree $\leqslant 2g$, then for $n \in \mathbb{N}$ the reduction of $h(x)y^{2n+1}\,dx$ (resp. $h(x)/y^{2n+1}\,dx$) becomes integral upon multiplication by $p^m$ with $m \geqslant \lfloor \log_p((2g+1)(n+1) - 2) \rfloor$ (resp. $m \geqslant \lfloor \log_p(2n+1) \rfloor$).

A further careful analysis [EDI 2003] shows we actually have to compute with $N$ digits instead of $B$ where
$$N = B + v_p(2g+1) + \left\lfloor \log_p\left(2g + 1 - \frac{2}{p}\right)\right\rfloor + \lfloor \log_p(2g-1) \rfloor,$$

and that it suffices to approximate $\mathcal{F}_p(y)^{-1}$ by
$$y^{-p} \sum_{0 \leqslant n < M} \binom{-1/2}{n} \left(\mathcal{F}_p(f(x)) - f(x)^p\right)^n y^{-2pn},$$

with $M$ the smallest integer such that $M - \lfloor \log_p(2M+1) \rfloor \geqslant N$. This finally leads to Algorithm 17.80.

---

**Algorithm 17.80** Kedlaya's point counting method for $p \geqslant 3$

INPUT: A hyperelliptic curve $\overline{\mathcal{C}}$ defined by $y^2 = \overline{f}(x)$ over finite field $\mathbb{F}_q$ with $q = p^d$ and $p \geqslant 3$.
OUTPUT: The zeta function $Z(\widetilde{\mathcal{C}}/F_q; T)$.

1. $B \leftarrow \left\lceil \log_p\left(2\binom{2g}{g}q^{g/2}\right)\right\rceil$
2. $N \leftarrow B + v_p(2g+1) + \left\lfloor \log_p\left(2g+1 - \frac{2}{p}\right)\right\rfloor + \lfloor \log_p(2g-1) \rfloor$
3. compute $M$ with $M - \lfloor \log_p(2M+1) \rfloor \geqslant N$
4. $S \leftarrow \left(y^{-p} \sum_{0 \leqslant n < M} \binom{-1/2}{n}\left(\mathcal{F}_p(f(x)) - f(x)^p\right)^n y^{-2pn}\right) \bmod p^N$  [using (17.34)]
5. **for** $i = 0$ **to** $2g - 1$ **do**
6. $\quad \sum_{j=0}^{2g-1} M[i][j] \frac{x^i}{y}\,dx \leftarrow$ Reduce cohomology$(px^{p(i+1)-1} S\,dx)$  [using 17.3.3.b]
7. $M_{\mathcal{F}_q} \leftarrow \Sigma^{d-1}(M) \cdots \Sigma(M) M \bmod p^N$
8. $\chi(T) \leftarrow$ characteristic polynomial of $M_{\mathcal{F}_q} \bmod p^B$
9. **for** $i = 0$ **to** $g$ **do**
10. $\quad$ **if** coefficient of $\chi$ of degree $2g - i > \binom{2g}{i}q^{i/2}$ **then**
11. $\quad\quad$ add $-p^B$ to the coefficient of $\chi$ of degree $2g - i$
12. $\quad$ coefficient of $\chi$ of degree $i \leftarrow q^{g-i} \times$ (coefficient of $\chi$ of degree $2g - i$)
13. **return** $Z(\widetilde{\mathcal{C}}/F_q; T) = \dfrac{T^{2g}\chi(1/T)}{(1-T)(1-qT)}$

## § 17.3 Overview of $p$-adic methods

A detailed complexity analysis of the above algorithm can be found in [KED 2001].

**Theorem 17.81 (Kedlaya)** There exists a deterministic algorithm to compute the zeta function of a smooth hyperelliptic curve of genus $g$ defined over a finite field $\mathbb{F}_{p^d}$ with $p \geqslant 3$ using $O(g^{4+\varepsilon}d^{3+\varepsilon})$ bit-operations and $O(g^3 d^3)$ space for $p$ fixed.

**Example 17.82** Let $\overline{C}$ be the genus 2 hyperelliptic curve over $\mathbb{F}_3$ defined by the equation

$$y^2 = x^5 + x^4 + 2x^3 + 2x + 2.$$

Algorithm 17.80 then computes the following intermediate results: $B = 4$, $N = 6$, $M = 9$ and $S = y^{-p} R$ where $R$ is given by

$$\begin{aligned}R \equiv{}& 1 + (-363x^4 + 96x^3 + 144x^2 - 6x + 207)\tau + (-123x^4 - 153x^3 - 21x^2 + 351x + 210)\tau^2 \\&+ (339x^4 - 228x^3 - 60x^2 - 204x + 186)\tau^3 + (-81x^4 + 54x^3 - 243x^2 - 243x + 27)\tau^4 \\&+ (-54x^4 - 162x^3 - 54x^2 - 54x + 162)\tau^5 + (351x^4 + 189x^3 + 189x^2 + 189x + 351)\tau^6 \\&+ (-243x^4 + 243x^3 - 108x^2 - 270x + 27)\tau^7 + (-135x^3 + 54x^2 + 81x - 108)\tau^8 \\&+ (216x^4 + 108x^3 - 297x^2 + 351x - 162)\tau^9 + (-243x^4 - 162x^3 - 324x^2 + 243x)\tau^{10} \\&+ (81x^4 - 243x^3 - 162x^2 + 162x - 81)\tau^{11} + (-162x^4 + 162x^3 + 324x^2 - 324x + 324)\tau^{12}\end{aligned}$$

with $\tau = y^{-2}$. The matrix $M_{\mathcal{F}_p}$ is given by

$$\begin{bmatrix} 27 & 39 & 30 & 108 \\ 129 & 36 & 27 & 126 \\ 204 & 186 & 12 & 138 \\ 46/3 & 76/3 & 41/3 & 169 \end{bmatrix}$$

and the characteristic polynomial is $\chi(T) \equiv T^4 + 80T^3 + T^2 + 78T + 9 \pmod{3^4}$, which gives

$$Z(\widetilde{C}/F_q; T) = \frac{9T^4 - 3T^3 + T^2 - T + 1}{(1-T)(1-3T)}.$$

**Remark 17.83** Kedlaya's algorithm does not apply when $p = 2$, since the nature of the problem changes: for $p \geqslant 3$, a hyperelliptic curve defines a Kummer extension, whereas for $p = 2$, it defines an Artin–Schreier extension. Denef and Vercauteren [DEVE 2002, DEVE 2005] extended Kedlaya's algorithm to characteristic 2 and showed that the average time and space complexity are the same as for Kedlaya's algorithm. The main difference is that the curve equation has to be lifted in a rather specific way. A ready to implement description of the algorithm can be found in [DEVE 2005].

**Remark 17.84** A related approach for Artin–Schreier covers based on Dwork–Reich cohomology was worked out by Lauder and Wan [LAWA 2002a, LAWA 2004]. For $p = 2$, their algorithm is slightly less general than the one of Denef and Vercauteren, and the complexity of the algorithm is not rigorously proven.

# Chapter 18

# Complex Multiplication

*Gerhard Frey and Tanja Lange*

## Contents in Brief

| | | |
|---|---|---|
| **18.1** | **CM for elliptic curves** | **456** |
| | Summary of background • Outline of the algorithm • Computation of class polynomials • Computation of norms • The algorithm • Experimental results | |
| **18.2** | **CM for curves of genus** 2 | **460** |
| | Summary of background • Outline of the algorithm • CM-types and period matrices • Computation of the class polynomials • Finding a curve • The algorithm | |
| **18.3** | **CM for larger genera** | **470** |
| | Strategy and difficulties in the general case • Hyperelliptic curves with automorphisms • The case of genus 3 | |

In this chapter we present a method for finding a curve and the group order of its Jacobian which can be seen as complementary to those in Sections 17.2 and 17.3. Instead of trying several random curves over a fixed finite field until a good one is found and determining the group order by computing the characteristic polynomial of the Frobenius endomorphism, we *start* with the endomorphism ring and vary the prime until one with a good group order is found. These checks can be computed relatively fast. Only the last step of actually computing the equation of the curve requires some effort, but at that time one knows already that the result is the desired one. On the other hand the curves one can construct are somewhat special as the running time depends on the discriminant of the CM-field and thus only small discriminants are possible.

The approach works in general for curves of arbitrary genus but the implementation has to be done for each genus separately. We first detail it for elliptic curves as it is easier to understand there. In genus $g = 2$ we can efficiently compute curves with the CM method. This is in contrast to the fact that point counting over fields of large characteristic as described in Section 17.2 is still rather inefficient and to date needs about one week to determine the order of the Jacobian of a genus two curve over the prime field with $p = 5 \times 10^{24} + 8503491$ [GASc 2004a].

For larger genera the constructions are possible in principle, but the difficulty is that hyperelliptic curves become very rare. We give some indications on what is possible and which difficulties have to be dealt with. For genus $g = 3$ we give examples of curves with additional automorphisms.

## 18.1 CM for elliptic curves

The theoretical background of this chapter is given in Chapter 5 for curves over the complex numbers. We briefly summarize what is necessary for the implementations and then concentrate on the algorithms.

### 18.1.1 Summary of background

Let $E$ be an elliptic curve over the complex numbers $\mathbb{C}$. Then it is isomorphic to $\mathbb{C}/\Lambda_E$ for some lattice $\Lambda_E$. We recall that an elliptic curve $E/\mathbb{C}$ has *complex multiplication* if its endomorphism ring is strictly bigger than $\mathbb{Z}$. Its lattice corresponds to an ideal $A$ of an order $\mathcal{O}$ in an imaginary quadratic field $K$. The $j$-invariant $j(A) = j_{E_A}$ is an *algebraic integer lying in the ring class field of* $\mathcal{O}$. Moreover $j_{E_A}$ depends only on the ideal class of $A$ in $\mathrm{Cl}(\mathcal{O})$. It satisfies a monic polynomial with integer coefficients whose zeroes are the $j$-invariants of the isomorphism classes of the elliptic curves corresponding to a set of representatives of ideal classes of $\mathcal{O}$.

For our purposes it is enough to restrict to the case that $\mathcal{O}$ is equal to the ring of integers $\mathcal{O}_K$ of $K = \mathbb{Q}(\sqrt{-d})$.

Then the minimal polynomial of $j(E)$ is the Hilbert class polynomial

$$H_d(X) = \prod_{r=1}^{h_d}(X - j(A_r)),$$

where $j(A_r)$ is the $j$-invariant of the elliptic curve corresponding to $A_r$, $h_d$ is the order of the ideal class group of $\mathcal{O}_K$, and $A_r$ are representatives of elements of the class group of $\mathcal{O}_K$.

The reduction of (a suitably chosen equation for) $E$ modulo a prime $\mathfrak{p}$ of the Hilbert class field $H$ is again an elliptic curve. Its $j$-invariant is a root of $H_d(X) \pmod{p}$, where $p \in \mathbb{Z}$ is the prime integer in $(\mathfrak{p})$.

So it is easy to find the $j$-invariant $j_\mathfrak{p}$ of the reduction of a model of $E$ defined over $H$ by factoring $H_d(X)$ over $\mathbb{F}_p$ if we can compute $H_d(X)$.

We get the most important example by choosing for $p$ a prime that splits completely in $H$. Then the reduced curve is defined over $\mathbb{F}_p$.

The important property is that one can easily compute the number of points on the reduced curve. Let $\omega \in \mathcal{O}_K$ be an element with $\omega\overline{\omega} = p$.

Then there is an elliptic curve $E_\mathfrak{p}$ defined over $\mathbb{F}_p$ with $j$-invariant $j_\mathfrak{p}$ having $p + 1 - (\omega + \overline{\omega})$ points; and $\omega$ and $\overline{\omega}$ are the eigenvalues of the Frobenius endomorphism on this curve. From this it follows that the curve is not supersingular.

Recall that for a given element $j$ an elliptic curve with $j$-invariant $j \neq 0, 12^3$ is isomorphic to

$$E_j : y^2 = x^3 - \frac{27j}{4(j-12^3)}x + \frac{27j}{4(j-12^3)}. \tag{18.1}$$

or to a quadratic twist of $E_j$.

### 18.1.2 Outline of the algorithm

From the background we derive a method to construct elliptic curves over finite prime fields or small extensions. In this exposition we shall restrict ourselves to the prime field case. Given a Hilbert class polynomial $H_d(X)$ of an imaginary quadratic field with not too large discriminant $d$ one can reduce it modulo primes $p$, which are the product of principal prime ideals in $\mathcal{O}_K$ and factor the result. For these primes, $H_d(X) \pmod{p}$ splits completely and one obtains $j_\mathfrak{p}$, from which one can reconstruct the elliptic curve $E_\mathfrak{p}$ defined over $\mathbb{F}_p$. To be more exact, one constructs one curve $E'_\mathfrak{p}$ with invariant $j_\mathfrak{p}$ and then distinguishes $E_\mathfrak{p}$ from its twists by checking the group order. This is performed by trial scalar multiplications with the group order and random points.

For use in cryptography one needs curves with almost prime order. To this aim one additionally requires that for $p = \omega\overline{\omega}$ at least one of the numbers $p + 1 \pm (\omega + \overline{\omega})$ is almost prime as this is the group order of $E_\mathfrak{p}$ or its quadratic twist. From theorems of analytic number theory it follows that by varying the prime $p$ one has a good chance to find elliptic curves with almost prime order after few trials. Likewise, one might be interested in curves with a smooth group order. In the sequel we assume that one wants to construct curves with a certain property $Pr$.

So first one chooses a squarefree natural number $d$ such that the class number $h_d \sim h_d^2$ of $K_d = \mathbb{Q}(\sqrt{-d})$ is not too large. Then one computes $H_d(X)$ or a variant (cf. Remark 18.3).

**Remark 18.1** Determining the class number $h_d$ and also the Hilbert class polynomial $H_d(X)$ is done as *precomputation* and thus does not belong to the actual algorithm. Furthermore, these data can be obtained by table lookup as all appropriate polynomials are precomputed and published [WENG].

Starting from $H_d$ we can use the one dimensional scheme $\text{Spec}(\mathcal{O}_K)$ as parameter space. So for each fixed $d$ we are in a situation that can be compared with the classical construction of discrete logarithm systems that uses the global group scheme $G_m$ (the multiplicative group) and then chooses the reduction modulo primes to find appropriate groups inside the multiplicative group of finite fields.

To compare the effectiveness of key generation one has to assume that $H_d(X)$ is known and then one has to find methods to compute both the equation and the group order of a corresponding elliptic curve. In detail one does the following: one chooses random prime numbers of the desired ground field size and solves the norm equation $p = \omega\overline{\omega}$ if possible. For such primes one factors $p + 1 \pm (\omega + \overline{\omega})$. If one of them satisfies property $Pr$, one computes the factorization of $H_d(X) \pmod{p}$. We take a factor $X - j_\mathfrak{p}$ and construct the elliptic curve $E_\mathfrak{p}$ by (18.1).

In the sequel of this section we study these steps in detail.

### 18.1.3 Computation of class polynomials

To compute $H_d$, one first needs a list of the $A_r$. This is obtained through the equivalence between the ideal classes of an algebraic number field with discriminant $d$ and the equivalence classes of primitive, positive definite binary quadratic forms of discriminant $d$.

**Definition 18.2** A quadratic form $ax^2 + bxy + cy^2$ is called a *reduced binary quadratic form* if it satisfies the conditions

- $|b| \leqslant a \leqslant c$
- $b \geqslant 0$ if $a = |b|$ or $a = c$
- $\gcd(a, b, c) = 1$.

Like in the case of the ideal class group of function fields (cf. Section 4.4.6) one shows that there is exactly one reduced binary quadratic form in each equivalence class, so one can enumerate all reduced binary quadratic forms of discriminant $d$ to obtain a complete system of representatives; see e.g., [ZAG 1981, Teil II, §10]. Then to each reduced quadratic form $ax^2 + bxy + cy^2$ corresponds the ideal $A = \mathbb{Z} + \tau\mathbb{Z}$, with

$$\tau = \frac{b + \sqrt{-d}}{2a}. \tag{18.2}$$

In that way one obtains a list of numbers $\tau_r$ with $A_r = \mathbb{Z} + \tau_r \mathbb{Z}$, and the $A_r$ form a complete system of representatives of the ideal classes of $K$.

The $j$-invariants $j(A_r)$ of the elliptic curves over $\mathbb{C}$ with lattice $A_r$ are calculated by

$$j(A_r)(q) = \frac{\left(1 + 240 \sum_{n=1}^{\infty} \sigma_3(n) q^n\right)^3}{q \prod_{n=0}^{\infty} (1-q^n)^{24}},$$

where $q = e^{2\pi i \tau_r}$ and $\sigma_3(n) = \sum_{t|n} t^3$.

Then

$$H_d(X) = \prod_{i=1}^{h_d} (X - j(A_r)) \in \mathbb{Z}[X].$$

All computations are done over $\mathbb{C}$ and, thus, it is only possible to compute approximate values for the $j(A_r)$. But since the coefficients of $H_d$ are rational integers, we find the exact polynomial if we compute with sufficient precision.

For a fixed discriminant $d$ of an imaginary quadratic field $K$ the complexity of computing $H_d(X)$ depends mainly on the size of the class number $h_d$, and so on the size of $\sqrt{d}$. It turns out that it is no great problem to deal with $d \leqslant 8 \times 10^6$. In this range we reach class numbers up to 5000. Because we need to factor $H_d(X)$ it makes no sense to choose larger class numbers because $\deg(H_d) = h_d$.

In practice we would propose to use a table of class polynomials, e.g., [WENG], where one finds many examples covering all CM-curves up to discriminant $d \leqslant 422500$, and choose a $d$ with a not too small class number.

**Remark 18.3** For theoretical reasons we have used the Hilbert class polynomial to explain the global part of the CM-method. In practice one should use Weber's polynomials whose coefficients have smaller absolute value. For details we refer to [ATMO 1993].

### 18.1.4 Computation of norms

To find suitable curves one chooses primes $p$ and factors $H_d$ modulo $p$. For this, $p$ has to split as $(p) = (\omega)(\overline{\omega})$ in $\mathcal{O}_K$. These are the primes for which we have integer solutions $x, y$ of the norm equation

$$x^2 + dy^2 = \varepsilon p, \quad \text{with } \varepsilon = \begin{cases} 1, & \text{if } d \equiv 1, 2 \pmod 4 \\ 4, & \text{if } d \equiv 3 \pmod 4. \end{cases} \tag{18.3}$$

A necessary condition for $p$ to split as above is that $-d$ is a square modulo $p$. Hence, one first computes the Legendre symbol $\left(\frac{-d}{p}\right)$ (cf. Section 2.3.4) and rejects the prime if the result is not 1.

In the positive case one uses Cornacchia's algorithm [COH 2000] to find a solution to $a^2 + db^2 = p$ from which one easily obtains the desired $x, y$.

---

**Algorithm 18.4** Cornacchia's algorithm

INPUT: A positive integer $d > 0$ and a prime $p$ such that $\left(\frac{-d}{p}\right) = 1$.
OUTPUT: Two integers $(a, b) \in \mathbb{Z}^2$ with $a^2 + db^2 = p$ if possible.

1. compute square root $p/2 < x_0 < p$ of $-d$, i.e., $x_0^2 \equiv -d \pmod p$
2. $x \leftarrow p, y \leftarrow x_0$ and $z \leftarrow \lfloor \sqrt{p} \rfloor$
3. **while** $y > z$ **do** $r \leftarrow x \bmod y, x \leftarrow y$ and $y \leftarrow r$     [Euclidean algorithm]
4. **if** $d \nmid p - y^2$ **or if** $c = (p - y^2)/d$ is not a square **then return** "no solution"
5. **else return** $(a, b) = (y, \sqrt{c})$

## 18.1.5 The algorithm

In the following we give the description of the algorithm using the theory from above. For more details see [SPA 1994].

The parameter $\varepsilon$ is chosen according to (18.3) and $\delta = +\sqrt{\varepsilon}$.

---

**Algorithm 18.5** Construction of elliptic curves via CM

INPUT: A squarefree integer $d \neq 1, 3$, parameters $\varepsilon$ and $\delta$, Hilbert class polynomial $H_d(X)$, desired size of $p$ and property $Pr$.
OUTPUT: A prime $p$ of the desired size, an elliptic curve $E/\mathbb{F}_p$ whose group order $|E(\mathbb{F}_p)|$ satisfies property $Pr$.

1. **repeat**
2.     **repeat** choose $p$ prime of desired size     [see remark below]
3.     **until** $\varepsilon p = x^2 + dy^2$ with $x, y \in \mathbb{Z}$
4.     $n_1 \leftarrow p + 1 - 2x/\delta$ and $n_2 \leftarrow p + 1 + 2x/\delta$
5. **until** $n_1$ **or** $n_2$ satisfies property $Pr$
6. compute a root $j$ of $H_d(X) \pmod{p}$
7. compute $E_j/\mathbb{F}_p$ from (18.1) and its twist $\widetilde{E}_j/\mathbb{F}_p$
8. **while true do**
9.     take $P \in_R E_j(\mathbb{F}_p)$ and compute $Q \leftarrow [n_1]P$
10.     **if** $Q = P_\infty$ **and** $[n_2]P \neq P_\infty$ **then return** $p$ and $E_j$
11.     **else if** $Q \neq P_\infty$ **then return** $p$ and $\widetilde{E}_j$

---

Examples of desired properties $Pr$ are that the group order contains a prime factor larger than a given bound or that the group order is smooth with a given smoothness bound. The latter is interesting for factorization methods (cf. Chapter 25) while the prior can be used as basis for a cryptosystem (cf. Chapter 23).

**Remarks 18.6**

(i) To simplify finding primes $p$ that split in $\mathcal{O}_K$ together with their decomposition, one replaces Lines 2 and 3 and chooses integers $x, y$ such that $x, \sqrt{dy}$ are of size $\sqrt{p}$ and tests whether $(x^2 + dy^2)/\varepsilon$ is a prime.

(ii) The factorization of $H_d(X) \pmod{p}$ needs the largest computational effort but it is done only once.

(iii) For $d = 1$, the splitting condition is $p \equiv 1 \pmod{4}$ and the $j$-invariant is equal to $12^3$, hence, $H_1(X) = X - 12^3$. There are 4 twists by the 4th roots of unity to consider.

(iv) For $d = 3$, the splitting condition is $p \equiv 1 \pmod{3}$ and the $j$-invariant is equal to zero, thus $H_3(X) = X$. There are 6 twists by the different 6th roots of unity.

## 18.1.6 Experimental results

We list the running times of the CM-method for elliptic curves as stated in [WEN 2001b, Appendix A]. Table 18.1 contains the timings needed on a 650 Mhz Pentium III computer to find 1000 prime numbers of size $2^{160}$, which satisfy a norm equation with respect to $K_d$ and lead to a group order

$|E(\mathbb{F}_p)| = c\ell$ with $c \leqslant 1000$ and $\ell$ prime. The discriminant of the imaginary quadratic field is denoted by $D = -\varepsilon d$. The entries are ordered with increasing class number, which ranges from 200 to 5800.

**Table 18.1** Time to find 1000 primes $p$ with $|E(\mathbb{F}_p)|$ of cofactor $c \leqslant 1000$.

| $D$ | Time (in s) | $D$ | Time (in s) |
| --- | --- | --- | --- |
| $-53444$ | 476.25 | $-973496$ | 437.32 |
| $-345124$ | 292.89 | $-2128136$ | 424.44 |
| $-17111$ | 734.06 | $-900311$ | 698.49 |
| $-19631$ | 733.88 | $-1139519$ | 807.23 |
| $-19031$ | 521.46 | $-2155919$ | 876.61 |
| $-56696$ | 293.71 | $-3300359$ | 836.26 |
| $-698472$ | 521.44 | $-4145951$ | 904.15 |
| $-98276$ | 293.99 | $-5154551$ | 817.52 |
| $-180164$ | 327.60 | $-6077111$ | 1013.19 |
| $-237236$ | 345.40 | $-7032119$ | 994.38 |
| $-326504$ | 365.75 | $-8282039$ | 928.72 |

Table 18.2 shows the timings for constructing an elliptic curve over $\mathbb{F}_p$ having complex multiplication by $\mathcal{O}_K$ for a fixed imaginary quadratic field $K$. The first column describes the imaginary quadratic field and the second gives the class number. For completeness, the third column contains the time needed to compute the class polynomial. In practice $H_d$ has been precomputed and stored once and for all, e.g., it can be obtained from the large data base [WENG].

The actual running time to construct the curve (given in the fourth column) contains only the factorization of the class polynomial at one prime $p$ and the time to determine the correct equation for $E$.

So the actual running time consists of the time to find a suitable prime $p$ as stated in Table 18.1 and the time stated in the fourth column.

## 18.2 CM for curves of genus 2

For genus $g = 2$ we can follow the lines we have explained in the beginning of this chapter if we replace elliptic curves by hyperelliptic curves of genus $g = 2$. Imaginary quadratic fields and their class field theory have to be replaced by CM fields of degree 4 and the Shimura–Taniyama version of class field theory. The invariant theory becomes more involved; we shall need 3 invariants in the place of the $j$-invariant of elliptic curves.

After a brief overview of the background we summarize the algorithm. Then we go into the details of the computations. We give a complete study starting from the choice of the CM field with computation of the period matrix and the computation of the class polynomials. These two steps are usually omitted in an actual implementation as the respective information can be obtained from tables available on the web. The important steps for the implementation are explained in Section 18.2.5, namely the choice of suitable primes and the computation of the curve equation.

**Table 18.2** Complexity of CM-method for $g = 1$, dependence on discriminant and class number.

| $D$ | $h_K$ | Time (in s) Precomp. | Curve | $D$ | $h_K$ | Time (in s) Precomp. | Curve |
|---|---|---|---|---|---|---|---|
| $-53444$ | 200 | 3 | 14 | $-345124$ | 200 | 4 | 9 |
| $-17111$ | 202 | 1 | 8 | $-19031$ | 203 | 1 | 9 |
| $-56696$ | 204 | 3 | 10 | $-395652$ | 204 | 16 | 10 |
| $-18119$ | 205 | 1 | 9 | $-19631$ | 208 | 2 | 14 |
| $-345624$ | 208 | 20 | 20 | $-690072$ | 208 | 23 | 9 |
| $-57944$ | 210 | 3 | 10 | $-58484$ | 210 | 6 | 11 |
| $-52664$ | 212 | 3 | 9 | $-159416$ | 212 | 3 | 9 |
| $-18191$ | 213 | 1 | 9 | $-55844$ | 216 | 3 | 10 |
| $-698472$ | 224 | 26 | 10 | $-698772$ | 228 | 91 | 10 |
| $-128456$ | 236 | 5 | 10 | $-158036$ | 242 | 11 | 9 |
| $-124004$ | 248 | 5 | 15 | $-78536$ | 252 | 4 | 10 |
| $-699752$ | 264 | 9 | 16 | $-113636$ | 284 | 7 | 16 |
| $-98276$ | 304 | 7 | 18 | $-132404$ | 332 | 24 | 23 |
| $-120056$ | 340 | 10 | 20 | $-144836$ | 352 | 13 | 20 |
| $-160676$ | 380 | 16 | 20 | $-168164$ | 384 | 18 | 19 |
| $-180164$ | 400 | 19 | 101 | $-185624$ | 402 | 32 | 76 |
| $-247796$ | 466 | 75 | 24 | $-248804$ | 468 | 31 | 24 |
| $-237236$ | 476 | 78 | 25 | $-283076$ | 520 | 43 | 35 |
| $-318776$ | 540 | 184 | 49 | $-326504$ | 578 | 60 | 39 |
| $-399944$ | 612 | 75 | 39 | $-434216$ | 630 | 84 | 42 |
| $-442196$ | 644 | 218 | 41 | $-450056$ | 676 | 103 | 44 |
| $-512984$ | 714 | 135 | 43 | $-607844$ | 752 | 161 | 142 |
| $-650744$ | 832 | 211 | 192 | $-727256$ | 866 | 246 | 170 |
| $-803864$ | 914 | 291 | 45 | $-914744$ | 972 | 367 | 48 |
| $-973496$ | 1044 | 460 | 69 | $-1202984$ | 1126 | 599 | 81 |
| $-1319876$ | 1188 | 728 | 93 | $-1435496$ | 1218 | 803 | 89 |
| $-1514036$ | 1250 | 2201 | 93 | $-1561544$ | 1280 | 966 | 98 |
| $-1617656$ | 1324 | 1134 | 99 | $-1890776$ | 1404 | 1467 | 102 |
| $-2128136$ | 1500 | 1724 | 99 | $-701399$ | 1581 | 566 | 92 |
| $-900311$ | 1626 | 626 | 94 | $-1139519$ | 2027 | 1306 | 109 |
| $-1238639$ | 2150 | 1595 | 176 | $-1614311$ | 2421 | 3465 | 193 |
| $-1884791$ | 2669 | 3407 | 211 | $-2155919$ | 2968 | 5373 | 223 |
| $-2336879$ | 3036 | 5883 | 212 | $-3300359$ | 3531 | 9458 | 250 |
| $-3190151$ | 3593 | 10034 | 272 | $-3312839$ | 3632 | 10424 | 269 |
| $-3524351$ | 3714 | 11585 | 262 | $-3983591$ | 3918 | 13694 | 293 |
| $-4145951$ | 4065 | 16008 | 281 | $-4305479$ | 4227 | 17515 | 479 |
| $-4972679$ | 4498 | 21830 | 507 | $-5154551$ | 4551 | 22698 | 521 |
| $-5652071$ | 4802 | 28634 | 501 | $-5892311$ | 4913 | 29785 | 550 |
| $-6077111$ | 5092 | 34459 | 509 | $-6606599$ | 5180 | 35831 | 529 |
| $-7032119$ | 5424 | 45254 | 615 | $-7651199$ | 5628 | 52417 | 589 |
| $-7741439$ | 5686 | 54300 | 641 | $-8282039$ | 5819 | 59305 | 668 |

## 18.2.1 Summary of background

For this section we refer to Section 5.1.6.b for background. Let $C$ be a hyperelliptic curve of genus 2 over the complex numbers $\mathbb{C}$. Then its Jacobian $J_C$ is isomorphic to $\mathbb{C}^2/\Lambda_C$ for some lattice $\Lambda_C$. We recall that $C$ has *complex multiplication* if the endomorphism ring of its Jacobian contains an order $\mathcal{O}$ of a field of degree 4 over $\mathbb{Q}$, which has to be a CM-field $K$, i.e., it is a totally imaginary quadratic extension of a totally real field $K_0$ of degree 2 over $\mathbb{Q}$. For our purposes it is enough to consider the case that $\mathcal{O} = \mathcal{O}_K$ is the ring of integers of $K$. The lattice $\Lambda_C$ is called the *period lattice of $C$* and can be chosen as an ideal $A$ of $\mathcal{O}_K$. The Jacobian $J_C$ is principally polarized. This is reflected by the period matrix $\Omega_C$. This matrix is determined by arithmetical data of $K$, and so the Igusa invariants $j_1, j_2, j_3$ (cf. Section 5.1.6.b) of the curve $C$ can be computed as complex numbers. Under suitable conditions they depend only on the ideal $A$ and the chosen polarization, which is reflected by the group of units of $\mathcal{O}_K$.

So, it is possible to determine a complete set $C_i$, $1 \leqslant i \leqslant s$, for some integer $s$, of representatives for isomorphism classes of curves whose Jacobian variety $J_{C_i}$ has endomorphism ring $\mathcal{O}_K$. This is done in terms of ideals $A_i$ and principal polarizations, which are also used to determine the Igusa invariants $j_k^{(i)}$, $k = 1, 2, 3$ of $C_i$. A consequence is that the polynomials

$$H_{K,k}(X) = \prod_{i=1}^{s}(X - j_k^{(i)}), \quad k = 1, 2, 3$$

have rational coefficients.

The computation of these polynomials corresponds to the computation of the Hilbert class polynomial $H_d(X)$ in the case of elliptic curves.

A slight complication is that the polynomials $H_{K,k}$ no longer have integer coefficients. But in practice, the denominators only have small prime divisors. Hence, we can reduce these polynomials modulo large enough primes $p$ and get $H_{K,k} \pmod{p}$ and the roots of these polynomials are the invariants $j_k$ of curves $C_i$ modulo $p$.

We get the most important example by choosing primes $p$ for which all $H_{K,k} \pmod{p}$ have at least one linear factor. This can be interpreted by class field theory in a slightly more complicated way than in the case of elliptic curves. Then the reduced curve is defined over $\mathbb{F}_p$ or over a quadratic extension. We consider only the first case.

In this case we find elements $\omega_1, \overline{\omega}_1 \in \mathcal{O}_K$ satisfying $\omega_1 \overline{\omega}_1 = p$. Up to negation and conjugation there can exist at most one second solution $\omega_2 \overline{\omega}_2 = p$, with

$$\omega_2 \notin \{\pm\omega_1, \pm\overline{\omega}_1, \pm\hat{\sigma}(\omega_1), \pm\hat{\sigma}(\overline{\omega}_1)\},$$

where $\hat{\sigma}$ is the extension of the real conjugation to an embedding of $K$ into $\mathbb{C}$. In this case put $W = \{\pm\omega_1, \pm\omega_2\}$, else $W = \{\pm\omega_1\}$. Then the group order of $J_{C_i \pmod{p}}(\mathbb{F}_p)$ is in

$$\{\chi_\omega(1) \mid w \in W\}, \tag{18.4}$$

where $\chi_\omega(T)$ is the characteristic polynomial of $\omega$.

In Section 5.1.6.b it is explained how to obtain the equation of $C_i \pmod{p}$ from its Igusa invariants. Note that this is not done by exploiting a simple formula as in the elliptic case.

## 18.2.2 Outline of the algorithm

The first step to construct curves $C$ of genus 2 with CM and known group order modulo a prime $p$ is to choose a suitable CM field of degree 4, e.g., the subfield $K_0$ has to have class number 1 and like for elliptic curves the class number of $K$ has to be moderately small. For further computations one has to determine the class group of $K$ and its fundamental units. One computes a complete set of representatives of the ideal classes and principal polarizations.

The next step is to compute the period matrix $\Omega_C$, cf. Definition 5.19 attached to the curve $C$ with lattice $\Lambda_C$ being a given ideal in $\mathcal{O}_K$ together with a polarization. Using $\Omega_C$ one determines the theta constants (5.2), which already provide a system of invariants over the complex numbers. An algebraic system of equation is given by the Igusa invariants $j_k$, which are explicit rational functions in the theta constants.

Doing this procedure for every $C_i$, one can compute the polynomials $H_{K,k}(X)$.

As for elliptic curves, these steps are done once and for all as a precomputation. Suitable CM fields and their arithmetical data and also the class polynomials $H_{K,k}$ can be obtained from tables.

Then one looks for a prime $p$ for which the norm equation $p = \omega\bar{\omega}$ has solutions in $\mathcal{O}_K$. At this point one computes the at most 4 numbers (18.4) that contain the information about the group order of $J_{C \pmod{p}}(\mathbb{F}_p)$. If at least one of them has a desired property one proceeds to compute the equation of the curve, otherwise one repeats this step with a different choice of $p$.

One factors the class polynomials modulo $p$ and determines the invariants $j_k^{(i)}$ of $C_i \pmod{p}$ as roots of $H_{K,k}(X) \pmod{p}$.

The last step is to determine an equation (up to a twist) for some $C := C_i \pmod{p}$ from the knowledge of its invariants. For this one uses either Gröbner basis techniques or the more efficient method of Mestre, which relies on the invariant theory of binary forms.

Finally, one needs to determine the group order of $J_C(\mathbb{F}_p)$ from the set of possible values (18.4). This is done by randomly choosing a point $P \in J_C(\mathbb{F}_p)$ and scalar multiplication by the candidate group orders. If one finds no matches one replaces $C$ by a twist and tries again. This immediately answers whether one has to take $C$ or a twist of $C$ to get the desired order of the group of rational points on the corresponding Jacobian variety.

### 18.2.3 CM-types and period matrices

Choose a squarefree $d \in \mathbb{N}$ such that $K_0 := \mathbb{Q}(\sqrt{d})$ has class number one. Choose $\alpha = a + b\sqrt{d}$ squarefree and totally positive, i.e., $a \pm b\sqrt{d} > 0$. Then $K := K_0(i\sqrt{\alpha})$ is a CM field of degree 4. We require that $K$ is not a Galois extension of $\mathbb{Q}$ with Galois group $\mathbb{Z}/2\mathbb{Z} \times \mathbb{Z}/2\mathbb{Z}$ (this ensures that the constructed Jacobian varieties are simple). We choose two non-conjugate distinct embeddings $\varphi_1, \varphi_2$ of $K$ into $\mathbb{C}$. The tuple $(K, \Phi) = (K, \{\varphi_1, \varphi_2\})$ is called *CM-type*. For an ideal $A$ we put

$$\Phi(A) = \left\{ \left(\varphi_1(\alpha), \varphi_2(\alpha)\right)^t, \alpha \in A \right\}.$$

By Theorem 5.58, $\mathbb{C}^2/\Phi(A)$ is an abelian variety with complex multiplications by $\mathcal{O}_K$. The final structure, abelian variety with polarization, is determined by Theorem 5.62. The following procedure uses this theorem and computes a complete set of representatives for isomorphism classes of principally polarized abelian varieties with endomorphism ring $\mathcal{O}_K$ of CM-type $(K, \Phi)$. We use the notation from Section 5.1.6.d.

The computation of a fundamental unit $\epsilon_0$ of $K_0$ is done with the help of a computer algebra system. Let $U^+$ be the group of units $\epsilon$ with $N_{K_0/\mathbb{Q}}(\epsilon) = 1$. Put

$$U_1 = \{\epsilon \in U^+ \mid \exists \gamma \in K \text{ with } N_{K/K_0}(\gamma) = \epsilon\}.$$

At this point one also determines the ideal classes of $\mathcal{O}_K$ that contain an ideal $A$ such that $A\bar{A} = \alpha\mathcal{O}_K$ with $\varphi_i(\alpha)$ real positive for $i = 1, 2$. We choose a complete set of representatives $A_1, \ldots, A_{h'_K}$ of these classes. We can assume that $A_j$ is of the form

$$\mathcal{O}_{K_0} + \tau_j \mathcal{O}_{K_0}, \tag{18.5}$$

where $\mathrm{Im}(\tau_j) > 0$ and if the norm of the fundamental unit $\mathrm{N}_{K_0/\mathbb{Q}}(\epsilon_0)$ is negative one also requires $\mathrm{N}_{K/K_0}(\tau_j)$ totally positive.

Let $\sigma$ be the real conjugation in $K_0$ and denote by $\sqrt{a}^+$ the positive square root of $a \in \mathbb{R}$. This gives rise to two embeddings of $K$ given by

$$\hat{\sigma}(i\sqrt{\alpha}^+) = i\sqrt{\alpha}^+ \quad \text{and} \quad \rho\hat{\sigma}(i\sqrt{\alpha}^+) = -i\sqrt{\alpha}^+.$$

We put $\psi = \rho\hat{\sigma}$.

Having these definitions, we make some observations concerning the CM-types:

1. $(K, \{1, \psi\})$ is a CM-type.
2. The abelian varieties obtained for a CM-type $(K, \Phi)$ are isomorphic to varieties obtained from either $(K, \{1, \psi\})$ or $(K, \{1, \overline{\psi}\})$. If $K$ is Galois over $\mathbb{Q}$ then $(K, \{1, \psi\})$ is sufficient.

For the following result we shall treat the two types $\{1, \psi\}$ and $\{1, \overline{\psi}\}$ in parallel and list for each $\tau_j$ a set $\mathcal{K}_j$ defining the period matrices of a complete set of isomorphism classes of principally polarized abelian varieties related to the ideal $A_j$.

**Definition 18.7** Let $\tau_j$ be as in (18.5).

If $K$ is Galois with Galois group $\mathbb{Z}/4\mathbb{Z}$ we define $\mathcal{K}_j$ by

- $\{(\tau_j, \tau_j^\varphi), (\epsilon_0\tau_j, (\epsilon_0\tau_j)^\varphi)\}$, if $\mathrm{N}_{K_0/\mathbb{Q}}(\epsilon_0) = 1, \epsilon_0 \in U_1, \mathrm{N}_{K/K_0}(\tau_j)$ totally positive,
- $\{(\tau_j, \tau_j^\varphi)\}$, if $\mathrm{N}_{K_0/\mathbb{Q}}(\epsilon_0) = 1, \epsilon_0 \in U^+ \smallsetminus U_1, \mathrm{N}_{K/K_0}(\tau_j)$ totally positive,
- $\{(\tau_j, \tau_j^\varphi), (\epsilon_0\tau_j, (\epsilon_0\tau_j)^{\overline{\varphi}})\}$, if $\mathrm{N}_{K_0/\mathbb{Q}}(\epsilon_0) = -1$,
- $\emptyset$, if $\mathrm{N}_{K/K_0}(\tau_j)$ not totally positive.

If $K$ is not Galois then we define $\mathcal{K}_j$ by

- $\{(\tau_j, \tau_j^\psi), (\epsilon_0\tau_j, (\epsilon_0\tau_j)^\psi)\}$, if $\mathrm{N}_{K_0/\mathbb{Q}}(\epsilon_0) = 1, \epsilon_0 \in U_1, \mathrm{N}_{K/K_0}(\tau_j)$ totally positive,
- $\{(\tau_j, \tau_j^\psi), (\epsilon_0\tau_j, (\epsilon_0\tau_j)^{\overline{\psi}})\}$, if $\mathrm{N}_{K_0/\mathbb{Q}}(\epsilon_0) = 1, \epsilon_0 \in U_1, \mathrm{N}_{K/K_0}(\tau_j)$ not totally positive,
- $\{(\tau_j, \tau_j^\psi)\}$, if $\mathrm{N}_{K_0/\mathbb{Q}}(\epsilon_0) = 1, \epsilon_0 \in U^+ \smallsetminus U_1, \mathrm{N}_{K/K_0}(\tau_j)$ totally positive,
- $\{(\tau_j, \tau_j^{\overline{\psi}})\}$, if $\mathrm{N}_{K_0/\mathbb{Q}}(\epsilon_0) = 1, \epsilon_0 \in U^+ \smallsetminus U_1, \mathrm{N}_{K/K_0}(\tau_j)$ not totally positive,
- $\{(\tau_j, \tau_j^\psi), (\epsilon_0\tau_j, (\epsilon_0\tau_j)^{\overline{\psi}})\}$, if $\mathrm{N}_{K_0/\mathbb{Q}}(\epsilon_0) = -1$.

By Spallek [SPA 1994] we have that a complete set of period matrices for all representatives of isomorphism classes of abelian varieties with complex multiplication with $\mathcal{O}_K$ is given by the following result.

**Theorem 18.8** Let $K$ be a CM field which is not a Galois extension of $\mathbb{Q}$ with Galois group $\mathbb{Z}/2\mathbb{Z} \times \mathbb{Z}/2\mathbb{Z}$ and let $\{\tau_1, \ldots, \tau_{h'_K}\}$ be as described above. Let $\mathcal{O}_{K_0} = \mathbb{Z} + w\mathbb{Z}$, with $w$ totally positive.

A complete set of period matrices for representatives of all isomorphism classes of simple principally polarized abelian varieties having complex multiplication by $\mathcal{O}_K$ is given by

$$\left\{\Omega_{(t,\tilde{t})} : (t, \tilde{t}) \in \bigcup_{j=1}^{h'_K} \mathcal{K}_j\right\},$$

where

$$\Omega_{t,\tilde{t}} = \frac{1}{w - w^\sigma} \begin{bmatrix} w^2 t - (w^\sigma)^2 \tilde{t} & wt - w^\sigma \tilde{t} \\ wt - w^\sigma \tilde{t} & t - \tilde{t} \end{bmatrix}.$$

### 18.2.4 Computation of the class polynomials

As in the previous section, we are dealing again with a precomputation. As explained in the background (cf. Chapter 5), the class polynomials are computed from the invariants of the curves over $\mathbb{C}$ represented by the period matrix $\Omega$. To get these invariants we first need to determine the theta constants. This needs to be done for each of the $s = |\mathcal{K}|$ isomorphism classes (cf. Section 18.2.3) to determine $j_k^{(i)}$, $k = 1, 2, 3$ and $i = 1, \ldots, s$. Here, we omit the upper index and explain how to compute one set of invariants.

#### 18.2.4.a Computation of the theta constants

We recall (5.2), the definition of the theta constants in terms of the period matrix $\Omega$ for genus $g = 2$:

$$\theta\begin{bmatrix}\delta\\ \epsilon\end{bmatrix}(z, \Omega) = \sum_{\mathbf{n} \in \mathbb{Z}^2} \exp\left(\pi i \left(\mathbf{n} + \frac{1}{2}\delta\right)^t \Omega \left(\mathbf{n} + \frac{1}{2}\delta\right) + 2\left(\mathbf{n} + \frac{1}{2}\delta\right)^t \left(z + \frac{1}{2}\epsilon\right)\right).$$

As explained in Chapter 5 we need only the even theta constants as the odd ones vanish. For genus 2 they are given by

$$\theta_1 := \theta\begin{bmatrix}\begin{pmatrix}0\\0\end{pmatrix}\\ \begin{pmatrix}0\\0\end{pmatrix}\end{bmatrix},\ \theta_2 := \theta\begin{bmatrix}\begin{pmatrix}0\\0\end{pmatrix}\\ \begin{pmatrix}1\\0\end{pmatrix}\end{bmatrix},\ \theta_3 := \theta\begin{bmatrix}\begin{pmatrix}0\\0\end{pmatrix}\\ \begin{pmatrix}0\\1\end{pmatrix}\end{bmatrix},\ \theta_4 := \theta\begin{bmatrix}\begin{pmatrix}0\\0\end{pmatrix}\\ \begin{pmatrix}1\\1\end{pmatrix}\end{bmatrix},\ \theta_5 := \theta\begin{bmatrix}\begin{pmatrix}1\\0\end{pmatrix}\\ \begin{pmatrix}0\\0\end{pmatrix}\end{bmatrix},$$

$$\theta_6 := \theta\begin{bmatrix}\begin{pmatrix}1\\0\end{pmatrix}\\ \begin{pmatrix}0\\1\end{pmatrix}\end{bmatrix},\ \theta_7 := \theta\begin{bmatrix}\begin{pmatrix}0\\1\end{pmatrix}\\ \begin{pmatrix}0\\0\end{pmatrix}\end{bmatrix},\ \theta_8 := \theta\begin{bmatrix}\begin{pmatrix}0\\1\end{pmatrix}\\ \begin{pmatrix}1\\0\end{pmatrix}\end{bmatrix},\ \theta_9 := \theta\begin{bmatrix}\begin{pmatrix}1\\1\end{pmatrix}\\ \begin{pmatrix}0\\0\end{pmatrix}\end{bmatrix},\ \theta_{10} := \theta\begin{bmatrix}\begin{pmatrix}1\\1\end{pmatrix}\\ \begin{pmatrix}1\\1\end{pmatrix}\end{bmatrix}.$$

The algorithmic problem is to evaluate the series with sufficiently high precision. The analysis can be found in [WEN 2003, Sect. 4].

#### 18.2.4.b Computation of Igusa invariants

We now show how to compute the Igusa invariants from the period matrix using the theta constants. To compute the Igusa invariants from the theta constants we introduce the values $h_4, h_{10}, h_{12}$, and $h_{16}$ related to modular forms of the respective weights. Then we express the invariants $I_2, I_4, I_6$, and $I_{10}$ in terms of the $h_i$ to define the Igusa invariants $j_i$ as in (5.3).

$$h_4 := \sum_{i=1}^{10} \theta_i^8,\ h_{10} := \prod_{i=1}^{10} \theta_i^2,$$

$$\begin{aligned}h_{12} :=\ & (\theta_1\theta_5\theta_2\theta_9\theta_6\theta_{10})^4 + (\theta_1\theta_2\theta_9\theta_6\theta_8\theta_3)^4 + (\theta_5\theta_9\theta_6\theta_8\theta_{10}\theta_7)^4 + (\theta_5\theta_2\theta_6\theta_8\theta_3\theta_7)^4 \\ & + (\theta_1\theta_5\theta_2\theta_{10}\theta_3\theta_7)^4 + (\theta_1\theta_9\theta_8\theta_{10}\theta_3\theta_7)^4 + (\theta_1\theta_5\theta_2\theta_8\theta_{10}\theta_4)^4 + (\theta_1\theta_5\theta_9\theta_8\theta_3\theta_4)^4 \\ & + (\theta_5\theta_9\theta_6\theta_{10}\theta_3\theta_4)^4 + (\theta_2\theta_6\theta_8\theta_{10}\theta_3\theta_4)^4 + (\theta_1\theta_2\theta_9\theta_6\theta_7\theta_4)^4 + (\theta_1\theta_5\theta_6\theta_8\theta_7\theta_4)^4 \\ & + (\theta_2\theta_9\theta_8\theta_{10}\theta_7\theta_4)^4 + (\theta_5\theta_2\theta_9\theta_3\theta_7\theta_4)^4 + (\theta_1\theta_6\theta_{10}\theta_3\theta_7\theta_4)^4,\end{aligned}$$

$$\begin{aligned}
h_{16} :=\ & \theta_8^4(\theta_1\theta_5\theta_2\theta_9\theta_6\theta_8\theta_{10})^4 + \theta_5^4(\theta_1\theta_5\theta_2\theta_9\theta_6\theta_8\theta_3)^4 + \theta_{10}^4(\theta_1\theta_2\theta_9\theta_6\theta_8\theta_{10}\theta_3)^4 \\
&+ \theta_3^4(\theta_1\theta_5\theta_2\theta_9\theta_6\theta_{10}\theta_3)^4 + \theta_1^4(\theta_1\theta_5\theta_9\theta_6\theta_8\theta_{10}\theta_7)^4 + \theta_2^4(\theta_5\theta_2\theta_9\theta_6\theta_8\theta_{10}\theta_7)^4 \\
&+ \theta_1^4(\theta_1\theta_5\theta_2\theta_6\theta_8\theta_3\theta_7)^4 + \theta_9^4(\theta_5\theta_2\theta_9\theta_6\theta_8\theta_3\theta_7)^4 + \theta_9^4(\theta_1\theta_5\theta_2\theta_9\theta_{10}\theta_3\theta_7)^4 \\
&+ \theta_6^4(\theta_1\theta_5\theta_2\theta_6\theta_{10}\theta_3\theta_7)^4 + \theta_5^4(\theta_1\theta_5\theta_9\theta_8\theta_{10}\theta_3\theta_7)^4 + \theta_2^4(\theta_1\theta_2\theta_9\theta_8\theta_{10}\theta_3\theta_7)^4 \\
&+ \theta_6^4(\theta_1\theta_9\theta_6\theta_8\theta_{10}\theta_3\theta_7)^4 + \theta_8^4(\theta_1\theta_5\theta_2\theta_8\theta_{10}\theta_3\theta_7)^4 + \theta_{10}^4(\theta_5\theta_2\theta_6\theta_8\theta_{10}\theta_3\theta_7)^4 \\
&+ \theta_3^4(\theta_5\theta_9\theta_6\theta_8\theta_{10}\theta_3\theta_7)^4 + \theta_7^4(\theta_1\theta_5\theta_2\theta_9\theta_6\theta_{10}\theta_7)^4 + \theta_7^4(\theta_1\theta_2\theta_9\theta_6\theta_8\theta_3\theta_7)^4 \\
&+ \theta_9^4(\theta_1\theta_5\theta_2\theta_9\theta_8\theta_{10}\theta_4)^4 + \theta_6^4(\theta_1\theta_5\theta_2\theta_6\theta_8\theta_{10}\theta_4)^4 + \theta_2^4(\theta_1\theta_5\theta_2\theta_9\theta_8\theta_3\theta_4)^4 \\
&+ \theta_6^4(\theta_1\theta_5\theta_9\theta_6\theta_8\theta_3\theta_4)^4 + \theta_1^4(\theta_1\theta_5\theta_9\theta_6\theta_{10}\theta_3\theta_4)^4 + \theta_2^4(\theta_5\theta_2\theta_9\theta_6\theta_{10}\theta_3\theta_4)^4 \\
&+ \theta_1^4(\theta_1\theta_2\theta_6\theta_8\theta_{10}\theta_3\theta_4)^4 + \theta_5^4(\theta_5\theta_2\theta_6\theta_8\theta_{10}\theta_3\theta_4)^4 + \theta_9^4(\theta_2\theta_9\theta_6\theta_8\theta_{10}\theta_3\theta_4)^4 \\
&+ \theta_8^4(\theta_5\theta_9\theta_6\theta_8\theta_{10}\theta_3\theta_4)^4 + \theta_{10}^4(\theta_1\theta_5\theta_9\theta_8\theta_{10}\theta_3\theta_4)^4 + \theta_3^4(\theta_1\theta_5\theta_2\theta_8\theta_{10}\theta_3\theta_4)^4 \\
&+ \theta_5^4(\theta_1\theta_5\theta_2\theta_9\theta_6\theta_7\theta_4)^4 + \theta_2^4(\theta_1\theta_5\theta_2\theta_6\theta_8\theta_7\theta_4)^4 + \theta_9^4(\theta_1\theta_5\theta_9\theta_6\theta_8\theta_7\theta_4)^4 \\
&+ \theta_8^4(\theta_1\theta_2\theta_9\theta_6\theta_8\theta_7\theta_4)^4 + \theta_1^4(\theta_1\theta_2\theta_9\theta_8\theta_{10}\theta_7\theta_4)^4 + \theta_5^4(\theta_5\theta_2\theta_9\theta_8\theta_{10}\theta_7\theta_4)^4 \\
&+ \theta_6^6(\theta_2\theta_9\theta_6\theta_8\theta_{10}\theta_7\theta_4)^4 + \theta_{10}^4(\theta_1\theta_2\theta_9\theta_6\theta_{10}\theta_7\theta_4)^4 + \theta_{10}^4(\theta_1\theta_5\theta_6\theta_8\theta_{10}\theta_7\theta_4)^4 \\
&+ \theta_1^4(\theta_1\theta_5\theta_2\theta_9\theta_3\theta_7\theta_4)^4 + \theta_6^4(\theta_5\theta_2\theta_9\theta_6\theta_3\theta_7\theta_4)^4 + \theta_8^4(\theta_5\theta_2\theta_9\theta_8\theta_3\theta_7\theta_4)^4 \\
&+ \theta_5^4(\theta_1\theta_5\theta_6\theta_{10}\theta_3\theta_7\theta_4)^4 + \theta_2^4(\theta_1\theta_2\theta_6\theta_{10}\theta_3\theta_7\theta_4)^4 + \theta_9^4(\theta_1\theta_9\theta_6\theta_{10}\theta_3\theta_7\theta_4)^4 \\
&+ \theta_8^4(\theta_1\theta_6\theta_8\theta_{10}\theta_3\theta_7\theta_4)^4 + \theta_{10}^4(\theta_5\theta_2\theta_9\theta_{10}\theta_3\theta_7\theta_4)^4 + \theta_3^4(\theta_1\theta_2\theta_9\theta_6\theta_3\theta_7\theta_4)^4 \\
&+ \theta_3^4(\theta_1\theta_5\theta_6\theta_8\theta_3\theta_7\theta_4)^4 + \theta_3^4(\theta_2\theta_9\theta_8\theta_{10}\theta_3\theta_7\theta_4)^4 + \theta_7^4(\theta_1\theta_5\theta_2\theta_8\theta_{10}\theta_7\theta_4)^4 \\
&+ \theta_7^4(\theta_1\theta_5\theta_9\theta_8\theta_3\theta_7\theta_4)^4 + \theta_7^4(\theta_5\theta_9\theta_6\theta_{10}\theta_3\theta_7\theta_4)^4 + \theta_7^4(\theta_2\theta_6\theta_8\theta_{10}\theta_3\theta_7\theta_4)^4 \\
&+ \theta_4^4(\theta_1\theta_5\theta_2\theta_9\theta_6\theta_{10}\theta_4)^4 + \theta_4^4(\theta_1\theta_2\theta_9\theta_6\theta_8\theta_3\theta_4)^4 + \theta_4^4(\theta_5\theta_9\theta_6\theta_8\theta_{10}\theta_7\theta_4)^4 \\
&+ \theta_4^4(\theta_5\theta_2\theta_6\theta_8\theta_3\theta_7\theta_4)^4 + \theta_4^4(\theta_1\theta_5\theta_2\theta_{10}\theta_3\theta_7\theta_4)^4 + \theta_4^4(\theta_1\theta_9\theta_8\theta_{10}\theta_3\theta_7\theta_4)^4.
\end{aligned}$$

This allows us to define:

$$I_2 := \frac{h_{12}}{h_{10}},\ I_4 := h_4,\ I_6 := \frac{h_{16}}{h_{10}},\ I_{10} := h_{10}$$

and we recall (5.3)

$$j_1 = \frac{I_2^5}{I_{10}},\ j_2 = \frac{I_2^3 I_4}{I_{10}}\ \text{and}\ j_3 = \frac{I_2^2 I_6}{I_{10}}.$$

### 18.2.4.c Computation of the class polynomial

The $3s$ invariants

$$j_1^{(i)},\ j_2^{(i)}\ \text{and}\ j_3^{(i)}, i = 1,\ldots,s$$

allow to compute the class polynomials

$$H_{K,1}(X) = \prod_{i=1}^{s}(X - j_1^{(i)}),\ H_{K,2}(X) = \prod_{i=1}^{s}(X - j_2^{(i)})\ \text{and}\ H_{K,3}(X) = \prod_{i=1}^{s}(X - j_3^{(i)}).$$

Unlike in the elliptic case the polynomials are no longer defined over the integers but over the rationals, hence we need to control the denominator. In all practical situations the denominators have no large prime factors and one can apply the continued fraction algorithm to the second highest coefficients of the polynomials to obtain possible denominators $d_{K,k}$. Often it is enough to multiply the polynomial with $d_{K,k}$ to obtain a polynomial over the integers. There are some exceptions, however, where the other coefficients have other factors or a higher power. Hence, one should check with care whether the coefficients times $d_{K,k}$ are close enough to integers and in case of doubt try to approximate the other coefficients, too.

In the sequel we take the integer polynomials $H'_{K,k}(X) \in \mathbb{Z}[X]$.

## 18.2.5 Finding a curve

This is the step in the CM method, which needs to be done for each curve one wants to construct. The previous sections have lead to the class polynomials $H_{K,k}$ corresponding to the CM field $K$ and the polynomials $H'_{K,k}$ defined over $\mathbb{Z}$. The aim here is to find suitable primes such that the $H'_{K,k}$ have a linear factor and that the cardinality of $\text{Pic}^0(C)$ for $C/\mathbb{F}_p$ has some desired property $Pr$, such as having a large prime factor or being smooth. Then we can use Mestre's algorithm to compute an equation of such a curve $C$.

### 18.2.5.a Choosing the characteristic $p$

In the algorithm we are required to find primes $p$ such that $(p) = (\omega)(\overline{\omega})$ with principal prime ideals $(\omega), (\overline{\omega})$.

In computational number theory there are algorithms to test whether this equation is solvable for a given $p$ and to compute solutions $\omega$ in the positive case (cf. [COH 2000]) and these algorithms are implemented in computer algebra systems like Pari [PARI].

Given the factors $\omega, \overline{\omega}$ we obtain the possible group orders of the Jacobian as in (18.4). The CM method is usually used in applications where one has special requirements $Pr$ and at this point one checks whether at least one of the candidate orders fulfills it.

If one does not want to use the heavy machinery of a computer algebra system, the inverse approach is interesting. One starts with an integer $\omega \in \mathcal{O}_K$, which has $N_{K/K_0}(\omega) \in \mathbb{Z}$. This condition can be phrased via quadratic polynomials in the coefficients of $\omega$ with respect to $\{1, w, \eta, w\eta\}$, where $\mathcal{O}_K = \mathcal{O}_{K_0} + \eta \mathcal{O}_{K_0}$.

We detail the case of $\eta = i\sqrt{a+bw}$, i.e., where the discriminant $D$ of $K_0$ is equivalent to $0$ modulo $4$. The other case is treated similarly and can be found in [WEN 2003, Sect. 8].

Let $\omega = c_1 + c_2 w + c_3 \eta + c_4 w \eta$. Then

$$\begin{aligned}N_{K/K_0}(\omega) &= (c_1 + c_2 w + c_3 \eta + c_4 w \eta)(c_1 + c_2 w - c_3 \eta - c_4 w \eta) \\ &= c_1^2 + c_2^2 w^2 + c_3^2 a + c_4^2 a w^2 + 2c_3 c_4 b w^2 + (2c_1 c_2 + c_3^2 b + c_4^2 b w^2 + 2c_3 c_4 a)w.\end{aligned}$$

The requirement $(p) = (\omega)(\overline{\omega})$ translates to

$$\begin{aligned}c_1^2 + c_2^2 w^2 + c_3^2 a + c_4^2 a w^2 + 2c_3 c_4 b w^2 &= p \\ 2c_1 c_2 + c_3^2 b + c_4^2 b w^2 + 2c_3 c_4 a &= 0.\end{aligned}$$

This leads to all possible solutions of the norm equation. We start with a random choice of $c_3, c_4$ such that they are coprime. From the second condition we immediately get the requirement

$$c_3^2 b + c_4^2 b w^2 \equiv 0 \pmod{2}$$

and have to make a different choice otherwise. By the same condition one has $2c_1 c_2 = -(c_3^2 b + c_4^2 b w^2 + 2c_3 c_4 a)$ and tries for different factors $c_1, c_2$ whether $c_1^2 + c_2^2 w^2 + c_3^2 a + c_4^2 a w^2 + 2c_3 c_4 b w^2$ is a prime. If not, one starts with a different choice of $c_3, c_4$.

**Remarks 18.9**

(i) Of course, one gets by this method (up to sign and complex conjugation) only one solution of the norm equation. It can happen that there is no curve of genus 2 over $\mathbb{F}_p$ for which the Jacobian has the expected group order. In this case one starts with a new choice of $\omega$.

(ii) In applications one needs primes of a certain size. To this end one chooses $c_3, c_4$ to have size $O(\sqrt{p})$.

(iii) In this method the most expensive step consists in the primality test (cf. Chapter 25). We note that primality tests or even factorization methods are usually needed in any case to test for the property $Pr$. One should use an early abort strategy and apply trial division first. With this method the time to find a suitable prime and its factors is usually shorter than for the method starting with candidate primes $p$.

### 18.2.5.b  Finding the invariants

By construction, the class polynomials $H_{K,k}$ have a linear factor modulo the primes generated in the previous paragraph.

One can also investigate the splitting behavior for other primes $p$. Factors of larger degree lead to curves defined over extension fields of $\mathbb{F}_p$. For our purposes we are interested in curves defined over a prime field, and hence refer to Weng [WEN 2003] for a complete study.

As already remarked, the denominators occurring in the polynomials $H_{K,k}$ are not divisible by large primes, and hence the invariants $j_k^{(i)}$ of the curves $C_i$ correspond in a canonical way to the roots of $H_{K,k}$ modulo $p$, which are the invariants of $C_i \pmod{p}$.

At this point a further difficulty compared to the case of elliptic curves occurs, this time caused by the more involved invariant theory. We do not know how to combine the 3 zeroes arising from the 3 polynomials $H_{K,k}, k = 1,2,3$ to form a triple of invariants. Hence, we have to do the steps in Mestre's algorithm and the final step for each of the at most $s^3$ triples whose entries are zeroes $j_1'^{(i_1)}, j_2'^{(i_2)}, j_3'^{(i_3)}$ of the polynomials $H_{K,k}$.

### 18.2.5.c  Mestre's algorithm

Let $(j_1, j_2, j_3) \in \mathbb{F}_p^3$ be a candidate triple of invariants of a curve $C_i \pmod{p}$. In Section 5.1.6.b we defined a conic $\mathcal{Q}_{j_1,j_2,j_3}(x_1, x_2, x_3)$ and a cubic $\mathcal{H}_{j_1,j_2,j_3}(x_1, x_2, x_3)$. As a first step one determines a rational point $(a_1, a_2, a_3)$ of $\mathcal{Q}_{j_1,j_2,j_3}$ by multivariate factorization methods. If there does not exist such a point one rejects the triple and tries with a new one. It can happen that for none of the triples found in Section 18.2.5.b the conic has a rational point. In this case one needs to start anew with a different choice of the prime $p$.

Now assume that $\mathcal{Q}_{j_1,j_2,j_3}(a_1, a_2, a_3) = 0$. Then one can parameterize the conic with

$$\mathcal{Q}_{j_1,j_2,j_3}\bigl(x_1(t), x_2(t), x_3(t)\bigr) = 0.$$

We find the points of intersection between $\mathcal{Q}_{j_1,j_2,j_3}$ and $\mathcal{H}_{j_1,j_2,j_3}$ by solving the equation

$$\mathcal{H}_{j_1,j_2,j_3}\bigl(x_1(t), x_2(t), x_3(t)\bigr) = 0.$$

In general, this leads to a polynomial $f(t)$ of degree 6 such that the roots $t_i$ give the 6 points of intersection over $\overline{\mathbb{F}}_p$.

**Remark 18.10** It is only possible to parameterize affine parts of the conic. If we assume that $\mathcal{Q}_{j_1,j_2,j_3}$ and $\mathcal{H}_{j_1,j_2,j_3}$ intersect only in this part we get $\deg(f) = 6$. Otherwise, there is another affine part on which the description works as described.

By Lemma 5.53 the curve $C$ with invariants $j_1, j_2, j_3$ is given, up to a twist, by the affine equation

$$C_a : y^2 = f(x).$$

### 18.2.5.d  The final step

Throughout this book we always assume that hyperelliptic curves have a $\mathbb{F}_p$-rational Weierstraß point. In our case that means that the polynomial $f(x)$ has a root $\xi$ over $\mathbb{F}_p$. By the transformation $\xi \to \infty$ we get a new affine equation $y^2 = \tilde{f}(x)$ with $\deg(\tilde{f}) = 5$.

## § 18.2 CM for curves of genus 2

Now we have to check whether the Jacobian of the constructed curve $C$ and of its twists has one of the at most 4 possible numbers in $\mathcal{N} = \{N_\omega = \chi_\omega(1) \mid \omega \in W\}$ as in (18.4) of $\mathbb{F}_p$-rational points. This is done as usual:

Choose a random divisor class $\overline{D} \in \text{Pic}_C^0$, check whether $[N]\overline{D} = 0$ for $N \in \mathcal{N}$. If this check fails for all $N_i$ one chooses a new triple $(j_1, j_2, j_3)$ or, if not available, another $p$.

In case of ambiguity one tries a different random divisor class $\overline{D} \in \text{Pic}_C^0$.

**Remark 18.11** The quadratic twist of $C$ is defined by

$$C_v : vy^2 = f(x),$$

where $v$ is not a square in $\mathbb{F}_p$. If $C$ has order $N_\omega$ then the twist $C_v$ has order $N_{-\omega}$ and it could be that $N_{-\omega}$ rather than $N_\omega$ has the desired property $Pr$.

### 18.2.6 The algorithm

In the following we give the complete description of the algorithm using the theory from above.

In Sections 18.2.3 and 18.2.4 we described the choice of the CM field and the computation of the theta constants and the class polynomials. In practice this is done as a precomputation step. So, we shall take the CM-field and the class polynomials as given in the following algorithm.

---

**Algorithm 18.12** Construction of genus 2 curves via CM

INPUT: A CM-field $K/K_0/\mathbb{Q}$, class polynomials $H_{K,k}(X)$, desired size of $p$ and property $Pr$.
OUTPUT: A prime $p$, a genus 2 curve $C/\mathbb{F}_p$ whose group order $N = |\text{Pic}_C^0(\mathbb{F}_p)|$ satisfies $Pr$.

1. **while true do**
2.     **repeat**
3.         **repeat** choose prime $p$ of desired size     [see remark below]
4.         **until** $p = \omega_1 \overline{\omega}_1$ with $\omega_1 \in \mathcal{O}_K$     [compute also $p = \omega_2 \overline{\omega}_2$ if possible]
5.         compute the set of possible group orders $\mathcal{N}$ as in (18.4)
6.     **until** at least one $N \in \mathcal{N}$ satisfies property $Pr$
7.     **for** $k = 1$ **to** $3$ compute all roots $J_k = \left\{ j_k^{(i)} \right\}$ of $H_{K,k}(X) \pmod{p}$
8.     **for** $(j_1, j_2, j_3) \in J_1 \times J_2 \times J_3$ **do**
9.         compute the conic $\mathcal{Q}_{j_1,j_2,j_3}(x_1, x_2, x_3)$
10.         **if** there exists a point $(a_1, a_2, a_3)$ with $\mathcal{Q}_{j_1,j_2,j_3}(a_1, a_2, a_3) = 0$ **then**
11.             compute $\mathcal{H}_{j_1,j_2,j_3}$ and the intersection polynomial $f(t)$
12.             **if** $f(t)$ has a root in $\mathbb{F}_p$ **then**
13.                 find the curve $C$ given by $y^2 = \tilde{f}(x)$ and $\deg(\tilde{f}) = 5$
14.                 deduce the twist $\widetilde{C}$ given by $dy^2 = \tilde{f}(x)$
15.                 choose a random divisor class $\overline{D} \in \text{Pic}_C^0(\mathbb{F}_p)$
16.                 use $\overline{D}$ to check if $N$ is the group order
17.                 **if** desired group order is assumed **then break**
18.     **return** the prime $p$, the curve $C$ and the order $N$

As for elliptic curves, examples of desired properties $Pr$ are that the group order contains a prime factor larger than a given bound or that the group order is smooth with a given smoothness bound. The latter is interesting for factorization methods (cf. Chapter 25), while the prior can be used as basis for a cryptosystem (cf. Section 23.4).

**Remarks 18.13**

(i) Lines 3 and 4 can be replaced as described in Section 18.2.5.

(ii) The largest complexity is introduced by the factorization of the $H_{K,k}$ and the following step, which is executed in the worst case for all possible triples.

## 18.3 CM for larger genera

We give a short outline of possibilities and difficulties that occur when one tries to construct hyperelliptic curves of genus $\geqslant 3$ by using the CM-method.

### 18.3.1 Strategy and difficulties in the general case

The CM-theory of Shimura and Taniyama is valid for all CM-fields $K$ of degree $2g$. Of course the arithmetic of $K$ becomes more complicated for increasing degrees but for moderate $g$ (e.g., $g \leqslant 5$) it is possible to determine the period matrices of a set or representatives of all *principally polarized abelian varieties* over the complex numbers with endomorphism ring equal to the ring of integers in $K$. The next step, to compute the theta constants attached to the matrices as complex numbers with high precision, causes only numerical difficulties that can be overcome if one uses enough effort.

The first theoretical problem arises when one has to decide which of the constructed abelian varieties are Jacobian varieties of curves. Due to a theorem of Weil this is always the case for $g = 3$; for $g \geqslant 4$ the question is unsolved (Schottky problem).

But we are interested in hyperelliptic curves and so we can say more. By results of Mumford and Poor one has a criterion that is easily verified: let $\Omega$ be the period matrix of a principally polarized abelian variety $\mathcal{A}$. A necessary and sufficient condition for $\mathcal{A}$ being the Jacobian variety of a hyperelliptic curve is that one of the even theta constants vanishes. So, one can determine the isomorphism classes (over $\mathbb{C}$) of all hyperelliptic curves whose Jacobian varieties have as ring of endomorphisms the integers of $K$.

Now the next complication arises. The invariant theory of such curves becomes more complicated and the structure of the invariant ring is not well understood at least when $g > 5$. So, we restrict ourselves to $g \leqslant 5$ (which is of course enough for practical applications). Then we can proceed as in Section 18.2.4 and compute class polynomials for a set of invariants that determine the isomorphism classes of the curves. These polynomials have rational coefficients, and the possible denominators are becoming larger but for primes $p$, large enough reduction theory works as in the case of $g \leqslant 2$.

By solving a norm equation with integers in $K$, one finds candidates for the Frobenius endomorphism modulo $p$ and so possible orders of the group of rational points of the Jacobian. Even the last step, to find the curve equation, can be done in reasonable time [WEB 1997].

So, in theory we can use the CM-method at least for hyperelliptic curves of genus $\leqslant 5$. But *in practice* we meet a big difficulty: if we do not impose additional conditions on $K$ we rarely find *hyperelliptic* curves.

The reason comes from the theory of the moduli space of curves of genus $g \geqslant 2$. The dimension of this space is $3g - 3$, and the locus of hyperelliptic curves has dimension $2g - 1$ (it has $2g + 2$ Weierstraß points containing (up to projective transformations) the points $P_\infty, (0,0), (1,0)$). So already for $g = 3$ the hyperelliptic curves lie in an algebraic set of codimension 1. An extensive search, cf. Weng [WEN 2001b], did not yield any example of a hyperelliptic curve of genus 3 and CM without additional automorphism. So only very special CM-fields $K$ can be chosen if we want to find hyperelliptic curves.

### 18.3.2 Hyperelliptic curves with automorphisms

Assume that $K$ contains the complex number $i$ with $i^2 = -1$. This means that $K = K_0(i)$ where $K_0$ is a totally real field of degree $g$ over $\mathbb{Q}$.

We get the following result.

**Lemma 18.14** Let $\mathcal{A}$ be a principally polarized abelian variety with endomorphism ring equal to the ring of integers of $K$. Assume that $\mathcal{A}$ is simple and that it is the Jacobian of a curve $C$.

Then $C$ is hyperelliptic and has an automorphism $\tau(i)$ of order 4.

**Proof.** By assumption $\mathcal{A}$ has an automorphism $\tau(i)$ of order 4. It is a general fact that $\tau(i)$ induces an automorphism of order 2 or 4 of $C$. In any case $C$ has an automorphism of order 2. Hence the function field of $C$ has a subfield of index 2. The Hurwitz genus formula yields that this subfield has genus $g_0 < g$. The Jacobian of the corresponding curve is a factor of $\mathcal{A}$ and so, since $\mathcal{A}$ is assumed to be simple, it is equal to $\{0\}$. Hence $g_0 = 0$ and $C$ is hyperelliptic.

Assume that the conditions of the lemma are satisfied and let $C$ be given as usual by

$$C_a : y^2 = x^{2g+1} + f_{2g}x^{2g} + \cdots + f_0.$$

Since $\tau(i)^2$ is equal to the hyperelliptic involution $w$ the automorphism $\tau(i)$ induces an automorphism of $\mathbb{C}(x)$ of order 2 and hence it maps $x$ to $-x$. Since $w(y) = -y$ it follows that $\tau(i)(y) = iy$.

This imposes conditions on the equation defining $C$. We must have

$$y^2 = x^{2g+1} + f_{2g}x^{2g} + \cdots + f_1 x + f_0 = x^{2g+1} - f_{2g}x^{2g} \pm \cdots + f_1 x - f_0$$

and so $f_i = 0$, for $i$ even. $\square$

**Corollary 18.15** Let $C$ be as in Lemma 18.14. Then $C$ is given by an affine equation

$$C_a : y^2 = x^{2g+1} + \sum_{j=0}^{g-1} f_{2j+1} x^{2j+1}.$$

**Remark 18.16** Until now, we have stated our results over the complex numbers. But of course Lemma 18.14 and its corollary are true over any field $K$ (of characteristic $\neq 2$) over which $\mathcal{A}$ and $\tau(i)$ are defined.

Assume now that $K = \mathbb{F}_q$ with $q$ odd and let $r$ be a nonsquare, i.e., $r \in \mathbb{F}_q \setminus \mathbb{F}_q^2$.

Since $C$ has an automorphism of order 4 it has a *quartic* twist $C^{(4)}$ of $C$ obtained from $C_a$ by the transformation $x \mapsto r^{1/2}x$, $y \mapsto r^{(2g+1)/4}$ defined over $\mathbb{F}_{q^4}$.

The equation of $C^{(4)}$ is

$$y^2 = x^{2g+1} + \frac{f_{2g-1}}{r} x^{2g-1} + \cdots + \frac{f_1}{r^g} x.$$

Hence either $f_{2g-1}$ or $f_{2g-1}/r$ is a square in $\mathbb{F}_q$. Now we can use a quadratic twist to get that $f_{2g-1} = 1$.

**Corollary 18.17** Assume that the conditions of Lemma 18.14 are satisfied and that $C$ and $\tau(i)$ are defined over $\mathbb{F}_q$. Then a twist of $C$ is given by an equation

$$y^2 = x^{2g+1} + x^{2g-1} + \sum_{j=2}^{g} a_{2j} x^{2g+1-2j}.$$

### 18.3.3 The case of genus 3

We now apply our results to the case that $K_0$ is totally real of degree 3 and class number 1. Because of the Theorem of Weil and Lemma 18.14 we know that every principally polarized abelian variety with endomorphism ring $\mathcal{O}_K$ is the Jacobian of a hyperelliptic curve.

In a precomputation we determine the period matrices and the theta constants of a set of representatives of the $\mathbb{C}$-isomorphism classes of such varieties. This is explained in detail in section 3 and 4 of [WEN 2001a].

As seen in the background there are 5 Shioda invariants that determine a curve up to isomorphism. Due to the automorphisms we have a symmetry and thus it is enough to compute only 2 invariants, $j_1$ and $j_3$. This is done either directly from the theta constants or via a Rosenhain model for $C$ (cf. [WEB 1997] and [WEN 2001a, Chap. 4]) over $\mathbb{C}$ for each of the representatives, and we use this to determine the class polynomials $H_{j_k}(X)$ for $k = 1$ and 3.

After this precomputation (which can be replaced by a look-up of tables) the algorithm proceeds as follows.

1. Choose a prime $p$, solve the norm equation

$$p = \omega\overline{\omega}$$

   with $\omega \in \mathcal{O}_K$ and test whether there is a candidate curve $C \pmod{p}$ for which the order of the rational points of its Jacobian has the desired property.

2. Compute the zeroes of $H_{j_k}(X)$ over $\mathbb{F}_p$.

3. At this step we use the extra structure we have.
   By Corollary 18.17 we know that the curve we are looking for is (up to a twist) given by

$$y^2 = x^7 + x^5 + ax^3 + bx; \ a, b \in \mathbb{F}_p.$$

The discriminant of $x^7 + x^5 + ax^3 + bx$ is

$$\begin{aligned}\Delta = &-17280a^5b^2 + 9216a^5b - 1024a^5 - 4352a^4b^3 + 512a^4b^2 + 9216a^4b^4 \\ &+ 62280a^6b - 13824a^5b^3 - 64a^3b^3 - 13824a^6 - 1024a^3b^3 + 512a^3b^5 - 46656a^7.\end{aligned}$$

Define

$$I_2 = -1/4a - 1/28b$$

and

$$I_4 = \frac{1}{288}(-504b - 120)(-504a - 120b^3) + \frac{1}{96}(-196a + 68b)^3.$$

Then we have the identities

$$j_1\Delta - I_2^5 = 0$$

and

$$j_3\Delta - I_2 I_4 = 0,$$

which give two polynomial equations for possible values of $a$ and $b$.
Take these equations with $(j = j_1, j' = j_3)$ a combination of the zeroes of $H_{j_1}(X)$ and $H_{j_3}(X)$ over $\mathbb{F}_p$ and compute the candidates $(a', b')$ for $(a, b)$.

4. Test whether $C' : y^2 = x^7 + x^5 + a'x^3 + b'x$ or one of its twist curves has a Jacobian $J_C$ with $J_C(\mathbb{F}_p)$ of desired order. If not, choose another pair $(a', b')$ and eventually another pair $(j, j')$.

**Remarks 18.18**

(i) Examples and a detailed discussion of the general case of hyperelliptic curves of genus 3 can be found in [WEN 2001a].

(ii) If one wants to use curves of genus 3 for cryptographic purposes, the examples constructed above may have some security deficiencies compared with random curves, for instance, caused by the existence of a nontrivial automorphism. But they are easily constructed and serve well if one wants to test, for example, efficiency of specific scalar multiplications.

On the other hand, Section 15.2 shows how to use endomorphisms to speed up scalar multiplication in $\operatorname{Pic}_C^0$.

(iii) One can drop the condition that $C$ is hyperelliptic. Then one finds many curves of genus 3 with complex multiplication. Another interesting subclass of such curves consists of Picard curves, which have an automorphism of order 3. They have as CM-field a field $K_0(\sqrt{-3})$. For more details see [KOWE 2005].

# Computation of Discrete Logarithms

# Chapter 19

# Generic Algorithms for Computing Discrete Logarithms

*Roberto M. Avanzi*

### Contents in Brief

| | | |
|---|---|---|
| **19.1** | **Introduction** | **478** |
| **19.2** | **Brute force** | **479** |
| **19.3** | **Chinese remaindering** | **479** |
| **19.4** | **Baby-step giant-step** | **480** |
| | Adaptive giant-step width • Search in intervals and parallelization • Congruence classes | |
| **19.5** | **Pollard's rho method** | **483** |
| | Cycle detection • Application to DL • More on random walks • Parallelization • Automorphisms of the group | |
| **19.6** | **Pollard's kangaroo method** | **491** |
| | The lambda method • Parallelization • Automorphisms of the group | |

This chapter is devoted to the generic methods for computing discrete logarithms in commutative groups and group orders. There are also applications where, apart from the group order, we have additional knowledge on the logarithm $\log_g(h)$, such as a certain amount of high or low order bits, or a probability distribution of the logarithm, or else the interval in which it lies is known beforehand. The last kind of information is exploited, for example, by the kangaroo methods (see Section 19.6): if we are interested in the order of an elliptic or hyperelliptic curve, this information by Hasse's bound (see Section 5.2.3) is an interval centered on the cardinality of the underlying field. In general, such knowledge (especially information about certain bits) can be used to reduce the running time for solving the DLP, so the designer of a DL system must take it into account [NGSH 2003]. The adaptation of the methods described here to the context where one has that particular information is often quite straightforward.

Our presentation has been influenced by Teske's excellent survey [TES 2001b].

## 19.1 Introduction

We begin our presentation with exponential time methods; we recall that this means that the complexity is exponential in the *logarithm* of the order. For factoring an integer $n$, this implies that these methods may take $n^C$ operations on integers of size comparable to $n$ for some positive constant $C$. For computing DLs, it means that they may require $|G|^C$ compositions on $G$.

The best methods that apply to *any* cyclic group provide an upper bound $C = 1/2$ — which is the reason they are also called *square root algorithms* — and are based on only a handful of ideas: brute force, the Chinese remainder theorem, the birthday paradox. In fact, most of the methods explained in this section are square root methods. This explains why subexponential time algorithms (despite their names) are in many cases better than square root methods! We first give a rigorous definition of a generic algorithm:

**Definition 19.1** An algorithm performing computations in groups is called a *generic algorithm* if the only computations it performs are:

- computation of the composition of two elements
- computation of the inverse of an element, and
- checking two elements for equality.

Such an algorithm is also said to operate in *black-box groups*.

Shoup [SHO 1997], generalizing a result of Nechaev [NEC 1994], has shown that if $p$ is the largest prime dividing the group order, a *generic* algorithm to solve the DLP with a probability bounded away from zero has to perform $\Omega(\sqrt{p})$ group operations. Let $G$ be a cyclic group of order $n$. Starting with the notion of an *oracle* that can be queried for the result of the three above generic group operations, Shoup computes the probability that an algorithm outputs the correct answer to a DLP in $G$ after $m$ oracle invocations. The numeration of elements of $G$ can be reinterpreted as an encoding of the $n$ (distinct) scalar multiples of $g$ that are given by a map $\sigma$ of $\mathbb{Z}/n\mathbb{Z}$ into a set $S$ of binary strings representing elements of $G$ uniquely. The problem $[t]g = h$ in $G$ is then rewritten as follows: Given $(\sigma(1), \sigma(t))$, find $t \in \mathbb{Z}/n\mathbb{Z}$. We cite Shoup's main result concerning generic DL algorithms.

**Theorem 19.2** [SHO 1997] Let $n$ be a positive integer whose largest prime divisor is $p$. Let $S$ be a set of binary strings of cardinality at least $n$. Let $\mathcal{A}$ be a generic algorithm for $\mathbb{Z}/n\mathbb{Z}$ on $S$ that makes $m$ oracle queries, and suppose that the encoding function $\sigma$ of $\mathbb{Z}/n\mathbb{Z}$ on $S$ is chosen randomly. The input to $\mathcal{A}$ is $(\sigma(1), \sigma(t))$ where $t \in \mathbb{Z}/n\mathbb{Z}$ is random. The output of $\mathcal{A}$ is $v \in \mathbb{Z}/n\mathbb{Z}$. Then the probability that $t = v$ is $O(m^2/p)$.

Therefore, for achieving a nonnegligible probability of success, one needs $O(\sqrt{p})$ oracle calls. The result is that a generic DL algorithm must perform $\Omega(\sqrt{p})$ group operations. The generic algorithms for the DLP that we shall describe provide a constructive proof of how the DLP can be solved in time $O(\sqrt{|G|})$ for any cyclic group $G$. The ideas underlying these methods are often used for other purposes too, for example, for factoring, as we shall see in Chapter 25, or to improve algorithms, which in principle are completely different. Generic methods like the ones described here are often superseded by better algorithms that have been designed for particular kinds of groups. Some of these methods are presented in Section 20 if still "generic" enough, and are then discussed in more detail for specific groups in the next chapters. *This is the reason why ideal group-based cryptosystems are based on primitives whose solution has square root complexity.* For a very interesting discussion of other types of information that can weaken the protocols, we refer to [MAWO 1998].

In the following, $G$ can always be assumed cyclic: in fact, since the discrete logarithm problem we want to solve is $[t]g = h$ with $h \in \langle g \rangle$, we can always replace $G$ with $\langle g \rangle$.

## 19.2 Brute force (exhaustive search)

Given a group $G$ generated by one of its elements $g$ and a second element $h \in G$, an integer $t$ such that $[t]g = h$ can be found by comparing $[t]g$ to $h$ for all $t$ with $0 \leqslant t < |G|$. This takes at most $\mathrm{ord}(g)$ additions in $G$ and is definitely not viable for large groups. For the integer factorization problem this is simply trial division: see Chapter 25. Usually this method is too slow, but, surprisingly, it has some important applications: due to its absence of overhead, it might even be the method of election for checking whether a given element splits completely over a small, given set of primes (smoothness test: see Section 20.1).

## 19.3 Chinese remaindering

Discrete logarithms in $G = \langle g \rangle$ can be computed easily if $n := \mathrm{ord}(g)$ has only small factors. More precisely, the complexity of computing discrete logarithms in a group of composite order $n$ is (from the point of view of computational complexity) bounded from above by the complexity of solving the DLP in a group whose order is the largest prime factor of $n$. This was first observed for cryptographic applications by Silver, Pohlig, and Hellman [POHE 1978].

Assume $n$ composite and let $p \mid n$. From $[t]g = h$ it follows that

$$[t \bmod p]\big(\big[\tfrac{n}{p}\big]g\big) = [t]\big(\big[\tfrac{n}{p}\big]g\big) = \big[\tfrac{n}{p}\big]h.$$

Thus, $t$ modulo each of the primes $p$ dividing $n$ can be found by solving the DLP in a cyclic group of order $p$. If $n$ is a product of distinct primes, then $t$ is recovered applying the Chinese remainder theorem (see Section 10.6.4).

If $n$ is not squarefree, the $p$-adic expansion can be used as described, for example, in [LELE 1990], to compute $t$ modulo the highest power of $p$ dividing $n$ for all primes $p$, and then the Chinese remainder theorem is employed. If, say, $p^k$ divides $n$, one first gets $t \bmod p$, then "lifts" this value to $t \bmod p^j$ for $j = 2, 3, \ldots, k$. For simplicity, suppose that $G$ is a *cyclic* group of order $p^k$ and that $[t]g = h$ with $t = t_0 + t_1 p + t_2 p^2 + \cdots + t_{k-1} p^{k-1}$ where $0 \leqslant t_j < p$. Now

$$[p^{k-1}]h = [t]\big([p^{k-1}]g\big) = [t_0]\big([p^{k-1}]g\big),$$

because the $p^k$-fold of any element of $G$ is $0_G$ (the unit of $G$), hence $t_0$ is found by computing a discrete logarithm in a the unique subgroup $G_0$ of $G$ of cardinality $p$. Similarly

$$[p^{k-2}]h = [t]\big([p^{k-2}]g\big) = [t_0 + t_1 p]\big([p^{k-2}]g\big)$$

whence

$$[p^{k-2}]h \oplus [-t_0]\big([p^{k-2}]g\big) = [t_1]\big([p^{k-1}]g\big) \ \text{ in } \ G_0.$$

In general

$$[p^{k-j}]h \oplus [-(t_0 + t_1 p + \cdots + t_{j-1} p^{j-1})]\big([p^{k-j}]g\big) = [t_j]\big([p^{k-1}]g\big) \ \text{ in } \ G_0$$

allowing us to find $t_j$ for $j = 3, 4, \ldots$, until $t$ is completely determined.

If $k$ is not too large (as in the cryptographic applications) we can precompute and store the scalar multiples $[p^j]g$ and $[p^j]h$ for $j = k-1, k-2, \ldots, 1$ in order to speed up the determination of $t_0, t_1, \ldots, t_{k-1}$.

If $p$ is itself not too large, there is the following efficient strategy for computing $t$: first precompute a table with the elements of $G_0$ (one is attained with high probability as $r := [p^{m-1}]x$ for a random $x \in G$, the other ones as multiples of $r$), then solve the above equations by means of table lookups.

The general setting for $p$-adic lifting is described in Chapter 12.

**Remark 19.3** The algorithms in the following sections are the "serious" generic methods to solve the DLP. Of course, in order to avoid this weakness one has to assume that the group order is known. In some cases, however, this is not the case. In particular, many algorithms can be adapted to this case simply by using them to solve $\log_g(1)$, i.e., to compute the order of $g$. In most cases simplifications and optimizations are just obvious.

## 19.4 Baby-step giant-step

The baby-step giant-step algorithm was first published by Shanks in [SHA 1971]. According to Nechaev [NEC 1994] it was known to Gelfond in 1962. See also [KNU 1997, Ex. 5.17]. The first application of this method was order computation: in fact Shanks used it to compute ideal class numbers of quadratic number fields. It can be used for discrete logarithm computation and we shall present it in this form, which is more general. Order computation is then done solving $[t]g = 0$ with for $t \neq 0$ (and of course with unknown order). The *baby-step giant-step* method is based on the following observation:

**Lemma 19.4** Let $n$ be a positive integer. Put $s := \lfloor \sqrt{n} \rfloor + 1$. Then for any $t$ with $0 \leqslant t < n$ there are integers $0 \leqslant U, V < s$ with $t = U + Vs$.

Suppose now $n = \text{ord}(g)$. Then $[t]g = h$ implies

$$h \oplus [-U]g = [Vs]g$$

for some $U, V, s$ as in Lemma 19.4.

---

**Algorithm 19.5** Shanks' baby-step giant-step algorithm

INPUT: A generator $g$ of a group $G$ of order $n$ and $h \in G$.
OUTPUT: An integer $t$ with $[t]g = h$.

1. $s \leftarrow \lfloor \sqrt{n} \rfloor + 1$
2. **for** $j = 0$ **to** $s$ store $(\beta_j \leftarrow h \oplus [-j]g, j)$ in a hash table  [the 'baby steps']
3. $i \leftarrow 0$ and $\gamma \leftarrow 0_G$  [$\gamma$ holds the 'giant steps' $[is]g$]
4. **while true do**
5.     **if** $(\gamma = \beta_j$ for some $j)$ **then return** $(is + j)$  [hash table lookup]
6.     $i \leftarrow i + 1$ and $\gamma \leftarrow \gamma \oplus [s]g$

---

This method is very important because it is the first generic deterministic method requiring at most $2\sqrt{n}$ compositions in $G$, thus matching the order of magnitude of Shoup's lower bound. Its main drawback is that it has to store $O(\sqrt{n})$ group elements.

What happens if we do not know the group order? We can work with an assumed bound $n$ on $t$ and compute a set of baby steps. Then we compute giant steps until the solution is found. The numbers of compositions required is about $\sqrt{n} + t/\sqrt{n} \leqslant 2\sqrt{n}$ if the bound is correct. This works even if $n < t$, but, when $t$ is much larger than $n$, we observe that $\sqrt{n} + t/\sqrt{n}$ becomes linear in $t$, turning this algorithm into a "sophisticated" brute force method. The same can be done to *compute the group order*: we just want to solve $t = \log_g(1)$ discarding the trivial solution $t = 0$. Usually we have some loose bounds on the cardinality of the group by construction or by its representation (for instance, if the elements of the group are represented by strings of at most $b$ bits, it can be assumed

## 19.4.1 Adaptive giant-step width

A different approach allows us to obtain results relative to the discrete logarithm itself rather than relative to an upper bound on it. It can also be used to compute unknown group orders without the complexity becoming possibly linear in the group order itself. Buchmann, Jacobson, and Teske [BUJA+ 1997] suggested starting with a conservative, moderate assumption on the order of magnitude of $t = \log_g(h)$ and doubling the giant step width at certain intervals. Their method is based on the following result:

**Lemma 19.6** [BUJA+ 1997, Lemma 2.1] For every integer $t > 1$ there are uniquely determined integers $k$, $c$ and $j$ such that

$$t = 2^{k+1}c + j \quad \text{with} \quad \left(2^{k-2} - \tfrac{1}{2}\right) \leqslant c < 2^{k+1},\ k \geqslant 0 \quad \text{and} \quad 1 \leqslant j \leqslant 2^{k+1}.$$

The lemma implies the correctness of the following algorithm.

---

**Algorithm 19.7** BJT variant of the baby-step giant-step algorithm

INPUT: A generator $g$ of a group $G$ of unknown order and $h \in G$ and an initial estimate $t$ of an integer $t > 1$ with $[t]g = h$.
OUTPUT: An integer $t$ with $[t]g = h$.

1. $k \leftarrow \lfloor \lg(\sqrt{t}) \rfloor$
2. **for** $j = 0$ **to** $2^{k+1} - 1$ store $(\beta_j \leftarrow h \oplus [-j]g, j)$ in a hash table [initial 'baby steps']
3. **while true do**
4.     **for** $c = \lfloor 2^{k-1} \rfloor$ **to** $2^{k+1} - 1$ **do**
5.         $\gamma \leftarrow [2^{k+1}c]g$     [the 'giant steps']
6.         **if** ($\gamma = \beta_j$ for some $j$ with $1 \leqslant j \leqslant 2^{k+1}$) **then return** $(2^{k+1}c + j)$.
7.     $k \leftarrow k + 1$
8.     **for** $j = 2^k$ **to** $2^{k+1} - 1$ insert $(\beta_j \leftarrow h \oplus [-j]g, j)$ in the baby step hash table

---

In Line 5, $\gamma$ should not be computed anew each time as $[2^{k+1}c]g$, as this would be too expensive. Instead, $[2^{k+1}c]g$ and $[4^k]g = [2^{k+1}2^{k-1}]g$ (where $2^{k-1}$ is the first value of $c$ unless $k$ is zero) are computed before the while loop and updated as $k$ increases. The values $[2^{k+1}c]g$ in the internal for loop are obtained upon adding repeatedly $[2^{k+1}c]g$ to a variable initially containing $[4^k]g$. As in the original baby-step giant-step method, the search in Line 6 should be performed using a hash table.

The complexity of this variant is $O(\sqrt{t})$.

Another approach is to increment the width of the giant steps after each baby step [TER 2000]. It is based on the following way of representing integers:

**Lemma 19.8** For every nonnegative integer $t$ there are uniquely determined integers $j$ and $k$ with $0 \leqslant k < j$ such that $t = T_{j+1} - k$, where $T_j$ is the $j$-th triangular number, i.e., $T_1 = 0$, $T_{n+1} = T_n + n$ for $n \geqslant 1$.

To compute $t = \log_g(h)$ we need to find the correct value of $j$ in the representation above. This is done alternating baby steps and giant steps as follows: first, we compute the baby step $\beta_0 =$

$h \oplus [0]g = h$ and the giant step $[T_1]g = [0]g = 0_G$; the baby steps are always stored, only the last giant step is kept in memory. For $j \geq 1$, the $j$-th iteration consists of computing the giant step $[T_{j+1}]g = [T_j]g \oplus [j]g$. If $j \geq 2$, we check whether $[T_{j+1}]h$ is in the baby step set for some $s$ with $0 \leq s < j$: if this is the case, then $t = T_{j+1} - s$, otherwise the baby step $\beta_j = h \oplus [j]g = h \oplus [j-1]g \oplus g$ is computed, $j$ increased by one and the next iteration performed.

---

**Algorithm 19.9** Terr's variant of the baby-step giant-step algorithm

INPUT: A generator $g$ of a group $G$ of unknown order and $h \in G$.
OUTPUT: An integer $t$ with $[t]g = h$.

1. $\beta_0 \leftarrow h, \gamma \leftarrow [T_1]g = 0_G, \delta \leftarrow 0_G$ and $j \leftarrow 0$
2. **while true do**
3.     $j \leftarrow j + 1$
4.     $\delta \leftarrow \delta \oplus g$ and $\gamma \leftarrow \gamma \oplus \delta$      $[\gamma = [T_{j+1}]g]$
5.     **if** $j \geq 2$ **then**
6.         **if** $(\gamma = \beta_s$ for some $s$ with $0 \leq s < j)$ **then**
7.             **return** $(T_{j+1} - s)$      $[T_{j+1} = j(j+1)/2]$
8.     $\beta_j \leftarrow \beta_{j-1} \oplus g$      [equivalently $\beta_j \leftarrow h \oplus \delta$]

---

This variant has complexity $O(\sqrt{t})$, exactly as in the previous variant, but the implied constants are smaller.

## 19.4.2 Search in intervals and parallelization

Sometimes it is known beforehand that the discrete logarithm $t$ or the group order lies in the interval $[a, b]$. This is in fact the situation for the group orders of elliptic and hyperelliptic curves. The modifications to Algorithm 19.5 to take advantage of this information are straightforward. The time and storage complexities of the modified algorithm become $O(\sqrt{b-a})$. Put $s = \lceil \sqrt{b-a} \rceil$. The baby steps $\beta_j$ are given by $h \oplus [-a - j]g$ for $0 \leq j < s$. The giant steps are $[is]g$ for $0 \leq i < s$. When a match $h \oplus [-a - j^*]g = [i^*s]g$ is found we get $t = a + i^*s + j^*$.

---

**Algorithm 19.10** Baby-step giant-step algorithm in an interval

INPUT: A generator $g$ of a group $G$, an element $h \in G$ and an interval $[a, b]$.
OUTPUT: An integer $t \in [a, b]$ with $[t]g = h$ or failure.

1. $s \leftarrow \lceil \sqrt{b-a} \rceil$
2. **for** $j = 0$ **to** $s - 1$ store $(\beta_j \leftarrow h \oplus [-a - j]g, j)$ in a hash table      [the 'baby steps']
3. $\gamma \leftarrow 0_G$      [$\gamma$ holds the 'giant steps' $[is]g$]
4. **for** $i = 0$ **to** $s - 1$ **do**
5.     **if** $(\gamma = \beta_j$ for some $j)$ **then return** $(a + is + j)$      [hash table lookup]
6.     $\gamma \leftarrow \gamma \oplus [s]g$
7. **return fail**

---

Upon application of the baby-step giant-step method on $m$ processors, a speedup by a factor of $\sqrt{m}$

is easily achieved. It suffices to divide the interval to be searched into $m$ approximately equally wide subintervals and to apply the algorithm independently on each of them. In order to get a better speedup one would presumably need a parallel machine with shared memory with unrestricted access. Furthermore it seems at least very intricate to combine parallelization with dynamically expanding giant step width. The conclusion is that the baby-step giant-step method cannot be parallelized in an efficient way.

### 19.4.3 Congruence classes

In other contexts the order is known to lie in one or more congruence classes modulo a given integer. For simplicity, assume that we know that $n \equiv n_0 \pmod{m}$. This can happen, for example, if we used a Schoof-like algorithm (cf. Section 17.2) to compute the order of the curve or of the Jacobian, but because of computational limitations we stop earlier and compute the order only modulo $m$. In this situation we usually also know the interval $[a, b]$ in which the order $t$ lies, where we can assume $a \equiv n_0 \pmod{m}$, and clearly $b - a > m$, otherwise the result is already uniquely determined. In this case we choose $s$ to be the smallest integer $\geqslant \sqrt{(b-a)/m}$ which is divisible by $m$. The baby steps $\beta_j$ are given by $h \oplus [-a - jm]g$ for $0 \leqslant j < s$. The giant steps are $[is]g$ for $0 \leqslant i < s$. When a match $h \oplus [-a - j^*m]g = [i^*s]g$ is found we get $t = a + i^*s + j^*m$. The algorithm is nearly identical to Algorithm 19.10. The complexity is $O(\sqrt{(b-a)/m})$.

The generalization to more congruence classes is just more complex from the technical point of view. The complexity of the corresponding method is $O(\sqrt{(b-a)r/m})$ where $r$ is the number of the given congruence classes modulo $m$.

## 19.5 Pollard's rho method

The methods which are presented here have been originally proposed for $\mathbb{F}_p^*$ (see [POL 1978]) and are based on the *birthday paradox*. Draw elements at random from a group $G$, always putting the element back after each draw: when an element is drawn that has already been drawn before we say we have a *match* or a *collision*.

**Theorem 19.11** If elements are drawn at random from $G = \langle g \rangle$ then the expected number of draws before a collision occurs is $\sqrt{\pi n/2}$ where $n = |G| = \text{ord}(g)$.

The basic idea is that we make a "random" *walk* $\{w_i\}_{i=0}^{\infty}$ in the considered group or semigroup until a match is found. If we remember how we computed the elements of the walk from $w_0$ then we hope to be able to recover some information about $w_0$. In order to do this we must pick elements in a deterministic way, but which in practice *behaves randomly with respect to the structure in which we are working*.

This can be achieved by a *random mapping* $\Phi$ of $G$ to itself, i.e., a mapping that is chosen uniformly at random from the set of the maps from $G$ to itself. Suppose then that $w_{i+1} = \Phi(w_i)$ where $\Phi$ is a random mapping. A pictorial description of the sequence $\{w_i\}_{i=0}^{\infty}$ is given by the Greek letter $\rho$: starting at the tail of the $\rho$ at some point the sequence will meet itself at some earlier point and loop from there on. In other words there exist positive integers $\mu$ and $\tau$ such that $w_i = w_{i+\tau}$ for all $i \geqslant \mu$. According to [HAR 1960], the expected values for $\mu$ and $\tau$ are around $\sqrt{\pi n/8}$. In practice, however, $\Phi$ is not a true random mapping, but such a behavior can be approached, as we shall see.

## 19.5.1 Cycle detection

Cycle detection algorithms in general do not exploit the group structure of $G$ in which the function $\Phi$ defines a random walk (the map $\Phi$, on the other hand, must be defined in terms of the group law). As a result, the methods described here in fact apply to any set $G$ on which an iterated map $\Phi$ is used to make random walks, and their utilization goes beyond DL and group order computations.

### 19.5.1.a Floyd's algorithm

It is not necessary to compare each new $w_i$ to all previous ones, which would make the method as slow as exhaustive search: for instance, according to *Floyd's cycle-finding algorithm* it suffices to compare $w_i$ to $w_{2i}$ for all $i$ (see [KNU 1997, §3.1, Ex. 6]). In fact, if $i$ is any multiple of $\tau$ that is larger than $\mu$, then $w_i = w_{2i}$, so any cycle will be detected. To avoid storing all the $w_i$ the algorithm is in fact implemented as follows, where only two intermediate values need to be stored.

---
**Algorithm 19.12** Floyd's cycle-finding algorithm

INPUT: An initial value $w_0$ and an iterating function $\Phi : G \to G$.
OUTPUT: An index $i > 0$ such that $\Phi^i(x) = \Phi^{2i}(x)$.

1. $x \leftarrow \Phi(w_0), y \leftarrow \Phi(x) = \Phi^2(w_0)$ and $i \leftarrow 1$
2. **while** $(x \neq y)$ **do**
3. $\quad i \leftarrow i+1, x \leftarrow \Phi(x)$ and $y \leftarrow \Phi^2(y)$
4. **return** $i$

---

Let the length of the prefix and of the loop be $\mu$ and $\tau$ respectively, as above. If $\tau \geqslant \mu$ the first collision will happen after $\tau$ iterations. If $\tau < \mu$ the match will be found after $\tau \lceil \frac{\mu}{\tau} \rceil$ iterations, which lies between $\mu$ and $2\mu$: the precise distribution depends on the group and on the mapping. Under the assumptions that $\Phi$ is a random mapping, and thus that the expected values of $\mu$ and $\tau$ are $\sqrt{\pi n/8}$ and with good probability relatively close, we estimate the expected number of iterations by $\frac{3}{2}\sqrt{\pi n/8} \approx 0.94\sqrt{n}$, and thus about $2.82\sqrt{n}$ evaluations of $\Phi$.

### 19.5.1.b Gosper's algorithm

Gosper's algorithm is item number 132 of the HAKMEM list [BEGO$^+$ 1972]. It has been designed to be used in the very specific context where it is difficult to recompute former values of the sequence $\Phi^k(w_0)$, because of the way the system is constructed. Such an example could be a pseudorandom number generator based on an iterative "black box" with a very limited interface. This context is quite far from that of order and discrete logarithm computations, but the beauty and ingenuity of the algorithm persuaded us that it should be popularized. It is a nice exercise to try to understand how it works and *why*.

Let $w_0$ be the initial value of the sequence, $L$ be an upper bound for the length of the cycle, and $T$ a table to keep $m := \lceil \lg L \rceil$ old values of $\Phi^k(w_0)$. Let $i$ be a counter of the number of times $\Phi$ has been applied. For each $i$ compare $\Phi^i(w_0)$ for equality with the first $s$ entries in the table $T$, where $s = \lceil \lg i \rceil$ (the number of bits necessary to represent $i$). If there is no match increment $i$ and (using this new value of the counter $i$) store $\Phi^i(w_0)$ into $T[r]$, where $r$ is the exponent to which 2 divides $i$ (the number of trailing zero bits in the binary representation of $i$). A match with the entry in position $e$ means the loop length is 1 more than the low $e + 2$ bits of $i - 2^{e+1}$. Note that if the bound $L$ is too small, then the algorithm will fail to find a loop. Otherwise, it will detect repetition before the third occurrence of any value, guaranteeing that the correct length of the loop is found and not a multiple of it.

---

**Algorithm 19.13** Gosper's cycle finding algorithm

INPUT: An initial value $w_0$, an iterating function $\Phi : G \to G$ and a bound $L$ on the period of the sequence $\{\Phi^i(w_0)\}$ that is to be found.
OUTPUT: Indexes $i \neq j > 0$ such that $\Phi^i(x) = \Phi^j(x)$.

1. $m \leftarrow \lceil \lg L \rceil$ and create the table $T[0, \ldots, m-1]$
2. $T[0] \leftarrow w_0, i \leftarrow 1$ and $z \leftarrow \Phi(w_0)$
3. **while** $(i \leqslant |G|)$ **do**
4.     $s \leftarrow \lceil \lg i \rceil$
5.     **for** $k = 0$ **to** $s - 1$ **do**
6.         **if** $(z = T[k])$ **then**
7.             $\tau \leftarrow 1 + \left((i - 2^{e+1}) \bmod 2^{e+2}\right)$
8.             **return** $(i, i + \tau)$
9.         **else**
10.             $i \leftarrow i + 1$
11.             $z \leftarrow \Phi(z)$
12.             let $r$ be the 2-adic valuation of $i$
13.             $T[r] \leftarrow z$
14. **return fail**

---

#### 19.5.1.c Brent's algorithm

An obvious problem with Floyd's cycle-finding algorithm lies in the fact that $w_i$ must be recomputed twice at least for large $i$ or, in other words, at every iteration step we have to evaluate $\Phi$ thrice. Brent [BRE 1980] improved on Floyd's cycle-finding algorithm. Brent's algorithm uses an auxiliary variable $z$, which holds $w_{\ell(i)-1}$, $\ell(i)$ being the largest power of two contained in the current index of the walk $i$, i.e., $\ell(i) = 2^{\lfloor \lg i \rfloor}$. For each newly computed step of the walk, we check whether $z = w_i$. Whenever the index $i$ equals a power of two minus one, we assign $z = w_i$. This is repeated until a match is found.

---

**Algorithm 19.14** Brent's cycle-finding algorithm

INPUT: An initial value $w_0$ and an iterating function $\Phi : G \to G$.
OUTPUT: Integers $i$ and $j$ such that $\Phi^i(w_0) = \Phi^j(w_0)$.

1. $z \leftarrow w_0, w \leftarrow w_0, i \leftarrow 0$ and $\ell \leftarrow 1$     [$\ell$ is always a power of two]
2. **while true do**
3.     $w \leftarrow \Phi(w)$ and $i \leftarrow i + 1$     [compute next step]
4.     **if** $w = z$ **then break**
5.     **if** $i \geqslant (2\ell - 1)$ **then** $z \leftarrow w$ and $\ell \leftarrow 2\ell$
6. **return** $(i, j \leftarrow \ell - 1)$

Under the assumption that $\Phi$ is a random mapping, the first match is *expected* to occur after $\sqrt{\pi n/2} \approx 1.25\sqrt{n}$ iterations. Brent analyzes his algorithm running time. He finds that the first match is *found* after an expected number of $\approx 1.9828\sqrt{n}$ iterations. This is a higher number of iterations than with Floyd's algorithm, in fact about twice as much, but in Floyd's algorithm $\Phi$ is evaluated three times per iteration, and only once in Brent's. The price for this reduction (of about one third) in evaluations of $\Phi$ before finding a match is a much higher number of comparisons, which can offset the advantages in some circumstances. The following observation yields a variant that performs much better in practice as it reduces the number of comparisons:

**Proposition 19.15** Let $i$ be the smallest index with $w_i = w_{\ell(i)-1}$. Then $i$ satisfies $\frac{3}{2}\ell(i) \leqslant i \leqslant 2\ell(i)$.

For a proof, see [COH 2000, Prop. 8.5.1].

We can now give the improved version of Brent's algorithm.

---

**Algorithm 19.16** Brent's improved cycle-finding algorithm

INPUT: An initial value $w_0$ and an iterating function $\Phi : G \to G$.
OUTPUT: Integers $i$ and $j$ such that $\Phi^i(w_0) = \Phi^j(w_0)$.

1. $z \leftarrow w_0 \; w \leftarrow w_0 \; i \leftarrow 0$ and $\ell \leftarrow 1$
2. **while true do**
3. $\quad w \leftarrow \Phi(w)$ and $i \leftarrow i+1$
4. $\quad$ **if** $w = z$ **then break**
5. $\quad$ **if** $i \geqslant (2\ell - 1)$ **then**
6. $\quad\quad z \leftarrow w$ and $\ell \leftarrow 2\ell$
7. $\quad\quad$ **while** $i < \left(\frac{3}{2}\ell - 1\right)$ **do** $w \leftarrow \Phi(w)$ and $i \leftarrow i+1$
8. **return** $(i, j \leftarrow \ell - 1)$

---

Variations of Brent's algorithm requiring slightly more storage and comparisons but less iterations can be found in [LESC 1984] and [TES 1998a]. The general idea is as follows: We have $r$ *cells* $A_1, \ldots, A_r$. Start with the element $w_0$ in each cell. At the $i$-th step of the algorithm, after we have computed $w_i$ we check whether this point lies in one of the $r$ cells. If it does, then we have found a match, otherwise suppose the element in $A_1$ is $w_j$. If $i \geqslant \alpha j$, for a fixed parameter $\alpha \geqslant 1$, shift the contents of $A_k$ to $A_{k-1}$ for $k = 2, \ldots, r$ and place $w_i$ into cell $A_r$. Teske found that with the choice of parameters $r = 8$ and $\alpha = 3$ the expected number of iterations before a match is detected is about $1.13$ times the expected number of iterations before a match occurs. Under the assumption that $\Phi$ is a random mapping, this is about $1.412\sqrt{n}$. The cells $A_1, \ldots, A_r$ should be implemented as a circular buffer. There are however better strategies, which we will describe now.

#### 19.5.1.d Sedgewick–Szymanski–Yao algorithm

Sedgewick, Szymanski, and Yao [SESZ[+] 1982] address the problem of optimizing worst-case performance with bounded memory. Their algorithm uses a table $T$ of size $m$, where $m$ is a free parameter. Another parameter $g$ is used.

Initially put $d = 1$. At the $i$-th iteration, if $i \bmod gd < d$, search for $w_i = \Phi(w_{i-1})$ in $T$. If $i$ is a multiple of $d$, store the pair $(w_i, i)$ in $T$. When there are no more free entries in $T$, before inserting another element double $d$ and erase from the table all entries $(w_j, j)$ where $j$ is not a multiple of the new value of $d$. The table $T$ should be implemented so as to asymptotically reduce the worst-case search time. (For example, as a balanced tree or using hashing.) Let $t_s$ be the time needed to perform one search in $T$, and let $t_\Phi$ be the time needed to evaluate $\Phi$ once. Then, the authors show, $g$ can be chosen as a function of $t_s, t_f$, and $M$, such that the algorithm's worst-case running time is $t_\Phi(\mu + \lambda)\big(1 + (t_s/Mt_\Phi)\big(1 + o(1)\big)\big)$. They also show that this worst-case performance is asymptotically optimal.

The next algorithm has worse worst-case performance, but better running time on average.

### 19.5.1.e Nivash's stack-based cycle detection

Nivash algorithm is introduced in [NIV 2004]. It is probably the most efficient known cycle detection algorithm on single processor machines.

**The basic method**

This algorithm requires that a total ordering $<$ be defined on the set $G$, and works as follows: keep a stack of pairs $(w_j, j)$, where, at all times, both the $j$'s and the $w_j$'s in the stack form strictly increasing sequences (with respect to the usual ordering of the natural integers and, for the $w_i \in G$, with respect to the ordering $<$). The stack is initially empty. At each step $i$, remove from the stack all entries $(w_j, j)$ where $w_j > w_i$. If a match $w_i = w_j$ is found with an element in the stack, the algorithm terminates successfully: the cycle length is equal to $i - j$. Otherwise, push $(w_i, i)$ on top of the stack, compute $w_{i+1} = \Phi(w_i)$, increase $i$ by one and continue.

We easily see that this algorithm always halts on the second occurrence of the element $z$ of the periodic part of the sequence that is minimal with respect to the given orderings. Let $w_{i_{\min}}$ be the first occurrence of $z$. This element is added to the stack the first time it appears, and is never removed. On the other hand, any other element in the cycle belongs to a pair that is greater than $(w_{i_{\min}}, i_{\min})$, so it will be removed (with probability that depends on the relative "magnitude" of the elements with respect to $w_{i_{\min}}$ according to the ordering in $G$, and at latest when $w_{i_{\min}}$ is encountered for the second time) before it has a chance to appear again.

Let $h$ be the number of evaluations of $\Phi$ of the algorithm before it terminates. Under the assumption that the $\Phi$ is a random mapping, the expected value of $h$ is $\mu + \tau\big(1 + \frac{1}{2}\big) = \frac{5}{2}\sqrt{\pi n/8} \approx 1.5666\sqrt{n}$. Under the same assumption, Nivash proves also that the expected size of the stack is $\ln h + O(1)$. Therefore, the algorithm only requires a probabilistic logarithmic amount of memory. Note that the search in the stack for the first element $w_j \geqslant w_i$ can be done via a binary search. The size of the stack can be estimated, as we just saw, and therefore it can be implemented by an array, handling the unlikely situation when it overflows by resizing and relocating it. There is therefore no need to explicitly erase or deallocate popped stack entries: they will simply be overwritten and only the stack size (i.e., the number of entries in it) needs to be kept. If the magnitudes of the elements of the sequence with respect to the ordering in $G$ present some regularities, Nivash suggests "randomizing" them by applying a fixed hash function to the values before comparing them.

**A partitioning technique**

The basic stack algorithm halts at a uniformly random point in the second loop through the sequence's cycle. In order to increase its probability of halting closer to the beginning of the second loop, which is especially important if loops may be quite long, while adding virtually no extra time penalty per step, we introduce the following partitioning technique: For some integer $m$, we divide $G$ into $m$ disjoint classes. This can be done, for example, according to the values of some bits in the internal representation of the elements of $G$. Ideally, the classes should have the same cardinality. We keep $m$ different stacks, one for each class of elements of $G$. Any new element in the sequence is compared only to the element in its corresponding stack, and pushed onto it; in particular any minimal element of a class, on its second occurrence, will collide with the first occurrence of the same element because they are pushed in the same stack. This method is called the *multi-stack* algorithm. The algorithm halts whenever the first of all the stacks detects a match. Each class contains its own cycle minimum, which is distributed uniformly and independently at random in the cycle. It follows that the expected number of evaluations of $\Phi$ before the algorithm halts is $\mu + \lambda\bigl(1 + 1/(m+1)\bigr)$. The size of each stack is about $\ln h - \ln m + O(1)$, hence the memory usage is roughly multiplied by $m$ with respect to the basic version.

---

**Algorithm 19.17** Nivash' cycle-finding algorithm

INPUT: An initial value $w_0$, an iterating function $\Phi : G \to G$, a number of stacks $K$ and a class function $\chi : G \to [0, \ldots, K-1]$.
OUTPUT: Integers $i$ and $j$ such that $\Phi^i(w_0) = \Phi^j(w_0)$.

1. create stacks $S[0], \ldots, S[K-1]$ containing pairs (element, index)
2. initialize stack indexes $p_0, \ldots, p_{K-1}$ to $-1$
3. $x \leftarrow w_0$
4. **while true**
5.     $\kappa \leftarrow \chi(x)$
6.     find smallest index $t \leqslant p_k$ such that element $(S[t]) \geqslant x$ or put $t = -1$
7.     **if** $t \neq -1$ **and** element$(S[t]) = x$ **then break**
8.     $i \leftarrow i+1$ and $p_k \leftarrow t+1$
9.     **if** $p_k$ is too large **then** resize $k$-th stack
10.     $S[p_k] \leftarrow (x, i)$
11.     $x \leftarrow \Phi(x)$
12. **return** $\bigl(i, \text{index}(S[t])\bigr)$

---

**Remark 19.18** In Line 1, the element of the pair is in $G$ and the index is a natural number. The initial depth of the stacks should be about $\frac{1}{2}\ln n$.

The multi-stack algorithm is especially useful on single processor computers with large memory. The *distinguished point technique* (see Section 19.5.4) can be used to find collisions early in the loop on multiple processors and, as Nivash himself observes, it offers in that situation a somewhat better time/memory trade-off. The multi-stack algorithm is also suited to the situations where the mapping $\Phi$ is not a random mapping and its cycles are short compared to the aperiodic part of the sequence. The (multi-)stack algorithm does not seem applicable to Pollard's rho factorization method (see Section 25.3.1).

### 19.5.2 Application to DL

The walk on $G$ is given by $\bigl\{w_i = [a_i]g \oplus [b_i]h\bigr\}_{i \geqslant 0}$ for known integers $a_i$ and $b_i$. A collision has the form $[a_i]g \oplus [b_i]h = [a_j]g \oplus [b_j]h$, so that

$$\log_g(h) = \frac{a_i - a_j}{b_j - b_i},$$

computed modulo the group order. According to [POL 1978], one way to create a 'random-ish' walk is the following. First, partition $G$ into three subsets $G_1$, $G_2$, and $G_3$ of approximately equal cardinality. This can be done fairly accurately and in a 'random' way exploiting the representation of the elements of $G$, for instance, using some fixed bits in the internal machine representation, in

## § 19.5 Pollard's rho method

other words *hashing* the elements of $G$. Take $w_0 = g$ (so $a_0 = 1$, $b_0 = 0$), and define $w_{i+1}$ as a function of $w_i$ as follows:

$$w_{i+1} = \Phi(w_i) := \begin{cases} h \oplus w_i & \text{if } w_i \in G_1 \\ [2]w_i & \text{if } w_i \in G_2 \\ g \oplus w_i & \text{if } w_i \in G_3 \end{cases} \Rightarrow (a_{i+1}, b_{i+1}) = \begin{cases} (a_i, b_i + 1) \\ (2a_i, 2b_i) \\ (a_i + 1, b_i) \end{cases} \text{respectively.}$$
(19.1)

Of course one can choose any other random mapping that allows an easy computation of the scalars $a_i$ and $b_i$.

### 19.5.3 More on random walks

As observed in [SASc 1985, TES 1998b] partitioning $G$ into only three sets does not in general lead to a truly random walk. This is reflected in the fact that the collision occurs on average later than expected. A truly random walk is difficult to achieve, but as the number of partitions is substantially increased, performance is improved and approaches the ideal one. One method is to use $r$-adding walks.

**Definition 19.19** Let $r > 1$ be a small integer and assume we have a partition $G_1 \cup \cdots \cup G_r$ of $G$ into subsets of approximately equal cardinality. For any $x \in G$ let $v(x)$ be the index with $x \in G_{v(x)}$. An *r-adding walk* is generated by an iterating function of the form $w_{i+1} = \Phi(w_i) = M_{v(w_i)} \oplus w_i$, where the elements $M_1, \ldots, M_r$ are of the form $M_s = [m_s]g \oplus [n_s]h$ with $m_s$ and $n_s$ chosen (possibly at random) in $[1, \ldots, n-2]$.

Usually $3 \leqslant r \leqslant 100$ and $v(x)$ is a hash function. From the definition it follows immediately that

$$w_i = [a_i]g \oplus [b_1]h \text{ with } a_{i+1} = a_i + m_{v(w_i)} \text{ and } b_{i+1} = b_i + n_{v(w_i)}.$$

These addings (whence the name) are defined modulo $n$. It is not strictly necessary to reduce $a_i$ and $b_i$ modulo $n$ — which is relatively inexpensive anyway — since their increase is only linear in the number of iterations and they will be therefore bounded by a constant times $\sqrt{n}$. This means that $r$-adding walks can be computed even if the group order is not known, by taking $m_s$ and $n_s$ bounded by some hopefully well-guessed large number. Since in some practical instances the group order is known to lie in an interval (for example, for elliptic and hyperelliptic curves), this can be done. Note that this would not be meaningful with mappings like (19.1) where the numbers $a_i$ and $b_i$ would grow exponentially with $i$. In particular, by taking $h = 0_G$ one can get a multiple of the logarithm $t = \log_g(0_G)$ which is at most off by a factor around $\sqrt{t}$, then one adjusts the result. This has been done by Sattler and Schnorr [SASc 1985] using 8-adding walks. Teske's investigations [TES 1998b] (the results have been slightly updated in [TES 2001a]) in elliptic curve (sub)groups allowed her to obtain matches after $\approx 1.452\sqrt{n}$ iterations using 20-adding walks. She also used a mix of 16 addings and 4 doublings, with similar performance.

### 19.5.4 Parallelization

The above methods can be run on $m$ processors in parallel, each starting with a different point $w_0^{(t)}$ for $t = 0, 1, \ldots, m - 1$. A speedup of a factor $\sqrt{m}$ can be expected.

A better parallelization is described in [QUDE 1990] — where it is actually used to find the first collisions in DES reported in the literature — and in [OOWI 1999]. The idea is to have a set $D$ of rarely occurring *distinguished points*.

There is a central server and $m$ client processors. The clients start with different points $w_0^{(t)}$ for $t = 0, 1, \ldots, m - 1$, which may be computed, in the DL case, as $[a_t]g \oplus [b_t]h$ for integers $a_t$ and $b_t$ which are different for each processor.

When a processor hits some element in $D$, it reports this fact to a central server, together with the element and the corresponding values of the scalars of $g$ and $h$. Then, the same processor starts afresh with a new trail. The starting points must of course always be different. Two strategies are possible:

1. Each client generates new starting points as $[a_t]g \oplus [b_t]h$ for integers $a_t$ and $b_t$ generated randomly, and the random number generator of each client is initialized with a different seed. The likelihood that the same point is reused is in practice negligible.

2. The server produces the unique starting points, to guarantee that they are all different. The points are generated either in packets, which are sent to each client when the client has used all the points in the previous batch (this can induce too much network traffic) or one by one (less network traffic, but the clients must wait until the server replies by delivering a new starting point after the finding of a distinguished point is reported).

In this way some collisions *between* trails can be detected. The central server applies a brute force approach to detect whether two of the distinguished points are in fact a match — usually aided by hashing. Nivash [NIV 2004] suggests using a partitioning technique (see Section 19.5.1).

One way of defining the set $D$ is to fix an integer $f$ and to define that $x \in D$ if and only if the $f$ least significant bits in the internal machine representation of $x$ are zero. This definition allows a fast test, and the size of $D$ can be easily monitored. By the theoretical analysis in [OOWI 1999] a speedup of a factor $m$ with $m$ processors can be expected.

The cardinality of $D$ is crucial for the performance. The more distinguished points there are, the easier it will be to detect collisions, but, on the other hand, the client processors will spend more time reporting the points to the server. Let $n = |G|$ and $\theta$ be the proportion of distinguished points. Further let $\tau_g$ and $\tau_r$ denote the costs of a group composition and of reporting the arrival on a distinguished point to the server, respectively. The expected running time (cf. [OOWI 1999]) on each processor is $\approx (\sqrt{\frac{\pi}{2}n}/m + 1/\theta)\tau_g + n_0\theta\tau_r$ where $n_0$ is the expected length of the trails (which shall also depend on $D$). We do not enter into details, but observe that Schulte–Geers [SCH 2000d] concludes that the cardinality of $D$ must be proportional to $\sqrt{|G|}$.

The DLP in subgroups of elliptic curves of about $2^{108}$ elements has been successfully solved by Pollard's parallelized rho algorithms distributed over the Internet. See http://www.certicom.com for the most recent successes. The extensive experimental work related to this challenge (see [ESSA+ 1998]) confirms the theoretically predicted linear speedup for the parallelized rho method. These implementations, whenever possible, also made use of automorphisms of the group (see the next section) to further speed up the computation.

## 19.5.5 Automorphisms of the group

In certain types of groups, a speedup up to a factor of $\sqrt{2}$ is obtained by means of the so-called inverse-point strategy. This strategy applies if the inverse of any group element can be computed very efficiently. Each group element is paired with its inverse: the iterating function must be defined with the property that if $\Phi(a) = b$ then $\Phi(-a) = -b$. If we use a distinguished point set $D$, this must fulfill the property that if $a \in D$ then also $-a \in D$. Further, we require from the representation of the group elements that it can be detected if the inverse of a previously submitted point is submitted. Of course the method would not work if such a collision would not help in revealing the discrete logarithm.

The inverse-point strategy has been successfully applied to groups of elliptic curves over finite fields [ESSA⁺ 1998, WIZU 1998]. It applies essentially unchanged to Jacobian varieties of hyperelliptic curves, but also to some groups that are outside the scope of the present book, like the XTR subgroup in the natural (i.e., not trace-based) representation [STA 2003, STLE 2003] (note that for the latter, which is a subgroup of a finite field, better algorithms belonging to the family of Index Calculus attacks (see Chapter 20) may apply). The fundamental fact here is that $a \mapsto -a$ is an automorphism of the elliptic curve. In fact, let $\psi$ be an automorphism of the group $G$ with the following properties:

- Almost all orbits of the elements of the group under $\psi$ have order $m$, say.
- It is easy (i.e., computationally negligible) to check whether two group elements belong to the same orbit.
- The automorphism acts like the multiplication by a known scalar $s$, say (coprime to the group order).

Usually, such a group automorphism is induced by an easily computable endomorphism of the algebraic variety underlying the considered group, said endomorphism satisfying a known polynomial equation. Suppose that the iterating function has been constructed in such a way that if $\Phi(a) = b$ then $\Phi(\psi(a)) = \psi(b)$. Then we can use such $\Phi$ to make the random walk and detect collisions between orbits, and not only between elements. Suppose we want to solve the DLP $[t]g = h$. If the random walk is given by the sequence $\{w_i = [a_i]g \oplus [b_i]h\}_{i \geqslant 0}$, then a collision $w_i = \psi^k(w_j)$ means that $w_i = [s^k]w_j$, hence

$$[a_i]g \oplus [b_i]h = [s^k a_j]g \oplus [s^k b_j]h$$

and the discrete logarithm is computed as

$$\log_g(h) = \frac{a_i - s^k a_j}{s^k b_j - b_i} \bmod |G|.$$

The theoretical speedup with this technique is up to a factor of $\sqrt{m}$. This claim is obvious, since the search for collisions is in fact made on the quotient of the group (as a set) modulo the equivalence relation induced by the orbits under the automorphism. The speedup is, however, in practice somewhat smaller, because the test for same orbit membership is not completely free, and it might be difficult to find an iterating function that acts almost like a random mapping on the orbits. Further difficulties (which are, however, in general not unsurmountable) include:

- If a distinguished point set is used, then it must be invariant under $\psi$ in order for the technique to be applicable
- If we want to use a (multi-)stack method a la Nivash, then the ordering should be on the group modulo the equivalence relation induced by $\psi$.

The Frobenius endomorphism and other automorphisms have been used for computing the DL in some classes of elliptic curves over fields of characteristic 2 [GALA⁺ 2000, WIZU 1998], such as Koblitz curves, and of hyperelliptic curves [DUGA⁺ 1999]. In particular, it is quite easy to set up an improved attack on elliptic and hyperelliptic Koblitz curves when the definition field of characteristic 2 is implemented using normal bases. In this particular case, distinguished point sets can be defined, for example, using the Hamming weight of some coordinates.

## 19.6 Pollard's kangaroo method

We have seen that there are variants of Shanks' baby-step giant-step method to solve the DLP or compute the order of an element when we know that the answer lies in a given interval, exploiting this information to get lower running times. We describe here a method due to Pollard [POL 1978, POL 2000] that achieves the same purpose while at the same time retaining the space efficiency advantages of Pollard's rho methods. The fundamental idea of this method is to have more random walks at the same time, and to watch when these collide. Since they are "deterministically" random as in Pollard's rho method (cf. Section 19.5), there is no need to check whether each element of each random walk collides with some element of the other walks: as with the rho method, checking for each possible collision would in fact slow down the approach and render it useless. Since the walks can be viewed as "jumping" inside the considered group, Pollard depicted them as if done by kangaroos jumping in that immense unknown expanse, which is the group where we want to solve the DLP. The first version of Pollard's kangaroo method, sometimes called the *lambda method*, is serial and works with a *tame kangaroo* $T$ and a *wild kangaroo* $W$ with respective starting points $w_0(T) = [b]g$ and $w_0(W) = h$. The first kangaroo is called tame because, in a figurative sense, we know where it starts, whereas we do not know where the wild kangaroo comes from. They will start jumping inside the group, and both will remember exactly the path they have followed. When a collision between the paths of the tame and the wild kangaroo happens, we say that the tame kangaroo has captured the wild one. If the paths of the two kangaroos cross, they will coincide after that event, because the next jump of each kangaroo is determined only by the current position. Usually, the tame kangaroo sets a trap after a certain number of jumps, then waits for the wild one: if the two paths cross, the wild kangaroo will fall in the trap. If the tame kangaroo successfully captures the wild kangaroo, then we can use the information from both about their travel to find "where" $h$ is, i.e., its discrete logarithm with respect to $g$.

### 19.6.1 The lambda method

Let $r > 1$ be an integer and $v : G \to \{1, \ldots, r\}$ be a hash function. The kangaroos follow $r$-adding walks of the form

$$w_{i+1}(K) = w_i(K) \oplus M\big(v\big(w_i(K)\big)\big), \text{ with } K = T \text{ or } W$$

with multipliers $M(j) = [s_j]g$ for *jump lengths* $s_j > 0$ of size $O(\sqrt{b-a})$ for all $j$. The *traveled distances* for each kangaroo

$$d_0(K) = 0 \text{ and } d_{i+1}(K) = d_i(K) + s_{v(w_i(K))}, \text{ with } i \in \mathbb{N}$$

are recorded. Note that $w_j(T) = [b + d_j(T)]g$ and $w_j(W) = h \oplus [d_j(W)]g$. The tame kangaroo is set off first and after a certain number of jumps, say $M$, it stops and installs a *trap* at the final spot $t_M$ (its distance from the start is then $d_M(T)$). Then $W$ is freed and starts jumping: after each jump of $W$ we check to see if it has fallen into the trap, i.e., if $w_M(T) = w_N(W)$ for some $N$. If this happens, a solution to $[t]g = h$ is computed immediately, namely $t = d_M(T) - d_N(W)$. After a certain number of steps the wild kangaroo is halted — because we may assume it has gone too far and is now in safe territory (in other words it has probably entered a cycle that does not include the trap) — and a new wild one starts jumping. Its starting point is $w_0(W) = hg^z$, with $z$ small and increasing with each new wild kangaroo. This fact about $z$ is important: we can imagine that the wild kangaroos start with "parallel" trails and we hope that at least one shall be caught. If $W$ falls into $T$'s trap, the paths of the two kangaroos have with high probability met earlier during the travel, from which point on the paths coincided. A graphical representation of this phenomenon resembles the Greek letter $\lambda$, whence the alternative name of this basic version of the kangaroo method.

Put $\xi = \sqrt{b-a}$. Van Oorschot and Wiener [OoWi 1999] show that the expected number of group operations is minimal if

1. The average of the $s_j$ is $\approx \xi/2$, and

2. The tame kangaroo $T$ makes about $0.7\xi$ jumps before installing the trap; and the maximal number of jumps done by each wild kangaroo is $\approx 2.7\xi$.

At that point either $W$ has fallen into $T$'s trap or is safe in the sense already described. The latter outcome happens on average 0.33 times, whereas the former has probability 0.75 after an expected number of $\approx 1.7\xi$ jumps. From this they get the expected value of $(0.7 + 0.33 \times 2.7 + 1.7)\xi \approx 3.3\sqrt{b-a}$ group operations. The algorithm needs storage only for the jump set and for the current positions of the two kangaroos. Using a set of distinguished points and slightly more storage one can obtain an algorithm with an expected total running time of $\approx 2\sqrt{b-a}$ group operations.

### 19.6.2 Parallelization

One can employ a distinguished point set $D$ to realize a parallel kangaroo algorithm in the same way as for the rho method [OoWI 1999].

Suppose then that we have $2m$ processors, each processor being the "home" to a kangaroo. There are two kangaroo *herds*, one consisting of $m$ tame kangaroos $\{T_1, \ldots, T_m\}$ and the other of $m$ wild ones $\{W_1, \ldots, W_m\}$. We use a global multiplier set, defined as in the serial variant, whose jump distances $s_j$ have mean value $\beta = \frac{m}{2}\sqrt{b-a}$. Let $\sigma \approx \frac{1}{m}\beta$ be an integer corresponding to the *spacing* between kangaroos of the same herd. The starting points of the kangaroos are given by

$$w_0(T_i) = \left[\tfrac{a+b}{2} + (i-1)\sigma\right]g \text{ and } w_0(W_i) = h \oplus [(i-1)\sigma]g, \text{ for } 1 \leqslant i \leqslant m/2.$$

Observe that the tame kangaroos start near the middle of the interval rather than its end $b$, as this leads to lowest average running times. The initial traveled distances are thus

$$d_0(T_i) = \frac{a+b}{2} + (i-1)\sigma \text{ and } d_0(W_i) = (i-1)\sigma, \text{ for } 1 \leqslant i \leqslant m/2.$$

Each kangaroo starts jumping and after each jump it is tested whether the kangaroo landed on a distinguished point, in which case a packet is sent to the central server consisting of: the distinguished point, the type (tame/wild) of kangaroo that has just landed there, and its traveled distance. The central server checks whether that point has already been previously submitted in order to detect a collision between a tame and a wild kangaroo, and otherwise stores it. Let $\tau_g$, respectively $\tau_r$, be the costs of a group composition and of reporting the arrival on a distinguished point, and $n_0$ the average length of trails before the algorithm terminates. Let, as usual, $\theta$ be the proportion of distinguished points. The expected running time (on each processor) is $\approx (\sqrt{b-a}/m + 1/\theta)\tau_g + n_0\theta\tau_r$. As for the rho method, the proportion of distinguished points must be chosen carefully.

With respect to the size of $D$, remarks similar to those of Section 19.5.4 apply here, in the sense that $|D|$ must be proportional to $\sqrt{b-a}$ (see [TES 2003, §6.4]).

Kangaroos of the same herd might collide, and such collisions are useless as they do not help recovering the discrete logarithm. Pollard [POL 2000] has developed a parallelized version where useless collisions cannot occur. He writes $m = A + B$ with $A$, $B$ coprime and $\approx m/2$. There are $A$ tame and $B$ wild kangaroos. The jump lengths for the multiplier set are multiples of $AB$ of the form $k_i AB$ where the $k_i$'s have average $\beta = \sqrt{(b-a)/(AB)}/2$. The starting points of the tame and wild kangaroos are

$$\left[\tfrac{a+b}{2} + (i-1)B\right]g \text{ with } 1 \leqslant i \leqslant A \text{ and } h \oplus [(j-1)A]g \text{ with } 1 \leqslant j \leqslant B,$$

respectively. Observe that the congruence

$$\frac{a+b}{2} + (i-1)B \equiv t + (j-1)A \pmod{AB}$$

has only one solution in $i$ and $j$ whence, for any residue class modulo $AB$, there is exactly one pair of kangaroos, one tame and one wild, which both travel in that class and thus can collide.

Pollard's variant avoids useless collisions at the price of preventing collisions from more than one tame/wild pair. Therefore the fact that the expected running time is about the same as for van Oorschot and Wiener's variant is not surprising. A big problem with Pollard's variant occurs with searches distributed over the Internet: if one of the players retires, the only possible tame/wild match might disappear. This makes the approach infeasible for such attacks. A running time analysis is found in [POL 2000].

The jump lengths $s_i$ and the jump length factors $k_i$ in the two versions should be chosen with some care. Randomly chosen integers in $\{1, 2, \ldots, \beta\}$ with no common factors are quite good, slightly better performance is obtained by picking different small powers of 2 including 1. Note further that kangaroos can enter in loops: in this case cycle recognition clearly does not help in recovering the discrete logarithms, so, ideally, this situation should be avoided. We can use a very inexpensive cycle-finding method like Brent's, or let the kangaroos start again with a new trail if after a certain number of jumps no distinguished point has been met, or use a distinguished point set that is dense enough to allow the server computer to detect most of the cycles by the clients. The exact values of these parameters must be chosen carefully for each case, by heuristic arguments or by doing some smaller experiments before the longer computation takes place.

### 19.6.3 Automorphisms of the group

Exactly as with Pollard's rho method we can use an efficiently computable automorphism $\psi$: all the kangaroos will in fact jump in the quotient set of $G$ modulo the equivalence relation determined by $\psi$. The expected speedup is about $\sqrt{k}$ where $k$ is the average size of the orbits.

# Chapter 20

# Index Calculus

*Roberto M. Avanzi and Nicolas Thériault*

**Contents in Brief**

| | | |
|---|---|---|
| 20.1 | **Introduction** | 495 |
| 20.2 | **Arithmetical formations** | 496 |
| | Examples of formations | |
| 20.3 | **The algorithm** | 498 |
| | On the relation search • Parallelization of the relation search • On the linear algebra • Filtering • Automorphisms of the group | |
| 20.4 | **An important example: finite fields** | 506 |
| 20.5 | **Large primes** | 507 |
| | One large prime • Two large primes • More large primes | |

## 20.1 Introduction

In view of Shoup's Theorem 19.2, methods that have lower time complexity than the methods explained in Chapter 19 cannot be generic. This applies not only to subexponential time methods, but also to methods whose complexity is still exponential and of the form $O(|G|^C)$ with $C < 1/2$.

In practice, we always work with a concrete representation of a given group, and many different types of groups share some common traits. As a consequence, there are methods that can be described in some generality for groups with additional properties and which attain much better complexity than the best generic algorithms. For some types of groups, these methods yield subexponential time algorithms, and for others, such as the Jacobians of hyperelliptic curves of moderate genus, the resulting methods are exponential with $C < 1/2$. These algorithms belong to the family of index calculus algorithms.

The idea behind index calculus is quite old: [ODL 1985] ascribed it to Western and Miller, but, as McCurley pointed out [MCC 1990], the idea goes back a few decades earlier to the work of Kraitchik [KRA 1922, KRA 1924]. Index calculus stems from the observation that if

$$\bigoplus_{i=1}^{r} [e_i]g_i = 1 \tag{20.1}$$

holds for some elements $g_i$ of a group $G$ (of order $N$) generated by an element $g$, and suitable

scalars $e_j$, then
$$\sum_{i=1}^{r} e_i \log_g(g_i) \equiv 0 \pmod{N} \, . \tag{20.2}$$

If we are able to obtain many equations of the form (20.1) with at least one of them involving an element for which the discrete logarithm is known, for example $g$ itself, and the number of the $g_i$'s is not too large, then we can solve the system (20.2) by linear algebra for the $\log_g(g_i)$'s. The set $\{g_1, \ldots, g_r\}$ is called the *factor base*. If we include an element $h$ in the factor base for which we know that $h = [t]g$, but $t$ itself is unknown, then we might hope to recover $t$.

Finding enough equations of the form (20.1) is equivalent to computing the structure of $G$ as $\mathbb{Z}$-module: If $\mathbb{Z}^r$ is the free abelian group generated by base elements $\{X_1, \ldots, X_r\}$ and $L$ is the lattice in $\mathbb{Z}^r$ generated by the relations $\prod_{i=1}^{r} X_i^{e_i} = 1$ corresponding to the equations (20.1), then

$$\begin{aligned} \Phi : \mathbb{Z}^k &\to G \\ (e_1, \ldots, e_k) &\mapsto [e_i]g_1 \oplus \cdots \oplus [e_k]g_k \end{aligned} \tag{20.3}$$

is a homomorphism with kernel $L$ so that $\mathbb{Z}^k/L \simeq G$.

Ideally, one has to prove that for a suitable choice of the factor base, given a randomly chosen element of $G$ there is a high probability that it can be written as a linear combination of a small number of elements of the factor base and with small coefficients. The index calculus methods achieve a subexponential complexity depending on the ability to efficiently generate such relations, which is often the dominant part of these algorithms.

Subexponential index calculus algorithms have been developed for a variety of discrete logarithm problems, for instance finite fields and Jacobians of hyperelliptic curves of large genus. Notable examples where they have *not* been made to work are elliptic curve discrete logarithms and discrete logarithms in Jacobians of genus 2 hyperelliptic curves.

## 20.2 Arithmetical formations

We follow here the presentation of Enge and Gaudry [ENGA 2002].

**Definition 20.1** [KNO 1975] Let $\mathcal{P}$ be a countable set, whose elements are called *primes*. An *additive arithmetical semigroup* is a free abelian monoid $\mathcal{M}$ over $\mathcal{P}$ together with an equivalence relation $\sim$ that is compatible with its composition law, such that $G$ is isomorphic to $\mathcal{M}/\sim$.

Each element $g$ of $G$ is represented by an unique element $\imath(g)$ of $\mathcal{M}$ such that the isomorphism $G \to \mathcal{M}/\sim$ is given by $g \mapsto \imath(g)/\sim$.

A *size map* is a homomorphism of monoids norm $: (\mathcal{M}, \oplus) \to (\mathbb{R}, +)$, and as such it is completely determined by its values at the primes of $\mathcal{M}$. We shall always assume that all primes $p \in \mathcal{P}$ have positive size. The size map is also applied to the elements of $G$ via $\imath$. The size of an element $g \in G$ (respectively $m \in \mathcal{M}$) is denoted by $|g|$ (respectively $|m|$).

Such a group $G$, together with the monoid $\mathcal{M}$, the equivalence relation $\sim$, the representation map $\imath$ and a size map — in other words the quintuple $(G, (\mathcal{M}, \cdot), \sim, \imath, |\cdot|)$ — is then called an *arithmetical formation*, or *formation* for short.

We assume that the elements of $G$ are represented by bit-strings associated to the corresponding elements of $\mathcal{M}$ and of length bounded by some constant $r$ and that all generic operations in $G$ (i.e., the operations listed in Definition 19.1) are performed in time polynomial in $r$.

**Definition 20.2** A *smoothness bound* $B$ is a positive integer and we denote by $\mathcal{M}_B$ (respectively $\mathcal{P}_B$) the set of elements of $\mathcal{M}$ (respectively $\mathcal{P}$) of size not larger than $B$.

## § 20.2 Arithmetical formations

We denote by $n_B$ the cardinality of $\mathcal{P}_B$ and by $n'_B$ the cardinality of $\mathcal{M}_B$. An element of $G$ is called $B$-smooth if the decomposition of its representation in $\mathcal{M}$ involves only primes in $\mathcal{P}_B$.

When possible, the factor base is defined by a smoothness bound and is denoted $\mathcal{P}_B$. In some cases, however, (see Sections 21.2.2 and 21.3), the factor base cannot be defined only in terms of a smoothness bound. In these cases, the factor base will be denoted $\mathcal{B}$.

In our context, we require that $n'_B$ is finite and of cardinality polynomial in $B$, and that the elements of $\mathcal{M}_B$ can be enumerated in a time polynomial in $B$ and linear in $n'_B$. We also require that an element $m \in \mathcal{M}$ can be tested for being prime in time polynomial in $|m|$ and linear in $n'_{|m|}$ (by trial division by all elements of norm smaller than $|m|$, for instance). Thus, $\mathcal{P}_B$ can be constructed in time polynomial in $B$ and quadratic in $n'_B$.

It should be possible to test elements of $G$ for $B$-smoothness and decompose them into primes in time $\tilde{O}(n'_B)$ (in practice, this can be done even faster).

### 20.2.1 Examples of formations

#### Prime fields

The multiplicative group $G = \mathbb{F}_p^*$ can be represented as $(\mathbb{Z}, \times)/\sim$ where $n_1 \sim n_2$ if and only if $n_1 \equiv n_2 \pmod{p}$ and $\mathcal{P}$ is the set of rational prime numbers. The size of an element of $G$ is the logarithm of the element of $\mathbb{N}$ by which it is represented. In practice, the bit length $\lceil \lg n \rceil$ of the integer $n$ is used.

#### Nonprime finite fields

The multiplicative group $G = \mathbb{F}_{p^d}^*$ with $d > 1$ can be represented by the polynomials of degree less than $d$ over $\mathbb{F}_p$. Let $f$ be a monic irreducible polynomial over $\mathbb{F}_p$ and let $\mathcal{M}$ be the multiplicative monoid of the ring of polynomials over $\mathbb{F}_p$ under the usual polynomial multiplication as composition. There exists a field isomorphism

$$\psi : \mathbb{F}_{p^d} \xrightarrow{\sim} \mathbb{F}_p[X]/\bigl(f(X)\bigr)$$

such that for each element $g \in \mathbb{F}_{p^d}$ there is a unique $u \in \mathbb{F}_p[X]$ of degree less than $d$ with $\psi(g) = U + (f)$. Write $\imath(g) = U$: the map $\imath$ can be restricted to a map $G \to \mathcal{M}$, because the inverse image of $0 \in \mathbb{F}_p[X]$ under $\imath$ consists of the zero of $\mathbb{F}_{p^d}$ alone. We can then represent $G$ as $\mathcal{M}/\sim$ where $U_1 \sim U_2$ if and only if $U_1 \equiv U_2 \pmod{f}$. The set $\mathcal{P}$ of primes consists of the set of monic irreducible polynomials over $\mathbb{F}_p$ together with a multiplicative generator for $\mathbb{F}_p^*$, embedded in the obvious way in the Cartesian product. The size of an element $g \in \mathbb{F}_{p^d}$ is now defined as $\deg \imath(g)$.

#### Jacobians of hyperelliptic curves

Let $C$ be a hyperelliptic curve of genus $g$ over a finite field $K$ of characteristic $p$, and consider the group $G = \operatorname{Pic}_C^0(K)$ of the divisor classes of $C$ over $K$. Here, $\mathcal{P}$ is the set of irreducible divisors. It is known that each element of $G$ can be represented by a $K$-rational divisor of degree at most $g$.

The set of primes is defined as the set of principal divisors whose effective divisors are irreducible over $K$. The latter can be single $K$-rational points, or sums of all Galois conjugates over $K$ of a non-$K$-rational point: In other words, if a divisor $D$ has Mumford representation $[u_D, v_D]$, $D$ is prime if and only if the polynomial $u_D$ is irreducible over $K$. The size of a divisor $D$ is the degree of $u_D$.

## 20.3 The algorithm

One possible way of describing the basic form of the index calculus method is the following one. There are in fact a few variants, but they differ only in minor details.

---

**Algorithm 20.3** Index calculus

INPUT: A group $G$ of order $N$, two elements $g, h \in G$ with $h \in \langle g \rangle$.
OUTPUT: An integer $t$ with $h = [t]g$.

---

1. **Construction of the factor base**
   Choose a smoothness bound $B$ and let the factor base be the set $\mathcal{P}_B = \{\pi_1, \ldots, \pi_{n_B}\}$ of the $B$-smooth primes of $G$.

2. **Produce relations**
   They are equalities of the form
   $$[a_i]g \oplus [b_i]h = \bigoplus_{j=1}^{n_B} [e_{i,j}]\pi_j, \text{ for } i = 1, 2, \ldots$$
   
   Put $c = n_B$. Construct a matrix $A$ with $c$ rows defined as the row vectors
   $$(e_{i,1}, e_{i,2}, \ldots, e_{i,c}), \text{ for } i = 1, 2, \ldots, c.$$
   
   Store the vectors $\mathbf{a} = (a_1, a_2, \ldots, a_c)$ and $\mathbf{b} = (b_1, b_2, \ldots, b_c)$.
   Put
   $$\mathbf{v} = (e_{c+1,1}, e_{c+1,2}, \ldots, e_{c+1,c}).$$
   
   Possibly, process the matrix and the vectors $\mathbf{a}$, $\mathbf{b}$ and $\mathbf{x}$ to make $c$ smaller.

3. **Linear algebra**
   Compute a solution to $\mathbf{x}A \equiv \mathbf{v} \pmod{N}$ or find an element $\mathbf{x}$ of the kernel of $A$, i.e., $\mathbf{x}A \equiv 0 \pmod{N}$.

4. **Extract solution**
   The matrix $A$ and the vectors $\mathbf{a}$, $\mathbf{b}$ satisfy by construction the following formal relation in $G$ (where the apex $t$ denotes transposition):
   $$(\mathbf{a}^t \ \mathbf{b}^t) \times \begin{bmatrix} g \\ h \end{bmatrix} = A \times \Pi \text{ where } (\mathbf{a}^t \ \mathbf{b}^t) = \begin{bmatrix} a_1 & b_1 \\ a_2 & b_2 \\ \vdots & \vdots \\ a_c & b_c \end{bmatrix} \text{ and } \Pi = \begin{bmatrix} \pi_1 \\ \pi_2 \\ \vdots \\ \pi_c \end{bmatrix}.$$
   
   Let $\mathbf{x} = (x_1, \ldots, x_c)$ be the vector found in Line 3. Let $\alpha$, respectively $\beta$ be equal to the inner product $\mathbf{x}\mathbf{a}^t$, respectively $\beta = \mathbf{x}\mathbf{b}^t$.
   **if** $\mathbf{x}A = 0$ **then**
   $\mathbf{x}A\Pi = 0$ and we obtain $(\alpha, \beta) \times \begin{bmatrix} g \\ h \end{bmatrix} = 0$, i.e., $[\alpha]g \oplus [\beta]h = 0$.
   Therefore $\log_g(h) = -\dfrac{\alpha}{\beta} \mod N$ provided that $\gcd(\beta, N) = 1$.
   **if** $\mathbf{x}A = \mathbf{v}$ **then**
   $[\alpha]g \oplus [\beta]h = \mathbf{x}A\Pi = \mathbf{v}\Pi = [a_{c+1}]g \oplus [b_{c+1}]h.$
   Therefore $\log_g(h) = -\dfrac{\alpha - a_{c+1}}{\beta - b_{c+1}} \mod N$

**Remark 20.4** If we want to solve a discrete logarithm problem in a proper subgroup $G_0$ of order $N_0$ of $G$, we can perform all the computations involving integers modulo $N$ (keeping track of $a_i$ and $b_i$, the linear algebra step, the final division) modulo $N_0$ instead — this is clear because the final result should be reduced modulo $N_0$ anyway — hence all the operations that take place in the ring $\mathbb{Z}/N\mathbb{Z}$ can be replaced with operations in the ring $\mathbb{Z}/N_0\mathbb{Z}$. If $N_0$ is, as in most applications, prime, then all the computations, including the linear algebra, take place in the finite field with $N_0$ elements.

When analyzing an index calculus variant, several problems have to be addressed. If $B$ is too small, the time to find relations will probably be too large; on the other hand, if $B$ is too large, the linear algebra step will be too expensive. The construction of the factor base has running time at most $\tilde{O}((n_B)^2) \leqslant \tilde{O}((n'_B)^2)$ (by enumerating the elements of norm at most $B$ and checking them for primality).

### 20.3.1 On the relation search

For the second step of the algorithm, we must take into account the probability of finding a relation. This probability is about $|G_B|/|G|$ where $G_B$ denotes the set of $B$-smooth elements of $G$. Candidate relations can be generated by random walks of the form $[a]g \oplus [b]h$ in the group $G$ and they all must be tested for smoothness, which takes an expected time $\tau_s$ per test. All random walk techniques seen in Section 19.5.3 can be of course adopted here.

Avanzi and Thériault [AvTh 2004] have a strategy that can be very efficient in many situations. It is often much faster to compute several group operations simultaneously than to compute them sequentially. For example, on elliptic and hyperelliptic curves using affine coordinates, Montgomery's trick (Section 11.1.3.c) can be used to perform the (independent) inversions in parallel. The obvious application is to perform several random walks simultaneously on a single processor, but several simultaneous group operations can be sped up further if one of the arguments is the same in all operations.

As in the $r$-adding walk method (Section 19.5.3), a set $\{M_1, \ldots, M_r\}$, with $M_s = [m_s]g \oplus [n_s]h$ fixed and $r$ registers $z_1, \ldots, z_r$, which are initialized, for example, as $z_i = M_i$ for all $i$. One index $s$ with $1 \leqslant s \leqslant r$ is picked, for example, by taking the value of a hash function from the group into the index set $\{1, 2, \ldots, r\}$ on the element $z_1$, and a single processor puts $w = z_s$ and computes $z_1 = M_1 \oplus w, z_2 = M_2 \oplus w, \ldots, z_r = M_r \oplus w$. These elements are then checked for smoothness, and, if the corresponding relation is smooth, it is added to the relation set. After one such step, if squarings (doublings) are cheaper than multiplications (additions), then all the elements $z_i$ can be squared (doubled) one or more times, and the results checked for smoothness. It is understood that these squarings (doublings) are also performed in parallel. Because of the seemingly random behavior of the smoothness property with respect to the group operations, this method will bring a noteworthy speedup (the factor is $r/r'$, where $r'$ is the time to compute $M_1 \oplus w, \ldots, M_r \oplus w$ for $1 \leqslant i \leqslant r$ relative to the time to compute just one element alone). If squarings or doublings are faster than multiplications or additions, then the speedup due to this strategy is even faster. If multiplying (respectively adding) and multiplying by the inverse (respectively subtracting) have similar costs, and the simultaneous computations of $ab$ and $ab^{-1}$ (respectively of $a+b$ and $a-b$) share some partial results, one can keep $2r$ registers $z_1, \ldots, z_{2r}$ and compute $z_1 = M_1 \oplus w, z_2 = M_2 \oplus w, \ldots, z_r = M_r \oplus w$ together with $z_{r+1} = M_1 \ominus w, z_{r+2} = M_2 \ominus w, \ldots, z_{2r} = M_r \ominus w$. By means of this we can compute more candidate relations in the same amount of time — and therefore find relations more quickly.

In general, regardless of how the relations are sought, one can choose to find $kc$ relations for a suitable $k = \tilde{O}(\lg c)$ and observe that there is an appreciable probability that $c-1$ relations are linearly independent [ENGA 2002]: This means that $\tilde{O}(n'_B|G|/|G_B|)$ candidates have to be tested. Another approach for generating relations is due to Dixon: All relations but one are found upon checking for smoothness multiples $[a]g$ of $g$, for several values of $a$, which can be chosen at random. The last relation is of the form $h \oplus [v]g$ for a suitable, possibly also randomly chosen, $v$. *This has the additional advantage that the relations involving only $g$ and not $h$ can be used later to compute the logarithms of other elements.*

### 20.3.2 Parallelization of the relation search

Relation search can be done in parallel by several machines, since the linear independence of the relation is not checked after each new relation is found. This has the advantage that, with $n$ machines, the total amount of time to find the desired number of relations is reduced by a factor $n$. Of course, at some point the relations must be sent to a central server, and time spent in network traffic must be taken into account too.

### 20.3.3 On the linear algebra

#### 20.3.3.a Complexity

In the third step of the algorithm, the linear dependency is found modulo prime divisors of $|G|$ (but usually this number is just a large prime). The matrix $A$ is sparse, because the elements of $G$ have bounded norm, so there are $O(\lg c) = O(\lg n'_B)$ (recall that $c$ is the dimension of the matrix) nonzero entries in each row for $c$ large enough. Sparse matrix techniques such as Wiedemann's or Lanczos' (described in the next sections) can be used: In this case, the running time of this step is $O(c^2 + c\omega)$, where $\omega$ is the total number of nonzero entries among all the $e_{i,j}$ and $a_i, b_i$, i.e., $\tilde{O}(c^2)$. The complexity of the algorithm, apart from the time required to factor the order of $G$ (which we assume to be known and factored) is then

$$\tilde{O}\left((n'_B)^2 + n'_B \frac{|G|}{|G_B|} \tau_s\right). \tag{20.4}$$

In particular, the smoothness bound $B$ and the complexity $\tau_s$ play an important role in determining the overall complexity of an index calculus algorithm for a specific type of group. These considerations lead to the following result:

**Theorem 20.5** (based on [ENGA 2002]): Assume that the following smoothness result holds for the group $G$: The bound $B$ can be chosen such that

$$n'_B = L_{|G|}\bigl(1/2, \rho + o(1)\bigr)$$

and

$$\frac{|G|}{|G_B|} = L_{|G|}\bigl(1/2, \sigma + o(1)\bigr)$$

for some constants $\rho, \sigma > 0$.

Also, suppose, that the smoothness test in $G$ can be done in time $\tilde{O}\bigl((n'_B)^\tau\bigr)$ for a constant exponent $\tau$. Then, solving the discrete logarithm problem in $G$ requires

$$\tilde{O}\bigl(L_{|G|}\bigl(1/2, \max\{2\rho, 1 + (1+\tau)\rho + \sigma\} + o(1)\bigr)\bigr)$$

operations in $G$.

### 20.3.3.b Methods

We now review the linear algebra methods that can be used for index calculus. Recall that we have a large matrix $A$ with $c$ columns and rows. The goal is to find an element of the kernel of $A$, i.e., a vector $\mathbf{x}$ such that $\mathbf{x}A = 0$ (more precisely $\mathbf{x}A \equiv 0 \pmod{|G|}$, but we assume that our group has prime order $N = p$, and therefore we obtain a linear system over $\mathbb{F}_p$). This is a very well-investigated problem and a complete treatment would definitely be outside the scope of this book. The matrix obtained is sparse, and in most cases the number of nonzero entries in the matrix is very small. We observe two very typical matrix types that arise in this context:

1. For the index calculus in extension fields (respectively prime fields), the number of entries per row is the number of factors in the polynomial over the base field that represents the field element (respectively the number of prime factors in the integer representing the field element).

   The "smaller" primes appear more often, and usually with higher multiplicities, hence the matrix is "denser" in the first columns. This is clearly true in the prime field case. In the extension field case we note that the case of binary fields has been investigated in more detail, and since the factor base must contains polynomials over $\mathbb{F}_2$ of several degrees, those of smaller degree will appear more often. Under these conditions, the well-known structured Gaussian elimination can reduce the size of the system by a considerable amount.

2. For hyperelliptic curves of small genus $g$ and the version of index calculus presented in Algorithm 20.3, the nonzero entries are scattered in a rather homogeneous way in the matrix. In this case, the structured Gaussian elimination has only a limited impact.

To complete the solution of the system, several algorithms can be used. For each type of discrete logarithm problem to be solved, a careful comparison of the algorithms by Lanczos and Wiedemann, along with their "block" variants, is necessary, and the best algorithm must be considered. This is not a trivial task. We will describe the basic versions of these two methods below. Further methods to be taken into account are the conjugate gradient method (which can be viewed as two pipelined Lanczos algorithms) and Lambert's variant (see [LAM 1996]).

The "block" variations try to minimize scattered memory reads and to replace them with consecutive reads as much as possible, and are also used in numerical analysis. They replace vector-matrix multiplications by multiplications of several vectors by the same matrix at once, thus also improving cache locality and in fact greatly reducing accesses to the main memory. We will not enter into details in their description, just as we will not deal with the parallelization of such methods. In the context of discrete logarithm computation, see [LAM 1996], [THO 2001], [THO 2003], and references therein for more details and literature on the subject.

### 20.3.3.c Wiedemann's method

This method [WIE 1986] solves a system $\mathbf{x}A = \mathbf{v}$ by building a Krylov subspace generated by the vectors

$$\mathbf{v}, \mathbf{v}A, \mathbf{v}A^2, \mathbf{v}A^3, \ldots \tag{20.5}$$

Such a sequence is eventually recurrent, which is evident from the Cayley–Hamilton formula, which says that a matrix satisfies its characteristic polynomial. Let the minimal polynomial for this sequence be $f(T)$. In order to determine $f(T)$, Wiedemann's method picks a random vector $\mathbf{u}$ and feeds the sequence

$$\mathbf{v}\mathbf{u}^t, \mathbf{v}A\mathbf{u}^t, \mathbf{v}A^2\mathbf{u}^t, \mathbf{v}A^3\mathbf{u}^t, \ldots \tag{20.6}$$

into the Berlekamp–Massey algorithm [MAS 1969], which determines the minimal polynomial of the sequence. When applied to the sequence (20.6) the polynomial resulting from the Berlekamp–

Massey algorithm must divide the minimal polynomial of the sequence (20.5) and also the minimal and characteristic polynomials of $A$. Hence by computing the minimal polynomials of sequences with various $\mathbf{u}$'s, some information on the factors of the minimal polynomial of the sequence (20.5) is discovered.

Under the assumption that $\mathbf{u}$ is randomly chosen, Wiedemann shows that it is quite likely that after a certain number of factors $f^{(i)}(T)$ of the minimal polynomial $f(T)$ have been found, $f(T)$ itself can be reconstructed as the least common multiple of these factors. More precisely, the success probability is 70% with just three factors.

In practice, the following accumulation procedure is used to determine $f(T)$:

1. Put $\mathbf{v}_0 = \mathbf{v}$ and let $f^{(0)}(T)$ be the minimal polynomial of the sequence

$$\mathbf{v}_0 \mathbf{u}_0^t, \ \mathbf{v}_0 A \mathbf{u}_0^t, \ \mathbf{v}_0 A^2 \mathbf{u}_0^t, \ \mathbf{v}_0 A^3 \mathbf{u}_0^t, \ldots$$

2. If $\mathbf{v}_0 f^{(0)}(A) \neq 0$, then put $\mathbf{v}_1 = \mathbf{v}_0 f_0(A)$, select another vector $\mathbf{u}_1$, compute the minimal polynomial $f^{(1)}(T)$ of the sequence

$$\mathbf{v}_1 \mathbf{u}_1^t, \ \mathbf{v}_1 A \mathbf{u}_1^t, \ \mathbf{v}_1 A^2 \mathbf{u}_1^t, \ \mathbf{v}_1 A^3 \mathbf{u}_1^t, \ldots$$

3. Repeating this, various factors $f^{(i)}(T)$, $i = 0, 1, 2, \ldots$ of $f(T)$ are found and $f(T)$ is just their product.

If the degree of $f^{(0)}(T)$ is smaller than that of $f(T)$, then the first pass, suitably pipelined with Berlekamp–Massey's algorithm, will terminate earlier. The computation of $f^{(1)}(T)$ will be faster because we have an easy bound on its degree, i.e., $c - \deg f^{(0)}(T)$, and the sequence $\{\mathbf{v}_1 A^j \mathbf{u}_1^t\}_{j \geqslant 0}$ must be at most twice that degree for the Berlekamp–Massey algorithm to determine $f^{(1)}(T)$ uniquely. Similar considerations and bounds hold for the successive factors $f^{(i)}(T)$.

Once $f(T)$ has been determined, the solution $\mathbf{x}$ or an element of the kernel of $A$ can be computed as follows:

- If the constant term of this minimal polynomial $f(T)$ is zero, $\mathbf{v}$ times $f(T)/T$ evaluated with $T = A$ yields an element of the kernel. This null vector is not trivial because of the minimality of $f(T)$.
- If $f(0) \neq 0$, then put $g(T) = \dfrac{f(T) - f(0)}{f(0) T}$. It is immediate to verify that $\mathbf{z} = -\mathbf{v} g(A)$.

Wiedemann's algorithm will, at best, be completed in a single pass, requiring $2c$ multiplications by $A$, where $c$ is the dimension of $A$, before the Berlekamp–Massey algorithm determines the (factor of the) minimal polynomial. The time required by the Berlekamp–Massey algorithm is negligible with respect to the time necessary to form the generating vectors of the Krylov subspace. The vectors $A^i b$ cannot be stored, so they must be recomputed in the last phase. Hence, at best $3c$ multiplications by $A$ are required. As a consequence of the remarks after the description of the accumulation approach for determining $f(T)$, the worst case cost is not much worse than the best case. Upon closer analysis, one sees that the Wiedemann method needs to store, apart from the matrix and the scalar quantities, only four vectors.

### 20.3.3.d Lanczos' method

Lanczos' method solves an equation $\mathbf{x}S = \mathbf{y}$ for $\mathbf{x}$ where $S$ is a symmetric matrix. To use it for solving $\mathbf{x}A = \mathbf{v}$ with $A$ nonsymmetric we set $S = AA^t$ and $\mathbf{y} = \mathbf{v}A^t$, then solve $\mathbf{x}S = \mathbf{y}$. The solution obtained may actually differ from the desired solution by an element of the kernel of $A^t$.

In applying this method to finding a vector in the kernel of $A$, as in the index calculus method, the matrix will have more columns (relations) than rows (primes). If it has full column rank $c$, then the solution of the system $\mathbf{x}S = 0$ with $S = AA^t$ will be a solution to the original system $Az = 0$,

If the rank is not full, this becomes unlikely, hence in practice one adds more relations to increase the likelihood that the column rank is in fact maximal.

Put

$$\mathbf{w}_0 = \mathbf{y}, \quad \mathbf{v}_1 = \mathbf{w}_0 S, \quad \mathbf{w}_1 = \mathbf{v}_1 - \frac{\langle \mathbf{v}_1, \mathbf{v}_1 \rangle}{\langle \mathbf{w}_0, \mathbf{v}_1 \rangle} \mathbf{w}_0$$

and define the following iterations

$$\mathbf{v}_{i+1} = \mathbf{w}_i S, \quad \mathbf{w}_{i+1} = \mathbf{v}_{i+1} - \frac{\langle \mathbf{v}_{i+1}, \mathbf{v}_{i+1} \rangle}{\langle \mathbf{w}_i, \mathbf{w}_{i+1} \rangle} \mathbf{w}_i - \frac{\langle \mathbf{v}_{i+1}, \mathbf{v}_1 \rangle}{\langle \mathbf{w}_{i-1}, \mathbf{v}_i \rangle} \mathbf{w}_{i-1}.$$

Note that the matrix $S$ is in general not sparse and it is therefore never computed explicitly: $\mathbf{v}_{i+1} = \mathbf{w}_i S$ is in fact computed by two multiplications of a vector by a sparse matrix as $\mathbf{v}_{i+1} = \mathbf{w}_i A A^t$.

The algorithm stops when $\mathbf{w}_j$ is found to be self-conjugate, i.e., $\langle \mathbf{w}_j, \mathbf{w}_j S \rangle = 0$: this expression appears in fact already as the first denominator of the above iterative formula.

If $\mathbf{w}_j = 0$, then a solution is

$$\mathbf{x} = \sum_{i=0}^{j-1} \frac{\langle \mathbf{w}_i, \mathbf{y} \rangle}{\langle \mathbf{w}_i, \mathbf{v}_{i+1} \rangle} \mathbf{w}_i$$

which is accumulated partially as the algorithm progresses.

Over the reals, if $S$ is positive definite and $\mathbf{w}_j$ is self-conjugate, then $\mathbf{w}_j$ must be 0, but this is not necessarily the case in finite characteristic where there exist nonzero self-conjugate vectors. If this happens, then the algorithm fails. When the Lanczos algorithm is used to find a solution modulo 2, such as in the number field sieve for factorization (see Section 25.3.4.g), Montgomery [MON 1995] proposed to keep subspaces instead of vectors, thus deriving a block version of the Lanczos algorithm. This has the dual advantage of greatly reducing the risk of zero denominators in the iteration above and of speeding up the algorithm by making use of the fact that most processors can operate on a block of elements modulo 2.

When $p$ is large, the zero denominator problem can be dealt with simply by restarting the algorithm with a different system.

The Lanczos algorithm uses seven vector variables, but in fact only the storage for six is required. Each iteration of the algorithm requires one matrix-vector product (which in the case of $S = AA^t$ costs as two matrix-vector products of the "simple" type used in Wiedemann's method — we multiply by $A$ and $A^t$ separately because it is faster than multiplying by $S$, which is, in general, too large to store anyway), and three inner product calculations (one calculation is preserved for the next iteration). Since the Lanczos method will terminate in at most $c$ iterations where $c$ is the dimension of $A$, this will be at most $2c$ "simple" matrix-vector products and $3c$ inner products. Compared with the Wiedemann method, the Lanczos method requires $c$ less matrix-vector products, $3c$ inner products where the Wiedemann algorithm required only one, and roughly 50% more storage. Since matrix-vector products will likely dominate the computation, if the storage is available, it would be easy to draw the conclusion that Lanczos' method seems preferable over Wiedemann's for finding solutions modulo large primes. In fact, if the system is very large, there might be big problems in constructing and storing the transpose matrix $A^t$. Therefore the use of Wiedemann's algorithm must be seriously considered in some circumstances.

### 20.3.4 Filtering

An important step that can be included between the relation search and the linear algebra is filtering. Its aim is to reduce the size of the linear system without losing the fact that it is overdetermined. This is particularly successful with the type of systems obtained in the index calculus of finite fields.

The most common filtering strategies can be divided into two types: removing useless and unnecessary relation, or *pruning*; and *merging* two (respectively $r$) relations to build one (respectively $r-1$) relation(s), removing one variable in the process.

Before entering into details, we should also briefly mention the removal of duplicate relations, which applies mainly to factorization problems. In factoring, relations are usually obtained by different means: two sieves are commonly used, the line-by-line siever and the lattice siever, and this may cause some overlap. Duplicate relations are removed using hashing techniques.

### 20.3.4.a Pruning

Suppose that a relation of the form $[a]g \oplus [b]h = \bigoplus_{j=1}^{n_B} [e_j]\pi_j$ is found, and after the relation collection stage we can determine that one of the primes $\pi_j$ appears in this relation with a nonzero coefficient, but its multiplicity is zero in all other relations. It is clear that this relation is useless for the purpose of solving the discrete logarithm, and can be purged from the system, decreasing dimension of the matrix and number of variables both by one. We also reduce the total weight of the matrix.

Since at some point during pruning, the system might have more relations than absolutely necessary to have a kernel of dimension bigger than zero, it may be possible to remove relations that contribute to the kernel. A natural approach to do this is to choose a variable that has a nonzero coefficient in only two relations (doublets) and remove one of these two relations. Since the variable now has nonzero coefficient in only one relation, that relation becomes useless and it is purged from the system, along with the variable. This is particularly successful if chains of doublets can be found, i.e., sequences of pairs of doublets that have nonzero coefficients in the same relation. Furthermore, closed cycles of doublets can be removed without affecting the dimension of the kernel.

In order to use this approach, a database with the total multiplicities of all the primes in the factor base is created as relations are collected.

### 20.3.4.b Merging

Merging is based on the structured Gaussian elimination as described in [POSM 1992]. This takes advantage of both the sparsity of the matrix, and (when possible) the "unbalanced" shape of its rows. It is in principle the usual Gaussian elimination, but it works from "right to left" in order to eliminate the elements that appear less frequently, i.e., the larger primes first, and is only done partially. This reduces the size of the system while attempting to preserve the sparse character of the matrix.

The present description of merging follows Avanzi and Thériault [AVTH 2004], who build upon the strategies developed for the number field sieve method for factorization, such as, for example, those described in [CAV 2000] and in the references therein.

Suppose we have two relations

$$[a_i]g \oplus [b_i]h = \bigoplus_{j=1}^{n_B} [e_{i,j}]\pi_j$$

for $i = 1, 2$, and, for simplicity, assume that $e_{1,1} = e_{2,1} = 1$, but that the coefficient of $\pi_1$ is zero in *all* other relations. If we replace the two relations with the single relation

$$[a_1 - a_2]g \oplus [b_1 - b_2]h = \bigoplus_{j=1}^{n_B} [e_{1,j} - e_{2,j}]\pi_j = \bigoplus_{j=2}^{n_B} [e_{1,j} - e_{2,j}]\pi_j$$

then, clearly, the prime $\pi_1$ does not appear in the system any longer, therefore we in fact reduce both the dimension of the matrix and the number of variables by one. It is clear that the first system is solvable if and only if the merged system is solvable, and finding an element of the kernel of the system before the merging is equivalent to finding an element of the kernel of the new system.

If the prime $\pi_1$ belongs to more than one relation we can, in the same way, merge the first relation to the other ones where $\pi_1$ appears, thus eliminating one variable and one relation once again.

Note that the total weight of the matrix usually increases when one relation is merged to more than one other relation. One should always pick the relation of smallest weight among those containing a given variable and merge that one to the other ones. The implementor of the filtering stage must take particular care to merge only if this actually *does not increase* the complexity of the linear algebra solution step that follows. Usually a given sparse linear algebra algorithm takes time $k_1 c^2 + k_2 c\omega$ where $\omega$ is the weight of the matrix and $k_1, k_2$ are time constants that depend on the algorithm and on the computing architecture chosen (in practice, due to the extreme complexity of today's computer architectures, the latter can be correctly determined only experimentally). This means that this quantity, at least in principle, should not increase after a merging step.

In principle, one determines a threshold $t$ and then merges the relations containing variables that appear at most $t$ times in the whole system. But there might be rather complicated networks of relations, where two relations are adjacent if they both contain a variable to be merged, and it is therefore not clear in which order they should be removed. This cannot be considered on a case-by-case basis, hence the most common situations have to be analyzed beforehand and "hardwired" in the code for a specific application, taking into account the impact on the total running time of the linear algebra step.

Note also that, due to the particular shape of the matrices arising from finite field discrete logarithm or integer factorization problems, filtering is usually much more effective in those cases than for the hyperelliptic curve index calculus method.

### 20.3.5 Automorphisms of the group

The presence of an automorphism $\psi$ for the group $G$ has a great impact on index calculus methods, as remarked by Gaudry in [GAU 2000b, GAU 2000a]. We restate his result in the context of index calculus methods for generic groups:

**Theorem 20.6** Let $G$ be an abelian group for which there exists an index calculus variant, with set of primes $\mathcal{P}$ and smoothness bound $B$, and which admits an automorphism $\psi$ of order $m$ with the following properties:

(i) Almost all orbits under $\psi$ of the elements of the group have order $m$. The automorphism $\psi$ acts transitively on the set of primes $\mathcal{P}_B$, also with almost all orbits of order $m$.

(ii) The computation of $\psi$ can be performed in reasonable time: a computation in polynomial time of $\psi$ and of the verification that two elements belong to the same orbit is enough for our purposes.

(iii) The automorphism acts like the multiplication by a known scalar $s$, say (coprime to the group order).

The automorphism can be exploited to improve the index calculus to make:

(i) The search for relations up to $m$ times faster.
(ii) The linear algebra phase up to $m^2$ times faster.

Gaudry's idea, originally formulated only for the Jacobians of hyperelliptic curves, is to include in the factor base only one element for each orbit under $\psi$. Let $\mathcal{P}_B$ be the original factor base, and let $\widehat{\mathcal{P}}_B$ be the factor base containing only one element per orbit. $\widehat{\mathcal{P}}_B$ has cardinality $\widehat{n}_B \approx n_B/m$, Let the elements of $\widehat{\mathcal{P}}_B$ be denoted by $\widehat{\pi}_i$.

The size of the factor base is then reduced by a factor $m$, and so is the number of relations to be found. The relations with respect to $\mathcal{P}_B$ are of the form

$$[a_i]g \oplus [b_i]h = \bigoplus_{j=1}^{n_B} [e_{i,j}]\pi_j.$$

Now for all $i$ we can write an element of $\mathcal{P}_B$ as $\pi_i = \psi^{\eta_i}(\widehat{\pi}_{\gamma_i})$ where $\widehat{\pi}_i$ is in the factor base $\widehat{\mathcal{P}}_B$ and $\eta_i$ and $\gamma_i$ are suitable integers. Hence

$$[a_i]g \oplus [b_i]h = \bigoplus_{j=1}^{n_B} [e_{i,j}s^{\eta_i}]\widehat{\pi}_{\gamma_j} = \bigoplus_{k=1}^{\widehat{n}_B} \Big[\sum_{j\,:\,\gamma_j=k} e_{i,j}s^{\eta_i}\Big]\widehat{\pi}_k.$$

The matrix $\widehat{A}$ of the resulting, reduced, linear algebra system is $m$ times smaller than $A$, so the result about the speedup of the linear algebra phase is immediate.

The important consequence of Theorem 20.6 is that in groups with automorphisms the index calculus can be sped up much more than Pollard's rho method. The theorem claims a speedup up to a factor $m^2$, but we expect it to be somewhat smaller for the following two reasons:

1. The nonzero elements of $A$ are, as integers, usually very small (almost always just $\pm 1$), whereas those of $\widehat{A}$ are often rather large numbers modulo $|G|$. Multiplications by small numbers can be done by repeated additions, but generic multiplications are more expensive, hence influence the performance of the linear algebra in the small system.

2. The automorphism $\psi$ introduces small overheads in the relation collection phase as well.

## 20.4 An important example: finite fields

We now outline a realization of the index calculus algorithm in the case $G$ is the group $\mathbb{F}_{p^d}^*$. The expected running times are given for $p \to \infty$ and $d$ fixed or for $p$ fixed and $d \to \infty$.

According to Definition 20.2 and the usual definition of rational primes, a positive integer is $B$-smooth if all its prime factors are $\leqslant B$.

**Lemma 20.7** [CAER+ 1983, BRU 1966] Let $\alpha, \beta, r, s \in \mathbb{R}_{>0}$ with $s < r \leqslant 1$. Then a random positive integer $\leqslant L_x(r, \alpha)$ is $L_x(s, \beta)$-smooth with probability $L_x(r-s, -\alpha(r-s)/\beta)$ for $x \to \infty$.

A polynomial in $\mathbb{F}_p[X]$ is $B$-smooth if it factors as a product of irreducible polynomials in $\mathbb{F}_p[X]$ of norm $\leqslant B$.

**Lemma 20.8** [ODL 1985] Let $\alpha, \beta, r, s \in \mathbb{R} > 0$ with $r > 1$ and $\frac{r}{100} < s < \frac{99r}{100}$. Then a random polynomial in $\mathbb{F}_p[X]$ of norm $\leqslant L_x(r, \alpha)$ is $L_x(s, \beta)$-smooth with probability

$$L_x(r-s, -\alpha(r-s)/\beta) \text{ for } x \to \infty.$$

Remember that if $d = 1$ the 'norm' of a field element is simply the integer representing the field element, and an element is 'prime' if that integer is prime. Also, if $d > 1$ the 'norm' of an element $g$ is given by $p^{\deg h}$ where $h$ is the polynomial representing $g$, and an element is 'prime' if its representation is an irreducible polynomial over $\mathbb{F}_p$.

We see now that the assumptions of Theorem 20.5 hold for the multiplicative group of finite fields (provided that $p$ and $d$ satisfy the conditions made at the opening of this section) so the index calculus approach yields subexponential algorithms for them.

Let $B$ be a smoothness bound, and let the factor base $S$ be the subset of $\mathbb{F}_{p^d}^*$ of primes of norm $\leqslant B$. We can now apply the index calculus Algorithm 20.3.

With $B = L_{(p^d-1)}(1/2, \sqrt{1/2})$ and Dixon's approach (cf. end of Section 20.3.1) the relation collection stage takes expected time $L_{p^d}(1/2, \sqrt{2})$. This follows from the smoothness probabilities given above, the running time of the elliptic curve factoring method if $d = 1$, and, if $d > 1$, the fact that polynomials over $\mathbb{F}_p$ of degree $k$ can be factored in expected time polynomial in $k$ and $\ln p$ [BER 1967, MCE 1969, CAM 1981, CAZA 1981, CAM 1983, SHO 1990, SHO 1991, GASH 1992]. Solving the system of linear equations takes the same expected running time $L_{p^d}(1/2, \sqrt{2})$ because the matrix is sparse. This results in an expected running time $L_{p^d}(1/2, \sqrt{2})$.

There are two important variants of the index calculus method for general finite fields $\mathbb{F}_{p^d}$. One is based on the number field sieve (cf. Section 25.3.4) and, given that $d < \sqrt{\ln p}$, it attains an expected running time $L_{p^d}(1/3, 1.923)$, similar to that of the number field sieve for factoring [SCH 2000b].

The other method, due to Coppersmith, is called the function field sieve and applies only to $d > (\ln p)^2$. It actually predates the number field sieve and was the first cryptanalytic method to break through the $L_x(1/2, c)$ barrier, but only for fields of small fixed characteristic, such as $\mathbb{F}_{2^d}$: For such fields it attains a complexity $L_{2^d}(1/3, 1.588)$. Observe that the constant 1.588 is substantially smaller than 1.923, making DL systems in groups $\mathbb{F}_{2^d}^*$ even less desirable than RSA systems. It has been applied to $\mathbb{F}_{2^{503}}$ [GOMC 1993] and $\mathbb{F}_{2^{607}}$ [THO 2003].

**Remark 20.9** For finite fields in the "gap" $\sqrt{\ln p} < d < (\ln p)^2$ there is currently no algorithm known with proved subexponential running time. The conjectured running time for some index calculus variants that apply to specific fields is $L_x(1/2, c)$. A survey of the number and function field sieve methods and on the current status concerning the gap can be found in [SCWE+ 1996].

All these variants can be seen in more generality from the unitary perspective of computations in Brauer groups of local and global fields and explicit class field theory. Under this optic these methods can probably be generalized further. See Nguyen's Ph.D. Thesis [NGU 2001] for details.

## 20.5 Large primes

The present section is devoted to the technique of using "large primes" in order to increase the amount of relations found. This technique mimics the one described in Section 25.3.4.e for factoring; Incidentally, the latter technique is described in a later chapter because of the ordering of the parts of the present book, even though from an historical perspective it predates the application to discrete logarithms in algebraic groups. Despite the fact that we usually cannot order the points and divisors on the curve in such a way that we can establish which of two points or divisors is "largest," we borrow the name of the method used for the integers and employ it also in the present context. It comes at the price of increasing the average weight of the relations, and thus of reducing the sparseness of the matrix.

### 20.5.1 One large prime

We now describe how to use a "large prime" to increase the amount of relations collected. Relations of the type

$$[a_i]g \oplus [b_i]h = \bigoplus_{j=1}^{n_B} [e_{i,j}]\pi_j + L_i, \quad (20.7)$$

where $L_i$ does not belong to the factor base, are collected at the same time as the $B$-smooth ones. The $L_i$ are larger than $B$, but are usually bounded in size, too. For example, the condition might be $L_i \in \mathcal{P}_{B_1}$ where $B_1$ is larger than $B$: that's why these $L_i$'s are called large primes. Relations like (20.7) are called almost-smooth relations (or 1-almost-smooth relations) or partial relations (1-partial relations), and are of course in far greater number than the $B$-smooth ones, called *full relations*. With this terminology, partial relations are used to construct full ones as follows: If two partial relations with the same large prime have been found, subtracting the second relation from the first (exactly as we did for merging) gives a smooth relation. Such relations are usually a bit heavier than the normal ones, and thus using large primes will create a matrix that is less sparse. On the other hand, the search becomes much faster, reducing the total running time.

Large prime matches are found using a hash table. When two relations are merged, one of them must be removed from the list of partial relations, otherwise we might merge both relations with a third one containing the same large prime, giving rise to three linearly dependent full relations.

Large prime variants have been used since the beginning of the development of factoring methods such as that of Morrison and Brillhart [MOBR 1975] (see Section 25.3.4) or in the first implementation of the quadratic sieve (see Section 25.3.4.b) by Gerver [GER 1983] as well as of index calculus methods (see Odlyzko's survey paper [ODL 1985]). In all these cases, relations obtained by merging partial relations with one large prime account for the majority of the relations found, and the usage of large primes brought a speedup of a factor 2 to 2.5.

## 20.5.2 Two large primes

A natural extension of the large prime idea is to use two or more large primes. Although the original concept of using two large primes can be traced back to ideas of Montgomery and Silverman from the mid 1980s, the classic paper on the subject is that of Lenstra and Manasse [LEMA 1994]. Merging in the single large prime case is a trivial task, but dealing with more than one large prime is more difficult.

One way of using two large primes is typical of factoring applications, and is related to the so-called "Waterloo variant" of searching smooth relations. The *Waterloo variant* has been developed by Blake et al. in [BLFU+ 1984] (see also [BLMU+ 1984]) and consists in the following: instead of testing values $b = g^a \bmod p$ for smoothness, the extended Euclidean algorithm is used to find integers $u, v, w$ with $w = ub + vp$ where both $u$ and $w$ are smooth. Instead of stopping when, as is more usual, $w = 1$ and $u, v$ are the Bezout coefficients with $u \equiv b^{-1} \pmod{p}$, the algorithm will be interrupted when $u \approx \sqrt{p}$, in which case $w \approx \sqrt{p}$ too. Then $u$ and $w$ are checked for smoothness: if they are both smooth we get a relation $b \equiv g^a \equiv wu^{-1} \pmod{p}$. It is more likely that two integers of size about $\sqrt{p}$ are both smooth than a single integer of size about $p$. Clearly, since we are actually building a single relation from two "halves," we can admit one (or more) large primes in $w$ independently from the large prime in $u$. This method has also been used by Thomé in his successful computation of a discrete logarithm in $\mathbb{F}_{2^{607}}$ [THO 2003] — but it does not seem adaptable to the solution of DLPs in groups arising from geometric constructions. Note that, by admitting negative exponents, we can consider either $b \equiv g^a \pmod{p}$ or $b^{-1} \equiv g^{-a} \pmod{p}$ for our purposes, according to the way we couple the large primes in "half relations" arising from different $b$'s. An analysis of the Waterloo variant can be found in Holt's PhD. Thesis [HOL 2003]. Another variant has been developed for the number field sieve with two factor bases [LELE 1993] and has not been adapted to the discrete logarithm case.

How are full relations built from partial relations with up to two large primes? During the relation search, a graph is built whose vertices are the large primes. Two vertices are connected by an edge if there exists at least one partial relation involving them both. There is a special vertex referred to as "0" (alternatively "1"), to which all primes that appear alone in a partial relation are connected.

For the purposes of factoring a cycle in the graph it is sufficient to "eliminate" the large primes by simply multiplying together all the partial relations corresponding to the edges, since each large prime will appear an even number of times in a cycle.

For solving DLPs, all cycles containing the special vertex "0" can be used to produce a full relation, but only about half of all other cycles are useful. Let us see why: Let

$$\rho_i = [a_i]g \oplus [b_i]h = L_i - \epsilon_i L_{i+1} + \{\text{smooth primes}\}$$

for $i = 1, 2, \ldots, n$ where the $L_i$ are large primes with $L_{n+1} = L_1$, and $\epsilon_i = \pm 1$ (this is always possible by suitably replacing some $\rho_i$'s with their opposites). All the large primes are pairwise different and, with the only possible exception of $L_1$, all different from the special prime "0". Then

$$\rho = \rho_1 + \epsilon_1 \rho_2 + \epsilon_1 \epsilon_2 \rho_3 + \cdots + \epsilon_1 \epsilon_2 \cdots \epsilon_{n-1} \rho_n$$

is a relation involving smooth primes and $(1 - \prod_{i=1}^{n} \epsilon_i) L_i$. If $L_i$ is "0," then $\rho$ is always a full relation; otherwise $\rho$ is a full relation if and only if $\prod_{i=1}^{n} \epsilon_i = 1$: Such a cycle is called an even cycle. This problem does not occur in factoring applications where all the multiplicities are considered modulo 2, hence all cycles produce full relations in these situations.

To find relations, one must therefore detect cycles containing "0" as well as even cycles in the connected components of the graph. Cycle detection involves only a relatively small overhead, but it must be implemented with care as the graph can be very large (for example, more than $10^8$ edges among $2 \times 10^9$ vertices) and managing it is not a trivial task.

Note, however, that long cycles may not be very effective because they tend to generate very heavy relations and using them can lead to a denser matrix.

### 20.5.3 More large primes

Here the landscapes for factoring and discrete logarithms start to be very different. Using more than two large primes (and up to 4) with the Waterloo variant of factorization is in fact just using up to two large primes for both "half relations." See [DOLE 1995] for a four large primes variant of the number field sieve. In general, and always for discrete logarithms, one needs to consider higher-dimensional generalizations of graphs, where the "edges" connect more than two vertices (still corresponding to primes), as in [HOL 2003], or follow Cavallar's [CAV 2000] method inspired from structured Gaussian elimination.

While the multi-large-prime schemes have been very efficient for factoring, it is not clear how more than two large primes can be used for computing discrete logarithms in groups arising from geometric constructions.

# Chapter 21

# Index Calculus for Hyperelliptic Curves

*Roberto M. Avanzi and Nicolas Thériault*

### Contents in Brief

| | | |
|---|---|---|
| **21.1** | **General algorithm** | 511 |
| | Hyperelliptic involution • Adleman–DeMarrais–Huang • Enge–Gaudry | |
| **21.2** | **Curves of small genus** | 516 |
| | Gaudry's algorithm • Refined factor base • Harvesting | |
| **21.3** | **Large prime methods** | 519 |
| | Single large prime • Double large primes | |

## 21.1 General algorithm

As stated in Section 20.2.1, the index calculus algorithm can be applied to compute discrete logarithms in the Jacobian of hyperelliptic curves. For groups of this type, the set of primes are *prime divisors* or, in terms of the ideal class group, *prime ideals*. These divisors can be single $\mathbb{F}_q$-rational points or the sums of all Galois conjugates over $\mathbb{F}_q$ of a non-$\mathbb{F}_q$-rational point: In other words, if a divisor $D$ has Mumford representation $[u(x), v(x)]$, $D$ is prime if and only if the polynomial $u(x)$ is irreducible over $\mathbb{F}_q$.

---

**Algorithm 21.1** Divisor decomposition

INPUT: A semi-reduced divisor $D = [u(x), v(x)]$.
OUTPUT: Prime divisors $P_1, \ldots, P_k$ and coefficients $e_1, \ldots, e_k$ such that $D = \sum_{j=1}^{k} e_j P_j$.

1. find the factorization $\prod_{i=1}^{k} u_i(x)^{e_i}$ of $u(x)$
2. $i \leftarrow 1$
3. **while** $i \leqslant k$ **do**
4.      $v_i \leftarrow v(x) \bmod u_i(x)$
5.      $P_i \leftarrow [u_i(x), v_i(x)]$

6.     $i \leftarrow i+1$
7.   **return** $\sum_{j=1}^{k} e_j P_j$

---

As is common for index calculus algorithms, the prime decomposition does not require a divisor to be reduced and two divisors in the same class will have two different decompositions.

**Remark 21.2** For the remainder of this chapter, once the prime decomposition of a divisor has been computed it will be assumed that both the Mumford representation and the prime decomposition can be used when required and the two representations will not be explicitly distinguished.

### 21.1.1  Hyperelliptic involution

As stated in Section 20.3.5, using group automorphisms can have a significant impact on the speed of the index calculus algorithm, although this impact is in the size of the constants involved and not in the asymptotic form of the resulting algorithm. In the case of hyperelliptic curves, the automorphism used is the hyperelliptic involution.

**Definition 21.3** Let $C$ be a hyperelliptic curve given by the equation $y^2 + h(x)y = f(x)$. The *hyperelliptic involution* of a divisor $D = [u(x), v(x)]$ is the divisor $[u(x), \tilde{v}(x)]$, denoted $-D$, where $\tilde{v}(x) := -v(x) - h(x) \pmod{u(x)}$.

Since the hyperelliptic involution is an automorphism of order two, it can be used to make the relation search twice as fast and the linear algebra four times faster (more if the sparse linear algebra algorithms of Section 20.3.3 cannot be used).

In order to take advantage of the hyperelliptic involution, we act as if for every prime $P \in \mathcal{P}_B$, its image under the involution were also in $\mathcal{P}_B$, but at most one copy of $P$ and $-P$ is actually included in $\mathcal{P}_B$. The prime decomposition is rewritten to reflect that fact, i.e., if $P$ is in the factor base and the prime decomposition of a divisor $D$ would call for the prime $(-P)$, then we replace $e(-P)$ by $(-e)P$.

### 21.1.2  Adleman–DeMarrais–Huang

With the exception of some special cases, the first general description of an index calculus algorithm for Jacobians of hyperelliptic curves is due to Adleman, DeMarrais, and Huang [ADDE$^+$ 1999].

The algorithm can be used for curves over any finite field, but describing in full generality the various sub-algorithms would only make things unnecessarily confusing, so only the case curves defined over fields of odd characteristic and defined by equations of the form $y^2 = f(x)$ will be described in this section.

#### Factor base

Just as the factor base is chosen such that it contains primes with a small norm when we are dealing with the index calculus algorithm for finite fields, we attach a notion of "size" to prime divisors. For a given smoothness bound $B$, the factor base will then be composed of all the prime divisors of degree at most $B$, where the degree of a prime divisor is the degree of its polynomial $u(x)$ in the Mumford representation $[u(x), v(x)]$.

**Definition 21.4** A divisor is said to be $B$-*smooth* if all the prime divisors in its decomposition have degree at most $B$.

## § 21.1 General algorithm

To find all the prime divisors of a given degree, it suffices to test all the irreducible polynomials $u(x)$ of that degree, checking to see if there exists a polynomial $v(x)$ satisfying $v(x)^2 \equiv f(x)$ (mod $u(x)$). If we also take advantage of the hyperelliptic involution, we get the following method to find all the elements of the factor base:

---
**Algorithm 21.5** Computing the factor base

INPUT: A hyperelliptic curve $C$ and a smoothness bound $B$
OUTPUT: A set $\mathcal{P}_B$ of $B$-smooth prime divisors

1. $\mathcal{P}_B \leftarrow \{\}$
2. **for** every irreducible monic polynomial $u(x) \in \mathbb{F}_q[x]$, $\deg u \leqslant B$ **do**
3.     **if** $u(x) \nmid f(x)$ and $f(x) \equiv$ square (mod $u(x)$) **then**
4.         find $v(x)$ such that $\deg v < \deg u$ and $v(x)^2 - f(x) \equiv 0$ (mod $u(x)$)
5.         $P \leftarrow [u(x), v(x)]$
6.         add $P$ to $\mathcal{P}_B$
7. **return** $\mathcal{P}_B$

---

### Relation search

Given a reduced divisor, the approach used by Adleman, DeMarrais, and Huang to obtain a smooth divisor consists of adding randomly chosen principal divisors (of degree at most $d$) to the reduced divisor until the semi-reduced divisor created is smooth. Because this can be applied separately to each of the prime divisors in the decomposition of a divisor, we consider only the case of unramified prime divisors and the divisor 0.

---
**Algorithm 21.6** Smoothing a prime divisor

INPUT: An unramified prime divisor $D = [u(x), v(x)]$, a factor base $\mathcal{P}_B$, and a parameter $d$.
OUTPUT: The prime decomposition of an equivalent $B$-smooth divisor $\widetilde{D} = [\tilde{u}(x), \tilde{v}(x)]$.

1. **if** $D$ is $B$-smooth **then**
2.     **return** $D$
3. **else**
4.     **repeat**
5.         **repeat**
6.             choose a random $A(x) \in \mathbb{F}_q[x]$ of degree $\leqslant d$
7.             $B(x) \leftarrow -A(x)v(x) \bmod u(x)$
8.         **until** $\gcd(A(x), B(x)) = 1$
9.         $\tilde{u}(x) \leftarrow (B(x)^2 - A(x)^2 f(x))/u(x)$
10.        $\tilde{v}(x) \leftarrow B(x)A(x)^{-1} \bmod \tilde{u}(x)$
11.        $\widetilde{D} \leftarrow [\tilde{u}(x), \tilde{v}(x)]$
12.        $\widetilde{D} \leftarrow \sum_{i=1}^{r} e_i P_i$       $\left[\text{decomposition of } \widetilde{D} \text{ into prime divisors}\right]$

13.       **until** $\widetilde{D}$ is $B$-smooth

14.       **return** $\widetilde{D}$

---

**Algorithm 21.7** Finding smooth principal divisors

INPUT: A factor base $\mathcal{P}_B$ and a parameter $d$.
OUTPUT: The prime decomposition of a $B$-smooth principal divisor $\widetilde{D} = [\tilde{u}(x), \tilde{v}(x)]$.

1. **repeat**
2.     **repeat**
3.         choose random $A(x), B(x) \in \mathbb{F}_q[x]$ of degree $\leqslant d$
4.     **until** $\gcd(A(x), B(x)) = 1$
5.     $\tilde{u}(x) \leftarrow B(x)^2 - A(x)^2 f(x)$
6.     $\tilde{v}(x) \leftarrow B(x) A(x)^{-1} \bmod \tilde{u}(x)$
7.     $\widetilde{D} \leftarrow [\tilde{u}(x), \tilde{v}(x)]$
8.     $\widetilde{D} \leftarrow \sum_{i=1}^{r} e_i P_i$               $\bigl[$decomposition of $\widetilde{D}$ into prime divisors$\bigr]$
9. **until** $\widetilde{D}$ is $B$-smooth
10. **return** $\widetilde{D}$

---

Since the smoothing algorithm does not work for ramified primes (i.e., primes of the form $[u(x), 0]$), it may be necessary to find an equivalent divisor that does not contain any ramified prime in its decomposition. This is done as follows:

---

**Algorithm 21.8** Decomposition into unramified primes

INPUT: A semi-reduced divisor $D = [u(x), v(x)]$.
OUTPUT: The decomposition of an equivalent divisor $\widetilde{D}$ into unramified primes.

1. $d(x) \leftarrow \gcd(u(x), f(x))$
2. **if** $d(x) = 1$ **then**
3.     find the decomposition of $[u(x), v(x)]$ into prime divisors $\sum_{i=1}^{r} e_i P_i$
4.     **return** $\widetilde{D} = \sum_{i=1}^{r} e_i P_i$
5. **else**
6.     $u_1(x) \leftarrow u(x)/d(x)$
7.     $v_1(x) \leftarrow v(x) \bmod u_1(x)$
8.     $u_2(x) \leftarrow \bigl(d(x)^2 - f(x)\bigr)/d(x)$
9.     $v_2(x) \leftarrow d(x) \bmod u_2(x)$
10.     find the decomposition of $[u_1(x), v_1(x)]$ into prime divisors $\sum_{i=1}^{r} e_i P_i$
11.     find the decomposition of $[u_2(x), v_2(x)]$ into prime divisors $\sum_{j=1}^{s} f_j Q_j$
12.     **return** $\widetilde{D} = \sum_{i=1}^{r} e_i P_i - \sum_{j=1}^{s} f_j Q_j$

## Algorithm

The index calculus algorithm presented by Adleman, DeMarrais, and Huang [ADDE+ 1999] differs somewhat from Algorithm 20.3. The idea behind the algorithm is to consider the kernel of the map between the free abelian group $\mathbb{Z}^n$ generated by the factor base and the Jacobian of the curve. Since this kernel is a lattice of $\mathbb{Z}^n$, working out the structure of the lattice makes it relatively easy to find the relationship between two smooth divisors in the Jacobian.

We first find smooth representations of the pair of divisors for which we want to compute the discrete logarithm. To obtain the structure of the lattice, we must find enough smooth principal divisors (which correspond to points in the lattice) and decompose the linear system obtained from these relations. It may not be necessary to completely work out the structure of the lattice, and the search for smooth principal divisors is ended once the discrete logarithm is found.

The resulting algorithm can be written as follows:

**Algorithm 21.9** Adleman–DeMarrais–Huang

INPUT: A hyperelliptic curve $C$, two divisors $g$ and $h$ with $h \in \langle g \rangle$, a smoothness bound $B$, and a parameter $d$.
OUTPUT: An integer $t$ with $[t]g = h$.

1.    compute the factor base $\mathcal{P}_B$                                       [use Algorithm 21.5]
2.    $n \leftarrow |\mathcal{P}_B|$
3.    $g \leftarrow \sum_{i=1}^r e_i P_i$                                           [decompose $g$ into unramified primes]
4.    $i \leftarrow 1$
5.    **while** $i \leqslant r$ **do**
6.        smooth the prime divisor $P_i$ and put the result in $\widetilde{P}_i$         [use Algorithm 21.6]
7.        $i \leftarrow i + 1$
8.    $\tilde{g} \leftarrow \sum_{i=1}^r e_i \widetilde{P}_i$
9.    write $\tilde{g}$ as a $1 \times n$ vector denoted by $v_g$
10.   $h \leftarrow \sum_{j=1}^s f_j Q_j$                                     [decompose $h$ into unramified primes]
11.   $j \leftarrow 1$
12.   **while** $j \leqslant s$ **do**
13.       smooth the prime divisor $Q_j$ and put the result in $\widetilde{Q}_j$        [use Algorithm 21.6]
14.       $j \leftarrow j + 1$
15.   $\tilde{h} \leftarrow \sum_{j=1}^s f_j \widetilde{Q}_f$
16.   write $\tilde{h}$ as a $1 \times n$ vector denoted by $v_h$
17.   $M \leftarrow 0 \times n$ matrix
18.   $j \leftarrow 0$
19.   **repeat**
20.       $j \leftarrow j + 1$
21.       find a smooth principal divisor $D_j$                         [use Algorithm 21.7]
22.       write $D_j$ as a $1 \times n$ vector denoted by $v_j$
23.       add $v_j$ to $M$ as the $j$-th row

24.   find matrices $L \in GL(j, \mathbb{Z})$ and $R \in GL(n, \mathbb{Z})$ such that $LMR = C$ where $C$ is a $m \times n$ matrix with all nonzero entries on the diagonal and with diagonal entries satisfying $c_i \mid c_{i+1}$ and $c_i > 0$

25.   $c \leftarrow \prod_{i=1}^{j} c_i$

26.   **if** $c > (\sqrt{q}+1)^{2g}$ **then**  [upper bound of the group order]

27.   $\tilde{v}_g \leftarrow v_g R$

28.   $\tilde{v}_h \leftarrow v_h R$

29.   **if** $\tilde{v}_h = t\tilde{v}_g$ for some $t \in \mathbb{Z}$ **then**

30.   **return** $t$

31.  **until** $t$ is found

By setting $B = \lfloor \log_q L_{q^{2g+1}}(1/2, 1/\sqrt{2k}) \rfloor$ and $d = \lfloor \log_q L_{q^{2g+1}}(1/2, \sqrt{2k}) \rfloor$ where the constant $k$ depends on the speed of the linear algebra, we get:

**Theorem 21.10** [ADDE+ 1999] Let $C$ be a hyperelliptic curve of genus $g$ over the finite field $\mathbb{F}_q$. If $\ln q \leqslant (2g+1)^{1-\epsilon}$, then there exists a constant $c \leqslant 2.181$ such that discrete logarithms in $J_C(\mathbb{F}_q)$ can be computed in expected time $L_{q^{2g+1}}(1/2, c)$.

The constant $c$ is in fact dependent on the value of $k$ coming from the linear algebra, with

$$c = \frac{k+1}{\sqrt{2k}}.$$

The value $c = 2.181$ is obtained when no sparse linear algebra arithmetic is assumed (with $k = 7.376$). Although the algorithms described in Section 20.3.3 cannot be applied in this context, sparse arithmetic might still be used, giving a value of $c = 4/\sqrt{6}$ (assuming $k = 3$ is possible).

### 21.1.3 Enge–Gaudry

The algorithm can be adapted to fit more closely with the index calculus as described in Chapter 20 (and used in the remainder of this chapter). Using results by Enge, Gaudry, and Stein [ENG 2002, ENGA 2002, ENST 2002] we get:

**Theorem 21.11** [ENGA 2002, Theorem 1] Let $C$ be a hyperelliptic curve of genus $g$ defined over $\mathbb{F}_q$ and let $\log_g$ denote the logarithm in base $g$. For $g/\log_g q > t$, the discrete logarithm in the divisor class group of $C$ can be computed with complexity bounded by

$$L_{q^g}\left(\frac{1}{2}, \sqrt{2}\left(\left(1 + \frac{1}{2t}\right)^{1/2} + \left(\frac{1}{2t}\right)^{1/2}\right)\right).$$

## 21.2 Curves of small genus

Although Algorithm 21.9 is very efficient when $g > \log_g q$, it must be modified in order to be applied for hyperelliptic curves of "small" genus, i.e., such that $g < \log_g q$.

In the remainder of this chapter, we will assume that $q > g!$ (which implies $g < \log_g q$) and running times can be viewed in terms of a fixed genus and a varying field size.

## § 21.2 Curves of small genus

### 21.2.1 Gaudry's algorithm

There are three major differences with the Algorithm 21.10 when adapting the index calculus algorithm to curves of small genus:

- The approach of Algorithm 20.3 is used.
- The only primes considered for the factor base are those coming from $\mathbb{F}_q$-rational points, i.e., we are looking for 1-smooth divisors.
- Rather than smoothing divisors, smooth relations are found by repeatedly looking at different reduced divisors until a smooth one is found.

Using these ideas, Gaudry [GAU 2000b] obtained an algorithm that is more efficient when the genus of the hyperelliptic curve is small. The second step of Algorithm 20.3 is done as follows:

---

**Algorithm 21.12** Relation search

INPUT: A hyperelliptic curve $C$, two divisors $g$ and $h$ with $h \in \langle g \rangle$, and a factor base $\mathcal{P}_1$.
OUTPUT: A system of $r$ 1-smooth divisors of the form $[\alpha_i]g \oplus [\beta_i]h$.

1.   choose random $\alpha, \beta \in \{1, \ldots, |J_C(\mathbb{F}_q)|\}$
2.   $D \leftarrow [\alpha]g \oplus [\beta]h$
3.   $i \leftarrow 1$
4.   **while** $i \leqslant r$ **do**
5.       $D \leftarrow \Phi(D)$
6.       update $\alpha$ and $\beta$
7.       decompose $D$ into prime divisors
8.       **if** $D$ is 1-smooth **then**
9.           $D_i \leftarrow D$
10.           $i \leftarrow i + 1$
11.   **return** $\{D_1, \ldots, D_r\}$

---

**Remark 21.13** To take into account the various linear algebra methods of Section 20.3.3, the desired number of smooth relations produced by Algorithm 21.12 (and by the remaining algorithms in this chapter) is denoted by $r$. In most cases, $r$ can be assumed to be close to the size of the factor base: If Wiedemann's method is used, we need $r = |\mathcal{B}| + 1$, while if Lanczos' method is preferred, $r$ may be chosen somewhat larger (to get a system with higher rank).

Since approximately 1 in $g!$ divisors in the Jacobian is 1-smooth, obtaining enough relations is quite fast and takes time $O\left(g^2 g! \, q^{1+\epsilon}\right)$. The other steps of Algorithm 20.3 are dominated by the linear algebra, which has running time $O\left(g^3 q^{2+\epsilon}\right)$. These running times combine to give the result:

**Theorem 21.14** [GAU 2000b] Let $C$ be a hyperelliptic curve of genus $g$ over the finite field $\mathbb{F}_q$. If $q > g!$ then discrete logarithms in $J_C(\mathbb{F}_q)$ can be computed in expected time $O(g^3 q^{2+\epsilon})$.

### 21.2.2 Refined factor base

Because the running time for Gaudry's algorithm is dominated by the cost of solving the linear algebra, a natural approach to improving the speed of the algorithm is to reduce the cost of the linear algebra part. Unless new sparse linear algebra algorithms are developed, the only option is to reduce the size of the system, which means reducing the size of the factor base (since we need at least as many relations as factor base elements).

To do this, we choose the factor base $\mathcal{B}$ as a subset of $\mathcal{P}_1$. Since reducing the size of the factor base makes it less likely that a random reduced divisor is smooth (or $\mathcal{B}$-*smooth*), this increases the cost of the relation search. If the factor base is reduced too much, the relation search will eventually become the dominant part of the algorithm as smooth relations become rarer. This brings us back to the usual situation for index calculus algorithms, where optimizing the running time is a balancing act between the relation search and the linear algebra.

By choosing the factor base $\mathcal{B}$ such that $|\mathcal{B}| = O(g^2 q^{g/(g+1)+\epsilon})$, we get the following result:

**Theorem 21.15** [THÉ 2003a, Theorem 2] Let $C$ be a hyperelliptic curve of genus $g$ over the finite field $\mathbb{F}_q$. If $q > g!$ then discrete logarithms in $J_C(\mathbb{F}_q)$ can be computed in expected time $O(g^5 q^{2 - \frac{2}{g+1} + \epsilon})$.

## 21.2.3 Harvesting

For the index calculus of hyperelliptic curves, Avanzi and Thériault [AVTH 2004] introduce a "drastic" extension of pruning they call harvesting. The idea behind harvesting is to start with a very overdetermined system, in fact of $k$ times as many relations as variables for $k$ relatively large (even $k \geqslant 100$ can be useful), and then remove as many relations as possible while at the same time reducing as much as possible the numbers of variables present in the system, in order to find a small subsystem that is still overdetermined.

How are we going to choose the equations that can be removed to obtain the smallest possible system after the filtering? We remove all relations that contain "rare" variables, i.e., the variables that appear less frequently with nonzero coefficients in the equations of the system. If this process is repeated until the system has only a few more relations than variables, we can reasonably expect that the result is almost as small as possible.

At the end, we can obtain a system that is still overdetermined, but with much less variables than the elements in the original factor base. Of course there might exist a smaller subsystem of the original system, but that subsystem cannot be found efficiently, hence the construction by an approximation method.

Such a harvesting step, just like pruning, has the nice side effect of decreasing the multiplicities of all other variables that appeared in the removed relations. This means that one can see if the new system can be harvested again, until it reaches the desired size or it can no longer be harvested.

After harvesting has been performed and the desired level of overdetermination has been reached, we can still perform merging. In fact, merging after harvesting will be, in comparison, *more efficient* than if it had been performed before.

Harvesting can also be viewed as a redefinition (reduction) of the factor base *a posteriori* (i.e., after the relation search has been done). Since harvesting requires more smooth relations but produces a smaller linear system, which leaves the relation search and linear algebra unbalanced (compared to an optimized algorithm with $k \sim 1$), it is only natural to readjust the original size of the factor base to re-balance the algorithm. If, for a given $k$, the reduction factor of harvesting is large enough, the running time for the new optimized algorithm will be smaller than the running time with $k \sim 1$.

With the approach of Section 21.2.2, harvesting can be used to reduce the running time of the index calculus algorithm by a non-negligible factor (although only the constant term in the asymptotic running time is affected). In Jacobians of genus 6 curves for example, a total saving of close to 45% can be obtained using $k = 100$. Harvesting can also be applied to the algorithm in Section 21.3.1 and Section 21.2.1 (although there is no re-balancing of the choice of the factor base in that case), but it is not yet known how well it can be adapted to the other algorithms of this chapter and of Chapter 20.

**Parallelization**

Although it is not clear how to perform pruning and merging in parallel with a non shared-memory computer, such as a cluster, some distribution is possible (and relatively easy) for harvesting.

We proceed by repeated steps as follows: Each node keeps track of its own part of the system and lets the server know how many times each variable is used. The server can then decide which variables will be removed from the system and tells the nodes. The nodes then look for the relations containing the variables removed (and remove those equations from the system) and update the number of times variables are used.

In order to reduce latency, variables are not removed one by one but in groups (still with the "rarest" variables first). Since removing "blocks" of variables together is likely to have a detrimental impact on the effectiveness of harvesting, one must be careful and choose these blocks as small as possible without getting too much latency.

Once the system is reduced to the point where harvesting is no longer possible or distributing the work is no longer advantageous, the remaining relations are collected and the filtering is completed on a single processor. If this is implemented carefully, the total amount of information transmitted is not much more than what would be transmitted to collect the full system (certainly much less than twice that amount) while most of the work is distributed among the nodes. A major advantage of this approach is the distribution of the system pre-harvesting, potentially making it possible to work with systems that would be too big to be handled by a single processor.

If the expansion factor $k$ is *very* large (and larger than the number of client computers involved in the filtering), then some harvesting may be done locally by each node (and independently of other nodes), reducing the number of relations but not the number of variables (since a variable removed on one node can still be used on another). The harvesting can then be done distributively on the reduced system or the relations can be collected together and the filtering (including harvesting) can be done on a single computer. This will reduce network traffic (as no information is communicated on the relations that were removed locally) but will also (presumably) reduce the effectiveness of the filtering.

## 21.3 Large prime methods

As was described in Section 20.5, using large primes to decrease the time required to find relations can have a significant impact on the running time of the index calculus algorithm.

The main difference when using large primes for Jacobians of hyperelliptic curves is in their definition. Whereas large primes are usually defined as all the primes not included in the factor base, in this situation we only consider all the primes in $\mathcal{P}_1$ not included in the factor base (in other words, we ignore large primes of degree greater than one). This restriction is not strictly speaking a necessary one, but more one of convenience.

In essence, this is because a large prime must appear in at least two different divisors found during the relation search before it can be of any use in building smooth relations. Although almost-smooth divisors with a large prime of degree at least 2 are more common than almost-smooth divisors with a linear large prime, the low probability that a specific prime of higher degree appears more than once far offsets the large number of these primes.

Even if all primes outside the factor base were to be considered, the number of smooth relations obtained due to the use of the primes of degree larger than one would be almost insignificant compared to what is obtained using only $\mathcal{P}_1$. In practice, restricting the definition of large primes does increase the running time of the relation search, but not in any meaningful way. The main reason to restrict the definition of large prime is to reduce the amount of memory required to run the algorithm, the set $\mathcal{P}_1$ being much more manageable than the set of all primes.

At the time this book was written, methods using relations with one or two large primes had been described, but no effective way had yet been found to use relations with more than two large primes. The algorithms described here will therefore be limited to single and double large prime variants.

### 21.3.1 Single large prime

In order to take advantage of the high number of almost-smooth divisors, we must find pairs of these divisors with the same large prime (up to sign, using the hyperelliptic involution). The approach relies on the birthday paradox, since the large primes in almost-smooth divisor are uniformly distributed among the set of linear large primes. In order to identify these pairs, we use a list $\mathcal{P}$ of the large primes encountered during the search and a list $\mathcal{R}$ of the almost-smooth divisors in which they appeared. To avoid producing redundant relations, only the first copy of a large prime is included in the list, and subsequent copies are used to produce smooth relations (using the corresponding divisors to cancel the large primes).

---

**Algorithm 21.16** Single large prime

INPUT: A hyperelliptic curve $C$, two divisors $g$ and $h$ with $h \in \langle g \rangle$, and a factor base $\mathcal{B}$.
OUTPUT: A system of $r$ $\mathcal{B}$-smooth divisors of the form $D_i = [\alpha_i]g \oplus [\beta_i]h$.

---

1.     $\mathcal{P} \leftarrow \{\}$ and $\mathcal{R} \leftarrow \{\}$
2.     choose random $\alpha, \beta \in \{1, \ldots, |J_C(\mathbb{F}_q)|\}$
3.     $D \leftarrow [\alpha]g \oplus [\beta]h$
4.     $i \leftarrow 1$
5.     **while** $i \leqslant r$ **do**
6.        $D \leftarrow \Phi(D)$, update $\alpha$ and $\beta$
7.        decompose $D$ into prime divisors
8.        **if** $D$ is $\mathcal{B}$-smooth **then**
9.            $D_i \leftarrow D$
10.           $i \leftarrow i + 1$
11.        **if** $D$ is almost-smooth with large prime $P$ **then**
12.            **if** $P = P_j \in \mathcal{P}$ **then**
13.                 $R \leftarrow$ almost-smooth divisor containing $P_j$
14.                 $D_i \leftarrow D \ominus R$
15.                 $i \leftarrow i + 1$
16.            **else if** $P = -P_j \in \mathcal{P}$ **then**
17.                 $R \leftarrow$ almost-smooth divisor containing $P_j$
18.                 $D_i \leftarrow D \oplus R$
19.                 $i \leftarrow i + 1$
20.            **else**
21.                 add $P$ to $\mathcal{P}$

22.           add $D$ to $\mathcal{R}$                      [almost-smooth divisor containing $P$]
23.    **return** $\{D_1, \ldots, D_r\}$

By choosing the factor base $\mathcal{B}$ such that $|\mathcal{B}| = O(g^2 q^{(g-\frac{1}{2})/(g+\frac{1}{2})+\epsilon})$, we get the following result:

**Theorem 21.17** [THÉ 2003a, Theorem 3] Let $C$ is a hyperelliptic curve of genus $g$ over the finite field $\mathbb{F}_q$. If $q > g!$ then discrete logarithms in $J_C(\mathbb{F}_q)$ can be computed in expected time $O(g^5 q^{2-\frac{4}{2g+1}+\epsilon})$.

In practice, smooth relations coming from smooth divisors are used, but they are not taken into account to obtain this result. Since only a small number of smooth divisors are found during the relation search, they do not have a significant impact on the time required for the relation search.

### 21.3.2 Double large primes

Since almost-smooth divisors can be used to produce relations so much faster, it is natural to also consider 2-almost-smooth divisors. The idea is to produce chains of 2-almost-smooth (and almost-smooth) divisors where two consecutive divisors have a common large prime. To each such chain, we associate a relation where the divisors of the chain are added to the first one in such a way that the common large primes between consecutive divisors are canceled.

The chain produces a smooth relation if all large primes are canceled. This can happen in two ways: The first and last divisors in the chain are almost-smooth, or the first and last divisors in the chain are 2-almost-smooth with a common large prime and the cancellation of the other large primes in the chain also cancels that one. It is important to notice that even if the divisors at the extremities of the chain are 2-almost-smooth with a common large prime, that large prime need not be canceled in the relation associated to that chain. On average, the large prime common to the first and last divisor will only cancel out in half of these chains.

In order to find chains of divisors producing smooth relations, we use a graph $G$ where the vertices are the large primes and an edge between two vertices correspond to a 2-almost-smooth divisor containing those two large primes. Since we also want to take advantage of almost-smooth divisors, the graph $G$ contains an extra vertex denoted 1 and an almost-smooth divisor is associated to the edge between its large prime and 1. The search for chains of divisors producing smooth relations can then be viewed as a search for cycles in the graph $G$ corresponding to the random walk.

The general algorithm for the relation search using divisors with up to two large primes can be given as follows:

**Algorithm 21.18** Double large primes

INPUT: A hyperelliptic curve $C$, two divisors $g$ and $h$ with $h \in \langle g \rangle$, and a factor base $\mathcal{B}$.
OUTPUT: A system of $r$ $\mathcal{B}$-smooth divisors of the form $D_i = [\alpha_i]g \oplus [\beta_i]h$.

1.    $G \leftarrow$ empty graph
2.    choose random $\alpha, \beta \in \{1, \ldots, |J_C(\mathbb{F}_q)|\}$
3.    $D \leftarrow [\alpha]g \oplus [\beta]h$
4.    $i \leftarrow 1$
5.    **while** $i \leqslant r$ **do**
6.        $D \leftarrow \Phi(D)$, update $\alpha$ and $\beta$

7.   decompose $D$ into prime divisors
8.   **if** $D$ is $\mathcal{B}$-smooth **then**
9.       $D_i \leftarrow D$
10.      $i \leftarrow i+1$
11.  **if** $D$ is almost-smooth or 2-almost-smooth **then**
12.      update $G$
13.      **if** a smooth divisor $S$ is created **then**
14.          $D_i \leftarrow S$
15.          $i \leftarrow i+1$
16.  **return** $\{D_1, \ldots, D_r\}$

Obviously, the graph must be updated in such a way that no redundant relations are created. For example, it may be possible to break down a cycle into distinct subcycles, but even if all the relations obtained from each subcycle and from the original cycle itself are smooth, it would be undesirable to use them all since they are clearly not linearly independent.

The main difference between the various double large prime algorithms is how the graph $G$ is updated, i.e., how edges are added to the graph. In the following sections, we will describe three ways of constructing the graph and produce only good relations. We will call these the *full graph*, *simplified graph*, and *concentric circles* method.

### 21.3.2.a  Full graph method

The first, and more natural, approach to building the graph is to use every almost-smooth and 2-almost-smooth divisor to add an edge to the graph, giving us the *full graph*.

To avoid producing redundant relations, edges that would close a cycle are used to produce relations but are not added to the graph, which means that the graph $G$ never actually contains any cycles. Since not all cycles produce smooth relations, the non-smooth relations that may be obtained from these cycles (which are in fact almost-smooth relations) are used to produce an edge between 1 and the non-canceled large prime.

**Algorithm 21.19** Full graph method

INPUT: A graph $G$, a set of relations $\mathcal{R} = \{R_e\}$ corresponding to the edges of $G$, and a new almost-smooth or 2-almost-smooth relation $R$.
OUTPUT: Updated $G$ and $\mathcal{R}$ and (if possible) a smooth relation $S$.

1.   **if** $R$ is almost-smooth **then**
2.       $P \leftarrow$ large prime in $R$
3.       **if** $P$ is connected to 1 **then**
4.           $(e_1, \ldots, e_i) \leftarrow$ path from 1 to $P$ in $G$
5.           use $R$ and $R_{e_1}, \ldots, R_{e_i}$ to cancel the large primes
6.           $S \leftarrow$ smooth divisor obtained
7.           leave $G$ and $\mathcal{R}$ untouched
8.       **else**
9.           add the edge $(1, P)$ to $G$

## § 21.3 Large prime methods

10.                   add $R_{(1,P)} = R$ to $\mathcal{R}$
11.      **if** $R$ is 2-almost-smooth **then**
12.          $P_1, P_2 \leftarrow$ large primes in $R$
13.          **if** $(P_1, P_2)$ would create a loop in $G$ **then**
14.              **if** the loop would contain the vertex 1 **then**
15.                  $(e_1, \ldots, e_i) \leftarrow$ path from 1 to $P_1$ in $G$
16.                  $(f_1, \ldots, f_j) \leftarrow$ path from 1 to $P_2$ in $G$
17.                  use $R$ and $R_{e_1}, \ldots, R_{e_i}$ and $R_{f_1}, \ldots, R_{f_j}$ to cancel the large primes
18.                  $S \leftarrow$ smooth divisor obtained
19.                  leave $G$ and $\mathcal{R}$ untouched
20.              **else**
21.                  $(e_1, \ldots, e_i) \leftarrow$ path from $P_1$ to $P_2$ in $G$
22.                  use $R$ and $R_{e_1}, \ldots, R_{e_i}$ to cancel the large primes other than $P_1$
23.                  **if** $P_1$ is also canceled **then**            [smooth relation]
24.                      $S \leftarrow$ smooth divisor obtained
25.                      leave $G$ and $\mathcal{R}$ untouched
26.                  **else**            [almost-smooth relation]
27.                      $R_{(1,P_1)} \leftarrow$ almost-smooth relation obtained
28.                      add the edge $(1, P_1)$ to $G$
29.                      add $R_{(1,P_1)}$ to $\mathcal{R}$
30.          **else**
31.              add the edge $(P_1, P_2)$ to $G$
32.              add $R_{(P_1,P_2)} = R$ to $\mathcal{R}$
33.   **return** $G, \mathcal{R}$ and if possible $S$

**Remark 21.20** This method is heuristically faster than the methods in next sections but its running time is (at the time of this writing) still unproven. Although the relation search is faster with the full graph method, there is no proven bound on the weight of the system generated using this method, hence the cost of solving the linear system is unclear. If the system can be proven to be sparse enough (and its weight is not significantly greater than with the other graph methods), this is the method that should be favored.

The other methods described in the next two sections look for cycles in specific subgraphs of $G$ instead of the entire full graph and limit themselves to the connected component of the vertex 1.

### 21.3.2.b  Simplified graph method

In order to make the analysis more accessible, Gaudry, Thériault, and Thomé [GATH[+] 2004] introduce a more restrictive approach to the graph construction. The idea is to discard edges that fall completely outside of the connected component of the graph containing the vertex 1.

**Algorithm 21.21** Simplified graph method

INPUT: A graph $G$, a set of relations $\mathcal{R} = \{R_e\}$ corresponding to the edges of $G$, and a new almost-smooth or 2-almost-smooth relation $R$.
OUTPUT: Updated $G$ and $\mathcal{R}$ and (if possible) a smooth relation $S$.

1. **if** $R$ is almost-smooth **then**
2.     $P \leftarrow$ large prime in $R$
3.     **if** $P$ is connected to 1 **then**
4.         $(e_1, \ldots, e_i) \leftarrow$ path from 1 to $P$ in $G$
5.         use $R$ and $R_{e_1}, \ldots, R_{e_i}$ to cancel the large primes
6.         $S \leftarrow$ smooth divisor obtained
7.         leave $G$ and $\mathcal{R}$ untouched
8.     **else**
9.         add the edge $(1, P)$ to $G$
10.         add $R_{(1,P)} = R$ to $\mathcal{R}$
11. **if** $R$ is 2-almost-smooth **then**
12.     $P_1, P_2 \leftarrow$ large primes in $R$
13.     **if** $P_1$ and $P_2$ are both connected to 1 **then**
14.         $(e_1, \ldots, e_i) \leftarrow$ path from 1 to $P_1$ in $G$
15.         $(f_1, \ldots, f_j) \leftarrow$ path from 1 to $P_2$ in $G$
16.         use $R$ and $R_{e_1}, \ldots, R_{e_i}$ and $R_{f_1}, \ldots, R_{f_j}$ to cancel the large primes
17.         $S \leftarrow$ smooth divisor obtained
18.         leave $G$ and $\mathcal{R}$ untouched
19.     **else if** one of $P_1$ or $P_2$ is connected to 1 **then**
20.         add the edge $(P_1, P_2)$ to $G$
21.         add $R_{(P_1,P_2)} = R$ to $\mathcal{R}$
22.     **else**
23.         leave $G$ and $\mathcal{R}$ untouched
24. **return** $G, \mathcal{R}$ and if possible $S$

This approach has the following advantages compared to the full graph method:

- The algorithm as proven bounds on both the relation search and the linear algebra.
- All the cycles found in the graph produce a smooth relation since they correspond to chains of divisors with first and last divisors almost-smooth.

Obviously the bound on the relation search with the simplified graph methods automatically gives an upper bound on the relation search for the full graph method. However, the bound on the linear algebra may not hold true for Algorithm 21.19, which is why Algorithm 21.21 is still important.

There are some significant disadvantages, however:

- Cycles in the full graph that are not connected to 1 cannot be found.

- When looking at an edge during the search, if neither of the vertices have already been connected to 1 then the edge is not used. But it can happen that the edge would be in the connected component of 1 in the full graph (because of edges that are found later in the search or other edges that were not used). This will most likely destroy some cycles.
- The relation search is slower (because of the previous points).

By choosing the factor base $\mathcal{P}_B$ such that $|\mathcal{P}_B| = O(g^2 q^{(g-1)/g+\epsilon})$, we get the following result:

**Theorem 21.22** [GATH$^+$ 2004, Theorem 2] Let $C$ be a hyperelliptic curve of genus $g$ over the finite field $\mathbb{F}_q$. If $q > g!$ then discrete logarithms in $J_C(\mathbb{F}_q)$ can be computed in expected time $O(g^5 q^{2-\frac{2}{g}+\epsilon})$.

**Remark 21.23** The $\epsilon$ in the exponents of Theorems 21.14, 21.15, 21.17, and 21.22 are due to $\ln q$ factor in the running times. These $\epsilon$ hide the fact that the double large prime algorithm contains an extra $\ln q$ factors compared to the three other variants.

### 21.3.2.c Concentric circles method

The approach described in this section is a modification of an algorithm by Nagao [NAG 2004].

Like the simplified graph method, we are looking for cycles in the connected component of the full graph containing 1. Instead of building the graph bit by bit as almost-smooth and 2-almost-smooth relations are found, the random walk is completely done first and then the graph is constructed.

To make it possible to keep the weight of the smooth relations low enough (and give a bound on the linear algebra), the random walk is prolonged so that the connected component of 1 contains more cycles than necessary. The shortest relations are found by looking at the large primes in concentric circles centered at 1, i.e., in terms of their distance to the vertex 1.

In order to reduce the total cost of the index calculus algorithm as much as possible, the factor base is again chosen with size $O(g^2 q^{(g-1)/g+\epsilon})$, giving the same running time of $O(g^5 q^{2(g-1)/g+\epsilon})$ as the simplified graph method (possibly with a slightly different constant). To make notation easier, three sets are constructed during the random walk:

- the set $\mathcal{F}$ contains the smooth divisors found,
- the set $\mathcal{G}$ contains the almost-smooth divisors, and
- the set $\mathcal{H}$ contains the 2-almost-smooth divisors.

For each concentric circle, we build a list $\mathcal{L}_n$ of the large primes at distance $n$ from the vertex 1 (and their corresponding almost-smooth relations) and a subset $\mathcal{H}_n$ of the divisors in $\mathcal{H}$ that have not yet been used to built smooth or almost-smooth relations.

---

**Algorithm 21.24** Concentric circles method

INPUT: Sets $\mathcal{F}, \mathcal{G}, \mathcal{H}$ of respectively smooth, almost-smooth, and 2-almost-smooth divisors.
OUTPUT: A system of $r$ smooth relations.

1.     $i \leftarrow 1$
2.     **while** $i \leqslant |\mathcal{F}|$ **and** $i \leqslant r$ **do**
3.         $D_i \leftarrow i$-th divisor in $\mathcal{F}$
4.     $\mathcal{L}_1 \leftarrow \{\}$
5.     $j \leftarrow 1$
6.     **while** $j \leqslant |\mathcal{G}|$ **and** $i \leqslant r$ **do**

7.     $R \leftarrow j$-th divisor in $\mathcal{G}$
8.     $P \leftarrow$ large prime in $R$
9.     **if** $\pm P \in \mathcal{L}_1$ **then**
10.         $R_0 \leftarrow$ almost-smooth relation for $\pm P$
11.         use $R$ and $R_1$ to cancel the large prime
12.         $D_i \leftarrow$ smooth divisor obtained
13.         $i \leftarrow i + 1$
14.         leave $\mathcal{L}_1$ untouched
15.     **else**
16.         almost-smooth divisor for $\pm P \leftarrow R$
17.         add $P$ to $\mathcal{L}_1$
18.     $j \leftarrow j + 1$
19. $\mathcal{H}_1 \leftarrow \mathcal{H}$
20. $n \leftarrow 1$
21. **while** $i \leqslant r$ **do**
22.     $\mathcal{L}_{n+1} \leftarrow \{\}$
23.     $\mathcal{H}_{n+1} \leftarrow \{\}$
24.     $j \leftarrow 1$
25.     **while** $j \leqslant |\mathcal{H}_n|$ **and** $i \leqslant r$ **do**
26.         $R \leftarrow j$-th divisor in $\mathcal{H}_n$
27.         $P_1, P_2 \leftarrow$ large primes in $R$
28.         **if** $\pm P_1$ and $\pm P_2 \in \mathcal{L}_n \cup \mathcal{L}_{n+1}$ **then**
29.             $R_1 \leftarrow$ almost-smooth relation for $\pm P_1$
30.             $R_2 \leftarrow$ almost-smooth relation for $\pm P_2$
31.             use $R, R_1$ and $R_2$ to cancel the large primes
32.             $D_i \leftarrow$ smooth divisor obtained
33.             $i \leftarrow i + 1$
34.             leave $\mathcal{L}_{n+1}$ and $\mathcal{H}_{n+1}$ untouched
35.         **else if** $\pm P_1 \in \mathcal{L}_n, \pm P_2 \notin \mathcal{L}_n \cup \mathcal{L}_{n+1}$ **then**
36.             $R_1 \leftarrow$ almost-smooth relation for $\pm P_1$
37.             use $R$ and $R_1$ to cancel $P_1$
38.             almost-smooth divisor for $\pm P_2 \leftarrow$ almost-smooth divisor obtained
39.             add $P_2$ to $\mathcal{L}_{n+1}$
40.             leave $\mathcal{H}_{n+1}$ untouched
41.         **else if** $\pm P_2 \in \mathcal{L}_n, \pm P_1 \notin \mathcal{L}_n \cup \mathcal{L}_{n+1}$ **then**
42.             $R_2 \leftarrow$ almost-smooth relation for $\pm P_2$
43.             use $R$ and $R_2$ to cancel $P_2$
44.             almost-smooth divisor for $\pm P_1 \leftarrow$ almost-smooth divisor obtained

| | | |
|---|---|---|
| 45. | | add $P_1$ to $\mathcal{L}_{n+1}$ |
| 46. | | leave $\mathcal{H}_{n+1}$ untouched |
| 47. | **else** | |
| 48. | | add $R$ to $\mathcal{H}_{n+1}$ |
| 49. | | leave $\mathcal{L}_{n+1}$ untouched |
| 50. | | $j \leftarrow j+1$ |
| 51. | $n \leftarrow n+1$ | |
| 52. | **return** $\{D_1, \ldots, D_r\}$ | |

## Parallelization

Contrary to Algorithms 21.19 and 21.21, Algorithm 21.24 can be easily adapted to work distributively even on non-shared memory computers. The sets $\mathcal{F}, \mathcal{G}$, and $\mathcal{H}$ are constructed independently on each node. At the beginning of every layer of concentric circles, each node has its own $\mathcal{H}_n$ and the server sends $\mathcal{L}_n$ and the corresponding partial relations to the nodes. Each node computes its $\mathcal{H}_{n+1}$ and its part of $\mathcal{L}_{n+1}$. All the partial $\mathcal{L}_{n+1}$'s and corresponding partial relations (as well as the smooth relations produced) are collected by the server and combined, keeping at most one copy of each large prime (and using supplementary appearances to produce more smooth relations).

# Chapter 22

# Transfer of Discrete Logarithms

### Gerhard Frey and Tanja Lange

**Contents in Brief**

| | | |
|---|---|---|
| **22.1** | **Transfer of discrete logarithms to $\mathbb{F}_q$-vector spaces** | 529 |
| **22.2** | **Transfer of discrete logarithms by pairings** | 530 |
| **22.3** | **Transfer of discrete logarithms by Weil descent** | 530 |
| | Summary of background • The GHS algorithm • Odd characteristic • Transfer via covers • Index calculus method via hyperplane sections | |

In this chapter we give three different methods of how to transfer the discrete logarithm problem on the Jacobian of a curve to the discrete logarithm problem on a different group where it might be easier to solve. The cryptographic implications on the security of the DL systems are discussed in Chapter 23.

We briefly show the transfer to the additive group of a finite field in the exceptional case where $\ell = p$. The next section shows how the pairings introduced in Chapter 6 can be used to transfer the discrete logarithm problem. Then we explain in detail how the methodology of Weil descent can be used to transfer the discrete logarithm problem from one variety to another.

## 22.1 Transfer of discrete logarithms to $\mathbb{F}_q$-vector spaces

Let $C/\mathbb{F}_q, q = p^d$ be a hyperelliptic curve of genus $g$ and assume that $p \mid |\text{Pic}_C^0|$. For this case, it is shown in Section 4.4.3.a that there exists a map from $\text{Pic}_C^0[p]$ to $\Omega^0(C)$, the $\mathbb{F}_q$-vector space of holomorphic differentials on $C$. This space is isomorphic to $\mathbb{F}_q^{2g-1}$. Even though computing the map involves some effort, the complexity of computing it is in $O(\lg q)$.

This means that the discrete logarithm problem in $\text{Pic}_C^0[p]$ can be transferred efficiently to $\mathbb{F}_q^{2g-1}$, where it can be solved by methods generalizing the Euclidean algorithm (for $g = 1$) and linear algebra techniques in general. These methods run in $O\big((2g-1)\lg(q)^k\big)$, where $k$ is a small constant.

Note that this transfer is only interesting if one considers the discrete logarithm problem in $\text{Pic}_C^0[p]$ and, hence, if $p$ is sufficiently large. For a random curve this situation will not occur.

The general case for this transfer is presented in [RÜC 1999], for elliptic curves this method was discovered by several authors [SAAR 1998, SEM 1998, SMA 1999a].

For applications this case can be easily identified. So far only destructive consequences of this transfer are known, namely breaking of the discrete logarithm problem in $\text{Pic}_C^0[p]$ by transfer to the easier problem.

## 22.2 Transfer of discrete logarithms by pairings

As usual, let $C/\mathbb{F}_q, q = p^d$ be a hyperelliptic curve of genus $g$. We consider the discrete logarithm problem in a subgroup of $\text{Pic}_C^0$ of prime order $\ell$. Let $k$ be such that $\ell \mid q^k - 1$.

In Chapter 6, the Tate–Lichtenbaum pairing was defined. For our purposes the following version is most useful. As in Chapter 16 we interpret the Tate–Lichtenbaum pairing as the pairing

$$T_\ell : J_C(\mathbb{F}_q)[\ell] \times J_C(\mathbb{F}_{q^k}) \to \mathbb{F}_{q^k}^*[\ell],$$

$$(\overline{D}, \overline{E}) \mapsto \left(f_{\overline{D}}(E)\right)^{\frac{q^k - 1}{\ell}}$$

where $\overline{D}$ and $\overline{E}$ are elements of $J_C(\mathbb{F}_q)[\ell]$ respectively $J_C(\mathbb{F}_{q^k})$ and $\mathbb{F}_{q^k}^*[\ell]$ are the $\ell$-th roots of unity. We let $\overline{D}$ and $\overline{E}$ be represented by the divisors $D$ and $E$ such that no point $P \in C$ occurs in both $E$ and $D$.

The original Tate–Lichtenbaum pairing is not degenerate. In our version this is expressed by the following fact: for randomly chosen $\overline{E} \in J_C(\mathbb{F}_{q^k})$ and $\overline{D} \neq 0$ we get $T_\ell(\overline{D}, \overline{E}) \neq 1$. Chapter 16 deals with the efficient implementation of this map showing that computing $T_\ell$ is polynomial in $q^k$.

The implications on the discrete logarithm problem are immediate. Consider the subgroup of order $\ell$ of $\text{Pic}_C^0$ and let $\overline{D}$ be the base point. The degree of extension $k$ can be obtained from $q$ and $\ell$. Choose $\overline{E}$ randomly in $J_C(\mathbb{F}_{q^k})$. One can expect that $T_\ell(\overline{D}, \overline{E}) = \zeta$, where $\zeta$ is an $\ell$-th root of unity. Then the discrete logarithm problem $\overline{F} = [n]\overline{D}$ can be transferred to $\mathbb{F}_{q^k}^*[\ell]$ by computing $T_\ell(\overline{D}, \overline{E})$ and $T_\ell(\overline{F}, \overline{E})$ and observing that

$$T_\ell(\overline{F}, \overline{E}) = T_\ell(\overline{D}, \overline{E})^n.$$

If $k$ is small, the pairing can be efficiently computed and the discrete logarithm problem can be solved in the group of $\ell$-th roots of unity in $\mathbb{F}_{q^k}$ by methods explained in the examples of Chapter 20. For elliptic curves this transfer was pointed out in [MEOK+ 1993]. The general case is considered in [FRRÜ 1994, FRMÜ+ 1999].

We like to stress that the transfer is possible in any case. Negative implications are that for small $k$ a cryptosystem built on $\text{Pic}_C^0$ can be attacked using the pairing (cf. Chapter 23) while the positive applications of the transfer are dealt with in Chapter 24. As shown in Section 24.2.2, the embedding degree $k$ is always small for supersingular curves.

## 22.3 Transfer of discrete logarithms by Weil descent

The principle of transfers by Weil descent is that one transfers the discrete logarithm problem from $J_C(\mathbb{F}_{q^k})$ to the discrete logarithm problem of its Weil descent and then uses additional structures on this higher dimensional variety. Here, we present 3 variants that have been proposed in the literature. The mathematical background is explained in Chapter 7.

Note that for security considerations it is not enough to exclude these cases as Weil descent is a far more general concept. It is applicable in composite fields. We start with a finite field $\mathbb{F}_{q^k}$, $q = p^d$, and assume $k > 1$.

With the main applications in mind (see Chapter 23) we are especially interested in the case that $\text{Pic}_C^0$ contains a large subgroup of prime order $\ell$. This is needed in some places and mentioned there, but we would like to stress here that this is no real restriction, just the most interesting case.

## 22.3.1 Summary of background

We recall that the Weil descent $W_{\mathbb{F}_{q^k}/\mathbb{F}_q}(\mathcal{A})(\mathbb{F}_{q^k})$ of an abelian variety $\mathcal{A}/\mathbb{F}_{q^k}$ has the property

$$W_{\mathbb{F}_{q^k}/\mathbb{F}_q}(\mathcal{A})(\mathbb{F}_q) = \mathcal{A}(\mathbb{F}_{q^k}).$$

By the construction (either by using affine pieces or by Galois theory cf. Chapter 7) this identification can be made explicit and realized computationally. So, obviously, the discrete logarithm problem can be transferred from $\mathcal{A}(\mathbb{F}_{q^k})$ to $W_{\mathbb{F}_{q^k}/\mathbb{F}_q}(\mathcal{A})(\mathbb{F}_q)$. It can happen that $W_{\mathbb{F}_{q^k}/\mathbb{F}_q}(\mathcal{A})(\mathbb{F}_q)$ is a factor of the Jacobian variety of a curve $D/\mathbb{F}_q$. Equivalently this means that we can embed $D$ into $W_{\mathbb{F}_{q^k}/\mathbb{F}_q}(\mathcal{A})$. Then the discrete logarithm problem in $W_{\mathbb{F}_{q^k}/\mathbb{F}_q}(\mathcal{A})(\mathbb{F}_q)$ is transferred to the discrete logarithm problem in the Jacobian of $D$.

This situation can be realized by covers for $\mathcal{A} = J_C$. A curve $D/\mathbb{F}_q$ is called a *cover* of $C/\mathbb{F}_{q^k}$ if there exists a nonconstant morphism

$$\psi : D_{\mathbb{F}_{q^k}} \to C$$

defined over $\mathbb{F}_{q^k}$. Under this condition the transfer can be made explicit by norm and conorm maps of divisor classes. As usual we denote by $\psi^*$ the induced map from $\operatorname{Pic}_C^0$ to $\operatorname{Pic}_{D_{\mathbb{F}_{q^k}}}^0$. It corresponds to the conorm map of divisors in the function fields $\psi^*\bigl(\mathbb{F}_{q^k}(C)\bigr) \subset \mathbb{F}_{q^k}(D)$.

Next we use the inclusion $\mathbb{F}_q(C) \subset \mathbb{F}_{q^k}(D)$ to define a correspondence map on divisor classes

$$\phi : \operatorname{Pic}_C^0 \to \operatorname{Pic}_D^0$$

given by

$$\phi := \operatorname{N}_{\mathbb{F}_{q^k}/\mathbb{F}_q} \circ \psi^*.$$

We assume that both $D$ and the cover map $\psi$ are explicitly given and that $\psi$ can be computed quickly.

Of course, the efficiency of the transfer depends heavily on the genus $g'$ of $D$. As $J_D$ needs to contain the considered subgroup of $W_{\mathbb{F}_{q^k}/\mathbb{F}_q}(\mathcal{A})(\mathbb{F}_q)$ the genus cannot be too small as by the Hasse–Weil Theorem 5.76 and its corollary we have $|J_D(\mathbb{F}_q)| = O(q^{g'})$. If the subgroup under consideration has size $O(q^{gk})$ we have

$$g' \gtrsim gk. \tag{22.1}$$

## 22.3.2 The GHS algorithm

In this section we want to explain the (until now) most successful applications of the idea to use Weil restriction to get a transfer of discrete logarithms. Building on [GASM 1999] Gaudry, Hess, and Smart [GAHE+ 2002b] developed it for $C$ an elliptic curve $E$ defined over binary fields, $q = 2^d$. For this reason we shall call it the *GHS algorithm*. This section contains material from [DIE 2003], [GAL 2001b],[GAHE+ 2002a], [HES 2003], [JAME+ 2001], [MAME+ 2002], [MEQU 2001], and [THÉ 2003b].

We describe the more general setting of starting with a hyperelliptic curve $C$ of genus $g$.

Let $C/\mathbb{F}_{q^k}$ be given by the usual equation

$$C : y^2 + h(x)y = f(x), \quad f(x), h(x) \in \mathbb{F}_{q^k}[x],$$

where $f$ is monic of degree $2g + 1$ and $\deg(h) \leqslant g$ (cf. 4.3).

**Remark 22.1** To use Artin–Schreier theory we make a change of variables and write
$$C : v^2 + v = \tilde{f}(x),$$
where $v = y/h(x)$ and $\tilde{f}(x) = f(x)/h^2(x)$. Note that in general $\tilde{f}(x)$ is not a polynomial. By abuse of notation, from now on we write the equation as
$$C : y^2 + y = f(x), f(x) \in \mathbb{F}_{q^k}(x).$$

The crucial ingredient for the algorithm is a careful choice of an extension of the Frobenius automorphism $\phi_q : \mathbb{F}_{q^k} \to \mathbb{F}_{q^k}$ to an isomorphism on $\mathbb{F}_{q^k}(C)$ into its algebraic closure. We extend $\phi_q$ to an automorphism of $\mathbb{F}_{q^k}(x)/\mathbb{F}_q$ by requiring $\phi_q(x) = x$ and $\phi_q(y) = y^{(1)}$ such that the following equation is satisfied
$$C^{\phi_q} : y^{(1)^2} + y^{(1)} = \phi_q\bigl(f(x)\bigr).$$

The field $\mathbb{F}_{q^k}(C^{\phi_q}) := \mathbb{F}_{q^k}(x)[T]/\bigl(T^2 + T - \phi_q(f(x))\bigr)$ is the function field of a hyperelliptic curve $C^{\phi_q}$ with rational subfield $\mathbb{F}_{q^k}(x)$.

We repeat this construction to get $\mathbb{F}_{q^k}(C^{\phi_q^i}) = \mathbb{F}_{q^k}\bigl((C^{\phi_q^{i-1}})^{\phi_q}\bigr)$ and note that $\mathbb{F}_{q^k}(C^{\phi_q^k}) = \mathbb{F}_{q^k}(C)$.

Take $F'$ as composite of the fields $\mathbb{F}_{q^k}(C), \mathbb{F}_{q^k}(C^{\phi_q}), \ldots, \mathbb{F}_{q^k}(C^{\phi_q^{k-1}})$. It is a Galois extension of $\mathbb{F}_{q^k}(x)$ with a two-elementary Galois group of order $2^m$ with $m \leqslant k$. The algebraic closure of $\mathbb{F}_{q^k}$ in $F'$ has degree at most 2. To simplify the exposition we assume from now on that the degree is equal to 1 and so $F'$ is the function field of an absolutely irreducible nonsingular curve $D'$ defined over $\mathbb{F}_{q^k}$. The inclusion $\mathbb{F}_{q^k}(C) \hookrightarrow F'$ induces a cover
$$\psi : D' \to C$$
of degree $2^m$.

We assume $k$ is either equal to 2 or odd. Then one has that the Frobenius automorphism $\phi_q$ on $\mathbb{F}_{q^k}(C)$ has an extension to an automorphism $\phi_q'$ of $F'$ of order $k$. Define $F = F'^{\phi_q'}$ as fixed field of $\phi_q'$.

**Lemma 22.2** The field $F$ is a function field in one variable with field of constants $\mathbb{F}_q$ containing $\mathbb{F}_q(x)$ as subfield of index $2^m$.

The projective curve $D$ corresponding to $F$ has the property that
$$D \cdot \mathbb{F}_{q^k} = D'.$$

Hence, we are in the situation described at the beginning and the cover $\psi$ induces a homomorphism
$$\phi : \mathrm{Pic}_C^0 \to \mathrm{Pic}_D^0, \phi = \mathrm{N}_{\mathbb{F}_{q^k}/\mathbb{F}_q} \circ \psi^*.$$

The map $\phi$ is useful if the subgroup we are interested in is mapped injectively to $\mathrm{Pic}_D^0$. For the most interesting cases Diem [DIE 2003, Theorem 1] shows that this requirement is satisfied.

**Proposition 22.3** Assume that $k$ is a prime and that $C$ cannot be obtained by constant field extension from a curve defined over $\mathbb{F}_q$. Then the kernel of $\phi$ contains only points of order 2.

This encourages us to go further and study explicitly the construction of $D$. In the sequel we assume $k$ to be prime. We give an affine part of $D$ as an intersection of hyperplanes. Therefore, we take a

§ 22.3 Transfer of discrete logarithms by Weil descent

basis $(\alpha_0, \alpha_1, \ldots, \alpha_{k-1})$ of $\mathbb{F}_{q^k}$ as $\mathbb{F}_q$-vector space, where we additionally require $\sum_{i=0}^{k-1} \alpha_i = 1$. We introduce the variables $x_j, y_j$ for $j = 0, \ldots, k-1$ by

$$x = \sum_{j=0}^{k-1} \alpha_j x_j \text{ and } y = \sum_{j=0}^{k-1} \alpha_j y_j.$$

By Proposition 7.1, we get a system of equations defining an open affine part $W_a$ of the Weil restriction $W_{\mathbb{F}_{q^k}/\mathbb{F}_q}(C)$ in the affine space $\mathbb{A}^{2k}$ by plugging in these expressions for $x, y$ into the equation defining $C$.

Define the hyperplanes

$$H_j : x_j = x_0$$

for $j = 1, \ldots, k-1$, then

$$D_a := W_a \cap \bigcap_{j=1}^{k-1} H_j$$

is an affine nonempty open part of the curve $D$. Thus $D$ can be computed in principle.

**Remark 22.4** Up to this point of discussion we did not use that $q$ is even. From now on this will become crucial.

We have defined the function fields $\mathbb{F}_{q^k}(C^{\phi_q^i})$ and their composite field $F'$. To determine $m$ we use Artin–Schreier theory and get that the Galois group of $F'/\mathbb{F}_{q^k}(x)$ is isomorphic to a subspace $U$ generated by the classes of $\{f, \phi_q(f), \ldots, \phi_q^{k-1}(f)\}$ of the $\mathbb{F}_2$-vector space $\mathbb{F}_{q^k}(x)/\mathcal{P}(\mathbb{F}_{q^k}(x))$, where $\mathcal{P} : z \mapsto \mathcal{P}(z) = z^2 + z$ is the Artin–Schreier operator. Thus $m$ is equal to $\dim_{\mathbb{F}_2}(U)$

$$m = \dim \left\{ \mathrm{span}_{\mathbb{F}_2} \{\overline{f}, \overline{\phi_q(f)}, \ldots, \overline{\phi_q^{k-1}(f)}\} \right\}.$$

The genus of the resulting curve will depend heavily on $m$. In fact we will show that for elliptic curves and some hyperelliptic curves one has $g' = g2^{m-1} - 1$ or $g' = g2^{m-1}$.

To compute $m$ under the additional assumption $\gcd(d, k) = 1$ we observe that $U$ is not only a vector space over $\mathbb{F}_2$ but even a $\mathbb{F}_2[\mathrm{Gal}(\mathbb{F}_{q^k}/\mathbb{F}_q)]$-module. The assumption that $C$ is not defined over $\mathbb{F}_q$ implies that $U$ is nontrivial. As $\phi_q(y) = y^{(1)}$ we get that $m$ depends on the length of the orbit of $y$.

Hence,

$$m = \kappa \varphi_2(k) \text{ or } m = \kappa \varphi_2(k) + 1 \tag{22.2}$$

with some natural number $\kappa = 1, \ldots, (k-1)/\varphi_2(k)$, where $\varphi_2(k)$ is the multiplicative order of 2 modulo $k$.

For *Mersenne primes*, i.e., primes $k$ of the form $2^a - 1$, $\varphi_2(k) = a$ is minimal. If $\kappa = 1$ then the genus of $D$ can be as low as $g2^{\varphi_2(k)} = gk$ which is minimal according to (22.1).

The Mersenne primes in the cryptographically important range are

$$3, 7, 31, \text{ and } 127.$$

For these primes some curves may allow $\kappa = 1$ and, hence, the transfer will lead to a probably easier problem. For $k = 3$ it is possible to find curves with $g' = 3g$.

A further class of primes $k$ leading to a small $\varphi_2(k)$ are the Fermat primes given by $k = 2^{2^a} + 1$. In the cryptographically important range they are

$$3, 5, 17, \text{ and } 257.$$

Explicitly, if $k = 5$ and $g = 1$, the minimal genus one can obtain is 7; if $k = 17$, the minimal genus is 127, and for $k = 257$, it is 32768. For the latter two cases, $g'$ is already quite high in relation to $k$.

In the literature [DIE 2001, GAHE+ 2002b, MEQU 2001, THÉ 2003b] one finds examples of curves $C$ and fields $\mathbb{F}_{q^k}$ with small $\kappa$ and $\varphi_2(k)$ and, hence, of examples where the transfer leads to a curve of relatively small genus.

The integer $m$ was determined such that $F'$ is defined by $\mathbb{F}_{q^k}(C), \ldots, \mathbb{F}_{q^k}(C^{\phi_q^{m-1}})$. An interesting special case is that $D$ is hyperelliptic. In this case one can use the highly optimized arithmetic in its divisor class group, and the index-calculus methods for computing the discrete logarithm in these groups are well analyzed. (But see Remark 22.8.) So it seems worthwhile to look for conditions which yield that $D$ is hyperelliptic.

We recall that we have to find a subfield $L$ of $F'$ of degree 2 such that $L$ is rational over $\mathbb{F}_{q^k}$.

We give a sufficient condition for this and show how the proof works. An additional benefit is that one can also easily compute the genus of $D'$ and hence of $D$ by using the explicit generation of the rational subfield. The field $F$ is then obtained as the fixed field of the Frobenius.

**Proposition 22.5** Let $C/\mathbb{F}_{q^k}$ with $k$ odd be a nonsingular hyperelliptic curve of genus $g$ given by an equation
$$C : y^2 + y = f(x) \tag{22.3}$$
with $f(x) = (\alpha x + \beta) + \dfrac{u(x)}{v(x)^2}$, $\beta \in \mathbb{F}_{q^k}$, $\alpha \in \mathbb{F}_{q^k} \setminus \mathbb{F}_q$, $u(x), v(x) \in \mathbb{F}_q[x]$, $\deg(u) \leqslant 2g+1$ and $\deg(v) \leqslant g$.

Then $D'$ is hyperelliptic of genus $g2^{m-1} - 1$ or $g2^{m-1}$.

**Remarks 22.6**

(i) Note that these conditions are always satisfied for elliptic curves. The original paper [GAHE+ 2002b] gives an explicit description of the equation that is a special case of the result given below.

(ii) Thériault [THÉ 2003b] shows which genus is assumed depending on conditions on $u$ and $v$.

The crucial step for establishing this result is the explicit construction of $L$. This field is constructed as composite of fields $L_i$ of index 2 in $\mathbb{F}_{q^k}(C) \cdot \mathbb{F}_{q^k}(C^{\phi_q^i})$. Since $\mathbb{F}_{q^k}(C^{\phi_q^i})$ is a quadratic extension of $\mathbb{F}_{q^k}(x)$ with equation
$$y^{(i)^2} + y^{(i)} = \phi_q^i(\alpha x + \beta) + \frac{u(x)}{v(x)^2}$$
we get by Artin–Schreier theory that $L_i$ is defined by
$$t_i^2 + t_i = (\alpha x + \beta) + \frac{u(x)}{v(x)^2} + \phi_q^i(\alpha x + \beta) + \frac{u(x)}{v(x)^2} = (\alpha x + \beta) + \phi_q^i(\alpha x + \beta)$$
for $t_i = y + y^{(i)}$ and, hence, has genus 0.

Taking the composite $L$ of these $L_1, \ldots, L_{m-1}$ and following the same procedure we obtain that $L$ is a rational extension of $\mathbb{F}_{q^k}$ with $[L : \mathbb{F}_{q^k}(x)] = 2^{m-1}$. This dependence can be made explicit by expressing $L = \mathbb{F}_{q^k}(c)$ with $x(c) = \lambda_{-1} + \sum_{i=0}^{m-1} \lambda_i c^{2^i}$, with $\lambda_i \in \mathbb{F}_{q^k}, \lambda_0, \lambda_{m-1} \neq 0$.

Inserting this expression for $x$ in the definition of $\mathbb{F}_{q^k}(C)$, we get an equation
$$y^2 + y = \tilde{f}(c)$$
from which one can read off the genus. As $k$ is assumed to be odd, the fixed field under the Frobenius has the same genus and one can obtain $F$ from $F'$.

§ 22.3 Transfer of discrete logarithms by Weil descent

For curves $C$ given by an equation of the type (22.3) we can explicitly give the equation of the resulting curve. Let $\min_\alpha(z)$ be the minimal polynomial of $\alpha$ in $\phi_q$, i.e., $\min_\alpha(\phi_q)(\alpha) = 0$ and one additionally requires $\min_\alpha(\phi_q)(1) = 0$. Furthermore, let $\mu \in \mathbb{F}_{q^k}$ be such that $\mathrm{Tr}_{\mathbb{F}_{q^k}/\mathbb{F}_q}(\mu) = 1$ and let $\mathbb{F}_{q^k} = \mathbb{F}_2(\theta)$. We now give the algorithmic description taken from [THÉ 2003b] of how to obtain $D$ explicitly.

---

**Algorithm 22.7** Weil descent of $C$ via GHS

---

INPUT: The curve $C : y^2 + h(x)y = f(x) + (\alpha x + \beta)h(x)^2$, of genus $g$ with $f(x), h(x) \in \mathbb{F}_q, \beta \in \mathbb{F}_{q^k}, \alpha \in \mathbb{F}_{q^k} \setminus \mathbb{F}_q, \deg(f) \leq 2g+1, \deg(h) \leq g$, and $\mathbb{F}_q = \mathbb{F}_{2^d}$.
OUTPUT: The curve $D : \tilde{y}^2 + \tilde{h}(c)\tilde{y} = \tilde{f}(c)$.

1. **for** $i = 0$ **to** $k$ compute $\phi_q^i(\alpha)$ as expression over $\mathbb{F}_2$
2. compute $\min_\alpha(z)$ and $m \leftarrow \deg(\min_\alpha(z))$
3. $\nu \leftarrow \big(\min_\alpha(\theta^{2d})/(\theta^2 - 1)\big)(\beta)$
4. **if** $\mathrm{Tr}_{\mathbb{F}_{q^k}/\mathbb{F}_2}(\beta) = 0$ **then** $\nu_\beta \leftarrow \nu$
5.     **else** $u(\phi_q) = \dfrac{\phi_q^k - 1}{\min_\alpha(\phi_q)}$ and $\nu_\beta \leftarrow \nu - \dfrac{\mathrm{Tr}_{\mathbb{F}_{q^k}/\mathbb{F}_2}(\beta)}{u(\phi_q)(1)}$
6. find $0 \leq i \leq k-1$ with $\mathrm{Tr}_{\mathbb{F}_{q^k}/\mathbb{F}_q}(\theta^i) \neq 0$ and set $\mu \leftarrow \dfrac{\theta^i}{\mathrm{Tr}_{\mathbb{F}_{q^k}/\mathbb{F}_q}(\theta^i)}$
7. **for** $i = 1$ **to** $m-1$ **do** $\gamma_i \leftarrow \phi_q^i(\alpha) - \alpha$ and $\delta_i \leftarrow \phi_q^i(\beta) - \beta$
8. **for** $i = 1$ **to** $m-1$ **do**
9.     $\epsilon_i \leftarrow \gamma_i$ and $\rho_i \leftarrow \delta_i$
10.     **for** $j = i+1$ **to** $m-1$ **do** $\gamma_j \leftarrow \left(\dfrac{\gamma_j}{\epsilon_i}\right)^{1/2} - \left(\dfrac{\gamma_j}{\epsilon_i}\right)$ and $\delta_j \leftarrow \delta_j - \dfrac{\rho_i \gamma_j}{\epsilon_i}$
11. $s_{m-1} \leftarrow c$
12. **for** $i = m-2$ **down to** $0$ **do**
13.     $s_i \leftarrow \dfrac{1}{\epsilon_{i+1}}\left(s_{i+1}^2 - s_{i+1} - \rho_{i+1}\right)$
14. express $s_0(c) = \lambda_{-1} + \sum \lambda_i c^{2^i}$
15. $\tilde{\lambda} \leftarrow \sum_{i=0}^{k-1} \phi_q^i(\mu) \sum_{j=0}^{i-1} \phi_q^j(\phi_q(\lambda_0)\nu_\beta)$ and $c \leftarrow \dfrac{1}{\lambda_0}(\tilde{c} - \tilde{\lambda})$
16. $x(\tilde{c}) \leftarrow s_0(\tilde{c})$ and $z(\tilde{c}) \leftarrow s_1(\tilde{c}), \tilde{h}(\tilde{c}) \leftarrow h(x(\tilde{c}))$
17. $\tilde{f}(\tilde{c}) \leftarrow f\big(x(\tilde{c})\big) + \big(\mathrm{Tr}_{\mathbb{F}_{q^k}/\mathbb{F}_q}(\mu^2\alpha)x(\tilde{c}) + \mathrm{Tr}_{\mathbb{F}_{q^k}/\mathbb{F}_q}(\mu^2\beta)\big)\tilde{h}(\tilde{c})^2$
18. $\tilde{f}(\tilde{c}) \leftarrow \tilde{f}(\tilde{c}) + \big(\sum_{i=0}^{k-1} \phi_q^i(\mu^2 - \mu)\sum_{j=0}^{i}\phi_q^j(s_1(\tilde{c}))\big)\tilde{h}(\tilde{c})^2$
19. **return** $D : \tilde{y}^2 - \tilde{h}(\tilde{c})\tilde{y} = \tilde{f}(\tilde{c})$

---

To actually transfer the discrete logarithm problem from $J_C(\mathbb{F}_{q^k})$ to $J_D(\mathbb{F}_q)$ one first uses $x = x(\tilde{c})$ to embed $\mathbb{F}_{q^k}(C)$ into $F' = L\mathbb{F}_{q^k}(C)$ and then the norm down to $\mathbb{F}_q(D)$.

**Remark 22.8** If one gives up the restriction that the resulting curve $D$ should be hyperelliptic, one does not need curves of the form (22.3). Nevertheless the transfer of the discrete logarithm problem can be very efficient. In fact, new results of Diem [DIE 2005] motivate that in many cases the index-calculus method will be more effective if $D$ is *not* hyperelliptic. The reason is that its efficiency depends on the degree of plane models of $D$, and this degree is $\leq g+1$ for generic curves but for hyperelliptic curves in canonical form it is $2g+1$ or $2g+2$.

Menezes and Qu [MEQU 2001] give a complete study of all fields $\mathbb{F}_{2^d}$ for primes $d \in [100, 600]$ stating the minimal $m$ that could occur for an elliptic curve over $\mathbb{F}_{2^d}$. Furthermore, they also consider the finite field $\mathbb{F}_{2^{155}}$, which had been proposed to offer computational advantages for the field arithmetic (cf. Chapter 11). The genus of the resulting curve $D$ is given by $2^m$ or $2^m - 1$. This study is extended in [MAME$^+$ 2002] to composite $d$ in the same interval.

Galbraith, Hess, and Smart [GAHE$^+$ 2002a] extend the GHS method to more curves over the same field by additionally applying isogenies of small degree. This shows that more curves allow $D$ to have comparably small genus but still the fraction is negligible (about $2^{33}$ out of the $2^{156}$ isogeny classes for elliptic curves over $\mathbb{F}_{2^{155}}$). Hess [HES 2003] considers Weil descent to nonhyperelliptic curves and bounds the genus of the resulting curve. Especially for the field $\mathbb{F}_{2^{155}}$ he shows that around $2^{123}$ isogeny classes of elliptic curves $E/\mathbb{F}_{2^{155}}$ lead to a curve $D$ such that the discrete logarithm in $J_D(\mathbb{F}_{2^5})$ can be computed faster than on $J_C(\mathbb{F}_{2^{155}})$.

In [METE$^+$ 2004] the authors consider Weil descent as a transfer of the discrete logarithm problem of an elliptic curve over a composite field $\mathbb{F}_{2^d}$ where $d$ is divisible by 4 or 5. Furthermore, they look at $\mathbb{F}_{2^{161}} = \mathbb{F}_{q^7}$ with $q = 2^{23}$.

**Remark 22.9** Hess [HES 2003] and Thériault [THÉ 2003b] study Weil descent on general Artin–Schreier curves defined by an affine equation of the form

$$y^p - y = f(x), f(x) \in \mathbb{F}_{q^k}(x) \text{ and } q = p^d.$$

In fact, most results stated here for $p = 2$ hold in more generality.

### 22.3.3 Odd characteristic

In principle the same approach can be taken in the case of odd characteristic. However, the resulting field $F'$ will rarely have a subfield rational over $\mathbb{F}_{q^k}$. The background of the generalization is presented in [DIE 2001] and [DIE 2003]. Thériault [THÉ 2003c] studies for which fields and types of equation one obtains that $D$ is of the same type as the original curve $C$.

For our exposition we concentrate on the case that $\mathbb{F}_q = \mathbb{F}_{p^d}$ with $p$ odd and that $C/\mathbb{F}_{q^k}$ is *hyperelliptic* of genus $g$ given by an affine equation of the form

$$C : y^2 = f(x), f(x) \in \mathbb{F}_{q^k}[x] \text{ and } \deg(f) = 2g + 1.$$

Like for even characteristic one needs a suitable extension of the Frobenius automorphism $\phi_q$ from $\mathbb{F}_{q^k}/\mathbb{F}_q$ to an isomorphism on $\mathbb{F}_{q^k}(C)$ into its algebraic closure. We extend $\phi_q$ to an automorphism of $\mathbb{F}_{q^k}(x)/\mathbb{F}_q$ by requiring $\phi_q(x) = x$ and $\phi_q(y) = y^{(1)}$ such that the following equation is satisfied

$$C^{\phi_q} : y^{(1)^2} = \phi_q\big(f(x)\big).$$

The field $\mathbb{F}_{q^k}(C^{\phi_q}) := \mathbb{F}_{q^k}(x)[T]/\big(T^2 - \phi_q(f(x))\big)$ is the function field of a hyperelliptic curve $C^{\phi_q}$ with rational subfield $\mathbb{F}_{q^k}(x)$. Let $F'$ be the composite of $\mathbb{F}_{q^k}(C), \mathbb{F}_{q^k}(C^{\phi_q}), \ldots, \mathbb{F}_{q^k}(C^{\phi_q^{k-1}})$. One obtains that

$$[F' : \mathbb{F}_{q^k}(x)] = 2^m \text{ with } m = \dim_{\mathbb{F}_2}(U),$$

where $U$ is an $\mathbb{F}_2$-vector space generated by the images of $\phi_q^i(f), 0 \leqslant i < k$ in $\mathbb{F}_{q^k}^*(x)/\mathbb{F}_{q^k}^*(x)^2$. In odd characteristic we are working in a Kummer extension and, hence, the Artin–Schreier operator is replaced accordingly by considering functions up to squares. It is possible to obtain upper and lower bounds on the genus $g'$ of the function field $F'$. We denote by $F$ the subfield of $F'$ fixed under the operation of $\phi_q$.

## § 22.3 Transfer of discrete logarithms by Weil descent

Let $r$ be the number of places in $\overline{\mathbb{F}}_q(x)$ that ramify in $\overline{\mathbb{F}}_q F$. Since all ramification orders are equal to 2, and hence the ramification is tame, the Hurwitz genus formula 4.110 yields

$$2g' - 2 = -2 \times 2^m + r \times 2^{m-1}$$

and so

$$g' = 2^{m-2}(r-4) + 1. \tag{22.4}$$

For prime $k$ and elliptic curves $E$ it is shown in [DIE 2003]:

- If $k = 2, 3, 5, 7$ there exists an elliptic curve $E$ over $\mathbb{F}_{q^k}$ such that $g = k$. If $k = 2$ or $3$, $E$ can additionally be chosen such that $F$ is hyperelliptic.
- If $k \geqslant 11$ and $q^k \geqslant 2^{160}$, then $q^g \geqslant 2^{5000}$ and thus the discrete logarithm problem in $J_E(\mathbb{F}_{q^k})$ is transferred to a far larger group.

It follows that the GHS algorithm in odd characteristic transfers the discrete logarithm problem to one that is harder to solve if the extension degree is different from $3, 5$, or $7$.

Extending this work, it is shown in [THÉ 2003b] that $k = 2$ and $k = 3$ are the only degrees of extension leading to a hyperelliptic curve $D$. Explicitly the following theorem is shown:

**Theorem 22.10**

1. Let

$$C : y^2 = (x-a)h(x) = (x-a)\sum_{i=0}^{2g} h_i x^i, h_i \in \mathbb{F}_q \text{ and } a \in \mathbb{F}_{q^2} \setminus \mathbb{F}_q.$$

   Then the Weil descent gives rise to a hyperelliptic curve of genus $2g$

$$D : v^2 = (a - \phi_q(a))^{2g+3} s(u) \sum_{i=0}^{2g} h_i \big(a(u - \phi_q(a))^2 - \phi_q(a)(u-a)^2\big)^i s(u)^{2g-i},$$

   where $s(u) = (u - \phi_q(a))^2 - (u-a)^2$.

2. Let

$$C : y^2 = (x-a)h(x), h(x) \in \mathbb{F}_q[x] \text{ and } a \in \mathbb{F}_{q^3} \setminus \mathbb{F}_q.$$

   Then the Weil descent gives rise to a hyperelliptic curve over $\mathbb{F}_q$ of genus $4g + 1$, which can be computed explicitly in the coefficients of $h$ and depending on the class of $q$ modulo 12.

3. Let

$$C : y^2 = (x-a)(x - \phi_q(a))h(x), h(x) \in \mathbb{F}_q[x] \text{ and } a \in \mathbb{F}_{q^3} \setminus \mathbb{F}_q.$$

   Then the Weil descent gives rise to a hyperelliptic curve over $\mathbb{F}_q$ of genus $4g - 1$, which can be computed explicitly in the coefficients of $h$ and depending on the class of $q$ modulo 12.

We refer to the paper for details and proofs. If one is willing to give up the hyperellipticity of the resulting curve, the transfer is useful for a larger class of curves. In [DIE 2003] one finds that for

$$C : y^2 = (x - a_1)(x - a_1^q)(x - a_2)(x - a_2^q) \cdots (x - a_{g+1})(x - a_{g+1}^q), \tag{22.5}$$

where the $a_1, a_1^q, \ldots, a_{g+1}, a_{g+1}^q \in \mathbb{F}_{q^3} \setminus \mathbb{F}_q$ are pairwise distinct, the resulting function field $F'/\mathbb{F}_{q^3}$ has genus $3g$. It does not contain a rational subfield but the fixed field of the Frobenius $\phi_q$ has degree 4 over $\mathbb{F}_q(x)$.

## 22.3.4 Transfer via covers

In Section 7.4.3 we have described a method of transferring the discrete logarithm from a subgroup of the Jacobian of a curve $C$ over $\mathbb{F}_{q^k}$ to the Jacobian of a curve $D$ defined over $\mathbb{F}_q$ by using covers in a more general way than is done in the GHS algorithm. In particular, it is not necessary that the curve $C$ is not defined over $\mathbb{F}_q$; the "twist," which ensures that the transfer map is injective on the interesting part of the divisor group is obtained by the *cover map*, which is not defined over $\mathbb{F}_q$.

Instead of following the Weil descent strategy as extended from the GHS method, we take a more theoretical approach and study which genera $g'$ could occur for $D/\mathbb{F}_q$, a cover of $C/\mathbb{F}_{q^k}$, depending on $k$ and $g$. It may then be possible to construct a curve $C$ that actually assumes these parameters. In fact, the example presented at the end of the previous section was constructed this way.

We first show how this method works in the GHS situation for curves defined over $\mathbb{F}_{q^k}$ that are not defined over a subfield. Then we consider transfers via covers for trace zero varieties. Here, it is important that the method applies also to curves defined over a subfield.

### 22.3.4.a The GHS situation

As in the previous section, we assume that the curve $C/\mathbb{F}_{q^k}$ is not defined over a subfield. We use the notions developed in the GHS method. Our main objective is to study systematically how the Galois group $\mathrm{Gal}(\overline{\mathbb{F}}_{q^k}/\mathbb{F}_q)$ operates on the branched places of $\overline{\mathbb{F}}_{q^k}F'/\overline{\mathbb{F}}_{q^k}(x)$, where like before $F'$ is the composite of $\mathbb{F}_{q^k}(C^{\phi_q^i})$, $0 \leqslant i < k$. As $C$ is not defined over a proper subfield of $\mathbb{F}_{q^k}$, at least for one root of $f(x)$ less than the whole orbit is contained in the branch points. Let $r \geqslant 2g+3$ be the number of branched places over the algebraic closure. The dimension $m$ is as defined above and can be explicitly obtained from the information on the orbits of the branch points.

**Example 22.11** To demonstrate the method we prove the claim stated about curve (22.5). The cycle pattern is given by

$$(110)(110)\ldots(110),$$

stating that $a_i$ and $\phi_q(a_i)$ are contained as branch points while $\phi_q^2(a_i)$ is not. Here, $r = 3g+3$ while the matrix

$$\begin{bmatrix} 1 & 0 & 1 \\ 1 & 1 & 0 \\ 0 & 1 & 1 \end{bmatrix}$$

obtained for each of the orbits has rank $m = 2$. Therefore, by (22.4) one obtains

$$g' = 2^{m-2}(r-4) + 1 = 2^0(3g+3-4) + 1 = 3g$$

as stated.

A systematic study of all possible factorization patterns depending on $k$ and $g$ results in the following table, which states minimal $r$ and $m$ and hence (lower) bounds on $g'$.

## § 22.3 Transfer of discrete logarithms by Weil descent

| $g$ | $k$ | Tuples | $r$ | $m$ | $g'$ |
|---|---|---|---|---|---|
| 1 | 2 | (1)(1)(1)(10) | 5 | 2 | 2 |
|   | 3 | (110)(110) | 6 | 2 | 3 |
|   | 4 | (1)(1110) | 5 | 4 | 5 |
|   | 5 | (11110) | 5 | 4 | 5 |
|   | 7 | (1110100) | 7 | 3 | 7 |
| 2 | 2 | (1)(1)(1)(1)(1)(10) | 7 | 2 | 4 |
|   | 3 | (110)(110)(110) | 9 | 2 | 6 |
|   | 4 | (1110)(1110) | 8 | 3 | 9 |
|   | 5 | (1)(111110) | 7 | 5 | 25 |
|   | 7 | (1111110) | 7 | 6 | 49 |
| 3 | 2 | (1)(1)(1)(1)(1)(1)(1)(10) | 9 | 2 | 6 |
|   | 3 | (110)(110)(110)(110) | 12 | 2 | 9 |
|   | 4 | (1)(1)(1)(1)(1)(1110) | 9 | 4 | 21 |
|   | 5 | (11110)(11110) | 10 | 4 | 25 |
|   | 7 | (1110100)(1110100) | 14 | 3 | 21 |

By these methods and extensions to the case of $k = 4$, [DIE 2001, DISC 2003] derive the following table. With the exception of the entry $g = 2, k = 4$ (marked by *) one finds curves $C$ attaining the bounds.

| $g\backslash k$ | 2 | 3 | 4 | 5 | 7 | 11 |
|---|---|---|---|---|---|---|
| 1 | 2 | 3 | 5 | 5 | 7 | $\geqslant 1793$ |
| 2 | 4 | 6 | 9* | 25 | 49 | $\geqslant 1793$ |
| 3 | 6 | 9 | 21 | 25 | 21 | $\geqslant 1793$ |

### 22.3.4.b  Transfer by covers for trace zero varieties

In Section 7.4.2 and Section 15.3 we introduced trace zero varieties of elliptic and hyperelliptic curves. Let $C$ be a hyperelliptic curve of genus $g$ defined over $\mathbb{F}_q$ and $k$ be a positive integer. The $\mathbb{F}_q$-rational points of the trace zero variety $J_{C,0}$ of the Jacobian of $C$ with respect to the extension $\mathbb{F}_{q^k}/\mathbb{F}_q$ is isomorphic to the kernel $G$ of the trace map from $\mathrm{Pic}^0_{C \cdot \mathbb{F}_{q^k}} \longrightarrow \mathrm{Pic}^0_C$, i.e.,

$$G := \{\overline{D} \in \mathrm{Pic}^0_{C \cdot \mathbb{F}_{q^k}} \mid \overline{D} + \phi_q(\overline{D}) + \cdots + \phi_q^{k-1}(\overline{D}) = 0\}.$$

For applications it is especially interesting to study the case that $k$ is small and $q$ is a prime $p$.

The difference to the GHS method lies in the fact that $C$ is defined over $\mathbb{F}_q$ and thus the original method fails. The way to overcome this problem is to construct a twist of $C$ that is defined over $\mathbb{F}_{q^k}$ but not over $\mathbb{F}_q$.

#### Construction of the twisted cover

Let $\psi : D \to C$ be a cover *defined over* $\mathbb{F}_q$, and suppose that $D$ has an automorphism $\tau$ of order $k$, such that $\psi \circ \tau \neq \psi$. Let $D^\tau$ denote the twist of $D$ over the extension $\mathbb{F}_{q^k}/\mathbb{F}_q$ with respect to $\tau$. (Note that $D \cdot \mathbb{F}_{q^k} \simeq D^\tau \cdot \mathbb{F}_{q^k}$.) In [DIE 2001, Theorem 9] it is proved that under some mild conditions the kernel of the map

$$J_{C,0}(\mathbb{F}_q) \hookrightarrow \mathrm{Pic}^0_{C \cdot \mathbb{F}_{q^k}} \xrightarrow{\psi^*} \mathrm{Pic}^0_{D^\tau} \xrightarrow{\mathrm{N}_{\mathbb{F}_{q^k}/\mathbb{F}_q}} \mathrm{Pic}^0_{D^\tau}$$

does not contain a subgroup of large prime order. If the genus of $D$ is not much bigger than $(k-1)$ times the genus of $C$, then index calculus algorithms on $\mathrm{Pic}^0_{D^\tau}$ are more efficient than square root algorithms on $G$ for solving the discrete logarithm problem.

To find such curves $D$ we again use the theory of covers of the projective line with given ramification type. For simplicity assume from now on that $k$ is a prime number. Assume that there is a rational function $f$ on $C$ of degree $k$. Take the Galois closure $F$ of the function field of $C$ over $\mathbb{F}_q(f)$. The corresponding curve $D$ covers $C$ and has the Galois group of $F/\mathbb{F}_q(f)$ as its group of automorphisms. By construction this group has cycles $\tau$ of length $k$ with $\tau \circ \psi \neq \psi$, as required.

So, we have to find suitable functions $f$ such that the genus of $D$ is small. The first step is to do this over $\overline{\mathbb{F}}_q$. One looks for tamely ramified covers and uses again the theory of Hurwitz spaces as in the classical case over the complex numbers. The background is delivered by the theory of *monodromy groups*. This allows us to study whether "geometrically" suitable covers exist. It remains to find the definition fields for these covers and, last but not least, to construct them explicitly. This is done in [DISC 2003].

### Results for small extension degrees

Here we consider small degrees of extension as proposed in Section 15.3 to give a complete study of which genera can occur depending on properties of the curve.

**Example 22.12 (Case $k=3$ and $g=2$)** Take $k=3$ and let $C$ be any curve of genus 2. Take a point $P \in C(\mathbb{F}_q)$, and take $f$ as function on $C$, which has a pole at $P$ of order 3, and which is holomorphic everywhere else. The Galois group is the symmetric group $S_3$, $D$ covers $C$ of degree 2, and the genus of $D$ is 6.

It is interesting to note that there is a kind of converse to this statement following from results in [HOW 2001, Theorem 3.3].

Let $C/\mathbb{F}_q$ be a curve of genus 2 and assume that 3 does not divide $\mathrm{Pic}^0_{C \cdot \mathbb{F}_{q^6}}$. Then by the cover approach the discrete logarithm in $J_{C,0}$ is transferred to discrete logarithms in Jacobians of curves of genus at least 6, except for very special cases.

Now we specialize the (arithmetical) conditions for $C$. Let $w : C \to C$ denote the hyperelliptic involution. If $\mathrm{Pic}^0_C$ has a divisor class $\overline{D}$ of order 3 with representative $P_1 + P_2 - 2P_\infty$, where $P_1$ and $P_2$ are in $C(\mathbb{F}_q)$, then $P_1 - w(P_2)$ has order 3 in $\mathrm{Pic}^0_C$. Hence, there is a function $f$ with divisor $3P_1 - 3w(P_2)$. Generically, this $f$ has the required ramification. The corresponding cover curve $D$ has genus 5.

By replacing $C$ by a quadratic twist one can do this construction in the case that the point $\overline{D}$ is defined over $\mathbb{F}_{q^6}$.

Diem and Scholten construct an explicit family of curves $C$, for which the covering curve $C$ of genus 5 is hyperelliptic.

Specializing even more and using Kummer theory one can construct in an explicit way a family of hyperelliptic curves of genus 2 such that the cover curve $D$ has genus 4.

**Example 22.13 (Elliptic curves)** We are now interested in covers of elliptic curves $E$, which can be used to transfer the discrete logarithm of the trace zero varieties to Jacobian varieties of small dimension. For $k=3$ the dimension of $E_0$ is 2 and so generic methods of computing discrete logarithms have complexity $O(q)$ (cf. Chapter 19), and, hence, a transfer cannot lead to a variety in which the DLP is easier to solve.

The first interesting case is $k=5$. One would like to get a cover curve of genus 4. By the described method this can be done *only* for the curve

$$E : y^2 = x^3 + 3165x - 31070.$$

Next assume $k=7$. Assume moreover that $q$ is prime to 6. Then one can show that there exists a cover $D$ of genus 8 for *all* elliptic curves $E$ defined over the algebraic closure of $\mathbb{F}_q$. Until now, no explicit example is known, but for many $q$ and many elliptic curves $E/\mathbb{F}_q$ this cover will be defined over $\mathbb{F}_q$. In these cases there is a very effective transfer of discrete logarithms on $E_0$, the trace zero variety of $E$ with respect to $\mathbb{F}_{q^7}$. This shows that extensions of degree 7 should be avoided in the design of discrete logarithm systems on elliptic curves.

### 22.3.5  Index calculus method via hyperplane sections

The previous sections used Weil descent to construct a curve $D$ such that the discrete logarithm problem on $J_C(\mathbb{F}_{q^k})$ can be transferred to the discrete logarithm problem in $J_D(\mathbb{F}_q)$. Such a transfer is useful if the resulting problem is easier to solve. Usually, index calculus methods (cf. Chapter 21) are used in $J_D(\mathbb{F}_q)$. If one wants to apply index-calculus methods in the group of rational points of elliptic curves over finite fields *without transfer*, one has to find an appropriate factor base. It seems to be impossible to do this by lifting techniques analogously to the case of discrete logarithms in finite fields. Diem [DIE 2004] and Gaudry [GAU 2004] found (independently) a first approach to overcome this difficulty in the case that the curve is defined over $\mathbb{F}_{q^k}$ with $k > 1$. They suggest to intersect the Weil descent of $J_C$ with the hyperplanes $x_i = 0$, where $i = 1, \ldots, k - 1$ in Gaudry's approach and $i \in I \subset \{0, 1, \ldots, k-1\}$ in Diem's. These elements constitute the factor base for the index calculus method.

To implement the index calculus algorithm with this factor base one needs an efficient method for membership testing. In the case of elliptic curves this is given by the summation polynomials introduced by Semaev [SEM 2004].

**Definition 22.14** Let $E : y^2 = x^3 + a_4 x + a_6$ be an elliptic curve over $\mathbb{F}_{q^k}$. The *summation polynomials* $f_n$ are defined by the following recurrence. The initial values for $n = 2$ and $n = 3$ are given by
$$f_2(x_1, x_2) = x_1 - x_2,$$
and
$$f_3(x_1, x_2, x_3) = (x_1 - x_2)^2 x_3^2 - 2\big((x_1 + x_2)(x_1 x_2 + a_4) + 2a_6\big)x_3 + \big((x_1 x_2 - a_4)^2 - 4a_6(x_1 + x_2)\big),$$
and for $n \geqslant 4$ and $1 \leqslant j \leqslant n - 3$ by
$$f_n(x_1, x_2, \ldots, x_n) = \mathrm{Res}_x\big(f_{n-j}(x_1, x_2, \ldots, x_{n-j-1}, x), f_{j+2}(x_{n-j}, \ldots, x_n, x)\big).$$

Semaev shows that $f_n$ is uniquely defined in spite of the redundancy in the definition. One can check that $f_n(x_1, x_2, \ldots, x_n) = 0$, with $x_i \in \overline{\mathbb{F}}_{q^k}$, if and only if there exists an $n$-tuple $(y_1, y_2, \ldots, y_n)$, with $y_i \in \overline{\mathbb{F}}_{q^k}$, such that $P_i = (x_i, y_i) \in E(\overline{\mathbb{F}}_{q^k})$ and
$$P_1 \oplus P_2 \oplus \cdots \oplus P_n = P_\infty.$$

The polynomials $f_n$ are symmetric and have degree $2^{n-2}$ in each variable.

The methods of Gaudry and Diem differ in the choice of the factor base. We describe both in the setting of elliptic curves as there the summation polynomials are already determined even though the method can be applied in more generality. Gaudry also gives the relation to the standard Weil descent method and shows that for hyperelliptic curves this approach corresponds to the standard index calculus method with factor base given by linear polynomials.

#### 22.3.5.a  Choice of the factor base by Gaudry

Gaudry's [GAU 2004] method is applicable for small $k \geqslant 3$ only, as the $(k+1)$-th summation polynomial is involved in a Gröbner basis computation and the complexity of this step depends heavily on the degree.

Let $E$ be an elliptic curve defined over $\mathbb{F}_{q^k} = \mathbb{F}_q(\theta)$ and not over a proper subfield and put

$$x = \sum_{i=0}^{k-1} x_i \theta^i \text{ and } y = \sum_{i=0}^{k-1} y_i \theta^i.$$

Assume that the intersection of an open affine part $W_a$ of the Weil restriction $W_{\mathbb{F}_{q^k}/\mathbb{F}_q}(E)(\mathbb{F}_q)$ with the hyperplanes $H_j : x_j = 0$ for $j = 1, \ldots, k-1$ is an irreducible curve (this is the case generically and can otherwise be reached by a different choice of the basis). Then one can choose the factor base

$$\mathcal{F} = \{P \in E(\mathbb{F}_{q^k}) \mid P = (x_P, y_P), x_P = x_{P,0} \in \mathbb{F}_q, y \in \mathbb{F}_{q^k}\}.$$

Like in the usual index calculus algorithm for computing the discrete logarithm of $Q \in \langle P \rangle$ one computes combinations

$$R = [m_P]P \oplus [m_Q]Q$$

with random integers $m_P, m_Q$, and hopes that $R$ can be expressed as sum of points in $\mathcal{F}$. Here we use that the arithmetic on $W_{\mathbb{F}_{q^k}/\mathbb{F}_q}(E)(\mathbb{F}_q)$ is inherited from $E$ and apply Weil descent only in the basic form of sorting the equations according to the powers of $\theta$.

Gaudry suggests to use only such points $R$ that allow a representation as

$$R = P_1 \oplus P_2 \oplus \cdots \oplus P_k,$$

with $P_i \in \mathcal{F}$. The existence of such a representation can be checked with the $(k+1)$-th summation polynomial which gives rise to an equation of the form

$$\sum_{i=0}^{k-1} \psi_i(x_{P_1,0}, x_{P_2,0}, \ldots, x_{P_k,0})\theta^i = 0,$$

where the $\psi_i$ depend on $R$. This is a system of $k$ equations in $k$ unknowns which all have to hold individually. Before applying Buchberger's algorithm one should symmetrize the equations. In case one finds a solution in the symmetric expressions one checks whether this in fact gives rise to a decomposition with $x_{P_i,0} \in \mathbb{F}_q$ and (finally) in $\mathcal{F}$. As soon as one has more equations than unknowns, sparse linear algebra techniques allow to compute the discrete logarithm. Note that each row has only $k$ nonzero entries.

**Remarks 22.15**

(i) The complexity of this algorithm is $O(q^{k-\frac{2}{k}})$, where the hidden constants depend (exponentially) on $k$. Gaudry proposes this method for $k = 3$ or $k = 4$ only and shows that for $E$ defined over $\mathbb{F}_{q^3}$, the discrete logarithm problem can be solved in time $O(q^{4/3}) \simeq O(q^{1.33})$. If $E$ is defined over $\mathbb{F}_{q^4}$, then the discrete log problem can be solved in time $O(q^{1.5})$. These complexities are no better than those resulting from Weil descent of curves over $\mathbb{F}_{q^k}$ as explained in the previous sections. But there are only a few curves leading to such a small genus of $D$ while Gaudry's method applies to *all* curves.

(ii) We point out that this algorithm does not work as described if $E$ is defined over a proper subfield $\mathbb{F}_{q^{k'}}$ as then all combinations of points in $\mathcal{F}$ lead to points in $E(\mathbb{F}_{q^{k'}})$ which is usually outside of the subgroup of large prime order one is working in. A way out is to make a different choice of the factor base by putting a different $x_i \neq 0$.

Like with the standard Weil descent approaches one can also use an isogenous curve defined over $\mathbb{F}_{q^k}$.

### 22.3.5.b  Choice of the factor base by Diem

Diem's version [DIE 2004] uses a larger factor base by observing that there is no need to allow only one nonzero coordinate of $x$. Let $c \leqslant k$ be an integer and put $m = \lceil \frac{k}{c} \rceil$ and $l = mc - k$. The $x$-coordinates of points in the factor base lie in an $m$-dimensional subspace of $\mathbb{F}_{q^k}$ over $\mathbb{F}_q$. We describe a simplified version and fix a polynomial basis $\mathbb{F}_{q^k} = \mathbb{F}_q(\theta)$ and a subset $I \subset \{0, 1, \ldots, k-1\}$ of cardinality $m$. We put

$$\mathcal{F} = \{P \in E(\mathbb{F}_{q^k}) \mid P = (x_P, y_P), x_P = \sum_{i \in I} x_{P,i} \theta^i, x_{P,i} \in \mathbb{F}_q, y_P \in \mathbb{F}_{q^k}\}.$$

To guarantee that a sufficiently large factor base can be obtained by this method in $O(q^m)$ field operations, Diem needs to make a random choice averaging over all bases of $m$-dimensional subspaces and put the fixed coordinates to some randomly chosen values. We refer to his paper for details.

For $c < k$ he also uses a second factor base $\mathcal{F}' \subset \mathcal{F}$ for which the $x$-coordinates lie in an $(m-1)$-dimensional space over $\mathbb{F}_q$. He looks for relations

$$R = [m_P]P \oplus [m_Q]Q = P_1 \oplus P_2 \oplus \cdots \oplus P_c,$$

such that $P_1, \ldots, P_{l+1} \in \mathcal{F}'$ and $P_{l+2}, \ldots, P_c \in \mathcal{F}$. To check for the existence of such a splitting, the $(c+1)$-th summation polynomial is used and $k$ equations are found. Unlike Gaudry he suggests to perform this Gröbner basis computation once symbolically involving the $k$ coefficients defining $x_R$. The task of checking whether a given $R$ allows such a decomposition is then simplified to checking whether the $k$ equations can be satisfied. To obtain a relation one finally needs to check whether the $x_{P_i}$ lead to a point defined over $\mathbb{F}_{q^k}$ or $\mathbb{F}_{q^{2k}}$. In the latter case the relation is rejected.

By the choice of $m$ and $l$ he has $k$ equations defining the $k-1$ variables which determine the elements in $\mathcal{F}'$ and $\mathcal{F}$ and this is used in the analysis of the algorithm (the case $c = k$ is already considered in the previous section). His paper considers the complexity of the algorithm for various choices of $c$ in relation to $k$.

The asymptotic result found by Diem (heuristic since one has to make a very plausible assumption) is stated in the following proposition.

**Proposition 22.16** Let $0 < a < b$ and $\epsilon > 0$ be fixed real numbers and put $D = \frac{4b+\epsilon}{a^{3/4}}$. Assume that $a \lg q \leqslant k \leqslant b \lg q$. Then there exists an algorithm which computes the discrete logarithm in $E(\mathbb{F}_{q^k})$ with $O\left(2^{D(k \lg q)^{3/4}}\right)$ field operations in $\mathbb{F}_q$, where the constants depend on $a, b, \epsilon$.

For the proof and more details we refer to [DIE 2004].

# VI

# Applications

# Chapter 23

# Algebraic Realizations of DL Systems

### Gerhard Frey and Tanja Lange

### Contents in Brief

| | |
|---|---|
| **23.1 Candidates for secure DL systems** | 547 |
| Groups with numeration and the DLP • Ideal class groups and divisor class groups • Examples: elliptic and hyperelliptic curves • Conclusion | |
| **23.2 Security of systems based on $\mathrm{Pic}_C^0$** | 554 |
| Security under index calculus attacks • Transfers by Galois theory | |
| **23.3 Efficient systems** | 557 |
| Choice of the finite field • Choice of genus and curve equation • Special choices of curves and scalar multiplication | |
| **23.4 Construction of systems** | 564 |
| Heuristics of class group orders • Finding groups of suitable size | |
| **23.5 Protocols** | 569 |
| System parameters • Protocols on $\mathrm{Pic}_C^0$ | |
| **23.6 Summary** | 571 |

We want to design a public-key cryptosystem that enables us to exchange keys, sign and authenticate documents and encrypt and decrypt (small) messages. The system should rely on simple protocols that are based on secure cryptographic primitives with a well understood mathematical background. The implementation rules should be clear and easy to understand.

By using the results obtained in the previous chapters, we hope to convince the reader that these criteria can be realized quite satisfyingly by systems based on the discrete logarithm in finite groups $G$ of prime order $\ell$.

The purpose of this chapter is to serve as a digest of the other chapters. To this aim we briefly state the main results and provide many references to the much more detailed descriptions in the book.

## 23.1 Candidates for secure DL systems

The protocols based on discrete logarithms are described and discussed in Chapter 1. It is obvious that the complexity of computing discrete logarithms in the chosen group is a key ingredient for the security of the system. For actual use in practice, the DLP is used as a *cryptographic primitive* in protocols. In this section we shall introduce examples for groups that are expected to be good candidates for this.

## 23.1.1 Groups with numeration and the DLP

For the convenience of the reader we shall repeat the essential notions from Chapter 1. Let $(G, \oplus)$ be a group of order $\ell$, where $\ell$ is a prime number. Let $P, Q \in G$ be two elements. The discrete logarithm in $G$ of $Q$ with respect to $P$ is a number $n = \log_P(Q)$ with

$$Q = [n]P = \underbrace{P \oplus P \oplus \cdots \oplus P}_{n \text{ times}}.$$

The number $n$ is determined modulo $\ell$. To compute $n$ (for randomly given $(P, Q)$) is the *discrete logarithm problem* in $G$ (the DLP in $G$).

To enable this computation for a computer, we have to assume that $G$ is given in a very concrete way. So, we shall assume that the elements of $G$ are given as bit-strings of length $O(\lg(\ell))$. The assumption used here is that $G$ is a group with *numeration*. For the exact definition of this notion we refer to [FRLA 2003].

Furthermore, for estimating the hardness of the discrete logarithm problem, the instantiation of $G$ is very important. The fact that the representation plays a key role for the complexity of the computations of discrete logarithms is demonstrated by two examples.

Up to isomorphisms there is only one group with $\ell$ elements.

As a first representation, we choose $(G, \oplus) = (\mathbb{Z}/\ell\mathbb{Z}, +)$ with natural numeration

$$\begin{aligned} f : \mathbb{Z}/\ell\mathbb{Z} &\to \{1, \ldots, \ell\} \\ m + \ell\mathbb{Z} &\mapsto r_m \text{ such that } r_m \equiv m \pmod{\ell}. \end{aligned}$$

The discrete logarithm of $m_1 + \ell\mathbb{Z}$ with respect to $m_2 + \ell\mathbb{Z}$ is computed in $O(\lg \ell)$ bit operations (cf. Chapter 10).

In Example 1.13 we choose another representation for such a group. We find $G$ embedded in the multiplicative group of the finite field $\mathbb{F}_p$ as the group of roots of unity of order $\ell$, where $p$ is a prime such that $\ell$ divides $p - 1$. In this case the complexity of the DLP is subexponential in $p$ (cf. Chapter 20).

We take this opportunity to recall our measure for the complexity of algorithms introduced in Chapter 1 and used many times.

Let $N$ be a natural number. Define

$$L_N(\alpha, c) := \exp\big((c + o(1))(\ln N)^\alpha (\ln \ln N)^{1-\alpha}\big)$$

with $0 \leqslant \alpha \leqslant 1$ and $c > 0$. $L_N(\alpha, c)$ interpolates between polynomial complexity for $\alpha = 0$ and exponential complexity for $\alpha = 1$. For $\alpha < 1$ the complexity is said to be *subexponential*.

So the second numeration makes $G$ to a group in which the DLP is much harder than in the first example but not optimal, i.e., not exponential in $\ell$. More details on computing discrete logarithms using the index calculus method can be found in Chapter 20. To make related cryptosystems secure one has to take $p$ rather large (the bit-size of $p$ should be at least 1024).

## 23.1.2 Ideal class groups and divisor class groups

All DL-based cryptosystems applied today use as groups the ideal classes of convenient (commutative) rings $\mathcal{O}$ with unit element and without zero divisors. Using a class group requires that in the scalar multiplication some notion of reduction of ideals is available and that the groups are chosen in a manner to guarantee that ideal classes can be (almost) uniquely represented by reduced ideals.

### 23.1.2.a Mathematical background

Let $\mathcal{O}$ be a commutative ring with unit element and without zero divisors. The quotient field $K$ of $\mathcal{O}$ consists of all fractions $f/g$ with $f, g \in \mathcal{O}$ with addition and multiplication defined by rules one is used to follow in $\mathbb{Q}$.

A fractional ideal is a set $A \subseteq K$ such that there exists an element $f \in \mathcal{O}$, so that $fA$ is an ideal in $\mathcal{O}$. The multiplication of fractional ideals is defined and associative. A first criterion to choose $\mathcal{O}$ is that the fractional ideals form a group called the *ideal group* $\mathrm{I}(\mathcal{O})$ of $\mathcal{O}$.

The group $\mathrm{I}(\mathcal{O})$ will have no elements of finite order. This changes if we introduce the following equivalence relation: two fractional ideals $A, A'$ are equivalent if there exists an element $f \in K$ such that $A = fA'$.

The resulting group is denoted by $\mathrm{Cl}(\mathcal{O})$, the *ideal class group* of $\mathcal{O}$. The neutral element in the class group consists of the group of *principal ideals* $\mathrm{Princ}(\mathcal{O})$ and so $\mathrm{Cl}(\mathcal{O}) = \mathrm{I}(\mathcal{O})/\mathrm{Princ}(\mathcal{O})$. The next criterion to choose $\mathcal{O}$ is that $\mathrm{Cl}(\mathcal{O})$ should have many elements of finite order. In fact, in all existing systems $\mathrm{Cl}(\mathcal{O})$ is a finite group.

This advantage has a price. As usual one computes in quotient structures like $\mathrm{Cl}(\mathcal{O})$ by using representatives (in our case, ideals) of the classes, composes these representatives, and then forms the class of the result. Since there are infinitely many elements in an ideal class, this does not lead to an algorithm if one does not have more information. There are two ways out of this difficulty.

1. It is possible to find a distinguished element in each ideal class (respectively a finite [small] subset of such elements).
2. It is possible to define "coordinates" and addition formulas directly for elements of $\mathrm{Cl}(\mathcal{O})$.

The first possibility can be used if we have efficient "reduction algorithms" that compute the distinguished element in ideal classes, and the second possibility can be realized if there is a geometric background of $\mathrm{Cl}(\mathcal{O})$.

Most interesting cases are those for which both methods can be used!

### 23.1.2.b Realization in number fields

Having in mind the requirements stated above, it is no wonder that the first suggestions for systems of discrete logarithms based on ideal classes came from number theory [BUWI 1988]. The highly developed "computational number theory" based on Minkowski's geometry of numbers made it possible to compute efficiently with ideal class groups of the ring of integers $\mathcal{O}_K$ of number fields $K$ that are finite algebraic extensions of $\mathbb{Q}$. An important special case is that $K$ is an imaginary quadratic field. In this case already Gauß developed a fast algorithm for computing with ideal classes. It relies on the identification of these classes with classes of binary quadratic forms. The distinguished ideals correspond to the uniquely determined reduced quadratic forms. A given form is transformed into a reduced one by an explicit algorithm using Euclid's algorithm which runs in polynomial time.

The disadvantage of these systems is that the *index calculus attack* is very effective, i.e., the algorithm based on the principles explained in Chapter 20 has only subexponential complexity. One uses prime ideals of $\mathcal{O}_K$ with small norm to build up a factor base for $\mathrm{Cl}(\mathcal{O})$.

### 23.1.2.c Realization in function fields

This motivates us to look for rings $\mathcal{O}$ having a similar simple structure as $\mathcal{O}_K$ but being more resistant against index calculus attacks. Since the beginning of arithmetic geometry in the last century, it is well-known and has been often exploited that the ring $\mathcal{O}_{C_a}$ of regular functions on a nonsingular irreducible affine curve $C_a$ defined over a finite field $\mathbb{F}_q$ is a Dedekind domain with finite ideal class group $\mathrm{Cl}(\mathcal{O}_{C_a})$ and that the arithmetic is analogous to the arithmetic of the ring of integers in number fields. The quotient field of $\mathcal{O}_{C_a}$ is the function field (of meromorphic functions) of $C_a$ and is denoted by $K(C_a)$.

These rings and their ideal class groups are explained in Section 4.4. As size of ideals one uses their degree, and the theorem of Minkowski about points in lattices is replaced by the Riemann–Roch theorem, which yields (amongst other results) that in every ideal class in $\mathrm{Cl}(\mathcal{O}_{C_a})$ there are ideals contained in $\mathcal{O}_{C_a}$ with small degree. (For a precise formulation see Section 4.4.6 and especially Theorem 4.143.) The reader familiar with the geometry of numbers will remark that the logarithm of the absolute value of number fields is replaced by the *genus $g$ of the curve $C_a$*, respectively of its function field $K(C_a)$ (cf. Definition 4.107). Already at this stage it follows that the group $\mathrm{Cl}(\mathcal{O}_{C_a})$ is at least as good as a candidate for groups, in which the group operation can be executed effectively, as the rings of integers in number fields.

But we can go further because of the geometrical background of $\mathcal{O}_{C_a}$. The prime ideals in $\mathcal{O}_{C_a}$ are closely related to points on $C_a$. To see this one takes the points $P$ on $C_a$ with coordinates in the algebraic closure $\overline{\mathbb{F}}_q$ of $\mathbb{F}_q$ *together* with the operation of the Galois group $G_{\mathbb{F}_q}$ of $\mathbb{F}_q$ (cf. Section 4.4.4). The resulting Galois orbits $G_{\mathbb{F}_q} \cdot P$ of points correspond one-to-one to the prime ideals (always taken differently from $\{0\}$) of $\mathcal{O}_{C_a}$ consisting of functions $f \in \mathcal{O}_{C_a}$ vanishing in $P$. By taking the order of vanishing of $f$ at $P$ we define a valuation $v_P$ on $\mathcal{O}_{C_a}$, and hence on $K(C_a)$, whose valuation ring contains $\mathcal{O}_{C_a}$. Its equivalence class is called a *prime divisor* $\mathfrak{p}$ of $K(C_a)$ corresponding to $G_{\mathbb{F}_q} \cdot P$.

One of the major advantages of the geometric theory of curves over finite fields is that it is very easy to compactify the affine curve $C_a$ by going to its projective closure. In principle this is done by homogenizing the equations defining $C_a$. The procedure is explained in Section 4.22. We find a projective curve $C$ containing the affine curve $C_a$ and the difference set of points consists of finitely many "points at infinity." The Galois orbits of these points define, as above, equivalence classes of valuations of $K(C_a)$ and hence divisors $\mathfrak{p}$. They correspond one-to-one to the equivalence classes of valuations (and hence to prime divisors) of $K(C_a)$ which do not contain $\mathcal{O}_{C_a}$.

Because of the compactness of $C$ the only functions that are regular at all points of $C$ are constants. To study the arithmetic of $C$ one has to introduce the divisor group of $C$ (see Section 4.4.2) replacing the ideal group of $\mathcal{O}_{C_a}$. Divisors can be identified with formal sums of points on $C$ with integers as coefficients. The role of principal ideals is taken by principal divisors (cf. Definition 4.102) and the resulting quotient group is the *divisor class group of degree 0* of $C$ denoted by $\mathrm{Pic}_C^0$. By construction, it is closely related to $\mathrm{Cl}(\mathcal{O}_{C_a})$ For instance, if there is only one point at infinity of $C$ then $\mathrm{Pic}_C^0$ is isomorphic to $\mathrm{Cl}(\mathcal{O}_{C_a})$ by Proposition 4.140. We shall assume from now on that this is satisfied. We denote by $P_\infty$ this *unique* point at infinity.

The group $\mathrm{Pic}_C^0$ is (in a canonical way) isomorphic to the group of rational points of the Jacobian variety $J_C$ of $C$ which is an abelian variety (cf. Definition 4.134) defined over $\mathbb{F}_q$ and intrinsically attached to $C$. Together with the abstract theory comes a very concrete way to construct $J_C$ (up to birational equivalence): a consequence of the theorem of Riemann–Roch is that in every divisor class of degree 0 of $C$, there is a divisor of the form $\sum_{i=1}^{r} P_i - rP\infty$ with $r \leqslant g$. Details are found in Section 4.4.4.

### 23.1.2.d Conclusion

Beginning with an affine curve $C_a$ and its ring of regular functions $\mathcal{O}_{C_a}$ we construct the associated projective curve $C$. Under the assumption that there is only a single point at infinity we can interpret $\mathrm{Pic}_C^0$, its divisor class group of degree 0, as both — as class group of ideals of the ring $\mathcal{O}_{C_a}$, for which we have an efficient reduction theory, and as an abelian variety. Hence, we have a compact way to represent its elements and we can introduce coordinates for ideal classes and expect algebraic formulas describing the group composition in these coordinates.

Since we have related the ideal class group of $\mathcal{O}_{C_a}$ to rational points of abelian varieties we can use their rich structure in general as described in Section 4.3 and especially over finite fields (see Section 5.2).

For the choice of a ring $\mathcal{O}_{C_a}$ or equivalently, an affine or projective curve $C$ we have a lot of freedom compared with the case that the chosen ring is a ring of integers in a number field. The parameters are

1. the characteristic $p$ of the base field $\mathbb{F}_p$,
2. the degree $d$ of the ground field $\mathbb{F}_q$ over $\mathbb{F}_p$,
3. the genus $g_C = g$ of the curve $C$ (resp. the function field $K(C)$).

The number of isomorphism classes of curves of genus $g = 1$ over $\mathbb{F}_{p^d}$ is about $p^d$ and for genus $g \geqslant 2$ it is of size $p^{d(3g-3)}$.

Even more important is that in the geometric case we have a much stronger relation between these parameters and the expected order of the divisor class group. By Theorem 5.76 of Hasse–Weil we get
$$|\text{Pic}_C^0| \sim p^{dg}.$$

So, if $\ell$ is the desired size of the group of prime order we want to embed into $\text{Pic}_C^0$, we have to take $d \lg p$ slightly larger than $\lg(\ell)/g$.

The explicit construction of the Jacobian variety represented as divisors of degree zero containing at most $g$ affine points yields that the number of bits needed to represent group elements is $O(dg \lg p) = O(\lg \ell)$.

### 23.1.3 Examples: elliptic and hyperelliptic curves

In this section we shall apply our general theory to two special families of curves called elliptic and hyperelliptic curves. The assumption we have made in Section 23.1.2.c that there is only a single point at infinity implies that the hyperelliptic curves have a rational Weierstraß point. In Section 4.4.2.b and Chapter 14 we have studied these curves in great detail. We shall repeat their definition and crucial properties for the convenience of the reader.

#### 23.1.3.a Elliptic curves

An elliptic curve $E$ defined over $\mathbb{F}_q$ is a projective absolute irreducible curve of genus 1 with a rational point $P_\infty$. It can be given by an affine Weierstraß equation
$$E_a : y^2 + a_1 xy + a_3 y = x^3 + a_2 x^2 + a_4 x + a_6.$$

If the characteristic is prime to 6 this equation can be transformed to
$$E_a : y^2 = x^3 + a_4 x + a_6, \text{ with } a_4, a_6 \in \mathbb{F}_q.$$

For normal forms and invariants of the curve we refer to Table 4.1.

The ring $\mathcal{O}_{E_a}$ is $\mathbb{F}_q[x,y]/(y^2 + a_1 xy + a_3 y - x^3 - a_2 x^2 - a_4 x - a_6)$. So it is a polynomial order in $\mathbb{F}_q(x,y)$ which has rank 2 over $\mathbb{F}_q[x]$.

By Riemann–Roch, we find in each ideal class of $\mathcal{O}_{E_a}$ a uniquely determined prime ideal $M_P$ of degree 1. It is generated by the two functions $x - x_1, y - y_1$ with $x_1, y_1 \in \mathbb{F}_q$ corresponding to the point $P = (x_1, y_1)$ of $E(\mathbb{F}_q)$. The Jacobian of $E$ is isomorphic to $E$ with zero element $P_\infty$. The point $P$ corresponds to the divisor class of $P - P_\infty$. Hence the $\mathbb{F}_q$-rational points of $E$ form in a natural way an abelian group. The addition law can be expressed either by polynomials in the homogeneous coordinates of points on $E$ or by rational functions (with the appropriate interpretation if the point $P_\infty$ is involved) in the affine coordinates $(x_1, y_1)$.

The explicit formulas can be found in great detail in Section 4.4.5 and Chapter 13.

### 23.1.3.b Hyperelliptic curves

A generalization of elliptic curves are hyperelliptic curves $C$. These are projective curves of genus $g > 1$. We assume that they have a rational Weierstraß point $P_\infty$. Then they are determined by affine equations

$$C_a : y^2 + h(x)y = f(x), \quad \text{with } h(x), f(x) \in \mathbb{F}_q[x], \tag{23.1}$$

where $\deg(h) \leq g$ and $\deg(f) \leq 2g+1$. Hence $\mathcal{O}_{C_a} = \mathbb{F}_q[x,y]/(y^2 + h(x)y - f(x))$ is a polynomial order of rank 2 in $\mathbb{F}_q(x,y)$.

By Riemann–Roch, we find in each ideal class of $\mathcal{O}_{C_a}$ an ideal lying in $\mathcal{O}_{C_a}$ of degree less than or equal to $g$. Again we get that $\mathrm{Cl}(\mathcal{O}_{C_a})$ is isomorphic to $\mathrm{Pic}^0_C$, the divisor class group of degree 0 of $C$. In the language of divisors we get that in each divisor class of degree 0 there is a divisor $D = \sum_{i=1}^r P_i - rP_\infty$ where $P_i$ are points on $C_a$, which are now not necessarily rational over $\mathbb{F}_q$. If we assume that $D$ is reduced (cf. Theorem 4.143) then $D$ is uniquely determined. For computations the representation by ideals is most convenient. It leads to the Mumford representation discussed already in Theorem 4.145, and which we repeat here to give a flavor of how to introduce "coordinates" for compact presentations of the classes.

**Theorem 23.1 (Mumford representation)**
Let $C$ be a genus $g$ hyperelliptic curve given as in (23.1). Each nontrivial group element $\overline{D} \in \mathrm{Pic}^0_C$ can be represented via a unique pair of polynomials $u(x)$ and $v(x)$, $u, v \in \mathbb{F}_q[x]$, where

(i) $u$ is monic,
(ii) $\deg v < \deg u \leq g$,
(iii) $u \mid v^2 + vh - f$.

Let $D \in \overline{D}$ be the unique reduced divisor, i.e., $D = \sum_{i=1}^r P_i - rP_\infty$, where $P_i \neq P_\infty, P_i \neq -P_j$ for $i \neq j$ and $r \leq g$. Put $P_i = (x_i, y_i)$. Then the divisor class of $D$ is represented by

$$u(x) = \prod_{i=1}^r (x - x_i)$$

and the property that if $P_i$ occurs $n_i$ times then

$$\left(\frac{d}{dx}\right)^j \left[v(x)^2 + v(x)h(x) - f(x)\right]_{|x=x_i} = 0, \quad \text{for } 0 \leq j \leq n_i - 1.$$

So the polynomials $[u, v]$ can be taken as coordinates of elements in $\mathrm{Pic}^0_C$. The group operations can be computed using this representation as given below in Cantor's algorithm. Moreover, it is explained in Section 14.1 how to express the group law in terms of the coefficients of $u$ and $v$.

---

**Algorithm 23.2** Cantor's algorithm

INPUT: Two divisor classes $\overline{D}_1 = [u_1, v_1]$ and $\overline{D}_2 = [u_2, v_2]$ on the curve $C : y^2 + h(x)y = f(x)$.
OUTPUT: The unique reduced divisor $D$ such that $\overline{D} = \overline{D}_1 \oplus \overline{D}_2$.

1. $d_1 \leftarrow \gcd(u_1, u_2)$      $[d_1 = e_1 u_1 + e_2 u_2]$
2. $d \leftarrow \gcd(d_1, v_1 + v_2 + h)$      $[d = c_1 d_1 + c_2(v_1 + v_2 + h)]$

3. $s_1 \leftarrow c_1 e_1$, $s_2 \leftarrow c_1 e_2$ and $s_3 \leftarrow c_2$
4. $u \leftarrow \dfrac{u_1 u_2}{d^2}$ and $v \leftarrow \dfrac{s_1 u_1 v_2 + s_2 u_2 v_1 + s_3(v_1 v_2 + f)}{d} \bmod u$
5. **repeat**
6. $\quad u' \leftarrow \dfrac{f - vh - v^2}{u}$ and $v' \leftarrow (-h - v) \bmod u'$
7. $\quad u \leftarrow u'$ and $v \leftarrow v'$
8. **until** $\deg u \leqslant g$
9. make $u$ monic
10. **return** $[u, v]$

## 23.1.4 Conclusion

To design DL-based cryptosystems we have the following results:

- In ideal class groups $\mathrm{Cl}(\mathcal{O})$ of orders $\mathcal{O}$ in number fields or function fields over finite fields one can perform the group operation in polynomial time.
- Let $\mathcal{O}$ be the ring of regular functions on an affine nonsingular curve $C_a$ of genus $g$ over a finite field $\mathbb{F}_q$. There is a close connection between $\mathrm{Cl}(\mathcal{O})$ and $\mathrm{Pic}_C^0$, the divisor class group of the projective curve associated to $C_a$. Moreover $\mathrm{Pic}_C^0$ is equal to the set of rational points of the Jacobian variety of $C$. So its order is of size $q^g$.
- If $C$ is an elliptic curve $E$ we have that

$$\mathrm{Pic}_E^0 \simeq \mathrm{Cl}\big(\mathbb{F}_q[x,y]/(y^2 + a_1 xy + a_3 y - x^3 - a_2 x^2 - a_4 x - a_6)\big).$$

The curve $E$ is isomorphic to its Jacobian variety with $P_\infty$ as zero element. The elements in $\mathrm{Pic}_E^0$ correspond one-to-one to the rational points $E(\mathbb{F}_q)$ and hence can be represented by bit strings of size $O(\lg |E(\mathbb{F}_q)|) = O(\lg q)$. The addition law described by explicit formulas can be found in Chapter 13 and is done in $O(\lg q)$ bit operations.

- If $C$ is a hyperelliptic curve of genus $g$ with an $\mathbb{F}_q$-rational Weierstraß point $P_\infty$ we have that

$$J_C(\mathbb{F}_q) \simeq \mathrm{Pic}_C^0 \simeq \mathrm{Cl}\big(\mathbb{F}_q[x,y]/(y^2 + h(x)y - f(x))\big),$$

with $h, f \in \mathbb{F}_q[x]$ and $\deg(h) \leqslant g$, $\deg(f) = 2g + 1$.

Hence, the points in $\mathrm{Pic}_C^0$ can be given by ideals in Mumford representation and the addition is done by Cantor's algorithm. So $\mathrm{Pic}_C^0$ is a group in which elements can be represented by bit strings of size $O(\lg |\mathrm{Pic}_C^0|)$ and addition is done in $O(\lg |\mathrm{Pic}_C^0|)$ bit operations.

**Remark 23.3** Obviously the arithmetical properties of elliptic and of hyperelliptic curves are completely analogous. For this reason we often interpret elliptic curves as "hyperelliptic curves of genus 1" without mentioning it.

## 23.2 Security of systems based on $\text{Pic}_C^0$

In the last section we have seen that ideal class groups attached to elliptic and hyperelliptic curves lead to groups in which elements have a compact representation and in which the composition law is computable in polynomial time.

In generic black-box groups of order $N$, the results by Shoup show that the discrete logarithm problem cannot be solved in less than $O(\sqrt{N})$. Hence, if only the generic algorithms (cf. Chapter 19) can be used, we can consider the discrete logarithm problem to be hard as they run in time $O(\sqrt{N})$.

There are also some results showing that the discrete logarithm problem and the Diffie–Hellman problem on elliptic curves have at least a certain complexity when attacked by means like Boolean functions or polynomials. However, the results given in [LAWI 2002, LAWI 2003] are much weaker than expected and can only give an indication on the hardness — only the opposite result would have catastrophic implications.

We now deal with security issues showing under which conditions there exist attacks that are stronger than the generic ones.

### 23.2.1 Security under index calculus attacks

We have stressed the similarity with the case of orders in imaginary quadratic fields and, hence, the disadvantage: namely, the possibility of computing the discrete logarithm by index calculus algorithms must be discussed.

In fact, one has seen in Chapter 21 that this type of algorithms works in certain ranges. In the sequel we briefly present these results.

We recall the result of Gaudry, Enge, and Stein [ENG 2002, ENGA 2002, ENST 2002] which is strong for large genus $g$.

**Theorem 23.4** For $g/\ln(q) > t$ the discrete logarithm in the divisor class group of a hyperelliptic curve of genus $g$ defined over $\mathbb{F}_q$ can be computed with complexity bounded by

$$L_{q^g}\left(\tfrac{1}{2}, \sqrt{2}\left(\left(1 + \tfrac{1}{2t}\right)^{1/2} + \left(\tfrac{1}{2t}\right)^{1/2}\right)\right).$$

The results of Gaudry [GAU 2000b] and more recently of Thériault [THÉ 2003a] and Gaudry, Thériault, and Thomé [GATH⁺ 2004] are serious for hyperelliptic curves of relatively small genus (in practice: $g \leqslant 9$).

There is an index calculus attack of complexity

$$O\left(g^5 q^{2 - \frac{2}{g} + \epsilon}\right)$$

with "reasonably small" constants and even for $g = 3$ and $4$ the security is reduced.

The explicit result for $g = 4$ is: for hyperelliptic curves $C$ of genus 4 defined over $\mathbb{F}_q$ there is an index calculus algorithm that computes the discrete logarithm in $\text{Pic}_C^0$ with complexity

$$O\left(q^{3/2+\epsilon}\right) = O\left(|\text{Pic}_C^0|^{0.375}\right).$$

This means that the discrete logarithm is considerably weaker than generically expected.

The explicit result for $g = 3$ is: for hyperelliptic curves $C$ of genus 3 defined over $\mathbb{F}_q$, there is an index calculus algorithm that computes the discrete logarithm in $\text{Pic}_C^0$ with complexity

$$O\left(q^{4/3+\epsilon}\right) = O\left(|\text{Pic}_C^0|^{0.44}\right).$$

§ 23.2 Security of systems based on $\mathrm{Pic}_C^0$

This shows that asymptotically genus 3 curves are weaker than elliptic curves and even for group sizes encountered in practice the security is reduced. However, the main application of genus three curves is over fields of 64 bits using one word on a 64-bit processor or two words on a 32-bit processor. This setting can easily offer group sizes of 192 bits. So even with some percent less security, one is above the usual security threshold of working in groups of size $2^{160}$.

We stress that these results hold for *all* curves of that genus, not only for hyperelliptic curves. The difference appears only in the constants that are smaller for hyperelliptic curves.

#### 23.2.1.a Conclusion

We can summarize our results.

- For curves $C$ of genus $g \geqslant 4$, a direct application of index calculus algorithms to $\mathrm{Cl}(\mathcal{O}_{C_a})$ gives a complexity of the DLP that is smaller than the generic one. Hence, orders related to curves of genus $g \geqslant 4$ or closely related abelian varieties should not be used as crypto primitives for public-key systems, or, if one has very good reasons for using them, one has to enlarge the group size considerably.
- The state of the art is: we have only three types of rings $\mathcal{O}$ which avoid serious index calculus attacks and for which addition in $\mathrm{Cl}(\mathcal{O})$ is fast enough. These are the *maximal orders belonging to curves of genus* 1, 2 *and* 3. Even for $g = 3$ one needs to take into account the group size to compare the complexities of the generic attacks and Thériault's large prime variant of the index calculus attack and the more recent double large prime variants (cf. Section 21.3).

### 23.2.2 Transfers by Galois theory

In the last section we have studied a "generic" attack to compute discrete logarithms based on ideal class groups and as a consequence we have to exclude all (not only hyperelliptic) curves of genus $g \geqslant 4$ from the candidate list. Now we want to show that special curves of genus $g \leqslant 3$ over special fields can deliver weak DL systems though no direct application of an index calculus algorithm can be used. The method is to transfer the discrete logarithm problem in the original group in polynomial time to a group in which index calculus algorithms are efficient. The transfer maps known today are treated in Chapter 22.

In the following sections we shall always work with a projective absolutely irreducible nonsingular curve $C$ defined over the finite field $\mathbb{F}_q$.

#### 23.2.2.a Pairings

The first method uses the duality theory of Jacobian varieties.

First look at the special case that $\ell$ divides $q$. Then we can transfer $\mathrm{Pic}_C^0$ in polynomial time into a subgroup of a vector space of dimension $g$ over $\mathbb{F}_q$, and in this group the discrete logarithm has complexity $O\big((2g-1)\lg(q)^k\big)$, where $k$ is a small constant. Hence, one has to avoid this case in all circumstances.

So assume now that $\ell$ is prime to $q$. In Chapter 6 one finds the definition and the mathematical background of the Tate pairing in the Lichtenbaum version. In Chapter 16 one finds algorithms to implement the pairing efficiently. The result to be kept in mind is Theorem 6.15. Together with its corollary it states: let $\mathbb{F}_q$ be the field with $q$ elements, $\ell$ a prime number prime to $q$ and $k \in \mathbb{N}$ be minimal with $\ell \mid (q^k - 1)$.

There is a bilinear map

$$T_\ell : \mathrm{Pic}_C^0[\ell] \times \mathrm{Pic}_{C \cdot \mathbb{F}_{q^k}}^0 \to \mathbb{F}_{q^k}^* / (\mathbb{F}_{q^k}^*)^\ell,$$

which is nondegenerate on the right side, i.e., for a random element in $\overline{D} \in \mathrm{Pic}^0_{C \cdot \mathbb{F}_{q^k}}$ the map

$$T_{\ell,\overline{D}} : \mathrm{Pic}^0_C[\ell] \to \mathbb{F}^*_{q^k}[\ell]$$
$$P \mapsto T_{\ell,\overline{D}}(P) := \bigl(T_\ell(P, \overline{D})\bigr)^{\frac{q^k-1}{\ell}}$$

is an injective homomorphism of groups, which has complexity $O(k \lg q)$.

As a consequence we get from the results in Chapter 20 the following proposition.

**Proposition 23.5** *The discrete logarithm in $\mathrm{Pic}^0_C$ has complexity $L_{q^k}\bigl(\frac{1}{2}, \sqrt{2}\bigr)$ and so is subexponential in $q^k$.*

**Example 23.6** Assume that $k \leqslant \ln q$. Then the complexity to compute the discrete logarithm in $\mathrm{Pic}^0_C$ is $L_q\bigl(\frac{1}{2}, \sqrt{2k}\bigr)$.

It is rather easy to avoid curves $C$ of genus $\leqslant 3$ for which a large prime $\ell$ divides $|\mathrm{Pic}^0_C|$ and for which at the same time the corresponding $k$ is relatively small. For elliptic curves we have proved in Section 6.4.2 a necessary and sufficient criterion for this situation. It turns out that for *supersingular elliptic curves* (cf. Definition 4.74) $k$ is always small, namely less than or equal to 6, while for ordinary random elliptic curves $k$ can be expected to be large. Analogous results hold for curves of genus $g \leqslant 3$. So, for a cryptographic system with discrete logarithms as crypto primitive, the divisor class groups $\mathrm{Pic}^0_C$ of supersingular curves are to be avoided — or, if there is a strong reason for using them, one has to choose the parameters in such a way that for given $k$ the complexity $L_q\bigl(\frac{1}{2}, \sqrt{2k}\bigr)$ is large enough. One of these strong reasons is motivated by the constructive aspects of the Tate pairing related to the bilinear structure on $\mathrm{Pic}^0_C$. A detailed discussion is found in Chapter 24.

### 23.2.2.b  Weil descent

In Section 23.1.2.d we have given a list for parameters of DL systems. By index calculus methods we have seen that we should restrict the genus to be less than or equal to 3. Because of duality we have to make sure that for the pair $(\mathbb{F}_q, \ell)$ the minimal number $k$ with $\ell \mid q^k - 1$ has to be large enough.

Now we come to the parameter $d$ with $q = p^d$. Assume that $d > 1$ and that $d_0 \mid d$ with $d_0 < d$. So we have a nontrivial operation of the Frobenius $\phi_{p^{d_0}}$ and we can use Weil descent (see Chapters 7 and 22). The results there give strong indications for the weakness of the discrete logarithm problem if either $d/d_0$ is small or if $d$ is a prime with the additional property that there is a small number $t$ with $2^t \equiv 1 \pmod{d}$. For instance $d$ is not allowed to be a Mersenne prime number. These assertions follow from the GHS algorithm (cf. Section 22.3.2) and its generalizations. But the reader should be aware that the algorithms described in Chapter 22 are only examples of possible attacks and, contrary to the attack by pairings, we do not have a clean criterion for a curve defined over nonprime fields with composite degree to be resistant against Weil descent attack.

The new ideas of Gaudry and Diem make the situation even worse. For instance the result of Gaudry in Section 22.3.5.a can be applied to *all* elliptic curves defined over extension fields of degree 4, and the result of Diem stated in Section 22.3.5.b at least gives a strong hint that for elliptic curves $d$ should be a large prime.

Taking this into account the choice of the ground field $\mathbb{F}_q$ for a DL system realized in $\mathrm{Pic}^0_C$ should be:

- a prime field, namely $\mathbb{F}_q = \mathbb{F}_p$, or
- an extension field $\mathbb{F}_q = \mathbb{F}_{p^d}$ with $p$ very small (usually $p = 2$) and $d$ a prime such that the multiplicative order of 2 modulo $p$ is large.

There should be a very good reason if one does not follow this rule, and then a careful discussion of the security has to be performed. We shall give one example for this.

**Example 23.7** Trace zero varieties introduced in Section 15.3 are constructed from curves defined over $\mathbb{F}_p$ which are considered over $\mathbb{F}_{p^d}$ for $d = 3$ or $d = 5$. The latter is only proposed for elliptic curves while the former construction could be used for elliptic and genus 2 curves.

As DL system one uses the group of divisor classes that have trace zero (hence the name), i.e.,

$$G = \left\{ \overline{D} \in \mathrm{Pic}_C^0(\mathbb{F}_{p^d}) \mid \overline{D} \oplus \phi_p(\overline{D}) \oplus \cdots \oplus \phi_p^{d-1}(\overline{D}) = 0 \right\}.$$

Geometrically this group is isomorphic to an abelian variety $\mathcal{G}$ of dimension $g(d-1)$ in the Weil descent of $C$, which shows $|G| \sim p^{g(d-1)}$. The advantages of this group come from efficient implementation and will be made clear in the following section. From the security considerations the choice of $(g, d) = (1, 3)$ has no known weakness. For good choices of parameters the other two possibilities that lead to groups of size $p^4$ offer better security than genus 4 curves or elliptic curves over fields $\mathbb{F}_{p^4}$, which would have the same group order. For $(g, d) = (2, 3)$ this comes from the fact that $\mathcal{G}$ can always be embedded in the Jacobian of a genus 6 curve over $\mathbb{F}_p$ and that there the index calculus algorithm runs in $O\bigl(q^{\frac{22}{13}+\epsilon}\bigr) = O\bigl(|G|^{\frac{11}{16}+\epsilon}\bigr)$. In this rough bound we did not include the constants. A more thorough study (cf. Section 22.3.4.b, [DISC 2003]) reveals that for low security applications, e.g., group sizes of 128 bits, the security is close to the generic one whereas for larger bit sizes one loses asymptotically about 17% of security.

### 23.2.2.c Conclusion

The results of Section 23.2.2 show that one has to be careful with the choice of the *pair* $(C, \mathbb{F}_q)$ if one wants to have instances in which the complexity of algorithms computing the discrete logarithm is $O(\ell^{1/2})$.

1. One has to choose $\ell$ so that it does not divide $q$ and so that the field $\mathbb{F}_q(\zeta_\ell)$ is an extension of $\mathbb{F}_q$ of sufficiently large degree $k$. For instance $k \geqslant 1000$ should be ensured. (Note that this does not mean that one needs to actually compute $k$ but that one checks for all $k' \leqslant k$ that $\ell \nmid q^{k'} - 1$.) This excludes especially the case that $C$ is supersingular.
2. One should take $\mathbb{F}_q$ either as prime field or as an extension of a field of small characteristic $p$ (e.g., $p = 2$), which has prime degree $d$ over $\mathbb{F}_p$. Moreover the number 2 should have large order modulo $d$, e.g., $d$ must not be a Mersenne or a Fermat prime number.

If one has strong reasons not to follow these directions, one has to make a careful analysis of the situation.

**Example 23.8** Assume the situation that one would like to find a group of order equivalent to $q^4$ as, e.g., the arithmetic in $\mathbb{F}_q$ is particularly suited to the hardware. In this case, one should not take curves of genus 4 over $\mathbb{F}_q$ or elliptic curves defined over $\mathbb{F}_{q^4}$, as due to the attacks described in Section 22.3.5.a the size of $q$ needs to be chosen much larger. One can do better with trace zero varieties of curves of genus 2 defined over $\mathbb{F}_q$ with respect to extensions of degree 3 (see Example 23.7).

## 23.3 Efficient systems

In the previous section we have shown that ideal class groups of function fields over finite fields allow us to obtain groups in which the discrete logarithm is supposed to be hard to compute, given that the genus is less than or equal to 3. The parameters $p, d$, and $g$ may then be chosen to accommodate fast computation of scalar multiples in the group provided that the group size $O(p^{dg})$ is of the correct size and none of the weaknesses mentioned before is introduced.

## 23.3.1  Choice of the finite field

From the point of view of finite field arithmetic, prime fields $\mathbb{F}_p$ and binary fields $\mathbb{F}_{2^d}$ are the most common choices, but for some implementations optimal extension fields and their generalizations might offer advantages, as they are ideally suited to the word size. For full details on arithmetic on finite fields we refer to Chapter 11, and the mathematical background can be found in Chapter 2. For special considerations for hardware implementation one should consult Chapter 26.

### 23.3.1.a  Prime fields

In prime fields $\mathbb{F}_p$, addition and subtraction are performed as in the integers and the result is reduced modulo the prime $p$. In general, much of the considerations for integer arithmetic (cf. Chapter 10) can be applied for the arithmetic modulo $p$.

For multiplication one uses the schoolbook method or Karatsuba's trick. Fast multiplication methods like FFT do not apply for the small size of $p$ we are considering here. Squaring has about the same complexity as multiplication and can either be implemented separately (which can lead to a speedup in trade-off for more code) or by reusing the multiplication routine. In general one should consider the multiplication separately from the reduction and take into account the effect that for computing $ab + cd$, for field elements $a, b, c, d$, one only needs one reduction instead of two.

Inversions and divisions are computed using the (binary) extended gcd.

To have fast arithmetic in $\mathbb{F}_p$ it is advisable to choose $p$ of low Hamming weight, ideally of the form $2^{wn} + c$, where $w$ is the word size and $c$ is small. This allows us to compute the modular reduction more efficiently.

Montgomery representation of elements in $\mathbb{F}_p$ speeds up the computations even further. The element $x$ is represented by $xR$, where $R$ is the smallest power of $2^w$ larger than $p$. This representation behaves well with respect to addition, multiplication, and inversion and the reductions are particularly simple. For full details we refer to Algorithms 11.3 and 11.9.

### 23.3.1.b  Extension fields

To represent extension fields one has two general methods, either using the multiplicative or the additive structure of $\mathbb{F}_{p^d}$. In the first case one uses a generator $g$ and represents each element as a power of $g$. This way, multiplications are very efficient but additions are problematic. For small fields one can use a lookup table but this gets inefficient very quickly.

So, for implementations in cryptographically relevant ranges, representations using the additive structure of $\mathbb{F}_{p^d}$ as $d$-dimensional vector space over $\mathbb{F}_p$ are favored. In this representation addition is done coefficient-wise.

There are two main trends for choosing the basis: either find an irreducible polynomial $m(X) \in \mathbb{F}_p[X]$ of degree $d$ and use $\left(1, \theta, \ldots, \theta^{d-1}\right)$ as basis, where $\theta$ is a root of $m(X)$ over $\mathbb{F}_{p^d}$, or choose a basis of the form $\left(\alpha, \alpha^p, \ldots, \alpha^{p^{d-1}}\right)$. The latter is called a *normal basis*; it has the advantage that the operation of the Frobenius automorphism $\phi_p$ is computed by a cyclic shift of the coefficients. As a drawback, multiplications are more complicated than in *polynomial basis* representation, where they are computed by a multiplication of polynomials followed by a reduction modulo $m(X)$.

We now give some details for binary fields $\mathbb{F}_{2^d}$ and optimal extension fields $\mathbb{F}_{p^d}$. Some applications discussed in Chapter 24 use the field $\mathbb{F}_{3^d}$. It shares many properties with $\mathbb{F}_{2^d}$. Hence, the mathematical background is covered in the following paragraph. Note that the basic arithmetic in

## Binary fields

The security considerations impose that $d$ is a large prime. So, in all applications $d$ is odd implying that $\mathrm{Tr}_{\mathbb{F}_{2^d}/\mathbb{F}_2}(1) = 1$. This means that some transformations on the curve equations, discussed later, are always possible. On the other hand this means that there is no type I optimal normal basis (cf. Section 11.2.2.b), i.e., if one represents the multiplication using a matrix, then it is not possible to get the most sparse type of matrix.

The use of normal basis representation can be useful if many more squarings than multiplications are needed or if one needs to compute square roots. For many of the coordinate systems on elliptic and hyperelliptic curves and ways of scalar multiplication discussed below, normal basis representation leads to slower implementations than cleverly chosen irreducible polynomials with a polynomial basis.

To be able to multiply in normal basis representation one is either given an explicit multiplication matrix or uses Gauß periods over extensions. Inversion is either done by transforming to a polynomial representation and then performing it there as described in the sequel or by using Lagrange's theorem. In the latter case one should apply an addition chain involving the Frobenius automorphism, as this map is particularly fast in normal basis representation.

We now turn our attention to polynomial basis representations. As for primes it is very useful if the irreducible polynomial is sparse, i.e., it has only few nonzero coefficients. Over $\mathbb{F}_2$ each binomial has either 0 or 1 as its root and hence it cannot be irreducible. For implementations irreducible trinomials are the best choice if they exist. Otherwise Section 11.2.1.b proposes to use redundant trinomials to obtain sparse polynomials for reduction.

Multiplication is performed as a multiplication followed by a reduction modulo $m(X)$. For the fields $\mathbb{F}_{2^d}$ a Montgomery representation of the fields exists as well. Also, for inversions the same tricks can be applied and usually they are computed as an extended gcd. For restricted devices inversions are usually too time and space-consuming. However, one should note that the ratio between multiplication and inversion is not as bad as in prime fields.

Even though in polynomial basis representation a squaring is more complicated than a cyclic shift, it is still a fast operation compared to multiplication as no mixed terms occur. A rough estimate is S = 0.1M, where S abbreviates a squaring and M stands for a multiplication.

## Optimal extension fields

For some platforms or applications other choices of $p$ and $d$ are more advantageous. For security reasons one chooses $d$ to be prime. The idea behind optimal extension fields and their generalization as *processor adapted finite fields* is to use a ground field $\mathbb{F}_p$ such that the elements of $\mathbb{F}_p$ fit within one word, with the consequence that they can be particularly efficiently handled. For the choice of $p$ the same considerations as in the prime field case apply.

To construct the field extension one tries to use an irreducible binomial $m(X) = X^d - a$, where optimally $a$ is small such that multiplications involving $a$ can be performed by a few modular additions. Hence, reduction modulo $m(X)$ can be efficiently computed.

Addition is trivial. To multiply, the reductions are simplified by the choices of $p$ and $m(X)$ and for small extension degrees $d$ like those needed for trace zero varieties, one can save a few multiplications over $\mathbb{F}_p$ by making a detailed look at the code. For $d = 3$ and 5 these formulas are included in Section 11.3.6.

For inversion one uses $\alpha^{p^d-1} = 1$ and expresses

$$\alpha^{-1} = \prod_{i=1}^{d-1} \alpha^{p^i} \bigg/ \prod_{j=0}^{d-1} \alpha^{p^j}.$$

Note that some of the powers can be rearranged to save even more operations. For $d = 5$ this reads

$$\alpha^{-1} = \frac{(\alpha^p \alpha^{p^2} (\alpha^p \alpha^{p^2})^{p^2})^{p^2}}{\alpha (\alpha^p \alpha^{p^2} (\alpha^p \alpha^{p^2})^{p^2})}.$$

For very small $d \leqslant 4$ an approach based on linear algebra is more efficient. We refer to Section 11.3.4.

### 23.3.2  Choice of genus and curve equation

In this section we assume that the scalar multiplications appearing in the DL-based protocols are carried out using doublings and additions, and, depending on the storage capacities, by using some precomputed points. The latter can be applied in windowing methods as explained in Chapter 9 to reduce the number of additions needed in a scalar multiplication. As the number of doublings remains unchanged such systems need especially cheap repeated doublings.

Clearly, Cantor's algorithm can be used to perform arithmetic in the ideal class group of arbitrary hyperelliptic curves. As soon as one fixes the genus of the curve one can derive explicit formulas from the general group operations, which are usually much faster in implementations. As the group size behaves like $p^{dg}$ a larger genus allows smaller finite fields. By the security considerations we are, however, limited to $g = 1$, 2, and 3.

#### 23.3.2.a  Elliptic curves

Elliptic curves are curves of genus 1. On them the ideal and divisor class group are isomorphic to the group of points and thus one can define addition and doubling in terms of the coordinates of points. The general addition formula is given in Section 13.1.1.

Over a binary field, each nonsupersingular curve can be given (after an isomorphic transformation) by an affine equation

$$y^2 + xy = x^3 + a_2 x^2 + a_6, \quad \text{where } a_2 \in \mathbb{F}_{2^d} \text{ and } a_6 \in \mathbb{F}_{2^d}^*.$$

As $d$ should be odd, we can even choose $a_2 \in \mathbb{F}_2$, cf. Remark 13.40.

In affine coordinates the addition formulas read as follows, where $-P = (x_1, x_1 + y_1)$.

**Addition**

Let $P = (x_1, y_1)$, $Q = (x_2, y_2)$ such that $P \neq \pm Q$ then $P \oplus Q = (x_3, y_3)$ is given by

$$x_3 = \lambda^2 + \lambda + x_1 + x_2 + a_2, \qquad y_3 = \lambda(x_1 + x_3) + x_3 + y_1, \qquad \lambda = \frac{y_1 + y_2}{x_1 + x_2}.$$

**Doubling**

Let $P = (x_1, y_1)$ then $[2]P = (x_3, y_3)$, where

$$x_3 = \lambda^2 + \lambda + a_2, \qquad y_3 = \lambda(x_1 + x_3) + x_3 + y_1, \qquad \lambda = x_1 + \frac{y_1}{x_1}.$$

Thus an addition and a doubling require exactly the same number of operations that is I + 2M + S. Here, we use the abbreviations I for inversion, M for multiplication, and S for squaring. We use $M_2$ to denote a multiplication by $a_2$. This is counted separately as for odd $d$ there always exists an isomorphic curve with $a_2 \in \mathbb{F}_2$ such that the multiplication is computed as an addition.

## § 23.3 Efficient systems

To avoid inversions one can use projective and different weighted projective coordinates. The following Table 23.1 lists the number of operations needed to double or add in different coordinate systems ($\mathcal{A}$ denotes affine coordinates, $\mathcal{P}$ projective coordinates, and $\mathcal{J}$ and $\mathcal{LD}$ stand for Jacobian and López–Dahab coordinates, respectively). For details on the arithmetic we refer to Section 13.3.

**Table 23.1** Operations required for addition and doubling.

| Doubling | | Addition | |
|---|---|---|---|
| Operation | Costs | Operation | Costs |
| $2\mathcal{P}$ | $7\text{M} + 4\text{S} + \text{M}_2$ | $\mathcal{J} + \mathcal{J}$ | $15\text{M} + 3\text{S} + \text{M}_2$ |
| $2\mathcal{J}$ | $5\text{M} + 5\text{S}$ | $\mathcal{P} + \mathcal{P}$ | $15\text{M} + 2\text{S} + \text{M}_2$ |
| $2\mathcal{LD}$ | $4\text{M} + 4\text{S} + \text{M}_2$ | $\mathcal{LD} + \mathcal{LD}$ | $13\text{M} + 4\text{S}$ |
| $2\mathcal{A} = \mathcal{P}$ | $5\text{M} + 2\text{S} + \text{M}_2$ | $\mathcal{P} + \mathcal{A} = \mathcal{P}$ | $11\text{M} + 2\text{S} + \text{M}_2$ |
| $2\mathcal{A} = \mathcal{LD}$ | $2\text{M} + 3\text{S} + \text{M}_2$ | $\mathcal{J} + \mathcal{A} = \mathcal{J}$ | $10\text{M} + 3\text{S} + \text{M}_2$ |
| $2\mathcal{A} = \mathcal{J}$ | $\text{M} + 2\text{S} + \text{M}_2$ | $\mathcal{LD} + \mathcal{A} = \mathcal{LD}$ | $8\text{M} + 5\text{S} + \text{M}_2$ |
| — | — | $\mathcal{A} + \mathcal{A} = \mathcal{LD}$ | $5\text{M} + 2\text{S} + \text{M}_2$ |
| $2\mathcal{A}$ | $\text{I} + 2\text{M} + \text{S}$ | $\mathcal{A} + \mathcal{A} = \mathcal{J}$ | $4\text{M} + \text{S} + \text{M}_2$ |
| $2\mathcal{A}'$ | $\text{I} + \text{M} + \text{S}$ | $\mathcal{A} + \mathcal{A} = \mathcal{A}'$ | $2\text{I} + 3\text{M} + \text{S}$ |
| $2\mathcal{A}' = \mathcal{A}$ | $\text{M} + 2\text{S}$ | $\mathcal{A} + \mathcal{A}$ | $\text{I} + 2\text{M} + \text{S}$ |

If inversions are affordable, affine coordinates are preferred as the total number of field operations is lowest. Otherwise, one should use López–Dahab coordinates, as for them doublings are fastest and additions, especially mixed addition $\mathcal{A} + \mathcal{LD} = \mathcal{LD}$, are cheap as well. For the binary elliptic curves proposed in the standards $a_6$ is chosen to be small such that multiplications by this constant are fast. Then the standard doubling formulas should be applied, while for random curves, and thus large and changing $a_6$, formulas (13.9) are preferred.

In odd characteristic for $p > 3$ one can make an isomorphic transform to get each elliptic curve represented by an affine equation of the form

$$y^2 = x^3 + a_4 x + a_6, \quad \text{with } a_4, a_6 \in \mathbb{F}_q,$$

such that $x^3 + a_4 x + a_6$ has only simple roots over $\overline{\mathbb{F}}_q$. The negative of $P = (x_1, y_1)$ is given by $-P = (x_1, -y_1)$.

### Addition

Let $P = (x_1, y_1)$, $Q = (x_2, y_2)$ such that $P \neq \pm Q$ and $P \oplus Q = (x_3, y_3)$. In this case, addition is given by

$$x_3 = \lambda^2 - x_1 - x_2, \qquad y_3 = \lambda(x_1 - x_3) - y_1, \qquad \lambda = \frac{y_1 - y_2}{x_1 - x_2}.$$

### Doubling

Let $[2]P = (x_3, y_3)$. Then

$$x_3 = \lambda^2 - 2x_1, \qquad y_3 = \lambda(x_1 - x_3) - y_1, \qquad \lambda = \frac{3x_1^2 + a_4}{2y_1}.$$

An addition needs $I + 2M + S$ and to compute a doubling $I + 2M + 2S$ are used.

As stated in the previous section, in odd characteristic inversions are very time-consuming compared to multiplications. Hence, it is even more important to consider inversion-free coordinate systems. The following Table 23.2 gives an overview of the number of field operations per group operation. In addition to the above abbreviations we use $\mathcal{J}^m$ for modified Jacobian coordinates and $\mathcal{J}^c$ for Chudnovsky Jacobian coordinates. For full details on the formulas, we refer to Section 13.2.

Table 23.2 *Operations required for addition and doubling.*

| Doubling | | Addition | |
|---|---|---|---|
| Operation | Costs | Operation | Costs |
| $2\mathcal{P}$ | $7M + 5S$ | $\mathcal{J}^m + \mathcal{J}^m$ | $13M + 6S$ |
| $2\mathcal{J}^c$ | $5M + 6S$ | $\mathcal{J}^m + \mathcal{J}^c = \mathcal{J}^m$ | $12M + 5S$ |
| $2\mathcal{J}$ | $4M + 6S$ | $\mathcal{J} + \mathcal{J}^c = \mathcal{J}^m$ | $12M + 5S$ |
| $2\mathcal{J}^m = \mathcal{J}^c$ | $4M + 5S$ | $\mathcal{J} + \mathcal{J}$ | $12M + 4S$ |
| $2\mathcal{J}^m$ | $4M + 4S$ | $\mathcal{P} + \mathcal{P}$ | $12M + 2S$ |
| $2\mathcal{A} = \mathcal{J}^c$ | $3M + 5S$ | $\mathcal{J}^c + \mathcal{J}^c = \mathcal{J}^m$ | $11M + 4S$ |
| $2\mathcal{J}^m = \mathcal{J}$ | $3M + 4S$ | $\mathcal{J}^c + \mathcal{J}^c$ | $11M + 3S$ |
| $2\mathcal{A} = \mathcal{J}^m$ | $3M + 4S$ | $\mathcal{J}^c + \mathcal{J} = \mathcal{J}$ | $11M + 3S$ |
| $2\mathcal{A} = \mathcal{J}$ | $2M + 4S$ | $\mathcal{J}^c + \mathcal{J}^c = \mathcal{J}$ | $10M + 2S$ |
| — | — | $\mathcal{J} + \mathcal{A} = \mathcal{J}^m$ | $9M + 5S$ |
| — | — | $\mathcal{J}^m + \mathcal{A} = \mathcal{J}^m$ | $9M + 5S$ |
| — | — | $\mathcal{J}^c + \mathcal{A} = \mathcal{J}^m$ | $8M + 4S$ |
| — | — | $\mathcal{J}^c + \mathcal{A} = \mathcal{J}^c$ | $8M + 3S$ |
| — | — | $\mathcal{J} + \mathcal{A} = \mathcal{J}$ | $8M + 3S$ |
| — | — | $\mathcal{J}^m + \mathcal{A} = \mathcal{J}$ | $8M + 3S$ |
| — | — | $\mathcal{A} + \mathcal{A} = \mathcal{J}^m$ | $5M + 4S$ |
| — | — | $\mathcal{A} + \mathcal{A} = \mathcal{J}^c$ | $5M + 3S$ |
| $2\mathcal{A}$ | $I + 2M + 2S$ | $\mathcal{A} + \mathcal{A}$ | $I + 2M + S$ |

### 23.3.2.b Curves of genus 2

For these curves the explicit formulas are more involved compared to elliptic curves. On the other hand the field one works in is of half the size for equal security as the group order is given by $O(p^{2d})$. In affine coordinates an addition needs $I + 22M + 3S$ and a doubling needs 2S more. We refer to Chapter 14 for the details on the group operation but mention that as for elliptic curves one has the choice between different coordinate systems including inversion-free systems. The choice of genus 2 curves leads to similar speed for scalar multiplication as elliptic curves. This can be seen from the implementation results listed in Table 14.13, p. 353. For each environment one needs to check which fields are faster to implement and also whether the longer code needed for the group operations for genus 2 curves is a problem. It is only possible to state the advantages or disadvantages of elliptic curves over genus 2 curves based on implementations. From a theoretical point of view their performance is similar.

Special choices of curves, like the family of binary curves

$$y^2 + xy = x^5 + f_3 x^3 + \varepsilon x^2 + f_0, \text{ with } f_3, f_0 \in \mathbb{F}_{2^d} \text{ and } \varepsilon \in \mathbb{F}_2,$$

over fields with odd $d$ allow us to obtain especially fast doubling formulas that outperform elliptic curves. There is no similar family of elliptic curves, and, hence, the richer structure of genus 2 curves pays off.

#### 23.3.2.c Curves of genus 3

So far explicit formulas for arithmetic on genus 3 curves were obtained in affine and projective coordinates. For general curves it seems that the number of operations is so large that the performance is worse than for elliptic and genus 2 curves. Furthermore, one needs to take into account the reduction of security due to index calculus attacks, such that the group order needs to be made even larger.

For special choices of curves like the binary curves

$$y^2 + y = f(x), \text{ with } f(x) \in \mathbb{F}_{2^d}[x]$$

particularly fast doublings can be designed without weakening security.

#### 23.3.2.d Conclusion

Over prime fields elliptic curves and genus 2 curves offer similar efficiency, while genus 3 curves are less efficient with the current explicit formulas. For completely general curves over binary fields the same observations hold.

If one is willing to trade off generality for higher speed, there exist genus 2 and 3 curves over $\mathbb{F}_{2^d}$ offering fast arithmetic. Comparable choices cannot be made for elliptic curves. We mention that these curves constitute families and thus they are not considered special curves.

### 23.3.3 Special choices of curves and scalar multiplication

Some curves have extra properties that can be used to compute scalar multiplication faster. This does not mean that the single group operations, i.e., additions and doublings, are sped up, but one uses a different procedure to compute the scalar multiples.

In Example 23.7 we introduced trace zero varieties. In general, subfield curves defined over $\mathbb{F}_p$ and considered over $\mathbb{F}_{p^d}$ offer the advantage that the Frobenius endomorphism $\phi_p$ operates on the points of the curve by raising each coordinate to the power of $p$. This can be used to obtain faster methods for scalar multiplication. In Chapter 15 we showed how to compute $[n]\overline{D}$ by using $\phi_p$. Instead of using a *double and add* algorithm, one applies a *Frobenius and add* algorithm, which leads to faster computations as $\phi_p$ can be computed efficiently.

It is also possible to use other endomorphisms of curves. A general curve will not have an efficiently computable endomorphism, but on specially chosen curves it can be used to speed up the scalar multiplication. The methods are described in Section 15.2.

In both cases one needs to be aware that the endomorphisms also speed up the attacks. The effects are discussed in Section 19.5.5. The use of special curves is particularly interesting for either large systems, where a central server has to perform a huge amount of encryptions and has no problems with a slightly larger group size (as long as the protocols run faster), or in lightweight cryptography applications, where the security needs are reduced but the devices cannot consume too much power and have limited storage. In this scenario curves with endomorphisms make the operations faster.

## 23.4 Construction of systems

Depending on the security needs one chooses the size $N$ of the group in which the discrete logarithm is to be used. The list of candidates for secure DL systems in Section 23.1.4 gives possible choices of parameters for the system. By considering the given environment and by looking at the results on efficiency in Section 23.3 one chooses candidate pairs $(\mathbb{F}_q, C)$ such that the expected group order of $\text{Pic}_C^0$ is of size $N$ and that the other needs are met. The final step is to look for primes $\ell \sim N$ for which $\mathbb{Z}/\ell\mathbb{Z}$ can be embedded into $\text{Pic}_C^0$ for a candidate curve $C$ in a bit-efficient way. Usually one has to do this by randomly choosing curves $C$ in a certain family $\mathcal{F}$ and then computing $\text{Pic}_C^0$.

So, one needs two ingredients: there have to be sufficiently many instances leading to an almost prime group order and one has to be able to compute the group order efficiently.

### 23.4.1 Heuristics of class group orders

First there has to be (at least) a heuristic prediction stating that with large probability the order of $\text{Pic}_C^0$ for $C \in \mathcal{F}$ is almost a prime, i.e., there is a number $c \leqslant B$ with $|\text{Pic}_C^0| = c\ell$ with $\ell$ a prime number. The number $c$ is called the cofactor, usually one sets the bound to $B \leqslant 1000$. A large choice of $B$ will raise the probability of finding a suitable curve $C$ but it will imply that we have to take the size of the ground field $\mathbb{F}_q$ (and hence the key length) larger, since we have

$$g \lg q \sim \lg c + \lg \ell.$$

The main sources for such heuristics are analytic number theory and analogous techniques over finite fields. Key words are Cohen–Lenstra heuristics about the behavior of class groups of global fields and Lang–Trotter conjectures about the distribution of traces of Frobenius automorphisms acting on torsion points of abelian varieties over number fields. Combined with sieving techniques one finds a positive probability for the existence of class numbers with only a few prime divisors. It is outside the scope of this book to give more details. For applications and statistics in the cases we are interested in we refer readers to [BAKO 1998] and [WEN 2001b]. For our purposes it is enough to state that in all families $\mathcal{F}$ that can be used in practice it will not take much time to find curves that can be used for DL systems.

In many cases one can impose $c = 1$. But one has to be cautious in special cases. There can be families $\mathcal{F}$ for which there is a number $c_0$ such that for each member, $C$, one has that $c_0$ divides $|\text{Pic}_C^0|$ and so $c$ has to be a multiple of $c_0$. For example take $\mathcal{F}$ as the set of ordinary binary elliptic curves. They are all of the form $y^2 + xy = x^3 + a_2 x^2 + a_6$ with $a_6 \neq 0$. Therefore, 2 always divides the cofactor as $(0, \sqrt{a_6})$ is a point of order 2 defined over the ground field. Another example is the family $\mathcal{F}$ consisting of curves $C$ of genus 3, which have an automorphism of order 4. Their class number is always even (cf. Section 18.3).

In other families it can occur that the class numbers are always divisible by different numbers bounded by $c_0$. For instance take $\mathcal{F}$ as a family of curves defined over a field $\mathbb{F}_{q_0}$ with $q_0 \mid q$. Then the class number of $C$ over $\mathbb{F}_q$ is divided by the class number of $C$ regarded as curve over $\mathbb{F}_{q_0}$ and so the cofactor $c$ has to be $\geqslant (\sqrt{q_0} - 1)^{2g}$. Examples of families $\mathcal{F}$ for which $c = 1$ is possible and for which one can (efficiently) determine the group order are random elliptic curves or hyperelliptic curves of genus 2 over fields $\mathbb{F}_q$ of odd characteristic, as well as elliptic curves over prime fields constructed by the method of complex multiplication (cf. Section 18), and not isogenous to a curve with $j$-invariant equal to 0 or $12^3$.

### 23.4.2 Finding groups of suitable size

In the following we shall give examples for families of curves $\mathcal{F}$ for which the (heuristic) results of the Section 23.4.1 predict that with a reasonable probability the number of elements in $\text{Pic}_C^0$ for $C \in \mathcal{F}$ is divisible by a large prime $\ell$ and for which it is possible to determine $|\text{Pic}_C^0|$ efficiently. To find a curve $C$ usable for DL systems one chooses a random curve in $\mathcal{F}$ and computes the order of $\text{Pic}_C^0$. If it is not of the right shape one chooses another random element in $\mathcal{F}$ and repeats the computation until finally a usable curve is found.

Once a suitable curve is found, one has to choose a base point in $\text{Pic}_C^0$ of order $\ell$.

#### 23.4.2.a Finding a curve

As said already the key ingredients for finding curves with divisor class group usable for DL systems are point counting algorithms, which are described in Chapter 17. Their common principle is that they compute the characteristic polynomial $\chi(\phi_q)_{\overline{C}}(T)$ of the Frobenius endomorphism $\phi_{q_C}$ acting on $J_C$ (cf. Section 4.3.6) and use the fact that $|\text{Pic}_C^0| = \chi(\phi_q)_C(1)$ (cf. Corollary 5.70).

**Curves defined over subfields**

Take as $\mathcal{F}$ the family of curves $C$ of genus $g$ over fields $\mathbb{F}_q$, which are defined over subfields $\mathbb{F}_{q_0}$. If $q_0$ is small, elementary methods like counting by enumeration (cf. Section 17.1.1) or square root algorithms (cf. Chapter 19) can be used. This is reasonable e.g., for $q_0^g \leqslant 10^{16}$.

For larger $q_0$ it may be necessary to replace these counting algorithms by more refined algorithms explained in the next paragraphs. Nevertheless this counting will be very fast compared to counting methods for curves defined over $\mathbb{F}_q$.

So, it is possible to compute the characteristic polynomial $\chi(\phi_{q_0})(T)$ of the Frobenius endomorphism $\phi_{q_0}$ over $\mathbb{F}_{q_0}$. As explained in Section 17.1.2 this can be used to compute the characteristic polynomial of $\phi_q$ and hence the order of $\text{Pic}_C^0$.

**Remark 23.9** This method is especially fast if $q_0$ is small. Moreover, the operation of the Frobenius endomorphism $\phi_{q_0}$ can be used to accelerate the computation of scalar multiples in $\text{Pic}_C^0$ considerably (cf. Section 15.1). The key word is "Koblitz curves."

**Example 23.10** The most famous Koblitz curves are

$$E_{a_2} : y^2 + xy = x^3 + a_2 x^2 + 1, \text{ with } a_2 = 0 \text{ or } 1$$

seen as curves over $\mathbb{F}_2$. The characteristic polynomial is given by

$$\chi_{a_2}(T) = T^2 - (-1)^{1-a_2} T + 2$$

and the number of points of $E(\mathbb{F}_{2^d})$ is given by (15.9).

The disadvantages are that if $q_0$ is very small there will be only a few curves in the cryptographically interesting range. At the same time the cofactors will be divisible by $\chi(\phi_{q_0})_C(1)$, which is of size equivalent to $q_0^g$. Moreover the degree of $\mathbb{F}_q$ over $\mathbb{F}_{q_0}$ should be a prime number because of the transfers related to Weil descent (cf. Section 22.3). This attack has to be considered for small extensions, too.

**Remark 23.11** If the degree of $\mathbb{F}_q$ over $\mathbb{F}_{q_0}$ is small ($\leqslant 5$) one is led to systems based on trace zero varieties (cf. Section 7.4.2), which can be interesting alternatives in special situations for fast arithmetic; see Section 15.3.

## Random elliptic curves

One chooses the size of the allowed cofactor $c$ and of the prime $\ell$, which should divide the order of the group of rational points of an elliptic curve $E$. Let $q$ be a power of a prime $p$ of size $c\ell$. As family $\mathcal{F}$ we take the set of elliptic curves defined over $\mathbb{F}_q$.

Let $E$ be a randomly chosen elliptic curve defined over $\mathbb{F}_q$. We can apply the SEA Algorithm 17.25 to count the elements of $E(\mathbb{F}_q)$ in $O\bigl(\lg(q)^{2\mu+2}\bigr)$ bit operations, where $\mu$ is a constant such that two $B$-bit integers can be multiplied in time $B^\mu$. The algorithm is fast enough to find cryptographically usable elliptic curves even with cofactor 1 in a short time.

## Random curves of genus 2

As above, the desired size of the group order is denoted by $N$. One chooses $q$ to be of size $N^{1/2}$. The family $\mathcal{F}$ consists of all curves of genus 2 defined over $\mathbb{F}_q$. In [GASC 2004a] one finds an algorithm combining $p$-adic, $\ell$-adic, and generic methods to count points on the Jacobian variety of random curves of genus 2 defined over $\mathbb{F}_q$, and to find such curves that can be used for DL systems. At the moment this is rather time-consuming — for a curve over a $\mathbb{F}_p$ with $p = 5 \times 10^{24} + 8503491$, the time reported in [GASC 2004a] is one week per curve — but it is to be expected that refinements will accelerate the algorithm in the near future.

## Curves over fields of small characteristic

Let $p$ be a small prime number, in practice $p = 2$ or $p = 3$. So, to reach cryptographic group sizes we need $q = p^d$ with $d$ large. (For security reasons $d$ should be a prime such that the multiplicative order of 2 modulo $d$ is large (cf. Section 22.3)). The family $\mathcal{F}$ consists of random curves $C$ of genus $g$ (with $1 \leqslant g \leqslant 3$) defined over $\mathbb{F}_q$. Hence $d \sim \log_g N/g$. To count points on random curves $C$ in $\mathcal{F}$ one uses $p$-adic methods as described in Section 17.3. For $p = 2$ the AGM method (cf. Section 17.3.2) generalizes Satoh's method for elliptic curves and constructs canonical liftings of Jacobian varieties of curves of genus 1, 2, and 3 in a most efficient way — provided that this curve is ordinary. For random curves this condition will be satisfied.

The AGM algorithm is very easy to implement, especially for elliptic curves. As an example, we state the algorithm for elliptic curves over $\mathbb{F}_2$. The idea is to use a recurrence to compute the trace $t$ of the Frobenius endomorphism.

---

**Algorithm 23.12** Elliptic curve AGM

INPUT: An elliptic curve $E : y^2 + xy = x^3 + a_6$ over $\mathbb{F}_{2^d}$ with $j(E) \neq 0$.
OUTPUT: The number of points on $E(\mathbb{F}_{2^d})$.

1. $L \leftarrow \lceil \frac{d}{2} \rceil + 3$      [$L$ is the precision to be used]
2. $a \leftarrow 1$ and $b \leftarrow (1 + 8e) \bmod 2^4$      [$e$ arbitrary lift of $a_6$]
3. **for** $i = 5$ **to** $L$ **do**
4.      $(a, b) \leftarrow \bigl((a+b)/2, \sqrt{ab}\bigr) \bmod 2^i$
5. $a_0 \leftarrow a$
6. **for** $i = 0$ **to** $d - 1$ **do**
7.      $(a, b) \leftarrow \bigl((a+b)/2, \sqrt{ab}\bigr) \bmod 2^L$
8. $t \leftarrow \dfrac{a_0}{a} \bmod 2^{L-1}$
9. **if** $t^2 > 2^{d+2}$ **then** $t \leftarrow t - 2^{L-1}$
10. **return** $2^d + 1 - t$

In Section 17.3.2.c one finds the algorithms for $g \geqslant 2$. The theoretical background is given in Section 17.3.

A universal method of computing the characteristic polynomial of the Frobenius endomorphism $\phi_q$ for arbitrary (small) $p$ and arbitrary hyperelliptic curves was developed for the first time by Kedlaya [KED 2001]. He uses the Monsky–Washnitzer cohomology, which computes the de Rham cohomology of a formal lifting of the curve. Though the background is rather involved, the algorithm is easily implemented. It is given in Section 17.3.3 as Algorithm 17.80.

If the AGM method can be used it is faster than algorithms based on Kedlaya's idea. The big advantage of Kedlaya's approach is that it is a universally applicable algorithm (which can be generalized to practically all curves). In any case, one can state that point counting for random hyperelliptic curves of genus 1, 2, and 3 over fields of characteristic 2, 3, and 5 in cryptographically relevant ranges can be done in a satisfying way.

**The CM method**

The methods described until now leave one gap open: it is difficult to compute the characteristic polynomial of the Frobenius endomorphism acting on the Jacobian variety of curves of genus 2 and 3 defined over fields $\mathbb{F}_q$ of large characteristic. To fill this gap, especially in the case that $q$ is a prime number $p$, one can use the theory of complex multiplication relying on the results of Taniyama and Shimura. Even for elliptic curves this method remains interesting, especially if one wants to find many curves to check heuristics or to construct an elliptic curve with prescribed group order [BRST 2004], e.g., if one wants to obtain a group order with low Hamming weight.

For the background of this theory, we refer to Section 5.1. In this method one begins with the choice of a ring of endomorphisms $\text{End}_C$ of an abelian variety that has complex multiplication and that is the Jacobian variety of a hyperelliptic curve $C$ of genus $g$. This ring $\text{End}_C$ has to be an order in a CM-field of degree $2g$. One characterizes the curve by its invariants, and to obtain the equation of the curve one computes the minimal polynomials over $\mathbb{Q}$ of the invariants, the *class polynomials*. For details on the computation of the class polynomial we refer to Section 18.1.3.

In practice both finding the ring $\text{End}_C$ and computing the class polynomials are to be regarded as part of a precomputation. For elliptic curves they can be taken from published lists (cf. [WENG]).

The family $\mathcal{F}$ consists of the curves $C_p$, which are obtained from $C$ by reduction modulo primes $p$. So, our space of parameters is now the set of prime numbers. Arithmetic geometry tells us that this can be regarded as the analogue of a curve, and so the richness of the family $\mathcal{F}$ is analogous to that of the family consisting of all curves of genus $g$ over a fixed finite field $\mathbb{F}_q$.

By class field theory it is possible to determine the order of $\text{Pic}^0_{C_p}$ by solving a norm equation in $\text{End}_C$ that can be done quickly. Only after having found a "good" prime $p$ one constructs the curve $C_p$ in an explicit way.

The algorithm for elliptic curves $E$ is found in Section 18.1.5. We repeat it here for this case.

---

**Algorithm 23.13** Construction of elliptic curves via CM

INPUT: A squarefree integer $d \neq 1, 3$, parameters $\varepsilon$ and $\delta$, Hilbert class polynomial $H_d(X)$, desired size of $p$ and $\ell$.

OUTPUT: A prime $p$ of the desired size, an elliptic curve $E/\mathbb{F}_p$ whose group order $|E(\mathbb{F}_p)|$ has a large prime factor $\ell$.

---

1. **repeat**
2.     **repeat** choose $p$ prime of desired size
3.         **until** $\varepsilon p = x^2 + dy^2$ with $x, y \in \mathbb{Z}$
4.         $n_1 \leftarrow p + 1 - 2x/\delta$ and $n_2 \leftarrow p + 1 + 2x/\delta$
5. **until** $n_1$ **or** $n_2$ has a large prime factor $\ell$

6. compute a root $j$ of $H_d(X) \pmod p$
7. compute $E_j/\mathbb{F}_p$ from (18.1) and its twist $\widetilde{E}_j/\mathbb{F}_p$
8. **while true do**
9.     take $P \in_R E_j(\mathbb{F}_p)$ and compute $Q \leftarrow [n_1]P$
10.     **if** $Q = P_\infty$ **and** $[n_2]P \neq P_\infty$ **then return** $p$ and $E_j$
11.     **else if** $Q \neq P_\infty$ **then return** $p$ and $\widetilde{E}_j$

---

The situation is more complicated for $g = 2$ but nevertheless the corresponding algorithm works very well. The details can be found in Section 18.2. For $g = 3$ one has to assume that $\mathrm{End}_C$ contains $\sqrt{-1}$ and hence $C$ has an automorphism of order 4, which implies that the cofactor is divisible by 2. So we get only special curves, and the situation is not totally satisfying. Nevertheless we can construct many hyperelliptic curves of genus 3, which can be used for DL systems.

### 23.4.2.b State of the art of point counting

We have seen that the computation of the order of $\mathrm{Pic}_C^0$ for curves $C$ of genus $g$ defined over fields $\mathbb{F}_q$ can be performed efficiently by not too complicated algorithms if

- The curve $C$ is already defined over a small subfield $\mathbb{F}_{q_0}$ of $\mathbb{F}_q$, or
- The genus $g$ is equal to 1, or
- The characteristic of $\mathbb{F}_q$ is small, or
- The genus of $C$ is 1 or 2, the field $\mathbb{F}_q$ is a prime field, and the curve $C$ is the reduction modulo $q$ of a curve $\widetilde{C}$ with complex multiplication over a given order $\mathrm{End}_C$ in a CM-field.

For random curves $C$ of genus 2 there is hope that one will find an efficient algorithm for point counting. The results of Gaudry and Schost [GASc 2004a] are very encouraging.

For genus $g = 3$ and $\mathbb{F}_q$ a prime field, we have to restrict ourselves, at least at the moment, to hyperelliptic curves that have an automorphism of order 4.

### 23.4.2.c Finding a base point

We assume now that $C$ is a hyperelliptic curve defined over $\mathbb{F}_q$ such that

$$|\mathrm{Pic}_C^0| = c\ell$$

and that $C$ is given by an affine equation

$$C_a : y^2 + h(x)y = f(x).$$

One chooses a random element $x_1 \in \mathbb{F}_q$ and tests whether the polynomial $y^2 + h(x_1)y - f(x_1)$ has a solution $y_1$ in $\mathbb{F}_q$. This will be the case with probability $1/2$. If the polynomial has no solution in $\mathbb{F}_q$ one chooses another $x_1$. After a few trials one has found a point $P = (x_1, y_1) \in C(\mathbb{F}_q)$. One uses the embedding of $C$ in its Jacobian. With probability $1 - 1/\ell$ the divisor class of $P - P_\infty$ has order divisible by $\ell$ and so $[c]\overline{(P - P_\infty)}$ has order $\ell$. To check this one verifies that $[c]\overline{(P - P_\infty)}$ is not the neutral element.

If this is the case we can take $\overline{D} = [c]\overline{(P - P_\infty)}$ as the base point of our DL system and embed $\mathbb{Z}/\ell\mathbb{Z}$ into $\mathrm{Pic}_C^0$ by mapping $n \mapsto [n]\overline{D}$ for $0 \leqslant n < \ell$.

If the trial multiplication fails, i.e., if $[c]\overline{(P - P_\infty)} = 0$, one repeats the procedure by choosing a new element $x_1$.

The expected time to find a base point by this method is $O(\lg \ell)$. For details concerning hyperelliptic curves implemented with explicit formulas we refer to Section 14.1.2.

## 23.5 Protocols

So far we have presented groups in which the discrete logarithm problem is assumed to be hard. In Chapter 1 we stated some protocols in Section 1.6 using only the structure of a DL system. We now briefly cover two issues: how to actually instantiate these protocols with groups based on $\text{Pic}_C^0$ of curves over finite fields; and which pitfalls one needs to be aware of.

We need to make a general remark. As of the time of writing this, only elliptic curves are in the standards. In this section we state system parameters and protocols for hyperelliptic curves by giving the generalizations from the elliptic curve setting.

### 23.5.1 System parameters

We now assume that the system parameters consisting of the ground field, the genus of the curve, and the curve equation are chosen in a manner that the DL problem should be hard to compute.

To use these parameters in a cryptosystem they need to be published. For the finite field this means that $q = p^d$ needs to be given but also the way the field elements are represented, i.e., for prime fields one needs to state whether the following parameters of curve and points refer to Montgomery or standard representation. In extension fields the type of basis and how to multiply in it need to be stated, e.g., for polynomial basis one gives the irreducible polynomial $m(X)$ of degree $d$ over $\mathbb{F}_p$.

Once the field elements can be represented, the equation of the curve, the base point, and also public keys can be stated. This information is sufficient for most applications — if all users behave well. In the following we sometimes assume that an attacker could get signatures on innocent-looking messages. In practice this is not too far fetched — there are service providers that sign all packages transmitted by them so that one could easily get signatures implying arbitrary group elements unless the following checks are implemented.

Standards additionally require that the group order $N = |\text{Pic}_C^0|$ is given together with the cofactor $c$ such that $N/c = \ell$ is prime, and also ask for a small cofactor. In signature schemes like Algorithm 1.18 one needs $\ell$, as the signature is an integer modulo $\ell$, and an inversion modulo $\ell$ is required when signing. We point out that there are inversion-free signature schemes that would not require knowledge of the group order but usually they are less efficient. However, what is worse is that a malicious user could ask for a signature on an element $\overline{D}$ that comes from a subgroup of $\text{Pic}_C^0$ of order dividing $c$. If $c$ has enough prime factors the attacker could determine the secret scalar modulo all these primes and recover a large part of the secret by using Chinese remaindering. This attack is called the *small subgroup attack*.

To avoid this attack one should check whether $\overline{D}$ has order $\ell$. This can be done by either actually checking $[\ell]\overline{D} = 0$ or by computing $[h]\overline{D}$ for $h = c/p_i$ for all prime divisors $p_i$ of $c$ and checking that the result is not 0. Both methods require that $\ell$ is prime, which could also be checked.

Browsing the algorithms for addition and doubling in $\text{Pic}_C^0$ one notices that not all curve coefficients are needed in the arithmetic. Hence, the same set of formulas could be used for different curves. The *invalid curve attack* works by requesting a signature on a group element of $\text{Pic}_{\overline{C}}^0$, where $\overline{C}$ has the same coefficients as $C$ on those positions appearing in the group operations and is different otherwise. If it is possible to find a curve $\overline{C}$ that offers less security, this could be used to find the secret key by attacking the DL problem in the easier group $\text{Pic}_{\overline{C}}^0$. In fact, it is very likely that there are curves $\overline{C}$ such that the group order contains many factors of moderate size, such that the DLP could be solved by using Chinese remaindering.

Hence, the protocols require to check whether $\overline{D}$ is actually a group element.

To sum up: the system parameters consist of the five entries

$$(\mathbb{F}_{p^d}, C, \overline{D}, N, c),$$

where we assume that $\mathbb{F}_{q^d}$ contains information on the field representation and $\overline{D}$ is the base point of the system. The system parameters are accepted if

1. all field elements are correctly represented,
2. $\overline{D} \neq 0 \in \text{Pic}_C^0$, and
3. $\overline{D}$ has order $\ell = N/c$.

User A has public key $\overline{D}_A = [a_A]\overline{D}$, which is included in her system parameters. To verify a signature issued by her or before sending her an encrypted message, B checks whether $\overline{D}_A \in \text{Pic}_C^0$.

Before signing a message that involves computation of $[a_A]\overline{E}$ for some $\overline{E}$, A needs to check whether $\overline{E} \in \text{Pic}_C^0$ and if so whether $\overline{E}$ has order $\ell$.

## 23.5.2 Protocols on $\text{Pic}_C^0$

For using elliptic curves, some special protocols were designed that make use of the representation of group elements by points. Applying compression techniques (cf. Sections 13.2.5 and 13.3.7) a point $P = (x_1, y_1)$ can be uniquely stored using its $x$-coordinate and a bit of $y$. As for each $x_1$ there are at most two $y_1$'s such that $(x_1, y_1)$ is on the curve and the points are the negatives of one another, one can also give up the uniqueness and only use $x_1$ to represent both $P$ and $-P$.

Hence, some reduction in the length of the key and signatures is possible such as is not possible in a generic group. In the following, we detail a signature algorithm to show the differences to the general purpose algorithm presented in Chapter 1. For other protocols similar observations hold.

The elliptic curve digital signature algorithm (ECDSA) [ANSI X9.62] is a signature scheme of ElGamal type as given in Section 1.6.3. We state the generalized version applicable to hyperelliptic curves in the following. For easier reference we call it *HECDSA*. We assume Mumford representation for elements in $\text{Pic}_C^0$ and assume that the finite field elements are ordered such that $0 \leqslant L(\alpha) < q$ is an integer uniquely assigned to $\alpha \in \mathbb{F}_q$ in an invertible way.

In Germany and Korea slightly different versions of the signature scheme are applied. The interested reader is referred to the standards and [HAME+ 2003].

---

**Algorithm 23.14** HECDSA – Signature generation

INPUT: The system parameters $(\mathbb{F}_{p^d}, C, \overline{D}, N, c)$, the private key $a_A$, a hash function $h$, and a message $m$.
OUTPUT: The signature $(U, s)$ on $m$.

1. **repeat**
2.     **repeat**
3.         $r \in_R [0, \ell - 1]$
4.         $\overline{E} \leftarrow [r]\overline{D}$   $[\overline{E} = [u_E, v_E]$ with $u_E = x^\nu + \sum_{i=0}^{\nu-1} u_i$ for some $\nu \leqslant g]$
5.         $U \leftarrow \sum_{i=0}^{\nu-1} L(u_i) q^i \bmod \ell$

6. **until** $U \neq 0$ **and** 1
7. $s \leftarrow \left(r^{-1}\left(h(m) - [a_\text{A}]U\right)\right) \bmod \ell$
8. **until** $s \neq 0$
9. **return** $(U, s)$

---

**Algorithm 23.15** HECDSA – Signature verification

INPUT: The system parameters $(\mathbb{F}_{p^d}, C, \overline{D}, N, c)$, the public key $\overline{D}_\text{A}$, a hash function $h$, the message $m$, and the possible signature $(U, s)$ on $m$.
OUTPUT: Acceptance or rejection of signature.

1. **if** $U$ **or** $s \notin [1, \ell - 1]$ **then return** "reject"
2. **else** $w \leftarrow s^{-1} \bmod \ell$
3. $u_1 \leftarrow \left(h(m)w\right) \bmod \ell$ and $u_2 \leftarrow (Uw) \bmod \ell$
4. $\overline{F} \leftarrow [u_1]\overline{D} \oplus [u_2]\overline{D}_\text{A}$ $\qquad\qquad\qquad\qquad\qquad\qquad [\overline{F} = [u_F, v_F]]$
5. **if** $\overline{F} = 0$ **then return** "reject"
6. **else** $U_1 \leftarrow \left(\sum_{i=0}^{\nu-1} L(u_{F,i})q^i\right) \bmod \ell$ $\quad [u_F = x^\nu + \sum_{i=0}^{\nu-1} u_{F,i}$ for some $\nu \leqslant g]$
7. **if** $U = U_F$ **then return** "accept" **else return** "reject"

---

The signature scheme works as specified as it is a special instance of Algorithm 1.18. Note, however, that the storage requirements are reduced as the first entry of the signature is not a group element but an integer modulo $\ell$.

As only the first part of the representation of $\overline{E}$ is used in the signature, the scheme is *malleable*, i.e., it is possible to obtain a valid signature on a different message by the observation that $\overline{E}$ and $-\overline{E}$ result in the same signature. This property was pointed out in [STPO[+] 2002] together with an attack that uses this property to choose the private key in a manner that two *a priori* chosen messages will result in the same signature.

An easy way to avoid this drawback is to apply a compression, (cf. Section 14.2) to $\overline{E}$, which means that $U$ depends uniquely on $\overline{E}$.

There are also encryption and key agreement protocols that take into account the special properties of elliptic curves. Usually it is easy to obtain hyperelliptic curve analogies. We do not include them in this book but refer the reader to the vast literature. A useful overview with many references is given in [HAME[+] 2003].

## 23.6 Summary

In Chapter 1 we have explained which purposes public-key cryptography can be used for as part of systems that provide data security. In Section 1.5 we have defined discrete logarithms in an abstract way and in Section 1.6 one finds protocols that use discrete logarithms as crypto primitives.

The purpose of the present chapter is to help to decide how to *realize* such systems in the most efficient way.

**Remark 23.16** In Chapter 1 we have discussed bilinear structures as additional structures of certain DL systems, as well as their use in protocols. Their realization is discussed in the next chapter.

To use DL systems in practice one first has to analyze which grade of security is needed. Then one inspects the offered types of groups and the hardness in the DLP in these groups. It could be that for specific practical reasons one is satisfied with subexponential complexity of DLP and chooses as the group a subgroup of prime order in the multiplicative group of a sufficiently large finite field $\mathbb{F}_q$. The system XTR is an example of an efficient realization of such a system (see Example 1.13). But in most cases one will want to have key sizes as small as possible with an easy scalable security, and so groups for which there are good reasons to believe in the exponential hardness of the DLP are the appropriate choice. In Section 23.2 we have discussed such groups; the results are stated in Sections 23.2.1.a and 23.2.2.c. The short answer is that the groups $\text{Pic}_C^0$ for curves of genus $1, 2, 3$ over prime fields, or over fields of order $2^d$ with $d$ a prime such that the multiplicative order of 2 modulo $\ell$ is large, are good candidates. A more subtle answer can be found by following the references in Section 23.2 leading to detailed discussions of the mathematical background and implementational details of attacks.

Roughly spoken, the proposed groups have the same level of security, require the same space for keys, and need the same time for scalar multiplication. But in concrete realizations, in particular the efficiency will depend heavily on the specific implementation and the computational environment. So, the discussion in Section 23.3 becomes relevant. One of the basic decisions is the choice of the ground field $\mathbb{F}_q$ and its arithmetic. Hints for this are to be found in Section 23.3.1 and references cited there. Since the group size $\ell$ is determined by the security level the choice of the bit size of the ground field is related to the genus $g$ of the curve $C$ by $\lg \ell \sim g \lg q$. So, the choice of $q$ determines $g$ and hence one has to check whether one finds a family of curves of genus $g$ over $\mathbb{F}_q$ that fits into the special situation one has.

The next criterion for the choice of $C$ is the efficiency of the group law in $\text{Pic}_C^0$. This is discussed in Sections 23.3.2 and 23.3.3. In many cases it will be sufficient to take random curves and standard versions of the scalar multiplication. But in special cases "the" best implementation will depend on special properties of the processors used, and adding additional structures like endomorphisms can result in significant accelerations. But then implementations (and choices) have to be done more carefully and the relevant background chapters have to be consulted.

Having gone through all choices one needs a pair $(C, \mathbb{F}_q)$ satisfying the properties one wants to have. At this stage one can either look for standard curves (especially for $g = 1$) or use the results described in Section 23.4. This part is one of the most attractive aspects of cryptography from the mathematical point of view and the algorithms related to point counting are relatively involved. On the other side, for all practical needs they are well documented and can be implemented without knowing the whole theoretical background. So in praxis it will be no problem to find suitable instances $(C, \mathbb{F}_q)$.

Now the last but not the least step has to be done: the embedding of the crypto primitive in the chosen protocol. In Section 23.5 it is explained how the generic protocols from Chapter 1 have to be transformed into protocols using $\text{Pic}_C^0$ as crypto primitive. This was done for signatures in Section 23.5.2, and it is not difficult to treat other protocols in a similar way.

# Chapter 24

# Pairing-Based Cryptography

*Sylvain Duquesne and Tanja Lange*

### Contents in Brief

**24.1 Protocols**   573
Multiparty key exchange • Identity-based cryptography • Short signatures

**24.2 Realization**   579
Supersingular elliptic curves • Supersingular hyperelliptic curves • Ordinary curves with small embedding degree • Performance • Hash functions on the Jacobian

Chapter 23 showed us how to build DL systems on the Jacobian of curves. In Chapter 1 we introduced DL systems with bilinear structure. In this chapter we first give more applications of this construction, namely the extension of the tripartite protocol given before to multiparty key exchange, identity-based cryptography, and short signatures. In recent years many systems using this extra structure have been proposed. We include some more references to further work in the respective sections, since giving a complete survey of all these schemes is completely out of the scope of this book. For a collection of results on pairings we refer to the "Pairing-Based Crypto Lounge" [BAR].

The second section is devoted to realizations of such systems. In Chapter 6 we gave the mathematical theory for the Tate–Lichtenbaum pairing and Chapter 16 provided algorithms for efficient evaluation of this pairing on elliptic curves and the Jacobian of hyperelliptic curves. There we assumed that the embedding degree (i.e., the degree $k$ of the extension field $\mathbb{F}_{q^k}$ to which the pairing maps), is small, so as to guarantee an efficiently computable map as required in a DL system with bilinear structure. In Section 24.2 we explain for which curves and fields these requirements can be satisfied and give constructions.

## 24.1 Protocols

For the convenience of the reader we repeat the definition of DL systems with bilinear structure. For the definition of general DL systems we refer to Chapters 1 and 23.

**Definition 24.1** Let $(G, \oplus)$ be a DL system, where $G$ is a group of prime order $\ell$ and such that there is a group $(G', \oplus')$ in which we can compute "as fast" as in $G$. Assume moreover that $(H, \boxplus)$ is another DL system and that a map

$$e : G \times G' \to H$$

satisfies the following requirements:

- the map is $e$ is computable in polynomial time (this includes the fact that the elements in $H$ should need only $O(\lg \ell)$ space),
- for all $n_1, n_2 \in \mathbb{N}$ and random elements $(P_1, P'_2) \in G \times G'$ we have

$$e([n_1]P_1, [n_2]P'_2) = [n_1 n_2] e(P_1, P'_2),$$

- the map $e(\cdot, \cdot)$ is nondegenerate in the first argument, so that for a random $P' \in G'$ we have $e(P_1, P') = e(P_2, P')$ if and only if $P_1 = P_2$.

We then call $(G, e)$ a *DL system with bilinear structure*.

If there exists such a bilinear map with $G = G'$ or an efficiently computable isomorphism $\psi : G \to G'$, the decision Diffie–Hellman problem becomes easy. Namely, given $(P, [a]P, [b]P, [r]P)$ we can compare $e([a]P, \psi([b]P)) = [ab]e(P, \psi(P))$ with $e([r]P, \psi(P)) = [r]e(P, \psi(P))$. As $H$ has the same order $\ell$ we have equality only if $ab \equiv r \pmod{\ell}$.

DL systems with bilinear structure are a special case of *Gap-DH groups*. These are groups in which the computational Diffie–Hellman problem (CDHP) is hard while the corresponding decision problem DDHP is easy. Joux and Nguyen [JoNG 2003] show that the two problems can be separated and give a realization via supersingular elliptic curves.

Supersingular elliptic curves were shown to have this additional structure (cf. Sections 24.2.1). For a long time this was considered a weakness since it makes the DLP in $G$ no harder than in $H$. Joux [JOU 2000] was the first to propose a positive application of bilinear maps, namely tripartite key exchange. For the extended version of the paper see [JOU 2004].

In this section we first repeat this protocol and then show how to extend it to more than three parties. As a second topic we introduce ID-based cryptography. We picked these examples as they are historically the first *constructive* applications of DL systems with bilinear structure. They motivate the study of suitable groups with respect to fast implementations of the bilinear map and also the investigation of their security. As a third application we describe how one can use DL systems with bilinear structure to build a signature scheme with short signatures.

Some protocols require $G = G'$, for the others we can usually save some computations of scalar multiples per participant if the groups are equal. In the remainder of this section we will sometimes assume that we are in this favorable case since it simplifies the description of the protocols. Furthermore we make the assumption that the DLP is hard in $G$ ($G'$) and $H$.

### 24.1.1 Multiparty key exchange

For the convenience of the reader we recall Joux's [JOU 2000, JOU 2004] tripartite key exchange protocol.

Let $P$ be a generator of $G$ and $P'$ one for $G'$. We state the algorithm for the viewpoint of user A; user B and C follow the same steps with the roles interchanged accordingly.

## Algorithm 24.2 Tripartite key exchange

INPUT: The public parameters $(G, \oplus, P, G', \oplus', P', H, \boxplus, e)$.
OUTPUT: The joint key $K \in H$ for A, B, C.

1. choose $a_A \in_R \mathbb{N}$          [$a_A$ is the secret key of A]
2. $(P_A, P'_A) \leftarrow ([a_A]P, [a_A]P')$
3. send $(P_A, P'_A)$ to B and C
4. receive $(P_B, P'_B)$ from B          [$P_B = [a_B]P$ and $P'_B = [a_B]P'$]
5. receive $(P_C, P'_C)$ from C          [$P_C = [a_C]P$ and $P'_C = [a_C]P'$]
6. **return** $K \leftarrow [a_A]\bigl(e(P_B, P'_C)\bigr)$

Applying this algorithm, A, B, and C obtain the same element $K$ of $H$ as

$$K = [a_A]\bigl(e(P_B, P'_C)\bigr) = [a_A a_B a_C]\bigl(e(P, P')\bigr) = [a_B]\bigl(e(P_A, P'_C)\bigr) = [a_C]\bigl(e(P_A, P'_B)\bigr).$$

The joint key could be computed by an eavesdropper if the *computational bilinear Diffie–Hellman problem* (CBDHP) were easy, namely the problem of computing $[a_A a_B a_C]\bigl(e(P, P')\bigr)$ given

- the points $P$ and $P'$
- the description of $G, G', H$, and $e$
- the transmitted values $[a_A]P, [a_B]P, [a_C]P \in G$ and $[a_A]P', [a_B]P', [a_C]P' \in G'$.

The security of protocols is usually based on decision problems. The *decision bilinear Diffie–Hellman problem* (DBDHP) is the problem of distinguishing

$$[a_A a_B a_C]\bigl(e(P, P')\bigr) \text{ from } [r]\bigl(e(P, P')\bigr)$$

under the same conditions as for the CBDHP.

To avoid a man-in-the-middle attack we usually assume that A, B, C participate in some public-key cryptosystem and can thus sign their messages. The receiver has to check the authenticity of the message before using it in the computation of the key. Some proposals use ideas from ID-based cryptography (see below) to achieve authenticity.

Assume that a group of more than three parties wants to agree on a joint key. If a *public-key infrastructure* exists, i.e., a directory listing public keys of all participants together with a certificate of their correctness, this can be solved by sending the encrypted key to each participant using Algorithm 1.16 or 1.17. The first proposal for such a group key exchange was given in [INTA+ 1982]. The Burmester–Desmedt schemes I and II [BUDE 1995, BUDE 1997] give extensions of the Diffie–Hellman key exchange to more participants based on ordinary DL systems. The first one is *contributory*, i.e., the contribution of each participant is reflected in the key while the second one is *non-contributory* but more efficient ($O(n)$ vs. $O(\lg n)$ for $n$ participants). Both protocols can be turned into authenticated key agreement protocols as described in [KAYU 2003, DELA+ 2004]. Note that this approach does not prevent attacks from malicious insiders as described in [JUVA 1996], e.g., participants not playing according to the rules — such as refusing to deliver messages or giving a valid signature on an incorrect message — can make the system insecure.

There exist many proposals combining multiparty key exchange with DL systems with bilinear structure. This allows us to gain efficiency by at least a constant factor as the basic block consists of three parties instead of two. The first authenticated scheme is proposed in [CHHW+ 2004] and improved in [DELA 2004b]. In [BADU+ 2004], Barua et al. propose a different arrangement of the participants which has some computational advantages, and to date the most efficient but noncontributory scheme is described in [DELA+ 2004].

Group key exchange is combined with ID-based cryptography (see below) in [CHHW+ 2004, DUWA+ 2003] to achieve authenticity. Hence, they keep the two-party DH protocol as a building block and apply the additional structure to get the sender's identity involved. This eases the verification of the messages since no extra public-key cryptosystem is involved, but it has the drawback that the long-term secret key of the ID-based system is involved in the protocol. Unless time stamps or other randomization techniques are used this allows for replay attacks.

## 24.1.2 Identity-based cryptography

The algorithms presented for general DL systems require the receiver of an encrypted message to have set up his public key in advance. To solve the problem of sending a ciphertext to a person who is not yet in the system leads to the concept of *ID-based cryptography*, [SHA 1984]. Here, the public key is derived in a deterministic way from the user's identity parameters.

In an ID-based system, the receiver then has to put in some effort to determine his private key. All systems proposed so far require a *trusted authority* (TA) to help in this process and it is TA who checks whether the entity is allowed to obtain the private key. The systems in Section 1.6.2 and Section 23.5 require the *sender* to make sure that he receives the latest public key together with a certificate that the key actually belongs to the intended receiver. Identity-based systems allow the sender to *choose* the public key of the recipient, thus putting the workload on the person interested in receiving the message.

In the usual DL systems, a third party was only needed to guarantee that the public key belongs to a certain entity by issuing a certificate of the public key. Such a *certification authority* (CA) never sees the user's private key. The drawback of ID-based systems proposed so far is that there the trusted authority *issues* the private keys of the participants, i.e., a malicious TA could decrypt all incoming messages and sign on behalf of any participant. We present ID-based systems using DL systems with bilinear structure — which up to now provide the only efficient examples. Burmester and Desmedt [BUDE 2004] study in detail advantages and disadvantages of ID-based cryptography.

We present the ID-based encryption protocol as proposed by Boneh and Franklin [BOFR 2001]. An extended version achieving chosen ciphertext security can be found in [BOFR 2003]. They base the protocol on a DL system with bilinear structure in the special case that $e : G \times G \to H$, i.e., the first two groups are equal.

We assume that each participant can be uniquely identified by a bit-string ID and that there is a map, usually taken to be a cryptographic hash function, $h_1 : \{0,1\}^* \to G$, such that one can associate a group element to every identity. We give implementation details for this hash function in Section 24.2.5 for the realization of $G$ as a subgroup of the Jacobian of some curve. Let the message space be given by bit-strings of fixed length $\mathcal{M} = \{0,1\}^n$. We also need a second hash function $h_2 : H \to \mathcal{M}$.

Before one can use the scheme, the trusted authority TA sets up the system by choosing a group $(G, \oplus, P)$ with bilinear map $e$ to $H$ and computing the public key $P_{\text{TA}} = [a_{\text{TA}}]P$. The system parameters $(G, \oplus, H, \boxplus, P, P_{\text{TA}}, e)$ are made public, the integer $a_{\text{TA}}$ is kept secret and serves as the *master key*.

## § 24.1 Protocols

**Algorithm 24.3** Identity-based encryption

INPUT: A message $m$, public parameters $(G, \oplus, H, \boxplus, P, P_{\text{TA}}, e)$, and the identity of the recipient ID.
OUTPUT: The ciphertext $(R, c)$.

1. choose $r \in_R \mathbb{N}$
2. $R \leftarrow [r]P$
3. $Q \leftarrow h_1(\text{ID})$
4. $S \leftarrow e(P_{\text{TA}}, Q)$                                                                              $[S \in H]$
5. $c \leftarrow m \text{ XOR } h_2([r]S)$
6. **return** $(R, c)$

The group element $Q = h_1(\text{ID})$ is the public key of the user with identity ID. As we assume the DLP in $G$ to be hard, it is not possible to compute the corresponding private key as $\log_P(Q)$. This is the point where TA enters the scene. After checking that the user is allowed to obtain his private key, TA performs the following algorithm:

**Algorithm 24.4** Private-key extraction

INPUT: The parameters $(G, \oplus, P)$, the master key $a_{\text{TA}}$, and the identity ID.
OUTPUT: The private key $A_{\text{ID}} \in G$ of ID.

1. $Q \leftarrow h_1(\text{ID})$
2. $A_{\text{ID}} \leftarrow [a_{\text{TA}}]Q$
3. **return** $A_{\text{ID}}$

Once this private key is obtained, TA is no longer needed and $A_{\text{ID}}$ is used as the private key. If he thinks that the key might be compromised, the sender can append the time to ID. Then a further interaction of ID with TA is needed. For the decryption, the bilinear structure is used again.

**Algorithm 24.5** Identity-based decryption

INPUT: The ciphertext $(R, c)$, the parameters $(G, \oplus, P, P_{\text{TA}}, e)$, and the private key $A_{\text{ID}}$.
OUTPUT: The message $m$.

1. $T \leftarrow e(R, A_{\text{ID}})$
2. **return** $m \leftarrow c \text{ XOR } h_2(T)$

The algorithm works as specified as

$$T = e(R, A_{\text{ID}}) = e\big([r]P, [a_{\text{TA}}]Q\big) = [ra_{\text{TA}}]\big(e(P, Q)\big) = [r]\big(e(P_{\text{TA}}, Q)\big) = [r]S.$$

Clearly, this cryptosystem is weak if one can solve the DLP in $G$ or $H$. The actual security is based on the intractability of the CBDHP. To prove security in the random oracle model one needs to assume the intractability of the DBDHP.

As TA poses a critical point of security, it is recommended to use a secret sharing scheme to store the master key $a_{TA}$. This was already noticed in [BOFR 2003] and considered further in [LIQU 2003, BAZH 2004, DELA 2004a].

For a large group of participants it is clearly not desirable that all participants need to contact TA to obtain their keys. Fortunately, the system allows a hierarchical structure as shown in [GESI 2002] and considered further in [DOYU 2003, BOBO$^+$ 2005].

### 24.1.3 Short signatures

In [OKPO 2001], Pointcheval and Okamoto show how a Gap-DH group can be used to design a signature scheme. Boneh, Lynn, and Shacham [BOLY$^+$ 2002, BOLY$^+$ 2004] give a realization using supersingular elliptic curves with the Tate–Lichtenbaum pairing as bilinear structure. In this realization the signature scheme has the advantage of offering shorter signature lengths compared to DSA and ECDSA (cf. Chapter 23). We present the basic scheme using a DL system with bilinear structure now and give details on the key length in Section 24.2.4.a. For simplicity we assume $G = G'$ but the scheme can be modified to allow different groups with a slightly modified security assumption (for a discussion of this *co-DHP* see [BOLY$^+$ 2002, BOLY$^+$ 2004]).

The system parameters consist of a DL system with bilinear structure $(G, \oplus, P, e)$. Furthermore, a hash function $h_1 : \{0,1\}^* \to G$, mapping from the message space in the group is required. For realizations of $h_1$ see Section 24.2.5.

Like in an ordinary DL system A's private key consists of a secret integer $a_A$, and the public key is $P_A = [a_A]P$ (cf. Section 1.6.2).

The signing procedure simplifies compared to Algorithm 1.18 and the signature consists only of one group element.

---
**Algorithm 24.6** Signature in short signature scheme

INPUT: A message $m$, system parameters $(G, \oplus, P)$, and the private key $a_A$.
OUTPUT: The signature $S$ on $m$.

1. $M \leftarrow h_1(m) \in G$
2. $S \leftarrow [a_A]M$
3. **return** $S$
---

If compression techniques exist in $G$, the signature can be even shorter. This will be the case for the realization with Jacobians of curves (cf. Section 24.2.4.a).

To verify the signature the receiver performs the following steps:

---
**Algorithm 24.7** Signature verification in short signature scheme

INPUT: The message $m$, the signature $S$, parameters $(G, \oplus, P, e)$, and the public key $P_A$.
OUTPUT: Acceptance or rejection of signature.

1. **if** $S \notin G$ **then return** "rejection"
2. **else** $M \leftarrow h_1(m)$
3. $\quad u \leftarrow e(P_A, M)$
4. $\quad v \leftarrow e(P, S)$
5. **if** $u = v$ **then return** "acceptance" **else return** "rejection"
---

Obviously, the signature scheme is valid since for a correct signature we have

$$u = e(P_A, M) = [a_A]e(P, M) = e(P, [a_A]M) = e(P, S) = v.$$

The larger workload lies on the side of the verifier, since he must compute two pairings, whereas the signer only needs to compute one scalar multiplication. But this meets the usual scenario encountered in practice that the sender should sign every message and the receiver can decide whether it is worth the effort checking the signature, or if he believes that it is correct right away.

**Remark 24.8** Obviously short signatures do not require a DL system with bilinear structure but use only the defining property of Gap-DH groups to compare $u$ and $v$. We described the system in this more restrictive setting, as curves lead to efficient instantiations allowing compression, and as until now curves with efficiently computable Tate–Lichtenbaum pairing constitute the only known examples of such groups.

## 24.2 Realization

We have seen in Section 6.4.1 that the Tate–Lichtenbaum pairing is a bilinear map from the Jacobian $J_C$ of a curve $C/\mathbb{F}_q$ to the multiplicative group of $\mathbb{F}_{q^k}$ for some $k$ depending on $C$ and $\mathbb{F}_q$. If $k$ is small, the pairing can be computed very efficiently, as explained in Chapter 16. The previous section shows how it can be used for cryptographic applications. In this section, we will focus on realizations of the Tate–Lichtenbaum pairing, namely how to use it in practice and how to create instances with small $k$.

Some systems require the bilinear map $e$ to map $G \times G$ to $H$, while for the Tate–Lichtenbaum pairing the inputs come from two different groups $e : G \times G' \to H$. Therefore, we study efficiently computable isomorphisms from $G$ to $G'$.

Let $C$ be a curve of genus $g$ defined over $\mathbb{F}_q$ with $q = p^d$ and $p$ prime. Let $J_C$ be its Jacobian. In this section, we are only interested in cryptographic applications, so we assume that we work in a cyclic subgroup $G$ of $J_C(\mathbb{F}_q)$ of prime order $\ell$. The Tate–Lichtenbaum pairing maps the points of order $\ell$ into some field extension of the base field $\mathbb{F}_q$ (cf. Section 16.1.1 for details). The degree of this extension is called the *embedding degree* and will always be denoted by $k$. In fact $k$ is the minimal number such that $\ell \mid q^k - 1$, and the Tate–Lichtenbaum pairing takes its values in $\mathbb{F}_{q^k}^* / (\mathbb{F}_{q^k}^*)^\ell$. To achieve a unique representation one usually raises the result to the power of $(q^k - 1)/\ell$ such that this modified pairing maps to the $\ell$-th roots of unity $\mu_\ell$ in $\mathbb{F}_{q^k}^*$.

The size of this embedding degree is very important. Indeed, as mentioned above, the DLP in $G$ cannot be harder than in $\mu_\ell$. The multiplicative group was considered as an example for index calculus methods in Section 20.4 There it was shown that the security is subexponential $L_{q^k}(1/2, \sqrt{2})$ for general fields and even $L_{q^k}(1/3, c)$ for the number and function field sieves as opposed to exponential security of $J_C$ for $g \leqslant 3$. Therefore the parameters have to be chosen in such a way that the DLP is hard in $J_C(\mathbb{F}_q)$ *and* in $\mu_\ell$. Due to the different levels of security, $k$ should not be too small to avoid using a large finite field. On the other hand, if $k$ is too large, the computation of the Tate–Lichtenbaum pairing becomes inefficient since a lot of computations are performed in $\mathbb{F}_{q^k}$ or the field $\mathbb{F}_q$ is too small such that the DLP in $J_C(\mathbb{F}_q)$ is easy.

By [BAKO 1998] a random elliptic curve has a large $k$ and this result generalizes to larger genera. In the next sections, we first detail implementations on supersingular elliptic curves as they are the most important examples of curves with small embedding degrees and were the first to be proposed for applications using pairings. We then consider supersingular hyperelliptic curves and conclude with constructions of nonsupersingular elliptic curves having a small embedding degree.

### 24.2.1 Supersingular elliptic curves

We now describe supersingular *elliptic* curves and their use for pairing-based cryptosystems. Supersingular abelian varieties are defined in Chapter 4 (cf. Definition 4.74). We recall some properties specialized to elliptic curves here (cf. Definition 13.14).

**Definition 24.9** Let $E$ be an elliptic curve defined over $\mathbb{F}_q$ with $q = p^d$. If $E[p^e] \simeq \{P_\infty\}$ for all $e$ the curve is called *supersingular*, otherwise it is called *ordinary*.

For us the most important property is that the trace $t = \mathrm{Tr}(\phi_q)$ of the Frobenius endomorphism satisfies $t \equiv 0 \pmod{p}$. By the Theorem of Hasse–Weil 13.28 we have $|t| \leqslant 2\sqrt{q}$. Hence, over prime fields $\mathbb{F}_p$ with $p \geqslant 5$, the condition implies $t = 0$ and the cardinality of $E(\mathbb{F}_p)$ satisfies $|E(\mathbb{F}_p)| = p + 1$. As $\ell$ divides $|E(\mathbb{F}_p)| = p + 1$, which in turn divides $(p^2 - 1)$, we see that for supersingular curves over large prime fields the embedding degree is bounded by 2. As outlined in Section 6.4.2, Menezes, Okamoto, and Vanstone [MEOK+ 1993] prove that the embedding degree for supersingular elliptic curves is always less than or equal to 6. In fact, we can even say more. It is proved in Section 6.4.2 that the upper bound on the embedding degree depends on the characteristic of the base field:

- in characteristic 2 we have $k \leqslant 4$,
- in characteristic 3 we have $k \leqslant 6$,
- over prime fields $\mathbb{F}_p$ with $p \geqslant 5$ we have $k \leqslant 2$,

and these bounds are attained.

This means that for supersingular elliptic curves over large prime fields we must work in a larger field than usual for curve-based cryptography. For current security parameters we should choose a ground field with around $2^{512}$ elements to obtain a satisfying level of security, since the finite field $\mathbb{F}_{q^k}$ needs to have at least $2^{1024}$ elements to adhere to the proposals. But this is to the detriment of efficiency, since we have to compute on some larger field for the same level of security (note that the current proposals assume elliptic curves with $2^{160}$ elements to offer sufficient security). Therefore, it is preferable to work in characteristic 2 or, even better, in characteristic 3 and on curves with the maximal possible $k$.

Once the curve is chosen and $k$ has been determined, the algorithms from Chapter 16, usually referred to as Miller's algorithms, can be used to implement the protocols from the previous section. Particular care should be taken when working in characteristic 3 as this is not very common and some specific algorithms and improvements to the computation of the pairing should be used, which we now consider.

#### 24.2.1.a Efficient Tate–Lichtenbaum pairing in characteristic 3

In characteristic 3 (i.e., $q = 3^d$), it is quite natural to use a triple and add algorithm instead of the standard double and add method used to describe Miller's algorithm as proposed in [GAHA+ 2002] and [BAKI+ 2002]. In the computation of the Tate–Lichtenbaum pairing we compute the $\ell$-fold of a point of order $\ell$ keeping some information along the way. In general, let $\sum_{i=0}^{l-1} n_i 3^i$ be the signed base 3 representation of the integer $n$ where $n_i \in \{0, \pm 1\}$ and $n_{l-1} = 1$. Let $P$ be some point on the elliptic curve; the computation of the scalar multiplication $[n]P$ proceeds as follows

## § 24.2 Realization

---

**Algorithm 24.10** Triple and add scalar multiplication

INPUT: An integer $n$ represented in signed base 3 format $\sum_{i=0}^{l-1} n_i 3^i$ and a point $P \in E(\mathbb{F}_{3^d})$.
OUTPUT: The scalar multiple $[n]P$.

1. $T \leftarrow P$
2. **for** $i = l-2$ **down to** $0$ **do**
3. $\quad T \leftarrow [3]T$
4. $\quad$ **if** $n_i = 1$ **then** $T \leftarrow T \oplus P$
5. $\quad$ **if** $n_i = -1$ **then** $T \leftarrow T \ominus P$
6. **return** $T$

---

**Remark 24.11** Obviously, all the algorithmic improvements explained in Chapter 9 can easily be applied to this basic algorithm.

To apply this algorithm efficiently, of course, we need the formulas for tripling. Note that a supersingular elliptic defined over $\mathbb{F}_{3^d}$ can always be represented by an equation of the form

$$y^2 = x^3 + a_4 x + a_6.$$

Let $P = (x_1, y_1) \in E(\mathbb{F}_{3^d})$, then using the division polynomials (cf. Section 4.4.5.a) one sees that $[3]P$ is given by

$$[3]P = \left( \frac{x_1^{3^3} + a_6^3}{a_4^4} + 2\frac{a_6}{a_4}, 2\frac{y_1^{3^3}}{a_4^6} \right). \tag{24.1}$$

As cubing is linear in characteristic 3, point tripling on curves of this form can be done in time $O(d)$ (and even $O(1)$ in normal basis representation), which is very efficient and corresponds to the linearity of point doubling for supersingular curves in characteristic 2 as discovered by Menezes and Vanstone [MEVA 1990]. Moreover this provides an inversion-free algorithm for tripling (assuming that $1/a_4$ is precomputed). In total, this leads to a triple and add algorithm, which is much more efficient than the standard double and add algorithm.

Unfortunately, an inversion is required to compute the pairings since we need to compute $1/y$ to obtain the equations of the lines (cf. Chapter 16 for details). Indeed, to triple the point $P = (x_1, y_1)$ we need the lines $L_1$ and $L_2$ involved in the doubling of $P$ and the lines $L_1'$ and $L_2'$ involved in the addition of $P$ and $[2]P$. These lines are given by the following equations [GAHA$^+$ 2002]:

$L_1 : y - \lambda x + (\lambda x_1 - y_1) = 0$ with $\lambda = 2\dfrac{a}{y_1}$ the slope of the tangent at $P$,

$L_2 : x - x_2 = 0$ with $x_2 = \lambda^2 + x_1$,

$L_1' : y - \mu x + (\mu x_1 - y_1) = 0$ with $\mu = \left( \dfrac{y_1^3}{a^2} - \lambda \right)$ the slope of the line through $P$ and $[2]P$,

$L_2' : x - x_3 = 0$ with $x_3 = \mu^2 - x_1 - x_2$.

As previously, let us write the integer $\ell$ in signed base 3 expansion, namely

$$\ell = \sum_{i=0}^{l-1} \ell_i 3^i \text{ with } \ell_i \in \{0, \pm 1\} \text{ and } \ell_{l-1} \neq 0. \tag{24.2}$$

The following algorithm is a version of Miller's algorithm 16.1.3.c in base 3 which is well adapted for base fields of characteristic 3.

**Algorithm 24.12** Tate–Lichtenbaum pairing in characteristic $3$

INPUT: An integer $\ell = \sum_{i=0}^{l-1} \ell_i 3^i$ as in (24.2) and the points $P \in E(\mathbb{F}_{3^d})[\ell]$, $Q \in E(\mathbb{F}_{3^{dk}})[\ell]$.
OUTPUT: The Tate–Lichtenbaum pairing $T_\ell(P, Q)$.

1. choose $S \in_R E(\mathbb{F}_{3^{dk}})$
2. $Q' \leftarrow Q \oplus S$
3. $P_2 \leftarrow [2]P$
4. $f_2 \leftarrow f(Q' - S)$            [$f$ is such that $\mathrm{div}(f) = 2P - P_2 - P_\infty$]
5. **if** $n_{l-1} = 1$ **then** $T \leftarrow P$ and $f_1 \leftarrow 1$
6. **else** $T \leftarrow P_2$ **and** $f_1 \leftarrow f_2$
7. **for** $i = l - 2$ **down to** $0$ **do**
8.      $T \leftarrow [3]T$            [use (24.1)]
9.      $f_1 \leftarrow f_1^3 \frac{L_1 L_1'}{L_2 L_2'}(Q' - S)$            [$L_1, L_1', L_2$ and $L_2'$ are the lines arising in $[3]T$]
10.      **if** $\ell_i = 1$ **then**
11.          $T \leftarrow T \oplus P$
12.          $f_1 \leftarrow f_1 \frac{L_1}{L_2}(Q' - S)$            [$L_1$ and $L_2$ are the lines arising in $T \oplus P$]
13.      **if** $\ell_i = -1$ **then**
14.          $T \leftarrow T \ominus P$
15.          $f_1 \leftarrow f_1 f_2 \frac{L_1}{L_2}(Q' - S)$            [$L_1$ and $L_2$ are the lines arising in $T \ominus P$]
16. **return** $f_1^{\frac{3^{dk}-1}{\ell}}$

**Remarks 24.13**

(i) Remember that $Q' - S$ is a divisor, so that $f(Q' - S) = \frac{f(Q')}{f(S)}$.

(ii) Algorithm 24.2 is just an equivalent form of Miller's basic algorithm, and all the improvements explained in Chapter 16 can of course be applied.

(iii) The final exponentiation can be omitted if one does not need a unique representative per class of $\mathbb{F}_{q^k}^* / (\mathbb{F}_{q^k}^*)^\ell$. For example, in Algorithm 24.6 one could first compute only the classes $\overline{u}$ and $\overline{v}$ and check whether $(\overline{u}/\overline{v})^{(q^k-1)/\ell} = \overline{1}$.

### 24.2.1.b Distortion maps

The main drawback of the Weil pairing is that $W_\ell(P, P)$ is always equal to 1. Hence, the pairing is degenerate if applied to the cyclic subgroup of order $\ell$ in both arguments. This is not the case with the Tate–Lichtenbaum pairing. Frey, Müller, and Rück state in [FRMÜ+ 1999] that if $\ell^2$ does not divide the cardinality of the elliptic curve over $\mathbb{F}_{q^k}$, the Tate–Lichtenbaum pairing applied to a point with itself yields a primitive $\ell$-th root of unity (see also Example 6.10).

Otherwise, i.e., if $\ell^2 \mid |E(\mathbb{F}_{q^k})|$, a modification is required to make the pairing nontrivial on the cyclic subgroup generated by $P \in E(\mathbb{F}_q)$. Verheul [VER 2001, VER 2004] suggests to make use of the fact that the endomorphism ring of a supersingular elliptic curve is an order in a quaternion algebra. In particular this implies that the ring is not commutative and, hence, there exists an endomorphism $\psi$ such that $P$ and $\psi(P)$ lie in disjoint cyclic groups of order $\ell$ in $E(\mathbb{F}_{q^k})$. He

## § 24.2 Realization

calls these endomorphisms *distortion maps*. They give the possibility of sending points from one $\ell$-torsion subgroup to another (different) one, over $\mathbb{F}_{q^k}$. In this way we define a new pairing by $T_\ell(P, \psi(Q))$. In most applications, $P$ and $Q$ are defined over $\mathbb{F}_q$ and lie in the same subgroup of order $\ell$ of $E(\mathbb{F}_q)$, and $\psi$ is a non-$\mathbb{F}_q$-rational endomorphism. We now give examples of such endomorphisms.

### 24.2.1.c Examples

In this part, we give some examples of supersingular elliptic curves in various characteristics that can be used to build pairing-based cryptosystems. These examples come from [JOU 2000, JOU 2004], [BAKI+ 2002], and [GAHA+ 2002]. One needs finite fields $\mathbb{F}_{q^k}$ of 1024 bits to achieve that the DLP in the multiplicative group $\mathbb{F}_{q^k}^*$ is as hard as the DLP of a 160-bit elliptic curve because there are subexponential algorithms solving the DLP in finite fields while no subexponential algorithm is known for solving the DLP on elliptic curves. The following types of supersingular elliptic curves come with a fixed embedding degree $k$ but allow to choose $q$. We state the needed size of $q$ to achieve that simultaneously the DLP on the curve and in $\mathbb{F}_{q^k}^*$ is hard. In fact as all these curves are supersingular, the finite fields are all chosen larger than necessary as $k$ is bounded by 6 (cf. Section 24.2.1).

Over prime fields $\mathbb{F}_p$ ($p \neq 2, 3$), there are two well-known families of supersingular elliptic curves. The first family is given by the equation

$$y^2 = x^3 + a_4 x \qquad (24.3)$$

when $p \equiv 3 \pmod{4}$. The latter condition ensures that $-1$ is a nonsquare in $\mathbb{F}_q$. As a distortion map we can thus choose $(x, y) \mapsto (-x, iy)$ with $i^2 = -1$ as this maps $P \in E(\mathbb{F}_q)$ to a point of the extension field $\mathbb{F}_{q^2}$ and hence to a linearly independent group.

The second family is given by the equation

$$y^2 = x^3 + a_6 \qquad (24.4)$$

when $p \equiv 2 \pmod{3}$. As a distortion map we can choose $(x, y) \mapsto (jx, y)$ with $j^3 = 1, j \neq 1$, which maps to $\mathbb{F}_{q^2}$ as there are no nontrivial cube roots of unity in $\mathbb{F}_q$. In [BOFR 2001, BOFR 2003], Boneh and Franklin suggested this setting with $a_6 = 1$ together with the Weil pairing to realize their ID-based encryption scheme.

In both cases, the order of the curve is equal to $p + 1$ and the embedding degree is equal to 2 (which is the maximal possible value in this case), so that $p$ must be chosen of the order of $2^{512}$.

However, as noticed by Boneh and Franklin [BOFR 2001] this does not mean that $\ell$ must be of this size. Indeed, for this level of security, it is sufficient to work on a subgroup of $E(\mathbb{F}_q)$ of approximately 160 bits. In fact, the cardinality of $E(\mathbb{F}_q)$ need not be almost prime as usual but can be of the form $p + 1 = c\ell$ with $\ell$ a 160-bit prime number and $c$ some large cofactor. Hence, the number of loops in Miller's algorithm is not increased but the computations need to be carried out over far too large a field.

It is possible to have an embedding degree equal to 3 in large characteristic using curves of the form (24.4) defined over $\mathbb{F}_{p^2}$ with $p$ prime and $a_4 \notin \mathbb{F}_p$ (in which case a 340-bit base field can be used).

Over $\mathbb{F}_{2^d}$, we can use the subfield curves given by the equation

$$y^2 + y = x^3 + x + a_6, \text{ with } a_6 = 0 \text{ or } 1$$

and the distortion map

$$(x, y) \mapsto (x + s^2, y + sx + t) \text{ with } s, t \in \mathbb{F}_{2^{4d}}, \ s^4 + s = 0 \text{ and } t^2 + t + s^6 + s^2 = 0.$$

These curves have $2^d + 1 \pm 2^{(d+1)/2}$ points and an embedding degree equal to 4 (which is the maximal value in characteristic 2). As a consequence a 256-bit base field is sufficient. Again, this is not satisfactory but better than in large characteristic and, of course, we can again use a subgroup of 160-bit prime order.

Furthermore, as shown in the first part of Section 13.3, doublings are particularly fast on this type of curve and the arithmetic profits from the coefficients being defined over the ground field.

Finally, in characteristic 3 we can use the curves given by the equations

$$y^2 = x^3 + x + a_6, \text{ with } a_6 = \pm 1$$

whose embedding degree is 6 and cardinality is $3^d + 1 \pm 3^{(d+1)/2}$ (the precise cardinality in terms of $d$ modulo 12 and $a_6$ is given in [JOU 2000]). In this case we can choose the distortion map

$$(x, y) \mapsto (-x + s, iy) \text{ with } s^3 + 2s + 2a_6 = 0 \text{ and } i^2 = -1.$$

For these curves the tripling in (24.1) is even simpler given by $[3]P = (x^{3^3}, -y^{3^3})$. The relatively large embedding degree allows at least a satisfactory efficiency with respect to the level of security obtained. Indeed, in this case it is sufficient to choose a 170-bit base field.

Moreover, for the examples in characteristic 2 and 3, it can be advantageous to use the particular structure of the extensions involved in these examples. In fact, Galbraith, Harrison, and Soldera explain in [GAHA+ 2002] how to work efficiently in $\mathbb{F}_{2^{4d}}$ and $\mathbb{F}_{3^{6d}}$ where $d$ is prime (cf. also Section 11.3).

Thanks to these results, it is possible to choose an elliptic curve with an embedding degree equal to 6 defined over some field of about $2^{170}$ elements, so that transferring the discrete logarithm problem gives one on a 1020-bit field, which is not easier to solve. Thus, using pairings in this case can be done without loss of security and (almost) without loss of performance.

Moreover, to increase the level of security, the growth in the group size for elliptic curves is much smaller than for finite fields, e.g., an elliptic curve of 160 bits is assumed to offer security similar to a finite field of 1024 bits, and when the finite field size is increased to 2048 bits, the size of the elliptic curve is only increased to 200-230 bits. Thus, larger embedding degrees are required to keep acceptable performance in both groups.

To get around this problem, we can either use hyperelliptic supersingular curves for genus $g \geqslant 3$, or try to find ordinary elliptic curves with larger embedding degrees. We treat these two possibilities in the next sections.

## 24.2.2 Supersingular hyperelliptic curves

Most of this section is based on [GAL 2001a]. We first recall that a hyperelliptic curve is said to be supersingular if its Jacobian is isogenous, over $\overline{\mathbb{F}}_q$, to a product of supersingular elliptic curves (cf. Definition 4.74).

As in the case for elliptic curves, there exists an upper bound, denoted by $k(g)$, on the embedding degree for supersingular curves depending only on the genus and not on the abelian variety [GAL 2001a] (cf. Section 6.4.2).

We will now give effective values for these bounds for higher genera (we have shown above that $k(1) = 6$). In fact, in terms of security, the interesting value is $k(g)/g$ and it is called the *security parameter*. It represents the logarithmic ratio between the size of the Jacobian and the size of the finite field $\mathbb{F}_{q^k}$, which are the important parameters for security. Using cyclotomic polynomials,

## § 24.2 Realization

Galbraith proves that

$$
\begin{aligned}
k(2) &= 12, \quad \bigl(k(2)/2 = 6\bigr), \\
k(3) &= 30, \quad \bigl(k(3)/3 = 10\bigr), \\
k(4) &= 60, \quad \bigl(k(4)/4 = 15\bigr).
\end{aligned}
$$

This shows that for hyperelliptic curves one can obtain larger embedding degrees, but it also shows that genus 2 supersingular curves do not solve the problem of the too-small embedding degree encountered with elliptic curves. On the contrary, higher dimensional abelian varieties seem to provide better ratios (for instance, 10 for dimension 3). Note that the bounds are upper bounds and that Galbraith shows that they are sharp, in the sense that they are attained, but it is not proved that they can be attained for the Jacobian of a hyperelliptic curve.

**Remark 24.14** In cryptosystems based on general abelian varieties $\mathcal{A}/\mathbb{F}_q$, the security depends on the cyclic subgroup of $\mathcal{A}(\mathbb{F}_q)$ of prime order. Thus, in cryptography, we are only interested in simple abelian varieties (i.e., those that do not decompose as products of lower dimensional abelian varieties). For this case, Rubin and Silverberg prove in [RISI 1997] that Galbraith's bounds are not attained for dimensions greater or equal to 3.

Just as for elliptic curves, the largest security parameters (or the largest embedding degree) achievable depend on the characteristic of the base field.

In the following table [RISI 1997], we give the highest security parameters attainable with simple supersingular abelian varieties for dimensions $g$ up to 6 defined over $\mathbb{F}_q$ with $q = p^d$ (the sign — means that there is no simple supersingular abelian variety in this case).

**Table 24.1** Security parameter $k(g)/g$ for simple supersingular abelian varieties.

| $g$ | 1 | 2 | 3 | 4 | 5 | 6 |
|---|---|---|---|---|---|---|
| $q$ arbitrary (Galbraith's bounds) | 6 | 6 | 10 | 15 | 24 | 35 |
| $q$ square | 3 | 3 | 3 | $\frac{15}{4}$ | $\frac{11}{5}$ | $\frac{7}{2}$ |
| $q$ nonsquare, $p > 3$ | 2 | 3 | $\frac{14}{3}$ | $\frac{15}{4}$ | $\frac{22}{5}$ | 7 |
| $q$ nonsquare, $p = 2$ | 4 | 6 | — | 5 | — | 6 |
| $q$ nonsquare, $p = 3$ | 6 | 2 | 6 | $\frac{15}{2}$ | — | 7 |

In dimension 2, the best bound obtained is the same as Galbraith's. It is interesting to note that it is attained in characteristic 2, which is more convenient for implementations than characteristic 3 (which is required in genus 1). As an example, Galbraith suggests the hyperelliptic curve $C/\mathbb{F}_2$ given by $C : y^2 + y = x^5 + x^3$, which has embedding degree equal to 12.

In higher dimensions, bounds for the security parameters are smaller for simple supersingular abelian varieties than for general ones. The largest security parameter is $\frac{15}{2} = 7.5$ and it is attained for a supersingular simple abelian variety of dimension 4 in characteristic 3. Rubin and Silverberg [RISI 1997] construct such an abelian variety by using the Weil restriction of scalars of an elliptic curve (which is explained in Chapter 7). For this example the field sizes can be chosen smaller than for supersingular elliptic curves.

### 24.2.3 Ordinary curves with small embedding degree

As seen in the previous sections, supersingular curves are not optimal for constructing cryptosystems based on the Tate–Lichtenbaum pairing, for at least two reasons.

The first reason is that supersingular elliptic curves (or higher genus curves and more generally abelian varieties) attain large embedding degrees only in small characteristic (e.g., $\text{char}(\mathbb{F}_q) = 3$ for elliptic curves and $\text{char}(\mathbb{F}_q) = 2$ for genus 2 curves). This implies in particular that for the finite field $\mathbb{F}_{q^k}$ the security estimates have to take into account Coppersmith's attack [COP 1984]. Furthermore, recent generalizations of the Weil descent approach [GAU 2004, DIE 2004] (cf. Section 22.3.5) make these fields with small characteristic more suspect to attacks.

The second reason is that larger embedding degrees will be necessary in the future in order to maintain an optimal ratio between the security in the Jacobian of the curve and the security in the finite field $\mathbb{F}_{q^k}$, and higher genus curves (or higher dimensional abelian varieties) do not really solve this problem as seen in Section 24.2.2.

In this section, we will explain how to solve these two problems by constructing ordinary elliptic curves defined over some prime field with a large embedding degree. This is not easy since the results of Balasubramanian and Koblitz [BAKO 1998] show that curves having a large prime order subgroup usually have a large extension degree. Luca and Shparlinski [LUSH 2005] generalize this study to arbitrary elliptic curves. Therefore, trying to find at random an elliptic curve suitable for pairing-based protocols (i.e., one with a moderately small embedding degree $k \sim 10$ and a large subgroup of prime order) is not appropriate. It is therefore quite natural to try to construct such curves using the CM-method explained in Chapter 18.

#### 24.2.3.a Constructing ordinary elliptic curves defined over $\mathbb{F}_p$ with $k \leqslant 6$

The first work in this direction is due to Miyaji, Nakabayashi, and Takano [MINA+ 2001] who describe how to construct ordinary elliptic curves with an embedding degree equal to 3, 4, or 6 having a large subgroup of prime order. Their method is quite restrictive since it is is necessary to find an elliptic curve with prime order. To allow a small cofactor, Scott and Barreto [SCBA 2004], following previous work with Lynn [BALY+ 2003, BALY+ 2004a], transformed these conditions into a condition on cyclotomic polynomials: let $E$ be an elliptic curve defined over $\mathbb{F}_q$ with trace $t$ and cardinality equal to $c\ell$ with $\ell$ prime (and $c$ small). Then, the subgroup of order $\ell$ in $E(\mathbb{F}_q)$ has embedding degree equal to $k$ if and only if $\ell \mid \Phi_k(t-1)$ and $\ell \nmid \Phi_i(t-1)$ for $i < k$, where $\Phi_k$ is the $k$-th cyclotomic polynomial.

In fact, this condition is just a combination of the two conditions $q + 1 - t \equiv 0 \pmod{\ell}$ and $q^k - 1 \equiv 0 \pmod{\ell}$. In this section, we are only interested in $k = 3, 4,$ or 6 so that only $\Phi_3, \Phi_4,$ and $\Phi_6$ are required. We have

$$\begin{aligned} \Phi_3(x) &= x^2 + x + 1, \\ \Phi_4(x) &= x^2 + 1, \\ \Phi_6(x) &= x^2 - x + 1. \end{aligned}$$

Unlike in Section 18.1, we here fix the ground field $\mathbb{F}_q$ and vary the discriminant of the quadratic number field. In the notation of (18.3) we put $D = \varepsilon d$. Recall that the complex multiplication method will find an elliptic curve with a given base field $\mathbb{F}_q$ and given trace $t$ if a solution can be found for the CM-equation

$$Dv^2 = 4q - t^2$$

for sufficiently small values of $D$. Introducing the new condition on the embedding degree yields the equation

$$Dv^2 = 4c\frac{\Phi_k(t-1)}{c'} - (t-2)^2$$

## § 24.2 Realization

using $q = c\ell + t - 1$ and $\Phi_k(t-1) = c'\ell$. In the cases we are interested in, the degree of the cyclotomic polynomial is always 2 so that this equation is quadratic. Setting $n_3 = 2c + c'$, $n_4 = c'$, $n_6 = -2c + c'$, $m = 4c - c'$, $r = c'mD$, $y = m(t-1) + n_k$ and $f_k = n_k^2 - m^2$, the CM-equation simplifies to the *generalized Pell equation*

$$y^2 - rv^2 = f_k.$$

Such Diophantine equations are a classical topic in number theory and can easily be solved (cf. [SMA 1998] or [COH] for details). For each solution, we must check that

- $t = \frac{y - n_k}{m} + 1$ is an integer
- $\ell = \frac{\Phi_k(t-1)}{c'}$ is prime, and
- $q = c\ell + t - 1$ is prime (or a power of a prime but, as noticed in [DUEN⁺ 2005], this seems hopeless).

These are restrictive conditions so that solutions are rare — especially since in addition $q$ and $\ell$ must be in a range interesting for cryptographic applications. Still, this leads to a probabilistic algorithm that finds an ordinary elliptic curve with a given embedding degree equal to 3, 4, or 6.

---

**Algorithm 24.15** Construction of elliptic curves with prescribed embedding degree

INPUT: An embedding degree $k = 3$, 4, or 6, the maximal cofactor $c_{\max}$, and the maximal discriminant for the CM-algorithm $D_{\max}$.
OUTPUT: A prime field $\mathbb{F}_q$ and an elliptic curve $E$ such that $|E(\mathbb{F}_q)| = c\ell$ with $c \leqslant c_{\max}$ and $\ell$ prime or "failure".

1.     $\lambda \leftarrow -2\lfloor k/2 \rfloor + 4$
2.     **for** $c = 1$ **to** $c_{\max}$ **do**
3.        **for** $c' = 1$ **to** $4c - 1$ **do**
4.           $n_k \leftarrow \lambda c + c'$
5.           $m \leftarrow 4c - c'$
6.           $f_k \leftarrow n_k^2 - m^2$
7.           **for** $D = 1$ **to** $D_{\max}$ squarefree **do**
8.              $r \leftarrow c'mD$
9.              **for** each solution of $y^2 - rv^2 = f_k$ **do**
10.                $t \leftarrow (y - n_k)/m + 1$
11.                $\ell \leftarrow \Phi_k(t-1)/c'$
12.                $q \leftarrow c\ell + t - 1$
13.                **if** $t \in \mathbb{Z}$ **and** $\ell$ is prime **and** $q$ is prime **then**
14.                   **return** $q$ and $CM(q, t, D)$
15.     **return** "failure"

---

### Remarks 24.16

(i) The input value of $c_{\max}$ must be chosen small for reasons of efficiency of the cryptographic application and $D_{\max}$ must not be too large to facilitate the CM-algorithm.

(ii) Scott and Barreto [SCBA 2004] give some congruential restrictions (on $c'$ and $D$) to eliminate some impossible solutions and speed up this algorithm. They also notice that some congruences give better results than others (for instance $D \equiv 3 \pmod 8$).

**Example 24.17** This algorithm allows Scott and Barreto to find several ordinary elliptic curves suitable for pairing-based cryptosystems such as

$$y^2 = x^3 - 3x + 259872266527491431103791444700778440496305560566$$

defined over the 160-bit prime field $\mathbb{F}_p$, where

$$p = 730996464809526906653170358426443036650700061957,$$

and which has a 160-bit prime order subgroup of size

$$\ell = 730996464809526906653171213409755627912276816323$$

and embedding degree $k = 6$.

They also find curves defined over fields of size 192, 224, and 256 bits with very small cofactors and embedding degrees equal to 6, cf. [SCBA 2004].

This algorithm cannot be generalized directly to higher values of the embedding degree. In fact, it uses the fact that the $k$-th cyclotomic polynomial for $k = 3, 4$, or 6 has degree equal to 2 so that the CM-equation can be solved. For higher values of $k$ the degree of the $k$-th cyclotomic polynomial is larger than or equal to 3 and we must use other methods.

### 24.2.3.b  Constructing ordinary elliptic curves with larger embedding degree

In [BALY$^+$ 2003, BALY$^+$ 2004a], the authors notice that the conditions required in the previous method are too restrictive for larger values of $k$ and the CM-equation cannot be reduced to Pell's equation and thus cannot easily be solved. They avoid these problems by

- relaxing the condition on the size of $c$, which can be of the size of $\ell$,
- choosing a value to serve as solution to the CM-equation, and then
- searching for parameters that yield a CM-equation with this particular solution.

In this way they produce examples of curves with embedding degree 7, 11, and 12.

However, since the condition on the size of $c$ has been omitted, the curves that are obtained have a subgroup of prime order $\ell$ of size only $\sqrt{q}$. This loss of efficiency is not very suitable for cryptographic purposes. In some applications such as short signatures [BOLY$^+$ 2002, BOLY$^+$ 2004] there are further reasons that require the same size for $\ell$ and $q$. Otherwise, the use of such curves results in an increase of length of ciphertexts or generated signatures.

Dupont, Enge, and Morain [DUEN$^+$ 2005] also find solutions of CM-equations for larger values of the embedding degree but they use a different approach. They choose the maximal value for the trace with respect to the Hasse–Weil bound so that the solution to the CM-equation must be small. They then use an exhaustive search on the remaining parameters. They also produce examples with large embedding degrees (up to 50) but their approach has the same drawback: the curves obtained cannot be used for efficient cryptosystems since $q$ has twice the number of bits as $\ell$. Note that so large degrees are not useful for the settings considered here; an embedding degree of $1024/160$, i.e., of 6 or 7, is optimal for current security requirements and in the near future, where the finite field $\mathbb{F}_{q^k}$ should have 2048 bits, $k = 10$ is a good choice.

More recently, Brezing and Weng obtained a better cofactor (i.e., a better ratio between the size of the field and the size of the subgroup of prime order) under certain conditions [BRWE 2004].

For instance they can find ordinary elliptic curves with an embedding degree equal to 8 such that $\frac{\ln q}{\ln \ell} = \frac{5}{4}$ so that a 200-bit prime field is required for 160-bit security.

Thus, by using CM-methods as explained above it is possible to construct ordinary elliptic curves with small embedding degrees. Of course, the main drawback of those methods is that ordinary elliptic curves have no distortion maps, which are required in many protocols, so that, up until now it is usually more practical to use supersingular curves.

Fortunately, as noticed in [BAKI$^+$ 2002] and mentioned above, some protocols or variants of protocols do not require such maps (i.e., can work with $G \neq G'$). This is for instance the case in the short signature protocol [BOLY$^+$ 2002, BOLY$^+$ 2004] or in the ID-based encryption [BOFR 2001] described in Section 24.1, since the second divisor can be chosen in $E(\mathbb{F}_{q^k}) \smallsetminus E(\mathbb{F}_q)$, thanks to the hash function (see next section for more details).

This approach has been partially generalized to hyperelliptic curves in [GAMC$^+$ 2004].

## 24.2.4 Performance

In the following we comment on additional performance issues, namely the signature length, and compare performance between supersingular and ordinary curves.

### 24.2.4.a Short signatures

The signature scheme based on Gap-DH systems was proposed as a scheme with *short* signatures. In [BOLY$^+$ 2002, BOLY$^+$ 2004] the authors suggest using a supersingular elliptic curve $E$ over $\mathbb{F}_{3^d}$ with maximal embedding degree $k = 6$. The signature then consists of a point on the curve. By Section 13.2.5 a point can be uniquely represented by the $x$-coordinate and one further bit. If compression is used the signature has a length of only $d \lg 3$. For the proposed parameters this results in a very much shorter signature than in DSA and of half the bit-size than in ECDSA since there, two integers modulo $\ell$ are required. As a drawback, the decompression requires some additional algorithms to recover the $y$-coordinate. The same considerations apply to hyperelliptic curves as well.

### 24.2.4.b Comparison of ordinary and supersingular elliptic curves

In [PASM$^+$ 2004], Page, Smart, and Vercauteren compare the efficiency of pairing-based cryptosystems using supersingular elliptic curves with those using ordinary elliptic curves constructed as explained above. As already noticed, the main drawback of ordinary curves is the absence of distortion maps. To cope with this, we can use hash functions taking their values in $E(\mathbb{F}_{q^k})$ as explained in the next section. In this manner, the hashed point is defined over $E(\mathbb{F}_{q^k})$, on which arithmetic is slower than on $E(\mathbb{F}_q)$. For instance in the short signature scheme explained in Section 24.1.3, a scalar multiplication of this hashed point is required. Thus if an ordinary curve is used, we must perform a scalar multiplication on $E(\mathbb{F}_{q^k})$, whereas if a supersingular curve is used, it is only necessary to perform a scalar multiplication on $E(\mathbb{F}_q)$ as the distortion map transforms the result to an element in $E(\mathbb{F}_{q^k})$.

On the other hand, ordinary elliptic curves are defined over a base field $\mathbb{F}_{q_l}$ of large characteristic, whereas supersingular curves interesting for cryptographic applications are defined over a base field $\mathbb{F}_{q_s}$ of small characteristic and algorithms for solving DL on finite fields are more efficient in small characteristic [COP 1984, JOLE 2002]. Studying the complexities in detail, the authors of [PASM$^+$ 2004] prove that, for the same level of security, we need

$$q_s = q_l^{1.7}.$$

It follows that for supersingular curves we need much larger base fields than for ordinary curves.

Taking this into account, the arithmetic on ordinary curves is more efficient than the one on supersingular curves for the same level of security.

Moreover, we have seen that embedding degrees larger than 6 can be obtained with ordinary elliptic curves, while this is not the case for supersingular ones.

Finally, ordinary elliptic curves are more efficient for pairing-based protocols, except for applications to signature schemes, for which a costly scalar multiplication is required. Nevertheless, we could swap the roles of $G$ and $G'$ to obtain a more efficient signature generation.

### 24.2.5 Hash functions on the Jacobian

Here we deal with the issue of defining a cryptographic hash function $h_1 : \{0,1\}^* \to G$, taking values in a subgroup $G$ of the Jacobian of some curve. Such a hash function is needed in ID-based cryptography (cf. Section 24.1.2).

In general, such hash functions are hard to build in practice. Therefore, Boneh and Franklin describe in [BOFR 2001] how to relax the requirements. Instead of hashing directly onto a Jacobian, one can hash onto some set $A \subseteq \{0,1\}^*$ (using standard hash functions) and then use an admissible encoding function to map $A$ onto the Jacobian. An encoding function is said to be *admissible* if it is computable, $c$-to-1 and samplable ([BOFR 2001]), where $c$ is the cofactor ($|J_C(\mathbb{F}_p)| = c\ell$). It is proved in [BOFR 2001] that this relaxation does not affect security.

**Example 24.18** Boneh and Franklin give an example of admissible encoding functions for the elliptic curve $y^2 = x^3 + 1$ defined over $\mathbb{F}_p$ (where $p \equiv 2 \pmod{3}$ is prime). They use a standard hash function from $\{0,1\}^*$ to $\mathbb{F}_p$. This allows us to obtain a $y$-coordinate $y_0$. Then the $x$-coordinate $x_0$ is obtained by
$$x_0 = (y_0^2 - 1)^{1/3} = (y_0^2 - 1)^{(2p-1)/3}.$$
Finally they compute the scalar multiplication $[c](x_0, y_0)$ to obtain a $c$-to-1 map.

Of course, this kind of map can be applied to other curves. Galbraith generalizes this example in [GAL 2001a] as follows: the identity bit-string is concatenated with a given padding string and then passed through a standard hash function taking its values in $\mathbb{F}_q$. This process is repeated using a deterministic sequence of padding strings until the output is the $x$-coordinate (or the first polynomial $u(x)$ in the higher genus case) of an $\mathbb{F}_q$-rational element of the Jacobian. It is then easy to find the remainder of the representation of this element using the decompression techniques (cf. Sections 13.2.5, 13.3.7, 14.2.1, and 14.2.2). Finally, this element is multiplied by $c$.

This method provides an $\mathbb{F}_q$-rational element of the Jacobian. Often an $\mathbb{F}_{q^k}$-rational element is required to ensure that the pairing is nontrivial. For supersingular curves, distortion maps can be used to transform the element obtained above to an $\mathbb{F}_{q^k}$-rational one. But for ordinary curves, this is not possible.

# Chapter 25

# Compositeness and Primality Testing Factoring

### Roberto M. Avanzi and Henri Cohen

### Contents in Brief

| | | |
|---|---|---|
| **25.1** | **Compositeness tests** <br> Trial division • Fermat tests • Rabin–Miller test • Lucas pseudoprime tests • BPSW tests | 592 |
| **25.2** | **Primality tests** <br> Introduction • Atkin–Morain ECPP test • APRCL Jacobi sum test • Theoretical considerations and the AKS test | 596 |
| **25.3** | **Factoring** <br> Pollard's rho method • Pollard's $p-1$ method • Factoring with elliptic curves • Fermat–Morrison–Brillhart approach | 601 |

Primality and factoring are essential in many aspects of public-key cryptography. RSA is still the most widely used public-key cryptosystem, and its fundamental idea is based on the difficulty of factoring compared to the simplicity of primality testing. In addition, in ECC and HECC we need to work in groups whose order is either prime or a small factor times a large prime, so once again primality is essential.

The simplest attempt to factor an integer $n$ is of course trial division. But the main breakthrough in primality and compositeness testing was made by Fermat thanks to his "little" theorem $a^{n-1} \equiv 1 \pmod{n}$ when $n$ is a prime not dividing $a$. Although this does not give a necessary and sufficient condition for primality, it can be modified so that the number of exceptions becomes vanishingly small. Such modifications are called *compositeness tests* since they will prove a number composite, but will never prove that a number is prime, only give a "moral certainty" of that fact. Deeper modifications lead to true primality tests, giving a proof of primality. Once a number $n$ is proven composite, we can either be content with that fact, or want to know the factors of $n$, and this is the whole subject of factoring.

There are two attitudes that one can have with respect to compositeness and primality testing. If we are content with the fact that a number is almost certainly, but not yet provably prime, we can stop there, and declare our $n$ to be an industrial-grade prime number. The disadvantage is that this may be false, and as a consequence it can induce either bugs or security holes in cryptosystems using that number. The big advantage is that the test is very fast and easy to implement, so for instance a dynamical generation of industrial-grade primes can be included in a smart card environment.

If we want higher security, or if we can spare the time, or simply if we want mathematical rigor, we must use a true primality test. These are much more complex than compositeness tests, both in the mathematics involved (although the mathematics of the AKS test is quite simple), and in the algorithmic description.

Finally, if we know that a number is composite, we may want to know its factors. This leads to the vast domain of factoring, which includes a wide variety of basic methods, and of variants of these methods.

Since this book is primarily targeted to elliptic and hyperelliptic curve cryptosystems, we will not give too much detail of the algorithms, but refer the reader to the abundant literature on the subject, for instance to [COH 2000].

Our focus is on generic methods. Techniques that only work for inputs of a certain form (such as Mersenne numbers) will not be considered here. Information about them can be found in any good book on computer algebra.

As we remarked in Chapter 19, when constructing a cryptosystem based on the DLP in a group, the order of the latter must be prime or almost prime. After point counting, the order must be checked for suitability, or, if a curve is constructed by means of complex multiplication, the possible orders must be checked for (almost) primality. Moreover, some of the methods presented here make use of the theory of elliptic and hyperelliptic curves, and are themselves applications of the ideas given in other chapters. In general, their implementation can take advantage of the techniques that we describe elsewhere in this book; see for example Chapters 9, 11, 13, and 14.

## 25.1 Compositeness tests

We begin with compositeness tests. These are tests that can prove that a number is composite or assess that it is a prime with a certain probability. This might seem strange, since a given integer is either prime or composite. It refers to the fact that a run of the test will possibly not be able to identify some composites as such. Running the test several times with different parameters — or using a combination of different tests — can increase the probability that a number identified as a probable prime is indeed prime, but will never give us complete certainty unless an explicit primality proof is given. For this purpose, primality tests have been devised, and we shall see them in Section 25.2.

### 25.1.1 Trial division

As already mentioned, the simplest method for factoring a number $n$, and in particular to detect whether it is prime or composite, is trial division. This can be done in several ways. We could divide $n$ by all integers up to $n^{1/2}$ until either a nontrivial factor is found or $n$ is declared prime because it is not divisible by any such integer. An evident improvement that gains a multiplicative factor is to divide only by 2, 3, and integers congruent to $\pm 1$ modulo 6, or more generally to use a *wheel*, in other words to use as divisors elements of congruence classes of integers coprime to some integer $W$, together with the prime divisors of $W$. The number $W$ should not be chosen too large since we have to store a table of $\varphi(W)$ integers, where $\varphi$ denotes Euler's totient function. Another possibility is to use a precomputed table of prime numbers up to some bound $B$. If $n \leqslant B^2$ this will be sufficient to factor $n$, otherwise the prime number table will at least tell us that the smallest prime factor of $n$ is larger than $B$. Once again $B$ should not be chosen too large since there are more efficient ways than trial division for removing small factors, for instance, Pollard's rho method.

## 25.1.2 Fermat tests

The fundamental remark, due in essence to Fermat, that allows us to distinguish between primes and composites without factoring is that an element $g$ of a finite group $G$ of cardinality $N$ satisfies $g^N = 1$, where 1 denotes the identity element of $G$. Applied, for instance, to $G = (\mathbb{Z}/n\mathbb{Z})^*$ for $n$ prime, it gives Fermat's little theorem

$$a^{n-1} \equiv 1 \pmod{n} \text{ when } \gcd(a, n) = 1.$$

If $n$ is not prime but satisfies this condition, we say that it is a *pseudoprime* to base $a$. If the condition is not satisfied, then we call $a$ a *Fermat witness* for the compositeness of $n$.

Although it is not a necessary and sufficient condition (for instance $2^{340} \equiv 1 \pmod{341}$ although $341 = 11 \times 31$ is not prime), the pseudoprimality test is already very useful for the following two reasons:

- The number of exceptions, although infinite, is quite small. In other words, if $n$ is a pseudoprime to base $a$ there is already a good chance that $n$ is in fact prime.
- The test is practical, in other words it is easy to compute $a^{n-1} \bmod n$ using any reasonable exponentiation method in $\mathbb{Z}/n\mathbb{Z}$, cf. Chapter 9. This is in marked contrast with other tests such as those based on Wilson's theorem $(n-1)! \equiv -1 \pmod{n}$ if and only if $n$ is prime, since it is not known how to compute $(n-1)! \bmod n$ in a reasonable amount of time (and this is probably impossible).

Pseudoprimes to base 2 are sometimes also called *Poulet numbers* or *Sarrus numbers*. There are 1091987405 primes less than $25 \times 10^9$ and 21853 nonprime pseudoprimes to base 2 – a relatively small number, but still too large for some applications.

To improve on this test (i.e., to reduce the number of exceptions), we can use several bases $a$. Unfortunately, even such a stronger test has infinitely many exceptions since there exist infinitely many nonprime $n$ such that $a^{n-1} \equiv 1 \pmod{n}$ for all $a$ coprime to $n$. Such a composite number $n$ without any Fermat witnesses is called a *Carmichael number*. It is easy to show that Carmichael numbers must satisfy very strong conditions: an integer $n$ is a Carmichael number if and only if it is squarefree and $p - 1$ divides $n - 1$ for all prime factors $p$ of $n$. As a consequence, Carmichael numbers are odd and have at least three prime factors. On the other hand, it has been quite difficult to show that there are infinitely many Carmichael numbers. In [ALGR$^+$ 1994] it has been proven that the number of Carmichael numbers up to $x$ is at least $x^{2/7}$ for $x$ large enough ($2/7 \approx 0.285$). The exponent has been improved to $\approx 0.293$ in [BAHA 1998] and, recently, to $\approx 0.332$ by Harman in [HAR 2005]. Erdős [ERD 1956] conjectured that there are $x^{1-o(1)}$ Carmichael numbers up to $x$ for $x$ large. Granville and Pomerance [GRPO 2002] give convincing arguments to support this conjecture and also their own that, for any given integer $k \geqslant 3$, there are $x^{1/k-o_k(1)}$ Carmichael numbers up to $x$ with exactly $k$ prime factors.

There are only 2163 Carmichael numbers less than $25 \times 10^9$ [POSE$^+$ 1980]. The Carmichael numbers under 100000 are

$$561, 1105, 1729, 2465, 2821, 6601, 8911, 10585,$$
$$15841, 29341, 41041, 46657, 52633, 62745, 63973, \text{ and } 75361.$$

Richard Pinch gives a table of Carmichael numbers up to $10^{17}$ on his web site [PINCH].

It is possible and important to use more sophisticated groups $G$ than $(\mathbb{Z}/n\mathbb{Z})^*$. These groups must have the following properties:

- The cardinality of $G$ should be equal to some simple function $f(n)$ when $n$ is prime, and rarely be a divisor of $f(n)$ if $n$ is not prime (or even better, the group law may not be defined everywhere if $n$ is not prime). In the example $G = (\mathbb{Z}/n\mathbb{Z})^*$ we have $f(n) = n - 1$.

- It must be easy to compute in $G$, and in particular to detect if an element is equal to the identity element 1 of $G$.

As examples of such groups we mention subgroups of finite fields and the group of rational points of an elliptic or (of the Jacobian of) a hyperelliptic curve over a finite field.

All these groups can also be used for factoring, but then we need some additional properties that we will mention below.

### 25.1.3 Rabin–Miller test

The Fermat test is very attractive because of its simplicity, but it is not able to identify the Carmichael numbers as composite. As Pomerance [POM 1990] writes: *Using the Fermat congruence is so simple that it seems a shame to give it up just because there are a few counter examples!* Hence improved tests have been devised that aim to reduce the number of incorrectly identified composites, but without sacrificing too much of the beauty of the Fermat test.

Certainly the most important of all compositeness tests improving upon the Fermat test is the Rabin–Miller test. It is based on the following additional property of a prime number $n$: if $x^2 \equiv 1 \pmod{n}$ then $x \equiv \pm 1 \pmod{n}$. Thus instead of testing only that $a^{n-1} \equiv 1 \pmod{n}$ we write $n - 1 = 2^t m$ with $m$ odd, and then it is easily seen that either $a^m \equiv 1 \pmod{n}$ or that there exists an integer $s$ with $0 \leqslant s \leqslant t - 1$ such that $a^{2^s m} \equiv -1 \pmod{n}$. If these conditions are satisfied we say that $n$ is a *strong pseudoprime to base* $a$.

If $a^{2^{s+1} m} \equiv 1 \pmod{n}$ and $b = a^{2^s m} \not\equiv \pm 1 \pmod{n}$ for some $0 \leqslant s \leqslant t - 1$, then $n$ must be composite, and $\gcd(b + 1, n)$ is a nontrivial factor of $n$. In fact, if $a$ is chosen uniformly at random in $[2, n - 2]$, the test furnishes a proper divisor of a Carmichael number with probability at least $1/2$.

As such, this simply gives a stronger test than the basic Fermat test. For instance, although 340 is a pseudoprime to base 2, it is not a strong pseudoprime because $2^{170} \equiv 1 \pmod{341}$ while $2^{85} \not\equiv \pm 1 \pmod{341}$. The smallest nonprime strong pseudoprime to base 2 is $n = 2047$. The following list [POSE$^+$ 1980] gives all composite numbers up to $25 \times 10^9$ that are simultaneously strong pseudoprime for bases 2, 3 and 5:

$$25326001, 161304001, 960946321, 1157839381, 3215031751,$$
$$3697278427, 5764643587, 6770862367, 14386156093,$$
$$15579919981, 18459366157, 19887974881, \text{ and } 21276028621.$$

The only number of the above that is also a strong pseudoprime to base 7 is 3215031751.

The basic result due to Rabin and Monier is that for a given nonprime $n$ the number of *false witnesses*, i.e., of $a$ between 1 and $n$ such that $n$ is a strong pseudoprime to base $a$ is at most equal to $n/4$, in fact at most equal to $\varphi(n)/4$, where $\varphi$ is Euler's totient function. Thus, by repeating the test a small number of times (for instance, 20), and assuming that the tests are independent (which they are not), we can say that $n$ has a very high probability of being prime, and so is an *industrial-grade prime*, according to a name coined by one of the authors. The above procedure is called the *Rabin–Miller test*.

Thus, after a small trial division search, one applies the Rabin–Miller test and the outcome is as follows: if the test has failed for some $a$ this gives a proof that $n$ is composite, usually without giving any clue to its factors. If it is necessary to know the factors of $n$, we must then use a factoring method. On the other hand if every test has succeeded, then there is a very high probability that $n$ is prime. Note however that in this case the tests will never prove that $n$ is prime. To prove primality, we must use a primality test. Such a test will have two possible outcomes: either it will give a proof that $n$ is prime, or it will give a proof that $n$ is composite, this last alternative being extremely unlikely since it is essentially ruled out by the Rabin–Miller test.

There are other similar strengthenings of the Fermat test, for instance, the Solovay–Strassen compositeness test, which checks whether

$$a^{(n-1)/2} \equiv \left(\frac{a}{n}\right) \pmod{n},$$

where $\left(\frac{a}{n}\right)$ is the Legendre–Kronecker symbol. This test is strictly weaker than the Rabin–Miller test since every strong pseudoprime to base $a$ passes this test, but the number of exceptions for a given nonprime $n$ can be as large as $\varphi(n)/2$ instead of $\varphi(n)/4$. Since the probability of error (that is, the probability that it fails when $n$ is composite) is higher than in the Rabin–Miller test, the same degree of confidence can be achieved by using more iterations of the test.

Slightly more complicated tests have been introduced that reduce the proportion to $1/4$ as in the Rabin–Miller test, but to little advantage: in practice the Rabin–Miller test efficiently distinguishes between prime and composite numbers, although if the number is prime a primality proof is always necessary afterwards, except if we are content with industrial-grade primes (for instance, when we need to generate large prime numbers in real time on a smart card).

### 25.1.4 Lucas pseudoprime tests

For a reference to the facts mentioned in this section, see [RIB 1996].

Let $P$ and $Q$ integers satisfying $D := P^2 - 4Q > 0$, and let $\rho = \frac{1}{2}(P + \sqrt{D})$, $\sigma = \frac{1}{2}(P - \sqrt{D})$ be the roots of the equation $x^2 - Px + Q = 0$. Consider the two sequences

$$U_k = \frac{\rho^k - \sigma^k}{\rho - \sigma} \quad \text{and} \quad V_k = \rho^k + \sigma^k.$$

The sequences $U_k$ and $V_k$ satisfy

$$\begin{cases} U_0 = 0, & U_1 = 1, \\ V_0 = 2, & V_1 = P, \end{cases} \quad \text{and, for } k \geq 0, \quad \begin{cases} U_{k+2} = PU_{k+1} - QU_k, \\ V_{k+2} = PV_{k+1} - QV_k. \end{cases}$$

The terms of the sequences can be very efficiently obtained by means of recurrence relations that permit the construction of a kind of double-and-add-like ladder for each value of the index $k$ consisting of $O(\lg k)$ steps. These relations are

$$\begin{cases} U_{2k} = U_k V_k, \\ V_{2k} = V_k^2 - 2Q^k, \end{cases} \quad \begin{cases} U_{2k+1} = U_{k+1} V_k - Q^k, \\ V_{2k+1} = V_{k+1} V_k - PQ^k. \end{cases}$$

We have the following result, which is exactly analogous to the one leading to the notion of strong pseudoprime.

**Theorem 25.1** Let $p$ be a prime number not dividing $2QD$, and write $p - \left(\frac{D}{p}\right) = 2^t m$ with $m$ odd. Then either $p \mid U_m$ or there exists $s$ with $0 \leq s \leq t-1$ such that $p \mid V_{2^s m}$.

A composite number $n$ relatively prime to $2QD$ and such that either $n \mid U_m$, or there exists $s$ with $0 \leq s \leq t-1$ such that $p \mid V_{2^s m}$, where $n - \left(\frac{D}{n}\right) = 2^t m$ with $m$ odd, is called a *strong Lucas pseudoprime* with respect to the parameters $P$ and $Q$.

A weaker notion is that of a *Lucas pseudoprime*, which is defined as a composite number $n$ that is coprime to $2QD$, and such that the Legendre symbol $\left(\frac{D}{n}\right) = 1$, and $n \mid U_{n+1}$. There are no even Lucas pseudoprimes [BRU 1994], and the first Lucas pseudoprimes are $705, 2465, 2737, 3745, \ldots$

There are several additional definitions, such as that of extra strong Lucas pseudoprime [MoJo], Frobenius pseudoprime, Perrin pseudoprime, and probably many more scattered in the literature. See, for example, [GRA 2001].

## 25.1.5 BPSW tests

Baillie and Wagstaff [BAWA 1980] and Pomerance et al. [POSE⁺ 1980, POM 1984] proposed a test (or rather a related set of tests), now known as the *BPSW probable prime test*, which is a combination of a strong pseudoprime test and a "true" Lucas test, i.e., with $\left(\frac{D}{n}\right) = -1$.

There are several variants, such as the following, modeled around the description in [POM 1984]:

1. Perform a strong pseudoprime test to base 2 on $n$.
   - If this test fails, declare $n$ composite and halt.
   - If this test succeeds, $n$ is probably prime: perform the next step.

2. In the sequence $5, -7, 9, -11, 13, \ldots$ find the first number $D$ for which the Legendre symbol $\left(\frac{D}{n}\right) = -1$, then perform a Lucas pseudoprime test with discriminant $D$ on $n$. Some authors perform here a strong Lucas pseudoprime test in place of the Lucas pseudoprime test. Marcel Martin uses a different sequence for finding $D$, namely $D = 5, 21, 45, \ldots, (2k+1)^2 - 4, \ldots$ which may lead to a faster search. This approach might have been suggested by Wei Dai, but the attribution is uncertain.
   - If this test fails, declare $n$ composite and halt.
   - If this test succeeds, $n$ is very probably prime.

Pomerance [POM 1984] originally offered a prize of $30 for discovery of a composite number that passes this test, but the offer was subsequently raised to $620 by Selfridge and Wagstaff. PRIMO [PRIMO] author Marcel Martin uses the BPSW test in his ECPP (cf. Section 25.2.2) software. Based on the fact that in more than 3 years of regular usage the ECPP part of his software never found a composite that passes the BPSW test, he estimates that no composite up to about 10000 digits (the current limit for ECPP) can fool the latter. Still, its exact reliability is not yet known.

Grantham [GRA 1998] provided a pseudoprimeness test (RQFT) with probability of error less than $1/7710$, and he pointed out that the lack of counter examples to the BPSW test was a quite strong hint that probability of error of the latter may be much lower.

Zhenxiang Zhang [ZHA 2002] has a variant of the BPSW test called the *one-parameter quadratic-base test* (OPQBT), that is passed by all primes $\geqslant 5$ and passed by an odd composite $n$ with a certain probability of error that is also very carefully analyzed.

The running time of the OPQBT is asymptotically 4 times that of a Rabin–Miller test in the worst cases, but only twice that of a Rabin–Miller test for most composites. This, together with the much higher reliability of the test, should result in a test that is faster that the Rabin–Miller test for a given, asymptotically very high, reliability level on large numbers.

## 25.2 Primality tests

This section is devoted to primality tests, i.e., to tests that can identify prime numbers with certainty.

### 25.2.1 Introduction

Note first that primality tests may be error-prone, since they may only give a one bit yes/no answer. Thus if there is any mistake, either in the algorithm or in the implementation, it will be difficult to detect, and the only remedy is to use another algorithm and another implementation on the same number. Thus, if possible, we will ask the primality test not only to give a proof of primality, but also a *primality certificate*, in other words some data that will allow anyone to prove the primality of the number for himself much more rapidly than the initial proof. Of the two main primality tests in use today, the ECPP test gives such a certificate, while the APRCL test does not, at least without considerable modifications.

The simplest primality test is the *Pocklington–Lehmer test*. Although completely superseded, it contains the basic ingredients of modern primality tests. It is based on the following result.

**Proposition 25.2** Let $n$ be a prime number. Then for every prime divisor $p$ of $n-1$, one can find an integer $a_p$ such that
$$a_p^{n-1} \equiv 1 \pmod{n} \text{ and } \gcd\left(a_p^{(n-1)/p} - 1, n\right) = 1.$$
Conversely, if this is satisfied and $n > 1$, then $n$ is prime.

The problem with using this proposition is that it is necessary to find the prime factors of $n-1$, hence essentially to factor $n-1$, which may be impossible in practice. Although this test can be improved in a number of ways, and combined with tests coming from other subgroups of finite fields (in other words, essentially Lucas-type tests), it is basically limited in scope because of the need to factor.

### 25.2.2 Atkin–Morain ECPP test

ECPP stands for elliptic curve primality proving. The initial method is described in [ATMO 1993]. See [FRKL+ 2004] for the latest version. This is currently the primality test of choice, and has for some time held the record for the size of general numbers that have been proved prime (special numbers such as Mersenne numbers can be proved prime using faster methods). In any case, for cryptographic purposes, ECPP and the other modern primality test APRCL are just as efficient, since a 1024 bit number, for example, can be proven prime in less than a minute by both algorithms.

The ECPP primality test is based on a result very similar to the one used for the Pocklington–Lehmer test, replacing the group $(\mathbb{Z}/n\mathbb{Z})^*$ by the group $E(\mathbb{Z}/n\mathbb{Z})$, where $E$ is a suitable elliptic curve.

**Proposition 25.3** Let $n > 1$ be an integer coprime to 6. Assume that we can find a plane elliptic curve $E$ defined over $\mathbb{Z}/n\mathbb{Z}$, a point $P \in E(\mathbb{Z}/n\mathbb{Z})$, and a positive integer $m$ satisfying the following conditions:

- $[m]P = P_\infty$, where $P_\infty$ is the identity element of the elliptic curve.
- There exists a prime divisor $q$ of $m$ such that $q > (n^{1/4}+1)^2$ and $[m/q]P = (X:Y:Z)$ in projective coordinates with $\gcd(Z,n) = 1$.

Then $n$ is prime.

The integer $m$ above is in practice going to be the cardinality of $E(\mathbb{Z}/n\mathbb{Z})$ (if $n$ is prime), but it is not necessary to know this in advance. This is similar to the Fermat–Pocklington–Lehmer test where one uses $n-1$ as a supposed cardinality of $(\mathbb{Z}/n\mathbb{Z})^*$, although this is not yet known.

There are two apparently quite difficult problems that we must solve to be able to use Proposition 25.3. The first one is to find a suitable integer $m$, hence essentially to compute the cardinality of $E(\mathbb{Z}/n\mathbb{Z})$. The second is once again to factor $m$ so as to find a suitable prime divisor $q$. A priori this seems to make life even more difficult than for the Pocklington–Lehmer test. However, this is not at all the case, because of the following two crucial remarks.

- Contrary to the Pocklington–Lehmer test we can vary the elliptic curve $E$, so that if we have some algorithm that can compute the cardinality $m$ of $E(\mathbb{Z}/n\mathbb{Z})$ (when $n$ is prime), we use trial division on $m$ to see whether there exists a large (pseudo)prime divisor $q$ of $m$, and if this does not seem to be the case or cannot be determined easily we simply choose another curve. This is the basis of the theoretical and nonpractical Goldwasser–Killian test.

- Although there do exist efficient (polynomial time) algorithms for computing $m$, it is much better to use the theory of complex multiplication to construct elliptic curves $E$ for which $m$ can be immediately computed. This is the basis of the Atkin–Morain ECPP test.

To use complex multiplication, cf. Chapter 18, we must find a negative discriminant $D$ (not necessarily fundamental) such that $n$ splits in the quadratic order of discriminant $D$ as a product of two elements $n = \pi\bar\pi$ or, equivalently, such that there exist integers $a$ and $b$ with $a^2 + |D|b^2 = 4n$. If this is the case, it is immediate to construct an elliptic curve $E$ over $\mathbb{Z}/n\mathbb{Z}$ with complex multiplication by the given order (see below), and we have the simple formula

$$|E(\mathbb{Z}/n\mathbb{Z})| = n + 1 - \pi - \bar\pi = n + 1 - a.$$

A necessary condition for $n$ to split as a product of two elements is $\left(\frac{D}{n}\right) = 1$. This will occur with probability approximately equal to $1/2$. This necessary condition will also be sufficient if the class number $h(D)$ of the order is equal to 1, and this occurs only for 13 values of $D$, namely $D = -3$, $-4, -7, -8, -11, -12, -16, -19, -27, -28, -43, -67$, or $-163$. Otherwise, if $\left(\frac{D}{n}\right) = 1$ there will be a probability close to $1/h(D)$ for $n$ to split as a product of two elements.

The basic Atkin–Morain test thus proceeds as follows. We choose discriminants $D$ in sequence of increasing $h(D)$, and for a given $D$ we test whether $\left(\frac{D}{n}\right) = 1$ and whether $4n$ is of the form $a^2 + |D|b^2$. This last condition can easily be checked using a Euclidean type algorithm called *Cornacchia's algorithm*, see [COH 2000]. Once a suitable $D$ has been found, as well as the corresponding integer $a$, we compute $m = n + 1 - a$. If $m$ can easily be factored and is found to have a pseudoprime divisor $q > (n^{1/4} + 1)^2$ we can apply Proposition 25.3 and prove (or very rarely disprove) that $n$ is prime. If we cannot easily factor $m$ we go on to the next value of $D$.

To finish the description of the test we explain how to construct the elliptic curve $E$, given a discriminant $D$ satisfying the above conditions. Assume first that $D \neq -3$ and $D \neq -4$. We first compute the elliptic $j$ invariants of the $h(D)$ classes of the order of discriminant $D$ as complex numbers, using one of the efficient formulas for $j$. We then form the monic polynomial $H_D(X)$ whose roots are these $h(D)$ values, in other words the class polynomial, see Chapter 18. The theory of complex multiplication tells us that $H_D(X)$ has integer coefficients, and gives an estimate on the size of these coefficients, so if we have used sufficient accuracy we simply round to the closest integer the coefficients of the polynomial that we have obtained so as to compute $H_D(X)$. Since we have chosen $D$ such that $n$ splits as a product of two elements, the theory of complex multiplication also tells us that the polynomial $H_D(X)$ will split modulo $n$ as a product of $h(D)$ linear factors, at least if $n$ is prime. If we call $j$ one of the $h(D)$ roots modulo $n$ the elliptic curve $E$ can be defined by the equation

$$y^2 = x^3 - 3cg^{2k}x + 2cg^{3k},$$

where $c = j/(j - 1728)$, $g$ is any quadratic nonresidue modulo $n$ and $k$ is 0 or 1. Although it seems that we obtain in this way many different elliptic curves, corresponding to the choice of the root $j$, of the nonresidue $g$ and of $k$, it is easily shown that there are only two nonisomorphic curves $E$, corresponding for instance to a fixed choice of $j$ and $g$ and the two possible values of $k$. The cardinality $|E(\mathbb{Z}/n\mathbb{Z})|$ of these curves is equal to $n + 1 - a$ or $n + 1 + a$, so we have two chances of obtaining a suitable curve for a given $D < -4$.

If $D = -3$ or $D = -4$ the (simpler) construction must be slightly modified. If $D = -4$, hence for $n \equiv 1 \pmod 4$, we have the four curves

$$y^2 = x^3 - g^k x,$$

where as before $g$ is any quadratic nonresidue modulo $n$ but now $0 \leqslant k \leqslant 3$. If $D = -3$, hence for $n \equiv 1 \pmod 3$, we have the six curves

$$y^2 = x^3 - g^k,$$

where $0 \leqslant k \leqslant 5$ and where $g$ is not only a quadratic nonresidue but also a cubic nonresidue, in other words must also satisfy

$$g^{(n-1)/3} \not\equiv 1 \pmod n.$$

**Remarks 25.4**

(i) The above method in fact leads to a recursive test: after having found a suitable elliptic curve $E$ and an integer $m$ having a large strong pseudoprime factor $q$, we must apply the method once again, but now with $n$ replaced by $q$, until we reach integers that are sufficiently small to be shown prime by some other method.

(ii) The ECPP test gives a primality certificate for the primality of $n$. Indeed, at each stage of the recursion it is sufficient to give the equation of the elliptic curve $E$, the integer $m$, and the prime $q$ (note that the choice of the point $P$ satisfying the hypotheses of Proposition 25.3 is unimportant and can be done very rapidly).

(iii) We have only described the basic Atkin–Morain test. In an actual implementation there are evidently many additional tricks that we will not give here.

### 25.2.3 APRCL Jacobi sum test

APRCL stands for the initials of the five authors of this test (Adleman, Pomerance, Rumely, Cohen, Lenstra), see [ADPO$^+$ 1983], [COLE 1984], [COLE 1987].

This test is based on a version of Fermat's theorem in cyclotomic fields. It is considerably more complicated to explain than ECPP so we will be content with a sketch.

Let $\chi$ be a Dirichlet character modulo some prime $q$ (which in practice will not be very large, at most $10^8$, say), and let $\zeta_q = \exp(2i\pi/q)$ be a primitive $q$-th root of unity in $\mathbb{C}$. Recall that the *Gauß sum* $\tau(\chi)$ is defined by

$$\tau(\chi) = \sum_{x \in \mathbb{Z}/q\mathbb{Z}} \chi(x) \zeta_q^x.$$

One of its basic properties is that when $\chi$ is not the unit character then $|\tau(\chi)| = q^{1/2}$. Furthermore if $\chi$ is a character of order 2 (in other words a real character) we even have $\tau(\chi) = (\chi(-1)q)^{1/2}$ for a suitable choice of the square root.

The basic idea of the APRCL test is the following proposition, stated purposely in vague terms.

**Proposition 25.5** *Let $n$ be prime, and let $\chi$ be a character of prime order $p$ with $p \mid q - 1$. In the ring $\mathbb{Z}[\zeta_q, \zeta_p]$ we have the congruence*

$$\tau(\chi)^n \equiv \chi(n)^{-n} \tau(\chi^n).$$

Conversely, if this congruence is true for sufficiently many characters $\chi$, and if some easily checked additional conditions are satisfied then $n$ is prime.

We note that if we choose $\chi(n) = \left(\frac{q}{n}\right)$ we obtain again the Solovay–Strassen test, which reduces to test if $q^{(n-1)/2} \equiv \left(\frac{q}{n}\right) \pmod{n}$.

We must solve two important problems. One is theoretical: we must make precise the "sufficiently many" and the additional conditions of Proposition 25.5. The second is practical: how does one test the congruence, since even if $q$ and $p$ are moderately small the ring $\mathbb{Z}[\zeta_q, \zeta_p]$ may be quite large. The first problem is (almost) solved by the following theorem.

**Theorem 25.6** Let $t > 1$ be an integer, and set $e(t) = \prod_{(q-1)|t} q$, where the product is over the primes $q$ such that $(q-1)$ divides $t$. For each $q \mid t$ and for each prime $p \mid (q-1)$ choose any character $\chi_{q,p}$ modulo $q$ of exact order $p$. Assume that the following conditions are satisfied:

- For each such pair $(p, q)$, we have the congruence
$$\tau(\chi_{p,q})^n \equiv \chi_{p,q}(n)^{-n} \tau(\chi_{p,q}^n).$$

- For each $p \mid t$ a technical condition $L_p$ is true.
- We have $\gcd(te(t), n) = 1$ and $e(t) > n^{1/2}$.
- For every $i$ with $0 < i < t$, we have $r_i = 1$, $r_i = n$, or $r_i \nmid n$, where $r_i$ denotes the remainder of the Euclidean division of $n^i$ by $e(t)$.

Then $n$ is prime.

The condition $L_p$ is slightly too technical to explain here. It is for instance satisfied if $p \neq 2$ and $n^{p-1} \not\equiv 1 \pmod{p^2}$.

The second problem is that of the practicality of the congruence test, since the ring $\mathbb{Z}[\zeta_q, \zeta_p]$ is large. The crucial ingredient here is to introduce Jacobi sums, which are closely linked to Gauß sums but belong to the much smaller ring $\mathbb{Z}[\zeta_p]$ (typically $p \leqslant 19$ even for $n$ having more than 1000 decimal digits). By definition, if $\chi_1$ and $\chi_2$ are two Dirichlet characters modulo the same $q$ the *Jacobi sum* $J(\chi_1, \chi_2)$ is defined by

$$J(\chi_1, \chi_2) = \sum_{x \in \mathbb{Z}/q\mathbb{Z}} \chi_1(x) \chi_2(1-x).$$

The crucial formula linking Jacobi and Gauß sums is the following: if $\chi_1$, $\chi_2$, and $\chi_1\chi_2$ are all three nontrivial characters then

$$J(\chi_1, \chi_2) = \frac{\tau(\chi_1)\tau(\chi_2)}{\tau(\chi_1\chi_2)}.$$

Thus if the $\tau(\chi_i)$ satisfy some congruence conditions, so does $J(\chi_1, \chi_2)$, in the smaller ring $\mathbb{Z}[\zeta_p]$. Conversely, using some additional group ring machinery it is possible to show that if $p \neq 2$ and $J(\chi, \chi)$ satisfies a suitable congruence in $\mathbb{Z}[\zeta_p]$, then $\tau(\chi)$ will satisfy the desired congruence in $\mathbb{Z}[\zeta_q, \zeta_p]$ (for $p = 2$ we must introduce more complicated Jacobi-type sums). It can thus be shown that all the congruences can in fact be done in $\mathbb{Z}[\zeta_p]$ using Jacobi sums, and this is why this test is called the Jacobi sum test. An evident additional gain is the following: if for instance $n \equiv 1 \pmod{p}$ (which occurs often since $p$ is very small), we can in fact replace $\zeta_p$ by a $p$-th root of 1 in $(\mathbb{Z}/n\mathbb{Z})^*$ (for instance by $a^{(n-1)/p}$ modulo $n$ for any $a$ for which this is not equal to 1), so the computations for that $p$ can be done in $\mathbb{Z}$.

## 25.2.4 Theoretical considerations and the AKS test

We begin with a few words concerning the running time of the two main practical algorithms for primality testing, the ECPP and the APRCL tests. The ECPP test has a heuristic (but unproved) running time that is polynomial in $\lg n$, of the order of $O(\lg n^{4+\varepsilon})$ for the most recent implementations (fastECPP). This is indeed what is observed in practice, but is not proved. Also the test is probabilistic.

In its practical version the APRCL test is also probabilistic, with a proven expected running time that is almost polynomial time, more precisely of $O(\lg n^{C \lg \lg \lg n})$. It seems that in practice the APRCL test is slower by a factor of 2 or 3 (although serious comparisons of professional-level implementations have never been made), but this is not at all because of its non-polynomial nature. Instead it is because of the higher complexity of the basic operations, in other words because of the size of the implicit $O$ constant. It is to be noted that there exists a deterministic version of APRCL with a similar running time, but which is less practical.

There have been two breakthroughs on the theoretical aspect of primality testing. The first one occured at the beginning of the 1980's: Adleman–Huang [ADHU 1992] proved that primality testing was in **RP**, in other words that there exists a probabilistic algorithm for primality proving that can be proved to run in expected polynomial time. Their quite sophisticated proof involves working in the Jacobians of curves of genus 2 over finite fields.

The most spectacular breakthrough took place in the summer of 2002 through the work of Agrawal–Kayal–Saxena [AGKA+ 2002], who proved that in fact primality is in **P**, in other words that there exists a deterministic polynomial time algorithm for primality testing. All the more remarkable is the fact that their algorithm is based on very simple considerations (in other words the underlying groups are not complicated) and not on curves or higher-dimensional objects.

Since their fundamental discovery, much progress has been done by experts in the field. Lenstra and Pomerance have for instance shown that modifications of the AKS ideas lead to a deterministic algorithm that runs in time $O(\lg n^{6+\varepsilon})$. Even though the initial goal was to remove probabilistic aspects, Bernstein [BER 2004a] has found a probabilistic variant of AKS that runs in $O(\lg n^{4+\varepsilon})$. This has been discovered independently by Mihăilescu and Avanzi [MIAV 2003], and builds upon improvements by Berrizbeitia [BER 2001b]. Qi Cheng has an interesting approach combining rounds of ECPP with rounds of AKS [CHE 2003]. For now it is not clear how promising this combination may be. The hope is to obtain a practical version, but for the moment we are far from that goal since the best implementations can handle numbers of 300 decimal digits, a far cry from the 10000 digits that ECPP can handle. Thus, since AKS is for the moment purely of theoretical interest, we do not describe it here but refer for instance to [MORAIN].

## 25.3 Factoring

Contrary to primality testing, factoring need not be rigorous in any way, since if an algorithm claims to output a factor of $n$ this can trivially be checked by division.

There are many more factoring methods than primality tests. This is due primarily to the fact that, contrary to primality, which is in a very satisfactory state, factoring is very difficult and almost no method is really superseded. For instance, even though Pollard's rho method, which we shall describe in a moment, is much slower than more sophisticated methods, it is still used to remove smallish prime factors from $n$, much faster than trial division. Thus a good factoring engine in a computer algebra system usually has a large combination of methods, and the implementer must choose carefully the sequence of methods to be used so as to optimize for speed.

### 25.3.1 Pollard's rho method

This method is remarkable by its simplicity (a few lines of code) and the fact that it is much more efficient than trial division for similar results. More precisely, while trial division requires (essentially) time $O(p)$ to find a prime factor $p$ of $n$, Pollard's rho method requires time approximately $O(p^{1/2})$, which of course makes a huge difference. It is based on the following idea. Let $f \in \mathbb{Z}[X]$ be a polynomial with integer coefficients. Consider the sequence $x_k$ defined by $x_{k+1} = f(x_k) \bmod n$, (in other words $x_{k+1}$ is obtained by reducing $f(x_k)$ modulo $n$), and $x_0$ being arbitrary. If $p$ is some (unknown) divisor of $n$ we also have $x_{k+1} \equiv f(x_k) \pmod{p}$, hence the sequence $x_k$ modulo $p$ can be considered as a sequence of iterates of the function $f$ in the ring $\mathbb{Z}/p\mathbb{Z}$ (note that $p$ need not be prime). Since this ring is finite, the sequence is necessarily ultimately periodic with some period $T$. Now it can be shown that if we average over all sequences of iterates of all functions $f$, the average of $T$ is $O(p^{1/2})$ (much more precise results are known, see for instance [COH 2000]). Thus if we assume that our polynomial $f$ behaves like a "random" map from $\mathbb{Z}/p\mathbb{Z}$ to itself, the period of our sequence $x_k$ modulo $p$ should be of the order $O(p^{1/2})$. This poses a number of questions which are all easily solved.

- *The choice of $f$.* Evidently linear polynomials will not be random. On the other hand there is no need to choose polynomials of degree greater or equal to 3. Thus we choose $f$ of degree 2, and to minimize computation time we simply choose $f(x) = x^2 + c$ for some $c \in \mathbb{Z}$. Evidently we must choose $c \neq 0$, but less evidently we must also choose $c \neq -2$ because $(y + 1/y)^2 - 2 = y^2 + 1/y^2$. All other choices seem to be acceptable.

- *The detection of the periodicity modulo $p$.* Since $p$ is unknown, the congruence $x_{k+T} \equiv x_k \pmod{p}$ cannot be tested as written. Instead, we simply test whether $\gcd(x_{k+T} - x_k, n) > 1$. If this is not the case then certainly $x_{k+T} \not\equiv x_k \pmod{p}$. On the other hand if $\gcd(x_{k+T} - x_k, n) > 1$ then we may not have discovered $p$ itself, but in any case we have found a divisor of $n$ strictly larger than 1, which with very high probability will be nontrivial, i.e., not equal to $n$ itself.

- *The detection of the condition $\gcd(x_{k+T} - x_k, n) > 1$, in other words cycle detection.* We have studied this subject in detail in Section 19.5.1. Recall that a simple solution is to test $\gcd(x_{2k} - x_k, n) > 1$, which will be true as soon as $k$ is simultaneously a multiple of $T$ and larger than the length of the nonperiodic part. Of course we may miss the first occurrence of the period, and make a computation that is a little longer than necessary, but this is largely compensated by the simplicity of the test. In practice one uses the slightly more elaborate but much faster cycle-detection method due to Brent and explained in Section 19.5.1.c, see [KNU 1997] and also [COH 2000].

- *The detection of the condition $\gcd(x_{2k} - x_k, n) > 1$ or of conditions of the same type.* We could of course store the $x_k$ as we go along. In fact it is much simpler, although wasteful in time, to have the two sequences $x_k$ and $y_k = x_{2k}$ considered as independent sequences, with the recursion $y_{k+1} = f(f(y_k)) \bmod n$. The storage requirements thus become $O(1)$. However, an easy but important improvement must also be made: since a gcd computation takes some time, it is much more efficient to group a large number of gcd computations together by multiplying modulo $n$ and doing a single gcd from time to time. For instance, setting $z_k = x_{2k} - x_k = y_k - x_k$, to test whether $z_0, \ldots, z_{20}$ are all coprime to $n$ (which they will be if $n$ is large), we simply compute recursively $Z = z_0 z_1 \cdots z_{20}$ modulo $n$ (which requires for each $z_k$ only one multiplication modulo $n$, not a gcd with $n$), and compute $\gcd(Z, n)$. If this is equal to 1, we have shown that all the $z_k$ are coprime to $n$. Otherwise (which happens only at the very end of the computation) we must backtrack our steps.

Because of its simplicity and importance, we give the algorithm explicitly, including Brent's cycle-detection method mentioned above adapted to our case.

### Algorithm 25.7 Pollard's rho with Brent's cycle detection

INPUT: A positive integer $n$.
OUTPUT: A nontrivial factor of $n$ or a failure message.

1. $y \leftarrow 2, x \leftarrow 2, x_1 \leftarrow 2, k \leftarrow 1, l \leftarrow 1, P \leftarrow 1$ and $c \leftarrow 0$
2. **while true do**
3.     $x \leftarrow x^2 + 1 \bmod n, P \leftarrow P \times (x_1 - x) \bmod n$ and $c \leftarrow c + 1$
4.     **if** $c = 20$ **then**
5.         **if** $\gcd(P, n) > 1$ **then break**
6.         $y \leftarrow x$ and $c \leftarrow 0$
7.     $k \leftarrow k - 1$
8.     **if** $k = 0$ **then**
9.         **if** $\gcd(P, n) > 1$ **then break**
10.        $x_1 \leftarrow x, k \leftarrow l$ and $l \leftarrow 2l$
11.        **for** $i = 1$ **to** $k$ **do**
12.            $x \leftarrow x^2 + 1 \bmod n$
13.            $y \leftarrow x$ and $c \leftarrow 0$
14. **repeat**
15.     $y \leftarrow y^2 + 1 \bmod n$ and $g \leftarrow \gcd(x_1 - y, n)$
16. **until** $g > 1$
17. **if** $g = n$ **then** the algorithm **return fail else return** $g$

**Example 25.8** If we choose $n = 49649$ the successive values of $x$ will be 2, 5, 26, 677, 11489, 30080, 3025, 15210, 29410, 12872, 9672, 8869, 15146, 22937, 25166, 4913, and we will discover that $\gcd(x_1 - x, n) = \gcd(15210 - 25166, 49649) = 131$ is a nontrivial factor of $n$ after 15 iterations (note that the last iteration leading to $x_k = 4913$ is unnecessary, but comes from the cycle-finding method).

## 25.3.2 Pollard's $p - 1$ method

This method is also due to Pollard [POL 1974]. By Fermat's little theorem $a^{p-1} \equiv 1 \pmod{p}$ for prime $p$ and any integer $a$ not divisible by $p$. It follows that if $k$ is any integer multiple of $p - 1$ we also have $a^k \equiv 1 \pmod{p}$. Furthermore, if $p \mid n$, then $p \mid \gcd(a^k - 1, n)$. Of course $p$ is unknown, but we can use this to find the prime factor $p$ of $n$ by computing $\gcd(a^k - 1, n)$ for an arbitrary integer $a \in [2, \ldots, n-1]$ coprime to $n$: this is checked by a gcd computation (and, even though unlikely, one might already find a factor of $n$). For this to work we have to find a multiple $k$ of $p - 1$, and the main problem is evidently that $p - 1$ is not known beforehand. Thus we hope that $n$ has a prime factor $p$ for which $p - 1$ consists of the product of some small primes. If this is the case we can choose for $k$ a product of many small prime powers. This number might be very large, but $a^k$ is only computed modulo $n$, and recursively without computing the number $k$ itself. If this does not work we choose $k$ having more prime power factors and try again, or, even better, we use a "second stage;" see [MON 1987] for implementation details.

The result is that a prime factor $p$ of $n$ can be found in time proportional to the largest prime factor of $p - 1$. There also exists a straightforward variant of this method using $p + 1$ instead of $p - 1$ [WIL 1982], which is the second cyclotomic polynomial. This is the reason that some cryptographic standards require RSA moduli consisting of products of primes $p$ for which $p - 1$ and $p + 1$ have large prime factors. However, there are similar methods using all cyclotomic polynomials: because of this and mainly because of the elliptic curve factoring method, which is only sensitive to the size of $p$ (see Section 25.3.3) and which has a much lower complexity, such precautions are superfluous [RISI 1997].

### 25.3.3 Factoring with elliptic curves

One can rephrase Pollard's $p - 1$ method in terms of $B$-smoothness: it attempts to find $p$ dividing $n$ for which $p - 1$ is $B$-smooth by computing $\gcd(a^k - 1, n)$ for $k$ equal to a product of the prime powers less than or equal to $B$. However, in almost all cases the largest prime in $p - 1$ is too large to make this practical.

The elliptic curve method (in short: ECM) due to H. W. Lenstra, Jr. [LEN 1987] is similar to Pollard's $p - 1$ method in the sense that it tries to exploit the smoothness of a group order. In Pollard's $p - 1$ method the groups are fixed: they are the groups $(\mathbb{Z}/p\mathbb{Z})^*$ for the primes $p$ dividing the number that we try to factor. The ECM randomizes the choice of the groups and of their orders. Such groups are, as the name of the method says, the groups of points of elliptic curves, therefore for their definitions and arithmetic we refer to the sections on elliptic curves mentioned above. However, unlike the presentation given in the sections mentioned previously, here it is necessary to consider curves defined over *rings* $\mathbb{Z}/n\mathbb{Z}$ having *zero divisors* corresponding to the factors of $n$, which "behave like fields" until some impossible division has to be performed, revealing a nontrivial factor of $n$. This is what Lenstra calls a "side-exit."

#### 25.3.3.a Elliptic curves over $\mathbb{Z}/n\mathbb{Z}$ with $n$ composite

Note that a complete theory of elliptic curves over $\mathbb{Z}/n\mathbb{Z}$ with $n$ composite exists. However, we do not need this theory here, but only some elementary remarks. Let $n$ be a composite number coprime to 6. An elliptic curve over $\mathbb{Z}/n\mathbb{Z}$ is defined by a pair $(a_4, a_6) \in (\mathbb{Z}/n\mathbb{Z})^2$ with $4a_4^3 + 27a_6^2 \in (\mathbb{Z}/n\mathbb{Z})^*$ as the "curve" with equation

$$E : y^2 = x^3 + a_4 x + a_6.$$

The set of points is, informally, the set $E(\mathbb{Z}/n\mathbb{Z})$ of pairs $(x_1, y_1) \in (\mathbb{Z}/n\mathbb{Z})^2$ which satisfy the above equation together with a point at infinity $P_\infty$. On that set we define, exactly as for the curves defined over fields, an addition that we call a *partial addition*. In other words, we define: $P \oplus P_\infty = P_\infty \oplus P = P$ for all $P$ on $E$. The *partial inverse* $-P$ of a point $P = (x_1, y_1)$ such that $P \oplus (-P) = P_\infty$ is the point $(x_1, -y_1)$. If $P, Q \in E(\mathbb{Z}/n\mathbb{Z}) \smallsetminus \{P_\infty\}$ with $Q = (x_2, y_2) \neq -P$, we set

$$\lambda = \begin{cases} (y_1 - y_2)/(x_1 - x_2) & \text{if } x_1 \neq x_2 \\ (3x_1^2 + a_4)/(2y_1) & \text{if } x_1 = x_2 \end{cases}$$

and $x_3 = \lambda^2 - (x_1 + x_2)$, and define $P \oplus Q = \big(x_3, \lambda(x_1 - x_3) - y_1\big)$. Since in $\mathbb{Z}/n\mathbb{Z}$ there exist zero divisors that may appear as denominators in these formulas, the composition law may *fail* to add points (this is why it is called a partial addition). Since in principle to invert a number $a$ modulo $n$ we have to compute its extended integer gcd with $n$, a number $a$ that corresponds to a nontrivial zero divisor in $\mathbb{Z}/n\mathbb{Z}$ will produce a nontrivial factor of $n$: one just has to put a "hook" in the modular inversion routine to detect this fact (instead of just halting the software).

Another, more mathematical way of considering the set of points of an elliptic curve over $\mathbb{Z}/n\mathbb{Z}$ for $n$ composite is by using projective coordinates. The set $E(\mathbb{Z}/n\mathbb{Z})$ is the disjoint union of three subsets: first the usual set of affine points having projective coordinates $(X : Y : Z)$ with $Z$ invertible modulo $n$, corresponding to the affine coordinates $(XZ^{-1}, YZ^{-1})$. Second, the point at infinity $P_\infty$ with projective coordinates $(0 : 1 : 0)$. Finally and most importantly for the ECM factoring method, the "interesting points" $(X : Y : Z)$ where $Z \neq 0$ and $Z$ is not invertible modulo $n$. If $n$ is prime there are no interesting points. On the other hand if $n$ is composite these points exist, and for any such point the gcd of $Z$ and $n$ is not equal to 1 or $n$ hence gives a nontrivial divisor of $n$. Thus the only purpose of the ECM method that we will see below is to perform computations that allow us after a while to land in the set of interesting points.

### 25.3.3.b  Elliptic curves over $\mathbb{Z}/n\mathbb{Z}$ modulo a prime

Let now $p$ be a prime dividing $n$. The elliptic curve $E$ over $\mathbb{Z}/n\mathbb{Z}$ and its set of points can be mapped onto an elliptic curve $E_p$ defined over $\mathbb{F}_p$ simply by reducing modulo $p$ the equation of $E$ and the coordinates of a point $P \in E(\mathbb{Z}/n\mathbb{Z})$, thus obtaining a point $P_p \in E_p(\mathbb{F}_p)$. Let $P, Q \in E(\mathbb{Z}/n\mathbb{Z})$. If the partial addition can successfully compute $P \oplus Q \in E(\mathbb{Z}/n\mathbb{Z})$, then it is trivial to verify that $P_p \oplus Q_p = (P \oplus Q)_p$. Note that the algorithm detects a prime factor $p$ of $n$ (by failure of inversion modulo $n$) if and only if it reaches a point $P \in E(\mathbb{Z}/n\mathbb{Z})$ for which the reduction modulo a factor $p$ yields $P_p = (P_\infty)_p$, the zero element in $E_p(\mathbb{F}_p)$, but $P \neq P_\infty$ in $E(\mathbb{Z}/n\mathbb{Z})$.

### 25.3.3.c  The algorithm

Recall also that by Hasse's theorem (cf. Corollary 5.76) we have $|E(\mathbb{F}_p)| = p + 1 - t$ for some integer $t$ with $|t| \leqslant 2\sqrt{p}$. If $E$ is randomly chosen among the elliptic curves over $\mathbb{F}_p$, it is reasonable to expect that $|E(\mathbb{F}_p)|$ behaves like a "random number close to $p+1$." Old results of Birch suggested that this is indeed plausible [BIR 1968] and we also have the following theorem of Lenstra:

**Theorem 25.9** [LEN 1987] *There exist effectively computable positive constants $c_1$ and $c_2$ such that for each prime $p \geqslant 5$ and for any subset $S$ of integers in the interval $[p + 1 - \sqrt{p}, p + 1 + \sqrt{p}]$, the probability $r_S$ that a random pair $(a_4, a_6) \in \mathbb{F}_p \times \mathbb{F}_p$ determines an elliptic curve $E : y^2 = x^3 + a_4 x + a_6$ with $|E(\mathbb{F}_p)| \in S$ is bounded as follows:*

$$c_1 \times \frac{|S| - 2}{2\lfloor \sqrt{p} \rfloor + 1} (\ln p)^{-1} \leqslant r_S \leqslant c_2 \times \frac{|S|}{2\lfloor \sqrt{p} \rfloor + 1} \ln p \, (\ln \ln p)^2.$$

Similar results for the case of elliptic curves over $\mathbb{F}_{2^m}$ can be deduced from the work of Waterhouse [WAT 1969] and Schoof [SCH 1987].

Furthermore, there are results about the density of smooth integers in intervals $[1, M]$ but not about those in $[M - \sqrt{M}, M + \sqrt{M}]$. Therefore the runtime arguments are based on the heuristic assumption that these densities are equal: this has been confirmed by experiments and by the fact that the ECM so far behaves as predicted. (In passing, we mention that Croot [CRO 2003] has some partial results on the density of smooth numbers in intervals $[M, M + H]$ where $H = c\sqrt{(M)}$ and the constant $c$ depends only on a lower bound on $x$.)

Pick $a_4, a_6 \in \mathbb{Z}/n\mathbb{Z}$ at random with $4a_4^3 + 27a_6^2$ coprime to $n$ (otherwise we might have been so lucky to already have found a nontrivial factor of $n$). With such $a_4$ and $a_6$, we define an elliptic curve $E$ over $\mathbb{Z}/n\mathbb{Z}$. In practice it is sufficient for $a_4$ to be a single precision random integer, which reduces the cost of operations on $E$. Moreover, there is no need to check if $\gcd(4a_4^3 + 27a_6^2, n) \neq 1$, as this is extremely unlikely unless $n$ has a small prime factor.

Thus an elliptic curve $E_p = E \bmod p$ over $\mathbb{F}_p$ is defined for each prime factor $p$ of $n$, and $|E_p(\mathbb{F}_p)|$ behaves (heuristically) as a random integer close to $p + 1$. Based on Lemma 20.7 with

$r = 1$, $\alpha = 1$, $s = 1/2$, and $\beta = \sqrt{1/2}$ we can assume that $|E_p(\mathbb{F}_p)| = L_p[1,1]$ is $L_p[1/2, \sqrt{1/2}]$-smooth with probability $L_p[1/2, -\sqrt{1/2}]$. Put $\xi = L_p[1/2, \sqrt{1/2}]$. Thus, for a fixed $p$, once every $\xi$ random elliptic curves over $\mathbb{Z}/n\mathbb{Z}$ one expects to find a curve $E$ for which $|E_p(\mathbb{F}_p)|$ is $\xi$-smooth: Assume that $E$ is such and let $k$ be the product of the primes $\leqslant \xi$ and some of their powers. Further, we let $P$ be a random element of $E(\mathbb{Z}/n\mathbb{Z})$. Note that since $n$ is composite, it is not easy to compute a square root modulo $n$, hence to find a point $P$, so in practice we proceed in reverse by first choosing $P = (x_1, y_1)$ at random modulo $n$, and setting $a_6 = y_1^2 - (x_1^3 + a_4 x_1)$. Let $k$ be a not too large product of enough prime powers so that the order of $P_p$ divides $k$, and attempt to compute $[k]P$ in $E(\mathbb{Z}/n\mathbb{Z})$ using the partial addition.

Suppose that for such a $k$ the computation of $[k]P$ is successful, i.e., it did not fail because of noninvertible denominators, which means that we find no nontrivial factors of $n$. Then $[k]P$ equals some $R \in E(\mathbb{Z}/n\mathbb{Z})$. Its reduction modulo $p$, i.e., $R_p \in E_p(\mathbb{F}_p)$ would have been also obtained by computing the elliptic curve scalar product $[k]P_p$ in $E_p(\mathbb{F}_p)$. Being $k$ a multiple of the order of $P_p$ we have $R_p = (P_\infty)_p \in E_p(\mathbb{F}_p)$. But $R_p = (P_\infty)_p$ if and only if $R = P_\infty$ whence $R_q = (P_\infty)_q$ for *any* prime $q$ dividing $n$. It follows that $k$ must be a multiple of the order of $P$ when taken modulo any prime dividing $n$. But $E$ was a random curve such that $|E_p(\mathbb{F}_p)|$ was $\xi$-smooth: It is extremely unlikely that for such a curve this would happen for all prime factors of $n$ simultaneously.

Thus it is much more likely that the attempt to compute $R$ fails: the partial addition breaks down during the computation of the scalar multiplication, producing a nontrivial factor of $n$ that is a multiple of $p$.

Since one in every $\xi$ elliptic curves over $\mathbb{Z}/n\mathbb{Z}$ can be expected to generate a factor, the total expected runtime is $\xi$ times the time required to compute $[k]P$ with $k$ as above, which in turn requires $\xi$ partial additions, that is a cost $O(M(\lg n)\xi)$ where $M(t)$ is the complexity of multiplications of $t$-bit numbers. The total heuristically expected running time is $O(\xi M(\lg n)\xi)$ equal to $O(M(\lg n)L_p[1/2, \sqrt{2}])$. This can be improved further using better scalar multiplication algorithms; see Chapter 9. In a nutshell: if it must fail, it will fail anyway. So it is better to do things as fast as possible to save time on all curves where it does not fail.

It follows that using the ECM small factors can be found faster than large factors. For $p \approx \sqrt{n}$, the worst case, the expected runtime becomes $L_n[1/2, 1]$. For composites without known properties, and, in particular, a smallest factor of unknown size, one generally starts off with a relatively small $k$ aimed at finding small factors. This $k$ is gradually increased for each new attempt, until a factor is found or until the factoring attempt is aborted. See [BOLE 1995] for implementation details.

The ECM can be easily turned into a parallel algorithm: any number of attempts can be run independently on, say $r$ processors in parallel, until one of them succeeds in finding a nontrivial factor. The expected speedup is of a factor $r$.

**Remark 25.10** In order to speed up the computation of the partial additions, one might want to use advanced implementations of the modular arithmetic. Suppose one wants to adopt the Montgomery representation, see Section 10.4, which works even for nonprime odd moduli. In this case, if the number $a$ is a multiple of a prime factor $p$ of the modulus $n$, also the Montgomery representation of $a$ will be an integer divisible by $p$, and the inversion algorithm will discover it. For the coward: since the quantities involved are rather large, one is also still safe from the performance point of view using the Barrett or the Quisquater reduction methods, cf. Section 10.4 again.

### 25.3.3.d  Using Montgomery's trick

The ECM requires one inversion modulo $n$ per partial addition or doubling, which takes approximately the same time as computing a gcd with $n$. This slows operations down, and one wants to avoid it. One possible improvement is the use of projective coordinates. At the end of the computation of $[k]P$ the gcd of the coordinates might yield a factor of $n$. The use of projective coordinates

## § 25.3 Factoring

requires about 12 multiplications (13 in the case of a doubling) and no inversions. There is, however, a usually much faster approach that makes use of the trick due to Montgomery described in Section 11.1.3.c, Algorithm 11.15. How can it be deployed to improve the ECM? The usual method would have required $i$ inversions (for instance: invocations of an extended Euclid algorithm) to do the job. But this method requires only one inversion and $3(i - 1)$ multiplications modulo $n$. Therefore it is superior as soon as one inversion is slower than three multiplications, and this is almost always the case. Let $\tau$ be the ratio between the timings of an inversion and of a multiplication, both modulo $n$. The computation of a partial addition requires the computation of $(x_1 - x_2)^{-1}$ if the points $P = (x_1, y_1)$ and $Q = (x_2, y_2)$ have distinct $x$-coordinates and of $(2y_1)^{-1}$ if $P = Q$, plus two multiplications modulo $n$ and some additions and subtractions. So if we work with $m$ curves in parallel — on a single processor in order to avoid slow intercommunication overheads — we get that the cost of a partial addition corresponds to $6 + \tau/m$ multiplications ($7 + \tau/m$ for a doubling). If $m$ is large enough ($m = 50$) this improvement can be remarkable (from a cost $6 + \tau$ to about 6). Note however that working with several curves simultaneously on one single processor means that the processor has to work with much more data, and the number of cache misses could lead to a performance penalty that might overcome the advantages. The quantity $m$ must be chosen with care. Moreover, it seems to be rather intricate to combine this idea with sophisticated scalar multiplication algorithms; see Chapter 9. However the combination with fixed-width non-sliding window methods seems to be relatively straightforward.

Remarkable successes of the ECM were the factorizations of the tenth and eleventh Fermat numbers.

A variant of the ECM suitable for the computation of discrete logarithms has never been published.

### 25.3.4 Fermat–Morrison–Brillhart approach

This approach is based on Fermat's factoring method of solving a congruence of squares modulo $n$: try to find integers $x$ and $y$ such that

$$x^2 \equiv y^2 \pmod{n} \text{ and } \gcd(xy, n) = 1. \tag{25.1}$$

It then follows that $n$ divides $x^2 - y^2 = (x - y)(x + y)$ so that $n = \gcd(n, x - y)(n/\gcd(n, x - y))$, which may yield a nontrivial factorization of $n$ with a probability of $1 - 2^{1-k}$ where $k$ is the number of distinct odd prime factors in $n$. Fermat's method to find $x$ and $y$ consists in trying $x = \lfloor\sqrt{n}\rfloor + 1, \lfloor\sqrt{n}\rfloor + 2, \ldots$ in succession, until an $x$ is found such that $x^2 - n$ is a perfect square. In this form, however, the method has the same worst case running time as trial division, but is faster if $n$ is a product of two primes that are close to each other.

**Remark 25.11** These methods do not work if $n$ is a prime power but this is checked easily. If $n = p^\ell$ where $p$ is prime and $\ell \geqslant 1$, then $n - 1 = p^\ell - 1$ is divisible by $p - 1$ and, by Fermat's little theorem, if $\gcd(a, n) = 1$, then $a^{n-1} \equiv 1 \pmod{p}$. Hence $p \mid \gcd(a^{n-1} - 1, n)$. Consequently, if we find an $a$ such that $\gcd(a^n - a, n) = 1$, then $n$ cannot be a prime power.

Morrison and Brillhart [MoBr 1975] proposed a faster way to find $x$ and $y$. Their approach consists in constructing $x$ and $y$ using identities modulo $n$ which are, supposedly, easier to solve. There are two phases. First, a set of relations is sought and, second, the relations are used to construct a solution to $x^2 \equiv y^2 \pmod{n}$. This is, essentially, the same scheme also used in the index calculus algorithms.

**Algorithm 25.12** Fermat–Morrison–Brillhart factoring algorithm

INPUT: A rational integer $n$.
OUTPUT: A nontrivial factor of $n$.

1. **Choose a factor base**
   Given a smoothness bound $B$, let the *factor base* $\mathcal{P}$ be the set of primes $\leq B$. The cardinality of $\mathcal{P}$ is $\pi(B)$.

2. **Collecting relations**
   Collect more than $|\mathcal{P}|$ integers $v$ such that $v^2 \bmod n$ is $B$-smooth:
   $$v^2 \bmod n = \left(\prod_{p \in \mathcal{P}} p^{e_{v,p}}\right).$$
   These congruences are called the *relations*.
   Let $\mathcal{V}$ be the resulting set of relations of cardinality $|\mathcal{V}| > |\mathcal{P}|$.

3. **Linear algebra**
   Each $v \in \mathcal{V}$ gives rise to a $|\mathcal{P}|$-dimensional vector $(e_{v,p})_{p \in \mathcal{P}}$. Since $|\mathcal{V}| > |\mathcal{P}|$, the vectors $\{(e_{v,p})_{p \in \mathcal{P}} : v \in \mathcal{V}\}$ are linearly dependent. This implies that there exist at least $|\mathcal{V}| - |\mathcal{P}|$ linearly independent subsets $\mathcal{S}$ of $\mathcal{V}$ for which $\sum_{v \in S} e_{v,p} = 2s_p$ is even for all $p \in \mathcal{P}$. These subsets $\mathcal{S}$ can, in principle, be found using Gaussian elimination modulo 2 on the matrix having the vectors $(e_{v,p})_{p \in \mathcal{P}}$ as rows.

4. **Compute a solution**
   For any subset $\mathcal{S}$ of $\mathcal{V}$ given by the linear algebra stage and the corresponding integer vector $(s_p)_{p \in \mathcal{P}}$, the integers
   $$x = \prod_{v \in S} v \bmod n \text{ and } y = \prod_{p \in \mathcal{P}} p^{s_p} \bmod n$$
   form a solution to the congruence $x^2 \equiv y^2 \pmod{n}$. Each of the $|\mathcal{V}| - |\mathcal{P}|$ independent subsets thus leads to an independent chance of at least 50% to produce a nontrivial factor of $n$.

We did not mention how to collect the relations in Line 2. Morrison and Brillhart found them using continued fractions. Dixon proposed a simpler (but slower) method: pick $v < n$ at random and keep the ones for which $v^2 \bmod n$ is $B$-smooth.

We do not make a complexity analysis here. In order to verify the candidates $v$ for $B$-smoothness in the relation collection stage the elliptic curve method can be used. Further, note that at most $\lg n$ entries are nonzero for each vector $(e_{v,p})_{p \in \mathcal{P}}$ so that sparse matrix techniques can be deployed in the linear algebra stage. This leads to a *provable* $L_n[1/2, \sqrt{2}]$ runtime for Dixon's algorithm. The Morrison-Brillhart method using continued fraction expansion performs somewhat better, but the running time is only heuristic.

### 25.3.4.a Continued fraction method

The *continued fraction method* (abbreviated CFRAC) is the method used originally by Morrison and Brillhart to factor the seventh Fermat number (39 digits) in 1970.

CFRAC looks for congruences $x^2 \equiv r \pmod{n}$ with small $r$, namely $r = O(\sqrt{n})$. For each such congruence it finds, it attempts to factor $r$ using the factor base. Where $r$ is smooth, the congruence is saved so it can be multiplied by other such congruences to form squares on both sides.

## § 25.3 Factoring

If $n$ is a perfect square, then it is easy to factor it, but we are excluding this case anyway in view of the remark opening the section. So $\sqrt{n}$ is irrational.

There exist infinitely many rational approximations $p/q$ of $\sqrt{n}$ such that

$$\left|\frac{p}{q} - \sqrt{n}\right| < \frac{1}{q^2},$$

which are obtained as the partial fractions (convergents) of the (infinite) continued fraction expansion for $\sqrt{n}$. If $p/q$ is any such an approximation and $\epsilon$ is such that $p/q = \sqrt{n} + \epsilon/q^2$ then

$$p^2 - nq^2 = (q\sqrt{n} + \epsilon/q)^2 - nq^2 = 2\epsilon\sqrt{n} + \epsilon^2/q^2.$$

Since $|\epsilon| < 1$, we get $|p^2 - nq^2| < 2\sqrt{n} + 1/q^2$, whence all such values of $p^2 - nq^2$ are $O(\sqrt{n})$ as desired.

**Remarks 25.13**

(i) Observe that the numerators and denominators $p$ and $q$ are given by recurrence formulas involving the coefficients of the continued fraction of $\sqrt{n}$. Since these continued fractions are periodic, one can compute $p$ and $q$ directly modulo $n$ instead of computing with unbounded integers. This trick was suggested by Montgomery.

(ii) Note that the factor base is actually only half as big as it could be: if $p' \mid (p^2 - nq^2)$ then $n$ must be a quadratic residue modulo $p'$ unless $p' \mid n$. Since $n$ is assumed not to be a perfect square, asymptotically only half of the primes have $n$ as a quadratic residue. Therefore, if $B$ is the smoothness bound, the cardinality of the factor base is $\approx \pi(B)/2$.

### 25.3.4.b The quadratic sieve

Much of the time in CFRAC is spent testing the residues $p^2 - nq^2$ for $B$-smoothness: factoring them essentially by brute force, but Pollard's $p-1$, rho methods and ECM might be also employed. The quadratic sieve, proposed by Pomerance [POM 1983, POM 1985], eliminates this burden almost completely. It also tries to find *small* quadratic residues: those found by the quadratic sieve are slightly larger than those found by CFRAC, but can, in practice, be tested much faster.

The method is based on the following observation: If $f(X) \in \mathbb{Z}[X]$ is a polynomial with integer coefficients and $p$ is a prime number, then $f(i) \equiv f(i+p) \pmod{p}$ for all $i$.

Suppose now that we want to check for smoothness the values of a polynomial $f$ at several consecutive values of $i$, say for $i \in [0, \ldots, L-1]$.

We present a first idea for doing it. Start by building a table of values of $f(i)$ with $0 \leq i < L$. For each prime $p$ in the factor base, and for each $i$ such that $f(i) \equiv 0 \pmod{p}$, we replace our tabulated value of $f(i)$ by $f(i)/p^{e(i)}$ where $p^{e(i)}$ is the largest power of $p$ dividing $f(i)$. This is done by computing the roots $r$ of $f(X)$ modulo $p$ that satisfy $0 \leq r < p$ and then dividing $f(r + kp)$ by the largest power of $p$ in it for all such roots and for all integers $k \geq 0$ such that $r + kp < L$. After processing all primes in our factor base, if any table entry is $\pm 1$, then the corresponding $f(i)$ was smooth. There are however too many evaluations and divisions. An alternative approach is the following

---

**Algorithm 25.14** Sieve method

---

INPUT: A polynomial $f(X) \in \mathbb{Z}[X]$, an integer $L$ and a smoothness bound $B$.
OUTPUT: A list $\mathcal{L}$ of values $i$ with $0 \leq i < L$ for which $f(i)$ is $B$-smooth.

1. **for** $i = 0$ **to** $L - 1$ **do** $s_i \leftarrow 0$         [the $s_i$ correspond to the $f(i)$]
2. **for** all $p \leq B$ **do**         [i.e., in the factor base]

3.  **for** all roots $r$ of $f(x)$ modulo $p$ with $0 \leqslant r < p$ **do**
4.     **for** all $k \geqslant 0$ with $r + kp < L$ **do**
5.        $s_{r+kp} \leftarrow s_{r+kp} + \ln p$
6.  **for** $i = 0$ **to** $L - 1$ **do**
7.     **if** $s_i$ is "close" to $\ln f(i)$ **then if** $f(i)$ is $B$-smooth **then** $\mathcal{L} \leftarrow \mathcal{L} \cup \{i\}$.
8.  **return** $\mathcal{L}$

In practice the sieve values $s_i$ and the values $\ln p$ are rounded, depending also on the base chosen (which might be 2 or a power of 2) and represented by small integers. We do not enter into the details how to check whether $s_i$ is close to $\ln f(i)$ but one can usually approximate it fairly well (for example by a linear function) for the range of values sieved. The values $s_i$ after Line 5 are called *residual logarithms*.

Let $v(i) = i + \lfloor \sqrt{n} \rfloor$ for small $i$, then $v(i)^2 \bmod n = (i + \lfloor \sqrt{n} \rfloor)^2 - n \approx 2i\sqrt{n}$. The above sieve method is applied to $f(i) := v(i)^2 - n$.

With $B = L_n[1/2, 1/2]$ and assuming that $(v(i)^2 \bmod n)$ behaves as a random number close to $\sqrt{n} = L_n[1, 1/2]$, based on Lemma 20.7 we might infer that it is $B$-smooth with probability $L_n[1/2, -1/2]$. This assumption is obviously incorrect: if an odd prime $p$ divides $(v(i)^2 \bmod n)$ but not $n$, then $(i + \lfloor \sqrt{n} \rfloor)^2 \equiv n \pmod{p}$ so that $n$ is a quadratic residue modulo $p$ (history repeats itself), so that the cardinality of the effective factor base is $\approx \pi(B)/2$. On the other hand, for each prime $p$ that might occur, one may expect two roots of $f(X)$ modulo $p$. The result is that the smoothness probabilities are very close to what one might predict naïvely.

Let then $\mathcal{P}$ be the chosen factor base. Since we have to find more than $|\mathcal{P}| = L_n[1/2, 1/2]$ relations,

$$\frac{L_n[1/2, 1/2]}{L_n[1/2, -1/2]} = L_n[1/2, 1] = O\left(e^{(1+o(1))\sqrt{\ln n \ln \ln n}}\right)$$

different $i$'s have to be sieved. This justifies the assumption that $i$ is small. Pomerance's analysis [POM 1983] shows that the total (heuristic) expected time of the quadratic sieve is $L_n[1/2, 1]$. See also Lenstra's survey in [LEN 2002].

Several ideas can be used to improve the performance of this algorithm, which will be explored in the next sections.

### 25.3.4.c Multiple polynomials

We have seen that $L_n[1/2, 1]$ different values $f(i)$ must be sieved. The effect of the larger $i$'s is noticeable because as $i$ gets larger, the proportion of $B$-smooth numbers decreases. Davis and Holdridge suggested the use of more polynomials, and a similar, more practical solution was proposed by Montgomery and other authors. We now consider these proposals.

Montgomery [MON 1994] proposes quadratic polynomials of the form $f(X) = a^2 X^2 + bX + c$ where $b^2 - 4a^2c = kn$ with $k = 1$ if $n \equiv 1 \pmod{4}$ and $k = 4$ if $n \equiv 3 \pmod{4}$. Observe that

$$f(X) = \left(aX + \frac{b}{2a}\right)^2 - \frac{b^2 - 4a^2c}{4a^2} \equiv \left(aX + \frac{b}{2a}\right)^2 \pmod{n}.$$

so if $a$, $b$, and $c$ are small enough these polynomials can be expected to be equally useful as $v(x)^2$. In order to make the values of $f(X)$ as small as possible on a sieving interval $\mathcal{I}$, say of length $2m$, we want first to center the interval at the minimum of the function $f(X)$, which is attained for $X = \xi := -b/(2a^2)$. Thus $\mathcal{I} = [\xi - m, \xi + m]$. We ask that $|f(\xi)| \approx |f(\xi + m)|$ subject to the condition $b^2 - 4a^2c = kn$. This leads us to discard the case when $f(\xi)$ and $f(\xi+m)$ have the same sign and thus in the case $f(\xi) < 0 < f(\xi+m)$ we pick

$$a^2 \approx \frac{\sqrt{kn/2}}{m}, b \approx 0 \text{ and } c \approx -m\sqrt{kn/8} < 0.$$

The largest polynomial value is about $|c|$, which is at most $m\sqrt{n/2}$. To sieve $L$ values then one can use $L/(2m)$ different polynomials, sieving $2m$ values per polynomial. Therefore the largest residue is about $O(m\sqrt{n})$ instead of $O(L\sqrt{n})$ as in the basic algorithm.

How to select $a$, $b$, and $c$? First, an odd prime value for $a$ (with $a \approx (kn/2)^{1/4}m^{-1/2}$) is selected such that $kn$ is a quadratic residue modulo $a$. Second, the equation $b_0^2 \equiv kn \pmod{a}$ is solved for $b_0$ and the equation $(b_0 + \ell a)^2 \equiv kn \pmod{a}^2$ for $\ell$. Third, we set $b = b_0 + \ell a$ or $b = b_0 + \ell a - a^2$, whichever has the same parity as $kn$. Last, put $c = (b^2 - kn)/(4a^2)$, which is by construction an integer and $b^2 - 4ac^2 = kn$.

Other authors (see, for example, Cohen [COH 2000]) suggest to take polynomials of the form $f(X) = aX^2 + 2bX + c$ with $a > 0, \delta := b^2 - ac > 0$ satisfying $n \mid \delta$. These polynomials are also equally useful as $v(x)^2$ since

$$af(X) = (aX + b)^2 - (b^2 - ac) \equiv (aX + b)^2 \pmod{n}.$$

As soon as the values of one get too large for $i$ large, a new polynomial is selected and sieving resumes again with $i = 0$.

By arguments similar to those for Montgomery's polynomials, we obtain that: $a$ could be chosen among the prime numbers close to $\sqrt{2n}/m$ with $\left(\frac{n}{a}\right) = 1$. Then $b$ should satisfy $b^2 \equiv n \pmod{a}$. The easiest way to do it works for about half of the primes: if $a \equiv 3 \pmod{4}$, pick $b = n^{(a+1)/4} \bmod a$. See Section 11.1.5 for more general square root algorithms. Finally, set $c = (b^2 - n)/a$. We have $\max_{i \in \mathcal{I}}\{|f(i)|\} \approx m\sqrt{n/2}$ as with Montgomery's choice.

The resulting method is called the multiple polynomial quadratic sieve, or MPQS. There are several ways this basic algorithm can be improved upon: we present here some of the most important and general ideas.

### 25.3.4.d  Self-initializing polynomials

The above description of the MPQS may lead us to think that we can try changing polynomials as often as possible in order to make the residues as small as possible. Apart from the fact that this might lead to collisions (i.e., small smooth residues found twice), this means, even with moderate switching, that for each polynomial all roots modulo all primes $\leqslant B$ have to be computed. Another very time-consuming part is the inversion of the leading coefficient $a^2$ (in Montgomery's version) or $a$ (in Cohen's version) modulo each prime in the factor base.

Self-initialization works choosing $a$ not prime, but a product of a few $t$ (about 10) medium-sized prime numbers $p$ with $\left(\frac{n}{p}\right) = 1$. The number of possible $b$, hence the number of possible polynomials with leading term $a$, is equal to the number of solutions of $b_0^2 \equiv kn \pmod{a}$ or $b^2 \equiv n \pmod{a}$ in Cohen's case (in Montgomery's case this is similar), which is equal to $2^{t-1}$. This procedure then leads to a speedup of a factor about $2^{t-1}$ in the most time-consuming part of the root computation during the sieve initialization phase. Faster root computation allows one to benefit by changing polynomials more frequently, hence sieving on smaller residues that are more likely to be smooth. The net result is about a factor of 2 speedup.

### 25.3.4.e  Large prime variations

The sieving procedure of Algorithm 25.14 looks for values of $i$ such that $f(i)$ is $B$-smooth. It is easily modified to also find values of $i$ for which $f(i)$ is a $B$-smooth number times one or more primes not much larger than $B$, by lowering the threshold used when inspecting logarithms after sieving. The extra prime in the factorization of $f(i)$ is called a *large prime*. If one finds two values of $i$ for which $f(i)$ has the same large prime, then the corresponding congruences, which are called *partial relations*, can be multiplied (or divided) together to obtain a new single relation, called a *full relation*, that can be used in the rest of the algorithm.

Combined partial relations make the matrix somewhat less sparse. It is a consequence of the birthday paradox that matches between large primes occur often enough to make this approach worthwhile: indeed, one large prime more than halves the sieving time. Two large primes again halve the sieving time and A. K. Lenstra claims that a third large prime has again the same effect, see [LEN 2002, § 4.2.5].

Large primes can be interpreted as a cheap way to extend the size of the factor base — cheap because they are not sieved with — but we cannot allow too many large primes. One reason is the decrease in sparseness of the matrix in the linear algebra stage. The other reason lies in the way they are recognized. After the sieving process, all powers of factor base elements are found. Therefore, allowing one large prime requires one primality test per relation after the sieving stage, until one is found, and then an additional comparison for all subsequent steps (in practice this primality test is not needed for a single large prime since those that are kept are less than $B^2$ and the residue is not divisible by any primes up to $B$, so it must be prime). Allowing a remainder of the form $\ell_1 \ell_2$ where $\ell_1$ and $\ell_2$ are large primes already requires the use of a fast special purpose factoring method to extract $\ell_1$ and $\ell_2$, unless one wants to find $\ell_1$ and $\ell_2$ first in separate occasions and only then checking if they appear – reducing drastically the efficiency of the whole idea. This is practical as long as the large primes fit in a computer word, so our special purpose factoring method has to split integers of size $2^{64}$. The use of Pollard's rho method and Lenstra's ECM have been reported for this purpose in the literature. The combination of relations having more than one large prime is an intricate task. The basic case of two large primes was done in [LEMA 1994] and later extended to four and applied to the number field sieve (see Section 25.3.4.g) and to the computation of discrete logarithms.

This variation is compatible with the use of multiple polynomials.

### 25.3.4.f  Small prime variation

During the sieving process (Algorithm 25.14) for collecting relations, the small primes and prime powers take quite a long time to process since about $1/p$ numbers are divisible by $p$. In addition, their contribution to the logarithms is smallest. So we do not sieve at all with prime powers less than a carefully chosen threshold, say 100. This makes it necessary to keep numbers whose residual logarithm is further away from zero than usual, but experience shows that this makes little difference: the important thing is to avoid ignoring some smooth numbers, at the expense of having to check and reject a few more that are not smooth. As a bonus, this might indirectly help in finding some large primes.

The small and large prime variations can also be used in a parallelized implementation of the MPQS.

### 25.3.4.g  The number field sieve

The number field sieve (NFS) is the fastest general purpose factoring algorithm that has been published until now.

It is based on an idea by Pollard in 1988 to factor numbers of the form $x^3 + k$ for small $k$. This method was quickly generalized to a factoring method for numbers of the form $x^d + k$, a method that is currently referred to as the special number field sieve (SNFS) to distinguish it from the method used for factoring general integers, called the general number field sieve (GNFS). It proved to be practical by factoring the ninth Fermat number $F_9 = 2^{2^9} + 1$. It was the first factoring algorithm with runtime substantially below $L_n[1/2, c]$ (for constant $c$) making cryptosystems based on the assumed hardness of the integer factorization problem at once less secure than thought. The heuristic expected runtime of the NFS is $L_n[1/3, \eta]$ where $\eta = (32/9)^{1/3}$ for the SFNS and $\eta = (64/9)^{1/3}$ for the GNFS. Coppersmith [COP 1993] has slightly improved the complexity of the GNFS to $\eta = \frac{1}{3}(92 + 26\sqrt{13})^{1/3}$ at least theoretically.

The NFS follows the Morrison–Brillhart approach of collecting relations involving smooth quantities, followed by linear algebra to find congruences among squares. Its better complexity compared with previous smoothness based methods is due to the fact that the numbers that are tested for smoothness are of order $n^{o(1)}$ for $n \to \infty$, as opposed to $n^c$ for some constant $c$ for the older methods. There are several reasons for the practicality of NFS. One is that recomputing roots is very fast. Another is that polynomial selection allows a lot more freedom and can result in great speedups. It also allows relatively easy use of more than two large primes, so that comparably small factor bases can be used during sieving [DOLE 1995], although this is not always the case (for instance, the GNFS RSA-130 factor base was 7 times the size of of the MPQS RSA-129 factor base). Initially there was skepticism as to whether this method would be practical, but those doubts have been dispelled by a series of remarkable improvements.

The *content* of a polynomial with integer coefficients is the gcd of its coefficients.

Suppose $n$ is a composite integer to be factored. It is easy to check whether $n$ is a prime number or a prime power (using primality tests), so we assume that $n$ is neither.

The algorithm has four main phases:

---

**Algorithm 25.15** The number field sieve

---

INPUT: A rational integer $n$.
OUTPUT: A nontrivial factor of $n$.

---

1. **Polynomial selection**
   Select two irreducible univariate polynomials $f(X)$ and $g(X)$ with content equal to one, which have at least one common root $m$ modulo $n$ – i.e., $f(m) \equiv g(m) \equiv 0 \pmod{n}$ – but have no common factors over $\mathbb{Q}$.
   Let $\alpha$ and $\beta$ denote complex roots of $f$ and $g$ respectively.

2. **Sieving**
   Find pairs $(a_i, b_i)$ with $\gcd(a_i, b_i) = 1$ and such that
   $$F(a_i, b_i) := b_i^{\deg(f)} f(a_i/b_i) \text{ and } G(a_i, b_i) := b_i^{\deg(g)} g(a_i/b_i)$$
   are both smooth with respect to a chosen factor base.

3. **Linear algebra**
   Use linear algebra to find a set $\mathcal{S}$ of indices such that the two products
   $$\prod_{s \in \mathcal{S}} (a_s - b_s \alpha) \text{ and } \prod_{s \in \mathcal{S}} (a_s - b_s \beta)$$
   are both squares of products of prime ideals in $\mathbb{Q}[\alpha]$ and $\mathbb{Q}[\beta]$, respectively.

4. **Square root**
   Using the set $\mathcal{S}$, try to find algebraic numbers $\gamma \in \mathbb{Q}(\alpha)$ and $\delta \in \mathbb{Q}(\beta)$ with
   $$\gamma^2 = \prod_{s \in \mathcal{S}} (a_s - b_s \alpha) \text{ and } \delta^2 = \prod_{s \in \mathcal{S}} (a_s - b_s \beta).$$
   Consider now the natural homomorphisms
   $$\phi_\alpha : \mathbb{Q}(\alpha) \to \mathbb{Z}/n\mathbb{Z}, \ \alpha \mapsto m \text{ and } \phi_\beta : \mathbb{Q}(\beta) \to \mathbb{Z}/n\mathbb{Z}, \ \beta \mapsto m,$$

where $m$ is the common root modulo $n$ of $f$ and $g$. The congruence

$$\phi_\alpha(\gamma)^2 = \phi_\alpha(\gamma^2) = \phi_\alpha\left(\prod_{s \in S}(a_s - b_s\alpha)\right) \equiv \prod_{s \in S}(a_s - b_s m) \equiv \phi_\beta(\delta)^2 \pmod{n}$$

has the form (25.1): the two sides will be coprime to $n$ if none of $F(a_i, b_i)$ and $G(a_i, b_i)$ for $i \in S$ share a factor with n.

**Remarks 25.16**

(i) *On the selection of polynomials.* A lot of research has been done in order to determine what a "good" polynomial is in terms of root properties and coefficient sizes. For a description of clever algorithms for finding such polynomials, see Murphy's PhD. Thesis on R. Brent's web page [BRENT]. On the other hand, when $n$ has a very special form the choice of polynomial is much simpler. For the 162-digit Cunningham number $(12^{151} - 1)/11$, which had been factored in 1994 by a group led by Montgomery, the polynomials $12X^5 - 1$ and $X - 12^{30}$ were chosen. For the ninth Fermat number, the polynomials $X - 2^{103}$ and $X^5 - 8$ were chosen, with common root $m = 2^{103}$.

In order to make arithmetic faster, one would want to pick $f$ and $g$ so that at least one has "small" integer coefficients.

One common method is the base $m$ method: for a small number $d$ (usually 4 or 5) put $m = \lfloor n^{1/d} \rfloor$ and define $f$ as the monic degree $d$ polynomial with the coefficients of the base $m$ expansion of $n$. Now $f(X)$ and $g(X) = X - m$ together with $m$ satisfy the requirements, unless $f$ is reducible, in which case it almost surely can yield a nontrivial factor of $n$.

(ii) *On the sieving phase.* In the classical sieving method we begin by choosing upper bounds for $a$ and $b$. Then we start sieving for $a$ with $b = 1$ fixed and, when finished with $a$, increase $b$ until we reach its bound. The estimates for the smoothness are done by approximated logarithms and the values $F(a, b)$, $G(a, b)$ computed from the remaining pairs can then verified by trial division [GOLE+ 1994]. Since we already computed the roots of $f$ and $g$ modulo each $p$, the first division by $p$ for each candidate is exact. At the end, the smooth value pairs are those that have been replaced in the meantime by a pair of $\pm 1$'s.

Here one can also use the small and large prime variations to speed up sieving.

The expressions $F(a, b)$, $G(a, b)$ are the norms of the algebraic numbers $a - b\alpha$ and $a - b\beta$, multiplied by the leading coefficients of $f$ and of $g$, respectively. These algebraic numbers generate ideals in $\mathbb{Q}(\alpha)$ and $\mathbb{Q}(\beta)$. In other words in Line 2 we sieve *ideals* and not just pairs of integers, by bounding their norms. What we have then is a set of smooth ideals. The principal ideals $(a - b\alpha)$ and $(a - b\beta)$ factor into products of prime ideals in the number fields $\mathbb{Q}(\alpha)$ and $\mathbb{Q}(\beta)$, respectively. All prime ideals appearing in these factorizations have small norm (since the norms are assumed to be smooth), so only a few different prime ideals can occur.

(iii) *On the square root computation.* There are various methods to find the square roots in Line 4 of the NFS, the most important being the methods of Couveignes, Montgomery and improvements to the latter by P. Nguyen. See [ELK 1996, NGU 1998].

The interested reader can find more information in one of the excellent surveys on the subject [LELE+ 1990, LELE 1993] and the book by Crandall and Pomerance [CRPO 2001].

# VII

# Realization of Discrete Logarithm Systems

# Chapter 26

# Fast Arithmetic in Hardware

*Kim Nguyen and Andrew Weigl*

### Contents in Brief

| | | |
|---|---|---|
| **26.1** | **Design of cryptographic coprocessors** <br> Design criterions | 618 |
| **26.2** | **Complement representations of signed numbers** | 620 |
| **26.3** | **The operation** $XY + Z$ <br> Multiplication using left shifts • Multiplication using right shifts | 622 |
| **26.4** | **Reducing the number of partial products** <br> Booth or signed digit encoding • Advanced recoding techniques | 625 |
| **26.5** | **Accumulation of partial products** <br> Full adders • Faster carry propagation • Analysis of carry propagation • Multi-operand operations | 627 |
| **26.6** | **Modular reduction in hardware** | 638 |
| **26.7** | **Finite fields of characteristic** $2$ <br> Polynomial basis • Normal basis | 641 |
| **26.8** | **Unified multipliers** | 644 |
| **26.9** | **Modular inversion in hardware** | 645 |

This chapter describes some aspects of the realization of arithmetic in dedicated hardware environments. It falls naturally into three main parts: the first one introduces some basic criteria applicable to the design of cryptographic hardware, highlighting some of the specific constraints to be observed for implementations in constrained hardware environments. The second one deals with the realization of multiplication as the main arithmetic operation in hardware, which can be seen as the foundation of nearly all public-key cryptosystems in use today. The motivation of this part is to introduce the reader to some of the peculiarities of the implementation of basic arithmetic in constrained environments. In the third part we touch on some topics more relevant to the preceding chapters, i.e., finite field arithmetic and modular inversion in hardware.

The main motivation for this dedicated treatment of arithmetic in hardware is that the hardware environments in which elliptic and hyperelliptic curve cryptosystems can be implemented can differ dramatically with respect to the computing power available: while a multiprecision arithmetical library on a PC can realize fast finite field arithmetic based on the arithmetic core operations provided by the CPU, on a smart card or an embedded device the standard CPU used will not offer sufficient support for this. Thus a dedicated piece of hardware will be added in order to offer support for these arithmetic operations. In order to make this support as efficiently as possible we need to analyze in detail the realization of the basic arithmetic operations in hardware.

This chapter can only give a brief introduction to the main challenges when implementing arithmetic operations in hardware. The interested reader should refer to the excellent books [KOR 2002] and [PAR 2000] for more details and variants of the implementation of the operations described here. Furthermore, these books also introduce important technical concepts that we cannot cover here (e.g., pipelining).

## 26.1 Design of cryptographic coprocessors

The main purpose of a cryptographic coprocessor is to accelerate the execution of cryptographic algorithms either on hardware with otherwise restricted computing power (e.g., on a security controller used for a smart card or an embedded device) or in situations where an especially high throughput is required (e.g., when generating digital signatures on a file server, etc.). In this section we want to describe some of the criteria that are important when designing a cryptographic coprocessor.

### 26.1.1 Design criterions

The main criteria influencing the design of cryptographic hardware come from three different areas:

1. algorithmic considerations
2. hardware considerations
3. application considerations

**Remarks 26.1**

(i) We first look at the algorithmic side. These considerations are of a more general nature and will be of interest to both application scenarios mentioned above.
The earliest cryptographic coprocessors were focused on supporting private-key algorithms (e.g., DES or later triple DES). In this case the complete flow of one complete DES operation is modeled in and executed by dedicated hardware.
The first cryptographic coprocessors that became available commercially focused on the support of the public-key system available at that time, i.e., RSA. It is plain to see that by offering hardware assistance for modular multiplication $XY \bmod N$ and modular squaring $X^2 \bmod N$, it is relatively easy to implement the modular exponentiation, which is the core operation of the RSA system.
For elliptic or hyperelliptic curves the situation is a little more complicated. This is due to the higher algorithmic complexity of the operations on an elliptic curve opposed to the case of RSA. For an elliptic curve coprocessor the main possibilities are:

- Supporting *atomic* finite field operations (e.g., multiplication, addition, subtraction, etc.).

- Supporting *elementary* operations on an elliptic curve or a Jacobian of a hyperelliptic curve (e.g., point addition, point doubling, etc.).

- Supporting *complex* operations on an elliptic curve or a Jacobian of a hyperelliptic curve (e.g., scalar multiplication or the computation of a digital signature).

(ii) All of these options have their own advantages and disadvantages:

- Supporting only *atomic* finite field operations leaves the implementer with the choice of different ways to actually implement the arithmetic on elliptic curves and Jacobians of hyperelliptic curves by choosing, for example, different coordinate systems, group operation formulas, and scalar multiplication methods. The implementer can make maximum use of special arithmetic properties of the underlying finite field, the curves or Jacobians used and of cryptographic algorithms to be implemented, which can lead to significant optimizations with respect to both performance and security of the implementation. However, this option also requires deep insights into the arithmetic and security of the objects involved.

- Supporting *elementary* operations on an elliptic curve narrows down the freedom of the implementer, but on the other hand allows a much easier, high-level-approach to the implementation of, for example, scalar multiplication or even more abstract cryptographic algorithms like digital signature algorithms. The flexibility to make use of special properties is restricted to the higher level of scalar multiplication. Special properties of the underlying finite field or of the elliptic curve can only be used if these are already built into the existing elementary operations provided by the hardware. This option requires a good understanding of the more high-level algorithms, but does not require an intimate knowledge of the arithmetic of the underlying curves or Jacobians.

- Supporting *complete* operations on an elliptic curve (e.g., scalar multiplication) allows us to quickly implement high-level cryptographic algorithms like a digital signature algorithm or a key exchange without requiring a deep knowledge of the underlying arithmetic or structures. However, the implementer is not able to make use of any special properties that are not already available.

(iii) Besides the specific algorithmic constraints imposed upon a cryptographic coprocessor, there are also technological and economical restraints one has to observe, and which also influence the design of the coprocessor:

- *Area consumption*: the area and thus the number of gates available for a complete smart card security controller is limited (normally less than $25\text{mm}^2$). Of these $25\text{mm}^2$ almost all is taken up by the different memory blocks. Thus the area available for the coprocessor is limited to a few $\text{mm}^2$. This imposes severe constraints on the complexity of operations that can be implemented in hardware.

  Furthermore, each additional $\text{mm}^2$ used will increase the price of the IC produced. Thus there are both technological and economical constraints resulting from the area consumption that will severely influence the architecture of the coprocessor.

- *Power consumption*: the energy available to a smart card or embedded device is usually strictly limited. In contactless mode of operation, the total power can decrease to only a few mW, which will be another severely limiting factor for the hardware design of a cryptographic coprocessor.

(iv) Finally, if there exists a specific application the hardware system is aimed at supporting, this particular application will impose specific requirements and a specific environment to be used. These will determine the algorithmic and technological constraints applicable and will, finally, determine the architecture of a suitable cryptographic coprocessor. A hardware unit designed for, say, generating fast elliptic curve-based digital signatures in a server could well consist of a large arithmetic unit capable of multiplying two 256-bit integers in one clock cycle, computing a modular inverse in a 256-bit sized finite field in only a few clock cycles and performing a complete hard-coded scalar multiplication on an elliptic curve in a few ms. This is due to the fact that neither restraints in terms of power nor due to area consumption will be applicable in this situation. A security controller designed for a smart card will more likely offer arithmetic support of considerably less computing power, or will make extensive use of special arithmetic properties to offer a performance increase.

However, the algorithmic requirements, for example, with respect to the number of digital signature computations performed in a given time, will also differ dramatically: while for a server application this could easily be something like several hundred per minute, a security controller will most likely be required to generate only one signature during a cryptographic algorithm, such as, an authentication scheme. Hence its more limited computing power will still be enough to satisfy the performance requirements imposed by the application.

We now turn to the technical part. We start by briefly recalling the representation of integers in different *complement representations*.

## 26.2 Complement representations of signed numbers

Recall from Section 10.1 that there are several techniques available to represent signed numbers in binary form. We will concentrate on the so-called complement representation. The name *complement* refers to the fact that the negative value $-x$ of a number $x$ is represented as the value $M - x$, where $M$ is a suitably chosen large complementation constant. If we want to represent numbers in the range of $[-N, +P]$ we need that $M \geqslant N + P + 1$, otherwise we would have an overlap between representations of positive and negative numbers. The choice of $M = N + P + 1$ will yield the most efficient version available.

**Remark 26.2** The advantages of complement representation are:

- addition can be performed by adding the respective unsigned representations modulo $M$
- subtraction can be performed by complementing the subtrahend and adding the complemented version.

The rules for addition in a complement number system can be summarized as follows:

Table 26.1 Operations in complement number systems.

| Operation | Computation mod $M$ | Result | Overflow if: |
|---|---|---|---|
| $(+x) + (+y)$ | $x + y$ | $x + y$ | $x + y > P$ |
| $(+x) + (-y)$ | $x + (M - y)$ | $x - y$ if $y \leqslant x$ | NA |
| | | $y - x$ if $y > x$ | |
| $(-x) + (+y)$ | $(M - x) + y$ | $y - x$ if $x \leqslant y$ | NA |
| | | $M - (x - y)$ if $x > y$ | |
| $(-x) + (-y)$ | $(M - x) + (M - y)$ | $M - (x - y)$ | $x + y > N$ |

In the most widely used case of ordinary binary representation, the choice of $M = 2^\ell$ is referred to as *two's complement* representation, generally known as *radix complement*.

## Remarks 26.3

(i) Recall from Section 10.1.2 that the computation of the two's complement can be accomplished very easily by inverting bitwise and adding 1.

(ii) Note that the choice of $M = 2^\ell$ makes computation modulo $M$ especially easy: if, for example, a carry is produced in addition, we simply ignore this. Since such a carry is worth $2^\ell$, we thus perform a reduction by $2^\ell$.

The range of numbers representable in the two's complement is given by $[-2^{\ell-1}, 2^{\ell-1} - 1]$. The zero element has the (unique) representation

$$\underbrace{0 \ldots 0}_{\ell \text{ times}}.$$

Another possible choice of $M$ is to set $M = 2^\ell - 1$. The representation obtained by this choice of $M$ is known as the one's complement or generally as a digit complement. The *one's complement* of a number $x$ can be obtained by bitwise complementation. This is due to the fact that

$$(2^\ell - 1) - x = x^{\text{compl}}.$$

## Remarks 26.4

(i) In contrast to the two's complement representation we do not need to add 1, therefore obtaining the one's complement is easier than obtaining the two's complement representation.

(ii) However, performing the reduction modulo $M$ is now slightly more complicated: when, for example, performing addition, we have to drop the carry-out corresponding to subtracting $2^\ell$ from the result. We also have to produce a carry-in at position 0 which will, together with the dropped carry-out, produce a reduction by $2^\ell - 1$. Note that in hardware the new carry-in could be produced by feeding the carry-out from position $\ell - 1$ into the carry-in at position 0. This technique is referred to as *end-around carry*.

The range of numbers in the one's complement is $[-(2^\ell - 1), 2^\ell - 1]$, which is symmetric in contrast to the range of numbers represented by the two's complement system. In contrast to the two's complement system, where zero had a unique representation, in the one's complement we obviously have two representations

$$\underbrace{0 \ldots 0}_{\ell \text{ times}} \text{ and } \underbrace{1 \ldots 1}_{\ell \text{ times}},$$

which are both valid.

A comparison of radix and digit complement is given in Table 26.2.

Which system is to be preferred? It turns out that the two's complement representation is the standard representation in almost all systems. The biggest disadvantage of two's complement might be the more complicated way of actually obtaining the complement representation of a number $x$.

However, the most interesting application of a complement representation will be the performance of subtraction by actually performing an addition.

In this application the addition of the two's complement of a number $X$ can be achieved by inverting all the bits of $X$ and setting the carry-in at position 0 to 1. Hence subtraction can be performed as an addition with building the two's complement on the fly.

We will assume that some sort of complement representation is used, thus we will not deal explicitly with the topic of subtraction in what follows.

**Table 26.2** Comparison of radix and digit complement.

| Property | Radix complement | Digit complement |
|---|---|---|
| Range of numbers represented | Asymmetric | Symmetric |
| Unique zero | Yes | No |
| Obtaining the complement | Bitwise complement, add 1 | Bitwise complement |
| Addition modulo $M$ | Drop carry-out | End-around carry |

## 26.3 The operation $XY + Z$

Contrary to the relatively high mathematical complexity of the topics of the previous chapters, the first question we want to address is a rather basic one: how to multiply two integers *of fixed size* efficiently using elementary operations? In Chapter 10 the main algorithms needed to implement *multiprecision* integer arithmetic were presented. Even in this presentation the core operation of multiplying two elements $u$ and $v$, both not larger than the base $b$, was assumed as given.

It is the aim of this section to precisely explain this operation, i.e., to explain the fundamentals of realizing the multiplication operation in hardware.

**Example 26.5** We consider the elementary multiplication method as it was described in Example 10.10. The two values to be multiplied are $u = (9712)_{10}$ and $v = (526)_{10}$. Recall that the main arithmetic operations used in order to compute $u \times v = 5108512$ were the computation of the three products $9712 \times 6, 9712 \times 20, 9712 \times 500$ and the addition of the intermediate results.

The first point of relevance for a hardware implementation is that the operation $9712 \times 20$ can be rewritten as $9712 \times 2 \times 10$, where in decimal representation the multiplication by 10 can be realized by a left shift of the decimal representation of $9712 \times 2$. The same argument applies for the multiplication $9712 \times 500$ where a double left shift needs to be performed.

Thus we obtain the following steps:

- first partial product is $9712 \times 6 = 58272$
- second partial product is $9712 \times 2 = 19424$
- third partial product is $9712 \times 5 = 48560$
- shift second partial product one position to the left $9712 \times 2 \times 10 = 194240$
- shift third partial product two positions to the left $9712 \times 5 \times 100 = 4856000$
- obtain first intermediate sum $9712 \times 6 + 9712 \times 20 = 252512$
- obtain second final result $252512 + 4856000 = 5108512$.

**Example 26.6** Consider the case of a $6 \times 6$ multiplication using the schoolbook method. The most straightforward way of arranging the partial products as given in the algorithm above has the graphical representation shown in Figure 26.3.

**Remarks 26.7**

(i) Taking a look at these two simple examples in a more abstract way we see that the process of multiplication falls into two main steps:
   - The first step generates a number of intermediate results, i.e., partial products derived from the two multiplicands $X$ and $Y$.

## § 26.3 The operation $XY + Z$

- In the second step, the final result is produced by adding all the intermediate partial products generated. This step is totally independent of the two multiplicands.

(ii) As surprising as it may be, this is exactly what happens in a hardware implementation of multiplication, and thus the two steps described above are the main challenges one has to overcome in order to implement an efficient multiplication method.

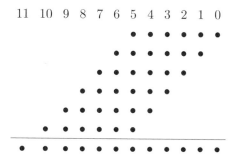

**Figure 26.3** Schoolbook multiplication of two 6-digits numbers.

**Goal 26.8** The main goal of this section is to show that the efficient realization of multiplication in hardware needs to solve the following main tasks:

- efficiently generate partial products
- efficiently compute the sum of several integers.

As it was pointed out before, the standard method of performing a multiplication of two single precision integers $X$ and $Y$ is a sequential algorithm generating partial products and accumulating these correctly shifted partial products in order to obtain the final result.

Now we give a more detailed explanation of this elementary method.

The correct alignment of the partial products can be accomplished using either left or right shifts. We will explain these two fundamental methods in the following two sections.

### 26.3.1 Multiplication using left shifts

The following algorithms describe a simple multiplication using left shifts.

**Algorithm 26.9** Sequential multiplication of two integers using left shift

INPUT: Single precision integers $X = (x_{\ell-1} \ldots x_1 x_0)_2$ and $Y = (y_{\ell-1} \ldots y_1 y_0)_2$.
OUTPUT: The product $P = XY = (p_{2\ell-1} \ldots p_1 p_0)_2$.

1. $P^{(0)} \leftarrow 0$
2. **for** $i = 0$ **to** $\ell - 1$ **do**
3.     $P^{(i+1)} \leftarrow (2P^{(i)} + x_{\ell-i-1} Y)$
4. **return** $P^{(\ell)}$

**Example 26.10** Consider the multiplication $XY$ where $X = (101)_2$ and $Y = (111)_2$. We obtain the following sequence:

| | | | | | | |
|---|---|---|---|---|---|---|
| $P^{(0)}$ | 0 | 0 | 0 | 0 | 0 | 0 |
| $2P^{(0)}$ | 0 | 0 | 0 | 0 | 0 | 0 |
| $x_2 Y$ | | | 1 | 1 | 1 | |
| $P^{(1)}$ | 0 | 0 | 0 | 1 | 1 | 1 |
| $2P^{(1)}$ | 0 | 0 | 1 | 1 | 1 | 0 |
| $x_1 Y$ | | | | 0 | 0 | 0 |
| $P^{(2)}$ | 0 | 0 | 1 | 1 | 1 | 0 |
| $2P^{(2)}$ | 0 | 1 | 1 | 1 | 0 | 0 |
| $x_0 Y$ | | | | 1 | 1 | 1 |
| $P^{(3)}$ | 1 | 0 | 0 | 0 | 1 | 1 |

## 26.3.2 Multiplication using right shifts

By using right shifts we obtain the following algorithm:

**Algorithm 26.11** Sequential multiplication of two integers using right shift

INPUT: Single precision integers $X = (x_{\ell-1} \ldots x_1 x_0)_2$ and $Y = (y_{\ell-1} \ldots y_1 y_0)_2$.
OUTPUT: The product $P = XY = (p_{2\ell-1} \ldots p_1 p_0)_2$.

1. $P^{(0)} \leftarrow 0$
2. $Y \leftarrow 2^\ell Y$
3. **for** $i = 0$ **to** $\ell - 1$ **do**
4. $\quad P^{(i+1)} \leftarrow (P^{(i)} + x_i Y) 2^{-1}$
5. **return** $P^{(\ell)}$

**Example 26.12** We again consider the multiplication $XY$ where $X = (101)_2$ and $Y = (111)_2$. We obtain the following sequence:

| | | | | | | |
|---|---|---|---|---|---|---|
| $P^{(0)}$ | 0 | 0 | 0 | 0 | 0 | 0 |
| $x_2 Y$ | 1 | 1 | 1 | | | |
| $2P^{(1)}$ | 1 | 1 | 1 | 0 | 0 | 0 |
| $P^{(1)}$ | 0 | 1 | 1 | 1 | 0 | 0 |
| $x_1 Y$ | 0 | 0 | 0 | | | |
| $2P^{(2)}$ | 0 | 1 | 1 | 1 | 0 | 0 |
| $P^{(2)}$ | 0 | 0 | 1 | 1 | 1 | 0 |
| $x_0 Y$ | 1 | 1 | 1 | | | |
| $2P^{(3)}$ | 1 | 0 | 0 | 0 | 1 | 1 |
| $P^{(3)}$ | | 1 | 0 | 0 | 0 | 1 | 1 |

## Remarks 26.13

(i) The right shift algorithm only uses $\ell$-bit addition, while in contrast to this the left shift algorithm has to use $2\ell$-bit addition in order to properly accumulate the partial products. Hence the right shift algorithm is more efficient to implement in hardware.

(ii) Inspection of the right shift algorithm yields that the output after $\ell$ steps is equal to $P^{(\ell)} = XY + P^{(0)}2^{-\ell}$. Therefore by setting $P^{(0)}$ to the value $2^{\ell}Z$ where $Z$ has $\ell$ bits, we see that we are able to compute $XY + Z$ with nearly no extra costs compared to the normal multiplication $XY$.

Since the multiplication operation falls into two main parts (generation and addition of partial products), the reduction of the number of partial products generated and the acceleration of the accumulation of these partial products or, preferably, a combination of these two ideas, will obviously lead to a speedup. We will discuss these ideas in the following sections.

## 26.4 Reducing the number of partial products

One way of reducing the numbers of partial products would be to examine more than one bit of the multiplier in one step, but this would require the generation of $Y$, $2Y$, $3Y$, ... which would increase the complexity of each single step.

There are several algorithms that reduce the number of partial products while still keeping the low complexity of the simple algorithm in each step. One of the first and most prominent is Booth's algorithm, also known as signed digit encoding.

### 26.4.1 Booth or signed digit encoding

The main observation is that fewer partial products can be generated for groups of either consecutive ones or zeroes. If a group of consecutive zeroes appears, there is no need for the multiplier to generate a new partial product, we simply have to shift the accumulated partial product one step to the right or left (according to which basic algorithm you use) for every 0 appearing in the multiplier. The procedure just described above is also referred to as recoding the multiplier in SD (signed digit) representation. The algorithm below gives the original formulation of Booth's algorithm.

---

**Algorithm 26.14** Recoding the multiplier in signed-digit representation

INPUT: A multiplier $X = (x_{\ell-1} \ldots x_1 x_0 x_{-1})_2$ with $x_{-1} = 0$ and $x_\ell = 0$.
OUTPUT: The recoded multiplier $Y = (y_{\ell-1} \ldots y_1 y_0)_s$ of $X$.

1. **for** $i = -1$ **to** $\ell - 1$ **do**
2.     **if** $x_{i+1} = 0$ **and** $x_i = 0$ **then** $y_{i+1} = 0$
3.     **if** $x_{i+1} = 1$ **and** $x_i = 1$ **then** $y_{i+1} = 0$
4.     **if** $x_{i+1} = 1$ **and** $x_i = 0$ **then** $y_{i+1} = \bar{1}$
5.     **if** $x_{i+1} = 0$ **and** $x_i = 1$ **then** $y_{i+1} = 1$
6. **return** $(y_{\ell-1} \ldots y_1 y_0)_s$

---

Note that given $x_{i+1}$ and $x_i$, the computation of $y_{i+1}$ can be represented by the formula

$$y_{i+1} = x_i - x_{i+1}.$$

**Remark 26.15** There are two major drawbacks of this SD recoding technique:

1. The number of add or subtract operations is variable, as is the number of shift operations between two consecutive add or subtract operations. This can pose a serious problem in a synchronous design where the steps of an algorithm have to be performed in a fixed specified order and where input values from intermediate steps are expected at a fixed point in time. In order to deal with varying computing time needed to generate intermediate values, one would need to store other values in internal memories known as latches in order to keep them available in all cases. This will lead to an increased power and area consumption and to a higher overall complexity of the hardware design and is thus best avoided.

2. The signed digit recoding can be very inefficient when isolated 1's occur.

**Example 26.16** Consider the binary number $X = (001010101)_2$. A multiplication with $X$ can be realized using four additions. Using the SD recoding algorithm given above we obtain

$$(001010101(0))_2 = (01\bar{1}1\bar{1}1\bar{1}1\bar{1})_s,$$

as recoding for $X$ implies that, by using the recoded multiplier, we need eight instead of four additions.

Inspecting more than two bits at a time can lead to significant performance gains (see a similar signed digit recoding technique introduced in Section 9 in order to speed up generic scalar multiplication in a group). The following section looks at two possible algorithms especially suitable for hardware implementation.

### 26.4.2 Advanced recoding techniques

Taking into consideration three consecutive bits at a time, the bits $x_{i+1}$ and $x_i$ are recoded into $y_{i+1}$ and $y_i$ while $x_{i-1}$ has to be inspected as well. We call this Radix-4 SD recoding:

---

**Algorithm 26.17** Recoding the multiplier in radix-4 SD representation

INPUT: A multiplier $X = (x_{\ell-1} \ldots x_1 x_0 x_{-1})_2$ where $x_{-1} = 0$.
OUTPUT: The recoded multiplier $Y = (y_{\ell-1} \ldots y_1 y_0)_s$ of $X$.

1. **for** $i = 0$ **to** $\ell - 1$ **do**
2.     **if** $x_{i+1} = 0$ **and** $x_i = 0$ **and** $x_{i-1} = 0$ **then** $y_{i+1} = 0$ and $y_i = 0$
3.     **if** $x_{i+1} = 0$ **and** $x_i = 1$ **and** $x_{i-1} = 0$ **then** $y_{i+1} = 0$ and $y_i = 1$
4.     **if** $x_{i+1} = 1$ **and** $x_i = 0$ **and** $x_{i-1} = 0$ **then** $y_{i+1} = \bar{1}$ and $y_i = 0$
5.     **if** $x_{i+1} = 1$ **and** $x_i = 1$ **and** $x_{i-1} = 0$ **then** $y_{i+1} = 0$ and $y_i = \bar{1}$
6.     **if** $x_{i+1} = 0$ **and** $x_i = 0$ **and** $x_{i-1} = 1$ **then** $y_{i+1} = 0$ and $y_i = 1$
7.     **if** $x_{i+1} = 0$ **and** $x_i = 1$ **and** $x_{i-1} = 1$ **then** $y_{i+1} = 1$ and $y_i = 0$
8.     **if** $x_{i+1} = 1$ **and** $x_i = 0$ **and** $x_{i-1} = 1$ **then** $y_{i+1} = 0$ and $y_i = \bar{1}$
9.     **if** $x_{i+1} = 1$ **and** $x_i = 1$ **and** $x_{i-1} = 1$ **then** $y_{i+1} = 0$ and $y_i = 0$
10. **return** $(y_{\ell-1} \ldots y_1 y_0)_s$

**Remarks 26.18**

(i) Again we note that a simple formula is the basis of this recoding technique; first computing
$$x_i + x_{i-1} - 2x_{i+1}$$
and then representing this result as a 2-bit binary number in SD representation gives $y_{i+1}$ and $y_i$ as required.

(ii) Using this algorithm, the occurrence of an isolated 1 or 0 no longer poses a problem, since if $x_i$ is an isolated 1, we have $y_i = 1$ and $y_{i+1} = 0$, hence there is only a single operation needed. Also, in the situation 101, we have $y_i = 0$ and $y_{i-1} = \bar{1}$, so again only one single operation is needed.

**Example 26.19** Let us again consider $(01010101(0))_2$. We see that this is recoded as $(01010101)_s$, hence the number of operations remains four. However, if we consider $(00101010(0))_2$ we see that this is recoded to $(010\bar{1}0\bar{1}\bar{1}0)_s$, which implies that the number of operations required goes up to four by the recoding process.

However, the number of patterns for which this phenomenon occurs is relatively small and the increase in the number of operations is minimal. Using the radix-4 SD representation it is possible to design a synchronous multiplier generating exactly $\ell/2$ partial products.

**Remark 26.20** Note that all the multiples required by this encoding scheme can be generated using simple shift operations on the multiplicand (in order to obtain $2Y$) and the two's complement representation (in order to obtain $-Y$ and $-2Y$).

**Remark 26.21** We can take this even further by changing the SD representation from radix-4 to radix-8, which needs only $\ell/3$ partial products.

Now we have to investigate 4 consecutive bits of the multiplier for encoding. Each partial product generated by this radix-8 encoded multiplier will come from the set $\{\pm 4Y, \pm 3Y, \pm 2Y, \pm Y, 0\}$. As in the case of radix-4 encoding, the elements $\pm 4Y$ and $\pm 2Y$ can be generated using shift operations and the two's complement. However, using these operations, it is not possible to generate the elements $\pm 3Y$ using these operations (therefore the element $3Y$ is sometimes referred to as the hard multiple). In order to generate it, a full carry propagate addition is required. Fortunately, we will not need the most general kind of adder (as described below), since an adder especially designed for the operation $Y + 2Y$ will suffice.

## 26.5 Accumulation of partial products

Once we have produced all the partial products needed for the multiplication operation, we have to accumulate them in order to obtain the final result. Obviously this fast accumulation will also speed up the overall multiplication process.

We start by giving an overview over the existing types of adders that can be used.

### 26.5.1 Full adders

A full adder (FA) is a logic circuit producing a sum bit $s_i$ and a carry bit $c_{i+1}$ from two input bits $x_i, y_i$ and an input carry bit $c_i$. The output bits $s_i$ and $c_{i+1}$ are given by the equations

$$s_i = x_i \text{ XOR } y_i \text{ XOR } c_i, \quad c_{i+1} = x_i \wedge y_i + c_i \wedge (x_i \vee y_i).$$

Here $x$ XOR $y$ refers to exclusive disjunction, $x \wedge y$ to logical conjunction AND and $x \vee y$ to logical disjunction OR. Addition of two $n$-bit operands can be accomplished using $n$ full adders by the following algorithm.

---
**Algorithm 26.22** Addition using full adders

INPUT: Operands $X = (x_{n-1} \ldots x_1 x_0)_2$ and $Y = (y_{n-1} \ldots y_1 y_0)_2$.
OUTPUT: The sum $X + Y = (s_{n-1} \ldots s_1 s_0)_2$ and a carrybit $c$.

1. **for** $i = 0$ **to** $n - 1$ **do**
2. $\quad c_0 \leftarrow 0$
3. $\quad s_i, c_{i+1} \leftarrow \text{FA}(x_i, y_i, c_i)$
4. **return** $(s_{n-1} \ldots s_1 s_0)_2$ and the carrybit $c_n$
---

**Remarks 26.23**

(i) Although we have all the bits of the operands available from the very beginning of the algorithm, the carries have to propagate from $j = 0$ until $j = i$ in order to produce the correct sum $s_i$ and carry $c_{i+1}$. Hence we have to wait until the carry ripples through all the stages of the algorithm before we can be sure that we have obtained the correct final result.

(ii) The adder at level $i = 0$ will always see the incoming carry $c_0 = 0$, so it can be replaced by a simpler adder called a half adder (HA), which produces $s_0$ and $c_1$ from

$$s_0 = x_0 \text{ XOR } y_0, \quad c_1 = x_0 \wedge y_0.$$

(iii) It might be necessary to keep the incoming carry $c_0$ as a variable, if we want to be able to perform not only addition but also subtraction using the same setup. Subtraction in two's complement is accomplished by complementing the subtrahend and then adding it to the minuend. Therefore we add the one's complement of the subtrahend and use a forced carry as input to the FA at level 0 by setting $c_0 = 1$.

(iv) A system of full adders imposes a carry propagation chain of length $n$ when adding two $n$-bit numbers. This is the main obstacle we have to overcome in order to achieve a significant speedup for addition.

Two ideas come to mind immediately: to reduce the time the carry needs for propagation and to change the system in such a way that enables us to see whether the carry propagation is finished instead of waiting for a fixed time of $n\Delta_{\text{FA}}$ where $\Delta_{\text{FA}}$ is the delay of a full adder.
We will develop these ideas in the next sections

### 26.5.2 Faster carry propagation

#### 26.5.2.a Carry-look-ahead adders

One of the most well-known techniques available to accelerate the carry propagation is the use of a *carry-look-ahead adder*.

The basic idea behind this type of adder is to simultaneously generate the carries at higher levels instead of waiting for the carry of level 0 to ripple through the system.

## Remarks 26.24

(i) We first note that in some situations the output carry $c_{i+1}$ is already known from the input bits $x_i$ and $y_i$, namely in the situation where $x_i = y_i = 1$. In this case the output carry $c_{i+1} = 1$ can be computed immediately.

(ii) If $x_i = 1$ and $y_i = 0$ or $x_i = 0$ and $y_i = 1$, then an incoming carry can only be propagated. It will not be annihilated, neither can an outgoing carry be generated.

(iii) Conversely, if $x_i = y_i = 0$, an incoming carry will be annihilated.

In order to be able to make use of this information, we define two more variables $G_i$ and $P_i$, which are called *generated carry* and *propagated carry*. They are defined by the following equations:
$$G_i = x_i y_i, \quad P_i = x_i + y_i.$$

Now the outgoing carry can be expressed as a function of $G_i$, $P_i$ and $c_i$ as follows:
$$c_{i+1} = x_i y_i + c_i(x_i + y_i) = G_i + c_i P_i.$$

Since also $c_i = G_{i-1} + c_{i-1} P_{i-1}$ we obtain
$$c_{i+1} = G_i + G_{i-1} P_i + c_{i-1} P_{i-1} P_i.$$

(iv) We can go on inductively until we obtain an expression for $c_{i+1}$, which is only dependent on $G_j, P_j$ for $j \leqslant i$ and $c_0$. Thus we are able to compute all the carries $c_i$ in parallel from the input bits $(x_{n-1} \ldots x_1 x_0)_2$ and $(y_{n-1} \ldots y_1 y_0)_2$.

**Example 26.25** Consider a 4-bit adder. Here the carries are:
$$\begin{aligned}
c_1 &= G_0 + c_0 P_0 \\
c_2 &= G_1 + G_0 P_1 + c_0 P_0 P_1 \\
c_3 &= G_2 + G_1 P_2 + G_0 P_1 P_2 + c_0 P_0 P_1 P_2 \\
c_4 &= G_3 + G_2 P_3 + G_1 P_2 P_3 + G_0 P_1 P_2 P_3 + c_0 P_0 P_1 P_2 P_3.
\end{aligned}$$

## Remarks 26.26

(i) What do we gain in comparison to a carry-ripple adder if we use a full $n$-stage carry-look-ahead adder?

- For each stage we have a delay of $\Delta_G$, which is needed in order to generate the $P_i$ and $G_i$. Assuming a two-level gate implementation of the formulas for the carries $c_i$ given above we need $2\Delta_G$ to generate all the carries $c_i$.
- Furthermore, we need another $2\Delta_G$ to generate the sum bits in a two-level implementation.
- Therefore we have a total time of $5\Delta_G$ for the whole addition regardless of the length $n$ of the operands.

(ii) Consulting the above example we see that for typical values like $n = 32$ we will need an enormous number of gates. Even worse, we will have $n+1$ gate inputs. Thus we have to balance the range of carry-look-ahead adders (and hence the complexity of the implementation we use) against the speedup factor we want to achieve.

### 26.5.2.b Carry-look-ahead adders arranged in groups

One of the techniques available to deal with the problems just described is the following one: we divide the stages of the total addition process into several groups, where each group will have a separate carry-look-ahead adder. The different groups are then connected in carry-ripple fashion. A natural way of doing this is to form groups of the same size.

**Remarks 26.27**

(i) A commonly used option is to divide the complete $n$-level structure into groups of size 4, hence there will be $n/4$ groups (assuming that $n$ is a multiple of 4, which is the case for all common word sizes).

(ii) The propagation of a carry through one group is $2\Delta_G$, once the inputs $P_i$, $G_i$ and $c_0$ are available. If $\Delta_G$ is needed in order to generate all $P_i$ and $G_i$, then $(n/4)2\Delta_G$ is spent on propagating the carry through all levels and finally $2\Delta_G$ is needed in order to compute the sum bits.

(iii) In total we have a timing of

$$\left(2\frac{n}{4}+3\right)\Delta_G = \left(\frac{n}{2}+3\right)\Delta_G,$$

which is about four times faster in delay than the delay of $2n\Delta_G$ needed in the case of a $n$-level carry-ripple adder.

**Remark 26.28** We can further improve on this situation if we also apply the idea of carry-look-ahead to the $n/4$ groups we used.

In order to achieve this we introduce a group-generated carry $G^*$ and a group-propagated carry $P^*$ by setting $G^* = 1$, if we have an outgoing carry (for the whole group) and $P^* = 1$, if we have an incoming carry that is propagated internally to an outgoing carry of the group.

If we assume a group size of four bits, we have the following expression:

$$G^* = G_3 + G_2P_3 + G_1P_2P_3 + G_0P_1P_2P_3, \quad P^* = P_0P_1P_2P_3.$$

Now we use this to combine several groups not in a carry-ripple but in a carry-look-ahead way, obtaining a so-called carry-look-ahead generator.

### 26.5.2.c  Conditional-sum adders and carry-select adders

While carry-look-ahead adders produce the actual incoming carry bits for all levels, we can also try to base a system on the idea of dealing with the case that the incoming carry is either set or not set simultaneously, which leads to the so-called *carry-select adders* or *conditional-sum adders*.

Assume that we generate two outputs for a set of $m$ input bits, whereby each output consists of a set of $m$ sum bits and one carry bit. Now one set will be computed based on the assumption that the incoming carry to the system will be set, while the other set is based upon the fact that it is not set. Of course, once we know the value of the incoming carry we do not need to compute the output of the adder; we just have to select the correct output from the two choices available. As before, we will form subgroups of $k$ operations.

**Example 26.29** To illustrate this, assume that we divide $2m$ input bits into two groups of size $k = m$ each. While the lower $m$ bits will be summed up directly, resulting in an $m$-bit sum and an outgoing carry $c$, the higher $m$ bits are summed up, generating two outputs, one corresponding to $c = 0$ and the other corresponding to $c = 1$.

An adder using this approach will be called a carry-select adder.

**Example 26.30** It is possible to take this approach even further and to divide the groups of size $k$ into further subgroups:

Assume for example that $n$ is a perfect power of two, then we could divide the $n$ bits into two groups of size $n/2$ each.

This could then be continued in an inductive fashion so that in the end we would reach a group size of 1. In this case we would need $k = \lg n$ steps.

## § 26.5 Accumulation of partial products

**Remark 26.31** Comparing the conditional-sum adder to a carry-save adder, both have about the same execution time. However, the design of a conditional-sum adder is considerably less modular than the carry-lookahead adders described above. They are therefore more to difficult to adapt to larger input sizes.

### 26.5.2.d Carry-skip adders

Yet another way of reducing carry propagation time is the *carry-skip adder*. The main observation here is that in certain stages of addition no carry propagation can occur, namely in the situation when we have $x_j \neq y_j$ or equivalently $P_j = x_j$ XOR $y_j = 1$. Thus, several consecutive steps can be skipped for carry propagation as long as we have $P_j = 1$.

How can we make use of this in an $n$-level addition? Again we divide this addition into several consecutive stages, where each stage is built in carry-ripple fashion. However, if we have $P_j = 1$ for all the internal stages, we allow the incoming carry to this group to skip all the internal levels and to generate a group-carry-out immediately.

Suppose we have a group consisting of the $k$ bits in position $j, j+1, \ldots, j+k-1$. We then have the incoming Group-Carry-in and an internal normal carry $c_{j+k}$ at level $j+k-1$. These allow the Group-Carry-out to be expressed as

$$\text{Group} - \text{Carry} - \text{Out} = c_{j+k} + \text{Group} - \text{Carry} - \text{In} \times (P_j \times P_{j+1} \times \cdots \times P_{j+k-1}).$$

The values of the $P_j$ can be computed simultaneously at the beginning of the process.

**Remark 26.32** What is the optimal value for the group-size $k$?

In order to compute this we assume that all groups will have the same size $k$ and that $n/k$ is an integer. We relate the time $t_r$ a carry needs to ripple through a single stage of a group to the time $t_s$ that is needed to skip a group of size $k$. The longest carry-propagation time one could imagine occurs when a carry is generated in the first level, skips all the following steps till the last one and then ripples through the last step.

The execution time for this comes down to

$$T = (k-1)t_r + (k-1)t_r + (n/k - 2)(t_s + t_b)$$

where $t_b$ denotes the time needed to combine the internal carry with the skipped carry. Assuming we have an implementation in which $t_r = t_s + t_b = 2\Delta_G$ for the gate delays, we obtain

$$T = (2k + n/k - 4) \times 2\Delta_G.$$

This expression is minimized if we choose $k$ to be equal to

$$k = \sqrt{n/2}.$$

### 26.5.3 Analysis of carry propagation

A natural question to ask is what the typical length of carry-propagation in an addition of two $n$-bit numbers is. More precisely, we call a *carry-chain* the sequence of positions starting with generation and ending with absorbing or annihilating the carry. Thus a carry-chain of length zero corresponds to no generation of a carry, while a carry-chain of length 1 corresponds to immediate absorbing of the carry in the next position.

**Table 26.4** Probabilities concerning carry propagation.

| Event | Probability |
|---|---|
| Generation of carry | 1/4 |
| Annihilation of carry | 1/4 |
| Propagation of carry | 1/2 |

**Remarks 26.33**

(i) Considering binary numbers with random values, the different cases of carry propagation are summarized in Table 26.4.

(ii) Now consider the probability that a carry generated at position $i$ is propagated up to position $j-1$ and annihilated at position $j$: it is given by $2^{-(j-i-1)} \times 1/2 = 2^{(j-i)}$. The expected length of a carry chain starting at position $i$ is therefore given by

$$\sum_{j=i+1}^{n-1}(j-i)2^{-(j-i)} + (n-1)2^{-(n-1-i)} = \sum_{l=1}^{n-1-i} l2^{-l} + (n-i)2^{-(n-1-i)}$$
$$= 2 - (n-i+1)2^{-(n-1-i)}$$
$$+ (n-i)^{-(n-1-i)}$$
$$= 2 - 2^{-(n-i-1)},$$

where the identity $\sum_{l=1}^{p} l2^{-l} = 2 - (p+2)2^{-p}$ was used. The term $(n-i)2^{-(n-1-i)}$ represents the fact that the carry has to stop at the last position $n$ (since we consider $n$ bit numbers).

(iii) So in the case that $i$ is much smaller than $n$ the expected length of the carry-chain is 2.

**Remark 26.34** Another question is what length the longest carry-chain occurring in the addition of two $n$-bit numbers on average will be. A celebrated result by Burks, Goldstine and von Neumann states that this value is given by $\lg n$ (see [BUGO$^+$ 1946]).

**Theorem 26.35** On average, the longest carry chain when adding $n$-bit numbers has length $\lg n$.

**Proof.** Let $\eta_n(h)$ denote the probability that the longest carry-chain in $n$-bit addition is of length at least $h$. The probability of the longest carry-chain being exactly of length $h$ is then given by $\eta_n(h) - \eta_n(h+1)$. We can use recursive techniques in order to compute $\eta_n(h)$.

How can a carry-chain of length at least $h$ be generated? There are two ways in which this can happen:

1. The least significant $n-1$ bits have a carry-chain of length at least $h$.
2. If the above condition is not met, the most significant $h$ bits, including the last bit, have a carry-chain of (exact) length $h$.

We thus obtain that

$$\eta_n(h) \leqslant \eta_{n-1}(h) + \frac{1}{4} \times 2^{-(h-1)} \tag{26.1}$$

where $\frac{1}{4}$ and $2^{-(h-1)}$ correspond respectively to the carry generation and the carry propagation of exactly $h-2$ steps. The inequality (26.1) in fact implies that $\eta_n(h) \leqslant \eta_{n-1}(h) + 2^{-(h+1)}$ holds.

## § 26.5 Accumulation of partial products

Assuming $\eta_i(h) = 0$ for $i < h$, we obtain

$$\begin{aligned} \eta_n(h) &= \sum_{i=h}^{n} (\eta_i(h) - \eta_{i-1}(h)) \\ &\leqslant (n - h + 1)2^{-(h+1)} \\ &\leqslant 2^{-(h+1)} n. \end{aligned}$$

Let $\lambda$ denote the expected length of the longest carry chain occurring. We have

$$\begin{aligned} \lambda &= \sum_{h=1}^{k} h(\eta_n(h) - \eta_n(h+1)) \\ &= (\eta_n(1) - \eta_n(2)) + 2(\eta_n(2) - \eta_n(3)) + \cdots + k(\eta_n(n) - 0) \\ &= \sum_{h=1}^{n} \eta_n(h). \end{aligned}$$

Now to further evaluate this sum we consider the first $\gamma = \lfloor \lg n \rfloor - 1$ terms and the remaining ones separately. We bound $\eta_n(h)$ by 1 for $h \leqslant \gamma$ and by $2^{-(h+1)}n$ for $h > \gamma$, thus obtaining

$$\begin{aligned} \lambda &= \sum_{h=1}^{n} \eta_n(h) \leqslant \sum_{h=1}^{\gamma} 1 + \sum_{h=\gamma+1}^{n} 2^{-(h+1)} n \\ &< \gamma + 2^{-(\gamma+1)} k. \end{aligned}$$

Now we write $\gamma = \lg n - 1 - \epsilon$ where $\epsilon = \lg n - \lfloor \lg n \rfloor$, and obtain

$$\lambda < \lg n - 1 - \epsilon + 2^\epsilon.$$

Here the transformation of the last terms follows simply from $2^{\lg n} = n$. Now since $0 \leqslant \epsilon < 1$ we have $2^\epsilon < 1 + \epsilon$, whence

$$\lambda < \lg n - 1 - \epsilon + 1 + \epsilon = \lg n$$

as claimed. □

**Remark 26.36** Experimental data verifies that the longest carry-chain in the worst case depends logarithmically on $n$. However, $\lg 1.25n$ seems to be a slightly better estimate [HEN 1961].

### 26.5.4 Multi-operand operations

Until now we only looked at the case of addition with two operands. However, having produced the partial products in a multiplication as outlined above we are faced with the problem of adding several operands.

One solution is to iterate several times the algorithms described above. But this implies that carry-propagation has to be done several times as well. If, for example we would like to add three $n$-bit inputs $X$, $Y$ and $Z$, we could for instance first add $X$ and $Y$ and then add the result $X + Y$ of this computation to $Z$, resulting to a double carry propagation.

### 26.5.4.a Carry-save adders

The idea behind carry-save adders is to use a redundant representation of the sum of the input vectors in the form $S + C$, where only one carry propagation is needed when adding $S$ and $C$ in the end.

**Remark 26.37** In the case of three inputs mentioned before, the easiest way of implementing a carry-save adder is to use a full adder with three 1-bit inputs $x$, $y$, and $z$, which generates the output values $s$ and $c$ given by

$$s = (x + y + z) \bmod 2 \quad \text{and} \quad c = \big((x + y + z) - s\big)/2.$$

Figure 26.5 illustrates how this principle applies to the case of three 6-bit inputs.

**Figure 26.5** A carry-save adder for three 6-bit inputs $X$, $Y$ and $Z$.

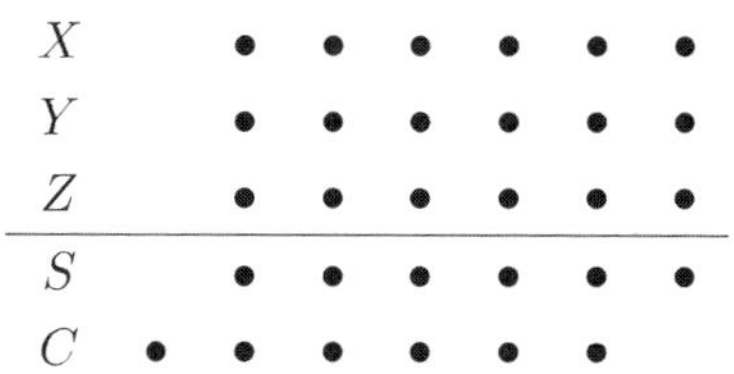

We refer to this principle as a $(3, 2)$-counter, since the output essentially represents the weighted binary representation of the number of 1's in the inputs.

A full $n$-bit carry-save adder consists of $n$ copies of a $(3, 2)$-counter acting in parallel without links between their carry outputs. In order to obtain the final result we need to add the $n$-bit sum vector and the carry vector together, using a standard two operand adder as described before.

Note that the carry output has to be shifted before it is added to the $n$-bit sum.

**Example 26.38** Consider $X = (100000)_2$, $Y = (010101)_2$ and $Z = (111111)_2$ with $X + Y + Z = (1110100)_2$. The carry-save adders generate the sum and carry bits as follows:

| $i$ | $X_i$ | $Y_i$ | $Z_i$ | Sum bit | Carry bit |
|---|---|---|---|---|---|
| 0 | 0 | 1 | 1 | 0 | 1 |
| 1 | 0 | 0 | 1 | 1 | 0 |
| 2 | 0 | 1 | 1 | 0 | 1 |
| 3 | 0 | 0 | 1 | 1 | 0 |
| 4 | 0 | 1 | 1 | 0 | 1 |
| 5 | 1 | 0 | 1 | 0 | 1 |

Hence the final adder obtains the result

| S |   | 0 | 0 | 1 | 0 | 1 | 0 |
|---|---|---|---|---|---|---|---|
| C | 1 | 1 | 0 | 1 | 0 | 1 | 0 |
|   | 1 | 1 | 1 | 0 | 1 | 0 | 0 |

implying that the value of the sum is $(110100)_2$ and that the carry bit is set.

§ 26.5 Accumulation of partial products

**Figure 26.6** Addition of four 4-bit operands using carry-save adders.

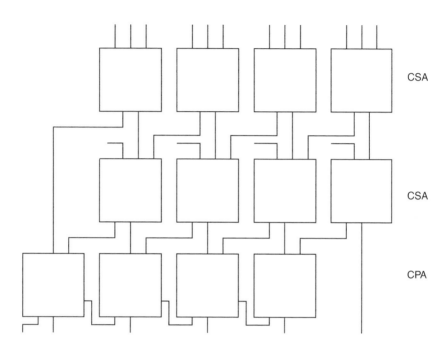

### 26.5.4.b Accumulation of partial products using carry-save adders

We will now explain how carry-save adders can be used to accumulate up to $n$ partial products generated in a $(n \times n)$-bit multiplication.

In general, we will have the following architecture: partial products will be generated as described above (see section 26.4) and will then be fed into a tree architecture, accumulating them in carry-save representation. Finally a carry-propagation adder will convert the result of the reduction tree from carry-save representation to normal binary representation.

**Example 26.39** In the most simple form we assume that all the $n$ multiples of the multiplicand are generated at once. Then we would use $n$-input carry-save adder to reduce the $n$ partial products to two operands for the final addition.

This could be accomplished using $(n-2)$ carry-save adder units and 1 carry-propagation adder.

Normally, one would expect that one has to use $(n-1)$ carry-save adders, however, note that the first counter compresses three partial products.

A typical arrangement for the addition of the four 4-bit numbers $x, y, z$, and $w$ with the 5-bit sum $S$ is shown in Figure 26.6.

There are much more efficient ways of setting up the carry-save adders, which result in a much higher performance. The most common solutions are the *Wallace tree* and *Dadda tree*.

**Remark 26.40** A $k$-input Wallace tree reduces its $k$ $n$ bit entries to two outputs of size

$$(n + \lg k - 1).$$

Figure 26.7 shows an example of a Wallace tree adding 7 numbers, each of size $n$ bits.

Note that some of the outputs of the carry-save adders have to be shifted before they are added again. By $[i, j]$ we denote that the $(j - i + 1)$ bit output is fed into the next block as bit $i$ up to and including bit $j$.

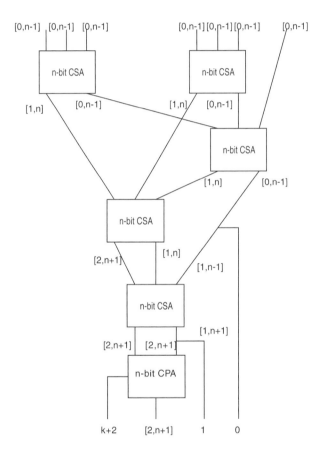

**Figure 26.7** A Wallace tree for the addition of seven $n$-bit numbers.

**Remarks 26.41**

(i) We have the following recurrence for the smallest possible height $h(k)$ of a $k$-input Wallace tree:
$$h(k) = 1 + h(\lceil 2k/3 \rceil)$$
which is easily deduced using the fact that each carry-save adder reduces the number of operands by a factor of $3/2$. By applying this recurrence we can obtain the exact height of an $k$-input Wallace tree, while setting $h(k) = 1 + h(2n/3)$ enables us to obtain a

§ 26.5 Accumulation of partial products

lower bound for $h(k)$ given by

$$h(k) \geqslant \log_{3/2}(k/2),$$

where equality holds for $k = 2, 3$.

(ii) Conversely we can also compute the maximal number of inputs $n(h)$ for a Wallace tree of given height. Again we obtain a recurrence relation for this given by

$$n(h) = \lfloor 3n(h-1)/2 \rfloor.$$

By removing the floor operation, we obtain an upper bound for $n(h)$ given by $n(h) \leqslant 2(3/2)^h$ and a lower bound $n(h) > 2(3/2)^{h-1}$.

(iii) The number of levels needed for a $k$ operand addition using Wallace trees is shown in the following table:

| Number of levels needed for given number of operands | | | | |
|---|---|---|---|---|
| Operands | 3 | 4 | $5 \leqslant k \leqslant 6$ | $7 \leqslant k \leqslant 9$ |
| Levels | 1 | 2 | 3 | 4 |
| $10 \leqslant k \leqslant 13$ | $14 \leqslant k \leqslant 19$ | $20 \leqslant k \leqslant 28$ | $29 \leqslant k \leqslant 42$ | $43 \leqslant k \leqslant 63$ |
| 5 | 6 | 7 | 8 | 9 |

(iv) The general philosophy in the Wallace adder is to reduce the number of operands as quickly as possible, therefore we apply $\lfloor m/3 \rfloor$ full adders to $m$ dots in one column. This means that we make the final carry-propagation adder as small as possible, which in turn will lead to the minimal possible delay at this stage.

Another philosophy is that of the Dadda tree. The Dadda tree aims at reducing the number of operands, thus using the fewest possible number of full and half adders. In general, this means that the carry-save adder tree will be simpler in the Dadda arrangement, while the final carry-propagation adder will be larger.

**Example 26.42** We consider the case of multiplication of two 4-bit numbers.

When using the conventional algorithm from Section 26.3.1, we would obtain the partial products in the following form:

```
6  5  4  3  2  1  0
            •  •  •  •
         •  •  •  •
      •  •  •  •
   •  •  •  •
```

We can rearrange these elements in the following way:

```
6  5  4  3  2  1  0
•  •  •  •  •  •  •
   •  •  •  •  •
      •  •  •
         •
```

**Table 26.8** Wallace tree for $4 \times 4$ multiplication.

| 1 | 2 | 3  | 4  | 3  | 2  | 1 |   |
|---|---|----|----|----|----|---|---|
|   |   | FA | FA | FA | HA |   |   |
|   | 1 | 3  | 2  | 3  | 2  | 1 | 1 |
|   |   | FA | HA | FA | HA |   |   |
|   | 2 | 2  | 2  | 2  | 1  | 1 | 1 |
|   |   | \multicolumn{4}{c}{4-bit adder} |   |   |
| 1 | 1 | 1  | 1  | 1  | 1  | 1 | 1 |

**Table 26.9** Dadda tree for $4 \times 4$ multiplication.

| 1 | 2 | 3  | 4  | 3  | 2  | 1 |   |
|---|---|----|----|----|----|---|---|
|   |   | FA | FA |    |    |   |   |
|   | 1 | 3  | 2  | 3  | 2  | 1 | 1 |
|   |   | FA | HA | HA | FA |   |   |
|   | 2 | 2  | 2  | 2  | 1  | 2 | 1 |
|   |   | \multicolumn{4}{c}{6-bit wide CP adder} |   |   |
| 1 | 1 | 1  | 1  | 1  | 1  | 1 | 1 |

The Wallace tree design for accumulating these partial products is shown in Table 26.8, where FA denotes a full adder, HA a half adder, and the integers represent the numbers of bits remaining in one column. In the Dadda tree the structure described in Table 26.9 is used.

Note that in the Wallace tree we use 5 full adders, 3 half adders, and a 4-bit carry-propagation adder. In the Dadda tree we use 4 full adders, 2 half adders and a 6-bit carry-propagation adder. See [KOR 2002] for more details on these examples.

**Remark 26.43** Since we gave an example for a 7-operand Wallace tree above, it is instructive to also show how this example may be applied in a $7 \times 7$ multiplication. This is shown in Figure 26.10.

Note that in contrast to the simple addition of 7 operands as shown in Figure 26.7, in this situation we have also to deal with shifted results due to the structure of the partial products generated. In total we have 14 positions for the product value. At the top level, the 7 partial products corresponding to positions $[0, 6]$, $[1, 7]$ up to $[6, 12]$, are fed into the carry-save adder tree.

In contrast to the standard addition of 7 operands of 7 bits each, we have to use a 10 bit wide carry-save adder and a 10-bit wide carry-propagation adder in the last two levels of the tree.

## 26.6 Modular reduction in hardware

Most of the modular reduction techniques described in Section 10.4 are directly applicable to hardware implementations. Note that, in general, interleaved multiplication and reduction methods as described in Algorithm 11.1 will be preferred for implementation in hardware since they guarantee that the intermediate result will not require significant register space to be stored.

Hardware implementations of modular reduction will favor techniques that allow an easy determination of the reduction factor for the intermediate results. For example, Montgomery multiplication, which is probably the most used algorithm for hardware implementations, (cf. Section 10.4.2) for

§ 26.6 Modular reduction in hardware

example, reduces this task to a simple lookup of the most significant word of the intermediate result, which can be be realized very efficiently in hardware.

As pointed out before, modular reduction implementations in hardware will most likely use an interleaved multiplication-reduction approach, since this guarantees that the bit size of the intermediate results can be stored in registers not much larger in size than the size of the input variables. One should note that there exist many different possibilities to realize a word-wise interleaved multiplication-reduction algorithm. All these are equivalent from an algorithmic point of view, but can differ greatly with respect to the suitability of implementation in hardware.

**Figure 26.10** Carry-save adder tree for $7 \times 7$ multiplication.

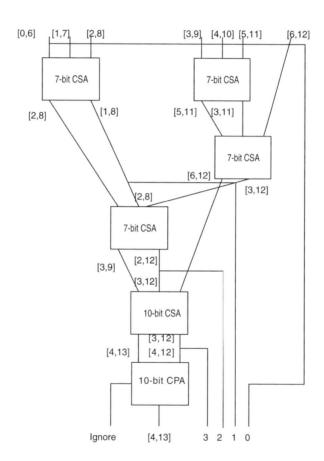

**Remark 26.44** Considering Montgomery reduction, Acar, Kaliski, and Koç have carefully analyzed the different ways of implementation in [KOAC⁺ 1996]. They classify algorithms according to whether multiplication and reduction are separated or interleaved, where the interleaving can be either coarse-grained or fine-grained, depending on how often multiplication and reduction alternate (i.e., after processing an array of words, or just one word). Furthermore, they consider two general forms of multiplication and reduction, one form being the operand scanning, where an outer loop

moves through words of one of the operands, while the other form considered is product scanning, where the loop moves through words of the product itself. This second aspect is independent from the first; moreover, it is also possible for multiplication to have one form and reduction to have the other form, even in the integrated approach.

Acar, Kaliski, and Koç come up with five variants and evaluate these according to the number of elementary multiplications, additions, and read- and write-operations, as well as the memory space used by the algorithm.

Based on implementations in C and assembler, they find that the so-called *coarsely integrated operand scanning method* has the best performance. This method is based on an interleaved multiplication-reduction algorithm that multiplies one word of the first operand with the complete second operand, then performs reduction on the intermediate result and also integrates the right shifting of the reduced intermediate result into the reduction process itself. However, they note that on processors with different arithmetic architecture, like digital signal processors, other algorithms can outperform this method. This is a typical phenomenon, showing again that, as mentioned before, there is no universal optimal choice of algorithms that will offer maximum performance on different types of hardware platforms.

**Remarks 26.45**

(i) To give the reader an idea of different types of optimizations, we briefly present two implementational options, each using different hardware and algorithmic options.

(ii) *Incomplete reduction.* In [YASA$^+$ 2002] the authors consider the use of incomplete reduction for modular arithmetic, i.e., when computing modulo $p$ the intermediate results are allowed to remain in a range of $[0, 2^m - 1]$, where $p < 2^m - 1$. This method is especially suited for cryptographic hardware that supports word-wise arithmetic operations, when the exponent $m$ is chosen in such a way that the elements of the interval $[0, 2^m - 1]$ fit exactly into one or more words. What is the benefit of this approach? Consider for example addition $x + y \bmod p$ of two reduced numbers $0 \leqslant x, y < p$. This requires the comparison of the intermediate result $s = x + y$ to $p$, if $s \geqslant p$ holds then the subtraction $s - p$ has to be performed. Now a comparison operation can be performed by bitwise operations on the most significant word of the intermediate result $s$, however, these operations can be quite costly to implement, if no use can be made of the cryptographic hardware, which is assumed to support a word-wise arithmetic. If one performs addition using incomplete reduction, the comparison between $x + y$ and $2^m - 1$ can be simply performed by using the word-wise oriented cryptographic hardware, since $2^m - 1$ by assumption fits exactly into a number of words. The authors also demonstrate that Montgomery multiplication can be easily adapted to the case of incomplete reduced numbers and that the two reduction algorithms differ only very slightly. Based on a C implementation of their findings, the authors report that an incomplete addition is 34–43% faster than the complete addition for primes $p$ of bit size of 161 to 256 bits. Similarly, the incomplete subtraction, which is an easy adaptation of the addition, is 17–23% faster than the complete subtraction. Since the two Montgomery reduction algorithms differ only very slightly, only a small speedup of 3–5% is reported for this.

(iii) *Special moduli.* It was reported before that special moduli can be used to obtain very efficient modular reduction (cf. Section 10.4.3). This can be used in both dedicated cryptographic hardware and systems that only use a standard controller without specific cryptographic functionality. Consider for example the case of Mersenne prime $p = 2^k - 1$. In Section 10.4.3, it was outlined that for $0 < x < p^2$ we can rewrite $x$ as $x = x_1 2^k + x_0$ and reduce this element modulo $p$ via the formula $x \equiv x_1 + x_0$

(mod $p$). Thus the modular reduction can be achieved by a simple modular addition. This operation is so simple that it can even be performed using the elementary addition provided by a standard micro controller. The techniques also described in Section 10.4.3 extend these reduction methods to primes of the more general form $p = 2^m + c$ with $|c|$ small. Again the main arithmetic operation used in these reduction techniques is addition and only a few multiplications, and thus can be realized even on a general purpose microcontroller. Again it should be noted that the benefit of the implementation of these special reduction techniques in hardware depends strongly on the application the hardware is intended to support. If this is a closed application (i.e., one in which all the system parameters are known in advance and can thus be chosen to allow the usage of special moduli), then the use of such specialized hardware will lead to a substantial performance gain and might even make the usage of general purpose hardware instead of specialized cryptographic hardware. However, if the hardware must support different moduli in an open application, where the system parameters like, for example, prime fields can vary, no automatic performance gain can be expected, since it cannot be guaranteed that all users in the system will actually make use of the special moduli supported by the hardware.

## 26.7 Finite fields of characteristic 2

While arithmetic modulo $p$ can be realized using normal long-integer arithmetic together with special reduction methods as described above, this does not hold for the case multiplication in a field of characteristic 2.

The main difference is that in normal long-integer arithmetic, multiplication can be realized as a combined shift-and-add operation involving carries, while addition in a field of characteristic 2 is equivalent to the binary logical XOR function, which implies that no carries occur.

Therefore, special multipliers have to be designed for the $\mathbb{F}_{2^d}$ situation.

**Remarks 26.46**

(i) The following two ways of representing an element of a finite field $\mathbb{F}_{2^d}$ (see also 11.2.1) are generally considered for hardware implementations:

- Using polynomial basis representation.
- Using normal basis representation.

(ii) From the point of hardware realization of multiplication, we can also distinguish between two different possibilities:

- bit parallel implementations
- bit serial implementations

(iii) Bit parallel implementations compute the product of two elements of $\mathbb{F}_{2^d}$ in one clock cycle. This method has the highest area consumption and will thus in general not be available in especially restricted environments. Therefore in the following only bit serial multipliers will be investigated.

## 26.7.1 Polynomial basis

Let two elements $a(X)$ and $b(X)$ in canonical basis be given as

$$a(X) = \sum_{i=0}^{d-1} a_i X^i \quad \text{and} \quad b(X) = \sum_{i=0}^{d-1} b_i X^i.$$

The standard method to multiply these two elements in hardware is to use the *Mastrovito multiplier*: this canonical basis multiplier is based on the fact that the $\mathbb{F}_{2^d}$ multiplication of $a(X)$ and $b(X)$ can be described as a matrix product

$$Z \times (b_0, b_1, \ldots, b_{d-1})^t,$$

where the matrix $Z = (z_{ij})_{1 \leqslant i,j \leqslant d}$ is a function of both the vector $(a_0, a_1, \ldots, a_{d-1})$ and the $(d-1) \times d$ basis reduction matrix $Q = (q_{ij})$ defined by

$$\left(X^d, X^{d+1}, \ldots, X^{2d-2}\right)^t = Q \times \left(1, X, X^2, \ldots, X^{d-1}\right)^t.$$

The values of $z_{ij}$ are defined to be

$$z_{ij} = \begin{cases} a_i & : \quad j=0, i=0,1,\ldots,d-1 \\ u(i-j)a_{i-j} + \sum_{t=0}^{j-1} q_{j-1-t,i} a_{m-1-t} & : \quad j=1,\ldots,d-1, i=0,1,\ldots,d-1 \end{cases}$$

where the step function $u$ is defined via

$$u(t) = \begin{cases} 1 & \text{if } t \geqslant 0 \\ 0 & \text{otherwise.} \end{cases}$$

**Remarks 26.47**

(i) The Mastrovito algorithm describes a way of directly implementing the matrix product $Z \times (b_0, b_1, \ldots, b_{d-1})^t$. It has a theoretical gate count of $d^2$ AND gates and at least $d^2 - 1$ XOR gates. Here the lower limit for the bound on the XOR gates is attained if the defining field polynomial $m(X)$ is a trinomial with certain coefficient configuration.

(ii) The delay is bounded from above by $T_A + 2\lceil \lg d \rceil T_X$ for an irreducible polynomial of the form $X^d + X + 1$.

In polynomial basis we can also use a bit-serial multiplier utilizing a shift-and-add, either with a least-significant- or a most-significant-bit-first approach. Algorithms that can be used for this are described in Section 11.2.2. Note that for a hardware implementation one will most likely choose an algorithm that interleaves multiplication and reduction, as described in Algorithm 11.1 for the case of modular reduction in odd characteristic. The main advantage of an interleaved algorithm is that the intermediate result one needs to store has much smaller size, thus preventing the usage of double sized registers.

**Remark 26.48** A generalization of this method is a least- or most-significant-digit-first approach. Choose a digit size $D$, then an element of $\mathbb{F}_{2^d}$ can be represented with $k_D$ digits, where $1 \leqslant k_D \leqslant \lceil d/D \rceil$. Now write

$$b(X) = \sum_{i=0}^{k_D-1} B_i X^{Di} \quad \text{where} \quad B_i(X) = \sum_{j=0}^{D-1} b_{Di+j} X^j.$$

The multiplication $a(X)b(X) \bmod m(X)$ can now be realized in the digit-approach by an algorithm similar to the one described in Algorithm 11.37. This will lead to especially efficient implementations if the digit size $D$ is chosen to be the width of the hardware multiplier used. In this case the operation $B_i(At^{Di})$ can be realized with a few calls to the internal hardware multiplier. Note that, in general, the possibility to use lookup tables as required in Algorithm 11.37 will lead to additional area consumption and therefore cannot be taken for granted.

---

**Algorithm 26.49** Least-significant-digit-first shift-and-add multiplication in $\mathbb{F}_{2^d}$

INPUT: An irreducible polynomial $m(X)$ over $\mathbb{F}_2$ of degree $d$, polynomials $a(X)$ and $b(X)$ of degree $d-1$ over $\mathbb{F}_2$.
OUTPUT: The product $r(X) = a(X)b(X) \bmod m(X)$.

1. $r(X) \leftarrow 0$
2. **for** $i = 0$ **to** $k_D - 1$
3. $\quad r(X) \leftarrow B_i(AX^{Di}) + r(X)$
4. $\quad r(X) \leftarrow r(X) \bmod m(X)$
5. $r(X) \leftarrow r(X) \bmod m(X)$

---

### 26.7.2 Normal basis

The standard multiplier for normal basis multiplication was proposed by Massey and Omura. A bit serial Massey–Omura multiplier uses the following main observation: let two elements $A$ and $B$ of the field $\mathbb{F}_{2^d}$ be given in normal basis representation

$$A = \sum_{i=0}^{d-1} a_i \alpha^{2^i} \text{ and } B = \sum_{i=0}^{d-1} b_i \alpha^{2^i}.$$

The product $C = A \times B$ has the representation

$$C = \sum_{i=0}^{d-1} c_i \alpha^{2^i}.$$

The Massey–Omura circuit has the property that it computes the coefficient $c_{d-1}$ from the inputs $(a_0, a_1, \ldots, a_{d-1})$ and $(b_0, b_1, \ldots, b_{d-1})$. It will produce the remaining coefficients $c_{d-2}, c_{d-3}, \ldots, c_1, c_0$, when the rotated input vectors

$$(a_{d-k}, a_{d-k+1}, \ldots, a_{d-1}, a_0, a_1, \ldots, a_{d-k-1})$$

and

$$(b_{d-k}, b_{d-k+1}, \ldots, b_{d-1}, b_0, b_1, \ldots, b_{d-k-1})$$

for $1 \leqslant k \leqslant d-1$ are used as inputs.

Thus using one Massey–Omura circuit we can compute the product of the two elements $A$ and $B$ in $d$ steps.

**Remark 26.50** A different approach is to use a canonical basis multiplier in order to perform a normal basis multiplication. It is described in [KOSU 1998].

The main observation here is that in some cases the basis $B_1 = (\alpha, \alpha^2, \alpha^{2^2}, \ldots, \alpha^{2^{d-1}})$ can be rewritten as $B_2 = (\alpha, \alpha^2, \alpha^3, \ldots, \alpha^d)$, which in turn implies that an element in normal basis representation can be rewritten in the shifted canonical basis $B_2$. Thus, instead of using a canonical basis multiplier, we first transform the operands from normal basis to canonical basis, then multiply in canonical basis and finally convert back. Of course, this is only worth doing if the transformations needed are not too expensive.

However, it turns out that the transformation can be obtained via a suitable permutation of the coefficients, thus the transformation can be easily implemented in hardware.

## 26.8 Unified multipliers

Since elliptic curve cryptography can be implemented both over fields of characteristic 2 and of characteristic larger than 2, it is desirable to support both these cases in hardware. Clearly, an optimal solution would be to have one multiplier unit, which is able to perform modular reduction for $\mathbb{F}_p$, as well as multiplication of polynomials over $\mathbb{F}_2$ modulo an irreducible polynomial $m(X)$ for $\mathbb{F}_{2^d}$ support.

There are several such unified multipliers described in literature.

**Remark 26.51** In 2000, Savas introduced a unified multiplier (see [SATE$^+$ 2000]), which uses Montgomery reduction for modular multiplication and an adaptation of Montgomery reduction for the $\mathbb{F}_{2^d}$ multiplication, which was introduced in [KOAC 1998].

Analogous to Montgomery reduction in the modular case, instead of computing $a \times b \in \mathbb{F}_{2^d}$ it is proposed to compute $a \times b \times r^{-1}$ for a special fixed element of $\mathbb{F}_{2^d}$. Let $m(X)$ be the defining irreducible polynomial of degree $d$ and $r(X) = X^d$, then $r$ is the element given by $r = r(X) \bmod m(X)$. As in the original method, $m(X)$ and $r(X)$ need to be prime to each other, hence $m(X)$ should be not divisible by $X$. Since $m(X)$ is irreducible over $\mathbb{F}_2$ this will always be the case. The relative primeness of $r(X)$ and $m(X)$ implies the existence of $r^{-1}(X)$ and $\widetilde{m}(X)$, such that

$$r(X)r^{-1}(X) + m(X)\widetilde{m}(X) = 1$$

via the extended Euclidean algorithm. Now the Montgomery product of $a$ and $b$ is defined to be

$$c(X) = a(X)b(X)r^{-1}(X) \bmod m(X).$$

A detailed description of the multiplication algorithm can be found in [KOAC 1998].

Note that we have to transfer elements $a(X)$ of $\mathbb{F}_{2^d}$ to the Montgomery representation, that is $a(X)X^{-d} \bmod m(X)$ in order to apply the Montgomery reduction as described.

This has the advantage that both multiplications with reduction essentially follow the same reduction algorithm. The implementation uses an array of word-size processing units organized in a pipeline. Therefore the architecture is highly scalable. The word size of the processing unit and the number of pipeline stages can be selected according to the desired area/performance relation.

**Remark 26.52** Goodman and Chandrasekaran [GOCH 2000] presented a domain specific reconfigurable cryptographic processor (DSRPC), which provides a unified multiplier by means of a single computation unit, equipped with data path cells that can be reconfigured on the fly. While modular arithmetic is performed using Montgomery reduction, the $\mathbb{F}_{2^d}$ multiplication is based on an iterated MSB-first approach.

**Remark 26.53** Großschädl [GRO 2001] introduced a unified multiplier that performs modular multiplication in a serial-parallel way, where the reduction is performed during multiplication by concurrent reduction of the intermediate result using an MSB-first approach. The $\mathbb{F}_{2^d}$ multiplication in this multiplier is simply achieved by setting all carry-bits of the intermediate result to 0 resulting in a shift-and-add reduction algorithm. Therefore, the area-cost of the modified multiplier is only slightly higher than that of the original modular multiplier. Furthermore, the transformation into Montgomery domain is not needed, and we also do not have to do precomputations.

## 26.9 Modular inversion in hardware

As it was pointed out before, the availability of a fast inversion algorithm is of great importance for the efficient implementation of elliptic and hyperelliptic curve cryptosystems. Software-based operations can normally rely on very fast inversion, which in turn implies that they will most likely use affine coordinates. However, the situation is quite different on hardware platforms that are used for embedded devices or smart cards. In the following we briefly discuss two main options.

If only the basic modular arithmetic operations (multiplication and squaring) are supported by hardware, the only way to realize modular inversion is to either implement the extended Euclidean algorithm on the main CPU, using some functionality of the cryptographic hardware, or to make use of Fermat's little theorem (see the introduction of Section 11.1.3). In most of these cases, the second solution will be better, since modular exponentiation is normally available anyway.

**Remarks 26.54**

(i) Hardware support for modular inversion in odd characteristic usually focuses on different variants of the Extended Euclidean algorithm as described in Section 11.2.4 for even characteristic and in Section 11.1.3 for odd characteristic.

(ii) We first deal with the odd characteristic case. The analysis of the main arithmetic operations used in this algorithm shows that the following operations are most important:

   (a) addition
   
   (b) subtraction
   
   (c) halving an even number
   
   (d) computing the absolute value $|A - B|$ of two odd numbers $A$ and $B$.

   While the first three operations are easy to implement (addition and subtraction are performed by the main arithmetic unit, halving of an even number can be realized very easily by a shift operation, and checking whether a number is even or odd can be accomplished by simply inspecting its least significant bit), the computation of the absolute value $|A - B|$ can be different to realize. This is mainly due to the fact that the absolute value is normally obtained by first computing $A - B$, then inspecting the sign of this result and finally computing the two's complement if the outcome of the subtraction was negative. Since the sign of the result will only be available after the end of subtraction, the two's complement building can only start once the subtraction operation is finished. Summarizing, we see that the cost for the comparison of two integers is mainly dominated by the cost of a subtraction. However, the computation of the absolute value will cost some extra cycles.
   
   If Montgomery multiplication is used, there are much more efficient inversion algorithms available that have already been described in Section 11.1.3.

(iii) In even characteristic, the same two main options as in the odd characteristic case are valid. The usage of the basic polynomial multiplication in order to compute the inverse of an element was described in Section 11.2.4.c. However, special properties of polynomial arithmetic in characteristic two allow especially efficient implementations of the inversion according to the extended Euclidean algorithm. This is mainly due to the fact that addition and subtraction are replaced by the simple XOR operation, which does not have carry propagation. Furthermore, the comparison of two elements can be accomplished by comparing single bits starting at the most significant bit.

The combination of these observations with specially designed hardware can lead to very fast implementations of the modular inversion in even characteristic.

(iv) Analogously to the design of unified multipliers able to perform modular arithmetic in both even and odd characteristic, designs for unified modular inversion also exist. These designs obviously rely on the choice of an inversion algorithm that can be easily adjusted to the case of even and odd characteristic. A good example of this is the Montgomery inversion algorithm as described in [KAL 1995], which was used by Savas and Koç to describe architectures for unified inversion in [SAKO 2002]. The main algorithmic ingredient here is that the comparison of two integers $u$ and $v$ (i.e., the decision whether $u > v$ holds) is replaced by the comparison of the bit-sizes of $u$ and $v$ respectively, as done in the characteristic two case. This replacement implies requiring the temporary variables of this algorithm to be either positive or negative (which in turn implies that a further bit has to be reserved to store the sign of these variables). However, as a result of this modification, identical hardware can now be used for the comparison.

(v) Inversion in optimal extension fields, as introduced in Section 11.3, relies on a combination of the computation of a modular inverse in a (small) prime field $\mathbb{F}_p$ and a modular exponentiation in an extension field $\mathbb{F}_{p^d}$. Due to the special nature of the prime $p$ this can be realized so efficiently that even without specific cryptographic hardware, fast inversion implementations are possible (see [WOBA+ 2000]). Their implementational results are based on an Infineon SLE44C24S, an 8051 derivative. For an OEF of approximately 135 bits field size their performance results are as follows: for a modular multiplication $5084\mu$s are needed, while an inversion is computed in $24489\mu$s. Thus they obtain a ratio multiplication/inversion below 5, which allows the usage of affine coordinates in the implemented elliptic curve operations. This rather low ratio is achieved by using an adaptation of the algorithm described in Section 11.3.4, where an addition chain is used for the main exponentiation $\alpha^{r-1}$, and the inversion $t^{-1}$ in the ground field $\mathbb{F}_p$ is performed via table lookup.

# Chapter 27

# Smart Cards

Bertrand Byramjee and Andrew Weigl

## Contents in Brief

| | | |
|---|---|---|
| 27.1 | **History** | 647 |
| 27.2 | **Smart card properties** | 648 |
| | Physical properties • Electrical properties • Memory • Environment and software | |
| 27.3 | **Smart card interfaces** | 659 |
| | Transmission protocols • Physical interfaces | |
| 27.4 | **Types of smart cards** | 664 |
| | Memory only cards (synchronous cards) • Microprocessor cards (asynchronous cards) | |

Although smart cards are now very common, this technology is still very new, with the first smart cards appearing in the 1970's. Since then, their evolution has been very rapid. Smart cards have advanced from simple memory cards to very efficient "microcomputers" with multiple applications.

Equipped with a microcontroller, these cards are able to store and protect information using cryptographic algorithms. They are also resistant to physical stresses such as twisting and bending. The physical structure of the smart card consist of a small plastic rectangle with a magnetic stripe, holograms, relief characters and an embedded chip. They are small, and easy to use and carry. The security and portability of smart cards provide a safe, reliable, convenient, and effective way to ensure secure transactions (banking, e-business, etc.), and to enable a broad range of applications. Thus, modern smart cards can actually be used in any system that needs security and authentication. They have been proven to be an ideal means of making high-level security available to everyone.

This chapter aims to present an overview of today's smart card technology and show the limitations that smart card manufacturers must take into account when implementing cryptographic algorithms, for example, elliptic or hyperelliptic curve algorithms, in a smart card environment.

## 27.1 History

In the beginning of the 1950's, the first plastic (PVC) cards appeared in the USA as a substitute for paper money. They were initially aimed at the rich and powerful, and were only accepted by prestigious hotels and restaurants. These cards were very simple with the owner's name printed in relief, and sometimes the handwritten signature was added. These cards provided a more convenient payment system than paper money. With the involvement of VISA™ and MasterCard™ in plastic money, credit cards spread rapidly around the world. Later a magnetic stripe was added to reduce fraud and to increase security. Confidential digitized data was stored on this stripe, but this information was accessible to anyone possessing the appropriate card reader.

Between 1970 and 1973 there was a significant development in plastic cards with the addition of microcircuits to the card. Many patents were filed during this time; the best known inventors include: J. Dethleff, K. Arimura, and R. Moreno. The term "smart card" was proposed by R. Bright. It was not until 1984 that the smart card was first put into commercial use by the French PTT (postal and telecom services) with their first telephone cards (smart cards with memory chips). In 1986, millions of these smart cards were sold in France and other countries. After telephone cards, the next big application was their use as banking cards. This development was more difficult because they contained more complicated chips that were able to compute cryptographic calculations. The French banks were the first to introduce this technology in 1984. A number of ISO standards were created to encourage interoperability of smart cards. By 1997, bank cards were widely used in France and Germany. The microcontrollers continued to advance and became more powerful with larger memory capacity. This allowed for sophisticated cryptographic algorithms, providing higher levels of security.

Nowadays, smart cards are present all over the world, and their use is likely to spread even further.

## 27.2 Smart card properties

Smart cards are physically similar to the classic embossed plastic cards. The older model cards are used as the base design for the newer smart cards. There are two different categories of smart cards: memory only cards, which are the cheapest and the simplest, and the microprocessor cards, which are more expensive, but have more applications and security features. The structure of smart cards is standardized by ISO, principally: ISO 7816 [ISO 1999a, ISO 1999b, ISO 1999c, ISO 1999d], and ISO 7810 [ISO 1995].

The following sections look at the different aspects of the smart card properties.

### 27.2.1 Physical properties

The most widely used smart card format, ID-1, is part of the 1985 ISO 7810 standard [ISO 1995]. Most smart cards are made from PVC (polyvinyl chloride), which is also used for credit cards. Some are made from ABS (acrylonitrile-butadiene-styrol), but they cannot be embossed; an example application is a mobile phone card.

The body of the card includes the following components: magnetic stripe, signature stripe, embossing, imprinting of personal data (picture, text, fingerprint), hologram, security printing, invisible authentication features (fluorescence, UV), and a microprocessor chip.

#### 27.2.1.a The chip module and its embedding

The chip module, also called the *micromodule*, is the thin gold contact on the left side of the smart card. This module needs to be firmly attached to the plastic of the card. Its purpose is to protect the card and the microprocessor chip. The contacts for contact-type smart cards can also be in the chip module.

Many embedding techniques have been tested and used with the aim to optimize overall card resilience to everyday physical and mechanical stresses (temperature abrasion, twisting, bending,

## § 27.2 Smart card properties

etc.) while keeping the production costs as low as possible.

### 27.2.1.b Contact and contactless cards

There are two main ways a smart card can communicate with the card terminal: through physical contact or by using a contactless connection. The contact cards were the first types of smart cards on the market. However, with new advances in microcircuit technology, contactless cards have become physically feasible.

#### Contact cards

This is currently the most common type of card. It communicates via a card reader where the information passes through the contacts. There are metal contacts inside the card reader and on the chip module of the smart card. The position and dimensions of these contacts (power supply, data transfer, etc.) are set in the ISO 7816-2 standard [ISO 1999b]. Another standard, AFNOR, is still in use by some cards in France, but is likely to disappear in the near future.

**Figure 27.1** Pin layout for contact smart cards.

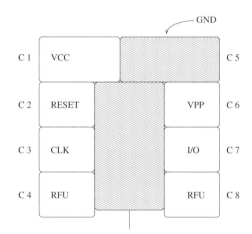

There are 8 contact areas $C_1, \ldots, C_8$:

$C_1$: Supply voltage, VCC,
$C_2$: Reset,
$C_3$: Clock, CLK,
$C_4$: Not in use, reserved for future use,
$C_5$: Ground, GND,
$C_6$: External voltage programming,
$C_7$: Input/Output for serial communication,
$C_8$: Not in use, reserved for future use.

#### Contactless cards

These cards contain special circuits, which allow data transmission over short distances without mechanical contact and without a direct supply of power. This technology is not new but is difficult to apply to smart cards. At the moment it is possible to incorporate a battery into the card, but it increases the size and cost of the card. Research is ongoing to reduce this problem.

Not only is there a problem supplying power to the smart card circuits, but data and clock signals also need to be transmitted between the card and the terminal. The technique of capacitive and inductive coupling, at this time, is the most suitable for smart cards and has been standardized in ISO/IEC 14443 [ISO 2000]. This standard presents a method for capacitive and inductive coupling where the card's conductive surfaces act as capacitor plates. One or several coupling loops are integrated into the card to receive energy from the terminal. A carrier frequency in the range 100-300 kHz is used, which allows very rapid transmission.

**Dual interface or "combi-cards"**

In the future it is likely that "combi-cards" will become more common. They combine the advantages of contact and contactless cards. In ISO/IEC 10536 the application is described as "slot or surface operation." Depending on the operation, the card must either be inserted in a slot to make contact or placed on a certain surface for contactless transaction. This type of card allows applications such as credit, debit, membership, and mass transit to be used on the same card.

## 27.2.2 Electrical properties

The electrical properties of a smart card depend on its embedded microcontroller, since this is the only component of the card with an electrical circuitry. The basic electrical requirements are defined by the ISO/IEC 7816-3 standard, Part 3: *Electronic signals and transmission protocols* [ISO 1999c]. Electrical characteristics and class indication for operating at 5 V, 3 V and 1.8 V are described within Amendment 1. Amendment 2, which describes an USB interface for smart cards, is currently under preparation. The GSM mobile telephone network (GSM 11.11) should be mentioned here, because it also contributes to the requirements in this area. Further modifications of the ISO/IEC 7816 standard are driven by the UMTS specification.

### 27.2.2.a Supply voltage

A smart card supply voltage is 5 V, with a maximum deviation of $\pm 10\%$. This voltage, which is the same as that used for conventional transistor-transistor-logic (TTL) circuits, is standard for all cards currently on the market. Since all modern cellular telephones are built with 1.8 V technology (GSM 11.18), modern smart cards are designed for a voltage range of 1.8-5 V $\pm 10\%$, which results in an effective voltage range of 1.6-5.5 V. They can be used in both, 1.8 V and 5 V terminals, to keep the advantage of simple and straightforward card usage.

### 27.2.2.b Supply current

The built-in microcontroller obtains its supply voltage via contact C1 (see Figure 27.1). According to the GSM 11.11 specification, the current may not exceed 10 mA, so the maximum power dissipation is 50 mW, with a supply voltage of 5 V and an assumed current consumption of 10 mA. Table 27.2 gives an overview of the actually defined maximum power consumption classes, specified by ISO 7816 and GSM.

The current consumption is directly proportional to the clock frequency used, so it is also possible to specify the current as a function of the clock frequency. State-of-the-art smart card microcontrollers use configurable internal clock frequencies for their processor and their arithmetic coprocessor. Hence, the current consumption is not only dependent on the external clock, but also on the given configuration of the microcontroller itself and the setting of the coprocessor. The coprocessor can be programmed to keep power consumption under a set value, for example, the GSM values.

## § 27.2 Smart card properties

**Table 27.2** Smart card power consumption specified by ISO 7816 and the GSM specifications.

| Specification | ISO 7816-3 | | GSM | | |
|---|---|---|---|---|---|
| Notation | Class A | Class B | GSM 11.11 | GSM 11.12 | GSM 11.18 |
| Supply voltage | 5 V | 3 V | 5 V | 3 V | 1.8 V |
| Supply current | 600 mA | 50 mA | 10 mA | 6 mA | 4 mA |
| Frequency | 5 MHz | 4 MHz | 5 MHz | 4 MHz | 4 MHz |
| Power consumption | 300 mW | 150 mW | 50 mW | 18 mW | 7.2 mW |

### 27.2.3 Memory

Smart cards can be divided into two main components: the processor (including coprocessor) and memory. Memory can again be divided into volatile and non-volatile memory. Table 27.3 shows the different types of volatile and non-volatile memory. Since a smart card needs to be able to function as an independent unit, most cards will be found with a combination of RAM, ROM, and EEPROM.

**Table 27.3** Types of memory found in smart cards.

| Memory types found in smart cards | |
|---|---|
| Volatile memory | Non-volatile memory |
| RAM | ROM |
| | PROM |
| | EPROM |
| | EEPROM |
| | Flash EEPROM |
| | FRAM |

#### 27.2.3.a Read-Only Memory (ROM)

ROMs are non-volatile memory that can be randomly accessed during reading. There is no limit to the number of times the memory can be read, but it can only be written during production. This type of memory requires no voltage to hold the information, so when the power is disconnected, the data is still retained. This is excellent memory for storing vital programs that the smart card needs to run, like the operating system and the diagnostic functions. The data is imprinted onto the chip by using lithographic techniques. ROM cells require the least amount of area per cell compared to other available types of memory.

#### 27.2.3.b Random Access Memory (RAM)

RAM is the work area for the smart card. It can quickly read and write data, and there is no limit to the number of writes a RAM cell can handle. However, since it is volatile memory, constant power needs to be supplied, or otherwise the contents will be lost. The method for accessing this memory is what gives it its name; *random access* means that the memory is selected and directly accessed without having to sequentially traverse the memory block.

In smart cards, the most common form of RAM is static RAM (SRAM), which, unlike dynamic RAM (DRAM), does not need to be periodically refreshed. SRAM has flip-flops as the basic component while DRAM uses capacitors with refresh circuitry.

Smart card chip designers try to keep the amount of RAM to a minimum, since it requires a large area per cell. Indeed, RAM cells require seventeen times more area than a ROM cell.

### 27.2.3.c  Programmable read-only memory (PROM)

Programmable read-only memory is similar to ROM in that once it has been written it cannot be rewritten. The difference is that the code does not need to be written with lithographic techniques. PROM has a serious drawback; access needs to be granted to the address, data, and control buses for the writing process. This leaves a security hole in the smart card that a hacker could use to read the data stored on the chip. PROM is not used in smart cards because of this vulnerability.

### 27.2.3.d  Erasable programmable read-only memory (EPROM)

An EPROM is essentially an $n$-channel MOSFET (metal-oxide-semiconductor field effect transistor) with an extra polysilicon gate called the *floating gate*. In Figure 27.4 the left curve has a relatively low $V_t$ and is normally chosen as state "1." This state is also called the *"preprogrammed"* state.

**Figure 27.4** Threshold voltage curves for programmed and preprogrammed state.

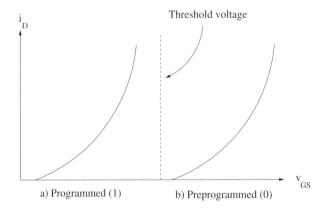

a) Programmed (1)     b) Preprogrammed (0)

A voltage needs to be applied between the drain and source to program the EPROM to the "0" state (see Figure 27.5). On the select gate a voltage of 17 V to 25 V needs to be applied. Since smart card controllers use a supply voltage between 3 and 5 V, a cascaded voltage-multiplier circuit, or charge pump, needs to be used to generate the required voltage levels.

The device acts as a regular $n$-channel enhancement MOSFET when there is no charge present on the floating gate. With the voltages present, a tapered $n$-type inversion layer is formed at the surface of the substrate. The drain-to-source voltage accelerates the electrons through the channel. The electric field formed by the voltage on the selected gate attracts the hot electrons (the accelerated electrons) towards the floating gate. At the floating gate the electrons collect, causing the gate to become negatively charged. This process continues until enough of a negative charge is formed on the floating gate to reduce the strength of the electric field to the point of not being able to accelerate any more hot electrons.

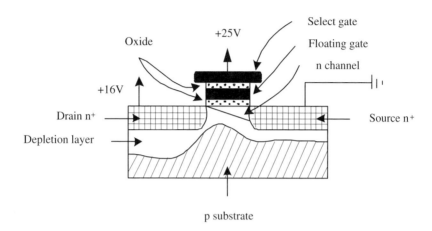

**Figure 27.5** EPROM during programming.

The negatively charged floating gate repels electrons away from the surface of the substrate. To compensate for the loss of electrons in the region, a larger select gate voltage is required to form an $n$-channel. This will shift the $i_D - v_{GS}$ characteristic graph upwards, as can be seen in Figure 27.4 [SESM 1991].

For the microcontroller to read the state of the EPROM, the unit only needs to apply a test $V_{GS}$ between the two $i_D - v_{GS}$ curves. If the current flows, the EPROM is in state "1" and if it does not flow then it is in state "0".

For smart cards, EPROM was used by the French PTT in their first telephone cards, since, at that time, it was the only ROM type memory available [RAEF 2000]. As with other ROM types, it does not require a supply voltage to retain the data. EPROM can be reprogrammed, but it first requires ultraviolet light to erase the old data. This method is not feasible for smart cards, so this technology has been abandoned for newer erasable ROMs.

### 27.2.3.e  Electrically erasable programmable read-only memory (EEPROM)

As with regular computers, sometimes data needs to be read, altered and then stored with the possibility that the voltage supply is disconnected. Computers use hard drives to store the data for longer periods of time, but smart cards do not have this option. Instead they use a type of ROM that can handle multiple writes. EPROM can only be erased with ultraviolet light, which makes it unsuitable as a multi-write memory. The solution is found with another type of ROM that can be electrically erased, the electrically erasable programmable read-only memory (EEPROM), see Table 27.7.

EEPROM operates similarly to the method described in Section 27.2.3.d. There are two main differences between EPROM and EEPROM. The first difference is how the electrons travel from the substrate to the floating oxide layer. The method described in Section 27.2.3.d uses hot electron injection, while standard EEPROM uses the tunnel effect (Fowler–Nordheim effect). A high positive voltage at the select gate causes electrons to migrate through the tunnel oxide to the floating gate, where they collect. Eventually, the floating gate becomes negatively charged.

The second difference between EPROM and EEPROM is how the data is erased. As stated earlier, EPROM requires ultraviolet light to reset its state. For EEPROM a negative voltage applied to the select gate forces the electrons from the floating gate back to the substrate. After this process, the EEPROM is classified again as discharged and the $V_t$ is low.

Similar to RAM and other types of ROM, EEPROM can be read an unlimited number of times. However, there is a limit to the number of writes that can be performed. The life expectancy is limited by the quality, type, and thickness of the tunnel oxide layer, which is the oxide layer between the floating gate and the substrate (see Figure 27.5). During production, the tunnel oxide is one of the first layers to be produced. As the rest of the production continues, it undergoes large thermal stresses that cause minute faults in the oxide layer. This allows the tunnel oxide to absorb electrons during the programming cycle, which are not returned to the substrate when the data is erased. The trapped electrons then collect at the channel between the drain and source. This process continues until enough electrons collect that they influence the threshold voltage to a greater degree than the floating gate. The threshold voltage then stays in one state, regardless of whether the floating gate is charged or not; the EEPROM is then useless.

### 27.2.3.f  Flash electrically erasable programmable read-only memory (flash EEPROM)

Flash EEPROM is a mixture of EEPROM and EPROM technology. It operates with hot electrons but uses an erase technique similar to EEPROM. Most manufacturers implement a combination of flash EEPROM and regular EEPROM onto their chips. Each memory type has its benefits when it comes to endurance and space requirements. A typical application for regular EEPROM is transient data or constants that may need to be occasionally changed. Flash EEPROM is better suited for program code, which may need to be upgraded or changed only a few times during the products, life cycle [BUR 1999].

EEPROM uses both, a read and a write transistor, while flash memory uses only one transistor. Also, each cell of the regular EEPROM requires signal routing and complex address-decoding logic, since each word has its own control signals. Flash memory, on the other hand, employs less complex word decoders that allow for more compact units. This leads to a trade-off in programming flexibility and write time.

With regular EEPROM any word can be altered by accessing that memory location, erasing the word, then writing in the new data. This is not possible with the word decoder of flash memory. Before any data can be changed, the whole array must be erased and then completely rewritten. This leads to a large difference in operating speed. EEPROM can be reprogrammed in a few milliseconds, whereas flash memory may take from a few microseconds to a second to reprogram.

Another key factor that makes regular EEPROM able to better handle transient data is the endurance of its cells. Flash EEPROM can only handle about 1000 write cycles, which is too short for most application requirements. Since application data does not change that often, it is perfect for it to be programmed onto flash EEPROM. It can be upgraded after manufacture giving it an advantage over ROM or PROM. Most smart cards now incorporate some flash EEPROM in their system.

### 27.2.3.g  Ferroelectric random access memory (FRAM)

Ferroelectric Random Access Memory (FRAM), a relatively new type of memory, has recently become available to the microprocessor manufacturers. The characteristics of FRAM are similar to those of both RAM and ROM. It is a non-volatile memory, but with a write speed considerably faster than flash or EEPROM memory.

Instead of using a floating oxide layer as the dielectric, FRAM employs a ferroelectric film. The composition of the film is usually either *PZT* $(Pb(Zr, Ti)O_3)$, *PLZT* $((Pb, La)(Zr, Ti)O_3)$, or *SBT*

## § 27.2 Smart card properties

$(SrBi_2Ta_2O_9)$. Figure 27.6 shows the crystalline structure of PZT. Polarization through electric fields has a dual advantage over the injecting of hot electrons or tunneling effect method, it is faster at writing and it requires less power (see Table 27.7 for the comparison between the different memory types).

**Figure 27.6** View of the crystalline structure of ferroelectric material.

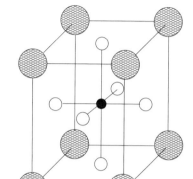

 Lead (Pb)

○ Oxygen (O)

● Zirconium / Titanium (Zr

**Table 27.7** Comparison of different memories [FUJITSU].

| Memory | SRAM | DRAM | EEPROM | FLASH | FRAM |
|---|---|---|---|---|---|
| Type | Volatile | Volatile | Non-volatile | Non-volatile | Non-volatile |
| Read cycle | 12 ns | 70 ns | 200 ns | 70 ns | 180 ns |
| Write cycle | 12 ns | 70 ns | 5 ms | 1 s | 180 ns |
| Data write method | Overwrite | Overwrite | Erase & Write | Erase & Write | Overwrite |
| Write Endurance | $\infty$ | $\infty$ | $10^4$ | $10^3$ | $10^{10}$ |

### 27.2.3.h  Memory management unit (MMU)

The physical memory is divided into memory pages with the assistance of the MMU. Inside the MMU a translation of the memory address occurs, and the visibility of the storage space is changed.

Each segment of the physical memory is placed into the logical memory. Figure 27.8 shows the behavior of the address translation.

**Figure 27.8** Memory segmentation.

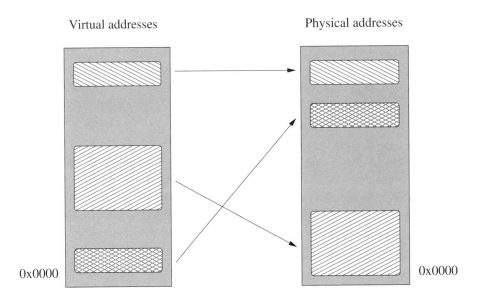

The size of one segment is programmable by the processor and can comprise the whole physical memory or only a small portion of it. It is only limited by the maximum number of allowed segmentations. The processor view of the memory may differ from the real memory alignment, since there is no direct connection between the processor and the memory.

In most cases, the MMU is not enabled after power-on or a global reset. At this point, the processor operates in "system" mode and the virtual addresses are equal to the physical addresses. All special function registers (SFR) of the MMU are accessible by the CPU; these registers perform the address translation.

When an application is executed from the code space the MMU is enabled and it is then in "user" mode. Direct access to the SFR is disabled to protect other applications. Thus, each application can only use its own memory region. This region contains the shared ROM for the program code, and a part of the RAM for the variable data (see Figure 27.9). Memory protection is distinctively important for the non-volatile data in the EEPROM, since that is where applications store sensitive data, and it is necessary to protect this information against spying or loss of power.

### 27.2.4 Environment and software

Smart card systems consist of two parts: the *host system*, residing in the terminal (or in a PC with a card reader) and the *card system*, inside the smart card. This section looks at the smart card software, including the system software and the user applications that run on the card system.

Developing a smart card application traditionally was a long and difficult process. Most smart card development tools were built by the smart card manufacturers using generic assembly lan-

§ 27.2 Smart card properties

guage tools and dedicated hardware emulators obtained from the silicon chip vendors. Therefore, developing smart card applications was limited to a group of highly skilled and specialized programmers who had intimate knowledge of the specific smart card hardware and software. They had to deal with very low-level communications protocols, memory management and other minute details, dictated by the specific hardware of the smart card. Upgrading software or moving applications to a different platform was particularly difficult or even impossible. Furthermore, because smart card applications were developed to run on proprietary platforms, applications from different service providers could not coexist and run on a single card. Lack of interoperability and limited card functions prevented a broader deployment of smart card applications. The development of the Java Card technology changed this situation. This language, designed by SUN technology, offers new possibilities for smart cards, as will be seen in the following sections.

**Figure 27.9** Example for the processor view.

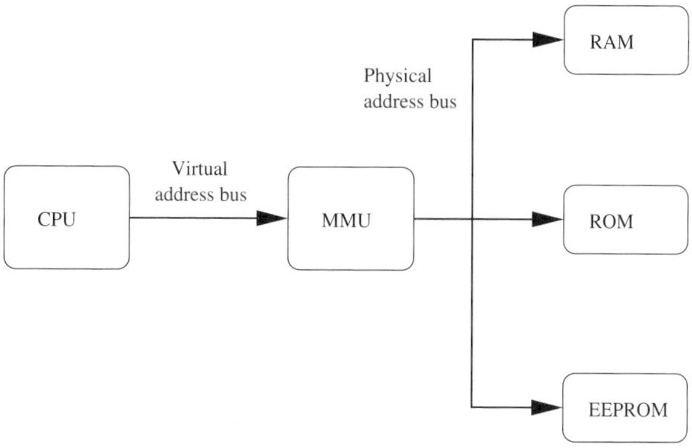

### 27.2.4.a Operating System (OS)

Each operating system depends on the manufacturer's philosophy. The conventional operating system is the one based on standard instructions, which is designed such that it can be implemented for any application on the component. Smart card operating systems support a collection of instructions that the user's applications can access.

Many operating systems have been designed for smart cards: Windows® for smart cards, which is no longer used, Multos, Java Card, etc. There is no real standard for a smart card operating system. The OS is strongly linked to the manufacturer, and standardizing it would affect the intellectual property of the manufacturer. The group SCOPE, directed by GlobalPlatform, tried to design a document specifying the main functionality that should be supported by a smart card OS. The ISO 7816-4 [ISO 1999d] standardizes a wide range of instructions in the format of the application protocol data unit, APDU. A smart card operating system may support some or all of these APDUs as well as the manufacturer's additions and extensions.

The operating system on a smart card has fewer operations to deal with than those present in standard personal computers. Its purpose is essentially to regulate the input and output (I/O), timers, exceptions, communications between the server and the terminal, memory management, and to securely load and run specific programs. In Figure 27.10, the OS is composed of the following elements:

- the BIOS: for the protocol communication T (0, 1, both, or others), the timers, the exceptions, the EEPROM management,
- the Java Card Virtual Machine (VM),
- some native applications used for access rights on memory.

The majority of applications are applets, native applications and the link with the Java Card VM. No external instruction is allowed to interfere with the operation of the card. System crashes or uncontrolled reactions due to a faulty instruction or as result of failed EEPROM sections must not occur under any circumstances.

**Figure 27.10** Architecture of the smart card.

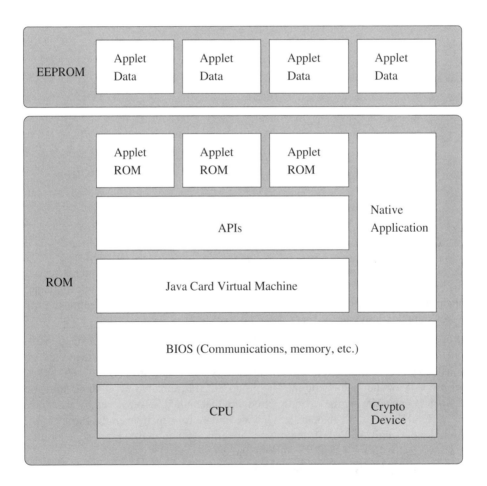

### 27.2.4.b Java Card

Java Card™ offers a way to overcome obstacles hindering smart card acceptance. It allows smart cards and other memory-constrained devices to run applications (called applets) written in the Java programming language. Essentially, Java Card technology defines a secure, portable, and multi-application smart card platform that incorporates many of the main advantages of the Java language. This type of platform has found wide acceptance and is currently being shipped in high volume. A Java Card platform-based smart card runs Java based applications in the form of byte-code. These are loaded into the memory of the smart card's microprocessor where they are run by the virtual machine (typically in the EEPROM). The advantages provided are:

- Multiple application management: multiple Java applications (electronic purse, authentication...) can reside on a single card. These applications can be upgraded with new or updated applets without the need of issuing a new or a different card.
- Security: the different applets present on the card are each separated by the applet firewall. It ensures their integrity and eliminates program tampering; the level of access to all methods and variables is strictly controlled.
- Hardware independence: Java Card technology is independent of the type of component used. It can run on any smart card (8, 16, or 32 bits) because the applets are written on top of the Java card platform.
- Compatibility: any card can run any application. Java Card technology is based on the Java Card international standard ISO 7816, applets can interoperate not only with all Java smart cards but also with existing card acceptance devices.
- High level development: applet developers do not have to deal with the details of microcontroller, with the exception of cryptographic algorithms.

The main elements in Java Card are:

- The Java Virtual Machine: necessary to compile the byte codes of the applet and to run the applications.
- The different applets that can be loaded in EEPROM or stored on ROM for some special applications.
- The APIs: these are the classic APIs of the functions used in the card, for example the cryptographic functions. These are designed and standardized by Sun to ensure that applets are portable.

For more detailed information, the Java Card environment is described in the book *Java Card Technology for Smart Cards: Architecture and Programmers Guide* [CHE 2000].

## 27.3 Smart card interfaces

### 27.3.1 Transmission protocols

This section explains the structure of transmission and specific control in half-duplex transmission protocols. Three types of protocols are defined in the ISO/IEC standards 7816 [ISO 1999a, ISO 1999b, ISO 1999c, ISO 1999d] and 14443 [ISO 2000, ISO 2000a, ISO 2000b, ISO 2000c, ISO 2000d]: one character protocol (T = 0), and two block protocols (T = 1 and T = CL). The designation T = CL is currently in the final stage of specification, which is why it does not appear in literature. Every smart card must support at least one of these protocol types, while terminals need to support all of them.

**Table 27.11** Transmission protocols [RAEF 2000].

| Protocol | Explanation | Specification |
|---|---|---|
| T = 0 | Asynchronous, half-duplex, byte-oriented | ISO/IEC 7816-3 |
| T = 1 | Asynchronous, half-duplex, byte-oriented | ISO/IEC 7816-3 |
| T = 2 | Asynchronous, half-duplex, byte-oriented | ISO/IEC 10536-4 |
| T = 3 | Full-duplex | – |
| T = 4 | Asynchronous, half-duplex, byte-oriented, extension of the T = 0 protocol | – |
| T = 5-13 | Reserved for future use | – |
| T = 14 | Reserved for national use (in Germany: Deutsche Telekom) | Proprietary standards |
| T = 15 | Reserved for future use and extensions | – |

Altogether, the ISO/IEC standards comprise of fifteen types of transmission protocols. Up to now, only three of them are fully specified while the rest is either still under definition, reserved for future use, or reserved for national use (see Table 27.11).

The half-duplex block transmission protocol T = CL addresses the special needs of contactless card environments and is not listed in Table 27.11. It sets out the requirements for contactless smart cards to operate in the vicinity of other contactless cards that conform to the ISO/IEC 10536 and ISO/IEC 15693 standards.

Which transmission protocol is to be used for subsequent communication between terminal and smart card is indicated within the answer-to-reset (ATR) interface bytes and shall always be T = 0 or T = 1 for cards with contacts. If the interface bytes are absent, T = 0 is assumed by default. For contactless cards only T = CL is designated. In Europe only Germany uses the T = 1 protocol, whereas the rest of Europe uses T = 0 protocol.

### 27.3.1.a Protocol T = 0

The half-duplex asynchronous transmission protocol T = 0 (cf. standards [ISO 1999a, ISO 1999b, ISO 1999c, ISO 1999d]) is the basic protocol for cards with contacts. It is the oldest transmission protocol and is designed for a minimum of technical requirements. For this reason, it is used in the present GSM technology for mobile phones. The protocol T = 0 is completely byte-oriented and it uses the following character frame format: a *character* consists of 10 consecutive bits:

- 1 start bit in state "Low,"
- 8 bits, which comprise the data byte,
- 1 even parity checking bit.

A character is the smallest data unit that can be exchanged. The interval between two consecutive character's leading edge start bits is comprised of the duration of the data plus a *guard time*. The minimum value of this guard time is 2 Elementary Time Units (ETU). Thus, for T = 0 the minimum duration between two consecutive start bits is 12 ETU. With the ATR parameter N, the guard time value can be changed. Its value represents the number of ETU to be added.

During the guard time, both communication partners are in receive mode (I/O line in state "High"). The transmitter checks the state of the I/O line by sampling the I/O port. In the event of an error-free transmission, the I/O line remains in the "High" state and the next character is expected after the guard time has expired.

If the card or the terminal detects a parity error in the received character, it sets the I/O line to the "Low" state for one or two ETU to indicate an error. Since the transmitter samples the I/O port during the guard time, it detects the "Low" state and identifies it as a parity error. After detection of the error signal, the sender repeats the character instead of transmitting the next one.

The error signal and character repetition procedure is mandatory for all cards offering the T = 0 protocol.

### 27.3.1.b Protocol T = 1

The half-duplex asynchronous transmission protocol T = 1 (cf. standards [ISO 1999a, ISO 1999b, ISO 1999c, ISO 1999d]) consists of block frames exchanged between the two communication partners. They convey either application data transparent to the protocol or transmission control data including transmission error handling. The protocol T = 1 is designed using the layering technique found in the Open System Interface (OSI) model [RAEF 2000], with particular attention paid to the minimization of interactions across boundaries.

A block is the smallest unit exchanged. It is defined as a sequence of bytes, whereby each byte is conveyed in a single character, see Section 27.3.1.a for the definition of a character. The protocol always starts with a block sent by the terminal and continues with alternating the right to send a block. The block structure allows us to check the received data before processing the conveyed data.

Within a block, the standard duration between the start bit leading edges of two consecutive characters is 12 ETU. The error signal and character repetition procedure, as defined for T = 0, does not take place in the T = 1 block protocol. Thus, the minimum delay between the leading edge of the start bits of two consecutive characters may be reduced from 12 ETU to 11 ETU. If the corresponding ATR interface parameter indicates N = 'FF' (N = 255), this leads to a guard time of only 1 ETU.

A block consists of the following three fields where the items in brackets indicate optional requirements (see Figure 27.12):

- The "Prologue" field consists of a "Node Address" byte (NAD), a "Protocol Control Byte" (PCB) and a "Length" byte (LEN).
- The "Information" field is optional. If present, it is comprised of 0 to 254 bytes (INF).
- The "Epilogue" field consists of one or two mandatory bytes for the "Error Detection Code" (EDC).

**Figure 27.12** Structure of a T = 1 block frame.

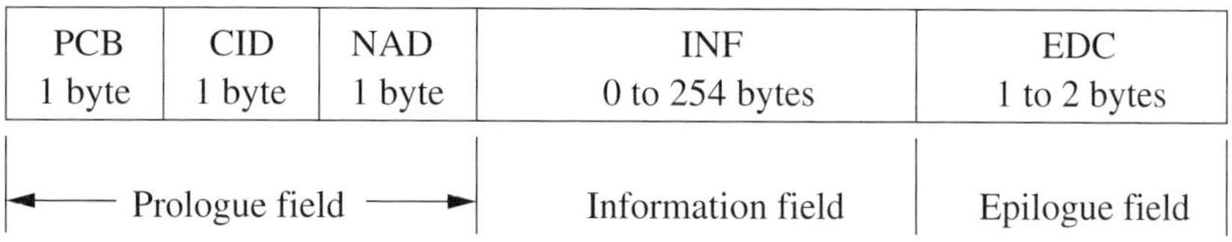

| PCB | CID | NAD | INF | EDC |
|---|---|---|---|---|
| 1 byte | 1 byte | 1 byte | 0 to 254 bytes | 1 to 2 bytes |

| Prologue field | Information field | Epilogue field |

## 27.3.1.c  Protocol T = CL

The half-duplex block transmission protocol T = CL controls the special needs of contactless smart card environments [ISO 2000, ISO 2000a, ISO 2000b, ISO 2000c, ISO 2000d] and supports both type A and type B frame formats (see Figure 27.13). An *extra guardtime* is inserted between the characters in type B frame, which is used to separate the single characters. As already annotated, this protocol is currently in the final stage of specification with the designation T = CL, which is the reason why it does not appear in current literature. This protocol is designed according to the principle of layering in the OSI reference model, with particular attention to the minimization of interactions across boundaries, which is similar to T = 1 (cf. Section 27.3.1.b).

**Figure 27.13**  Type A and type B frame formats.

Type A

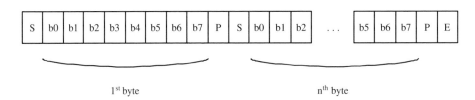

Type B

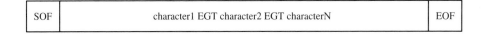

Another similarity with T = 1 is the method of transferring the character frame formats. The protocol T = CL also exchanges block frames between terminal and smart card. A block is the smallest data unit that can be exchanged, and it may be used to convey application data or transmission control data, including transmission error handling. This is one reason why in T = CL the error signal and character repetition procedure, as specified for the T = 0 protocol, does not take place. Another reason is that T = CL uses different frame formats than the T = 0 or the T = 1 protocols.

**Figure 27.14**  Character and next character transmission.

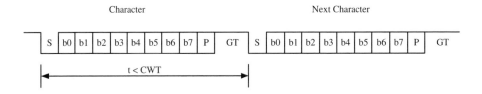

The protocol starts with a first block sent by the terminal and continues with alternating the right to send a block. As shown in Figure 27.14, a block consists of the following three fields, where the items in brackets indicate optional requirements:

- The "Prologue" field consists of up to three bytes, where the first byte (PCB) is mandatory and the two other bytes (CID and NAD) are optional.
- The "Information" field is optional. If present, it comprises 11 to 253 bytes (INF).
- The "Epilogue" field consists of two mandatory bytes for the "Error Detection Code" (EDC).

**Figure 27.15** Structure of a T = CL block frame.

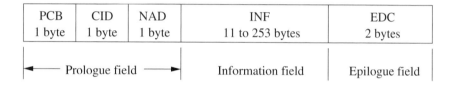

## 27.3.2 Physical interfaces

There exist different interfaces in the smart card environment. The 'universal asynchronous receiver transmitter' (UART) is the most used interface for data exchange. There also exist the 'universal serial bus' (USB) and, for special cases, radio frequency waves. The UART and the USB are investigated in the following sections.

### 27.3.2.a UART

The UART enables hardware supported data transfer at the serial I/O line. In former smart card architectures, the software, in conjunction with the OS, exclusively handled the process of transmitting and receiving data. This was sufficient as long as the desired baud rate did not exceed approximately 111 Kb/s. However, with higher rates, the received bits could not always be identified as valid, since the verification time was too short. An error-free communication was then no longer possible.

Today's smart card applications are becoming more complex. Apart from the increasing baud rate mentioned above, the security requirements, like cryptographic algorithms, need CPU time for their computation. Another factor is the possibility of several applications running simultaneously during a card session. These developments result in an increasing effort by the CPU, OS, and software to manage all requests in a satisfactory manner.

Using a UART is an efficient solution to increase the speed of data exchange without loading the CPU (see Figure 27.16). The hardware and software, for the UART implementation and control, are of a manageable size. The UART decouples the CPU from direct I/O data handling during communication. In this way, high baud rate can be achieved without decreasing the speed of the running applications. The UART autonomously organizes the transfer via the I/O lines, and, when necessary, prepares the data for the CPU. Also, there is a UART extension for direct memory access (DMA) operation. The function allows the UART to exchange the data with the memory without assistance of the microprocessor.

**Figure 27.16** Example for an UART environment.

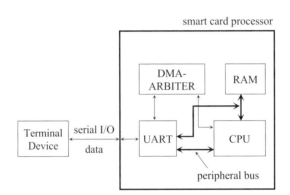

### 27.3.2.b  USB

The universal serial bus (USB) specification [USB] is a standardized peripheral connection developed in 1995 by leading companies in the PC industry. The major goal of USB is to define an external expansion bus, which makes adding of peripherals to a PC as easy as possible. In the smart card world, the USB interface serves as a high speed connection between smart card chip and external peripheral components.

With an implemented USB interface, the same smart card chip can be used for communication (card and reader).

The employment of a standard smart card chip on the reader side leads to a cheaper system, since developing one chip for both sides reduces the costs. This results in a shorter development period. A standard USB interface consists of three components: the Interface Logic, the Serial Interface Engine, and the USB Transceiver. The transceiver (see Figure 27.17) transforms the differential signal at ports D+ and D- into traditional digital logic.

## 27.4  Types of smart cards

### 27.4.1  Memory only cards (synchronous cards)

This is the first type of card to be widely used. The prepaid telephone cards mentioned in the introduction are an example of this type of card. The data required for the applications are stored in the EEPROM memory (EPROM for the first cards). In the simplest case the cards use memory that can only be written once, and after use, the memory is deleted and made inoperable (the Thomson ST1200 SGS, introduced in 1983 for the French telephone card, worked in this way). The addition of a security logic device allows more control over the memory access. There now exist more complex memory cards, which can perform simple encryption.

## § 27.4 Types of smart cards

**Figure 27.17** Example for an USB interface.

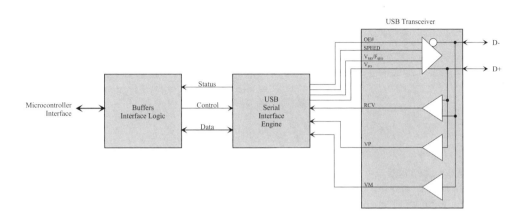

These types of cards are easy to use, the electronics are simple, the chip is small, and the price is low. However, memory space and flexibility are limited, and they are not adapted to security applications.

### 27.4.2 Microprocessor cards (asynchronous cards)

These cards are equipped with an "intelligent circuit": a processor connected to memory blocks capable of carrying out complex calculations. The added functionality of the microprocessor allows for higher security and application choices. However, as a result, these cards are larger and more complex. It is possible to connect other devices to the microprocessor for communication, special operations or security. Figure 27.18 shows many of the possible components that can be added to the microprocessor card. In smart cards, there are many different types of microprocessors. All of them function as a secured unit, protected from unauthorized access.

**Figure 27.18** Components of the microprocessor.

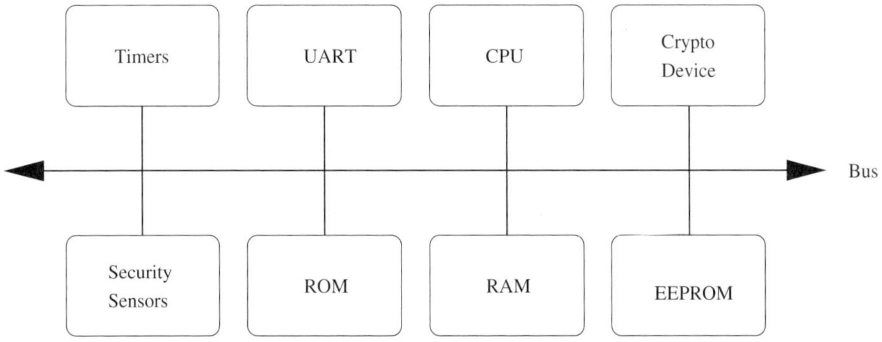

**Table 27.19** Characteristics of CISC and RISC based processors.

| CISC | RISC |
|---|---|
| Extensive instruction set | Small instruction set |
| Complex and efficient machine instructions | Simple instructions |
| Advanced instructions microencoded | Hardwired machine instructions |
| Extensive addressing capabilities for memory operations | Few addressing modes |
| Few registers | Many registers |

All microprocessors (and most computers) employ the principle of *stored program digital computer*. This means that data and instructions, which are stored in a memory area, must first be loaded into registers. Then the central processing unit, CPU, operates on these registers and places the results back into the memory areas.

The CPUs used in smart cards are usually built around proven modules from other applications. Many CPUs are based on the CISC (complex instruction set computer) architecture, which requires several clock cycles per instruction. However, CPUs based on the RISC (reduced instruction set computer) architecture are becoming more common. Table 27.19 shows the different characteristics between the CISC and RISC type processors. Many current CISC type processors are based on either one of the two main families: the Intel 8051 or the Motorola 6805 family. Manufacturers, such as Philips, Infineon, and ARM, take the base design of either a CISC or RISC processor and add their own functionality as needed.

The processing speed of the smart card is controlled by a clock circuit normally set to 5 MHz. Modern smart card processors use clock multipliers (by two or four) to increase this operating clock speed for internal calculations. Using clock multipliers, smart cards are able to operate at speeds between 20 and 48 MHz.

The area occupied by the microprocessor on the chip has a big influence on its manufacturing costs and its resistance to bending and shearing forces. Therefore, effort is made to reduce the chip's size as much as possible. The chip's surface area must be less than 25 mm$^2$. Using current chip technology, 0.25 or 0.30 $\mu$m, this means that the microprocessor contains between 150,000 and 200,000 transistors. Future microprocessors will be produced using newer 0.18 $\mu$m technology.

To provide additional functionality to the smart card, manufacturers add specialized processors and coprocessors to perform only specific tasks. The next section takes a closer look at coprocessors in smart cards.

### 27.4.2.a Coprocessors

Coprocessors are used on the majority of current chips for special operations and to optimize standard operations. Among those used for cryptography are:

- a coprocessor for DES encryption/decryption,
- a random number generator coprocessor: allows the use of random values in algorithms,
- an arithmetic coprocessor: dedicated to arithmetic operations (modular operations) on long integers; for example, 160-bit or longer integers.

An arithmetic coprocessor element is essential for asymmetric cryptography algorithms such as DSA, ECDSA, etc. Now there are more coprocessors that are not only optimized for RSA operations but also for elliptic and hyperelliptic curves operations. The first ECC cards started to appear around 2001 and 2004 the first smart card using HEC was produced.

Adding such coprocessors has a significant impact on the cost of the chip, increasing it by as much as a factor of ten. This being the case, one may wonder why with increasingly powerful processors it continues to be necessary to add coprocessors. But at the same time, cryptographic algorithms require longer keys to keep them secure, so coprocessors are likely to remain necessary for high performance cards.

# Chapter 28

# Practical Attacks on Smart Cards

Bertrand Byramjee, Jean-Christophe Courrège, and Benoît Feix

### Contents in Brief

| | |
|---|---|
| **28.1 Introduction** | 669 |
| **28.2 Invasive attacks** | 670 |
| Gaining access to the chip • Reconstitution of the layers • Reading the memories • Probing • FIB and test engineers scheme flaws | |
| **28.3 Non-invasive attacks** | 673 |
| Timing attacks • Power consumption analysis • Electromagnetic radiation attacks • Differential fault analysis (DFA) and fault injection attacks | |

## 28.1 Introduction

From the very first use of cryptography, people have always tried to decrypt enciphered messages in order to gain access to the sensitive and hidden information. We have seen, in the mathematical part of this book, different signature or cipher schemes based on the use of elliptic or hyperelliptic curves. These are considered to be secure from a theoretical point of view if the parameters are well chosen.

But, as far as a cryptosystem can be theoretically secure against actual cryptanalysis, in real implementations (for instance, in a smart card), it faces other threats than the mathematical ones; in particular *side-channel attacks*, SCA for short. As we have seen in the previous hardware Chapter 27, a smart card is very complex with many constraints such as speed and storage limitations. Side-channel attacks exploit the behavior of the chip while computing. Analyzing power consumption, Electromagnetic Emissions, calculation time, reactions to perturbations and fault injections, can reveal information on secret keys present in memory.

In this chapter, we are first going to deal with invasive attacks, which can be considered the most detectable because they partially modify (or destroy) the product. Then we will discuss non-invasive attacks, which can be performed on elliptic and hyperelliptic curves including: timing attacks, power analysis, electromagnetic analysis, and differential fault analysis.

## 28.2 Invasive attacks

The techniques in the following discussion originate from the failure analysis domain, but have also been used to attack electronic devices such as smart cards. We are not going to discuss in detail the most sophisticated types of attacks due to cost constraints and the complexities involved. For more details on these techniques we refer the reader to [MFA 2004]. Equipping a laboratory is very expensive, typically costing several million US$. However, price is less important as it is possible to rent a full laboratory station. Depending on the complexity of the work involved and the available knowledge of the chip being targeted, it can take many days or even weeks of work in a very specialized laboratory (see [CAMB] and [KÖKU 1999] for more details). Moreover, the owner of the card should notice these attacks in most cases because of the (partial) destruction of the smart card, and warn his provider. Nevertheless, it is important to bear in mind that such attacks could fatally undermine an entire security system. For instance, shared master keys should not be present into a card; in that case only the secret proper to the card attacked could be extracted.

### 28.2.1 Gaining access to the chip

Removal of the chip module from the card in order to connect it in a test package is quite easy; with a sharp knife, simply cut away the plastic behind the chip until the epoxy resin becomes visible. A smarter solution is to heat the card. This softens the glue of the chip which can then easily be removed by bending the card.

Once the chip module has been separated from the card, any remaining resin must be removed. A number of techniques are available depending upon the nature of the coating and of the targeted result (e.g., keep the module functional or not).

For instance, dropping fuming nitric acid completely dissolves the resin after a while. The process can be accelerated by heating the nitric acid. Then, the chip is plunged into acetone in order to wash it, followed by a short bath in deionized water and isopropanol.

Functional tests have proven that nitric acid damages neither the chip nor the EEPROM content. This is important as protected information is stored in non-volatile memory.

The chip is now ready for reconnection into another package, with fine aluminum wire, in a manual bonding machine. Now that access to the chip has been gained, analysis of the microelectronic characteristics can begin.

### 28.2.2 Reconstitution of the layers

The next step of the invasive attack involves reverse engineering the chip layout. This type of attack can be sufficient to provide direct access to sensitive data in the memory. It can also give rise to more complex scenarios such as micro-probing of critical signals or chip reconfiguration. Reverse engineering mainly consists of removing layers sequentially. The oxide layers can be selectively removed using a plasma machine, or chemically (for instance, fluoric acid). Metal layers are also chemically etched (for example, chloric acid for aluminum lines). Polishing techniques can also be used.

Images of each layer can be made using an optical or electronic microscope. Then by combining these images it is possible to regenerate the layout of the device (or at least a part of the device). Once the different layers and interconnections between them have been rebuilt, it becomes possible to regenerate the electronic schemes of the chip and so to explain how it really works.

In Figure 28.1, on the left, there is a top view of a chip with all the metal layers (3 layers). On the right, this is the same chip, but after removal of the upper metal layers.

However, even if a complete reverse engineer of a circuit is possible, the increasing complexity of devices can make this quite unrealistic.

**Figure 28.1** Reverse engineering, layer observation.

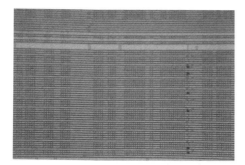 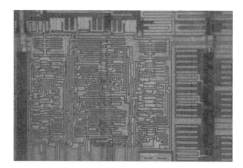

### 28.2.3 Reading the memories

In the previous section we explained that it is possible to determine the electric design of the circuit. In the same way, non-volatile memory contents can be "directly" read. Nevertheless, the remaining effort depends upon the kind of memory or the scrambling of the memory plan. In a ROM, for example, reading of the bits' values can be performed directly by optical observation of the first metal layer in the case of hard-wired memory. For diffused memory it will be necessary to reach the bulk level (lift off all layers until reaching the active one). Then, additional selective etching techniques could be necessary to highlight the diffusion area. In case of canal-implanted ROMs, dopant-selective crystallographic etching techniques or AFM (atomic force microscope) would be needed, cf. [BEC 1998].

In Figure 28.2, on the left, there is a view of the ROM before chemical treatment. On the right, this is the same ROM, but after chemical operation. The dark spots indicate the transistors set to transmit 0 volt by ion implantation (the bit is equal to 0). The bright spots indicate the transistors set to transmit 5 volt (the bit is equal to 1).

Other non-volatile memories such as EEPROM, Flash, and FRAM can, theoretically, be read by similar techniques. Nevertheless, because of the nature of the data coding, it is much more difficult to achieve the sample preparation without losing part of the memory content.

Finally, in most smart card products, the memories are fully scrambled and sometimes encrypted. So even if it is possible to access the memory, it is also necessary to completely rebuild the scrambling/ciphering process.

### 28.2.4 Probing

Previous sections deal with directly obtaining information through reverse engineering. But these analyses are intrinsically useful for additional attack paths such as the probing of internal signals. Indeed, if it is possible to observe data flow into internal buses then one can gain direct access to sensitive data. Different kinds of techniques can be used to observe the internal signals: galvanic probing (using a probe station cf. Figure 28.3); voltage contrast on a scanning electron microscope (SEM) or even light emission during transistor flips.

Figure 28.2 Crystallographic attack.

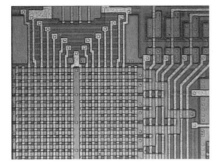 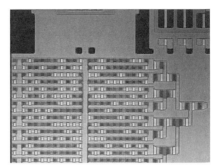

Figure 28.3 Microprobing station.

## 28.2.5 FIB and test engineers scheme flaws

Attackers also have a lot of interest in *focused ion beam* (FIB) workstations. This useful tool is a vacuum chamber with an ion gun (usually gallium), which is comparable to a scanning electron microscope. A FIB can generate images of the surface of the chip down to a few nm resolution and operates circuit reconfiguration with the same resolution. The first basic usage consists in increasing the "testability" of the circuit by managing access to deep internal signals. The FIB could be used to add test pads (probing) on critical signals. But the FIB can also be used to modify the internal behavior of the device by changing the internal connections. Finally, FIB is used to facilitate physical analysis of the device by making localized cross sections.

Figure 28.4 shows some typical FIB modification results. On the left, the image shows the creation of a connection between the second metal layer and the third one with a platinum strap. On the right the image has been annotated to explain what has been done in this FIB manipulation.

**Figure 28.4** FIB circuit modifications.

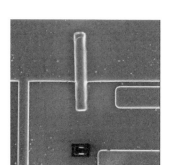
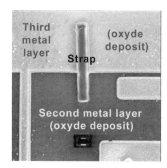

It is clear that such equipment presents a security threat to smart cards. It is possible, for instance, to disable security sensors. In particular it can be used to bypass some layout protections such as active shields.

It can also be used to rebuild a security fuse that has been blown after test during manufacture. We have seen above that some invasive attacks can be performed in order to access sensitive data.

Nevertheless they are very expensive in terms of time, resources, and equipment. In addition, technological advances are enabling the design of more and more complex and compact circuits, increasing the difficulty of such attacks as a side effect.

## 28.3 Non-invasive attacks

Non-invasive attacks destroy neither the chip nor damage the packaging around it. So, they are less likely to be detected by the owner of the card. They are extremely dangerous for this reason and also because the equipment needed to perform them is relatively inexpensive. Smart cards are nowadays utilized in many security domains and applications: bank cards, mobile communication and secure access, etc. Thus, groups of attackers or illegal organizations could create laboratories in order to process such attacks on the different products that are widely available within our society.

### 28.3.1 Timing attacks

Most of the programs and the code naturally designed contains conditional branching operations. Therefore, algorithm implementations have no constant time and thanks to these variations, it is possible to retrieve information. That is why the running time of an algorithm can constitute a side channel and give information on data operated during computations, which is the principle of timing attacks. In 1996, taking advantage of the variation in the algorithm execution time as a result of input changes, Kocher performed the first timing attacks on real PC implementations of several cryptographic algorithms (RSA, DSA, etc), cf. [KOC 1996].

The first timing attack on smart card was performed in 1998 by Dhem et al. in [DHKO$^+$ 2000]. Take for example, the following algorithm computing a reduction modulo the secret integer $N$:

---

**Algorithm 28.1** Modular reduction
---
INPUT: Two integers $x$ and $N$ of size respectively $\ell_1$ and $\ell_2$ bits with $\ell_1 > \ell_2$.
OUTPUT: The integer $x \bmod N$.

1. $t \leftarrow 2^{\ell_1 - \ell_2} \times N$
2. **for** $i = 0$ **to** $(\ell_1 - \ell_2)$ **do**
3.     **if** $x \geqslant t$ **then** $x \leftarrow x - t$
4.     $t \leftarrow t/2$
5. **return** $x$

---

As one can see in this algorithm, if $x$ is not greater than the integer $t$, Line 3 is not computed. Where $x \geqslant t$ the calculation takes longer and it is possible to deduce information on the value of the number $x$, particularly, the bit at the $i$th step. The attacker will analyze the running times of different computations of known or chosen messages in order to recover the modulus $N$.

Although it is an efficient attack, it's quite easy to thwart by implementing constant time algorithms, for instance by doing dummy computations.

For example, the following reduction algorithm will avoid timing attacks:

---

**Algorithm 28.2** Modular reduction against timing attacks
---
INPUT: Two integers $x$ and $N$ of size respectively $\ell_1$ and $\ell_2$ bits with $\ell_1 > \ell_2$.
OUTPUT: The integer $x \bmod N$.

1. $t \leftarrow 2^{\ell_1 - \ell_2} \times N$
2. **for** $i = 0$ **to** $(\ell_1 - \ell_2)$ **do**
3.     **if** $x \geqslant t$ **then** $x \leftarrow x - t$
4.     **else** $u \leftarrow x - t$     [dummy computation]
5.     $t \leftarrow t/2$
6. **return** $x$

---

In this reduction algorithm, the subtraction is always computed, but the result is discarded in the variable $u$ if the computation is not relevant for the reduction.

Moreover, nowadays, smart cards can prevent such attacks thanks to hardware countermeasures: dummy cycles, internal clocks, etc. To summarize, the cryptographic programmer has to implement algorithms in such a way that their execution times are independent of the input data. So, a timing attack proof algorithm is constant in time and, therefore, always longer than an efficient algorithm.

This type of attack has not been detailed here but they are not the most used against smart cards. For further details, we refer the interested reader Koeune's Thesis [KOE 2001]. Moreover, almost no timing attacks against ECC or HEC have been published yet, mainly because power analysis attacks, which have been deeply studied, are more powerful. Moreover, power consumption of the smart card gives more information about the secret key than the running time and can be more difficult to thwart. In [KAKI⁺ 2004], the authors give an example of attacking hyperelliptic curves. Nevertheless, the attack is not really practical as only the general case is implemented in hyperelliptic curves on smart cards, whereas the attack is based on degenerated cases.

### 28.3.2 Power consumption analysis

Kocher, Jaffe, and Jun first presented the idea of using the measure of power consumption of an electronic device to retrieve information about the secret keys inside a tamper-proof device. In [KOJA⁺ 1999], Kocher et al. described two methods for extracting the keys, which they named simple power analysis (SPA) and differential power analysis (DPA).

An electronic device such as a smart card is made of thousands of logical gates that switch differently depending on the complexity of the operations executed. These commutations create power consumption for a few nanoseconds. Thus the current consumption is dependent on the operations of its different peripherals: CPU, cryptographic accelerators, buses and memories, etc. In particular, during cryptographic computations, for the same instruction, the current consumption changes if the value of registers and data processed are different. Simply monitoring the power consumption, eventually followed by a statistical treatment, one can expect to deduce information on sensitive data when they are manipulated. With some experience and knowledge on the cryptographic algorithms, such analysis can be applied to many smart cards.

To mount these attacks the necessary equipment is a numerical oscilloscope, a computer, and a modified card reader to communicate with the card. For the most complicated attacks (such as DPA) other softwares are also necessary: to acquire the consumption curves, to treat the curves (signal processing), and process the attack (see Figure 28.5). So, nowadays the setup required to attack recent devices is still affordable for small organizations.

**Figure 28.5** Setup for power analysis.

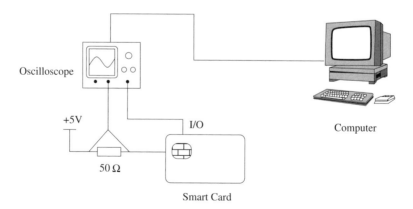

#### 28.3.2.a Simple power analysis

SPA needs only a single observation of the current consumption curve. The attacker can find information just by looking carefully at the curve representing the execution of a cryptographic algorithm. This is carried out by a detailed analysis of the curve. The power consumption curve of a smart card is different according to the executed instruction and data manipulated. For instance, a multiply instruction executed by the CPU needs more clock cycles than a XOR operation, or in a circular rotation operation, where the value of the carry could be observed.

As the implementation variants are limited and mainly known for a given algorithm, an attacker can deduce the structure of the implementation (from the sequence of double and add, the attacker can find which coordinate system is used during the computation of $Q = [n]P$ on an elliptic curve). With a straightforward implementation of a cryptographic algorithm, he can perhaps determine lots of details about it, for instance, to see precisely when the key is used and finally to retrieve all the bits of the key.

For example, we can analyze the scalar multiplication of $[n]P$ for a smart card using the double and add algorithm. If $n$ is the secret exponent, an attacker monitoring the power consumption of the smart card can guess the bits of $n$ with a single curve. Moreover, with some experience it is easy to know if the trace is a double or an addition. Furthermore, an addition should consume more than a double (at least for projective coordinates). In any case, as it is not possible in the double and add algorithm for two additions to be consecutive, it must be two doubles. So the attacker can guess when a bit of $n$ is a zero or a one and with the whole consumption curve discover all the secret exponents.

This general technique can be performed against implementations of elliptic curves scalar multiplication based on the double and add algorithm with affine or projective coordinates. In a more general way, every algorithm that treats the bits scanned of the secret parameter differently if it equals zero or one, is vulnerable to simple power analysis. In the same way, it is obvious that SPA can also be efficient against implementations of hyperelliptic curves.

Techniques to counteract SPA are quite simple and generally implemented nowadays by programmers. For instance a solution is to avoid conditional branching operations or to compute the same operations in all cases. Unfortunately, that mean slowing down the overall execution time and a compromise must be found between timing and security.

In the following the double-and-add-always algorithm is protected against a SPA attack:

---
**Algorithm 28.3** Double and add always
---
INPUT: A point $P$ and an integer $n = (n_{\ell-1} \ldots n_0)_2$.
OUTPUT: The point $Q = [n]P$.

1. $Q \leftarrow P$
2. **for** $i = (\ell - 2)$ **down to** $0$ **do**
3. $\quad Q_0 \leftarrow [2]Q$
4. $\quad Q_1 \leftarrow Q_0 \oplus P$
5. $\quad Q \leftarrow Q_{\ell_i}$
6. **return** $Q$
---

Another generic countermeasure is the use of the Montgomery ladder Algorithm, [JOYE 2000]. This can be viewed as a variant of the double and add algorithm, since point doubling and addition are performed for any input bit. Some other elliptic forms have been proposed, like the Jacobi form [LISM 2001] and the Hessian form [JOQU 2001], and as we have seen in the mathematical part, they have the same formulas for adding and doubling. Unfortunately, they are excluded from recommended curves by standards, such as NIST or Certicom curves, as their order is a multiple of 4 or 3, respectively.

Manufacturers also try to make SPA more difficult by including hardware countermeasures. We can quote the most frequently used: noise generators, dummy cycles, clock jitters, power filtering, variable frequency oscillators, and trying to make instructions cycles consuming identically.

### 28.3.2.b  Differential power analysis

As we said above, DPA was first introduced by Kocher et al. Nowadays any symmetric or asymmetric algorithm implementation can be threatened by DPA including computations based upon elliptic and hyperelliptic curves.

Today most chips are designed in CMOS technology. The power consumption depends mainly on the number of cells switching at a given time. Consumption for an instruction varies depending on the Hamming weight of the data manipulated. For instance, consider a multiplication operation between two registers. The power consumption needed for this operation is different if most of the bits equal zero compared to the case where most of the bits are equal to one. Thus for the same operation the chip power consumption varies according to the data manipulated, for instance a XOR between 0x33 and 0x01 will not have the same power consumption as a XOR between 0x33 and 0xFF.

DPA uses statistics to amplify and reveal these differences that cannot be easily exploited and seen with SPA. To be performed DPA requires several hundreds or even thousands of power consumption traces corresponding to the computations of the algorithm in the smart card. The aim is to retrieve information on the key used at each of these executions. Therefore other software has to be developed for signal processing and computing the differential traces. Of course, DPA works only if the key or the secret parameter the attacker wants to retrieve is constant during the monitoring of the power consumption traces.

### Principle of an attack on an algorithm $F$

Let $F(K, M)$ be a cryptographic algorithm, with a key $K$ on a message $M$.

An attacker collects many consumption curves $C_i$ of the execution of $F(K, M_i)$ on the smart card for $k$ (random or chosen) known messages $M_i$ for $i = 1, \ldots, k$. Then the attacker wants to guess bits of the secret key used. He makes a supposition on the bits and uses these to generate some intermediate values of the algorithm. Then he supposes that at an instant $t$ in the algorithm (corresponding to certain cycles of operation) the data will vary depending on the value of some bits of $K$: for instance $\ell$ bits of $K$. At this instant $t$ the value computed in the algorithm is $D(M_i, K_{1\ldots\ell})$ with $D(.,.)$ a known function. With this function the attacker wants to separate the curves in two sets $G_0$ and $G_1$ where basically:

- $G_0 = \{C_i \mid$ at instant $t$ the power consumption is low $(D(M_i, K_{1\ldots\ell})$ and has one (or several) bit(s) to zero)$\}$
- $G_1 = \{C_i \mid$ at instant $t$ the power consumption is high $(D(M_i, K_{1\ldots\ell})$ and has one (or several) bit(s) to one)$\}$

Then by subtracting the means of the two groups for each supposition we obtain $2^\ell$ differential traces $T_j$, with $j = 0, \ldots, 2^\ell - 1$. For the correct guess $h$ the trace $T_h$ will show many peaks of consumption. Indeed only in this case the separation will create two groups of different average consumptions. Basically one can choose a bit of $D(M_i, K_{1\ldots\ell})$ to make the selection if it equals to zero ($G_0$) or not ($G_1$) but many more evolved methods exist to select the curves into sets and to process the attack.

To perform such an attack we need to know the value of inputs and the structure of the algorithm implemented.

We observe that in ECDSA or in HECDSA, the scalar multiplications are not vulnerable to DPA, indeed at each signature step it takes a new random scalar to undertake it. However, DPA can be mounted on elliptic curves; an example is the attack against the EC ElGamal type encryption by Coron [COR 1999]. The attacker knows (or guesses in SPA) which algorithm is used to perform the scalar multiplication. Let the double and add be the algorithm used by the smart card.

**Figure 28.6** DPA example on DES: correct supposition (left) and wrong supposition (right).

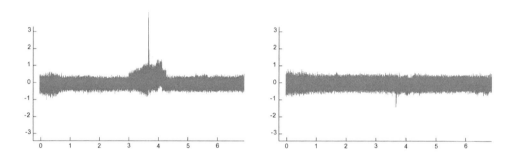

Let $P_1, \ldots, P_j, \ldots, P_k$ be $k$ random points on the curve. The attacker knows for instance the first bit of the exponent, generally it equals 1. Then he wants to guess the next one. Either it equals to 1 and then the card computes $C_j = [3]P_j$ or not and then $[3]P_j$ will never be calculated. So, the attacker wants to verify its supposition. He collects $k$ curves $C_1, \ldots, C_k$ corresponding to the computations $[n]P_1, \ldots, [n]P_k$ by the card. He divides the power consumption curves $C_1, \ldots, C_k$ collected into two groups $G_0$ and $G_1$ according to a chosen (or many) bit(s) of $[3]P_j$ (most significant or least significant for instance) or to the hamming weight of a byte of $[3]P_j$. If the difference between the means of the two groups is not zero then we observe peaks of consumption and we know that $[3]P_j$ has been computed. This means that the second bit of $n$ equals one.

Where Figure 28.7 shows a significant peak (it reveals at the same time the moment in the algorithm when $[3]P_j$ is computed or used in the next computation), the bit is equal to one. The attacker then guesses the third bit of $n$: if this bit is equal to one, $[7]P_j$ is computed and not otherwise. By repeating this attack until the last bit of $n$ he can recover the whole secret key.

The same attack can be used against the double and add always algorithm (see Algorithm 28.3) to find the second bit of $n$, indeed the attacker does not exploit the value $[3]P$ that is always computed but the value $[6]P$ that is computed in the next step only if the second bit is one.

Similarly the Montgomery ladder algorithm can be attacked in DPA; either the second bit is one and the value $[7]P$ will be computed by the smart card, otherwise it is equal to zero and the value $[5]P$ will be computed. Differential traces are computed for the value $[7]P$ and $[5]P$ and the trace showing peaks will indicate the right supposition, cf. Figure 28.8.

To speed up the attack, instead of guessing the key bit per bit, the attacker can guess $k$ bits but must check $2^k$ hypotheses. It is also the case if sliding window methods are implemented.

§ 28.3 Non-invasive attacks    679

**Figure 28.7** Means for a correct ($[3]P$) and a wrong guess on the $2$-nd bit of $n$ on the double and add.

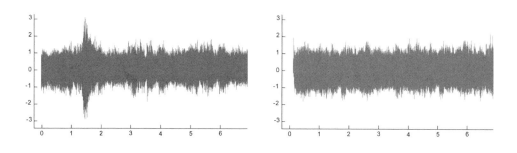

**Figure 28.8** Means for a correct ($[7]P$) and a wrong guess ($[5]P$) on the $2$-nd bit of $n$ on Montgomery ladder.

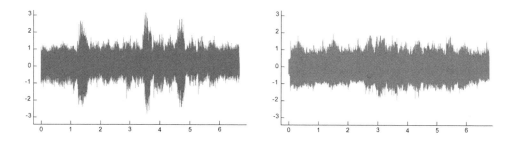

**Countermeasures**

As previously, when we studied SPA, we can identify two types of countermeasures against DPA: hardware protections and software countermeasures.

Since DPA is based on the ability of the attacker to predict one temporary variable of an algorithm by guessing a few bits of the secret key, a solution for preventing DPA is to randomize the value of the temporary variables of the algorithm with a random number. For elliptic curves the possibilities are numerous: randomizing the scalar $n$ and randomizing the coordinates (projective, Jacobian, etc.), blinding of the message [COR 1999], random elliptic curve, or field isomorphism [JOTY 2002].

Hardware countermeasures are included in the chip design. We presented some of them in the SPA paragraph, the effects are mainly desynchronization of the curves and more noise in the signal, so reducing the efficiency of the attack. However this kind of countermeasure might be sometimes bypassed by signal processing techniques and by increasing the number of curves and messages. Then more powerful countermeasures are appearing: dual rails encoding should theoretically eliminate the correlation between the hamming weight of the data and the power consumption but such designs increase the size of the chip and cannot always be used. Moreover the final chip can be still sensitive to DPA for manufacturing reasons. Manufacturers can also replace synchronous technology by an asynchronous one, in this case the consumption curve may be more difficult to exploit.

Statistics have been studied to improve DPA and some techniques exist to decrease the number of power consumption curves needed to perform the attack, see [COKO$^+$ 2001] and [BEKN 2003]. In [BRCL$^+$ 2004] the authors present correlation power analysis attacks that exploit the correlation factor and a linear model of consumption. These methods have been proven both theoretically and practically as more efficient than classical DPA, but they must only be considered as an optimization of DPA.

### 28.3.2.c High order power analysis

Kocher et al. [KOJA$^+$ 1999] also present an enhanced type of DPA called *high order DPA*, HODPA for short. We briefly introduce it in this paragraph. The idea is that instead of studying a single moment $t$ in the algorithm corresponding to an interval $T_1$ of points in the consumption curve in DPA, one has to study many moments $t_1, \ldots, t_n$ and then $n$ intervals $T_1, \ldots, T_n$ of points. Basically, for a second order DPA ($n = 2$), the attacker predicts a few bits at these two different places corresponding to two different intermediate values $V_1$ and $V_2$ of the computation of the algorithm. If those bits are "masked" with the same value, (i.e., XOR with this value), this "mask" can be *eliminated* if a supposition is made on a combination of $V_1$ and $V_2$ value (XOR for instance). Then the attacker constructs for each power curve a new one equal to $T_1 - T_2$ (for instance) and performs the DPA with those new curves. Indeed, as seen by Messerges in [MES 2000a], "masking" a value is not sufficient to protect DPA attacks. To be protected against high order DPA, several masks have to be used, instead of just one.

Even if HODPA is generally used against symmetric implementations, where countermeasures are often based on masking methods, a theoretical second order DPA attack has been recently presented in [JOY 2004] against the *Additive Randomization* countermeasure. However, high order DPA is complex to set up and not so easy to use against most of the asymmetric algorithm countermeasures.

### 28.3.2.d Goubin's refined power analysis attack

Here we will see how difficult it can be to secure elliptic curves and, even more, hyperelliptic curves against power analysis. Indeed Goubin [GOU 2003], shows that even an algorithm protected against SPA and DPA with the random (projective coordinates, random elliptic curves isomorphism, or field

## § 28.3 Non-invasive attacks

isomorphism) countermeasures can still be vulnerable to DPA. We will only present the elliptic case as the hyperelliptic case does not differ a lot. We will just give some remarks regarding that case.

The idea of the attack is to find a divisor that, given a correct guess of a scalar bit during the scalar multiplication, leads to the neutral divisor or at least one of its coordinates leads to 0. It, therefore, leads to a significant difference in the consumption traces.

We will describe the attack on the protected double and add always algorithm, introduced in Section 28.3.2.a, but the result would be similar with an algorithm that, for instance, uses Montgomery ladder method; for further details we refer the reader to [GOU 2003]. Let us assume that the attacker knows the most significant bits $n_{\ell-1} \ldots n_{i+1}$ of the scalar multiplier and that the elliptic curve $E(\mathbb{F}_p)$ contains a "special" point $P_0$, we mean in that sense that one of its coordinates $x$ or $y$ is 0.

In Algorithm 28.3, knowing a point $P$, the point $Q$ at the end of the $(\ell - i - 1)$-th step of the loop is the following:

$$Q = \left[ \sum_{j=i+1}^{\ell-1} n_j 2^{j-i} + n_i \right] P$$

During the next step of the loop, depending on the value of $n_i$, we have the two following cases:

- if $n_i = 0$,  $Q_0 = \left[ \sum_{j=i+1}^{\ell-1} n_j 2^{j-i+1} \right] P$ and $Q_1 = \left[ \sum_{j=i+1}^{\ell-1} n_j 2^{j-i+1} + 1 \right] P$.

- if $n_i = 1$,  $Q_0 = \left[ \sum_{j=i+1}^{\ell-1} n_j 2^{j-i+1} + 2 \right] P$ and $Q_1 = \left[ \sum_{j=i+1}^{\ell-1} n_j 2^{j-i+1} + 3 \right] P$.

Depending on the value of $n_i$, we construct the point $P_1$ in the following way.
If

$$\left( \sum_{j=i+1}^{\ell-1} n_j 2^{j-i+1} + 1 + 2n_i \right)$$

is coprime to $|E(\mathbb{F}_p)|$, which means no restriction, since $|E(\mathbb{F}_p)|$ must be a prime number or a prime number times a cofactor less than or equal to 4 for most standardized elliptic curves, then

$$P_1 = \left[ \left( \sum_{j=i+1}^{\ell-1} n_j 2^{j-i+1} + 1 + 2n_i \right)^{-1} \mod |E(\mathbb{F}_p)| \right] P_0.$$

Let us now assume that one of the DPA countermeasures discussed previously has been implemented. In that case the computation in the card of $[n]P_1$ will at the step $(\ell - i)$ lead to a point with one coordinate equal to zero. We collect $k$ curves $C_1, \ldots, C_k$ of these computations, these consumption curves are different for repeated computations of $[n]P$ because of the randomization of the countermeasure. Then we compute the average trace

$$T_{P_1} = \frac{1}{k} \sum_{j=1}^{k} C_j.$$

If the guess for $n_i$ is correct then in all curves $C_j$ one coordinate at the step $(\ell - i)$ equals 0 and then we can observe notable peaks due to its computation. In the same way, we can recursively recover the remaining bits $n_{i-1}, \ldots, n_0$.

**Remarks 28.4**

(i) For elliptic curves, special points do exist [GOU 2003].

(ii) For hyperelliptic curves, these kind of divisors also exist, but in a different way. First, we can achieve that attack with a divisor with 0 coordinates, which means that the degree of the divisor is inferior or equal to $g + 1$, $g$ being the genus of the curve. This can be well exploited, as in a smart card. All different cases cannot be implemented, and just the more common case is actually in the card. Secondly, this attack can also work in the general case, if we choose a degree $g$ divisor, having in Mumford representation its first polynomial with some 0 coefficients. Let us say, for a genus 2 curve, $D = [u(x), v(x)] = [x^2 + u_1 + u_0, v_1 x + v_0]$, with $u_1 = 0$ or $u_0 = 0$ (choosing carefully the divisor, we even can have $v_1 = 0$ or $v_0 = 0$).

We stress here to the reader that this idea has recently been extended by Akishita and Takagi in [AKTA 2003], to other points that do not contain a 0 among their coordinates but lead to 0 in the intermediate computation of the double or addition of a point.

As Goubin's attack is based on the prediction of the scalar multiplier and knowing the base point, a solution for preventing this attack can be to randomize the secret scalar or make it impossible to choose the base point. So, the first two countermeasures of Coron [COR 1999] can be applied, in addition to the third one (and for hyperelliptic curves also): randomizing the private exponent and blinding the base point. A better solution has been proposed by Smart using isogenies for elliptic curves in [SMA 2003], but it is much slower to implement on a smart card.

### 28.3.3 Electromagnetic radiation attacks

Electromagnetic analysis (EMA) is quite similar to power analysis but we analyze here the electromagnetic emanations of the chip instead of its power consumption.

The threat through EM radiation leakage has been well known for a long time in visual display units and the use of EM against smart cards is not more complicated. At the beginning of this century, some declassified U.S. Government Tempest documents [NSA] and preliminary claims by researchers at international cryptographic conference rump sessions (such as Eurocrpypt 2000) made public the threat of EM radiation.

The current consumed by the chip creates electromagnetic fields that can be measured with a specific probe. Thus one can monitor the radiation emitted by a precise part of the chip on a numerical oscilloscope and analyze the signal obtained.

Such analysis allows isolating and observing the activity of particular areas of the chip, depending on the position of the probe on its surface. Thus countermeasures against power analysis could be bypassed by EMA. For instance, if an attacker wants to analyze elliptic curve computations he can place its probe on the crypto accelerator to analyze its operations and consumption. In practice, however, this is not so easy. Better results could be obtained when the probe is placed on specific power lines. Reverse engineering and a scanning of the chip can also give information on the best place to put the probe.

The equipment needed for these attacks is not much more expensive than PA. The attacker needs one magnetic or electric probe, an amplifier and a system with precision movement. A careful study that has shown efficient results [GAMO+ 2001] has been made with very small (a few mm) but simple hand made probes. These consist of small solenoids made of coiled copper wire.

To refine the analysis of the EM radiations, one may need AM or FM demodulators which can be very expensive [AGAR+ 2003]. Techniques used in power analysis can be adapted to EMA: simple electromagnetic analysis (SEMA); differential electromagnetic analysis (DEMA).

However, cards secured against power analysis can be vulnerable to EMA. For instance, comparisons of bytes that were not visible in the power curve could be seen in EMA; noise generators or dummy operations effects can be eliminated and then an EM curve contains more information than the power curve. Thus DEMA can be successful when DPA has failed. Accesses to memories can be observed, perhaps a dummy addition in the double-and-add-always algorithm can be seen because the memory treatment is different.

The DPA software countermeasures that aim at preventing an attacker to predict intermediate values are still efficient against DEMA. However, hardware countermeasures against SPA can be inefficient against EMA and then specific countermeasures need to be added. It can take the form of shielding to contain or blur the emanation, by including a metal layer on top of the other layers, or by generating noisy fields.

One can observe that EMA is not really a non-invasive attack, as is the DPA, but a semi-invasive attack. Indeed to place the probe on the chip the card must be de-packaged. Keep in mind that the efficiency of these attacks is dependent upon the equipment, the quality of the EM probe, and the amplifier.

### 28.3.4 Differential fault analysis (DFA) and fault injection attacks

The effects of faults and perturbations on electronic devices have been observed since the 1970s. The first faults were accidentally obtained because the chips were used in particular contexts such as in the aerospace domain. Later we observed that fault injections and perturbations into a component during computations could lead to information leakages and reveal sensitive data.

The first attack, known as the *Bellcore attack*, presented by researchers from Bellcore, was published in [BODE$^+$ 1997]. The authors use fault injection during a RSA CRT computation to recover the two prime factors that compose the public modulus. A few months later in [BISH 1997], Biham and Shamir adapted this idea on symmetric algorithms. Basically, any kind of algorithm can be threatened by fault attacks. Firstly considered as non-exploitable, these attacks are today a real threat for smart cards. They must nowadays be considered very seriously and countermeasures must be added into products to protect them. These attacks are being increasingly studied because the possibilities have not been fully exploited today.

These attacks rely on generating errors in the chip during the execution of the algorithm. Then, the comparison between the genuine outputs and the outputs generated with errors gives information on the key. For instance, for RSA using speedup based on Chinese remainder theorem, a key can be retrieved either with the knowledge of a faulty signature and a correct one, or with the knowledge of a message and its erroneous signature. For an elliptic curve, the method is different: faults in the base point or in the definition field lead to solving the discrete logarithm problem in easier subgroups than in the original elliptic curve group.

There are several techniques available to perform a fault attack. The most common types are presented below:

- *Glitch attacks* induce glitches on one of the contacts of the card: the VCC pad, the Reset pad or the clock. A brief injection of power during computations can perturb a component behavior so that it will not return the right value. The equipment necessary for such an attack can be made of a pulse generator or a function and waveform generator or a pattern generator. Nevertheless, for more precise attacks it can be necessary to use a very precise generator together with highly sophisticated synchronization systems in order to perturb only chosen instructions during computation.

- *Light attacks* have been presented by Anderson and Skorobogatov in [SKAN 2003]. They used a simple camera flash placed on a microscope to focus light on the surface of the chip and then perturb it. More generally the principle is to use the energy of a light emission to perturb the silicon of a chip. Indeed, because silicon is very sensitive to light, such an attack will create parasitic currents sufficient to perturb the electrical behavior of the circuit. Contrary to glitch attacks, light attacks are semi-invasive because the card must be opened to allow light to reach the surface of the chip. The two main ways to process light attacks are the use of a camera flash, and for more precise and powerful attacks one can use a laser.

To illustrate these types of attacks, based upon [BIME$^+$ 2000], we present two ways to exploit flaws in implementing hyperelliptic curves.

For further details on fault attacks, we refer the interested reader to [BACH$^+$ 2004, GITH 2004, CIE 2003].

### 28.3.4.a Flaw in input points

This attack is a little bit different as it does not need to generate any faults. Indeed, the flaw exploits the fact that faulty points can be interpreted in order to find the key if the device does not check that the input points belong to the curve. Here, we will treat the example of a hyperelliptic curve following [BIME$^+$ 2000].

Let $\mathcal{H}$ be a genus 2 hyperelliptic curve, on $\mathbb{F}$ in the following equation:

$$\mathcal{H}: y^2 + h(x)y = f(x) \tag{28.1}$$

with:

$$h(x) = h_2 x^2 + h_1 x + h_0, \ h \in \mathbb{F}[x]$$

and

$$f(x) = x^5 + f_4 x^4 + f_3 x^3 + f_2 x^2 + f_1 x + f_0, \ f \in \mathbb{F}[x] \tag{28.2}$$

We will denote $J_{\mathcal{H}}$ the Jacobian of the hyperelliptic curve $\mathcal{H}$ and $D = P_1 + P_2 - 2P_\infty$ a divisor of $J_{\mathcal{H}}$. We can assume that our divisor is of degree 2.

As is well-known, not all the coefficients for elliptic curves are used in the formulas of addition and doubling. The same holds for hyperelliptic curves; in that case $f_0$ and $f_1$ are not used in explicit formulas, as we can see in Chapter 14.

We choose $D' = P'_1 + P'_2 - 2P_\infty$, which is not a divisor of $J_{\mathcal{H}}$, but one of $J_{\mathcal{H}'}$(with $\mathcal{H}'$ given by (28.3)) such that

$$f'_1 = \frac{c_1 - c_2}{x'_1 - x'_2}, \ f'_0 = \frac{c_2 x'_1 - c_1 x'_2}{x'_1 - x'_2}$$

where, for $i = 1, 2$

$$c_i = y'^2_i + h_2 x'^2_i y'_i + h_1 x'_i y'_i + h_0 y'_i - x'^5_i - f_4 x'^4_i - f_3 x'^3_i - f_2 x'^2_i.$$

So, $\mathcal{H}'$ has the following equation:

$$\mathcal{H}': y^2 + h(x)y = f'(x) \tag{28.3}$$

with

$$f'(x) = x^5 + f_4 x^4 + f_3 x^3 + f_2 x^2 + f'_1 x + f'_0$$

We calculate the cardinality of $J_{\mathcal{H}'}$, and denote by $r$ a small divisor of $|J_{\mathcal{H}'}|$. Let $D'_1 = \frac{|J_{\mathcal{H}}|}{r} D'$, it has order $r$. We know that with $D'$ as an input, the output $[n]D'$ is still in $J_{\mathcal{H}'}$. Therefore, we change the discrete logarithm problem from a secure hyperelliptic curve into the discrete logarithm problem in a subgroup of order $r$. We found a value $n'$ such as $n \equiv n' \pmod{r}$. Then, thanks to the Chinese remainder theorem cf. Section 10.6.4, we repeat that operation with other input divisors until we have enough $r$ such that $\prod r > \mathrm{ord}(D)$.

Note that this algorithm is even more efficient if, instead of choosing the divisor $D'$, we first chose the hyperelliptic curve $\mathcal{H}$ and then compute the divisor. In the case of an elliptic curve, it is very efficient as we can construct an elliptic curve. For hyperelliptic curves, it is still quite difficult, and no method is known to compute the cardinality in the general case. Algorithms are still under development cf. [GASC 2004a] and [MACH+ 2002].

### 28.3.4.b Flaw in output points

This time, creating a fault during the computation, we can recover the key, if the device does not check that output points belong to the curves.

Actually, we can apply exactly the same method as previously described. Assuming that the input is correct, we succeed in provoking a fault, just before the beginning of the computation of $[n]D$. Indeed, we assume that the correct input $D$ has been provided, but while computing, the device has taken a divisor $D'$ that only differs from $D$ in one bit. Therefore, reduce the problem the same way as in Section 28.3.4.a, even if we do not know the divisor $D'$. Indeed, we first find a hyperelliptic curve $\mathcal{H}'$, thanks to $[n]D'$ and then recover $D'$ as there are only a few possibilities.

The authors in [BIME+ 2000] also explain how to perform DFA attacks even if we don't know the position of the fault during the scalar computation. They basically calculate two different outputs, one correct and one faulty. Then they guess at which step of the loop of a double and add algorithm the fault occurs further simulating the computation until the output point is produced. Doing that, they recover MSB or LSB block bits — depending on the choice of the algorithm — of the secret key $n$. So, recursively, they recover the remaining bits with other random faults.

Even though the way to find the secret scalar $n$ is very different from the previous attack, it does exploit the same characteristic: it uses faulty output points.

Countermeasures against fault injection attacks are not always as easy as it seems to implement and can be time-consuming. Moreover, improvements on the fault injection techniques could succeed in bypassing insufficient countermeasures. Semaphores for critical parts of the programs are sometimes used, for instance, verifying that the final result computed belongs to the curve.

# Chapter 29

# Mathematical Countermeasures against Side-Channel Attacks

*Tanja Lange*

### Contents in Brief

| | |
|---|---|
| **29.1 Countermeasures against simple SCA** | 688 |
| Dummy arithmetic instructions • Unified addition formulas • Montgomery arithmetic | |
| **29.2 Countermeasures against differential SCA** | 697 |
| Implementation of DSCA • Scalar randomization • Randomization of group elements • Randomization of the curve equation | |
| **29.3 Countermeasures against Goubin type attacks** | 703 |
| **29.4 Countermeasures against higher order differential SCA** | 704 |
| **29.5 Countermeasures against timing attacks** | 705 |
| **29.6 Countermeasures against fault attacks** | 705 |
| Countermeasures against simple fault analysis • Countermeasures against differential fault analysis • Conclusion on fault induction | |
| **29.7 Countermeasures for special curves** | 709 |
| Countermeasures against SSCA on Koblitz curves • Countermeasures against DSCA on Koblitz curves • Countermeasures for GLV curves | |

This chapter has been influenced by Avanzi's report [AVA 2005c], which provides an excellent overview of side-channel attacks on curves and their countermeasures — including a historical perspective. Also Joye's chapter [JOY 2005] in [BLSE+ 2005] has been a source of inspiration. For the most recent research on SCA one should consider the proceedings of the CHES workshop series. A good overview with many links is the side-channel lounge [SCA LOUNGE].

In Chapter 28 attacks against *implementations* of cryptosystems were introduced. For restricted devices like smart cards it is possible for an attacker to derive side-channel information on the operations performed. This additional information can be the timing of the total operation or (more precise) the power consumption at different time points during the execution of the algorithm. For the introduction to side-channel attacks (SCA), we refer to Chapter 28. Here, we concentrate on software countermeasures, i.e., different methods of implementing the same group operations in such a manner that the information obtained from the side-channels is useless. Obviously, these countermeasures apply only to non-invasive attacks, but these are the most likely to occur in practice. In applications, software and hardware countermeasures complement each other.

In this chapter we provide alternative ways of performing group operations and of computing scalar multiplications on the Jacobian of elliptic and hyperelliptic curves in order to avoid side-channel attacks. The main ideas of these *mathematical countermeasures* are as follows.

- Against *simple side-channel attacks* (cf. Section 28.3.2.a, Section 29.1) make the information uniform, i.e., independent of the operation performed.
- To avoid *differential side-channel attacks* (cf. Section 28.3.2.b, Section 29.2) insert randomness, e.g., by modifying the scalar or by changing the internal representation.
- To avoid *Goubin type attacks* (cf. Section 29.3) insert randomness but additionally one needs to ensure that there is no set of elements unchanged under these methods.
- To detect and thwart *fault attacks* (cf. Section 28.3.4, Section 29.6) one should check the in- and output elements and also make sure that error messages leak no information.

In Chapters 13 and 14 we have described the most efficient ways of performing group operations on the Jacobian of elliptic and hyperelliptic curves but on devices like smart cards that are used in hostile environments, *security* becomes the main issue. The solutions we present in the sequel try to achieve security while keeping a certain level of efficiency. To avoid simple side-channel attacks the easiest method described already in Chapter 28 consists of performing the double and always add Algorithm 28.3, which strongly decreases the efficiency but makes the side-channel information useless. Our countermeasures will be more efficient than this direct one. Furthermore, combinations of different side-channel attacks have to be taken into account, for instance the double and always add algorithm is vulnerable against fault attacks and if the same scalar is used multiple times it does not provide any security against differential attacks.

We like to stress that no perfect countermeasures exist. The situation is similar to classical (mathematical) attacks. One has to make an estimate of the attacker's resources and abilities and design the countermeasures in a way that the device can resist at least these attacks.

For the time being there is no sound theory of side-channel cryptanalysis as this area is still relatively new. Hence, the implementor should carefully choose the most efficient countermeasure the constraints like performance and chip area allow.

This chapter is organized as follows: we first present countermeasures against simple side-channel attacks, then deal with differential and timing attacks and consider fault attacks. Finally we briefly state additional approaches for special curves like Koblitz curves.

For the scalar multiplication we assume that the result is computed by a sequence of additions, subtractions, and doublings by using addition chains. For full details and the windowing methods we refer to Chapter 9. We fix the following notation: the secret scalar is denoted by $n$ and one uses an expansion of length $l$ given by $(n_{l-1} \ldots n_1 n_0)_b$ to some base $b$. Furthermore, as we are working in the Jacobian of a curve, we assume that an addition and a subtraction need about the same time unless we work on a very low atomic level. To give the number of field operations needed for the group operations we use the notation S to denote a squaring, M for a multiplication and I for an inversion.

## 29.1 Countermeasures against simple SCA

*simple side-channel attacks (SSCAs)* obtain information from a single scalar multiplication by observing leaked information. We stress that also short term secrets like the nonces in signatures (cf. Sections 1.6.3 and 23.5.2) need to be protected as their knowledge is sufficient to obtain the long-term secret key.

To harden a cryptographic primitive against simple side-channel attacks one makes the observable information independent of the secret scalar. The attacker only sees a fixed sequence of operations that cannot be linked to the bits of $n$ being processed. This can be achieved by one of the following three approaches: insert dummy arithmetic instructions, use indistinguishable or unified addition and doubling formulas, or apply Montgomery's ladder for scalar multiplication. Furthermore, these strategies can be applied on different levels. We like to point out that some approaches belong to more than one category and that this distinction is sometimes too strong. We explain the method in that section where we deem it to be most suitable and give references in the other ones.

### 29.1.1 Dummy arithmetic instructions

The double and always add Algorithm 28.3 already mentioned in the introduction and in the previous chapter inserts dummy group additions such that after each group doubling one performs one addition. This is one example of inserting dummy operations on the top level. We also show how to insert dummy field operations to make the observable information uniform and thus predictable.

#### 29.1.1.a Dummy group operations

The methods described here are universal and work for any group no matter how different addition and doubling are. However, the big drawback is that much more operations are needed — on average $l/3$ additions or subtractions are used in a non-shielded implementation using an NAF of $n$, which compares to the fixed number of $l$ additions needed with the double and always add algorithm. For the scalar multiplication methods used here we refer to Chapter 9.

Obviously, this drawback is less dramatic when applied together with a (fixed) window scalar multiplication algorithm. A width $w$ expansion of $n$ is given by

$$n = \sum_{i=0}^{\lceil l/w \rceil - 1} n_i 2^{wi}, \text{ where } n_i \in [0, 2^w - 1].$$

The likelihood of zero coefficients $n_i$ is much lower than in a binary expansion and the absolute number of dummy group additions (inserted only if $n_i = 0$) is accordingly also much lower. On the other hand, these methods need storage, which is often a concern on the small devices we consider in this context.

For curve-based cryptography where the negations are efficiently computable one might try to keep the advantages of signed expansions. Hitchcock and Montague [HiMo 2002] provide a method transforming an NAF into a sequence consisting of the fixed blocks DBL, DBL, ADD, which is achieved by inserting dummy additions and doublings. Compared to the unprotected NAF this method needs 1.5 times as long in the worst case. On average 44% additions are saved compared to the double and always add algorithm while 11% extra doublings are needed. The advantage of this method is that no storage is needed except for the dummy operations. If the negation could be detected, one more group element is needed to store the negative of the base. Alternatively one could perform a negation before each addition that might be a dummy negation.

Actually a similar but more efficient idea can be found in [Gie 2001]. The NAF representation of $n$ is grouped into double-bits from $00, 01, 0\bar{1}, 10, \bar{1}0$. For each double-bit they perform 2 doublings and one addition — which is not used only in the case of the double-bit $00$. Thus, compared with the NAF they started with, they have the same number of doublings but for every two coefficients they perform an addition instead of one only for one third of the coefficients. This leads to a density of $1/2$, which is much better compared to density 1 obtained above, using the obvious method and does not introduce a larger complexity except for a bit of bookkeeping. We like to mention that the computations of the two doublings and one addition can be interleaved to need fewer inversions in affine coordinates or to save some multiplications in other coordinate systems. This way this method differs from using a sliding windowing method with inserted dummy operations.

As dummy operations can be detected by fault induction attacks (cf. Section 29.6), Möller proposed in [MÖL 2001a, MÖL 2001b] to use windowing methods with coefficients from $[1, 2^w - 1] \cup \{-2^w\}$ or $[1, 2^{w-1}] \cup [-2^{w-1} - 1, -1] \cup \{-2^w\}$. Okeya and Takagi [OKTA 2003] achieve expansions without zero coefficients requiring storage of $2^{w-1}$ elements. We like to point out that these methods do not use dummy operations, but we decided to mention them in this context as the efficiency is similar to the combination of dummy operations and windows of width $w$ except that one avoids the drawback of having (possibly detectable) dummy operations. If only the positive scalar multiples are stored one needs to pay attention that the negation cannot be detected. This can be done by always inserting a negation that might be a dummy negation in case of a positive coefficient.

For elliptic curves, both methods can be combined with fast repeated doublings (cf. Chapter 13). For hyperelliptic curves it can be expected that such formulas will be published in the near future. In [IZTA 2002a], Izu and Takagi report on a parallel implementation resistant against simple power analysis using Möller's coefficients. They additionally apply countermeasures against DPA attacks.

### 29.1.1.b Dummy field operations

One can also introduce dummy operations on the low-level of field operations inside the group operations and achieve that the side-channel information is identical for group additions and doublings. As each group operation consists of several field operations, it can be expected that this approach leads to faster shielded algorithms compared to the ones of the previous section. On the other hand we need to point out that this way the total number of group operations is leaked by the time the scalar multiplication takes. If the length of the scalar and, hence, the number of doublings is known, this means that the Hamming weight of the used expansion is obtained by the attacker. For extreme low or high weight a brute force attack is possible. For attacks using this leakage we refer to [CAKO+ 2003]. Note that this is less problematic if the method is combined with sliding windowing techniques, as they tend to unify the Hamming weight of the scalar. Additionally, scalar randomization techniques can be used to thwart differential attacks, and they change the Hamming weight, too.

We now state the explicit formulas for elliptic curves first over binary fields and then in odd characteristic. For hyperelliptic curves we only give the references.

For nonsupersingular binary elliptic curves (cf. Section 13.3.1.a) both a doubling and a group addition require $I + 2M + S$ but in a straightforward implementation they can nevertheless be distinguished by the different sequences of operations. Let $P = (x_1, y_1), Q = (x_2, y_2)$ be points on an elliptic curve $E/\mathbb{F}_{2^d}$ given by

$$E : y^2 + xy = x^3 + a_2 x^2 + a_6. \tag{29.1}$$

The formulas for addition and doubling differ mainly in the slope $\lambda$ which is given by

$$\lambda = \begin{cases} \dfrac{y_1 + y_2}{x_1 + x_2} & \text{if } P \neq \pm Q, \\ x_1 + \dfrac{y_1}{x_1} & \text{if } P = Q. \end{cases}$$

Then $R = (x_3, y_3) = P \oplus Q$ reads $x_3 = \lambda^2 + \lambda + a + x_1 + x_2$ and $y_3 = \lambda(x_2 + x_3) + x_1 + y_2$. Based on these observations, in [CHCI+ 2004] the following algorithm is provided, that allows us to compute an addition with the *same sequence of field operations* as a doubling by inserting only two field additions. This is a very cheap countermeasure using dummy operations as field additions

§ 29.1 *Countermeasures against simple SCA*

are the cheapest operations in a finite field. The input registers $T_1, T_2, T_3, T_4$ contain the $x$- and $y$-coordinates of the points $P, Q \in E(\mathbb{F}_{2^d})$ and the left column is used for $P \neq \pm Q$ while the right one performs a doubling $P = Q$.

---

**Algorithm 29.1** Atomic addition-doubling formulas

INPUT: The points $P = (T_1, T_2)$ and $Q = (T_3, T_4)$ on $E(\mathbb{F}_{2^d})$.
OUTPUT: The points $P \oplus Q$ or $[2]P$.

| Addition: $P \leftarrow P \oplus Q$ | | | Doubling: $P \leftarrow [2]P$ | | |
|---|---|---|---|---|---|
| 1. | $T_1 \leftarrow T_1 + T_3$ | $[x_1 + x_2]$ | 1. | $T_6 \leftarrow T_1 + T_3$ | [fake] |
| 2. | $T_2 \leftarrow T_2 + T_4$ | $[y_1 + y_2]$ | 2. | $T_6 \leftarrow T_3 + T_6$ | $[x_1]$ |
| 3. | $T_5 \leftarrow T_2/T_1$ | $[\lambda]$ | 3. | $T_5 \leftarrow T_2/T_1$ | $[y_1/x_1]$ |
| 4. | $T_1 \leftarrow T_1 + T_5$ | | 4. | $T_5 \leftarrow T_1 + T_5$ | $[\lambda]$ |
| 5. | $T_6 \leftarrow T_5^2$ | $[\lambda^2]$ | 5. | $T_1 \leftarrow T_5^2$ | $[\lambda^2]$ |
| 6. | $T_6 \leftarrow T_6 + a_2$ | $[\lambda^2 + a_2]$ | 6. | $T_1 \leftarrow T_1 + a_2$ | $[\lambda^2 + a_2]$ |
| 7. | $T_1 \leftarrow T_1 + T_6$ | $[x_3]$ | 7. | $T_1 \leftarrow T_1 + T_5$ | $[x_3]$ |
| 8. | $T_2 \leftarrow T_1 + T_4$ | $[x_3 + y_2]$ | 8. | $T_2 \leftarrow T_1 + T_2$ | $[x_3 + y_1]$ |
| 9. | $T_6 \leftarrow T_1 + T_3$ | $[x_2 + x_3]$ | 9. | $T_6 \leftarrow T_1 + T_6$ | $[x_1 + x_3]$ |
| 10. | $T_5 \leftarrow T_5 \times T_6$ | | 10. | $T_5 \leftarrow T_5 \times T_6$ | |
| 11. | $T_2 \leftarrow T_2 + T_5$ | $[y_3]$ | 11. | $T_2 \leftarrow T_2 + T_5$ | $[y_3]$ |
| 12. | **return** $(T_1, T_2)$ | | 12. | **return** $(T_1, T_2)$ | |

---

The choice of the numbers of the extra registers was done in such a way that with a simple assignment of two variables depending on the bit of the scalar, both group addition and doubling can be implemented with the same algorithm using no if/else conditional branching.

The big advantage of this idea is that only cheap dummy operations are introduced and that the scalar multiplication can make use of windowing techniques to have a sparse representation. The above algorithm can be modified to deal with subtractions, too, by inserting one more addition at the beginning to change $y_2$ to $y_2 + x_2$ if necessary.

For elliptic curves over fields of odd characteristic, a direct application of this method would imply that one needs to insert a dummy squaring in each addition as an addition needs $I + 2M + S$ while a doubling needs $I + 2M + 2S$. First of all, squarings take a non-negligible effort and furthermore, for the environments we consider in this chapter, inversion-free coordinate systems are more useful. To circumvent this difficulty, [CHCI+ 2004] introduce the concept of *side-channel atomicity*. Instead of trying to make the group operations look identical, they split the operations into identical blocks, each consisting of one field multiplication (which could also be a squaring), one field addition, one field negation, and one further field addition. An attacker can only observe a sequence of identical blocks and cannot link them to the operation that is performed. For elliptic curves over $\mathbb{F}_{p^d}$ with $p$ odd, [CHCI+ 2004] implement a curve doubling in Jacobian coordinates using 10 blocks and an addition in 16 blocks. The computational overhead involved in this countermeasure is almost negligible, introducing only a few further field additions and negations but no multiplications.

The security is based on the assumption that real and dummy operations cannot be distinguished and that multiplications cannot be told from squarings, which is often a valid assumption in odd characteristic. (This might not hold if higher order differential attacks can be used, cf. Section 29.4.) When the scalar multiplication is computed with these addition and doubling algorithms, it becomes a computation of a series of atomic blocks and the side-channel information becomes uniform. This difference is visualized in the following two pictures. Figure 29.1 shows the view of the attacker

**Figure 29.1** Attacker's view on algorithm.

**Figure 29.2** Actual operations inside blocks.

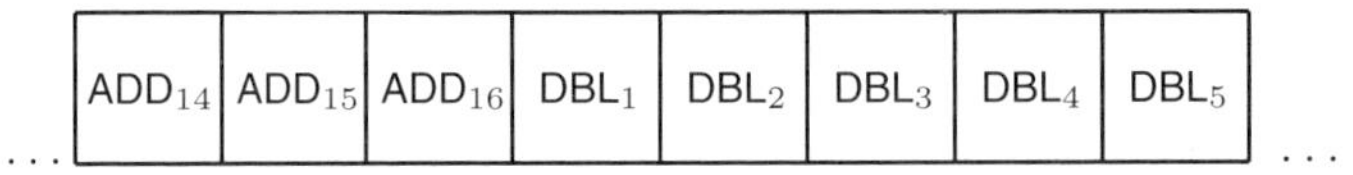

who only sees a uniform sequence of operations. In Figure 29.2 one sees that the first three steps belonged to an addition while the following ones show the beginning of a doubling.

We do not state the original formulas in atomic blocks for odd characteristic but state a simplified version in Algorithms 29.2 and 29.3. Unlike in the original version, we only list a $*$ to denote a dummy operation while in [CHCI$^+$ 2004] the authors state the access to the registers by a uniform algorithm in a compact representation given by a matrix. In total 10 registers are used to hold the variables for doubling and addition and the dummy operations are performed by different registers.

Our tables come from Mishra's approach [MIS 2004a] to combine side-channel atomicity and pipelined implementations on two processors to achieve a high throughput. Also, he considers mixed additions, i.e., additions in which one point is in affine coordinates while the other is in Jacobian. In Section 13.2.2 this was identified to be the fastest system for unprotected implementations and using the cheap countermeasure of side-channel atomicity, Mishra obtains a very efficient pipelined implementation, which we detail in the following.

We now list the algorithms for doubling in Jacobian coordinates and mixed addition. Like before the sequence of field operations is M, A, Neg, A, where Neg denotes a field negation. A $*$ denotes a dummy operation of the respective type. Note that no dummy multiplications appear. The value of the current variable is given in brackets. The intermediate point is stored in Jacobian coordinates in $(T_6 : T_7 : T_8)$ and the result is written back to these registers. The notation refers to the abbreviations used in Section 13.2.1.c.

---

**Algorithm 29.2** Elliptic curve doubling in atomic blocks

INPUT: The point $P_i = (X_i : Y_i : Z_i) = (T_6 : T_7 : T_8)$.
OUTPUT: The point $[2]P_i = (X_{i+1} : Y_{i+1} : Z_{i+1}) = (T_6 : T_7 : T_8)$.

| | | | | |
|---|---|---|---|---|
| $\Delta_1$. | $R_1 \leftarrow T_8 \times T_8$ $[Z_i^2]$ | $*$ | $*$ | $*$ |
| $\Delta_2$. | $R_1 \leftarrow R_1 \times R_1$ $[Z_i^4]$ | $*$ | $*$ | $*$ |
| $\Delta_3$. | $R_1 \leftarrow a_4 \times R_1$ $[a_4 Z_i^4]$ | $*$ | $*$ | $*$ |
| $\Delta_4$. | $R_2 \leftarrow T_6 \times T_6$ $[X_i^2]$ | $R_3 \leftarrow R_2 + R_2$ $[2X_i^2]$ | $*$ | $R_2 \leftarrow R_3 + R_2$ $[3X_i^2]$ |

| | | | | |
|---|---|---|---|---|
| $\Delta_5$. | $T_8 \leftarrow T_7 \times T_8$ | $T_8 \leftarrow T_8 + T_8$ | $*$ | $R_1 \leftarrow R_1 + R_2$ |
| | $[Y_i Z_i]$ | $[Z_{i+1}]$ | | $[B]$ |
| $\Delta_6$. | $R_4 \leftarrow T_7 \times T_7$ | $R_2 \leftarrow R_4 + R_4$ | $*$ | $*$ |
| | $[Y_i^2]$ | $[2Y_i^2]$ | | |
| $\Delta_7$. | $R_4 \leftarrow T_6 \times R_2$ | $R_4 \leftarrow R_4 + R_4$ | $R_4 \leftarrow -R_4$ | $R_5 \leftarrow R_4 + R_4$ |
| | $[2X_i Y_i^2]$ | $[A]$ | $[-A]$ | $[-2A]$ |
| $\Delta_8$. | $R_3 \leftarrow R_1 \times R_1$ | $T_6 \leftarrow R_3 + R_5$ | $*$ | $R_4 \leftarrow T_6 + R_4$ |
| | $[B^2]$ | $[X_{i+1}]$ | | $[X_{i+1} - A]$ |
| $\Delta_9$. | $R_2 \leftarrow R_2 \times R_2$ | $R_2 \leftarrow R_2 + R_2$ | $*$ | $*$ |
| | $[4Y_i^4]$ | $[8Y_i^4]$ | | |
| $\Delta_{10}$. | $T_7 \leftarrow R_1 \times R_4$ | $T_7 \leftarrow T_7 + R_2$ | $T_7 \leftarrow -T_7$ | $*$ |
| | $[B(X_{i+1} - A)]$ | $[-Y_{i+1}]$ | $[Y_{i+1}]$ | |

**Algorithm 29.3** Elliptic curve addition in atomic blocks

INPUT: The points $P = (T_x, T_y)$ and $P_i = (X_i : Y_i : Z_i) = (T_6 : T_7 : T_8)$.
OUTPUT: The point $P + P_i = (X_{i+1} : Y_{i+1} : Z_{i+1}) = (T_6 : T_7 : T_8)$.

| | | | | |
|---|---|---|---|---|
| $\Gamma_1$. | $R_1 \leftarrow T_8 \times T_8$ | $*$ | $*$ | $*$ |
| | $[Z_i^2]$ | | | |
| $\Gamma_2$. | $R_2 \leftarrow T_x \times R_1$ | $*$ | $R_2 = -R_2$ | $*$ |
| | $[A]$ | | $[-A]$ | |
| $\Gamma_3$. | $R_3 \leftarrow T_y \times T_8$ | $*$ | $*$ | $*$ |
| | $[YZ_i]$ | | | |
| $\Gamma_4$. | $R_3 \leftarrow R_3 \times R_1$ | $R_1 \leftarrow R_2 + T_6$ | $R_1 \leftarrow -R_1$ | $*$ |
| | $[C]$ | $[E]$ | $[-E]$ | |
| $\Gamma_5$. | $T_8 \leftarrow R_1 \times T_8$ | $*$ | $*$ | $*$ |
| | $[Z_{i+1}]$ | | | |
| $\Gamma_6$. | $R_4 \leftarrow R_1 \times R_1$ | $*$ | $*$ | $*$ |
| | $[E^2]$ | | | |
| $\Gamma_7$. | $R_2 \leftarrow R_2 \times R_4$ | $R_5 \leftarrow R_2 + R_2$ | $*$ | $*$ |
| | $[-AE^2]$ | $[-2AE^2]$ | | |
| $\Gamma_8$. | $R_1 \leftarrow R_4 \times R_1$ | $R_1 \leftarrow R_1 + R_5$ | $R_3 \leftarrow -R_3$ | $R_5 \leftarrow R_3 + T_7$ |
| | $[-E^3]$ | $[-E^3 - 2AE^2]$ | $[-C]$ | $[F]$ |
| $\Gamma_9$. | $T_6 \leftarrow R_5 \times R_5$ | $T_6 \leftarrow T_6 + R_1$ | $*$ | $R_2 \leftarrow T_6 + R_2$ |
| | $[F^2]$ | $[X_{i+1}]$ | | $[X_{i+1} - AE^2]$ |
| $\Gamma_{10}$. | $R_2 \leftarrow R_5 \times R_2$ | $*$ | $R_1 \leftarrow -R_1$ | $*$ |
| | $[-F(X_{i+1} - AE^2)]$ | | $[E^3]$ | |
| $\Gamma_{11}$. | $T_7 \leftarrow R_3 \times R_4$ | $T_7 \leftarrow T_7 + R_2$ | $*$ | $*$ |
| | $[-CE^3]$ | $[Y_{i+1}]$ | | |

If, instead, a subtraction should be performed, i.e., one wants to add the negative $(-T_x, T_y)$, this can be incorporated as $R_2 = -T_x$ in $\Gamma_1$ and by replacing the first step of $\Gamma_2$ by $R_2 = R_2 \times R_1$.

**Table 29.3** EC-operations in the pipeline.

| Time | DBL-DBL PS1 | DBL-DBL PS2 | DBL-ADD PS1 | DBL-ADD PS2 | ADD-DBL PS1 | ADD-DBL PS2 |
|---|---|---|---|---|---|---|
| $k$ | $\vdots$ | $\vdots$ | $\vdots$ | $\vdots$ | $\vdots$ | $\vdots$ |
| $k+1$ | $\Delta_1^{(i)}$ | — | $\Delta_1^{(i)}$ | — | $\Gamma_1^{(i)}$ | — |
| $k+2$ | $\Delta_2^{(i)}$ | — | $\Delta_2^{(i)}$ | — | $\Gamma_2^{(i)}$ | — |
| $k+3$ | $\Delta_3^{(i)}$ | — | $\Delta_3^{(i)}$ | — | $\Gamma_3^{(i)}$ | — |
| $k+4$ | $\Delta_4^{(i)}$ | — | $\Delta_4^{(i)}$ | — | $\Gamma_4^{(i)}$ | — |
| $k+5$ | $\Delta_5^{(i)}$ | — | $\Delta_5^{(i)}$ | — | $\Gamma_5^{(i)}$ | — |
| $k+6$ | $\Delta_1^{(i+1)}$ | $\Delta_6^{(i)}$ | $\Gamma_1^{(i+1)}$ | $\Delta_6^{(i)}$ | $\Delta_1^{(i+1)}$ | $\Gamma_6^{(i)}$ |
| $k+7$ | $\Delta_2^{(i+1)}$ | $\Delta_7^{(i)}$ | $\Gamma_2^{(i+1)}$ | $\Delta_7^{(i)}$ | $\Delta_2^{(i+1)}$ | $\Gamma_7^{(i)}$ |
| $k+8$ | $\Delta_3^{(i+1)}$ | $\Delta_8^{(i)}$ | $\Gamma_3^{(i+1)}$ | $\Delta_8^{(i)}$ | $\Delta_3^{(i+1)}$ | $\Gamma_8^{(i)}$ |
| $k+9$ | $\Delta_4^{(i+1)}$ | $\Delta_9^{(i)}$ | $\Gamma_4^{(i+1)}$ | $\Delta_9^{(i)}$ | $*$ | $\Gamma_9^{(i)}$ |
| $k+10$ | $*$ | $\Delta_{10}^{(i)}$ | $\Gamma_5^{(i+1)}$ | $\Delta_{10}^{(i)}$ | $\Delta_4^{(i+1)}$ | $\Gamma_{10}^{(i)}$ |
| $k+11$ | $\Delta_5^{(i+1)}$ | $*$ | $\vdots$ | $\Gamma_6^{(i+1)}$ | $*$ | $\Gamma_{11}^{(i)}$ |
| $k+12$ | $\vdots$ | $\Delta_6^{(i+1)}$ | $\vdots$ | $\Gamma_7^{(i+1)}$ | $\Delta_5^{(i+1)}$ | $*$ |

Table 29.3 shows the pipelining for all 3 possible cases — 2 doublings in a row, a doubling followed by a mixed addition and a mixed addition followed by a doubling. This scheduling introduces complete dummy blocks in the wait stages, but this is the usual drawback one needs to take into account when using more than one processor. As in a scalar multiplication each addition is followed by a doubling, the last two columns always appear together. This implies that each group operation consumes 6 atomic blocks, i.e., a scalar multiplication using an NAF of the scalar needs on average 8 atomic blocks per bit of the scalar.

Additionally, this method can be combined efficiently with windowing methods as the SCA resistance is implied by the atomic block structure [MIS 2004b]. The generalization to genus 2 curves was done in [LAMI 2004] in which the authors provide the expressions for performing parallel doubling or addition in atomic blocks. In even characteristic affine coordinates are chosen and due to the large similarity of addition and doubling, each group operation is made one big atomic block. In odd characteristic new coordinates (cf. Section 14.4.2) are used and it turns out that the atomic blocks proposed for implementing elliptic curves remain optimal. These formulas also apply to nonparallel implementations and achieve the lowest number of registers.

Countermeasures inserting dummy operations are always vulnerable to fault attacks (cf. Section 29.6). If the faults can be induced very precisely it may be possible to distinguish real from dummy operations, as a fault in the latter does not influence the result. Nevertheless, these attacks are less likely for the quick dummy operations appearing in the atomicity approach, but they are a real threat in the double and always add algorithm.

### 29.1.2 Unified addition formulas

Using unified formulas means that in the scalar multiplication algorithm the same set of formulas is invoked with different inputs. The operations are valid for both addition and doubling. To avoid leakage from the if/else instruction one should use arithmetic operations to schedule which of the inner paths is taken in the few places where the formulas still differ. We point out that these differences will not be between doubling and addition.

So far such formulas have been obtained only for elliptic curves. There is some ongoing research for hyperelliptic curves but so far nothing is published. Hence, we concentrate on elliptic curves for the remainder of this section.

In the previous section we have shown how to make addition and doubling on a binary curve use the same sequence of field operations, such that the side-channel information looks the same. There this was done by inserting dummy field additions. Here we achieve this aim *without* dummy operations. We first deal with general elliptic curves over arbitrary fields and then consider countermeasures for more special types of curves.

### 29.1.2.a  Unified formulas for general curves

Here we consider elliptic curves given by

$$E : y^2 + a_1 xy + a_3 y = x^3 + a_2 x^2 + a_4 x + a_6,$$

for $a_i \in \mathbb{F}_q$, for some prime power $q$. By writing the slope $\lambda$ in a different manner one can also obtain really unified addition formulas needing identical computations for addition and doubling. Building on [BRJO 2003] it is shown in [BRDÉ$^+$ 2004] that in arbitrary characteristic we have:

$$\begin{aligned}
\lambda &= \frac{y_2 - y_1}{x_2 - x_1} = \frac{y_2 - y_1}{x_2 - x_1} \frac{y_1 + y_2 + a_1 x_2 + a_3}{y_1 + y_2 + a_1 x_2 + a_3} \\
&= \frac{x_1^2 + x_1 x_2 + x_2^2 - a_1 y_1 + a_2(x_1 + x_2) + a_4}{y_1 + y_2 + a_1 x_2 + a_3} \\
&= \frac{x_1^2 + x_1 x_2 + x_2^2 - a_1 y_1 + a_2(x_1 + x_2) + a_4 + (y_1 - y_2)m}{y_1 + y_2 + a_1 x_2 + a_3 + (x_1 - x_2)m} \\
&= \frac{x_1^2 + x_1 x_2 + x_2^2 - a_1 y_2 + a_2(x_1 + x_2) + a_4 + (y_2 - y_1)\widetilde{m}}{y_1 + y_2 + a_1 x_1 + a_3 + (x_2 - x_1)\widetilde{m}},
\end{aligned}$$

where $m = m(x_1, y_1; x_2, y_2)$ is an arbitrary polynomial and $\widetilde{m} = m(x_2, y_2; x_1, y_1)$. The very same expression holds also for doublings. We remark that it is necessary to introduce both $m$ and $\widetilde{m}$ as the first version [BRJO 2003], corresponding to the second line, was vulnerable to the exceptional procedure attack [IZTA 2003b] (cf. Section 29.3). Namely, the algorithm would fail if $y_1 + y_2 + a_1 x_2 + a_3 = 0$, i.e., if the $y$-coordinate of $P$ equaled the $y$-coordinate of $-Q$.

By defining $\lambda_m$ as in the last line whenever $y_1 + y_2 + a_1 x_2 + a_3 + (x_1 - x_2)m = 0$ and as in the one to last line otherwise, this expression is valid for *all* points for appropriately chosen $m$. For nonsupersingular binary elliptic curves as given in (29.1) and curves over fields of odd characteristic, one can choose $m = 1$.

In affine coordinates this means that $\lambda_m$ can be evaluated with I+5M including 2M with the constants $a_1$ and $a_2$. In odd characteristic these constants can be chosen to be zero and in characteristic 2 one has $a_1 = 1$. In characteristic 2, affine coordinates are often the fastest method of computation as inversions are not much more expensive than multiplications. In odd characteristic, one tries to avoid inversions and moves to projective or Jacobian coordinates. The same ideas can be applied to the differently defined slope $\lambda_m$ given above. For isomorphic transformations and ways to derive inversion-free formulas we refer to Sections 13.2 and 13.3.

In particular, the inversion-free systems in odd characteristic suffer from the loss of speed due to the countermeasure. If the curve has points of order 2 or 3, other representations of the curve leading to faster unified group operations can be designed.

### 29.1.2.b Unified formulas for Hessian curves

In Section 13.1.5.b Hessian coordinates were introduced. Let $q \equiv 2 \pmod{3}$ and consider an elliptic curve $E$ over $\mathbb{F}_q$ with an $\mathbb{F}_q$-rational point of order 3. The *Hessian form* is given by

$$H : X^3 + Y^3 + Z^3 = cXYZ$$

for some $c \in \mathbb{F}_q$.

Joye and Quisquater [JoQu 2001] suggested using Hessian curves to avoid simple side-channel attacks. It is possible to compose the scalar multiplication of additions only as

$$[2](X_1 : Y_1 : Z_1) = (Z_1 : X_1 : Y_1) \oplus (Y_1 : Z_1 : X_1)$$

and $(Z_1 : X_1 : Y_1) \neq (Y_1 : Z_1 : X_1)$. An addition needs $12M + 6S$, which compares favorably to the general unified formulas presented before. Hence, the use of unified addition formulas is immediate. Additionally, Hessian curves allow for fast parallel implementation [SMA 2001].

### 29.1.2.c Elliptic curves in Jacobi model

A further set of unified formulas is obtained in [LiSm 2001] by viewing the curve as intersection of two quadrics. They need that $E(\mathbb{F}_q)$ contains a copy of $\mathbb{Z}/2\mathbb{Z} \times \mathbb{Z}/2\mathbb{Z}$. Their formulas were improved in [BiJo 2003] and the approach was generalized to curves having one affine point of order two only. The extended Jacobi form of an elliptic curve is given by

$$E_J : Y^2 = \epsilon X^4 - 2\delta X^2 Z^2 + Z^4.$$

The sum of $P = (X_1 : Y_1 : Z_1)$ and $Q = (X_2 : Y_2 : Z_2)$ is given by $(X_3 : Y_3 : Z_3)$ with

$$\begin{aligned} X_3 &= X_1 Y_2 Z_1 + X_2 Y_1 Z_2, \; Z_3 = (Z_1 Z_1)^2 - \epsilon(X_1 X_2)^2, \\ Y_3 &= (Z_3 + \epsilon(X_1 X_2)^2)(Y_1 Y_2 - 2\delta X_1 X_2 Z_1 Z_2) + 2\epsilon X_1 X_2 Z_1 Z_2 (X_1^2 Z_2^2 + X_2^2 Z_1^2), \end{aligned}$$

with the only exception $Q = -P = (-X_1 : Y_1 : Z_1)$. Hence, the formula holds also for $P = Q$.

Olson [OLS 2004] generalized the approach in [BiJo 2003] by showing how to obtain an isomorphic projective quartic curve with weighted inversion-free coordinates, in which doubling and addition need the same formulas. This can be done for each elliptic curve but is highly inefficient in the general case requiring $31M$. If the curve contains a point of order 2, i.e., with $Y = 0$, his transformation yields the curve considered above. Another case leading to a simple equation is a curve containing a point with $X = 0$ as this results in a quartic

$$W^2 = S^4 - \beta S - a_4/4$$

and the unified formulas are efficient.

## 29.1.3 Montgomery arithmetic

The Montgomery ladder 9.5 was introduced in Chapter 9 with a reference to side-channel attacks. In fact this idea thwarts simple side-channel attacks as for each bit of the scalar a doubling and an addition are performed. The difference to the double and always add approach is that the additions are *used additions*; hence, the method is not subject to fault attacks.

We briefly recall the algorithm here in the setting of $J_C$ the Jacobian of an elliptic or hyperelliptic curve $C$ and $\overline{D} \in J_C$. Then we consider issues specific to elliptic and hyperelliptic curves.

**Algorithm 29.4** Montgomery's ladder

INPUT: An element $\overline{D} \in J_C$ and a positive integer $n = (n_{l-1} \ldots n_0)_2$.
OUTPUT: The element $[n]\overline{D} \in J_C$.

1. $\overline{D}_1 \leftarrow \overline{D}$ and $\overline{D}_2 \leftarrow [2]\overline{D}$
2. **for** $i = l - 2$ **down to** $0$ **do**
3.     **if** $n_i = 0$ **then**
4.         $\overline{D}_1 \leftarrow [2]\overline{D}_1$ and $\overline{D}_2 \leftarrow \overline{D}_1 \oplus \overline{D}_2$
5.     **else**
6.         $\overline{D}_1 \leftarrow \overline{D}_1 \oplus \overline{D}_2$ and $\overline{D}_2 \leftarrow [2]\overline{D}_2$
7. **return** $\overline{D}_1$

This algorithm can be applied universally for any group. However, it suffers from the additional group operations like the double and always add method and additionally doubles the storage requirements. On the other hand, fault attacks cannot be applied, which made [IZTA 2002b] use this method against SPA.

For elliptic curves and hyperelliptic curves of genus 2 we have shown in Chapters 13 and 14 that one does not need the $Y$ coordinate, respectively the second polynomials $v(x)$, to carry out the scalar multiplication with Algorithm 29.4 and that this coordinate can be recovered uniquely.

We do not repeat the formulas here and refer to Sections 13.2.3 and 13.3.4 for Montgomery coordinates on elliptic curves in odd and even characteristic. In the latter case they can be evaluated efficiently for arbitrary curves but for odd characteristic they are more efficient for curves having a group order divisible by 4. We present both cases in the section on efficient arithmetic.

For genus 2 curves different approaches exist depending on whether the group order is divisible by 2. Like in the case of elliptic curves a larger rational 2-torsion part facilitates the arithmetic. For the more efficient case, due to Duquesne [DUQ 2004], the formulas are given in Section 14.4.4.

Lange's [LAN 2004a] approach is valid for arbitrary curves over fields of odd characteristic but it is less efficient. Both methods are better than the double and always add approach but far less efficient than the introduction of dummy field operations considered in Section 29.1.1.b.

The drawback of the Montgomery method is that one cannot use precomputations to speed up the scalar multiplications. On the other hand, Montgomery arithmetic can be parallelized efficiently even for small devices as reported in [FIGI[+] 2002]. As fault attacks cannot be applied the implementation is hedged against non-differential attacks.

## 29.2 Countermeasures against differential SCA

*Differential side-channel attacks (DSCAs)* are usually applied only if the simple side-channel attacks do not succeed, e.g., if the leaked information cannot be linked directly to the bits of the secret scalar. If one has access to side-channel information of several scalar multiplications involving the *same secret scalar* and *different group elements*, differential techniques can be applied. Note that we consider averaging over several measurements of the *same* scalar multiplication, i.e., the same scalar and same base point, as an enhanced simple side-channel attack and not as a differential one. Furthermore, this situation is very unlikely to occur in practice.

To mount a differential attack one needs to be able to simulate the operations in the attacked device. In particular one needs to know the internal representations as the operations are likely to depend on this. The common strategy in all countermeasures is therefore to introduce randomness in the representation of the points, of the curve or of the scalar such that the simulation cannot be achieved. In practice, these countermeasures are often combined.

The next sections deal with these methods. We like to mention that these countermeasures are only needed if a long-term secret is involved. If the scalar is changed at each iteration of the protocol they are not needed. However, still one needs to make sure that the implementation does not leak information on the bits of the nonces in signature schemes, as this can be used in a lattice based attack as shown in [BOVE 1996, HOSM 2001, NGSH 2003].

### 29.2.1 Implementation of DSCA

Suppose that by applying the countermeasures from the previous section the implementation is secured against SSCA. DSCA on a double and add scalar multiplication algorithm computing $[n]\overline{D}$, where $n$ is a secret scalar used for several executions works as follows:

Let $n = (n_{l-1}n_{l-2}\ldots n_0)_2$, and suppose that the first digits $n_{l-1}n_{l-2}\ldots n_{j+1}$ are known. The attacker wants to find $n_j$. He proceeds as follows:

1. The attacker first makes a *guess*: $n_j = 0$ or $1$.
2. He chooses several inputs $\overline{D}_1,\ldots,\overline{D}_t$ and computes $\overline{E}_i = \big[\sum_{k=j}^{l-1} n_k 2^{k-j}\big]\overline{D}_i$.
3. He picks a Boolean selection function $B$ to construct two index sets

$$\mathcal{S}_t = \{i : B(\overline{E}_i) = \texttt{true}\} \text{ and } \mathcal{S}_f = \{i : B(\overline{E}_i) = \texttt{false}\}.$$

For instance, $B$ might be the least significant bit of the representation of $\overline{E}_i$ in case this influences the computation.

4. He puts $SC_i = SC_i(t)$ to be the side-channel information obtained from the computation of $[n]\overline{D}_i$. This is a function of the time $t$. In the case where electromagnetic emission is used, this function is in fact also a function of space, for example, of a point on the surface of the card: namely $SC_i = SC_i(t; x, y)$.
5. Let $\langle SC_i \rangle_{i \in \mathcal{S}}$ denote the average of the functions $SC_i$ for the $i \in \mathcal{S}$. If the guess of $n_j$ was incorrect, then one expects

$$\langle SC_i \rangle_{i \in \mathcal{S}_t} - \langle SC_i \rangle_{i \in \mathcal{S}_f} \approx 0$$

i.e., the two sets are uncorrelated as the selection function was not related to the operation actually carried out. On the other hand, if the guess of $n_j$ was correct then

$$\langle SC_i \rangle_{i \in \mathcal{S}_t} - \langle SC_i \rangle_{i \in \mathcal{S}_f}$$

as a function of time (and possibly space) will present spikes, i.e., deviations from zero, cf. Figures 28.7 and 28.8.

This means that the cases of correlation and uncorrelation should be easy to distinguish, provided that the selection function captures the properties of the algorithm and enough samples are obtained. Once $n_j$ is obtained, the attack is used on the next digit.

Note that in practice the attacker does not choose the $\overline{D}_i$ but waits for further operations involving the private key or issues elements $\overline{D}_i$ on which $[n]\overline{D}_i$ is computed. Usually the attacker has no influence on the choice of the $\overline{D}_i$ so they are assumed to be random.

Additionally we like to mention that this method does not require making fresh measurements after each bit $n_j$ obtained but can work as an offline attack. In a short time many samples of side-channel leakage are recorded and stored. Then they allow us to recover the secret key bit-by-bit by following the above steps for each $n_j, l-2 \leqslant j \leqslant 0$ in an offline computation. This has the drawback that the complete side-channel traces need to be recorded, increasing the storage requirement — but it might be harder to notice the attacker.

### 29.2.2 Scalar randomization

The idea of this countermeasure is to change the representation of the scalar (bit-pattern or windowing expansion) such that it is not linked between the different executions of the algorithm. This can be achieved by either changing the scalar itself or by only changing the addition chain used for the scalar multiplication.

#### 29.2.2.a Varying the scalar

In all applications the group order $\ell$ is assumed to be known. We have $[n]\overline{D} = [n + i\ell]\overline{D}$ for any integer $i$. Random choices of $i$ should randomize the binary pattern of the applied scalar and thus the representations during the execution. In [OKSA 2000] it has been shown that this randomization leaves a bias in the least significant bits of the scalar. For elliptic curves, Möller [MÖL 2001a] combines this method with an idea of Clavier and Joye, and suggests the computation of $[n]\overline{D} = [n + k_1 + k_2 \ell]\overline{D} \ominus [k_1]\overline{D}$, where $k_1$ and $k_2$ are two suitably sized random numbers. In the context of group-based cryptography this method is called *Coron's first countermeasure* [COR 1999]. It induces a performance penalty depending linearly on the bit lengths of the random quantities $k$, respectively $k_1$ and $k_2$.

These countermeasures can be combined with any of those described in the following section to provide stronger defense.

#### 29.2.2.b Varying the representation

The insertion of random decisions during the execution of the scalar multiplication algorithm also helps in preventing side-channel analysis. In general, these decisions amount to randomly choosing among several different redundant weighted binary representations of the same integer [HAMO 2002, OSAI 2001]. Such methods must be used with care, and indeed their soundness has been questioned [WAL 2004], sometimes under the assumption that no SSCA-countermeasures are implemented [OKSA 2002, OKSA 2003].

Liardet and Smart [LISM 2001] propose not only unified addition formulas as explained in the previous section but also a countermeasure against DSCA. Assuming the indistinguishability of addition and doubling, they present three general-purpose randomized signed window methods for performing SCA-resistant scalar multiplication.

However, their argument depends heavily on the unified group operations. Under the hypothesis that additions are distinguishable from doublings, Walter [WAL 2002a] shows that repeated use of the same secret key with the algorithm of Liardet and Smart is insecure. Another randomized $2^k$-ary method is given in [ITYA$^+$ 2002].

It is also possible to apply a variant of MIST [WAL 2001, WAL 2002b]. The basic idea is to randomly choose between several different representations on a mixed base number system and at the same time select small addition chains for the different parts of the computation that are composed from similar sub-blocks of group operations. Ideally, such a technique would provide both SSCA and DSCA resistance. However, especially in the presence of long keys, this method alone can also leak some information on the exponent [WAL 2003].

## 29.2.3 Randomization of group elements

The methods described in this section leave the scalar invariant but change the representation of the base group element. Hence, it is no longer possible to simulate the attack as the internal representation is unknown.

### 29.2.3.a Blinding of group elements

In any group it is possible to apply Coron's second countermeasure [COR 1999]. Let $n$ be the long-term secret. To implement this countermeasure one chooses a group element $\overline{E} \in \text{Pic}_C^0$ from a list stating pairs $(\overline{E}, [n]\overline{E})$ for various $\overline{E}$ and computes $[n]\overline{D} = [n](\overline{D} \oplus \overline{E}) \ominus [n]\overline{E}$. So, the computation is hidden due to the blinding addend $\overline{E}$.

Besides the precomputations this method only adds an addition and a doubling per scalar multiplication. The list should be large enough and kept secret, to avoid brute force attacks. The element $\overline{E}$ should be changed at each execution.

### 29.2.3.b Randomization by redundant coefficients

A particularity of cryptography based on curves is that the groups come with a natural way of redundant representation. In inversion-free coordinate systems like projective or Jacobian coordinates a point is not uniquely represented. In Jacobian coordinates the points $(X_1 : Y_1 : Z_1)$ and $(\lambda^2 X_1 : \lambda^3 Y_1 : \lambda Z_1)$ are equal for any $\lambda \in \mathbb{F}_q^*$. This means that it is possible to change the representation of a point by $4M + S$ and even less in projective coordinates. The same approach can be used in all inversion-free coordinate systems [JOTY 2002] and it applies to elliptic curves as well as to the divisor class group of higher genus curves [AVA 2004b]. In addition to those results, we mention that in the meantime also projective formulas for genus 3 curves were published. Hence, this method can be applied there as well. The randomization can also be applied additionally at random places during the scalar multiplication.

Note that the final inversion to obtain affine coordinates and to change between the systems needs to be performed in the secure environment. Otherwise, i.e., if the triple is the direct result of a scalar multiplication without further randomization, [NAST+ 2004] show that the previous operations (doublings or additions) can be distinguished.

It is important that this attack was found, however, in practice one would either provide the result in affine coordinates (which is standard in transmission to reduce the bandwidth) or perform a further random blinding with a new choice of $\lambda$.

However, there are some coordinates that are not changed by multiplication by $\lambda$. Namely if the respective field element is 0 then it is unchanged by this method. Goubin's attack [GOU 2003] makes use of this observation together with the fact that a zero coordinate leads to shorter computation times in most operations. We like to point out that this method is not restricted to zero-coordinates but it can also work if just some coefficients in the representation of $\mathbb{F}_q$ as vector space over $\mathbb{F}_p$ are zero. We discuss this in more detail in the following Section 29.3.

## 29.2.4 Randomization of the curve equation

So far we implicitly identified isomorphic affine equations for curves and different representations of the finite fields. However, it is possible to use the isomorphisms as secret maps to prevent the attacker from simulating the scalar multiplication.

### 29.2.4.a  Isomorphic field representations

Using a different irreducible polynomial of degree $d$ leads to an isomorphic description of $\mathbb{F}_{p^d}$. This method does not depend on the curve and, hence, [JOTY 2002] covers this idea for general curves. There are a few drawbacks attached to this method: computations in the isomorphic fields might be considerably slower than usual as less efficient field representations (e.g., perhaps no binomials) are used. Furthermore, the field isomorphism is often only given by a $(d \times d)$ matrix over $\mathbb{F}_p$.

### 29.2.4.b  Isomorphic curve equations

In Sections 13.1.5 and 14.1.3 we introduced isomorphic transformations on the curve equation. The resulting curve is given by a different equation and the representation of the group elements is changed, too. If the isomorphism is secret the attacker has no information on the representation of the group elements, and hence, cannot simulate the computation. This countermeasure was proposed for elliptic curves in [JOTY 2002] and generalized to hyperelliptic curves in [AVA 2004b]. We present the version from the latter reference in the setting of hyperelliptic curves.

Let $C$ and $\widetilde{C}$ be two hyperelliptic curves of genus $g \geqslant 1$ over a finite field $\mathbb{F}_q$. Suppose that $\psi : C \to \widetilde{C}$ is an $\mathbb{F}_q$-isomorphism that is easily extended to an $\mathbb{F}_q$-isomorphism of the divisor class groups $\psi : \mathrm{Pic}^0_C \to \mathrm{Pic}^0_{\widetilde{C}}$. Let us further assume that $\psi$, and the inverse $\psi^{-1}$, are computable in a reasonable amount of time, i.e., small with respect to the time of a scalar multiplication. We do not require *a priori* the computation time of $\psi$ to be negligible with respect to a single group operation. Then, instead of computing $\overline{E} = [n]\overline{D}$ in $\mathrm{Pic}^0_C$ we perform:

$$\overline{E} = \psi^{-1}\bigl([n]\,\psi(\overline{D})\bigr)$$

so that the bulk of the computation is done in $\mathrm{Pic}^0_{\widetilde{C}}$. We visualize it in the following *commutative diagram*.

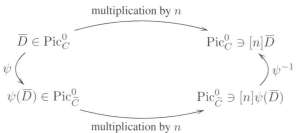

and we follow it along the longer path. The countermeasure is effective if the representations of the images under $\psi$ of the curve coefficients and of the elements of $\mathrm{Pic}^0_C$ are unpredictably different from those of their sources. This can be achieved by using randomly chosen isomorphisms $\psi$, which boils down to multiplying all the quantities involved in a computation with "random" numbers.

In Section 14.1.3 we gave examples of isomorphisms and in the following sections, optimal equations for each isomorphism class of curves were obtained. Applying random isomorphisms means that this optimal equation cannot be used. We briefly recapitulate the effects of isomorphisms on hyperelliptic curves, implying elliptic curves.

Let $C, \widetilde{C}$ be two hyperelliptic curves of genus $g$ defined by *Weierstraß equations*

$$C \; : \; y^2 + h(x)y - f(x) = 0 \qquad (29.2)$$

$$\widetilde{C} \; : \; y^2 + \tilde{h}(x)y - \tilde{f}(x) = 0 \qquad (29.3)$$

over $\mathbb{F}_q$, where $f, \tilde{f}$ are monic polynomials of degree $2g+1$ in $x$ and $h(x), \tilde{h}(x)$ are polynomials in $x$ of degree at most $g$. All $\mathbb{F}_q$-isomorphisms of curves $\psi : C \to \widetilde{C}$ are of the type

$$\phi \; : \; (x, y) \mapsto \bigl(s^{-2}x + b,\, s^{-(2g+1)}y + A(x)\bigr) \qquad (29.4)$$

for some $s \in \mathbb{F}_q^*$, $b \in \mathbb{F}_q$ and a polynomial $A(x) \in \mathbb{F}_q[x]$ of degree at most $g$. Upon substituting $s^{-2}x + b$ and $s^{-(2g+1)}y + A(x)$ in place of $x$ and $y$ in equation (29.3) and comparing with (29.2) we obtain

$$\begin{cases} h(x) = s^{2g+1}\bigl(\tilde{h}(s^{-2}x + b) + 2\,A(x)\bigr) \\ f(x) = s^{2(2g+1)}\bigl(\tilde{f}(s^{-2}x + b) - A(x)^2 - \tilde{h}(s^{-2}x + b)A(x)\bigr) \end{cases} \quad (29.5)$$

whose inversion is

$$\begin{cases} \tilde{h}(x) = s^{-(2g+1)}h(\hat{x}) - 2A(\hat{x}) \\ \tilde{f}(x) = s^{-2(2g+1)}f(\hat{x}) + s^{-(2g+1)}h(\hat{x})A(\hat{x}) - A(\hat{x})^2 \end{cases} \quad (29.6)$$

where $\hat{x} = s^2(x - b)$.

Avanzi [AVA 2004b] gives a detailed study on this countermeasure for hyperelliptic curves including elliptic curves. Hence, we only state these more general results in the sequel. For elliptic curves we refer to the original paper [JOTY 2002].

- For hyperelliptic curves in odd characteristic we can restrict the isomorphisms to the type

$$\psi \;:\; (x, y) \mapsto \bigl(s^{-2}x, s^{-(2g+1)}y\bigr), \quad (29.7)$$

with $h(x) = \tilde{h}(x) = 0$. This gives a fast countermeasure that effectively multiplies all quantities involved in the computations of the group operations by powers of the randomly chosen nonzero parameter $s$.

Also, here we need to mention that zero coefficients remain zero when applying this randomization. Additional care needs to be taken to avoid Goubin type attacks (cf. Section 29.3).

- For genus 2 hyperelliptic curves in characteristic 2, one loses performance compared to the optimal formulas stated in Chapter 14. On the other hand, the general formulas can be applied in any case and thus one has to analyze on a case-by-case basis which costs are acceptable. For example, in Proposition 14.37 a unique representative per isomorphism class is given while the restricted parameter $\varepsilon$ does not influence the performance and it is not necessary to map $f_1$ to 0. Hence, different choices are possible without affecting the performance.

- For genus 3 curves in even characteristic similar observations hold. Additionally, the type of curve given by an equation $y^2 + h_0 y = f(x)$ can be randomized by

$$\psi \;:\; (x, y) \mapsto \bigl(s^{-2}x, s^{-(2g+1)}y\bigr) \;, \quad (29.8)$$

leaving the shape of the curve unchanged and varying only $h_0$ on the left-hand side; thus the group operations remain particularly fast.

**Remarks 29.5**

(i) Clearly, using more general isomorphisms, e.g., including a constant term, Goubin type attacks can be easily avoided on the cost of a less efficient map.

(ii) Note that it is also possible to use isogenies of low degree, which are efficiently computable as the DLP is usually stated in a prime order subgroup and thus the group structure of the interesting part is not altered by isogenies. In particular, when searching for maps that avoid Goubin type attacks one might come back to the more general isogenies.

## 29.3 Countermeasures against Goubin type attacks

Goubin observed that some elements remain unchanged under some of the randomization techniques presented in the previous section, e.g., as in the example we already mentioned, the randomization by redundant coefficients leaves zero coordinates unchanged.

This can be used in an *active attack*. Assume that an attacker can observe the computation of $[n]\overline{D}_i$ for elements $\overline{D}_i \in \text{Pic}_C^0$ chosen by himself, and assume that the $\overline{D}_i$ were chosen such that for each $\overline{D}_i$ the result of $[3]\overline{D}_i$ has a zero coordinate while $[4]\overline{D}_i$ does not. As a zero coordinate is invariant under the randomization by redundant coefficients, all side-channel traces should show the particularities of a zero coordinate if $[3]\overline{D}$ is used in the scalar multiplication, i.e., if the second most significant bit is set in a left-to-right algorithm, and be normal otherwise.

In full generality, we have the following approach: let $H \subset \text{Pic}_C^0$ be a subset of elements having some property that makes their processing detectable by side-channel analysis (for example, zeroes in the internal representation) and that are invariant under a given randomization procedure $\mathcal{R}$.

The set $H$ is called the set of *special* group elements. It is used in an attack as follows:

Suppose that the most significant bits $n_{l-1}, n_{l-2}, \ldots, n_{j+1}$ of the secret scalar $n$ are known and that we want to discover the next bit $n_j$.

1. Make a *guess*: $n_j = 0$ or 1.
2. Set up a *chosen message attack*: choose a number of elements $\overline{E}_i \in H$ and determine the corresponding $\overline{D}_i$ such that $[(n_r n_{r-1} \ldots n_{j+1} n_j)_2]\overline{D}_i = \overline{E}_i$ and additionally check that $[(n_r n_{r-1} \ldots n_{j+1} \bar{n}_j)_2]\overline{D}_i \notin H$.
3. Then, statistical correlation of the side-channel traces corresponding to the computations of $[n]E_i$ may reveal if the guess was correct even if the randomization procedure $\mathcal{R}$ is used.

As already mentioned, such sets $H$ exist:

- The set of points with a zero coordinate of an elliptic curve.
- The set of divisors classes on a hyperelliptic curve with a zero coordinate in the unique representation. As we use Mumford representation, $\overline{D} = [u(x), v(x)]$, the set of special elements could consist of those divisor classes for which $\deg(u) < g$. If explicit formulas are used such elements require special routines and thus should be easy to detect.

In both cases, the probability that a random element is in $H$ is $O(q^{-1})$. So, one would not hit such an element by accident and, hence, the check whether $[(n_r n_{r-1} \ldots n_{j+1} \bar{n}_j)_2]\overline{D}_i \notin H$ should almost always hold.

The sets $H$ defined above are clearly preserved by multiplicative divisor randomization, field isomorphisms, and curve isomorphisms (cf. Sections 29.2.3.b, 29.2.4.a, and 29.2.4.b). The side-channel trace correlation may reveal if the guess was correct in the presence of such randomization procedures.

For elliptic curves over prime fields $\mathbb{F}_p$ given by $y^2 = x^3 + a_4 x + a_6$, one can avoid this attack by choosing $x^3 + a_4 x + a_6$ irreducible over $\mathbb{F}_p$ and such that $a_6$ is not a square. Then no point with a zero coordinate exists. In general, for elliptic curves, this attack is not too serious, and can be avoided easily, as shown by Smart [SMA 2003].

The initial attacks were described for elliptic curves only, but the generalization to divisor class groups of hyperelliptic curves has been done by Avanzi [AVA 2004b]. In the same paper he also provides a generalization of the countermeasures.

In fact, on a genus $g$ curve ($g > 1$), let $H$ be the set of divisor classes $\overline{D} = [u(x), v(x)]$ such that $\deg(u) < g$. After the computation of an intermediate result $\overline{E} = [t]\overline{D} \in H$, the scalar multiplication algorithm will double it. To do it, the formulas for the most common case (cf. Chapter 14) will not work — and even if Cantor's algorithm 14.7 is used to implement the group arithmetic, there will be less reduction steps than in the general case. The consequences are easily detectable differences in power consumption and even timing. The timing differences are the leaked information used in the attack described in [KAKI+ 2004], which we can thus consider as a *Goubin-type timing attack*.

An approach to thwart Goubin's attack could use very general curve isomorphisms that do not respect the short Weierstraß form using $b$, $A(x) \neq 0$ to randomize also the vanishing coefficients of the divisors: this has the disadvantage of requiring more general and slower formulas for the operations.

However, as shown by Avanzi [AVA 2004a, AVA 2004b], the methods of randomizing the scalar and blinding the group elements (cf. Section 29.2.2 and 29.2.3.a) can be effective against Goubin's attack.

Under Goubin type attacks we also subsume the exceptional procedure attack [IZTA 2003b]. It is also an active attack in which the attacker chooses the input $\overline{D}$, on which $[n]\overline{D}$ is computed, in such a way that the computation fails for one assignment of the next unknown bit and does not for the other. By recording the error messages or observing the algorithm to choose an exceptional procedure, the attacker can learn subsequent bits. This attack is feasible, e.g., if the group order is not prime by submitting a point of small order such that the computation would hit the point at infinity. However, the standards suggest choosing curves with small cofactor only (cf. Chapter 23) such that the attack could discover only a very small portion of $n$. Furthermore, checks that the submitted point is in fact on the curve and of the given prime group order are imposed by the protocols (cf. Section 23.5).

This attack could also be mounted against the initial version of unified addition formulas (given in [BRJO 2003]) as they proposed only one set of formulas that would fail for only a very few cases. Clearly, these cases would never occur in a random scalar multiplication, but assuming an active attacker such instances can be found even if the group order is prime. Note that the unified formulas presented in Section 29.1.2.a no longer suffer from this problem. Furthermore, the countermeasures of blinding and scalar randomization are also effective against these attacks.

## 29.4 Countermeasures against higher order differential SCA

In the DSCAs explained so far the attacker monitors leaked signals and calculates the individual statistical properties of the signals at each sample time. In a higher order DPA attack, the attacker calculates joint statistical properties of power consumption at *multiple* sample times *within* the power signals.

More formally, an $k$-th order DSCA is defined as a DSCA that makes use of $k$ different samples in the power consumption signal that correspond to $k$ different intermediate values calculated during the execution of an algorithm.

The attacks described so far are *first order* DSCAs. The idea behind higher order DSCA had already been defined in [KOJA+ 1999]. A description of a second-order DPA can be found in [MES 2000a].

A possible scenario in the setting of curve-based cryptography is that the implementer chooses a windowing method such that the leakage of zero coefficients does not seem to be a real concern, e.g., a fixed or sliding window of width $w = 4$, or he applies a method such that no zero coefficients can occur (cf. Section 29.1.1.a). At each addition, one adds one of few (e.g., 16) possible precomputed multiples of the base group element $\overline{D}$ to the intermediate value. The higher order DSCA method deems to determine when a precomputed multiple is reused. This might be observable from the side-channel information, as half of the input is fixed while the other can be assumed to vary randomly.

This method might also be used to distinguish multiplications from squarings in odd characteristic. Implementers should be aware of this when using the algorithms presented in Section 29.1.1.b; for implementations of RSA these issues were studied in [SCH 2000a] and following papers.

In [SASc+ 2004] it is shown how such an attack can be used for elliptic curves over $\mathbb{F}_p$ where the field operations are implemented in Montgomery representation. They use the visibility of the final reduction in the multiplication to distinguish the computations of $x^2, xy$, and $3xy$ to attack curves in Jacobian coordinates given by the (especially efficient) equation $y^2 = x^3 - 3x + a_6$.

In order to avoid the detection of reuse of a precomputed point or prevent the observation of the final reduction via side-channel attacks, it is also advisable to use a redundant, i.e., inversion-free, representation of the group elements and to randomize the representation of a precomputed point after each time it is used. This is trivial to implement, and comparably inexpensive.

SSCA-countermeasures will also make it more difficult for the attacker to guess which portions of single traces are to be correlated. This shall force him, ultimately, to adopt a brute force strategy, i.e., to try to correlate all possible sub-traces: the amount of possible combinations will increase exponentially with the length of the scalar multiplication and thereby make $k$-th order attacks computationally infeasible.

The same holds true for address bit DPA in which the attacker obtains information about reuse of locations by using the side-channel information. A further countermeasure is to randomly change the assignment of storage locations during the execution of the algorithm.

## 29.5  Countermeasures against timing attacks

Timing attacks were the first side-channel attacks ever described [KOC 1996]. The attacker is only assumed to be able to measure the time needed per complete scalar multiplication and not to obtain a more detailed side-channel information.

If the attacker is allowed to issue group elements $\overline{D}$ on which $[n]\overline{D}$ is computed, he might choose them in such a way that, depending on a certain bit, the complete scalar multiplication differs in the computation time. In other words, the timings of the part of the whole computation that processes the bit $n_j$ must be *biased* in the step obtaining the $j$-th bit.

Timing attacks have been shown to work on real life smart cards in [DHKO+ 2000] for an implementation of RSA, and the same methods can be used also for curve-based cryptography.

The countermeasures are in principle the same as for differential side-channel analysis. Randomizing the representation of $\overline{D}$ and of $n$ should make it infeasible to obtain information from the total computation time as the internal representation is not known.

## 29.6  Countermeasures against fault attacks

Attacks based on fault induction essentially force the device to perform erroneous instructions — e.g., by changing some bits in the internal memory. They have been announced officially in 1996 in a Bellcore press release, followed by a paper by Boneh, DeMillo, and Lipton [BODE+ 1997].

Originally, the attack was presented for RSA and usually the scenario assumes that the attacker has access to the result of the computation $[n]\overline{D}$ and that he might request this computation for the same input $\overline{D}$ more than once. Hence, like for DSCA this attack does not apply to the randomly chosen nonces in the protocol but only to the long term secret key, e.g., used in signing.

First of all, inserting errors during the execution of an algorithm allows us to determine if the hit operation was a dummy or a real one by comparing the outputs of the original computation of $[n]\overline{D}$ to the output after the insertion of a fault. The obvious countermeasure is to check the output by a check computation.

Unfortunately for curve-based cryptography, checking the result amounts to performing the whole computation twice or in parallel requiring either double time or size. (This is in contrast to RSA, where usually one of the operations is chosen to be cheap.) Hence, a more thorough look at possible effects is necessary. We follow Avanzi [AVA 2005c] and distinguish between *simple fault analysis (SFA)* and *differential fault analysis (DFA)* depending on whether the attack makes use of a single execution of the algorithm or of many in combination with statistical analysis.

### 29.6.1 Countermeasures against simple fault analysis

At Crypto2000, Biehl, Meyer, and Müller [BIME+ 2000] presented three different types of attacks on ECC that can be used to derive information about the secret key if bit errors can be inserted into the elliptic curve computations in a tamper-proof device. They also estimate the effectiveness of the attacks using a software simulation.

Their methods require *very precise placement and timing* of the faults. Generalizing the approach of [LILE 1997], their first two techniques depend on the ability to change the coordinates of a point on the curve at the beginning of the scalar multiplication. This can work because the elliptic curve formulas do not use all the coefficients of the equation of the curve to perform their operations (cf. Chapter 13); e.g., in affine coordinates and a finite field $\mathbb{F}_q$ of odd characteristic the coefficient $a_6$ in the curve equation $E : y^2 = x^3 + a_4 x + a_6$ is not used in either doubling or addition. Therefore, if the input point is changed it will no longer lie on $E$ but has good chance of lying on $\widetilde{E} = y^2 + a_4 x + \widetilde{a}_6$ for some $\widetilde{a}_6$. The implemented addition and doubling formulas will compute the multiple of the changed point on the changed curve. The largest prime factor of $\widetilde{E}(\mathbb{F}_q)$ will most likely be smaller than the group order $E(\mathbb{F}_q)$ such that solving the DLP on $\widetilde{E}$ is easier (cf. Section 19.3 for invalid curve attack).

To actually have a DLP to solve we assume that the attacker obtains the output $[n]\widetilde{P}$ from which he can determine $\widetilde{a}_6$ as $a_4$ is known to him. Furthermore, we assume that the attacker has some knowledge of how the fault was induced, e.g., he could place it such that only the $y$-coordinate is changed while $x$ is fixed. Given $a_4, x$ and $\widetilde{a}_6$ he gets two candidates for the $y$-coordinates. By solving the DLP for one of them he obtains either $n$ or $\ell - n$ from which the DL is derived by checking.

Note that this scenario holds also for hyperelliptic curves and for all other coordinate systems, as usually not all curve coefficients appear in the explicit group operations and a divisor class group in a random Jacobian is likely to have no big prime factor.

The countermeasure consists of checking whether the resulting point satisfies the curve equation. Note that the standards require this check for the input point (cf. Section 23.5) and this comparably cheap procedure also at the end of the computation.

Furthermore, the attack is not possible if the result is not available to the attacker. On the other hand, holding back the result just in case of a detected fault allows an attacker to find dummy instructions.

### 29.6.2 Countermeasures against differential fault analysis

We now introduce a class of fault attacks that need more than one scalar multiplication.

#### 29.6.2.a Faults changing the curve

Here we assume the faults can be placed at exact timings but their effects on $\overline{D}$ cannot be controlled. Hence, most likely one works on a different curve after the fault is induced.

## § 29.6 Countermeasures against fault attacks

The method of detecting dummy operations mentioned in the introduction belongs to this class of attacks. For the purpose of illustration we describe an attack on the right to left scalar multiplication Algorithm 29.6, which we repeat here for easier reference.

---

**Algorithm 29.6** Right to left binary exponentiation

INPUT: An element $\overline{D} \in G$ and a nonnegative integer $n = (n_{l-1} \ldots n_0)_2$ with $n_{l-1} = 1$.
OUTPUT: The element $[n]\overline{D} \in G$.

1. $\overline{E} \leftarrow 0, \overline{F} \leftarrow \overline{D}$ and $i \leftarrow 0$
2. **while** $i \leqslant l - 2$ **do**
3.     **if** $n_i = 1$ **then** $\overline{E} \leftarrow \overline{E} \oplus \overline{F}$
4.     $\overline{F} \leftarrow [2]\overline{F}$
5.     $i \leftarrow i + 1$
6. $\overline{E} \leftarrow \overline{E} \oplus \overline{F}$
7. **return** $\overline{E}$

---

The group element $\overline{F}$ contains $[2^i]\overline{D}$ in round $i$. Note that in Line 3 a dummy addition could be performed to avoid SSCA attacks.

Assume that we know the binary length $l$ of the unknown multiplier $n$ (note that an attacker can easily guess this length). First use the device to compute $[n]\overline{D}$ without inducing faults to have a value for comparison. Then change the content of $\overline{F}$ in the next to last round. If the result is changed then $n_{l-2} \neq 0$. Proceeding to inserting an error in the previous round and repeating this process, the attacker can read out all bits of $n$.

The paper [BiME+ 2000] also contains an approach in which faults are introduced during the scalar multiplication in a more sophisticated way. Assume that the attacker can, at any prescribed iteration, say flip just one bit of the variable $\overline{E}$. (The case where the variable $\overline{F}$ is modified is handled in a similar way.) Assume also that the scalar is fixed and not randomized, and that we know how the internal variables are represented. Then, essentially, the bits of the key can be recovered in small blocks as follows:

1. Perform a normal scalar multiplication $\overline{E} = [n]\overline{D}$ with a given input.
2. Repeat the computation of $[n]\overline{D}$, but this time induce a bit flip in a register $m$ steps before the end of the scalar multiplication, giving a result $\overline{E}'$. Of course all computations before the fault in the two cases will be equal, and all those involving the processing of the $m$ most significant bits of the unknown scalar $n$ in the faulted computation will have been changed with respect to those of the reference computation — the very first difference consisting of just a single bit.
3. For all possible $m$ bit integers $x$, "reverse" the correct computation and the faulted one, i.e., determine $\overline{E} \ominus [x 2^{l-m}]\overline{D}$ and $\overline{E}' \ominus [x 2^{l-m}]\overline{D}$.
4. If pairs of results that differ in only one bit are found, then the correct and faulted register values are now determined together with the $m$ most significant bits of the scalar.
5. If one bit of $\overline{F}$ (which is supposed to contain a copy of the input base point $\overline{D}$) was flipped, and not one of $\overline{E}$, the computations need to be done not only for all possible combinations of $m$ bits, but also for all possible single bit faults induced in $\overline{E}$. There exist $O(\lg q)$ different possibilities over $\mathbb{F}_q$.
6. Iteration of the above process yields all the bits of the secret scalar, $m$ bits at a time.

The attack works essentially unchanged if the binary representation of the scalar is replaced with any other deterministic recoding (NAF, $m$-ary, $w$-NAF, etc.).

In [BIME$^+$ 2000] the authors deal with more sophisticated attacks, such as attacks where the faults are induced but it is not possible to determine precisely *when* – such attacks are reasonable if clock randomization or randomized processor instructions are implemented on the card.

All such attacks have been originally presented for elliptic curves but are in fact entirely generic and apply to implementations of hyperelliptic curves, trace zero varieties and other geometric objects. Countermeasures are obvious: checking at regular intervals to see whether the intermediate result is on the curve, and restarting the computation (possibly from a backup value) if something has gone wrong; never allowing wrong results to leave the device; randomizing the scalar.

#### 29.6.2.b Faults preserving the curve

The countermeasures described so far would not apply if the resulting points happened to be valid points of the curve. In [BLOT$^+$ 2004] the authors analyze the scenario that the attacker is able to insert such precise faults that it is possible to change the sign of an integer modulo $p$. This might be unrealistic in practice but offers the interesting scenario that the attack would not be noticed through the above countermeasures. They show how one could still obtain the secret scalar $n$ from sign change fault attacks.

Apart from the attack, they also propose a countermeasure generalizing Shamir's idea [SHA 1999] for RSA to ECC, which works similar to performing the computation twice — but choosing the second group to be a small group over $\mathbb{F}_{p'}$ for $p' < q$ which facilitates this task enormously.

Their paper considers only elliptic curves; generalizations to larger genera are possible. For the explanation we restrict ourselves to elliptic curves.

Let the original curve $E$ be defined over $\mathbb{F}_p$ and let $p_0$ be a small prime. Consider an appropriate lift of the equation of $E$ to the integers and reduce it modulo $N = p_0 p$. The effects on the speed of working on the reduced curve $\overline{E}$ are the same as choosing a larger modulus. Let $P_N$ be an arbitrary lift of the base point $P$ reduced modulo $N$ and let $P_{p_0}$ be its reduction modulo $p_0$.

Instead of computing $[n]P$ directly, one performs $Q_N = [n]P_N$ and $Q_{p_0} = [n]P_{p_0}$. At the end, one checks whether the reduction of $Q_N$ modulo $p_0$ equals the result $Q_{p_0}$. If this does not hold, then at least one fault has been induced (unless one has the rare case that $n$ is a multiple of $|\overline{E}_{p_0}|$) and the computation should be started anew. Otherwise $[n]P$ is obtained as the reduction of $Q_N$ modulo $p$.

This countermeasure induces the small overhead of working with a larger modulus and performing the scalar multiplication twice. However, this is a comparably small overhead and the probability that the attacker can introduce a fault in both computations at the same place is rather low. On the other hand, inserting faults that change the sign, i.e., keep the point on the curve, are technically hard to achieve.

### 29.6.3 Conclusion on fault induction

For this type of attack it is even more important to analyze the capability of the attacker. In particular, the fact that a restart of the computation after an unsuccessful check does tell the attacker that the operation at which the fault was induced was not a dummy operation, needs to be taken into account.

If faults or intrusions have been detected, then the computation has to be redone, restarted with the state before the fault, or aborted without output. This can lead to *observable* differences in the behavior of the system: in [JOQU$^+$ 2002] the case of an RSA system has been investigated. The observable behavior can consist of the success/failure to encipher/decipher, but, for example, also in the total timing for the cryptographic operations, or in the power trace/EM emission of the device.

In other words, an (apparently) improved cryptosystem may actually leak useful observable information. If this behavior can be observed for several different faults induced in different moments of a scalar multiplication with fixed scalar, then an attacker may be able to guess the secret key. Accordingly we strongly advise randomizing the scalar.

## 29.7 Countermeasures for special curves

The curves considered in Chapter 15 allow fast scalar multiplications by using endomorphisms of the curve. On Koblitz curves (cf. Section 15.1) the Frobenius endomorphism $\phi_q$ operates by raising each coefficient of the representation to the power of $q$. On GLV curves (cf. Section 15.2.1) other efficiently computable endomorphisms are applied. For such curves it is possible to design countermeasures against side-channel attacks that are more efficient than for general curves.

In the sequel we first report on countermeasures for Koblitz curves as there is a vast literature proposing methods against SSCA and DSCA. As the Frobenius endomorphism is usually used to speed up the scalar multiplication (cf. Section 15.1) we now assume that $n$ is given by its $\tau$-adic expansion $n = \sum_{i=0}^{l-1} n_i \tau^i$ for $n_i \in R$ for some set of coefficients $R$. Finally we report on countermeasures for GLV curves.

### 29.7.1 Countermeasures against SSCA on Koblitz curves

In general, the SSCA countermeasures of inserting dummy group operations apply also to $\tau$-adic instead of binary expansions. However, avoiding simple side-channel attacks on Koblitz curves by resorting to the Frobenius and always add method would reduce much of the advantages of Koblitz curves, as many expensive additions are inserted. Likewise, there is no use in obtaining unified group operations, as the advantage of using Koblitz curves is that the Frobenius endomorphism is a cheap operation and thus its cost cannot and should not be made equal to that of an addition.

To find the most appropriate countermeasure it is necessary to specify how the computations are done.

#### 29.7.1.a Normal basis situation

If the implementation is using a normal basis representation, e.g., in hardware based implementations, computing $\phi_q^i(\overline{D})$ requires the same time for each exponent $i$. Therefore, a countermeasure against SSCA is not needed as the attacker automatically sees the deterministic sequence of operations consisting of one application of the Frobenius endomorphism and one addition. For genus $g > 1$ or $q > 2$ one table lookup is used before the addition. Note that this reveals the $\tau$-adic Hamming weight of the representation as the number of additions is observable.

#### 29.7.1.b Power of $\phi_q^i$ can be detected

We now detail methods to counteract SSCAs if it is possible to determine the number of Frobenius operations between two additions. The first generalizations of side-channel attacks to elliptic Koblitz curves were obtained in [HAS 2000]. The methods are stated for elliptic curves but hold the same for arbitrary genus curves.

First of all, it is possible to design an algorithm that resists SSCA and DSCA at the same time. As the Frobenius endomorphism can be applied almost for free, it is possible to insert dummy applications of $\phi_q$. We repeat Algorithm 3 of [SMGE 2003] in the general setting of arbitrary Koblitz curves. We need a random nonce $j$, by which we mean a binary sequence of length $l + 1$ if the expansion of $n$ has length $l$. By the logical complement of the expansion of $n$, we mean a

$\tau$-adic expansion that has a 0 whenever the expansion of $n$ has a nonzero coefficient, and which contains a random coefficient from the set of coefficients otherwise. By abuse of notation we also use $n$ to denote its expansion and $n'$ for the logical complement. The logical complement $j'$ of $j$ is the binary sequence such that $j + j'$ is the *all one string*.

---

**Algorithm 29.7** SSCA and DSCA by random assignment

INPUT: An integer $n = \sum_{i=0}^{l-1} r_i \tau^i$, a divisor $\overline{D}$, and precomputed multiples $[r_i]\overline{D}$.
OUTPUT: The divisor $\overline{E} = [n]\overline{D}$.

1. generate random nonce $j$ of length $l + 1$
2. $\text{kval}[j_l] \leftarrow n$ and $\text{kval}[j'_l] \leftarrow n'$
3. $\overline{E}[0] \leftarrow 0$ and $\overline{E}[1] \leftarrow 0$
4. **for** $i = l - 1$ **to** 0 **do**
5.      $\overline{E}[j_i] \leftarrow \phi_q(\overline{E}[j_i])$
6.      $\text{bit} \leftarrow \text{kval}_i[j_i]$
7.      **if** $(\text{bit} \neq 0)$ **then** $\overline{E}[j_i] \leftarrow \overline{E}[j_i] \oplus [\text{bit}]\overline{D}$
8.      $\overline{E}[j'_i] \leftarrow \phi_q(\overline{E}[j'_i])$
9.      $\text{bit} \leftarrow \text{kval}_i[j'_i]$
10.     **if** $(\text{bit} \neq 0)$ **then** $\overline{E}[j'_i] \leftarrow \overline{E}[j'_i] \oplus [\text{bit}]\overline{D}$
11. **return** $\overline{E}[j_l]$

---

Note that at each step $\phi_q$ was applied twice while only one of the additions took place and the order of used and dummy additions is randomized due to $j$. As in the double-and-always-add method the density of the expansion is constantly 1.

The same idea works also for the windowing method with fixed window width by redefining the logical complement of $n$, which has the advantage of needing fewer additions on the cost of larger storage requirements.

Furthermore, one needs to take into account that in even characteristic negating can be observed from the side-channel information, such that it might be necessary to store both $[i]\overline{D}$ and $[-i]\overline{D}$. This observation also holds true for the cases considered next.

**Binary elliptic Koblitz curves**

We are working on the original Koblitz curves $g = 1, q = 2$ and no precomputations are used. For an unshielded implementation it is possible to use a $\tau$-adic NAF expansion that has density $1/3$.

From the NAF property one knows that each addition is followed by at least two applications of $\phi_2$ as there are no two consecutive nonzero coefficients. One can use a direct generalization of the idea presented in Section 29.1.1.a by using a fixed pattern of operations. This way the scalar multiplication would consist of applying $\phi_2$ twice followed by an addition, where some of the additions and some of the Frobenius operations are actually dummy operations.

Note that dummy Frobenius operations appear if the number of zeroes between two nonzero coefficients is even whereas dummy additions are inserted whenever two adjacent zero-coefficients appear in the expansion.

As for Koblitz curves $\phi_2$ is much cheaper than a group addition, and it is advisable to prescribe more (dummy) Frobenius operations to reduce the number of dummy additions. Depending on the length of the scalar, we propose to let each addition be followed by 3 or 4 applications of $\phi_2$. To thwart fault attacks, the dummy Frobenius operations could be placed randomly among the 3 resp.

## Generalized Koblitz curves

In the case of larger genus or characteristic the density of the $\tau$-adic expansion is much closer to 1. So, inserting some dummy additions does not constitute a major drawback. To additionally take into account DSCA one can randomize the times where the dummy operations are executed and the places where they are stored. Some clever ways for this are detailed in [SMGE 2003].

It is also possible to directly generalize the use of windowing methods in which $w$ operations of the Frobenius endomorphism are followed by an addition, which might be a dummy addition. Here, the use of fixed windows is advantageous.

An analogue of Möller's method of deriving expansions with no zero coefficients can be obtained as well. In a more sophisticated way the following adaptation of Hasan [HAS 2000] to arbitrary Koblitz curves $C/\mathbb{F}_q$ works (for the details on Koblitz curves we refer to Section 15.1). Note that we consider $C$ over an extension field $\mathbb{F}_{q^k}$.

Using $\tau^k - 1 = 0$ in the subgroup under consideration we reduce the length of the expansion of $n$ allowing larger coefficients. For large $k$ the coefficients will all be of a size less than $q^g - 1$. Put $r_{\min} < 0$ the minimal and $r_{\max}$ the maximal coefficient of this enlarged set of coefficients. Then for all coefficients $r_i$, $0 \leqslant i \leqslant k-1$ of the expansion $r_i - r_{\min} + 1$ is an integer $> 0$. We precompute the multiples $\overline{D}, [2]\overline{D}, \ldots, [r_{\max} - r_{\min} + 1]\overline{D}$ in advance. Then $\sum_{i=0}^{k-1}[r_i - r_{\min} + 1]\phi_q^i(\overline{D})$ still computes $[n]\overline{D}$ as $(-r_{\min} + 1)(\phi_q^k - 1)/(\phi_q - 1)(\overline{D}) = 0$. But now we perform a table lookup and non-dummy addition for each of the $d$ coefficients. The density is still 1 but this trick avoids fault attacks.

### 29.7.2 Countermeasures against DSCA on Koblitz curves

To counteract differential attacks all methods introduce randomness in the computation. Several of the methods introduced in Section 29.2 have direct analogies for Koblitz curves, which can be applied much more efficiently as the doublings are replaced by cheap applications of $\phi_q$.

#### 29.7.2.a Varying the scalar or the representation

Coron's countermeasure [COR 1999] of adding a random multiple of the group order to the multiplier $n$ before computing $[n]\overline{D}$ does not help at all for Koblitz curves. Note that in the process of computing the $\tau$-adic expansion we first reduce modulo $(\tau^k - 1)/(\tau - 1)$ and that the group order is always a multiple of this in $\mathbb{Z}[\tau]$ (at least when it is almost prime as we suppose). This means that $n$ plus any multiple of the group order is always reduced to the *same* element in $\mathbb{Z}[\tau]$. This holds for all hyperelliptic curves independent of the genus. To find efficient countermeasures we again need to distinguish whether the number of applications of $\phi_q$ can be determined from the side-channel information or not.

**Normal basis situation**

As before, we assume that the side-channel traces of $\phi_q^i$ and $\phi_q^j$ are indistinguishable.

Hasan [HAS 2000] deals especially with elliptic Koblitz curves over $\mathbb{F}_2$ but the method works for any genus and ground field. Use the relation $\tau^k - 1 = 0$ to reduce the length of the expansion to have fixed length $k$ allowing larger coefficients. We can rotate the expansion using

$$r_{k-1}\tau^{k-1} + r_{k-2}\tau^{k-2} + \cdots + r_1\tau + r_0 = \tau^t(r_{t-1}\tau^{k-1} + r_{t-2}\tau^{k-2} + \cdots + r_t k + 1\tau + r_t)$$

### Power of $\phi_q^i$ can be detected

The exponent splitting algorithm of [SMGE 2003, Algorithm 2] works efficiently if the final addition is implemented in such a manner that no information on the $\overline{E}_i$ is leaked. We present their idea in the general setting of arbitrary Koblitz curves. They suggest splitting the expansion into $s$ groups of $t$ coefficients, where $t = \lceil k/s \rceil$. For each subsequence

$$u_i = (r_{(i+1)t-1} r_{(i+1)t-2} \ldots r_{it+1} r_{it}), \quad \text{for } 0 \leqslant i \leqslant s-1$$

the computation of $\overline{E}_i = [u_i]\overline{D}$ is done in a random order for the $i$. Finally the result is obtained as

$$\overline{E} = \sum_{i=0}^{s-1} \phi_q^{it}(\overline{E}_i).$$

Adding random multiples of $(\tau^k - 1)/(\tau - 1)$ does not change the content of the integer but changes the representation. Note, however, that this makes the coefficients larger and that each coefficient is increased by one.

The method described in Algorithm 29.7 is actually a DSCA countermeasure that also works against SSCA.

Joye and Tymen [JOTY 2002] consider elliptic Koblitz curves in the case that one would be able to tell the power of the Frobenius. Instead of reducing $n$ modulo $\tau^k - 1$ before computing the expansion, they reduce it modulo $\rho(\tau^k - 1)$, where $\rho$ is a random element of $\mathbb{Z}[\tau]$ of bounded norm. If for the complex norm N we have $\lg N(\rho) \sim 40$ then the expansion will be of length 200 instead of 160, which does not seem to be too bad as the density of the expansion is unchanged. Besides, one does not need to introduce further routines except for choosing a random $\rho$ as only reduction modulo an element of $\mathbb{Z}[\tau]$ and the algorithm to compute the expansion are needed.

In [LAN 2001a] it is proposed to first compute $n \bmod (\tau^k - 1)/(\tau - 1)$ as in the unshielded algorithm and then to add a random multiple $\alpha(\tau^k - 1)/(\tau - 1)$, where $\alpha$ is of norm less than some $K$. The result is similar to what is obtained above and $n$ is equally hidden in the full length of the element to expand. The important advantage is that we can make use of the precomputed values for $(\tau^k - 1)/(\tau - 1)$ and $(\tau - 1)/(\tau^k - 1)$ in $\mathbb{Z}[\tau]$ and only need to compute the product $\alpha(\tau^k - 1)/(\tau - 1)$ in $\mathbb{Z}[\tau]$ additionally. Using this for $K$ of about 40 bits also results in expansions of length 200 but the computation of the expansion is faster.

In the case of higher genus curves we can use the same approach, except that finding elements of bounded norm is not as immediate as for the elliptic case. However, slightly reducing the space for these random elements $\alpha$ we can choose those where the coefficients $\alpha = c_0 + \cdots + c_{2g-1}\tau^{2g-1}$ satisfy

$$\left( \sum_{j=0}^{2g-1} |c_j| \sqrt{q}^j \right)^2 \leqslant 2^{L-2}/(\sqrt{q} - 1)^2;$$

this means that we can choose all $c_i$ at random under the condition that $|c_j| \leqslant 2^{(L+2)/2}/(q^g - 1)$, where $L$ is the chosen parameter to guarantee enough security. The expansions will be longer by $+L$. If we use a new random $\alpha$ each time the algorithm is invoked, the information of the expansion of the multiplier should be well hidden.

#### 29.7.2.b Other countermeasures

As in the general case, one can also use randomization of group elements together with the Frobenius operation. In inversion-free coordinate systems some more representing field elements need to be raised to the power of $q$, but this remains much cheaper than an ordinary scalar multiplication.

Clearly, blinding methods also work universally in the setting of Koblitz curves, but we are not aware of an improvement making use of the possibilities Koblitz curves offer.

It is possible to use an isomorphic field representation, but usually this comes at the penalty of less efficient arithmetic. Concerning curve isomorphisms: they need to be defined over the small field $\mathbb{F}_q$ as otherwise the advantage of Koblitz curves cannot be used — and there are just very few isomorphic curves, e.g., for the case $g = 1, q = 2$ there are 4 curves in each isomorphism class. A random choice among such a small number does not considerably increase the security for the cost of far less efficient group operations.

### 29.7.3 Countermeasures for GLV curves

We now deal with side-channel attacks on GLV curves. Recall that these are curves with an endomorphism $\psi$ and that to compute $[n]\overline{D}$ one splits $n = n_0 + n_1 s_\psi$, where $\psi(\overline{D}) = [s_\psi]\overline{D}$ for all $\overline{D}$ such that $n_0, n_1$ are of size $O(\sqrt{n})$ only.

To avoid SSCA, all methods described in Section 29.1 can be applied as they only make use of the scalar multiplication. One needs to take into account that the computation of $[n_0]\overline{D} \oplus [n_1]\psi(\overline{D})$ applies a joint sparse form together with the Straus–Shamir trick of simultaneous doubling (cf. Section 9.1.5). Thus the initial density of the expansion is $1/2$ asymptotically. The effects and losses due to the SSCA countermeasures need to be slightly adjusted but basically that issue is solved.

To counteract DSCA, Ciet, Quisquater, and Sica [CIQU[+] 2002] give three approaches on how to randomize the decomposition of the scalar in the GLV scalar multiplication Algorithm 15.41. In the original GLV method one first computes a short basis $v_0, v_1$ of the lattice $\mathbb{Z}[s_\psi]$. Denote by $f$ the map

$$\mathbb{Z} \times \mathbb{Z} \to \mathbb{Z}/\ell\mathbb{Z}$$
$$(i, j) \mapsto f(i, j) \equiv i + j s_\psi \pmod{\ell}.$$

One has that $f(v_0) = f(v_1) = 0$.

Then one solves the linear system of equations over $\mathbb{Q}$ to get $(n, 0) = k_0 v_0 + k_1 v_1$ with $k_i \in \mathbb{Q}$. Let $k_i'$ be the integer closest to $k_i$, then $(n, 0) - k_0' v_0 + k_1' v_1 = (n_0, n_1)$ is a short vector in the lattice and $f(n_0, n_1) \equiv n \pmod{\ell}$.

In [CIQU[+] 2002], Ciet et al. basically show how to obtain slightly longer vectors $(n_0, n_1)$ in a random manner.

In order to do this, the easiest way is to add scalar multiples of $v_1$ and $v_2$ to $(n_0, n_1)$. The result will still be a comparably short vector and it operates like $n$ on all $\overline{D}$. This countermeasure induces an overhead as the resulting $(n_0', n_1')$ has larger coefficients but it is rather easy to control the effects and the increase of security.

To reach a higher order of randomization they propose not only to randomize the splitting of the scalar in base $\{1, \psi\}$ but also to randomize the basis such that the resulting splitting $n_0', n_1'$ of $n$ refers to $[n_0']\psi_0(\overline{D}) \oplus [n_1']\psi_1(\overline{D})$ for some random basis $\{\psi_0, \psi_1\}$.

To get the random basis, they apply a random invertible matrix

$$A = \begin{bmatrix} \alpha & \beta \\ \gamma & \delta \end{bmatrix}$$

to $(1, \psi)$ and then apply the usual GLV method including the computation of the short vectors $v_0', v_1'$ in the changed basis. The only difference is that one does not search for a rational approximation of $(n, 0)$ but needs to express this first with respect to the changed basis. Define $s_{\psi_0}$ by $\psi_0(\overline{D}) = [s_{\psi_0}]\overline{D}$ and let $s_{\psi_0}^{-1}$ be the inverse of $s_{\psi_0}$ modulo $\ell$. Then the modified GLV method proceeds as follows:

**Algorithm 29.8** Randomized GLV method

INPUT: An integer $n$, the endomorphism $\psi$, the order $s_\psi$, and short basis $v_0, v_1$ for $\{1, \psi\}$.
OUTPUT: The elements $\psi_0, \psi_1$ and $n'_0, n'_1$ such that $[n]\overline{D} = [n'_0]\psi_0(\overline{D}) \oplus [n'_1]\psi_1(\overline{D})$.

1. choose random invertible matrix $A$
2. $(\psi_0, \psi_1)^t \leftarrow A(1, \psi)^t$ and $s_{\psi_0} \leftarrow (\alpha + \beta s_\psi) \bmod \ell$
3. $v'_0 \leftarrow \delta v_0 - \beta v_1$ and $v'_1 \leftarrow -\gamma v_0 + \alpha v_1$     [compute short basis $v'_0, v'_1$ for $\psi'_0, \psi'_1$]
4. $s^{-1}_{\psi_0} \leftarrow (s^{-1}_{\psi_0}) \bmod \ell$
5. find $n'_0, n'_1$ for $v'_0, v'_1$ like in original GLV for $(ns^{-1}_{\psi_0}, 0)$     [use Algorithm 15.41]
6. **return** $\psi_0, \psi_1, n'_0$ and $n'_1$

**Remarks 29.9**

(i) The integers $n'_0, n'_1$ obtained by this method are at most $2R$ times larger as with the original GLV method if $\alpha, \beta, \gamma, \delta$ are bounded by $R$.

(ii) If $\det(A) = 1$, one can use $v_0, v_1$ as short vectors, also with respect to the new basis. A special effect is that the maximal size of $n'_i$ is at most twice as large as that of $n_i$ independent of the bound $R$. Note however, that the matrices of determinant 1 are very special.

(iii) It is possible to combine both ideas and use this changed basis and an affine translation of the resulting $n'_0, n'_1$ to achieve higher randomization.

So far the GLV method on hyperelliptic curves is applied only to endomorphisms having a minimal polynomial of degree 2 instead of the maximal degree of $2g$. Hence, the same method also applies literally to the hyperelliptic GLV curves proposed so far, and it can be adjusted also for more general endomorphisms.

# Chapter 30

# *Random Numbers Generation and Testing*

*Tanja Lange, David Lubicz, and Andrew Weigl*

### Contents in Brief

| | | |
|---|---|---|
| 30.1 | **Definition of a random sequence** | 715 |
| 30.2 | **Random number generators** | 717 |
| | History • Properties of random number generators • Types of random number generators • Popular random number generators | |
| 30.3 | **Testing of random number generators** | 722 |
| 30.4 | **Testing a device** | 722 |
| 30.5 | **Statistical (empirical) tests** | 723 |
| 30.6 | **Some examples of statistical models on $\Sigma^n$** | 725 |
| 30.7 | **Hypothesis testings and random sequences** | 726 |
| 30.8 | **Empirical test examples for binary sequences** | 727 |
| | Random walk • Runs • Autocorrelation | |
| 30.9 | **Pseudorandom number generators** | 729 |
| | Relevant measures • Pseudorandom number generators from curves • Other applications | |

## 30.1 Definition of a random sequence

What exactly are random numbers? Is number "5" random? In this section as well as in Section 30.4 and Section 30.5 we closely follow the exposition of [LUBICZ]. Let $\Sigma = \{0, 1\}$ and $\Sigma^*$ be the set of sequences of countable infinite length with coefficients in the alphabet $\Sigma$. An element of $u \in \Sigma^*$ can be written as a sequence of 0 and 1:

$$u = u_0 u_1 u_2 u_3 u_4 u_5 \ldots,$$

with $u_i \in \{0, 1\}$. For $n \in \mathbb{N}$, the set of finite binary sequences of length $n$ is denoted by $\Sigma^n$. An element $u \in \Sigma^n$ can be written as:

$$u = u_0 u_1 u_2 \ldots u_{n-1}.$$

The objective of this paragraph is to define among all the elements of $\Sigma^*$ those that are random.

Let $W^k$ be the map from $\Sigma^*$ in the set of sequences with coefficients in $\Sigma^k$, which associates to $u \in \Sigma^*$ the unique sequence such that:

$$u = w_0 \parallel w_1 \parallel \ldots \parallel w_q \parallel \ldots$$

with $\parallel$ the concatenation and $w_i \in \Sigma^k$.

In the following, a sequence of events is defined as a sequence $(u_n)_{n \in \mathbb{N}}$ with values in a set $\Omega$ which will always be finite. The probability denoted by

$$P_e[(u_n) = x]$$

is the empirical probability that an event is equal to $x$ if the following limit exists

$$\lim_{k \to \infty} \frac{S_k(x)}{k}, \tag{30.1}$$

with $S_k = |\{n \leqslant k \mid u_n = x\}|$. If $(w_n)$ is a sequence of words of $\Sigma^k$, then $E\big((w_n)\big)$ denotes the *Shannon entropy function* of $(w_n)$, defined by

$$E\big((w_n)\big) = \sum_{x \in \Sigma^k} P_e[(w_n) = x] \ln\big(1/P_e[(w_n) = x]\big).$$

The definition from [KNU 1997] can now be stated.

**Definition 30.1** A sequence $(u_n) \in \Sigma^*$ is *l-distributed* for $l \in \mathbb{N}^*$, if $E\big(W^l((u_n))\big) = l$ or that for all $x \in \Sigma^l$, $P_e[W^l((u_n)) = x] = \left(\frac{1}{2}\right)^l$. A sequence $(u_n) \in \Sigma^*$ is then $\infty$-*distributed* if it is $l$-distributed for all $l \in \mathbb{N}^*$.

Temporarily, it can be stated that a sequence is random if it is $\infty$-distributed. In particular, if $(u_n)$ is a random sequence, then $W^k((u_n))$ is an equidistributed sequence of words of $\Sigma^k$. If a random subsequence of length $k$ is picked from a random sequence, then the probability of selecting a given subsequence is the same for all words in $\Sigma^k$. This illustrates well the intuitive idea of a random phenomenon. A consequence of this is that it is impossible to precisely define what is a finite random sequence.

The link between the statistical tests and the preceding definition of a random sequence can be shown by rewriting the preceding definition in the terms of probability theory. For that, let $(\Omega, \mathcal{A}, P)$ be a probability space, which is defined by $\Omega$, a set that is finite, endowed by the discrete sigma-algebra, i.e., the one generated by all the elements of $\Omega$ and a positive measure $P$ on $\mathcal{A}$ equidistributed and of total weight 1. For this paragraph, $\Omega$ will be $\Sigma^n$, the set of binary sequences of length $n$. The probability space is then denoted by $(\Sigma^n, \mathcal{A}^n, P^n)$.

A random variable is a map $X : \Omega \mapsto \mathbb{R}$. This endows $\mathbb{R}$ with a structure of measured space, and the induced measure is indicated by the abuse of notation $P_X$. The function, which maps $x \in \mathbb{R}$ to $P[X = x] = P(X^{-1}(x))$ is called the *law of X*. This gives the following alternative definition of a random sequence, which is just a reformulation of Definition 30.1.

**Definition 30.2** A sequence $(u_n) \in \Sigma^*$ is *random* if and only if for all random variables from $\Sigma^k$ endowed with the equidistributed law of probability to $\mathbb{R}$ and for all $x \in \mathbb{R}$ there is

$$P_e[X\big(W^k((u_n))\big) = x] = P[X = x].$$

In other words, the empiric law determined by the sequence $X(u)$ follows the theoretical law induced by the random variable on $\mathbb{R}$ by the equidistributed probability law of $\Sigma^k$. This definition gives a general principle that underlies statistical tests in order to assess if a sequence is random: some random variables are defined on the sets $\Sigma^k$, $k$ being an integer endowed with the equidistributed probability. This gives a law on $\mathbb{R}$ that is able to be computed or approximately computed thanks to the results from the probability theory. Most of the time, this law will use a Gaussian or a $\chi^2$ distribution. This law is then compared, for example, using a test of Kolmogorov–Smirnov, to the empiric law, obtained from the limit in (30.1), which is approximated with a computation on a sample finite sequence.

The problem is that the preceding general principle is asymptotic by nature: as by definition all the sequences of fixed length $l$ have the same probability to occur in a random sequence. Without any further hypothesis, it is not possible to distinguish a random sequence from a nonrandom sequence only having a finite subsequence. It is important to remember two main ideas: an infinite sequence can be associated with a probability distribution on the space of finite sequences of length $l$ and a property for all random sequences of length $l$ is that they have a uniform distribution.

As noted in [KNU 1997], the definition of a random sequence that has been stated does not catch all the properties that may be expected from a random sequence. For instance, let $u \in \Sigma^*$ be a $\infty$-distributed sequence and let $u^0$ be the sequence deduced from $u$ by forcing to zero the bits of rank $n^2$, $n \geqslant 2$. Then it is easy to see that the sequence $u^0$ is also $\infty$-distributed and is not random, because the value of some of its bits can be easily predicted *a priori*. However, even if the definition does not catch the unpredictability notion that is expected from a random sequence, it is enough for the purpose of statistical tests.

The next section will take a closer look at generating random sequences and the testing to see if these generators are operating properly.

## 30.2 Random number generators

### 30.2.1 History

Progress in generating random number sequences has been significant. However, people are still trying to figure out new methods for producing fast, cryptographically secure random bits. Before the first table of random numbers was published in 1927, researchers had to work with very slow and simple random number generators (RNG), like tossing a coin or rolling a dice. Needless to say, these methods were very time consuming. It was not until 1927 when Tippetts published a table of 40,000 numbers derived from the census reports that people had access to a large sequence of random numbers.

This lack of a ready source of random number sequences led people to try and create more efficient means of producing random numbers. In 1939, the first mechanical random number machine was created by Kendell and Babington–Smith. Their machine was used to generate a table of 100,000 numbers, which was later published for further use. The practice of using random number machines to generate tables of random numbers continued with the publishing of 1,000,000 digits by the Rand Corporation. Their generator could be best described as an electronic roulette wheel. The first version produced sequences with a statistical biases. The Rand Corp had to optimize and fix their machine, but even after this the new sequences showed a slight statistical bias. However, the random sequences were deemed to be "good enough."

Even though tables provided researchers with a larger selection of random numbers, this method still had its drawbacks. It required large amounts of memory, since each random number had to be preloaded into memory, and it took a long time to input the data. At this point, RNG research branched into two paths: the algorithmic approach and the sampling of physical systems. The algorithmic approach looked into producing random numbers by using the computer's arithmetic operations, and this led to the creation of deterministic random number generators or pseudorandom number generators. Sampling of physical systems, however, looked at how to create statistically acceptable sequences from natural random sources. These random number generators are called "true" random number generators, since they are based on a truly random source.

**Remark 30.3** A detailed time for the random number machine can be found in [RIT].

### 30.2.2 Properties of random number generators

When looking at a random number generator, how is it possible to determine if it is a source of random numbers? Four properties distinguish a random number generator from just an ordinary number generator. The best way to illustrate these properties is to examine a simple random number generator. One of the most recognized and used RNG is the coin toss; if the coin is assumed to be "fair."

By giving the coin a "0" and "1" for each side, it can be used to generate a random binary sequence. One of the first properties noticed is that the result from each toss is not affected, in any way, by the previous tosses. This means that if ten ones are tossed in a row, the probability of tossing an eleventh one is still 50%. This example illustrates the property of *independence*: previous results do not affect future results.

Random number generators can be designed to produce any range of values, or *distribution*. When examining the output of common RNGs, the values usually fall into a uniform distribution, which means that they have an equal probability of obtaining any of the values in the specified range. This distribution does not need to be uniform; for some simulations a designer may wish to produce a random sequence following a normal or other distribution. For cryptographic applications it is important that the distribution is uniform. Using a nonuniform distribution allows a hacker to concentrate on a smaller group of numbers to attack the system.

There are physical and computational limits to the size of numbers that an RNG can create. These limitations impose a natural boundary on the RNG and once it has reached these limits, the RNG repeats its output. This defines the *period* of the RNG. A well designed RNG will only be bound by the hardware limits. If the RNG is designed without taking care, there can be multiple sequence groups that the RNG could produce, with each group less than the ideal period.

The size of random sequences required is dependent upon the desired application. Cryptographic applications require relatively small sequences, in the range of 1024 bits depending on the algorithm, whereas simulations require extremely large sequences. A good example is the Monte Carlo simulation, which may require random sequences up to a billion bit bits in length, or even more. Therefore, RNGs need to be very *efficient* and must quickly generate numbers.

The next sections examine the different properties of three classes of random number generators: pseudo, true, and cryptographic random number generators. Each has its own unique requirements and restrictions.

### 30.2.3 Types of random number generators

#### 30.2.3.a Pseudorandom number generators

As mentioned in the history of RNGs (cf. Section 30.2.1), development of random number generators branched with the advent of computers. Researchers looked for methods to create large random sequences by using algorithms. Using such algorithms, they were able to make sequences, which mimic the properties of "true" random generators. Since they were created with a deterministic

## § 30.2 Random number generators

equation, they could not be called "truly" random. This led to a new class of generators, called *pseudorandom number generators* (PRNGs).

Compared to true random number generators, PRNGs are easier to implement in both hardware and software, and they also produce large sequences very quickly. In [ECU 1998, ECU 2001], the PRNG is described as a structure of the form $(X, x_0, f, f_t, f_o, Z)$ where $X$ is the finite set of states with a distribution of $\delta$. The element $x_0 \in X$ is called the initial state or seed. Using the transition function $f_t$ and the output function $f_o$ as shown in Algorithm 30.4 a pseudorandom sequence can be generated, $(z_0, \ldots, z_n)$ with $z_i \in Z$ and $Z = [0, 1)$ as the output set.

---
**Algorithm 30.4** A pseudorandom number generator

INPUT: An integer $n$ and a seed $x_0$.
OUTPUT: A pseudorandom sequence $(z_0, \ldots, z_n)$ with $z_i \in Z$.

1. **for** $i = 0$ **to** $n$ **do**
2. $\quad x_{i+1} \leftarrow f_t(x_i)$
3. $\quad z_i \leftarrow f_o(x_i)$
4. **return** $(z_0, \ldots, z_n)$

---

The benefit of the PRNG is its ability to quickly produce large sequences of statistically random numbers. This is very important for running simulations when input data may require millions or even billions of random values. Caution must be taken when using pseudorandom number generators for cryptographic applications. Attacks have been published that are able to reveal the secret generator values for some types of pseudorandom generators, which then enables a hacker to accurately reproduce the sequence. Cryptographic secure RNGs will be looked at in Section 30.2.3.c.

### 30.2.3.b True random number generators

A computer algorithm can only create pseudorandom sequences. However, there exist a variety of phenomenons related to a computation that are nondeterministic. Some examples are noise generated by a transistor, a dual oscillator, air turbulence in a hard drive, or capturing user input on the computer. Whatever the source of natural entropy, the data need to be digitized and converted into a working space, often a binary sequence. True random number generators provide a source of random numbers that is impossible to predict (nondeterministic), but at the cost of the sequence generation speed. Therefore, these generators are generally suitable for cryptographic applications but unsuitable for simulations. The use of natural entropy is a good source of randomness, but care must still be taken to examine the sequence for other weaknesses: correlation or superposition of regular structures. To overcome these weaknesses, RNG sources are mathematically altered to mask weaknesses in the digitized analogue signal. Table 30.1 shows the characteristics of both pseudo- and true random number generators.

**Table 30.1** Characteristics of pseudo- and true random number generators.

| True RNG | Pseudo-RNG |
|---|---|
| Physical random source | Deterministic algorithm |
| Slow | Fast |
| Hard to implement | Easy to implement |

### 30.2.3.c Cryptographic random number generators

Cryptography has taken on a new importance as more personal and financial information is available in digital form. The strength of encrypted messages depends on many factors, one of which is the random number sequence used in key generation. Many people believe that the random number generator, provided with their compiler or mathematical package, is good enough. However, research has shown that they are very insecure for cryptographic applications. An example of an insecure RNG is where an attacker, who knows the pseudorandom algorithm and has a generated sequence, can take this information and calculate future values. With these values the attacker can calculate a secret key.

Cryptographic random number generators have an added property compared to other generators. They need to be *unpredictable*, given knowledge of the algorithm and previously generated bits.

These properties can be found in both pseudo- and true random number generators. Often the most efficient method of creating secure cryptographic random number sequences is by using a combination of the two types of generators.

## 30.2.4 Popular random number generators

This section describes three common random number generators, but there are many more available [FIPS 140-2, MEOO+ 1996, RUK 2001, KNU 1997, ENT 1998]. Care must be taken to select the correct generator for the required application.

### 30.2.4.a Linear congruential generator

The *linear congruential generator (LCG)* is a classic pseudorandom number generator and has been published in many journals and books [KNU 1997, CAR 1994, ENT 1998]. The LCG can be fully described using the following formula:

$$X_n = (aX_{n-1} + c) \bmod m$$

with $a$ the multiplier, $c$ the increment and $m$ the modulus. Care has to be taken when selecting the constants since it is very easy to create a poor random generator. This generator is so popular because it is simple to implement in both software and hardware after having selected the constants. Another benefit of this algorithm is its low memory requirement, since only the last value and the secret constants are required to calculate a new value. Knuth [KNU 1997] dedicates a large portion of the chapter on LCGs to the selection of each constant.

Table 30.2 is a list of popular linear congruential generators. The constants used and the quality of the generator are shown along with the generator's name. Two noteworthy LCGs are the RANDU and the ANSI-C generators, which can still be found in many mathematical packages and compilers. Both generators have been extensively researched and it was found that their quality is very poor. Park and Miller [PAMI 1998] describe the RANDU as:

> "RANDU represents a flawed generator with no significant redeeming features. It does not have a full period and it has some distinctly non-random characteristics."

As for the ANSI-C generator, it was found to be very nonrandom at lower bits.

### 30.2.4.b Blum–Blum–Shub generator (computationally perfect PRNG)

The Blum–Blum–Shub (BBS) generator is an example of a class of provably secure random number generators. It works under the complexity theory assumption that $\mathbf{P} \neq \mathbf{NP}$. The BBS generator was first published in 1986 by Blum et al. [BLBL+ 1986], where they showed that a quadratic residue of the form:

$$X_{n+1} = X_n^2 \bmod m$$

is very easy to calculate in the forwards direction. However, the backwards calculation of finding the square root of a number modulo $m$, when $m$ is large, is very difficult. The modulus is $m = p_1 p_2$, where $p_1$ and $p_2$ are large Blum prime numbers. Blum primes are prime numbers, satisfying:

$$p \equiv 3 \pmod 4$$

as $-1$ is not a square modulo $p$.

The BBS generator is targeted towards cryptographic applications, since it is not a permutation generator, which means the period of the generator is not necessarily $m - 1$. This makes the BBS generator not suitable for stochastic simulations.

Table 30.2 Popular LCGs.

| Generator | Constants | | | seed | Good/Poor |
|---|---|---|---|---|---|
| | $a$ | $c$ | $m$ | | |
| RANDU | 65539 | 0 | $2^{31}$ | | Poor |
| ANSI-C | 1103515245 | 12345 | $2^{31}$ | 12345 | Poor |
| Minimum Standard [PAMI 1998] | 16807 | 0 | $2^{31} - 1$ | | Good |

*Note: Good and bad and generators are rated on how well they pass empirical tests.*

### 30.2.4.c Cryptographic RNG (hardware RNG)

All previous examples of random number generators used deterministic algorithms. These generators statistically act like true RNGs but in fact are not. In order to be thought of as a true random number generator, the source of bits needs to be nondeterministic, which is usually achieved by sampling a natural stochastic process. There are many sources of natural randomness, including measuring radioactive decay, thermal noise, or noise generated by a reversed biased diode.

The problem with nondeterministic random sources is the possible presence of biasing, which means that ones or zeroes occur more often. A variety of methods have been developed to reduce the effect of biasing. A few common methods include the XORing of successive bits using the von Neumann algorithm [DAV 2000], or XORing the nondeterministic bit stream with the bits from a cryptographically secure random number generator (see Figure 30.3).

Hardware random number generators tend to be slower than their pseudorandom counterparts. However, for cryptographic applications, which may need only a few thousand bits, this is usually not a factor. For applications that need many random digits, hardware random generators are generally too slow.

**Remark 30.5** There are many implementations of hardware cryptographic random number generators [DAV 2000, INTEL 8051, CR 2003].

**Figure 30.3** Cryptographic hardware design.

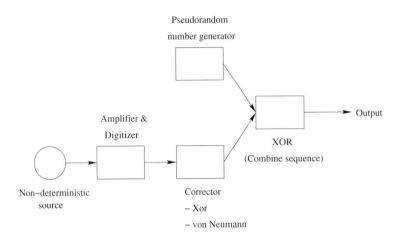

## 30.3 Testing of random number generators

There are two methods for testing a random number generator. One is to treat the generator as a black-box and only examine a portion of the resulting sequence, this is called *empirical testing*. The other method is to open the box and examine *a priori* the internal structure. This type of testing is called *theoretical testing*. Both empirical and theoretical tests use statistical tests, but they differ in the length of a sequence they examine. For theoretical tests, the full period of the generator is used; therefore, they detect *global nonrandomness*. Not all statistical tests are suitable for this type of testing.

Empirical testing is used to detect *local nonrandomness*. It is used to examine subsequences of length a lot less than the full period. Often these tests are used during the operation of the RNG to determine if the generator is still functioning properly, or as a quick test of a newly selected randomness generator. When selecting a RNG for an application, if possible, it is best to use both theoretical and empirical testing. This then helps to avoid both local and global abnormalities.

## 30.4 Testing a device

This section presents a definition of the mathematical objects that represent the device under test. A source $S_T$ is the mapping from a parameter space $T$ in the set $\Sigma^*$ to a binary sequence of infinite length with either discrete or continuous parameter space. In the case of a physical generator $T$ there can be a set of continuous variables that describes the state of the RNG (temperature of the circuit, position of each of the bits). For a LFSR, $T$ is the discrete space describing the initialization vector, the polynomial of retroaction, and the filtration function.

For an infinite binary sequence there can be associated for all $n \in \mathbb{N}^*$ a probability distribution on $\Sigma^n$ given by the definition of the empiric probability of $W^k((u_n))$. In particular, a source defines a map from the set of parameters $T$ to the set of probability distributions on $\Sigma^n$ for all $n$. This justifies the following definition:

**Definition 30.6** Let $T$ be a set of parameters, the *statistical model* on $\Sigma^n$ is the data for all $n \in \mathbb{N}^*$ with a probability distribution denoted by $P_t^n$ for $t \in T$ on the set $\Sigma^n$.

In practice, the set of parameters can take into account the normal operation of the source as well as flaws. It is possible that the source can produce sequences with good statistical properties for some values of the parameter in $T$ and poor statistical properties for the other values of $T$. For instance, a physical random generator can be built so that the output bits have a bias $p$ independent of the preceding draws. It outputs "1" with a probability of $p$ and a "0" with a probability of $q = 1 - p$. A hard to control production process may influence the parameter $p$. Therefore, a means is needed to assess the generator and reject any source that has a parameter $p$ too far from $\frac{1}{2}$.

## 30.5 Statistical (empirical) tests

Often it is not possible or feasible to look at the physical structure of the random number generator; for example, when the RNG needs to be tested before each operation. The only method to determine, to any degree of certainty, if the device is producing statistically independent and symmetrically distributed binary digits, is to examine a sample sequence of a given length $n$. In [MAU 1992] the idea is presented where a *statistical* or *empirical test* $\mathcal{T}$ is an algorithm that has as input a binary sample sequence and produces as output an "accept" or "reject" decision

$$\mathcal{T} : B^n \to \{\text{"accept", "reject"}\} \tag{30.2}$$

where $B$ is a binary set of $\{0, 1\}$. Using this function, all the possible binary sequences $x$ of length $n$, $x^n = x_1, \ldots, x_n$ are divided into two subsets

$$A_{\mathcal{T}} = \{s^n \mid \mathcal{T}(s^n) = \text{"accept"}\} \subseteq B^n \tag{30.3}$$

and

$$R_{\mathcal{T}} = \{s^n \mid \mathcal{T}(s^n) = \text{"reject"}\} \subseteq B^n \tag{30.4}$$

with $A_{\mathcal{T}}$ being the set of accepted or "random" sequences and $R_{\mathcal{T}}$ being the set of rejected or "nonrandom" sequences.

### Hypothesis testing

The method used to determine whether a device is operating properly, as a binary symmetric source, or is malfunctioning, is to test a parameter using the theory of hypothesis testing. The first step of this testing method is to calculate a test parameter by comparing the estimated parameters from a sample sequence for the given statistical model to the parameters for a binary stationary source. The sample is then accepted or rejected by comparing the test parameter to a probability distribution from a binary symmetric source.

**Remark 30.7** Randomness is a property of the device being tested, not of the finite sequence.

The researcher wishes to test the hypothesis that the device's parameter follows the parameter of the theoretical distribution. For hypothesis testing, the *null hypothesis*, $H_0$, is the claim that the sequence is acceptable as random, while the alternative hypothesis, $H_a$, states that the sequence is rejected. This hypothesis is in a general form and can take on a wide variety of parameters. One example is the examining of the population mean of the sample sequence and comparing it to the distribution of the mean for a binary symmetric sequence, $\mu_0$. The hypothesis can then be written as follows:

$$H_0 : \mu = \mu_0 \qquad H_a : \mu \neq \mu_0.$$

In order to decide between $H_0$ and $H_a$, the researcher needs to first determine the error threshold or *significance level* $\alpha$. This level indicates the probability the researcher is willing to take in rejecting a true $H_0$. For a significance level of $\alpha = 0.001$, the probability is that one sequence in a thousand will be rejected when in fact it should be accepted. This level is also called a *Type I error*.

**Figure 30.4** Parameters $\alpha$ and $\beta$ for a statistical test.

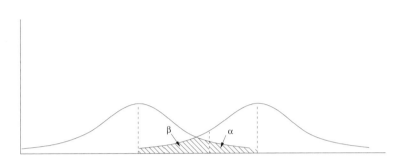

The next step in hypothesis testing is to calculate the statistical test. This step is dependent on the data under study. From the previous example, using the mean, the statistical test can be calculated by examining the sample mean, $\bar{x}$; the sample variance, $s^2$; the theoretical mean from a truly random sequence, $\mu_0$; the theoretical variance, $\sigma^2$; and the sample size, $n$. The statistical test is then as follows:

$$|Z| = \left| \frac{\bar{x} - \mu_0}{\frac{\sigma}{\sqrt{n}}} \right| > Z_{\frac{\alpha}{2}}$$

The rejection region works by examining the sample mean and determining whether there are too many standard deviations, more than $Z_{\frac{\alpha}{2}}$, from $\mu_0$. The rejection region can be seen in Figure 30.4 and if the statistical test falls in this region, then the null hypothesis is rejected in favor of the alternative hypothesis.

Often empirical tests described in the literature use a value called the *P-value*, to determine whether the sample sequence should be rejected or not. The significance level, as described in the last paragraph, is the boundary value between acceptance and rejection of the null hypothesis:

$$P > \alpha, \; H_0 \text{ is accepted}$$

$$P \leqslant \alpha, \; H_0 \text{ is rejected.}$$

Hypothesis testing can have two possible conclusions; the test accepts $H_0$ or it accepts $H_a$. As can be seen in Table 30.5, there are two possible errors that may arise. The Type I error has already been discussed and it is the significance level of the test. Type II error $\beta$ is the probability that the device is random, when it is not. The goal of the statistical test is to minimize the possibility of both types of errors. When dealing with statistical tests, the researcher is often able to set the sample size and one of the two types of errors, usually the Type I error. Setting the two points produces a $\beta$ as small as possible. It is not possible to determine the $\beta$ probability, which means that it is only possible to draw a firm conclusion about the Type I error. However, if the statistical test does not fall inside the rejection region, it can be stated that *there is insufficient evidence to reject $H_0$*. The null hypothesis is not affirmatively accepted, since there is a lack of information about the Type II error.

## § 30.6 Some examples of statistical models on $\Sigma^n$

**Table 30.5** Type I and II errors.

|  | Decision | |
| --- | --- | --- |
|  | Reject $H_0$ | Do not reject $H_0$ |
| $H_0$ true | Type I error | Correct |
| $H_0$ false | Correct | Type II error |

## 30.6 Some examples of statistical models on $\Sigma^n$

This paragraph presents some statistical models currently used (sometimes in an implicit way) in the definition of random sequence tests. Further information can be found in [MAU 1992, LUBICZ].

A random variable $X$ is said to be binary if its values are in the set $B = \{0, 1\}$. In that case, the distribution of probability defined on $B$ is given by a unique parameter called the bias of $X$, which is by definition $P[X = 1]$. Let $X_1, \ldots, X_n, \ldots$ be a sequence of binary independent random variables. They define a distribution of probability on $\Sigma^n$. When all these random variables have the same bias, the previous distribution depends only on the parameter $p$.

This model describes a binary memoryless source (BMS) that outputs independent random variables with a bias $p$. As stated, a BMS defines a distribution of probability on the sets $\Sigma^n$ depending on the parameter $p$, and is therefore a statistical model on $\Sigma^n$. A particular case of a BMS is the binary symmetric memoryless canal, which corresponds to the parameter $p = \frac{1}{2}$.

Another model is the Source Transition (ST) that outputs a sequence of binary random variables $X_1, \ldots, X_n, \ldots$ of parameter $\frac{1}{2}$ such that $P[X_i + X_{i+1} = 1] = p$ and $P[X_i + X_{i+1} = 0] = 1 - p$ for $i \in \mathbb{N}$.

Generally, a source can produce a sequence of binary random variables $X_1, \ldots, X_n, \ldots$ such that the conditional probability of $X_n$ given $X_1, X_2, \ldots, X_{n-1}$ depends only on the last $m$ bits, i.e., such that

$$P_{X_n|X_{n-1}\ldots X_1}(x_n|x_{n-1}\ldots x_1) = P_{X_n|X_{n-1}\ldots X_{n-m}}(x_n|x_{n-1}\ldots x_{n-m}). \tag{30.5}$$

The least $m$ satisfying this preceding property is called the memory of the source $S$ and $\Sigma_n = [X_n - 1, \ldots X_{n-m}]$ is the state at the time $n$. Therefore taking the sequence $(X_n)_{n \in \mathbb{N}}$ is equivalent to consider an initial state $\Sigma_{m-1}$, represented by the trivial random variables $[X_m, \ldots, X_0]$ (their weight being totally concentrated either on 0 or 1) as well as a distribution of probability for the transition of states $P_{\Sigma_n|\Sigma_{n-1}}$ for all $n$ greater than $m$. If this last probability is independent of $n$, then the source is classified as stationary. So, a stationary source is completely described by its initial state and $P_{\Sigma_{m+1}|\Sigma_m}$.

The set of states is a particular case of a Markov chain, with the restriction that each state can have only two successors. If this Markov chain is ergodic, the limit of the distribution of probability on the set of states converge towards a limit. Let the integers between 0 and $2^{m-1}$ represent the set of possible states of the sources, then the Chapman–Kolmogorov equations give:

$$\lim_{n \to +\infty} P_{\Sigma_n}(j) = p_j$$

where the $p_j$ are the solution of a system of $2^m$ equations:

$$\sum_{j=0}^{2^m-1} p_j = 0, \tag{30.6}$$

$$p_j = \sum_{k=0}^{2^m-1} P_{\Sigma_2|\Sigma_1} p_k, \text{ for } 0 \leqslant j \leqslant 2^m - 2. \tag{30.7}$$

There are two interesting points to consider with the statistical model of ergodic stationary sources:

- this model seems to be the most general of the models presented. In particular, it contains the BMS and ST models.
- this model has been extensively studied in the field of information theory. In particular, it is possible to compute its entropy.

## 30.7 Hypothesis testings and random sequences

In the previous sections it was stated that all the statistical models that can be used to carry out statistical tests on a binary sequence describe, for a given parameter value, a distribution that is verified if the random variables are Bernoulli with a parameter equal to $\frac{1}{2}$. From [LUBICZ], the link between the theory of hypothesis testings and random sequences is given as follows: a statistical model is adapted to the device that is under test, an $H_0$ chosen so that the values of the parameters of the model are verified if the random input variables are Bernoulli with a parameter of $\frac{1}{2}$, and as alternative hypothesis there is a large deviation from this $\frac{1}{2}$ value.

For example, if it is known that the statistical model of the device is a BMS, the monobit frequency test can be used on its own: this is the best test associated to this model. It may happen that the statistical model is more general and includes several different tests. For instance, the BMS is contained in the general model of a stationary ergodic source with a certain amount of memory. In this case, the advantage of the more specific test is that it is more powerful. However, it may not discover deviations in the parameters that it does not control. Therefore, it is important to first use the more specific tests and then the more general ones. It amounts to restraining the variance, in some direction, of the parameter space.

In general, the use of the techniques of hypothesis tests in order to verify the random quality of a source is characterized by:

- the choice of a statistical model based on the operation of the device
- the use of only a small number of tests (one or maybe two) that are associated with the statistical model.

It should be pointed out that this general technique does not describe the set of available procedures in order to test a random number generator. It is apparent that it is difficult to attach a statistical model to some tests that are widely published and recommended. Moreover, in the available test suites it is quite common to use many different tests. In practice, it is often difficult to prove that a certain physical device corresponds to a given statistical model apart from very general models, which then leads to tests of very poor quality.

In the case where no statistical model is available, it is possible to use the property that the estimators computed by the tests are consistent. Then, under the assumption of the Bernoulli distribution with a parameter equal to $\frac{1}{2}$ (BSS), the property that the sequence is $\infty$-distributed can be checked by the convergence in probability of certain estimators. Therefore, it is possible to use a group of several tests, so that each of them, with a given probability, outputs a pass for a random sequence. It should be noted that it is not easy to compute the rejection rate of a full test suite, because the estimators of different tests are often extremely nonindependent. This rate can, nevertheless, be estimated by stochastic simulations.

The reader should keep in mind that if the device is not provided with a statistical model and if the statistical tests cannot be interpreted with respect to the cryptographic use of the random sequence, the rejection zone selected by the statistical tests is totally arbitrary. If we have a statistical model, the rejection zone is chosen to contain most of the weight of probability when the device is faulty. But, if we do not know this statistical model, it may happen, on the contrary, that the rejection zone contains sequences with a low probability of appearance: this means that the probability of passing the test is higher when the device is faulty. In this respect, a statistical test is nothing but a convenient way to choose a certain proportion of sequences in the set of all binary sequences of a given length. In particular, if the tests do not pass, it is difficult to pronounce with any degree of certainty that there is no systematic interpretation of the result of the tests.

It is also important to realize that a random test may undermine cryptographic security in some applications. The problem is that, if a statistical test is used to filter the flux of a random generator, it introduces a bias that is very easy to detect by using the same test. A practical example of this is given to draw the reader's attention to this topic.

**Example 30.8** The user may want to cypher the content of a hard drive by using a strong symmetric encryption function. It may be required that an intruder, who does not posses the secret key, is not able to distinguish the written sectors on the hard drive from the blank ones. One way to implement this functionality is to consider the symmetric encryption function as a pseudorandom function. Therefore, a random number generator can be used to write random noise on nonwritten sectors of the harddrive. If the output of this random number generator is filtered by a statistical test with, for instance, a rejection rate of 1%, this means that 1% of the sequences of a given length will never appear in the nonwritten sectors of the harddrive, but will be present in the written sectors. This allows an attacker to find the distinguishing point between the written and nonwritten sectors very easily.

## 30.8 Empirical test examples for binary sequences

### 30.8.1 Random walk

The last example of counting the number of ones in the 100-bit sequence is an example of an empirical test based on the random walk. The random walk, $Y_n$, is the sum of independent Bernoulli random variables, $X_i$. It can be written:

$$Y_n = \sum_{i=1}^{n} X_i \qquad (30.8)$$

Using the Central Limit Theorem and the De Moivre–Laplace theorem, a binomial sum, normalized by $\sqrt{n}$, follows a normal distribution, if the sample size $n$ is large. This can be written as:

$$\lim_{n \to \infty} P\left(\frac{Y_n}{\sqrt{n}} \leqslant y\right) = \frac{1}{\sqrt{2\pi}} \int_{-\infty}^{y} e^{-\frac{h^2}{2}} dh = f(y). \qquad (30.9)$$

This theory is the basis for one of the simplest but most important statistical tests, the *frequency (monobit) test*. The null hypothesis for this test states that a sequence of independent, identically distributed Bernoulli variables will have a probability:

$$P(X_i = 1) = \frac{1}{2}. \qquad (30.10)$$

As already mentioned in previous sections, this statistical test is based on the model for a binary memoryless source.

Another implementation of the random walk is a variation on the previous frequency test called the *frequency block test*. This test performs multiple frequency tests on smaller, equally distributed subsequence blocks of the main sample sequence. This allows the detection of localized deviations from randomness. The sample sequence is divided into $n$ sets of $m$ bits. The number of ones in each $m$ sequence is counted, $\pi_i$. A test characteristic is then calculated by using the following formula:

$$X = 4m \sum_{i=1}^{n} \left( \frac{\pi_i}{m} - \frac{1}{2} \right)^2. \tag{30.11}$$

## 30.8.2 Runs

Tests based on the runs property observe the trend of a sequence. There are many variations of the runs test; a run can be treated as a grouping of ones and zeroes in a binary sequence, or it could also refer to the grouping of the sequence into subsequences, which end when the trend changes from ascending to descending or vice versa. These are only two examples of the many different forms in which a run can be defined.

For binary sequences, the FIPS 140-2 standard provides a *runs test*, which is a one parameter test, contained in a model of a stationary ergodic source with a memory of 6 bits. The runs test counts the length of runs of one and zero in the sequence and then compares the results to a precalculated range. The test inspects a sequence to see if the oscillations between the zeroes and ones are too fast or too slow.

As published in [MEO0+ 1996], the statistical test is calculated by first counting the length, $i$, of each run for both zeroes, $G$, and ones, $B$. A sequence of length $n$ will have an expected number of runs, $e_i$, for each length $i$. The following formulas show how to calculate $e_i$ and the test characteristic:

$$e_i = \frac{(n - i + 3)}{2^{i+2}}, \tag{30.12}$$

$$X = \sum_{i=1}^{k} \frac{(B_i - e_i)^2}{e_i} + \sum_{i=1}^{k} \frac{(G_i - e_i)^2}{e_i}. \tag{30.13}$$

Repeating bits are stored and counted until a change is noticed. If the run has a length of greater than six, then it is treated as a run of length six. If using the FIPS 140-2 suggested sequence length of 20,000 bits, then Table 30.6 can be used as acceptance ranges. Should any of the run's count fall outside of the acceptable range, then the sequence is rejected.

The *turning point test* is another type of run test and it can be found in [KAN 1993]. This test counts the number of turning points (peaks and troughs) in a sequence. To calculate the statistical test, the number of samples tested needs to be large. This allows for the assumption of a normal distribution with a mean of $\mu = \frac{2}{3}(n-2)$ and a variance of $\sigma^2 = \frac{(16n-29)}{90}$. The test characteristic can then be calculated with the following:

$$X = \left| \frac{\bar{x} - \mu}{\sigma} \right|. \tag{30.14}$$

## 30.8.3 Autocorrelation

Visually, it is possible to detect regular waveforms as nonrandom. How can this property be automated for randomness testing in applications? One method is to compare the signal with a shift copy of itself, which is the autocorrelation function. A random sequence will have very little correlation with any copy of itself.

Table 30.6 FIPS 140-2 intervals for runs test.

| Length of run ($i$) | Required interval |
|---|---|
| 1 | 2343–2657 |
| 2 | 1135–1365 |
| 3 | 542–708 |
| 4 | 251–373 |
| 5 | 111–201 |
| 6+ | 111–201 |

The *autocorrelation test*, as described in [MEOO+ 1996], checks for the correlation between the current sequence and a shifted version. The statistical model for this test is the source transition model. It operates by taking a sample sequence and XORing with a $d$ delayed version of itself. With a large sample, $n$, and $n - d \geqslant 10$ the statistical test can again be assumed to follow a normal distribution. The test characteristic is calculated using the following formulas:

$$A(d) = \sum_{i=0}^{n-d-1} s_i \text{ XOR } s_{i+d} \tag{30.15}$$

$$X = 2\left(\frac{A(d) - \frac{n-d}{2}}{\sqrt{n-d}}\right). \tag{30.16}$$

**Remark 30.9** The empirical tests presented here are only a small fraction of what is available in literature. Three randomness test suites have been created to help evaluate the selected random number generators:

- NIST statistical test suite [RUSO+],
- The diehard battery of strigent statistical randomness tests [MAR],
- ENT: A pseudorandom number sequence test program [ENT 1998].

## 30.9 Pseudorandom number generators

The previous part introduced physical mechanisms to build true unbiased random number generators and detailed test suites such generators have to pass. In this section pseudorandom number generators based on curves are considered. A *pseudorandom number generator (PRNG)* is a *deterministic* algorithm taking as input some (truly random) seed and producing a sequence of numbers that satisfies certain statistical properties. Here "statistical properties" are meant in a really broad sense. An important requirement, called *unpredictability*, is that one should not be able to compute further elements given a substring of the sequence. Another requirement, called *indistinguishability*, is that the sequence should resemble a truly random sequence, i.e., one should not be able to distinguish a sequence generated by a PRNG from a random sequence.

It is important to note that there is a whole wealth of other pseudorandom number generators based on much cheaper devices, e.g., on shift register sequences. Some popular ones are briefly described in Section 30.2.4. In this section only applications of elliptic and hyperelliptic curves are considered. One of the advantages of these generators is that one can base their security on well-studied problems like the ECDLP or the ECDHP. On the other hand, their use as building blocks for pseudorandom number generators is less studied and thus it might bear further disadvantages besides being rather complicated. Some of the following algorithms are generalizations from PRNGs over finite fields. Just as with the normal DLP, the advantage of using curves is the smaller field size for the same level of security.

This section studies sequences of random points on elliptic curves or of random divisor classes on hyperelliptic curves. Let $E : y^2 = x^3 + a_4 x + a_6$ be an elliptic curve over a finite field of odd characteristic $\mathbb{F}_q$. Assume that $x^3 + a_4 x + a_6$ has a root $\alpha \in \mathbb{F}_q$. Then $P = (0, \alpha)$ is as random as any other point even if it is clearly not a *random point* in our usual understanding as, for example, the multiples of $P$ alternate between $P$ and $P_\infty$. Therefore, like in the previous section, measures are needed to determine the (pseudo-)randomness of such sequences. For applications to generate random numbers one will usually only use the $x$-coordinates of the points.

This study also serves a second purpose — if the sequence of powers of an element would turn out to be biased (for instance some bits were fixed for certain multiples) the discrete logarithm problem would be weaker than assumed as some bits of the exponent would leak.

Furthermore, there are many more results on pseudorandom sequences from curves published in the area of stream ciphers [NIXI 2001, NIE 2003] that will not be studied within the scope of this book.

### 30.9.1 Relevant measures

This section gives some basic definitions concerning pseudorandom sequences. In particular, it defines expressions to measure their quality. First, note that a sequence over a finite field for which a sequence element depends deterministically on the previous ones needs to have a finite period, i.e., there exist some integers $N$ and $t$ such that $s(i + t) = s(N + i + t)$ for all $i \geqslant 0$. The integer $N$ is called the *period* and $t$ is called the *preperiod*; in the applications studied here one usually has $t = 0$.

**Definition 30.10** Let $S = \{s(0), s(1), \ldots, s(N-1)\}$ be a sequence of elements of $\mathbb{F}_q$ and let $\alpha \in \mathbb{F}_q^*$. The *balance of $S$ with respect to $\alpha$* is defined in the following way:

$$B_S(\alpha) = \frac{1}{N} \sum_{i=0}^{N-1} \zeta_p^{\mathrm{Tr}_{\mathbb{F}_q/\mathbb{F}_p}(\alpha s(i))},$$

where $\zeta_p = \exp(2\pi i/p)$ denotes a primitive $p$-th root of unity.

Furthermore, the *balance* is defined to be

$$B_S = \max_{\alpha \in \mathbb{F}_q^*} \{|B_S(\alpha)|\}.$$

So the balance is the largest absolute value of character sums with nontrivial additive characters of order $p$. As the the $p$-th roots of unity sum up to zero $\sum_{i=0}^{p-1} \zeta_p^i = 0$, the balance evaluates to zero if each element of $\mathbb{F}_p$ appears equally often as trace $\mathrm{Tr}_{\mathbb{F}_q/\mathbb{F}_p}(\alpha s(i))$ for some element of the sequence. If some elements occur more often than others, the balance gives a measure of the bias.

The autocorrelation was already introduced at the end of the previous section. The following definition provides a more concrete formula.

## § 30.9 Pseudorandom number generators

**Definition 30.11** Let $\{s(0), s(1), \ldots, s(N-1)\}$ be a sequence $S$ of period $N$ defined over the finite field $\mathbb{F}_q$. Furthermore, let $\alpha, \beta \in \mathbb{F}_q^*$.

The *autocorrelation with respect to $\alpha$ and $\beta$* of a sequence is defined as follows:

$$C_S(d, \alpha, \beta) = \frac{1}{N} \sum_{i=0}^{N-1} \zeta_p^{\mathrm{Tr}_{\mathbb{F}_q/\mathbb{F}_p}(\alpha s(i+d) - \beta s(i))},$$

with $0 \leqslant d < N$.

Note that in the above definition $i + d$ should be read modulo $N$. Also, note that for sequences over $\mathbb{F}_2$ this definition amounts to

$$C_S(d) = \frac{1}{N} \sum_{i=0}^{N-1} (-1)^{s(i+d)+s(i)},$$

which is the usual definition of the autocorrelation (see, e.g., [MEOO+ 1996, Chapter 5, Section 4] and the previous section).

Another useful object is the crosscorrelation of two sequences. It is defined as follows:

**Definition 30.12** Let $S = \{s(i)\}$ and $T = \{t(i)\}$ be two sequences defined over $\mathbb{F}_q$ of the same period $N$. Let $\alpha, \beta \in \mathbb{F}_q^*$. We define the *crosscorrelation of $S$ and $T$ with respect to $\alpha$ and $\beta$* by

$$C_{S,T}(d, \alpha, \beta) = \frac{1}{N} \sum_{i=0}^{N-1} \zeta_p^{\mathrm{Tr}_{\mathbb{F}_q/\mathbb{F}_p}(\alpha s(i+d) - \beta t(i))}.$$

The problem is to find a family of sequences $\Sigma = \{S_i | i \in I\}$ such that for each choice of $i, j \in I$ the crosscorrelation $C_{S_i, S_j}(d, \alpha, \beta)$ is small.

The discrepancy is usually defined for sequences taking their values in the interval $[0, 1)$. If the sequence $S$ assumes values in $\mathbb{F}_p$ this can be achieved by using $\{0, \ldots, p-1\}$ as a set of representatives of the field elements and dividing each element by $p$. If an extension field $\mathbb{F}_q = \mathbb{F}_{p^d} = \mathbb{F}_p(\theta)$ is used with a polynomial basis representation, one associates the number $a = \sum_{i=0}^{d-1} a_i p^i$, with $a_i \in \{0, \ldots, p-1\}$ to $\alpha = \sum_{i=0}^{d-1} a_i \theta^i, a_i \in \mathbb{F}_p$ and normalizes by dividing by $p^d$. One can also consider the distribution of the coordinate tuples $(x, y)$. Generalizations to hyperelliptic curves seem possible but have not been proposed yet.

**Definition 30.13** Let $S$ be normalized to the interval $[0, 1]$ and consider a subsequence of $t$ elements. Denote by $\Delta(S)_t$ the *discrepancy of the sequence $S$*, given by

$$\Delta(S)_t = \sup_{[\alpha, \beta] \subseteq [0,1]} \left| \frac{N(\alpha, \beta)}{t} - (\beta - \alpha) \right|,$$

where $N(\alpha, \beta)$ is the number of elements $s(i)$, $i = 0, \ldots, t-1$, which hit the interval $[\alpha, \beta]$.

For periodic sequences it is common to choose $t$ equal to the period $N$. In this case we use the notation $\Delta(S)$ to denote the discrepancy of $S$.

The discrepancy provides a measure for the distribution of the elements. For a good sequence, the number of elements hitting an interval should be proportional to the size of the interval, i.e., the discrepancy should be small.

It is also possible to define the discrepancy of multidimensional distributions $(s(i), s(i+1), \ldots, s(i+n-1))$ and in fact one needs to consider these measures to ensure the strength of pseudorandom number generators (see e.g., [NISH 1999]).

A further indicator of the randomness is that there is no low-order linear recursion among the outputs.

**Definition 30.14** The *linear complexity* $\mathcal{L}(S)$ of an infinite sequence $S = s(i), i = 0, 1, \ldots$ over a ring $R$ is the length of the shortest linear recurrence relation, i.e., $\mathcal{L}(S) = l$ if and only if

$$s(i+l) = a_{l-1}s(i+l-1) + \cdots + a_0 s(i), \text{ for } i = t, t+1, \ldots,$$

with $a_0, \ldots, a_{l-1} \in R$, which is satisfied by this sequence and $l$ is minimal with this property. Here, $t$ is the preperiod and thus usually $t = 0$ in our examples.

## 30.9.2 Pseudorandom number generators from curves

Throughout this section $E$ denotes an elliptic curve and $P \in E$ is a point of order $\ell$. In the sequel, different ways to construct sequences $S$ of points are presented. They can be used to construct sequences of field elements by considering only $x(P) \in \mathbb{F}_q$, the $x$-coordinate of $P$, maybe after discarding some bits. The references to balance, correlation and discrepancy assume this conversion to be done. One can also consider the distribution of the coordinate tuples $(x, y)$.

### 30.9.2.a The linear congruential generator

In 1994, Hallgren [HAL 1994] proposed a *linear congruential generator* from an elliptic curve. Gong, Berson, and Stinson [GOBE$^+$ 2000] study binary sequences derived from this generator where they use the trace to map to $\mathbb{F}_2$. This study has been extended in [BEDO 2002, ELSH 2002, GOLA 2002, HESH 2005, KOSH 2000].

The linear congruential generator builds a sequence of points on $E$ by the rule $s(0) = P_0$ for some point $P_0 \in E$ and $s(i) = P \oplus s(i-1) = [i]P \oplus P_0$.

Obviously, for this generator one can obtain the next element in the sequence given the two previous elements, as this reveals the difference $P$. To avoid this problem, one additionally uses a function $f \in \mathbb{F}_q(E)$ from the function field of $E$, i.e., a function that can be evaluated at points of the curve. If the function is kept secret and subsequently evaluated at the points $s(i)$ one obtains a sequence of field elements that (presumably) cannot be reconstructed. To scramble the results even further and to allow a better analysis, most authors propose to apply the trace $\text{Tr}_{\mathbb{F}_q/\mathbb{F}_p}$ to the resulting field elements, mapping them to the smaller field $\mathbb{F}_p$.

Clearly the distribution of the sequence depends a lot on the function $f$, i.e., the most striking example being a constant function $f$.

The following part briefly outlines the first construction of [GOBE$^+$ 2000] and states some results (the proofs can be found in the original papers). The authors suggest using a binary supersingular curve

$$E : y^2 + y = x^3 + a_4 x + a_6 \text{ over } \mathbb{F}_q = \mathbb{F}_{2^d} \text{ with } d \text{ odd},$$

and let $P = P_0$ be a generator of $E(\mathbb{F}_q)$ with $\text{ord}(P) = |E(\mathbb{F}_q)| = v + 1$. The sequence is an interleaved sequence given by

$$s(2i) = \text{Tr}\bigl(x([i]P)\bigr), \ s(2i+1) = \text{Tr}\bigl(y([i]P)\bigr), \text{ for } 1 \leqslant i \leqslant v$$

where $x(P)$ and $y(P)$ are the $x$- and $y$-coordinate of $P$, respectively.

For this sequence they prove that the period is maximal, namely equal to $2v$.

For the more special curve $E_1 : y^2 + y = x^3$ they use $|E_1(\mathbb{F}_{2^d})| = 2^d + 1$ and for odd $d = 2m+1$ they show that the number of of 1's and 0's in $S$ is equal to $2^d \pm 2^m$, respectively. Hence, the bias is small but not negligible.

Beelen and Doumen [BEDO 2002] consider curves of arbitrary genus over arbitrary finite fields. The following is a simplified version for elliptic curves. The full version can be found in the paper (cf. also [KOSH 2000]).

They define a whole family of PRNGs by using functions $f \in \mathbb{F}_{q^k}(E)$ inside the trace map. As $\mathrm{Tr}(a^p - a) = 0$ for all $a \in \mathbb{F}_{q^k}$, one can change the function $f$ to $f - h^p + h$ for arbitrary $h \in \mathbb{F}_{q^k}(E)$ without changing the value. If $(f - h^p + h)(Q)$ is not defined for any such $h$, put $(f - h^p + h)(Q) = 0$ in the definition below.

**Definition 30.15** Let $E$ be an elliptic curve defined over $\mathbb{F}_{q^k}$. Let $P$ be a generator of a prime order subgroup of $E(\mathbb{F}_{q^k})$ and denote its order by $\ell$. Let $f \in \mathbb{F}_{q^k}(E)$. The sequence

$$S^{\mathrm{AS}}(f, P) = \{s(i)\}_{0 \leqslant i < \ell}$$

is defined by

$$s(i) = \mathrm{Tr}_{\mathbb{F}_{q^k}/\mathbb{F}_q}\bigl(f([i]P)\bigr).$$

The weighted degree $\mathrm{wdeg}(f)$ of a function $f \in \mathbb{F}_q(E)$ is defined by $\mathrm{wdeg}(x) = 2$ and $\mathrm{wdeg}(y) = 3$. One obtains the following result on the balance.

**Theorem 30.16** Let $E$ be an elliptic curve defined over the finite field $\mathbb{F}_{q^k}$ given by a Weierstraß equation. Let $f$ be a polynomial in the coordinate functions $x$ and $y$, such that the degree in $y$ is $\leqslant 1$. Furthermore, let $P$ be a generator of a prime order subgroup of $E(\mathbb{F}_{q^k})$ and let $\ell = \mathrm{ord}(P)$. Suppose that $p$ does not divide $\mathrm{wdeg}(f)$. Then

$$B_{S^{\mathrm{AS}}(f,P)} \leqslant \bigl(1 + (1 + \mathrm{wdeg}(f))\sqrt{q^k}\bigr)/\ell.$$

For the autocorrelation and crosscorrelation the following results hold.

**Theorem 30.17** Let $E$ be an elliptic curve defined over the field $\mathbb{F}_{q^k}$ given by a Weierstraß equation. Let $f$ be a polynomial in the two coordinate functions $x$ and $y$, such that $\deg_y(f) \leqslant 1$. Choose $\alpha, \beta \in \mathbb{F}_q^*$. Further, choose a generator $P$ of a cyclic prime order subgroup $E(\mathbb{F}_{q^k})$ and a number $d$ satisfying $1 \leqslant d < \ell$ with $\ell = |\langle P \rangle|$. Suppose that the characteristic does not divide $\mathrm{wdeg}(f)$. Then

$$|C_{S^{\mathrm{AS}}(f,P)}(d, \alpha, \beta)| \leqslant \bigl(2 + 2(1 + \mathrm{wdeg}(f))\sqrt{q^k}\bigr)/\ell.$$

Let $f_1$ and $f_2$ be two polynomials in the coordinate functions $x$ and $y$ such that $\deg_y(f_i) \leqslant 1$ for $i = 1, 2$, and such that for all $(\alpha, \beta) \in \mathbb{F}_{q^k}^2 \setminus \{(0,0)\}$ one has $p \nmid \mathrm{wdeg}(\alpha f_1 - \beta f_2)$. Write $S_1 = S^{\mathrm{AS}}(f_1, P)$ and $S_2 = S^{\mathrm{AS}}(f_2, P)$. For all $\alpha, \beta \in \mathbb{F}_q^*$ and $0 \leqslant d < \ell$ one has

$$|C_{S_1, S_2}(d, \alpha, \beta)| \leqslant \bigl(2 + (2 + \mathrm{wdeg}(f_1) + \mathrm{wdeg}(f_2))\sqrt{q^k}\bigr)/\ell,$$

unless $d = 0$ and $\alpha f_1 = \beta f_2$.

**Remark 30.18** Beelen and Doumen [BEDO 2002] perform the same study also for the case of multiplicative characters $\chi$ instead of using the trace. Their paper also contains a study on the period and balance of the sequence by making use of linear recurrence relations on the points of $E$.

### 30.9.2.b  The elliptic curve power generator

The analogue of the power generator was introduced by Lange and Shparlinski [LASH 2005a].

**Definition 30.19** Let $P \in E(\mathbb{F}_q)$ be a point of order $\ell$ on a nonsupersingular elliptic curve $E$. The *elliptic curve power generator* uses an integer $e$, with $\gcd(e, \ell) = 1$ and considers the sequence $S^{\mathrm{PG}}$ defined by:

$$s(0) = P, \ s(i) = [e]s(i-1) = [e^i]P.$$

Determining $e$, which is the obvious way to compute the next element of the sequence, corresponds to solving the discrete logarithm problem. Given a part of the sequence, one needs to solve a problem related to the Diffie–Hellman problem in order to compute further elements of the sequence. The statistical properties like discrepancy are studied in [LASH 2005a].

It is easy to see that the period of this sequence equals $T = \mathrm{ord}_\ell(e)$, the multiplicative order of $e$ modulo $\ell$.

**Theorem 30.20** Let $E$ be a non-supersingular elliptic curve defined over $\mathbb{F}_q$ and let $P$ be a point of order $\ell$. Then, for any integer $\nu \geqslant 1$, the bound

$$\Delta_e(S^{\mathrm{PG}}) \ll T^{1-(3\nu+2)/2\nu(\nu+2)} t^{(\nu+1)/\nu(\nu+2)} q^{1/4(\nu+2)} (\ln p + 1)^\gamma$$

holds for the discrepancy of the sequence $S^{\mathrm{PG}}$.

From the discrepancy one obtains the following result on the linear complexity.

**Theorem 30.21** Let $E$ be a non-supersingular elliptic curve defined over $\mathbb{F}_q$. Then the bound

$$\mathcal{L}(S^{\mathrm{PG}}) \gg T\ell^{-2/3}$$

holds, where $T$ is the multiplicative order of $e$ modulo $\ell$.

#### 30.9.2.c  The Naor–Reingold generator

The following PRNG is similar to the power generator, but one uses a vector $(e_0, \ldots, e_{n-1})$ of secret integers instead of only one secret scalar $e$. For elliptic curves it was proposed by Shparlinski [SHP 2000] and studied further in [SISH 2001]. For simplicity the definition and results are restricted to elliptic curves over prime fields.

**Definition 30.22** Let the bit representation of $i$ be given by $(i_{n-1} \ldots i_0)_2$ and let $e_0, \ldots, e_{n-1}$ be nonzero integers and $P$ be a point of prime order $\ell$ of a nonsupersingular elliptic curve $E/\mathbb{F}_p$.

The $i$-th element of the sequence $S^{\mathrm{NR}}$ produced by the *Naor–Reingold generator* is given by

$$s(i) = [e_0^{i_0} \ldots e_{n-1}^{i_{n-1}}]P,$$

for $0 \leqslant i < \ell$.

**Example 30.23** Let $n = 4$ and consider the vector $(2, 5, 3, 4)$. Then

$$\begin{aligned}
s(0) &= 2^0 5^0 3^0 4^0 P = P, \\
s(1) &= 2^0 5^0 3^0 4^1 P = [4]P, \\
s(2) &= 2^0 5^0 3^1 4^0 P = [3]P, \\
&\vdots \\
s(15) &= 2^1 5^1 3^1 4^1 P = [120]P = [6]P.
\end{aligned}$$

In the publications about the Naor–Reingold PRNG, usually only the $x$-coordinates of the resulting sequence of points are considered. Using bounds on character sums one can show that as soon as the order $\ell$ of $P$ is sufficiently large, namely $\ell > p^{1/2+\epsilon}$, the discrepancy is small. In more detail, Shparlinski [SHP 2000] proves the following result:

**Theorem 30.24** For any $\delta > 0$ and a random vector $(e_0, \ldots, e_{n-1})$ chosen uniformly from $(\mathbb{F}_\ell^*)^n$ the bound
$$\Delta(S^{\mathrm{NR}}) \ll \delta^{-1} B(n, \ell, p) \lg^2 p$$
holds, where
$$B(n, \ell, p) = 2^{-n/2} + 3^{n/2} 2^{-n} \ell^{-1/2} p^{1/4} + n^{1/2} \ell^{-1/2} + \ell^{-1} p^{1/2}$$
with probability at least $1 - \delta$.

For the linear complexity [SISH 2001] provides the following result.

**Theorem 30.25** Suppose that $\gamma > 0$ and $n$ are chosen to satisfy
$$n \geqslant (2 + \gamma) \lg \ell.$$
Then for any $\delta > 0$ and sufficiently large $\ell$, the linear complexity $L(S^{\mathrm{NR}})$ of the sequence of the $x$-coordinates of the points satisfies
$$L(S^{\mathrm{NR}}) \gg \min\{l^{1/3-\delta}, l^{\gamma-3\delta} \lg^{-2} l\}$$
for all except $O\big((\ell-1)^{n-\delta}\big)$ vectors $(e_0, \ldots, e_{n-1}) \in (\mathbb{F}_\ell^*)^n$.

### 30.9.3 Other applications

This section briefly highlights two other applications of elliptic and hyperelliptic curves related to randomness.

To hide that a message is sent, Kaliski [KAL 1986] uses that a given $x_P \in \mathbb{F}_q$ is the $x$-coordinate of a point on $E/\mathbb{F}_q$ or on its twist $\widetilde{E}/\mathbb{F}_q$ (cf. Remark 13.17). He proposes to modify the usual encryption schemes (cf. Algorithms 1.16 and 1.17) by using both curves. The public key of A consists of $P_A \in E(\mathbb{F}_q)$ *and* $\widetilde{P}_A \in \widetilde{E}(\mathbb{F}_q)$. The sender randomly selects one of $P_A, \widetilde{P}_A$ and uses it as public key in the normal protocol. The ciphertext consists of the compressed representation, i.e., the $x$-coordinate of the result and one bit of the $y$-coordinate, as described in Sections 13.2.5 and 13.3.7. As the result corresponds to an arbitrary point on either $E$ or $\widetilde{E}$, it cannot be distinguished from a random bit-string and therefore an eavesdropper does not learn that a message was sent. Upon receiving the message, the receiver determines the curve on which a point with this $x$-coordinate exists and recovers the $y$-coordinate by decompression. This idea was studied further and brought to implementation in [BAI 2003].

Section 15.1 introduced Koblitz curves as curves defined over a very small finite field, which are then considered over a large extension field. The first curves proposed by Koblitz were even defined over $\mathbb{F}_2$. Up to isogenies, there are only two curves
$$E_{a_2} : y^2 + xy = x^3 + a_2 x^2 + 1, \quad a_2 \in \{0, 1\}.$$

For applications in DL systems, these curves offer the advantage that to compute scalar multiples of a point one can make use of the Frobenius endomorphism. This avoids doublings and thus increases the efficiency considerably. As described in Section 15.1.3 one can use an alternative setup to obtain random scalar multiples of a point $P$ by picking random $\tau$-adic expansions.

Obviously, this can be used in the setting of pseudorandom sequences as well. The sequence of such points is shown to be almost uniformly distributed in [LASH 2005b].

# References

*Numbers in the margin specify the pages where citations occur*

[AdDe+ 1999]  L. M. Adleman, J. DeMarrais, & M.-D. Huang, *A subexponential algorithm for discrete logarithms over hyperelliptic curves of large genus over* $GF(q)$, Theoret. Comput. Sci. **226** (1999), 7–18. [512, 515, 516]

[AdHu 1992]  L. M. Adleman & M.-D. Huang, *Primality testing and Abelian varieties over finite fields*, Lecture Notes in Math., vol. 1512, Springer-Verlag, Berlin, 1992. [601]

[AdHu 1996]  _____, *Counting rational points on curves and abelian varieties over finite fields*, Algorithmic Number Theory Symposium – ANTS II, Lecture Notes in Comput. Sci., vol. 1122, Springer-Verlag, Berlin, 1996, 1–16. [422]

[AdHu 2001]  _____, *Counting points on curves and abelian varieties over finite fields*, J. Symbolic Comput. **32** (2001), 171–189. [422]

[AdPo+ 1983]  L. Adleman, C. Pomerance, & R. Rumely, *On distinguishing prime numbers from composite numbers*, Ann. of Math. **117** (1983), 173–206. [599]

[AgAr+ 2003]  D. Agrawal, B. Archambeault, J. R. Rao, & P. Rohatgi, *The EM Side-Channel(s)*, Cryptographic Hardware and Embedded Systems – CHES 2002, Lecture Notes in Comput. Sci., vol. 2523, Springer-Verlag, Berlin, 2003, 29. [682]

[AgKa+ 2002]  M. Agrawal, N. Kayal, & N. Saxena, *PRIMES is in P*, preprint, date Aug. 6th, 2002. [601]
http://www.cse.iitk.ac.in/primality.pdf

[AkTa 2003]  T. Akishita & T. Takagi, *Zero-value point attacks on elliptic curve cryptosystem*, Information Security Conference – ISC 2003, Lecture Notes in Comput. Sci., vol. 2851, Springer-Verlag, Berlin, 2003, 218–233. [682]

[AlGr+ 1994]  R. Alford, A. Granville, & C. Pomerance, *There are infinitely many Carmichael numbers*, Ann. of Math. **139** (1994), 703–722. [593]

[AlMa+ 2002]  E. Al-Daoud, R. Mahmod, M. Rushdan, & A. Kilicman, *A new addition formula for elliptic curves over* $GF(2^n)$, IEEE Trans. on Computers **51** N°8 (2002), 972–975. [293]

[AnAn+ 1999]  I. Anshel, M. Anshel, & D. Goldfeld, *An algebraic method for public key cryptography*, Math. Res. Lett. **6** (1999), 287–291. [15]

[Ansi X9.62]  ANSI X9.62-1999, *Public key cryptography for the financial services industry: The elliptic curve digital signature algorithm (ECDSA)*, 1999. [13, 570]

[Apecs]  I. McConnell, *Maple programs*. [267]
ftp://ftp.math.mcgill.ca/pub/apecs

[Atk 1988]  A. O. L. Atkin, *The number of points on an elliptic curve modulo a prime*, 1988. E-mail on the Number Theory Mailing List. [414, 421]

[Atk 1991]  _____, *The number of points on an elliptic curve modulo a prime*, 1991. E-mail on the Number Theory Mailing List. [414, 415]

[AtMo 1993]  A. O. L. Atkin & F. Morain, *Elliptic curves and primality proving*, Math. Comp. **61** (1993), 29–68. [458, 597]

[Ava 2002] R. M. Avanzi, *On multi-exponentiation in cryptography*, Tech. Report 154, AREHCC, 2002.
http://citeseer.nj.nec.com/545130.html [155]

[Ava 2004a] _____, *Aspects of hyperelliptic curves over large prime fields in software implementations*, Cryptographic Hardware and Embedded Systems – CHES 2004, Lecture Notes in Comput. Sci., vol. 3156, Springer-Verlag, 2004, 148–162. [267, 352, 704]

[Ava 2004b] _____, *Countermeasures against Differential Power Analysis for hyperelliptic curve cryptosystems*, Cryptographic Hardware and Embedded Systems – CHES 2003, Lecture Notes in Comput. Sci., vol. 2779, Springer-Verlag, Berlin, 2004, 366–381. [700–704]

[Ava 2005a] _____, *A note on the signed sliding window integer recoding and a left-to-right analogue*, Selected Areas in Cryptography – SAC 2004, Lecture Notes in Comput. Sci., vol. 3357, Springer-Verlag, Berlin, 2005, 130–143. [153, 154]

[Ava 2005b] _____, *On the complexity of certain multi-exponentiation techniques in cryptography*, J. Cryptology (2005), to appear. [155]

[Ava 2005c] _____, *Side channel attacks on implementations of curve-based cryptographic primitives*, preprint, extended version of AREHCC-report, 2005.
http://eprint.iacr.org/2005/017/ [687, 706]

[AvCe 2005] R. M. Avanzi & E. Cesena, *Trace zero varieties over binary fields for cryptography*, preprint, 2005. [383]

[AvCi$^+$ 2004] R. M. Avanzi, M. Ciet, & F. Sica, *Faster scalar multiplication on Koblitz curves combining point halving with the Frobenius endomorphism*, Public Key Cryptography – PKC 2004, Lecture Notes in Comput. Sci., vol. 2947, Springer-Verlag, 2004, 28–40. [301, 365]

[AvHe$^+$ 2004] R. M. Avanzi, C. Heuberger, & H. Prodinger, *Scalar multiplication on Koblitz curves using the Frobenius endomorphism and its combination with point halving: extensions and mathematical analysis*, preprint, 2004.
http://finanz.math.tu-graz.ac.at/~cheub/publications/tauext.pdf [359, 365]

[AvLa 2005] R. M. Avanzi & T. Lange, *Cryptographic applications of trace zero varieties*, preprint, 2005. [13, 383, 386]

[AvMi 2004] R. M. Avanzi & P. Mihăilescu, *Generic efficient arithmetic algorithms for PAFFs (Processor Adequate Finite Fields) and related algebraic structures*, Selected Areas in Cryptography – SAC 2003, Lecture Notes in Comput. Sci., vol. 3006, Springer-Verlag, Berlin, 2004, 320–334. [182, 230, 236]

[AvTh 2004] R. M. Avanzi & N. Thériault, *Random walks and filtering strategies for index calculus*, Manuscripts, 2004. [499, 504, 518]

[BaCh$^+$ 2004] H. Bar-El, H. Choukri, D. Naccache, M. Tunstall, & C. Whelan, *The sorcerer's apprentice guide to fault attacks*, Workshop on Fault Diagnosis and Tolerance in Cryptography – FDTC 2004, 2004.
http://www.elet.polimi.it/res/FDTC04/Naccache.pdf [684]

[BaDu$^+$ 2004] R. Barua, R. Dutta, & P. Sarkar, *Provably secure authenticated tree based group key agreement protocol using pairing*, preprint, 2004.
http://eprint.iacr.org/2004/90/ [576]

[BaEn$^+$ 2002] A. Basiri, A. Enge, J.-C. Faugère, & N. Gürel, *The arithmetic of Jacobian groups of superelliptic cubics*, Tech. report, INRIA – RR-4618, 2002. [352]

[BaEn$^+$ 2004] _____, *Implementing the arithmetic of $C_{3,4}$ curves*, Algorithmic Number Theory Symposium – ANTS VI, Lecture Notes in Comput. Sci., vol. 3076, Springer-Verlag, Berlin, 2004, 87–101. [352]

[BaHa 1998] R. C. Baker & G. Harman, *Shifted primes without large prime factors*, Acta Arith. **83** N°4 (1998), 331–361. [593]

[BAI 2003]    H. BAIER, *A fast Java implementation of a provably secure pseudo random bit generator based on the elliptic curve discrete logarithm problem*, Tech. Report TI 7/03, University of Darmstadt, 2003. [735]

[BaKi⁺ 2002]  P. S. L. M. BARRETO, H. Y. KIM, B. LYNN, & M. SCOTT, *Efficient algorithms for pairing-based cryptosystems*, Advances in Cryptology – Crypto 2002, Lecture Notes in Comput. Sci., vol. 2442, Springer-Verlag, Berlin, 2002, 354–368. [389, 580, 583, 589]

[BaKo 1998]   R. BALASUBRAMANIAN & N. KOBLITZ, *The improbability that an elliptic curve has a sub-exponential discrete log problem under the Menezes–Okamoto–Vanstone algorithm*, J. Cryptology **11** (1998), 141–145. [395, 564, 579, 586]

[BaLy⁺ 2003]  P. S. L. M. BARRETO, B. LYNN, & M. SCOTT, *Constructing elliptic curves with prescribed embedding degrees*, Security in Communication Networks – SCN 2002, Lecture Notes in Comput. Sci., vol. 2576, Springer-Verlag, Berlin, 2003, 257–267. [586, 588]

[BaLy⁺ 2004a] _____, *Efficient implementation of pairing-based cryptosystems*, J. Cryptology **17** (2004), 321–334. [586, 588]

[BaLy⁺ 2004b] _____, *On the selection of pairing-friendly groups*, Selected Areas in Cryptography – SAC 2003, Lecture Notes in Comput. Sci., vol. 3006, Springer-Verlag, Berlin, 2004, 17–25. [389]

[BaPa 1998]   D. V. BAILEY & C. PAAR, *Optimal extension fields for fast arithmetic in public key algorithms*, Advances in Cryptology – Crypto 1998, Lecture Notes in Comput. Sci., vol. 1462, Springer-Verlag, Berlin, 1998. [229]

[BAR]         P. S. L. M. BARRETO, *The pairing-based crypto lounge*. http://planeta.terra.com.br/informatica/paulobarreto/pblounge.html [389, 573]

[BAR 1987]    P. BARRETT, *Implementing the Rivest Shamir and Adleman public key encryption algorithm on a standard digital signal processor*, Advances in Cryptology – Crypto 1986, Lecture Notes in Comput. Sci., vol. 263, Springer-Verlag, Berlin, 1987, 311–323. [179]

[BaWa 1980]   R. BAILLIE & S. S. WAGSTAFF, JR., *Lucas pseudoprimes*, Math. Comp. **35** (1980), 1391–1417. [596]

[BaZh 2004]   J. BAEK & Y. ZHENG, *Identity-based threshold decryption*, Public Key Cryptography – PKC 2004, Lecture Notes in Comput. Sci., no. 2947, Springer-Verlag, 2004, 262–276. [578]

[BeBe⁺ 1989]  F. BERGERON, J. BERSTEL, S. BRLEK, & C. DUBOC, *Addition chains using continued fractions*, J. Algorithms **10** N°3 (1989), 403–412. [161, 166]

[BeBe⁺ 1994]  F. BERGERON, J. BERSTEL, & S. BRLEK, *Efficient computation of addition chains*, J. Théor. Nombres Bordeaux **6** (1994), 21–38. [162]

[BEC 1998]    F. BECK, *Integrated circuit failure analysis – a guide to preparation techniques*, John Wiley & Sons, Ltd., 1998. [671]

[BeDo 2002]   P. H. T. BEELEN & J. M. DOUMEN, *Pseudorandom sequences from elliptic curves*, Finite fields with applications to coding theory, cryptography and related areas, Springer-Verlag, 2002, 37–52. [732, 733]

[BeGe⁺ 1991]  T. BETH, W. GEISELMANN, & F. MEYER, *Finding (good) normal basis in finite fields*, International Symposium on Symbolic and Algebraic Computations – ISSAC 1991, ACM Press, Bonn, 1991, 173–178. [221]

[BeGo⁺ 1972]  M. BEELER, R. W. GOSPER, & R. SCHROEPPEL, *HAKMEM*, Memo 239, Massachusetts Institute of Technology Artificial Intelligence Laboratory, February 1972. [484]

[BeKn 2003]   R. BEVAN & E. W. KNUDSEN, *Ways to enhance differential power analysis*, Information Security and Cryptology – ICISC 2002, Lecture Notes in Comput. Sci., vol. 2587, Springer-Verlag, 2003, 327–342. [680]

[BeMc⁺ 1978]  E. R. BERLEKAMP, R. J. MCELIECE, & H. C. VAN TILBORG, *On the inherent intractability of certain coding problems*, IEEE Trans. Inform. Theory **24** N°3 (1978), 384–386. [15]

[BER 1967]   E. R. BERLEKAMP, *Factoring polynomials over finite fields*, Bell System Tech. J. **46** [507]
(1967), 1853–1859.

[BER 1974]   P. BERTHELOT, *Cohomologie cristalline des schémas de caractéristique $p > 0$*, Lecture [136]
Notes in Math., vol. 407, Springer-Verlag, Berlin, 1974.

[BER 1982]   E. R. BERLEKAMP, *Bit-serial Reed–Solomon encoder*, IEEE Trans. Inform. Theory [35]
**IT-28** (1982), 869–874.

[BER 1986]   P. BERTHELOT, *Géométrie rigide et cohomologie des variétés algébriques de carac- [136]
téristique $p$*, Mém. Soc. Math. France (N.S.) $N^\circ 23$ (1986), 3, 7–32, Introductions aux
cohomologies $p$-adiques (Luminy, 1984).

[BER 1998]   D. J. BERNSTEIN, *Detecting perfect powers in essentially linear time*, Math. Comp. **67** [198, 199]
$N^\circ 223$ (1998), 1253–1283.

[BER 2001a]   _____, *Multidigit multiplication for mathematicians*, 2001. [174]
http://cr.yp.to/papers.html

[BER 2001b]   P. BERRIZBEITIA, *Sharpening "primes in P" for a large family of numbers*, preprint, [601]
2001.
http://lanl.arxiv.org/abs/math.NT/0211334

[BER 2002]   D. J. BERNSTEIN, *Pippenger's exponentiation algorithm*, 2002. preprint. [146, 155, 159,
http://cr.yp.to/papers.html                                                                                           166]

[BER 2004a]   _____, *Proving primality in essentially quartic random time*, preprint, 2004. [601]
http://cr.yp.to/papers.html

[BER 2004b]   _____, *Scaled remainder trees*, preprint, 2004. [184]
http://cr.yp.to/papers.html

[BIGNUM]   J.-C. HERVÉ, B. SERPETTE, & J. VUILLEMIN, *BigNum: A portable and efficient* [169]
*package for arbitrary-precision arithmetic*, Tech. report, Digital Paris Research Laboratory, 1989, available via e-mail from librarian@decprl.dec.com.

[BIJO 2003]   O. BILLET & M. JOYE, *The Jacobi model of an elliptic curve and Side-Channel Anal-* [696]
*ysis*, Applicable Algebra, Algebraic Algorithms and Error-Correcting Codes – AAECC
2003, Lecture Notes in Comput. Sci., vol. 2643, Springer-Verlag, Berlin, 2003, 34–42.

[BIME$^+$ 2000]   I. BIEHL, B. MEYER, & V. MÜLLER, *Differential fault attacks on elliptic curve cryp-* [684, 685,
*tosystems*, Advances in Cryptology – Crypto 2000, Lecture Notes in Comput. Sci., vol.   706–708]
1880, Springer-Verlag, Berlin, 2000, 131–146.

[BIR 1968]   B. J. BIRCH, *How the number of points of an elliptic curve over a fixed prime field* [605]
*varies*, J. London Math. Soc. **43** (1968), 57–60.

[BISH 1997]   E. BIHAM & A. SHAMIR, *Differential fault analysis of secret key cryptosystems*, Ad- [683]
vances in Cryptology – Crypto 1997, Lecture Notes in Comput. Sci., vol. 1294, Springer-
Verlag, 1997, 513–525.

[BLA 2002]   G. BLADY, *Die Weil-Restriktion elliptischer Kurven in der Kryptographie*, Master's the- [383]
sis, Universität-Gesamthochschule Essen, 2002.

[BLBL$^+$ 1986]   L. BLUM, M. BLUM, & M. SHUB, *A simple unpredictable pseudo-random number* [721]
*generator*, SIAM J. Comput. **15** (1986), 364–383.

[BLFL]   D. BLEICHENBACHER & A. FLAMMENKAMP, *An efficient algorithm for computing* [158]
*shortest addition chains*.
http://www.uni-bielefeld.de/~achim/ac.dvi

[BLFU$^+$ 1984]   I. F. BLAKE, R. FUJI-HARA, R. C. MULLIN, & S. A. VANSTONE, *Computing* [508]
*logarithms in finite fields of characteristic two*, SIAM J. Algebraic Discrete Methods **5**
$N^\circ 2$ (1984), 276–285.

[BlGa+ 1994a] I. F. BLAKE, S. GAO, & R. J. LAMBERT, *Constructive problems for irreducible polynomials over finite fields*, Proceedings of the 1993 Information Theory and Applications Conference, Lecture Notes in Comput. Sci., vol. 793, Springer-Verlag, Berlin, 1994, 1–23. [217]

[BlGa+ 1994b] I. F. BLAKE, S. GAO, & R. C. MULLIN, *Normal and self dual normal bases from factorization of $cx^{q+1} + dx^q - ax - b$*, SIAM J. Discrete Math. **7** N°3 (1994), 499–512. [35]

[BlGa+ 1996] I. F. BLAKE, S. GAO, & R. J. LAMBERT, *Construction and distribution problems for irreducible trinomials over finite fields*, Applications of Finite Fields, Oxford University Press, New York, 1996, 19–32. [217]

[BlMu+ 1984] I. F. BLAKE, R. C. MULLIN, & S. A. VANSTONE, *Computing logarithms in $\mathbb{F}_{2^n}$*, Advances in Cryptology – Crypto 1984, Lecture Notes in Comput. Sci., vol. 196, Springer-Verlag, 1984, 73–82. [508]

[BlMu+ 2004] I. F. BLAKE, K. MURTY, & G. XU, *Refinements of Miller's algorithm for computing Weil/Tate pairing*, preprint, 2004. http://eprint.iacr.org/2004/065/ [401]

[BlOt+ 2004] J. BLÖMER, M. OTTO, & J.-P. SEIFERT, *Sign change fault attacks on elliptic curve cryptosystems*, preprint, 2004. http://eprint.iacr.org/2004/227/ [708]

[BlRo+ 1998] I. F. BLAKE, R. M. ROTH, & G. SEROUSSI, *Efficient arithmetic in $GF(2^n)$ through palindromic representation*, Tech. Report HPL-98-134, Hewlett–Packard, August 1998. http://www.hpl.hp.com/techreports/98/HPL-98-134.pdf [218, 221]

[BlSe+ 1999] I. F. BLAKE, G. SEROUSSI, & N. P. SMART, *Elliptic curves in cryptography*, London Mathematical Society Lecture Note Series, vol. 265, Cambridge University Press, Cambridge, 1999. [197]

[BlSe+ 2005] ———, *Advances in elliptic curve cryptography*, London Mathematical Society Lecture Note Series, vol. 317, Cambridge University Press, Cambridge, 2005. [687]

[BoBo+ 2005] D. BONEH, X. BOYEN, & E.-J. GOH, *Hierarchical Identity Based encryption with constant size ciphertext*, preprint, 2005. http://eprint.iacr.org/2005/015/ [578]

[BoCo 1990] J. BOS & M. J. COSTER, *Addition chain heuristics*, Advances in Cryptology – Crypto 1989, Lecture Notes in Comput. Sci., vol. 435, Springer-Verlag, Berlin, 1990, 400–407. [162, 163]

[BoDe+ 1997] D. BONEH, R. DEMILLO, & R. LIPTON, *On the importance of checking cryptographic protocols faults*, Advances in Cryptology – Eurocrypt 1997, Lecture Notes in Comput. Sci., vol. 1233, Springer-Verlag, Berlin, 1997, 37–51. [683, 705]

[BoDi+ 2004] I. BOUW, C. DIEM, & J. SCHOLTEN, *Ordinary elliptic curves of high rank over $\overline{\mathbb{F}}_p$ with constant $j$-invariant*, Manuscripta Math. **114** (2004), 487–501. [131]

[BoFr 2001] D. BONEH & M. FRANKLIN, *Identity based encryption from the Weil pairing*, Advances in Cryptology – Crypto 2001, Lecture Notes in Comput. Sci., vol. 2139, Springer-Verlag, Berlin, 2001, 213–229. [395, 576, 583, 589, 590]

[BoFr 2003] ———, *Identity based encryption from the Weil pairing*, SIAM J. Comput. **32** N°3 (2003), 586–615. [576, 578, 583]

[BoGa+ 2004] A. BOSTAN, P. GAUDRY, & É. SCHOST, *Linear recurrences with polynomial coefficients and application to integer factorization and Cartier–Manin operator*, Proceedings of Fq7, Lecture Notes in Comput. Sci., vol. 2948, Springer-Verlag, Berlin, 2004, 40–58. [412]

[BoGo+ 1994] A. BOSSALAERS, R. GOVAERTS, & J. VANDEWALLE, *Comparison of three modular reduction functions*, Advances in Cryptology – Crypto 1993, Lecture Notes in Comput. Sci., vol. 773, Springer-Verlag, Berlin, 1994, 175–186. [179, 182]

[BoLe 1995] W. BOSMA & A. K. LENSTRA, *An implementation of the elliptic curve integer factorization method*, Computational algebra and number theory (W. BOSMA & A. VAN DER POORTEN, eds.), Kluwer Academic Publishers, 1995. [606]

[BoLy+ 2002]   D. Boneh, B. Lynn, & H. Shacham, *Short signatures from the Weil pairing*, [578, 588, 589]
Advances in Cryptology – Asiacrypt 2001, Lecture Notes in Comput. Sci., vol. 2248,
Springer-Verlag, Berlin, 2002, 514–532.

[BoLy+ 2004]   _____, *Short signatures from the Weil pairing*, J. Cryptology **17** (2004), 297–319. [578, 588, 589]

[Boo 1951]   A. D. Booth, *A signed binary multiplication technique*, Quarterly J. Mech. Appl. Math. [151]
**4** (1951), 236–240.

[Bos 2001]   W. Bosma, *Signed bits and fast exponentiation*, J. Théor. Nombres Bordeaux **13** (2001), [151]
27–41.

[BoVe 1996]   D. Boneh & R. Venkatesan, *Hardness of computing the most significant bits of se-* [376, 698]
*cret keys in Diffie–Hellman and related schemes*, Advances in Cryptology – Crypto 1996,
Lecture Notes in Comput. Sci., vol. 1109, Springer-Verlag, Berlin, 1996, 129–142.

[Bra 1939]   A. Brauer, *On addition chains*, Bull. Amer. Math. Soc. **45** (1939), 736–739. [148, 158]

[BrBr 1996]   G. Brassand & P. Bratley, *Fundamentals of Algorithms*, Prentice-Hall, Inc., [4]
Englewood Cliffs NJ, 1996, first published as *Algorithmics — Theory & Practice*, 1988.

[BrCl+ 2004]   É. Brier, C. Clavier, & F. Olivier, *Correlation power analysis with a leakage* [680]
*model*, Cryptographic Hardware and Embedded Systems – CHES 2004, Lecture Notes in
Comput. Sci., vol. 3156, Springer-Verlag, Berlin, 2004, 16–29.

[BrCu+ 1993]   H. Brunner, A. Curiger, & M. Hofstetter, *On computing multiplicative in-* [223]
*verses in $GF(2m)$*, IEEE Trans. on Computers **42** N°8 (1993), 1010–1015.

[BrDé+ 2004]   É. Brier, I. Déchène, & M. Joye, *Unified point addition formulæ for elliptic* [695]
*curve cryptosystems*, Embedded Cryptographic Hardware: Methodologies & Architectures, Nova Science Publishers, 2004.

[Bre 1980]   R. P. Brent, *An improved Monte Carlo factorization algorithm*, BIT **20** (1980), 176– [485]
184. The paper can be obtained as a series of .gif bitmaps from [Brent].

[Brent]   _____, *homepage*, Oxford University Computing Laboratory. [614, 742]
http://web.comlab.ox.ac.uk/oucl/work/richard.brent

[BrGo+ 1993]   E. F. Brickell, D. M. Gordon, K. S. McCurley, & D. B. Wilson, *Fast* [165]
*exponentiation with precomputation*, Advances in Cryptology – Eurocrypt 1992, Lecture
Notes in Comput. Sci., vol. 658, Springer-Verlag, Berlin, 1993, 200–207.

[BrJo 2002]   É. Brier & M. Joye, *Weierstraß elliptic curves and side channels attacks*, Public Key [286]
Cryptography – PKC 2002, Lecture Notes in Comput. Sci., vol. 2274, Springer-Verlag,
2002, 335–345.

[BrJo 2003]   _____, *Fast point multiplication on elliptic curves through isogenies*, Applicable Alge- [282, 695, 704]
bra, Algebraic Algorithms and Error-Correcting Codes – AAECC 2003, Lecture Notes in
Comput. Sci., vol. 2643, Springer-Verlag, Berlin, 2003, 43–50.

[BrKu 1978]   R. P. Brent & H. T. Kung, *Fast algorithms for manipulating formal power series*, [225]
J. Assoc. Comput. Mach. **25** N°4 (1978), 581–595.

[BrKu 1983]   _____, *Systolic VLSI arrays for linear-time GCD computation*, VLSI 1983, Elsevier [205, 223]
Science Publishers B. V., 1983, 145–154.

[BrMy+ 2001]   E. Brown, B. T. Myers, & J. A. Solinas, *Elliptic curves with compact param-* [382, 383]
*eters*, Combinatorics and Optimization Research Report CORR 2001-68, University of
Waterloo, 2001.
http://www.cacr.math.uwaterloo.ca/techreports/2001/corr2001-68.ps

[BrSt 2004]   R. Bröker & P. Stevenhagen, *Elliptic curves with a given number of points*, Al- [567]
gorithmic Number Theory Symposium – ANTS VI, vol. 3076, Springer-Verlag, Berlin,
2004, 117–131.

[Bru 1966]   N. G. de Bruijn, *On the number of positive integers $\leqslant x$ and free of prime factors* [506]
*$> y$, II*, Indag. Math. **38** (1966), 239–247.

[Bru 1994] P. S. Bruckman, *Lucas pseudoprimes are odd*, Fib. Quart. **32** (1994), 155–157.

[BrWe 2004] F. Brezing & A. Weng, *Elliptic curves suitable for pairing based cryptography*, preprint, 2004.
http://eprint.iacr.org/2003/143/

[BrZi 2003] R. P. Brent & P. Zimmermann, *Random number generators with period divisible by a Mersenne prime*, Computational Science and its Applications – ICCSA 2003, vol. 2667, Springer-Verlag, Berlin, 2003, 1–10.

[BuDe 1995] M. Burmester & Y. Desmedt, *A secure and efficient conference key distribution system*, Advances in Cryptology – Eurocrypt 1994, Lecture Notes in Comput. Sci., vol. 950, Springer-Verlag, Berlin, 1995, 275–286.

[BuDe 1997] ———, *Efficient and secure conference key distribution*, Proceedings of the 1996 Workshop on Security Protocols, Lecture Notes in Comput. Sci., vol. 1189, Springer-Verlag, Berlin, 1997, 119–130.

[BuDe 2004] ———, *Identity based key infrastructures*, Proceedings of the IFIP 2004 World Computer Congress, Kluwer Academic Publishers, 2004.

[BuGo$^+$ 1946] A. W. Burks, H. H. Goldstine, & J. von Neumann, *Preliminary discussion of the logical design of an electronic computing instrument*, Tech. Report Princeton, NJ, Institute for Advanced Study, 1946.

[BuJa$^+$ 1997] J. Buchmann, M. J. Jacobson, Jr., & E. Teske, *On some computational problems in finite abelian groups*, Math. Comp. **66** (1997), 1663–1687.

[Bur 1999] D. Bursky, *Flash and EEPROM storage boost 8-bit mcu flexibility*, Electronic Design **47** N°5 (1999).

[BuWi 1988] J. Buchmann & H. C. Williams, *A key-exchange system based on imaginary quadratic fields*, J. Cryptology **1** N°2 (1988), 107–118.

[BuZi 1998] C. Burnikel & J. Ziegler, *Fast recursive division*, Tech. Report MPI-I-98-1-022, Max Planck Institut für Informatik, October 1998.
http://data.mpi-sb.mpg.de/internet/reports.nsf/

[ByDu 2004] B. Byramjee & S. Duquesne, *Classification of genus 2 curves over $\mathbb{F}_{2^n}$ and optimization of their arithmetic*, preprint, 2004.
http://eprint.iacr.org/2004/107/

[CaEr$^+$ 1983] E. R. Canfield, P. Erdős, & C. Pomerance, *On a problem of Oppenheim concerning factorizatio numerorum*, J. Number Theory **17** (1983), 1–28.

[CaFl 1996] J. W. S. Cassels & E. V. Flynn, *Prolegomena to a middlebrow arithmetic of curves of genus 2*, London Mathematical Society Lecture Note Series, vol. 230, Cambridge University Press, 1996.

[CaKo$^+$ 2003] J. Cathalo, F. Koeune, & J.-J. Quisquater, *A new type of timing attack: applications to GPS*, Cryptographic Hardware and Embedded Systems – CHES 2003, Lecture Notes in Comput. Sci., vol. 2779, Springer-Verlag, Berlin, 2003, 291–303.

[Cam 1981] P. Camion, *Factorisation des polynômes de $\mathbb{F}_q[X]$*, Tech. Report RR-0093, INRIA, September 1981, in French.

[Cam 1983] ———, *Improving an algorithm for factoring polynomials over a finite field and constructing large irreducible polynomials*, IEEE Trans. Inform. Theory **29** N°3 (1983), 378–385.

[Camb] Cambridge University, TAMPER Lab homepage.
http://www.cl.cam.ac.uk/Research/Security/tamper/

[Can 1987] D. G. Cantor, *Computing in the Jacobian of a hyperelliptic curve*, Math. Comp. **48** (1987), 95–101.

[CAR 1932] L. CARLITZ, *The arithmetic of a polynomial in a Galois field*, Amer. J. of Math. **54** (1932), 39–50. [37]

[CAR 1994] E. F. CARTER, *The generation and application of random numbers*, Forth Dimensions **XVI** N° 1 & 2 (1994). [720]

[CAR 2003] R. CARLS, *Generalized AGM sequences and approximation of canonical lifts*, September 2003. [139, 440]

[CAS 1991] J. W. S. CASSELS, *Lectures on elliptic curves*, Cambridge University Press, New York, 1991. [270, 275]

[CAV 2000] S. CAVALLAR, *Strategies in filtering in the Number Field Sieve*, Algorithmic Number Theory Symposium – ANTS IV, Lecture Notes in Comput. Sci., no. 1838, Springer-Verlag, 2000, 209–232. [504, 509]

[CAZA 1981] D. G. CANTOR & H. ZASSENHAUS, *A new algorithm for factoring polynomials over finite fields*, Math. Comp. **36** N° 154 (1981), 587–592. [507]

[CES 2005] E. CESENA, *Varietá a traccia zero su campi binari: Applicazioni crittografiche*, Master's thesis, Universitá degli Studi di Milano, 2005. [383]

[CHCH 1999] C.-Y. CHEN & C.-C. CHANG, *Fast modular multiplication algorithm for calculating the product AB modulo N*, Inform. Process. Lett. **72** (1999), 77–81. [202]

[CHCI$^+$ 2004] B. CHEVALLIER-MAMES, M. CIET, & M. JOYE, *Low-cost solutions for preventing simple Side-Channel Analysis: Side-Channel Atomicity*, IEEE Trans. on Computers **53** (2004), 760–768. [690–692]

[CHCO$^+$ 1991] L. S. CHARLAP, R. COLEY, & D. P. ROBBINS, *Enumeration of rational points on elliptic curves over finite fields*, Draft, 1991. [422]

[CHE 2000] Z. CHEN, *Java card technology for smart cards: Architecture and programmers guide*, Addison-Wesley Publishing Company, Reading, MA, 2000. [659]

[CHE 2003] Q. CHENG, *Primality proving via one round ECPP and one iteration in AKS*, preprint, 2003. [601]

[CHHW$^+$ 2004] K. Y. CHOI, J. Y. HWANG, & D. H. LEE, *Efficient ID-based group key agreement with bilinear maps*, Public Key Cryptography – PKC 2004, Lecture Notes in Comput. Sci., vol. 2947, Springer-Verlag, 2004, 130–144. [576]

[CHJU 2003] J. H. CHEON & B. JUN, *A polynomial time algorithm for the braid Diffie–Hellman conjugacy problem*, Advances in Cryptology – Crypto 2003, Lecture Notes in Comput. Sci., vol. 2729, IACR and Springer-Verlag, 2003, 212–225. [15]

[CHYU 2002] Y. CHOIE & D. YUN, *Isomorphism classes of hyperelliptic curves of genus 2 over $\mathbb{F}_q$*, Australasian Conference on Information Security and Privacy – ACISP 2002, Lecture Notes in Comput. Sci., vol. 2384, Springer-Verlag, Berlin, 2002, 190–202. [334, 336]

[CIE 2003] M. CIET, *Aspects of fast and secure arithmetics for elliptic curve cryptography*, PhD. Thesis, Université Catholique de Louvain, 2003. [684]

[CIJO$^+$ 2003] M. CIET, M. JOYE, K. LAUTER, & P. L. MONTGOMERY, *Trading inversions for multiplications in elliptic curve cryptography*, preprint, 2003.
http://eprint.iacr.org/2003/257/ [281, 292]

[CILA$^+$ 2003] M. CIET, T. LANGE, F. SICA, & J.-J. QUISQUATER, *Improved algorithms for efficient arithmetic on elliptic curves using fast endomorphisms*, Advances in Cryptology – Eurocrypt 2003, Lecture Notes in Comput. Sci., vol. 2656, Springer-Verlag, 2003, 388–400. [365, 381, 382]

[CIQU$^+$ 2002] M. CIET, J.-J. QUISQUATER, & F. SICA, *Preventing differential analysis in GLV elliptic curve scalar multiplication*, Cryptographic Hardware and Embedded Systems – CHES 2002, Lecture Notes in Comput. Sci., vol. 2523, Springer-Verlag, Berlin, 2002, 540–550. [713]

[CoFi+ 2001] N. COURTOIS, M. FINIASZ, & N. SENDRIER, *How to achieve a McEliece-based digital signature scheme*, Advances in Cryptology – Asiacrypt 2001, Lecture Notes in Comput. Sci., no. 2248, Springer-Verlag, 2001, 157–174. [15]

[COH] H. COHEN, *Diophantine Equations, p-adic Numbers and L-functions*, Springer-Verlag, to appear. [587]

[COH 2000] ———, *A course in Computational Algebraic Number Theory*, Graduate Texts in Mathematics, vol. 138, Springer-Verlag, Berlin, 2000, fourth edition. [95, 149, 170, 190, 191, 194, 195, 198, 209, 377, 410, 458, 467, 486, 592, 598, 602, 611]

[COH 2005] ———, *Analysis of the flexible window powering algorithm*, J. Cryptology **18** N°1 (2005), 63–76. [153, 154]

[CoJo+ 1992] M. J. COSTER, A. JOUX, B. A. LAMACCHIA, A. M. ODLYZKO, C.-P. SCHNORR, & J. STERN, *Improved low-density subset sum algorithms*, Comput. Complexity **2** (1992), 111–128. [376]

[CoKo+ 2001] J.-S. CORON, P. KOCHER, & D. NACCACHE, *Statistics and secret leakage*, Financial Cryptography – FC 2000, Lecture Notes in Comput. Sci., vol. 1962, Springer-Verlag, 2001, 157–173. [680]

[COL 1969] G. E. COLLINS, *Computing multiplicative inverses in $GF(p)$*, Math. Comp. **23** (1969), 197–200. [205]

[COL 1980] ———, *Lecture notes on arithmetic algorithms*, 1980. University of Wisconsin. [192]

[COLE 1984] H. COHEN & H. W. LENSTRA, JR., *Primality testing and Jacobi sums*, Math. Comp. **42** (1984), 297–330. [599]

[COLE 1987] H. COHEN & A. K. LENSTRA, *Implementation of a new primality test*, Math. Comp. **48** (1987), 103–121. [599]

[COMI+ 1997] H. COHEN, A. MIYAJI, & T. ONO, *Efficient elliptic curve exponentiation*, Information and Communication Security – ICICS 1997, Lecture Notes in Comput. Sci., vol. 1334, Springer-Verlag, Berlin, 1997, 282–290. [153]

[COMI+ 1998] ———, *Efficient elliptic curve exponentiation using mixed coordinates*, Advances in Cryptology – Asiacrypt 1998, Lecture Notes in Comput. Sci., vol. 1514, Springer-Verlag, Berlin, 1998, 51–65. [267, 280, 282, 283, 285, 296, 321, 327, 328]

[CON 2005] S. CONTINI, *FactorWorld: General purpose factoring records*, 2005. http://www.crypto-world.com/FactorRecords.html [7]

[COP 1984] D. COPPERSMITH, *Fast evaluation of logarithms in fields of characteristic two*, IEEE Trans. Inform. Theory **30** N°4 (1984), 587–594. [215, 586, 589]

[COP 1993] ———, *Modifications to the Number Field Sieve*, J. Cryptology **6** (1993), 169–180. [613]

[COR 1999] J.-S. CORON, *Resistance against differential power analysis for elliptic curve cryptosystems*, Cryptographic Hardware and Embedded Systems – CHES 1999, Lecture Notes in Comput. Sci., vol. 1717, Springer-Verlag, Berlin, 1999, 392–302. [678, 680, 682, 699, 700, 711]

[COSH 1997] D. COPPERSMITH & A. SHAMIR, *Lattice attacks on NTRU*, Advances in Cryptology – Eurocrypt 1997, Lecture Notes in Comput. Sci., vol. 1233, Springer-Verlag, Berlin, 1997, 52–61. [15]

[COSTER] M. J. COSTER, *homepage*. http://www.coster.demon.nl/ [162]

[COU 1996] J. M. COUVEIGNES, *Computing l-isogenies with the p-torsion*, Algorithmic Number Theory Symposium – ANTS II, Lecture Notes in Comput. Sci., vol. 1122, Springer-Verlag, 1996, 59–65. [421]

[CR 2003] CRYPTOGRAPHY RESEARCH, INC., *Evaluation of VIA C3 Nehemiah Random Number Generator*, Tech. report, 2003. http://www.cryptography.com/resources/whitepapers/VIA_rng.pdf [721]

[Cra 1992]  R. Crandall, *Method and apparatus for public key exchange in a cryptographic system*, United States Patent 5,159,632, Date: Oct. 27th 1992. [182]

[Cro 2003]  E. S. Croot III, *Smooth numbers in short intervals*, preprint, 2003. [605]
http://www.math.gatech.edu/~ecroot/papers.html

[CrPo 2001]  R. Crandall & C. Pomerance, *Prime numbers, a computational perspective*, Springer-Verlag, Berlin, 2001. [170, 177, 182, 207, 614]

[Dav 2000]  R. Davies, *Hardware random number generators*, New Zealand Statistics Conference, 2000. [721]

[Del 1974]  P. Deligne, *La conjecture de Weil. I*, Inst. Hautes Études Sci. Publ. Math. **43** (1974), 273–307. [135]

[DeLa 2004a]  Y. Desmedt & T. Lange, *Pairing based threshold cryptography improving on Libert, Quisquater, Baek, and Zheng*, preprint, 2004. [578]

[DeLa 2004b]  _____, *Pairing variants of Burmester–Desmedt I and Katz–Yung*, preprint, 2004. [576]

[DeLa+ 2004]  Y. Desmedt, T. Lange, & M. Burmester, *Exponential improvement on Katz–Yung's constant round authenticated group key exchange and tripartite variants*, preprint, 2004. [575, 576]

[Deu 1941]  M. Deuring, *Die Typen der Multiplikatorenringe elliptischer Funktionenkörper*, Abh. Math. Sem. Hansischen Univ. **14** (1941), 197–272. [138, 423]

[DeVe 2002]  J. Denef & F. Vercauteren, *An extension of Kedlaya's algorithm to Artin–Schreier curves in characteristic* 2, Algorithmic Number Theory Symposium – ANTS V, Lecture Notes in Comput. Sci., vol. 2369, Springer-Verlag, Berlin, 2002, 308–323. [453]

[DeVe 2005]  _____, *An extension of Kedlaya's algorithm to hyperelliptic curves in characteristic* 2, J. Cryptology (2005), to appear. [453]

[Dhe 1998]  J.-F. Dhem, *Design of an efficient public key cryptographic library for RISC-based smart cards*, PhD. Thesis, Faculté des sciences appliquées, Laboratoire de microélectronique, Université catholique de Louvain-la-Neuve, Belgique, 1998. [203, 204]
http://users.belgacom.net/dhem/these/these_public.pdf

[DhKo+ 2000]  J.-F. Dhem, F. Koeune, P.-A. Leroux, P. Mestre, J.-J. Quisquater, & J.-L. Willems, *A practical implementation of the timing attack*, Smart Card Research and Advanced Application – CARDIS 1998, Lecture Notes in Comput. Sci., vol. 1820, Springer-Verlag, 2000, 167–182. [674, 705]

[Die 2001]  C. Diem, *A study on theoretical and practical aspects of Weil-restriction of varieties*, PhD. Thesis, Universität Gesamthochschule Essen, 2001. [125, 383, 534, 536, 539]

[Die 2003]  _____, *The GHS-attack in odd characteristic*, J. Ramanujan Math. Soc. **18** N°1 (2003), 1–32. [531, 532, 536, 537]

[Die 2004]  _____, *On the discrete logarithm problem in elliptic curves over non-prime finite fields*, 2004. preprint. [541, 543, 586]

[Die 2005]  _____, *Index calculus in class groups of plane curves of small degree*, preprint, 2005. [535]
http://eprint.iacr.org/2005/119/

[DiHe 1976]  W. Diffie & M. E. Hellman, *New directions in cryptography*, IEEE Trans. Inform. Theory **22** N°6 (1976), 644–654. [xxix, 10]

[DiSc 2003]  C. Diem & J. Scholten, *Cover Attacks – A report for the AREHCC project*, 2003. [383, 539, 540, 557]
http://www.arehcc.com

[Doc 2005]  C. Doche, *Redundant trinomials for finite fields of characteristic* 2, Australasian Conference on Information Security and Privacy – ACISP 2005, Lecture Notes in Comput. Sci., vol. 3574, Springer-Verlag, Berlin, 2005, 122–133. [217]

[Doche]  _____, *homepage*. [217]
http://www.math.u-bordeaux.fr/~cdoche/

[DoLe 1995] B. DODSON & A. K. LENSTRA, *NFS with four large primes: an explosive experiment*, Advances in Cryptology – Crypto 1995, Lecture Notes in Comput. Sci., vol. 963, Springer-Verlag, Berlin, 1995, 372–385. [509, 613]

[DoLe+ 1981] P. DOWNEY, B. LEONG, & R. SETHI, *Computing sequences with addition chains*, SIAM J. Comput. **10** (1981), 638–646. [159]

[DoYu 2003] Y. DODIS & M. YUNG, *Exposure-resilience for free: Hierarchical ID-based encryption case*, IEEE Security in Storage 2003, 2003, 45–52. [578]

[DuEn+ 2005] R. DUPONT, A. ENGE, & F. MORAIN, *Building curves with arbitrary small MOV degree over finite prime fields*, J. Cryptology **18** N°2 (2005), 79–89. [587, 588]

[DuGa+ 1999] I. DUURSMA, P. GAUDRY, & F. MORAIN, *Speeding up the discrete log computation on curves with automorphisms*, Advances in Cryptology – Asiacrypt 1999, Lecture Notes in Comput. Sci., vol. 1716, Springer-Verlag, Berlin, 1999, 103–121. [491]

[DuKa 1990] S. R. DUSSÉ & B. S. KALISKI, JR., *A cryptographic library for the Motorola DSP56000*, Advances in Cryptology – Eurocrypt 1990, Lecture Notes in Comput. Sci., vol. 478, Springer-Verlag, Berlin, 1990, 230–244. [181]

[DuQ 2004] S. DUQUESNE, *Montgomery scalar multiplication for genus 2 curves*, Algorithmic Number Theory Symposium – ANTS VI, Lecture Notes in Comput. Sci., vol. 3076, Springer-Verlag, 2004, 153–168. [328, 334, 697]

[DuWa+ 2003] X. DU, Y. WANG, J. GE, & Y. WANG, *An improved ID-based authenticated group key agreement scheme*, preprint, 2003.
http://eprint.iacr.org/2003/260/ [576]

[Dwo 1960] B. DWORK, *On the rationality of the zeta function of an algebraic variety*, Amer. J. Math. **82** (1960), 631–648. [135, 138, 422]

[Ecu 1998] P. L'ECUYER, *Uniform random number generators*, Proceedings of the 1998 Winter Simulation Conference (1998), 97–104. [719]

[Ecu 2001] ———, *Software for uniform random number generation: Distinguishing the good and the bad*, Proceedings of the 2001 Winter Simulation Conference (2001), 95–105. [719]

[Edi 2003] B. EDIXHOVEN, *Point counting after Kedlaya*, EIDMA-Stieltjes graduate course Leiden, 2003. [452]

[EiLa+ 2003] K. EISENTRÄGER, K. LAUTER, & P. L. MONTGOMERY, *Fast elliptic curve arithmetic and improved Weil pairing evaluation*, Topics in Cryptology – CT-RSA 2003, Lecture Notes in Comput. Sci., vol. 2612, Springer-Verlag, Berlin, 2003, 343–354. [281, 292]

[ElG 1985] T. ELGAMAL, *A public key cryptosystem and a signature scheme based on discrete logarithms*, Advances in Cryptology – Crypto 1984, Lecture Notes in Comput. Sci., vol. 196, Springer-Verlag, Berlin, 1985, 10–18. [154]

[Elk 1991] N. D. ELKIES, *Explicit isogenies*, Draft, 1991. [414, 419]

[Elk 1996] R. ELKENBRACHT-HUIZING, *An implementation of the Number Field Sieve*, Experiment. Math. **5** N°3 (1996), 231–253. [614]

[ElSh 2002] E. EL MAHASSNI & I. E. SHPARLINSKI, *On the uniformity of distribution of congruential generators over elliptic curves*, Sequences and their Applications – SETA 2001, Discrete Mathematics and Theoretical Computer Science, Springer-Verlag, 2002, 257–264. [732]

[Eng 2002] A. ENGE, *Computing discrete logarithms in high-genus hyperelliptic Jacobians in provably subexponential time*, Math. Comp. **71** N°238 (2002), 729–742. [516, 554]

[EnGa 2002] A. ENGE & P. GAUDRY, *A general framework for subexponential discrete logarithm algorithms*, Acta Arith. **102** N°1 (2002), 83–103. [496, 499, 500, 516, 554]

[EnSt 2002] A. ENGE & A. STEIN, *Smooth ideals in hyperelliptic function fields*, Math. Comp. **71** (2002), 1219–1230. [516, 554]

[ENT 1998] K. ENTACHER, *Bad subsequences of well-known linear congruential pseudorandom number generators*, ACM Transactions on Modeling and Computer Simulation **8** N°1 (1998), 61–70. [720, 729]

[ERD 1956] P. ERDŐS, *On pseudoprimes and Carmichael numbers*, Publ. Math. Debrecen **4** (1956), 201–206. [593]

[EsSa+ 1998] A. E. ESCOTT, J. C. SAGER, A. P. L. SELKIRK, & D. TSAPAKIDIS, *Attacking elliptic curve cryptosystems using the parallel Pollard rho method*, CryptoBytes (The technical newsletter of RSA laboratories) **4** N°2 (1998), 15–19.
http://www.rsa.com/rsalabs/pubs/cryptobytes [490, 491]

[FaJo 2003] J.-C. FAUGÈRE & A. JOUX, *Algebraic cryptanalysis of Hidden Field Equations (HFE) using Gröbner bases*, Advances in Cryptology – Crypto 2003, Lecture Notes in Comput. Sci., vol. 2729, IACR and Springer-Verlag, 2003, 44–60. [15]

[FaWa 2004] X. FAN & Y. WANG, *Inversion-free arithmetic on genus 3 hyperelliptic curves*, preprint, 2004.
http://eprint.iacr.org/2004/223/ [348]

[FeGa+ 1999] S. FEISEL, J. VON ZUR GATHEN, & M. A. SHOKROLLAHI, *Normal bases via general Gauß periods*, Math. Comp. **68** N°225 (1999), 271–290. [35]

[FeMa+ 1996] R. FERREIRA, R. MALZAHN, P. MARISSEN, J.-J. QUISQUATER, & T. WILLE, *FAME: A 3rd generation coprocessor for optimising public key cryptosystems in smart card applications*, Smart Card Research and Advanced Application – CARDIS 1996, Stichting Mathematisch Centrum, CWI, Amsterdam, 1996. [204]

[FiGi+ 2002] W. FISCHER, C. GIRAUD, E. W. KNUDSEN, & J.P. SEIFERT, *Parallel scalar multiplication on general elliptic curves over $\mathbb{F}_p$ hedged against non-differential side-channel attacks*, preprint, January 2002.
http://eprint.iacr.org/2002/007/ [288, 697]

[FIPS 140-2] FIPS 140-2, *Security requirements for cryptographic modules*, Federal Information Processing Standards Publication 140-2, 1999.
http://csrc.nist.gov [720]

[FIPS 186-2] FIPS 186-2, *Digital signature standard*, Federal Information Processing Standards Publication 186-2, 2000.
http://csrc.nist.gov [183, 215]

[FIPS 197] FIPS 197, *Advanced encryption standard (AES)*, Federal Information Processing Standards Publication 197, 2001.
http://csrc.nist.gov [2]

[FLOy 2004] S. FLON & R. OYONO, *Fast arithmetic on Jacobians of Picard curves*, Public Key Cryptography – PKC 2004, Lecture Notes in Comput. Sci., vol. 2947, Springer-Verlag, Berlin, 2004, 55–68. [352]

[FLOy+ 2004] S. FLON, R. OYONO, & C. RITZENTHALER, *Fast addition on non-hyperelliptic genus 3 curves*, preprint, 2004.
http://eprint.iacr.org/2004/118/ [352]

[FLSA 1997] P. FLAJOLET & B. SALVY, *The SIGSAM challenges: Symbolic asymptotics in practice*, SIGSAM Bulletin **31** N°4 (1997), 36–47. [412]

[FLY] E. V. FLYNN, *Formulas for the Kummer surface of a genus 2 curve*.
ftp://ftp.liv.ac.uk/pub/genus2/kummer [330]

[FLY 1993] ———, *The group law on the Jacobian of a curve of genus 2*, J. Reine Angew. Math. **439** (1993), 45–69. [329]

[FoGa+ 2000] M. FOUQUET, P. GAUDRY, & R. HARLEY, *An extension of Satoh's algorithm and its implementation*, J. Ramanujan Math. Soc. **15** N°4 (2000), 281–318. [432]

| | | |
|---|---|---|
| [FRE 1998] | G. FREY, *How to disguise an elliptic curve*, Talk at Waterloo workshop on the ECDLP, 1998.<br>http://cacr.math.uwaterloo.ca/conferences/1998/ecc98/slides.html | [125, 383] |
| [FRE 2001] | ———, *Applications of arithmetical geometry to cryptographic constructions*, Proceedings of the 1998 Finite Fields and Applications Conference, Springer, Berlin, 2001, 128–161. | [131, 383] |
| [FREELIP] | A. K. LENSTRA & P. LEYLAND, *Free version of the LIP package*, 1996. | [169] |
| [FRI 2001] | H. R. FRIUM, *The group law on elliptic curves on Hesse form*, Sixth International Conference on Finite Fields and Applications, Springer-Verlag, Berlin, 2001. See also the technical report CORR 2001-09.<br>http://www.cacr.math.uwaterloo.ca/techreports/2001/corr2001-09.ps | [275, 276] |
| [FrKl$^+$ 2004] | J. FRANKE, T. KLEINJUNG, F. MORAIN, & T. WIRTH, *Proving the primality of very large numbers with fast ECPP*, Algorithmic Number Theory Symposium – ANTS VI, vol. 3076, Springer-Verlag, Berlin, 2004, 194–207. | [597] |
| [FrLa 2003] | G. FREY & T. LANGE, *Mathematical background of public key cryptography*, Tech. Report 10, IEM Essen, 2003, To appear in Séminaires et Congrès. | [548] |
| [FrMü$^+$ 1999] | G. FREY, M. MÜLLER, & H.-G. RÜCK, *The Tate pairing and the discrete logarithm applied to elliptic curve cryptosystems*, IEEE Trans. Inform. Theory **45** N°5 (1999), 1717–1719. | [395, 396, 530, 582] |
| [FrRü 1994] | G. FREY & H.-G. RÜCK, *A remark concerning m-divisibility and the discrete logarithm problem in the divisor class group of curves*, Math. Comp. **62** (1994), 865–874. | [336, 530] |
| [FrTa 1991] | A. FRÖHLICH & M. TAYLOR, *Algebraic number theory*, Cambridge Studies in Adv. Math., vol. 27, Cambridge Univ. Press, 1991. | [19] |
| [FUJITSU] | FUJITSU LIMITED, *Fram guide book*.<br>http://www.fujitsu.com/global/services/microelectronics/technical/ | [655] |
| [FUL 1969] | W. FULTON, *Algebraic curves: An introduction to algebraic geometry*, Benjamin, 1969. | [45] |
| [GaGa$^+$ 2000] | S. GAO, J. VON ZUR GATHEN, D. PANARIO, & V. SHOUP, *Algorithms for exponentiation in finite fields*, J. Symbolic Comput. **29** (2000), 879–889. | [35, 226, 227] |
| [GaGe 1996] | J. VON ZUR GATHEN & J. GERHARD, *Arithmetic and factorization of polynomials over $\mathbb{F}_2$*, International Symposium on Symbolic and Algebraic Computation – ISSAC 1996, 1–9. | [220] |
| [GaGe 1999] | ———, *Modern computer algebra*, Cambridge University Press, 1999. | [3] |
| [GaHa 2000] | P. GAUDRY & R. HARLEY, *Counting points on hyperelliptic curves over finite fields*, Algorithmic Number Theory Symposium – ANTS IV, vol. 1838, Springer-Verlag, Berlin, 2000, 313–332. | [422] |
| [GaHa$^+$ 2002] | S. D. GALBRAITH, K. HARRISON, & D. SOLDERA, *Implementing the Tate pairing*, Algorithmic Number Theory Symposium – ANTS V, Lecture Notes in Comput. Sci., vol. 2369, Springer-Verlag, Berlin, 2002, 324–337. | [389, 393, 400, 401, 580, 581, 583, 584] |
| [GaHe$^+$ 2002a] | S. GALBRAITH, F. HESS, & N. SMART, *Extending the GHS Weil-descent attack*, Advances in Cryptology – Eurocrypt 2002, Lecture Notes in Comput. Sci., vol. 2332, Springer-Verlag, Berlin, 2002, 29–44. | [531, 536] |
| [GaHe$^+$ 2002b] | P. GAUDRY, F. HESS, & N. P. SMART, *Constructive and destructive facets of Weil descent on elliptic curves*, J. Cryptology **15** N°1 (2002), 19–46. | [531, 534] |
| [GAL 2001a] | S. D. GALBRAITH, *Supersingular curves in cryptography*, Advances in Cryptology – Asiacrypt 2001, Lecture Notes in Comput. Sci., vol. 2248, Springer-Verlag, Berlin, 2001, 495–513. | [124, 336, 584, 590] |
| [GAL 2001b] | ———, *Weil descent of Jacobians*, Workshop on Coding and Cryptography, 2001, Electronic Notes in Discrete Mathematics, vol. 6, Elsevier Science Publishers, 2001. | [531] |

[GaLa+ 2000] R. P. Gallant, R. J. Lambert, & S. A. Vanstone, *Improving the parallelized Pollard lambda search on anomalous binary curves*, Math. Comp. **69** (2000), 1699–1705. [491]

[GaLa+ 2001] _____, *Faster point multiplication on elliptic curves with efficient endomorphisms*, Advances in Cryptology – Crypto 2001, Lecture Notes in Comput. Sci., vol. 2139, Springer-Verlag, Berlin, 2001, 190–200. [376–378]

[GaLe 1992] S. Gao & H. W. Lenstra, Jr., *Optimal normal bases*, Des. Codes Cryptogr. **2** (1992), 315–323. [217]

[GaMc 2000] S. D. Galbraith & J. McKee, *The probability that the number of points on an elliptic curve over a finite field is a prime*, J. London Math. Soc. (2) **62** N°3 (2000), 671–684. [272]

[GaMc+ 2004] S. D. Galbraith, J. McKee, & P. Valenca, *Ordinary abelian varieties having small embedding degree*, preprint, 2004.
http://eprint.iacr.org/2004/365/ [589]

[GaMo+ 2001] K. Gandolfi, C. Mourtel, & F. Olivier, *Electronic analysis: concrete results*, Cryptographic Hardware and Embedded Systems – CHES 2001, Lecture Notes in Comput. Sci., vol. 2162, Springer-Verlag, Berlin, 2001, 251–261. [682]

[GaNö 2005] J. von zur Gathen & M. Nöcker, *Polynomial and normal bases for finite fields*, J. Cryptology (2005), to appear. [214, 215, 220]

[Gao 2001] S. Gao, *Abelian groups, Gauß periods and normal bases*, Finite Fields Appl. **7** N°1 (2001), 148–164. [35]

[Gar 1959] H. Garner, *The residue number system*, IRE Transactions on Electronic Computers **EC-8** (1959), 140–147. [197]

[GaSc 2004a] P. Gaudry & É. Schost, *Construction of secure random curves of genus 2 over prime fields*, Advances in Cryptology – Eurocrypt 2004, Lecture Notes in Comput. Sci., vol. 3027, Springer-Verlag, 2004, 239–256. [422, 455, 566, 568, 685]

[GaSc 2004b] _____, *A low-memory parallel version of Matsuo, Chao, and Tsujii's algorithm*, Algorithmic Number Theory Symposium – ANTS VI, Lecture Notes in Comput. Sci., vol. 3076, Springer-Verlag, Berlin, 2004, 208–222. [411]

[GaSc 2005] _____, *Modular equations for hyperelliptic curves*, Math. Comp. **74** N°249 (2005), 429–454 (electronic). [422]

[GaSh 1992] J. von zur Gathen & V. Shoup, *Computing Frobenius maps and factoring polynomials (extended abstract)*, ACM Symposium on Theory of Computing, 1992, 97–105. [507]

[GaSm 1999] S. D. Galbraith & N. P. Smart, *A cryptographic application of Weil descent*, Proceedings of the 1999 Cryptography and Coding Conference, Lecture Notes in Comput. Sci., vol. 1746, Springer-Verlag, Berlin, 1999, 191–200. A version is available as HP Technical report HPL-1999-70. [531]

[GaTh+ 2004] P. Gaudry, N. Thériault, & E. Thomé, *A double large prime variation for small genus hyperelliptic index calculus*, preprint, 2004.
http://eprint.iacr.org/2004/153/ [523, 525, 554]

[Gau 1973] C. F. Gauss, *Werke*, Georg Olms Verlag, 1973, in German. [434]

[Gau 2000a] P. Gaudry, *Algorithmique des courbes hyperelliptiques et applications à la cryptologie*, PhD. Thesis, École polytechnique, 2000.
http://www.lix.polytechnique.fr/Labo/Pierrick.Gaudry/publis/ [505]

[Gau 2000b] _____, *An algorithm for solving the discrete log problem on hyperelliptic curves*, Advances in Cryptology – Eurocrypt 2000, vol. 1807, Springer-Verlag, Berlin, 2000, 19–34. [505, 517, 554]

[Gau 2002] _____, *A comparison and a combination of SST and AGM algorithms for counting points of elliptic curves in characteristic* 2, Advances in Cryptology – Asiacrypt 2002, Lecture Notes in Comput. Sci., vol. 2501, Springer-Verlag, Berlin, 2002, 311–327. [433, 441]

[Gau 2004] _____, *Index calculus for abelian varieties and the elliptic curve discrete logarithm problem*, preprint, 2004. [541, 586]
http://eprint.iacr.org/2004/073/

[Ger 1983] J. L. Gerver, *Factoring large numbers with a quadratic sieve*, Math. Comp. **41** (1983), 287–294. [508]

[GeSi 2002] C. Gentry & A. Silverberg, *Hierarchical ID-based encryption*, Advances in Cryptology – Asiacrypt 2002, Lecture Notes in Comput. Sci., no. 2501, Springer-Verlag, 2002, 548–566. [578]

[GeSm 2003] K. Geissler & N. P. Smart, *Computing the $M = UU^t$ integer matrix decomposition*, Proceedings of the 2003 Cryptography and Coding Conference, Lecture Notes in Comput. Sci., vol. 2898, Springer-Verlag, 2003, 223–233. [15]

[Gie 2001] E.-G. Giessmann, *Ein schneller Algorithmus zur Punktevervielfachung, der gegen Seitenkanalattacken resistent ist*, talk at Workshop über Theoretische und praktische Aspekte von Kryptographie mit Elliptischen Kurven, Berlin, 2001. [689, 711]

[GiTh 2004] C. Giraud & H. Thiebeauld, *A survey on fault attacks*, Smart Card Research and Advanced Application – CARDIS 2004, Kluwer Academic Publishers, 2004, 159–176. [684]

[GMP] Free Software Foundation, *GNU MP library, version 4.1.4*, 2004. [169, 176]
http://www.swox.com/gmp/

[GoBe+ 2000] G. Gong, T. A. Berson, & D. R. Stinson, *Elliptic curve pseudorandom sequence generators*, Selected Areas in Cryptography – SAC 1999, Lecture Notes in Comput. Sci., vol. 1758, Springer-Verlag, Berlin, 2000, 34–48. [732]

[GoCh 2000] J. Goodman & A. Chandrasekaran, *An energy efficient reconfigurable public key cryptography processor architecture*, Cryptographic Hardware and Embedded Systems – CHES 2000, Lecture Notes in Comput. Sci., vol. 1965, Springer-Verlag, Berlin, 2000, 174–191. [644]

[GoHa+ 1996] D. Gollman, Y. Han, & C. Mitchell, *Redundant integer representations and fast exponentiation*, Des. Codes Cryptogr. **7** (1996), 135–151. [160]

[GoLa 2002] G. Gong & C. C. Y. Lam, *Recursive sequences over elliptic curves*, Sequences and their Applications – SETA 2001, Discrete Mathematics and Theoretical Computer Science, Springer-Verlag, 2002, 182–196. [732]

[GoLe+ 1994] R. A. Golliver, A. K. Lenstra, & K. S. McCurley, *Lattice sieving and trial division*, Algorithmic Number Theory Symposium – ANTS I, Lecture Notes in Comput. Sci., vol. 877, Springer-Verlag, Berlin, 1994, 18–27. [614]

[GoMa+ 2005] M. Gonda, K. Matsuo, K. Aoki, J. Chao, & S. Tsuji, *Improvements of addition algorithm on genus 3 hyperelliptic curves and their implementations*, IEICE Trans. Fundamentals **E88-A** N°1 (2005), 89–96. [348]

[GoMc 1993] D. M. Gordon & K. S. McCurley, *Massively parallel computation of discrete logarithms*, Advances in Cryptology – Crypto 1992, Lecture Notes in Comput. Sci., vol. 740, Springer-Verlag, Berlin, 1993, 312–323. [215, 507]

[Gor 1989] J. Gordon, *Fast multiplicative inverse in modular arithmetic*, Proceedings of the 1986 Cryptography and Coding Conference, Oxford University Press, New York, 1989, 269–279. [192]

[Gor 1998] D. M. Gordon, *A survey of fast exponentiation methods*, J. Algorithms **27** N°1 (1998), 129–146. [146]

[Gou 2003] L. Goubin, *A refined power analysis attack on elliptic curve cryptosystems*, Public Key Cryptography – PKC 2003, Lecture Notes in Comput. Sci., vol. 2567, Springer-Verlag, Berlin, 2003, 199–210. [680–682, 700]

[Gra 1998] J. Grantham, *A probable prime test with high confidence*, J. Number Theory **72** (1998), 32–47, MR 2000e:11160. [596]

[GRA 2001] ———, *Frobenius pseudoprimes*, Math. Comp. **70** (2001), 873–891. [595]

[GRA 2004] T. GRANLUND, *The GNU Multiple Precision arithmetic library, version 4.1.4*, 2004. http://www.swox.com/gmp [176, 178]

[GRHA 1978] P. GRIFFITHS & J. HARRIS, *Principles of algebraic geometry*, John Wiley & Sons, Ltd., 1978. [88, 90]

[GRHE+ 2004] P. GRABNER, C. HEUBERGER, & H. PRODINGER, *Distribution results for low-weight binary representations for pairs of integers*, Theoret. Comput. Sci. **319** (2004), 307–331. [156]

[GRKN+ 1994] R. L. GRAHAM, D. E. KNUTH, & O. PATASHNIK, *Concrete Mathematics, 2nd edition*, Addison-Wesley Publishing Company, Reading, MA, 1994, first edition 1989. [4]

[GRO 1968] A. GROTHENDIECK, *Crystals and the de Rham cohomology of schemes*, Dix Exposés sur la Cohomologie des Schémas, North-Holland, Amsterdam, 1968, 306–358. [136]

[GRO 2001] J. GROSSSCHÄDL, *A bit-serial unified multiplier architecture for finite fields $GF(p)$ and $GF(2^m)$*, Cryptographic Hardware and Embedded Systems – CHES 2001, Lecture Notes in Comput. Sci., Springer-Verlag, 2001, 202–219. [644]

[GRPO 2002] A. GRANVILLE & C. POMERANCE, *Two contradictory conjectures concerning Carmichael numbers*, Math. Comp. **71** (2002), 883–908. [593]

[GUKA+ 2004] C. GUYOT, K. KAVEH, & V. M. PATANKAR, *Explicit algorithm for the arithmetic on the hyperelliptic Jacobians of genus 3*, J. Ramanujan Math. Soc. **19** (2004), 119–159. [348]

[GÜLA+ 2000] C. GÜNTER, T. LANGE, & A. STEIN, *Speeding up the arithmetic on Koblitz curves of genus two*, Selected Areas in Cryptography – SAC 2000, Lecture Notes in Comput. Sci., vol. 2012, Springer-Verlag, Berlin, 2000, 106–117. [367, 368, 374]

[GUPA 1997] J. GUAJARDO & C. PAAR, *Efficient algorithms for elliptic curve cryptosystems*, Advances in Cryptology – Crypto 97, Lecture Notes in Comput. Sci., vol. 1294, Springer-Verlag, Berlin, 1997, 342–356. [296]

[HAL 1994] S. HALLGREN, *Linear congruential generators over elliptic curves*, Tech. Report CS-94-143, Carnegie Mellon Univ., 1994. [732]

[HALó+ 2000] D. HANKERSON, J. LÓPEZ, & A. J. MENEZES, *Software implementation of elliptic curve cryptography over binary fields*, Cryptographic Hardware and Embedded Systems – CHES 2000, Lecture Notes in Comput. Sci., vol. 1965, Springer-Verlag, Berlin, 2000, 1–24. [267, 291]

[HAME+ 2003] D. HANKERSON, A. J. MENEZES, & S. A. VANSTONE, *Guide to elliptic curve cryptography*, Springer-Verlag, Berlin, 2003. [172, 215, 219, 223, 231, 234, 267, 301, 570, 571]

[HAMO 2002] J. HA & S. MOON, *Randomized signed-scalar multiplication of ECC to resist power attacks*, Cryptographic Hardware and Embedded Systems – CHES 2002, Lecture Notes in Comput. Sci., vol. 2523, Springer-Verlag, Berlin, 2002, 551–563. [699]

[HAN 1959] W. HANSEN, *Zum Scholz–Brauerschen Problem*, J. Reine Angew. Math. **202** (1959), 129–136, in German. [158]

[HAQU+ 2002] G. HANROT, M. QUERCIA, & P. ZIMMERMANN, *Chronométrages d'algorithmes multiprécision*, janvier 2002. http://www.pauillac.inria.fr/~quercia/papers/mesures2.tar [187, 188]

[HAQU+ 2004] ———, *The middle product algorithm, I*, Appl. Algebra Engrg. Comm. Comput. **14** N°6 (2004), 415–438. [188]

[HAR 1960] B. HARRIS, *Probability distributions related to random mappings*, Ann. of Math. Statistics **31** (1960), 1045–1062. [483]

[HAR 1977] R. HARTSHORNE, *Algebraic Geometry*, Graduate Texts in Mathematics, vol. 52, Springer-Verlag, 1977. [50]

[HAR 2000]   R. HARLEY, *Fast arithmetic on genus 2 curves*, 2000.
See http://cristal.inria.fr/~harley/hyper for C source code and further explanations.

[HAR 2002a]   _____, *Algorithmes avancés pour l'arithmétique des courbes (Advanced algorithms for arithmetic on curves)*, PhD. Thesis, Université Paris 7, 2002.

[HAR 2002b]   _____, *Asymptotically optimal p-adic point-counting*, e-mail to NMBRTHRY list, December 2002.

[HAR 2005]   G. HARMAN, *On the number of Carmichael numbers up to $x$*, Bull. London Math. Soc. (2005), to appear.

[HAS 2000]   M. A. HASAN, *Power analysis attacks and algorithmic approaches to their countermeasures for Koblitz curve cryptosystems*, Cryptographic Hardware and Embedded Systems – CHES 2000, Lecture Notes in Comput. Sci., vol. 1965, Springer-Verlag, Berlin, 2000, 93–108.

[HEN 1908]   K. HENSEL, *Theorie der algebraischen Zahlen*, Leipzig, 1908.

[HEN 1961]   H. C. HENDRICKSON, *Fast high-accuracy binary parallel addition*, IRE Trans. Electronic Computers **10** (1961), 465–468.

[HES 2003]   F. HESS, *The GHS attack revisited*, Advances in Cryptology – Eurocrypt 2003, Lecture Notes in Comput. Sci., vol. 2656, Springer-Verlag, Berlin, 2003, 374–387.

[HES 2004]   _____, *A note on the Tate pairing of curves over finite fields*, Arch. Math. (Basel) **82** (2004), 28–32.

[HeSe$^+$ 2001]   F. HESS, G. SEROUSSI, & N. P. SMART, *Two topics in hyperelliptic cryptography*, Selected Areas in Cryptography – SAC 2000, Lecture Notes in Comput. Sci., vol. 2259, Springer-Verlag, Berlin, 2001, 181–189.

[HeSh 2005]   F. HESS & I. E. SHPARLINSKI, *On the linear complexity and multidimensional distribution of congruential generators over elliptic curves*, Des. Codes Cryptogr. **35** (2005), 111–117.

[HiMo 2002]   Y. HITCHCOCK & P. MONTAGUE, *A new elliptic curve scalar multiplication algorithm to resist simple power analysis*, Australasian Conference on Information Security and Privacy – ACISP 2002, Lecture Notes in Comput. Sci., vol. 2384, Springer-Verlag, Berlin, 2002, 214–225.

[HiTa 2000]   A. HIGUCHI & N. TAKAGI, *A fast addition algorithm for elliptic curve arithmetic in $GF(2^n)$ using projective coordinates*, Inform. Process. Lett. **76** (2000), 101–103.

[HoHo$^+$ 2003]   J. HOFFSTEIN, N. HOWGRAWE-GRAHAM, J. PIPHER, J. H. SILVERMAN, & W. WHYTE, *NTRUSign: Digital Signatures Using the NTRU Lattice*, Topics in Cryptology – CT-RSA 2003, Lecture Notes in Comput. Sci., vol. 2612, Springer-Verlag, Berlin, 2003, 122–140.

[HOL 2003]   A. J. HOLT, *On computing Discrete Logarithms: Large prime(s) variants*, PhD. Thesis, University of Bath, 2003.

[HoOh$^+$ 1996]   S.-M. HONG, S.-Y. OH, & H. YOON, *New modular multiplication algorithms for fast modular exponentiation*, Advances in Cryptology – Eurocrypt 1996, Lecture Notes in Comput. Sci., vol. 1070, Springer-Verlag, Berlin, 1996, 166–177.

[HoPi$^+$ 1998]   J. HOFFSTEIN, J. PIPHER, & J. H. SILVERMAN, *NTRU: a ring-based public key cryptosystem*, Algorithmic Number Theory Symposium – ANTS III, Lecture Notes in Comput. Sci., vol. 1423, Springer-Verlag, Berlin, 1998, 267–288.

[HoSm 2001]   N. G. HOWGRAVE-GRAHAM & N. P. SMART, *Lattice attacks on digital signature schemes*, Des. Codes Cryptogr. **23** (2001), 283–290.

[HOW 2001]   E. HOWE, *Isogeny classes of abelian varieties with no principal polarizations*, Moduli of abelian varieties (Texel Island, 1999), Progr. Math., vol. 195, Birkhäuser, Basel, 2001, 203–216.

[HuIe 1998]    M.-D. Huang & D. Ierardi, *Counting points on curves over finite fields*, J. Symbolic Comput. **25** N°1 (1998), 1–21. [422]

[HuPa 1998]    X. Huang & V. Y. Pan, *Fast rectangular matrix multiplication and applications*, J. Complexity **14** N°2 (1998), 257–299. [251]

[IbKi 1975]    O. H. Ibarra & C. E. Kim, *Fast approximation algorithms for the knapsack and sum of subsets problems*, J. of the ACM **22** (1975), 463–468. [14]

[Igu 1960]     J.-I. Igusa, *Arithmetic variety of moduli for genus two*, Ann. of Math. (2) **72** (1960), 612–649. [101]

[IiMa⁺ 2002]   T. Iijima, K. Matsuo, J. Chao, & S. Tsujii, *Construction of Frobenius maps of twists elliptic curves and its application to elliptic scalar multiplication*, Symposium on Cryptography and Information Security – SCIS 2002. [378]

[InTa⁺ 1982]   I. Ingemarsson, D. T. Tang, & C. W. Wong, *A conference key distribution system*, IEEE Trans. Inform. Theory **28** (1982), 714–720. [575]

[Intel 8051]   Literature Department Intel Corporation, *Microcontroller handbook*. [721]

[ISO 1995]     ISO group, *International Standard ISO 7810 Identification cards — Physical characteristics*, Tech. report, ISO/IEC Copyright Office, 1995. [648]

[ISO 1999a]    _____, *Part 1: Physical characteristics, International Standard ISO/IEC 7816: Identification cards — Integrated circuit(s) cards with contacts*, Tech. report, ISO/IEC Copyright Office, 1995-99. [648, 659–661]

[ISO 1999b]    _____, *Part 2: Dimensions and location of the contacts, International Standard ISO/IEC 7816: Identification cards — Integrated circuit(s) cards with contacts*, Tech. report, ISO/IEC Copyright Office, 1995-99. [648, 649, 659–661]

[ISO 1999c]    _____, *Part 3: Electronic signals and transmission protocols, International Standard ISO/IEC 7816: Identification cards — Integrated circuit(s) cards with contacts*, Tech. report, ISO/IEC Copyright Office, 1995-99. [648, 650, 659–661]

[ISO 1999d]    _____, *Part 4: Interindustry commands for interchange, International Standard ISO/IEC 7816: Identification cards — Integrated circuit(s) cards with contacts*, Tech. report, ISO/IEC Copyright Office, 1995-99. [648, 657, 659–661]

[ISO 2000]     _____, *International Standard ISO/IEC 14443: Identification cards — Contactless integrated circuit(s) cards — Proximity cards*, Tech. report, ISO/IEC Copyright Office, 2000. [650, 659, 662]

[ISO 2000a]    _____, *Part 1: Physical characteristics, International Standard ISO/IEC 14443: Identification cards — Contactless integrated circuit(s) cards - Proximity cards*, Tech. report, ISO/IEC Copyright Office, 2000. [659, 662]

[ISO 2000b]    _____, *Part 2: Radio frequency power and signal interface, International Standard ISO/IEC 14443: Identification cards — Contactless integrated circuit(s) cards — Proximity cards*, Tech. report, ISO/IEC Copyright Office, 2000. [659, 662]

[ISO 2000c]    _____, *Part 3: Initialization and anticollision, International Standard ISO/IEC 14443: Identification cards — Contactless integrated circuit(s) cards — Proximity cards*, Tech. report, ISO/IEC Copyright Office, 2000. [659, 662]

[ISO 2000d]    _____, *Part 4: Transmission protocol, International Standard ISO/IEC 14443: Identification cards — Contactless integrated circuit(s) cards — Proximity cards*, Tech. report, ISO/IEC Copyright Office, 2000. [659, 662]

[ItTs 1988]    T. Itoh & S. Tsujii, *A fast algorithm for computing multiplicative inverses in $GF(2^m)$ using normal bases*, Inform. and Comp. **78** N°3 (1988), 171–177. [225, 234]

[ItTs 1989]    _____, *Structure of parallel multipliers for a class of fields $GF(2^m)$*, Inform. and Comp. **83** (1989), 21–40. [217]

[ItYa⁺ 2002]   K. Itoh, J. Yajima, M. Takaneka, & N. Torii, *DPA countermeasures by improving the window method*, Cryptographic Hardware and Embedded Systems – CHES 2002, Lecture Notes in Comput. Sci., vol. 2523, Springer-Verlag, Berlin, 2002, 303–317. [699]

# References

[IzTa 2002a]   T. Izu & T. Takagi, *Fast elliptic curve multiplication with SIMD operations*, Information and Communication Security – ICICS 2002, Lecture Notes in Comput. Sci., vol. 2513, Springer-Verlag, 2002, 217–230. [288, 690]

[IzTa 2002b]   _____, *A fast parallel elliptic curve multiplication resistant against Side-Channel Attacks*, Public Key Cryptography – PKC 2002, Lecture Notes in Comput. Sci, vol. 2274, Springer-Verlag, Berlin, 2002, 280–296. [697]

[IzTa 2003a]   _____, *Efficient computations of the Tate pairing for the large MOV degrees*, Information Security and Cryptology – ICISC 2002, Lecture Notes in Comput. Sci., vol. 2587, Springer-Verlag, Berlin, 2003, 283–297. [401]

[IzTa 2003b]   _____, *Exceptional procedure attack on elliptic curve cryptosystems*, Public Key Cryptography – PKC 2003, Lecture Notes in Comput. Sci, vol. 2567, Springer-Verlag, Berlin, 2003, 224–239. [695, 704]

[JaMe+ 2001]   M. J. Jacobson, Jr., A. J. Menezes, & A. Stein, *Solving elliptic curve discrete logarithm problems using Weil descent*, J. Ramanujan Math. Soc. **16** (2001), 231–260. [531]

[Jeb 1993a]   T. Jebelean, *An algorithm for exact division*, J. Symbolic Comput. **15** N°2 (1993), 169–180. [189, 190]

[Jeb 1993b]   _____, *Improving the multiprecision Euclidean algorithm*, Design and Implementation of Symbolic Computation Systems – DISCO 1993, Lecture Notes in Comput. Sci., vol. 722, Springer-Verlag, Berlin, 1993, 45–58. [192]

[JoLe 2002]   A. Joux & R. Lercier, *The function field sieve is quite special*, Algorithmic Number Theory Symposium – ANTS V, Lecture Notes in Comput. Sci., vol. 2369, Springer-Verlag, 2002, 431–445. [589]

[JoNg 2003]   A. Joux & K. Nguyen, *Separating decision Diffie–Hellman from Diffie–Hellman in cryptographic groups*, J. Cryptology **16** (2003), 239–247. [574]

[JoQu 2001]   M. Joye & J.-J. Quisquater, *Hessian elliptic curves and side-channel attacks*, Cryptographic Hardware and Embedded Systems – CHES 2001, Lecture Notes in Comput. Sci., vol. 2162, Springer-Verlag, Berlin, 2001, 402. [275, 276, 676, 696]

[JoQu+ 2002]   M. Joye, J.-J. Quisquater, S.-M. Yen, & M. Yung, *Observability analysis – detecting when improved cryptosystems fail*, Topics in Cryptology – CT-RSA 2002, Lecture Notes in Comput. Sci., vol. 2271, Springer-Verlag, Berlin, 2002, 17–29. [708]

[JoTy 2002]   M. Joye & C. Tymen, *Protections against differential analysis for elliptic curve cryptography – an algebraic approach*, Cryptographic Hardware and Embedded Systems – CHES 2001, Lecture Notes in Comput. Sci., vol. 2162, Springer-Verlag, Berlin, 2002, 377–390. [680, 700–702, 712]

[Jou 2000]   A. Joux, *A one round protocol for tripartite Diffie–Hellman*, Algorithmic Number Theory Symposium – ANTS IV, Lecture Notes in Comput. Sci., vol. 1838, Springer-Verlag, 2000, 385–394. [13, 574, 583, 584]

[Jou 2004]   _____, *A one round protocol for tripartite Diffie–Hellman*, J. Cryptology **17** (2004), 263–276. [574, 583]

[Joy 2004]   M. Joye, *Smart-card implementation of elliptic curve cryptography and DPA-type attacks*, Smart Card Research and Advanced Application – CARDIS 2004, Kluwer Academic Publishers, 2004, 114–125. [680]

[Joy 2005]   _____, *Defenses against side-channel analysis*, Advances in Elliptic Curve Cryptography (I. F. Blake, G. Seroussi, & N. P. Smart, eds.), Cambridge University Press, 2005. [687]

[JoYe 2000]   M. Joye & S. M. Yen, *Optimal left-to-right binary signed-digit recoding*, IEEE Trans. on Computers **49** N°7 (2000), 740–748. [151, 152, 401, 676]

[JuMe+ 1990]   D. Jungnickel, A. J. Menezes, & S. A. Vanstone, *On the number of self dual basis of $GF(q^m)$ over $GF(q)$*, Proc. Amer. Math. Soc. **109** (1990), 23–29. [35]

[Jun 1993] D. Jungnickel, *Finite fields*, B.I.-Wissenschaftsverlag, Manheim, Leipzig, Wien, Zürich, 1993. [201, 230]

[JuVa 1996] M. Just & S. Vaudenay, *Authenticated multi-party key agreement*, Advances in Cryptology – Asiacrypt 1996, Lecture Notes in Comput. Sci., vol. 1163, Springer-Verlag, 1996, 36–49. [575]

[KaKi+ 2004] M. Katagi, I. Kitamura, T. Akishita, & T. Takagi, *Novel efficient implementations of hyperelliptic curve cryptosystems using degenerate divisors*, Workshop on Information Security Applications – WISA 2004, Lecture Notes in Comput. Sci., vol. 3325, Springer-Verlag, Berlin, 2004, 347–361. [675, 704]

[Kal 1986] B. S. Kaliski, Jr., *A pseudorandom bit generator based on elliptic logarithms*, Advances in Cryptology – Crypto 1986, Lecture Notes in Comput. Sci., vol. 293, Springer-Verlag, Berlin, 1986, 84–103. [735]

[Kal 1995] _____, *The Montgomery inverse and its applications*, IEEE Trans. on Computers **44** N°8 (1995), 1064–1065. [207, 646]

[KaLu 1982] G. C. Kato & S. Lubkin, *Zeta matrices of elliptic curves*, J. Number Theory **15** N°3 (1982), 318–330. [138, 423]

[Kam 1991] W. Kampkötter, *Explizite Gleichungen für Jacobische Varietäten hyperelliptischer Kurven*, PhD. Thesis, Universität Gesamthochschule Essen, 1991. [104, 422]

[KaMa 1997] A. H. Karp & P. Markstein, *High-precision division and square root*, ACM Trans. Math. Software **23** N°4 (1997), 561–589. [249]

[Kan 1993] G. Kanji, *100 statistical tests*, Sage Publications, 1993. [728]

[KaOf 1962] A. A. Karatsuba & Yu. Ofman, *Multiplication of multiplace numbers on automata*, Dokl. Acad. Nauk SSSR **145** N°2 (1962), 293–294. [176]

[KaOf 1963] _____, *Multiplication of multidigit numbers on automata*, Soviet Physics-Doklady **7** (1963), 595–596. [244]

[Kar 1995] A. A. Karatsuba, *The complexity of computations*, Trudy Mat. Inst. Steklov. translated in Proc. Steklov Inst. Math. **211** (1995), 186–202. [176]

[KaYu 2003] J. Katz & M. Yung, *Scalable protocols for authenticated group key exchange*, Advances in Cryptology – Crypto 2003, Lecture Notes in Comput. Sci., vol. 2729, Springer-Verlag, 2003, 110–125. [575]

[Ked 2001] K. S. Kedlaya, *Counting points on hyperelliptic curves using Monsky–Washnitzer cohomology*, J. Ramanujan Math. Soc. **16** (2001), 323–338. [138, 261, 433, 449, 452, 453, 567]

[Kin 2001] B. King, *An improved implementation of elliptic curves over $GF(2)$ when using projective point arithmetic*, Selected Areas in Cryptography – SAC 2001, Lecture Notes in Comput. Sci., vol. 2259, Springer-Verlag, Berlin, 2001, 134–150. [221]

[KiPa+ 1999] A. Kipnis, J. Patarin, & L. Goubin, *Unbalanced oil and vinegar signature schemes*, Advances in Cryptology – Eurocrypt 1999, Lecture Notes in Comput. Sci., vol. 1592, Springer-Verlag, 1999, 206–222. Extended version: [KiPa+ 2003]. [15, 756]

[KiPa+ 2002] H. Y. Kim, J. Y. Park, J. H. Cheon, J. H. Park, J. H. Kim, & S. G. Hahn, *Fast elliptic curve point counting using Gaussian Normal Basis*, Algorithmic Number Theory Symposium – ANTS V, Lecture Notes in Comput. Sci., vol. 2369, Springer-Verlag, 2002, 292–307. [240, 433]

[KiPa+ 2003] A. Kipnis, J. Patarin, & L. Goubin, *Unbalanced oil and vinegar signature schemes*, extended version of [KiPa+ 1999], 2003. http://citeseer.nj.nec.com/231623.html [15, 756]

[Kno 1975] J. Knopfmacher, *Abstract analytic number theory*, North–Holland Mathematical Library, vol. 12, North–Holland, Amsterdam, 1975. [496]

[KNPA 1981]  D. E. KNUTH & C. H. PAPADIMITRIOU, *Duality in addition chains*, Bull. Eur. Assoc. Theoret. Comput. Sci. **13** (1981), 2–4. [159]

[KNU 1981]  D. E. KNUTH, *The art of computer programming. Vol. 2, Seminumerical algorithms*, second ed., Addison-Wesley Publishing Company, Reading, MA, 1981, Addison-Wesley Series in Computer Science and Information Processing. [165]

[KNU 1997]  _____, *The art of computer programming. Vol. 2, Seminumerical algorithms*, third ed., Addison-Wesley Publishing Company, Reading, MA, 1997, Addison-Wesley Series in Computer Science and Information Processing. [146, 160, 162, 170, 177, 185, 226, 480, 484, 602, 716, 717, 720]

[KNU 1999]  E. W. KNUDSEN, *Elliptic scalar multiplication using point halving*, Advances in Cryptology – Asiacrypt 1999, Lecture Notes in Comput. Sci., vol. 1716, Springer-Verlag, Berlin, 1999, 135–149. [229, 299, 300]

[KOAC 1998]  Ç. K. KOÇ & T. ACAR, *Montgomery multiplication in $GF(2^k)$*, Des. Codes Cryptogr. **14** N°1 (1998), 57–69. [218, 644]

[KOAC$^+$ 1996]  Ç. K. KOÇ, T. ACAR, & B. S. KALISKI, JR., *Analyzing and comparing Montgomery multiplication algorithms*, IEEE Micro **16** N°3 (1996), 26–33. [205, 639]

[KOB 1989]  N. KOBLITZ, *Hyperelliptic cryptosystems*, J. Cryptology **1** (1989), 139–150. [308, 368]

[KOB 1992]  _____, *CM-curves with good cryptographic properties*, Advances in Cryptology – Crypto 1991, Lecture Notes in Comput. Sci., vol. 576, Springer-Verlag, Berlin, 1992, 279–287. [356, 358, 375]

[KOB 1994]  _____, *A course in Number Theory and Cryptography*, Graduate Texts in Mathematics, vol. 114, Springer-Verlag, 1994, second edition. [212]

[KOC 1996]  P. KOCHER, *Timings attacks on implementations of Diffie–Hellman, RSA, DSS and other systems*, Advances in Cryptology – Crypto 1996, Lecture Notes in Comput. Sci., vol. 1109, Springer-Verlag, Berlin, 1996, 104–113. [674, 705]

[KOE 2001]  F. KOEUNE, *Careful design and integration of cryptographic primitives, with contribution to timing attack, padding schemes and random number generators*, PhD. Thesis, Katholieke Universiteit Leuven, 2001.
http://www.dice.ucl.ac.be/~fkoeune/thesis.ps.gz [674]

[KOH 2003]  D. R. KOHEL, *The $AGM$-$X_0(N)$ Heegner point lifting algorithm and elliptic curve point counting*, Advances in Cryptology Proceedings – Asiacrypt 2003, Lecture Notes in Comput. Sci., vol. 2894, Springer-Verlag, Berlin, 2003, 124–136. [442]

[KOJA$^+$ 1999]  P. KOCHER, J. JAFFE, & B. JUN, *Differential power analysis*, Advances in Cryptology – Crypto 1999, Lecture Notes in Comput. Sci., vol. 1666, Springer-Verlag, Berlin, 1999, 388–397. [675, 680, 704]

[KÖKU 1999]  O. KÖMMERLING & M. G. KUHN, *Design principles for tamper-resistant smartcard processors*, Proceedings of the 1999 USENIX Workshop on Smartcard Technology, 1999, 9–20. [670]

[KOLE$^+$ 2000]  K. H. KO, S. J. LEE, J. H. CHEON, J. W. HAN, J. KANG, & C. PARK, *New public key cryptosystem using braid groups*, Advances in Cryptology – Crypto 2000, Lecture Notes in Comput. Sci., vol. 1880, Springer-Verlag, Berlin, 2000, 166–183. [15]

[KOMO$^+$ 1999]  T. KOBAYASHI, H. MORITA, K. KOBAYASHI, & F. HOSHINO, *Fast elliptic curve algorithm combining Frobenius map and table reference to adapt to higher characteristic*, Theory and Application of Cryptographic Techniques, 1999, 176–189. [237]

[KOR 2002]  I. KOREN, *Computer arithmetic algorithms*, A. K. Peters, 2002. [618, 638]

[KOSH 2000]  D. R. KOHEL & I. E. SHPARLINSKI, *Exponential sums and group generators for elliptic curves over finite fields*, Algorithmic Number Theory Symposium – ANTS IV, Lecture Notes in Comput. Sci., vol. 1838, Springer-Verlag, Berlin, 2000, 395–404. [732]

[KOSU 1998]  Ç. K. KOÇ & B. SUNAR, *Low-complexity bit-parallel canonical and normal basis multipliers for a class of finite fields*, IEEE Trans. on Computers **47** N°3 (1998), 353–356. [643]

[KoTs 1993] K. KOYAMA & Y. TSURUOKA, *Speeding up elliptic cryptosystems by using a signed binary window method*, Advances in Cryptology – Crypto 1992, Lecture Notes in Comput. Sci., vol. 740, Springer-Verlag, Berlin, 1993, 345–357. [152, 302]

[KoWe 2005] K. KOIKE & A. WENG, *Construction of CM-Picard curves*, Math. Comp. **74** N°249 (2005), 499–518. [473]

[KRA 1922] M. KRAITCHIK, *Théorie des nombres*, vol. 1, Gauthier-Villars, 1922. [495]

[KRA 1924] _____, *Recherches sur la théorie des nombres*, Gauthier-Villars, 1924. [495]

[KRI 1997] U. KRIEGER, *Anwendung hyperelliptischer Kurven in der Kryptographie*, Master's thesis, Universität Gesamthochschule Essen, 1997. [313]

[KrJe 1996] W. KRANDICK & T. JEBELEAN, *Bidirectional exact integer division*, J. Symbolic Comput. **21** N°4–6 (1996), 441–445. [190]

[KuGo$^+$ 2002] J. KUROKI, M. GONDA, K. MATSUO, J. CHAO, & S. TSUJI, *Fast genus three hyperelliptic curve cryptosystems*, Symposium on Cryptography and Information Security – SCIS 2002, 503–507.
http://lab.iisec.ac.jp/~matsuo_lab/pub/pdf/8b-2_1244.pdf [348]

[KÜH 1902] H. KÜHNE, *Eine Wechselbeziehung zwischen Funktionen mehrerer Unbestimmten, die zu Reciprocitätsgesetzen führt*, J. Reine Angew. Math. **124** (1902), 121–133. [37]

[KuYa 1998] N. KUNIHIRO & H. YAMAMOTO, *Window and extended window methods for addition chain and addition-subtraction chain*, IEICE Trans. Fundamentals **E81-A** N°1 (1998), 72–81. [160]

[LAG 1973] J. L. LAGRANGE, *Œuvres*, Georg Olms Verlag, 1973, in French. [434]

[LAM 1996] R. J. LAMBERT, *Computational aspects of Discrete Logarithms*, PhD. Thesis, University of Waterloo, Ontario, Canada, 1996. [501]

[LaMi 2004] T. LANGE & P. K. MISHRA, *SCA resistant parallel explicit formula for addition and doubling of divisors in the Jacobian of hyperelliptic curves of genus 2*, preprint, 2004. [694]

[LAN 1973] S. LANG, *Elliptic functions*, Addison-Wesley Publishing Company, Reading, MA, 1973. [99, 123]

[LAN 1982] _____, *Introduction to algebraic and abelian functions*, 2nd edition ed., Springer-Verlag, Berlin, 1982. [100, 106]

[LAN 2001a] T. LANGE, *Efficient arithmetic on hyperelliptic curves*, PhD. Thesis, Universität-Gesamthochschule Essen, 2001. [313, 314, 367, 374, 376, 383, 410, 712]

[LAN 2001b] _____, *Efficient arithmetic on hyperelliptic Koblitz curves*, Tech. Report 2-2001, Universität-Gesamthochschule Essen, 2001. [367]

[LAN 2001c] _____, *Hyperelliptic curves allowing fast arithmetic*, webpage, 2001.
http://www.ruhr-uni-bochum.de/itsc/tanja/KoblitzC.html [368, 371]

[LAN 2002a] S. LANG, *Algebra*, Graduate Texts in Mathematics, vol. 211, Springer-Verlag, Berlin, 2002, third edition. [19, 24, 92]

[LAN 2002b] T. LANGE, *Efficient arithmetic on genus 2 hyperelliptic curves over finite fields via explicit formulae*, preprint, 2002.
http://eprint.iacr.org/2002/121/ [314]

[LAN 2002c] _____, *Inversion-free arithmetic on genus 2 hyperelliptic curves*, preprint, 2002.
http://eprint.iacr.org/2002/147/ [321]

[LAN 2002d] _____, *Weighted coordinates on genus 2 hyperelliptic curves*, preprint, 2002.
http://eprint.iacr.org/2002/153/ [323, 324, 342]

[LAN 2004a] _____, *Montgomery addition for genus two curves*, Algorithmic Number Theory Symposium – ANTS VI, Lecture Notes in Comput. Sci., vol. 3076, Springer-Verlag, 2004, 309–317. [328, 329, 697]

[LAN 2004b] ———, *A note on López–Dahab coordinates*, preprint, 2004. [294]
http://eprint.iacr.org/2004/323/

[LAN 2004c] ———, *Trace zero subvarieties of genus 2 curves for cryptosystems*, J. Ramanujan Math. [131, 383, 385, 386]
Soc. **19** N°1 (2004), 15–33.

[LAN 2005a] ———, *Arithmetic on binary genus 2 curves suitable for small devices*, preprint, 2005. [313, 346, 347]

[LAN 2005b] ———, *Formulae for arithmetic on genus 2 hyperelliptic curves*, Appl. Algebra Engrg. [339, 342, 346]
Comm. Comput. **15** N°5 (2005), 295–328.

[LAN 2005c] ———, *Koblitz curve cryptosystems*, Finite Fields Appl. **11** N°2 (2005), 220–229. [367, 370, 376]

[LaRu 1985] H. LANGE & W. RUPPERT, *Complete systems of addition laws on abelian varieties*, [56]
Invent. Math. **79** (1985), 603–610.

[LaSh 2005a] T. LANGE & I. E. SHPARLINSKI, *Certain exponential sums and random walks on* [733, 734]
*elliptic curves*, Canad. J. Math. **57** (2005), 338–350.

[LaSh 2005b] ———, *Collisions in fast generation of ideal classes and points on hyperelliptic and* [376, 735]
*elliptic curves*, Appl. Algebra Engrg. Comm. Comput. **15** N°5 (2005), 329–337.

[LaSt 2005] T. LANGE & M. STEVENS, *Efficient doubling for genus two curves over binary fields*, [334, 336, 337, 352]
Selected Areas in Cryptography – SAC 2004, Lecture Notes in Comput. Sci., vol. 3357,
Springer-Verlag, Berlin, 2005, 170–181.

[Lau 2004] A. G. B. LAUDER, *Deformation theory and the computation of zeta functions*, Proc. [138]
London Math. Soc. (3) **88** N°3 (2004), 565–602.

[LaWa 2002a] A. G. B. LAUDER & D. WAN, *Computing zeta functions of Artin–Schreier curves* [138, 453]
*over finite fields*, LMS J. Comput. Math. **5** (2002), 34–55 (electronic).

[LaWa 2002b] ———, *Counting points on varieties over finite fields of small characteristic*, Algorith- [138]
mic Number Theory: Lattices, Number Fields, Curves and Cryptography, Mathematical
Sciences Research Institute Publications, 2002. Proceedings of an MSRI workshop. To
appear.

[LaWa 2004] ———, *Computing zeta functions of Artin-Schreier curves over finite fields II*, J. Com- [138, 453]
plexity **20** N°2-3 (2004), 331–349.

[LaWi 2002] T. LANGE & A. WINTERHOF, *Polynomial interpolation of the elliptic curve and XTR* [554]
*discrete logarithm*, Computing and Combinatorics – COCOON 2002, Lecture Notes in
Comput. Sci., vol. 2387, Springer-Verlag, Berlin, 2002, 137–143.

[LaWi 2003] ———, *Interpolation of the elliptic curve Diffie–Hellman mapping*, Applicable Alge- [554]
bra, Algebraic Algorithms and Error-Correcting Codes – AAECC 2003, Lecture Notes in
Comput. Sci., Springer-Verlag, Berlin, 2003, 51–60.

[Leh 1938] D. H. LEHMER, *Euclid's algorithm for large numbers*, Amer. Math. Monthly **45** (1938), [192]
227–233.

[LeLe 1990] A. K. LENSTRA & H. W. LENSTRA, JR., *Algorithms in number theory*, Handbook of [479]
theoretical computer science, Volume A, algorithms and complexity (J. VAN LEEUWEN,
ed.), Elsevier, Amsterdam, 1990.

[LeLe 1993] A. K. LENSTRA & H. W. LENSTRA, JR. (eds.), *The development of the Number* [508, 614]
*Field Sieve*, Lecture Notes in Math., vol. 1554, Springer-Verlag, Berlin, 1993.

[LeLe+ 1982] A. K. LENSTRA, H. W. LENSTRA, JR., & L. LOVÁSZ, *Factoring polynomials with* [448]
*rational coefficients.*, Math. Ann. **261** (1982), 513–534.

[LeLe+ 1990] A. K. LENSTRA, H. W. LENSTRA, JR., M. S. MANASSE, & J. M. POLLARD, [614]
*The Number Field Sieve*, Symposium on the Theory of Computing – STOC 1990, ACM,
1990, 564–572.

[LeLu 2003] R. LERCIER & D. LUBICZ, *Counting points on elliptic curves over finite fields of small* [253, 434]
*characteristic in quasi quadratic time*, Advances in Cryptology – Eurocrypt 2003, Lecture
Notes in Comput. Sci., vol. 2656, Springer-Verlag, 2003, 360–373.

[LeMa 1994] A. K. Lenstra & M. S. Manasse, *Factoring with two large primes*, Math. Comp. **63** (1994), 77–82.

[Len 1987] H. W. Lenstra, Jr., *Factoring integers with elliptic curves*, Ann. of Math. (2) **126** N°3 (1987), 649–673.

[Len 1999] A. K. Lenstra, *Efficient identity based parameter selection for elliptic curve cryptosystems*, Australasian Conference on Information Security and Privacy – ACISP 1999, Lecture Notes in Comput. Sci., vol. 1587, Springer-Verlag, 1999, 294–302.

[Len 2002] _____, *Computational methods in public key cryptology*, 2002.
http://www.win.tue.nl/~klenstra/notes.ps

[Ler 1996] R. Lercier, *Computing isogenies in $GF(2^n)$*, Algorithmic Number Theory Symposium – ANTS II, Lecture Notes in Comput. Sci., vol. 1122, Springer-Verlag, 1996, 197–212.

[Ler 1997] _____, *Algorithmique des courbes elliptiques dans les corps finis*, PhD. Thesis, École polytechnique, 1997.
http://www.medicis.polytechnique.fr/~lercier/preprints/these.pdf

[LeSc 1984] H. W. Lenstra, Jr. & C. P. Schnorr, *A Monte Carlo factoring algorithm with linear storage*, Math. Comp. **43** N°167 (1984), 289–311.

[LeVe 2000] A. K. Lenstra & E. R. Verheul, *The XTR public key system*, Advances in Cryptology – Crypto 2000, Lecture Notes in Comput. Sci., vol. 1880, Springer-Verlag, Berlin, 2000, 1–19.

[LeZi 1978] A. Lempel & J. Ziv, *Compression of individual sequences via variable rate coding*, IEEE Trans. Inform. Theory **IT-24** N°5 (1978), 530–536.

[Lic 1969] S. Lichtenbaum, *Duality theorems for curves over p-adic fields*, Invent. Math. **7** (1969), 120–136.

[LiDe+ 1994] Y. X. Li, R. H. Deng, & X. M. Wang, *On the equivalence of McEliece's and Niederreiter's public-key cryptosystems*, IEEE Trans. Inform. Theory **40** N°1 (1994), 271–273.

[Lidia] Lidia, *A C++ library for computational number theory, version 2.1.3*, 2004.
http://www.informatik.tu-darmstadt.de/TI/LiDIA/

[LiLe 1994] C. H. Lim & P. J. Lee, *More flexible exponentiation with precomputation*, Advances in Cryptology – Crypto 1994, Lecture Notes in Comput. Sci., vol. 839, Springer-Verlag, Berlin, 1994, 95–107.

[LiLe 1997] _____, *A key recovery attack on discrete log-based schemes using a prime order subgroup*, Advances in Cryptology – Crypto 1997, Lecture Notes in Comput. Sci., vol. 1294, Springer-Verlag, 1997, 249–263.

[LiNi 1997] R. Lidl & H. Niederreiter, *Finite fields*, second ed., Cambridge University Press, 1997.

[LiQu 2003] B. Libert & J.-J. Quisquater, *Efficient revocation and threshold pairing based cryptosystems*, Principles of Distributed Computing – PODC 2003, ACM, 2003, 163–171.

[LiSm 2001] P.-Y. Liardet & N. P. Smart, *Preventing SPA/DPA in ECC systems using the Jacobi form*, Cryptographic Hardware and Embedded Systems – CHES 2001, Lecture Notes in Comput. Sci., vol. 2162, Springer-Verlag, Berlin, 2001, 391.

[LóDa 1998] J. López & R. Dahab, *Improved algorithms for elliptic curve arithmetic in $GF(2^n)$*, Tech. Report IC-98-39, Relatório Técnico, October 1998.

[LóDa 1999] _____, *Fast multiplication on elliptic curves over $GF(2^m)$ without precomputation*, Cryptographic Hardware and Embedded Systems – CHES 1999), Lecture Notes in Comput. Sci., vol. 1717, Springer-Verlag, Berlin, 1999, 316–327.

[LóDa 2000a] _____, *High-speed software multiplication in $\mathbb{F}_{2^m}$*, Progress in Cryptology – Indocrypt 2000, Lecture Notes in Comput. Sci., vol. 1977, Springer-Verlag, Berlin, 2000, 203–212.

[LóDa 2000b] _____, *An overview of elliptic curve cryptography*, Tech. Report IC-00-10, Relatório Técnico, May 2000. [295]

[Lor 1996] D. Lorenzini, *An invitation to arithmetic geometry*, Graduate studies in mathematics, vol. 9, AMS, 1996. [45]

[Lub 1968] S. Lubkin, *A p-adic proof of Weil's conjectures*, Ann. of Math. (2), 105-194; ibid. (2) **87** (1968), 195–255. [136]

[Lubicz] D. Lubicz, *homepage*, Sur les tests statistiques de générateurs aléatoires. http://www.math.u-bordeaux.fr/~lubicz/ [715, 725, 726]

[LuSe+ 1964] J. Lubin, J.-P. Serre, & J. Tate, *Elliptic curves and formal groups*, July 1964. Lecture notes prepared in connection with the seminars held at the Summer Institute on Algebraic Geometry, Whitney State, Woods Hole, Massachusets. http://ma.utexas.edu/users/voloch/lst.html [138, 423]

[LuSh 2005] F. Luca & I. E. Shparlinski, *On the exponent of the group of points on elliptic curves in extension fields*, Intern. Math. Research Notices **23** (2005), 1391–1409. [586]

[MaCh+ 2001] K. Matsuo, J. Chao, & S. Tsujii, *Fast genus two hyperelliptic curve cryptosysytems*, Tech. Report ISEC2001-23, IEICE, 2001. [314]

[MaCh+ 2002] _____, *An improved baby-step giant-step algorithm for point counting of hyperelliptic curves over finite fields*, Algorithmic Number Theory Symposium – ANTS V, Lecture Notes in Comput. Sci., vol. 2369, Springer-Verlag, 2002, 461–474. [411, 685]

[Mad 2003] M. S. Madsen, *A general framework for p-adic point counting and application to elliptic curves in Legendre form*, preprint, 2003. [442]

[Magma] *The Magma computational algebra system for algebra, number theory and geometry, version 2.11-14*, April 2005. http://magma.maths.usyd.edu.au/magma/ [201, 267]

[MaIm 1988] T. Matsumoto & H. Imai, *Public quadratic polynomial-tuples for efficient signature verification and message-encryption*, Advances in Cryptology – Eurocrypt 1988, Lecture Notes in Comput. Sci., vol. 330, Springer-Verlag, 1988, 419–445. [15]

[MaMe+ 2002] U. M. Maurer, A. J. Menezes, & E. Teske, *Analysis of the GHS Weil descent attack on the ECDLP over characteristic two finite fields of composite degree*, LMS J. Comput. Math. **5** (2002), 127–174. [531, 536]

[Man 1961] Ju. I. Manin, *The Hasse-Witt matrix of an algebraic curve*, Izv. Akad. Nauk. SSSR Ser. Mat. **25** (1961), 153–172. [411]

[Man 1962] _____, *On the theory of Abelian varieties over a field of finite characteristic*, Izv. Akad. Nauk. SSSR Ser. Mat. **26** (1962), 281–292. [412]

[MaNa 2002] D. Maisner & E. Nart, *Abelian surfaces over finite fields as Jacobians*, Experiment. Math. **11** (2002), 321–337, with an appendix by E. Howe. [113]

[Mar] G. Marsaglia, *The diehard battery of stringent statistical randomness tests*. http://random.com.hr/products/random/manual/html/Diehard.html [729]

[Mas 1969] J. L. Massey, *Shift–register synthesis and BCH decoding*, IEEE Trans. Inform. Theory **IT-15** (1969), 122–127. [501]

[Mau 1992] U. M. Maurer, *A universal statistical test for random number generators*, J. Cryptology **5** N°2 (1992), 89–105. [723, 725]

[MaWo 1998] U. M. Maurer & S. Wolf, *Lower bounds on generic algorithms in groups*, Advances in Cryptology – Eurocrypt 1998, Lecture Notes in Comput. Sci., vol. 1403, Springer-Verlag, Berlin, 1998, 72–84. [478]

[MaWo 1999] _____, *The relationship between breaking the Diffie–Hellman protocol and computing Discrete Logarithms*, SIAM J. Comput. **28** N°5 (1999), 1689–1721. [10]

[McC 1990] K. S. McCurley, *The discrete logarithm problem*, Cryptography and computational number theory (C. Pomerance, ed.), Proc. Symp. Appl. Math., vol. 42, AMS, 1990, 49–74. [495]

[McE 1969] R. J. McEliece, *Factorization of polynomials over finite fields*, Math. Comp. **23** (1969), 861–867. [507]

[McE 1978] ———, *A public-key cryptosystem based on algebraic coding theory*, Deep Space Network Progress Report 42-44, Jet Propulsion Lab., California Institute of Technology (1978), 114–116. [15]

[MeBu+ 2004] G. Meurice de Dormale, P. Bulens, & J.-J. Quisquater, *Efficient modular division implementation: ECC over $GF(p)$ affine coordinates application*, Field-Programmable Logic and Applications – FPL 2004, Lecture Notes in Comput. Sci., vol. 3203, Springer-Verlag, Berlin, 2004, 231–240. [206]

[MeOk+ 1993] A. J. Menezes, T. Okamoto, & S. A. Vanstone, *Reducing elliptic curve logarithms to a finite field*, IEEE Trans. on Inform. Theory **39** (1993), 1639–1646. [395, 530, 580]

[MeOo+ 1996] A. J. Menezes, P. van Oorschot, & S. A. Vanstone, *The Handbook of Applied Cryptography*, CRC Press, Inc., 1996.
http://www.cacr.math.uwaterloo.ca/hac/ [xxix, 5, 10, 12, 13, 146, 170, 180, 183, 197, 720, 728, 729, 731]

[MeQu 2001] A. J. Menezes & M. Qu, *Analysis of the Weil descent attack of Gaudry, Hess and Smart*, Topics in cryptology – CT-RSA 2001, Lecture Notes in Comput. Sci., vol. 2020, Springer-Verlag, Berlin, 2001, 308–318. [531, 534, 535]

[Mes 1991] J.-F. Mestre, *Construction des courbes de genre 2 à partir de leurs modules*, Prog. Math., Birkhäuser **94** (1991), 313–334. [102]

[Mes 2000a] T. S. Messerges, *Using second order power analysis to attack DPA resistant software*, Cryptographic Hardware and Embedded Systems – CHES 2000, Lecture Notes in Comput. Sci., vol. 1965, Springer-Verlag, Berlin, 2000, 238–251. [680, 704]

[Mes 2000b] J.-F. Mestre, *Lettre adressée à Gaudry et Harley*, December 2000. In French.
http://www.math.jussieu.fr/~mestre/ [138, 433, 434]

[Mes 2002] ———, *Applications de l'AGM au calcul du nombre de points d'une courbe de genre 1 ou 2 sur $\mathbb{F}_{2^n}$*. Talk given to the Séminaire de Cryptographie de l'Université de Rennes, March 2002.
http://www.maths.univ-rennes1.fr/crypto/2001-02/Mestre2203.html [434, 442]

[MeSt 1993] W. Meier & O. Staffelbach, *Efficient multiplication on certain nonsupersingular elliptic curves*, Advances in Cryptology – Crypto 1992, Lecture Notes in Comput. Sci., vol. 740, Springer-Verlag, Berlin, 1993, 333–344. [359, 360, 362]

[MeTe+ 2004] A. J. Menezes, E. Teske, & A. Weng, *Weak fields for ECC*, Topics in Cryptology – CT-RSA 2004, Lecture Notes in Comput. Sci., vol. 2964, Springer-Verlag, 2004, 366–386. [536]

[MeVa 1990] A. J. Menezes & S. A. Vanstone, *The implementation of elliptic curve cryptosystems*, Advances in Cryptology – Auscrypt 1990, Lecture Notes in Comput. Sci., vol. 453, Springer-Verlag, Berlin, 1990, 2–13. [356, 581]

[MFA 2004] Microelectronics Failure Analysis Desk Reference Fifth Edition EDFAS (Electronic Device Failure Analysis Society) and ASM International editors, 2004. [670]

[MiAv 2003] P. Mihăilescu & R. M. Avanzi, *Efficient "quasi"-deterministic primality test improving AKS*, preprint, March 2003. [601]

[MiDo+ 2002] Y. Miyamoto, H. Doi, K. Matsuo, J. Chao, & S. Tsuji, *A fast addition algorithm of genus two hyperelliptic curve*, Symposium on Cryptography and Information Security – SCIS 2002, 497–502. In Japanese. [314, 321]

[Mih 1997] P. Mihăilescu, *Optimal Galois field bases which are not normal*, 1997. Presented at the Workshop on Fast Software Encryption in Haifa. [229]

[MIH 2000] _____, *Medium Galois fields, their bases and arithmetic*, 2000. Preprint. [231]
http://grouper.ieee.org/groups/1363/P1363a/contributions/Medium.pdf

[MIL 1986] V. S. MILLER, *Short programs for functions on curves*, 1986. IBM, Thomas J. Watson Research Center. [122, 392]
http://crypto.stanford.edu/miller/

[MIL 2004] _____, *The Weil pairing, and its efficient calculation*, J. Cryptology **17** (2004), 235–261. [122, 392]

[MiNa+ 2001] A. MIYAJI, M. NAKABAYASHI, & S. TAKANO, *New explicit conditions of elliptic curve traces for FR-reduction*, IEICE Trans. Fundamentals **E84-A** N°5 (2001), 1234–1243. [586]

[MIS 2004a] P. K. MISHRA, *Pipelined computation of scalar multiplication in elliptic curve cryptosystems*, Cryptographic Hardware and Embedded Systems – CHES 2004, Lecture Notes in Comput. Sci., vol. 3156, Springer-Verlag, Berlin, 2004, 328–342. [302, 692]

[MIS 2004b] _____, *Scalar multiplication in elliptic curve cryptosystems: Pipelining with precomputations*, preprint, 2004. [694]
http://eprint.iacr.org/2004/191/

[MOBO 1972] R. T. MOENCK & A. B. BORODIN, *Fast modular transforms via division*, Conf. Record, IEEE 13th Annual Symp. on Switching and Automata Theory, IEEE Press, 1972, 90–96. [184]

[MOBR 1975] M. A. MORRISON & J. BRILLHART, *A method of factorization and the factorization of $F_7$*, Math. Comp. **29** (1975), 183–205. [508, 607]

[MOE 1973] R. T. MOENCK, *Fast computation of GCDs*, Proceedings of the 5th Annual ACM Symposium on the Theory of Computing, 1973, 142–151. [263]

[MOJO] Z. MO & J. P. JONES, *A new primality test using Lucas sequences*, preprint. [595]

[MÖL 2001a] B. MÖLLER, *Securing elliptic curve point multiplication against Side-Channel Attacks*, Information Security Conference – ISC 2001, Lecture Notes in Comput. Sci., vol. 2200, Springer-Verlag, 2001, 324–334. See also [MÖL 2001b]. [690, 699]

[MÖL 2001b] _____, *Securing elliptic curve point multiplication against Side-Channel Attacks*, Tech. report, TU Darmstadt, 2001, Addendum "Efficiency Improvement" added 2001-08-27/2001-08-29. Errata added 2001-12-21. [690, 763]
http://www.informatik.tu-darmstadt.de/TI/Mitarbeiter/moeller/

[MÖL 2003] _____, *Improved techniques for fast exponentiation*, Information Security and Cryptology – ICISC 2002, Lecture Notes in Comput. Sci., vol. 2587, Springer-Verlag, 2003, 298–312. [154]

[MON 1968] P. MONSKY, *Formal cohomology. II: the cohomology sequence of a pair*, Ann. of Math. (2) **88** (1968), 218–238. [136, 139]

[MON 1970] _____, *p-adic analysis and zeta functions*, Lectures in Mathematics, Department of Mathematics Kyoto University, vol. 4, Kinokuniya Book-Store Co. Ltd., Tokyo, 1970. [139]

[MON 1971] _____, *Formal cohomology. III: fixed point theorems*, Ann. of Math. (2) **93** (1971), 315–343. [136, 139]

[MON 1985] P. L. MONTGOMERY, *Modular multiplication without trial division*, Math. Comp. **44** N°170 (1985), 519–521. [180, 207]

[MON 1987] _____, *Speeding the Pollard and elliptic curve methods of factorization*, Math. Comp. **48** N°177 (1987), 243–264. [285, 603]

[MON 1994] _____, *A survey of modern integer factorization algorithms*, CWI Quarterly **7** N°4 (1994), 337–366. [610]

[MON 1995] _____, *A block Lanczos algorithm for finding dependencies over GF(2)*, Advances in Cryptology – Eurocrypt 1995, Lecture Notes in Comput. Sci., no. 921, Springer-Verlag, 1995, 106–120. [503]

[MoOl 1990]   F. Morain & J. Olivos, *Speeding up the computations on an elliptic curve using addition-subtraction chains*, Inform. Theory Appl. **24** (1990), 531–543. [151, 153]

[Mor 1995]   F. Morain, *Calcul du nombre de points sur une courbe elliptique dans un corps fini : aspects algorithmiques*, J. Théor. Nombres Bordeaux **7** (1995), 255–282. [416, 420]

[Morain]   _____, homepage. [601]
http://www.lix.polytechnique.fr/~morain/

[MoWa 1968]   P. Monsky & G. Washnitzer, *Formal cohomology. I*, Ann. of Math. (2) **88** (1968), 181–217. [136, 139]

[Mül 1995]   V. Müller, *Ein Algorithmus zur Bestimmung der Punktanzahl elliptisher Kurven über endlichen Körpen der Charakteristik größer drei*, PhD. Thesis, Technische Fakultät der Universität des Saarlandes, February 1995. [416, 420]

[Mül 1998]   _____, *Fast multiplication on elliptic curves over small fields of characteristic two*, J. Cryptology **11** (1998), 219–234. [367, 370]

[Mum 1966]   D. Mumford, *On the equations defining Abelian varieties I-III*, Invent. Math. **1** (1966), 287–354. [56]

[Mum 1974]   _____, *Abelian varieties*, Oxford University Press, New York, 1974. [59, 60, 62, 63, 92, 93, 111, 115]

[MuOn+ 1989]   R. C. Mullin, I. M. Onyszchuk, S. A. Vanstone, & R. M. Wilson, *Optimal normal bases in $GF(p^n)$*, Discrete Appl. Math. **22** (1989), 149–161. [217, 221]

[MuSm+ 2004]   A. Muzereau, N. P. Smart, & F. Vercauteren, *The equivalence between the DHP and DLP for elliptic curves used in practical applications*, LMS J. Comp. Math. **7** (2004), 50–72. [10]

[MuSt 2004]   J. A. Muir & D. R. Stinson, *Minimality and other properties of the width-$w$ nonadjacent form*, Combinatorics and Optimization Research Report CORR 2004-08, University of Waterloo, 2004. [153]
http://www.cacr.math.uwaterloo.ca/techreports/2004/corr2004-08.ps

[MuSt 2005]   _____, *New minimal weight representations for left-to-right window methods*, Topics in cryptology – CT-RSA 2005, Lecture Notes in Comput. Sci., vol. 3376, Springer-Verlag, Berlin, 2005, 366–383. [154]

[Nag 2004]   K.-I. Nagao, *Improvement of Thériault algorithm of index calculus for Jacobian of hyperelliptic curves of small genus*, preprint, 2004. [525]
http://eprint.iacr.org/2004/161/

[NaSt+ 2004]   D. Naccache, J. Stern, & N. P. Smart, *Projective coordinates leak*, Advances in Cryptology – Eurocrypt 2004, Lecture Notes in Comput. Sci., vol. 3027, Springer-Verlag, Berlin, 2004, 257–267. [700]

[Nau 1999]   N. Naumann, *Weil-Restriktion abelscher Varietäten*, Master's thesis, University Essen, 1999. [383, 386]

[Nec 1994]   V. I. Nechaev, *On the complexity of a deterministic algorithm for a discrete logarithm*, Mat. Zametki **55** N°2 (1994), 91–101, 189, Russian. Translation in Math. Notes 55 (1994), no. 1-2, 165–172. [478, 480]

[NeMa 2002]   N. Nedjah & L. de Macedo Mourelle, *Minimal addition chain for efficient modular exponentiation using genetic algorithms*, Proceedings of the 2002 Industrial and Engineering, Applications of Artificial Intelligence and Expert Systems Conference, Lecture Notes in Comput. Sci., vol. 2358, Springer-Verlag, Berlin, 2002, 88–98. [163]

[NgSh 2003]   P. Q. Nguyen & I. E. Shparlinski, *The insecurity of the elliptic curve digital signature algorithm with partially known nonces*, Des. Codes Cryptogr. **30** (2003), 201–217. [376, 477, 698]

[NgSt 1999]   P. Q. Nguyen & J. Stern, *The hardness of the hidden subset sum problem and its cryptographic implications*, Advances in Cryptology – Crypto 1999, Lecture Notes in Comput. Sci., vol. 1666, Springer-Verlag, Berlin, 1999, 31–46. [376]

[NGU 1998] P. Q. NGUYEN, *A Montgomery-like square root for the Number Field Sieve*, Algorithmic Number Theory Symposium – ANTS III, Lecture Notes in Comput. Sci., vol. 1423, Springer-Verlag, Berlin, 1998, 151–168. [614]

[NGU 2001] K. NGUYEN, *Explicit arithmetic of Brauer groups, ray class fields and index calculus*, PhD. Thesis, Universität Gesamthochschule Essen, 2001. [119, 121, 507]

[NIE 1986] H. NIEDERREITER, *Knapsack-type cryptosystems and algebraic coding theory*, Prob. Contr. Inform. Theory **15** N°2 (1986), 157–166. [15]

[NIE 2003] _____, *Linear complexity and related complexity measures for sequences*, Progress in Cryptology – Indocrypt 2003, Lecture Notes in Comput. Sci., vol. 2904, Springer-Verlag, 2003, 1–17. [730]

[NISH 1999] H. NIEDERREITER & I. E. SHPARLINSKI, *On the distribution and lattice structure of nonlinear congruential pseudorandom numbers*, Finite Fields Appl. **5** (1999), 246–253. [731]

[NIST] National Institute of Standard and Technology, *Recommended elliptic curves for federal government use*, 1999.
http://www.csrc.nist.gov/encryption/ [286]

[NIV 2004] G. NIVASH, *Cycle detection using a stack*, Inform. Process. Lett. **90** N°3 (2004), 135–140. [487, 490]

[NIXI 2001] H. NIEDERREITER & C. XING, *Rational points on curves over finite fields. Theory and Applications*, London Mathematical Society Lecture Note Series, vol. 285, Cambridge University Press, 2001. [730]

[NÖC 1996] M. NÖCKER, *Exponentiation in finite fields: theory and practice*, Diplomarbeit im Fach Informatik, Universität Paderborn, 1996. [224, 226]

[NÖC 2001] _____, *Data structures for parallel exponentiation in finite fields*, PhD. Thesis, Universität Paderborn, 2001. [35]

[NSA] NSA TEMPEST SERIES.
http://cryptome.org/nsa-tempest.htm [682]

[NTL] V. SHOUP, *NTL: A Library for doing Number Theory, version 5.4*, 2005.
http://www.shoup.net/ [201]

[NTRU] NTRU CRYPTOLAB.
http://www.ntru.com/cryptolab/index.htm [14]

[O'CO 2001] L. J. O'CONNOR, *On string replacement exponentiation*, Des. Codes Cryptogr. **23** (2001), 173–183. [160]

[ODL 1985] A. M. ODLYZKO, *Discrete logarithms in finite fields and their cryptographic significance*, Advances in Cryptology – Eurocrypt 1984, Lecture Notes in Comput. Sci., vol. 209, Springer-Verlag, Berlin, 1985, 224–314. [215, 495, 506, 508]

[ODL 1990] _____, *The rise and fall of knapsack cryptosystems*, Cryptology and computational number theory (Boulder, CO, 1989) (Providence, RI), Amer. Math. Soc., 1990, 75–88. [14]

[OKPO 2001] T. OKAMOTO & D. POINTCHEVAL, *The gap-problems: a new class of problems for the security of cryptographic schemes*, Public Key Cryptography – PKC 2001, Lecture Notes in Comput. Sci., vol. 1992, Springer-Verlag, 2001, 104–118. [578]

[OKSA 2000] K. OKEYA & K. SAKURAI, *Power analysis breaks elliptic curve cryptosystems even secure against the timing attack*, Progress in Cryptology – Indocrypt 2000, Lecture Notes in Comput. Sci., vol. 1977, Springer-Verlag, Berlin, 2000, 178–190. [699]

[OKSA 2001] _____, *Efficient elliptic curve cryptosystems from a scalar multiplication algorithm with recovery of the y-coordinate on a Montgomery-form elliptic curve*, Cryptographic Hardware and Embedded Systems – CHES 2001, Lecture Notes in Comput. Sci., vol. 2162, Springer-Verlag, Berlin, 2001, 126–141. [286, 298]

[OKSA 2002] _____, *On insecurity of the Side Channel Attack countermeasure using addition-subtraction chains under distinguishability between addition and doubling*, Australasian Conference on Information Security and Privacy – ACISP 2002, Lecture Notes in Comput. Sci., vol. 2384, Springer-Verlag, Berlin, 2002, 420–435. [699]

[OKSA 2003] _____, *A simple power attack on a randomized addition-subtraction chains method for elliptic curve cryptosystems*, IEICE Transactions **E86-A** (2003), 1171–1180. [699]

[OKTA 2003] K. OKEYA & T. TAKAGI, *The width-w NAF method provides small memory and fast elliptic scalar multiplication secure against side channel attacks*, Topics in Cryptology – CT-RSA 2003, Lecture Notes in Comput. Sci., vol. 2612, Springer-Verlag, Berlin, 2003, 328–342. [690]

[OLI 1981] J. OLIVOS, *On vectorial addition chains*, Journal of algorithms **2** (1981), 13–21. [159]

[OLS 2004] L. D. OLSON, *Side-Channel Attacks in ECC: A general technique for varying the parametrization of the elliptic curve*, Cryptographic Hardware and Embedded Systems – CHES 2004, Lecture Notes in Comput. Sci., vol. 3156, Springer-Verlag, Berlin, 2004, 220–229. [696]

[OMMA 1986] J. OMURA & J. L. MASSEY, *Computational method and apparatus for finite field arithmetic*, United States Patent 4, 587, 627, Date: May 6th 1986. [35, 220]

[OOTS 1986] F. OORT & S. TSUTOMU, *The canonical lifting of an ordinary Jacobian variety need not be a Jacobian variety*, J. Math. Soc. Japan **38** N°3 (1986), 427–437. [139]

[OOWI 1999] P. VAN OORSCHOT & M. J. WIENER, *Parallel collision search with cryptanalytic applications*, J. Cryptology **12** (1999), 1–28. [489, 490, 492, 493]

[OSAI 2001] E. OSWALD & M. AIGNER, *Randomized addition-subtraction chains as a countermeasure against power attacks*, Cryptographic Hardware and Embedded Systems – CHES 2001, Lecture Notes in Comput. Sci., vol. 2162, Springer-Verlag, Berlin, 2001, 39–50. [699]

[PAJE$^+$ 2002a] Y.-H. PARK, S. JEONG, C. KIM, & J. LIM, *An alternate decomposition of an integer for faster point multiplication on certain elliptic curves*, Public Key Cryptography – PKC 2002, Lecture Notes in Comput. Sci., vol. 2274, Springer-Verlag, Berlin, 2002, 323–334. [380]

[PAJE$^+$ 2002b] Y.-H. PARK, S. JEONG, & J. LIM, *Speeding up point multiplication on hyperelliptic curves with efficiently-computable endomorphisms*, Advances in Cryptology – Eurocrypt 2002, Lecture Notes in Comput. Sci., vol. 2332, Springer-Verlag, 2002, 197–208. [380]

[PAMI 1998] S. K. PARK & K. W. MILLER, *Random number generators: Good ones are hard to find*, Comm. ACM **31** N°10 (1998), 1192–1201. [720, 721]

[PAP 1994] C. H. PAPADIMITRIOU, *Computational complexity*, Addison-Wesley Publishing Company, Reading, MA, 1994. [2]

[PAR 2000] B. PARHAMI, *Computer arithmetic – algorithms and hardware design*, Oxford University Press, New York, 2000. [618]

[PARI] The PARI Group, Bordeaux, *PARI/GP*, version 2.1.6, 2004. http://pari.math.u-bordeaux.fr/ [267, 467]

[PASM$^+$ 2004] D. PAGE, N. P. SMART, & F. VERCAUTEREN, *A comparison of MNT curves and supersingular curves*, preprint, 2004. http://eprint.iacr.org/2004/165/ [589]

[PAST 1973] M. S. PATERSON & L. J. STOCKMEYER, *On the number of nonscalar multiplications necessary to evaluate polynomials*, SIAM J. Comput. **2** (1973), 60–66. [251, 253, 260, 261]

[PAT 1996] J. PATARIN, *Hidden Fields Equations (HFE) and Isomorphisms of Polynomials (IP): two new families of asymmetric algorithms*, Advances in Cryptology – Eurocrypt 1996, Lecture Notes in Comput. Sci., vol. 1070, Springer-Verlag, Berlin, 1996, 33–48. [15]

[PEL 2002] J. PELZL, *Fast hyperelliptic curve cryptosystems for embedded processors*, Master's thesis, Ruhr-University of Bochum, 2002. [348]

| | | |
|---|---|---|
| [PeWo+ 2004] | J. Pelzl, T. Wollinger, & C. Paar, *Special hyperelliptic curve cryptosystems of genus two: Efficient arithmetic and fast implementation*, Embedded Cryptographic Hardware: Design and Security, Nova Science Publishers, 2004. | [334] |
| [Pil 1990] | J. Pila, *Frobenius maps of abelian varieties and finding roots of unity in finite fields*, Math. Comp. **55** N°192 (1990), 745–763. | [137, 422] |
| [Pinch] | R. Pinch, *homepage*.<br>http://www.chalcedon.demon.co.uk/rgep.html | [593] |
| [Pip 1979] | N. Pippenger, *The minimum number of edges in graphs with prescribed paths*, Math. Systems Theory **12** (1979), 325–346. | [166] |
| [Pip 1980] | \_\_\_\_\_, *On the evaluation of powers and monomials*, SIAM J. Comput. **9** (1980), 230–250. | [166] |
| [PKCS] | Public Key Cryptography Standards, *PKCS #1 v1.5: RSA encryption standard*, 1993.<br>http://www.rsasecurity.com/rsalabs/pkcs | [1] |
| [PoHe 1978] | S. Pohlig & M. Hellmann, *An improved algorithm for computing logarithms over $GF(p)$ and its cryptographic significance*, IEEE Trans. Inform. Theory **IT-24** (1978), 106–110. | [479] |
| [Pol 1974] | J. M. Pollard, *Theorems on factorization and primality testing*, Proceedings of the Cambridge philosophical society **76** (1974), 521–528. | [603] |
| [Pol 1978] | \_\_\_\_\_, *Monte Carlo methods for index computation (mod p)*, Math. Comp. **32** (1978), 918–924. | [483, 488, 492] |
| [Pol 2000] | \_\_\_\_\_, *Kangaroos, monopoly and discrete logarithms*, J. Cryptology **13** (2000), 437–447. | [492–494] |
| [Pom 1983] | C. Pomerance, *Analysis and comparison of some integer factoring algorithms*, Computational methods in number theory (H. W. Lenstra, Jr. & R. Tijdeman, eds.), Math. Centre Tracts, no. 154, 155, Mathematisch Centrum, Amsterdam, 1983, 89–139. | [609, 610] |
| [Pom 1984] | \_\_\_\_\_, *Are there counter-examples to the Baillie-PSW primality test?*, Dopo Le Parole aangeboden aan Dr. A.K. Lenstra (H. W. Lenstra, Jr., J. K. Lenstra, & P. van Emde Boas, eds.), Amsterdam, 1984.<br>http://www.pseudoprime.com/dopo.ps | [596] |
| [Pom 1985] | \_\_\_\_\_, *The quadratic sieve factoring algorithm*, Advances in Cryptology – Eurocrypt 1984, Lecture Notes in Comput. Sci., vol. 209, Springer-Verlag, Berlin, 1985, 169–182. | [609] |
| [Pom 1990] | \_\_\_\_\_, *Factoring*, Cryptology and Computational Number Theory, Proceedings of Symposia in Applied Mathematics, vol. 42, American Mathematical Society, 1990, 27–47. | [594] |
| [PoSe+ 1980] | C. Pomerance, J. L. Selfridge, & S. S. Wagstaff, Jr., *The pseudoprimes to $25 \cdot 10^9$*, Math. Comp. **35** N°151 (1980), 1003–1026. | [593, 594, 596] |
| [PoSm 1992] | C. Pomerance & J. W. Smith, *Reduction of huge, sparse matrices over finite fields via created catastrophes*, Experiment. Math. **1** N°2 (1992), 89–94. | [504] |
| [PRIMO] | M. Martin, *PRIMO – Primality Proving*.<br>http://www.ellipsa.net | [596] |
| [Put 1986] | M. van der Put, *The cohomology of Monsky and Washnitzer*, Mém. Soc. Math. France **23** (1986), 33–60. | [139] |
| [QuDe 1990] | J.-J. Quisquater & J.-P. Delescaille, *How easy is collision search? Application to DES*, Advances in Cryptology – Eurocrypt 1989, Lecture Notes in Comput. Sci., no. 434, Springer-Verlag, 1990, 429–434. | [489] |
| [Qui 1990] | J.-J. Quisquater, *Procédé de codage selon la méthode dite RSA, par un microcontrôleur et dispositifs utilisant ce procédé*, Demande de brevet français. N° de dépôt 90 02274, Date: Feb. 23rd 1990. | [187, 203] |

[Qui 1992] _____, *Encoding system according to the so-called RSA method, by means of a microcontroller and arrangement implementing this system*, U.S. Patent # 5,166,978, Date: Nov. 24th 1992. [187, 203]

[QuWa+ 1991] J.-J. Quisquater, D. de Waleffe, & J.-P. Bournas, *A chip with fast RSA capability*, Proceedings of Smart Card 2000, Elsevier Science Publishers, Amsterdam, 1991, 199–205. [204]

[Rab 1980] M. O. Rabin, *Probabilistic algorithms in finite fields*, SIAM J. Comput. N°9 (1980), 273–280. [214]

[RaEf 2000] W. Rankl & W. Effing, *Smart card handbook*, 2nd ed., John Wiley & Sons, Ltd., 2000. [653, 660, 661]

[Rei 1962] G. Reitwiesner, *Binary arithmetic*, Adv. Comput. **1** (1962), 231–308. [151]

[Rib 1996] P. Ribenboim, *Lucas pseudoprimes*, p. 129, Springer, 1996. [595]

[RiSh+ 1978] R. L. Rivest, A. Shamir, & L. Adleman, *A method for obtaining digital signatures and public key cryptosystems*, Comm. ACM **21** (1978), 120–126. [7]

[RiSi 1997] R. L. Rivest & R. D. Silverman, *Are "strong" primes needed for RSA?*, preprint, 1997.
http://eprint.iacr.org/2001/007/ [585, 604]

[Rit] T. Ritter, *Random number machines: A literature survey*.
http://www.ciphersbyritter.com/RES/RNGMACH.HTM [718]

[Roo 1995] P. de Rooij, *Efficient exponentiation using precomputation and vector addition chains*, Advances in Cryptology – Eurocrypt 1994, Lecture Notes in Comput. Sci., vol. 950, Springer-Verlag, Berlin, 1995, 389–399. [166]

[RSA] *The new RSA factoring challenge.*
http://www.rsasecurity.com/rsalabs/node.asp?id=2092 [7]

[Rüc 1999] H.-G. Rück, *On the discrete logarithm problem in the divisor class group of curves*, Math. Comp. **68** (1999), 805–806. [77, 529]

[Ruk 2001] A. Rukhin, *Testing randomness: A suite of statistical procedures*, Theory Probab. Appl. **45** N°1 (2001), 111–132. [720]

[RuSo+] A. Rukhin, J. Soto, J. Nechvatal, M. Smid, E. Barker, S. Leigh, M. Levenson, M. Vangel, D. Banks, A. Heckert, J. Dray, & S. Vo, *A statistical test suite for random and pseudorandom number generators for cryptographic Applications*, NIST Special Publication 800-22.
http://csrc.nist.gov/rng/ [729]

[SaAr 1998] T. Satoh & K. Araki, *Fermat quotients and the polynomial time discrete log algorithm for anomalous elliptic curves*, Comm. Math. Univ. Sancti Pauli **47** (1998), 81–92. [529]

[Sah 1975] S. Sahni, *Approximate algorithms for the 0/1 knapsack problem*, J. ACM **22** (1975), 115–124. [14]

[SaKo 2002] E. Savaş & Ç. K. Koç, *Architectures for unified field inversion with applications in elliptic curve cryptography*, International Conference on Electronics, Circuits and Systems – ICECS 2002, vol. 2, 2002, 1155–1158. [646]

[SaSc 1985] J. Sattler & C. P. Schnorr, *Generating random walks in groups*, Ann. Univ. Sci. Budapest. Sect. Comput. **6** (1985), 65–79. [489]

[SaSc+ 2004] H. Sato, D. Schepers, & T. Takagi, *Exact analysis of Montgomery multiplication*, Progress in Cryptology – Indocrypt 2004, vol. 3348, Springer-Verlag, Berlin, 2004, 290–304. [705]

[SaSk+ 2003] T. Satoh, B. Skjernaa, & Y. Taguchi, *Fast computation of canonical lifts of elliptic curves and its application to point counting*, Finite Fields Appl. **9** (2003), 89–101. [258, 262, 433]

[Sat 2000]  T. Satoh, *The canonical lift of an ordinary elliptic curve over a finite field and its point counting*, J. Ramanujan Math. Soc. **15** N°4 (2000), 247–270.

[SaTe⁺ 2000]  E. Savaş, A. F. Tenca, & Ç. K. Koç, *A scalable and unified multiplier architecture for finite fields $GF(p)$ and $GF(2^m)$*, Cryptographic Hardware and Embedded Systems – CHES 2000, Lecture Notes in Comput. Sci., vol. 1965, Springer-Verlag, Berlin, 2000, 277–292.

[SCA Lounge]  Ecrypt-Vampire, *The side-channel cryptanalysis lounge*.
http://www.crypto.ruhr-uni-bochum.de/en_sclounge.html

[ScBa 2004]  M. Scott & P. S. L. M. Barreto, *Compressed pairings*, Advances in Cryptology – Crypto 2004, Lecture Notes in Comput. Sci., vol. 3152, Springer-Verlag, Berlin, 2004, 140–156.

[Sch 1927]  F. K. Schmidt, *Zur Zahlentheorie in Körpern der Charakteristik p. (Vorläufige Mitteilung.)*, Sitz.-Ber. phys. med. Soz. Erlangen **58/59** (1926/1927), 159–172.

[Sch 1974]  B. Schoeneberg, *Elliptic modular functions*, Die Grundlehren der mathematischen Wissenschaften in Einzeldarstellungen, vol. 203, Springer-Verlag, 1974.

[Sch 1975]  A. Schönhage, *A lower bound on the length of addition chains*, Theoret. Comput. Sci. **1** (1975), 1–12.

[Sch 1985]  R. Schoof, *Elliptic curves over finite fields and the computation of square roots mod p*, Math. Comp. **44** (1985), 483–494.

[Sch 1987]  _____, *Nonsingular plane cubic curves*, J. Combin. Theory Ser. A **46** N°2 (1987), 183–211.

[Sch 1995]  _____, *Counting points on elliptic curves over finite fields*, J. Théor. Nombres Bordeaux **7** (1995), 219–254.

[Sch 1996]  B. Schneier, *Applied cryptography: protocols, algorithms and source code in C*, John Wiley & Sons, Ltd., New York, 1996, second edition.

[Sch 2000a]  W. Schindler, *A timing attack against RSA with the Chinese Remainder Theorem*, Cryptographic Hardware and Embedded Systems – CHES 2000, Lecture Notes in Comput. Sci., vol. 1965, Springer-Verlag, Berlin, 2000, 109–124.

[Sch 2000b]  O. Schirokauer, *Using number fields to compute logarithms in finite fields*, Math. Comp. **69** N°231 (2000), 1267–1283.

[Sch 2000c]  R. Schroeppel, *Elliptic curves: Twice as fast!*, 2000. Presentation at the Crypto 2000 Rump Session.

[Sch 2000d]  E. Schulte-Geers, *Collision search in a random mapping: some asymptotic results*, Talk at ECC 2000, the fourth Workshop on Elliptic Curve Cryptography, Essen, Germany, 2000.
http://www.cacr.math.uwaterloo.ca/conferences/2000/ecc2000/

[ScOr⁺ 1995]  R. Schroeppel, H. Orman, & S. O'Malley, *Fast key exchange with elliptic curve systems*, Tech. report, Department of Computer Science. The University of Arizona, 1995.

[ScSt 1971]  A. Schönhage & V. Strassen, *Schnelle Multiplikation grosser Zahlen*, Computing (Arch. Elektron. Rechnen) **7** (1971), 281–292.

[ScWe⁺ 1996]  O. Schirokauer, D. Weber, & Th. F. Denny, *Discrete logarithms: the effectiveness of the index calculus method*, Algorithmic Number Theory Symposium – ANTS II, Lecture Notes in Comput. Sci., vol. 1122, Springer-Verlag, Berlin, 1996, 337–361.

[Sec]  Standards for Efficient Cryptography, *Elliptic curve cryptography Ver.0.5*, 1999.
http://www.secg.org/drafts.htm

[SeLe 1980]  G. Seroussi & A. Lempel, *Factorization of symmetric matrices and trace-orthogonal bases in finite fields*, SIAM J. Comput. **9** N°4 (1980), 758–767.

[Sem 1983] I. Semba, *Systematic method for determining the number of multiplications required to compute $x^m$, where $m$ is a positive integer*, J. Inform. Process. **6** N°1 (1983), 31–33. [166]

[Sem 1998] I. Semaev, *Evaluation of discrete logarithms in a group of p-torsion points of an elliptic curve in characteristic p*, Math. Comp. **67** (1998), 353–356. [529]

[Sem 2004] ———, *Summation polynomials and the discrete logarithm problem on elliptic curves*, preprint, 2004.
http://eprint.iacr.org/2004/031/ [541]

[Ser 1958] J.-P. Serre, *Sur la topologie des variétés algébriques en caractéristique p*, Symposium internacional de topología algebraica – International symposium on algebraic topology (México City 1956), Universidad Nacional Autónoma de México and UNESCO, 1958, 24–53. [76]

[Ser 1970] ———, *Cours d'arithmétique*, Presses Universitaires de France, 1970. [416]

[Ser 1979] ———, *Local Fields*, Graduate Texts in Mathematics, vol. 67, Springer-Verlag, 1979. [39]

[Ser 1998] G. Seroussi, *Table of low-weight binary irreducible polynomials*, Tech. Report HPL-98-135, Hewlett–Packard, August 1998.
http://www.hpl.hp.com/techreports/98/HPL-98-135.pdf [214, 217]

[SeSm 1991] A. S. Sedra & K. C. Smith, *Microelectronic circuits*, Oxford University Press, New York, 1991. [653]

[SeSz$^+$ 1982] R. Sedgewick, T. G. Szymanski, & A. C. Yao, *The complexity of finding cycles in periodic functions*, SIAM J. Comput. **11** N°2 (1982), 376–390. [486]

[SGA 4] *Théorie des topos et cohomologie étale des schémas.*, Springer-Verlag, Berlin, 1973, Séminaire de Géométrie Algébrique du Bois-Marie 1963–1964 (SGA 4), Dirigé par M. Artin, A. Grothendieck et J. L. Verdier. Lecture Notes in Math., Vol. 269, 270, 305. [136]

[Sha 1971] D. Shanks, *Class number, a theory of factorization and genera*, Proc. Symp. Pure Math. **20** (1971), 415–440. [480]

[Sha 1984] A. Shamir, *Identity-based cryptosystems and signature schemes*, Advances in Cryptology – Crypto 1984, Lecture Notes in Comput. Sci., vol. 196, Springer-Verlag, 1984, 47–53. [576]

[Sha 1999] ———, *Method and apparatus for protecting public key schemes from timing and fault attacks*, United States Patent 5991415, 1999. [708]

[Shi 1967] T. Shioda, *On the graded ring of invariants of binary octavics*, Amer. J. Math. **89** (1967), 1022–1046. [101]

[Shi 1998] G. Shimura, *Abelian Varieties with complex multiplication and modular functions*, revised ed., Princeton University Press, 1998. [107]

[Sho] V. Shoup, *A computational introduction to number theory and algebra*.
http://www.shoup.net/ntb/ntb-b5.pdf [201]

[Sho 1990] ———, *On the deterministic complexity of factoring polynomials over finite fields*, Inform. Process. Lett. **33** N°5 (1990), 261–267. [507]

[Sho 1991] ———, *A fast deterministic algorithm for factoring polynomials over finite fields of small characteristic*, International Symposium on Symbolic and Algebraic Computations – ISSAC 1991, ACM Press, Bonn, 1991, 14–21. [507]

[Sho 1994a] ———, *Exponentiation in $GF(2^n)$ using fewer polynomial multiplications*, preprint, 1994. [226]

[Sho 1994b] ———, *Fast construction of irreducible polynomials over finite fields*, J. Symbolic Comput. **17** N°5 (1994), 371–391. [214]

[Sho 1997] ———, *Lower bounds for discrete logarithms and related problems*, Advances in Cryptology – Eurocrypt 1997, Lecture Notes in Comput. Sci., vol. 1233, Springer-Verlag, Berlin, 1997, 256–266. [478]

| | | |
|---|---|---|
| [SHP 1999] | I. E. SHPARLINSKI, *Finite fields: Theory and computation*, Kluwer Academic Publishers, Dordrecht/Boston/London, 1999. | [35, 201] |
| [SHP 2000] | _____, *On the Naor–Reingold pseudo-random function from elliptic curves*, Appl. Algebra Engrg. Comm. Comput. **11** (2000), 27–34. | [734] |
| [SHP 2003] | _____, *Cryptographic applications of analytic number theory*, Progress in Computer Science and Applied Logic, vol. 22, Birkhäuser Verlag, Basel, 2003, Complexity lower bounds and pseudorandomness. | [4] |
| [SHTA 1961] | G. SHIMURA & Y. TANIYAMA, *Complex multiplication of abelian varieties and its applications to number theory*, Publications of the Mathematical Society of Japan, vol. 6, the Mathematical Society of Japan, Tokyo, 1961. | [104, 105] |
| [SICI$^+$ 2002] | F. SICA, M. CIET, & J.-J. QUISQUATER, *Analysis of the Gallant-Lambert-Vanstone method based on efficient endomorphisms: Elliptic and hyperelliptic curves*, Selected Areas in Cryptography – SAC 2002, Lecture Notes in Comput. Sci., vol. 2595, Springer-Verlag, 2002, 21–36. | [377, 379, 380] |
| [SID 1994] | V. M. SIDEL'NIKOV, *A public-key cryptosystem based on Reed–Muller codes*, Discr. Math. Appli. **4** N°3 (1994), 191–207. | [15] |
| [SIL 1986] | J. H. SILVERMAN, *The Arithmetic of Elliptic Curves*, Graduate Texts in Mathematics, vol. 106, Springer-Verlag, Berlin, 1986. | [45, 72, 95, 96, 99, 115, 424, 432] |
| [SIL 1994] | _____, *Advanced topics in the arithmetic of elliptic curves*, Springer-Verlag, Berlin, 1994. | [98, 416] |
| [SIL 1999] | _____, *Fast multiplication in finite fields $GF(2^n)$*, Cryptographic Hardware and Embedded Systems – CHES 1999, Lecture Notes in Comput. Sci., vol. 1717, Springer-Verlag, Berlin, 1999, 122–134. | [217] |
| [SIMATH] | *Simath, a computer algebra system for number theoretic applications.* http://tnt.math.metro-u.ac.jp/simath | [267] |
| [SIRU 2004] | A. SILVERBERG & K. RUBIN, *Algebraic tori in cryptography*, High Primes and Misdemeanours: lectures in honour of the 60th birthday of Hugh Cowie Williams, Fields Institute Communications, AMS, 2004. | [56] |
| [SISH 2001] | J. H. SILVERMAN & I. E. SHPARLINSKI, *On the linear complexity of the Naor–Reingold pseudo-random function from elliptic curves*, Des. Codes Cryptogr. **243** (2001), 279–289. | [734, 735] |
| [SKAN 2003] | S. P. SKOROBOGATOV & R. J. ANDERSON, *Optical fault induction attacks*, Cryptographic Hardware and Embedded Systems – CHES 2002, Lecture Notes in Comput. Sci., vol. 2523, Springer-Verlag, Berlin, 2003, 281–290. | [684] |
| [SKJ 2003] | B. SKJERNAA, *Satoh's algorithm in characteristic* 2, Math. Comp. **72** N°241 (2003), 477–487. | [432] |
| [SMA 1998] | N. P. SMART, *The algorithmic resolution of Diophantine equations*, London Mathematical Society Student Texts, vol. 41, Cambridge University Press, 1998. | [587] |
| [SMA 1999a] | _____, *The Discrete Logarithm problem on elliptic curves of trace one*, J. Cryptology **12** (1999), 193–196. | [529] |
| [SMA 1999b] | _____, *Elliptic curve cryptosystems over small fields of odd characteristic*, J. Cryptology **12** (1999), 141–151. | [367, 370] |
| [SMA 2001] | _____, *The Hessian form of an elliptic curve*, Cryptographic Hardware and Embedded Systems – CHES 2001, Lecture Notes in Comput. Sci., vol. 2162, Springer-Verlag, 2001, 118–125. | [275, 276, 288, 696] |
| [SMA 2003] | _____, *An analysis of Goubin refined power analysis attack*, Cryptographic Hardware and Embedded Systems – CHES 2003, Lecture Notes in Comput. Sci., vol. 2779, Springer-Verlag, Berlin, 2003, 281–290. | [682, 703] |

[SMGE 2003] E. SMITH & C. GEBOTYS, *SCA countermeasures for ECC over binary fields on a VLIW DSP core*, Combinatorics and Optimization Research Report CORR 2003-06, University of Waterloo, 2003.
http://www.cacr.math.uwaterloo.ca/techreports/2003/cacr2003-06.pdf

[SMSK 1995] P. SMITH & C. SKINNER, *A public key cryptosystem and a digital signature system based on the Lucas function analogue to discrete logarithms*, Advances in Cryptology – Asiacrypt 1994, Lecture Notes in Comput. Sci., vol. 917, Springer-Verlag, Berlin, 1995, 357–364.

[SOL 1997] J. A. SOLINAS, *An improved algorithm for arithmetic on a family of elliptic curves*, Advances in Cryptology – Crypto 1997, Lecture Notes in Comput. Sci., vol. 1294, Springer-Verlag, Berlin, 1997, 357–371.

[SOL 1999a] ———, *Generalized Mersenne numbers*, Combinatorics and Optimization Research Report CORR 99-39, University of Waterloo, 1999.
http://www.cacr.math.uwaterloo.ca/techreports/1999/corr99-39.ps

[SOL 1999b] ———, *Improved algorithms for arithmetic on anomalous binary curves*, Combinatorics and Optimization Research Report CORR 99-46, University of Waterloo, 1999, updated and corrected version of [SOL 1997].
http://www.cacr.math.uwaterloo.ca/techreports/1999/corr99-46.ps

[SOL 2000] ———, *Efficient arithmetic on Koblitz curves*, Des. Codes Cryptogr. **19** (2000), 195–249.

[SOL 2001] ———, *Low-weight binary representations for pairs of integers*, Combinatorics and Optimization Research Report CORR 2001-41, University of Waterloo, 2001.
http://www.cacr.math.uwaterloo.ca/techreports/2001/corr2001-41.ps

[SPA 1994] A. M. SPALLEK, *Kurven vom Geschlecht 2 und ihre Anwendung in Public-Key-Kryptosystemen*, PhD. Thesis, Universität Gesamthochschule Essen, 1994.

[STA 2003] M. STAM, *Speeding up subgroup cryptosystems*, PhD. Thesis, Technische Universiteit Eindhoven, 2003.

[STA 2004] C. STAHLKE, *Point compression on Jacobians of hyperelliptic curves over $\mathbb{F}_q$*, preprint, 2004.
http://eprint.iacr.org/2004/030/

[STE 1910] E. STEINITZ, *Algebraische Theorie der Körper*, J. Reine Angew. Math. **137** (1910), 167–309.

[STI 1979] H. STICHTENOTH, *Die Hasse–Witt–Invariante eines Kongruenzfunktionenkörpers*, Arch. Math. (Basel) **33** N°4 (1979/80), 357–360.

[STI 1993] ———, *Algebraic function fields and codes*, Springer-Verlag, Berlin, 1993.

[STI 1995] D. STINSON, *Cryptography – theory and practice*, CRC Press, Inc., 1995.

[STLE 2003] M. STAM & A. K. LENSTRA, *Efficient subgroup exponentiation in quadratic and sixth degree extensions*, Cryptographic Hardware and Embedded Systems – CHES 2002, Lecture Notes in Comput. Sci., vol. 2523, Springer-Verlag, 2003, 317–332.

[STPO+ 2002] J. STERN, D. POINTCHEVAL, J. MALONE-LEE, & N. P. SMART, *Flaws in applying proof methodologies to signature schemes*, Advances in Cryptology – Crypto 2002, Lecture Notes in Comput. Sci., vol. 2442, Springer-Verlag, Berlin, 2002, 93–110.

[STR 1964] E. G. STRAUS, *Addition chains of vectors (problem 5125)*, Amer. Math. Monthly **70** (1964), 806–808.

[STXI 1995] H. STICHTENOTH & C. P. XING, *On the structure of the divisor class group of a class of curves over finite fields*, Arch. Math. (Basel) **65** N°2 (1995), 141–150.

[SUMA+ 2002] H. SUGIZAKI, K. MATSUO, J. CHAO, & S. TSUJII, *An Extension of Harley algorithm addition algorithm for hyperelliptic curves over finite fields of characteristic two*, Tech. Report ISEC2002-9(2002-5), IEICE, 2002.

[SWA 1962]   R. G. SWAN, *Factorization of polynomials over finite fields*, Pacific J. Math. **12** (1962), 1099–1106. [214]

[TAK 1998]   N. TAKAGI, *A VLSI algorithm for modular division based on the binary GCD algorithm*, IEICE Trans. Fundamentals **E81-A** N°5 (1998), 724–728. [206]

[TAK 2002]   M. TAKAHASHI, *Improving Harley algorithms for Jacobians of genus 2 hyperelliptic curves*, Symposium on Cryptography and Information Security – SCIS 2002. In Japanese. [314, 317]

[TAK 2004]   K. TAKASHIMA, *New families of hyperelliptic curves with efficient Gallant-Lambert-Vanstone method*, Preproceedings ICISC 2004. [381]

[TAT 1958]   J. TATE, *WC-groups over p-adic fields*, Séminaire Bourbaki; 10e année: 1957/1958. Textes des conférences; Exposés 152 à 168; 2e éd. corrigée, Exposé 156, vol. 13, Secrétariat mathématique, Paris, 1958. [118]

[TAT 1966]   _____, *Endomorphisms of abelian varieties over finite fields*, Invent. Math. **2** (1966), 134–144. [112, 448]

[TER 2000]   D. C. TERR, *A modification of Shanks' baby-step giant-step algorithm*, Math. Comp. **69** N°230 (2000), 767–773. [481]

[TES 1998a]   E. TESKE, *A space efficient algorithm for group structure computation*, Math. Comp. **67** N°224 (1998), 1637–1663. [486]

[TES 1998b]   _____, *Speeding up Pollard's rho method for computing discrete logarithms*, Algorithmic Number Theory Symposium – ANTS III, Lecture Notes in Comput. Sci., no. 1423, Springer-Verlag, Berlin, 1998, 541–554. [489]

[TES 2001a]   _____, *On random walks for Pollard's rho method*, Math. Comp. **70** N°234 (2001), 809–825. [489]

[TES 2001b]   _____, *Square-root algorithms for the Discrete Logarithm Problem (a survey)*, Public-Key Cryptography and Computational Number Theory, Walter de Gruyter, 2001, 283–301. [477]

[TES 2003]   _____, *Computing discrete logarithms with the parallelized kangaroo method*, Discrete Appl. Math. **130** N°3 (2003), 61–82. [493]

[THÉ 2003a]   N. THÉRIAULT, *Index calculus attack for hyperelliptic curves of small genus*, Advances in Cryptology – Asiacrypt 2003, Lecture Notes in Comput. Sci., vol. 2894, Springer-Verlag, Berlin, 2003, 75–92. [518, 521, 554]

[THÉ 2003b]   _____, *Weil descent attack for Artin–Schreier curves*, preprint, 2003. http://www.cacr.math.uwaterloo.ca/~ntheriault/weildescent.pdf [531, 534–537]

[THÉ 2003c]   _____, *Weil descent attack for Kummer extensions*, J. Ramanujan Math. Soc. **18** (2003), 281–312. [536]

[ThKe+ 1986]   J. J. THOMAS, J. M. KELLER, & G. N. LARSEN, *The calculation of multiplicative inverses over GF(p) efficiently where p is a Mersenne prime*, IEEE Trans. on Computers **35** N°5 (1986), 478–482. [206]

[THO 2001]   E. THOMÉ, *Computation of Discrete Logarithms in $\mathbb{F}_{2^{607}}$*, Advances in Cryptology – Asiacrypt 2001, Lecture Notes in Comput. Sci., vol. 2248, Springer-Verlag, 2001, 107–124. [501]

[THO 2003]   _____, *Algorithmes de calcul des Logarithmes Discrets dans les corps finis*, PhD. Thesis, École Polytechnique, Palaiseau, September 2003. [501, 507, 508]

[THU 1999]   E. G. THURBER, *Efficient generation of minimal length addition chains*, SIAM J. Comput. **28** N°4 (1999), 1247–1263. [158]

[TrZh+ 1997]   J. TROMP, L. ZHANG, & Y. ZHAO, *Small weight bases for Hamming codes*, Theoret. Comput. Sci. **181** N°2 (1997), 337–345. [217]

[TUN 1968]   B. P. TUNSTALL, *Synthesis of noiseless compression codes*, PhD. Thesis, Georgia Inst. Tech. Atlanta, 1968. [160]

[USB]     USB, *Universal serial bus specification revision 2.0*, April 27 2000.   [664]
          http://www.usb.org

[VÉL 1971]   J. VÉLU, *Isogénies entre courbes elliptiques*, C. R. Acad. Sci. Paris Sér. A-B **273** (1971),   [415, 427]
             A238–A241.

[VEPR⁺ 2001]  F. VERCAUTEREN, B. PRENEEL, & J. VANDEWALLE, *A memory efficient version of*   [433, 437]
              *Satoh's algorithm*, Advances in Cryptology – Eurocrypt 2001, Lecture Notes in Comput.
              Sci., vol. 2045, Springer-Verlag, Berlin, 2001, 1–13.

[VER 2001]   E. VERHEUL, *Evidence that XTR is more secure than supersingular elliptic curves cryp-*   [582]
             *tosystems*, Advances in Cryptology – Eurocrypt 2001, Lecture Notes in Comput. Sci., vol.
             2045, Springer-Verlag, Berlin, 2001, 195–210.

[VER 2004]   _____, *Evidence that XTR is more secure than supersingular elliptic curves cryptosys-*   [582]
             *tems*, J. Cryptology **17** (2004), 277–296.

[WAL 2001]   C. D. WALTER, *MIST: An efficient, randomized exponentiation algorithm for resisting*   [699]
             *power analysis*, Topics in Cryptology – CT-RSA 2002, Lecture Notes in Comput. Sci.,
             vol. 2271, Springer-Verlag, Berlin, 2001, 53–66.

[WAL 2002a]  _____, *Breaking the Liardet–Smart randomized exponentiation algorithm*, Smart Card   [699]
             Research and Advanced Applications – CARDIS 2002, Usenix Association, 2002, 59–68.

[WAL 2002b]  _____, *Some security aspects of the MIST randomized exponentiation algorithm*, Cryp-   [699]
             tographic Hardware and Embedded Systems – CHES 2002, Lecture Notes in Comput.
             Sci., vol. 2523, Springer-Verlag, Berlin, 2002, 276–290.

[WAL 2003]   _____, *Seeing through MIST given a small fraction of an RSA private key*, Topics in   [699]
             Cryptology – CT-RSA 2003, Lecture Notes in Comput. Sci., vol. 2612, Springer-Verlag,
             Berlin, 2003, 391–402.

[WAL 2004]   _____, *Security constraints on the Oswald–Aigner exponentiation algorithm*, Topics in   [699]
             Cryptology – CT-RSA 2004, Lecture Notes in Comput. Sci., vol. 2964, Springer-Verlag,
             Berlin, 2004, 208–221.

[WAT 1969]   W. WATERHOUSE, *Abelian varieties over finite fields*, Ann. Sci. École Norm. Sup. $4^e$   [278, 605]
             série **2** (1969), 521–560.

[WEB 1997]   H. J. WEBER, *Hyperelliptic simple factors of $J_0(N)$ with dimension at least* 3, Exper-   [104, 470, 472]
             iment. Math. **6** (1997), 273–287.

[WEI 1948]   A. WEIL, *Sur les courbes algébriques et les variétés qui s'en déduisent*, Publ. Inst. Math.   [135]
             Univ. Strasbourg **7** (1945), Hermann et Cie., Paris, 1948.

[WEI 1949]   _____, *Numbers of solutions of equations in finite fields*, Bull. Amer. Math. Soc. **55**   [134]
             (1949), 497–508.

[WEI 1957]   _____, *Zum Beweis des Torellischen Satzes*, Nachr. Akad. Wiss. Göttingen, Math. Phys.   [101]
             Klasse (1957), 33–53.

[WEI 2001]   A. WEIMERSKIRCH, *The application of the Mordell–Weil group to cryptographic sys-*   [383]
             *tems*, Master's thesis, Worchester polytechnic institute, 2001.

[WEI 2005]   E. W. WEISSTEIN, *RSA-200 factored*, 2005.   [7]
             http://mathworld.wolfram.com/news/2005-05-10/rsa-200/

[WEN 2001a]  A. WENG, *Hyperelliptic CM-curves of genus* 3, J. Ramanujan Math. Soc. **16** (2001),   [104, 108, 472, 473]
             339–372.

[WEN 2001b]  _____, *Konstruktion kryptographisch geeigneter Kurven mit komplexer Multiplikation*,   [106, 459, 471, 564]
             PhD. Thesis, Universität Gesamthochschule Essen, 2001.

[WEN 2003]   _____, *Constructing hyperelliptic curves of genus 2 suitable for cryptography*, Math.   [101, 465, 467, 468]
             Comp. **72** N°241 (2003), 435–458.

[WENG]      _____, *A table of class polynomials covering all CM-curves up to discriminant* 422500.   [457, 458, 460, 567]
            http://www.exp-math.uni-essen.de/zahlentheorie/classpol/class.html

[WIE 1986] D. H. WIEDEMANN, *Solving sparse linear equations over finite fields*, IEEE Trans. Inform. Theory **IT-32** N°1 (1986), 54–62. [501]

[WIL 1982] H. C. WILLIAMS, *A p+1 method of factoring*, Math. Comp. **39** N°159 (1982), 225–234. [604]

[WIZU 1998] M. WIENER & R. ZUCCHERATO, *Faster attacks on elliptic curve cryptosystems*, Selected Areas in Cryptography – SAC 1998, Lecture Notes in Comput. Sci., vol. 1556, Springer-Verlag, Berlin, 1998, 190–200. [491]

[WOBA+ 2000] A. WOODBURY, D. BAILEY, & C. PAAR, *Elliptic curve cryptography on smart cards without coprocessor*, Smart Card Research and Advanced Application – CARDIS 2000, IFIP Conference Proceedings, vol. 180, Kluwer Academic Publishers, 2000, 71–92. [229, 646]

[WOL 2004] T. WOLLINGER, *Software and hardware implementation of hyperelliptic curve cryptosystems*, PhD. Thesis, Ruhr-University of Bochum, 2004. [348, 352]

[WUHA+ 2002] H. WU, M. A. HASAN, I. F. BLAKE, & S. GAO, *Finite field multiplier using redundant representation*, IEEE Trans. on Computers **51** N°11 (2002), 1306–1316. [217]

[WUWU+ 2004] C.-H. WU, C.-M. WU, M.-D. SHIEH, & Y.-T. HWANG, *High-speed, low-complexity systolic designs of novel iterative division algorithms in $GF(2m)$*, IEEE Trans. on Computers **53** N°3 (2004), 375–380. [223]

[YAC 1998] Y. YACOBI, *Fast exponentiation using data compression*, SIAM J. Comput. **28** N°2 (1998), 700–703. [160]

[YAN 2001] T. YANIK, *New methods for finite field arithmetic*, PhD. Thesis, Oregon State University, 2001.
http://islab.oregonstate.edu/papers/01Yanik.pdf [202]

[YAO 1976] A. C. YAO, *On the evaluation of powers*, SIAM J. Comput. **5** (1976), 100–103. [158, 165]

[YASA+ 2002] T. YANIK, E. SAVAŞ, & Ç. K. KOÇ, *Incomplete reduction in modular arithmetic*, IEEE Micro **149** N°2 (2002), 46–52. [202, 640]

[ZAG 1981] D. ZAGIER, *Zetafunktionen und quadratische Zahlkörper*, Springer-Verlag, Berlin, 1981, Eine Einführung in die höhere Zahlentheorie. [An introduction to higher number theory], Hochschultext. [University Text]. [457]

[ZASA 1976] O. ZARISKI & P. SAMUEL, *Commutative algebra. vol. II.*, Springer, 1976. [45, 48, 67, 75, 78]

[ZEN] F. CHABAUD & R. LERCIER, *ZEN, version 3.0*, 2001.
http://zenfact.sourceforge.net/ [201]

[ZHA 2002] Z. ZHANG, *A one-parameter quadratic-base version of the Baillie-PSW probable prime test*, Math. Comp. **71** N°240 (2002), 1669–1734. [596]

[ZILE 1977] J. ZIV & A. LEMPEL, *A universal algorithm for data compression*, IEEE Trans. Inform. Theory **IT-23** (1977), 337–343. [160]

[ZIM 2001] P. ZIMMERMANN, *Arithmétique en précision arbitraire*, Tech. Report 4272, INRIA Lorraine, September 2001.
http://www.loria.fr/~zimmerma/papers/RR4272.ps.gz [172, 187]

# Notation Index

**Symbols**

$(G, \oplus)$: group $G$ with operation $\oplus$, 8
$(K, \Phi)$: CM-type, 105, 463
$(X_0 : X_1 : \cdots : X_n)$: projective point in $\mathbb{P}^n/K$, 46
$(X : Y : Z)$: projective point in $\mathbb{P}^2/K$, 270
$(u_{n-1} \ldots u_0)_2$: binary expansion of $u \in \mathbb{N}$, 170
$(u_{n-1} \ldots u_0)_b$: expansion of $u \in \mathbb{N}$ in base $b$, 170
$(n_{\ell-1} \ldots n_0)_s$: signed-digit expansion of $n \in \mathbb{N}$, 151
$(n_{\ell-1} \ldots n_0)_{\text{NAF}}$: expansion of $n$ in non-adjacent form, 151
$(n_{\ell-1} \ldots n_0)_{\text{NAF}_w}$: expansion of $n \in \mathbb{N}$ in width-$w$ non-adjacent form, 153
$(r_{l-1} \ldots r_0)_\tau$: $\tau$-adic expansion of $\eta \in \mathbb{Z}[\tau]$, 358, 370
$(r_{l-1} \ldots r_0)_{\tau\text{NAF}}$: expansion of $\eta \in \mathbb{Z}[\tau]$ in $\tau$-adic non-adjacent form, 359
$(r_{l-1} \ldots r_0)_{\tau\text{NAF}_w}$: expansion of $\eta \in \mathbb{Z}[\tau]$ in width-$w$ $\tau$-adic non-adjacent form, 363
$\begin{pmatrix} n_{0,\ell} \ldots n_{0,0} \\ n_{1,\ell} \ldots n_{1,0} \end{pmatrix}_{\text{JSF}}$: expansions of $n_0, n_1 \in \mathbb{N}$ in joint sparse form, 155
$\begin{pmatrix} r_{0,l+2} \ldots r_{0,0} \\ r_{1,l+2} \ldots r_{1,0} \end{pmatrix}_{\tau\text{JSF}}$: expansions of $\eta_0, \eta_1 \in \mathbb{Z}[\tau]$ in $\tau$-adic joint sparse form, 365
$\left(\frac{a}{p}\right)$: Legendre symbol of the integer $a$ modulo the prime $p$, 36
$\left(\frac{a}{b}\right)$: Kronecker–Jacobi symbol of the integers $a$ and $b$, 36
$\left(\frac{f(X)}{m(X)}\right)$: Legendre–Kronecker–Jacobi symbol of $f(X)$ and $m(X) \in \mathbb{F}_p[X]$, 36
$(r_1, r_2)$: signature of the number field $K/\mathbb{Q}$, 29
$(v_0, \ldots, v_s)$: addition chain computing the integer $v_s$, 157
$[\frac{1}{2}]P$: halving of the point $P$ on the elliptic curve $E$, 299
$[2]P$: doubling of the point $P$ lying on the elliptic curve $E$, 270
$[2]\overline{D}$: doubling of the divisor class $\overline{D}$, 307
$[G : H]$: index of the subgroup $H$ in the group $G$, 20
$[L : K]$: degree of the extension $L/K$, 25
$[n]$: scalar multiplication by $n$, i.e., $(n-1)$-fold application of the addition $\oplus$, 60
$[u(x), v(x)]$: divisor class in Mumford representation, 83, 307
$[x]$: Montgomery representation of the integer $x$, 180
$\langle x \rangle$: subgroup generated by $x$, 20
$\langle SCi \rangle_{i \in \mathcal{S}}$: average of the functions $SC_i$ for the $i \in \mathcal{S}$, 698
$\Delta_E$: discriminant of the elliptic curve $E$, 71
$\Delta_G$: delay at each stage in a full $n$-stage carry-look-ahead adder, 629
$\Lambda_C$: period lattice of curve $C$, 91, 462
$\Lambda_E$: period lattice of elliptic curve $E$, 95
$\Lambda_\tau$: period lattice of elliptic curve $E_\tau$ for $\tau \in \mathcal{H}$, 97
$\Omega^0(K(C))$: set of holomorphic differentials of $C$, 76
$\Omega$: period matrix of the abelian variety $\mathcal{A}$, 93

$\Omega_C$: period matrix of $\Lambda_C$ called period matrix of $C$, 93, 100
$\chi(\phi_q)_{J_C}(T)$: characteristic polynomial of the Frobenius endomorphism $\phi_q$ on $C$ and of $J_C$, 110, 310
$\chi(\varphi)_\mathcal{A}(T)$: characteristic polynomial of endomorphism $\varphi$ on abelian variety $\mathcal{A}$, 62
$\chi_E(T)$: characteristic polynomial of the Frobenius endomorphism of an elliptic curve $E$, 278
$\chi_{a_2}(T)$: characteristic polynomial of the Frobenius endomorphism on $E_{a_2}$, 356
$\chi_\psi(T)$: characteristic polynomial of the endomorphism $\psi$ of a GLV curve, 377
$\delta_\mathcal{N}$: density of the multiplication matrix $T_\mathcal{N}$, 221
$\eta_n(h)$: probability that the longest carry-chain in $n$-bit addition is of length at least $h$, 632
$\Phi_k(x)$: $k$-th cyclotomic polynomial, 586
$\Phi_\ell(X,Y)$: $\ell$-th modular polynomial, 414
$\Phi_{\ell,X}(X,Y)$: partial derivative of $\Phi_\ell(X,Y)$ with respect to $X$, 419
$\Phi_{\ell,Y}(X,Y)$: partial derivative of $\Phi_\ell(X,Y)$ with respect to $Y$, 419
$\Phi_{\ell,XX}(X,Y)$: second partial derivative of $\Phi_\ell(X,Y)$ with respect to $X$, 420
$\Phi_{\ell,XY}(X,Y)$: second partial derivative of $\Phi_\ell(X,Y)$ with respect to $X$ and $Y$, 420
$\Phi_{\ell,YY}(X,Y)$: second partial derivative of $\Phi_\ell(X,Y)$ with respect to $Y$, 420
$\Phi_\ell^c(X,Y)$: $\ell$-th canonical modular polynomial, 419
$\phi_p$: absolute Frobenius automorphism of a finite field, 33
$\phi_q$: relative Frobenius automorphism of a finite field, 33
$\phi_p$: Frobenius morphism from the variety $V$ to $\phi_p(V)$, 53
$\psi_n$: $n$-th division polynomial, 80
$\Sigma$: Frobenius substitution of $\mathbb{Z}_q$, 240
$\Sigma$: $\{0,1\}$, 715
$\Sigma^*$: set of sequences of countable infinite length with coefficients in $\Sigma$, 715
$\Sigma_{F/K}$: set of places of $F/K$, 65
$\Sigma^n$: set of finite sequences of length $n$ with coefficients in $\Sigma$, 715
$\tau, \overline{\tau}$: complex roots of $\chi_E(\phi_q)$, 279
$\tau_1, \ldots, \tau_{2g}$: complex roots of $\chi(\phi_q)_C(T)$ for curve of genus $g$, 311
$\Theta^{(\Omega)}$: Riemann theta divisor associated to the period matrix $\Omega$, 100
0xA: 10 in hexadecimal notation, 290
$\infty$: point at infinity $(0:1)$, 50
$|u|_b$: length of the expansion of $u$ in base $b$, 170
$|x|_p$: $p$-adic norm of $x \in \mathbb{Q}_p$, 41
$|x|_K$: norm of $x$ belonging to the complete discrete valuation field $K$, 41
$\lfloor \lambda \rceil$: rounded value of $\lambda \in \mathbb{Q}$ given by $\lfloor \lambda + 1/2 \rfloor$, 360
$\lceil \lambda \rfloor$: rounded value of $\lambda \in \mathbb{Q}$ of smallest absolute value, 360
$\lfloor \lambda \rceil_\tau$: rounded value of $\lambda \in \mathbb{Z}[\tau]$, 360
$\Delta(S)$: discrepancy of sequence $S \in [0,1]^\mathbb{N}$, 731
$\overline{K}$: algebraic closure of the field $K$, 27
$\overline{u}$: bitwise complement of the single $u$, 172
$\overline{x}$: complement of the bit $x$, 170
$\varphi^*$: induced $K$-algebra morphism $\varphi^* : K[W] \to K[V]$ by morphism $\varphi \in \mathrm{Mor}_K(V,W)$, 52
$\hat{\psi}$: dual isogeny of the isogeny $\psi$ of an elliptic curve, 277
$\hat{\psi}$: dual isogeny of the isogeny $\psi$ of the Jacobian of a hyperelliptic curve, 309
$\mu(j)$: Möbius function of the integer $j$, 33
$\nu(n)$: Hamming weight of the integer $n$, 147
$\varphi(N)$: Euler totient function of the integer $N$, 23
$\pi(x)$: number of primes less than $x$, 6
$\tau(\chi)$: Gauß sum of the character $\chi$, 599

Notation Index

$\theta(z, \Omega)$: Riemann theta function associated to the period matrix $\Omega$, 100
$\theta \begin{bmatrix} \delta \\ \epsilon \end{bmatrix}(z, \Omega)$: theta characteristics associated to the period matrix $\Omega$, 100
$[x][y]$: Montgomery multiplication of the integers $x$ and $y$ in Montgomery representation, 204
$\langle \cdot, \cdot \rangle_L$: Lichtenbaum pairing, 120
$\langle \cdot, \cdot \rangle_{T,n}$: Tate pairing, 117
$P \oplus Q$: addition of the points $P$ and $Q$, 79, 270
$\overline{D} \oplus \overline{E}$: addition of divisor classes $\overline{D}$ and $\overline{E}$, 307
$v \oplus j$: composition of the addition chain $v$ and of the integer $j$, 161
$v \otimes w$: composition of the addition chains $v$ and $w$, 161
$x \vee y$: disjunction of the bits $x$ and $y$, 170
$u \vee v$: bitwise disjunction of the singles $u$ and $v$, 172
$x \wedge y$: conjunction of the bits $x$ and $y$, 170
$u \wedge v$: bitwise conjunction of the singles $u$ and $v$, 172
$u \ll t$: left shift of $t$ bits of the single $u$, 172
$u \gg t$: right shift of $t$ bits of the single $u$, 172
$\|$: concatenation, 5
$\Omega_C$: period matrix of $\Lambda_C$ called period matrix of $C$, 462
$\mathcal{A} \sim \mathcal{B}$: isogenous abelian varieties $\mathcal{A}$ and $\mathcal{B}$, 58
$\mathcal{A} \simeq \mathcal{B}$: isomorphic abelian varieties $\mathcal{A}$ and $\mathcal{B}$, 58
$x \equiv y \pmod{n}$: $x$ and $y$ are congruent modulo $n$, 21
$x \in_R E$: choose $x$ at random in the set $E$, 10

# A

$A$: an elementary addition, 692
$\mathcal{A}$: abelian variety, 56
$\mathcal{A}$: affine coordinates, 280, 291, 316, 336
$a_1, a_2, \ldots, a_{2g}$: coefficients of $\chi(\phi_q)_C(T)$ for curve of genus $g$, 135
$a_1, a_2, a_3, a_4, a_6$: coefficients of elliptic curve in Weierstraß form, 69, 268
$\mathcal{A}[n]$: set of $n$-torsion points of an abelian variety $\mathcal{A}$, 60
$\mathbb{A}^n$: affine space of dimension $n$ over $K$, 47
$\mathbb{A}^n(L)$: set of $L$-rational affine points, 47
$A^\dagger$: dagger ring, 140
$\mathrm{AGM}(a_0, b_0)$: Arithmetic-Geometric-Mean of $a_0$ and $b_0$, 434

# B

$B$: a smoothness bound, 496
$\mathfrak{B}$: a basis of the vector space $\mathfrak{D}_0(\mathcal{J}_c, \mathbb{Q}_q)$, 443
$b_2$: $a_1^2 + 4a_2$, 70
$b_4$: $2a_4 + a_1 a_3$, 70
$b_6$: $a_3^2 + 4a_6$, 70
$b_8$: $a_1^2 a_6 - a_1 a_3 a_4 + 4a_2 a_6 + a_2 a_3^2 - a_4^2$, 72
$B_S$: balance of sequence $S \in \mathbb{F}_q^{\mathbb{N}}$, 730
$B_S(\alpha)$: balance of sequence $S \in \mathbb{F}_q^{\mathbb{N}}$ with respect to $\alpha \in \mathbb{F}_q^*$, 730

# C

$\mathfrak{C}$: Cartier operator, 411
$\widetilde{C}$: desingularization of $C$, 64
$c_4$: $b_2^2 - 24 b_4$, 70

$c_6$: $-b_2^3 + 36b_2b_4 - 216b_6$, 70
$C_S$: autocorrelation of sequence $S \in \mathbb{F}_q^{\mathbb{N}}$, 731
$C_{S,T}$: crosscorrelation of sequences $S$ and $T \in \mathbb{F}_q^{\mathbb{N}}$, 731
$C_{ab}$: $C_{ab}$ curve, 82
$C_{an}$: analytic curve, 89
$\mathrm{char}(R)$: characteristic of the ring $R$, 22
$\mathrm{Cl}(\mathcal{O})$: ideal class group of $\mathcal{O}$, 81
$\mathrm{Cl}_K$: class group of $\mathcal{O}_K$, 30

# D

$\overline{D}$: divisor class of $D$, 76
$\mathfrak{D}_0(\mathcal{A}_c, \mathbb{Q}_q)$: space of holomorphic differential forms of degree 1 on $\mathcal{A}_c$ defined over $\mathbb{Q}_q$, 139
$\deg(L/K)$: degree of the extension $L/K$, 25
$\deg(D)$: degree of a divisor $D$, 66
$\deg_s(L/K)$: degree of separability of $L$ over $K$, 27
$df$: differential of $f \in K(C)$, 75
$\dim(V)$: dimension of the variety $V$, 49
$D \in \mathrm{Div}_C$: divisor on $C$, 66
$\mathrm{div}(f)$: principal divisor of $f \in K(C)^*$, 67
$\mathrm{div}_{an}(f)$: analytic divisor of the meromorphic function $f$, 88
$\mathrm{Div}_C$: divisor group of the curve $C$, 66
$\mathrm{Div}_C^0$: group of degree zero divisors of the curve $C$, 67, 76, 306

# E

$E$: elliptic curve, $E : y^2 + a_1xy + a_3y = x^3 + a_2x^2 + a_4x + a_6$, 69, 268
$\mathcal{E}$: canonical lift over $\mathbb{Q}_q$ of an ordinary elliptic curve $E$ defined over $\mathbb{F}_q$, 423
$E((w_n))$: Shannon entropy function of the sequence $(w_n)$ of words of $\Sigma^k$, 716
$E(K)$: $K$-rational points of the elliptic curve $E$, 272
$e(P_1, P_2)$: a bilinear map, 9
$E[n]$: set of $n$-torsion points of an elliptic curve $E$, 273
$E_{a_2}$: binary Koblitz curve $y^2 + xy = x^3 + a_2x^2 + 1$, 356
$E_M$: elliptic curve in Montgomery form, 285
$\widetilde{E}_v$: quadratic twist of the elliptic curve $E$ by $v$, 71, 274
$\mathrm{End}_K(\mathcal{A})$: set of endomorphisms of the abelian variety $\mathcal{A}$, 59
$\mathrm{End}_K(\mathcal{A})^0$: $\mathrm{End}_K(\mathcal{A}) \otimes_\mathbb{Z} \mathbb{Q}$, 60
$\mathrm{End}_K(E)$: set of endomorphisms of the elliptic curve $E$ defined over $K$, 277
$\mathrm{End}_K(J_C)$: set of endomorphisms of the Jacobian of the hyperelliptic curve $C$ defined over $K$, 310
$\exp(x)$: $p$-adic exponential of $x \in \mathbb{Z}_q$, 259

# F

$\mathcal{F}_p$: lift of the Frobenius isogeny $\phi_p$ of an ordinary elliptic curve defined over $\mathbb{F}_q$, 423
$\mathcal{F}_q$: lift in $\mathrm{End}(\mathcal{E})$ of the Frobenius endomorphism $\phi_q$, 423
$\mathbb{F}_q$: the unique finite field of order $q$, 31

# G

$g$: genus, 68, 304
$G_B$: set of $B$-smooth elements of the group $G$, 499
$G_L$: absolute Galois group of $L$, 45

Notation Index

$G_{L/K}$: Galois group of $L$ over $K$, 28
$G_n(\Lambda)$: Eisenstein series of weight $n$ associated to the lattice $\Lambda$, 96
$\mathrm{Gal}(L/K)$: Galois group of $L$ over $K$, 28
$\gcd(x,N)$: greatest common divisor of the integers $x$ and $N$, 190
$GF(q)$: the unique finite field of order $q$, 31

# H

$\mathcal{H}$: Poincaré upper half plane $\{\tau \in \mathbb{C} \mid \Im m(\tau) > 0)\}$, 97
$h(K)$: class number of the number field $K$, 30
$h_d$: class number of number field $\mathbb{Q}(\sqrt{-d})$, 99
$H_d(x)$: Hilbert class polynomial for the discriminant $d$, 99, 456
$H_{K,k}(X)$: class polynomial for a hyperelliptic curve of genus 2 and 3, 107, 462
$\mathbb{H}_g$: Siegel upper half plane, 100
$H^i(\overline{X}, \mathbb{Q}_\ell)$: $\ell$-adic cohomology group, 136
$H^i_{DR}(M)$: de Rham cohomology group of the compact complex manifold $M$, 134
$H^n(G_K, M)$: quotient of the group of $n$-cocycles modulo the subgroup of $n$-coboundaries, given the $G_K$-module $M$, 116
$\mathrm{Hom}_K(\mathcal{A},\mathcal{B})$: set of homomorphisms from the abelian variety $\mathcal{A}$ to the abelian variety $\mathcal{B}$, 57
$\mathrm{Hom}_K(\mathcal{A},\mathcal{B})^0$: $\mathrm{Hom}_K(\mathcal{A},\mathcal{B}) \otimes_\mathbb{Z} \mathbb{Q}$, 57

# I

I: an elementary inversion, 280
$I_2, I_4, I_6, I_{10}$: absolute invariants of a hyperelliptic of genus 2, 101
$\mathrm{INV}(u)$: Montgomery inverse of the integer $u$, 207

# J

$\mathcal{J}$: Jacobian coordinates, 282, 292
$\mathcal{J}^c$: Chudnovsky Jacobian coordinates, 282
$\mathcal{J}^m$: modified Jacobian coordinates, 282
$\mathcal{J}^s$: simplified Chudnovsky Jacobian coordinates, 401
$j_E$: absolute invariant or $j$-invariant of the elliptic curve $E$, 71
$j(\tau)$: $j$-function of $\tau \in \mathcal{H}$, 97
$j(q)$: $j$-function of $q = e^{2\pi i \tau}$, 98
$j_1, j_2, j_3$: Igusa invariants of a hyperelliptic curve of genus 2, 101, 462
$j'_1, j'_2, j'_3$: Mestre's invariants of a hyperelliptic curve of genus 2, 102
$j_1, j_3, j_5, j_7, j_9$: Shioda invariants of a hyperelliptic curve of genus 3, 104
$J(\chi_1, \chi_2)$: Jacobi sum of the characters $\chi_1$ and $\chi_2$, 600
$J_C$: Jacobian variety of the curve $C$, 78
$J_C(\mathbb{F}_q)[n]$: $\mathbb{F}_q$-rational $n$-torsion points on Jacobian of curve $C$, 111
$J_C[n]$: set of $n$-torsion points of Jacobian of $C$, 309
$J_K$: group of fractional ideals of the field $K$, 30

# K

$\mathrm{K}(n)$: complexity to multiply two $n$-word integers with Karatsuba method, 176
$K(P)$: field of definition of $P$, 46
$K(V)$: function field of the variety $V$, 51
$\ker \psi$: kernel of a group homomorphism, 21
$\ker(\varphi)^0$: connected component of the unity of $\ker(\varphi)$ for $\varphi \in \mathrm{Hom}_K(\mathcal{A},\mathcal{B})$, 58

## L

$L(D)$: $\{f \in K(C) \mid \operatorname{div}(f) \geqslant -D\}$, 67
$\ell(D)$: $\dim_K(L(D))$, 67
$\ell(n)$: length of the shortest addition chain computing the integer $n$, 157
$\mathcal{L}(S)$: linear complexity of sequence $S$ with coefficients in a ring, 732
$L(T)$: $L$-polynomial of a curve, 135
$L/K$: $L$ is an extension field of $K$, 25
$L^H$: invariant elements of the field $L$ under the action of the group $H$, 28
$L_N(\alpha, c)$: complexity of algorithm depending on $N$, 4
$L_N(\alpha)$: $L_N(\alpha, 1/2)$, 4
$\mathcal{LD}$: López–Dahab coordinates, 293
$\lg x$: binary logarithm of $x$, 3
$\varprojlim A_i$: inverse limit of $(A_i, \{p_{ij}\}_{j \in I})$, 40
$\ln x$: natural logarithm of $x$, 3
$\log(x)$: $p$-adic logarithm of $x \in \mathbb{Z}_q$, 258
$\log_a b$: logarithm of $b$ to base $a$, 3
$\log_P(Q)$: discrete logarithm of $Q$ to the base of $P$, 8

## M

$\mathcal{M}$: additive arithmetical semigroup, 496
M: an elementary multiplication, 149
$\mathrm{M}(n)$: complexity to multiply two $n$-word integers, 174
$\mathcal{M}_B$: set of elements of $\mathcal{M}$ of size not larger than $B$, 496
$\mathfrak{m}_P$: local ring of the point $P$, 64
$\mathfrak{m}_v$: maximal ideal of valuation $v$, 65
$x \bmod n$: canonical representative of $x$ modulo $n$, 21
$x \bmod s\, n$: centered representative of $x$ modulo $n$, 21
$\operatorname{Mor}_K(V, W)$: set of morphisms from the variety $V$ to the variety $W$, 52

## N

$\mathcal{N}$: new coordinates, 323, 342
$n_\psi$: norm of the endomorphism $\psi$ of a GLV curve, 377
$n_B$: cardinality of $\mathcal{P}_B$, 497
$n'_B$: cardinality of $\mathcal{M}_B$, 497
$N_k$: number of $\mathbb{F}_{q^k}$-rational points on an algebraic variety defined over $\mathbb{F}_q$, 112, 134
$\mathrm{N}_{L/K}(x)$: norm of $x \in L$ over $K$, 26
$n_P(f)$: order of vanishing of the meromorphic function $f$ at the point $P$, 88
Neg: an elementary negation, 692
**NP**: set of decision problems whose "yes" answer can be verified in polynomial time, 4

## O

$O(f(N_1, \ldots, N_s))$: big-$O$ of $f$, 3
$o(f(N_1, \ldots, N_s))$: small-$o$ of $f$, 3
$\tilde{O}(g(n))$: $O(g(n) \lg^k g(n))$, 3
$\mathcal{O}_K$: integer ring of the field $K$, 29
$\mathcal{O}_P$: set of rational functions that are regular at the point $P$, 64
$\mathcal{O}_P/\mathfrak{m}_P$: residue field of the point $P$, 64
$\mathcal{O}_S$: ring of regular functions on $S$, 64

$\mathcal{O}_v$: local ring of valuation $v$, 65

# P

**P**: set of decision problems solvable in polynomial time, 4
$\mathcal{P}$: a countable set whose elements are called primes, 496
$\mathcal{P}$: projective coordinates, 281, 292, 321, 341, 346
$\mathfrak{p}$: place of $K(C)$, 65
$P_\infty$: point at infinity, 269, 306
$\mathcal{P}_B$: set of $B$-smooth prime divisors, 513
$P_K$: subgroup of fractional principal ideals of the field $K$, 30
$\mathbb{P}^n(L)$: set of $L$-rational points, 46
$\mathbb{P}^n/K$: $n$-dimensional projective space over $K$, 46
$\wp(z, \Lambda)$: Weierstraß $\wp$-function associated to the lattice $\Lambda$, 96
$\mathrm{Pic}_C^0$: divisor class group of degree zero or Picard group of the curve $C$, 76, 306
$\mathrm{Pic}_{C \cdot L}^0$: Picard group of base change $\mathrm{Pic}_{C \cdot L}^0 = (\mathrm{Pic}_{C \cdot \overline{K}}^0)^{G_L}$, 77
$\mathrm{Princ}_C$: group of principal divisors of the curve $C$, 67

# Q

$\mathbb{Q}_q$: unramified extension of $\mathbb{Q}_p$, 43

# R

$\mathcal{R}$: recent coordinates, 346
$R^*$: multiplicative group of the invertible elements of the ring $R$, 23
$R(N)$: reciprocal integer of the integer $N$, 178
$\mathrm{REDC}(u)$: Montgomery reduction of the integer $u$, 180

# S

S: an elementary squaring, 149
$S^{\mathrm{AS}}$: pseudorandom generator from elliptic curves, 733
$S^{\mathrm{NR}}$: Naor–Reingold pseudorandom generator, 734
$S^{\mathrm{PG}}(f, P)$: elliptic curve power pseudorandom generator, 733
$SC_i$: side-channel information obtained from the computation of $[n]\overline{D}_i$, 698
$\mathrm{Sp}(2g, \mathbb{Z})$: symplectic group over $\mathbb{Z}$, 100

# T

$t_\psi$: trace of the endomorphism $\psi$ of a GLV curve, 377
$T_\mathcal{N}$: multiplication matrix of a normal basis, 220
$T_{d,N}$: time to multiply two polynomials of degree less than $d$ in $(\mathbb{Z}/p^N\mathbb{Z})[X]$, 243
$T_f(N)$: time to evaluate $f \in \mathbb{Z}_q[X]$ at precision $N$, 247
$T_\ell(\mathcal{A})$: $\ell$-adic Tate module of $\mathcal{A}$, 61
$T_n$: Tate–Lichtenbaum pairing, 122
$t_P$: translation by point $P$, 57
$\mathrm{Tr}(\phi_q)$: trace of the Frobenius endomorphism on $C$, 110
$\mathrm{Tr}_{L/K}(x)$: trace of $x \in L$ over $K$, 26

# U

$U^+$: totally positive units of real subfield $K_0$ of CM-field $K$, 106
$U_k$: Lucas sequence, 357, 595

$U_i$: affine open parts, standard covering, 47

## V

$V_I$: $\{P \in \mathbb{P}^n(\overline{K}) \mid f(P) = 0,\ \forall f \in I\}$ for a homogeneous ideal $I$, 47
$V_I$: $\{P \in \mathbb{A}^n(\overline{K}) \mid f(P) = 0,\ \forall f \in I\}$, for ideal $I$, 47
$V_k$: Lucas sequence, 357, 595
$v_K$: discrete valuation of the complete discrete valuation field $K$, 41
$V_L$: variety $V/K$ considered as $V_L/L$, 51
$v_P$: valuation at the point $P$ on $\mathcal{O}_P$, 65
$v_p(x)$: $p$-adic valuation, 41
$\overline{V}_p$: Verschiebung of $\phi_p$, 61
$\overline{V}_q$: Verschiebung of $\phi_q$, 427
$\mathcal{V}_q$: lift of Verschiebung on $\mathfrak{B}$, 443

## W

$W_n$: Weil pairing, 117
$w(K)$: order of the group of roots of unity in $\mathcal{O}_K$, 31
$W_{L/K}(V)$: Weil descent of the variety $V$ defined over $L$, 126

## X

$x^n$: $x$ to the power $n$, 145
$x$ XOR $y$: exclusive disjunction of the bits $x$ and $y$, 170
$u$ XOR $v$: bitwise exclusive disjunction of the singles $u$ and $v$, 172

## Z

$Z(n)$: Zech's logarithm of $\gamma^n$ where $\langle \gamma \rangle = \mathbb{F}_q^*$, 33
$Z(X/\mathbb{F}_q; T)$: zeta function of the algebraic variety $X$ defined over $\mathbb{F}_q$, 134
$\mathbb{Z}/n\mathbb{Z}$: quotient group of $\mathbb{Z}$ modulo $n$, 20
$\mathbb{Z}_p$: ring of $p$-adic integers, 40
$\mathbb{Z}_q$: ring of integers of $\mathbb{Q}_q$, 43

# General Index

**Symbols**

$2^k$-ary method, **148**, 165, 356

**A**

ABC, *see* anomalous binary curve
Abel–Jacobi map, 91
Abel–Jacobi theorem, 91
abelian group, 20
abelian variety, 56
    canonical lift, 138
    complex multiplication, 63, 456
    endomorphism, 59
    homomorphism, 57
    isogeny, 58
    isogeny over $\mathbb{C}$, 94
    isomorphism, 58
    $\ell$-adic Tate module, 61
    ordinary, 60
    $p$-rank, 61
    period matrix, 93
    principally polarized, 93
    simple, 60
    supersingular, 61
    Verschiebung, 61
    *see also* Jacobian variety
absolute invariant of an elliptic curve, 71, 72, 97
absolute ramification index, 43
absolutely irreducible variety, 48
addition
    binary field, 218
    hardware
        carry-chain, 632
        carry-look-ahead adder, **628**, 629
        carry-save adder, 634
        carry-select adder, 630
        conditional-sum adder, 630
        Dadda tree, **635**, 638
        full adder, 627
        generated carry, 629
        half adder, 628
        propagated carry, 629
        Wallace tree, **635**, 638
    integers, 173
    optimal extension field, 231
    prime field, 202
addition chain, **157**, 157–164, 224, 233, 237, 559, 646, 699
    dichotomic strategy, 161
    dyadic strategy, 162
    factor method, 162
    Fermat's strategy, 162
    power tree method, 160
    Thurber's algorithm, 158
    total strategy, 162
    Tunstall method, 160
    *see also* addition-subtraction chain, addition sequence, and vectorial addition chain
addition law
    elliptic curve, 79, 270
    hyperelliptic curve, 306
addition sequence, **158**, 162
addition-subtraction chain, 158
additive arithmetical semigroup, 496
Adleman–DeMarrais–Huang algorithm, 512–516
admissible change of variables, 274
admissible encoding function, 590
affine coordinates
    elliptic curve, 280–281, 283, 284, 291–292, 294–295, 297
    hyperelliptic curve, 316, 336
affine space, **47**, 49
    closed set, 47
    open set, 47
affine variety, 48
AGM algorithm, 434–449
    elliptic curve, 441
    generalized sequence, 445
    hyperelliptic curve, 446
    iteration, 434

univariate sequence, 440
AKS test, 600
algebraic closure, 27
algebraic coding theory, 15
algebraic element, 25
algebraic extension of a field, 25
algebraic group, 55
algebraic integer, 29
algebraic number, 29
    conjugate, 27
    integral over $\mathbb{Z}$, 29
    minimal polynomial, 25
all one polynomial, 215
almost irreducible trinomial, 217
almost Montgomery inverse, 208
analytic variety, 88
anomalous basis, 217
anomalous binary curve (ABC), *see* Koblitz curve
anomalous curve, 76
ANSI-C generator, 720
abcs, 267
APRCL Jacobi sum test, 599
arithmetic
    binary field, 213–229
    elliptic curve, 280–302
        Hessian form, 696
        Jacobi model, 696
        supersingular, 290
        triple and add, 581
        unified formulas, 694
    hardware, 617–646
    hyperelliptic curve, 303–352
        genus 2, 313–347
        genus 3, 348–352
    optimal extension field, 229–237
    $p$-adic numbers, 239–263
    prime field, 201–213
    special curve, 355–387
        elliptic Koblitz curve, 356–367
        generalized Koblitz curve, 367–376
        GLV curve, 377–380
        trace zero variety, 383–387
Arithmetic-Geometric-Mean, 434–435
    *see also* AGM algorithm
arithmetical formation, **496**, 496–497
    example
        Jacobian of hyperelliptic curves, 497
        nonprime finite field, 497
        prime field, 497
    prime, 496
    size map, 496
    smoothness bound, 496
Artin–Schreier
    cover, 453
    curve, 536
    equation, **252**, 252–256, 434
        Harley's algorithm, 255
        Lercier–Lubicz' algorithm, 253, 434
    extension, 453
    operator, **533**, 536
    root, 254, 255
    theory, 532, 534
associative law, 20
asymmetric Diffie–Hellman encryption, 10
asynchronous card, 665
Atkin prime, **414**, 415
    *see also* SEA algorithm
atomic force microscope, 671
attack
    Chinese remaindering, 479
    Frey–Rück, 530
    invalid curve, 569, 706
    man-in-the-middle, 10
    MOV, 530
    Rück, 529
    Silver–Pohlig–Hellman, *see* Chinese remaindering
    small subgroup, 569
    *see also* baby-step giant-step algorithm, index calculus, pairing, Pollard's kangaroo, Pollard's rho, side-channel attacks, transfer, and Weil descent
autocorrelation of a sequence, 731
autocorrelation test, 729

## B

$B$-smooth
    divisor, 512–514
    number, 604, 608–611
baby-step giant-step algorithm, 155, 226, **480**, 480–483
    class number computation, 480
    DL computation, 376, 482
    point counting, 411, 416
balance of a sequence, 730
Barrett method, **179**, 182, 202, 203, 210, 606
    reciprocal integer, 178

base $b$ representation, 170
base change, 57, 66, 77
basis
    complementary, 35
    dual, 35
    normal, **34**, 224, 228
        self-complementary, 35
        optimal normal basis, 217–218, **221**, 220–221
    polynomial, **34**, 218–220, 225–226
        anomalous, 217
        ghost bit basis, 217
        pentanomial, 214
        redundant trinomial, **217**, 228
        sedimentary polynomial, 215
        sparse, **34**, 214, 216
        trinomial, 214
    self-dual, 35
    vector space, 24
Bellcore attack, 683
    *see also* side-channel attacks
Berlekamp–Massey algorithm, 501
BGMW algorithm, *see* Yao's exponentiation method
BH curve, 381
big endian, 171
BigNum, 169
binary extended gcd, 195
binary field
    arithmetic, 213–229
        exponentiation, *see* exponentiation in a binary field
        inversion, *see* inversion in a binary field
        multiplication, *see* multiplication in a binary field
        square root, 228–229
        squaring, 221
    hardware, 641–645
        Massey–Omura multiplier, 643
        Mastrovito multiplier, 642
        normal basis, 643–645
        polynomial basis, 642–643
        unified multipliers, 644–645
    quadratic equation, 228
    representation, *see* normal basis, optimal normal basis, and polynomial basis
    trace computation, 35, 229
binary quadratic form, 457
birational map, 53
birationally equivalent varieties, 53

birthday paradox, 478, **483**, 520, 612
bit, 170
bitwise complement, 172
bitwise conjunction, 172
bitwise disjunction, 172
bitwise exclusive disjunction, 172
black-box group, **478**, 554
Blum–Blum–Shub generator, 720
Booth recoding technique, **151**, 625
borrow bit, 172
BPSW test, 596
Braid groups, 15
Brent–Kung modular composition, 225
byte, 170

# C

$C_{ab}$-curve, 82, 352
canonical lift
    abelian variety, 138
    elliptic curve, 423
    hyperelliptic curve, 442
Cantor's algorithm, 308, 552
cardinality
    elliptic curve, 278–279
        admissible, 278
        cofactor, 272
        Hasse–Weil theorem, 278
        subfield curve, 279
        trace of the Frobenius, 278
        twist, 279
    heuristics, 564
    hyperelliptic curve, 310–311
        divisor class group, 111
        Hasse–Weil theorem, 310
    *see also* point counting and complex multiplication
Carmichael number, 593
carry bit, 172
carry-chain, 632
carry-look-ahead adder, **628**, 629
carry-save adder, 634
carry-select adder, 630
Cartier operator, 411
CBDHP, *see* computational bilinear Diffie–Hellman problem
CDHP, *see* computational Diffie–Hellman problem
Čebotarev's density theorem, 236
centered representative, 21

certification authority, 576
CFRAC, *see* continued fraction method
character of a finite field, 35
characteristic of a field, 23
characteristic of a ring, 22
characteristic polynomial of an endomorphism, 62
Chinese remainder theorem (CRT), **22**, 196–197, 199, 316, 413, 685
Chinese remaindering attack, 479
Chudnovsky Jacobian coordinates, 282–284
class group, 30, 81, 84, 306
class modulo $n$, 20
class number, **30**, 457, 480, 598
class polynomial, 107, 462, 463, 465, 466, 472
CM-field, 105, 107, 462
CM-type, 63, 105, 463
cohomology
    crystalline, 136, 451
    de Rham, **134**, 140, 567
    Dwork–Reich, 138, 453
    Monsky–Washnitzer, 136, **139**, 451, 567
    rigid, 136
combi-card, 650
commutative group, 20
commutative law, 19
compact curve, 382
complement, **170**, 172
    one's ∼, 621
    two's ∼, **171**, 620
complementary basis, 35
complete discrete valuation field, 41
complex homomorphism, 29
complex multiplication, 63
    abelian variety, 63, 456
    elliptic curve, 98, 277, 456–460
        computation of norms, 458
        construction, 459
    Hilbert class field, 99
    Hilbert class polynomial, 99, 456
    hyperelliptic curve, 104, 310
        computation of theta constant, 465
        construction, 469
        genus 2, 460–470
        genus 3, 470
        genus $\geqslant 3$, 470–473
        Mestre's algorithm, 468
        with endomorphisms, 471
complexity, **2**, 3
    exponential, 4, 548
    polynomial, 4, 548
    space, 3
    subexponential, 4, 548
    time, 3
compositeness test, **591**, 592–596
    BPSW, 596
    Fermat, 593
    Lucas pseudoprime, 595
    OPQBT, 596
    pseudoprime, 593
    Rabin–Miller, **594**, 596
    Solovay–Strassen, 595
    trial division, 592
    *see also* primality test
composition law, 19
    associative, 20
    commutative, 19
    distributive, 21
compression
    elliptic curve, 288–289, 302
    hyperelliptic curve, 311–313
computation of norms, 458
computation problem, 3
computational bilinear Diffie–Hellman problem (CBDHP), 575
computational Diffie–Hellman problem (CDHP), 10, 574
computer algebra system
    Magma, 201, 267
    Maple, 267
    PARI/GP, 267, 467
    SIMATH, 267
concentric circles method, 525–527
conditional-sum adder, 630
congruence, 21
conjugate of an algebraic number, 27
conjunction, **170**, 172
construction of DL systems, 565
construction of ordinary elliptic curves with small embedding degree, 586
contact card, 649
contactless card, 649
content of a polynomial, 613
continued fraction method, 608
coordinate ring of a variety, 51
coordinates
    elliptic curve
        affine, 280–281, 283, 284, 291–292, 294–295, 297

Chudnovsky Jacobian, 282–284
  comparison, 283, 296
  generalized projective, 271
  Jacobian, 282, 292–295, 297
  López–Dahab, 293–295, 297
  mixed, 283–285, 296–298
  modified Jacobian, 282–284
  Montgomery, 285–288, 696
  projective, 281, 283, 284, 292, 294–295, 297
  simplified Chudnovsky Jacobian coordinates, 401
homogeneous, 46, 69
hyperelliptic curve
  affine, 316, 336
  comparison, 325, 344
  mixed, 325–328, 344–345
  new, 323, 342
  projective, 321, 341, 346
  recent, 346
coprime ideals, 22
Cornacchia's algorithm, **458**, 598
Coron's first countermeasure, 682, 699
correlation power analysis, 680
countermeasures
  against differential fault analysis, 706–708
  against differential side-channel attacks, 697–702
    Koblitz curves, 711–713
  against fault attacks, 705–709
  against Goubin type attacks, 703–704
  against higher order differential side-channel attacks, 704–705
  against side-channel attacks
    GLV curves, 713–714
    special curves, 709–714
  against simple fault analysis, 706
  against simple side-channel attacks, 688–697
    Koblitz curves, 709–711
  against timing attacks, 705
  dummy operations, 689
    atomic blocks, 691
    field operations, 690
    group operations, 689
  Montgomery arithmetic, 696
  randomization
    curve equation, 700
    group elements, 700

    scalar, 699
cover map, 538
cover method
  trace zero variety, 539
  twisted cover, 539
crosscorrelation of two sequences, 731
CRT, *see* Chinese remainder theorem
cryptographic primitive, 1, 8, 547, 569
crystalline cohomology, 136, 451
curve, 49
  anomalous, 76
  BH, 381
  $C_{ab}$, 82, 352
  desingularization, 64
  divisor, 66
  divisor class group $\mathrm{Pic}_C^0$ of $C$ of degree zero, 76, 81
  divisor group, 66, 306
  genus, 68, 304
  Hasse–Weil theorem, 110, 112, 278, 310
  ideal class group, 81–83
  imaginary quadratic, 83, 304
  Jacobian variety, **78**, 77–80
  nonsingular, 64, 65
  period lattice, 91, 95, 462
  period matrix, 93
  Picard, 83, 352, 473
  Picard group, 76
  Serre bound, 112
  smooth, 64, 65
  uniformizer at point $P$, 65
  *see also* elliptic curve and hyperelliptic curve
cyclic group, 20

## D

Dadda tree, **635**, 638
dagger ring, **140**, 451
DBDHP, *see* decision bilinear Diffie–Hellman problem
DDHP, *see* decision Diffie–Hellman problem
de Rham cohomology, **134**, 140, 567
decision bilinear Diffie–Hellman problem (DBDHP), 575
decision Diffie–Hellman problem (DDHP), 10, 574
decision problem, 3
Dedekind ring, 30
defining element, 28

degree
    divisor, 66
    extension field, 25
    isogeny, 59, 277, 309
    separability, 27
    transcendence, 26
dehomogenization, 49
density of a normal basis, 221
desingularization of a curve, 64
differential, 75
    electromagnetic analysis, 682
    fault analysis, 683–685
    power analysis, 677–678
    side-channel attacks, 697, 698
differential holomorphic, 76
differential over $\mathbb{C}$, 89
Diffie–Hellman (DH) protocol, 10
digit, 170
dimension
    analytic variety, 88
    variety, 49
    vector space, 24
directed family of groups, 39
directed set, 39
discrepancy of a sequence, 731
discrete logarithm
    computation, 477–543
        baby-step giant-step, 376, 482
        generic algorithm, **478**, 554
        GHS algorithm, 131, 531, 538
        square root algorithms, 477–494
        see also index calculus, Pollard's rho, and Pollard's lambda
    transfer, 9, 529
        by cover method for trace zero varieties, 539
        by pairings, 530, 555
        by Weil descent, 530–543, 556
        odd characteristic, 536
        to $\mathbb{F}_q$-vector spaces, 529
        via covers, 538
discrete logarithm problem (DLP), 8
discrete logarithm system (DLS), 8, **547**, 571
    choice of genus and curve equation, 560
    choice of secure candidates, 547
    choice of the finite field, 558
    construction of systems, 565
    function field, 549
    generic, 8
    index calculus, 554

number field, 549
    summary of elliptic curves, 551
    summary of hyperelliptic curves, 552
    transfers, see transfer of DLP
    with bilinear structure, 9, 573, 574
discrete Newton iteration, 179
discrete valuation, 41
discriminant of an elliptic curve, 71, 268
disjunction, **170**, 172
distortion map, 582
distributive law, 21
division
    binary field, see inversion in a binary field
    integers, 184–190
        exact, **189**, 204
        Karatsuba method, 188
        normalization, 186
        Quisquater's normalization, **187**, 202
        recursive method, 188
        schoolbook method, 185
    $p$-adic numbers, 245
    prime field, see inversion in a prime field
division ideal, 60
division polynomial of an elliptic curve, 60, 80, 413
divisor, 66
    $B$-smooth, 512
    class group
        Mumford representation, 84, 306
        of order equal to $\mathrm{char}(K)$, 76
        $\mathrm{Pic}_C^0$ of $C$ of degree zero, 76, 81, 306
        relation between ideal class group and divisor, 81, 306
    degree, 66
    effective, 66
    group of a curve, 66, 306
    prime, 66
    principal, 67, 306, 550
    random class, 307
    rational function, 67, 306, 550
    reduced, 84, 307, 552
DL, see discrete logarithm
DLP, see discrete logarithm problem
DLS, see discrete logarithm system
dominant rational map, 53
double precision integer, 171
DSCA, see differential side-channel attacks
dual basis, 35
dual isogeny

elliptic curve, 277
    hyperelliptic curve, 309
dummy operations, 689
    atomic blocks, 691
    field operations, 690
    group operations, 689
Dwork–Reich cohomology, 138, 453

**E**

early abort strategy, 468
    *see also* SEA algorithm
ECDSA, *see* elliptic curve digital signature algorithm
ECM, *see* elliptic curve method
ECPP test, **597**, 601
EEPROM, 653
effective divisor, 66
eigenvalues of the Frobenius endomorphism, 110, 111, 279, 311, 356
Eisenstein polynomial, 43
Eisenstein series, 96
electromagnetic analysis, 682–683
ElGamal
    encryption, 10, 11
    signature, 12
Elkies prime, **414**, 416, 419
    *see also* SEA algorithm
elliptic curve, 68–72, 95–100, 268, 551–552
    absolute invariant, 71, 72, 97, 268
    addition in atomic blocks, 690, 693
    addition law, 79, 270
    admissible change of variables, 274
    arithmetic, 267–302, 355–387
        binary field, 289–302
        prime field, 280–289
        *see also* special curve
    canonical lift, 423
    cardinality, 278–279
        admissible, 278
        Hasse–Weil interval, 112
        Hasse–Weil theorem, 110, 112, 278
        subfield curve, 279
        trace of the Frobenius, 278
        twist, 279
    characteristic polynomial of the Frobenius, 278
    compact curve, 382
    comparison of coordinate systems, 283, 296

complex multiplication, 98, 277, 456–460
    construction, 459
compression, 288–289, 302
coordinates, *see* coordinates for elliptic curves
discriminant, 71, 268
distortion map, 582
division polynomial, 60, 80, 413
doubling
    followed by addition, 281, 292
    in atomic blocks, 690, 692
    repeated, 295
dual isogeny, 277
eigenvalues of the Frobenius endomorphism, 279
endomorphism, 277–278
endomorphism ring, 98
Frobenius endomorphism, 277
GLV curve, 377–380
group law, 79, 270
group structure of the $\mathbb{F}_q$-rational points, 272
halving, **299**, 299–302, 365
Hessian form, 696
Hilbert class field, 99
Hilbert class polynomial, 99, 456
    computation, 457
isogeny, **277**, 282
    point counting, 415, 419, 420, 424, 435, 439
isomorphism, 71, 273–276
    Hessian form, 275–276
    Jacobi model, **275**, 696
    Legendre form, 275
    twist, **71**, 99, **274**, 279, 378, 459, 568, 735
$j$-function, **97**, 417
$j$-invariant, 71, 72, 97, 268
Koblitz curve, 356–367
library and software
    apecs, 267
    Magma, 267
    PARI/GP, 267
    SIMATH, 267
minimal 2-torsion, 299
Montgomery form, **285**, 286, 288
nonsingular, 268
nonsupersingular, **273**, 289, 290, 584
    comparison with supersingular, 589

use in pairings, 586–588
  with small embedding degree, 586
 ordinary, see nonsupersingular
 pairing, 122, 123, 390–392, 396, 397
   embedding degree, 123, 390, 573
   Tate–Lichtenbaum, 396
   Tate–Lichtenbaum in characteristic 3, 580
 period lattice, 95, 456
 point at infinity, 269
 rational point, 272
 Serre bound, 112
 short Weierstraß equation, 70, 73, 268
 singular point, 268
 smooth, 268
 supersingular, 123, **273**, 279, 556, 574, 580, 583
   arithmetic, 289
   comparison with nonsupersingular, 589
 torsion point, 273, 413, 414
 trace of the Frobenius endomorphism, 110, 413, 426
 trace zero variety, 383–387
 triple and add, 581
 tripling, 580
 unified formulas, 694
   Hessian form, 696
   Jacobi model, 696
 Vélu's formulas, 415
 Verschiebung, 427, 443
 Weierstraß equation, 69, 73, 268
 Weierstraß point, 83
elliptic curve digital signature algorithm (ECDSA), 13, 570
elliptic curve method (ECM), 7, **604**, 604–607, 612
 partial addition on an elliptic curve, 604
 partial inverse on an elliptic curve, 604
elliptic curve power generator, 733
elliptic function, 95
embedding degree, 123, 390, 573, 585
 construction of ordinary elliptic curves, 586
end-around carry, 621
endomorphism
 abelian variety, 59
 curve with identity-based parameters, 382
 elliptic curve, 277–278
 hyperelliptic curve, 310
 ring of an elliptic curve, 98

 *see also* Frobenius endomorphism and scalar multiplication
Enge–Gaudry's theorem, 516
EPROM, 652
Euclid extended gcd, 191
Euclidean exponentiation method, 166
Euler totient function, 23
exact differential, **141**, 450
exceptional procedure attack, *see* Goubin type attack
exclusive disjunction, **170**, 172
exponent, 145
exponent recoding, 151
 $\nu$-adic expansion, 381
 $\tau$-adic expansion, 358, 368
   length reduction, 359–362, 371–373
 $\tau$JSF, 365
 $\tau$NAF, 358
 $\tau$NAF$_w$, 363
 Booth technique, 151
 JSF, 155
 Koyama-Tsuruoka, 152
 NAF, 151
 NAF$_w$, 153
 signed fractional window method, 154
exponential complexity, 4, 548
exponentiation, 145–168
 $2^k$-ary method, **148**, 165, 356
 addition chain, **164**, 157–164, 224, 233, 237, 559, 646, 699
 base point, 145
 BGMW algorithm, *see* Yao's method
 binary field, 225–227
   Brent–Kung composition, 225
   Shoup algorithm, 227
 Euclidean method, 166
 exponent, 145
 fixed-base comb method, 167
 left-to-right binary method, **146**, 210, 253, 261
 Montgomery, 210
 Montgomery's ladder, **148**, 697
 multi-exponentiation, 154–157, 164, 377
 optimal extension field, 231–233
 Pippenger's algorithm, 166
 prime field, 209–210
 right-to-left binary method, 146
 sliding window, **150**, 163

square and multiply, *see* left-to-right binary method
Yacobi's method, 160
Yao's method, **165**, 227
*see also* scalar multiplication
extension field, 25

## F

factor base
   factoring, 608
   index calculus, 496
factoring, 601–614
   continued fraction method, 608
   elliptic curve method, 7, **604**, 604–607
   factor base, 608
   Fermat–Morrison–Brillhart approach, 607–614
   general number field sieve, 612
   multiple polynomial quadratic sieve, 611
   number field sieve, 7, **613**, 612–614
   Pollard's $p-1$ method, 603
   Pollard's rho method, 601–603
   quadratic sieve, 508, 609–610
   special number field sieve, 612
false witness, 594
fastECPP, 601
fault injection attack, 683–685
Fermat prime, **533**, 557
Fermat test, 593
Fermat witness, 593
Fermat's little theorem, **23**, 599, 603, 607, 645
Fermat–Morrison–Brillhart factoring method, 607–614
FFT multiplication, 177
field, 23
   algebraic closure, 27
   automorphism, 27
   CM-field, 105, 107, 462
   CM-type, 63, 105, 463
   complete discrete valuation, 41
   extension, 25
     algebraic, 25
     defining element, 28
     degree, 25
     degree of separability, 27
     finite, 25
     Galois, 28
     monogenic, 26
     normal, 27
     purely transcendental, 26
     separable, 27
     transcendental, 26
   fraction ∼ of a ring, 23
   function ∼ of a variety, 51
   homomorphism, 23
   isomorphism, 25
   norm, 26
   of constants of a function field, 51
   of definition of $P$, 46
   perfect, 28
   trace, 26
   *see also* finite field, number field, and $p$-adic field
finite extension field, 25
finite field, **31**, 31–37
   library and software
     Lidia, 201
     Magma, 201
     NTL, 201
     ZEN, 201
   multiplicative group, 32
   norm, 33
   Rabin irreducibility test, 214
   representation, 33–35
   trace, 33
   *see also* binary field, prime field, and optimal extension field
finitely generated ideal, 22
fixed-base comb exponentiation method, 167
flash EEPROM, 654
focused ion beam, 672
formation, *see* arithmetical formation
fractional ideal, 30, 549
   inverse, 30
fractional $\mathcal{O}$-ideal, 81
FRAM, 654
FreeLip, 169
frequency block test, 728
Frey–Rück attack, 530
Frobenius automorphism
   absolute, 33
   relative, 33
Frobenius endomorphism, 59, 109, 355
   elliptic curve
     characteristic polynomial, 278
     eigenvalues, 279, 356
     Koblitz, 356
     trace, 278, 413, 426, 564

hyperelliptic curve, 109, 310
   characteristic polynomial, 310
   eigenvalues, 311
   Koblitz, 367
   trace, 564
variety
   characteristic polynomial, 109–110
   eigenvalues, 110, 111
   trace, 110, 111
Frobenius morphism, 53, 59, 109, 133
   see also Frobenius endomorphism
Frobenius substitution, **43**, 240, 250–252, 423
full adder, 627
full-graph method, 522–523
function field, 51
   ideal class group, 81, 84, 306
   variety, 51
fundamental domain, 416
fundamental unit, **31**, 463

# G

Galois extension, 28
Galois field, see finite field
Galois group, 28
Gap-Diffie–Hellman group, 10, 574
Gauß period, 34
   general ~, 35
Gaudry's algorithm, 517
   see also index calculus
Gauß sum, 599
Gaussian normal basis, **35**, 240
   $p$-adic field, **240**, 252, 261, 433
   see also optimal normal basis
gcd
   integer, 190–197
      binary extended, 195
      Euclid extended, 191
      generalized binary, 195
      Lehmer extended, 192
   polynomial, 222
general number field sieve (GNFS), 612
generalized binary gcd, 195
generalized projective coordinates, 271
generated carry, 629
generating set, 24
generic algorithm, **478**, 554
genus, 68, 304
   Hurwitz formula, 68
   hyperelliptic curve, 73, 304

ghost bit basis, 217
GHS algorithm, 131, 531, 538
   Artin–Schreier curve, 536
   magic number, 532
   see also Weil descent
glitch attack, 683
GLV curve, 377–380
   basis of the endomorphism ring, 378
   combination with Koblitz curves, 381
   computation of expansion, 379
   countermeasures against side-channel
      attacks, 713–714
   elliptic curve, 377
   generalizations, 380–381
   hyperelliptic curve, 380
GMP, 169, 176
GNFS, see general number field sieve
Goldwasser–Killian test, 598
Goubin's refined power analysis, 680–682
greatest common divisor, see gcd
group, **19**, 19–21
   abelian, 20
   action, 21
   cyclic, 20
   Galois, 28
   homomorphism, 21
      kernel, 21
   index, 20
   inverse, 20
   normal, 20
   order, 20
   quotient, 20
   subgroup, 20
   unit element, 20

# H

half adder, 628
halve and add scalar multiplication algorithm, 301
halving map, 299
Hamming weight, **147**, 158, 234, 393, 491, 558, 677
hardware
   addition, see addition in hardware
   binary field, see binary field in hardware
   inversion, 645–646
   modular reduction, see modular reduction in hardware
   multiplication, see multiplication in hardware

hardware random number generator, 721
hash function, 12
hash function on the Jacobian, 590
Hasse–Weil interval, 112, 410
Hasse–Weil theorem, 110, 112, 278, 310
HECDSA, *see* hyperelliptic curve digital signature algorithm
Hensel odd division, 180
    *see also* Montgomery reduction
Hessian form of an elliptic curve, 275–276
heuristics of class number, 564
hidden field equations (HFE), 15
higher order differential power analysis, 680
Hilbert class field, 99
Hilbert class polynomial, 99, 456
    computation, 457
holomorphic differential, 76
homogeneous
    coordinates, 46, 69
        *see also* projective coordinates
    ideal, 47
    polynomial, 46
homomorphism
    abelian variety, 57
    complex, 29
    connected component of the unity of $\ker(\varphi)$, 58
    field, 23
    group, 21
        kernel, 21
    real, 29
homothetic lattices, 97
Horner-like scheme, 227
Hurwitz genus formula, 68
hyperelliptic curve, 73, **303**, 552–553
    addition in atomic blocks, 694
    addition law, 304, 308
    arithmetic
        genus 2, 313–347
            genus 2 in even characteristic, 334–347
            genus 2 in odd characteristic, 320–334
        genus 3, 348–352
    BH curve, 381
    canonical lift, 442
    Cantor's algorithm, 308, 552
    cardinality, 310–311
        Hasse–Weil interval, 410
        Hasse–Weil theorem, 110, 112, 310

characteristic polynomial of the Frobenius, 310
class polynomial, 107, 462, 463, 465, 466, 472
    computation, 462, 472
comparison of coordinate systems, 325, 344
complex multiplication, 104, 310, 460–473
    computation of theta constant, 465
    construction, 469
    curve with automorphisms, 471
    Mestre's algorithm, 468
compression, 311–313
computation of Igusa invariants, 465
coordinates, *see* coordinates for hyperelliptic curves
divisor class group $\text{Pic}_C^0$ of $C$ of degree zero, 306
doubling in atomic blocks, 694
dual isogeny, 309
eigenvalues of the Frobenius endomorphism, 311
endomorphism, 310
Frobenius endomorphism, 109, 310
genus, 73, 304
group law, 304
group structure of the $\mathbb{F}_q$-rational points $J_C(\mathbb{F}_q)[n]$, 111
ideal class group, 83–85, 306
Igusa invariants, 101, 107, 462
index calculus method, 511
isogeny, 309
isomorphism, 74, 308
Koblitz curve, 367–376
Mestre's invariants, 102, 468
Montgomery like form, 329
nonsingular, 304
$p$-rank of a hyperelliptic curve, 310
pairing, 122, 390–392, 394, 398
    embedding degree, 123, 390, 573
    Tate–Lichtenbaum, 398
period lattice, 462
period matrix, 100, 462, 463
Picard group, 306
random divisor class, 307
Riemann theta divisor, 100
Rosenhain model, 104, 472
Serre bound, 112
Shioda invariants, 104, 472

smooth, 304
supersingular, **310**, 584, 585
   arithmetic, 340
theta characteristic, 100, 444
theta constants, 100, 444, 463
torsion point, 309
trace zero variety, 383–387
Weierstraß equation, 74, 83, 304
Weierstraß point, 73, 304
hyperelliptic curve digital signature algorithm (HECDSA)
   generation, 570
   verification, 570
hyperelliptic involution, 73, 512
hypersurface, 47

# I

ID-based
   cryptography, 576–578
   decryption, 577
   encryption, 576
   parameters, 382
ideal, 21
   above $p\mathbb{Z}$, 31
   coprime, 22
   finitely generated, 22
   fractional, 30, 549
   homogeneous, 47
   inert, 31
   maximal, 22
   prime, 22
   principal, 22
   product, 30
   ramified, 31
   split, 31
ideal class group, 81–83
   hyperelliptic curve, 83–85, 306
   Mumford representation, 84, 306
   relation with divisor class group, 81, 306
identity-based, *see* ID-based
Igusa invariants, 101, 107, 462
   computation, 465
imaginary quadratic curve, 83, 304
incompletely reduced number, 202
index calculus, 495–527, 554–555
   arithmetical formation, 496–497
   automorphism of the group, 505–506
   factor base, 496
   filtering, 503–505

      merging, 504–505
      pruning, 504
   finite field, 506–507
   hyperelliptic curve, 511–527
      Adleman–DeMarrais–Huang algorithm, 512–516
      $B$-smooth divisor, 512
      concentric circles method, 525–527
      double large primes, 521–522
      Enge–Gaudry's theorem, 516
      factor base, 512, 515, 517
      full-graph method, 522–523
      Gaudry's algorithm, 517
      general algorithm, 511–512
      harvesting, 518–519
      hyperelliptic involution, 512
      refined factor base, 517–518
      relation search, 513–514
      simplified graph method, 523–525
      single large prime, 520–521
   large primes, 507–509
      1 large prime, 507–508
      2 ~, 508–509
      more ~, 509
   linear algebra, 500–503
      *see also* Lanczos' method and Wiedemann's method
   prime, 496
   relation search, 499–500
      parallelization, 500
   smoothness bound, 496
   via hyperplane sections, 541
   Waterloo variant, 508
index of a group, 20
indistinguishability, 729
induced morphism, 52
industrial-grade prime, 591
inert ideal, 31
inertia degree, 43
integer arithmetic, 169–199
   addition, 173
   concatenation, 187
   division, *see* division of integers
   exact division, **189**, 204
   gcd, *see* gcd of integers
   integer square root, 198
   multiplication, *see* multiplication of integers
   reduction, *see* modular reduction
   square root, 198

squaring, *see* squaring of integers
subtraction, 173
integer factorization problem, 1, 7, 479
integer ring of a number field, 29
integral basis, 29
integral domain, 21
integrally closed ring, 30
interleaved multiplication-reduction, 203
invalid curve attack, 569, 706
invasive attacks, 670–673
inverse limit, 40
inverse of a fractional ideal, 30
inverse of a group element, 20
inverse of a ring element, 23
inverse square root of $p$-adic numbers, 248
inversion
    binary field, 222–225
        binary method, 223
        extended gcd, 222
        Itoh and Tsujii algorithm, 225
        Lagrange's theorem, 224
    hardware, 645–646
    integers, *see* division of integers
    optimal extension field, 233–234, 237
    $p$-adic numbers, 247
    prime field, 205–209
        almost Montgomery inverse, 208
        Montgomery inverse, 208
        plus-minus method, 206
        simultaneous, **209**, 283, 296, 327
        Thomas et al. method, 207
involution of a hyperelliptic curve, 73, 512
irreducible subset of a topological space, 48
isogeny
    abelian variety, 58
        over $\mathbb{C}$, 94
    degree, 59, 277
    dual, 277
    elliptic curve, **277**, 282
        point counting, 415, 419, 420, 424, 435, 439
        Vélu's formulas, 415
    hyperelliptic curve, 309
    purely inseparable, 59
    separable, 59
isomorphism
    abelian variety, 58
    elliptic curve, 71, 273–276
        admissible change of variables, 274
        Hessian form, 275–276

        Jacobi model, **275**, 696
        Legendre form, 275
    field, 25
    hyperelliptic curve, 74, 308
    variety, 52
Itoh and Tsujii inversion, 225

## J

$j$-function, **97**, 417
$j$-invariant, 71, 72, 97, 268
Jacobi criterion, 64
Jacobi model, **275**, 696
Jacobi sum, 600
Jacobian coordinates, 282, 292–295, 297
    *see also* Chudnovsky Jacobian coordinates, modified Jacobian coordinates, and simplified Chudnovsky Jacobian coordinates
Jacobian variety, **78**, 77–80
Java Card, 659
Jebelean exact division, **189**, 204
joint Hamming weight, 156
joint sparse form, *see* JSF
JSF, 155
    simple, 156
    $\tau$JSF, 365

## K

Karatsuba
    integer division, 188
    integer multiplication, **176**, 202
    integer squaring, 178
    polynomial multiplication, 220, 227, 236, 244, 317
Kedlaya's algorithm, 449
kernel of a group homomorphism, 21
key exchange
    contributory, 575
    multiparty, 574
    noncontributory, 575
key generation, 11
Knapsack problem, 14
Koblitz curve
    $\tau$-adic expansion, 358, 368
    alternative generation of $\tau$-adic expansion, 375
    combination with GLV method, 381
    countermeasures against

differential side-channel attacks, 711–713
simple side-channel attacks, 709–711
elliptic curve, 356–367
  $\tau$JSF, 365
  length reduction of the $\tau$-adic expansion, 359–362
  main subgroup membership, 359
  $\tau$NAF, 358
  $\tau$NAF$_w$, 363
hyperelliptic curve, 367–376
  length reduction of the $\tau$-adic expansion, 371–373
Koyama–Tsuruoka recoding, 152
Kronecker relation, 425
Kronecker–Jacobi symbol, 36
Kummer surface, 329

**L**

$\ell$-adic Tate module, 61
$L$-polynomial of a curve, **135**, 407
$L$-rational points, 46
Lanczos' method, 500–503, 517
LCG, *see* linear congruential generator
least significant digit, 170
left-to-right binary method, **146**, 210, 253, 261
Legendre form, 275
Legendre symbol, **36**, 210
Legendre–Kronecker–Jacobi symbol, **36**, 235
Lehmer extended gcd, 192
Lempel–Ziv compression algorithm, 160
Lercier–Lubicz' algorithm, 253, 434
library and software
  aPecs, 267
  BigNum, 169
  FreeLip, 169
  GMP, 169, 176
  Lidia, 201
  Magma, 201, 267
  Maple, 267
  NTL, 201
  PARI/GP, 267, 467
  SIMATH, 267
  ZEN, 201
Lidia, 201
lift of an element in a valuation ring, 41
lift of $p$-adic numbers
  Hensel, **250**, 249–250, 428

Newton, 246–249
  generalized, **257**, 447
  Teichmüller, 241, **257**, 258
light attack, 684
linear complexity of a sequence, 732
linear congruential generator (LCG), 720
linearly independent vectors, 24
little endian, 171
local ring of a point, 64
local Tate pairing, 118
López–Dahab coordinates, 293–295, 297
Lubin–Serre–Tate theorem, 423
Lucas pseudoprime, 595
Lucas pseudoprime test, 595
Lucas sequence, 357

**M**

Möbius function, 33
magic number, 532
Magma, 201, 267
man-in-the-middle attack, 10
Maple, 267
  aPecs, 267
Massey–Omura multiplier, 643
Mastrovito multiplier, 642
match and sort, 421
  *see also* SEA algorithm
maximal ideal, 22
maximal order, 30
memory management unit, 655
memory only card, *see* synchronous card
Mersenne prime, **182**, 207, 533, 556, 640
  pseudo-~, 230
Mestre's algorithm for complex multiplication, 468
Mestre's invariants, 102, 468
micromodule, 648
microprocessor card, *see* asynchronous card
Miller's pairing computation algorithm, 122, 392
minimal 2-torsion for an elliptic curve, 299
minimal polynomial of an algebraic element, 25
mixed coordinates
  elliptic curve, 283–285, 296–298
  hyperelliptic curve, 325–328, 344–345
modified Jacobian coordinates, 282–284
modular function, 97
modular polynomial, 416–419

canonical, 418
classical, 418
Kronecker relation, 425
modular reduction, 178–184
    Barrett method, **179**, 182, 202, 203, 210, 606
    hardware, 638–641
        incomplete reduction, 640
        special moduli, 640
    modulo several primes, 184
        remainder tree, 184
        scaled remainder tree, 184
    modulo special integers, **183**, 182–184
        Mersenne prime, 182
        NIST prime, 183
    Montgomery reduction, **181**, 180–182
    Montgomery representation, 180
    residue number system arithmetic, 197
    *see also* interleaved multiplication-reduction
module, 24
monobit test, 726
Monsky–Washnitzer cohomology, 136, **139**, 451, 567
Montgomery
    almost inverse, 208
    exponentiation, 210
    inversion, 208
    multiplication, 204
    reduction, **181**, 180–182
        coarsely integrated operand scanning method, 640
        hardware, 639–640
    representation, 180
Montgomery coordinates on an elliptic curve, 285–288, 696
Montgomery form of an elliptic curve, **285**, 286, 288
Montgomery like form of hyperelliptic curve, 329
Montgomery's ladder, **148**, 287, 298, 328, 331, 401, 676, 697
morphism
    affine variety, 52
    Frobenius, 53, 59, 109, 133
    *see also* Frobenius endomorphism
    from $\mathbb{A}^n$ to $\mathbb{A}^1$, 52
    from $\mathbb{A}^n$ to $\mathbb{A}^m$, 52
    from $V \subset \mathbb{A}^n$ to a variety $W \subset \mathbb{A}^m$, 52
    induced, 52

projective variety, 55
most significant digit, 170
MOV attack, 530
MPQS, *see* multiple polynomial quadratic sieve
multi-exponentiation, 154–157, 164, 377
multi-stack algorithm, 487
multiple polynomial quadratic sieve (MPQS), 611
multiplication
    binary field, 218–221
        optimal normal basis, 220–221
        polynomial basis, 218–220
    hardware
        using left shift, 623
        using right shift, 624
    integers, 174–177
        FFT, 177
        Karatsuba method, **176**, 202
        schoolbook method, 174
        Toom–Cook, 177
    optimal extension field, 231, 236–237
    $p$-adic numbers, 244
    prime field, 202–205
        interleaved with reduction, 203
        Montgomery method, 204
    *see also* squaring
multiplication matrix of a normal basis, 220
    density, 221
multiplicative group, 32
multiplier recoding
    Booth method, 625
    radix-4 signed digit, 626
multiprecision, 171
multiprecision library
    BigNum, 169
    FreeLip, 169
    GMP, 169, 176
Mumford representation, 84, 306

# N

$n$-word integer, 171
NAF, 151
NAF$_w$, 153
Naor–Reingold generator, 734
new coordinates for hyperelliptic curves of genus 2, 323, 342
Newton–Girard formula, 229, 408
NFS, *see* number field sieve

Noetherian ring, 30
non-adjacent form, *see* NAF
non-invasive attacks, 673–685
   *see also* side-channel attacks
nonempty affine part of a variety, 50
nonsingular curve, 64, 65
   elliptic curve, 268
   hyperelliptic curve, 304
nonsingular point, 64, 65, 268, 304
nonsupersingular elliptic curve, 273, 584
norm
   algebraic number, 29
   endomorphism, 377
   field, 26
   finite field, 33
   $p$-adic number, 41, 261–263
normal basis, **34**, 224, 228
   density, 221
   hardware, 643–645
   multiplication matrix, 220
   self-complementary, 35
   *see also* optimal normal basis
normal element, **34**, 218
normal extension, 27
normal group, 20
normalized valuation of a place, 65
NTL, 201
NTRU encryption system, 14
null hypothesis, 723
number field, **29**, 29–31
   class number, **30**, 457, 480, 598
   fundamental unit, **31**, 463
   ideal class group, 30
   ideal decomposition, 31
   integer ring, 29
   maximal order, 30
   norm, 29
   order, 30
   signature, 29
   totally complex, 29
   totally real, 29
   trace, 29
number field sieve (NFS), 7, **613**, 612–614

## O

$O$-notation, 3
$o$-notation, 3
OEF, *see* optimal extension field
ONB, *see* optimal normal basis

one's complement, 621
one-parameter quadratic-base test (OPQBT), 596
one-way function, 5
OPQBT, *see* one-parameter quadratic-base test
optimal extension field (OEF)
   arithmetic, 229–237
      exponentiation, 231–233
      inversion, 233–234, 237
      multiplication, 231, 236–237
      specific improvements, 235–237
      square root, 234–235
      trace, 235
   type I, 230
   type II, 230
optimal normal basis (ONB), 217–218, **221**, 220–221
   density, 221
   multiplication matrix, 220
   palindromic representation, 221
   type I, 217
   type II, 218
   *see also* Gaussian normal basis
oracle, 10, **478**
   random model, 577
order of a group, 20
order of a number field, 30
order of an element
   finite, 20
   infinite, 20
ordinary
   abelian variety, 60
   curve for pairings, 586–588
   elliptic curve, **273**, 289, 290, 584

## P

$p$-adic exponential, 259
$p$-adic field
   absolute ramification index, 43
   inertia degree, 43
   unramified extension, **43**, 136, 138, 239, 428, 436
$p$-adic integer ring, 40
$p$-adic logarithm, 258
$p$-adic norm, 41
$p$-adic numbers
   arithmetic, 239–263
      fast division with remainder, 245

inverse, 247
inverse square root, 248
multiplication, 244
square root, 249
exponential, 259
Frobenius substitution, **43**, 240, 250–252, 423
generalized Newton lifting, **257**, 447
Hensel lifting, **250**, 249–250, 428
logarithm, 258
Newton lifting, 246–249
norm, 261–263
representation
   Gaussian normal basis, **240**, 252, 261, 433
   sparse modulus, 240
   Teichmüller modulus, 240
Teichmüller lift, 241, 256, **257**, 258
Teichmüller modulus increment, 242
trace, 260
$p$-adic valuation, 41
$p$-rank
   abelian variety, 61
   hyperelliptic curve, 310
padding, 170
pairing
   comparison of ordinary and supersingular curves, 589
   computation in characteristic 3, 580
   distortion map, 582
   elliptic curve, 123, 396, 397
   embedding degree, 123, 390, 573, 585
   hyperelliptic curve, 398
   improvements for elliptic curves, 400
   local Tate, 118
   Miller's algorithm, 122, 392
   ordinary curve, 586–588
   over local fields, 118
   security parameter, 585
   short signature protocol, 578, 589
   supersingular curve, 580, 584
   Tate, 116
   Tate–Lichtenbaum, 122, 390–392, 394
      comparison with Weil pairing, 395
      efficient computation, 400
      elimination of divisions, 400
      elliptic curve, 396
      hyperelliptic curve, 398
      subfield computations, 401
   transfer of DLP, 530, 555

Weil, 115
   comparison with Tate–Lichtenbaum pairing, 395
palindromic representation, 221
PARI/GP, 267, 467
pentanomial, 214
perfect field, 28
period lattice of a curve, 91, 95, 462
period matrix
   abelian variety, 93
   curve, 93
   hyperelliptic curve, 100, 462, 463
period of a pseudorandom sequence, 730
Picard curve, 83, 352, 473
Picard group
   curve, 76
   hyperelliptic curve, 306
PID, *see* principal ideal domain
Pila's algorithm, 422
Pippenger's exponentiation algorithm, 166
place of a function field, 65
plus-minus inversion method, 206
Pocklington–Lehmer test, 597
Poincaré upper half plane, 97, **416**
point at infinity, 50, 74, 269–271, 306
point counting
   AGM, *see* AGM algorithm
   Cartier–Manin method, 411
   enumeration, 407
   Kedlaya's algorithm, 449
   $\ell$-adic methods, 413–422
   $p$-adic methods, 422–453
   Pila's algorithm, 422
   Satoh's algorithm, **430**, 423–434, 437
   Schoof's algorithm, 413
   SEA, *see* SEA algorithm
   square root algorithms, 410–411
   subfield curve, 409
point halving, **299**, 299–302, 365
points at infinity, 50, 550
Pollard's kangaroo, 491–494
   automorphism, 494
   lambda method, 492–493
   parallelization, 493–494
   tame kangaroo, 492
   trap, 492
   wild kangaroo, 492
Pollard's $p - 1$ factoring method, 603
Pollard's rho, 483–491
   automorphisms, 490–491

collision, 483
    distinguished point technique, 488, **489**
    factoring method, 601–603
    for discrete logarithms, 488–489
    match, 483
    multi-stack algorithm, 487
    parallelization, 489–490
    random mapping, **483**, 486, 489
polynomial basis, **34**, 218–220, 225–226
    all one polynomial, 215
    anomalous, 217
    ghost bit basis, 217
    hardware, 642–643
    pentanomial, 214
    redundant, 34
    redundant trinomial, **217**, 228
    sedimentary, 215
    sparse, **34**, 214, 216
    trinomial, 214
polynomial complexity, 4, 548
Poulet number, 593
power consumption analysis, 675
power tree method, 160
powering, *see* exponentiation
precision of an integer, 170
preperiod of a pseudorandom sequence, 730
primality certificate, 597
primality test, 596–601
    AKS, 600
    APRCL Jacobi sum, 599
    Atkin–Morain ECPP, **597**, 601
    `fastECPP`, 601
    Goldwasser–Killian, 598
    Pocklington–Lehmer, 597
    primality certificate, 597
    *see also* compositeness test
prime divisor, 66
prime field, 32
    arithmetic, 201–213
    exponentiation, *see* exponentiation in a prime field
    inversion, *see* inversion in a prime field
    multiplication, *see* multiplication in a prime field
    quadratic nonresidue, 36
    quadratic residue, 36
    reduction, *see* modular reduction
    representative of a class, 202
    square root, *see* square root in a prime field
    squaring, 205
prime ideal, 22
prime number theorem, 6
prime of an arithmetical formation, 496
primitive element, **32**, 215, 217, 230
primitive polynomial, 34
principal divisor, 67, 306, 550
principal ideal, 22
principal ideal domain (PID), 22
principally polarized abelian variety, 93
probing, 671
projective closure of an affine set, 50
projective coordinates
    elliptic curve, 281, 283, 284, 292, 294–295, 297
    hyperelliptic curve, 321, 341, 346
    *see also* homogeneous coordinates
projective point, 46, 270, 271
projective space, **46**, 49
    closed set, 46
    open set, 46
    standard covering, 50
projective variety, 48
PROM, 652
propagated carry, 629
protocols, 9, 569–571
    ECDSA, 570
    HECDSA, 570
    ID-based
        cryptography, 576–578
        decryption, 577
        encryption, 576
    multiparty key exchange, 574
    short signature, 578, 589
pseudoprime, 593
    Lucas, 595
    strong, 594
    strong Lucas, 595
pseudorandom number generator, 718–719, 729–735
public-key cryptography, 5
public-key infrastructure, 11, 575
pure transcendental extension, 26
purely inseparable isogeny, 59

## Q

quadratic nonresidue, 36
quadratic reciprocity law
    Legendre symbol, 36

Legendre–Kronecker–Jacobi symbol, 37
quadratic residue, 36
quadratic sieve, 508, 609–610
quadratic twist, **71**, 99, **274**, 279, 378, 459, 568, 735
quotient group, 20

# R

Rabin polynomial irreducibility test, 214
Rabin–Miller compositeness test, **594**, 596
radix, 170
radix complement, 620
RAM, 651
ramified ideal, 31
random mapping, **483**, 486, 489
random oracle model, 577
random walk, 484, 489
    $r$-adding walk, 489, 492
randomization
    curve equation, 700
    group elements, 700
    scalar, 699
randomized GLV method, 713
randomness
    $\infty$-distributed sequence, 716
    autocorrelation of a sequence, 731
    autocorrelation test, 729
    balance of a sequence, 730
    crosscorrelation of two sequences, 731
    discrepancy of a sequence, 731
    frequency block test, 728
    hardware random number generator, 721
    indistinguishability, 729
    $l$-distributed sequence, 716
    linear complexity of a sequence, 732
    monobit test, 726
    null hypothesis, 723
    period of a pseudorandom sequence, 730
    preperiod of a pseudorandom sequence, 730
    pseudorandom number generator, 718–719, 729–735
        Blum–Blum–Shub generator, 720
        elliptic curve power generator, 733
        linear congruential generator, 720
        Naor–Reingold generator, 734
        RANDU, 720
    random number generator, 717–721
        ANSI-C, 720
    random sequence, 716
    random walk, 727
    runs test, 728
    seed, 490, **719**
    Shannon entropy function, 716
    statistical model, 723
    statistical tests, 723–724
    true random number generator, 718
    turning point test, 728
    unpredictability, 729
RANDU generator, 720
rational function, 53
    divisor, 67, 306, 550
    pole, 67
    regular at point $P$, 54
    zero, 67
rational map, 53
    birational, 53
    dominant, 53
    regular at point $P$, 54
rational point
    elliptic curve, 272
    hyperelliptic curve, 306
    variety, 46
real homomorphism, 29
recent coordinates for hyperelliptic curves of genus 2, 346
reciprocal integer, 178
recursive integer division, 188
recursive middle product, 188
reduced divisor, 84, 307, 552
redundant trinomial basis, **217**, 228
refined factor base, 517–518
register, **170**, 172
remainder tree, 184
representation
    finite field, 33–35
    integer in base $b$, 170
    *see also* basis
representative of a congruence class, 202
    canonical, 21
    centered, 21
    incompletely reduced number, 202
residual logarithm, 610
residue field, 41
    of a point, 64
residue number system arithmetic, 197
reverse engineering, 670
Riemann form, 93
Riemann theta divisor, 100

Riemann–Roch theorem, 68
right-to-left binary method, 146
rigid cohomology, 136
ring, **21**, 21–23
    characteristic, 22
    commutative, 21
    Dedekind, 30
    ideal, 21
    integrally closed, 30
    inverse, 23
    invertible element, 23
    Noetherian, 30
    $p$-adic integers, 40
    regular functions, 64
    unit, 23
ROM, 651
Rosenhain model, 104, 472
RSA, 7
RSA assumption, 7
RSA problem, 7
Rück attack, 529
runs test, 728

## S

Sarrus number, 593
Satoh's algorithm, **430**, 423–434, 437
SCA, *see* side-channel attacks
scalar, 24
scalar multiplication
    curve with endomorphism, 376–383
    elliptic curve
        halve and add, 301
    GLV curve
        elliptic curve, 377–380
        hyperelliptic curve, 380–381
    Koblitz curve
        elliptic curve, 356–367
        hyperelliptic curve, 367–376
    Montgomery's ladder, **287**, 298, 328, 331, 401, 676
    sliding window, **271**, 284, 356, 678, 704
    trace zero variety, 383–387
    triple and add, 581
    tripling, 580
    *see also* arithmetic, elliptic curve, hyperelliptic curve, exponentiation, and special curve
scalar randomization, 699
scalar restriction, 125

scaled remainder tree, 184
scanning electron microscope, 672
Schoof's algorithm, 413
Schoof–Elkies–Atkin's algorithm, *see* SEA algorithm
SEA algorithm, **421**, 414–422, 566
    Atkin prime, **414**, 415
    canonical modular polynomial, 418
    classical modular polynomial, 418
    early abort, 468
    Elkies prime, **414**, 416, 419
    match and sort, 421
security parameter, 585
sedimentary polynomial, 215
seed, 490, **719**
self-complementary normal basis, 35
self-dual basis, 35
separable extension of a field, 27
separable isogeny, 59
Serre bound, 112
Shannon entropy function, 716
shift
    integer multiplication using left ~, 623
    integer multiplication using right ~, 624
    left, 172
    right, 172
Shioda invariants, 104, 472
short signature, 578, 589
Shoup exponentiation algorithm, 227
side-channel atomicity, 691
    elliptic curve
        addition, 690, 693
        doubling, 690, 692
    hyperelliptic curve
        addition, 694
        doubling, 694
side-channel attacks (SCA), 285, 288, 328, 673–685
    Bellcore attack, 683
    correlation power analysis, 680
    differential electromagnetic analysis, 682
    differential fault analysis, 683–685
    differential power analysis, 677–678
    electromagnetic analysis, 682–683
    fault injection, 683–685
    glitch attack, 683
    Goubin's refined power analysis, 680–682

higher order differential power analysis, 680
light attack, 684
power consumption analysis, 675
simple power analysis, 675–677
timing attack, 673–675
*see also* countermeasures against ~
signature of a number field, 29
signature scheme, 12
signature verification, 13
signed fractional window method, 154
signed-binary representation, 151
signed-digit representation, 151
signed-magnitude representation, 171
Silver–Pohlig–Hellman attack, *see* Chinese remaindering attack
SIMATH, 267
simple abelian variety, 60
simple joint sparse form, 156
simple power analysis, 675–677
simple side-channel attacks (SSCA), 688
*see also* side-channel attacks and countermeasures against ~
simplified Chudnovsky Jacobian coordinates, 401
simplified graph method, 523–525
simultaneous inversion modulo $p$, **209**, 283, 296, 327
single precision integer, 171
singular point, 64, 65, 268, 304
size map, 496
sliding window
  exponentiation algorithm, **150**, 163
  scalar multiplication algorithm, **271**, 284, 356, 678, 704
small subgroup attack, 569
smart card
  asynchronous card, 665
  card system, 656
  combi-card, 650
  contact card, 649
  contactless card, 649
  coprocessor, 666
  electrical properties, 650
  floating gate, 652
  host system, 656
  invasive attacks, 670–673
  memory, 651–656
    EEPROM, 653
    EPROM, 652
    flash EEPROM, 654
    FRAM, 654
    PROM, 652
    RAM, 651
    ROM, 651
  memory management unit, 655
  memory only card, *see* synchronous card
  metal layers, 670
  micromodule, 648
  microprocessor card, *see* asynchronous card
  non-invasive attacks, 673–685
    *see also* side-channel attacks
  operating system, 657
    Java Card, 659
    Multos, 657
    Windows, 657
  physical properties, 648
  preprogrammed state, 652
  probing, 671
  reverse engineering, 670
  synchronous card, 664
  transmission protocol, 659–663
  UART, 663
  USB, 664
smooth
  curve, 64, 65
  elliptic curve, 268
  hyperelliptic curve, 304
smoothness bound, 496
SNFS, *see* special number field sieve
Solovay–Strassen compositeness test, 595
space complexity, 3
sparse modulus representation, 240
sparse polynomial basis, **34**, 214, 216
special curve, 355–387
  countermeasures against side-channel attacks, 709–714
  *see also* GLV curve, Koblitz curve, and trace zero variety
special number field sieve (SNFS), 612
splitting field, 27
splitting ideal, 31
square and multiply, *see* left-to-right binary method
square root
  binary field, 228–229
  integers, 198
  optimal extension field, 231, 234–235
  $p$-adic numbers, 249

prime field, 212, 213
    Tonnelli and Shanks square root algorithm, 212
squaring
    binary field, 221
    integers
        Karatsuba method, 178
        schoolbook method, 177
    prime field, 205
SSCA, *see* simple side-channel attacks
standard covering of a projective space, 50
star addition chain, **157**, 225
statistical model, 723
Straus–Shamir's trick, **155**, 387, 713
    *see also* multi-exponentiation
strong Lucas pseudoprime, 595
strong pseudoprime, 594
subexponential complexity, 4, 548
subgroup, 20
subgroup generated by an element, 20
subtraction
    binary field, 218
    integers, 173
    optimal extension field, 231
    prime field, 202
summation polynomial, 541
supersingular
    abelian variety, 61
        security parameter, 585
    elliptic curve, 123, **273**, 279, 289, 556, 574, 580, 583
        attack via pairing, 530
        distortion map, 582
        security parameter, 585
        use in pairings, 580
    hyperelliptic curve, **310**, 584, 585
        arithmetic, 340
        security parameter, 585
        use in pairings, 584
        Jacobian, 310
symmetric key cryptography, 2
synchronous card, 664
system parameter, 569

# T

$\tau$-adic expansion
    alternative generation, 375
    comparison, 374
    elliptic curve

computation, 358
length reduction, 359–362
windowing methods, 363
hyperelliptic curve
    computation, 368
    length reduction, 371–373
*see also* $\tau$NAF, $\tau$NAF$_w$, and $\tau$JSF
$\tau$-adic joint sparse form, *see* $\tau$JSF
$\tau$-adic non-adjacent form, *see* $\tau$NAF
$\tau$JSF, 365
$\tau$NAF, 358
$\tau$NAF$_w$, 363
Tate pairing, 116, 117, 119
    *see also* Tate–Lichtenbaum paring
Tate–Lichtenbaum pairing, 122, 390, 392
    characteristic 3, 580
    comparison with Weil pairing, 395
    computation, 391, 394
    efficient computation, 400
    elimination of divisions, 400
    elliptic curve, 123, 396, 397
    hyperelliptic curve, 398
    ordinary curve, 586–588
    subfield computations, 401
    supersingular curve, 580, 584
Teichmüller lift, 256
Teichmüller modulus increment, 242
Teichmüller modulus representation, 240
theta characteristic, 100, 444
theta constant, 100, 444, 463
    computation, 465
    even, 100
    odd, 100
Thomae formula, 446
Thurber's algorithm for addition chain, 158
time complexity, 3
timing attack, 673–675
Tonnelli and Shanks square root algorithm, 212
Toom–Cook multiplication, 177
torsion point
    abelian variety, 60
    elliptic curve, 273, 413, 414
        minimal 2-torsion, 299
    hyperelliptic curve, 309
torus, 92
totally complex number field, 29
totally real number field, 29
trace
    algebraic number, 29

binary field, 35
  efficient computation, 229
endomorphism, 377
field, 26
finite field, 33
Frobenius endomorphism, 110, 111, 413, 426, 564
$p$-adic number, 260
trace zero variety, 130, 383–387, 539, 557
  arithmetic, 385–387
  background, 384
transcendence degree, 26
transcendental element, 26
transcendental extension, 26
transfer of DLP, 9, 529, 555–557
  by pairings, 530, 555
  by Weil descent, 530–543, 556
    in odd characteristic, 536
  to $\mathbb{F}_q$-vector spaces, 529
  via covers, 538
translation by point $P$, 57
trial division, 592
trinomial, 214
tripartite key exchange, 14
triple and add scalar multiplication algorithm, 581
tripling on an elliptic curve, 580
true random number generator, 718
trusted authority, 576
turning point test, 728
twist of an elliptic curve, **71**, 99, **274**, 279, 378, 459, 568, 735
  see also isomorphism of elliptic curves
twisted cover, 539
two's complement, **171**, 620

## U

unified arithmetic for elliptic curves, 694
  Hessian form, 696
  Jacobi model, 696
unified multipliers, 644–645
uniformizer for the curve $C$ at the point $P$, 65
uniformizing element, **41**, 44
unit element of a group, 20
unit of a ring, 23
universal asynchronous receiver transmitter (UART), 663
universal serial bus (USB), 664
unpredictability, 729

unramified extension of a $p$-adic field, **43**, 136, 138, 239, 428, 436
USB, see universal serial bus

## V

valuation at a point, 65
valuation ring, 41
variety
  absolutely irreducible, 48
  birational equivalence, 53
  coordinate ring, 51
  dimension, 49
  function field, 51
vector space, 24
  basis, 24
  dimension, 24
  generating set, 24
  linearly independent vectors, 24
  scalar, 24
  vector, 24
vectorial addition chain, **158**, 164
Vélu's formulas, 415
Verschiebung
  abelian variety, 61
  elliptic curve, 427, 443

## W

Wallace tree, **635**, 638
Waterloo variant, 508
weak completion, see dagger ring
Weierstraß equation
  elliptic curve, 69, 73, 268
    short, 70, 73, 268
  hyperelliptic curve, 74, 83, 304
Weierstraß $\wp$-function, 96
Weierstraß point, 73, 83, 304
Weil descent, 90, 125, 127, 530–543, 556
  in odd characteristic, 536
  index calculus
    via hyperplane sections, 541
  over $\mathbb{C}$, 89
  trace zero variety, 539
  transfer of DLP, 530–543, 556
  via GHS, 131, 531, 538
Weil pairing, 115
  comparison with Tate–Lichtenbaum pairing, 395
  see also Tate paring and Tate–Lichtenbaum paring

width-$w$ $\tau$-adic non-adjacent form, *see* $\tau\text{NAF}_w$
width-$w$ non-adjacent form, *see* $\text{NAF}_w$
Wiedemann's method, 500–503, 517
Witt vector, 44

## X

XOR, *see* exclusive disjunction

## Y

Yacobi's exponentiation method, 160
Yao's exponentiation method, **165**, 227

## Z

Zariski topology, 46, 47
Zech's logarithm, 33
ZEN, 201
zeta function, **134**, 408, 422, 451